ESSENTIALS OF NEURAL SCIENCE AND BEHAVIOR

ESSENTIALS OF NEURAL SCIENCE AND BEHAVIOR

EDITED BY

Eric R. Kandel
James H. Schwartz
Thomas M. Jessell

CENTER FOR NEUROBIOLOGY AND BEHAVIOR
COLLEGE OF PHYSICIANS & SURGEONS OF COLUMBIA UNIVERSITY
and
THE HOWARD HUGHES MEDICAL INSTITUTE

PRENTICE HALL INTERNATIONAL, INC.

96 97 98 / 10 9 8 7 6 5 4 3 2

Prentice Hall International (UK) Limited, *London*
Prentice Hall of Australia Pty. Limited, *Sydney*
Prentice Hall Canada, Inc., *Toronto*
Prentice Hall Hispanoamericana, S.A., *Mexico*
Prentice Hall of India Private Limited, *New Delhi*
Prentice Hall of Japan, Inc., *Tokyo*
Simon & Schuster Asia Pte. Ltd., *Singapore*
Editora Prentice Hall do Brasil Ltda., *Rio de Janeiro*
Prentice Hall, *Englewood Cliffs, New Jersey*

ISBN 0-8385-2247-5

Library of Congress Cataloging-in-Publication Data

Essentials of neural science and behavior/edited by Eric R. Kandel,
James H. Schwartz, Thomas M. Jessell.
p. cm.
1. Neurosciences. I. Kandel, Eric R. II. Schwartz, James H.
(James Harris), 1932- . III. Jessell, Thomas M.
[DNLM: 1. Nervous Systems—physiology. 2. Behavior. WL 102 E78
1995]
QP355.2.E87 1995
612.8—dc20
DNLM/DLC
for Library of Congress 94-42332 CIP

Senior Medical Editor: John Dolan
Editorial Assistant: Debbie Feher
Executive Director of Production: Alan Winick
Manager of Production: John Williams
Production Assistants: Todd Miller, Jennifer Sinsavich
Manufacturing Buyer: Shirley Dahlgren
Cover Production: Michael J. Kelly
Art Production: Hudson River Studio

PRINTED IN THE UNITED STATES OF AMERICA

Contents in Brief

List of Authors vi
Contents vii
Box Features xiii
Preface xv
Acknowledgments xix
How to Use This Book xxi

Section I
Introduction 1

1 Brain and Behavior 5
2 Nerve Cells and Behavior 21

Section II
Cell Biology, Anatomy, and Development of the Nervous System 41

3 The Neuron 45
4 Neuronal Proteins 57
5 The Nervous System 71
6 Development of the Nervous System 89

Section III
Signaling Within Nerve Cells 111

7 Ion Channels 115
8 Membrane Potential 133
9 Local Signaling: Passive Electrical Properties of the Neuron 149
10 Propagated Signaling: The Action Potential 161

Section IV
Signaling Between Nerve Cells 179

11 An Introduction to Synaptic Transmission 183
12 Transmission at the Nerve–Muscle Synapse 197
13 Synaptic Integration 219
14 Modulation of Synaptic Transmission: Second-Messenger Systems 243
15 Transmitter Release 269
16 Neurotransmitters 293
17 A Clinical Example: Myasthenia Gravis 307

Section V
Cognitive Neural Science 317

18 From Nerve Cells to Cognition 321
19 Cognition and the Cortex 347

Section VI
Perception 365

20 The Sensory Systems 369
21 Construction of the Visual Image 387
22 Visual Processing by the Retina 407
23 Perception of Form and Motion 425
24 Color 453
25 Sensory Experience and the Formation of Visual Circuits 469

Section VII
Action 485

26 An Introduction to Movement 489
27 Muscles and Muscle Receptors 501
28 Spinal Reflexes 515
29 Voluntary Movement 529

Section VIII
Genes, Emotions, and Instincts 551

30 Genes and Behavior 555
31 Sex and the Brain 579
32 Emotional States 595
33 Motivation 613

Section IX
Language, Learning, and Memory 629

34 Language 633
35 Learning and Memory 651
36 Cellular Mechanisms of Learning and Memory 667

Figure Sources 695
Glossary 701
Index 717

List of Authors

Chapter	Author(s)
1	Eric Kandel
2	Eric Kandel
3	James Schwartz
4	James Schwartz
5	Tom Jessell
6	Tom Jessell
7	John Koester & Steven Siegelbaum
8	John Koester
9	John Koester & Steven Siegelbaum
10	John Koester
11	Eric Kandel & Steven Siegelbaum
12	Eric Kandel & Steven Siegelbaum
13	Eric Kandel
14	James Schwartz & Eric Kandel
15	Eric Kandel
16	James Schwartz
17	Lewis P. Rowland
18	Eric Kandel & Irving Kupfermann
19	Irving Kupfermann
20	Jack Martin & Tom Jessell
21	Eric Kandel
22	Marc Tessier-Lavigne
23	Eric Kandel & Carol Mason
24	Marc Tessier-Lavigne & Peter Gouras
25	Eric Kandel & Tom Jessell
26	Claude Ghez & Jim Gordon
27	Claude Ghez & Jim Gordon
28	Claude Ghez & Jim Gordon
29	Claude Ghez & Jim Gordon
30	Ralph Greenspan, Eric Kandel & Tom Jessell
31	Dennis Kelly & Tom Jessell
32	Eric Kandel & Irving Kupfermann
33	Irving Kupfermann & James Schwartz
34	Eric Kandel
35	Irving Kupfermann & Eric Kandel
36	Eric Kandel

Contents

Box Features xiii
Preface xv
Acknowledgments xix
How to Use This Book xxi

Section I
Introduction 1

1 Brain and Behavior 5
Two Alternative Views Describe the Relationship Between Brain and Behavior 6
Regions of the Brain Are Specialized for Different Functions 8
Language and Other Cognitive Functions Are Localized Within the Cerebral Cortex 9
Mental Processes Are Represented in the Brain by Their Elementary Operations 16

2 Nerve Cells and Behavior 21
The Nervous System Has Two Classes of Cells 23
Nerve Cells Are the Signaling Units for Behavioral Responses 29
Signaling Is Organized in the Same Way in All Nerve Cells 31
Functional Differences Among Nerve Cells Are Most Evident on the Molecular Level 39
Complexity of Interconnection Allows Relatively Similar Nerve Cells to Convey Unique Information 39

Section II
Cell Biology, Anatomy, and Development of the Nervous System 41

3 The Neuron 45
The Neurons that Mediate the Stretch Reflex Illustrate the Cytology of Nerve Cells 47
The Axons of Both Sensory and Motor Neurons Are Ensheathed in Myelin 51
A Major Function of the Neuron's Cell Body Is the Synthesis of Macromolecules 53
An Overall View 55

4 Neuronal Proteins 57
Messenger RNA Gives Rise to Three Classes of Proteins 58
Membranes and Secretory Proteins Are Actively Transported in the Neuron 62
Fibrillar Proteins of the Cytoskeleton Are Responsible for the Shape of Neurons 65
An Overall View 67

5 The Nervous System 71
The Nervous System Has Peripheral and Central Components 77
The Central Nervous System Consists of Seven Main Regions 77
The Cerebral Cortex Is Divided into Four Functional Lobes 81
Interaction of the Sensory, Motor, and Motivational Systems Is Essential for Even Simple Behavior 83
The Anatomical organization of Each Major Functional System Follows Four Principles 84
An Overall View 87

6 Development of the Nervous System 89
Neural Cell Identity Is Controlled by Cell Lineage and Inductive Interactions 91
Axonal Pathways Are Formed in Response to Local Guidance Cues 99
Synapse Formation at the Neuromuscular Junction Involves Inductive Interactions Between the Motor Neuron and the Muscle Cell 104
Survival of Neurons Is Regulated by Interactions with Their Targets 107
An Overall View 107

Section III
Signaling Within Nerve Cells 111

7 Ion Channels 115
Ion Channels Are Proteins that Span the Cell Membrane 116
Ion Channels Can Be Investigated by Both Functional and Structural Methods 118
Ion Channels in All Cells Share Several Characteristics 124
An Overall View 131

8 Membrane Potential 133

The Resting Membrane Potential Results from the Separation of Charge Across the Cell Membrane 134
The Resting Membrane Potential Is Determined by Resting Ion Channels 134
The Balance Ion that Gives Rise to the Resting Membrane Potential Is Abolished During the Action Potential 139
The Contribution of Different Ions to the Resting Membrane Potential Can Be Quantified by the Goldman Equation 140
The Functional Properties of the Neuron Can Be Represented as an Electrical Equivalent Circuit 140
An Overall View 144
Postscript: An Equation for the Resting Membrane Potential Can Be Derived from an Equivalent Circuit 144

9 Local Signaling: Passive Electrical Properties of the Neuron 149

Membrane Resistance Affects the Magnitude of Electrical Signals 150
Membrane Capacitance Prolongs the Time Course of Electrical Signals 150
Membrane and Axoplasmic Resistance Affect the Efficiency of Signal Conduction 153
Passive Membrane Properties and Axon Diameter Affect the Velocity of Action Potential Propagation 156
An Overall View 159

10 Propagated Signaling: The Action Potential 161

The Action Potential Is Generated by the Flow of Ions Through Voltage-Gated Channels 161
The Action Potential Can Be Reconstructed from the Known Electrical Properties of the Neuron 168
Variations in the Action Potential in Different Neurons Are Accounted for by Variations on the Basic Theme of the Hodgkin–Huxley Theory 170
Voltage-Gated Channels Have Characteristic Molecular Properties 171
An Overall View 177

Section IV
Signaling Between Nerve Cells 179

11 An Introduction to Synaptic Transmission 183

Synapses Are Either Electrical or Chemical 184
Electrical Synapses Provide Instantaneous Signal Transmission 185
Chemical Synapses Can Simplify Signals 189
An Overall View 194

12 Transmission at the Nerve–Muscle Synapse 197

The Neuromuscular Junction Permits Study of Directly Gated Transmission 198
Synaptic Excitation at the Nerve–Muscle Synapse Involves Transmitter-Gated Ion Channels 200
The Ion Channel at the Muscle End-Plate Is Permeable to Both Sodium and Potassium 203
Patch-Clamp Experiments Reveal the Nature of Current Flow Through Single Ion Channels 204
The Nicotinic Acetylcholine Receptor–Channel Is a Membrane-Spanning Protein 209
Transmitter-Gated Channels Differ from Voltage-Gated Channels 210
An Overall View 212
Postscript: The End-Plate Current Can Be Calculated from an Equivalent Circuit 213

13 Synaptic Integration 219

A Central Neuron Receives Both Excitatory and Inhibitory Signals 221
Excitatory and Inhibitory Signals Are Integrated into a Single Response by the Cell 222
Synapses onto a Single Central Neuron Are Grouped According to Function 224
Excitatory Synaptic Action Is Mediated by Transmitter-Gated Channels Selective for Sodium and Potassium 227
Inhibitory Synaptic Action Is Usually Mediated by Receptor–Channels Selective for Chloride 232
Excitatory and Inhibitory Synapses Have Distinctive Ultrastructures 234
Synaptic Receptors for Glutamate, GABA, and Glycine Are Transmembrane Proteins 236
Transmitter-Gated, Voltage-Gated, and Gap-Junction Ion Channels Share Certain Structural Features 237
The Signals Produced by Transmitter-Gated and Voltage-Gated Channels Have Features in Common 239
An Overall View 239

14 Modulation of Synaptic Transmission: Second-Messenger Systems 243

Second-Messenger Pathways Share a Common Molecular Logic 245
Second-Messenger Pathways Can Interact with One Another 255
Second Messengers Often Act Through Protein Phosphorylation to Open or Close Ion Channels 257
Second Messengers and G-Proteins Can Sometimes Act Directly on Ion Channels 258
Second Messengers Can Alter the Properties of Transmitter Receptors: Desensitization 260
Second Messengers Can Endow Synaptic Transmission with Long-Lasting Consequences 261
An Overall View 266

15 Transmitter Release 269
Transmitter Release Is Not Controlled by Sodium Influx or Potassium Efflux 269
Transmitter Release Is Triggered by Calcium Influx 271
Transmitter Is Released in Quantal Units 273
Each Quantum of Transmitter Is Stored in a Synaptic Vesicle 276
Transmitter Is Discharged from Synaptic Vesicles by Exocytosis at the Active Zone 276
The Docking of Synaptic Vesicles, Fusion, and Exocytosis Are Controlled by Calcium Influx 282
Synaptic Vesicles Are Recycled 285
The Number of Transmitter Vesicles Released by an Action Potential Is Modulated by Calcium Influx 286
An Overall View 290

16 Neurotransmitters 293
Chemical Messengers Must Fulfill Four Criteria to Be Considered Transmitters 293
There Are a Small Number of Small-Molecule Transmitter Substances 294
There Are Many Neuroactive Peptides 298
Peptides and Small-Molecule Transmitters Differ in Several Ways 301
Peptides and Small-Molecule Transmitters Can Coexist and Be Co-released 301
Removal of Transmitter from the Synaptic Cleft Terminates Synaptic Transmission 302
An Overall View 303

17 A Clinical Example: Myasthenia Gravis 307
Myasthenia Gravis Affects Transmission at the Nerve–Muscle Synapse 307
Antibodies to the Acetylcholine Receptor Cause the Physiological Abnormality 310
Myasthenia Gravis Is More Than One Disease 314
An Overall View 315

Section V
Cognitive Neural Science 317

18 From Nerve Cells to Cognition 321
The Major Goal of Cognitive Neural Science Is to Study Internal Representations of Mental Events 322
Cognitive Neural Science Relies on Five Main Approaches 323
The Brain Has an Orderly Representation of Personal Space 324
The Internal Representation of Personal Space Is Modifiable by Experience 329
The Internal Representation of Personal Space Can Be Studied at the Cellular Level: Each Central Neuron Has a Specific Receptor Field 335
Real As Well As Imagined and Remembered Space Is Represented in the Posterior Parietal Association Areas 339
An Overall View 345

19 Cognition and the Cortex 347
The Three Association Areas Are Involved in Different Cognitive Functions 348
The Frontal Association Areas Are Involved in Motor Strategies and Motor Planning 351
The Parietal Association Areas Are Involved in Higher Sensory Functions and Language 354
The Temporal Association Areas are Involved in Memory and Emotional Behavior 354
The Two Hemispheres Have Different Cognitive Capabilities 355
Cognitive Functions Can Now Be Simulated by Artificial Neural Networks that Employ Parallel Distributed Processing 359
An Overall View 362

Section VI
Perception 365

20 The Sensory Systems 369
Sensory Information Underlies Motor Control and Arousal As Well As Perception 371
Modality, Intensity, Duration, and Location Are the Major Attributes of Sensation 371
All Sensory Systems Have a Common Plan 372
Stimulus Information Is Encoded at the Entrance to the Nervous System 375
The Different Modalities Place Specific Demands on the Common Neural Architecture of Sensory Systems 383
An Overall View 385

21 Construction of the Visual Image 387
Visual Perception Is a Creative Process 388
Three Parallel Pathways Process Information for Depth and Form, Motion, and Color 393
Attention Focuses Visual Perception by Facilitating Coordination Between Separate Visual Pathways 401
The Analysis of Visual Attention May Provide Important Clues About Conscious Awareness 403
An Overall View 404

22 Visual Processing by the Retina 407
The Retina Contains the Eye's Receptive Sheet 408
Phototransduction Results from a Cascade of Biochemical Events in the Photoreceptors 410

Photoreceptors Slowly Adapt to Changes in Light Intensity 414
The Output of the Retina Is Conveyed by the Ganglion Cells 414
Signals from Photoreceptors to Ganglion Cells Are Relayed Through a Network of Interneurons 420
An Overall View 423

23 Perception of Form and Motion 425
The Retinal Image Is an Inversion of the Visual Field 425
The Retina Projects to the Lateral Geniculate Nucleus 428
Information About Contrast from the Retina Is Not Altered Significantly in the Lateral Geniculate Nucleus 431
Form Is Analyzed in the Primary Visual Cortex by Cells Whose Receptive Fields Have Linear Features 433
The Primary Visual Cortex Is Organized into Columns and Layers 440
Beyond the Primary Visual Cortex: The Representation of Faces and Other Complex Forms Occurs in the Inferotemporal Cortex 445
Motion in the Visual Field Is Analyzed by a Special Neural System 446
Visual Attention Can Now Be Studied on the Cellular Level 449
An Overall View 450

24 Color 453
Three Separate Cone Systems Respond to Different Parts of the Visible Spectrum 455
Color Discrimination Requires at Least Two Types of Photoreceptors with Different Spectral Sensitivities 456
Color Opponency, Simultaneous Color Contrast, and Color Constancy Are Key Features of Color Vision 459
Color Blindness Can Be Caused by Genetic Defects in Photoreceptors or by Retinal Disease 464
An Overall View 467

25 Sensory Experience and the Formation of Visual Circuits 469
The Development of Visual Perception Requires Sensory Experience 470
The Development of Ocular Dominance Columns Serves as a Model for Understanding the Fine Tuning of Visual Circuitry by Activity 471
Different Regions of the Brain Have Different Critical Periods of Development 482
There Is an Early Critical Period in the Development of Social Competence 482
An Overall View 483

Section VII
Action 485

26 An Introduction to Movement 489
The Psychophysics of Movement 490
The Motor Systems Generate Three Types of Movement 491
The Spinal Cord, Brain Stem, and Motor Cortex Represent Three Levels of Motor Control 492
The Spinal Cord Contains the Cell Bodies of Motor Neurons 493
The Brain Stem Modulates Motor Neurons and Interneurons in the Spinal Cord Through Two Systems 494
The Motor Cortex Acts on Spinal Motor Neurons Directly via the Corticospinal Tract and Indirectly Through Brain Stem Pathways 496
The Motor Areas of the Cerebral Cortex Are Organized Somatotopically 498
The Cerebellum and Basal Ganglia Control the Cortical and Brain Stem Motor Systems 499
An Overall View 499

27 Muscles and Their Receptors 501
A Motor Unit Consists of a Single Motor Neuron and the Muscle Fibers It Innervates 501
The Nervous System Grades the Force of Muscle Contraction in Two Ways 502
Muscles Contain Specialized Receptors that Sense Different Features of the State of the Muscle 506
The Central Nervous System Controls Sensitivity of the Muscle Spindles Through the Gamma Motor Neurons 510
An Overall View 512

28 Spinal Reflexes 515
The Stretch Reflex Is a Simple Model of a Stereotyped Reflex 516
Most Spinal Reflexes Are Mediated by Polysynaptic Circuits that Allow the Reflex to Be Modified 518
Muscle Action Around a Joint Is Coordinated by Inhibitory Interneurons 518
Complex Reflexes that Serve Protective and Postural Functions Are Initiated by Stimulation of the Skin 520
The Main Features of Walking Movements Are Controlled by the Spinal Cord 523
An Overall View 526

29 Voluntary Movement 529
Neurons in the Primary Motor Cortex Encode the Force and Direction of Voluntary Movements 530
Premotor Cortical Areas Prepare the Motor Systems for Movement 534

The Cerebellum Regulates Movement Indirectly 536
The Basal Ganglia Integrate Information from Diverse Areas of Cortex 544
An Overall View 549

Section VIII
Genes, Emotions, and Instincts 551

30 Genes and Behavior 555
Is There a Genetic Component to Human Behavior? 556
What Components of Behavior Are Inherited? 558
How Do Genes Organize Behavior? 565
An Overall View 576

31 Sex and the Brain 579
A Single Gene Switches the Initial Development of the Gonad from Female to Male 580
Gonadal Hormones from Both the Mothor and the Male Fetus Regulate Continued Development 580
Perinatal Hormones Impose A Permanent Sex-Specific Blueprint on the Developing Nervous System 582
The Brain Can Be Masculinized Not Only by Male Hormones But Also by Many Other Compounds 584
Sexually Differentiated Brains Have Different Physiological Properties and Behavioral Tendencies 586
A Wide Range of Behaviors Is Influenced by Sex Differences in the Organization of the Brain 590
An Overall View 592

32 Emotional States 595
A Theory of Emotion Must Explain the Relationship of Cognitive and Physiological States 596
The Hypothalamus Is a Critical Subcortical Structure in the Regulation of Emotion 597
The Search for Cortical and Subcortical Representation of Emotions Has Led to the Amygdala 607
An Overall View 611

33 Motivation 613
Motivation Is an Inferred Internal State Postulated to Explain the Variability of Behavioral Responses 614
Homeostatic Processes Such as Temperature Regulation, Feeding, and Thirst Correspond to Motivational States 614
Temperature Regulation Involves Integration of Autonomic, Endocrine, and Skeletomotor Responses 616
Feeding Behavior Is Regulated by a Great Variety of Mechanisms 618
Thirst is Regulated by Tissue Osmolality and Vascular Volume 623
Motivational States Can Be Regulated by Factors Other Than Tissue Needs 624
Intracranial Stimulation Can Simulate Motivational States and Reinforce Behavior 625
The Mesolimbic Dopamine Pathways Important for Reinforcement Are Also Acted on by Some Drugs of Abuse 625
An Overall View 627

Section IX
Language, Learning, and Memory 629

34 Language 633
Language Is Distinctive from All Other Forms of Communication 634
Animals Models of Human Language Have Been Largely Unsatisfactory 635
What Is the Origin of Human Language? 636
Is the Capability for Language Innate or Learned? 637
Aphasias Are Disorders of Language that Also Interfere with Other Cognitive Functions 639
Certain Affective Components of Language Are Impaired by Damage to the Right Hemisphere 644
Some Disorders of Reading and Writing Can Be Localized 644
An Overall View 648

35 Learning and Memory 651
Memory Functions Can Be Localized to Specific Regions of the Brain 651
Memory Is Not Unitary and Can Be Classified as Implicit or Explicit on the Basis of How Information Is Stored and Recalled 653
Implicit Forms of Learning Can Be Nonassociative or Associative 658
The Relationship Between Implicit and Explicit Forms of Memory in Learning 663
Certain Implicit Forms of Memory Involve the Amygdala and Cerebellum 663
The Neural Basis of Memory Can Be Summarized in Two Generalizations 664
An Overall View 666

36 Cellular Mechanisms of Learning and Memory 667

Simple Forms of Implicit Learning Lead to Changes in the Effectiveness of Synaptic Transmission 668
Long-Term Memory Requires the Synthesis of New Proteins and the Growth of New Synaptic Connections 671
Classical Conditioning Involves an Associative Enhancement of Presynaptic Facilitation That Is Dependent on Activity 677
Storage of Explicit Memory In Mammals Involves Long-Term Potentiation in the Hippocampus 680
Is There a Molecular Alphabet for Learning? 685
The Somatotopic Map in the Brain Is Modifiable by Experience 688
Neuronal Changes Associated with Learning Provide Insights into Psychiatric Disorders 690
An Overall View 692

Figure Sources 695
Glossary 701
Index 717

Box Features

Box 1-1	*The Central Nervous System*	10
Box 4-1	*Proteins of the Nerve Cell*	60
Box 4-2	*Dynamics of Polymerization*	68
Box 5-1	*Computerized Tomography*	72
Box 7-1	*Recording Current Flow from Single Ion Channels*	120
Box 8-1	*Recording the Membrane Potential*	135
Box 10-1	*Voltage-Clamp Technique*	164
Box 10-2	*Calculation of Membrane Conductances from Voltage-Clamp Data*	167
Box 12-1	*Reversal Potential of the Excitatory Postsynaptic Potential*	206
Box 14-1	*Regulation of Gene Expression*	263
Box 15-1	*Calculating the Probability of Transmitter Release*	277
Box 15-2	*Freeze-Fracture Technique*	278
Box 16-1	*Histochemical Detection of Chemical Messengers Within Neurons*	304
Box 21-1	*Movement Agnosia: A Case*	399
Box 21-2	*Prosopagnosia: A Case*	400
Box 22-1	*The Dark Current*	415
Box 22-2	*Calcium and Light Adaptation*	416
Box 22-3	*The Center–Surround Receptive Field of Bipolar Cells*	421
Box 24-1	*The Cone Pigments*	457
Box 24-2	*A Life Without Color Vision: The Case of the Colorblind Painter*	465
Box 30-1	*Genetic Systems for Studying Behavior: Their Pluses and Minuses*	556
Box 30-2	*Genetic Polymorphisms*	562
Box 30-3	*Manipulating Genes in Transgenic Flies and Mice*	566
Box 30-4	*Making Targeted Mutations in Mice*	568
Box 34-1	*The Common Aphasias*	640
Box 35-1	*The Transformation of Explicit Memories*	656
Box 35-2	*Measurements of Learning*	658
Box 36-1	*Single Genes, Long-Term Potentiation, and Spatial Learning*	686

Preface

The past decade has seen a great interest in the biology of mental processes. We predict that this interest will only increase. It therefore is likely that understanding the brain will be to the 21st century what the study of the gene has been to the 20th century and what understanding the cell was to the 19th century. As neural science is assuming a more central role within both biology and psychology, it is even now being taught in many undergraduate schools. In the future, an introduction to the biological basis of the mind will probably play a significant role in the core curriculum of undergraduate liberal arts institutions because neurobiology is a natural bridge between the humanities and the natural sciences.

In our larger and more detailed textbook, *Principles of Neural Science*, we outline the basic principles of brain and behavior for advanced undergraduate students, medical students, and graduate students interested in studying the biology of the brain. The increased interest in neural science among undergraduates has now encouraged us to write an introductory textbook based on the approach we used in *Principles of Neural Science*. The present book is designed to be a primary textbook for a variety of undergraduate courses, including those that require only an elementary knowledge of biology. By emphasizing the interrelationship of neurobiology and cognitive science, we have specifically sought to provide a textbook for introductory courses in the biology of behavior.

To facilitate the beginning student's use of the book, we have provided three additional teaching aids. First, a complete and detailed glossary of key terms in the book gives students a handy summary in one place of the fundamental concepts of nervous system function and behavior. Second, an interactive computer program allows the student to gain direct familiarity with electrical signaling in neurons. Finally, a workbook written by several outstanding teachers of neural science allows students to test their mastery of the material in the textbook.

When we began work on *Principles of Neural Science* in 1979, it was clear that the decade of the 1980s would see a greater integration at the molecular level of cellular neurobiology with the rest of biology. We encouraged this unification by emphasizing the molecular aspects of neural function and by drawing the outlines of a neural science that connects molecular studies with studies of the behavioral actions of the nervous system. Less heralded during the 1980s than the advances in understanding the molecular action of the nervous system was the progress in cognitive neuroscience. But nevertheless this progress has been extremely important because, in the long run, it promises to unify psychology and neurobiology.

The most challenging aspect of this unification centers on the highest cognitive functions. We therefore believe that the task for the next several decades is to begin to understand the biological basis of consciousness and the mental processes by which we perceive, act, learn, and remember. Are these processes localized to specific regions of the brain, or do they represent a collective and emergent property of the whole brain? If various mental processes can be localized, what rules relate the anatomy and physiology of a region to its specific function? Can these rules be understood better by examining the regions as a whole or by studying its individual nerve cells? How do genes contribute to behavior, and how is gene expression in nerve cells regulated by development and learning? How does experience alter the way we perceive subsequent events?

In the present textbook we emphasize the importance of neural science for cognitive psychology in two ways. First, we believe important and surprising inroads to the logic of cognitive processes will come from molecular neurobiology, from understanding genes and proteins. We therefore think it important to introduce the beginning student to the details of the molecular mechanisms of nerve cell signaling. Because behavior is determined by the precise pattern of interconnections between neurons, the study of how these connections form during development will tell us much about the rules that relate neuronal circuits to behavioral acts. Finally, the use of genetically modified animals opens up, for the first time, a molecular approach to cognition by examining how altering one gene at a time affects behavior: learning, memory, perception, and attention.

Second, in addition to molecular neuroscience other approaches will, however, be required to decipher the special computational power of the brain. We therefore also emphasize in the textbook behavioral and clinical approaches from systems and computational neuroscience, neurology, psychiatry, and neuropsychology. To give but one example, recent developments in brain imaging have made it possible for psychologists to study the internal representation of mental processes directly, often in living human subjects. As a result, higher mental functions need no longer be inferred from behavioral observations, and the study of cognition can be carried out noninvasively in human subjects and in parallel with comparable studies in experimental animals.

From this convergence of molecular neural science and cognitive psychology, we are gaining a new perspective on behavior. We have also seen that the concept of mentation does not suffer by discussing issues in terms of neural science. Contrary to some expectations, biological analysis is unlikely to diminish our fascination with mentation or to make mentation trivial by reduction. Rather, as we try to show in this text, neurobiology expands our vision by allowing us to perceive previously unanticipated relationships between biological and psychological phenomena.

Acknowledgments

Many colleagues read portions of the manuscript critically. We are particularly indebted to Ronald Calabrese of Emory University, one of the premier teachers of neural science to undergraduates, who read and critically commented upon the entire book. In addition, the following colleagues have made constructive comments on specific chapters: David Amaral; Elizabeth Bates; Michael Brownstein; Maxwell Cowan; Antonio Damasio; Hanna Damasio; Michael Davis; Howard Eichenbaum; Richard S. Frackowick; Christopher Frith; Michela Gallagher; Apostolos Georgopoulos; Amiram Grinvald; Miles Herkenham; Larry Katz; Joseph LeDoux; Keir Pearson; Michael Posner; Donald Reis; Paul Sawchenko; Arthur Shimamura; Gerald Smith; Larry Squire; Larry Swanson; and Semir Zeki. Lise Eliot, Ted Abel, and Mark Barad read and commented on the Glossary.

We are again indebted to Seta Izmirly who single-handedly, and with great thoughtfulness and intelligence, coordinated all aspects of the production of this book at Columbia. Howard Beckman edited the manuscript with his usual high standards for style of presentation and clarity of content, and brought to this book a level of enthusiasm that inspired all of us. We also were extremely fortunate to recruit for this book Sarah Mack, who assumed responsibility for the art program. Our colleague Jane Dodd assessed the art for scientific accuracy and, together with Sarah Mack, developed a consistent and didactic approach to the use of color in the figures. In these tasks they were ably assisted by Charles Lam. Finally, Andy Krawetz organized and oversaw the typing of the many versions of the manuscript, and together with Harriet Ayers typed the final draft.

How to Use This Book

This book is designed to be used as a comprehensive textbook for a variety of introductory college courses in neurobiology, physiology, psychology, and cognitive neural sciences. The book can be used for either a two-semester or one-semester course. A student manual is available to facilitate students' mastering of the material.

For a two-semester course, Chapters 1 through 17 should be covered in the first semester. These chapters describe the signaling properties of nerve cells. Chapters 18 through 36, which are concerned with systems and cognitive neural science, should be covered in the second semester.

For a one-semester course, we would recommend one of two approaches. For courses emphasizing the cellular biology of neurons, students should read Chapters 1 through 17; and Chapters 18, 25 and 36. For courses emphasizing behavior and cognitive psychology, the students should read Chapters 1 and 2, 5, 6, 10 and 11, and Chapters 18 through 36.

ESSENTIALS OF NEURAL SCIENCE AND BEHAVIOR

I

AN OVERALL VIEW

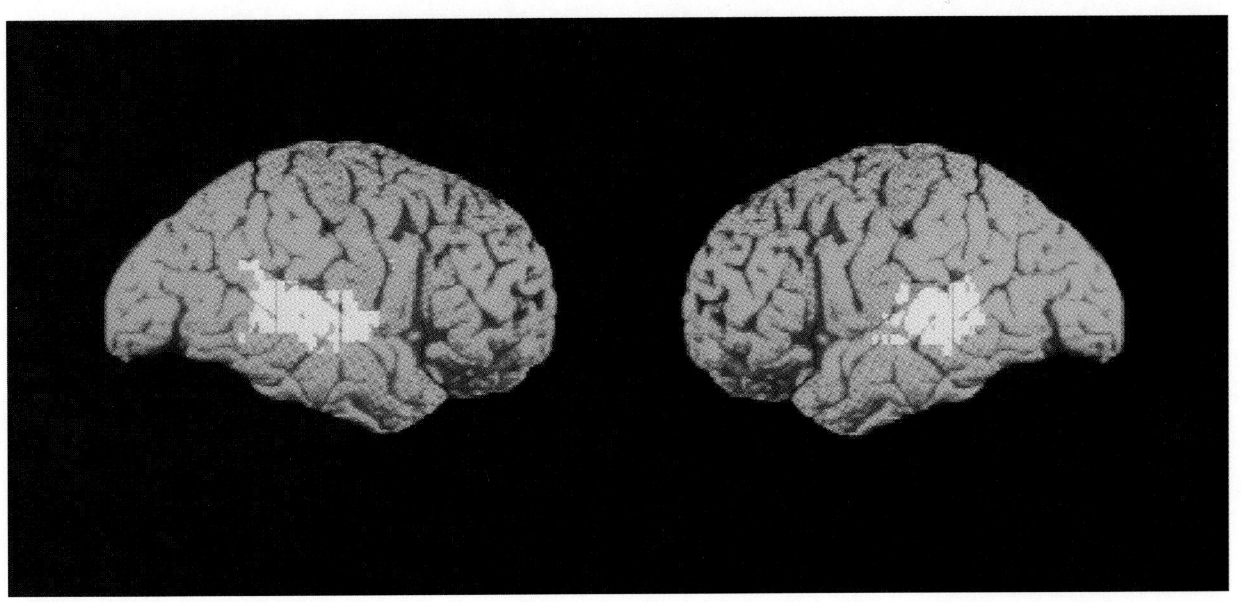

About the image on the preceding page:

A computer image of a living human brain while it listens to sentences. As the subject listens, the primary auditory and association cortices (including Wernicke's area) in both cerebral hemispheres are activated. This activation is detected by increases in blood flow that correlate significantly with the listening test. Blood flow is measured quantitatively using three-dimensional positron emission tomography (PET) after injection of water containing the unstable isotope ^{15}O. The areas of the brain activated are projected on computer images of the lateral aspects of the cortex obtained by magnetic resonance imaging (MRI). The right hemisphere is here depicted on the left and the left hemisphere on the right. (Courtesy of Drs. David Silbersweig and Emily Stern and colleagues at the MRC Cyclotron Unit, Hammersmith Hospital, London, and New York Hospital.)

About the hieroglyphs on the next page:

The ancient Egyptian hieroglyphics for the word "brain" (*'yś*) first appeared in a Seventeenth Century B.C. papyrus (the Edwin Smith Surgical Papyrus) describing the symptoms, diagnosis, and prognosis of two patients wounded in the head. This writing is the earliest known reference to the brain anywhere in the human record.

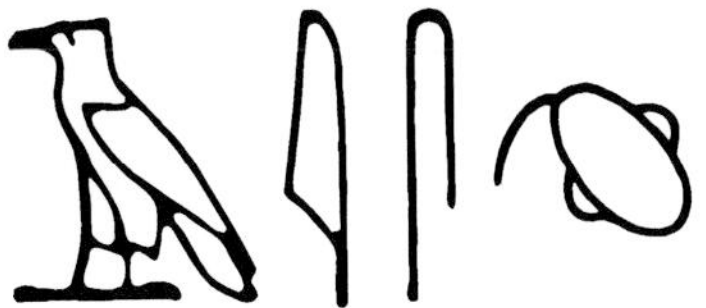

Introduction

> The microscope and telescope opened vast domains of unexpected scientific discovery. Now that new imaging methods can visualize the brain systems . . . a similar opportunity may be available for human cognition. . . .
>
> MICHAEL POSNER
> *Seeing the Mind*

The primary challenge of neural science is to understand how the brain produces the remarkable individuality of human action. This book illustrates how the neurobiological study of behavior spans the distance between molecules and mind—how the molecules responsible for the activities of nerve cells relate to the complexity of the mental processes.

The brain is a precise network of more than 100 billion discrete nerve cells, interconnected in systems that produce our perception of the external world, fix our attention, and control the machinery of action. Our first step in understanding the mind, therefore, is to understand how neurons become organized into signaling pathways and how individual nerve cells of the brain talk with one another by means of synaptic transmission. In later chapters we shall see how alterations in individual genes affect signaling between nerve cells, and how alterations in signaling affect behavior.

Neural science has emerged over the last century from studies of the nervous system in a variety of classical disciplines. Today, new techniques provide the means of directly linking the molecular dynamics of individual nerve cells with representations of perceptual and motor acts in the brain, and relating these internal mechanisms to observable behavior. New imaging techniques, for example, allow us to observe the human brain in action—to identify specific regions associated with thinking and feeling. In its ability to bridge molecular biology and cognitive studies, neural science makes it possible to begin to explore the biology of human potential, to understand what makes us what we are.

1

Brain and Behavior

Two Alternative Views Describe the Relationship Between Brain and Behavior

Regions of the Brain Are Specialized for Different Functions

Language and Other Cognitive Functions Are Localized Within the Cerebral Cortex

Mental Processes Are Represented in the Brain by Their Elementary Operations

Perhaps the last frontier of science—its ultimate challenge—is to understand the biological basis of consciousness and the mental processes by which we perceive, act, learn, and remember. In the last two decades a remarkable unity has emerged among the biological sciences, a unity that has resulted in the merger of neural science with the rest of cell and molecular biology. The next and even more challenging step in this progressive unification, which we outline in this book, is the merger of the study of behavior—the science of the mind—with neural science, the science of the brain.

The central tenet of this unification is that what we commonly call mind is a range of functions carried out by the brain. The actions of the brain underlie all behavior, not only relatively simple motor behaviors such as walking and eating, but all the complex cognitive actions that we associate with specifically human behavior, such as thinking, speaking, and creating works of art. According to this view, behavioral disorders that characterize psychiatric illness are disturbances of brain function.

The task of neural science is to provide explanations of behavior in terms of activities of the brain, to explain how millions of individual nerve cells in the brain operate to produce behavior and how, in turn, these cells are influenced by the environment, including the behavior of other people. The progress of neural science in explaining behavior is a major theme of this book.

Are mental processes localized to specific regions of the brain, or do they represent a collective and emergent property of the whole brain? If various mental processes can be localized to different brain regions, what rules relate the anatomy and physiology of a region to its specific function in perception, thought, or movement? Can these rules be understood better by examining the region as a whole or by studying its individual nerve cells? In this chapter we consider to what degree mental functions are located in specific regions of the brain. We also want to know to what degree any such localized mental processes can be understood in terms of the properties of specific nerve cells and their interconnections.

To answer these questions, we look at how modern neural science approaches one of the most elaborate cognitive behaviors—language. We focus on the cerebral cortex, the part of the brain that has developed most in recent primate evolution and is concerned with higher aspects of human behavior. Here we see how the brain is organized into large groups of neurons and how, by understanding the functions of groups of neurons, highly complex behaviors can be traced to specific regions of the brain. In the next chapter we consider how the nervous system works at the cellular level, using the example of a simple reflex behavior to illustrate how sensory signals are transformed into motor acts.

Two Alternative Views Describe the Relationship Between Brain and Behavior

The current view of nerve cells, the brain, and behavior has emerged over the last century from five major experimental disciplines: anatomy, embryology, physiology, pharmacology, and psychology.

Before the invention of the compound microscope in the eighteenth century, nervous tissue was thought to be glandular in function—an idea that can be traced to antiquity and Galen's proposal that nerves are ducts conveying fluid secreted by the brain and spinal cord to the periphery of the body. The microscope made clear the true cellular structure of nervous tissue. Nevertheless, nervous tissue did not become the subject of a precise science until the late 1800s, when the investigations of the Italian physician Camillo Golgi and the Spanish histologist Santiago Ramón y Cajal described in detail the structure of nerve cells.

Golgi developed a silver staining method that under the microscope revealed the entire structure of the neuron, including its cell body and its two major types of projections (or processes), the dendrites and the axon. Ramón y Cajal used this silver staining technique to label individual cells, thus showing that nervous tissue is not a continuous mass but a network of discrete cells. In the course of this work he developed some of the key conceptual insights and much of the early evidence for the *neuron doctrine*—the principle that individual neurons are the elementary signaling elements of the nervous system.

Additional experimental support for the neuron doctrine was provided by the American embryologist Ross Harrison, who devised tissue culture methods to show directly that the dendrites and the axon are continuous with the cell body and grow out from it. Following Ramón y Cajal's lead, Harrison showed that the tip of the axon gives rise to the growth cone, which leads the developing axon to its targets (muscle or other nerve cells).

Physiological investigation of the nervous system began in the late eighteenth century when the Italian physician and physicist Luigi Galvani discovered that living excitable muscle and nerve cells produce electricity. During the nineteenth century the foundations of electrophysiology were laid by three German physiologists—Emil DuBois-Reymond, Johannes Müller, and Hermann von Helmholtz—who discovered that the electrical activity of one nerve cell affects the activity of another cell in predictable ways.

The impact of pharmacology on our understanding of the nervous system and behavior started at the end of the nineteenth century, when Claude Bernard in France, Paul Ehrlich in Germany, and John Langley in England demonstrated that drugs interact with specific receptors on cells. This discovery became the basis of the all-important study of the chemical nature of communication between nerve cells.

The psychological investigation of behavior has its roots in the beginnings of Western science in classical Greek philosophy. Many key issues in the modern investigation of behavior, particularly in the area of perception, were formulated in the writings of René Descartes (died 1650), John Locke (died 1704), and David Hume (died 1776). In the mid-nineteenth century, Charles Darwin's investigations on evolution set the stage for the systematic observation of actions and behavior. This new approach gave rise to experimental psychology, the study of human and animal behavior under controlled conditions, and to ethology, the study of animal behavior in the wild.

Attempts to bring together biological and psychological concepts in the study of behavior occurred as early as the end of the eighteenth century when the German physician and neuroanatomist Franz Joseph Gall (Figure 1–1) proposed that discrete regions of the cerebral cortex control specific functions. Gall asserted that the brain does not act as a unitary organ but is divided into at least 35 organs (others were added later), each corresponding to a specific mental faculty. He thought that even the most abstract of human behaviors, such as generosity and secretiveness, occur in discrete areas of the brain. Gall further proposed that the center for each mental function increased in size as a result of use, much as the size of a muscle is increased by exercise. As each center grew, it was thought to cause the overlying skull to protrude, creating a pattern of bumps and ridges on the skull that indicated which regions of the brain were most developed (Figure 1–2). By correlating the personality of individuals with the bumps on their skulls, Gall sought to develop an anatomical basis for describing character traits. This anatomical personology Gall later called *phrenology*.

Gall's ideas were subjected to experimental analysis by Pierre Flourens in France in the late 1820s. Flourens attempted to isolate the contributions of different parts of the nervous system to behavior by removing from the brains of experimental animals the functional centers identified by Gall. From these experiments Flourens concluded that specific sites in the brain are not solely responsible for specific behaviors, but that all regions of the brain—especially the cerebral hemispheres of the forebrain—participate in every mental function. He proposed that any part of the cerebral hemisphere is able to perform all the functions of the hemisphere. Injury to a specific area of the cerebral hemisphere would therefore affect all higher functions equally.

In 1823 Flourens wrote: "All perceptions, all volitions occupy the same seat in these (cerebral) organs; the faculty of perceiving, of conceiving, of

Figure 1–1 Franz Joseph Gall (1757–1828).

A German physician and neuroanatomist, Gall practiced and lectured in Vienna. His studies convinced him that the brain is the organ of the mind, and that the cerebral cortex is not homogenous but contains distinctive centers that control specific mental functions. Gall was therefore the first proponent of *localization*. Because he was not an experimentalist, he misidentified the function of most parts of the cortex, relying on his belief that the location of a given mental faculty could be ascertained by examining the surface of the skulls of people well endowed with that function (*cranioscopy*). Because of his materialistic ideas about the mind, he was forced to leave Vienna in 1805. He was welcomed in Paris, where he continued his neuroanatomical studies and with his followers founded phrenology, an attempt to correlate personality with features of the skull. This bronze medal by J. J. Barre was struck to commemorate his death; on its reverse is written, "To the creator of brain physiology." (Courtesy of J. H. Schwartz.)

willing merely constitutes therefore a faculty which is essentially one." The rapid and fairly general acceptance of this belief (later called the *aggregate field* view of the brain) was based only partly on Flourens's experimental work. It also represented a reaction against a strict materialistic view of mind—the idea that mind is completely biological.

The aggregate field view prevailed until the middle of the nineteenth century, when it was first seriously challenged by the British neurologist J. Hughlings Jackson. In clinical studies of focal epilepsy, a disease characterized by convulsions that begin in one part of the body, Jackson showed that different motor and sensory processes are localized in different parts of the cerebral cortex. These studies were later elaborated systematically at the end of the nineteenth and the beginning of the twentieth century by the German neurologist Karl Wernicke, by the English physiologist Charles Sherrington, and by Ramón y Cajal into an opposing view of brain function called *cellular connectionism*. According to this view, individual neurons are the signaling units of the brain; they are generally arranged in functional groups and connect to one another in a precise fashion. Wernicke's work in particular showed that different behaviors are mediated by different brain regions that are interconnected by discrete neural pathways.

The history of the dispute between the proponents of the aggregate field and the cellular connection views of cortical function can best be illustrated by an analysis of how the brain manages language, the highest and certainly the most characteristic human mental function. Before we consider the relevant clinical and anatomical studies concerned with the localization of language, let us briefly survey the structure of the brain.

Regions of the Brain Are Specialized for Different Functions

The central nervous system, which is bilateral and essentially symmetrical, consists of seven main parts: the spinal cord, the medulla oblongata, the pons, the cerebellum, the midbrain, the diencephalon, and the cerebral hemispheres (Box 1–1). Imaging techniques developed in the last several years have made it possible to visualize these structures in the living human brain (Figure 1–3). Each of these brain regions has been shown, through a variety of experimental methods, to have specific functions. As a result, the idea that different regions are specialized for different functions is now accepted as one of the cornerstones of brain science.

One of the reasons this conclusion eluded investigators for so many years lies in another organizational principle of the nervous system known as *parallel processing*. As we shall see below, each major sensory, motor, and other integrative function is commonly subserved by more than one

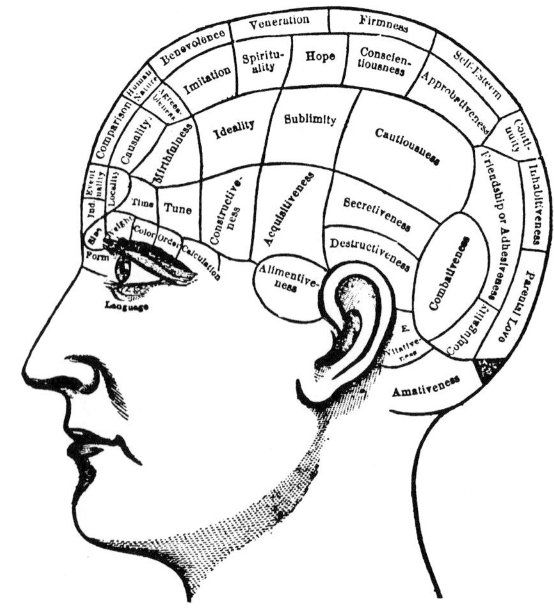

Figure 1–2 According to the nineteenth century doctrine of phrenology, complex traits such as combativeness and hope are carried out at specific sites in the brain, which expand as the traits develop. This enlargement of local areas of the brain was thought to produce characteristic bumps and ridges on the overlying skull, from which an individual's character could be determined. This map, taken from a drawing of the early 1800s, distinguishes 35 intellectual and emotional faculties in distinct areas of the skull and the cerebral cortex underneath.

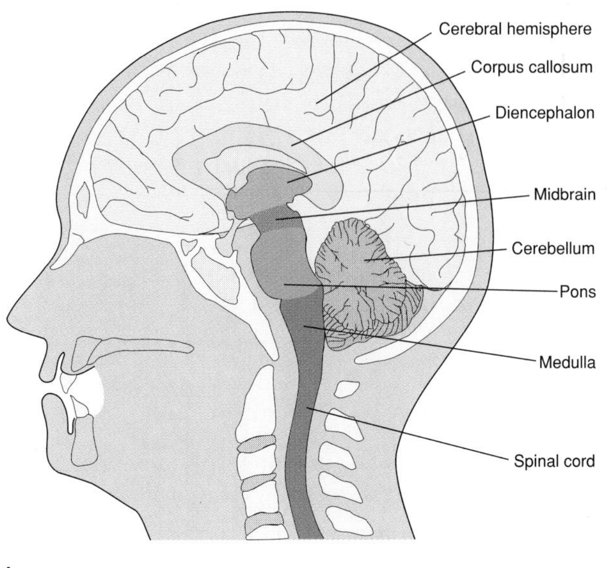

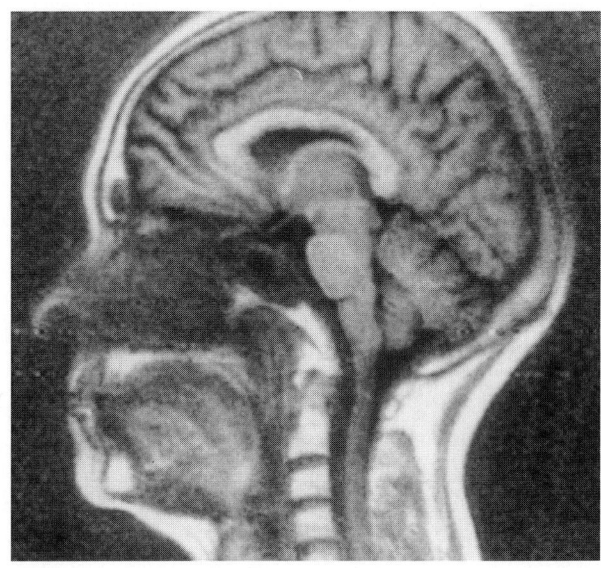

Figure 1–3 The main divisions of the brain are clearly visible when the brain is cut down the midline between the two hemispheres.

A. This schematic drawing shows the position of major structures of the brain in relation to external landmarks. Students of brain anatomy quickly learn to distinguish internal landmarks, such as the corpus callosum, a large bundle of nerve fibers that connects the left and right hemispheres.

B. The same section drawn in **A** appears here in a magnetic resonance image of the living brain.

neural pathway. When one region or pathway is damaged, others often are able to compensate partially for the loss, thereby obscuring the behavioral evidence for localization. However, the precision with which certain higher functions are actually localized emerges clearly from a consideration of language, to which we now turn.

Language and Other Cognitive Functions Are Localized Within the Cerebral Cortex

Brain functions relating to language are located primarily in the cerebral cortex, which wraps around and covers the cerebral hemispheres like the bark on a tree. In each of the brain's two hemispheres the overlying cortex is divided into four anatomically distinct lobes: frontal, parietal, occipital, and temporal (Box 1–1). Originally named for the specific bones of the skull that cover them, the lobes have specialized functions. The frontal lobe is largely concerned with the planning for future action and with the control of movement; the parietal lobe with tactile sensation and body image; the occipital lobe with vision; and the temporal lobe with hearing as well as aspects of learning, memory, and emotion. Each lobe has several characteristic convolutions or infoldings; this is an evolutionary strategy to increase valuable surface area within a restricted space. The crests of the convolutions are called *gyri*. The intervening grooves are called *sulci*. The more prominent gyri and sulci are the same in all individuals and have specific names. For example, the precentral gyrus (concerned with motor function) is separated from the postcentral gyrus (concerned with sensory function) by the central sulcus (Figure 1–5).

The cerebral cortex has two important organizational features. First, each hemisphere is concerned primarily with sensory and motor processes on the opposite or *contralateral* side of the body. Sensory information that enters the spinal cord from the left side of the body crosses over to the right side of the nervous system before being conveyed to the cerebral cortex. Similarly, the motor areas in one hemisphere of the brain exert control over the movements of the opposite half of the body. Second, although the hemispheres appear to be similar in humans, they are not completely symmetrical in structure—nor are they equivalent in function.

Much of what we know about the localization of language comes from the study of aphasia, a category of language disorders resulting from medical problems. Aphasia is found most often in patients who have suffered a stroke (the obstruction or rup-

BOX 1–1 The Central Nervous System

The central nervous system has seven main parts, illustrated in Figure 1–4.

1. The **spinal cord**, the most caudal part of the central nervous system, receives and processes sensory information from the skin, joints, and muscles of the limbs and trunk, and controls movement of the limbs and trunk. It is subdivided into cervical, thoracic, lumbar, and sacral regions. The spinal cord continues rostrally as the *brain stem*, which conveys information to and from the spinal cord and brain. The brain stem contains several distinct clusters of cell bodies, the *cranial nerve nuclei*. Some of these nuclei receive information from the skin and muscles of the head; others control motor output to muscles of the face, neck, and eyes. Still others are specialized for information from the special senses: for hearing, balance, and taste. The brain stem also regulates levels of arousal and awareness through the diffusely organized *reticular formation*. The brain stem consists of three parts: the medulla, pons, and midbrain.
2. The **medulla oblongata**, which lies directly above the spinal cord, includes several centers responsible for such vital autonomic functions as digestion, breathing, and the control of heart rate.
3. The **pons**, which lies above the medulla, conveys information about movement from the cerebral hemisphere to the cerebellum.
4. The **cerebellum** lies behind the pons and is connected to the brain stem by several major fiber tracts called *peduncles*. The cerebellum modulates the force and range of movement and is involved in the learning of motor skills.
5. The **midbrain**, which lies rostral to the pons, controls many sensory and motor functions, including eye movement and the coordination of visual and auditory reflexes.
6. The **diencephalon** lies rostral to the midbrain and contains two structures. One, the *thalamus*, processes most of the information reaching the cerebral cortex from the rest of the central nervous system. The other, the *hypothalamus*, regulates autonomic, endocrine, and visceral function.
7. The **cerebral hemispheres** consist of the *cerebral cortex* and three deep-lying structures: the *basal ganglia*, the *hippocampus*, and the *amygdaloid nucleus*. The basal ganglia participate in regulating motor performance; the hippocampus is involved with aspects of memory storage; and the amygdaloid nucleus coordinates autonomic and endocrine responses in conjunction with emotional states.

Overlying both hemispheres is the heavily wrinkled layer of the *cerebral cortex*, which is divided into four lobes: frontal, parietal, temporal, and occipital. The brain is also commonly divided into three broader regions: the *hindbrain* (the medulla, pons, and cerebellum), the *midbrain*, and the *forebrain* (diencephalon and cerebral hemispheres). The hindbrain and midbrain (excluding the cerebellum) constitute the brain stem.

The anatomy of the central nervous system is described in more detail in Chapter 5.

ture of a blood vessel supplying a portion of the cerebral hemisphere). Many of the important discoveries in the study of aphasia occurred in rapid succession during the last half of the nineteenth century. Taken together, these advances form one of the most exciting chapters in the study of human behavior, because they offered the first insight into the biological basis of a complex mental function.

The first advance occurred in 1861 when the French neurologist Pierre Paul Broca (Figure 1–6) described the case of a patient who could understand language but could not speak. The patient had no conventional motor problems with his

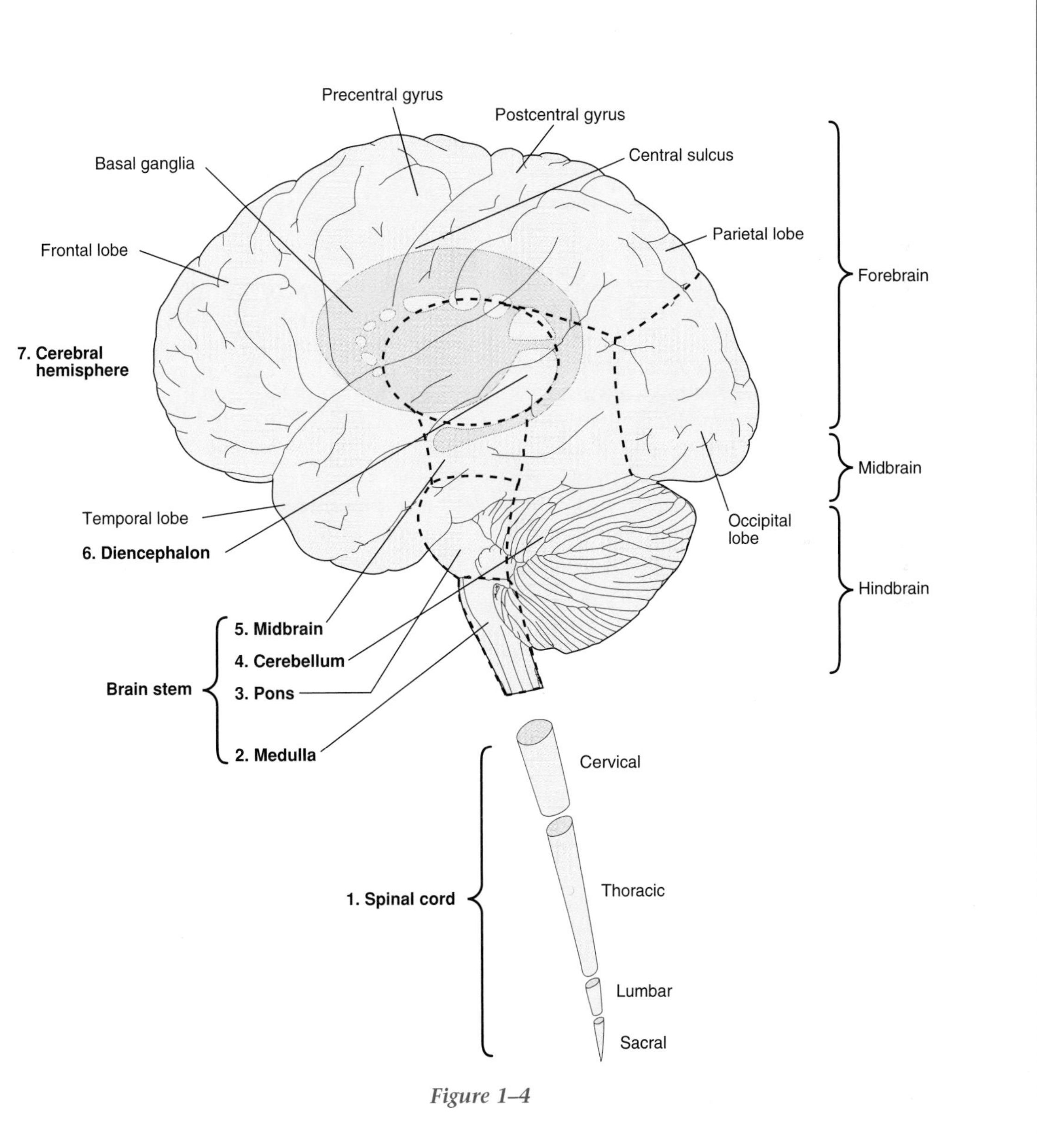

Figure 1–4

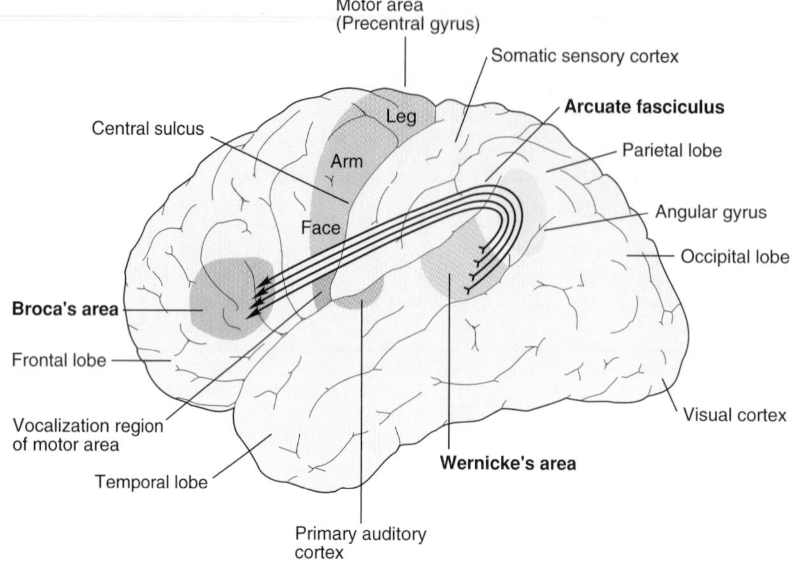

Figure 1–5 The major areas of the cerebral cortex involved in language are shown in this lateral view of the left hemisphere. *Wernicke's area* processes the auditory input for language and is important to the understanding of speech. It lies near the primary auditory cortex and the angular gyrus, which combines auditory input with information from other senses. *Broca's area* controls the production of speech. It lies near the region of the motor area that controls the mouth and tongue movements that form words. Wernicke's area communicates with Broca's area by a fiber tract, the *arcuate fasciculus*. (Adapted from Geschwind, 1979.)

tongue, mouth, or vocal cords that would affect speech. He could utter isolated words and sing a melody without difficulty but could not speak grammatically or in full sentences, nor could he express ideas in writing. Examination of the patient's brain after death revealed a lesion in the posterior region of the frontal lobe (an area now called *Broca's area*; Figure 1–5). Broca studied eight similar patients, all of whom showed lesions in this region. In each case the lesion was located in the left cerebral hemisphere. This discovery led Broca to announce, in 1864, one of the most famous principles of brain function: *"Nous parlons avec l'hémisphère gauche!"* ("We speak with the left hemisphere!")

Broca's work stimulated a search for the cortical sites of other specific behavioral functions—a search that was soon rewarded. In 1870, in Germany, the physiologist Gustav Fritsch and the psychiatrist Eduard Hitzig galvanized the scientific community with their discovery that electrical stimulation of certain regions of a dog's brain produced characteristic movements of the limbs. They found that individual movements are controlled by small, quite discrete regions of the cortex; moreover, these movements of a limb were produced by stimulating the precentral gyrus in the contralateral motor cortex (Figure 1–4). Thus, in human beings the right hand, commonly used for writing and skilled movements, is controlled by the same left hemisphere that controls speech. In most people, therefore, the left hemisphere is regarded as being *dominant*.

The next step was taken in 1876 by the German neurologist Carl Wernicke (Figure 1–7). At the age of 26 Wernicke published a now classic paper, "The Symptom Complex of Aphasia: A Psychological Study on an Anatomical Basis." In this paper he described a new type of aphasia, involving an impairment of comprehension rather than execution (a *receptive* as opposed to an *expressive* malfunction). Whereas Broca's patients could understand but not speak, Wernicke's patient could speak but not understand language—not even his own words. Wernicke found that this new type of aphasia was produced by lesions in a site different from that described by Broca: in the posterior part of the temporal lobe where it joins the parietal and occipital lobes (Figure 1–5).

In addition to making this discovery, Wernicke proposed a theory of language that attempted to reconcile and extend the two existing theories of brain function. Phrenologists argued that the cortex is a mosaic of specific functions and that even abstract mental attributes are located in and represented by single, functionally specific cortical areas. The opposing school of thought, viewing the brain as an aggregate field, argued that mental functions are not localized to specific brain regions but that each function is represented diffusely throughout the cortex.

Based on his discoveries and those of Broca, Fritsch, and Hitzig, Wernicke proposed that only the most basic mental functions, those concerned with simple perceptual and motor activities, are localized to single areas of the cortex, and that more complex intellectual functions result from interconnections between several functional sites. In placing the principle of localized function within a connectionist framework, Wernicke appreciated that different components of a single behavior are processed in different regions of the brain. He thus advanced the first evidence for the idea of *distributed processing,* which is now central to our understanding of brain function.

Wernicke postulated that language involves separate motor and sensory programs, each governed by separate cortical regions. He proposed that the motor program, governing the mouth movements involved in speech, is located in Broca's area, which is suitably situated immediately in front of the motor area that controls the mouth, tongue, palate, and vocal cords (Figure 1–5). The sensory program, governing word perception, was attributed to the temporal lobe area that he had discovered (now called Wernicke's area). This area is also suitably placed, being surrounded by the auditory cortex as well as by areas that integrate auditory, visual, and somatic sensation into complex perceptions (areas known as *association cortex*).

Thus, Wernicke formulated a coherent model for the organization of language that, with certain modifications we shall learn about later, is still useful today. According to this model, the initial auditory or visual perceptions of language are formed in separate sensory areas of the cortex specialized

Figure 1–6 Pierre Paul Broca (1824–1880).

Broca was educated in Paris as a medical doctor and was an intern at the Hotel-Dieu during the revolution of 1848. He became a surgeon, winning a prize from the French Academy of Medicine for his cytological studies on cancer. He next examined several infectious diseases before turning to the new science of human anthropology. As professor of both surgery and anthropology in Paris, he was the first to describe the skulls of Cro-Magnon and Aurignacian (paleolithic) man using craniometry with statistical standardization. Broca is best remembered for his presentation, beginning in 1861, of a set of cases of expressive aphasia. On autopsy, the brains of these patients showed a small lesion in the left frontal cortex, now called Broca's area. The portrait in oil shows Broca at about the time he localized the area for speech; it hangs in the town hall of Sainte-Foy-la-Grande, his birthplace.

Figure 1–7 Carl Wernicke (1848–1905).

A German neurologist and psychologist, Wernicke trained in Vienna at Theodor Meynert's Institute of Cerebral Neurology. He was 26 years old when he published his monograph entitled *The Symptom Complex of Aphasia: A Psychological Study on an Anatomical Basis*, in which he described a patient who could speak but not understand. Profoundly influenced by Meynert's division of the cortex into sensory and motor areas that are associated by nerve tracks, he called this type of aphasia *sensory*, in contrast to Broca's, which he termed *motor* aphasia. Wernicke's interest extended beyond merely describing new forms of aphasia. He proposed an ambitious model of how the brain processes language, a model that is still used today, although with some modifications.

for auditory or visual information. The neural representations of these perceptions are then conveyed to an association area of cortex specialized for both visual and auditory information (the angular gyrus). Here, spoken or written words are transformed into a common neural representation, a code shared by both speech and writing. From the angular gyrus this code is conveyed to Wernicke's area, where it is recognized as language and associated with meaning. Without that association, the ability to comprehend language is lost. The common neural code is then relayed from Wernicke's to Broca's area, where it is transformed from a sensory (auditory or visual) representation into a motor representation that can lead to spoken or written language.

When the last-stage transformation from sensory to motor representation cannot take place, the ability to express language (either as spoken words or in writing) is lost. With this reasoning, Wernicke predicted a new type of aphasia, later to be demonstrated clinically. This form of aphasia is produced by a type of lesion different from those in Broca's and Wernicke's aphasias: the receptive and motor speech zones are spared, but the fiber pathways that connect them are destroyed (Figure 1–5). The resulting disconnection syndrome, now called *conduction aphasia*, is characterized by an incorrect use of words (paraphasia). Patients with conduction aphasia can understand words that are heard and seen. They also have no motor difficulties in speech. They cannot, however, speak correctly; they omit parts of some words or substitute incorrect sounds. Painfully aware of their own errors, they are unable to correct them.

Inspired in part by Wernicke, a new school of cortical localization arose in Germany at the beginning of the twentieth century led by the anatomist Korbinian Brodmann. This school sought to distinguish different functional areas of the cerebral cortex based on the structure of cells and the characteristic arrangement of these cells into layers. Using this *cytoarchitectonic* method, Brodmann distinguished 52 functionally distinct areas in the human cerebral cortex (Figure 1–8).

Thus, by the beginning of the twentieth century there was compelling functional and anatomical evidence for many discrete areas in the cortex, some of which could be assigned a specific role in certain behaviors. Nevertheless, the aggregate field view of the brain, not cellular connectionism, dominated experimental thinking and clinical practice during the first half of this century. This surprising state of affairs owed much to the arguments of prominent neural scientists, including such major figures as the British neurologist Henry Head, the German neuropsychologist Kurt Goldstein, the Russian behavioral physiologist Ivan Pavlov, and the Americans Jacques Loeb and Karl Lashley, who continued to advocate the aggregate field view.

Most influential of this group was Lashley, Professor of Psychology at Harvard. Lashley was deeply skeptical of the cortical subdivisions determined by the cytoarchitectonic approach. "The 'ideal' architectonic map is nearly worthless," Lashley wrote. "The areal subdivisions are in large part anatomically meaningless, and . . . misleading as to the presumptive functional divisions of the cortex." Lashley's skepticism was reinforced by his attempts, in the tradition of Flourens's work, to find a specific locus of learning by studying the effects of various brain lesions on the ability of rats to learn to run a maze. Instead of finding a specific learning center, Lashley found that the severity of the learning defect produced by brain lesions seemed to depend on the extent of the damage, not on its precise location. This observation and his disappointment with cortical cytoarchitectonics led Lashley—and, after him, many other psychologists—to conclude that learning and other mental functions have no special locus in the brain and consequently cannot be related to specific collections of neurons. Lashley therefore reformulated the aggregate field view in a theory of brain function called *mass action*, which further belittled the importance of individual neurons, of specific neuronal connections, and of discrete, functionally specific regions of the brain. According to this view, brain mass, not its neuronal components, is important to brain function.

Applying this logic to aphasia, Head and Goldstein argued that disorders of language cannot be attributed to lesions at specific sites, but could result from injury to almost any cortical area. They asserted that cortical damage, regardless of site, caused the patient to regress from a rich, abstract language to the impoverished language that is characteristic of aphasia.

The experimental findings of Lashley and Head have gradually been reinterpreted. A variety of studies have demonstrated that maze learning, the task used by Lashley, is unsuitable for studying localization of function because it involves so

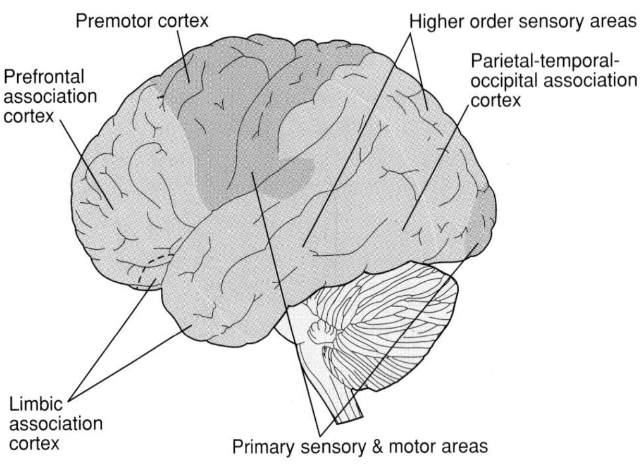

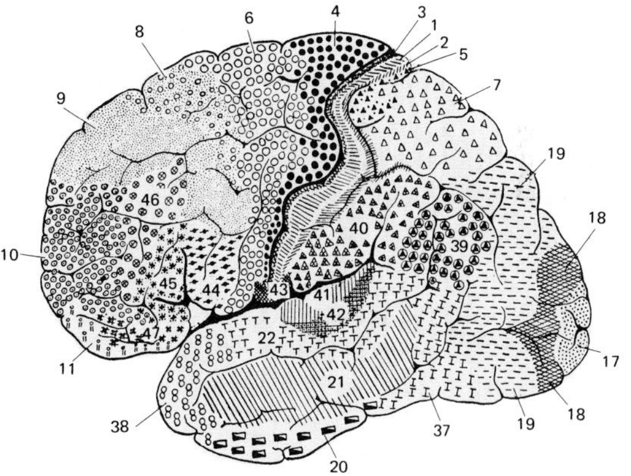

Figure 1–8 A schematic drawing of the lateral surface of the human brain (**top**) shows the major regions of the primary sensory and motor cortices as well as the higher-order motor and sensory cortices, and the three association cortices. This scheme is a simplified version based on the work of Korbinian Brodmann (**bottom**), who in the early part of the 20th century divided the cerebral cortex into 52 discrete areas on the basis of distinctive nerve-cell structure and characteristic arrangements of cell layers.

Several of the areas numbered by Brodmann have been found to controll a specific brain function. For instance, three distinct areas (Brodmann's areas 3, 1, and 2) make up the primary somatosensory cortex concerned with perceiving bodily sensation. Area 4 is the motor cortex responsible for the nerve signals that cause voluntary movement. Area 17 is the primary visual cortex, which receives signals from the eyes and relays them to other areas for further processing. Areas 41 and 42 house the primary auditory cortex. Areas that are not visible from the outer surface of the cortex are not shown in the drawing.

many complex motor and sensory capabilities. Deprived of one sensory capability (such as vision), an animal can still learn with another (for instance, by following tactile or olfactory cues).

In addition, the evidence for localization of function has been greatly strengthened. Beginning in the late 1930s, Edgar Adrian in England and Wade Marshall and Philip Bard in the United States discovered that tactile stimuli elicit electrical activity in discrete regions of the cerebral cortex. Shortly after, Jerzy Rose and Clinton Woolsey, and others after them, rigorously reexamined the concept of the architectonic area. Together, these studies established that cortical areas could be defined unambiguously according to several independent criteria, including cell type and cell layering, input and output connections, and—most important—physiological function. As we shall see in later chapters, recent studies suggest that regional specialization is a key principle of cortical organization and that the brain is divided into even more functional regions than Brodmann had identified.

By combining studies of brain localization with progressively more sophisticated observations of behavior, it has been possible to learn a great deal about the localization of mental functions in the brain. For example, in the late 1950s Wilder Penfield used small electrodes to stimulate the cortex of awake patients during brain surgery for epilepsy (carried out under local anesthesia). Penfield was testing the cortex specifically for areas that produce disorders of language, to ensure that the epilepsy surgery would not damage the patient's communication skills. On the basis of verbal reports from his conscious patients, he confirmed directly, in the living brain, the language areas that Broca and Wernicke had described.

Until recently, almost everything we knew about the anatomical organization of language came from clinical studies of patients who had suffered lesions of the brain. These studies have now been extended to normal individuals by Michael Posner and Marcus Raichle and their colleagues using positron emission tomography (PET) scanning. PET is a noninvasive imaging technique for visualizing the local changes in cerebral blood flow and metabolism that accompany mental activities, such as reading, speaking, and thinking. Posner and his colleagues discovered that the neural inputs for language production and understanding are processed by more than one pathway.

Recall that according to Wernicke both visual and auditory information are transformed into a shared auditory representation of language. This information is then conveyed to Wernicke's area, where it becomes associated with meaning before being transformed in Broca's area into output as written or spoken language. Posner and his colleagues asked: Is the neural representation of a word that is read also carried along auditory pathways before it can be associated with a meaning? Or can visual information be transferred directly to Broca's area? Using PET imaging, they determined how individual words are coded in the brain when the words are read or heard. They found that when words are heard, Wernicke's area becomes active, but when words are seen but not heard or spoken, there is no activation of Wernicke's area (Figure 1–9). The visual information from the occipital cortex appears to be conveyed directly to Broca's area without first being transformed into an auditory representation in the posterior temporal cortex. From these findings Posner and his colleagues concluded that different brain pathways and sensory codes are used in the perception of words presented visually or orally. Moreover, they proposed that these pathways have independent access to higher-order regions concerned with the assignment of meaning and the expression of language.

Not only are active reading and listening processed separately, but also when a subject merely thinks about the *meaning* of a word, without recourse to other inputs, a still different area in the left frontal cortex becomes active (Figure 1–9D). Thus, the processing of language is both *serial* and *parallel*. As we shall see later, a similar conclusion has also been reached from studies of behaviors other than language. These studies demonstrate that information processing requires that individual cortical areas be appropriately interconnected and that they respond to, and therefore code for, only certain aspects of specific sensory stimuli or motor movement and not for others.

Mental Processes Are Represented in the Brain by Their Elementary Operations

Why has the evidence for localization, which seems so obvious and compelling in retrospect, been rejected so often in the past?

First, the phrenologists introduced the idea of localization in an extreme form and without adequate evidence. They thought of each region of the

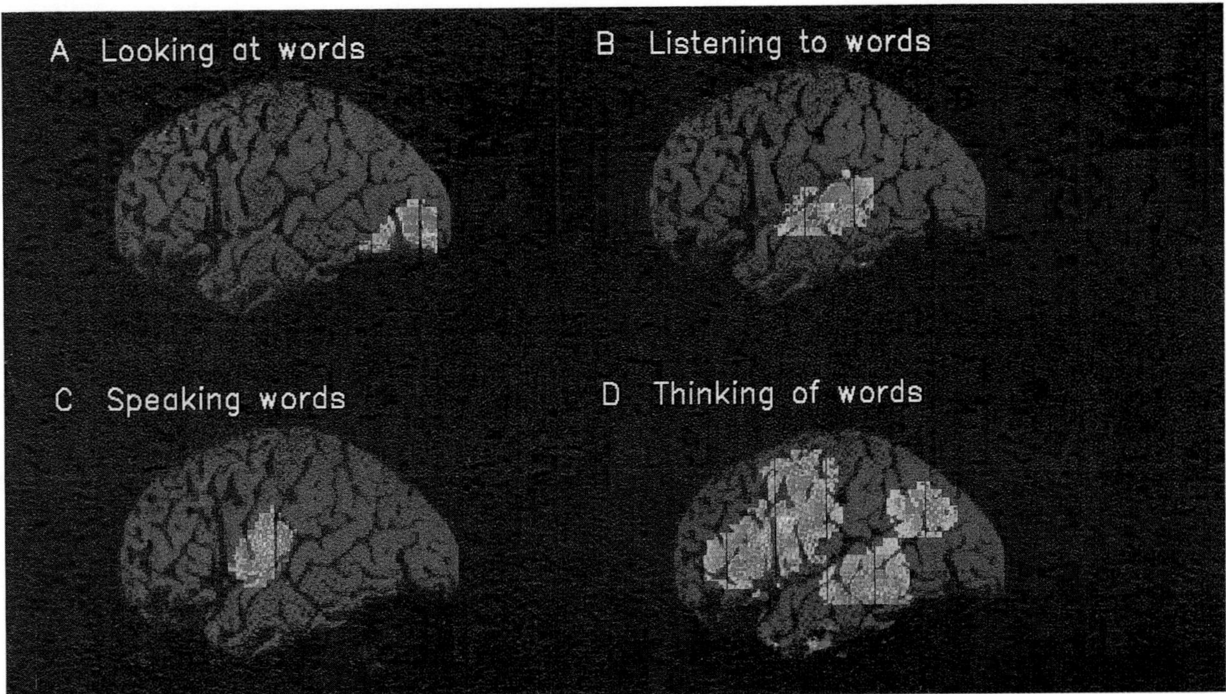

Figure 1–9 Specific regions of the cortex involved in the recognition of a spoken or written word can be identified with PET scanning. Each of the four images of the human brain shown here (lateral views of the left hemisphere) actually represents the averaged brain activity of several normal subjects. In PET images white represents the areas of highest activity, yellow the areas of quite high activity, and blue the areas of minimal activity. The "input" component of language (reading or hearing a word) activates the regions of the brain shown in A and B. The "output" component (speech or thought) activates the regions shown in C and D. (Courtesy of Cathy Price, MRC Cyclotron Unit, Hammersmith Hospital, London, England.)

A. The reading of a single word produces a response both in the primary visual cortex and in the visual association cortex. (see Figures 1–5 and 1–8)

B. Hearing a word activates an entirely different set of areas in the temporal cortex and at the junction of the temporal-parietal cortex. This shows that the brain uses discrete auditory and visual pathways for processing language, not a single common pathway. (To control for irrelevant differences, the same words were used in both the reading and listening tests.)

C. Speaking a word activates the supplementary motor area of the medial frontal cortex. Subjects were asked to repeat a word presented either through earphones or on a screen. Broca's area is activated whether the word is presented orally or visually. Thus, both visual and auditory pathways converge on Broca's area, the common site for the motor output of speech.

D. Thinking, such as analyzing the meaning of a word, activates the frontal cortex. To identify the regions active during thinking, the subjects were asked to respond to the word "brain" with an appropriate verb (for example, "to think"). Broca's and Wernicke's areas are also involved.

cerebral cortex as an independent mental organ dedicated to a distinct, complex mental function (much as the pancreas and the liver are independent digestive organs). The subsequent rebuttal by Flourens and the ensuing dialectic between proponents of the aggregate field view (against localization) and the cellular connectionists (for localization) were responses to a theory of localization that, although correct in a general sense, was extreme in principle and wrong in detail. The concept of localization that ultimately emerged, and that has prevailed, is much more complex than Gall (or even Wernicke) had envisioned.

Discrete local regions in the brain are not responsible for complex faculties of mind, but rather perform *elementary operations*. More elaborate faculties are made possible by the serial and parallel interconnections of several brain regions. As a result, damage to a single area need not lead to the disappearance of an entire faculty as many earlier neurologists had predicted. Even if the function initially disappears, it may in time partially return because undamaged parts of the brain can reorganize to some extent to perform the lost function. Thus, mental processes are not usually represented by a series of links in a single chain, for in such an arrangement the entire process is disrupted when one link is disrupted. Rather, mental processes are composed of several components represented by many neural pathways, like a variety of railroad lines that all feed into the same terminal. The malfunction of a single spot on one pathway affects the information carried by that pathway, but this need not interfere permanently with the performance of the system as a whole. The remaining parts of the system can modify their performance and accommodate extra traffic following the breakdown of a line.

Models of localized function were slow to be accepted because there was, and to some extent still is, great difficulty in demonstrating which components of a mental operation are represented by a particular pathway or region of the brain. There also is the additional difficulty of analyzing mental operations into their component parts. Only during the last decade, with the convergence of modern cognitive psychology and the brain sciences, have we begun to appreciate that *all* mental functions are divisible into subfunctions. We experience mental processes as instantaneous, unified operations. Intuitively, therefore, we think of individual mental processes—perceiving, thinking, learning, remembering—as continuous and indivisible. Actually these processes are composed of several independent information-processing components, and even the simplest cognitive task requires the coordination of several distinct brain areas.

To illustrate this point, consider how we store and recall the representation of objects and people —or even of the simplest event in our environment. Our impression is that we store each piece of knowledge—each object or fact about the world— as one unified representation that can be recalled by sensory stimuli or even by the imagination alone. For example, we feel that our knowledge about our grandmother is stored in one unified representation as grandmother, a representation that is equally accessible to us whether we see this person, hear her voice, or simply think about her. Elizabeth Warrington and her colleagues have found, however, that this belief is not supported by the facts. Knowledge is not stored as general representations, but is subdivided into distinct categories. Accordingly, selected lesions in the association areas of the left temporal lobe can lead to the loss of one specific category of knowledge—to a loss of knowledge of living things, especially people, without loss of knowledge of inanimate objects. Moreover, these categories are further subdivided by sensory modality. Thus, a small lesion in the left temporal lobe can destroy recognition of the names of living things without affecting visual recognition.

The most astonishing example of the combinatory structure of mental processes is the finding that our very sense of ourselves as a *self*—a coherent being—depends on neural connections between distinct operations carried out independently in the two cerebral hemispheres. In some epileptic patients the corpus callosum, a fiber tract connecting the two cerebral hemispheres, is severed to treat severe epileptic seizures. As a result of this surgical separation, each hemisphere carries an independent awareness of the self. For example, each hemisphere responds to tactile stimuli applied to the opposite (contralateral) hand, but not to stimuli applied to the same-side (ipsilateral) hand. Thus, when identical objects are placed in both hands of such patients, the object in the left hand cannot be compared with the object in the right hand, because that object can be identified only by the left hemisphere, which is no longer in communication with the right hemisphere. Even more dramatic is the demonstration that in most of these patients the right hemisphere cannot understand language that is well understood by the isolated left hemisphere. As a result, opposing commands can be given selectively to each hemisphere!

As we may suspect from these examples, perhaps the primary reason it has taken so long to appreciate that mental activities are localized within the brain is that we are dealing here with some of the deepest riddles in biology: the neural representation of consciousness and self awareness. It is important to appreciate that we have only begun to understand how complex behavior is represented in the brain. To study the relationship between a mental process and specific regions of the brain, we must be able to identify the components and properties of the behavior that we are attempting to explain. Yet, of all behaviors, higher mental processes are the most difficult to describe and measure objectively. In addition, the brain is immensely complex anatomically, and the structure and interconnections of many of its parts are still not fully understood. To analyze how a specific mental activity is represented we need to discern which aspects of a mental activity are represented in which regions of the brain and how they are represented.

Today, the application of brain imaging techniques in cognitive psychology is providing important information about the regions of the brain involved in specific complex behaviors, and how these behaviors can be broken down into simpler mental operations in specific interconnected brain regions. Such results have generated new excitement in neural science today, an excitement based on the conviction that we at last have the proper conceptual and methodological tools—cognitive psychology, brain imaging techniques, and new anatomical methods—to explore the organ of the mind. With these tools and this conviction come the optimism that the principles underlying the biology of mental function will in time be understood.

Selected Readings

Churchland, P. S., and Sejnowski, T. J. 1992. The Computational Brain. Cambridge, Mass.: MIT Press.

Cooter, R. 1984. The Cultural Meaning of Popular Science: Phrenology and the Organization of Consent in Nineteenth-Century Britain. Cambridge, England: Cambridge University Press.

Dennett, D. C. 1991. Consciousness Explained. Boston: Little, Brown.

Ferrier, D. 1890. The Croonian Lectures on Cerebral Localisation. London: Smith, Elder.

Geschwind, N. 1974. Selected Papers on Language and the Brain. Dordrecht, Holland: Reidel.

Harrington, A. 1987. Medicine, Mind, and the Double Brain: A Study in Nineteenth-Century Thought. Princeton, N.J.: Princeton University Press.

Harrison, R. G. 1935. On the origin and development of the nervous system studied by the methods of experimental embryology. Proc. R. Soc. Lond. [Biol.] 118:155–196.

Jackson, J. H. 1884. The Croonian Lectures on Evolution and Dissolution of the Nervous System. Br. Med. J. 1:591–593; 660–663; 703–707.

Kosslyn, S. M. 1988. Aspects of a cognitive neuroscience of mental imagery. Science 240:1621–1626.

Marshall, J. C. 1988. Cognitive neurophysiology: The lifeblood of language. Nature 331:560–561.

Marshall, J. C. 1988. Cognitive neuropsychology: Sensation and semantics. Nature 334:378.

Posner, M. I., and Raichle, M. E. 1994. Images of Mind. New York: Scientific American Library.

Shepherd, G. M. 1991. Foundations of the Neuron Doctrine. New York: Oxford University Press.

Sperry, R. W. 1968. Mental unity following surgical disconnection of the cerebral hemispheres. Harvey Lect. 62:293–323.

Young, R. M. 1970. Mind, Brain and Adaptation in the Nineteenth Century. Oxford: Clarendon Press.

2

Nerve Cells and Behavior

The Nervous System Has Two Classes of Cells

Nerve Cells

Glial Cells

Nerve Cells Are the Signaling Units for Behavioral Responses

Signaling Is Organized in the Same Way in All Nerve Cells

The Input Component Produces Graded Local Signals

The Trigger Component Makes the Decision to Generate an Action Potential

The Conductile Component Propagates an All-or-None Action Potential

The Output Component Releases Transmitter

The Signal Is Transformed as It Passes from One Component to the Next

Functional Differences Among Nerve Cells Are Most Evident on the Molecular Level

Complexity of Interconnection Allows Relatively Similar Nerve Cells to Convey Unique Information

All animals, including humans, obtain information about their environment through a variety of sensory receptors. The information from these receptors is transformed by the brain into perceptions or into commands for movement. These remarkable tasks are accomplished using only nerve cells and the connections between them.

The basic units of the brain, the nerve cells, are quite simple. The brain is able to produce highly complex behavior because it contains an extraordinarily large number of nerve cells—about 100 billion—that communicate with each other through specific interconnections. Despite the large number, nerve cells share many common features. A key discovery in the understanding of the brain is that the potential for complex behavior does not depend so much on the variety of nerve cells as on the number of nerve cells and their precise connections with each other and with sensory receptors and muscle.

Since few organizational principles in the brain give rise to considerable complexity of behavior, it is possible to learn a great deal about behavior by paying attention to how the nervous system solves four general problems:

1. How neurons produce stereotyped signals that allow communication between parts of one cell.
2. How neurons are interconnected to allow communication between cells.
3. How different patterns of interconnection give rise to different types of behavior.

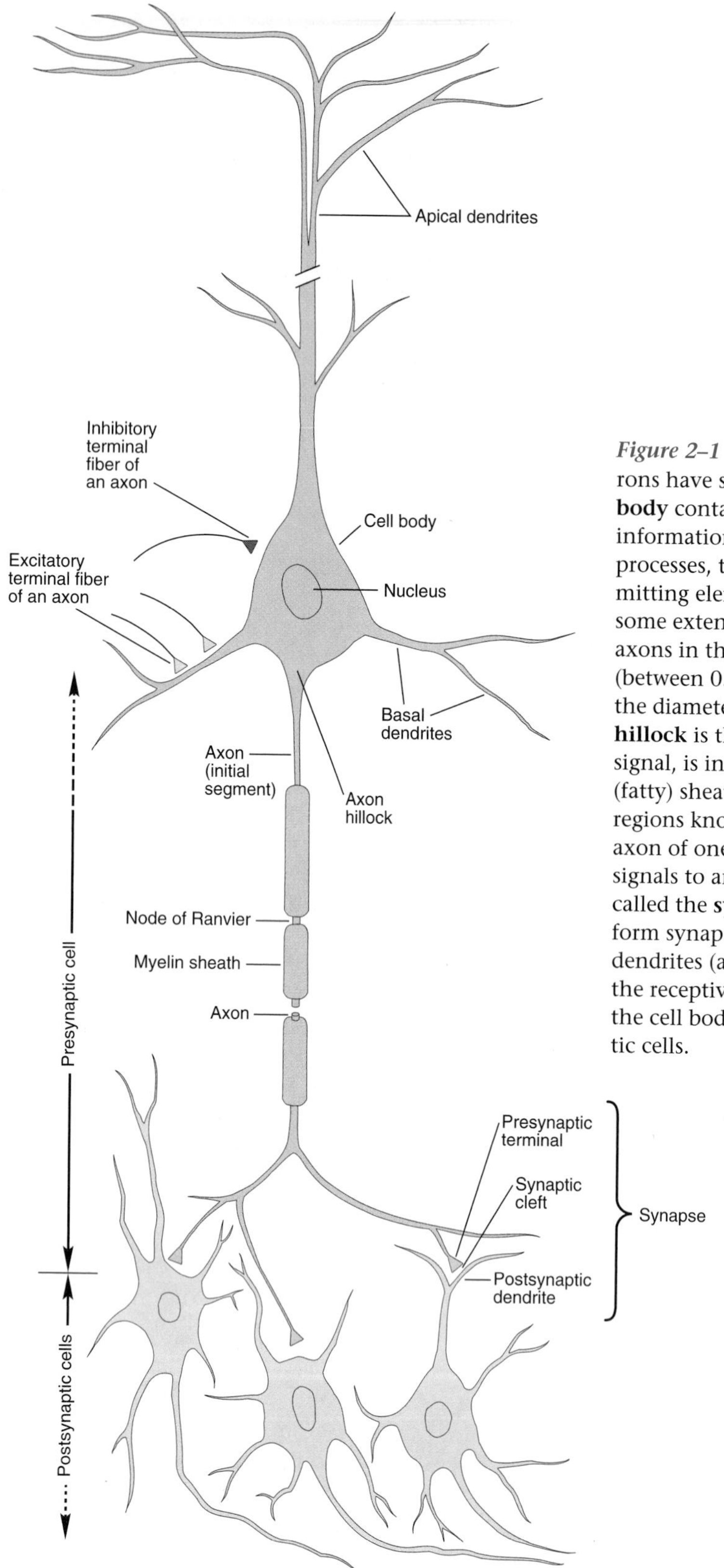

Figure 2–1 In the vertebrate nervous system most neurons have several main features in common. The **cell body** contains the nucleus, the storehouse of genetic information. The cell body gives rise to two types of cell processes, the **dendrites** and the **axon**. Axons, the transmitting element of neurons, can vary greatly in length; some extend more than 1 m within the body. Most axons in the central nervous system are very thin (between 0.2 and 20 μm in diameter) as compared with the diameter of the cell body (50 μm or more). The **axon hillock** is the site at which the action potential, the cell's signal, is initiated. Many axons are insulated by a myelin (fatty) sheath that is interrupted at regular intervals by regions known as the **nodes of Ranvier**. Branches of the axon of one neuron (the presynaptic neuron) transmit signals to another neuron (the postsynaptic cell) at a site called the **synapse**. The branches of a single axon may form synapses with as many as 1000 other neurons. The dendrites (apical and basal) make up a major portion of the receptive surface of the neuron and, together with the cell body, receive the synaptic input from presynaptic cells.

4. How behavior is modified in light of experience (synaptic plasticity).

In this chapter we introduce the basic features of neuronal signaling by considering some structural and functional properties of neurons and their surrounding support cells. We examine how the interconnections between nerve cells produce a simple behavior, the *knee jerk*. We then look briefly at the location and function of the various signaling mechanisms, and consider how signaling is transformed within the neural circuit so as to produce behavior.

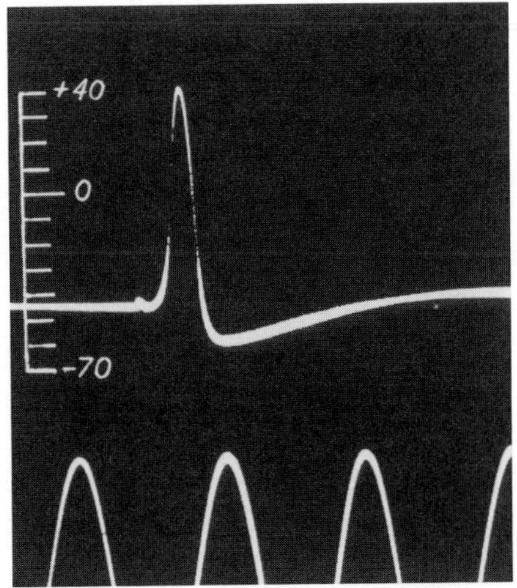

Figure 2–2 The action potential and resting potential as recorded with a capillary pipette across the membrane of the squid giant axon in a bathing solution of seawater. Time marker on the horizontal axis is 500 Hz. The vertical scale indicates the potential of the internal electrode in millivolts; the sea water outside is taken as zero potential. (From Hodgkin and Huxley, 1939.)

The Nervous System Has Two Classes of Cells

There are two distinct classes of cells in the nervous system: nerve cells (or neurons) and glial cells (or glia).

Nerve Cells

A typical neuron has four morphologically defined regions: cell body, dendrites, axon, and presynaptic terminals (Figure 2–1). As we shall see later, each of these regions has a distinct function in the generation of signals.

The cell body (*soma*) is the metabolic center of the cell. It contains the nucleus, which stores the genes of the cell, and the rough and smooth endoplasmic reticulum, which synthesizes the proteins of the cell. The cell body usually gives rise to two types of outgrowths (or processes), the *dendrites* and the *axon*. Most neurons have several dendrites; these branch out in treelike fashion and serve as the main apparatus for receiving signals from other nerve cells. In contrast, the cell body typically gives rise to only one axon, a tubular process that grows out from a specialized region of the cell body called the *axon hillock*. Axon diameters range from 0.2 to 20 μm.

The axon is the main conducting unit of the neuron, capable of conveying electrical signals along distances that range from as short as 0.1 mm to as long as 2 m. Many axons split into several branches, thereby conveying information to different targets.

The electrical signals propagated along the axon, called *action potentials*, are rapid and transient all-or-none nerve impulses, with an amplitude of 100 millivolts (mV) and a duration of about 1 ms (Figure 2–2). Action potentials are initi-ated at the axon hillock and are conducted along the axon, without failure or distortion, at rates between 1 and 100 meters per second. The amplitude of the action potential remains constant throughout the axon because the all-or-none impulse is continually regenerated as it travels down the axon.

Action potentials—the signals used by the brain to receive, analyze, and convey information—are highly stereotyped throughout the nervous system. They are, of course, initiated by a variety of physical events in the environment that impinge on our bodies—mechanical contact, odorants, light, or pressure waves. However, the signals that convey visual information are identical to those that carry information on, say, odors. Here we encounter one of the key principles of brain action. The *information* conveyed by an action potential is determined not by the form of the signal, but by the pathway the signal travels in the brain. It is up to the brain to analyze and interpret *patterns* of incoming electrical signals and in this way to create our everyday sensations of sight and sound.

To ensure high-speed conduction of action potentials, large axons are surrounded by a fatty

Golgi

Ramón y Cajal

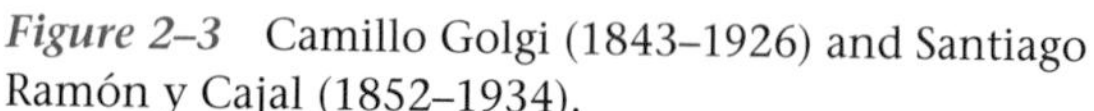

Figure 2–3 Camillo Golgi (1843–1926) and Santiago Ramón y Cajal (1852–1934).

Golgi studied medicine at the University of Pavia, specializing in microscopic anatomy. His chief interest was the cytology of the nervous system. He discovered a cytoplasmic network within nerve cells, which is known to be present in all cells and is called the Golgi apparatus. In the 1860s he developed a silver impregnation method that stains only a small number of nerve cells in a microscopic field, thus making the cytology of the nervous system simpler to examine. Although both Golgi and Ramón y Cajal used similar staining methods, they came to opposite conclusions. Golgi believed that neurons form a reticulum, and that dendrites are nutritive in function. Ramón y Cajal thought that nerve cells are distinct units with a receptive end (dendrites), a conducting portion (the axon), and a transmitting end (nerve terminals). Golgi shared the Nobel Prize with Ramón y Cajal in 1906, and was a senator in Rome in 1900.

As a boy, the Spanish neuroanatomist Ramón y Cajal wanted to be a painter. At the age of 15, however, his father, a surgeon, began teaching him anatomy, using bones unearthed from an ancient cemetery. He became a physician in 1873, served in a Cuban army hospital, and then joined the Faculty of Medicine at Zaragosa and began histological research. He recognized the usefulness of Golgi's silver impregnation method for neurons (published in 1885) and improved it. He realized that "an exact knowledge of the structure of the brain was of supreme interest for constructing a rational psychology," and devoted the rest of his career to the nervous system. Ramón y Cajal showed that the nervous system is not a continuous reticulum (a theory favored by several histologists, including Golgi) but is made up of discrete nerve cells. He provided anatomical evidence that the nerve impulse is transmitted by contact, which led to Sherrington's formulation of the synapse. He also discovered and described the growing axon process and called it the growth cone. Finally, he formulated the theory of dynamic polarization: "Every neuron possesses a receptor apparatus, the body and dendritic propagations, an apparatus of emission, the axon, and an apparatus for distribution, the terminal arborization of the nerve fiber." In 1892, now world famous, he became professor of histology and pathological anatomy at Madrid.

insulating sheath called *myelin*. The sheath is interrupted at regular intervals by the *nodes of Ranvier*, and it is at these uninsulated sites along the axon that the action potential becomes regenerated. We shall learn more about myelination in Chapter 3 and about the action potential in Chapter 10.

Near its end the axon divides into fine branches that make contact with other neurons. The point of contact is known as a *synapse*. The cell transmitting a signal is called the *presynaptic cell*, and the cell receiving the signal is the *postsynaptic cell*. Specialized swellings on the axon's branches serve as the transmitting site in the presynaptic cell. These *presynaptic terminals* do not communicate anatomically with the postsynaptic cell. Instead, the two cells are separated at the synapse by a space, the *synaptic cleft*. Most axons terminate near a postsynaptic neuron's dendrites, but communication may also occur at the cell body or, less often, at the initial segment or terminal portion of the axon of the postsynaptic cell.

As we saw in Chapter 1, Ramón y Cajal (Figure 2–3) provided much of the evidence for the neuron doctrine, which holds that neurons are the basic signaling units of the nervous system and that each neuron is a discrete cell whose several processes arise from its cell body. In retrospect, it is hard to appreciate how difficult it was to persuade scientists of this elementary idea. After the anatomists Jacob Schleiden and Theodor Schwann put forward the cell theory in the early 1830s, the idea that cells are the structural units of all living matter became central dogma. For years, however, most anatomists believed that the cell theory did not apply to the brain. Unlike other tissues, whose cells are simple in shape and fit into a single field of the light microscope, the cells of the nervous system are large and have complex shapes, with dendrites and axons that appear to extend endlessly; these processes were thought to be unrelated to the cell body.

The coherent structure of the neuron did not become clear until late in the nineteenth century, when Ramón y Cajal began to use the silver impregnation method that had been introduced by Golgi (Figure 2–3). This method, which is still used today, has two advantages. First, for reasons that are still not understood, the silver solution stains, in a random manner, only about one percent of the cells in any particular region of the brain, making it possible to study a single nerve cell in isolation from its neighbors. Second, the neurons that do take up the stain are delineated in their entire extent, including the cell body, axon, and full dendritic tree.

Ramón y Cajal applied Golgi's method to the embryonic nervous systems of many animals, including the human brain. By carefully examining the structure of nerve cells and the types of contacts between cells in histological sections of almost every region of the nervous system, Ramón y Cajal was able to describe the differences between classes of nerve cells and to delineate the precise connections between many of them.

In addition to the fundamental principles of the neuron doctrine, Ramón y Cajal grasped two other principles that proved particularly important. First, the *principle of dynamic polarization* states that electrical signals flow in a predictable and consistent direction, and only in that direction, within a nerve cell. The direction runs from the receiving sites of the neuron (usually the dendrites and cell body) to the trigger zone at the axon hillock. There the action potential is initiated and propagated unidirectionally along the axon to the presynaptic release sites in the axon terminal. Although neurons vary in shape and function, most follow this rule of information flow. Later in this chapter we describe a generalized physiological model of this principle.

Second, the *principle of connectional specificity* entails two important considerations. First, there is no cytoplasmic continuity between nerve cells; even at the synapse the presynaptic terminal is separated from the postsynaptic cell (by the synaptic cleft). Second, nerve cells do not communicate indiscriminately with one another or form random networks. Rather, each cell communicates with certain postsynaptic target cells but not with others and always at specialized points of synaptic contact.

Taken together, the principles of dynamic polarization and connectional specificity form the cellular basis of the modern connectionist approach to the brain discussed in Chapter 1. In addition, Ramón y Cajal and the neuroanatomists who followed him found the one feature that most distinguishes one neuron from another: its shape, and, specifically, the number and form of its processes. On the basis of the number of processes that arise from the cell body, neurons are classified into three large groups: unipolar, bipolar, and multipolar.

Unipolar cells are the simplest type of neuron. They have a single primary process, usually with

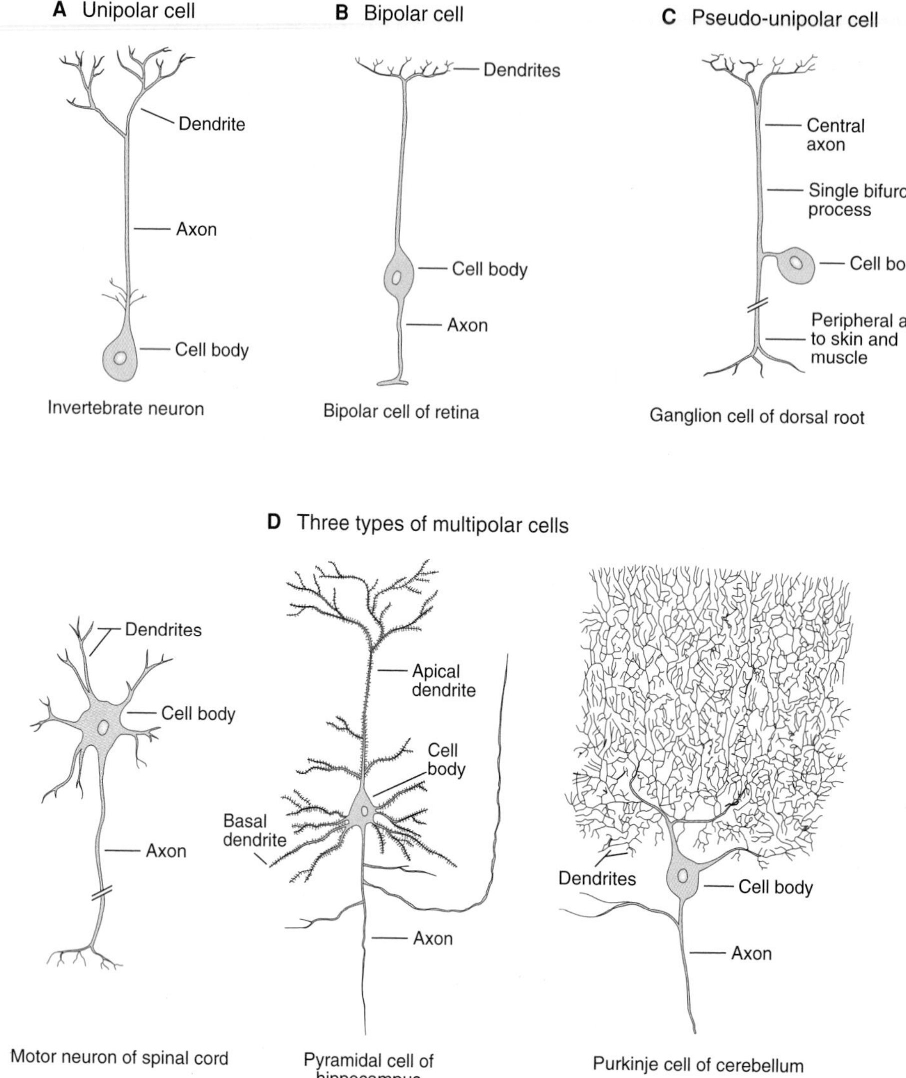

Figure 2–4 Neurons can be classified as unipolar, bipolar, or multipolar according to the number of processes that originate from the cell body.

A. Unipolar cells have a single process, different segments of which serve as receptive surfaces or releasing terminals. Unipolar cells are characteristic of the invertebrate nervous system.

B. Bipolar cells have two processes that are functionally specialized: the dendrite carries information to the cell, and the axon transmits information to other cells.

C. Certain neurons that carry sensory information to the spinal cord belong to a subclass of bipolar cells called pseudo-unipolar cells. As such cells develop, the two processes of the embryonic bipolar cell become fused and emerge from the cell body as a single process. This process then splits into two processes, both of which function as axons, one going peripherally to skin or muscle, the other going centrally to the spinal cord.

D. Multipolar cells have an axon and many dendrites. They are the most common type of neuron in the mammalian nervous system. Three examples illustrate the large diversity of shape and organization of multipolar cells. **Spinal motor neurons** innervate skeletal muscle fibers. **Pyramidal cells** have a roughly triangular cell body; dendrites emerge from both the apex (the apical dendrite) and the base (the basal dendrite). Pyramidal cells are found in the hippocampus and throughout the cerebral cortex. **Purkinje cells** of the cerebellum are characterized by their rich and extensive dendritic tree in one plane. Such a structure permits an enormous synaptic input. (Adapted from Ramón y Cajal, 1933.)

many branches. One branch serves as the axon; other branches function as dendritic receiving structures. Unipolar cells have no dendrites emerging from the soma. Unipolar cells predominate in the nervous systems of invertebrates (Figure 2–4A); they also occur in vertebrates in certain ganglia of the autonomic nervous system.

Bipolar neurons have an oval-shaped soma that gives rise to two processes: a dendrite that conveys information from the periphery to the soma, and an axon that carries information from the soma to the central nervous system (Figure 2–4B). Many bipolar neurons are sensory, such as the bipolar cells of the retina and olfactory epithelium. The sensory cells that carry information about touch, pressure, and pain to the spinal cord are special examples of bipolar cells. They initially develop as bipolar cells, but then the two processes fuse to form a single axon that emerges from the cell body and splits into two. One segment runs to the periphery (to sensory receptors in the skin, joints, and muscle), the other to the spinal cord. As a result, these sensory cells are called *pseudo-unipolar* (Figure 2–4C).

Multipolar neurons predominate in the nervous system of vertebrates. These cells have a single axon and one or more dendrites that typically emerge from all parts of the cell body (Figure 2–4D). The size and shape of multipolar cells vary greatly. In particular, multipolar cells differ in the number and length of their dendrites and the length of their axons. In most such cells the number and extent of the dendrites correlate with the number of synaptic contacts that other neurons make onto it. A spinal motor cell, whose dendrites are moderate in both number and extent, receives about 10,000 contacts—2000 on the cell body and 8000 on the dendrites. The dendritic tree of a Purkinje cell in the cerebellum is much larger, as well it might be—it receives approximately 150,000 contacts!

Neurons can be classified by their function into three major groups: sensory, motor, and interneuronal. *Sensory* (or *afferent*) neurons carry information into the nervous system both for perception and for motor coordination.[1] *Motor neurons* carry commands to muscles and glands. Interneurons constitute by far the largest class, consisting of all the cells in the nervous system that are not specifically sensory or motor. *Relay or projection interneurons* have long axons and relay information over great distances, from one brain region to another. *Local interneurons* have short axons and process information within local circuits.

Glial Cells

Nerve cell bodies and axons are surrounded by glial cells. There are between 10 and 50 times more glial cells than neurons in the central nervous system of vertebrates. Although their name comes from the Greek word for glue, glial cells do not actually hold other cells together. Nor, as far as is known, are they essential for processing information. However, glial cells are thought to have several vital roles.

1. They serve as supporting elements, providing firmness and structure to the brain. They also separate and occasionally insulate groups of neurons from each other.
2. Two types of glial cells produce myelin, the insulating sheath that covers most large axons.
3. Some glial cells are scavengers, removing debris after injury or neuronal death.
4. Glial cells buffer and maintain the concentration of potassium ions in the extracellular space; some also take up and remove chemical transmitters released by neurons during synaptic transmission.
5. During development of the brain, certain classes of glial cells guide the migration of neurons and direct the outgrowth of axons.
6. Certain glial cells help with the forming of a special, impermeable lining in the capillaries and venules of the brain, creating a *blood–brain barrier* that prevents toxic substances in the blood from entering the brain.
7. There is suggestive evidence that some glial cells play a role in nourishing nerve cells, although this has been difficult to demonstrate conclusively.

There are three predominant types of glial cells in the vertebrate nervous system: oligodendrocytes, Schwann cells, and astrocytes. *Oligodendrocytes* and *Schwann cells* are small cells with relative-

[1]Some primary sensory neurons are also commonly called afferent neurons, and we use these two terms interchangeably in this chapter. The term *afferent* (carried *toward* the nervous system) applies to all information reaching the central nervous system from the periphery, whether or not this information leads to sensation. The term *sensory* should, strictly speaking, be applied only to afferent input that leads to a perception.

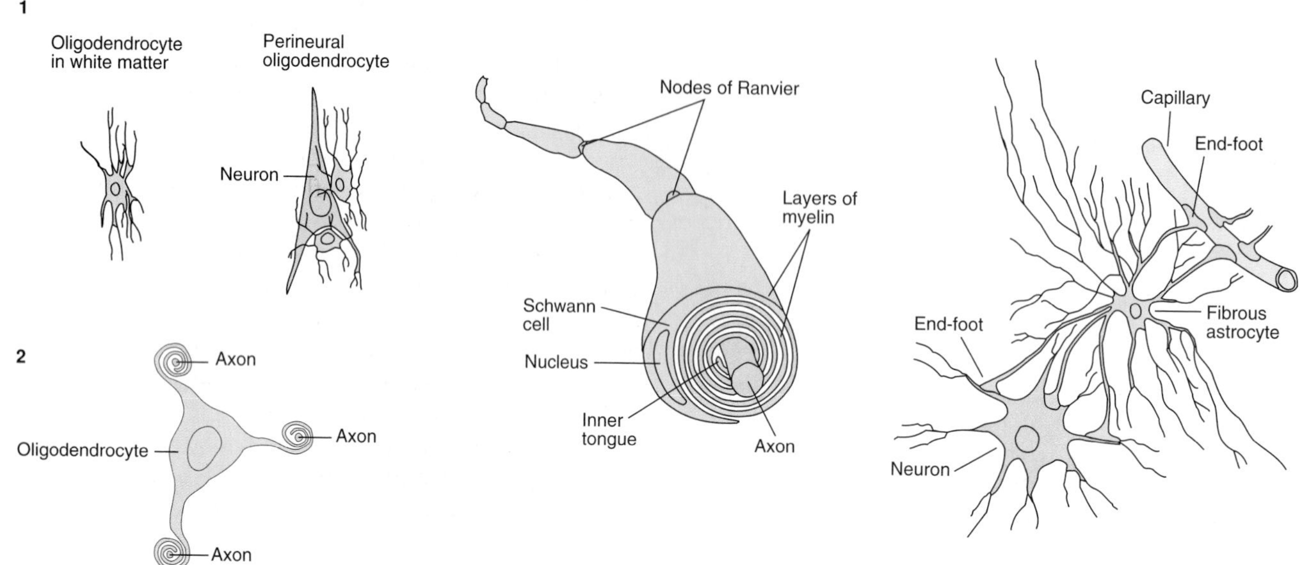

Figure 2–5 The principal types of glial cells in the nervous system are the astrocytes and oligodendrocytes in the central nervous system and the Schwann cells in the peripheral nervous system.

A. Oligodendrocytes are small cells with many processes. **1.** In white matter (left) they participate in myelination; in gray matter (right) they surround the cell bodies of neurons, providing needed support. **2.** A single oligodendrocyte forms myelin sheaths around many axons by wrapping its plasma membrane around the axons. (Adapted from Penfield, 1932.)

B. Schwann cells furnish the myelin sheaths that insulate axons in the peripheral nervous system. Each of several Schwann cells, positioned along the length of a single axon, forms a segment of myelin sheath about 1 mm long. The sheath assumes its form as the inner tongue of the Schwann cell turns around the axon several times, wrapping it in concentric layers of membrane. The intervals between segments of myelin are known as the nodes of Ranvier. In living cells the layers of myelin are more compact than shown here. (Adapted from Alberts et al., 1989.)

C. Astrocytes, the most numerous of the glial cells, are characterized by their starlike shape and by the broad endfeet on their processes. Because these endfeet put the astrocyte into contact with both capillaries and neurons, this type of cell is thought to have a nutritive function. Astrocytes are also important in helping with the formation of the blood–brain barrier.

ly few processes. These cells perform the important function of insulating axons by wrapping their membranous processes concentrically around the axon in a tight spiral, forming a myelin sheath. Oligodendrocytes, which occur in the central nervous system, may each envelop several axons, 15 on average (Figure 2–5A). Schwann cells, which occur in the peripheral nervous system, each envelop only one axon (Figure 2–5B). Oligodendrocytes and Schwann cells also differ to some degree in their chemical makeup.

Astrocytes, the third major class of glial cells, are the most numerous. They have irregular, roughly star-shaped cell bodies and often relatively long processes. Some of these form *endfeet* on the surface of the nerve, brain, and spinal cord. Other astrocytes contact blood vessels and cause the endothelial (lining) cells to form tight junctions, as in the blood–brain barrier (Figure 2–5C).

When neurons fire repeatedly, potassium ions accumulate in the extracellular space. This excess extracellular potassium could depolarize nerve cells in the vicinity. Because of their high potassium permeability, astrocytes can take in the excess potassium and store it so as to protect the neighboring neurons from depolarization. Although the electrical properties of some glial cells can be altered by changes in potassium ion concentration outside the cell, and although many glia have a variety of ion channels in their plasma membranes, there is no evidence that glia are directly involved in electrical signaling. Signaling is the function of nerve cells.

Nerve Cells Are the Signaling Units for Behavioral Responses

The critical signaling functions of the brain—the processing of sensory information, the programming of motor and emotional responses, learning and memory—are all carried out by interconnected sets of neurons. We examine in general terms how these interconnections produce a behavior by considering a simple stretch reflex, the knee jerk.

The anatomical components of the knee jerk are shown in Figure 2–6. The tendon of the quadriceps femoris, an extensor muscle that moves the lower leg, is attached to the tibia by way of the patellar tendon. Tapping the patellar tendon just below the kneecap (patella) pulls (stretches) the quadriceps femoris. This initiates a reflex contraction of the quadriceps femoris and a simultaneous relaxation of the opposing flexor muscles, the hamstrings. This stretch reflex thus changes the position of the lower leg by increasing the tension of a selected group of muscles. Stretch reflexes also maintain muscle tone, a background level of tension. (The regulation of muscle action by the nervous system is discussed in Section VII of the book.)

Stretch reflexes, like the knee jerk, are special types of *spinal reflexes*, behaviors that are mediated by neural circuits that are entirely contained within the spinal cord. As we shall see later in the book, such spinal circuits relieve the brain of having to micromanage elementary actions. Stretch reflexes are mediated by *monosynaptic circuits*, in which the

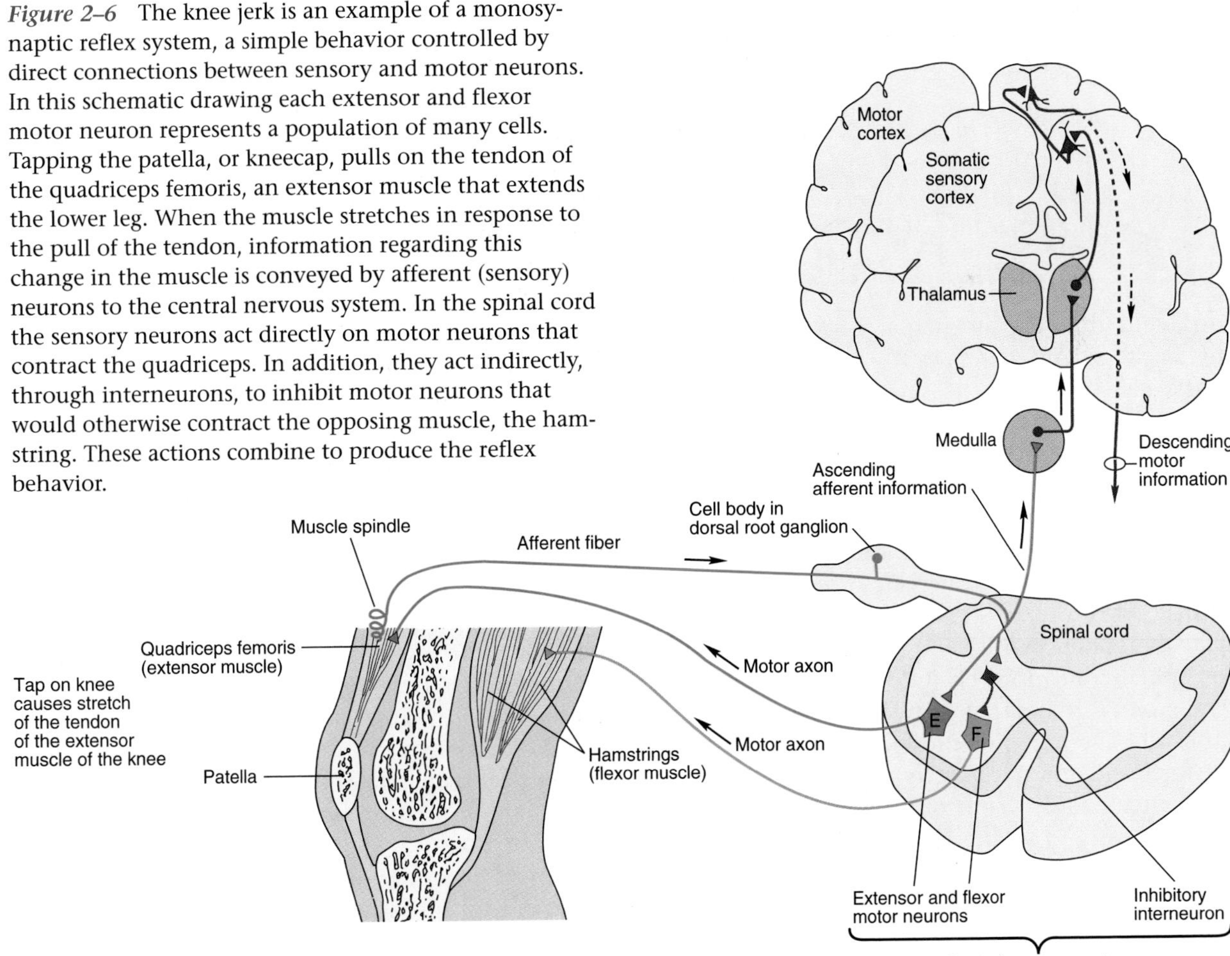

Figure 2–6 The knee jerk is an example of a monosynaptic reflex system, a simple behavior controlled by direct connections between sensory and motor neurons. In this schematic drawing each extensor and flexor motor neuron represents a population of many cells. Tapping the patella, or kneecap, pulls on the tendon of the quadriceps femoris, an extensor muscle that extends the lower leg. When the muscle stretches in response to the pull of the tendon, information regarding this change in the muscle is conveyed by afferent (sensory) neurons to the central nervous system. In the spinal cord the sensory neurons act directly on motor neurons that contract the quadriceps. In addition, they act indirectly, through interneurons, to inhibit motor neurons that would otherwise contract the opposing muscle, the hamstring. These actions combine to produce the reflex behavior.

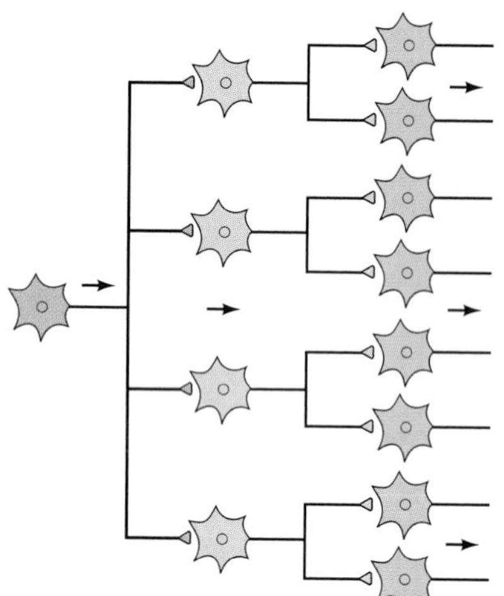

B Convergence

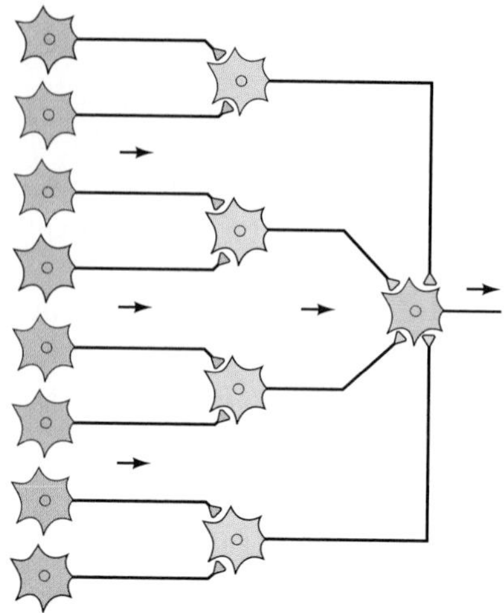

Figure 2–7 The divergence and convergence of neuronal connections represent a key principle in the organization of the brain. In the sensory systems the receptor neurons at the input stage usually branch out and make multiple connections with neurons that represent the second stage of information processing (**A**). Subsequent connections diverge even more. By contrast, motor neurons are targets of progressively converging connections (**B**). With convergence, the target cell receives the sum of information from many presynaptic cells.

sensory and the motor neurons are connected directly, without the intervention of an interneuron. (Most spinal reflexes, however, use *polysynaptic* circuits, which include one or more interneurons, and thus are amenable to modification by higher processing centers in the brain.)

The cell bodies of the sensory neurons in our example reflex are clustered near the spinal cord in a *dorsal root ganglion* (Figure 2–6). They are pseudo-unipolar cells, a special type of bipolar cell (see Figure 2–4C). One branch of the cell's axon goes out to the muscle and the other runs into the spinal cord. The branch that innervates the muscle makes contact with receptors in the muscle, called *muscle spindles*, that are sensitive to stretch. The branch in the spinal cord forms excitatory connections with two groups of neurons: the motor neurons that innervate the extensor muscles and control their contraction, and local interneurons that inhibit the motor neurons controlling the *opposing* flexor muscles.

The stretching of a single muscle activates several hundred sensory neurons, each of which activates between 100 and 150 motor neurons. This type of connection is called *neuronal divergence* and is especially common in the input stages of the nervous system. By distributing its signals to many target cells, a single neuron can exert widespread influence (Figure 2–7A). For example, the central axons of the sensory neurons involved in a stretch reflex also contact projection interneurons that transmit information about the local neural activity to higher regions of the brain concerned with movement. In contrast, many sensory cells terminate on a single motor cell (there are usually 5 to 10 times more sensory neurons than motor neurons). This type of connection, called *convergence*, is common at the output of the nervous system. By receiving signals from multiple neurons, the target cell is able to integrate diverse information from many sources (Figure 2–7B).

Thus, a simple stretch reflex involves three kinds of neural information.

1. Sensory information is conveyed from the body surface to the central nervous system (the spinal cord).

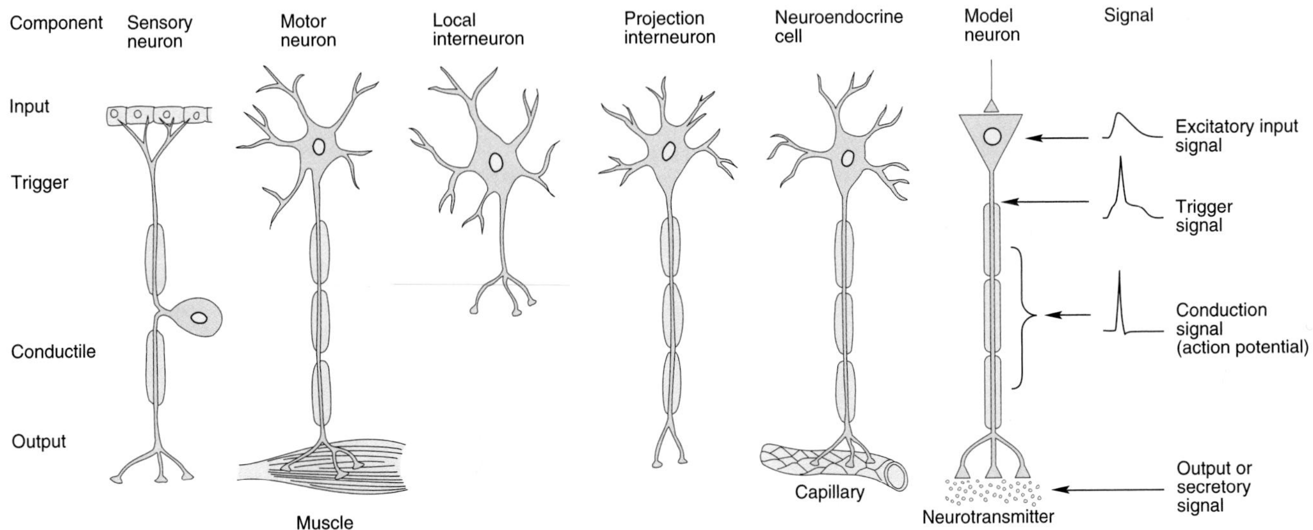

Figure 2–8 Most neurons, regardless of type, have four functional regions in common: an input component, a trigger component, a conductile component, and an output component. Thus, the functional organization of neurons in general can be represented by a *model neuron*. Each component produces a characteristic signal: the input, trigger, and conductile signals are all electrical, and the output signal consists of the release of a chemical transmitter into the synaptic cleft. Not all neurons share all these features; for example, local interneurons often lack a conductile component.

2. Complementary motor commands are issued from the central nervous system to opposing sets of muscles.
3. Information about this behavior is transmitted to other parts of the central nervous system with an interest in the behavior.

In our example a transient imbalance of the body produces sensory information that is conveyed to motor cells, which convey commands to the muscles to contract so that balance will be restored — all in far less time than it takes to say "the stretch reflex."

Signaling Is Organized in the Same Way in All Nerve Cells

To produce a behavior, each participating sensory and motor nerve cell generates, in sequence, four types of signals at different sites within the cell: an input signal, an integration (trigger) signal, a conducting signal, and an output signal. Regardless of size, shape, transmitter biochemistry, or behavioral function, almost all neurons can be described by a generalized model neuron that has four functional regions: a local input (receptive) component, an integrative (trigger) component, a conductile (signaling) component, and an output (secretory) component (Figure 2–8). This model neuron represents the physiological evidence for Ramón y Cajal's principle of dynamic polarization. To understand how the four types of signals are produced one after another, one should first be familiar with the electrical properties of the cell membrane.

Neurons maintain an electrical charge difference of about 65 mV across their external or plasma membrane. This difference is called the *resting membrane potential*. It results, as we shall see below, from an unequal distribution of sodium, potassium, and chloride ions and organic anions across the membrane of cells and the selective permeability of the membrane to potassium. These two factors act in concert to leave the inside of the nerve cell membrane negatively charged in relation to the outside. Because the outside of the membrane is arbitrarily defined as zero, we say the resting membrane potential is –65 mV. In different nerve

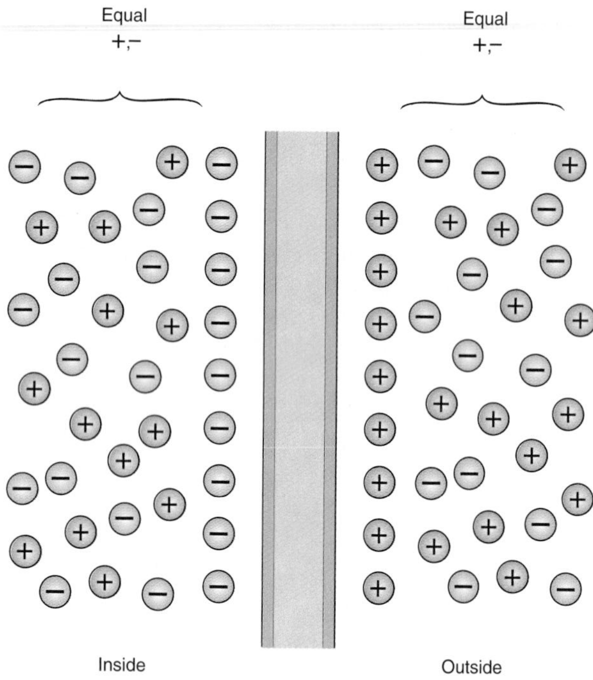

Figure 2–9 The membrane potential results from a separation of positive and negative charges across the cell membrane. The excess of positive charges outside and negative charges inside the membrane of a nerve cell at rest represents a small fraction of the total number of ions inside and outside the cell.

cells the resting membrane potential may range from about –40 to –80 mV. In muscle cells the resting potential is higher still, about –90 mV.

The unequal distribution of ions is maintained by a specific membrane protein that serves as a pump, transporting sodium out of the cell and potassium into it. This *sodium–potassium pump,* which we shall learn more about in Chapter 8, keeps the sodium concentration low within the cell (about 10 times lower than outside) and the potassium concentration high (about 20 times higher than outside). The resting membrane potential results from two properties of the cell: (1) the concentration gradients established by the sodium–potassium pump, and (2) the resting membrane's high permeability to potassium and relatively low permeability to sodium. Because of its high concentration inside the cell, potassium tends to be driven out of the cell under the influence of the concentration gradient. As potassium, a positively charged ion, leaks out of the cell, it leaves behind a cloud of unneutralized negative charge on the inside surface of the membrane, which makes the membrane more negative on the inside (by about 65 mV) than on the outside (Figure 2–9).

Excitable cells, such as nerve and muscle cells, are different from other cells in that their resting membrane potential can be significantly altered and therefore can serve as a signaling mechanism. A reduction in membrane potential of a nerve cell by 10 mV (to –55 mV) initiates an action potential, which transiently nullifies and then reverses the membrane potential. The action potential is a small electrical change that propagates along the axon during this reversal of the membrane's permeability, and it is the action potential that is the conducting signal of the neuron. During the action potential the nerve cell membrane suddenly becomes highly permeable to sodium. After a delay, the membrane returns to its resting state, with its normal separation of charges and high permeability to potassium. We shall learn more about the mechanisms underlying the resting and action potentials in Chapters 8 and 10.

As we saw earlier, the action potential is regularly regenerated (at the nodes of Ranvier) as it travels down the axon. As a result, at its destination the amplitude of the action potential is the same as when the signal was initiated. Nerve cells also produce local signals, such as receptor potentials and synaptic potentials, which typically decay within a distance of a few millimeters. Both classes of signals—action potentials and local signals—result from perturbations of the membrane, which cause the membrane potential to increase or decrease with respect to the resting potential. The resting membrane potential therefore provides the baseline against which all other signals, local as well as propagated, are expressed. An increase in membrane potential (e.g., from about –65 to –75 mV) is called *hyperpolarization*; a reduction (e.g., from –65 to –55 mV) is called *depolarization*. As we shall see later, hyperpolarization makes it less likely that a cell will generate a transmitter signal (the action potential) and is therefore *inhibitory*. Depolarization increases a cell's ability to generate an action potential and is therefore *excitatory*.

Using the sensory and motor neurons involved in the simple knee jerk as an example, we now examine how neural information is generated and transformed both within and between neurons by the four cell components essential for signaling.

The Input Component Produces Graded Local Signals

In most neurons in the resting state no current flows from one part of the neuron to another, so the resting potential is the same throughout the cell. In sensory neurons current flow typically is initiated at a specialized region of the neuron, called the *receptive surface,* where certain protein molecules are sensitive to sensory stimuli.

In our example reflex the specialized receptor proteins form ion channels, pores in the membrane through which sodium and potassium flow. These channels open when the cell is stretched, as we shall learn in Chapters 7 and 19. Ionic current flows through the open channels, producing a change in the resting potential of the cell membrane. The change in membrane potential, or *receptor potential,* represents an input signal. The receptor potential can vary in both amplitude and duration. The larger or longer-lasting the stretch of the muscle, the larger and longer-lasting are the resulting receptor potentials (Figure 2–10A). Most receptor potentials are depolarizing (excitatory). Hyperpolarizing (inhibitory) receptor potentials are found in the visual system, as we shall learn in Chapter 20.

The receptor potential is the first representation of stretch to be coded in the nervous system, but it alone would not cause any signals to appear in the rest of the system. This is because the proteins that transduce sensory stimuli into receptor potentials are restricted to the receptive surface of the sensory neurons and the receptor potential spreads only passively along the axon. The receptor potential thus decreases in amplitude with distance and cannot be conveyed much farther than 1 or 2 mm. At about 1 mm along the axon the amplitude of the signal is only about one-third what it was at the site of generation. The receptor potential is thus a purely *local* signal within the neuron. To be conveyed successfully to the rest of the nervous system, a signal must be amplified—it must be regenerated.

In motor neurons (and interneurons) an input signal occurs when the membrane potential in one cell (the postsynaptic neuron) is perturbed by the output of another cell (the presynaptic neuron). This occurs when the presynaptic neuron (in our example, a sensory neuron) releases a chemical transmitter at the synapse between the two cells. The transmitter then interacts with receptor molecules at the surface of the postsynaptic cell. Finally, the receptor molecules transduce chemical potential energy into an electrical signal: the *synaptic potential.* Like the receptor potential in sensory neurons, the synaptic potential is graded; its amplitude is a function of the amount of chemical transmitter released and its duration is determined by the amount of time the transmitter is active (the period of release). The synaptic potential can be either depolarizing or hyperpolarizing, depending on the type of receptor molecule.

Receptor molecules for inhibitory and excitatory chemical neurotransmission are typically found at different sites in the neuron. Inhibitory synapses are usually located on the cell body of the neuron, whereas excitatory synapses are usually located on the dendrites or on dendritic specializations called *spines.* Although synapses are not found along the main portion of the axon in most neurons, they are not unusual at the presynaptic terminals. Synaptic potentials, like receptor potentials, spread passively from one region of the neuron to another. The features of receptor and synaptic potentials are summarized in Table 2–1.

Table 2–1 Features of Local Signals (Receptor and Synaptic Potentials) and Propagated Signals (Action Potentials)

	Amplitude	*Duration*	*Summation*	*Effect of Signal*	*Type of Propagation*
Local Signals					
Receptor potentials	Small (0.1–10 mV)	Brief (5–100 ms)	Graded	Hyperpolarizing or depolarizing	Passive
Synaptic potentials	Small (0.1–10 mV)	Brief to long (5 ms–20 min)	Graded	Hyperpolarizing or depolarizing	Passive
Propagated Signals					
Action potentials	Large (70–110 mV)	Brief (1–10 ms)	All-or-none	Depolarizing	Active

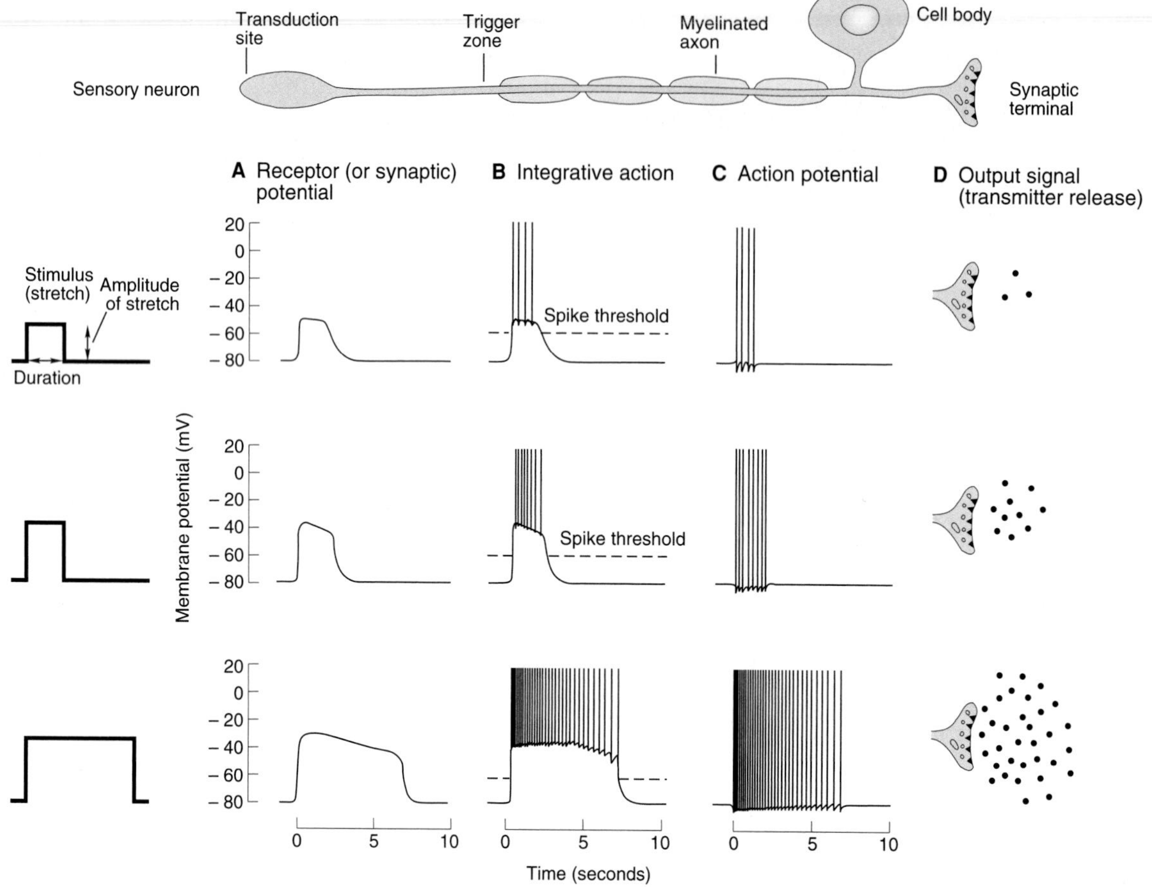

Figure 2–10 A sensory neuron transforms a physical stimulus (in our example, a stretching) into electrical activity in the cell. Each of four components of the neuron produces a characteristic signal.

A. The input signal is graded in amplitude and duration, proportional to the amplitude and duration of the stimulus.

B. The trigger signal transforms the input signal into action potentials that will be propagated along the axon. An action potential is generated only if the receptor potential (in sensory neurons) or synaptic potential (in motor neurons) is greater than a certain *spike threshold.* Once the input signal surpasses this threshold, any further increase in amplitude of the input signal increases the *frequency* with which the action potentials are generated, not their amplitude. Thus, the graded nature of input signals is translated into a frequency code of action potentials at the trigger zone. The *duration* of the input signal determines the number of action potentials.

C. Action potentials are all-or-none. Every action potential has the same amplitude and duration and, thus, the same wave form on an oscilloscope. Since action potentials are conducted without fail along the full length of the axon, the information in the signal is represented only by the frequency and number of spikes, not by the amplitude. The greater the amplitude of the stimulus, the greater the frequency of spikes. The greater the duration of the stimulus, the longer the burst of potentials and therefore the greater the number of spikes.

D. When the action potential reaches the synaptic terminal, the cell releases a chemical neurotransmitter that serves as the output signal. The total number of action potentials per unit time determines exactly how much neurotransmitter will be released by the cell.

The Trigger Component Makes the Decision to Generate an Action Potential

Action potentials, the conducting signals of neurons, are generated by a sudden influx of sodium through voltage-sensitive sodium channels. These channels are specialized protein molecules that are inserted through the lipid bilayer of the membrane, similar to the specialized receptor proteins described earlier. When the cell membrane is depolarized, the change in membrane potential opens these channels, allowing sodium to flow down its concentration gradient from outside to inside the cell. These voltage-sensitive sodium channels are located near the input region of the neuron, but not in the region itself. In sensory neurons the highest density of sodium channels occurs at the first node of Ranvier in the myelinated axon; in motor cells or interneurons the highest density occurs at the axon hillock, where the axon emerges from the cell body.

The initial segment of the axon has not only the highest density of voltage-sensitive sodium channels, but also the lowest threshold for generating an action potential. Thus, as an input signal spreads passively to the axon hillock it will, if it exceeds the threshold, give rise to one or more action potentials. It is here that the activity of all receptor (or synaptic) potentials is summed and where the neuron decides, depending on the magnitude of the summed activity, whether to fire off an all-or-none signal. Consequently, this region in the axon is the site of impulse initiation or the *trigger zone.*

Some neurons have an additional trigger zone in the dendrites, where the threshold for an action potential is also relatively low. Dendritic trigger zones serve to amplify the effectiveness of synapses distant from the cell body. The action potentials produced at these dendritic trigger zones then discharge the final common trigger zone in the initial segment of the axon. Many cell bodies also have the capability of generating action potentials, but the threshold of the cell body is usually higher than that of the initial segment of the axon.

The Conductile Component Propagates an All-or-None Action Potential

The action potential is the conducting signal of the neuron. Unlike the graded input signals (receptor or synaptic potentials), the conducting signal is *all-or-none.* This means that stimuli below the threshold do not produce a signal, whereas all stimuli above the threshold produce the same signal. However much the stimuli may vary in intensity or time, the amplitude and duration of the signal are always the same. Moreover, unlike receptor and synaptic potentials, which spread passively and thus decrease in amplitude with distance, the action potential does not decay as it travels from the initial segment of the axon to the terminal of the target neuron—a distance that can be up to 3 meters. Since the action potential is a large depolarizing signal, up to 110 mV in amplitude, the interior of the axon membrane momentarily becomes positive with respect to the outside. This signal, which often lasts only 1 ms, can travel at rates of up to 100 meters per second (Table 2–1).

The remarkable feature of action potentials is that they are highly stereotyped throughout the nervous system, varying only subtly (although in some cases importantly) from one nerve cell to another. This feature was demonstrated in the 1920s by Edgar Adrian (Figure 2–11), an English physiologist and one of the first to study the nervous system on the cellular level. Adrian found that all action potentials, wherever they occur in the nervous system, have a similar shape or wave form on the oscilloscope (see Figure 2–10). Indeed, the voltage signals of action potentials carried into the nervous system by a sensory axon often are indistinguishable from those carried out of the nervous system to the muscles by a motor axon.

Only two features of the conducting signal convey information: the number of action potentials and the time intervals between them. As Adrian put it in 1928, summarizing his work on sensory fibers, "all impulses are very much alike, whether the message is destined to arouse the sensation of light, of touch, or of pain; if they are crowded together the sensation is intense, if they are separated by long intervals the sensation is correspondingly feeble." Thus, what determines the intensity of sensation or speed of movement is not the magnitude or duration of individual action potentials, but their *frequency.* Likewise, the duration of a sensation or movement is determined by the period during which action potentials are generated.

This important feature of signaling in the nervous systems points to one of the deep questions on the organization of the brain. If the signaling mechanisms are stereotyped and do not reflect properties of the stimulus, how do neural messages carry specific behavioral messages? How is infor-

mation about the image of a bee distinguished from information about a bee sting, and how do both of these signals differ from messages that send commands for voluntary movement? As we have already seen, and will learn to appreciate even more in later chapters, *the message of an action potential is determined entirely by the neural pathway in which it is carried.* The pathways activated by receptor cells in the retina responding to light are completely different from those activated by sensory cells in the skin that respond to touch. The function of the signal—be it visual or tactile, sensory or motor—is determined not by the signal itself, but by the specific pathway along which it travels.

The Output Component Releases Transmitter

When the action potential reaches the terminal region of the neuron, it stimulates the release of packets of chemical transmitter. Transmitters can be small molecules, such as L-glutamate or acetylcholine, or they can be peptides like enkephalin. These transmitter molecules are held in subcellular organelles called *vesicles,* which are loaded into specialized release sites in the presynaptic terminals called *active zones.* The vesicles release their contents at these sites by fusing with the surface membrane, a process known as *exocytosis.* (The

Figure 2–11 Edgar Douglas Adrian (1889–1977)

Adrian was the first cellular neurophysiologist. Born in London and educated at Trinity College, Cambridge, where he studied medicine, Adrian became a resident at the National Hospital for Nervous Diseases at Queens Square, London, and then a medical officer during the First World War. After the war he returned to Cambridge and research. His first discovery was that nerve impulses are propagated in an all-or-nothing fashion (1925). He next developed techniques for dissecting and recording from single sensory fibers. This led him to the discovery that there is a simple relationship between sensation and impulse frequency in an individual sensory axon (1928): "The simplicity of the relation is at once very natural and very surprising. It means that our mind receives all the information which can be got out of the messages from those receptors which are in touch with it, but it means also that the mental correlate is a very close copy of the physical events in the sensory nerves." He later recorded from single motor fibers. In 1932 Adrian shared the Nobel Prize with Charles Sherrington.

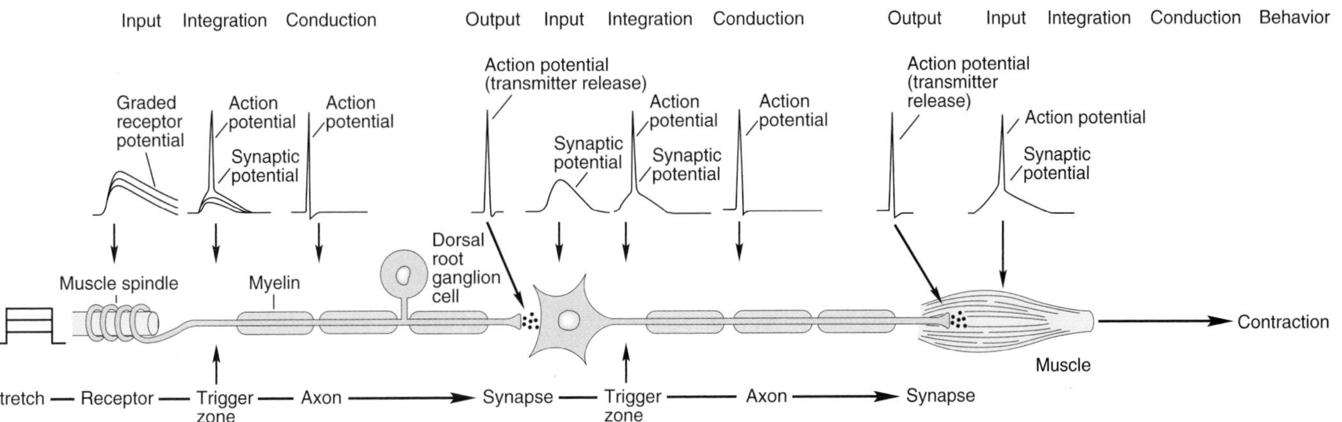

Figure 2–12 The sequence of signals that produces a reflex action. The stretching of a muscle produces a *receptor potential* in the terminal fibers of the sensory neuron (the dorsal root ganglion cell). The amplitude of the receptor potential is proportional to the intensity of the stretch. This potential then spreads passively to the trigger zone, at the first node of Ranvier. If the receptor potential is sufficiently large, it triggers an *action potential* at the trigger zone; the action potential then propagates actively and without change along the axon to the terminal region. At the terminal the action potential leads to an output signal: the release of a chemical neurotransmitter. The transmitter diffuses across the synaptic cleft and interacts with receptor molecules on the external membranes of the motor neurons that innervate the stretched muscle. This interaction initiates a synaptic potential in the motor cell. The synaptic potential then spreads passively to the trigger zone of the motor neuron axon, where it may initiate an action potential that propagates actively to the terminal of the motor neuron. At the terminal the action potential causes transmitter release, which triggers a synaptic potential in the muscle. This signal produces an action potential in the muscle, causing contraction of the muscle fiber.

mechanisms for neurotransmitter release are described in detail in Chapter 15.)

The release of chemical transmitter serves as the output signal. Like the input signal, the output signal is graded. The amount of transmitter released is determined by the amplitude of the depolarization of the nerve terminals, which in turn is determined by the number and frequency of the action potentials in the axon (see Figure 2–10D). The transmitter released by the presynaptic neuron diffuses across the synaptic cleft to receptor molecules located in the membrane of the postsynaptic cell. The binding of the transmitter to the receptor molecules causes the postsynaptic cell to generate a synaptic potential, which will be excitatory or inhibitory depending on the type of receptors in the postsynaptic cell.

The Signal Is Transformed as It Passes from One Component to the Next

A critical feature of neuronal signaling is that the information is *transformed* as it passes from one component of the neuron to the next. The information is transformed even more elaborately as it passes from one neuron to the next. In the stretch reflex we can see aspects of these transformations in their most elementary form (Figure 2–12).

When a muscle is stretched, the particular features of the stimulus—its amplitude and duration—are reflected in the graded amplitude and duration of the receptor potential in the sensory neuron. If the receptor potential exceeds the threshold for action potentials in that cell, the graded signal is transformed at the trigger zone into an all-or-none signal. The all-or-none character of the action potential guarantees that the signal will be propagated faithfully to the terminals of the neuron. The larger the amplitude of the receptor potential above threshold, the greater the depolarization and consequently the greater the frequency of the action potentials in the axon; likewise, the duration of the input signal determines the number of action potentials. (Several action potentials together are often called a *train* of action potentials.) This *digital* information—the action potentials—is conveyed along the entire extent of the axon without failure. At the presynaptic terminals

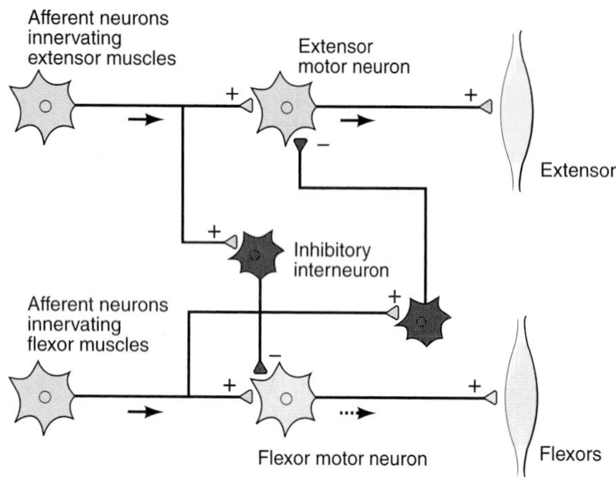

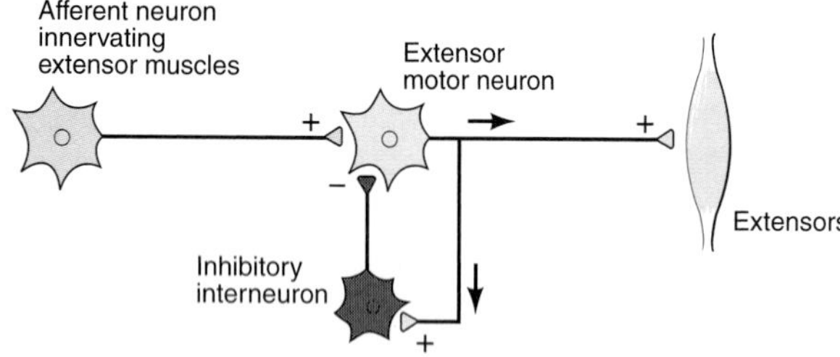

Figure 2–13 Inhibitory interneurons can produce either feed-forward or feedback inhibition.

A. Feed-forward inhibition is common in monosynaptic reflex systems, such as the knee-jerk reflex (see Figure 2–6). Afferent neurons from extensor muscles excite not only the extensor motor neurons, but also inhibitory neurons that prevent the firing of the motor cells in the opposing flexor muscles. Feed-forward inhibition enhances the effect of the active pathway by suppressing the activity of other, opposing pathways.

B. Negative feedback is a self-regulating mechanism. The effect is to dampen activity within the stimulated pathway and prevent it from exceeding a certain critical maximum. Here, the extensor motor neurons act on inhibitory interneurons, which feed back and reduce the probability of firing in the extensor motor neurons themselves.

of the sensory neuron the frequency of action potentials determines how much transmitter will be released. In this way the *digital* signal, the action potential, is transformed with repitition into an *analog* signal (a graded amount of transmitter).

These stages of transformation have their counterparts in the motor neuron. The transmitter released by a sensory neuron interacts with receptor molecules on a motor neuron to initiate a graded synaptic potential, which spreads to the initial segment of the motor axon. There, if it reaches a critical threshold, it can initiate an action potential, which propagates without fail to the motor cell's terminals, where it causes transmitter release. The neurotransmitter in turn triggers a synaptic potential in the muscle. This synaptic potential in the muscle fiber then produces an action potential that leads to the final transformation of this reflex: muscle contraction and the generation of movement. The sequence of signal transformations from sensory to motor neuron to muscle is illustrated in Figure 2–12.

The stretch reflex is a simple behavior, produced by two classes of neurons connected to each other at excitatory synapses. Not all the important signals in the brain are excitatory, however. In fact, fully half of all neurons have the function of inhibiting the firing of certain signals. Inhibitory neurons release transmitter that hyperpolarizes the membrane potential of the postsynaptic cell and thus reduces the likelihood of firing. The sensory neurons in the knee-jerk reflex have both excitatory and inhibitory connections. Excitatory connections with the extensor muscles of the leg cause these muscles to contract, while inhibitory connections with certain interneurons prevent the antagonist flexor muscles from being brought into action. This circuit is a form of *feed-forward inhibition* (Figure 2–13A). Feed-forward inhibition in the knee-jerk reflex is reciprocal, ensuring that the flexor and extensor pathways inhibit each other.

Neurons can also have inhibitory connections that provide *feedback inhibition*. For example, an active neuron may have excitatory connections with both a target cell and an inhibitory interneuron that has its own feedback connection with the active neuron. In this way, signals from the active neuron simultaneously excite the target neuron and the inhibitory interneuron, which thus is able to limit the ability of the active neuron to excite its target (Figure 2–13B). We will encounter many examples of feed-forward and feedback inhibition when we examine more complex behaviors in later chapters.

Functional Differences Among Nerve Cells Are Most Evident on the Molecular Level

The model we have outlined here, although applicable to the vast majority of neurons, is a simplification and is not accurate in detail for all neurons. For example, some neurons do not generate action potentials. Typically, these are local interneurons that lack a conductile component—they have no axon or only a very short one. In these neurons the input signals are summed and spread passively to the terminal region, where they directly affect transmitter release. Other cells do not have a steady resting potential; consequently, they are spontaneously active.

Even cells that have similar organization can differ in important details at the molecular level. For example, different neurons express different combinations of ion channels in their membranes. As we shall learn in Chapters 7 and 10, a diversity of ion channels provides a neuron with a variety of thresholds, excitability properties, and firing patterns. Thus, neurons with different ion channels can encode the same class of synaptic potential into different patterns of firing and thereby can encode and convey different signals.

Neurons also differ in their chemical transmitters and receptors. Indeed, drugs that act on the brain often do so by modifying synaptic transmission. These differences have physiological importance for day-to-day functioning; they also account for the fact that a disease may strike one class of neurons but not others. Certain diseases strike only motor neurons (for example, amyotrophic lateral sclerosis and poliomyelitis), while others, such as tabes dorsalis, affect primarily sensory neurons. Parkinson's disease, a motor disorder, damages a local population of interneurons that use dopamine as a chemical transmitter. In Chapter 17 we shall see how research in selective disorders of motor neuron functioning have provided important insights into the brain. Some diseases are selective even within the neuron: some affect only the receptive elements, others the cell body, and still others the axon. Indeed, because there are so many cell types and each type has distinct components at the molecular level, the nervous system is attacked by a greater number of diseases, both neurological and psychiatric, than any other organ in the body.

Despite these differences, the basic properties of electrical signaling are similar among all nerve cells. Given the large number of different nerve cells in the brain, this simplicity is fortunate for those who study the brain. If we understand in detail the molecular mechanisms that produce signaling in any one kind of cell, we shall be well along the way to understanding these mechanisms in many other kinds of nerve cells.

Complexity of Interconnection Allows Relatively Similar Nerve Cells to Convey Unique Information

We have seen how a limited number of nerve cells can interact to produce a simple behavior. But can more complex behaviors be related so specifically to individual neurons? In invertebrate animals, and in some lower vertebrates, a single (command) cell can initiate a complex behavioral sequence. However, as far as we know, in the human brain no such complex functions are initiated by a single neuron. Rather, every behavior is generated by many cells. The neural mediation of behavior is subdivided into three discrete aspects: sensory input, intermediate processing, and motor output. Each of these components of a behavior is mediated by a distinct group of neurons, and a single component of behavior often recruits several parallel neuronal groups. The deployment of several groups of neurons or several pathways to convey similar information is called *parallel processing*. As an evolutionary strategy for building a more powerful brain, parallel processing makes sense: it increases both the richness and the reliability of function within the central nervous system.

Nevertheless, as we have seen, a key strategy used by the nervous system is localization of function. Specific aspects of information processing are restricted to particular regions within the brain. For example, information for each of the senses is processed in distinct regions where the sensory connections represent precisely a map of the appropriate surface of the body—the skin, tendons and joints, the retina, the basilar membrane of the cochlea (hearing), or the olfactory epithelium (smell). Muscles and movements are also represented in an orderly arrangement of connections.

Thus, the brain contains at least two types of maps or representations: one for sensory perceptions and the other for motor commands. The two types of maps are interconnected in ways that we do not as yet fully understand.

The neurons that make up these representations, whether motor, sensory, or interneuronal, do not differ greatly in their electrical properties. Neurons with similar properties have different functions because of the connections they make in the nervous system. These connections, established during development of the brain, determine the cell's role in behavior. Our understanding of how information is processed and how components of a mental process are represented in the brain is limited to a few regions. In these regions, however, it is clear that the functions or logical operations of a representation can only be understood by defining the flow of information through the connections of the network.

The importance of both abundant and highly specific connections is now recognized by scientists attempting to construct computer models of brain function. Scientists working in this field, a branch of computer science known as *artificial intelligence*, first used serial processing models to simulate the higher-level cognitive processes of the brain—processes such as pattern recognition, the acquisition of new information, memory, and motor performance. They soon realized that although these serial models solved many problems rather well, including such difficult tasks as playing chess, they performed poorly and slowly on other computations that the brain does rapidly and well, such as the almost immediate recognition of faces or the comprehension of speech.

As a result, most computational neurobiologists who model neural function have turned from serial systems to systems with parallel and distributed components, which they call *connectionistic models*. In these models computational elements distributed throughout the system process related information simultaneously. The preliminary insights that have emerged from such models are often consistent with physiological studies. Individual elements in the model do not transmit large amounts of information. *It is the complexity of the connections between the many elements, not that of individual components, that make complex information processing possible.* Individual neurons can carry out important computations because they are wired together in organized and diverse ways. As we shall see in the second part of the book, this wiring is modifiable during development, and later through learning. It is this plasticity in the relationships between relatively stereotyped units in the nervous system that endows us with individuality.

Selected Readings

Adrian, E. D. 1928. The Basis of Sensation: The Action of the Sense Organs. London: Christophers.

Barres, B. A. 1991. New roles for glia. J. Neurosci. 11:3685–3694.

Churchland, P. S., and Sejnowski, T. J. 1992. The Computational Brain. Cambridge, Mass: MIT Press.

Jones, E. G. 1988. The nervous tissue. In L. Weiss (ed.), Cell and Tissue Biology: A Textbook of Histology, 6th ed. Baltimore: Urban & Schwarzenberg, pp. 277–351.

Katz, B. 1966. Nerve, Muscle, and Synapse. New York: McGraw-Hill.

Posner, M. I. (ed.) 1989. Foundations of Cognitive Science. Cambridge, Mass: MIT Press.

Ramón y Cajal, S. 1852–1937. Recollections of My Life. E. H. Craigie (trans.) Philadelphia: American Philosophical Society. Republished 1989. Cambridge, Mass.: MIT Press.

II

CELL BIOLOGY, ANATOMY, AND DEVELOPMENT OF THE NERVOUS SYSTEM

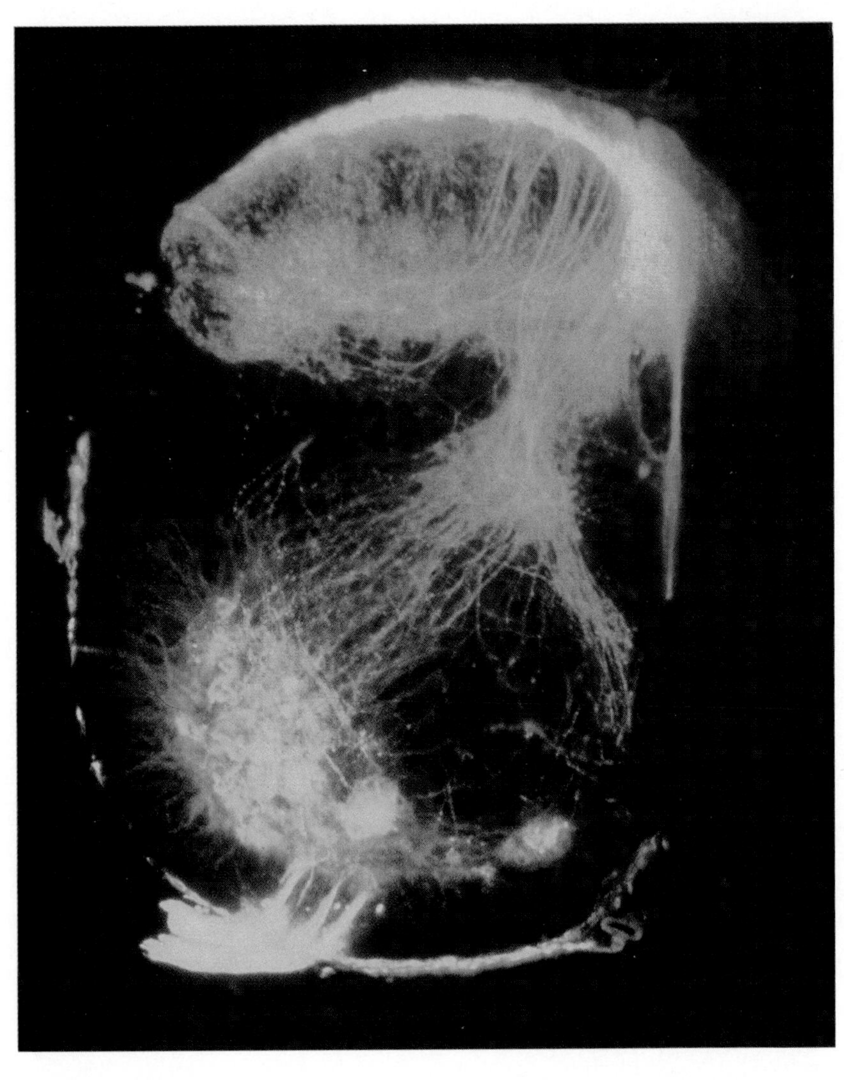

About the image on the preceding page:

The micrograph shows the projection of primary sensory afferents (orange) to the dendrites of motor neurons (green) in the embryonic rat spinal cord. (Courtesy of W. Snider, Washington University, St. Louis, Missouri.)

Introduction

The great diversity of nerve cells—the fundamental units from which nervous systems are assembled—is derived from one basic plan. Three features of this plan give nerve cells the unique ability to communicate precisely, rapidly, and over long distances with one another and with target cells such as muscles and gland cells. First, the neuron is polarized, having receptive dendrites, a biosynthetic cell body, and axons bearing synaptic terminals. Second, the neuron's excitability is due to its external membrane, which contains specialized proteins—ion channels and receptors—that allow specific inorganic ions to pass rapidly into the cell. Finally, the neuron's cell body contains intracellular proteins that provide the neuron with its specialized secretory properties.

The ways in which neural circuits are built up from the connections of single nerve cells have been traced by modern neuroanatomical labeling techniques. These techniques have shown, for example, that nerve fibers from the periphery typically make connections in the central nervous system that maintain the spatial relationship of inputs from the periphery. This topographical organization of inputs is maintained throughout the brain and is the information on the spatial relationships of different stimuli.

Developmental studies of the nervous system aim to elucidate how neural cells acquire specific identities and how the patterns of neuronal connections are established and maintained. The nervous system develops in a series of ordered steps, with each neural structure developing at a characteristic time. Each neuron connects with only a selected subset of potential target cells, and connections are formed only at specific regions of the surface membrane of target cells.

The total genetic information available to an animal—perhaps 10^5 genes in mammals—is not enough to specify directly all the neuronal interconnections that are made, perhaps as many as 10^{15}. Instead, epigenetic events, both embryonic and environmental, activate specific genes in a combinatorial manner at different times during development. Intercellular signals consisting of many diffusible factors and surface molecules are an important epigenetic influence in the embryo. The external environment provides nutritive factors, sensory and social experiences, and learning, and these affect the nervous system by changing neural activity. Thus, optimal timing of innate and environmental factors is critical for proper differentiation of each nerve cell and for the development of the entire nervous system and its ability to produce behavior.

3

The Neuron

The Neurons that Mediate the Stretch Reflex Illustrate the Cytology of Nerve Cells

The Sensory Neuron

The Motor Neuron

The Axons of Both Sensory and Motor Neurons Are Ensheathed in Myelin

A Major Function of the Neuron's Cell Body Is the Synthesis of Macromolecules

The cells of the nervous system are more varied than those in any other part of the body. Neurons differ from one another in many ways; nevertheless, they all share features that distinguish them from liver cells, fibroblasts, and cells in other tissues. For example, neurons typically contain distinct regions—they exhibit regional polarization, the cytological basis of Ramón y Cajal's principle of functional or dynamic polarization discussed in Chapter 2. The cell body, which contains the nucleus and the organelles for making ribonucleic acid (RNA) and protein, is only one of the important regions of the neuron, and in most neurons it contains less than a tenth of the cell's total volume. The remaining cell volume is contained in the cell's communicating processes, the dendrites and the axon that originate from the cell body (Chapter 2). These regions, which carry out different functions, can be separated from one another by relatively great distances. Thus, the axons of some nerve cells exceed a meter in length.

Neurons also differ from most other cells in being excitable—that is, they undergo rapid shifts in electrical potential because of the flow of ions into and out of the cells. This excitability results from specialized proteins in the neuron's membranes (the ion channels and pumps to be described in later chapters).

The great diversity of neurons is well illustrated in the cerebellum, a part of the brain that is important in controlling motor behavior. The cerebellum contains five types of nerve cells, each with its own

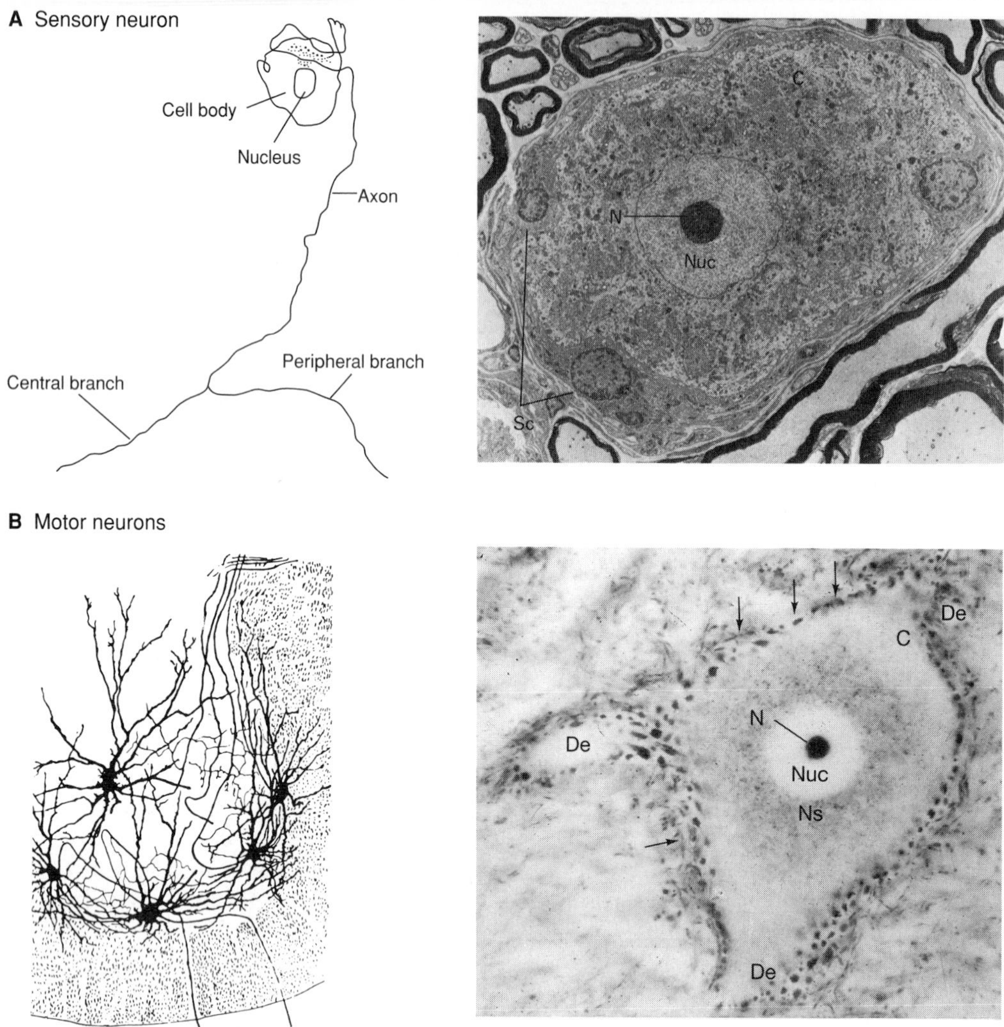

Figure 3–1 A sensory (dorsal root ganglion) cell and a spinal motor neuron form a monosynaptic circuit that controls the knee-jerk stretch reflex.

A. Left: The cell body of the dorsal root ganglion cell contains a prominent nucleus. The axon typically is quite convoluted before it bifurcates into a central and a peripheral branch. (From Dogiel, 1908.) **Right:** Detail of the nucleus (**Nuc**) is shown in this low-power electron micrograph. A prominent nucleolus (**N**) can be seen. The cell body of the neuron is surrounded by Schwann cells (**Sc**), the type of glial cells found in the peripheral nervous system. (Courtesy of R. E. Coggeshall and F. Mandriota.)

B. Left: Five spinal motor neurons in the ventral horn of a kitten show the many dendritic processes that typically branch from the cell bodies of spinal motor neurons. (From Ramón y Cajal, 1909.) **Right:** Detail of the cell body of a motor neuron is shown in this photomicrograph. An enormous number of nerve endings from presynaptic neurons (**arrows**) are visible. These terminals, called synaptic boutons, appear as knoblike enlargements on the cell membrane. Three dendrites (**De**) are also shown. The nucleus and its nucleolus are surrounded by Nissl substance (**Ns**), clumps of ribosomes associated with the membrane of the endoplasmic reticulum. The synaptic boutons (**arrows**) are prominent in this micrograph because the tissue is specially impregnated with silver. (Courtesy of G. L. Rasmussen.)

functions and its own distinct morphology. The Purkinje cells in the cerebellum are among the largest neurons in the vertebrate nervous system. Their cell bodies are 80 μm in diameter and their dendrites branch extensively over relatively long distances to receive diverse inputs. By contrast, the cell bodies of the granule neurons of the cerebellum are only 6 to 8 μm in diameter and consist of a nucleus surrounded by the thinnest shell of cytoplasm; the dendritic processes of these cells do not extend as far from the cell body.

Diversity among cells is the result of differentiation during development, as discussed in Chapter 6. By the genes it expresses, each type of cell synthesizes only certain macromolecules—enzymes, structural proteins, membrane constituents, and secretory products—and avoids making others. In essence, each cell is the macromolecules that it makes. Nevertheless, not all constituents of a neuron are specialized. Many molecules are common to all cells in the body; some are characteristic of all neurons, others of large classes of neurons, and still others are restricted to only a few nerve cells. Thus, each neuron consists of a combination of specific and general molecules.

In this chapter and the next we consider the cytology of the nerve cell, describing both distinctive and general constituents. Although the details of neuronal cytology might be illustrated by different nerve cells that exhibit particular features in a striking manner, we choose to illustrate these features with the two types of neurons that mediate the simple behavior discussed in Chapter 2, the stretch reflex operating in the knee jerk. In this way the relationship between structure and function should be more easily appreciated.

The monosynaptic component of the stretch reflex consists of only two types of neurons: large sensory neurons of the dorsal root ganglion that are connected to muscle spindles, and the motor neurons in the spinal cord that cause the thigh muscle to contract (see Figure 2–5). These two types of nerve cells differ in function and structure as well as in certain macromolecular components. They also display some cytological and biochemical features that are typical of other neurons. In addition, they have many parts that are common to all cells of the body.

The Neurons That Mediate the Stretch Reflex Illustrate the Cytology of Nerve Cells

The Sensory Neuron

The cell bodies of the primary afferent (sensory) fibers are located in dorsal root ganglia that are situated immediately adjacent to the spinal cord. The cell bodies are round with large diameters, about 60 to 120 μm (Figure 3–1). As we saw in Chapter 2, these dorsal root ganglion cells are classified as pseudo-unipolar neurons because they give rise to only one process, an axon that bifurcates into two branches a short distance from the cell body (Figure 3–1). The peripheral branch projects to the muscle; the central branch extends to the spinal cord, where it synapses on the motor neurons that control the reflex (Figure 3–2).

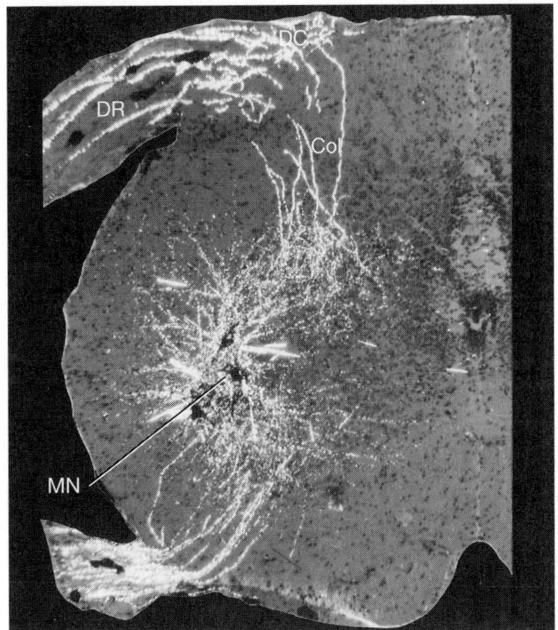

Figure 3–2 This micrograph shows the connections between sensory neurons of the triceps muscle and motor neurons in the brachial spinal cord of a bullfrog. The triceps nerve was labeled with horseradish peroxidase. The sensory axons enter the spinal cord through the dorsal root (**DR**) and then run longitudinally in the dorsal columns (**DC**). Collaterals (**Col**) descend from the dorsal columns to the spinal gray matter, where they arborize and make synaptic contact with the dendrites of brachial motor neurons (**MN**). Dorsal is up, lateral is to the left. × 50. (Courtesy of E. Frank.)

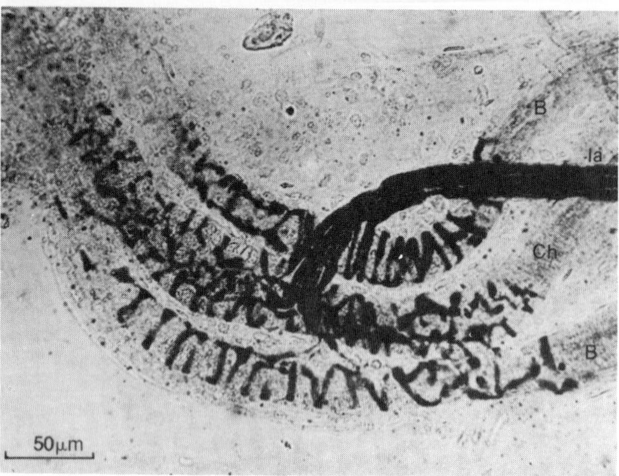

Figure 3–3 The primary sensory nerve endings in the cat soleus muscle are shown in this photomicrograph. The endings of the 1a (primary) afferent axon coil around specialized muscle fibers within the muscle spindle. This structure, the sensory organ for stretch, is a major component of the knee-jerk reflex. (From Boyd and Smith, 1984.)

At its receiving end the peripheral branch of the sensory axon coils around a fine, specialized muscle fiber that lies within the *muscle spindle*, a receptor that is sensitive to stretch (Figure 3–3). From the muscle the peripheral axon travels in the femoral nerve to the cell body in the dorsal root ganglion in the lumbosacral region of the spinal cord. The sensory axon is 14 to 18 μm in diameter and is coated with a white insulating sheath of *myelin* 8 to 10 μm thick. (We examine the myelination of axons in some detail later in the chapter.) This sheath is regularly interrupted along the length of the axon by gaps called nodes of Ranvier (Figure 3–4). At these gaps the plasma membrane of the axon, called the *axolemma*, is exposed to the extracellular space for about 0.5 μm. As will be explained in Chapter 10, this segmental arrangement of myelin greatly increases the speed at which the nerve impulse is conducted along the axon, because the impulse can jump from one node to the next.

The Motor Neuron

The sensory axon projects directly to two kinds of motor neurons: those that innervate the same muscle from which the sensory fiber emerges, and those that innervate synergistic muscles (other muscles that cooperate in stretching the knee joint). Both types of motor neurons are located in the anterior horn of the spinal cord (see Figures 2–5 and 3–1B). Motor neurons have large cell bodies up to 80 μm in diameter, whose nucleus is distinctive because of its large size and prominent nucleolus (Figure 3–1B, right). Unlike dorsal root ganglion cells, which lack dendrites, motor neurons have extensive dendritic trees (Figure 3–1B, left). As was discussed in Chapter 2, the function of the dendrites is to receive synaptic input from other neurons; often this is accomplished at specialized short extensions of the dendrite, which are called *spines*.

Most of the protein made in neurons is synthesized in the cell body (by subcellular organelles that are discussed in the next chapter); some synthesis also occurs in dendrites, but not in axons. In dendrites the organelles required for protein synthesis are frequently situated just beneath a spine. The cytoskeleton of the dendrite differs to some degree from that of the axon. Its molecular composition is similar to that of the cytoskeleton of the cell body (the molecular components of the cytoskeleton are also described in the next chapter).

The number of primary or first-order dendrites of a motor neuron ranges from 7 to 18, and each one ramifies four to six times—usually by bifurcating, but sometimes by even more extensive branching. Each primary dendrite gives rise to about 10 or more terminal branches, and so the total number of terminal dendritic branches per cell is commonly greater than 100. The average length of a dendrite from the cell body to its termination is about 20 cell body diameters, but some branches are twice as long (the mean path length is about 1.5 mm). The branches project radially, so that the dendritic tree of a single motor neuron can extend within the spinal cord over an area about 2 to 3 mm in diameter. Such extensive dendritic branching permits a cell to receive many inputs.

Although each motor neuron has many dendrites, it gives rise to only one axon, which originates from a specialized region of the cell body called the *axon hillock* (Figure 3–5). Together, the axon hillock and the initial segment of axon extend the length of about one cell body diameter, at which point the axon becomes ensheathed in myelin. As described in Chapter 2, the axon hillock

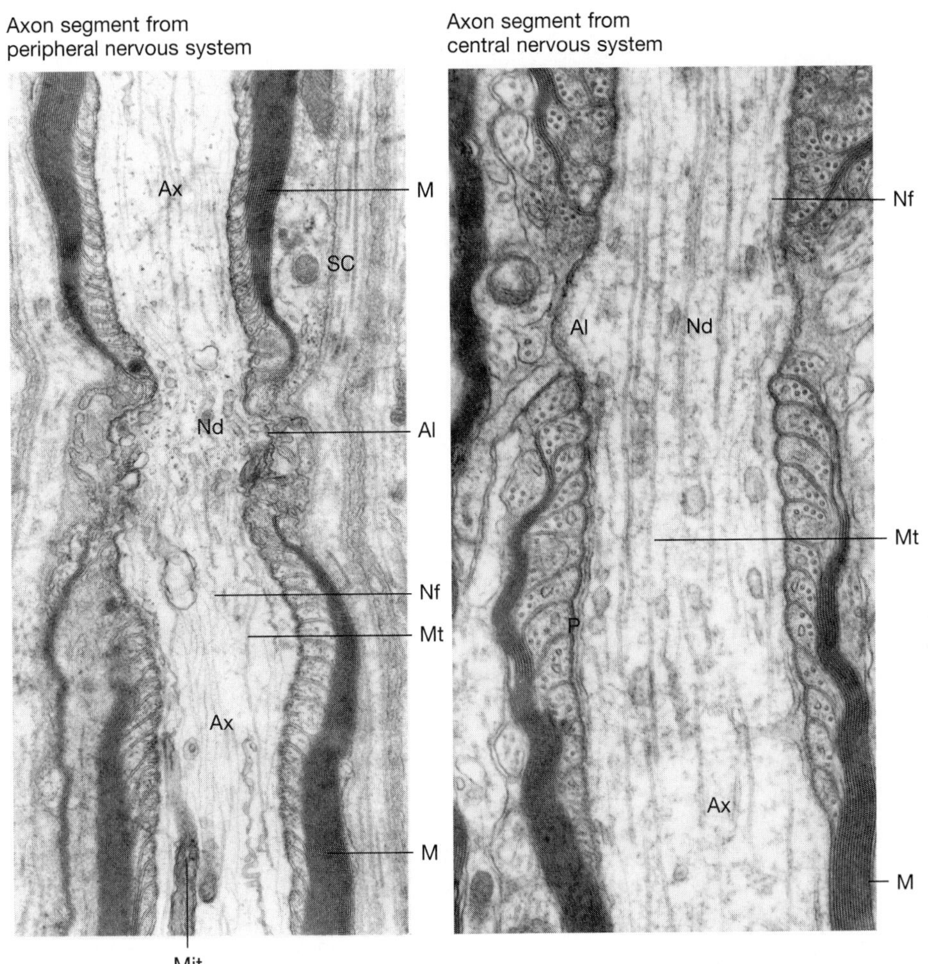

Figure 3–4 The insulating myelin sheath of the axon has regularly spaced gaps called the nodes of Ranvier. These electron micrographs show axon segments from the peripheral nervous system (**left**) and the central nervous system (**right**) in the region of a node. The axon (**Ax**) runs from the top to the bottom in both micrographs. The axon is coated with many layers of myelin (**M**), which is periodically terminated at the nodes (**Nd**). In the peripheral nervous system the myelin is produced by Schwann cells (**SC**), and in the central nervous system by oligodendrocytes. At the node the axolemma (**Al**) is exposed. The elements of the cytoskeleton that can be seen within the axon are microtubules (**Mt**) and neurofilaments (**Nf**). Mitochondria (**Mit**) are also seen. (From Peters, Palay and Webster, 1991.)

and initial segment of the axon function as a trigger zone that integrates the many incoming signals from other cells and initiates the signal that the neuron sends to the muscle. Close to the cell body, the motor axon itself gives off one to five collateral branches (Figure 3–5). These branches are called *recurrent* because, as a rule, they synapse on inhibitory interneurons whose axons project back to the motor neurons.

About half the surface area of the axon hillock and cell body and three quarters of the dendritic membrane are covered by knoblike enlargements, *synaptic boutons*, that are the axon terminals of other neurons (Figure 3–1B, right). The motor neuron receives several types of signals: excitatory input from the primary sensory neurons, excitatory and inhibitory inputs from interneurons that control motor behavior, and feedback inhibition

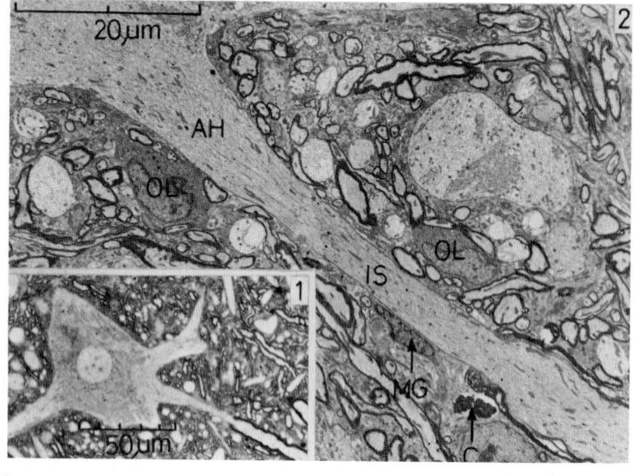

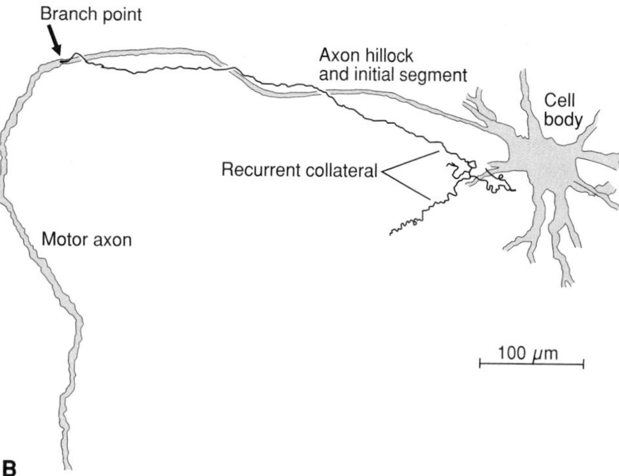

Figure 3–5 The axon of a spinal motor neuron branches to make synaptic contact with several interneurons and, rarely, a recurrent (feedback) connection on the motor neuron.

A. An electron micrograph (**2**) shows the cell body, axon hillock (**AH**), initial segment (**IS**), and the first part of the myelinated portion of a cat's spinal motor neuron. Glial cells (**OL**, oligodendrocytes; **MG**, microglial cell) surround the initial part of the axon. C is a cross section of a capillary. The inset (**1**) shows two dendrites emerging from opposite sides of the cell body. (From Conradi, 1969.)

B. The axons of motor neurons typically give off from one to five recurrent branches that usually make synaptic contact with inhibitory interneurons. In rare instances an axonal branch (a recurrent collateral) makes direct contact with its own cell body. (Courtesy R. E. Burke.)

from inhibitory interneurons. All of these synaptic inputs are tallied together by mechanisms that are described in Chapter 13. The resultant change in membrane potential is monitored at the trigger zone, where the cell membrane is rich in voltage-gated sodium channels. When these channels open, the trigger zone propagates an action potential.

One striking difference between the motor neuron and the sensory neuron is the location of their synaptic inputs. The sensory neuron has few if any presynaptic boutons on its cell body or on its axons within the dorsal root ganglion. In contrast, the motor neuron receives most of its primary and modifying inputs from muscle spindles on the cell body and dendrites. The sensory neuron receives primary input from muscle spindles and modifying inputs from the spinal cord. (These modifying connections often have the function of causing presynaptic inhibition, as will be discussed in Chapter 15.) Examination of many individual motor neurons shows that almost all of the presynaptic boutons are located on the dendritic branches; only 5% are located on the cell body. The synaptic input to the motor neuron is arranged in an orderly fashion. Most inhibitory synapses are close to the cell body, while excitatory ones are further out along dendrites. Each motor neuron receives two to six contacts from a single sensory neuron, and each sensory neuron contacts 500 to 1000 motor neurons. The neurotransmitter used by the primary sensory neuron has not been identified with certainty, but much evidence indicates that it is the amino acid L-glutamate.

The axon of the motor neuron, about 20 μm in diameter, leaves the spinal cord in the ventral root. In our example of the stretch reflex, the axon leaves the lumbosacral region of the spinal cord to join the femoral nerve. Thus, the motor axon travels along the same peripheral path as the sensory fiber from the muscle. When the motor neuron enters the muscle, it ramifies into many branches that become increasingly thinner, reaching a diameter of only a few micrometers. Eventually, each branch loses its myelin sheath and runs along the surface of a muscle fiber where it makes synaptic contacts called *neuromuscular junctions*. At these synapses the motor neuron releases the neurotransmitter acetylcholine. The neuromuscular junction is the most completely characterized and best understood of all synapses and is described in detail in Chapter 12.

In summary, the sensory and motor neurons that mediate the stretch reflex use similar signaling mechanisms but differ in their appearance, in their location in the nervous system, and in how their axons and dendrites are distributed. All of these cytological features have important behavioral consequences. In addition, the two types of cells use different neurotransmitters (although both transmitters are excitatory in function) and receive markedly different kinds of input. Synaptic transmission by the motor neuron, which occurs through the release of acetylcholine, requires not only a specialized enzyme, but also at least one special membrane protein that is not made in the sensory cell or in other types of neurons: a specific transporter protein for choline, which is an essential precursor of the transmitter.

The Axons of Both Sensory and Motor Neurons Are Ensheathed in Myelin

The signal-conducting processes of both sensory and motor neurons are ensheathed in myelin along much of their length (see Figure 3–4). The myelin sheath is arranged in concentric layers (Figures 3–6) consisting of repeating bimolecular layers of lipids interspersed between protein layers. Biochemical analysis shows that myelin has a composition similar to that of plasma membranes, consisting of 70% lipid and 30% protein, with a high concentration of cholesterol and phospholipid.

Our understanding of how the myelin sheath is formed helps explain both the orderly fine structure and the biochemical composition of myelin. During development, before myelination takes place, the sensory cell axon lies along a peripheral nerve in a trough formed by a series of glia called *Schwann cells*. Schwann cells line up along the axon with intervals between them that will eventually become the nodes of Ranvier. Next, the external cell membrane of each Schwann cell surrounds a single axon and forms a double-membrane structure called the *mesaxon*, which then elongates and spirals around the axon in concentric layers. During this ensheathing process, the cytoplasm of the Schwann cell appears to be squeezed out. The Schwann cell's processes then condense into the compact lamellae (thin layers) of the mature myelin sheath. Because the primary sensory axon in the femoral nerve is about 0.5 m long and the distance between nodes is 1 to 1.5 mm, it can be estimated that some 300 to 500 nodes of Ranvier occur along a primary afferent fiber between the thigh muscle and the dorsal root ganglion, where the cell body lies. And since each internodal segment of myelin is formed by a single Schwann cell, as many as 500 Schwann cells participate in the myelination of a single peripheral sensory axon.

The central axonal branch of dorsal root ganglion cells in the spinal cord is also myelinated, as are the axons of motor neurons. The myelin in the central nervous system differs somewhat from its peripheral counterpart, however, because the glial cell responsible for elaborating central myelin is the *oligodendrocyte*, which typically ensheaths several axon processes. Schwann cells and oligodendrocytes differ developmentally and biochemically. Whereas the genes in Schwann cells that encode myelin are turned on by the presence of axons, the genes in oligodendrocytes that encode myelin appear to be activated by the presence of astrocytes, the other major glial cell type in the central nervous system.

Early during myelination of peripheral nerves the Schwann cell expresses a protein called myelin-associated glycoprotein (MAG). This protein is situated primarily at the margin of the mature myelin sheath, just adjacent to the axon. MAG belongs to a superfamily of proteins that is related to the immunoglobulins and includes several important cell surface proteins thought to be involved in the ability of cells to recognize one another. (Other examples of such cell recognition proteins are found in the major histocompatibility complex of antigens, T-cell surface antigens, and the neural cell adhesion molecule, NCAM.) Its early expression, subcellular location, and structural similarity to other surface recognition proteins suggest that MAG is an adhesion molecule important for the initiation of the myelination process.

Myelin in both the central and the peripheral nervous systems contains the same group of proteins, the *myelin basic proteins*. This group of proteins consists of at least seven related proteins that are produced from a single gene by alternative RNA splicing. The myelin basic proteins are capable of eliciting a strong immune response. When injected into animals, myelin basic proteins cause experimental allergic encephalomyelitis, which is characterized by local inflammation and by destruction of the myelin sheaths (*demyelination*) in the central nervous system. This experimental disease has

A Myelination in the peripheral nervous system

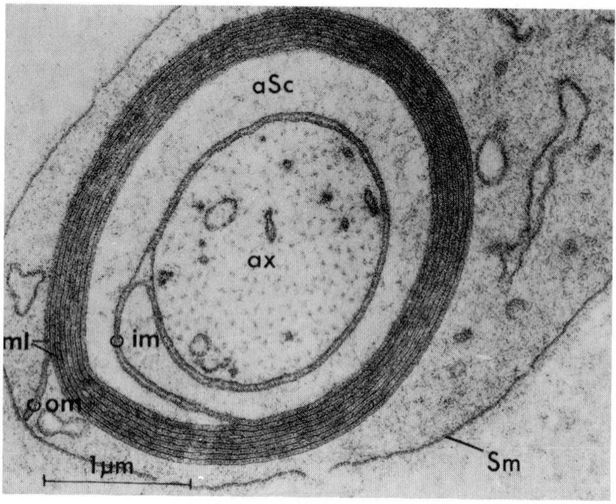

B Myelination in the central nervous system

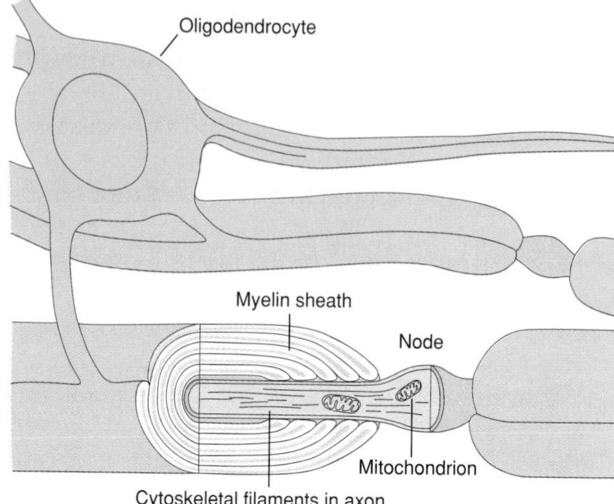

C Development of myelin sheath in the peripheral nervous system

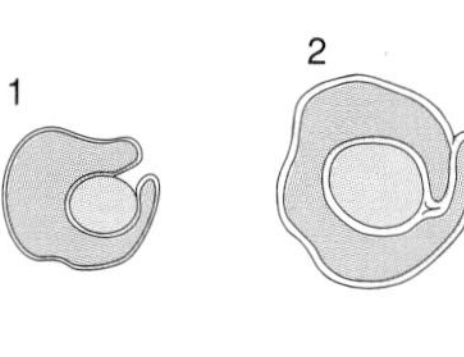

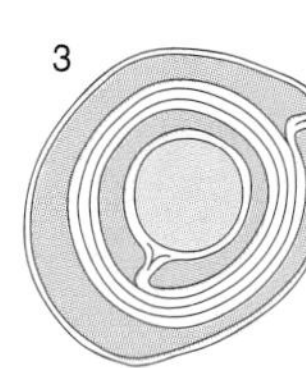

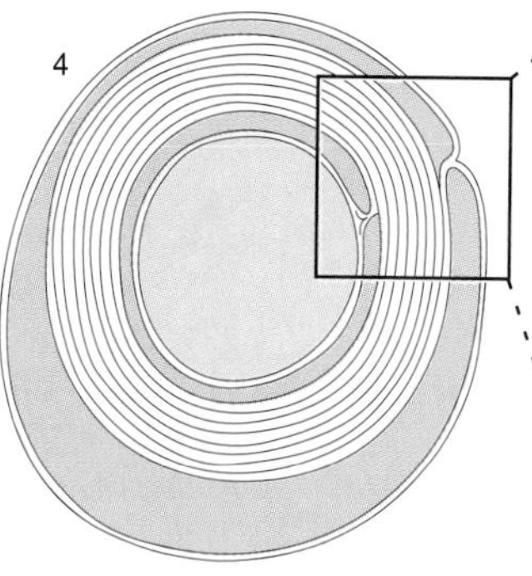

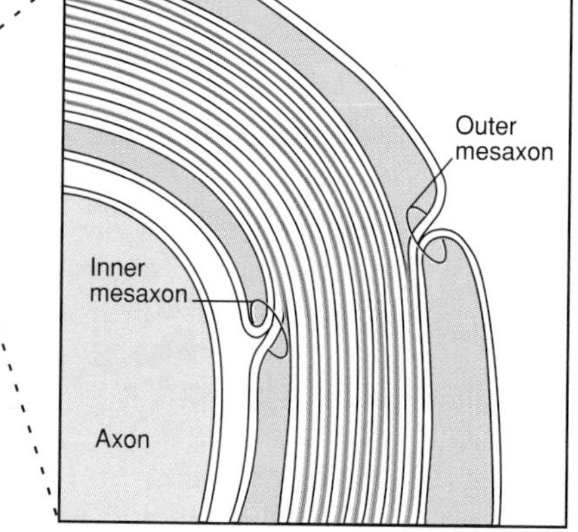

Figure 3–6 The axons of both motor and sensory neurons are insulated by a myelin sheath.

A. An electron micrograph of a transverse section through an axon (**ax**) in the sciatic nerve of a mouse. The spiraling lamellae of the myelin sheath (**ml**) start at a structure called the inner mesaxon (**im**). The spiraling sheath is still developing and is seen arising from the surface membrane (**Sm**) of the Schwann cell, which is continuous with the outer mesaxon (**om**). The Schwann cell cytoplasm (**aSc**) is still present next to the axon; eventually it is squeezed out and the sheath becomes compact. (From Dyck et al., 1984.)

B. An axon in the central nervous system receives its myelin sheath from an oligodendrocyte. (Adapted from Bunge, 1968.)

C. A peripheral nerve fiber is myelinated by a Schwann cell. (Redrawn from Williams et al., 1989.)

been used by some investigators as a model for multiple sclerosis, a relatively common disease in humans. Patients with multiple sclerosis and other demyelinating diseases (for example, Guillain–Barré syndrome) have problems with sensation and moving, because the demyelination of axons slows down transmission of nerve signals (the action potential), and they therefore have problems with sensory perception and proper motor coordination. These diseases can have devastating effects on behavior.

Another disease affecting the myelin sheath is found in mice with the *shiverer* (or *shi*) mutation. In *shiverer* mice myelination in the central nervous system is greatly deficient and abnormal. These mice experience coarse tremors and frequent convulsions, and tend to die young. The mutation that causes this disease is a deletion of five of the six exons of the gene for myelin basic proteins on chromosome 18. The mutation is recessive; a mouse develops the disease only if it has inherited the defective gene from *both* parents. *Shiverer* mice contain only about 10% of the myelin basic proteins found in normal mice. When the wild-type (nonmutant) gene is injected into fertilized eggs of the *shiverer* mutant, the resulting transgenic mice express the wild-type gene at the right time during development (during the first month after birth) and produce about 20% of the normal amounts of myelin basic proteins. Electron microscopic examination reveals much improvement in the myelination of central neurons in the transgenic mice. Thus, these experiments are early examples of successful gene therapy. As shown in Figure 3–7, the introduction of the wild-type gene partially restores these animals to health: although they still have occasional tremors, they do not experience convulsions and they have a normal life span.

A Major Function of the Neuron's Cell Body Is the Synthesis of Macromolecules

One of the important differences between sensory and motor neurons is the role of the cell body in the transmission of synaptic signals. In motor neurons the cell body receives synaptic input (especially inhibitory input), whereas in sensory neurons it does not. In normal circumstances an action potential in the peripheral branch of the sensory neuron axon is transmitted directly to the central branch. What, then, is the function of the sensory neuron's cell body?

The answer to this question was suggested by Augustus Waller in the mid-nineteenth century. Waller cut the various roots and nerves of the spinal cord and observed which fibers degenerated as a result. From the patterns of degeneration, Waller concluded that the cell body of the dorsal root ganglion maintains the vitality of the axon and dendrites attached to it. In a lecture delivered to the Royal Institution of Great Britain in 1861, he said: "A nerve-cell would be to its effluent nerve fibers what a fountain is to the rivulet which trickles from it—a *centre of nutritive energy*." For the most part this nourishment is provided in the form of proteins.

A Normal mouse has abundant myelination

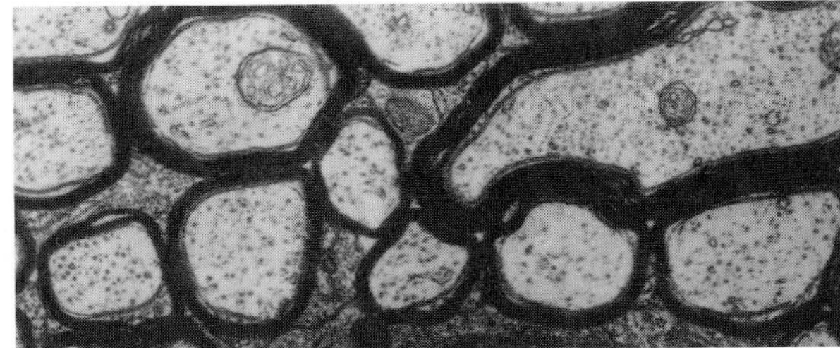

B *Shiverer* mutant has scant myelination

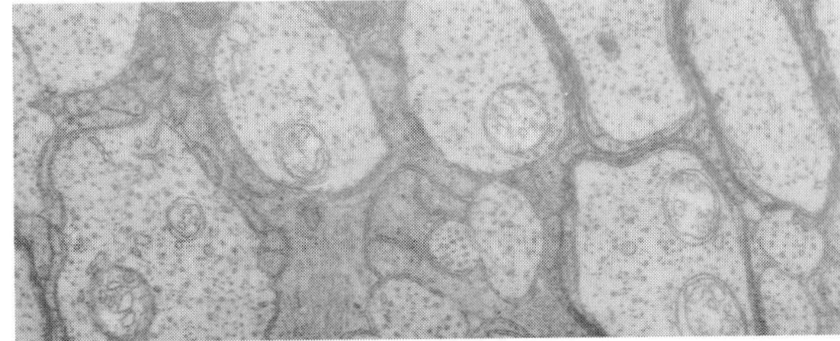

C Transfected normal gene improves myelination

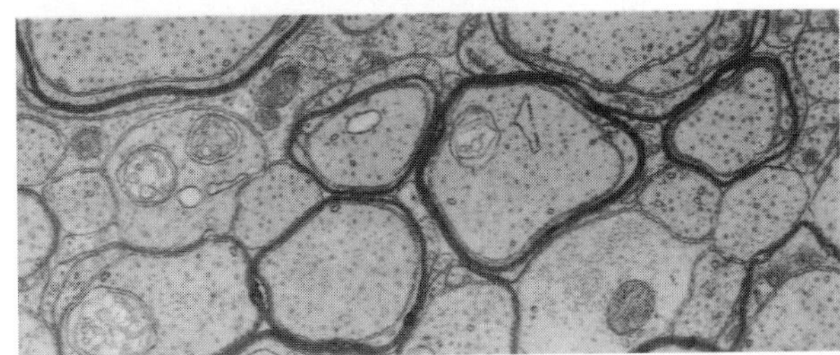

D

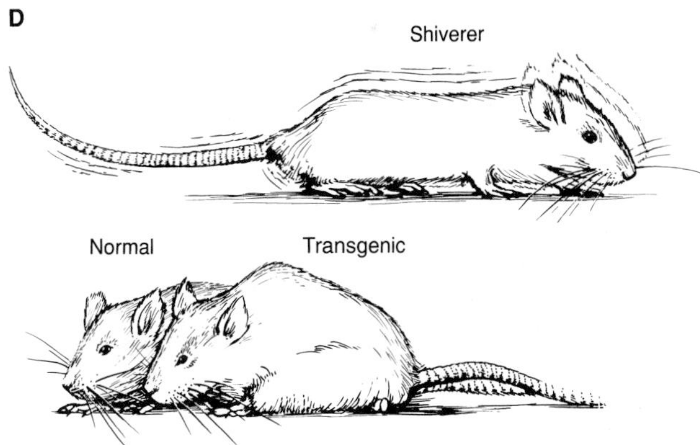

Figure 3–7 A genetic disorder of myelination in mice (*shiverer* mutant) can be partially cured by transfection of the normal gene that encodes myelin basic protein. Electron micrographs show the state of myelination in the optic nerve (**A**) a *shiverer* mutant, a normal mouse (**B**), and a *shiverer* mutant transfected with the gene for myelin basic protein (**C**). (From Readhead et al., 1987.) Myelination is incomplete in the *shiverer* mutant and greatly improved in the transfected animal. As a result of the incomplete myelination, the *shiverer* mutant exhibits poor posture and evident weakness, whereas a normal mouse and a transfected *shiverer* mutant look perky (**D**).

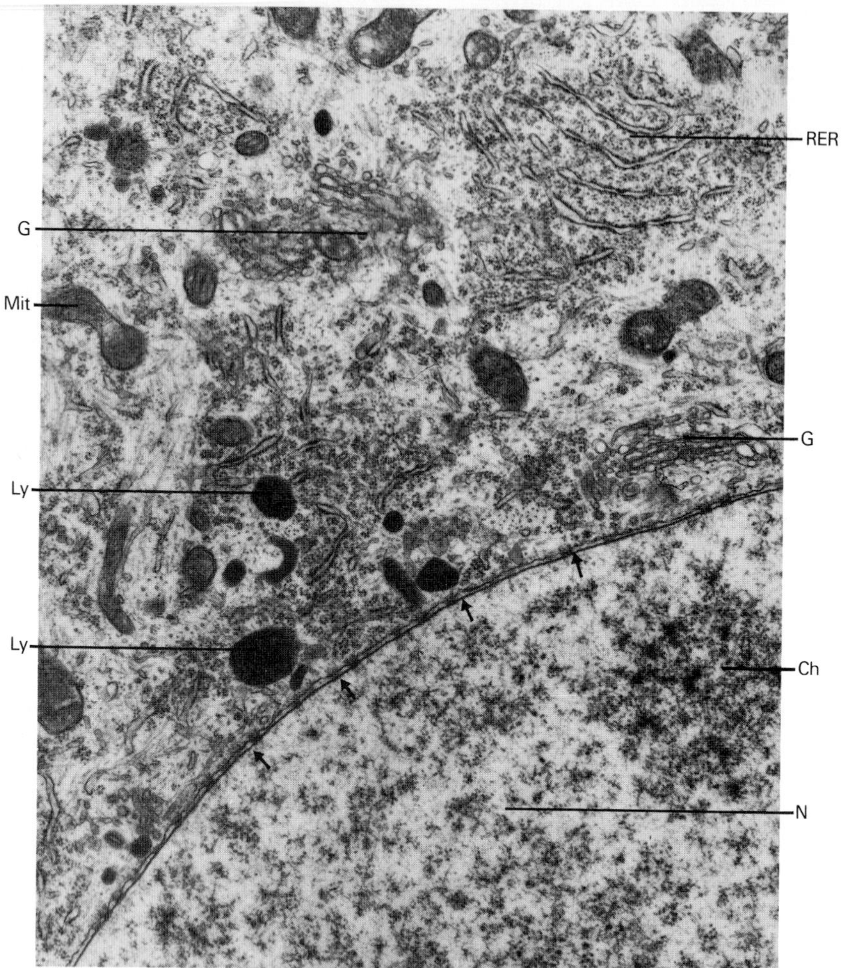

Figure 3–8 Some of the components of a spinal motor neuron that participate in the synthesis of macromolecules. The nucleus (**N**), containing masses of chromatin (**Ch**), is bounded by a double-layered membrane, the nuclear envelope, which contains many nuclear pores (**arrows**). The mRNA leaves the nucleus through these pores and attaches to polyribosomes that either remain free in the cytoplasm or attach to the membranes of the endoplasmic reticulum to form the granular or rough endoplasmic reticulum (**RER**). Several parts of the Golgi apparatus (**G**) are seen. Also present in the cytoplasm are lysosomes (**Ly**) and mitochondria (**Mit**). (From Peters, Palay, and Webster, 1991.)

From modern cell biology we now know that information for the synthesis of proteins is encoded in the deoxyribonucleic acid (DNA) of the chromosomes within the cell's nucleus. In all cell types there are two important ways in which this information can be processed: (1) the genetic information is passed from parent cell to daughter cell during cell division (heredity), and (2) a selected portion of the genetic information in each cell is transcribed into RNA and translated into proteins (gene expression). In almost all mature nerve cells, including the two cells we have been discussing in this chapter, cell division has stopped, and the chromosomes function only in gene expression.

Because mature neurons cannot divide, the chromosomes are not arranged in compact structures, but exist in a relatively uncoiled state. Thus, the cell nucleus, even when viewed in the electron microscope, has a rather amorphous appearance, except for a prominent spherical body called the *nucleolus* (see Figure 3–1). The nucleolus contains the specific portion of DNA that encodes the RNA (rRNA) of future ribosomes. During development this part of the genetic material is reduplicated many times and is especially prominent in secretory cells, like neurons, that make large quantities of proteins.

In addition to the ribosomal genes, many other genes in the nucleus are also actively transcribed into the precursors of messenger RNA (mRNA). These are then selectively processed or spliced to form mature mRNA. The mRNA is transported out of the double-layered nuclear envelope through pores, which are arranged in regular rows (Figure 3–8). Because the pores in adjacent rows are staggered, the pattern of pores appears as a roughly hexagonal array in tangential views of the nuclear membrane. The outer leaflet of the nuclear envelope is continuous with the highly folded membrane of the endoplasmic reticulum, an extensive system of sheets, sacs, and tubules that extends throughout the cytoplasm around the nucleus. This structure is described more fully in the next chapter.

Although most of the genetic information for the synthesis of proteins is encoded in the cell's nucleus, a small amount is contained in circular DNA molecules within mitochondria. It is in these organelles, each of which is about the size of a bacterial cell, that the energy generated by the metabolism of sugars and fats is transformed into adenosine triphosphate (ATP), the molecule most often used to drive biochemical reactions within the cell. The sequence of the 16,569 nucleotides in the human mitochondrial genome has been determined; it encodes information for mitochondrial transfer RNAs (tRNAs) and rRNAs (which differ from those in the rest of the cell) and for a few of the mitochondrion's proteins. The rest of the mitochondrion's proteins are encoded by genes in nuclear chromosomes, synthesized on cytoplasmic ribosomes, and then taken up into the mitochondrion, as discussed in the next chapter.

An Overall View

As in other cells, the neuron's DNA contains the genetic information for encoding proteins and the complex apparatus for synthesizing them. In common with other cells, neurons have mitochondria and enzymes both for intermediary metabolism (the major pathways that convert carbohydrates and other substances into usable energy) and for the synthesis of small molecules. Since nerve cells are excitable, some membrane constituents are the same ones found in all excitable tissue, but many components are highly specialized and unique to specific classes of nerve cells. Thus, neurons contain specific transmitter substances, special ion channels, membrane transport mechanisms, or receptors for neurotransmitters. Our understanding of neural function ultimately depends on identifying and characterizing these molecules, both general and neuron specific. In the next chapter we examine how proteins are synthesized and processed in nerve cells.

Selected Readings

Baldissera, F., Hultborn, H., and Illert, M. 1981. Integration in spinal neuronal systems. In V. B. Brooks (ed.), Handbook of Physiology, Section 1: The Nervous System, Vol. II: Motor Control, Part 1. Bethesda, Md.: American Physiological Society, pp. 509–595.

Burke, R. E. 1990. Spinal cord: Ventral horn. In G. M. Shepherd (ed.), The Synaptic Organization of the Brain, 3rd ed. New York: Oxford University Press, pp. 88–132.

Jones, E. G. 1988. The nervous tissue. In L. Weiss (ed.), Cell and Tissue Biology: A Textbook of Histology, 6th ed. Baltimore: Urban & Schwarzenberg, pp. 277–351.

Peters, A., Palay, S. L., and Webster, H. deF. 1991. The Fine Structure of the Nervous System: Neurons and Their Supporting Cells, 3rd ed. New York: Oxford University Press.

Siegel, G. J., Agranoff, B. W., Albers, R. W., and Molinoff, P. B. (eds.) 1994. Basic Neurochemistry: Molecular, Cellular, and Medical Aspects, 5th ed. New York: Raven Press.

Williams, P. L., Warwick, R., Dyson, M., and Bannister, L. H. (eds.) 1989. Gray's Anatomy, 37th ed. Edinburgh: Churchill Livingstone, pp. 859–919.

4

Neuronal Proteins

Messenger RNA Gives Rise to Three Classes of Proteins

Membranes and Secretory Proteins Are Actively Transported in the Neuron

Fast Axonal Transport

Slow Axoplasmic Flow

Fibrillar Proteins of the Cytoskeleton Are Responsible for the Shape of Neurons

A greater portion of the total genetic information encoded in DNA is expressed in the brain than in any other organ in the body. About 200,000 distinct sequences of messenger RNA are thought to be expressed—10 to 20 times more than in the kidney or liver. In part, this diversity results from the greater number and variety of cell types in the brain. But many neurobiologists also believe that each of the brain's 100 billion nerve cells actually expresses a greater amount of its genetic information than does a liver or kidney cell. What proteins are encoded by these mRNAs in the neuron, and where are they synthesized?

In this chapter we discuss the synthesis and distribution of proteins in general, concentrating on proteins that are important for the release of neurotransmitters. Distribution of proteins depends on the filamentous proteins that make up the cytoskeleton. We therefore describe these filamentous proteins in some detail. In addition to their role in moving other proteins among the various regions of the neuron, the proteins of the cytoskeleton also determine the shape of the neuron. During development and the cell cycle the cytoskeleton is altered dramatically; dynamic changes in the axonal cytoskeleton underlie the extension of the axons to target cells. Changes in synaptic strength that occur in the mature nervous system during long-term memory involve the formation of new synaptic connections, and the filamentous molecules of the cytoskeleton provide the foundation for these new contacts.

Messenger RNA Gives Rise to Three Classes of Proteins

There are three distinct membrane systems within a cell. All three membrane systems constitute separate compartments within the neuron, are made up of different proteins, and serve separate functions within the cell. The most extensive is the *major membrane system of the cell,* which extends from the nucleus to the plasma (external) membrane. This membrane system contains many parts: (1) the nuclear envelope; (2) the continuation of this membrane in a series of channels and tubes throughout the cell known as the endoplasmic reticulum; (3) the membranes of the Golgi apparatus (which packages and distributes molecules within the cell); (4) the secretory granules; (5) the endosomes; and (6) the plasma membrane surrounding the entire neuron (Figures 4–1 and 4–2 and see Box 7–1). Lysosomes (which break down proteins and lipids) constitute the second membrane system. The third distinct and independent membrane system belongs to the mitochondria.

These membrane systems are all embedded in the cytosol, a gel consisting of water-soluble proteins and a variety of insoluble filaments that form the cytoskeleton. (The cytoskeleton, which determines cell shape, is discussed at the end of this chapter.)

With the exception of the few proteins encoded by the mitochondrial genome, essentially all of the macromolecules of a neuron are made in the cell body from mRNA originating in the nucleus. Like other cells, each nerve cell makes only three classes of proteins:

1. Proteins that are synthesized in the cytosol and remain there.
2. Proteins that are synthesized in the cytosol but are later incorporated into the nucleus and mitochondria.
3. Proteins that are synthesized in association with the cell membrane systems. These include at least three subcategories:
   a. *Proteins* that remain attached to the membranes of the endoplasmic reticulum, the Golgi apparatus, and the vesicles that later pinch off from the Golgi apparatus. These are of three types: *membrane-spanning, anchored,* and *associated.* Some proteins are loosely associated with the membrane through weak

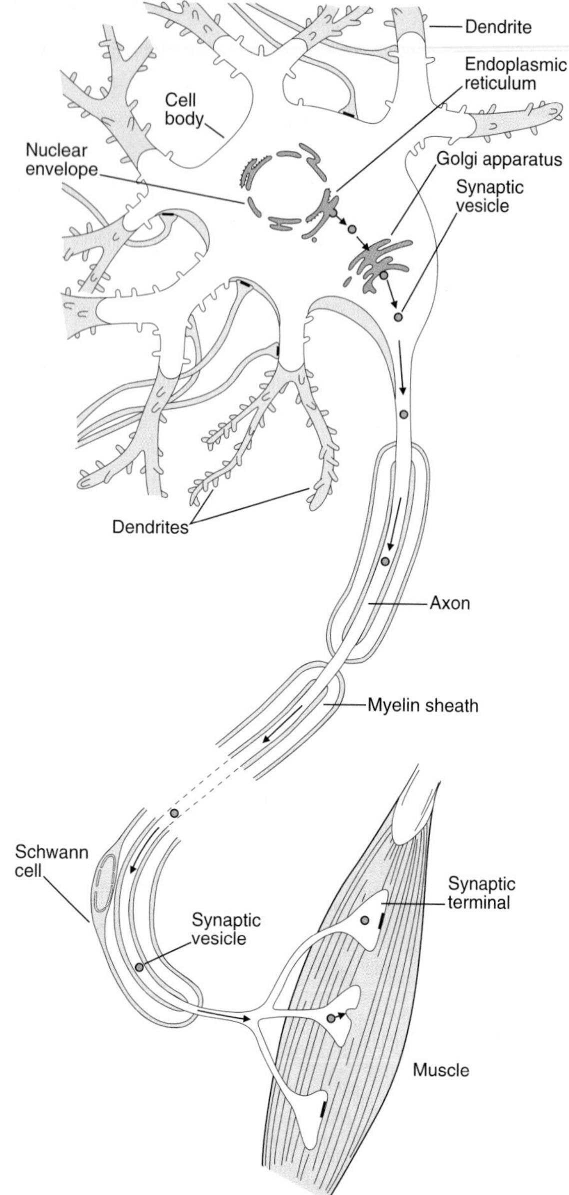

Figure 4–1 The topological continuity of the cell's major membrane system. The figure shows the cell body of a spinal neuron containing the nucleus surrounded by the nuclear envelope, which is continuous with the endoplasmic reticulum. The space between the two membranes that constitute the nuclear envelope is continuous with the extracellular space. Vesicles, which pinch off from the endoplasmic reticulum, shuttle to the *cis* face of the Golgi apparatus. These vesicles maintain the topological relationship between the inside of the cell (the cytosol) and the space within the membrane system. Various kinds of vesicles bud off from the *trans* face of the Golgi apparatus. In this diagram a precursor of a synaptic vesicle is shown making its way down the axon. At the nerve ending the synaptic vesicle fuses with the external (synaptic) membrane. During this exocytosis the inside of the vesicle's membrane will face the extracellular space. (From Williams et al., 1989.)

protein–protein or protein–lipid interactions. Others are anchored by covalent bonds to membrane constituents. Both anchored and associated proteins typically are easier to extract than membrane-spanning proteins, because they do not traverse the lipid bilayer.[1]

b. Proteins that remain within the lumen of the endoplasmic reticulum or Golgi sacs but are not attached to the membrane.

c. Proteins that are transported by means of vesicles that pinch off from the Golgi apparatus and are distributed in secretory vesicles or other organelles, for example, lysosomes. The proteins that are destined to become secretory products are of particular interest because they serve in the release of neurotransmitters.

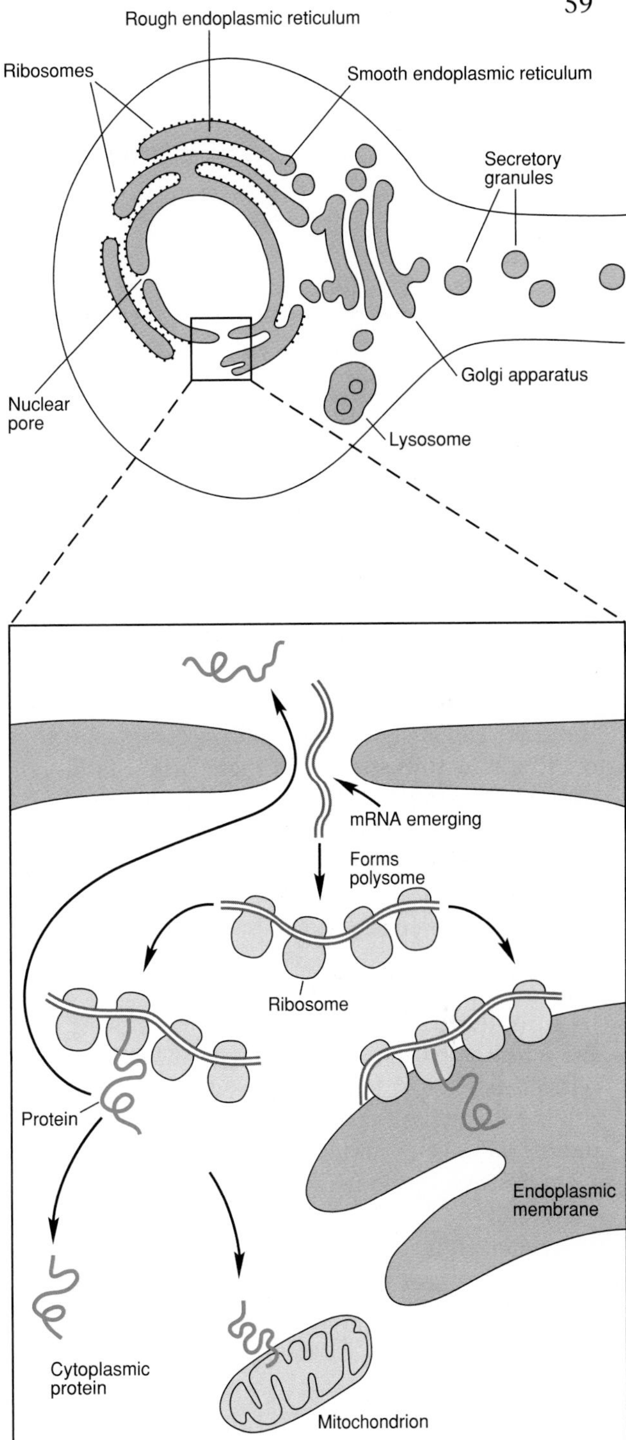

Figure 4–2 The organelles responsible for the synthesis and processing of proteins. The enlargement of a region of a nuclear pore shows mRNA molecules emerging from the nucleus through a pore and attaching to ribosomes to form polysomes. Three classes of proteins are formed as a result. The class of protein depends on the fate of the polysome, which is determined by information encoded in the mRNA.

The mRNAs that encode the proteins of the first two classes (cytosolic proteins and the proteins imported into the nucleus and mitochondria) are translated on free polyribosomes (polysomes). The mRNAs that encode the third class form polysomes that become attached to the flattened sheets of the endoplasmic reticulum (Figure 4–2). What determines the class to which a particular protein will belong? The mRNAs contain information that, when translated into polypeptide sequences, targets the new protein to its final destination.

As in other cells, most of the protein formed in neurons goes toward making up the cytosol. Nevertheless, secretory products also account for a large proportion of the macromolecules synthesized. Moreover, although most of the macromolecules made by nerve cells do not appear to differ from those made by other cells, there are cytosolic, cell membrane, and secretory proteins that are specific to groups of neurons and even to single cells. The three classes of neuronal proteins are described in further detail in Box 4–1.

[1]Membrane-spanning proteins are also called *integral proteins*, a term that was first introduced as an experimental definition for molecules that cannot be extracted from membranes by dilute aqueous salt solutions buffered at a mildly alkaline pH (7–8), but rather require strong *chaotropic* conditions that disrupt the lipid bilayer (for example, detergents, extremely alkaline pH, or high concentrations of certain salts). In more recent usage the term is reserved for a protein molecule whose polypeptide chain traverses the lipid bilayer at least once. All other proteins associated with membrane were called *peripheral*. Many proteins are now known to be linked covalently (anchored) to membrane constituents (as is discussed later in this chapter) and are frequently also called integral proteins.

BOX 4–1 *Proteins of the Nerve Cell*

Nerve cells produce three classes of proteins, each with distinctive characteristics.

Cytosolic Proteins

Cytosolic proteins comprise the two most abundant groups of proteins in the cell: (1) the *fibrillar elements* that make up the cytoskeleton (neurofilaments, tubulins, and actins and their associated proteins, which together account for at least 25% of the total protein in the neuron), and (2) the numerous *enzymes* that catalyze the metabolic reactions of the cell. After they are synthesized in the cell body, both the soluble components of the cytoplasm and the fibrillar elements of the cytoskeleton move into the dendrites and axons of the neuron by slow axoplasmic flow.

Messenger RNA molecules for cytosolic proteins associate with ribosomes to form free polyribosomes or *polysomes* in the neuron's cytoplasm. It is on this structure that the genetic information of the mRNA is translated into protein.

Other functional groups may be added to a protein. An important example is the phosphorylation of serine, threonine, or tyrosine residues by special enzymes, as discussed in Chapter 14. Phosphorylation is reversible and is probably the most common mechanism for altering the biochemical activity of proteins in all cells. Another modification is the addition of many molecules of the small protein *ubiquitin* to certain groups of lysine residues. Adding ubiquitin is a way to tag proteins for degradation by special proteases, an important mechanism the cell uses to turn over its proteins.

Nuclear and Mitochondrial Proteins

The nuclear and mitochondrial proteins encoded by the cell's nucleus are also formed on free polysomes. Soon after synthesis these are targeted to their proper organelle by a mechanism called *posttranslational importation.* Specific receptors around nuclear pores bind and translocate the recently synthesized polypeptide chains. These nuclear receptors recognize structural features of the proteins, allowing the proteins to pass through nuclear pores. Similar membrane receptors insert or translocate the proteins into mitochondria. Although nuclear and mitochondrial proteins represent only a small fraction of the total protein produced by the neuron, they illustrate an important feature of cell biology. This group of proteins reaches its destination in the cell *after* synthesis has been completed on free ribosomes. In contrast, most membrane and secretory proteins are incorporated by cotranslational transfer, which takes place *while* the protein is being synthesized.

Cell Membrane Proteins and Secretory Products

Messenger RNAs encoding proteins that are destined to become secretory products or parts of the cell's major membrane system are formed on polysomes that attach to the endoplasmic reticulum (Figure 4–2). These sheets of membrane, when studded with ribosomes, have a granular appearance in the electron microscope and are therefore called *rough endoplasmic reticulum* (Figure 4–3). Ribosomal RNA in the rough endoplasmic reticulum stains intensely with basic dyes (toluidine blue, cresyl violet, and methylene blue). Under the light microscope this basophilic material is called *Nissl substance* after the histologist who first described changes in the intensity and distribution of staining in neurons after their axons are cut. A large portion of the endoplasmic reticulum lacks attached ribosomes and this portion is called *smooth endoplasmic reticulum.* Attainment of proper secondary structure in the endoplasmic reticulum is fostered by special enzymes called *chaperonins.*

Fate of Major Membrane Proteins

The organelles that contribute to the major membrane system of the cell (nuclear membrane, endoplasmic reticulum, Golgi apparatus, plasma membrane, secretory granules, and endosomes) each have biochemically distinct membranes. These differences result from the fates of the proteins described here.

Cell membrane and secretory proteins, unlike those made in the cytosol, are extensively modified after translation. For example, secretory products typically are synthesized as parts of longer chains of precursor polypeptides, which then undergo sequential and specific tailoring. This process begins in the endoplasmic reticulum and continues in the Golgi apparatus. Indeed, modification can continue even within finished organelles. These biochemical steps result in the precise and orderly segregation of membrane constituents within the neuron. Membrane components leave the Golgi apparatus in a variety of vesicles. Especially important to the neuron are secretory granules or synaptic vesicles, and their precursors. In all cells the membranous and secretory material is conveyed to the plasma membrane and the extracellular space by one of two pathways—the *constitutive pathway* and the *regulated pathway* (see Chapter 15).

Production of smaller peptides from a larger polypeptide can have several physiological consequences. For example, the final orientation of the protein in the membrane depends on just how a particular signal sequence is removed. Another consequence is the masking of a potential activity that would be undesirable within the cell. The production of large proteins also permits amplification or diversification of secreted peptide products, a feature that is particularly important for hormones and neuroactive peptides (see Chapter 14).

Motor neuron in spinal cord

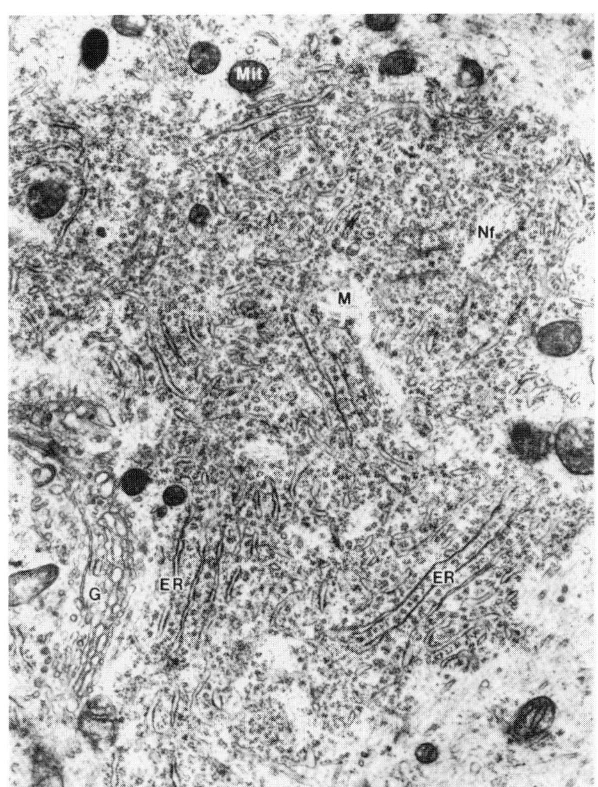

Dorsal root ganglion cell (sensory neuron)

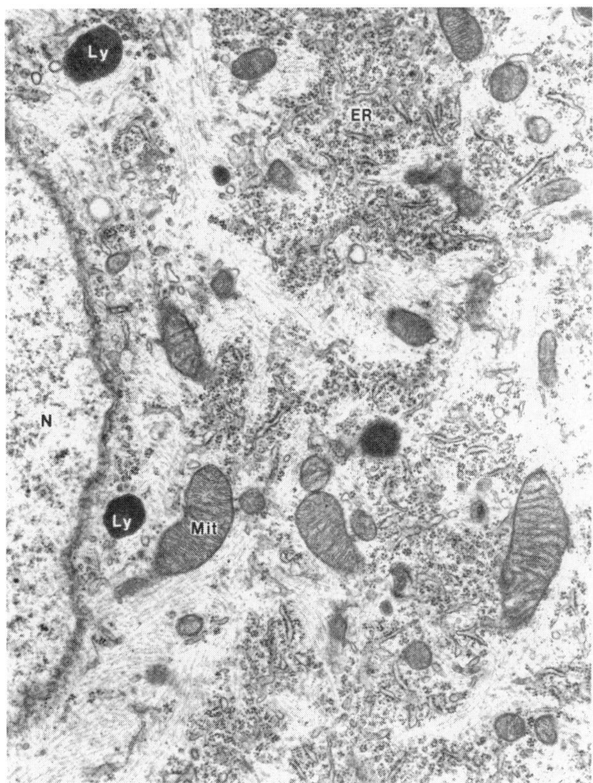

Figure 4–3 Most proteins in the neuron are synthesized by organelles in the cell body from mRNA originating in the nucleus. Messenger RNA enters the cytoplasm through the double-layered nuclear envelope that surrounds the nucleus (**N**). There they form polyribosomes, some of which synthesize proteins in the cytoplasm. One class of proteins remains in the cytoplasm, and another class is transported after they are synthesized into mitochondria (**Mit**). A third major class of proteins is formed after the polysomes attach to the membrane of the endoplasmic reticulum (**ER**). In the light microscope this is called Nissl substance. The motor neuron and the sensory neuron have similar kinds of organelles. The Golgi apparatus (**G**), in which membrane and secretory proteins are further processed is shown in the micrograph of the motor neuron. Some of the newly synthesized proteins leave the Golgi apparatus in vesicles that move down the axon to the synapses; other membrane proteins are incorporated into lysosomes (**Ly**). (From Peters, Palay, and Webster, 1991.)

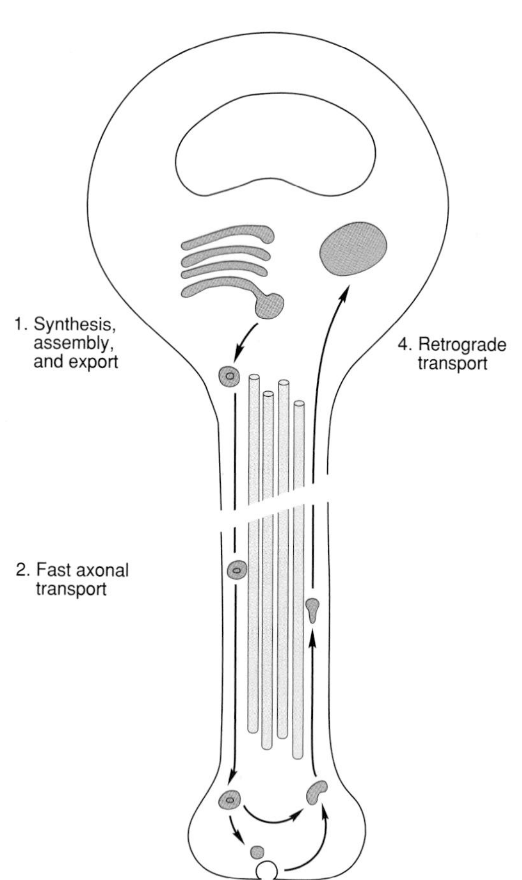

Figure 4–4 Synaptic vesicles and other membranous organelles involved in synaptic transmission at the nerve terminal are returned to the cell body for recycling after they are used at the synapse. **1.** Proteins and lipids are synthesized and incorporated into membranes within the endoplasmic reticulum and Golgi apparatus in the neuron's cell body. **2.** Organelles are then assembled from these components and exported from the cell body into the axon, where they are rapidly moved toward terminals by anterograde axonal transport. **3.** Synaptic vesicles and their precursors reach the neuron's terminals, where they participate in the release of transmitter substances by exocytosis. At random, a small proportion of the membrane becomes degraded, and this material is returned to the cell body by fast retrograde axonal transport. **4.** The degraded membrane is partly recycled; its residue is progressively accumulated in large, end-stage lysosomes that are characteristic of neuronal cell bodies.

Membranes and Secretory Proteins Are Actively Transported in the Neuron

Although the secretory process in neurons is formally similar to that in other cells, in practice it is quite different because of the extremely extended structure (polarity) of the nerve cell. Typically, cell bodies and nerve terminals are at considerable distances from each other. For example, a spinal motor neuron that innervates muscles around the knee joint perhaps half a meter away requires a special transport system to bring newly formed membrane and secretory products from the Golgi apparatus in the cell body to the end of the axon.

There are three ways by which constituents move within the axon: two types of *fast axonal transport* (anterograde and retrograde) and *slow axoplasmic flow*. Essentially all newly synthesized membranous organelles within axons and dendrites (except in the regions of primary dendrites nearest the cell body) are exported from the cell body by anterograde axonal transport. In adult warm-blooded animals the transported organelles move at a rate greater than 400 mm (nearly half a meter) per day. A large proportion of this material consists of synaptic vesicles or their precursors on their way to the terminals (Figure 4–4).

At the nerve terminals the vesicle membranes are recycled many times for reuse in synaptic transmission. Membrane is constantly being replaced by new components arriving from the cell body. At a compensating rate, existing membrane components are returned from nerve terminals to the cell body, where they are either degraded or reused (Figure 4–4).

Fast Axonal Transport

Direct microscopic analysis can be used to examine the movement of particles in a variety of nerve fibers. Continuous direct observation using video-enhanced light microscopy reveals that large particles move in a stop-and-go (saltatory) fashion in both directions, away from and toward the cell body.

Proteins synthesized in dorsal root ganglion cell bodies can be labeled by radioactive amino acids injected into the ganglion. Transport is then assessed by measuring the amount of radioactivity in uniform sequential segments along a nerve. Measuring the distribution of labeled protein along

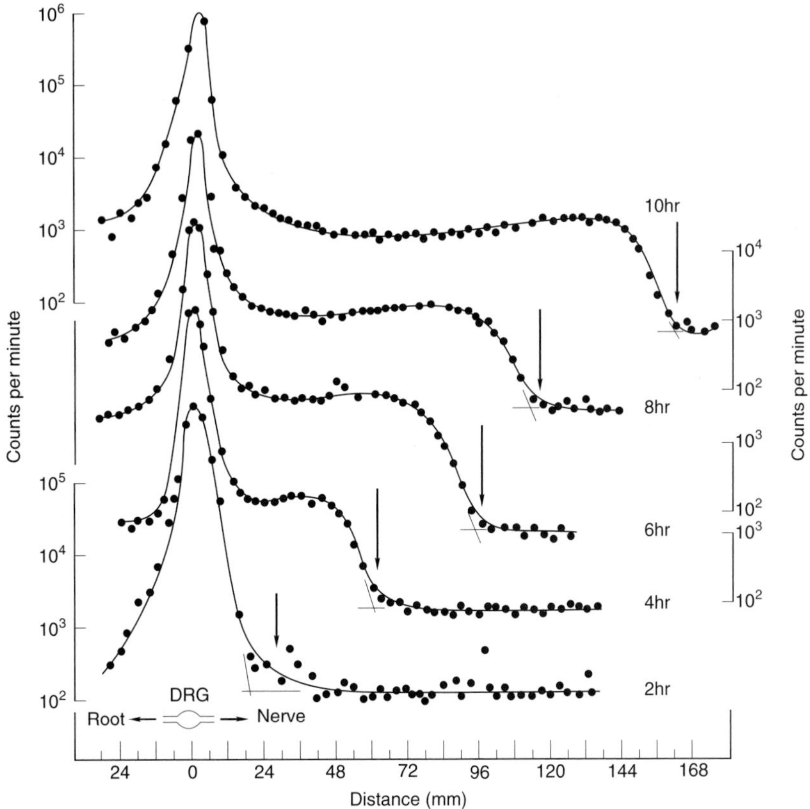

Figure 4–5 The distribution of radioactive proteins along the sciatic nerve of the cat at various times after injection of [^{3}H]leucine into dorsal root ganglia (DRG) in the lumbar region of the spinal cord. The radioactive leucine is incorporated into protein in the cell bodies within the ganglion. (In this figure only radioactivity incorporated into protein is shown.) To display transport curves from various times (2, 4, 6, 8, and 10 hours after the injection) in one figure, several ordinate scales (in logarithmic units) had to be used. Large amounts of labeled protein stay in the ganglion cell bodies. With time, protein moves out along axons in the sciatic nerve. Since the advancing front of the labeled proteins (**arrows**) is displayed progressively farther from the cell body with time, the velocity of transport can be calculated from the distances displaced. From experiments of this kind, Sidney Ochs found that the rate of fast transport is constant at 410 mm per day at body temperature. (Adapted from Ochs, 1972.)

the nerve at various times after the injection yields a transport profile (Figure 4–5). Studies using this approach show that fast anterograde transport depends critically on the production of ATP for energy, is not affected by inhibitors of protein synthesis (once the label is incorporated), and in fact is independent of the cell body, since it occurs even in nerves that have been severed from their cell bodies in the ganglion.

Fast anterograde transport depends on one or more of the filaments that make up the neuron's cytoskeleton (which is discussed later in this chapter). Microtubules provide an essentially stationary track on which specific organelles move in a saltatory fashion. Evidence for this idea is that colchicine and vinblastin, drugs that disrupt microtubules and block the process of mitosis (which is known to depend on microtubules), also interfere with fast transport. Current work has implicated one or more microtubule-associated ATPases (enzymes that catalyze the breakdown of ATP). The motor molecule for anterograde movement is *kinesin*, an ATPase large enough to be seen in the electron microscope. Kinesins form cross-bridges between microtubules and the moving membranous organelles; in a closeup view, kinesins have the appearance of little feet walking along the microtubules (Figure 4–6).

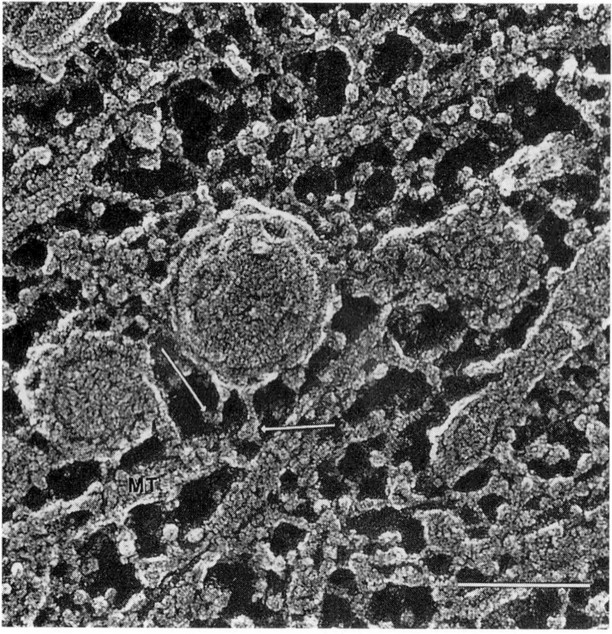

A

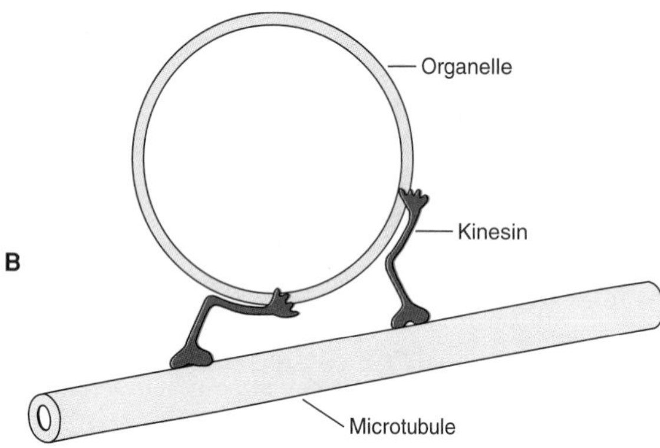

B

Figure 4–6 Structures with the morphology of kinesin molecules appear to crosslink membrane-bound organelles to microtubules.

A. Many rod-shaped structures bridge between organelles (large round structures) and microtubules (**MT**) in this quick-freeze, deep-etched electron micrograph from rat spinal cord. Several of these rod-shaped cross-bridges have globular ends that appear to contact the microtubules (**arrows**). Bar = 100 nm.

B. Model for how kinesin moves organelles along microtubules. Kinesin contains a pair of globular heads that bind to microtubules and a fan-shaped tail that binds the organelle to be moved. A hinge region is present near the center of the kinesin molecule. The similarities between kinesin and muscle myosin suggest that movement is produced by the sliding of the kinesin molecules along microtubular tracks. (From Hirokawa et al., 1989.)

Rapid transport also occurs in the retrograde direction, returning materials from nerve endings to the cell body either for degradation or for restoration and reuse. These materials are packaged in large membrane-bound organelles that are part of the lysosomal system. The rate of fast retrograde transport is about one-half to two-thirds that of fast anterograde transport. As in fast anterograde transport, particles move along microtubules. The motor molecule for fast retrograde transport is a form of *dynein*; like kinesin in the fast anterograde system, this is a microtubule-associated ATPase (MAP-1C).

Although transport in the retrograde direction serves a necessary scavenging function, the movement of materials from nerve endings back to the cell body also has other important functions. There is strong evidence from developmental neurobiology that retrograde transport has a role in informing the cell body about events that occur at the distant ends of axonal processes. For example, nerve growth factor, a protein released by postsynaptic cells and taken up at the terminals for transport to the cell body, stimulates growth of the neuron. Retrograde transport does not always work to the benefit of the cell. Some viruses (for example, poliovirus and herpesvirus) and toxins reach the central nervous system by ascending from peripheral nerve terminals to cell bodies by fast retrograde transport.

Slow Axoplasmic Flow

Whereas subcellular organelles are moved out along the axon by fast transport, the cytosol (cytoskeletal elements and soluble proteins) is transported by slow axoplasmic flow. Axoplasmic flow is somewhat more complex than fast anterograde axonal transport. It consists of at least two kinetic components that can be distinguished by their relative rates of movement and by the proteins that each transports. The slower of the two components moves at a rate of 0.2 to 2.5 mm per day and carries the subunits that make up neurofilaments and two of the subunits that make up the microtubule. These fibrous proteins constitute about 75% of the total protein moved by the slower component.

The faster component of axoplasmic flow is about twice as fast as the other component. Its protein composition is more complex; except for actin, which constitutes 2% to 4% of the protein carried by this component, all the other proteins are present in much smaller amounts. This component also transports the enzymes of intermediary metabolism that are formed on free ribosomes. It should be obvious that axoplasmic flow does not occur in the retrograde direction.

Fibrillar Proteins of the Cytoskeleton Are Responsible for the Shape of Neurons

The bulk of the axoplasm, including the more threadlike (fibrillar) elements of the cytoskeleton, moves by slow transport. As we have seen, the proteins that constitute the cytoskeleton mediate the movement of organelles from one region of the cell to another and serve to anchor membrane constituents, for example, receptors, at appropriate locations on the cell's surface. They also determine the shape of the neuron. The organization of the cytoskeleton in an axon is revealed in the freeze-etched preparation shown in Figure 4–7. Three types of fibrillar elements of varying thickness are the chief constituents of the cytoskeleton of neurons: microtubules, neurofilaments, and microfilaments together with their associated proteins. Each of the filaments is a polymer—a macromolecule consisting of a varying number of repeating monomeric subunits.

Microtubules, the thickest of the neuron's cytoskeletal fibers, are long macromolecules usually constructed of 13 protofilaments packed in a tubular array with an outside diameter of 25 to 28 nm (Figure 4–8A). Each monomer binds two guanosine triphosphate (GTP) molecules, or one GTP and one GDP molecule. In the axon the monomers that constitute the polymer are oriented longitudinally, with their polarity always in the same direction. This arrangement is presumably important for the directional specificities of the two forms of fast axonal transport, anterograde and retrograde. Although axonal or dendritic microtubules can be as long as 0.1 mm, they usually do not extend the full length of the axon or dendrite, and are not continuous with microtubules in the cell body.

Neurofilaments, 10 nm in diameter, typically are the most abundant fibrillar components in axons and serve as the bones of the cytoskeleton (Figure 4–8B). The discovery of neurofilaments, which are also referred to as neurofibrils, was important in verifying the neuron theory because these are the elements that retain silver nitrate, the stain first applied by Golgi and later used extensively by Ramón y Cajal. Neurofilaments are related to the intermediate filaments of other cell types, all of which belong to a family of proteins called *cytokeratins*. Neurofilaments are essentially totally polymerized in the cell: like hair, to which they are related, there is hardly any physiological condition under which these proteins can exist in solution.

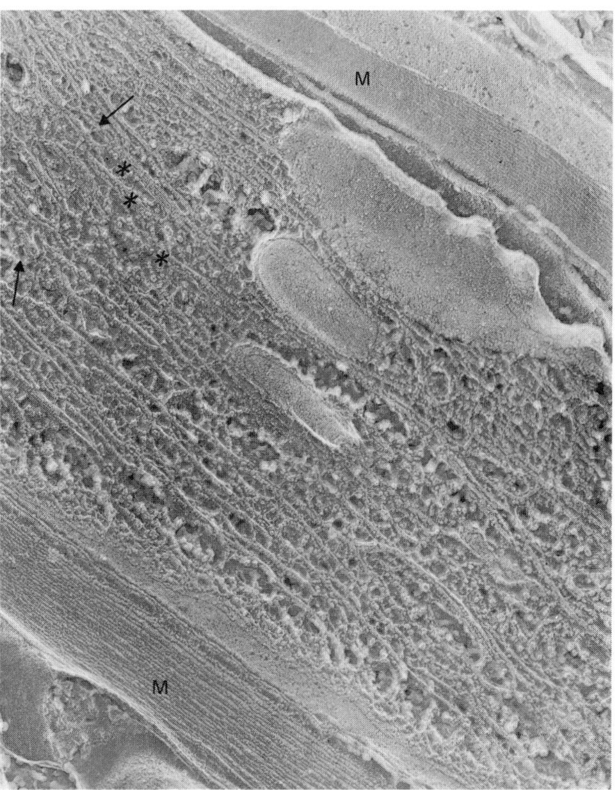

Figure 4–7 The distribution of fibrillar elements of the cytoskeleton is not uniform in the axon. Some regions are enriched in microtubules and others with neurofilaments. The regions rich in microtubules (**stars**) appear to contain the vesicular particles that are moved by axonal transport. At the top of the figure are two sausage-shaped organelles that probably correspond to components moving by retrograde axonal transport. These organelles are in a microtubule domain of the axoplasm (bracketed by **arrows** at left) that passes obliquely through the plane of fracture and consequently has a very irregular outline. **M**, myelin sheath. × 105,000. (Courtesy of B. Schnapp and T. Reese.)

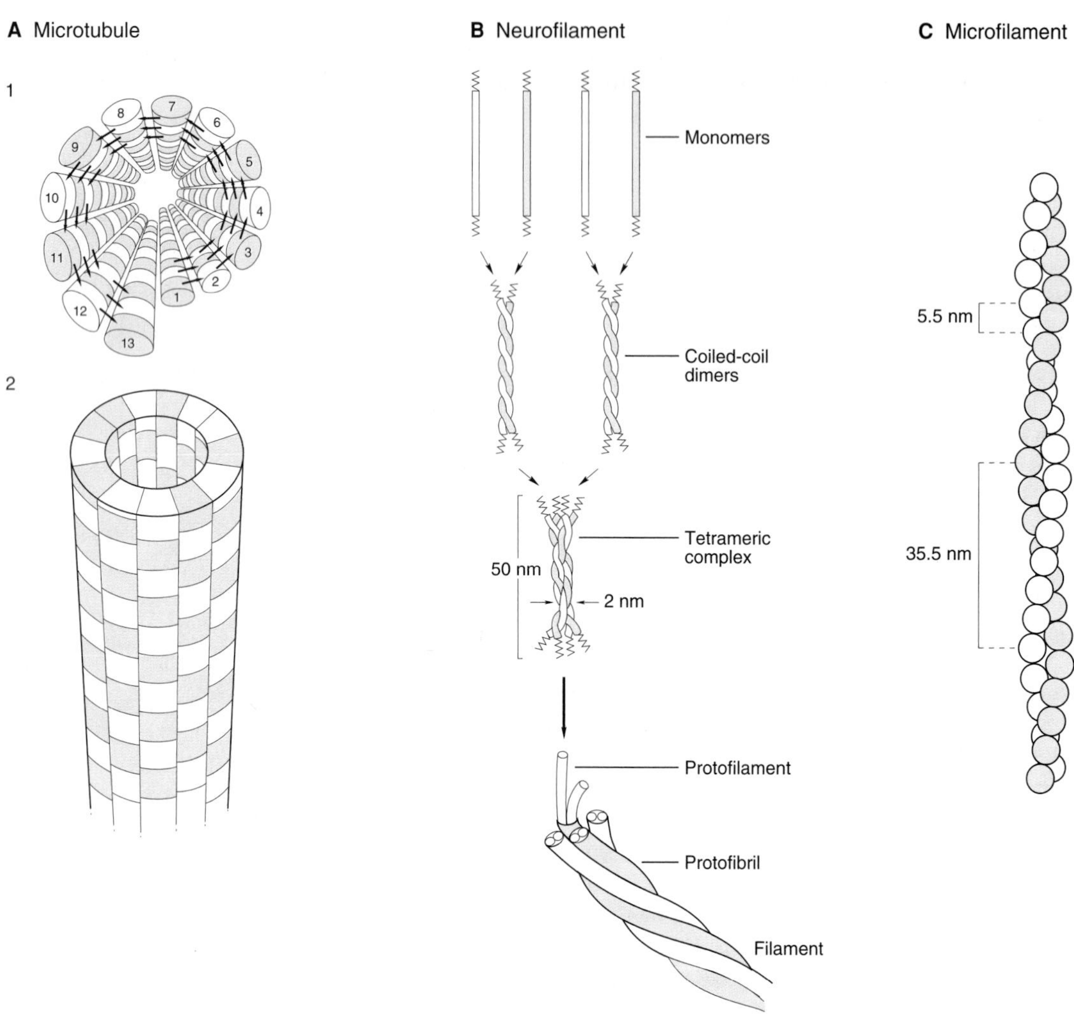

Figure 4–8 Atlas of fibrillary structures.

A. Microtubules, the largest-diameter fibers (25 nm), are helical cylinders composed of 13 protofilaments each 5 nm in width. Protofilaments are linearly arranged pairs of alternating α- and β-tubulin subunits (each subunit has a molecular weight of about 50,000). A tubulin molecule is a heterodimer consisting of one α- and one β-tubulin subunit. **1.** In this view up a microtubule the arrows indicate the direction of the right-handed helix. **2.** Alternating α- and β-subunits are shown in this side view.

B. Neurofilaments are built with fibers that twist around each other to produce coils of increasing thickness. The thinnest units are monomers that form coiled-coil heterodimers. These dimers form a tetrameric complex that becomes the protofilament. Two protofilaments become a protofibril, and three protofibrils are helically twisted to form the 10 nm neurofilament. (From Bershadsky and Vasiliev, 1988.)

C. Microfilaments, the smallest-diameter fibers (about 7 nm), are composed of two strands of polymerized globular (**G**) actin monomers arranged in a helix. The length of each monomer unit is 5.5 nm; 35.5 nm is the distance of one turn of the helix. Several isoforms of G-actin are encoded by families of actin genes. In mammals there are at least six different (but closely related) actins, two of which are classified as nonmuscle, cellular, or β and γ. Each variant is encoded by a separate gene. Microfilaments are polar structures, as the globular monomers actually are asymmetric. These monomers polymerize tip to tail.

They also are oriented along the length of the axon. On average, there are 3 to 10 times more neurofilaments than microtubules in an axon; indeed, some small axons have few, if any, microtubules. In Alzheimer's disease and some other degenerative disorders these proteins appear to be modified, forming a characteristic lesion called the *neurofibrillary tangle*.

Microfilaments, 3 to 5 nm in diameter, are the thinnest of the three types of fibers that make up the cytoskeleton (Figure 4–8C). Like the thin filaments of muscle, microfilaments are polar polymers of globular actin monomers (each bearing a molecule of ATP or ADP) wound into a two-stranded helix. Actins are a major constituent of all cells (perhaps the most abundant animal protein in nature). Much of the actin in neurons is associated with the plasma membrane; in the dendrites of cortical neurons it is concentrated at the dendritic spines, specialized sites where most synapses occur. Some axonal microfilaments are oriented longitudinally. As in other cells, these filaments are attached to the plasma membrane through several associated proteins linked to actin. (These structural proteins are not considered integral membrane proteins because their polypeptide chains do not extend through the lipid bilayer.) Microfilaments are also able to interact with proteins in the matrix outside the cell through their association with a family of transmembrane proteins called *integrins*.

No cell of the body is fixed in shape. This is especially evident during development and cell division, but continues throughout the life of the animal. Nerve cells, which do not divide, nonetheless also change their shape. Particularly important are shape changes in the nerve cell's processes, its dendrites and axons. The dynamic state of the cytoskeleton underlies these changes—the fibers of the neuronal cytoskeleton are continually expanding or contracting in length. Microfilaments and microtubules are in a state of continuous flux in the neuron. At any one time about half of the total actin in neurons may exist as unpolymerized monomers. The state of actin within the cell is controlled by several different binding proteins that block polymerization by capping the growing end of the filament or by severing the filament, or both. Other binding proteins crosslink or bundle microfilaments. The dynamics of polymerization varies with each type of fiber (Box 4–2).

An Overall View

Genetic information from the nucleus is transcribed into mRNA and carried through nuclear pores into the cytoplasm, where it is translated into one of the three major classes of proteins: (1) cytosolic proteins, (2) nuclear and mitochondrial proteins, and (3) proteins for the major membrane system of the neuron. Each of these classes of macromolecules has distinctive functions in the physiology of the neuron.

Cytosolic proteins, which are distributed throughout the neuron primarily by slow axoplasmic flow, include cytosolic enzymes and the fibrillar elements of the cytoskeleton that determine the shape of the cell. These enzymes, which are used both for the production of energy and for special biosynthetic pathways, consume or transform the many low-molecular-weight substances in the cell.

Proteins destined for the nucleus include the important enzymes that synthesize DNA and RNA and the various transcription factors that regulate gene expression. The primary function of mitochondria, which are also distributed throughout the neuron by one component of axoplasmic flow, is to generate ATP, the major molecule by which cellular energy is transferred or spent.

Finally, membrane and secretory proteins, which are moved along axons and dendrites by fast axonal transport, participate in the signaling function of the neuron and its interaction with the environment—through secretion by exocytosis and through maintenance of the plasma membrane by recycling of membrane and endocytosis.

Proteins in each of the three classes have properties that, while not *directly* relevant to axonal conduction and synaptic transmission, contribute indirectly. Thus, cytosolic enzymes catalyze the synthesis of the small-molecule transmitter substances; the energy of the ATP formed in mitochondria is needed in the mechanism of synaptic transmission. Furthermore, one of the chief products of the class of cell membrane and secretory proteins in neurons is the synaptic vesicle and, in some neurons, its neurosecretory contents. Finally, this class includes most of the integral membrane proteins that are destined to become ion channels and receptors.

BOX 4–2 Dynamics of Polymerization

The simplest way all filamentous proteins polymerize is *self-assembly*. Self-assembly requires no input of energy but instead depends on an equilibrium between the concentrations of monomers and the polymer, which is more stable than are the free monomers. For example, the neurofilament polymer is essentially stable in the cell and, thermodynamically, is at a lower energy level than the free monomers and protofilaments (Figure 4–9A).

The polymerization of microfilaments and microtubules also depends on the equilibrium conditions imposed by self-assembly, but in addition consumes the energy of a nucleotide triphosphate (ATP or GTP) during polymerization. In microfilaments the hydrolysis of ATP to ADP permits the molecule to polymerize at one end and depolymerize at the other. This process of directional polymerization is called *treadmilling* (Figure 4–9B). Rapid growth or disappearance of microtubules plays a crucial role in many cellular processes, for example, in the movement of chromosomes during cell division and in the nervous system during the growth and extension of axons and dendrites.

In microtubules each monomer of tubulin can act as a GTPase (an enzyme that hydrolyzes GTP to GDP and inorganic phosphate), but hydrolyzes only one of the two GTP molecules bound. This process is called *dynamic instability* (Figure 4–9C). In addition to self-assembly and treadmilling, microtubules can also undergo rapid alterations in length. At the growing ends of microtubules the GTP bound to the recently added monomers is not yet hydrolyzed, and these monomers do not readily depolymerize. This situation favors continued elongation. In the midregion of the polymer each monomer binds one GTP and one GDP (only one of the two bound GTPs can be hydrolyzed). Depolymerization is dependent on the nature of the nucleotides at the ends of the polymer; the rate of polymerization is related to the length of the caps bearing monomers that bind unhydrolyzed GTP. If for any reason polymerization is physically blocked or delayed, more of the GTP at the capped ends is hydrolyzed, and the polymer shortens rapidly. Alternatively, microtubules can be stabilized by capping the ends of the polymer.

A True equilibrium

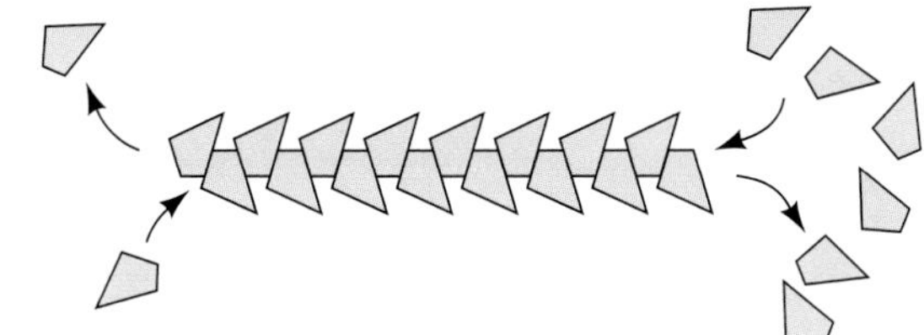

B Treadmilling

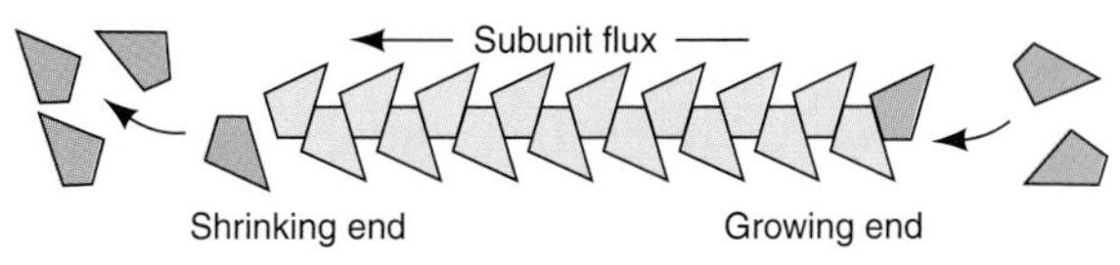

C Dynamic instability

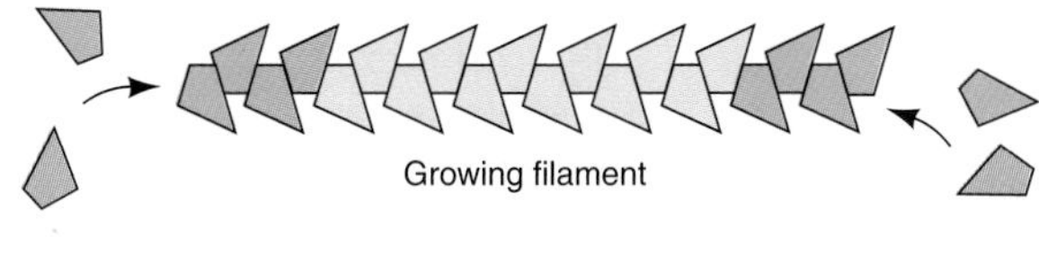

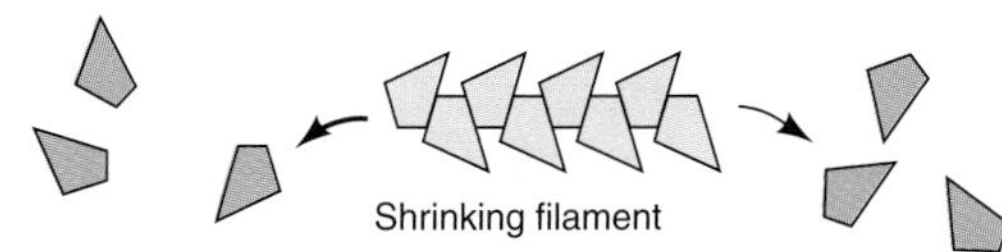

Figure 4–9 Three states of polymerization. At the concentrations present within the neuron, all three fibers (microtubules, neurofilaments, and microfilaments) are more stable than the free monomers from which they are formed.

A. A true equilibrium between the polymer and monomers. Above a critical concentration, more monomers will exist incorporated into polymers than as free molecules. Nevertheless, as with all other chemical equilibria, the components will be in a dynamic state, monomers adding and dissociating from both ends of the polymer.

B. Addition of monomers to one end of the polymer with dissociation at the other (treadmilling or directional polymerization) can be imposed on the true equilibrium described in **A** if some source of input energy is made. Energy is put into the system when a nucleotide triphosphate is hydrolyzed to a nucleotide diphosphate. The **shaded** monomers in the diagram signify molecules with bound nucleotide triphosphates.

C. In addition to true equilibrium and treadmilling, microtubules and microfilaments can also change in length abruptly. Again the energy needed for these changes, which are called *dynamic instability*, is provided in the form of the nucleotide triphosphates. Dynamic instability is a transient (nonequilibrium) process that depends on the amounts of monomers that have bound nucleotide triphosphates at the ends of the polymer. Any condition that increases the number of monomers with unhydrolyzed nucleotide triphosphates at the ends of the polymer (called *caps*) will increase the rate of polymerization by stabilizing its growth. Any condition that decreases the number of monomers with nucleotide triphosphates in the caps leads to rapid shortening of the polymer.

Selected Readings

Alberts, B., Bray, D., Lewis, J., Raff, M., Roberts, K., and Watson, J. D. 1994. Molecular Biology of the Cell, 3rd ed. New York: Garland.

Bershadsky, A. D., and Vasiliev, J. M. 1988. Cytoskeleton. New York: Plenum.

Darnell, J., Lodish, H., and Baltimore, D. 1990. Molecular Cell Biology, 2nd ed. New York: Scientific American Books.

Evans, W. H., and Graham, J. M. 1989. Membrane Structure and Function. Oxford: IRL Press.

Fawcett, D. W. 1981. The Cell, 2nd ed. Philadelphia: Saunders.

Grafstein, B., and Forman, D. S. 1980. Intracellular transport in neurons. Physiol. Rev. 60:1167–1283.

Holtzman, E. 1989. Lysosomes. New York: Plenum.

Kelly, R. B. 1993. A question of endosomes. Nature 364:487–488.

Kelly, R. B. 1993. Storage and release of neurotransmitters. Cell 72/Neuron 10 (Suppl.):43–53.

Kreis, T., and Vale, R. (eds.) 1993. Guidebook to the Cytoskeletal and Motor Proteins. Oxford: Oxford University Press.

Spudich, J. A. (ed.) 1989. Molecular Genetic Approaches to Protein Structure and Function: Applications to Cell and Developmental Biology. New York: Liss.

Warner, F. D., and McIntosh, J. R. (eds.) 1989. Cell Movement, Vol. 2: Kinesin, Dynein, and Microtubule Dynamics. New York: Liss.

Warner, F. D., Satir, P., and Gibbons, I. R. (eds.) 1989. Cell Movement, Vol. 1: The Dynein ATPases. New York: Liss.

5

The Nervous System

The Nervous System Has Peripheral and Central Components

The Central Nervous System Consists of Seven Main Regions

The Cerebral Cortex Is Divided into Four Functional Lobes

Interaction of the Sensory, Motor, and Motivational Systems Is Essential for Even Simple Behavior

The Anatomical Organization of Each Major Functional System Follows Four Principles

- *Each System Contains Relay Centers*
- *Each System Is Composed of Several Distinct Pathways*
- *Each Pathway Is Topographically Organized*
- *Most Pathways Cross the Body's Midline*

The architecture of the nervous system, although complex, follows a relatively simple set of functional, organizational, and developmental principles. Taken together, these principles bring order to the myriad details of brain anatomy. In this chapter we first review the gross anatomical organization of the nervous system and then consider how the *functional systems* for perception, motor coordination, and motivation interact during a simple behavioral act. We then discuss four general principles that underlie the anatomical organization of these functional systems. In Chapter 6 we examine how in development the specific wiring of the brain can be influenced by both genetic and environmental factors.

Until quite recently descriptions of the anatomy of the nervous system relied on postmortem dissection. Modern radiography has changed that in a very important way by permitting the nervous system to be seen in living humans. Two recent imaging techniques in particular have revolutionized the study of the regional anatomy of the brain. Positron emission tomography (PET) and magnetic resonance imaging (MRI) are used primarily by clinicians to locate lesions of the brain with remarkable accuracy and safety, circumventing invasive procedures that interfere with normal function. Neuroscientists, however, also use these techniques to observe the brain in action—during thought, perception, and voluntary actions (Box 5–1).

Box 5–1 Computerized Tomography

Computer-assisted tomography (CAT), like conventional radiography, produces X-ray images of various structures—bones, soft tissues, and fluid-filled vessels. The CAT scan, however, is able to reveal subtle differences and fine structures in tissue that cannot be detected in conventional X-ray images (Figure 5–1). Structures within the brain can be distinguished: the thalamus, the basal ganglia, the gray and white matter of the cerebral cortex, and the ventricles (Figure 5–2). Moreover, the CAT scan generates an image of a single plane or section of tissue (hence the term *tomography* from the Greek for "cut"). Because it reveals anatomical detail, computer-assisted tomography has greatly expanded the clinician's capacity for diagnosis. One limitation, however, is that the views of the brain produced by computer-assisted tomography are necessarily *static*—CAT scans make it possible to explore the structure but not the function of the brain.

Positron emission tomography (PET) does provide images of live, real-time brain functioning, and thus has revolutionized the study of human cognitive processes and psychiatric and neurological disease. Positron emission tomography combines the principles of computer-assisted tomography with those of radioisotope imaging. Both CAT and PET images are derived from data gathered by a detector that rotates around the patient's head. However, whereas CAT images reflect the *absorption* of radiation (specifically X-rays) by tissue, PET images are produced by the *emission* of radiation by an injected or inhaled isotope in tissue.

Because it is able to measure differential rates of glucose metabolism in the active brain, PET scanning has proven to be a powerful tool in locating sites or structures in the brain that are particularly active during a specific behavior. This technique takes advantage of the fact that all neurons must metabolize glucose to obtain energy. When, for purposes of a PET scan, an analog of glucose (2-deoxyglucose) is introduced into the cerebral bloodstream, the neurons take up this molecule and begin to treat it like glucose; that is, they add a phosphorus compound to it. Unlike glucose 6-phosphate, however, deoxyglucose 6-phosphate cannot be further metabolized. Nor can it leave the area, because it is now too large to cross the cell membrane. Thus, deoxyglucose 6-phosphate accumulates within the active brain cells. By adding a positron-emitting isotope that binds very specifically to the trapped molecules of deoxyglucose 6-phosphate, the isotope emissions identify the areas of glucose metabolism, and hence activity, in the brain.

The PET scans in Figure 5–3 show the rate of glucose metabolism in the brain of a normal person at rest. These images illustrate not only functioning components but also the underlying nervous structure, because all neural structures use glucose. White matter uses much less glucose than gray matter, and different regions of gray matter have distinctive patterns of glucose metabolism. Figure 5–4 shows PET scans of glucose metabolism in a normal brain during visual stimulation. When the person views a complex scene, activity in the primary visual cortex increases, and higher-order visual cortical areas become active as well.

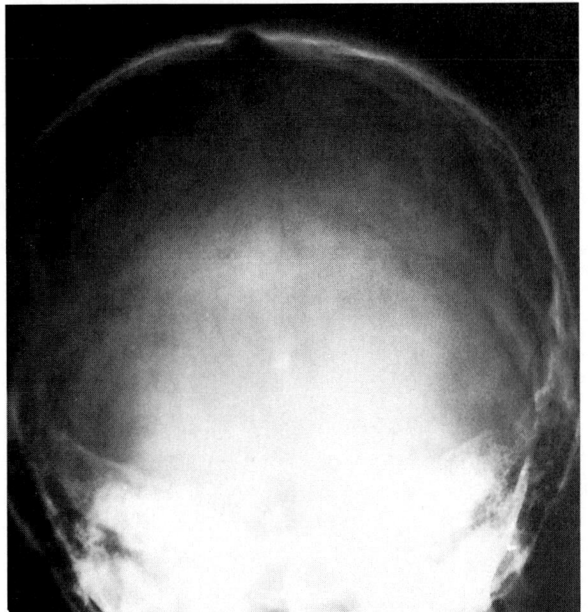

A Radiograph

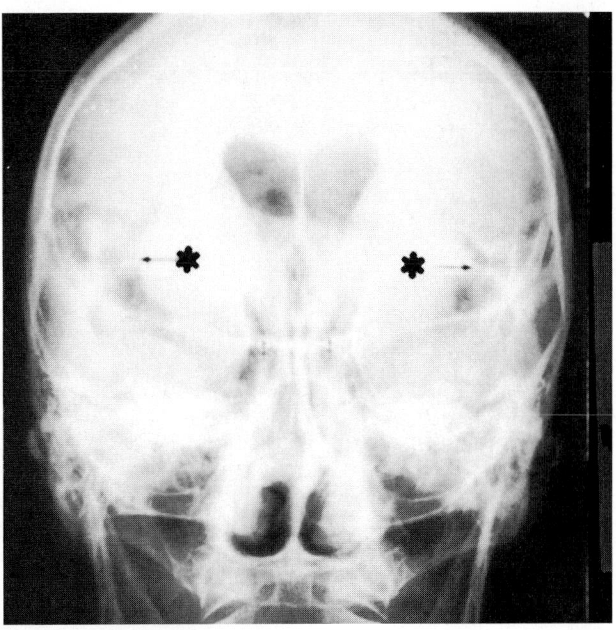

B Pneumoencephalogram

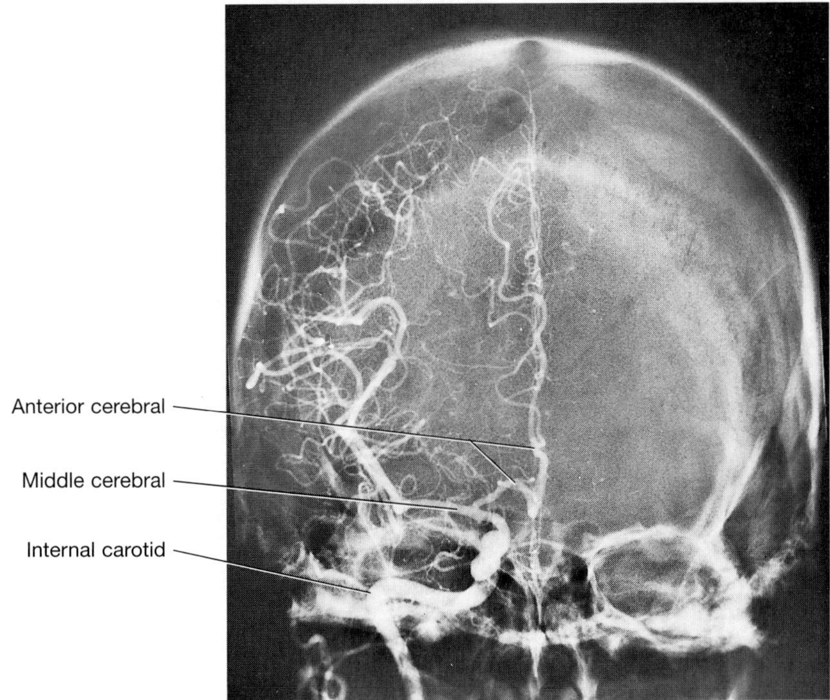

C Angiogram

Figure 5–1 Traditional radiographic methods have contributed much to our understanding of the anatomy of the nervous system.

A. In a radiograph of the skull X-rays are absorbed by bones and other tissues that absorb calcium. All these structures then appear light on X-ray film. X-rays cannot detect the gray matter or white matter.

B. In a pneumoencephalogram the cerebrospinal fluid in the ventricles of the brain is replaced with air. As a result, the ventricles appear dark on X-ray film. The arrows with asterisks indicate air in the lateral sulcus. (Courtesy of Dr. Robert McMasters.)

C. In an angiogram a radiopaque (X-ray-absorbing) material is injected into the cerebral bloodstream. X-ray film then reveals the cerebral blood vessels, here presented in a frontal view.

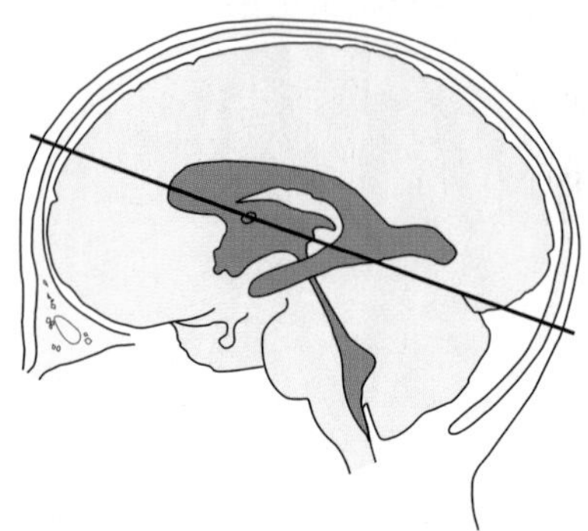

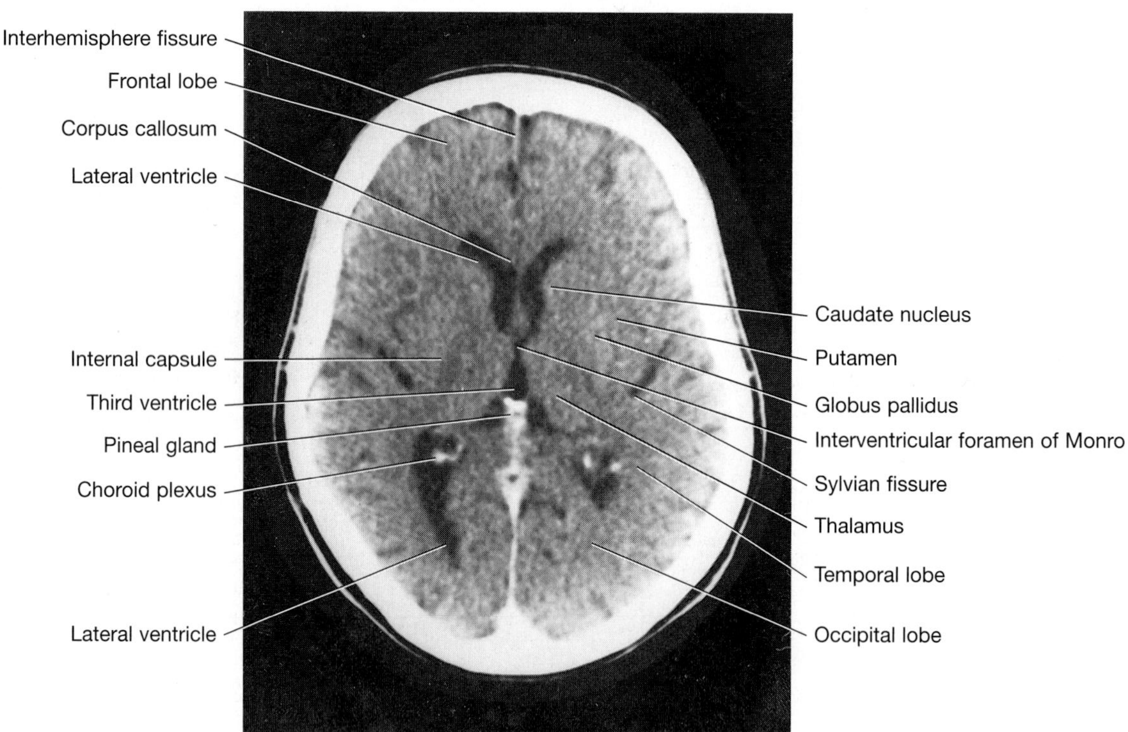

Figure 5–2 Computerized tomography is able to distinguish gray matter (neurons) and white matter (glial cells) in the brain. The imaging plane in this CAT scan through the cerebral hemisphere and diencephalon is parallel to an imaginary line between the eye and the ear canal.

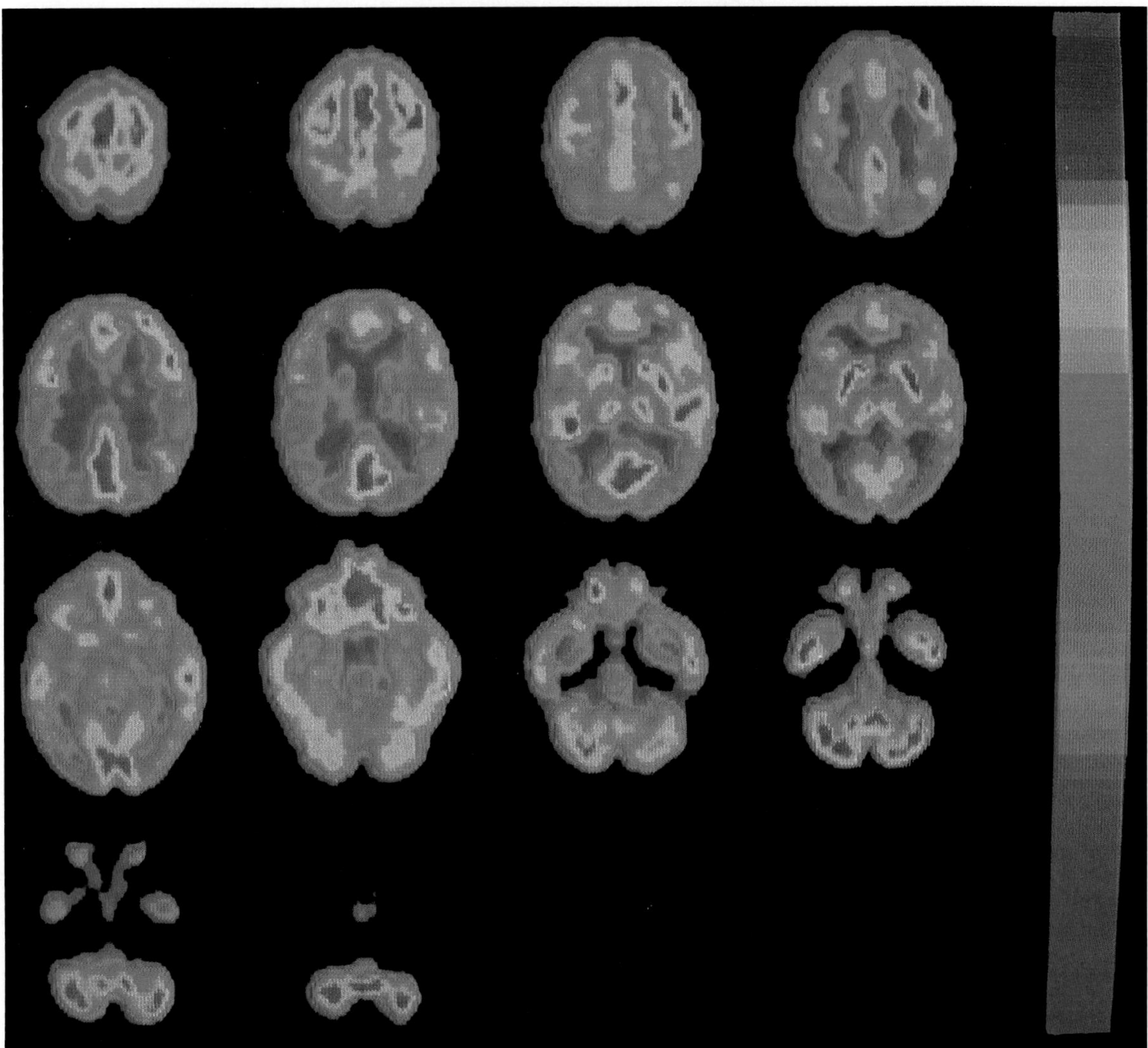

Figure 5–3 This series of PET scans discloses local patterns of glucose metabolism in a normal brain at rest. The 14 consecutive sections of the brain are each 8 mm apart, from dorsal (**top**) to ventral (**bottom**) levels. Gray matter, which contains the cell bodies and dendrites of neurons as well as the regions of synaptic contact, is metabolically more active than white matter, which contains the myelinated axons. The areas of gray matter that are especially active are in the cerebral cortex, cerebellum, basal ganglia, and thalamus. (Courtesy of Drs. Michael E. Phelps and John C. Mazziotta.)

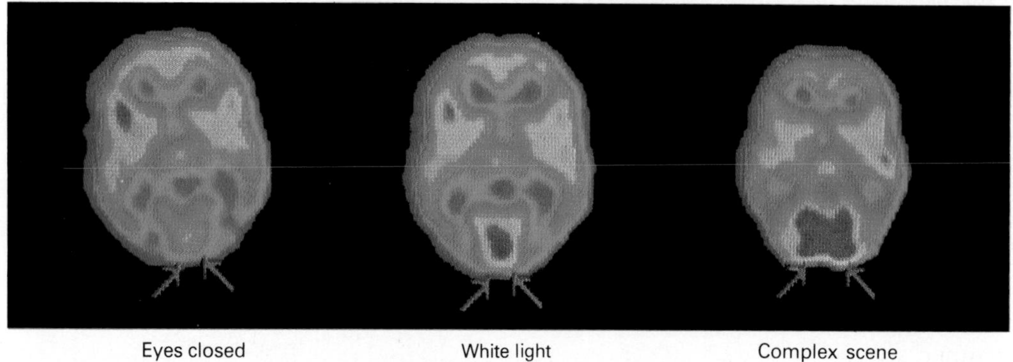

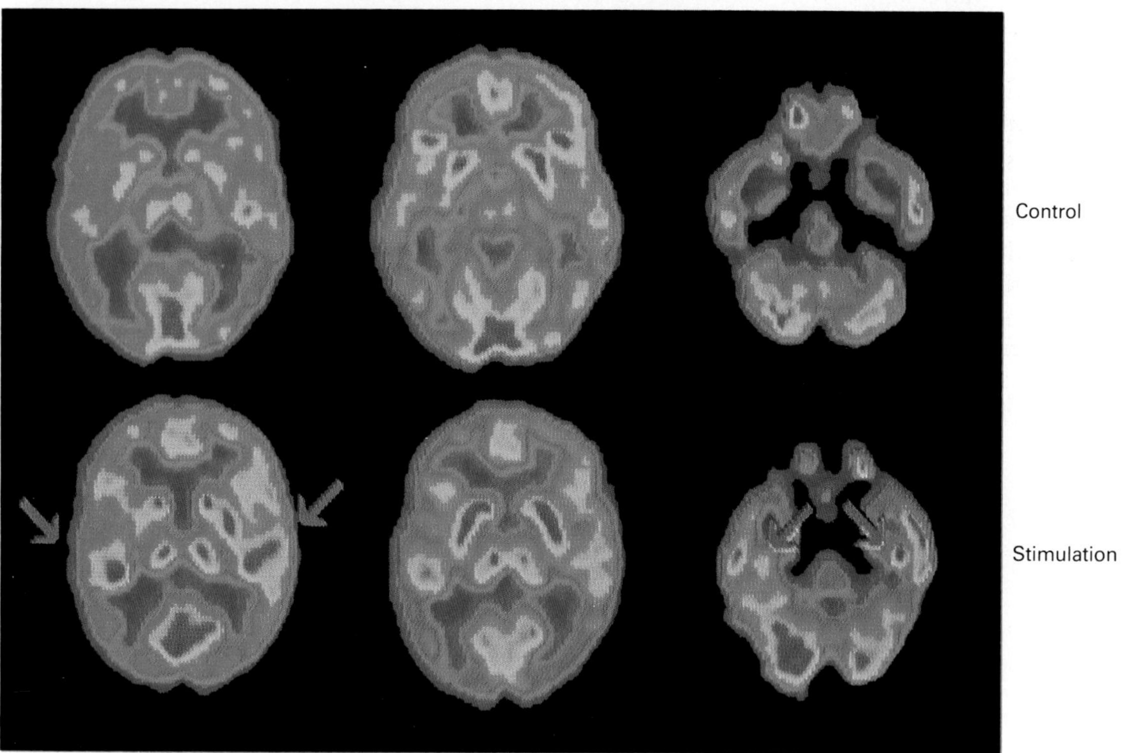

Figure 5–4 These PET scans show that different brain regions are activated by visual stimuli of different degrees of complexity. The simplest type of illumination, white light, causes the primary visual cortex to be active (**center**). Rates of activity are clearly higher than when the eyes are closed (**left**). However, the higher-order visual cortex becomes active only when the subject views a complex scene (**right**). (Courtesy of Drs. Michael E. Phelps and John C. Mazziotta.)

The Nervous System Has Peripheral and Central Components

The nervous system has two components: the *central nervous system*, composed of the brain and the spinal cord, and the *peripheral nervous system*, composed of the groups of neurons termed *ganglia* and *peripheral nerves* that lie outside the brain and spinal cord. The two systems are anatomically separate but functionally interconnected.

The peripheral nervous system has somatic and autonomic divisions. The *somatic division* provides the central nervous system with sensory information about muscle and limb position and about the environment outside the body. This division includes sensory neurons of the dorsal root and cranial ganglia that innervate the skin, muscles, and joints. Somatic motor neurons, which innervate skeletal muscle, have axons that project to the periphery; these axons are often considered part of the somatic division, even though the cell bodies are located in the central nervous system.

The *autonomic division* is the motor system for the viscera, the smooth muscles of the body, and the exocrine glands, and is commonly called the *autonomic motor system*. It consists of three spatially segregated subdivisions: the sympathetic, the parasympathetic, and enteric nervous systems. The *sympathetic system* participates in the response of the body to stress, whereas the *parasympathetic system* acts to conserve the body's resources and restore the equilibrium of the resting state. The *enteric nervous system* controls the function of smooth muscle of the gut. These components of the autonomic nervous system are described in Chapter 32.

The central nervous system is organized along two major axes that establish themselves very early in development—a longitudinal rostral-to-caudal axis (roughly, "head to tail") and a dorsal-to-ventral axis ("back to belly"). Lower vertebrates maintain the orientation of these axes throughout life. In primates the nervous system develops differently. The longitudinal axis flexes into a sharp curve as the brain begins to form, so that rostral and caudal areas of the central nervous system are no longer found at two opposite ends of a straight line. For this reason the terms used to indicate the relative position of organs of the body refer to different directions when applied to structures of the nervous system above and below the junction of the midbrain and diencephalon (Figure 5–5).

The Central Nervous System Consists of Seven Main Regions

The central nervous system of the adult can be divided into seven anatomical regions, each of which develops from a distinct region of the neural tube (Table 5–1). The seven major regions are (1) the spinal cord, (2) the medulla, (3) the pons, (4) the cerebellum, (5) the midbrain, (6) the diencephalon, and (7) the cerebral hemispheres (Figure 5–6). Each of the seven divisions is bilaterally paired.

Table 5-1 Main Subdivisions of the Embryonic Central Nervous System and Mature Adult Forms

Three-Vesicle Stage	*Five-Vesicle Stage*	*Mature Structures Derived from Vesicle*	*Related Cavity*
1. Forebrain	1a. Telencephalon (endbrain)	1. Cerebral cortex, basal ganglia, hippocampal formation, amygdala, olfactory bulb	Lateral ventricles
	1b. Diencephalon	2. Thalamus, hypothalamus, subthalamus, epithalamus, retinas, optic nerves and tracts	Third ventricle
2. Midbrain	2. Mesencephalon (midbrain)	3. Midbrain	Cerebral aqueduct
3. Hindbrain	3a. Metencephalon (afterbrain)	4. Pons	Fourth ventricle
		5. Cerebellum	Fourth ventricle
	3b. Myelencephalon (medullary brain)	6. Medulla oblongata	Fourth ventricle
4. Caudal part of neural tube	4. Caudal part of neural tube	7. Spinal cord	Central canal

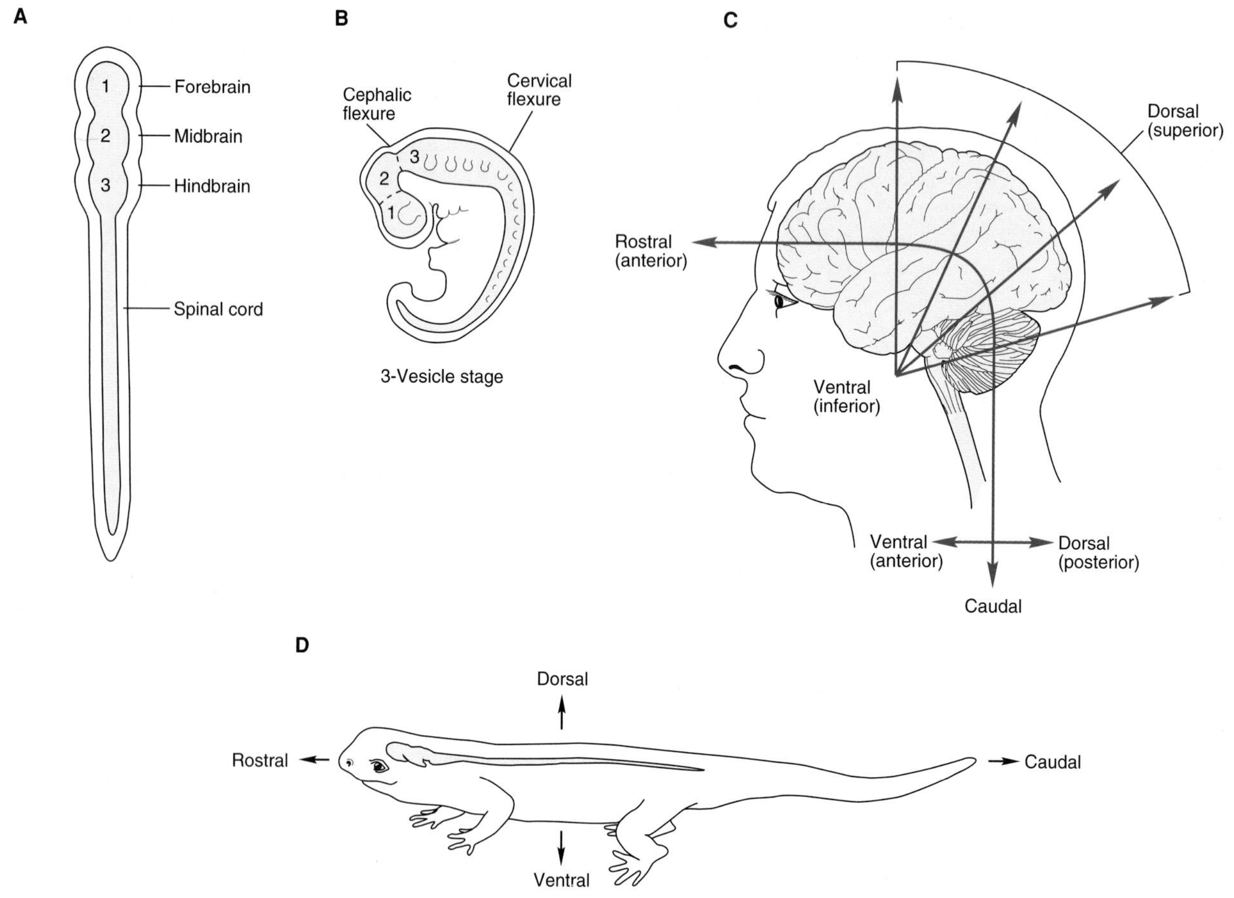

Figure 5–5 The human nervous system develops from the neural tube. The rostral neural tube gives rise to the brain and the most candal neural tube to the spinal cord (**A**). As the human nervous system develops it flexes at the junction between the midbrain and diencephalon (**B**). Thus, in the cerebral hemisphere and upper brain stem (**C**), the relative positions denoted by the terms rostral, caudal, dorsal, and ventral are different from those in the spinal cord. In the spinal cord, rostral means toward the head, caudal means toward the coccyx, ventral toward the belly, and dorsal toward the back. Above the diencephalon, however, rostral means toward the nose, caudal toward the back of the head, ventral toward the jaw, and dorsal toward the top of the skull. The terms *superior* (instead of dorsal) and *inferior* (instead of ventral) are also common. At all levels of the nervous system the neuraxis is the longitudinal (rostral-to-caudal) axis. In contrast, in lower vertebrates (**D**) the central nervous system is organized in a straight line.

1. The *spinal cord* is the most caudal part of the central nervous system. It extends from the base of the skull through the first lumbar vertebra and, thus, does not run the entire length of the spinal column. The spinal cord receives sensory information from the skin, joints, and muscles of the trunk and limbs and, in turn, contains the motor neurons responsible for both voluntary and reflex movements. It also receives sensory information from the internal organs and has clusters of neurons that control many visceral functions.

The spinal cord has a clear pattern of segmentation, evidenced in humans by 31 pairs of *spinal nerves*. The spinal nerves are peripheral nerves formed by the joining of the dorsal and the ventral roots. The *dorsal roots* carry sensory information into the spinal cord from the muscles, skin, and viscera. The *ventral roots* carry outgoing motor axons that innervate muscles, as well as axons of the sympathetic and parasympathetic systems. Within the spinal cord both the sensory cell groups that receive input from the periphery and the motor cell groups that control specific groups of muscle fibers are clustered together and not intermixed randomly. In addition to these cell groups, the spinal cord contains ascending pathways through which sensory information reaches the brain and descending pathways that relay motor commands from the brain to motor neurons.

Three of the next four divisions of the central nervous system—the medulla, the pons, and the midbrain—form a continuous structure termed the *brain stem*. The brain stem is located rostral to the spinal cord (i.e., further toward the head). It receives sensory information from skin and joints in the head, neck, and face, and contains the motor neurons that control the muscles of the head and neck. It is also concerned with specialized senses, such as hearing, taste, and balance. The sensory input and motor output of the brain stem are carried by 12 pairs of *cranial nerves*. The brain stem also contains ascending and descending pathways that carry sensory and motor information to and from higher brain regions. In addition, a network of neurons in the brain stem, known as the *reticular formation*, mediates aspects of alertness and attention.

2. The *medulla* is a rostral extension of the spinal cord and resembles the spinal cord in both organization and function. Together with the pons (below) it participates in regulating blood pressure and respiration.

3. The *pons* lies rostral to the medulla and appears as a protuberance from the ventral surface of the brain stem. Named by the Latin word for *bridge*, the pons contains a large number of neurons that relay information from the cerebral hemispheres to the cerebellum.

4. The *cerebellum* lies dorsal to the pons and the medulla and bulges out toward the back of the head, wrapping around the brain stem. It has a pleated surface and is divided into several lobes, each with its own function. The cerebellum receives sensory input from the spinal cord, motor information from the cerebral cortex, and input about balance from the vestibular organs of the inner ear. The convergence of all this input enables the cerebellum to coordinate the planning, timing, and patterning of skeletal muscle during movement. The cerebellum also plays a role in maintaining posture and in coordinating head and eye movements. Many of the motor functions of the pons and cerebellum are closely related.

5. The *midbrain*, the smallest brain stem component, lies rostral to the pons. Several regions of the midbrain play a dominant role in the direct control of eye movements, whereas others are involved in motor control of skeletal muscles. The midbrain also is an essential relay station of auditory and visual signals.

6. The thalamus and the hypothalamus together form the *diencephalon*, or between-brain, so called because it lies between the midbrain and the cerebral hemispheres. The thalamus processes and distributes almost all sensory and motor information going to the cerebral cortex. It is also thought to regulate levels of awareness and emotional aspects of sensation. The *hypothalamus* lies ventral to the thalamus and regulates the autonomic nervous system and the secretion of hormones by the pituitary gland. The hypothalamus has extensive afferent and efferent connections with the thalamus, the midbrain, and some cortical areas that receive information from the autonomic nervous system.

7. The *cerebral hemispheres* form by far the largest

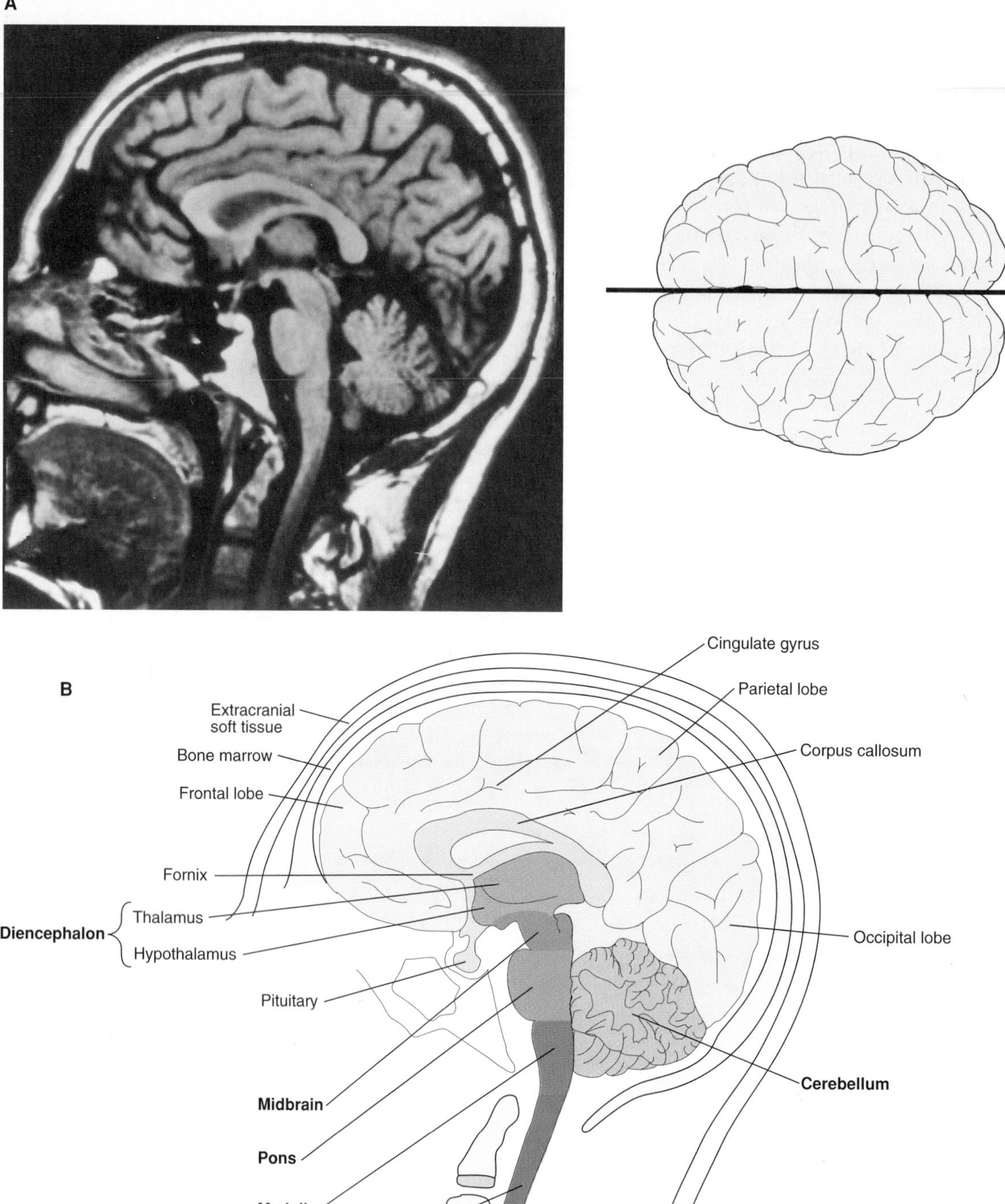

Figure 5–6 **A.** The central nervous system has seven major anatomical regions: spinal cord, medulla, pons, cerebellum, midbrain, diencephalon, and cerebral hemispheres (not shown). These can be observed in a MRI scan of a midsagittal section (**A**) through the cerebral hemispheres, corpus callosum, brain stem, and spinal cord, which reveals all major regions of the central nervous system as well as components of the ventricular system. Whereas dense bone is not seen on MRI, marrow is. The diagram (**B**) shows the detail visible in the MRI scan. The cingulate gyrus, a prominent gyrus on the medial surface, overlies the corpus callosum and fornix. These three structures each have a C shape. The cingulate gyrus and fornix are both part of the limbic system. The corpus callosum contains the axons of neurons that interconnect the two cerebral hemispheres. Each hemisphere is divided into four lobes: frontal, parietal, temporal (not visible in this drawing), and occipital (see Figure 5–8).

region of the brain. They consist of the cerebral cortex, the underlying white matter (largely myelinated axons and glial cells), and three deep-lying clusters of functionally related neurons, termed *nuclei*: the basal ganglia, the hippocampal formation, and the amygdala. The cerebral hemispheres are divided by a deep fissure that runs from front to back. For the most part, the right and left hemispheres constitute mirror images of each other. Though each hemisphere has specialized functions, both are jointly concerned with perceptual, cognitive, and higher motor functions as well as emotion and memory.

The Cerebral Cortex Is Divided into Four Functional Lobes

The cerebral cortex is the highly wrinkled surface of the cerebral hemisphere. Its peculiar shape arose during evolution of the primate brain as the volume of the cerebral cortex increased more rapidly than did the cranium. The disparity has resulted both in the crumpling of the cortical surface and in the folding of the structure as a whole (Figure 5–7).

The surface convolutions consist of grooves (*sulci*) that separate elevated regions (*gyr*). Smaller sulci vary among individuals, but larger ones are more consistent in position and thus can be used as landmarks to divide the cortex into four lobes, named after their overlying cranial bones: *frontal, parietal, temporal*, and *occipital* (Figure 5–8).

There are two other subdivisions of the cortex that are organized in a similar manner. The *insular cortex* occupies the medial wall of the lateral sulcus and is not visible on the surface of the brain. The

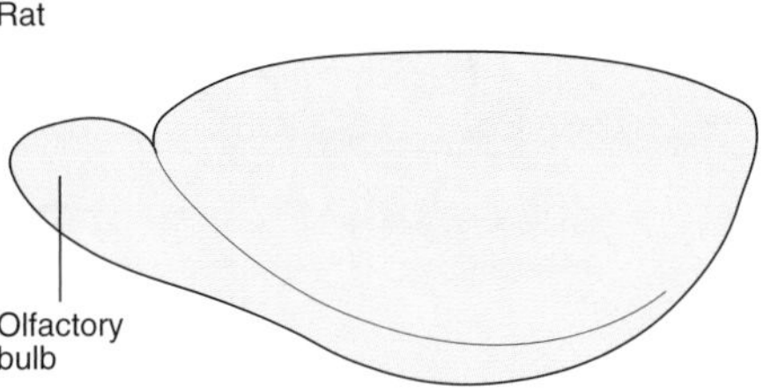

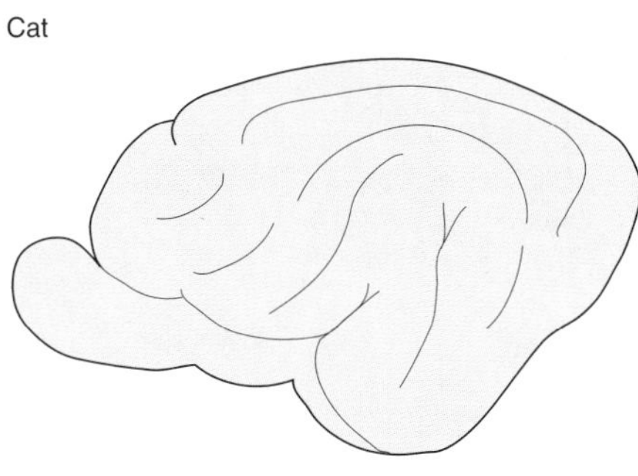

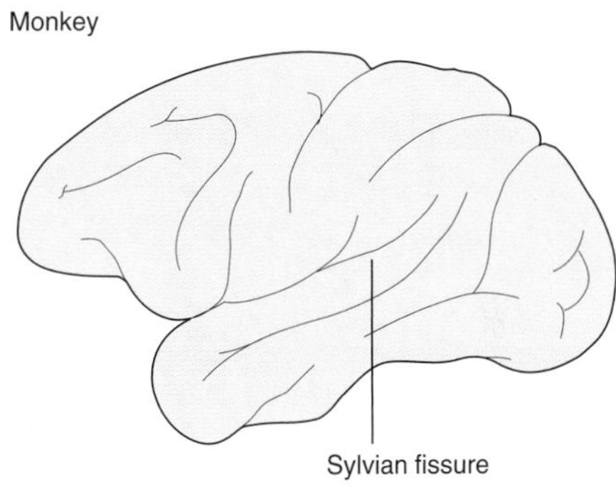

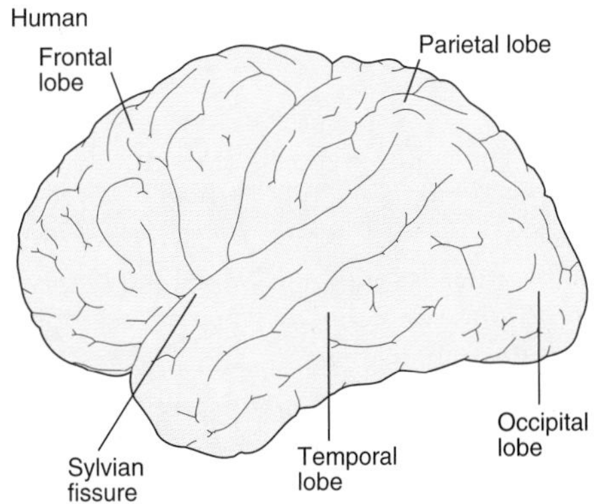

Figure 5–7 The folding of the cerebral hemisphere into a horseshoe shape during evolution has created a bulging region known as the temporal lobe. In the rat the cerebral hemisphere is not folded. In the cat the folding becomes more evident; the caudal end of the cerebral hemisphere extends slightly downward. In the monkey and in humans the folding is pronounced; the cerebral hemisphere curves from the frontal lobe toward the temporal lobe, which extends forward around the brain stem. The site of folding is an anatomical landmark called the Sylvian fissure. (Modified from Nauta and Feirtag, 1986.)

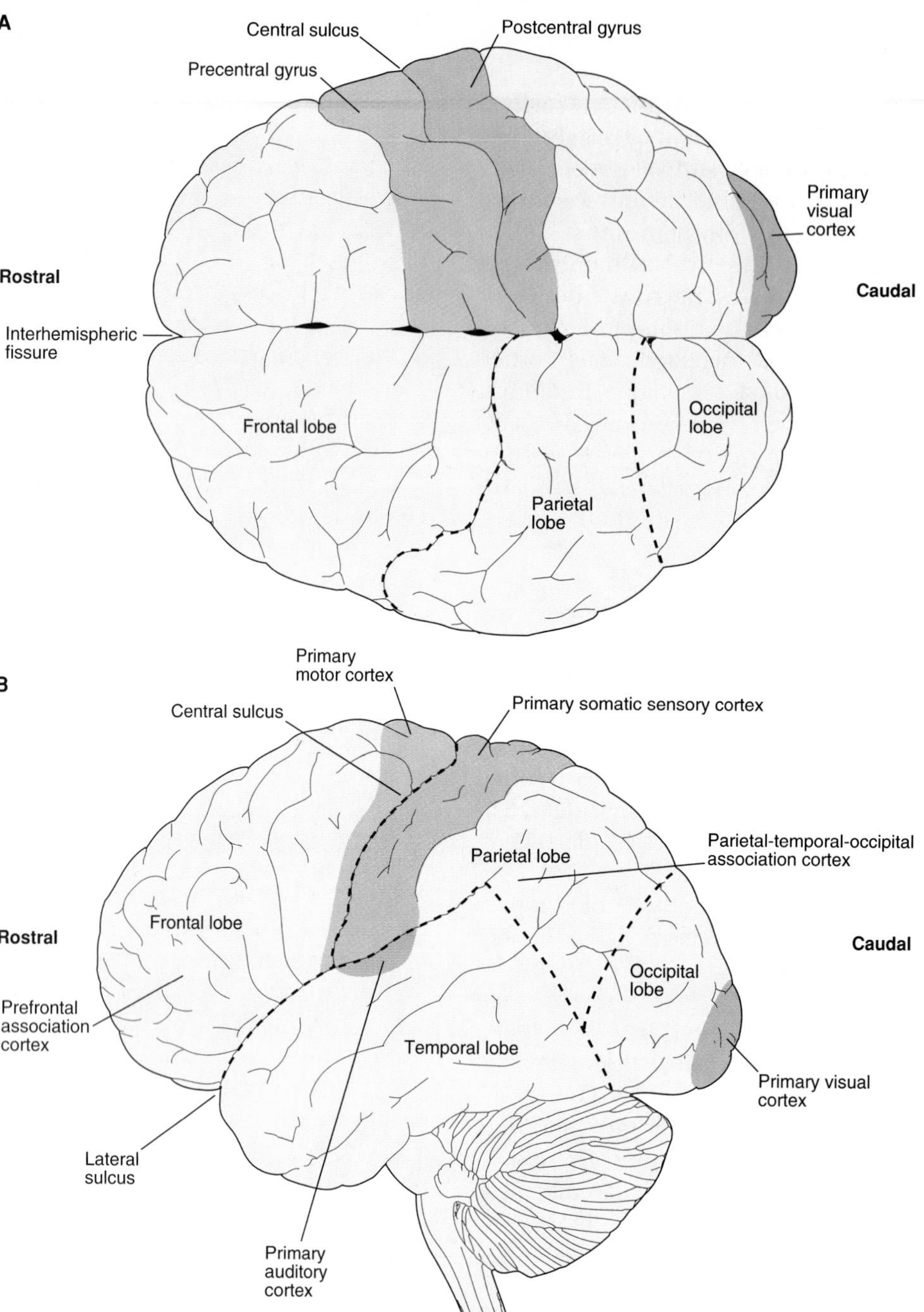

Figure 5–8 The major divisions of the human cerebral cortex are the frontal, parietal, temporal, and occipital lobes. The cortex of each lobe is thrown into folds or *gyri*, separated by grooves called *sulci*. The boundaries between the lobes are defined somewhat arbitrarily along the lines of the major sulci.

A. The separation of the cerebral hemispheres by a fissure is clear in this dorsal (overhead) view of the brain. The central sulcus defines the border between the frontal and parietal lobes. The precentral gyrus, which contains the motor cortex, lies in the frontal lobe; the postcentral gyrus, which contains the somatic sensory cortex, lies in the parietal lobe. The occipital lobe, at the caudal end of the hemisphere, contains the visual cortex. The temporal lobe, which lies ventrally, is not visible in this view of the brain.

B. The primary sensory and motor areas and the various association areas are shown in this lateral view of the left hemisphere. The primary auditory cortex, near the junction of the temporal and parietal lobes, lies within the lateral sulcus and is hidden from view. Two large association areas are visible: the prefrontal association cortex and the parietal–temporal–occipital association cortex. The most prominent cleft visible in a lateral view of the brain is the lateral sulcus, which separates the temporal lobe from the frontal and parietal lobes.

limbic lobe is not really a distinct area but consists of the medial portions of the frontal, parietal, and temporal lobes that form a continuous band of cortex overlying the rostral brain stem and diencephalon. This band is considered a unit because its neurons form complex circuits that play an important role in learning, memory, and emotions.

Many areas of the cerebral cortex are concerned primarily with processing either sensory information or motor commands. These areas are known as primary, secondary, or tertiary (sensory or motor) areas, depending on the level of information processing they carry out. For example, the *primary motor cortex* mediates voluntary movements of the limbs and trunk; it is called primary because it contains neurons that project directly to the spinal cord to activate the somatic motor neurons. The primary sensory areas receive information from peripheral receptors with only a few synaptic relays interposed. The *primary visual cortex* is located at the caudal pole of the occipital lobe, just above the cerebellum. The *primary auditory cortex* lies in the temporal lobe, near the language areas (in the left hemisphere). The *primary somatic sensory cortex* is found on the postcentral gyrus; just beyond the gyrus is the primary motor cortex.

Surrounding the primary areas are the higher-order (secondary and tertiary) sensory and motor areas. These areas process the complex aspects of a single sense or of information related to motor function. Higher-order sensory areas integrate information coming from the primary sensory areas. In contrast, higher-order motor areas send complex information required for a motor act to the primary motor cortex.

Three other large regions of cortex, called association areas, surround the primary, secondary, and tertiary areas. In primates the association areas constitute a great majority of the cortex. Their function is mainly to integrate diverse information for purposeful action, and they are involved to different degrees in the control of three major brain functions: perception, movement, and motivation. The *parietal–temporal–occipital association* cortex occupies the interface between the three lobes for which it is named (Figure 5–8B). It is concerned with higher perceptual functions related to somatic sensation, hearing, and vision—that is, the primary sensory inputs to the parietal, temporal, and occipital lobes, respectively. Information from these different senses is combined in the association cortex to form complex perceptions. The *prefrontal association cortex* occupies most of the rostral part of the frontal lobe; one important function of this area is the planning of voluntary movement. The *limbic association cortex* is found in portions of the parietal, frontal, and temporal lobes; it is devoted mainly to motivation, emotion, and memory. The organization and function of the association areas of the cortex are considered in greater detail in Chapter 19 in connection with cognition.

Finally, three deep-lying structures are part of the cerebral hemispheres: the basal ganglia, the hippocampus, and the amygdala. The *basal ganglia* are groups of neurons that have an important role in the regulation of movement and also contribute to cognition. They receive input from all four lobes of the cerebral cortex, but they send out signals only to the frontal cortex, via the thalamus.

The *hippocampus* and *amygdala* are part of the limbic system (Chapter 32). The hippocampus plays an important role in memory; the amygdala coordinates the actions of the autonomic and endocrine systems and is involved in emotions. The pathways that control the emotional quality of sensation or of motor behavior are not yet understood completely, but damage to the limbic system (or the autonomic nervous system) affects emotional expression. By means of direct connections with the hypothalamus, the limbic system modulates the activity of the autonomic nervous system, which regulates the body's basic physiology; in this way the visceral responses (such as blood pressure, heart rate, and pupillary size) are coordinated with emotional state. The limbic system exercises a major control over the endocrine systems by regulating the release of hormones from the hypothalamus.

Interaction of the Sensory, Motor, and Motivational Systems Is Essential for Even Simple Behavior

A simple purposeful behavior such as throwing a ball requires several kinds of sensory information: information about the motion of the ball, about the impact of the ball in the hand, and about the position of the arms, legs, and trunk in space.

Sensory information is fed to association areas of the cortex where the movement is planned. From there, information is transmitted to the motor system, which generates commands for movements involved in anticipating, catching, and holding the ball. These motor commands from the brain must be targeted to the correct muscles in the back, shoulder, arm, and hand, and they must be timed to coordinate the contraction and relaxation of opposing muscle groups. They must also regulate body posture as a whole. Finally, to regulate performance, the motor systems require continuous information about changes in muscle tension.

Whereas the sensory and motor systems are important in actually throwing the ball, the decision to initiate and complete the behavior is regulated by the motivational system (Chapter 33). This system exerts an unmistakable influence on the motor output to skeletal muscles: how well the ball is caught may depend on whether the catcher is excited, bored, or distracted.

Neurons of the motivational system influence voluntary movement by forming connections with neurons in the somatic motor system in the brain. The motivational system also influences behavior through its action on the autonomic nervous system. Thus, the same system that modulates the activity of the skeletal motor system also controls the physiological signs of excitement, such as sweating and heart rate.

The main modulator of the autonomic motor system is the hypothalamus. The hypothalamus sends out descending fibers that regulate sympathetic and parasympathetic nuclei in the spinal cord and brain stem. The hypothalamus is also involved in the regulation of hormone release and is influenced by the levels of insulin and glucose in the bloodstream. Thus, in its role as central governor of the autonomic motor system, the hypothalamus integrates information concerning a variety of behaviorally important stimuli. The role of the hypothalamus in motivation and emotion is discussed in detail in Chapters 32 and 33.

The interconnection of the major motor and sensory systems that process sensory information from the arm and control the arm muscles is shown in Figure 5–9. The interactions of these systems with the motivational system are summarized in Figure 5–10. In later chapters we shall see how separate neural pathways in the three major systems work together to produce appropriate motor responses to sensory stimuli.

The Anatomical Organization of Each Major Functional System Follows Four Principles

Each System Contains Relay Centers

The major functional systems of the brain and spinal cord—the sensory, motor, and motivational systems—are interconnected by *relay nuclei*. These nuclei are not simply relay connections for distributing signals to different sites. They are also important information processing centers where neural information is *modified* by interactions between the neurons within a nucleus as well as by synaptic inputs from higher centers in the system.

Relay nuclei typically contain several types of neurons, two of which are particularly important. (1) *Local interneurons* have axons that are confined to the area of the relay nucleus itself; they mediate local excitatory and inhibitory synaptic interactions. (2) *Projection* (or *principal*) *interneurons* transmit the output of the nucleus; these neurons have long axons that leave the nucleus to communicate with cells in other areas of the central nervous system.

Synaptic relay nuclei are found throughout the spinal cord and the brain. Perhaps the most prominent relay structure is the thalamus, a collection of many functionally distinct nuclei. *Almost all the sensory information that reaches the cerebral cortex is first processed in the thalamus*. The cerebral cortex, in turn, sends recurrent axons back to the thalamus.

Each System Is Composed of Several Distinct Pathways

The sensory, motor, and motivational systems each have anatomically and functionally distinct subsystems that perform specialized tasks. The visual system, for example, has separate pathways for analyzing the attributes of objects and for tracking motion. These pathways can also work together, as in the spotting of moving objects. Likewise, the somatic sensory system has anatomically separate pathways for touch and pain. The motor system, too, consists of separate specialized pathways, running from the highest centers of information processing in the brain to the spinal cord. For example,

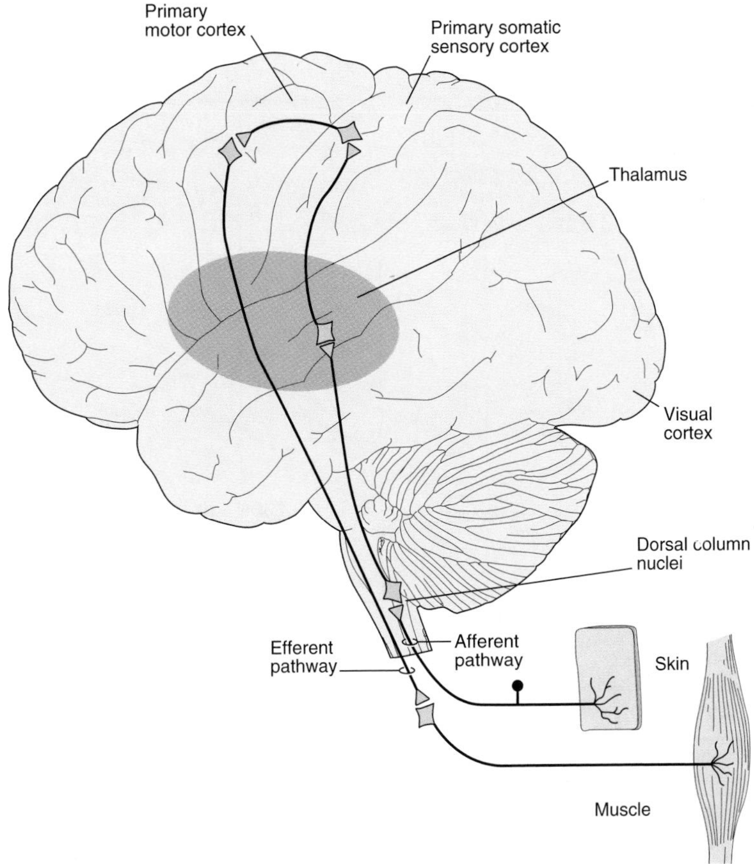

Figure 5–9 The major sensory systems and the motor system cooperate to carry out most physical actions. Sensory input from the skin, eyes, and other organs of perception is transmitted to the brain (the afferent pathway). The inputs ascend through the spinal cord to a synaptic relay in the brain stem (the dorsal column nuclei). They then proceed to a synaptic relay in the thalamus and eventually reach the primary somatic sensory cortex. Using this information, the brain issues commands to motor neurons (the efferent pathway). The motor pathway descends from the primary motor cortex through the brain stem to the motor neurons of the spinal cord, and from there to the muscle.

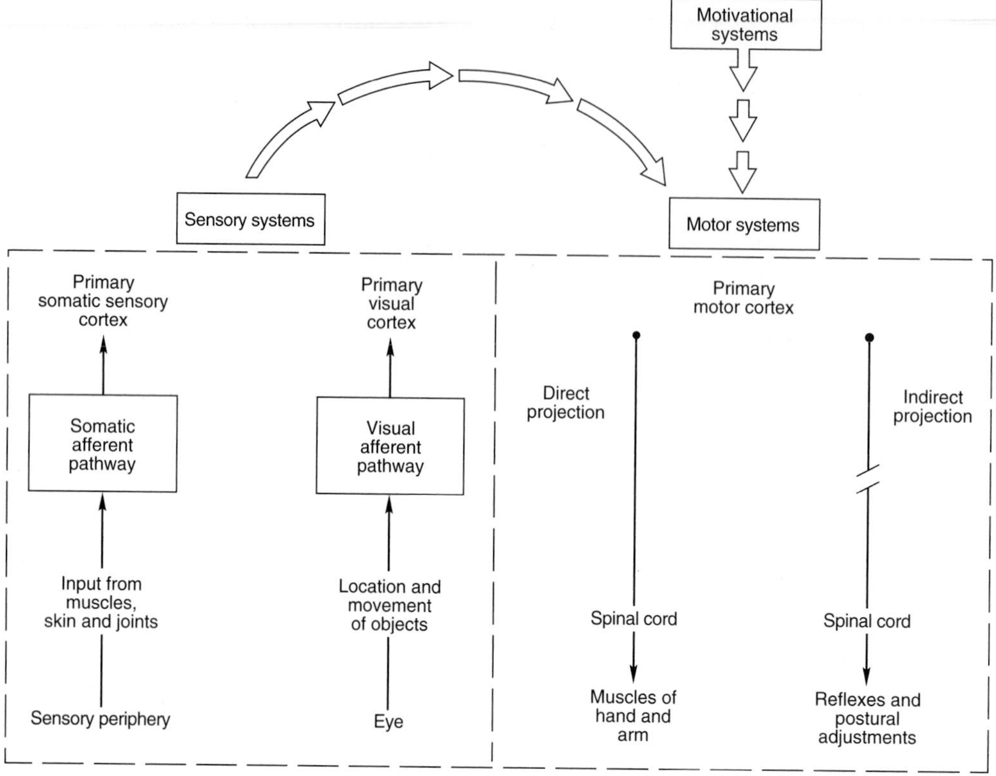

Figure 5–10 Most physical actions involve all three major functional systems of the brain—the sensory, motor, and limbic (motivational) systems. In catching a ball, for example, information about the movement of the ball and its eventual impact in the hand is relayed from the periphery of the body all the way to the primary sensory areas in the cerebral cortex. These areas provide input to the primary motor cortex through cortical connections and through multisynaptic pathways involving the basal ganglia, the cerebellum, and the thalamus. The motivational system also sends information to the motor cortex. The direct and fast-acting projections from the motor cortex regulate the activity of motor neurons that innervate the muscles of the hand and arm involved in the fine control of movement. The indirect and slower-acting projections play an important role in the overall regulation of body posture; these indirect pathways include synaptic relays (represented by the break in the arrow).

the pyramidal tract controls precise movements of the fingers and hand, while other motor pathways control overall body posture and regulate spinal reflexes.

Each Pathway Is Topographically Organized

The most striking feature of the sensory systems is that the spatial arrangement of the receptors in the peripheral sensory organs—the retina, the cochlea of the inner ear, or the skin—is preserved in point-to-point or *topographic connections* in the sensory pathways throughout the central nervous system. For example, neighboring groups of cells in the retina project on neighboring groups of cells in the thalamus, which in turn project on neighboring regions of the visual cortex. In this way an orderly *neural map* of the visual field is retained at each successive level of processing in the brain. Not all parts of the visual field are represented equally in this map. The central region of the retina, the area of greatest visual acuity, has a disproportionately large representation in the cortex because of the larger number of neurons and synaptic connections required to process the detailed information from this area.

Likewise, the body surface is represented by a neural map in the somatosensory cortex. As in the visual map, not all areas of the skin are represented equally. Regions that are particularly important for sensory discrimination and thus densely innervated, such as fingertips and lips, have more massive connections in the cortex and thus occupy the largest areas of the cortical map of the body. The auditory system has a unique neural map for sound. The neural code for each frequency of sound excites a distinct set of neurons within the relay nucleus. Thus the neural "sound map" is organized by tone rather than by the relationships in space of sounds.

In the motor pathways, neurons that regulate adjacent body parts are clustered together, thus forming a motor map, which is particularly distinct in the primary motor cortex. As in the sensory maps, the motor map is not a one-to-one representation of the body. Some parts of the body, particularly the hands, are more finely controlled than others and so are disproportionately represented in the cortex.

In addition to offering clues as to how the brain experiences the outside world, these central sensory and motor maps are useful in clinical medicine. Damage to a particular subdivision of a pathway produces characteristic deficits in motor or sensory function; hence, familiarity with the maps permits the neurologist to localize lesions in the central nervous system with precision.

Most Pathways Cross the Body's Midline

A striking, and as yet unexplained, feature of the central nervous system is that many neural pathways not only are bilaterally symmetrical but also cross over to the opposite (contralateral) side of the brain or spinal cord. As a result, sensory events on one side of the body are processed by the cerebral hemisphere on the opposite side; likewise, voluntary movement on one side of the body is controlled by the opposite-side hemisphere. Pathways cross at different anatomical levels in different systems. For example, the pathway for pain crosses in the spinal cord, whereas the direct motor pathway from the motor cortex to the spinal cord crosses in the medulla. Crossings of this kind within the brain stem and spinal cord are called *decussations* (from the Latin *decussare*, "to form a cross").

Structures that contain only decussating axons are termed *commissures*. By far the largest commissure, and indeed the largest fiber bundle in the brain, is the corpus callosum, which connects the two cerebral hemispheres (see Figure 5–6).

The crossing pathway in the human visual system is slightly more complicated. About half of the axons from each retina cross to the opposite side of the brain, while the remaining axons terminate on the same side. The crossing of axons from the retina takes place in the optic chiasm, where the left and right optic nerves meet. Axons are redistributed in the chiasm so that each half of the brain receives *all* the fibers that mediate sight from the opposite half of the visual field, just as somatic sensation on one side of the body is represented in the opposite half of the brain.

An Overall View

The nervous system is divided into two components: the central nervous system, consisting of the brain and the spinal cord, and the peripheral nervous system, composed of ganglia and peripheral nerves. The peripheral nervous system relays information to the central nervous system and executes motor commands generated in the brain and spinal cord. The simplest action involves the integrated activity of several distinct sensory, motor, and motivational pathways in the central nervous system. Each pathway contains relay nuclei and each nucleus has several functional subdivisions. In addition, most neurons are precisely arranged on the basis of their function, and many pathways cross from one side of the nervous system to the other. These basic principles govern the organization of the nervous system from the spinal cord through the brain stem to the highest levels of the cerebral cortex.

In the same way that the detailed structure of proteins reveals important principles of protein function, knowledge of neuroanatomy, seemingly a static science, can provide profound insight into how the nervous system functions. Many of the prevailing ideas about the dynamic mechanisms involved in the development of connectivity in the nervous system were forecast a century ago by Ramón y Cajal on the basis of Golgi-stained images of neurons in histological specimens. Today, much of our understanding of higher brain function depends on refined mapping of neuronal circuits with new anatomical and imaging techniques.

Indeed, many of the established properties of neuronal connectivity were first discovered by classical anatomy methods. Golgi staining first showed the existence of two major classes of nerve cells in the brain: projection neurons, whose axons connect the major regions of the nervous system, and local interneurons, which integrate information within specific nuclei of the brain. Next, tracing techniques demonstrated the considerable convergence and divergence of projections between brain regions. Convergent pathways permit a given region of the brain to integrate the input it receives from different sensory systems. Divergent pathways permit small groups of cells to exert widespread influence on many different brain regions.

The introduction of electron microscopic methods to neuroanatomy in the 1950s revealed the structure of synapses and illustrated that different classes of neurons form synapses with quite different features. Some synaptic terminals are located on dendrites, others on axon terminals, and still others on the soma of the postsynaptic cell. The location of the synapse of the neuronal surface affects the function of the cell in almost as critical a way as the organization of neuronal connections.

Modern neuroanatomical labeling techniques have also defined the principles by which neural circuits are organized. For example, the topographic organization of projections from one brain region to the next—maintaining the spatial relationship of inputs from the periphery in neighboring groups of neurons in the brain—ensures efficient coding of spatial information within the brain.

Modern imaging techniques have revolutionized the study of higher brain functions and placed neurology and psychiatry within reach of the methods of cell biology. The introduction of positron emission tomography (PET) and magnetic resonance imaging (MRI) has made the functional neuroanatomy of the human brain accessible during behavioral experiments. As a consequence we now have a much clearer idea of the brain regions involved in many complex cognitive functions.

Selected Readings

Andreasen, N. C. 1988. Brain Imaging: Applications in psychiatry. Science 239:1381–1388.

Barr, M. L., and Kiernan, J. A. 1988. The Human Nervous System: An Anatomical Viewpoint, 5th ed. Philadelphia: Lippincott.

Brodal, A. 1981. Neurological Anatomy in Relation to Clinical Medicine, 3rd ed. New York: Oxford University Press.

Brownell, G. L., Budinger, T. F., Lauterbur, P. C., and McGeer, P. L. 1982. Positron tomography and nuclear magnetic resonance imaging. Science 215:619–626.

Edelman, R. R. 1990. Magnetic resonance imaging of the nervous system. Discuss. Neurosci. 7:11–63.

Heimer, L. 1983. The Human Brain and Spinal Cord: Functional Neuroanatomy and Dissection Guide. New York: Springer.

Martin, J. H. 1989. Neuroanatomy: Text and Atlas. New York: Elsevier.

Moonen, C. T. W., van Zijl, P. C. M., Frank, J. A., Le Bihan, D., and Becker, E. D. 1990. Functional magnetic resonance imaging in medicine and physiology. Science 250:53–61.

Nauta, W. J. H., and Feirtag, D. 1986. Fundamental Neuroanatomy. New York: Freeman.

Oldendorf, W., and Oldendorf, W., Jr. 1991. MRI Primer. New York: Raven Press.

Oldendorf, W. H. 1980. The Quest for an Image of Brain: Computerized Tomography in the Perspective of Past and Future Imaging Methods. New York: Raven Press.

Posner, M. I., and Raichle, M. E. 1994. Images of Mind. New York: Scientific American Library.

Pykett, I. L. 1982. NMR imaging in medicine. Sci. Am. 246(5):78–88.

Valk, J., and van der Knaap, M. S. 1989. Magnetic Resonance of Myelin, Myelination, and Myelin Disorders. Berlin: Springer.

6

Development of the Nervous System

Neural Cell Identity Is Controlled by Cell Lineage and Inductive Interactions

- *Cell Lineage Is a Factor in the Fate of Nerve Cells*
- *Signaling Between Neighboring Cells Can Control Neuronal Identity*
- *Signals from the Mesoderm Induce the Neural Plate*
- *The Brain and Spinal Cord Become Segmented by Different Mechanisms*
- *Diffusible Growth Factors Control Nerve Cell Differentiation*

Axonal Pathways Are Formed in Response to Local Guidance Cues

- *The Migration of Neurons Establishes the Basic Plan of the Central Nervous System*
- *The Birthday of a Cortical Neuron Defines Its Eventual Position and Properties*
- *Some Neurons Migrate on a Scaffold of Glial Cells*
- *Growth Cones Guide the Axon to the Target*
- *Chemical Cues Guide the Developing Axon*

Synapse Formation at the Neuromuscular Junction Involves Inductive Interactions Between the Motor Neuron and the Muscle Cell

- *The Innervation of Muscle Can Change the Distribution and Stability of Receptors*
- *The Functional Properties of Receptors Change After Innervation*
- *Some Synapses Are Eliminated During Development*

Survival of Neurons Is Regulated by Interactions with Their Targets

Behavior is dependent on the formation of specific interconnections between classes of nerve cells with specialized functions. Thus, a comprehensive neurobiological approach to understanding behavior must consider the process by which neurons connect with specific target cells to form functional pathways. Studies of the development of the nervous system aim to elucidate how neural cells acquire specific identities and how patterns of neuronal connections are established and maintained.

Although the genetic information available to an animal is immense—perhaps 100,000 genes in mammals—it is not sufficient to specify all the neuronal interconnections that are made—perhaps as many as 10^{14}. The development of the nervous system must therefore involve *epigenetic* processes that activate specific genes in a combinatorial manner at different times during development.

Epigenetic influences that control neural cell differentiation originate both from within the embryo and from the external environment. Influences within the embryo include intercellular signals mediated by diffusible factors and cell surface molecules. The external environment provides nutritive factors, sensory and social experience, and learning, the effects of which can be seen in changes in neural activity. Thus, the appropriately timed actions of a complex array of distinct factors are critical for the proper differentiation of an individual neural cell.

Establishment of the mature pattern of neuronal connections is a gradual process that can be considered to occur in six major stages. First, a uniform population of neural precursor cells is induced from ectodermal cells by signals from the mesoderm. Second, these precursor cells begin to diversify, giving rise to glial cells and to immature neurons. Third, immature neurons migrate from germinal zones to their final position. Fourth, neurons extend axons, which project to the vicinity of their eventual targets. Fifth, axons form synaptic connections with selected target cells. Finally, some of the initial synaptic contacts are modified to generate the mature pattern of neural connections.

The nervous system has a greater variety of cell types than any other organ of the body. Nevertheless, the problems inherent in understanding the developmental mechanisms underlying the immense diversity of neural cell types are simply an extension of the central question in development: How does a single cell, the fertilized egg, give rise to each of the many differentiated cell types in the entire organism?

Modern insights into cellular differentiation began with studies by Theodor Boveri and Edmund Wilson around the turn of the century, and derive in large part from the work of Jacques Monod and François Jacob in the late 1950s. Monod and Jacob proposed that cell differentiation is achieved by the activation of specific sets of genes, with each distinct cell type expressing a different set of genes. We now know that the activation of specific genes within individual cells is controlled by nuclear proteins that bind to specific DNA sequences, thereby regulating the transcription of genes (see Box 12–1). These transcriptional regulatory proteins are themselves controlled by other signals that may be inherited from ancestor cells or provided by neighboring cells. The mechanisms controlling transcriptional regulatory proteins define two major mechanisms by which cell differentiation is regulated: cell lineage and cell–cell interactions.

The fate of some nerve cells depends exclusively on lineage, the program of divisions that the cell undergoes. Lineage-dependent programs of cell differentiation are controlled by cytoplasmic or nuclear proteins that are inherited by the progeny of a dividing precursor cell. At each division not all protein or RNA molecules need be partitioned equally in the descendants, so that daughter cells may inherit distinct molecules. Differences in molecular composition may therefore direct the differentiation of some cell types.

The differentiation of most nerve cells, however, does not depend exclusively on cell lineage but rather on signals from other cells in the environment. The molecules that mediate these signals can be secreted or they can be cell surface molecules. Many of the receptors for these signaling molecules are membrane proteins that transduce signals across the plasma membrane, triggering intracellular second-messenger pathways that directly or indirectly regulate the activity of transcription factors. Other receptors for diffusible signals, for example, steroid hormones, are located within the nucleus and themselves regulate gene transcription.

Thus, the different strategies of neural cell differentiation discussed in this chapter depend on whether transcriptional regulatory proteins are regulated by signals intrinsic to the cell or by signals from surrounding cells. Since no single organism, vertebrate or invertebrate, has so far presented a complete picture of all the known signaling mechanisms underlying neural development, it is necessary to illustrate different strategies with examples from different species.

We begin by examining how the development of a neural cell can be controlled by its program of division (i.e., its lineage), using the example of the nematode worm *Caenorhabditis elegans*. We then describe several examples of neural cell differentiation that depend on signals outside the cell. We discuss the control of neuronal identity in the eye of the fruit fly *Drosophila*, where the position of cells defines the local signals to which the cells are exposed, and thus their identity. We then examine external signals that induce neural tissue in vertebrates and establish the segmental organization of the vertebrate nervous system. The role of diffusible growth factors in the control of cell identity is then discussed with reference to two vertebrate nerve cell types, neural crest cells and glial cells.

We then consider the next steps involved in neuronal differentiation, focusing our attention on vertebrates. In particular, we examine how axons are guided to their targets and how they establish functional synaptic connections once they reach their target.

The pattern of connections established early in development does not represent the mature pattern of neuronal connections. Subsequent neuronal activity plays a crucial role in determining whether the synapse becomes stable or nonfunctional. The role of neuronal activity in the stabilization and rearrangement of synapses during development of the visual system is discussed in considerable detail in Chapter 25.

Neural Cell Identity Is Controlled by Cell Lineage and Inductive Interactions

Cell Lineage Is a Factor in the Fate of Nerve Cells

Every cell in the nervous system has its own developmental history, which can be traced through successive cell divisions from the fertilized egg. The division of individual cells or groups of cells can be examined under the microscope for long periods of time, sometimes throughout the entire program of cell division, until the cells stop dividing. Complete histories for the precursors of certain neural cells have been pieced together in several invertebrate species, and with this information it is possible to determine whether the program of divisions that a cell undergoes is important in defining its fate.

The best-studied neural cell lineage is that of the nematode *Caenorhabditis elegans*. The lineage of every somatic cell of this round worm has been mapped in detail and is invariant. Each animal has 302 neurons and 56 glial support cells, which develop from precursor cells spaced out along the body of the worm (Figure 6–1).

The degree to which this invariant program of cell division is governed by events within the cells themselves can be assessed by monitoring the fate of a single neuronal precursor after deleting one or several of its neighbors. Specific cells in *C. elegans* can be killed by focusing the beam of a laser at the nucleus of the cell. The fate of cells destined to give rise to neurons often is not affected by the absence of a neighbor or by the appearance of new neighbors. In such cases it is possible to conclude that cell lineage rather than cell–cell interactions determines the fate of the cell. Sometimes, however, deletion of neighboring cells does change the fate of particular cells. Even in animals such as *C. elegans* whose cell lineage is invariant, signals from other cells can have a role in regulating cell identity.

Signaling Between Neighboring Cells Can Control Neuronal Identity

Cell–cell interactions play a critical role in the differentiation of most types of neurons. Some of the signals that control the fate of neuronal precursor cells act over extremely short distances, influencing one cell but not its immediate neighbors. For example, the identity of each photoreceptor in the compound eye of the fruit fly, *Drosophila*, is determined by signals received from its neighboring cell. The compound eye of *Drosophila* consists of a highly ordered array of identical units called *ommatidia* (Figure 6–2). Each ommatidium consists of 20 cells, of which 8 are specialized photoreceptor neurons. These photoreceptors, labeled R1 through R8, fall into three distinct groups on the basis of the projections of their axons and the light-sensitive proteins (opsins) they produce.

The molecular analysis of how these photoreceptor neurons acquire their unique characteristics has focused on the last photoreceptor to develop, the R7 cell, as genetic screens had identified mutations in which the R7 photoreceptor is missing. The first of these mutations to be studied was called *sevenless*. In normal *Drosophila* the R7 cell is responsible for the response of the fly to ultraviolet light. In the mutant the cell that normally is destined to become the photoreceptor R7 gives rise instead to a cone cell, a glial-like cell. The *sevenless* gene is required in the prospective R7 cell itself and not in the neighboring cells. Molecular cloning of the protein encoded by the *sevenless* genes revealed it to be a transmembrane receptor protein with tyrosine kinase enzymatic function. Taken together, these two findings suggested that the *sevenless* gene encodes a receptor for a signal transmitted by neighboring photoreceptor cells.

Once the receptor function of the Sevenless protein was deduced, it became important to know which cells transmit the signal that triggers the differentiation of the R7 photoreceptor. By use of genetic screens a second gene was identified which, when mutated, leads to the loss of the R7 photoreceptor. This second gene, named *boss* (*b*ride *o*f *s*evenle*ss*), is required only in the R8 photoreceptor. Since both the *sevenless* and *boss* mutations produce identical defects, it was likely that both genes are involved in the same signaling pathway. Indeed, biochemical studies have shown

A Overall plan of *C. elegans*

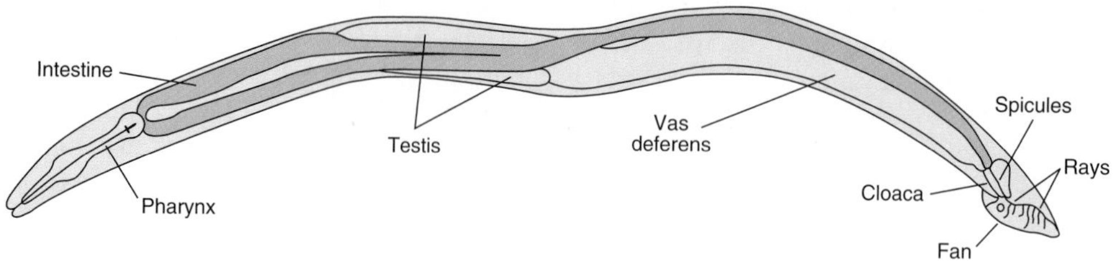

B An invariant cell lineage leads to motor neurons and hypodermal cells

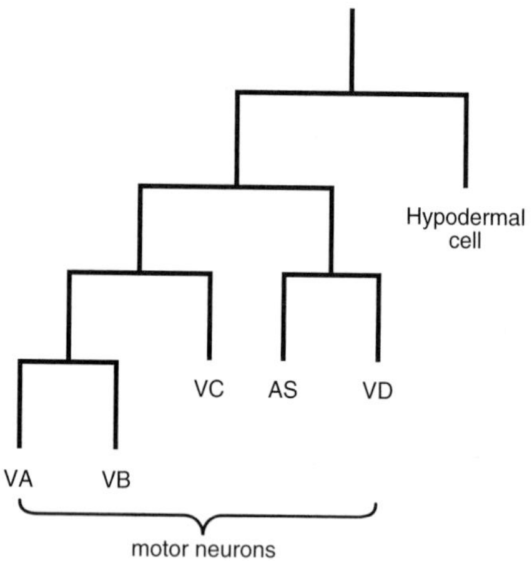

Figure 6–1 Neurons in the nematode *Caenorhabditis elegans* are generated by an invariant program of cell divisions.

A. Schematic diagram showing the appearance of an adult hermaphrodite *C. elegans*.

B. Specific classes of neurons in *C. elegans* derive from an invariant pattern of cell divisions. The lineage that gives rise to motor neurons also generates hypodermal cells.

C. Equivalent classes of neurons at different positions along the body axis are generated by minor variations in the basic program of cell divisions shown in **B**. The death of cells is marked by **X**. Each of the lineage branches shown in **C** represents a variation of the basic program of cell divisions shown in **B**.

C A single sublineage is repeated with variation in progeny cells

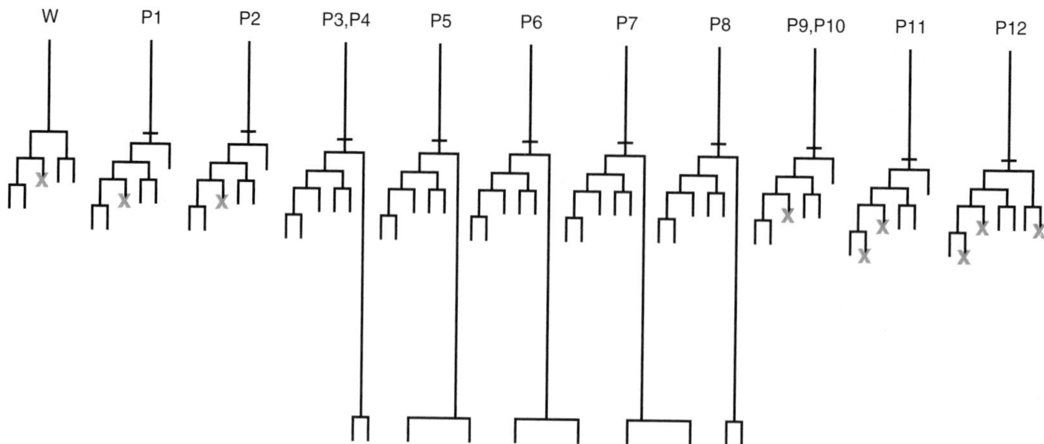

that the Boss protein is produced by the R8 cell and functions as the ligand for the Sevenless receptor protein on the surface of the prospective R7 cell (Figure 6–3).

The development of the R7 photoreceptor provides a clear example of the way in which cell–cell interactions can determine the identity of an individual neuron. It also illustrates the importance of cell surface molecules as signals in the determination of neural cell identity. Moreover, the Sevenless protein and the intracellular proteins it activates are strikingly similar to proteins in vertebrates, sug-

A Mature *Drosophila* eye

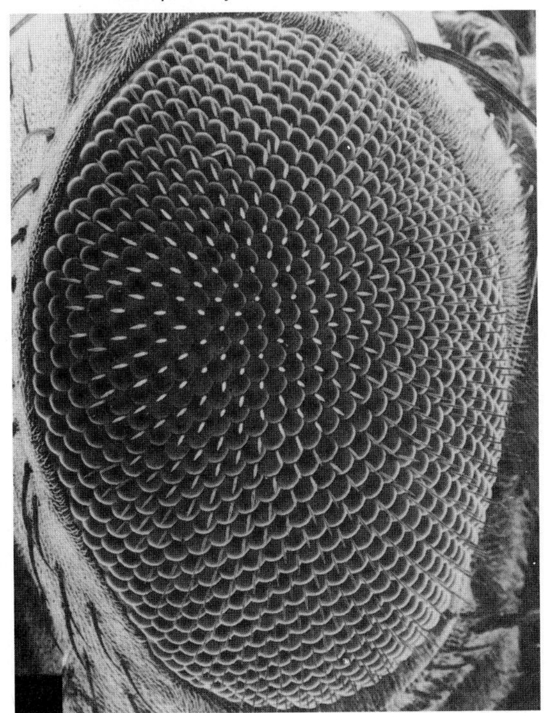

B Position of morphogenetic furrow

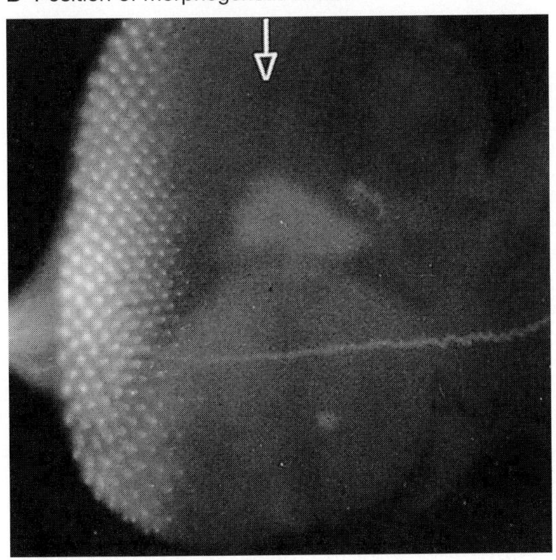

C Clusters form in wake of furrow

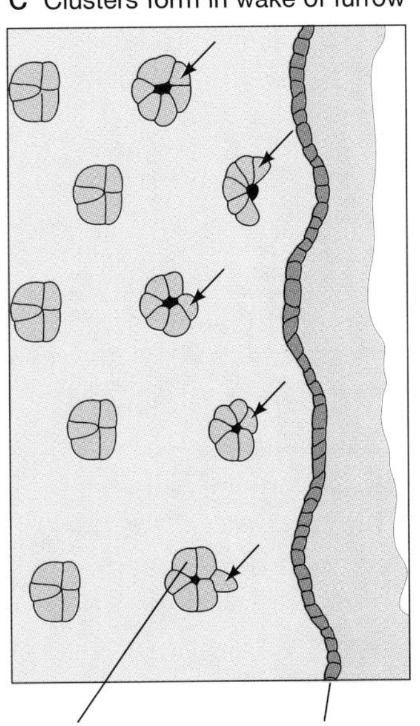

D Clusters give rise to ommatidia

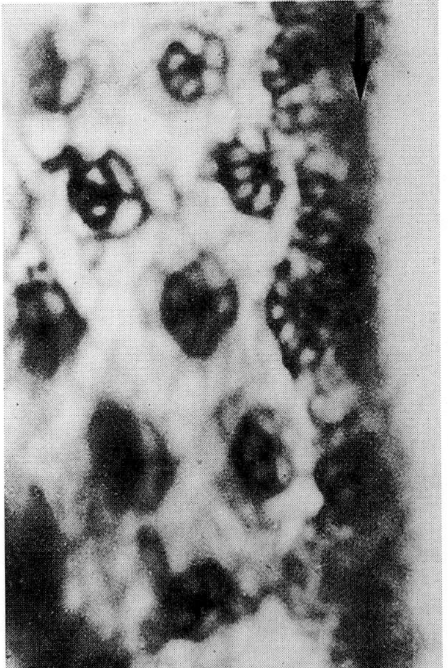

Figure 6–2 Development of the compound eye in the fruitfly *Drosophila melanogaster.*

A. The eye of the adult *Drosophila* consists of a regular array of identical units, or ommatidia. Each ommatidium consists of 20 cells. (From Ready et al., 1976.)

B. During the larval stage of development a wave of cell division and cell rearrangement sweeps across the eye disc. The wave is visible here as a furrow (indicated by an **arrow**), which represents the movement of cells in the epithelial (membranous) sheet of the eye disc. (From Banerjee and Zipursky, 1990.)

C. As the morphogenetic furrow passes across the eye disc, cells in the epithelial layer begin to form clusters in its wake. (From Tomlinson, 1988.)

D. This electron micrograph clearly shows cell clusters that represent the beginning of each ommatidium.

A Sequential recruitment of photoreceptor cells during assembly of ommatidium

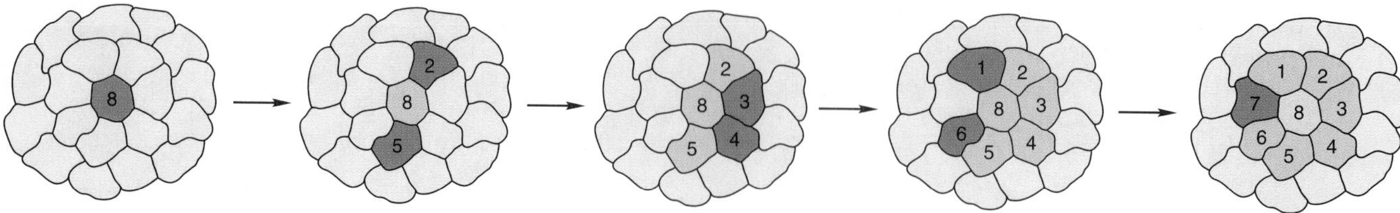

B Mutations that affect the fate of the R7 photoreceptor

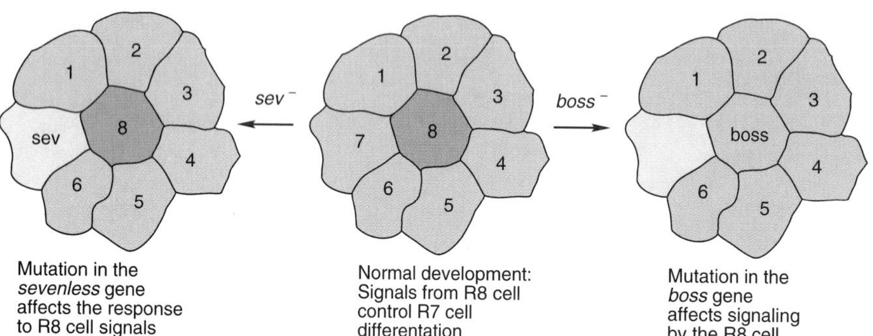

C Localization of the sevenless protein

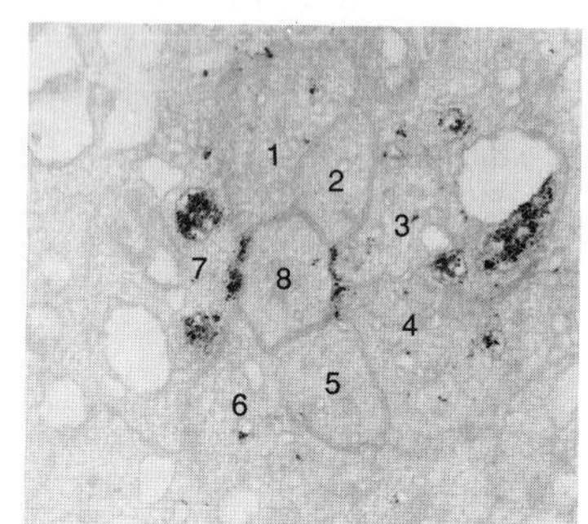

Figure 6–3 Differentiation of a photoreceptor cell in *Drosophila* depends on a signal from its neighboring cell.

A. Photoreceptor neurons develop in a strict temporal sequence. The first photoreceptor cell to express neuronal properties is the R8 cell. This is followed by the simultaneous differentiation of R2 and R5. Shortly thereafter, R3 and R4 begin to differentiate, followed by R1 and R6. The final cell to be added to the developing cluster is R7. (Adapted from Banerjee and Zipursky, 1990.)

B. Normal development of the R7 cell (**center**) depends on cell interactions that are controlled by at least two genes: the *sevenless* gene in the R7 cell and the *boss* gene in the R8 cell. Mutations in the *sevenless* gene (**left**) block the ability of the prospective R7 cell to respond to signals from the R8 cell. Mutations in the *boss* gene interfere with successful signaling by the R8 cell. The *boss* gene product may be a ligand for the Sevenless protein or may be required in the R8 cell for expression of the Sevenless ligand. (Based on Banerjee and Zipursky, 1990; Tomlinson, 1988.)

C. Immunocytochemistry locates the Sevenless protein at the junction of the R7 and R8 cells. Even though this protein is expressed in other photoreceptor types, it is essential for the development of the R7 cell only. Activation of the Sevenless tyrosine kinase by the Boss protein triggers an intracellular cascade of protein kinase activity, involving the GTP-binding protein Ras, the serine/threonine kinase Raf, and several other kinases. Strikingly, this entire signaling cascade is highly conserved in other cell types that respond to inductive signals in a wide variety of vertebrate and invertebrate organisms. (Provided by A. Tomlinson.)

gesting that signaling mechanisms involved in neural development are conserved in vertebrates and invertebrates.

This idea is supported also by the striking conservation between the intracellular signaling pathway triggered by activation of the Sevenless tyrosine kinase and the activation of receptor tyrosine kinases in other signaling systems in both invertebrates and vertebrates. The intracellular signaling initiated by activation of Sevenless and other receptor tyrosine kinases involves the Ras guanine nucleotide-binding protein, the Raf serine/threonine kinase, and the activation of the *Drosophila* homologs of the protein kinases Mek and MAP kinase. The targets of MAP kinase appear to be two transcription factors, the products of the *pointed* and *yan* genes.

Signals from the Mesoderm Induce the Neural Plate

How do vertebrate cells acquire neural properties? The entire vertebrate nervous system develops from the ectoderm, the outermost layer of the early embryo. It first appears as a thickened columnar epithelium known as the *neural plate*. (Ectodermal cells that do not acquire neural properties give rise to the skin.) Soon after the neural plate forms, it develops different regional identities along its anteroposterior axis. At the same time, the neural plate folds to form a tubular structure, the neural tube. During this process, cells in the anterior part of the neural tube begin to form the primitive forebrain and midbrain; cells in the posterior part form the hindbrain and spinal cord.

What controls the formation and regional identity of the neural plate? In 1924 Hans Spemann and Hilde Mangold discovered that the differentiation of the neural plate from uncommitted ectoderm is induced by cells in the adjacent mesoderm. Spemann and Mangold transplanted cells from a region of the embryo destined to form a portion of mesoderm into a ventral region that normally gives rise to skin. The transplanted cells were obtained from a pigmented embryo and grafted into an unpigmented host for ease of identification. The transplanted cells developed normally, differentiating into mesodermal tissues such as the notochord and somites. At the same time, however, the transplantation changed the fate of the *surrounding* ectodermal cells, which formed a second neural tube. Spemann and Mangold transplanted other regions of the early embryo, but none was able to induce a second nervous system. This finding suggested that during normal development the nervous system is induced by cells of a special region, a region Spemann called the *organizer*. Recent assays of neural induction in frog embryos have confirmed Spemann's findings. Moreover, two proteins with neural-inducing activity in frog embryos, noggin and follistatin, have been identified. (Figure 6–4).

The Brain and Spinal Cord Become Segmented by Different Mechanisms

One striking feature of the anatomy of the mature vertebrate nervous system is the segmental organization of the spinal cord and hindbrain. The axons of sensory and motor neurons enter and leave the spinal cord at regular intervals as the *dorsal roots* and *ventral roots*. The dorsal root ganglia and the sympathetic ganglia also are arranged in a segmental order. The events that lead to segmentation in the spinal cord and hindbrain differ in fundamental ways.

In higher vertebrates the segmental organization of sensory and motor axons in the spinal cord is imposed by the adjacent mesodermal tissues. When the mesoderm first breaks up into blocks or somites, the axons of motor neurons project into the anterior half of each somite. In contrast, segmentation of the hindbrain, and perhaps also the midbrain and forebrain, is thought to result from cell interactions that are intrinsic to the neural tube. The segmental form of the developing hindbrain can easily be observed in conspicuous dorsal swellings termed *rhombomeres*. The segregation of cells into rhombomeres may be responsible for the segmental organization of individual cranial motor nerve nuclei and other developing hindbrain neurons. Segmentation of the hindbrain may also contribute to the patterning of nonneural tissues in the periphery—neural crest cells that migrate out of specific segments of the hindbrain appear to shape the mesoderm that eventually gives rise to the muscles and bone of the head. Strikingly, the segmental organization of the hindbrain correlates with domains of expression of certain transcription factors, including homeobox genes, a class of transcription factors that contain a conserved DNA binding domain. Many of these genes represent vertebrate counterparts of genes that control regional identity along the anteroposterior axis of the *Drosophila* embryo. Some of the genes that subdivide the fly and human nervous systems may therefore be conserved in structure and function throughout evolution.

Diffusible Growth Factors Control Nerve Cell Differentiation

After their induction by signals from the organizer region, cells within the neural plate can differentiate into neurons or glial cells. We discuss here only two of many examples of the role of cell–cell interactions in differentiating neuronal and glial cells in vertebrates—first the neural crest cells and then the glial cells of the optic nerve.

Neural crest cells. The neural crest is a transient and migratory group of cells that emerges from the dorsal region of the neural tube and rapidly disperses along different pathways (Figure 6–5). Groups of cells migrate to different peripheral locations, where they coalesce to form the neurons and Schwann cells of the sensory and autonomic ner-

A Mesodermal induction depends on signals from vegetal cells (stage 8 embryo)

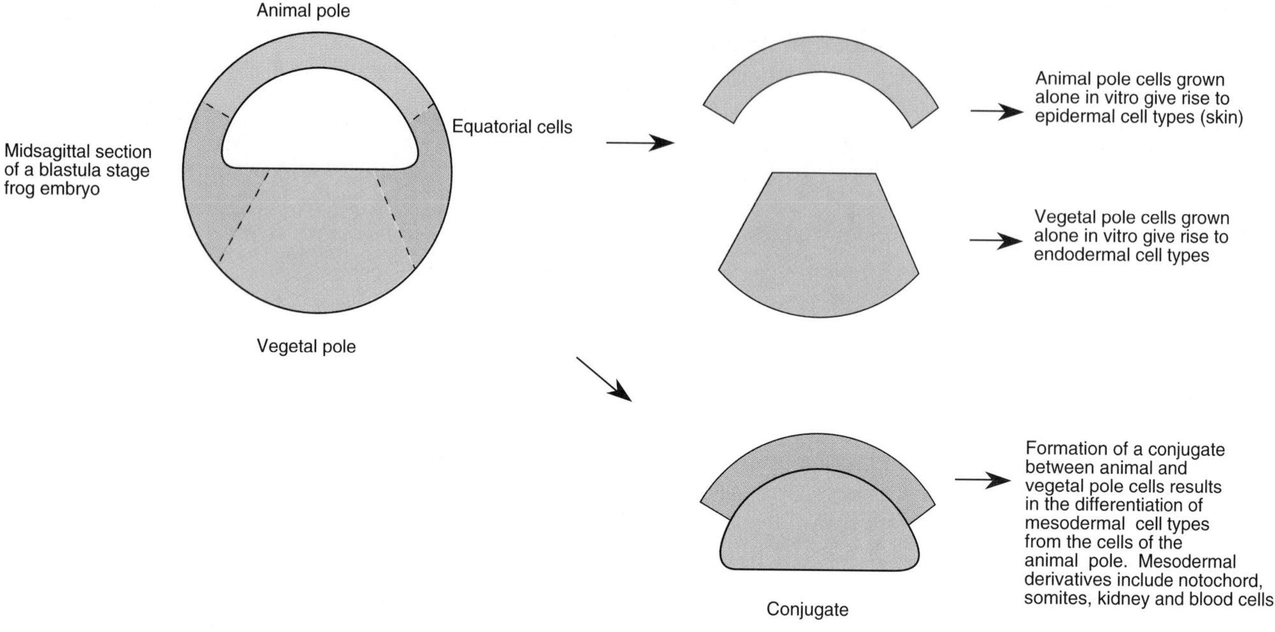

B Neural induction depends on signals from the organizer region (stage 9-10 embryo)

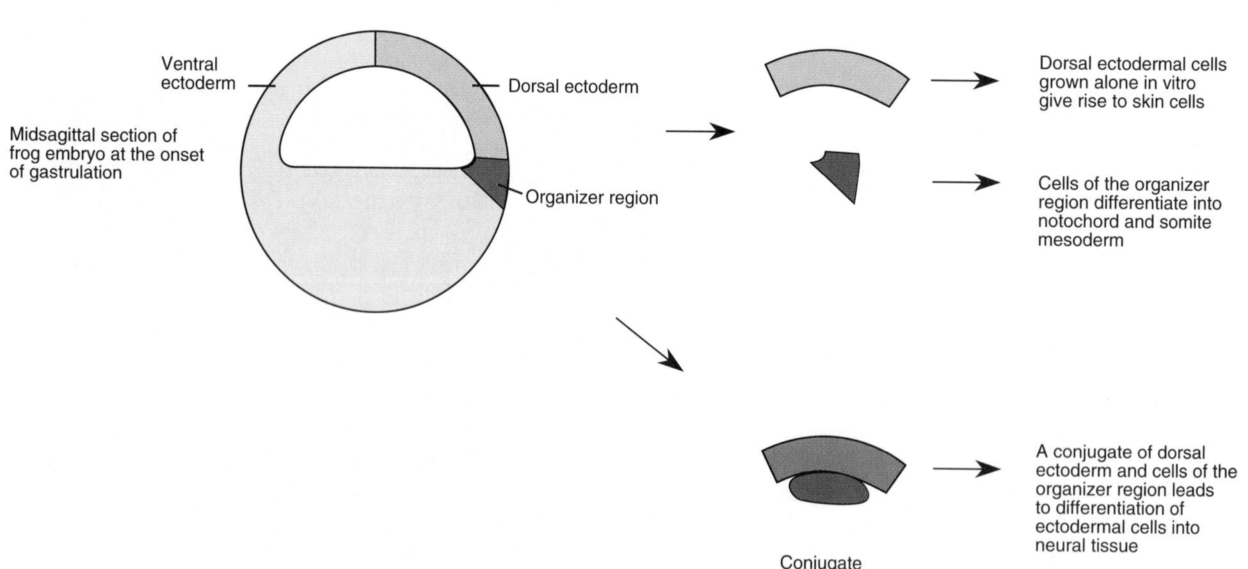

Figure 6–4 Inductive interactions lead to the formation of neural tissue. Neural induction has been studied most intensively in amphibian embryos.

A. This simplified scheme illustrates the sequence of the signals involved in mesodermal induction. The diagram shows a midsagittal section of a frog embryo at the blastula stage (stage 8). The equatorial region of the embryo normally gives rise to mesoderm. If ectoderm from the animal pole is placed alone *in vitro*, it forms an epidermal derivative (skin) is formed. If the animal pole tissue is placed in contact with vegetal cells, then the animal pole tissue is induced to differentiate into mesoderm of several different types: notochord, somites, kidney, blood cells. Peptide growth factors, in particular a transforming growth factor (TGF) β family member, activin, and fibroblast growth factor (FGF), can mimic the effect of vegetal cells in inducing mesodermal differentiation in animal pole ectoderm.

B. The sequence of induction of neural tissue is shown in this midsagittal view of a frog embryo at the onset of gastrulation (stages 9–10). Dorsal ectoderm normally gives rise to the neural plate, but will differentiate into skin when removed from the embryo. The organizer region normally gives rise to the notochord and somites, and these cells maintain the same fate *in vitro*. When the organizer region is placed in contact with the dorsal ectoderm, neural tissues develop from the ectoderm. The signals that induce neural differentiation are therefore likely to originate from cells of the organizer region.

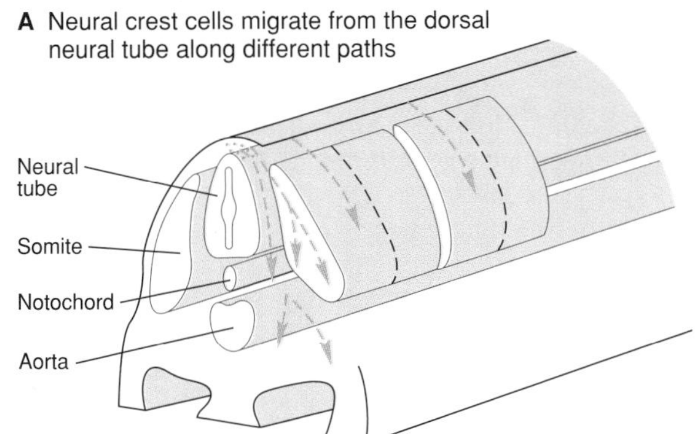

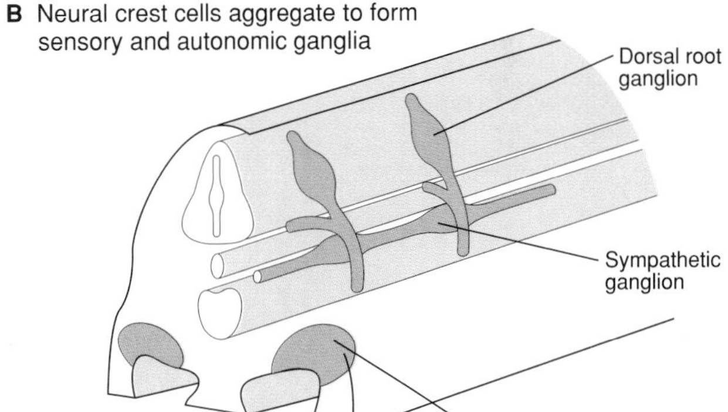

C Migration of neural crest cells in progress

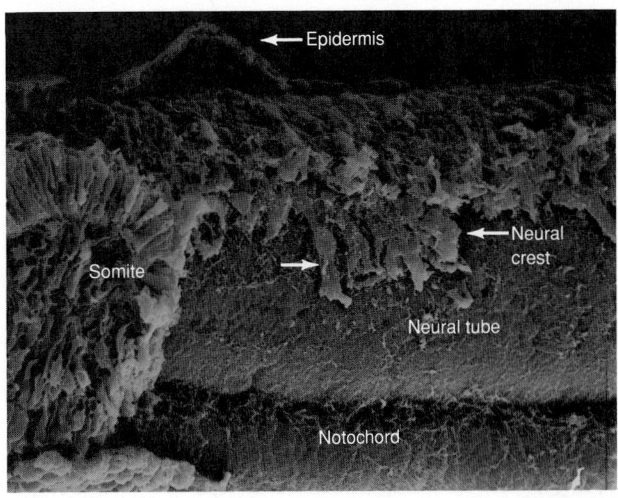

Figure 6–5 Neural crest cells migrate during early development.

A. The cells that take the superficial pathway, just beneath the ectoderm, will form pigment cells of the skin. Those that take the deep pathway via the somites will form sensory ganglia, sympathetic ganglia, and parts of the adrenal gland. This diagram shows a cross section through the middle part of the trunk. Migration occurs on both sides of the neural tube; only one side is shown here.

B. The sympathetic and dorsal root ganglia mature after migration from the neural crest is complete.

C. This scanning electron micrograph shows neural crest cells migrating away from the dorsal surface of the neural tube of a chick embryo. (Courtesy of K. Tosney.)

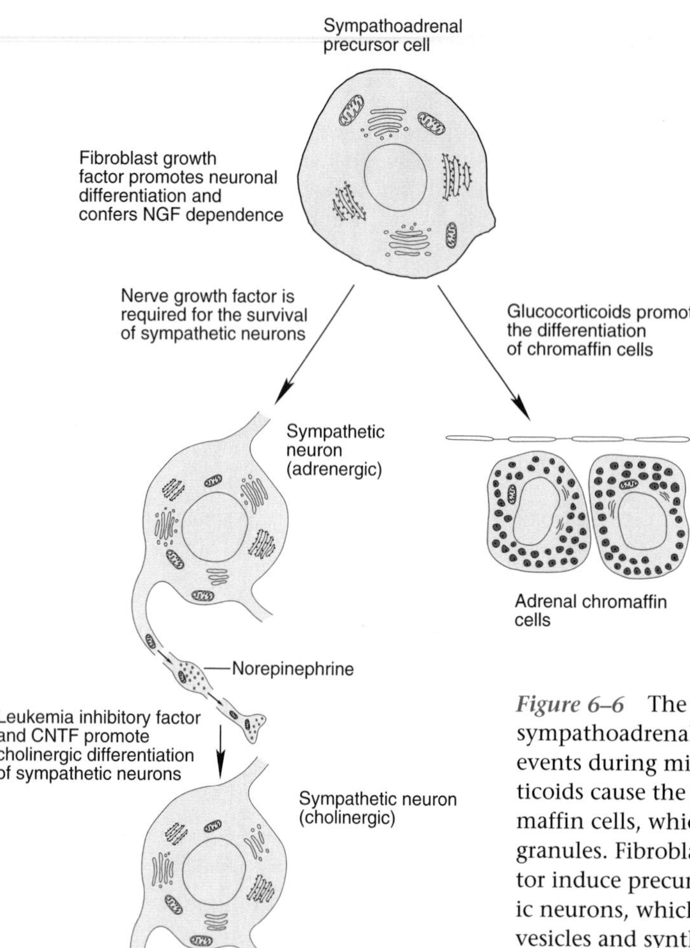

Figure 6–6 The differentiation of a precursor cell in the sympathoadrenal lineage is controlled by a series of events during migration from the neural crest. Glucocorticoids cause the precursor cell to differentiate into chromaffin cells, which have large (150–350 nm) dense-core granules. Fibroblast growth factor and nerve growth factor induce precursor cells to differentiate into sympathetic neurons, which possess smaller (50 nm) electron-dense vesicles and synthesize norepinephrine as transmitter. If these neurons are cultured in the presence of a conditioned medium containing leukemia inhibitory factor, they acquire cholinergic properties, synthesize acetylcholine, and contain small (30–50 nm) electron-translucent vesicles. (Adapted from Doupe et al., 1985.)

vous systems, the chromaffin cells of the adrenal medulla, the melanocytes of the skin, and mesenchymal tissues of the face and skull.

Early neural crest cells have the potential of becoming any one of a wide variety of cell types. The developmental fate of a cell is dependent on the signals it receives from the environment through which it migrates, and the developmental options of the cell become increasingly restricted by changes in the environment of the cell during migration.

Some of the molecules that control the fate of neural crest cells have been identified in the sublineage of cells that give rise to the sympathetic nervous system and adrenal medulla (the sympathoadrenal lineage). This lineage comprises the major catecholaminergic descendants of the neural crest: sympathetic neurons, chromaffin cells, and cells in the sympathetic ganglia defined on the basis of intense catecholamine histofluorescence, called *small intensely fluorescent (SIF) cells*. Precursors of these distinct cell classes can be isolated from the embryonic adrenal medulla or sympathetic ganglia. When grown in culture, it is possible to control the fate of these progenitor cells by varying the culture conditions (Figure 6–6).

The differentiation of adrenal progenitor cells into chromaffin cells is dependent on the presence of glucocorticoid hormones. Differentiation of chromaffin cells from neural crest precursors is probably triggered by the migration of these cells into the adrenal gland, where they are exposed to the high levels of glucocorticoids synthesized by the adrenal cortex. Glucocorticoids activate nuclear receptors that function directly as transcription factors. Neural crest cells that form sympathetic ganglia follow one of two fates—they can become SIF cells or sympathetic neurons. As with chromaffin cells, the decision to become an SIF cell may also be dependent on glucocorticoids in the blood supply, since *in vitro* low concentrations of glucocorticoids promote the appearance of SIF cells.

Glial cells. The development of glial cells in the optic nerve of the rat is one of the better understood examples of the cell–cell interactions that influence cell identity in the vertebrate nervous system. There are three distinct classes of glial cells in cultures of the rat optic nerve: oligodendrocytes and two types of astrocytes. Each class can be distinguished by its expression of specific intracellular and surface proteins. Because optic nerve cultures are relatively simple, the lineage and cellular interactions that control glial cell diversification can be readily analyzed. Oligodendrocytes (O) and type 2 astrocytes (2A) differentiate postnatally from a common precursor called the O–2A progenitor cell. Type 1 astrocytes differentiate from a distinct precursor cell.

Growth factors appear to play a crucial role in the decision of O–2A cells to differentiate into type 2 astrocytes or oligodendrocytes. When cultured in the absence of growth factors, O–2A cells rapidly differentiate into oligodendrocytes. Type 1 astrocytes release growth factors that induce the proliferation of O–2A progenitor cells and participate in their differentiation into type 2 astrocytes. A protein that promotes O–2A cell proliferation is platelet-derived growth factor. Astrocyte differentiation can also be induced, at least transiently, by ciliary neurotrophic factor (CNTF).

Axonal Pathways Are Formed in Response to Local Guidance Cues

The Migration of Neurons Establishes the Basic Plan of the Central Nervous System

A characteristic feature of many neuronal precursors (neuroblasts) and neurons in vertebrates is that they migrate from the site at which they begin to differentiate. As noted earlier, neurons in the peripheral nervous system derive from neural crest cells that migrate extensively throughout the body before completing their differentiation into autonomic, sensory, and enteric neurons. Similarly, in the central nervous system the eventual position of many different classes of neurons is achieved by the migration of neuroblasts from the site of their proliferation in the ventricular zones of the neuroepithelium.

Different classes of neuroblasts migrate at different stages, some migrating before and some after they have extended their axons. For example, motor neurons in the ventral horn of the spinal cord migrate from the ventricular zone of the neural tube to form the motor column before they send an axon out into the periphery. Other neurons, such as granule cells of the cerebellum, extend axons for considerable distances, and only at a relatively late stage in the maturation of the neuron does the cell body migrate from the external granular layer to its final settling place in the granule cell layer.

The migration of neuronal precursors serves a dual function. It has a role in establishing the identity of some neurons and it may define the functional properties and future connections of the neuron.

The Birthday of a Cortical Neuron Defines Its Eventual Position and Properties

In many regions of the central nervous system neurons are arranged in orderly layers. For example, neurons in the cerebral cortex that have different morphologies and connections are organized into well-defined layers. Large pyramidal-shaped projection neurons are located in layer 5, while smaller stellate neurons are found in layer 4. Yet all of the different neurons that eventually populate the several layers of the cerebral cortex derive from neuroblasts that originate in the *ventricular zone*, the region that lies near the ventricles of the brain.

The layering of cortical neurons appears to be associated with the birthdays of these neurons. The term *birthday* is used to indicate the time at which a dividing precursor undergoes its final round of cell division to give rise to a postmitotic neuron. Neuronal birthdays can be determined by applying pulses of [^{3}H]thymidine to developing neuroblasts.

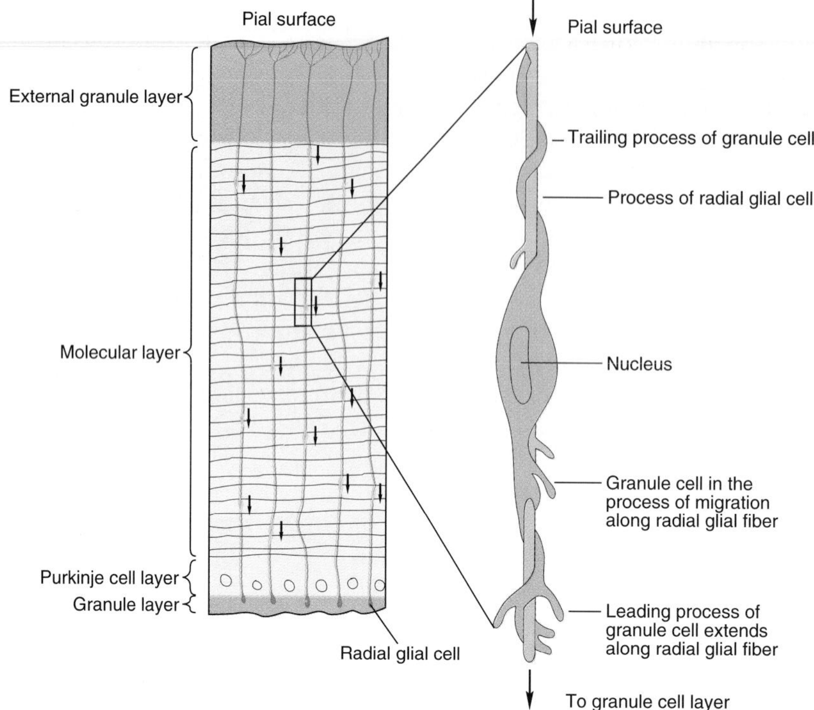

Figure 6–7 Some developing neurons crawl along glial cell fibers to their permanent location in the nervous system. Here, granule cells migrate through the molecular and Purkinje cell layers of the cortex along the processes of radial glial cells, which reach from the granule layer to the pial surface. The cell bodies of the radial glial cells lie near the junction of the Purkinje and granule cell layers. (Adapted from Rakic, 1971.)

The [^{3}H]thymidine is incorporated by precursor neuroblasts that are in the DNA replication (S) phase of the cell cycle. Daughter cells that arise from the next mitotic division are heavily labeled; in those cells that continue to divide, the label becomes diluted. Thus, heavily labeled cells are those that are born a short time after the pulse of [^{3}H]thymidine. This technique has been used to show that neurons born at early stages of cortical development end up in the deepest cortical layers, while those born at later times end up in progressively more superficial layers. Neurons born at later times must therefore migrate past neurons that have already reached their final position in the cortex. Thus, the organization of cortical layers is achieved by an inside-out sequence of neuronal differentiation. Similar processes of cell migration are also seen in many other layered structures in the brain.

The mechanisms by which neurons in different cortical layers come to acquire different properties have not been resolved. The identity and function of a neuron could be defined by its birthday. Alternatively, interactions within the local environment could be more important. Studies of neuronal cell migration in the *reeler* mutant mouse have provided some evidence that cell birthday is important in defining the eventual properties and projections of neurons in the cerebral cortex. In *reeler* mice the normal inside-out layering of cortical neurons is inverted; neurons born at early times end up in the most superficial layers, whereas neurons born at later stages end up in deeper layers. Although these neurons are located in inappropriate positions, they still appear to acquire their normal morphology and connections, leading to an inversion in the functional arrangement of cortical neurons. These observations suggest that the position occupied by a neuron in the cortex is less critical in determining its final identity and connections than its birthday or other events in its developmental history.

The time at which the fate of cortical neurons is fixed has been examined by transplanting cortical neuronal precursors destined for outer layers (layers 5 and 6) into newborn animals in the ventricular zone where the surrounding cells migrate to inner layers (layers 2 and 3). Many of the transplanted neurons migrated to layers 5 and 6 and developed axonal projections appropriate for their birthday. The stage of the cell cycle of the transplanted precursors affects the fate of the cells. Progenitor cells transplanted into older hosts when they are in the S-phase migrate to layers 2 and 3. Those transplanted late in the cell cycle or after their final cell division migrate to layers 5 and 6 and form axonal connections appropriate for their birthday. Thus, commitment occurs prior to migration; the cortical layer for which a neuron is destined is determined just before the cell is born. These principles of cortical cell development differ from those of neural crest cells, whose fate is determined more by their migratory route and final location than by cell birthday.

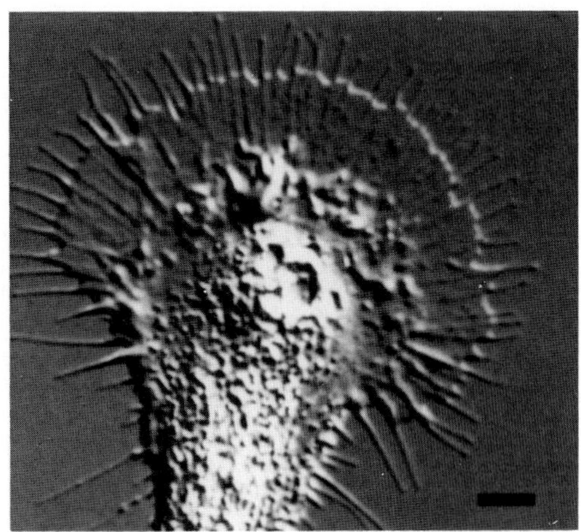

Figure 6–8 A growth cone from a neuron isolated from the marine snail *Aplysia californica* is visible under differential interference contrast optics. Numerous slender projections of ectoplasm, or filopodia, extend from the growth cone. (From Forscher and Smith, 1988.)

Some Neurons Migrate on a Scaffold of Glial Cells

In many regions of the developing brain the migration of neurons is dependent on radial glial cells, specialized glial cells that span the neuroepithelium. Neurons in the cerebral cortex, for example, use radial glia to guide their migration from the ventricular zone to their final destinations (Figure 6–7). However, radial glial fibers are absent in many regions of the nervous system. In these regions extracellular matrix molecules may serve a similar purpose.

Growth Cones Guide the Axon to the Target

Once a neuron has migrated to its final position, and sometimes even before, it begins to extend an axon at the end of which is a specialized sensory and motor apparatus, the structure called the *growth cone*. The importance of growth cones was first recognized by Ramón y Cajal, who likened them to battering rams, forcing their way through densely packed neuroepithelium. Several fingerlike extensions, the *filopodia*, project from growth cones (Figure 6–8). Filopodia are highly motile, continually extending and retracting. As the growth cone and axon extend, the surface area of the neuronal membrane increases. New membrane is synthesized in the cell body, packaged into vesicles, and transported along microtubules that extend into the body of the growth cone. Once in the growth cone, these vesicles are incorporated into the surface membrane by fusion.

Chemical Cues Guide the Developing Axon

For more than a century neurobiologists have been intrigued by the puzzle of how axons of developing neurons reach their targets. Ramón y Cajal's early descriptions of developing axons left the impression that growth cones move in an ordered and directed manner, but experimental support for the specificity of axonal pathfinding was not available until the 1940s and 1950s. At that time Roger Sperry (Figure 6–9) began to examine the visual system of the newt, a cold-blooded vertebrate. Ñeuronal regeneration occurs readily in the newt, in contrast to the situation in mammals. After cutting the optic nerve Sperry was able to study how retinal neurons regenerate axons and reestablish connections with their target cells.

In one key set of experiments the optic nerve was cut and the eye inverted 180 degrees in its orbit. The axons regenerated and the visual system continued to function, but with a difference—

Figure 6–9 Roger Sperry.
Sperry discovered the function of the corpus collosum. After receiving his Ph.D. at the University of Chicago, where he trained with the embryologist Paul Weiss, Sperry first worked in developmental neurobiology, where he demonstrated the precision of wiring in the brain by following the regeneration, regrowth, and restoration of function on the visual system of the frog after section of the optic nerve. He next turned his attention to the functioning of the corpus callosum, the fiber bundle that connects the two hemispheres of the brain. By first studying experimental animals and then human patients in whom the corpus callosum had been sectioned surgically to treat epilepsy, he showed that the two hemispheres carried out different cognitive functions. Sperry shared the 1981 Nobel Prize in Physiology and Medicine with David Hubel and Torsten Wiesel. (Photo courtesy of V. Hamburger.)

when tested behaviorally, the animals behaved as if their visual world had been inverted. For example, visual stimuli presented above and to the left of the animal evoked a motor response directed to the lower right of the visual field. Although such a response was clearly ineffectual, the animals were unable to learn to correct their responses even after many trials. Sperry concluded that the axons of retinal ganglion neurons had grown back to their original target in the optic tectum, even though the regenerated connections were behaviorally inappropriate.

Sperry and others later obtained anatomical evidence to support this conclusion, and on this basis proposed that the specific regeneration of axons to their original targets depended on chemical affinities that existed between individual neurons. The basic idea of this *chemoaffinity hypothesis* is that individual neurons acquire distinctive molecular markers, recognition molecules, early in development. The establishment of appropriate connections between two neurons would thus depend on the correct matching of molecules on the surfaces of the pre- and postsynaptic neurons. These seminal studies suggested that the guidance of regenerating axons is directed and involves the recognition of guidance cues in the axon's local environment.

Sperry's studies have been extended to the analysis of initial axon pathfinding during development (Figure 6–10). In the visual system there is evidence that developing retinal axons can recognize molecules that are distributed in a graded manner along the anterior-posterior axis of the tectum, much as Sperry had predicted. For example, when axons from the anterior part of the retina are confronted *in vitro* with alternating stripes of anterior and posterior membranes, the axons grow only toward the anterior membrane stripes, a choice that reflects their normal target *in vivo*. Retinal axons choose anterior membranes not because they contain some growth-promoting factor, but because the posterior membranes have a higher concentration of a repellent molecule.

Another good example of early and specific axon pathfinding in vertebrate embryos is the selection of peripheral pathways by the axons of developing motor neurons. The axons of different classes of motor neurons destined for different muscle targets emerge from the spinal cord intermingled in

A Axon regeneration in the mature goldfish retina

Remove half of retina; sever optic nerve

Retina

Dorsal

Anterior

Posterior

Ventral

Dorsal

Anterior

Posterior

Ventral

Axons from remaining half of retina regenerate and project to normal positions

Tectum

Anterior

Posterior

Anterior

Posterior

B Projection of retinal axons to the tectum in the developing *Xenopus* embryo

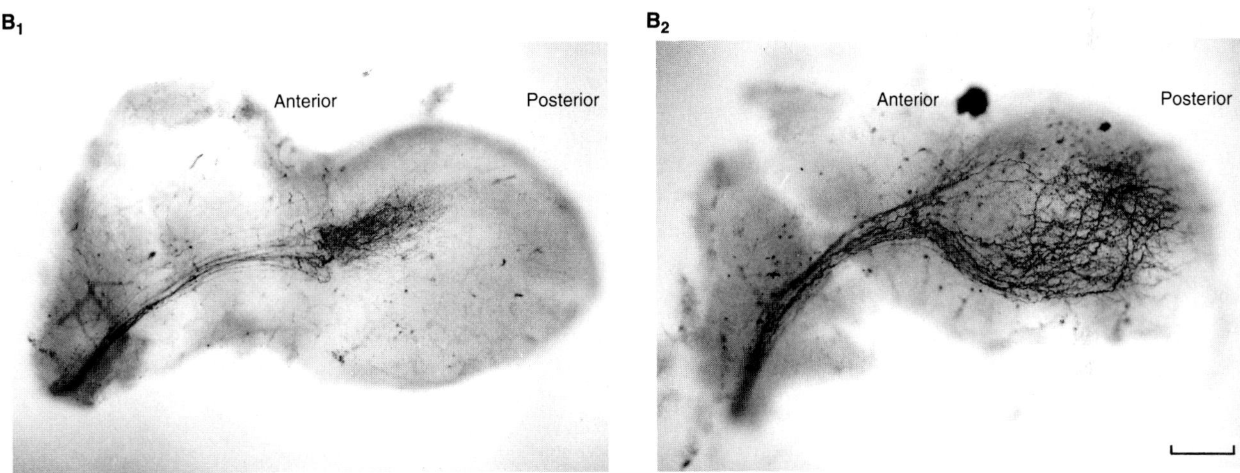

Figure 6–10 The axons of retinal ganglion neurons project to appropriate positions on the roof of the midbrain (tectum) during regeneration as well as in early development.

A. In adult goldfish the anterior or posterior half of the retina was experimentally removed, and the optic nerve was cut. The regenerated axons were observed several weeks later by means of silver staining. Just as in normal development, posterior retinal axons projected to the anterior tectum, and anterior retinal axons projected to the posterior tectum. (From Sperry, 1963.)

B. In *Xenopus*, the specificity with which retinal ganglion axons project to their target neurons in the tectum during development is clear from a simple labeling experiment. Small regions of the retina were labeled with horseradish peroxidase, which is transported by retinal ganglion axons as they grow toward their targets. Material injected into the posterior retina (**left**) shows up in axon terminals in the anterior tectum, whereas injections into the anterior retina (**right**) label terminals in the posterior tectum. (From Fujisawa et al., 1982.)

different spinal nerve roots. However, when the motor axons reach the base of the developing limb, axons from the same class of motor neurons sort themselves out to form nerve branches that contain only axons destined for the same muscle target. The growth cones of different classes of motor neurons must therefore recognize specific cues within the limb. These cues can also guide the correct projection of motor axons that, for purposes of experiment, have been forced to enter the limb in an inappropriate position (Figure 6–11). The fact that many distinct sets of motor axons can emerge from a mixed bundle of fibers implies a high degree of selectivity in the recognition of these guidance cues.

What is the nature of these guidance cues? Some cues for developing axons involve adhesive cell–cell contacts between the growth cone and specialized molecules, *cell adhesion molecules*, on the surface of neighboring cells or in the extracellular matrix. Many of the glycoproteins involved in the adhesion of neural cells belong to one of three major structural families. The first is the immunoglobulin superfamily. The second family comprises a group of structurally related glycoproteins called *cadherins*, of which N-cadherin is a prominent member expressed within the nervous system. The third family consists of a large family of glycoproteins called *integrins*. The integrins mediate interactions between the cell surface and molecules in the extracellular matrix, such as laminin and fibronectin. Each integrin consists of two subunits that together confer the binding properties of the molecule.

Developing axons may also be attracted toward chemical factors that are released by target cells. Growth that is oriented in this way is termed *chemotropism* (or *chemotaxis*). As noted earlier, axons may also be guided, in a negative fashion, by molecules that repel growth cones.

Synapse Formation at the Neuromuscular Junction Involves Inductive Interactions Between the Motor Neuron and the Muscle Cell

The formation of synapses is a key event in establishing functional neuronal connections. Our understanding of synapse formation is most advanced at the synapse between motor neurons and skeletal muscle—the best studied of all the vertebrate synapses. Many of the steps in the formation, stabilization, and modification of synaptic connections are probably controlled in similar ways at synapses between other classes of neurons.

When a developing motor axon first approaches its target—a skeletal muscle fiber—neither the axon nor the muscle cell is well equipped to transmit signals across a synapse. The growth cone of the axon does not resemble a mature presynaptic nerve terminal, and the postsynaptic muscle mass has not yet cleaved to form individual muscles. Nevertheless, a primitive form of synaptic transmission exists from the moment the axon reaches the muscle target. This primitive form of synaptic transmission requires two important features. First, the presynaptic axon must be capable of releasing its neurotransmitter acetylcholine (ACh) before it can make contact with its postsynaptic target muscle. Second, the postsynaptic muscle membrane must be capable of responding to ACh before it can be contacted by the motor neuron.

Soon after the motor neuron and muscle fiber make contact, specialized receptor zones (the end-plates) begin to form in the muscle, and the amplitude of the postsynaptic potentials increases dramatically. This increase in transmission appears to involve changes in both the pre- and postsynaptic components of the synapse over a protracted period of time. The development of communication between a motor neuron and a muscle involves changes in the distribution and stability of ACh receptors, changes in the functional properties of the receptors, and an increase in the number of nerve–muscle contacts.

The Innervation of Muscle Can Change the Distribution and Stability of Receptors

Before innervation, ACh receptors are distributed relatively uniformly over the surface of muscle fibers. When the axon of the motor neuron reaches the muscle, however, the density of ACh receptors at the sites of innervation increases and the density of receptors at extrasynaptic sites (that is, outside of the synapses) decreases. First, the receptors that are already present in the muscle membrane are redistributed. These receptors diffuse within the plane of the membrane and become immobilized at the synaptic site. Second, there is an increase in the synthesis of new receptors, which are inserted into the muscle membrane at or

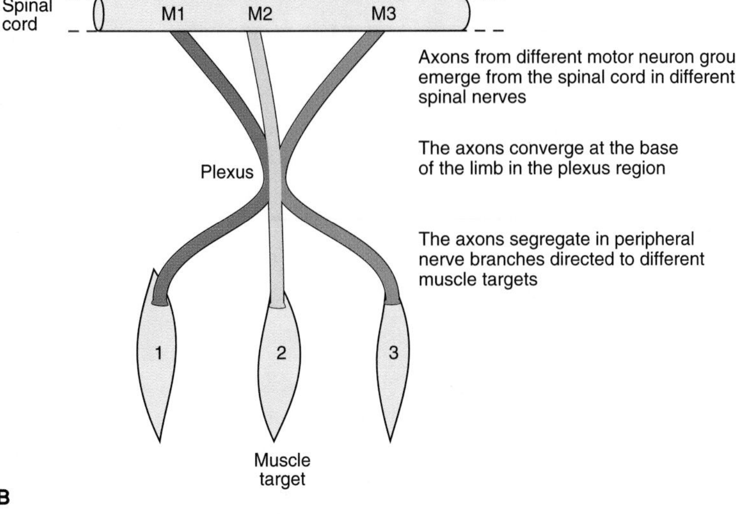

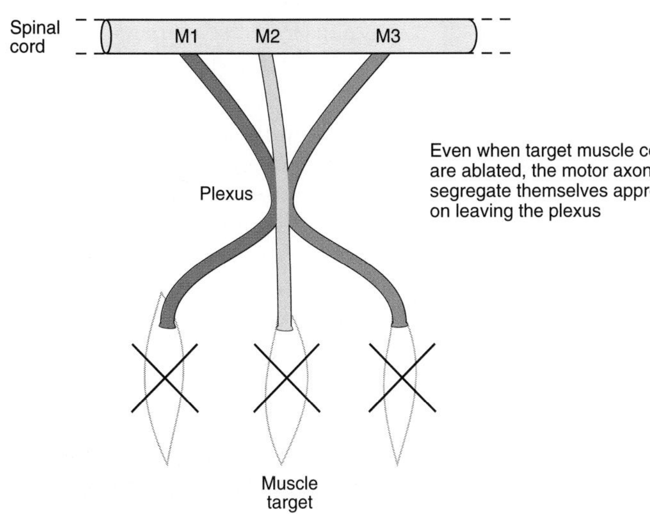

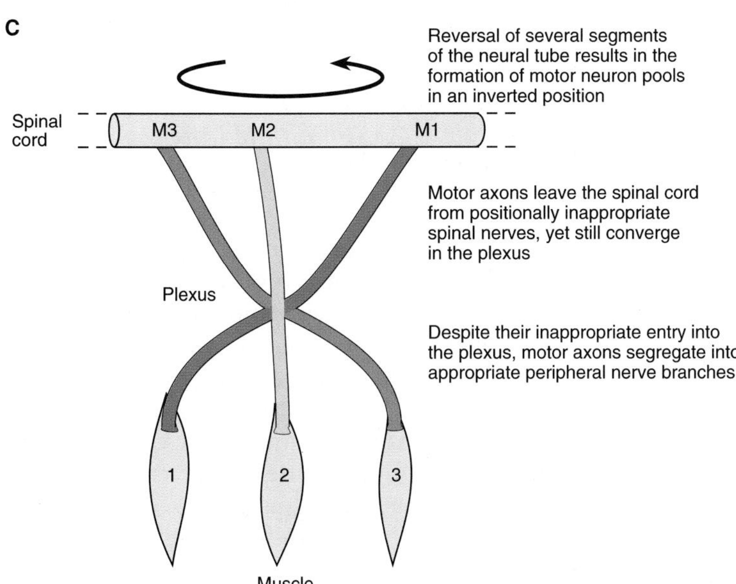

Figure 6–11 Motor axons are guided to their targets by chemical cues in the plexus.

A. During normal development different classes of motor neurons in the spinal cord (M1, M2, M3) project to different muscle target groups. The axons of each motor neuron group leave the spinal cord through a different spinal nerve, converge in the plexus region at the base of the limb, and then segregate into nerve branches that are directed to particular muscle targets.

B. Motor axons are sorted in the plexus even when the muscle targets have been removed, indicating that guidance cues in the plexus region are responsible for the sortings.

C. Rearrangement of the order of motor neuron groups in the spinal cord causes the axons to emerge from spinal nerves in inappropriate locations. Despite this, axons converge in the plexus and segregate into peripheral nerve branches that are directed to their normal muscle targets. Thus the growth cones of motor axons must be highly specific in the recognition of cues in the plexus region.

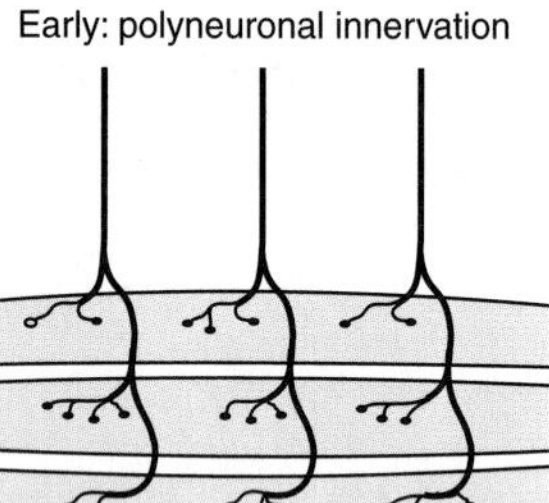

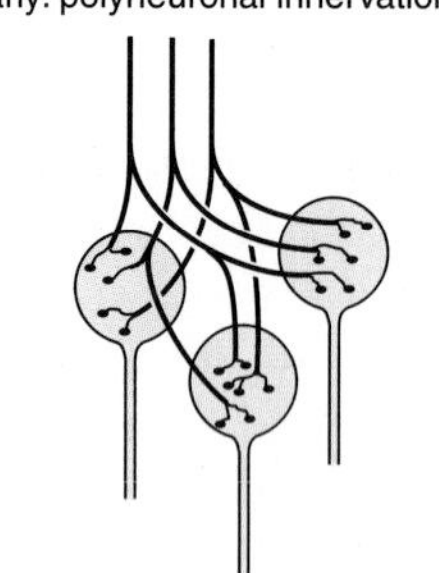

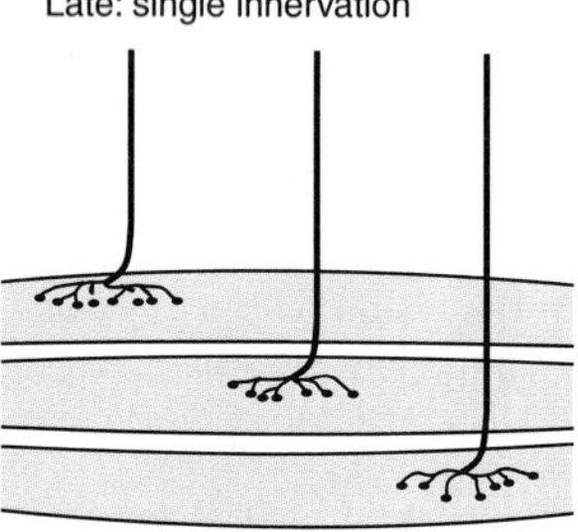

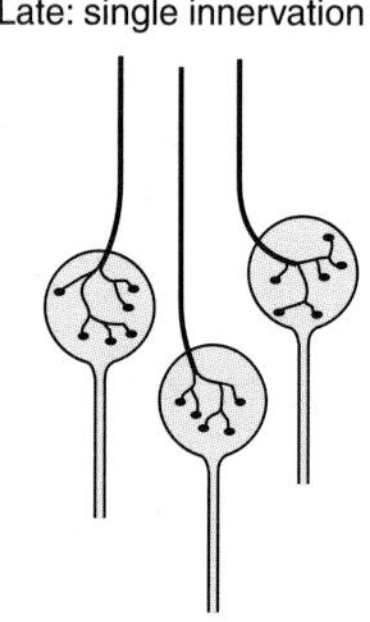

Figure 6–12 Synaptic connections in the peripheral nervous system of mammals are rearranged during the first few weeks of postnatal life. In muscles (**left**) and ganglia comprising neurons without dendrites (**right**) each axon innervates more target cells at birth than in maturity. The number of neurons innervating each target cell decreases with development, but the size and complexity of the remaining terminal increase with time. (Based on Purves and Lichtman, 1980; Purves, 1988.)

near the synaptic site. By the time these changes are complete, the density of ACh receptors at the endplate (up to 20,000 molecules per square micrometer) is several thousandfold greater than at extrasynaptic regions. These results indicate that the nerve terminal controls both the synthesis and the distribution of receptors on the postsynaptic muscle fiber.

The influence of the nerve on the distribution of receptors is realized by means of proteins released from the presynaptic nerve terminal. One such protein, agrin, causes the clustering of preexisting receptors, whereas a second protein, ARIA, increases the total number of ACh receptors on the muscle surface.

At the same time that ACh receptors begin to cluster at the synapse, extrasynaptic receptors disappear. The loss of ACh receptors is controlled by an electrical mechanism, rather than a chemical one. If mature muscle loses the axons that innervate it, ACh receptors soon begin to appear at extrasynaptic sites. These extrasynaptic receptors can be suppressed if the muscle is activated by direct electrical stimulation.

The Functional Properties of Receptors Change After Innervation

Besides influencing the distribution of receptors, the motor nerve terminal also triggers many other changes in the receptors. Acetylcholine receptors at synapses lose their ability to diffuse in the plane of the membrane and gradually become immobilized at the site of the synapse. Thus, ACh receptors at synapses also have a much longer half-life than those outside synapses. Moreover, ACh receptors actually change their electrical properties once the skeletal muscle is innervated. In embryonic rat muscle the ACh receptor channels have a relatively small conductance, but remain open for long periods when ACh is bound. In more mature muscle the junctional ACh receptors have a significantly larger conductance, but remain open for much shorter periods, usually only about 1 millisecond. The number of slow, embryonic channels decreases gradually after innervation. The fact that the nerve, and not some other feature, brings about this change in channel properties can be demonstrated by denervating mature muscle, which prevents fast channels from developing. The change itself is effected through a change in the subunit composition of the ACh receptor.

Some Synapses Are Eliminated During Development

At the mature neuromuscular junction a single muscle fiber is typically innervated by a single motor axon; that is, there is one endplate per muscle fiber. At earlier stages of neuromuscular development, however, each muscle fiber is innervated by several different motor axons. During development all but one of these axons is eliminated, and the one that remains becomes more complex over the same period; it acquires more endplate area and more release sites. Thus, the elimination of synapses actually increases synaptic input to the muscle fiber (Figure 6–12).

The elimination of synapses during development may involve *competition* among different axons that innervate the same target cell. That is, a motor neuron may lose its connections to some muscle fibers because other motor axons are more successful in maintaining input to those fibers. This reasoning has been borne out in experiments. When some of the competing motor axons are removed, each remaining motor axon continues to innervate several muscle fibers.

The mechanisms of the competition among neurons are not yet fully understood, but appear to involve access to *trophic factors* that are required to maintain synaptic contacts (see the next section). Moreover, during what is known as a *critical period* in early development, sensory experience itself is an important factor in strengthening some synaptic contacts and eliminating others. In Chapter 24 we take a closer look at how early experience shapes the brain's circuitry for vision.

Survival of Neurons Is Regulated by Interactions with Their Targets

A surprising finding in the study of development is that many neurons die in the normal course of development. Extensive studies have established that the death of neurons is a normal and widespread occurrence during embryonic development, often resulting in the loss of up to *half* of all the neurons initially generated. In addition, experimentally induced changes in the size of a neuronal target affect the survival of postmitotic (fully differentiated) neurons, but do not affect the population of neuronal precursor cells. This process of overproduction followed by drastic reduction occurs in almost all regions of the central and peripheral nervous systems and is usually referred to as *programmed neuronal death*.

Studies of programmed cell death in *Caenorhabditis elegans* have identified several genes that appear to have a role in a cell death program. Strikingly, one of these genes, *ced*-9, is related to a mammalian protein, bcl-2. Overexpression of the bcl-2 protein can suppress the death of sensory and sympathetic neurons that normally occurs after the removal of neurotrophic factors. Thus, vertebrates and invertebrates may possess a common cell death program.

Neurotrophic factors do have a role in the survival of neurons and may act by suppressing a cell death program in postmitotic cells. Several neurotrophic factors have now been identified, in particular nerve growth factor, brain-derived neurotrophic factor, neurotrophin 3, and ciliary neurotrophic factor, each of which supports the survival of distinct groups of neurons.

Two key sets of experiments have established the physiological role of nerve growth factor (NGF), discovered by Rita Levi-Montalcini (Figure 6–13), and other related trophic factors in the survival of sensory and sympathetic neurons. In one set of experiments, injection of antibodies against NGF into newborn mice and rats caused the sympathetic ganglia to disappear almost completely (Figure 6–14). The same antibodies administered at earlier stages of development also reduced the number of dorsal root ganglion neurons. Thus, both sympathetic and sensory neurons depend on this factor for their survival. Moreover, administering large amounts of NGF to embryonic or newborn animals prevents the natural death of many sensory and sympathetic neurons.

Other neurons that appear to depend on their targets for survival—for example, parasympathetic neurons, spinal motor neurons, and sensory neurons that derive from ectodermal placodes rather than from the neural crest—are supported by other related neurotrophic factors.

An Overall View

In all developing nervous systems cell differentiation depends on a series of signals that control the transcription of specific genes. These signals can be regulated by factors that are inherited by the cell or provided by other cells in the local environment. The relative contribution of these two basic programs of differentiation varies among species and between different types of neurons in the same embryo. The development of the nematode *Caenorhabditis elegans* is based in part on autonomous cell development, whereas vertebrate embryos rely more on local cell interactions to direct developmental fate.

Some of the cellular and molecular mechanisms that underlie neural cell differentiation have now been established. Not surprisingly, many of the genes that control the fate of invertebrate cells in a cell-autonomous manner are nuclear proteins that regulate the transcription of other genes. In contrast, local cell interactions in the nervous system involve diffusible signaling molecules and membrane receptors. Many of these signaling molecules

Figure 6–13 Rita Levi-Montalcini
Levi-Montalcini isolated the first-known growth factor in the nervous system, nerve growth factor (NGF), which she showed to be essential for the survival of sympathetic neurons. Levi-Montalcini received her medical degree from the University of Turin in Italy in 1938, where she trained in developmental biology with Guiseppe Levi. During World War II she hid from the fascists but continued doing research in the attic of a private home. It was here that she began to work on apoptosis, the overproduction of neurons during development, and programmed cell death. In 1948 she joined Viktor Hamburger, one of the pioneers of developmental neurobiology, at Washington University in St. Louis. There, together with Stanley Cohen she began work that led to the discovery of NGF. In 1986 she shared the Nobel Prize in Physiology and Medicine with Stanley Cohen. (Photo courtesy of R. Levi-Montalcini.)

are not restricted to the nervous system but have critical roles in the differentiation of other organs. Similarly, the receptors for these signaling molecules, for example transmembrane tyrosine kinases, are familiar proteins found in many different eukaryotic cell types. Thus, the molecules that control the differentiation and fate of cells in the nervous system are identical or closely related to the molecules that regulate other aspects of embryonic development.

Once the identity of a neuron has been established, the cell begins to project an axon and form complex but precise connections with other cells. The growth of developing vertebrate axons to their targets is dependent on guidance cues from the cells and extracellular matrices with which the growth cones come in contact. Molecules that mediate neural adhesion and recognition have been isolated and characterized. Many of these proteins belong to multigene families, whose other members serve similar recognition and adhesive functions in nonneural cells. Chemotropism and repulsion have also been established as mechanisms of guidance.

The ability of an axon to reach its appropriate target is essential for the survival of the neuron; if the target is absent, the neuron is likely to atrophy and die. There is now considerable evidence that target cells supply neurotrophic factors that nurture presynaptic neurons during the formation of functional synaptic connections. The matching of the number of presynaptic neurons with the size of the target organ may also be controlled by the availability of trophic factors.

The formation of synaptic contacts involves a complex series of interactions between the nerve terminal and its postsynaptic target. In many cases the presynaptic terminal plays a critical role in organizing the postsynaptic membrane to ensure that synaptic transmission becomes functional. For example, the developing presynaptic terminal regulates the number and distribution of transmitter receptors and other molecules on the postsynaptic membrane. At the same time, the postsynaptic cell may also regulate the differentiation of the presynaptic terminal.

In some regions of the nervous system the initial synaptic contacts between cells are accurate and stable, providing evidence for a high degree of initial specificity in cell recognition. In other regions, however, initial contacts are dramatically re-

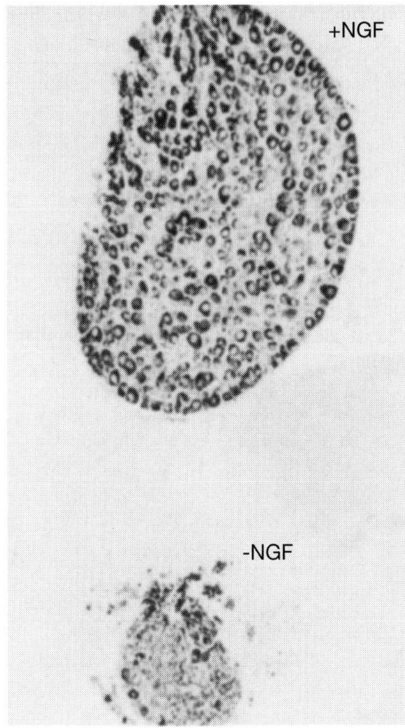

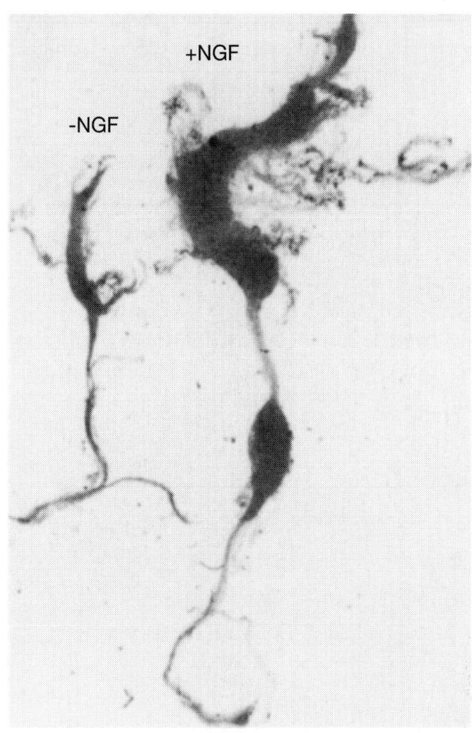

Figure 6–14 An antiserum against nerve growth factor (NGF) seriously impairs the development of the sympathetic ganglia in rodents. (From Levi-Montalcini, 1972.)

A. The superior cervical ganglion from a 9-day-old mouse injected daily since birth with NGF antiserum (–NGF, **below**) shows marked atrophy, with obvious loss of nerve cells. The superior cervical ganglion from normal 9-day-old mouse is shown (**above**) for comparison.

B. At 20 days of age the sympathetic chain ganglia of control mice (**right**) and of experimental mice injected since birth with NGF antiserum (**left**) are markedly different. The sympathetic chain from the experimental mouse shows gross atrophy.

arranged and many synapses are eventually eliminated during development. Thus, events that occur early in the development of the nervous system influence the final pattern of connections between nerve cells.

Selected Readings

Anderson, D. J. 1989. The neural crest cell lineage problem: Neuropoiesis? Neuron 3:1–12.

Banerjee, U., and Zipursky, S. L. 1990. The role of cell–cell interaction in the development of the *Drosophila* visual system. Neuron 4:177–187.

Chao, M. V. 1992. Neurotrophin receptors: A window into neuronal differentiation. Neuron 9:583–593.

Dodd, J., and Jessell, T. M. 1988. Axon guidance and the patterning of neuronal projections in vertebrates. Science 242:692–699.

Doupe, A J., Landis, S. C., and Patterson, P. H. 1985. Environmental influences in the development of neural crest derivatives: Glucocorticoids, growth factors, and chromaffin cell plasticity. J. Neurosci. 5:2119–2142.

Forscher, P., and Smith, S. J. 1988. Actions of cytochalasins on the organization of actin filaments and microtubules in a neuronal growth cone. J. Cell Biol. 107:1505–1516.

Frank, E., and Wenner, P. 1993. Environmental specification of neuronal connectivity. Neuron 10:779–785.

Fujisawa, H., Tani, N., Watanabe, K., and Ibata, Y. 1982. Branching of regenerating retinal axons and preferential selection of appropriate branches for specific neuronal connections in the newt. Dev. Biol. 90:43–57.

Goodman, C. S., and Shatz, C. J. 1993. Developmental mechanisms that generate precise patterns of neuronal connectivity. Cell 72/Neuron 10 (Suppl.):77–98.

Hall, Z. W., and Sanes, J. R. 1993. Synaptic structure and development: The neuromuscular junction. Cell 72/Neuron 10 (Suppl.):99–121.

Hamburger, V. 1988. The Heritage of Experimental Embryology. Hans Spemann and the Organizer. New York: Oxford University Press.

Hunt, R. K., and Cowan, W. M. 1990. The chemoaffinity hypothesis: An appreciation of Roger W. Sperry's contributions to developmental biology. In C. Trevarthen (ed.), Brain Circuits and Functions of the Mind. Cambridge, England: Cambridge University Press, pp. 19–74.

Hynes, R. O., and Lander, A. D. 1992. Contact and adhesive specificities in the associations, migrations, and targeting of cells and axons. Cell 68:303–322.

Jessell, T. M., and Melton, D. A. 1992. Diffusible factors in vertebrate embryonic induction. Cell 68:257–270.

Keynes, R. J., and Cook, G. M. W. 1992. Repellent cues in axon guidance. Curr. Opin. Neurobiol. 2:55–59.

Kreis, T. and Vale, R. (eds.). 1993. Guidebook to the Extracellular Matrix and Adhesion Proteins. Oxford: Oxford University Press.

Le Douarin, N. M. 1986. Cell line segregation during peripheral nervous system ontogeny. Science 231:1515–1522.

Levi-Montalcini, R. 1972. The morphological effects of immunosympathectomy. In G. Steiner and E. Schönbaum (eds.), Immunosympathectomy. Amsterdam: Elsevier, pp. 55–78.

Lillien, L. E., and Raff. M. C. 1990. Differentiation signals in the CNS: Type-2 astrocyte development in vitro as a model system. Neuron 5:111–119.

McConnell, S. K. 1989. The determination of neuronal fate in the cerebral cortex. Trends Neurosci. 12:342–349.

O'Leary, D. D. M., and Koester, S. E. 1993. Development of projection neuron types, axon pathways, and patterned connections of the mammalian cortex. Neuron 10:991–1006.

Oppenheim, R. W. 1991. Cell death during development of the nervous system. Annu. Rev. Neurosci. 14:453–501.

Patterson, P. H., and Nawa, H. 1993. Neuronal differentiation factors/cytokines and synaptic plasticity. Cell 72/Neuron 10 (Suppl.):123–137.

Purves, D. 1988. Body and Brain. A Trophic Theory of Neural Connections. Cambridge, Mass.: Harvard University Press.

Purves, D., and Lichtman, J. W. 1980. Elimination of synapses in the developing nervous system. Science 210:153–157.

Rakic, P. 1971. Neuron–glia relationship during granule cell migration in developing cerebellar cortex. A Golgi and electronmicroscopic study in *Macacus rhesus*. J. Comp. Neurol. 141:283–312.

Ready, D. F., Hanson, T. E., and Benzer, S. 1976. Development of the *Drosophila* retina, a neurocrystalline lattice. Dev. Biol. 53:217–240.

Schlessinger, J., and Ullrich, A. 1992. Growth factor signaling by receptor tyrosine kinases. Neuron 9:383–391.

Schuetze, S. M., and Role, L. W. 1987. Developmental regulation of nicotinic acetylcholine receptors. Annu. Rev. Neurosci. 10:403–457.

Solomon, F. 1992. Neuronal cytoskeleton and growth. Curr. Opin. Neurobiol. 2:613–617.

Tessier-Lavigne, M., and Placzek, M. 1991. Target attraction: Are developing axons guided by chemotropism? Trends Neurosci. 14:303–310.

Tomlinson, A. 1988. Cellular interactions in the developing *Drosophila* eye. Development 104:183–193.

III

SIGNALING WITHIN NERVE CELLS

```
5'---CAGCUAUCAGCUGUCGCUGAGACAGGUGGCAUAAGAGUGGAACAGAGAGUUGAAAAGGCAGGAAACUGGCUUAUCUCUUCACUAGAAAAGAGCUGAACACAGAAGUCCAGAAGAU
                      -240                -220                -200                -180                -160

                                                                                        -20
                                                                        Met Ile Leu Cys Ser Tyr Trp His Val Gly Leu Val
CUAACAAGUUCAUCGUUUAGUUAUUAGAAGUGGCAGAUUUGCUUGAAAAGCCAAUUAUUGAAAGCUGAAGA AUG AUU CUG UGC AGU UAU UGG CAU GUA GGG UUG GUG
 -140               -120                -100                 -80                        -60                         -40

        -10                                 -1  1                                   10
Leu Leu Leu Phe Ser Cys Cys Gly Leu Val Leu Gly Ser Glu His Glu Thr Arg Leu Val Ala Asn Leu Leu Glu Asn Tyr Asn Lys Val
CUA CUG UUA UUU UCG UGU UGU GGU CUG GUA CUA GGU UCU GAA CAU GAA ACA CGU UUG GUU GCU AAU UUA UUA GAA AAU UAU AAC AAG GUG
                    -20                     -1  1                       20                          40

    20                                      30                                      40
Ile Arg Pro Val Glu His His Thr His Phe Val Asp Ile Thr Val Gly Leu Gln Leu Ile Gln Leu Ile Ser Val Asp Glu Val Asn Gln
AUU CGU CCA GUG GAG CAU CAC ACC CAC UUU GUA GAU AUU ACA GUG GGG CUA CAG CUG AUA CAA CUC AUC AGU GUG GAU GAA GUA AAU CAA
    60                          80                         100                     120                         140

    50                                      60                                      70
Ile Val Glu Thr Asn Val Arg Leu Arg Gln Gln Trp Ile Asp Val Arg Leu Arg Trp Asn Pro Ala Asp Tyr Gly Gly Ile Lys Lys Ile
AUU GUG GAA ACA AAU GUG CGC CUA AGG CAG CAA UGG AUU GAU GUG AGG CUU CGC UGG AAU CCA GCC GAU UAU GGU GGA AUU AAA AAG AUC
                    160                     180                         200                         220

    80                                      90                                      100
Arg Leu Pro Ser Asp Asp Val Trp Leu Pro Asp Leu Val Leu Tyr Asn Asn Ala Asp Gly Asp Phe Ala Ile Val His Met Thr Lys Leu
AGA CUG CCU UCU GAU GAU GUU UGG CUG CCA GAU UUA GUU CUG UAC AAC AAU GCU GAU GGU GAU UUU GCC AUU GUU CAC AUG ACC AAA CUG
    240                         260                         280                     300                         320

    110                                     120                                     130
Leu Leu Asp Tyr Thr Gly Lys Ile Met Trp Thr Pro Pro Ala Ile Phe Lys Ser Tyr Cys Glu Ile Ile Val Thr His Phe Pro Phe Asp
CUU UUG GAU UAU ACG GGA AAA AUA AUG UGG ACA CCU CCA GCA AUC UUC AAA AGC UAU UGU GAA AUU AUU GUA ACA CAU UUC CCA UUU GAU
                    340                     360                         380                         400

    140                                     150                                     160
Gln Gln Asn Cys Thr Met Lys Leu Gly Ile Trp Thr Tyr Asp Gly Thr Lys Val Ser Ile Ser Pro Glu Ser Asp Arg Pro Asp Leu Ser
CAA CAA AAU UGC ACU AUG AAG UUG GGA AUC UGG ACG UAC GAU GGG ACA AAA GUU UCC AUA UCC CCG GAA AGU GAC CGU CCG GAU CUG AGU
    420                         440                         460                     480                         500

    170                                     180                                     190
Thr Phe Met Glu Ser Gly Glu Trp Val Met Lys Asp Tyr Arg Gly Trp Lys His Trp Val Tyr Tyr Thr Cys Cys Pro Asp Thr Pro Tyr
ACA UUU AUG GAA AGU GGA GAG UGG GUA AUG AAA GAU UAU CGU GGA UGG AAG CAC UGG GUG UAU UAU ACC UGC UGU CCU GAC ACU CCU UAC
                    520                     540                         560                         580

    200                                     210                                     220
Leu Asp Ile Thr Tyr His Phe Ile Met Gln Arg Ile Pro Leu Tyr Phe Val Val Asn Val Ile Ile Pro Cys Leu Leu Phe Ser Phe Leu
CUG GAU AUC ACC UAC CAU UUU AUC AUG CAG CGU AUU CCU CUU UAU UUU GUU GUG AAU GUC AUC AUU CCU UGU CUG CUU UUU UCA UUU UUA
    600                         620                         640                     660                         680

    230                                     240                                     250
Thr Gly Leu Val Phe Tyr Leu Pro Thr Asp Ser Gly Glu Lys Met Thr Leu Ser Ile Ser Val Leu Leu Ser Leu Thr Val Phe Leu Leu
ACU GGA UUA GUA UUU UAC UUA CCA ACU GAU UCA GGU GAG AAG AUG ACU UUG AGU AUU UCC GUU UUG CUG UCU CUG ACU GUG UUC CUU CUG
                    700                     720                         740                         760

    260                                     270                                     280
Val Ile Val Glu Leu Ile Pro Ser Thr Ser Ser Ala Val Pro Leu Ile Gly Lys Tyr Met Leu Phe Thr Met Ile Phe Val Ile Ser Ser
GUU AUU GUU GAG CUG AUC CCC UCA ACU UCC AGC GCU GUG CCU UUG AUU GGC AAA UAC AUG CUU UUU ACA AUG AUU UUU GUC AUC AGU UCA
    780                         800                         820                     840                         860

    290                                     300                                     310
Ile Ile Ile Thr Val Val Val Ile Asn Thr His His Arg Ser Pro Ser Thr His Thr Met Pro Gln Trp Val Arg Lys Ile Phe Ile Asp
AUC AUC AUU ACU GUU GUU GUA AUU AAU ACU CAC CAU CGC UCU CCA AGU ACA CAU ACA AUG CCA CAA UGG GUA CGA AAG AUC UUU AUU GAU
                    880                     900                         920                         940

    320                                     330                                     340
Thr Ile Pro Asn Val Met Phe Phe Ser Thr Met Lys Arg Ala Ser Lys Glu Lys Gln Glu Asn Lys Ile Phe Ala Asp Asp Ile Asp Ile
ACU AUA CCC AAU GUU AUG UUU UUC UCA ACA AUG AAA CGA GCU UCU AAG GAA AAG CAA GAA AAU AAG AUA UUU GCU GAU GAC AUU GAU AUC
    960                         980                         1,000                   1,020                       1,040

    350                                     360                                     370
Ser Asp Ile Ser Gly Lys Gln Val Thr Gly Glu Val Ile Phe Gln Thr Pro Leu Ile Lys Asn Pro Asp Val Lys Ser Ala Ile Glu Gly
UCU GAC AUU UCU GGA AAG CAA GUG ACA GGA GAA GUA AUU UUU CAA ACA CCU CUC AUU AAA AAU CCA GAU GUC AAA AGU GCU AUU GAG GGA
                    1,060                   1,080                       1,100                       1,120

    380                                     390                                     400
Val Lys Tyr Ile Ala Glu His Met Lys Ser Asp Glu Glu Ser Ser Asn Ala Ala Glu Glu Trp Lys Tyr Val Ala Met Val Ile Asp His
GUC AAA UAU AUU GCA GAG CAC AUG AAG UCU GAU GAG GAA UCA AGC AAU GCU GCA GAG GAA UGG AAA UAU GUU GCA AUG GUG AUU GAU CAC
    1,140                       1,160                       1,180                   1,200                       1,220

    410                                     420                                     430
Ile Leu Leu Cys Val Phe Met Leu Ile Cys Ile Ile Gly Thr Val Ser Val Phe Ala Gly Arg Leu Ile Glu Leu Ser Gln Glu Gly
AUU CUG CUG UGU GUC UUC AUG CUG AUU UGU AUA AUU GGU ACA GUU AGC GUG UUU GCU GGC CGU CUC AUU GAA CUC AGU CAA GAG GGC UAA
                    1,240                   1,260                       1,280                       1,300

AUCUUCAUUGUGAGCAAAAAAGGCAAUACUGGAAUAAGGGAUGGAUAUCACUCCACAGAAAAGAUGUGUGGGUUUAGUGUUGCAAUUGUAGUCUGUUUUAUGAGAUAUAUAGUUUGCUUU
    1,320               1,340               1,360               1,380               1,400               1,420

GUUUUACAAUGAAAUGUACUUAAGGUAUUUGAAUAUGUAAAAAAAAGUAAUGAAAUAACAGUAAGUGAAAAAUGUUAUUAUGCAAGUACCUGAAACGUGUAAUAAGUGGAACAACUUUUU
    1,440               1,460               1,480               1,500               1,520               1,540

AAUACAUUACAUAAAAGUAAGCAAAAAAUAAGUUUAACAAAUUAUGAGGGUAGUCAUUUGAAAUGUAACAGAGAAAUGAAAAUUAUUAGAAAUAUAAACAGUAUAUAUUAAGUUAAACAA
    1,560               1,580               1,600               1,620               1,640               1,660

AGUUAAUCCAUUCUUUUAUAUCCAAAGUUUGUAUUAUACAUUUAGAAGUGUAGUUCUAUUGUAUAAUUUAAAGUAUGUUUUACAGAUCAUUAAUAAAUAUUCAAUGCAUUACU---3'
    1,680               1,700               1,720               1,740               1,760               1,780
```

About the image on the preceding page:

The complete nucleotide base sequence of the messenger RNA that encodes the α subunit of the nicotinic acetylcholine receptor was first presented in the early 1980s. Shosaku Numa and his colleagues also cloned and sequenced the β, γ, and δ subunits at that time. Since then, the sequences of many ion channels and other receptors critical to neuronal signaling have been determined. This information has been used to construct specific and detailed models of the structure and function of these proteins.

Introduction

From modern cell biology we learn that complex biological systems—not only the higher metazoan animals but also relatively primitive organisms—are built from similar, repeating modules. In all biological systems the basic unit of these modules is the cell. Thus, understanding the cellular mechanisms of the vertebrate brain is an important step in understanding the architecture of the brain's function. Behavior is determined not merely by how the brain is wired, but also by the cytology, biophysics, and biochemistry of the nerve cells that make up the brain.

Although they share many functions with other cells of the body, nerve cells are unique in their ability to communicate precisely and rapidly over long distances with one another and with nonneural target cells, such as muscle and gland cells. This signaling ability depends on rapid changes in the electrical potential difference across the cell membrane. Specialized proteins in the cell membrane form channels that allow specific inorganic ions to pass through the membrane at astonishing rates, providing the rapid current fluxes that underlie the changes in membrane potential.

Each channel is selective for a particular ion species, and each is responsive to a particular type of external stimulus. Thus the current flow into and out of the individual nerve cell is regulated by a complex array of ion channels. In large part, the great variety of functions carried out by different brain modules is due to the variety and distribution of these membrane channels. In this part of the book we examine the basic properties of ion channels, the types of channels, the relationship between ion channels and the electrical properties of the nerve cell, and the dynamics of the channels during signaling between cells.

7

Ion Channels

Ion Channels Are Proteins that Span the Cell Membrane

Ion Channels Can Be Investigated by Both Functional and Structural Methods

The Current in a Single Channel Can Be Measured

The Structure of Ion Channels Is Inferred from Biochemical and Molecular Biological Studies

Ion Channels in All Cells Share Several Characteristics

The Flux of Ions Through the Ion Channel Is Passive

The Opening and Closing of a Channel Involve Conformational Changes

Variants of Each Type of Ion Channel Are Found in Different Tissues

Genes that Encode Ion Channels Can Be Grouped into Families

Neuronal signaling depends on rapid changes in the electrical potential difference across nerve cell membranes. During an action potential the membrane potential changes quickly, up to 500 volts per second. These rapid changes are made possible by *ion channels*, a class of integral proteins that traverse the cell membrane and are found in all cells throughout the body. These channels have three important properties: (1) they conduct ions, (2) they recognize and select among specific ions, and (3) they open and close in response to specific electrical, mechanical, or chemical signals.

Ion channels in nerve and muscle cells conduct ions across the cell membrane at extremely rapid rates—up to 100,000,000 ions *per second* may pass through a single channel, thereby providing a significant flow of ionic current. This current flow causes the rapid changes in membrane potential needed to generate an action potential, as will be discussed in Chapter 10. The high rate of ionic flow in channels is comparable to the rate of the fastest enzymes, catalyase and carbonic anhydrase, which have rates that are diffusion limited. (The turnover rate of most enzymes is considerably slower, ranging from 10 to 10^3 per second.)

In addition to their high rate of conductance, ion channels allow only a few types of ions to pass. For example, the membrane potential of nerve cells at rest is largely determined by channels that are selectively permeable to K^+ (these channels are typically a hundredfold more permeable to K^+ than to Na^+). During the action potential, however, ion channels selective for Na^+ are activated (these channels are 10- to 20-fold more permeable to Na^+ than to K^+). Thus a key to the great versatility of neuronal signaling is the activation of *different classes* of ion channels, each of which is selective for specific ions.

Finally, the channels involved in neuronal signaling are also *gated*: they actively open and close in response to various stimuli. There are, however, many *resting channels* that are usually open; these contribute significantly to the resting potential, as will be discussed in Chapter 8. Channels are regulated in three ways: by voltage (voltage-gated channels), by chemical transmitters (ligand-gated channels, so-called because the molecule of transmitter binds to the receptor), and by pressure or stretch (mechanically gated channels). An individual channel is usually most sensitive to only one type of influence.

In this chapter we consider four questions: Why do nerve cells have channels? How can channels conduct ions at such high rates and be selective? How are channels gated? And how are the properties of these channels modified by various intrinsic and extrinsic conditions? In Chapters 8 and 9 we consider how resting channels contribute to the resting potential and the passive electrical properties of the neuron. In Chapter 10 we describe how voltage-gated channels generate the action potential. In Chapters 11 to 14 we examine how transmitter-gated channels produce synaptic potentials.

Ion Channels Are Proteins that Span the Cell Membrane

To appreciate why nerve cells need channels, we need to understand the nature of the plasma membrane and the physical chemistry of ions in solution. The plasma membrane of all cells, including nerve cells, is about 6 to 8 nm thick and consists of a mosaic of lipids and proteins. The surface of the membrane is formed by a double layer of phospholipids. Embedded within this continuous lipid sheet are proteins, including ion channels.

The lipids of the membrane do not mix with water—they are hydrophobic. In contrast, the ions within the cell and those outside strongly attract water molecules—they are hydrophilic (Figure 7–1). The attraction between ions and water results because water molecules are *dipolar*. Charge is separated within the molecule, although the net charge on a water molecule is zero. The oxygen atom in a

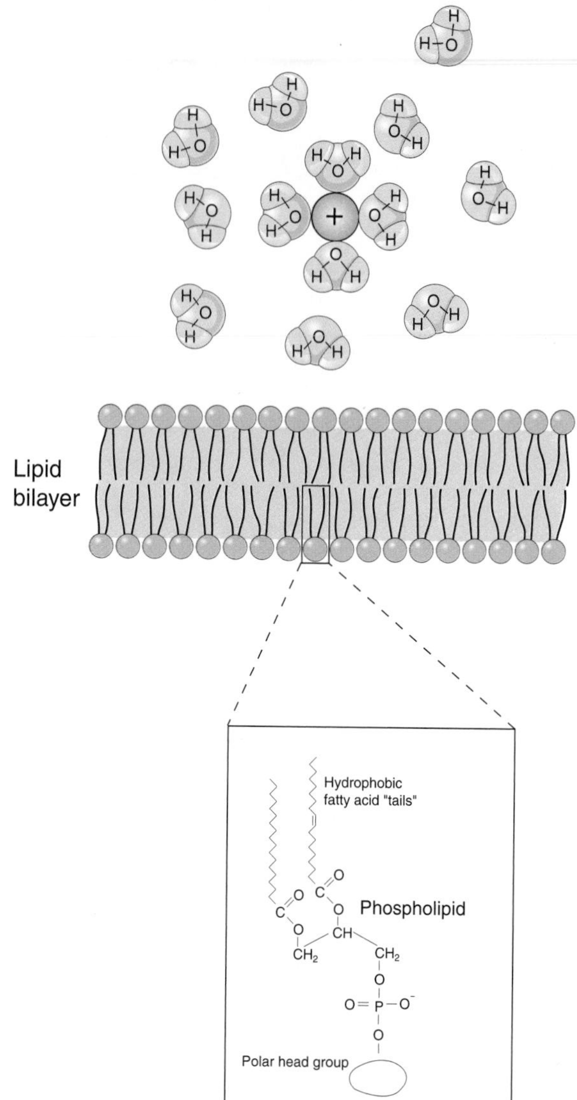

Figure 7–1 Ions in solution are surrounded by a cloud of water molecules (waters of hydration) that are attracted by the net charge of the ion. This cloud is carried along by the ion as it diffuses through a solution, adding to the ion's effective size. The energetics of leaving this polar environment for the nonpolar environment of the lipid bilayer of the cell membrane are extremely unfavorable to the ion, and therefore it is improbable that the ion will enter the lipid bilayer. Phospholipids and glycoproteins form the self-sealing lipid bilayers that are the basis for all cellular membrane. In phospholipids two of the –OH groups in glycerol are linked to fatty acids while the third –OH group is linked to phosphoric acid. The phosphate is further linked to one of a variety of small polar head groups (alcohols). Phospholipids have a hydrophilic head and a hydrophobic tail. The hydrophobic tails join to exclude water and ions, and the polar hydrophilic heads face the aqueous environment of the extracellular fluid and cytoplasm.

water molecule tends to attract electrons and so bears a small net negative charge, and the hydrogen atoms tend to lose electrons and therefore carry a small net positive charge. As a result of this distribution of charge, cations (which are positively charged) are strongly attracted electrostatically to the oxygen atom of water, and anions (negatively charged) are attracted to the hydrogen atoms. By the same token, ions attract water; in fact they become surrounded by electrostatically bound *waters of hydration* (Figure 7–1).

For an ion to move away from water into the noncharged hydrocarbon tails of the lipid bilayer in the membrane, a large amount of energy must be supplied to overcome the attractive forces between the ion and the surrounding water molecules. For this reason it is extremely unlikely that an ion will move from solution into the lipid bilayer, and therefore the bilayer itself is almost completely impermeable to ions. Ions cross the membrane only through specialized pores or openings in the membrane, such as ion channels, where, as we shall see, the energetics favor ion movement.

The fact that ion channels are made up of protein and are not simply holes in the lipid membrane has been known with certainty for only about 15 years. The idea of ion channels, however, dates to the end of the nineteenth century. At that time physiologists knew that cells are permeable to many small solutes, including some ions, despite the barrier of the cell membrane. To explain the flow of water across biological membranes (osmosis), Ernst Brucke proposed that membranes contain channels that would work like pores, allowing water to flow across membranes but excluding larger solutes. Later, William Bayliss suggested that a water-filled channel would permit ions to cross the cell membrane, since the ions would not need to be stripped of their waters of hydration.

The idea that ions move through channels leads to a question: How can a water-filled channel conduct at high rates and yet be selective? For instance, how does a channel allow K^+ to pass while excluding Na^+ ions? This function cannot be based solely on ionic diameter, because K^+, with its crystal radius of around 0.133 nm, is actually larger than Na^+ (crystal radius of 0.095 nm). As we have seen, however, ions in solution are surrounded by waters of hydration. Thus, the ease with which an ion moves in solution (its mobility or *diffusion constant*) does not simply depend on the size of the ions, but on the size of the ion *and* the shell of water surrounding it. The smaller the ion, the more highly localized is its charge and the stronger is its electric field; smaller ions such as Na^+ have stronger effective electric fields than larger ions like K^+. As a result, smaller ions ions exert a stronger attraction on water. Thus, as Na^+ moves through solution, its extra electrostatic attraction for water tends to slow it down relative to K^+, so that, by virtue of its larger water shell, Na^+ behaves as if it is larger than K^+. In fact, there is a precise relation between the size of an ion and its mobility in solution: the smaller the ion, the lower its mobility. We therefore can construct a model of a channel that selects K^+ rather than Na^+ simply on the basis of the interaction of the two ion species with water in a water-filled channel (Figure 7–2).

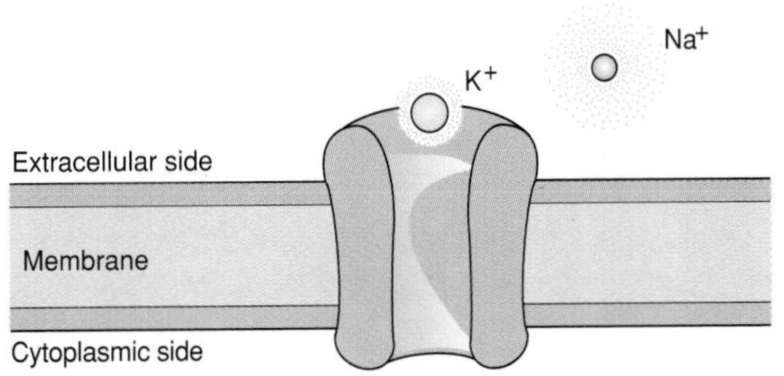

Figure 7–2 Most voltage-gated channels allow only one or another type of ion to pass through. The channel shown here is selective for K^+ ions. The model assumes that channels are water-filled pores that conduct ions by diffusion. Although the K^+ ion is larger than the Na^+ ion, its effective diameter in solution is smaller because its local field strength is less intense, causing it to attract a smaller cloud of water molecules. Thus, the K^+ channel could select by excluding hydrated ions whose diameter is larger than the pore.

This model explains how a channel can select K^+ and exclude Na^+, but it presents us with the converse problem: explaining how a channel could select Na^+ and exclude K^+. This problem led many physiologists in the 1930s and 1940s to abandon the channel theory in favor of a new idea: ions cross cell membranes by first binding to a specific carrier protein that then transports the ion through the membrane. In this carrier model, selectivity is based on the chemical binding between the ion and the carrier protein, not on the mobility of the ion in

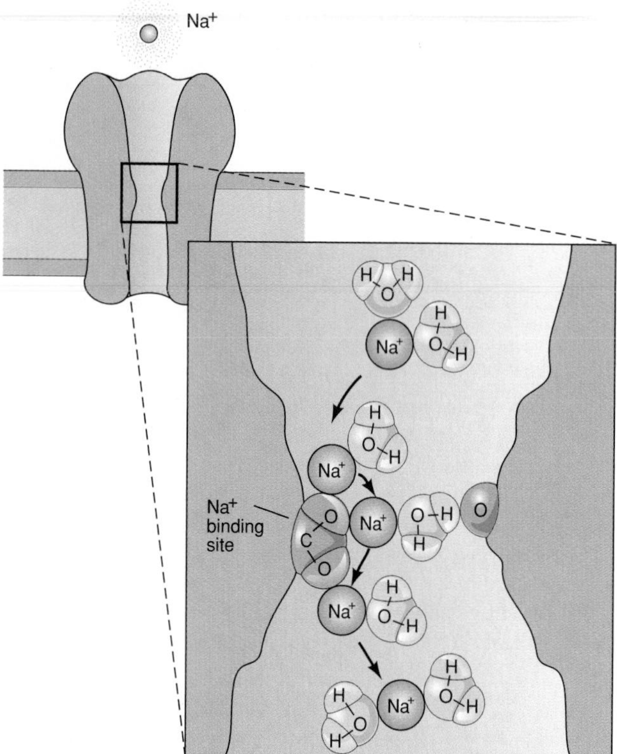

Figure 7–3 Sodium channels have a selectivity filter somewhere along the length of the channel, with a site that weakly binds Na^+ ions. According to the hypothesis developed by Bertil Hille and colleagues, a Na^+ ion binds transiently at an active site as it moves through the filter. At the binding site the positive charge of the ion is stabilized by a hydrophilic negatively charged amino acid residue lining the channel, and also by a water molecule that is attracted to a second polar amino acid residue lining the other side of the channel wall. It is thought that the larger K^+ ion is unable to move through the filter because it is not as effectively stabilized by its electrostatic interactions with the negative charge at the filter. (From Hille, 1984.)

solution. We now know that ions can cross membranes by means of transporters, the sodium–potassium pump being a well-characterized example (see Chapter 8).

Nevertheless, many observations on ion conductance across the cell membrane do not fit the carrier model. One of the most telling pieces of evidence is the rate of transfer across membranes. This was first examined in the early 1970s, in acetylcholine-activated ion channels located in the cell membrane of skeletal muscle at the synapse between nerve and muscle (see Chapter 12). Using measurements of membrane current noise (small statistical fluctuations in the mean ionic current induced by acetylcholine), it has been inferred that a single acetylcholine-activated receptor channel can transport 10 million ions per second. In contrast, the Na^+–K^+ pump can transport at most a thousand ions per second. If the acetylcholine receptor acted as a carrier, it would have to shuttle an ion across the membrane in 0.1 microsecond (one ten-millionth of a second), an implausibly fast rate. Therefore, the acetylcholine receptor (and other ligand-gated receptors) must conduct ions through a channel. Later measurements on many voltage-gated current pathways selective for K^+, Na^+, and Ca^{2+} demonstrated similarly large conductances, indicating that they too are channels.

But we are still left with the original problem: What makes a channel selective? To explain selectivity, the pore theory was extended by the proposal that channels have narrow regions that act as molecular sieves (Figure 7–3). At this *selectivity filter* an ion sheds most of its waters of hydration and in their place forms weak chemical bonds (electrostatic interactions) with polar (charged) amino acid residues that line the walls of the channel. Since the shedding of the waters of hydration is energetically unfavorable for an ion, the ion will traverse a channel only if the energy of interaction with the selectivity filter compensates for the loss of waters of hydration. Ions traversing the channel are normally bound to the selectivity filter for only a short time (less than 1 microsecond), after which the electrostatic and diffusional forces propel the ion through the channel. In some channels the pore diameter is large enough to accommodate several water molecules. An ion traversing such a channel need not be stripped completely of its water shell.

Ion Channels Can Be Investigated by Both Functional and Structural Methods

To understand fully how channels work, we ultimately will need the sort of three-dimensional structural information that has proven so informative in the study of enzymes and other soluble proteins. So far, X-ray crystallographic and other structural analyses have not been generally applied to integral membrane proteins, such as ion channels, because their hydrophobic regions make them difficult to crystallize. However, two other powerful methods, single-channel recording and gene cloning, have taught us a good deal.

The Current in a Single Channel Can Be Measured

Before it became possible to resolve the small amount of current that flows through a single ion channel in biological membranes, channel function was studied in artificial lipid bilayers. Functional lipid bilayers are formed by painting a thin drop of phospholipid over a small hole in a nonconducting barrier that separates two salt solutions. Although lipid membranes have a very high resistance to ions, ion conductance across the membrane increases dramatically when certain bacterial proteins are added to the salt solution. Early studies with a 15-amino-acid peptide, gramicidin A, were especially informative. Gramicidin A has two special features: the stereochemistry of the hydrophobic residues alternates (L, D, L, D, etc.) and the C and N termini are modified (and uncharged). Because of these structural features the peptide forms a channel that spans the membrane. Application of low concentrations of gramicidin A brings about small steplike changes in current flow across the membrane. These brief pulses of current reflect the all-or-none opening and closing of the single ion channel formed by the peptide.

The channel behaves as a simple resistor because the unitary current varies with membrane potential in a linear manner (Figure 7–4). The amplitude

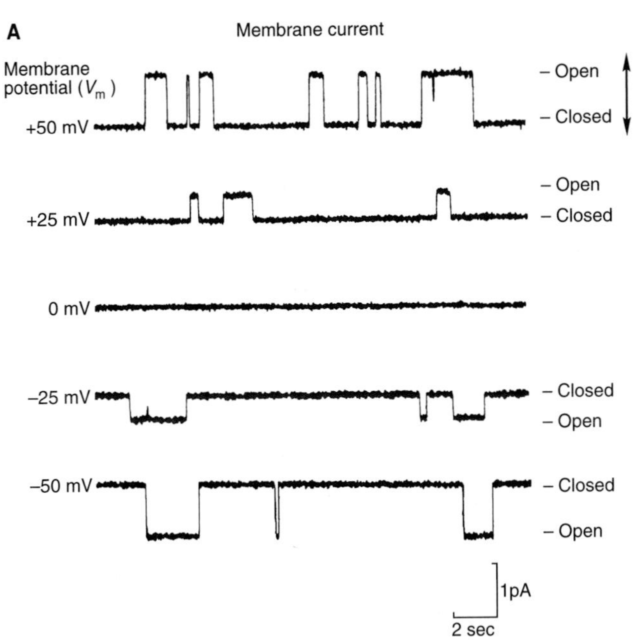

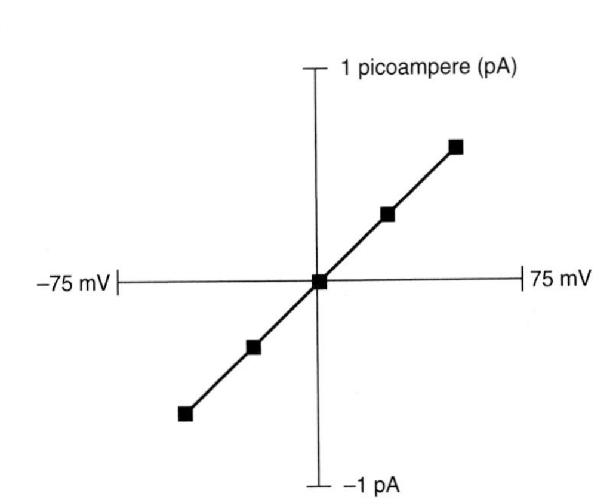

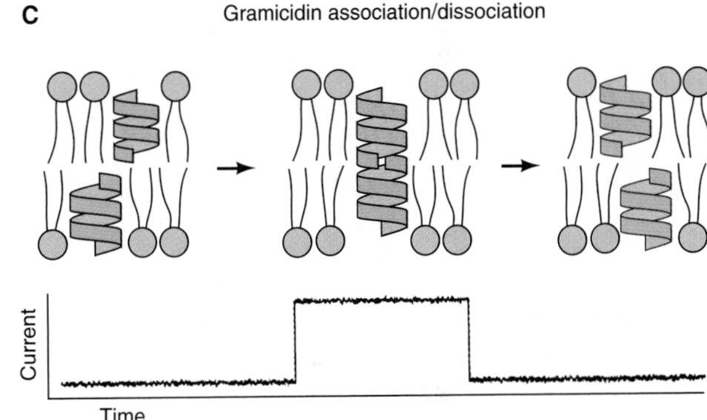

Figure 7–4 Characteristics of the current in a single ion channel. The data presented here are from a channel formed by the addition of gramicidin A molecules to the solution bathing an artificial lipid bilayer.

A. The channel opens and closes in an all-or-none fashion, resulting in brief current pulses through the membrane. If the electrical potential (V_m) across the membrane is varied, the current through the channel (i) changes proportionally. V_m is measured in millivolts (mV); i is measured in picoamperes (pA).

B. A plot of the current through the channel versus the potential difference across the membrane reveals that the current is linearly related to the driving force; in other words, the channel behaves as an electrical resistor that follows Ohm's law ($I = V/R$ or $I = gV$). (Data courtesy of Olaf Anderson and Lyndon Providence.)

C. Proposed structure of the gramicidin A channel. A functional channel is formed by end-to-end dimerization of two gramicidin peptides. (From Sawyer et al. , 1989.)

BOX 7–1 Recording Current Flow from Single Ion Channels

The patch-clamp technique is a refinement of voltage clamping (Box 8–1) that was developed in 1976 by Erwin Neher and Bert Sakmann to record current flow from single ion channels. A small fire-polished glass micropipette with a tip diameter of around 1 μm is pressed against the membrane of a skeletal muscle fiber that has been treated with proteolytic enzymes to remove connective tissue from the muscle surface. The pipette is filled with a salt solution resembling that normally found in the extracellular fluid. A metal electrode in contact with the electrolyte in the micropipette connects it to a special electrical circuit that measures the current flowing through channels in the membrane under the pipette tip.

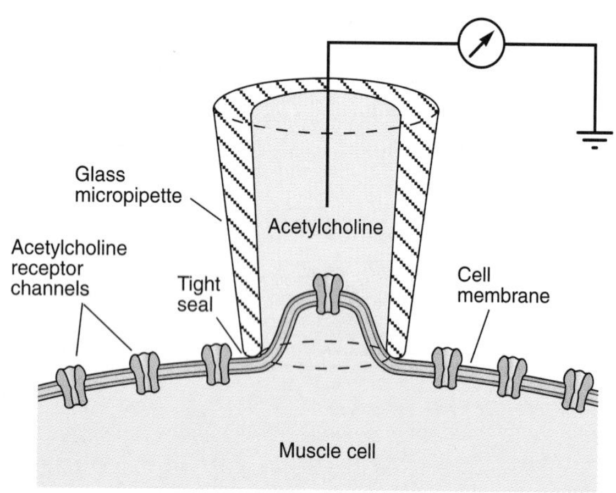

Figure 7–5A Patch-clamp setup. A pipette containing acetylcholine is used to record transmitter-gated channels in skeletal muscle. (Adapted from Alberts et al., 1989.)

In 1980 Neher discovered that applying a small amount of suction to the patch pipette greatly increased the tightness of the seal between the pipette and the membrane. The result was a seal with extremely high resistance between the inside and the outside of the pipette. The seal lowered the electronic noise and extended the utility of the technique to the whole range of channels involved in electrical excitability, including those with small conductance. Since this discovery, Neher and Sakmann, and many others, have used the patch-clamp technique to study, in a variety of neurons and other cells, all three major classes of ion channels: voltage gated, transmitter gated, and mechanically gated.

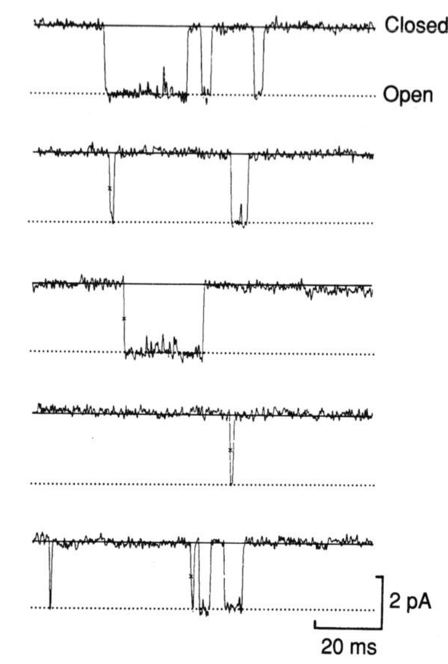

Figure 7–5B Record of the current flowing through a single ion channel as the channel switches between closed and open states. (Courtesy of B. Sakmann.)

Independently Christopher Miller developed a method for incorporating channels from biological membranes into planar lipid bilayers. With this technique, biological membranes are first homogenized in a laboratory blender; centrifugation then separates out a portion composed only of membrane vesicles. Under appropriate ionic conditions, these membrane vesicles fuse with a planar lipid membrane, incorporating any ion channel in the vesicle into the planar membrane. This technique has two experimental advantages. First, it allows recording from ion channels in regions of cells that are inaccessible to patch clamp; for example, Miller has successfully studied a K^+ channel isolated from the internal membrane of skeletal muscle (the sarcoplasmic reticulum). Second, it allows researchers to study how the *composition* of the membrane lipids influences channel function.

of the single-channel current can thus be obtained from Ohm's law, $i = V/R$, where i is the current through the single channel, V is the voltage across the channel, and R is the resistance of the open channel. The slope of the relation between i and V yields a value of R for a single open channel of around 8×10^{10} ohms (Figure 7–4B). In dealing with ion channels it is more useful to speak of the reciprocal of resistance or *conductance* ($\gamma = 1/R$), as this provides an electrical measure of ion permeability. Thus, Ohm's law can be expressed as $i = \gamma V$. The conductance of the gramicidin A channel is around 12×10^{-12} siemens or 12 picosiemens (pS), where 1 siemen = 1/ohm.

The insights into basic channel properties provided by artificial systems were later confirmed in biological membranes by the patch-clamp technique (Box 7–1). A glass micropipette containing acetylcholine—the neurotransmitter that activates channels in the membrane of skeletal muscle—was pressed tightly against a frog muscle membrane. Small unitary current pulses representing the opening and closing of single acetylcholine-activated ion channels were recorded from the area of the membrane under the pipette tip. As in the case of gramicidin A channels, the relation between current and voltage in these acetylcholine-activated channels proved to be linear, with a single-channel conductance of around 25 pS.

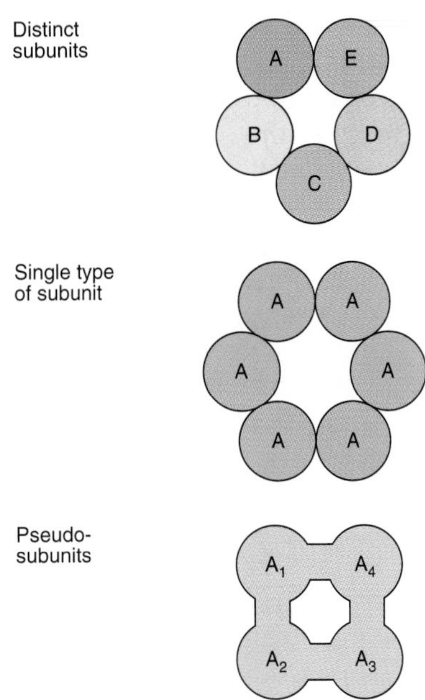

Figure 7–6 Ion channels can be constructed as heteromultimers from distinct subunits (**top**), as homo-oligmers from a single type of subunit (**middle**), or from a single polypeptide chain organized into repeating motifs, where each motif is the equivalent of one subunit (**bottom**).

The Structure of Ion Channels Is Inferred from Biochemical and Molecular Biological Studies

What do ion channels look like? How does the channel protein span the membrane? What happens to the structure of the channel when it opens and closes? Where along the length of the channel do drugs and transmitters bind? Biochemical and molecular biological approaches have resulted in considerable progress in understanding channel structure and function. Ion channels are large membrane-spanning proteins with carbohydrate groups (*glycoproteins*) attached to their surface. All channels have a central aqueous pore that spans the entire width of the membrane. Many channels are made up of two or more subunits, which may be identical or distinct (Figure 7–6).

The genes for most of the major classes of ion channels have now been cloned and sequenced. The primary amino acid sequence of the channel, as inferred from the DNA sequence, has been used to create models of the structure of different channel proteins. These models rely on computer programs to predict regions of secondary structure, such as the arrangement of the amino acid residues into α-helices and β-sheets that are likely to correspond to membrane-spanning domains of the channel (Figure 7–7). The predictions, in turn, are based on existing information from proteins whose actual structure is known from electron and X-ray diffraction analysis.

The membrane protein whose structure was first well understood is bacteriorhodopsin, a photopigment in the cell membrane of *Halobacterium*. (Photopigment converts sunlight into electrochemical energy.) Bacteriorhodopsin contains regions with charged (hydrophilic) amino acids and regions with uncharged (hydrophobic) amino acids. There are, in all, seven hydrophobic regions. Each of these is about 15 to 20 amino acids long and spans the membrane in the form of α-helices.

These membrane-spanning regions are in turn linked by six hydrophilic loops—three that extend just outside the cell, and three within.

Additional insight into channel structure and function can be obtained by comparing the primary amino acid sequences of the same type of channel from different species—for example, voltage-gated Na^+ channels from humans, rat, and fly—and identifying the regions that show high degrees of sequence similarity. Those regions that have been highly conserved through evolution are likely to be important in maintaining the effective structure and function of the channel. Further insight into the relationship between structure and function can be obtained from sequence similarities among different, but related, channels. These regions are likely to serve a common biophysical function in different channels. For example, all voltage-gated channels have a certain membrane-spanning domain that contains positively charged amino acids (lysine or arginine) spaced at every third position along the α-helix. The fact that this motif is observed in all voltage-gated channels, but not in chemically gated channels, lends support to the view that this charged region may play an important role in voltage gating (see Chapter 10).

Once a structure for a channel has been proposed, it can be tested in several ways. One way is to raise antibodies against synthetic peptides corresponding to different hydrophilic regions in the protein sequence. Using immunocytochemistry, one can then determine whether the antibody binds to the extracellular or cytoplasmic surface of the membrane, thus defining whether a particular region of the channel is extracellular or intracellular.

A second way uses genetic engineering to produce so-called chimeric channels, which are made up of various parts encoded by the genes of different species. This technique takes advantage of the fact that channels in different species have somewhat different properties. For example, the bovine acetylcholine-gated receptor channel has a slightly greater single-channel conductance than the same channel in the electric fish. By comparing the properties of a chimeric channel to those of the two original channels, it is possible to assess which regions of the channel are involved in which functions. This technique has been used to identify a specific membrane-spanning segment of the acetylcholine-gated channel as the region that

Figure 7–7 The secondary structure of a membrane-spanning protein can be determined from its amino acid sequence through a hydropathy plot.

A. The nicotinic acetylcholine receptor in the cell membrane is a cylindrical molecule composed of five similar membrane-spanning subunits: α, γ, α, β, δ. The two α subunits are identical.

B. The receptor is opened out to show the similar topology of the subunits, each of which has an extensive N-terminal extracellular domain and four transmembrane segments. Since each subunit is a single continuous polypeptide chain, these transmembrane segments are connected by alternating intracellular (cytoplasmic) and extracellular loops; the intracellular loop between M3 and M4 is the longest. These topological features are determined from a computer-generated hydropathy plot of an amino acid sequence.

C. This hydropathy plot is for a short segment of the human γ subunit that includes some of the N-terminal extracellular domain and M1 (residues 205–270). This part of the molecule was selected from the complete sequence because it shows the difference between the hydrophilic extracellular domain and the hydrophobic M1, and because this rather short sequence contains at least one of each of the 20 naturally occurring amino acids. At the right of the figure is a line representing the peptide chain with the amino acid residue numbers, starting with 205 at the top and ending with 270 at the bottom. To the right of this line the residues are represented by their single-letter codes. The structures of all 20 amino acids are shown at the right of this sequence, starting with the N-terminal (top): **E**, glutamate; **W**, tryptophan; **H**, histidine; **R**, arginine; **K**, lysine; **G**, glycine; **T**, threonine; **D**, aspartate; **Q**, glutamine; **F**, phenylalanine; **P**, proline, **I**, isoleucine; **N**, asparagine; **V**, valine; **C**, cysteine; **L**, leucine; **S**, serine; **M**, methionine; **Y**, tyrosine; **A**, alanine. The plot is constructed by assigning each amino acid a number representing its aqueous solubility (hydrophilicity is positive, hydrophobicity is negative). A running average of hydropathy is plotted for the entire sequence: Each point represents the average values for a window of four amino acids plotted at the position of the residue at the midpoint of the window. The topology of the polypeptide can be inferred from the resulting plot. Here the N-terminal hydrophilic domain extends into the aqueous extracellular space; the hydrophobic segment agent crosses through the membrane.

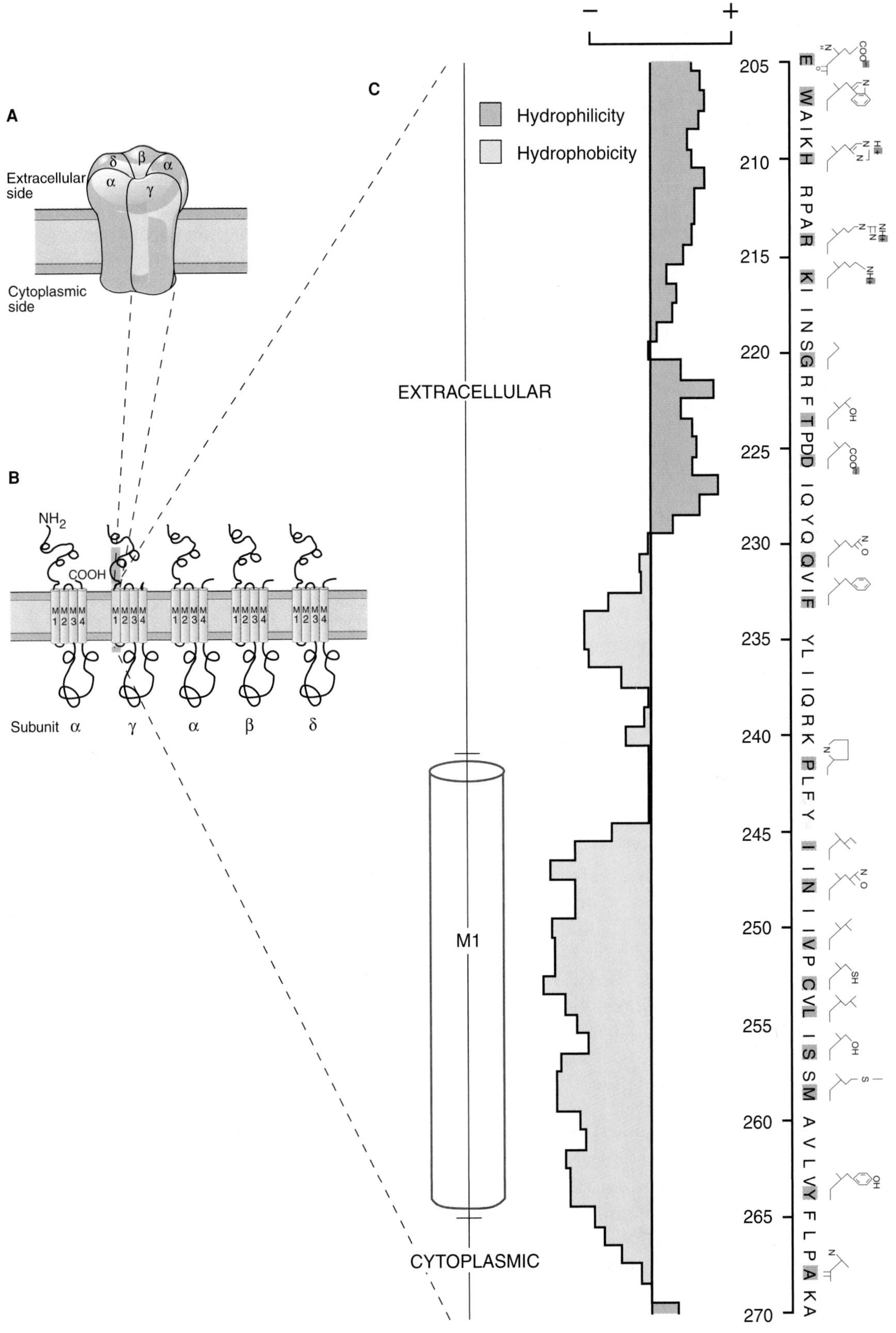
A
Extracellular side
Cytoplasmic side
B
NH2
COOH
Subunit α γ α β δ
C
Hydrophilicity
Hydrophobicity
EXTRACELLULAR
M1
CYTOPLASMIC
205
210
215
220
225
230
235
240
245
250
255
260
265
270
E W A I K H R P A R K I I N S G R F T P D D I Q Y Q Q V I F Y L I I Q R K P L F Y I I N I I V P C V L I S S M A V L V Y F L P A K A

forms the lining of the pore (see Chapter 12). Finally, the roles of different amino acid residues or stretches of residues can also be tested using site-directed mutagenesis, a type of genetic engineering in which specific amino acid residues are substituted or deleted.

Ion Channels in All Cells Share Several Characteristics

All cells make use of local intercellular signaling processes, but only nerve and muscle cells are specialized for rapid signaling over long distances. Although nerve and muscle cells have a particularly rich variety and high density of membrane ion channels, their channels do not appear to differ fundamentally from those in other cells in the body. Thus, the general properties of ion channels described here are found in a wide variety of cell types.

The Flux of Ions Through the Ion Channel Is Passive

The flux of ions through ion channels is passive, requiring no expenditure of metabolic energy by the channels. The direction and eventual equilibrium for this flux are determined not by the channel itself, but rather by the electrostatic and diffusional driving forces across the membrane.

Most ion channels select the types of ions that they allow to cross the membrane. By discriminating between ions on the basis of ionic charge, a channel allows either cations or anions to permeate. Some types of cation-selective channels do not select among the cations that are usually present in extracellular fluid—they pass Na^+, K^+, Ca^{2+}, and Mg^{2+} almost indiscriminantly. However, most cation-selective channels are permeable primarily to a single type of ion, whether it is Na^+, K^+, or Ca^{2+}. Most types of anion-selective channels are highly selective also; they conduct only one physiological ion, Cl^-.

The kinetic properties of ion permeation are best described by the channel's conductance, which is determined by measuring the current (i.e., ion flux) that flows through the open channel in response to a given *electrochemical driving force*. The net electrochemical driving force is determined by two factors—the electrical potential difference across the membrane and the concentration gradient of the permeant ions across the membrane. Changing either one can change the net driving force (see Chapter 8).

As we have seen, in some channels the current flow varies linearly with driving force—that is, the channels behave as simple resistors. In others the current flow is a nonlinear function of driving force. This type of channel behaves as a rectifier—it conducts ions more readily in one direction than in the other. Whereas the conductance of a resistor-like channel is constant—it is the same at all voltages—the conductance of a rectifying channel is variable, and must be determined by plotting current versus voltage over the entire physiological range of membrane potential (Figure 7–8).

The rate of ion flux (i.e., current) through a channel depends on the concentration of the ions in the surrounding solution. At low concentrations the current increases almost linearly with concentration. At higher concentrations the current tends to reach a point beyond which it no longer increases as the concentration is raised. At this point the current is said to *saturate*.

This saturation effect is consistent with the idea that ion permeation involves binding of ions to specific polar sites within the pore of the channel, rather than obeying the laws of electrochemical diffusion in free solution. A simple electrodiffusion model would predict that the ionic current should continue to increase as long as the ionic concentration also increases—the more charge carriers in solution, the greater the current flow.

The relation between current and ionic concentration for a wide range of ion channels is well described by a simple one-to-one binding equation, suggesting that a single ion binds to the channel during permeation. The ionic concentration at which current flow reaches half its maximum defines the *dissociation constant* for ion binding in the channel. One striking feature of such current–concentration plots is that the dissociation constant is typically quite high, around 100 m*M*, indicating weak binding, as compared with the typical interaction between enzymes and substrates (0.05–0.5 μM), for example. This weak interaction indicates that the bonds between the ion and the channel are rapidly formed and broken. In fact, an ion typically stays bound in the channel for less than 1 microsecond. The rapid off-rate for ion

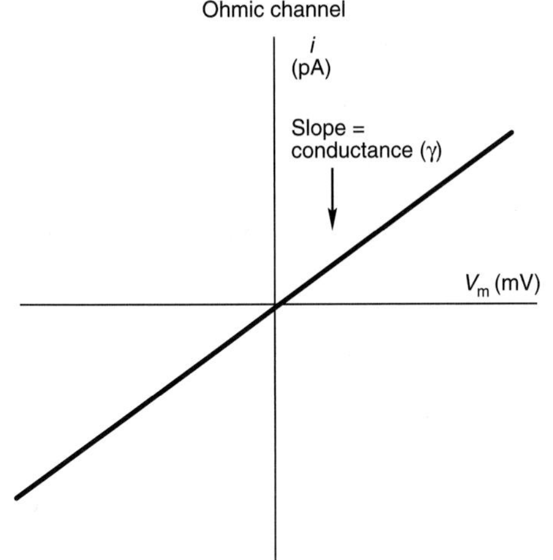

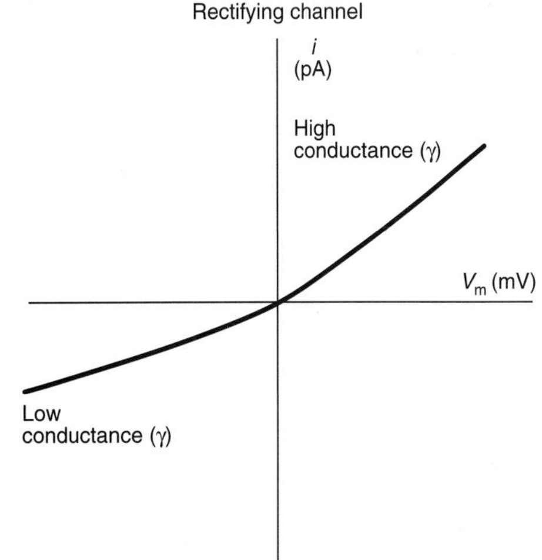

Figure 7–8 In many ion channels the relation between current flow (*i*) through the open channel and the applied membrane voltage (V_m) is linear, as illustrated in the plot at the **left**. Such channels are said to be "ohmic," as they follow Ohm's law, $i = V_m/R$ or $\gamma\ V_m$. In other channels the relation between current and membrane potential is nonlinear, as shown on the **right**. This kind of channel is said to "rectify," in the sense that it tends to conduct ions more readily in one direction (here positive current) than in the other.

binding ensures that channels achieve the very high conduction rates needed to rapidly change the membrane potential during signaling.

Ion channels are also susceptible to occlusion by various free ions or molecules in the cytoplasm or extracellular fluid. Passage through the channel can be blocked by particles that bind either to the mouth of the aqueous pore or somewhere within the pore. If the inhibitor is an ionized molecule that binds to a site within the pore, the charged inhibitor molecule will be influenced by the membrane electric field as it enters the channel. For example, if a positively charged blocker enters the channel from outside the membrane, anything that makes the membrane potential more negative will drive the blocker into the channel, increasing the inhibition. While blocking molecules are typically toxins or drugs that originate outside the cell, some are common ions present in the cell under normal conditions, such as Mg^{2+}, Ca^{2+}, and Na^+.

The Opening and Closing of a Channel Involve Conformational Changes

In all ion channels so far studied, each channel protein has two or more conformational states that are relatively stable. Each of these stable conformations represents a different functional state. For example, each ion channel has at least one open state and one or two closed states. The transition of a channel between different states is called *gating*.

Relatively little is known about the molecular mechanisms of gating, other than that they involve a temporary change in the channel's structure. Although the picture of a gate swinging open and shut is a convenient image, it probably is accurate only for certain channels (for example, the inactivation of Na^+ and K^+ channels, which we consider in Chapter 10). More commonly, channel gating involves *widespread* changes in the channel's conformation. For example, evidence from high-resolution electron microscopy and image analysis of the gap-junction type of ion channel (which we consider in Chapter 11) suggests that the opening and closing of this channel involve a concerted twisting and tilting of the six subunits that make up the channel. Three general physical models of channel gating are illustrated in Figure 7–9.

Because the primary function of ion channels in neurons is to mediate the flow of transient electrical signals, several specialized mechanisms have evolved to control the amount of time a channel spends in each of its different conformations. Some channels are regulated by the binding of chemical ligands; these ligands may be neurotransmitters or hormones in the extracellular environment that bind to the extracellular side of the channel, or they may be second messengers within the cell that are activated by transmitters and which act on the cytoplasmic side of the channel. Other ion channels are influenced either by changes in membrane potential or by mechanical stretch of the membrane (Figure 7–10). Under the influence of these regulators, channels can enter one of three functional states: closed and activatable (*resting*), open (*active*), closed and nonactivatable (*refractory*). Some channels are regulated by more than one of these three types of regulators.

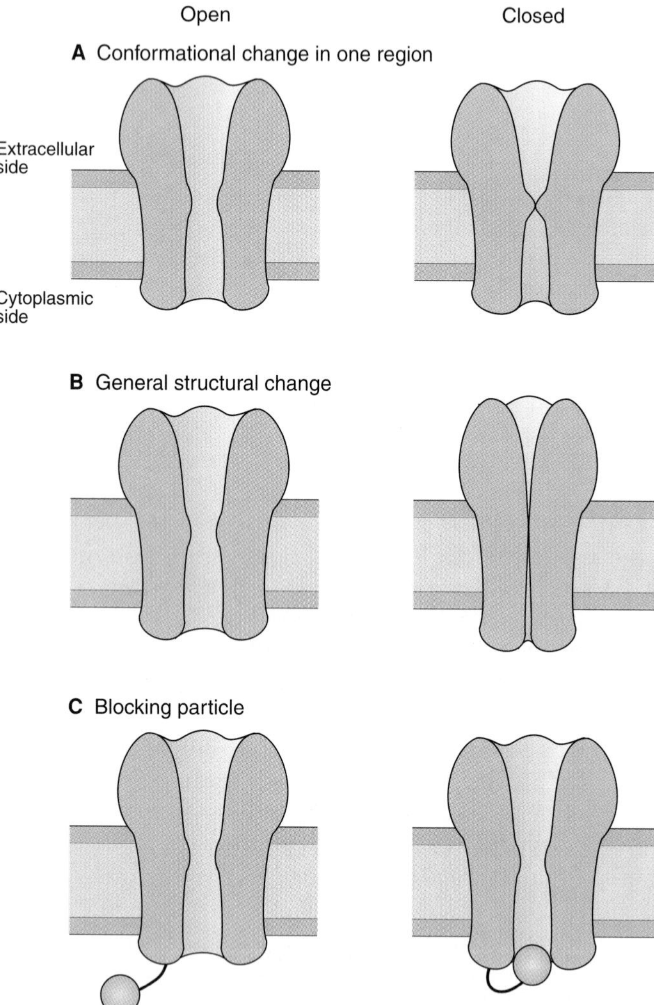

Figure 7–9 Three different physical models for the opening and closing of ion channels.

A. A conformational change occurs in one region of the channel.

B. A generalized conformational change occurs along the length of the channel.

C. A blocking particle swings into and out of the channel mouth.

How does a given stimulus, such as a change in voltage or the binding of transmitter, change the conformation of a channel? For voltage-gated channels, such as the Na^+ channel, opening and closing are associated with a movement of a region of charge in the channel through the electric field of the membrane. Changes in the membrane voltage tend to move this charged region back and forth through the electric field and, thus, drive the channel between closed and open states. For transmitter-gated channels, the energy released by the transmitter as it binds to the receptor site on the channel leads to channel opening. For mechanically activated channels the energy associated with membrane stretch is thought to be transferred to the channel through the cytoskeleton.

The signals that gate the channel also control the *rate* of transition between the open and closed states of a channel. For voltage-gated channels, the rates are steeply dependent on membrane potential. Although the time scale can vary from a microsecond to a minute, the transition tends to require a few milliseconds on average. Thus, once a channel opens, it stays open for a few milliseconds before closing, and after closing it stays closed for a few milliseconds before opening again. Once a transition between an open and closed state begins, it proceeds virtually instantaneously (in less than 10 microseconds, the present limits of experimental measurements), giving rise to abrupt, all-or-none steplike changes in current through the channel as it goes from a fully closed to a fully open state.

Transmitter-gated and voltage-gated channels enter refractory states through different processes. Transmitter-gated channels can enter the refractory state when they receive prolonged exposure to the ligand. This process, called *desensitization*, is discussed in Chapter 14. At present, desensitization of ion channels is not completely understood. In some channels it appears to be an intrinsic property of the channel, while in others it is due to phosphorylation of the channel molecule by a protein kinase. As for voltage-gated channels, many, but not all, can enter a refractory state following activation. This process is termed *inactivation*. In voltage-gated Na^+ and K^+ channels, inactivation is thought to be due to a conformational change, controlled by a subunit or region of the channel separate from that which controls activation. For example, by applying certain proteolytic enzymes within the cell, it is possible to eliminate the ability of voltage-gated Na^+ channels to inactivate, without affecting the channel's ability to be activated. In contrast, the inactivation of certain voltage-

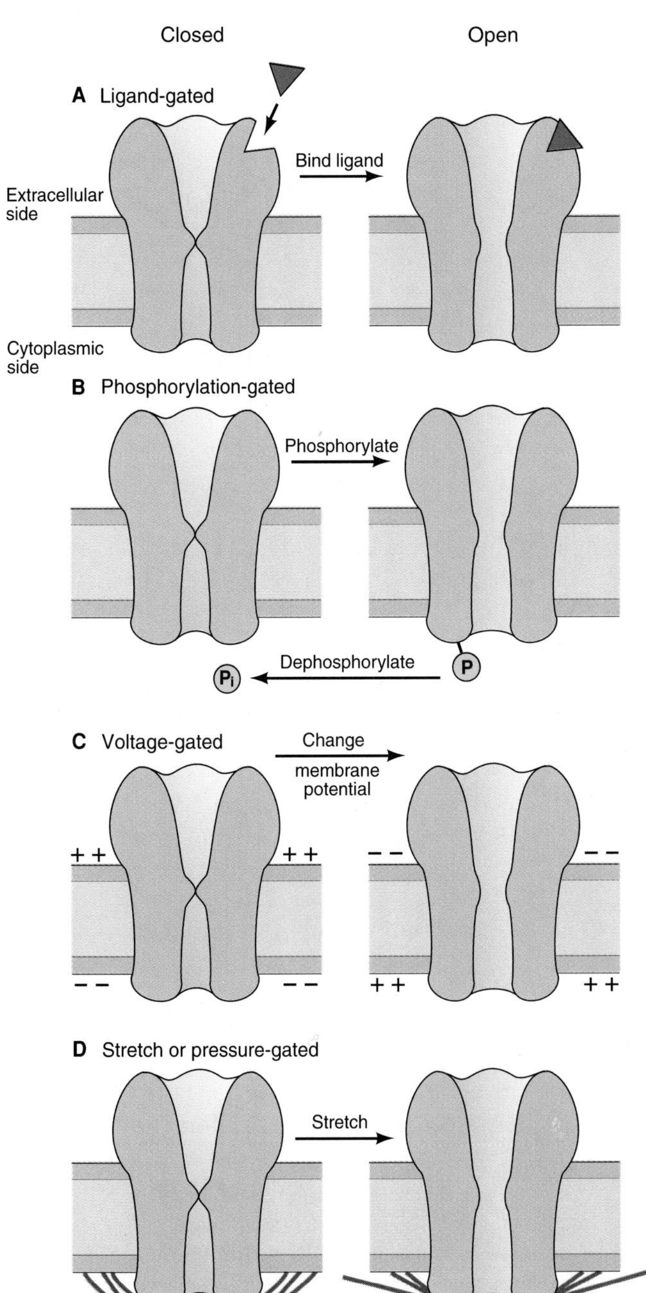

Figure 7–10 Several types of stimuli control the opening and closing of ion channels.

A. Ligand-gated channels open when the ligand binds to its receptor. The energy from ligand binding drives the channel toward an open state.

B. Protein phosphorylation and dephosphorylation regulate the opening and closing of some channels. The energy for channel opening comes from the transfer of the high-energy phosphate, P_i.

C. Changes in membrane voltage can open and close some channels. The energy for channel gating comes from a change in the electrical potential difference across the membrane, which causes a conformational change by acting on a component of the channel that has a net charge.

D. Channels can be activated by stretch or pressure. The energy for gating may come from mechanical forces that are passed along to the channel through the cytoskeleton.

gated Ca^{2+} channels is thought to be a consequence only of Ca^{2+} influx. In this case an increase in internal Ca^{2+} concentration inactivates the Ca^{2+} channel either directly, by binding to a control site on the inside of the channel, or indirectly, by activating an intracellular enzyme that inactivates the channel by protein dephosphorylation (Figure 7–11).

Exogenous factors, such as drugs and toxins, can affect the gating control sites of an ion channel.

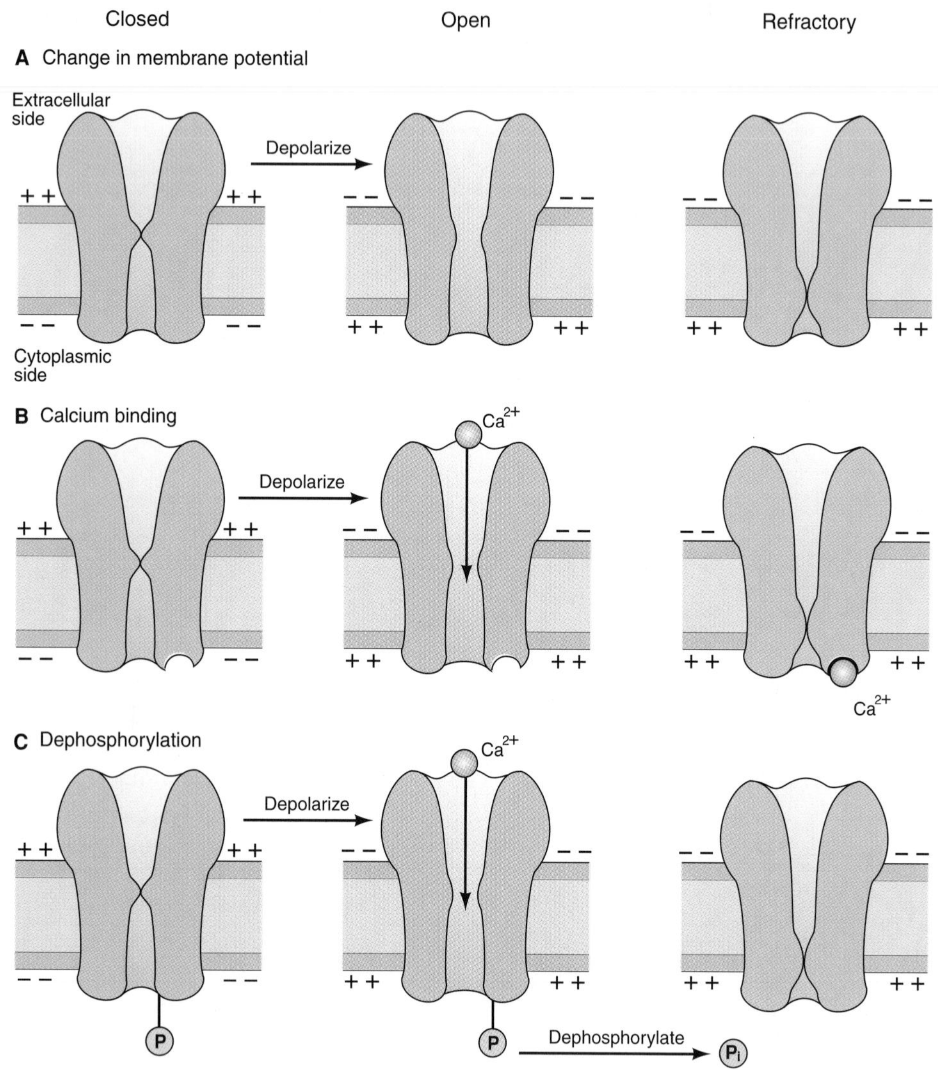

Figure 7–11 Three mechanisms by which channels become closed and nonactivatable (the refractory state).

A. Many voltage-gated channels can be inactivated by the change in membrane potential that occurs in going from a resting state to a transient open state. They recover from the refractory state and return to the resting state only after the membrane potential is restored to its original value.

B. When voltage-dependent Ca^{2+} channels are opened in response to depolarization, the internal Ca^{2+} level rises. The internal Ca^{2+} may then inactivate the channel that permitted its entry by binding to a specific recognition site.

C. An increase in internal Ca^{2+} concentration in Ca^{2+} channels may produce inactivation through dephosphorylation of the channel. At high concentrations, Ca^{2+} may even produce an irreversible, nonspecific inactivation of the channel due to the recruitment of protein-splitting enzymes activated by the Ca^{2+} ions.

Most of these agents tend to close the channel; a few tend to open it. Some compounds bind to the same site at which the gating ligand normally binds, and thereby prevent the activator from exerting its usual effect. This binding can be low energy and reversible, as in the blockade of the nicotinic acetylcholine-gated channel in skeletal muscle by curare, a South American arrow poison. Or it can be high energy and not reversible, as in the blockade of the same channel by the snake venom α-bungarotoxin. Other exogenous substances act in a noncompetitive manner and affect the normal gating mechanism without directly interacting with a ligand binding site. This type of indirect effect on the gating mechanism works not only on ligand gating, but also on gating controlled by voltage or stretch (Figure 7–12).

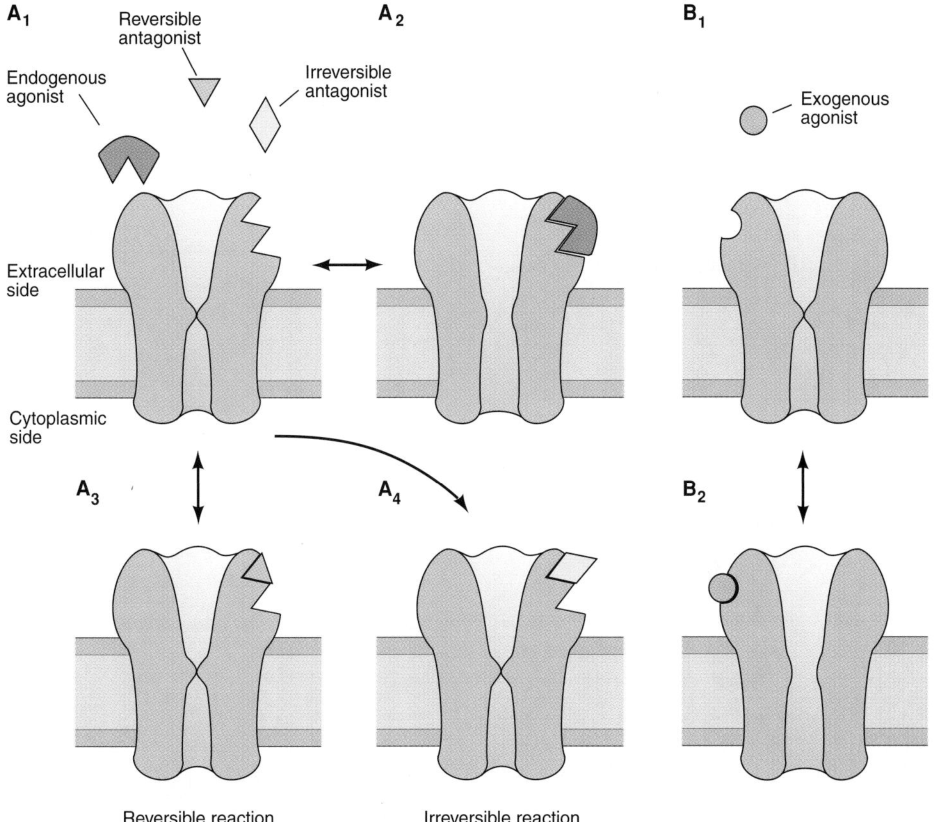

Figure 7–12 The binding of exogenous ligands, such as drugs, can bias an ion channel to either an open or a closed state by a variety of mechanisms.

A. In channels that are normally opened by the binding of an endogenous ligand (**1**, **2**), a drug or toxin may block the binding of the activator by either a reversible (**3**) or a nonreversible (**4**) reaction.

B. Some exogenous regulators can bias a channel to the *open* state by binding to a regulatory site.

Variants of Each Type of Ion Channel Are Found in Different Tissues

More than a dozen basic types of channels are known to exist in neurons, and each type includes several closely related forms (isoforms) that differ in their rate of opening or closing, their conductance, and their sensitivity to different activators. This variability is produced either by differential expression of two or more closely related genes or by alternative splicing of mRNA transcribed from the same gene. As with isozymes of a particular enzyme, variants of a channel type are expressed at different developmental stages, in different cell types, and even in different regions within a cell (Figure 7–13). These subtle variations in structure and function are presumed to adapt channels to specific functions. The rich variety of ion channels in different types of cells may make it possible to develop drugs that can activate or block channels in selected regions of the nervous system. Such drugs would, in principle, have maximum therapeutic effectiveness with a minimum of side effects.

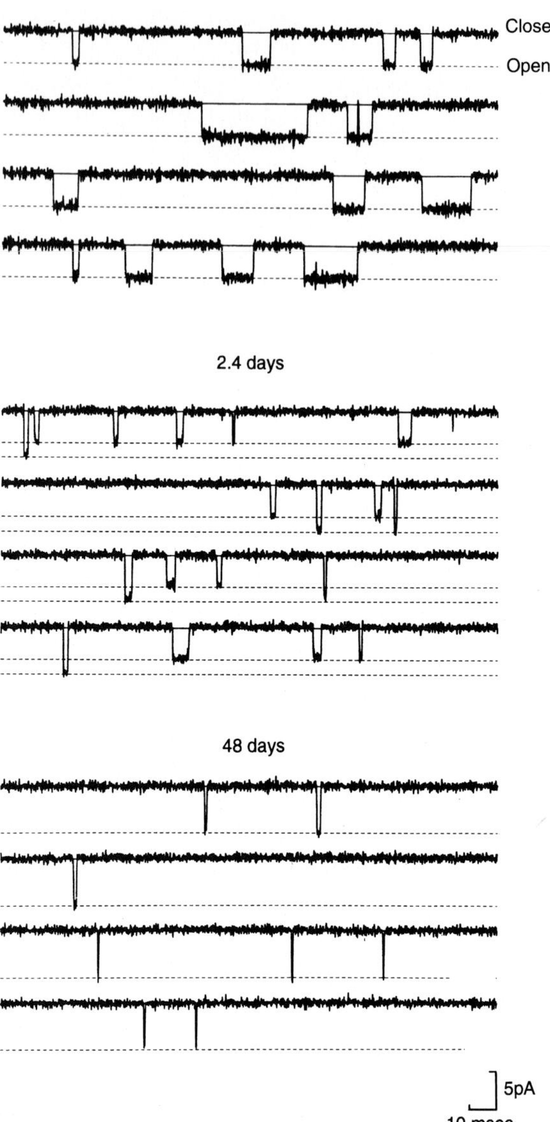

Figure 7–13 The functional properties of ion channels can change over the course of development. These examples of conductance in individual acetylcholine-activated channels were recorded from frog skeletal muscle at early (1.1 day), intermediate (2.4 days), and late (48 days) stages of development. In immature muscle the single channels have a small conductance and a relatively long open time. In mature muscle the channel conductance is larger, and the average open time is shorter. At intermediate stages of development the population of channels is mixed, with both long and short openings and both large and small classes of conductance. (From Owens and Kullberg, 1989.)

Genes that Encode Ion Channels Can Be Grouped into Families

Most of the ion channels described to date are encoded by genes that can be grouped into three families. Members of each gene family have substantially similar amino acid sequences. Each family is thought to have evolved from a common ancestral gene by gene duplication and divergence.

One family comprises all the genes that encode voltage-gated ion channels, selective for Ca^{2+}, Na^+, or K^+. A second family includes genes for ligand-gated ion channels that are activated by acetylcholine, γ-aminobutyric acid (GABA), glycine, or glutamate. Like the family of voltage-gated ion channels, the members of the ligand-gated family can differ from each other in their ion selectivity. The third gene family includes the genes coding for gap-junction channels, specialized channels that bridge the cytoplasm of two cells at electrical synapses (see Chapter 11). Because the genes for only a few ion channels have been sequenced, it remains to be seen how many other channel families exist.

An Overall View

Ion channels regulate the flow of ions across the membrane in all cells. In nerve and muscle cells they are important for controlling the rapid changes in membrane potential associated with the action potential and with the postsynaptic potentials of target cells. In addition, the influx of Ca^{2+} ions controlled by these channels can alter many metabolic processes within cells, leading to the activation of various enzymes and other proteins. As will be described in Chapter 15, Ca^{2+} influx also acts as a trigger for the release of neurotransmitter.

Channels differ from one another in their ion selectivity and in the factors that control their opening and closing, or gating. Ion selectivity is achieved through physical–chemical interaction between the ion and various amino acid residues that line the walls of the channel pore. Gating involves a change of the channel's conformation in response to various external stimuli, including voltage, ligands, and stretch or pressure.

Two methodological advances in the past several years have greatly increased our understanding of channel function. First, the patch-clamp technique has made it possible to measure directly the current flow through a single open channel. Second, gene cloning and sequencing have determined the primary amino acid sequences of many ion channel genes. From these results, many of the channels described so far can be grouped into three gene families: voltage-gated channels (including channels selective for Na^+, K^+, and Ca^{2+}), ligand-gated channels, and gap-junction channels.

The activity of channels can be modified by cellular metabolic reactions, including protein phosphorylation; by various ions that act as blockers; and by toxins, poisons, and drugs. Channels are also important targets in various diseases. Certain autoimmune neurological disorders, such as myasthenia gravis and the Lambert–Eaton syndrome (which we discuss in Chapter 17), result from the actions of specific antibodies that interfere with channel function. Hyperkalemic periodic paralysis involves a genetic defect in voltage-gated sodium channels. With our increasing understanding of channel structure and function it seems likely that other diseases of channel function will soon be identified. Detailed knowledge of the genetic basis of channel structure and function may one day make it possible to devise new pharmacological therapies for certain neurological and psychiatric disorders.

Selected Readings

Anderson, O. S., and Koeppe, R. E. II. 1992. Molecular determinants of channel function. Physical Rev. 72 (Suppl): S89–S158.

Catterall, W. A. 1993. Structure and function of voltage-gated ion channels. Trends Neurosci. 16:500–506.

Hille, B. 1992. Ionic Channels of Excitable Membranes, 2nd ed. Sunderland, Mass.: Sinauer.

Miller, C. 1987. How ion channel proteins work. In L. K. Kaczmarek and I. B. Levitan (eds.), Neuromodulation: The Biological Control of Neuronal Excitability. New York: Oxford University Press, pp. 39–63.

Miller, C. 1989. Genetic manipulation of ion channels: A new approach to structure and mechanism. Neuron 2:1195–1205.

8

Membrane Potential

The Resting Membrane Potential Results from the Separation of Charge Across the Cell Membrane

The Resting Membrane Potential Is Determined by Resting Ion Channels

Resting Channels in Glial Cells Are Selective Only for Potassium

Resting Channels in Nerve Cells Are Selective for Several Ion Species

Passive Flux of Sodium and Potassium Is Balanced by Active Pumping of the Ions

Chloride Ions May Be Passively Distributed

The Ionic Balance that Gives Rise to the Resting Membrane Potential Is Abolished During the Action Potential

The Contribution of Different Ions to the Resting Membrane Potential Can Be Quantified by the Goldman Equation

The Functional Properties of the Neuron Can Be Represented as an Electrical Equivalent Circuit

Each Ion Channel Acts as a Conductor and Battery

An Equivalent Circuit Model of the Membrane Includes Batteries, Conductors, a Capacitor, and a Current Generator

Postscript: An Equation for the Resting Membrane Potential Can Be Derived for an Equivalent Circuit

The flow of information within and between neurons is conveyed by electrical and chemical signals. Electrical signaling is particularly important for transferring information rapidly and over long distances. Electrical signals—receptor potentials, synaptic potentials, and action potentials—are all produced by transient changes in the current flow into and out of the cell that drive the electrical potential across the cell membrane away from its resting value.

Current flow into and out of the cell is controlled by ion channels embedded in the cell membrane. There are two types of ion channels in membranes—resting and gated. Resting channels normally remain open and are not influenced significantly by extrinsic factors such as the potential across the membrane. They are primarily important in maintaining the resting membrane potential—the electrical potential across the membrane in the absence of signaling. Gated channels, in contrast, open and close in response to various signals. Most gated channels are closed when the membrane is at rest, and their opening is regulated by the three factors that we considered in the last chapter—changes in membrane potential, ligand binding, or stretch of the membrane.

To understand how transient electrical signals are generated we must appreciate the mechanisms underlying the resting membrane potential. In this chapter we first discuss how resting ion channels establish and maintain the resting potential. We then briefly describe the mechanism by which the resting potential can be perturbed, giving rise to transient electrical signals such as the action potential. In Chapter 9 we consider the passive properties of neurons—their resistive and capacitative characteristics—and how they contribute to local signaling within the neuron. Finally, in Chapter 10 we examine the voltage-gated Na^+, K^+, and Ca^{2+} channels that underlie the action potential.

The Resting Membrane Potential Results from the Separation of Charge Across the Cell Membrane

Every neuron has a thin cloud of positive and negative ions spread over the inner and outer surfaces of the cell membrane. At rest a nerve cell has an excess of positive charges on the outside of the membrane and an excess of negative charges on the inside. This separation of charge is maintained because ions cannot move freely through the lipid bilayer of the membrane, as explained in Chapter 7. The charge separation gives rise to a difference of electrical potential, or voltage, across the membrane, called the *membrane potential* (Box 8–1). The membrane potential (V_m) is defined as

$$V_m = V_{in} - V_{out} \tag{1}$$

where V_{in} is the potential on the inside of the cell and V_{out} the potential on the outside.

The membrane potential of a cell at rest is called the *resting membrane potential.* Since, by convention, the potential outside the cell is defined as zero, the resting potential (V_R) is negative. Its usual range is –60 to –70 mV. All electrical signaling results from brief changes from the resting membrane potential due to alterations in the flow of electrical current across the cell membrane.

The current that flows into and out of the cell is carried by ions, both positively charged (cations) and negatively charged (anions), that move through the ion channels in the cell membrane. The direction of current flow is conventionally defined as the direction of *net* movement of *positive* charge. Thus, in an ionic solution cations move in the direction of the current, anions in the opposite direction. Whenever there is a net flow of cations or anions into or out of the cell, the charge separation across the resting membrane is disturbed, thus altering the polarization of the membrane. A reduction of charge separation, leading to a less negative membrane potential, is called *depolarization.* An increase in charge separation, leading to a more negative membrane potential, is called *hyperpolarization.* Passive responses of the membrane to current flow, which do not lead to the opening of gated ion channels, are called *electrotonic potentials.* Hyperpolarizing responses are passive, as are small depolarizations. However, at a critical level of depolarization, called the *threshold,* the cell responds actively with the opening of voltage-gated ion channels sufficient to produce an all-or-none *action potential* (see Box 8–1).

We begin examining the membrane potential by analyzing how the passive flux of individual ion species through resting membrane channels generates the resting potential. We shall then be able to understand how the selective gating of different types of ion channels generates the action potential, as well as the receptor and synaptic potentials.

The Resting Membrane Potential Is Determined by Resting Ion Channels

No single ion species is distributed equally on the two sides of a nerve cell membrane. Of the four most abundant types of ions found on either side of the cell membrane, Na^+ and Cl^- are more concentrated outside the cell, and K^+ and organic anions (A^-) are more concentrated inside. The organic anions are primarily amino acids and proteins. Table 8–1 shows the distribution of these ions inside and outside one particularly well-studied nerve cell process, the giant axon of the squid, whose blood has a salt concentration similar

Table 8–1 Distribution of the Major Ions Across a Neuronal Membrane at Rest: The Giant Axon of the Squid

Ion Species	*Cytoplasm* (mM)	*Extracellular Fluid* (mM)	*Equilibrium Potential** (mV)
K^+	400	20	–75
Na^+	50	440	+55
Cl^-	52	560	–60
A^- (organic ions)	385	—	—

*The membrane potential at which there is no net flux of the ion species across the cell membrane.

BOX 8–1 Recording the Membrane Potential

Reliable techniques for recording the electrical potential across cell membranes were developed in the late 1940s. These techniques allow accurate recordings of both the resting and the action potentials.

Glass pipettes filled with a concentrated salt solution serve as electrodes and are placed on either side of the cell membrane. Wires inserted into the back ends of the pipettes are connected via an amplifier to an oscilloscope, which displays the amplitude of the membrane potential in volts.

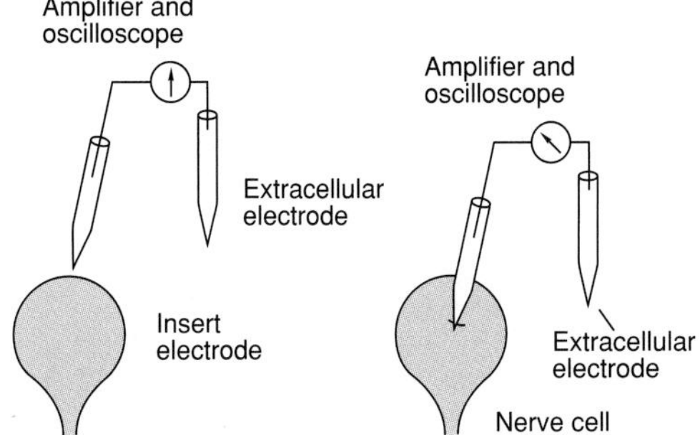

Figure 8–1A

When both electrodes are outside the cell, no electrical potential difference is recorded. But as soon as one electrode is inserted into the cell, the oscilloscope displays a steady level, the resting membrane potential. In most nerve cells at rest the membrane potential is around –65 mV.

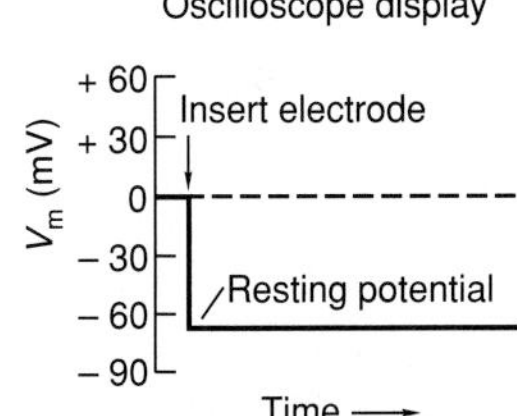

Figure 8–1B

The membrane potential can be experimentally changed using a current generator connected to a second pair of electrodes—one intracellular and one extracellular. When the intracellular electrode is made positive with respect to the extracellular one, a pulse of current from the generator will cause positive current to flow into the neuron from the intracellular electrode. As a result, the inside of the membrane becomes more positive while the outside of the membrane becomes more negative. This progressive *decrease* in the normal separation of charge is called *depolarization.*

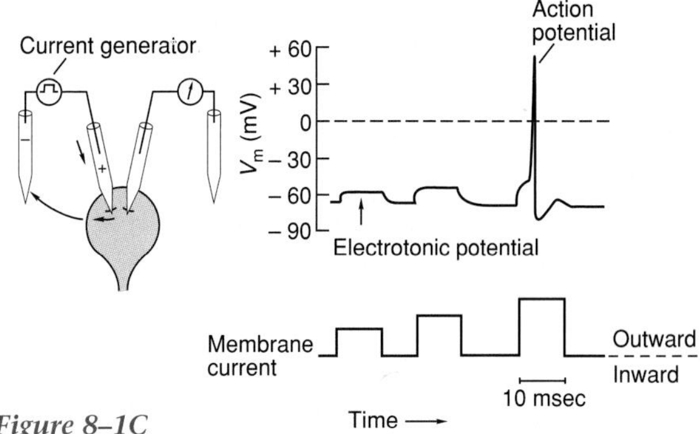

Figure 8–1C

Small depolarizing current pulses evoke purely electrotonic (passive) potentials in the cell—the size of the change in potential is proportional to the size of the current pulses. However, depolarizing current eventually drives the membrane potential to a critical level, the *threshold,* which triggers an active response—the regenerative opening of voltage-gated ion channels. The opening of these channels leads to the action potential, which differs from electrotonic potentials not only in the way in which it is generated but also in magnitude and duration.

Reversing the direction of current flow—making the intracellular electrode negative with respect to the extracellular electrode—makes the membrane potential more negative. This *increase* in charge separation is called *hyperpolarization.*

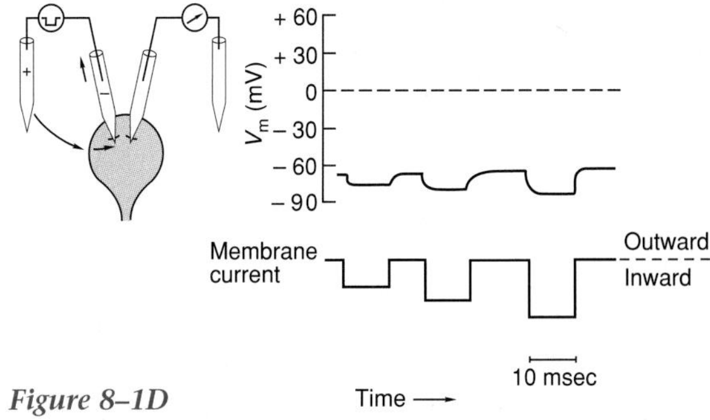

Figure 8–1D

The responses of the cell to hyperpolarization are usually purely electrotonic—as the size of the current pulse increases, the hyperpolarization simply increases proportionately. Hyperpolarization does not usually trigger an active response in the cell.

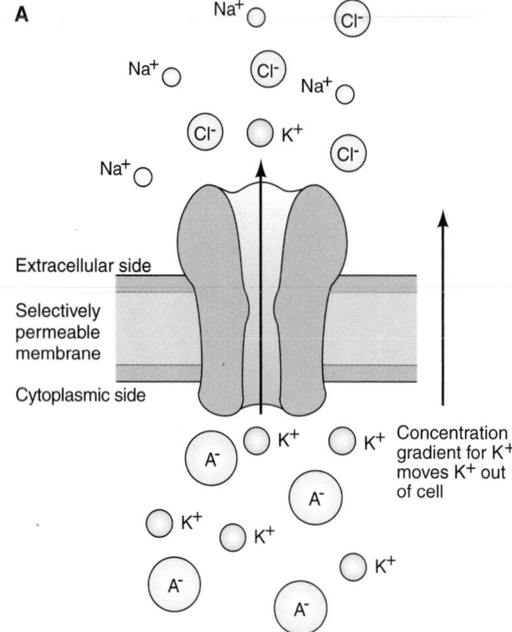

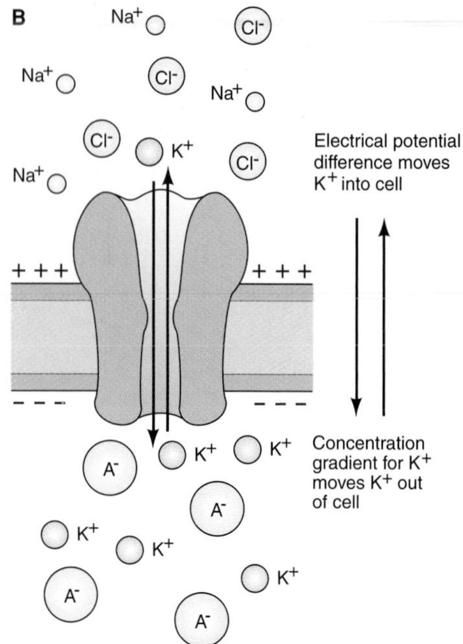

Figure 8–2 The flux of K^+ across the glial cell membrane is determined by both the K^+ concentration gradient and the electrical potential across the membrane.

A. Glial cells at rest are permeable only to K^+ and thus the resting potential is generated by the efflux of K^+ down its concentration gradient.

B. The continued efflux of K^+ builds up an excess of positive charge on the outside of the cell and leaves behind on the inside an excess of negative charge. This buildup of charge generates an electrical field that impedes the further efflux of K^+, so that eventually an equilibrium is reached, at which the electrical and chemical driving forces are equal and opposite.

to that of sea water. Although the absolute values of the concentrations for vertebrate nerve cells are two- to threefold lower than those for the squid giant axon, the concentration *gradients* (the ratio of external ion concentration to internal ion concentration) are about the same.

The unequal distribution of ions raises two important questions. First, how do ionic gradients contribute to the resting membrane potential? Second, how are they maintained? What prevents the ionic gradients from dissipating by diffusion of ions across the membrane through the passive (resting) channels? These two questions are interrelated, and we answer them by considering two examples of membrane permeability: the resting membrane of glial cells, which is permeable to only one species of ions, and the resting membrane of nerve cells, which is permeable to three. For the purposes of this discussion we consider only the resting ion channels, which are usually open.

Resting Channels in Glial Cells Are Selective Only for Potassium

A membrane's overall selectivity for individual ion species is determined by the relative proportions of the different types of ion channels that it contains. The simplest case is that of the glial cell. Most resting channels in the glial cell membrane are permeable only to K^+. As a result, the glial cell membrane at rest is almost exclusively permeable to K^+ ions. A glial cell has a high concentration of K^+ and negatively charged organic anions on the inside and a high concentration of Na^+ and Cl^- on the outside.

How do these ionic gradients generate the membrane potential of the glial cell? Because K^+ ions are present at a high concentration inside the cell and because glial cells are selectively permeable to them, K^+ ions tend to diffuse from inside to outside of the cell, down their concentration gradient (Figure 8–2A). As a result, the outside of the membrane accumulates a positive charge (due to the slight excess of K^+) and the inside a negative charge (because of the deficit of K^+ and the resulting preponderance of anions). Since opposite charges attract each other, the positive charge on the outside and the negative charge on the inside collect locally on either side of the membrane.

The diffusion of K^+ out of the cell is self-limiting, however. The separation of charge resulting from the diffusion of K^+ gives rise to an electrical potential difference: positive outside, negative inside. The more K^+ continues to flow out, the more charge will be separated and the greater the potential difference will become. The electrical force due to the buildup of this potential opposes the force of the concentration gradient and impedes the further outward movement of the K^+ ions. At a certain potential the electrical force due to the charge separation becomes equal to the oppositely directed chemical force (the concentration gradient) and no further net movement of K^+ occurs. This potential is called the *potassium equilibrium potential.* At the potassium equilibrium potential the K^+ ions inside and outside the cell are in equilibrium: the outward movement of K^+ driven by its concentration gradient is balanced by an equal and opposite inward movement of K^+ driven by the electrical potential difference (Figure 8–2B).

In a cell permeable only to K^+ ions, the K^+ equilibrium potential determines the resting membrane potential, which in most glial cells is about –75 mV. The membrane potential at which K^+ ions are in equilibrium across the membrane can be calculated from an equation derived in 1888 from basic thermodynamic principles by the German physical chemist Walter Nernst:

$$E_K = \frac{RT}{ZF} \ln \frac{[K^+]_o}{[K^+]_i} \tag{2}$$

where E_K is the K^+ equilibrium potential (or K^+ *Nernst potential*), R is the gas constant, T the temperature in degrees Kelvin, Z the valence of K^+, F the Faraday constant, and $[K^+]_o$ and $[K^+]_i$ are the concentrations of K^+ on the outside and inside of the cell. To be precise, chemical activities should be used rather than concentrations. At 25° C, RT/F is 25 mV. The constant for converting from natural logarithms to base 10 logarithms is 2.3. Thus, the Nernst equation is usually written

$$E_A = \frac{58}{Z} \log \frac{[A]_o}{[A]_i} \tag{3}$$

when applied to any ion A. For K^+, $Z = +1$ and the concentrations inside and outside the cell are as in Table 8–1 (for the squid giant axon):

$$E_K = \frac{58}{1} \log \frac{[20]}{[400]} = -75 \text{ mV}. \tag{4}$$

Other ions are not important in determining the resting potential of glial cells because their flux is insignificant relative to K^+. The Nernst equation, however, can be used to find the equilibrium potential of any ion that is present on both sides of a membrane permeable to that ion. (The Na^+, K^+, and Cl^- equilibrium potentials for the distributions of ions across the squid axon are given in Table 8–1.) Although metabolic energy is not used in the diffusion of ions through resting channels, it is required to generate and maintain the ionic concentration gradients.

Resting Channels in Nerve Cells Are Selective for Several Ion Species

Measurements of the resting membrane potential with intracellular electrodes and flux studies using radioactive tracers have verified that, unlike glial cells, nerve cells at rest are permeable to Na^+ and Cl^- ions in addition to K^+ ions. Of the abundant ion species in nerve cells, only the large organic anions (A^-), such as amino acids and proteins, are unable to permeate the cell membrane. How can three concentration gradients (for Na^+, K^+, and Cl^-) be maintained across the membrane of a single cell, and how do these three gradients interact to determine the cell's resting membrane potential?

To answer these questions, it will be easiest to examine first only the diffusion of K^+ and Na^+. Let us return to the simple example of a cell having only K^+ channels, with concentration gradients of K^+, Na^+, Cl^-, and A^- as shown in Table 8–1. Under these conditions the resting membrane potential, V_R, is determined solely by the K^+ concentration gradient. Now consider what happens if a few resting Na^+ channels are added to the membrane, making it slightly permeable to Na^+. Two forces act on Na^+ to drive it into the cell. First, Na^+ is more concentrated outside than inside and therefore tends to flow into the cell down its concentration gradient. Second, Na^+ is driven into the cell by the negative electrical potential difference across the membrane.

The influx of Na^+ depolarizes the cell through the influx of positive charge. However, this depolarization varies the membrane potential only slightly from the K^+ equilibrium potential and does not come close to the Na^+ equilibrium potential of +55 mV. This is because there are many more resting K^+ channels than Na^+ channels in the membrane. As soon as the membrane potential begins to depolarize from the value of the K^+ equilibrium potential, K^+ flux will no longer be in equilibrium across the membrane. The reduction in the negative electrical force driving K^+ into the cell results in a net efflux of K^+ from the cell, tending to counteract the Na^+ influx.

The more the membrane potential is depolarized and differs from the K^+ equilibrium potential, the greater is the electrochemical force driving K^+ out of the cell and, consequently, the greater the K^+ efflux. Eventually, the membrane potential reaches a new resting potential at which the outward movement of K^+ just balances the inward movement of Na^+ (Figure 8–3). This balance point (usually –60 mV) is far from the Na^+ equilibrium potential (+55 mV) and is only slightly more positive than the equilibrium potential for K^+ (–75 mV).

To understand how this balance point is determined, bear in mind that the flux of an ion across a cell membrane is the product of its *electrochemical driving force* (the sum of the electrical driving force and the chemical driving force due to the concentration gradient) and the *conductance* of the membrane to the ion. A cell has relatively few resting Na^+ channels, so at rest the conductance to Na^+ is quite low. Thus, despite the large chemical and electrical forces driving Na^+ into the cell, the influx of Na^+ is small. In contrast, since there are many resting K^+ channels, the membrane conductance to K^+ is relatively large. As a result, the small net outward force acting on K^+ at the resting membrane potential is enough to produce a K^+ efflux equal to the Na^+ influx.

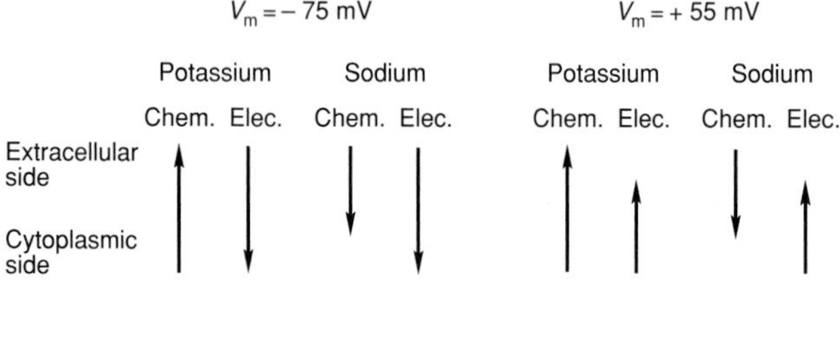

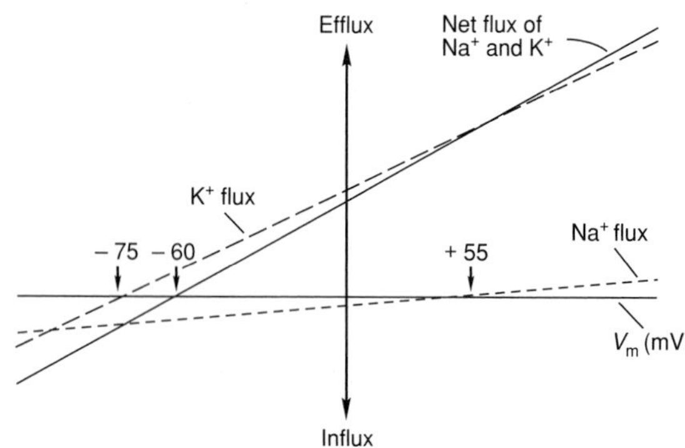

Figure 8–3 The resting potential of a cell with resting Na^+ and K^+ channels is defined as the potential at which K^+ efflux is equal to Na^+ influx. In the illustration the direction and amplitude of the chemical and electrical driving forces acting on Na^+ and K^+ are shown by arrows for two different values of membrane potential (V_m). The length of an arrow represents the relative amplitude of the force. (For a given change in V_m the changes in driving force are the same for Na^+ and K^+.) These forces result in the flux curves shown for each ion (**broken lines**) and the net flux curve for Na^+ and K^+ combined (**solid line**). The difference in the slopes of the Na^+ and K^+ flux curves reflects the fact that the resting membrane is more permeable to K^+ than to Na^+. The shapes of the curves for Na^+ and K^+ flux in this plot are simplified considerably. In a typical neuron these curves become quite nonlinear as voltage-gated channels begin to open at values of V_m more positive than about –50 mV, as described in Chapter 10.

Passive Flux of Sodium and Potassium Is Balanced by Active Pumping of the Ions

For a cell to have a steady resting membrane potential, the charge separation across the membrane must be constant: the influx of positive charge must be balanced by the efflux of positive charge. If these fluxes were not equal, the charge separation across the membrane, and thus the membrane potential, would vary continually. As we have seen, the passive movement of K^+ out of the cell through resting channels balances the passive movement of Na^+ into the cell. However, these steady ion leaks cannot be allowed to continue unopposed for any appreciable length of time because the (Na^+ and K^+) ionic gradients would eventually run down, reducing the resting membrane potential.

Dissipation of ionic gradients is prevented by the *sodium–potassium pump,* which moves Na^+ and K^+ *against* their net electrochemical gradients: it extrudes Na^+ from the cell while taking in K^+. The pump therefore requires energy on which to run. The energy comes from the hydrolysis of ATP. Thus, at the resting membrane potential the cell is not in equilibrium, but rather in a *steady state*: metabolic energy must be used to maintain the ionic gradients across the membrane. The sodium–potassium pump itself is a large membrane-spanning protein with catalytic binding sites for Na^+, K^+, and ATP. The sites for Na^+ and ATP are located on its intracellular surface, and the sites for K^+ on its extracellular surface.

The pump extrudes three Na^+ ions for every two K^+ ions it brings in, and thus is said to be electrogenic; that is, it generates a net outward ionic current. This outward flux of positive charge tends to hyperpolarize the membrane to a new resting potential. The greater the hyperpolarization, the greater the inward electrochemical force driving Na^+ into the cell, and the smaller the force driving K^+ out. Thus, there is a net inward flux of positive charge through the resting channels. At the new resting potential the net passive inward flux through the ion channels exactly counterbalances the active outward flux driven by the pump. This balance occurs when three Na^+ ions diffuse in for every two K^+ ions that diffuse out. To achieve the ratio of three Na^+ in for two K^+ out, the resting potential for a cell with an electrogenic pump settles at a value that is typically a few millivolts more negative than would be expected for a cell with an electroneutral pump.

Chloride Ions May Be Passively Distributed

So far we have ignored the contribution of Cl^- to the generation of the resting potential, even though all nerve cells have resting Cl^- channels. This simplification is valid, however, for those cells that do not have a chloride pump. In such cells the resting potential is ultimately determined by K^+ and Na^+ fluxes, because their intracellular concentrations are fixed by the sodium–potassium pump. The Cl^- concentration inside the cell is free to change, because it is affected only by passive forces (electrical potential and concentration gradient). In a cell with no chloride pump, therefore, Cl^- ions tend toward equilibrium across the membrane.

In nerve cells that do have a chloride pump the active transport is directed outward, so that the outside-to-inside ratio of Cl^- is greater than would result from passive diffusion alone. The effect of increasing the Cl^- gradient is to make the equilibrium point for Cl^- ions more negative than the resting membrane potential. This difference results in a steady inward leak of Cl^- that is balanced by active extrusion of Cl^- by the pump.

The Ionic Balance that Gives Rise to the Resting Membrane Potential Is Abolished During the Action Potential

In the nerve cell at rest the steady Na^+ influx is balanced by a steady K^+ efflux, so that the membrane potential is constant. This balance changes, however, when the membrane is depolarized past the threshold for generating an action potential. A transient depolarization past threshold, for example, produced by an excitatory synaptic input, causes voltage-gated sodium channels to open rapidly. The resultant increase in membrane permeability to Na^+ causes Na^+ influx to exceed the K^+ efflux. This net influx of positive charge causes further depolarization. The increase in depolarization causes more voltage-gated Na^+ channels to open, resulting in a greater influx of positively charged Na^+, which accelerates the depolarization still further.

This regenerative, positive feedback cycle develops explosively, driving the membrane potential toward the Na^+ equilibrium potential of +55 mV:

$$E_{Na} = \frac{RT}{F} \ln \frac{Na_o}{Na_i} = +55 \text{ mV}. \qquad (5)$$

However, the membrane potential never quite reaches that point because K^+ efflux continues throughout the depolarization. A slight diffusion of Cl^- into the cell also counteracts the depolarizing tendency of the Na^+ influx. Nevertheless, so many voltage-gated Na^+ channels open during the rising phase of the action potential that the cell's permeability to Na^+ is much greater than that to Cl^- or K^+. At the peak of the action potential, the membrane potential approaches the Na^+ equilibrium potential, just as at rest (when permeability to K^+ is predominant) the membrane potential tends to approach the K^+ equilibrium potential.

The membrane potential would remain at this large positive value near the Na^+ equilibrium potential indefinitely but for two processes that repolarize the membrane, terminating the action potential. First, as the depolarization continues, the population of voltage-gated Na^+ channels gradually becomes inactivated (see Chapter 7). Second, the delayed opening of the voltage-gated K^+ channels causes the K^+ efflux to increase gradually. This delayed increase in K^+ efflux combines with a decrease in Na^+ influx to produce a net efflux of positive charge from the cell, which continues until the cell has repolarized to its resting membrane potential.

The Contribution of Different Ions to the Resting Membrane Potential Can Be Quantified by the Goldman Equation

Although Na^+ and K^+ fluxes set the value of the resting potential, V_m is not equal to either E_K or E_{Na}, but lies between them. As a general rule, when V_m is determined by two or more species of ions, the influence of each species is determined both by its concentrations inside and outside the cell and by the permeability of the membrane to that ion. This relationship is given quantitatively by the Goldman equation:

$$V_m = \frac{RT}{F} \ln \frac{P_K[K^+]_o + P_{Na}[Na^+]_o + P_{Cl}[Cl^-]_i}{P_K[K^+]_i + P_{Na}[Na^+]_i + P_{Cl}[Cl^-]_o} \qquad (6)$$

This equation applies only when V_m is not changing. It states that the greater the concentration of a particular ion species and the greater the permeability of the membrane to that species, the greater the role that species has in determining the membrane potential. In the limiting case, when permeability to one ion is exceptionally high, the Goldman equation reduces to the Nernst equation for that ion. For example, if $P_K \gg P_{Cl}, P_{Na}$, as in glial cells, the equation becomes

$$V_m \approx \frac{RT}{F} \ln \frac{[K^+]_o}{[K^+]_i} \qquad (7)$$

In 1949 Alan Hodgkin and Bernard Katz first applied the Goldman equation systematically to analyze changes in membrane potential evoked by altering external ion concentrations in the squid giant axon. They measured the variation of V_m while changing extracellular concentrations of Na^+, Cl^-, and K^+. Their results showed that if V_m is measured shortly after the concentration change, before the internal ionic concentrations are altered, $[K^+]_o$ has a strong effect on the resting potential, $[Cl^-]_o$ has a moderate effect, and $[Na^+]_o$ has little effect. Their data could be fit accurately by the Goldman equation using the following permeability ratios for the membrane at rest:

$$P_K:P_{Na}:P_{Cl} = 1:0.04:0.45.$$

For the membrane at the peak of the action potential,[1] however, the variation of V_m with external ionic concentrations could be fit best by assuming a quite different set of permeability ratios:

$$P_K:P_{Na}:P_{Cl} = 1:20:0.45.$$

For this set of permeabilities (P_{Na}, P_K, P_{Cl}) the Goldman equation approaches the Nernst equation for Na^+:

$$V_m \approx \frac{RT}{F} \ln \frac{[Na^+]_o}{[Na^+]_i} = +55 \text{ mV}. \qquad (8)$$

Thus at the peak of the action potential, when the membrane is much more permeable to Na^+ than to any other ion, V_m approaches E_{Na}, the Nernst potential for Na^+. However, the finite permeability to K^+ and Cl^- prevents V_m from quite reaching E_{Na}.

The Functional Properties of the Neuron Can Be Represented as an Electrical Equivalent Circuit

A simple mathematical model derived from electrical circuits is helpful for describing the three critical features used by the nerve cell for electrical signaling—the ion channels, the concentration gradients of relevant ions, and the ability of the membrane to store charge. In this model, called an *equivalent circuit*, all of the important functional properties of the neuron are represented by an electrical circuit consisting only of conductors (resistors), batteries, and capacitors. This model provides an intuitive understanding as well as a quantitative description of how current flow due to the movement of ions generates signals in nerve cells.

[1]At the peak of the action potential there is an instant in time when V_m is not changing so the Goldman equation is applicable.

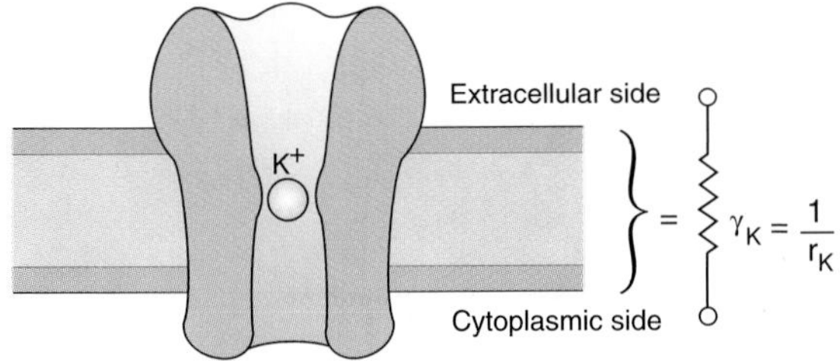

Figure 8–4 A single K^+ channel can be represented as a conductor or resistor (conductance, γ_K, is the inverse of resistance, r_K).

The first step in developing the model is to relate the membrane's discrete physical properties to its electrical properties.[2]

Each Ion Channel Acts as a Conductor and Battery

As described in Chapter 7, the lipid bilayer of the membrane is a poor conductor of ionic current as it is not permeable to ions. Even a large potential difference will produce practically no current flow across a pure lipid bilayer. Consider the cell body of a typical spinal motor neuron, which has a membrane area of about 10^{-4} cm^2. If that membrane were composed solely of lipid bilayer, its electrical *conductance* would be only about 1 pS (1×10^{-12} S, where 1 S = 1/ohm). But because thousands of resting ion channels are embedded in the membrane, ions constantly diffuse across it, so that its actual resting conductance is about 40,000 times greater, or about 40 nS (40×10^{-9} S).

In the equivalent circuit model each K^+ channel can be represented as a resistor (r_K), so that the current flow through K^+ channels (I_K) can be calculated using Ohm's law ($I_K = V/r_K$). However, electrophysiologists typically describe channels in terms of their conductance (e.g., γ_K) because it provides a direct measure of how efficiently the channel can conduct ions. Since conductance is inversely proportional to resistance, the conductance of a single channel to current flow can also be used in the equivalent circuit (Figure 8–4):

$$i = V \times \gamma_K \qquad (9)$$

The conductance of a membrane (or single channel) to an ion is somewhat similar to the permeability of the membrane (or channel) to the ion in that both measure the ease with which the ion crosses the membrane. However, the permeability (in units of centimeters per second) measures an intrinsic property of the membrane, whereas the conductance is an electrical measurement (in units of amperes per volts) that depends on both the permeability of the membrane and the ionic concentrations in the bath. Thus, a membrane with a high permeability to an ion will still have zero conductance to that ion if the ion is not present in the bath solution. In practical terms we use permeability for the Goldman equation and conductance for electrical measurements and equivalent circuits.

Each open ion channel contributes to the generation of an electrical potential difference across the membrane. For example, K^+, which is present at a higher concentration inside the cell, tends to diffuse out of the resting cell through resting channels selective for K^+. This diffusion leads to a net separation of charge across the membrane, resulting in an electrical potential difference. A source of electrical potential is called an *electromotive force*, and an electromotive force generated by a difference in chemical potentials is called a *battery*. We may therefore represent the electrical potential across each K^+ channel as a battery (Figure 8–5).

[2]The basic principles of electrical circuit theory are reviewed in a separately published companion workbook for students.

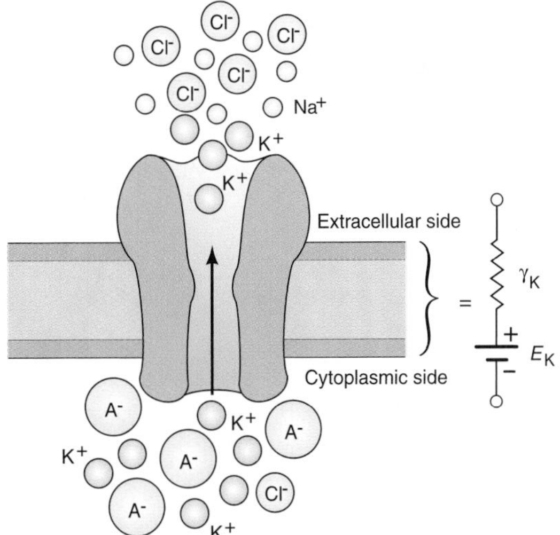

Figure 8–5 A channel selectively permeable to K^+ ions gives rise to an electromotive force, with a value equal to the K^+ Nernst potential. This can be represented by a battery, E_K. In this circuit the battery is in series with a conductor, γ_K, representing the conductance of the K^+ channel.

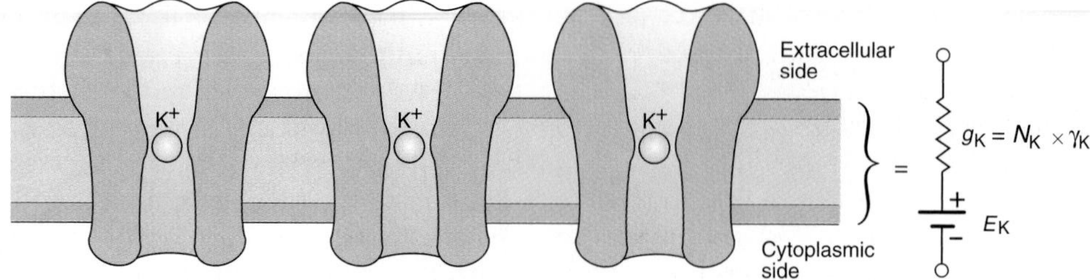

Figure 8–6 All of the passive K^+ channels in a nerve membrane can be lumped into a single equivalent electrical structure: a battery (E_K) in series with a conductor (g_K). The conductance $g_K = N_K \times \gamma_K$, where N is the number of passive K^+ channels and γ_K is the conductance of a single K^+ channel.

The potential generated by this battery is equal to E_K, which is typically about –75 mV.

Of course a cell has many passive K^+ channels in the membrane, and all of these channels can be combined into a single equivalent circuit, consisting of a conductor in series with a battery (Figure 8–6). In this equivalent structure the total conductance of all the K^+ channels (g_K), that is, the K^+ conductance of the cell membrane in its resting state, is equal to the number of passive K^+ channels (N_K) multiplied by the conductance of an individual K^+ channel (γ_K):

$$g_K = N_K \times \gamma_K. \quad (10)$$

Since the battery in this equivalent circuit of all the passive K^+ channels depends solely on the concentration gradient for K^+ and is independent of the number of K^+ channels, its value is simply E_K (Figure 8–6).

An Equivalent Circuit Model of the Membrane Includes Batteries, Conductors, a Capacitor, and a Current Generator

As we have seen, the entire population of passive K^+ channels can be represented by a single conductor in series with a single battery. By analogy, all the passive Cl^- channels can be represented by a similar combination, as can the passive Na^+ channels (Figure 8–7). These three types of channels account for the bulk of the passive ionic pathways through the membrane in the cell at rest.

We can incorporate these electrical representations of the total population of passive Na^+, K^+, and Cl^- channels into a simple equivalent circuit of a neuron to calculate the membrane potential. To construct this circuit we need only connect the elements representing each type of channel at their two ends by elements representing the extracellu-

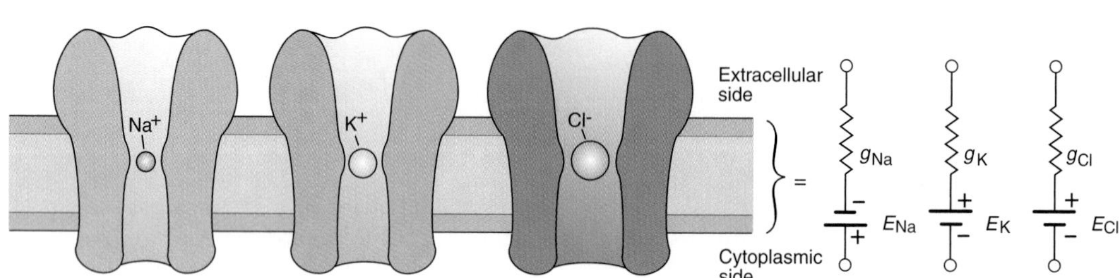

Figure 8–7 Each population of ion channels selective for Na^+, K^+, or Cl^- can be represented by a battery in series with a conductor.

lar fluid and cytoplasm. (These channels are, of course, in parallel with the conductance of the lipid bilayer. But because the conductance of the bilayer is so much lower than that of the ion channel pathways, virtually all transmembrane current flows through the channels, and the negligible conductance of the bilayer can be ignored.) Since the extracellular fluid and cytoplasm are both excellent conductors (because they have relatively large cross-sectional areas and many ions available to carry charge) they can each be approximated by a *short circuit*—a conductor with zero resistance (Figure 8–8).

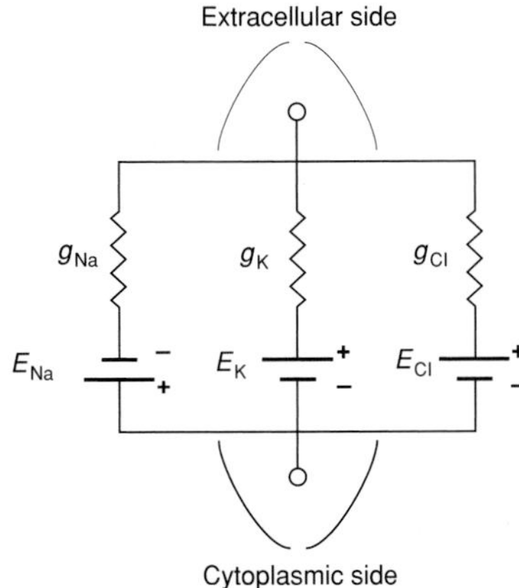

Figure 8–8 The passive current flow in a neuron can be modeled by an electrical equivalent circuit that includes elements representing the ion-selective membrane channels and the short-circuit pathways provided by the cytoplasm and extracellular fluid.

As described above, steady fluxes of Na^+ and K^+ ions through the passive membrane channels are exactly counterbalanced by active ion fluxes driven by the sodium–potassium pump, which extrudes Na^+ ions from the cell and pumps in K^+ ions. This ATP-dependent pump, which keeps the ionic batteries charged, can be added to the equivalent circuit in the form of a *current generator* (Figure 8–9).

In addition to electromotive force and conductance, the third important passive electrical property of the neuron is *capacitance*. In general, an electrical capacitor is defined as two conducting materials separated by an insulating material. For the neuron the conducting materials are the cytoplasm and the extracellular fluid; the insulating material is the cell membrane, specifically the lipid bilayer. Because the bilayer is penetrated by ion channels, the membrane is a leaky capacitor. Nevertheless, since the density of ion channels is low, the capacitor portion of the membrane occupies at least 100 times the area of all the ion channels combined. Membrane capacitance is included in the equivalent circuit in Figure 8–9.

The fundamental property of a capacitor is the ability to store charges of opposite sign on its two surfaces. The electrical potential difference across a capacitor, V, is expressed as:

$$V = \frac{Q}{C} \tag{11}$$

where Q is the excess of positive or negative charges on each side of the capacitor and C is the capacitance. Capacitance is measured in units of

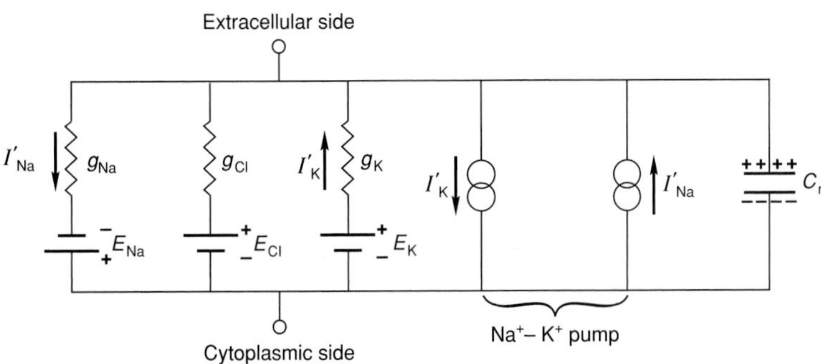

Figure 8–9 Under steady-state conditions the passive Na^+ and K^+ currents are balanced by active Na^+ and K^+ fluxes (I'_{Na} and I'_K) driven by the sodium–potassium pump. The lipid bilayer endows the membrane with electrical capacitance (C_m).

farads (F), where a charge separation of 1 coulomb (C) across a 1F capacitor produces a potential difference of 1 V.

A typical value of membrane capacitance for a nerve cell is about 1 μF/cm^2 of membrane area. The excess of positive and negative charges separated by the membrane of a spherical cell body with a diameter of 50 μm and a resting potential of –60 mV is 29×10^6 ions. Although this number may seem large, it represents only a tiny fraction (1/200,000th) of the total number of positive or negative charges within the cytoplasm. The bulk of the cytoplasm and the bulk of the extracellular fluid are electroneutral.

An Overall View

The lipid bilayer, which is virtually impermeant to ions, is an insulator separating two solutions, the cytoplasm and the extracellular fluid. Ions can leak across the lipid bilayer only by diffusing through the ion channels. When the cell is at rest, these passive ionic fluxes into and out of the cell are balanced, so that the charge separation across the membrane remains constant and the membrane potential remains at its resting value.

The value of the resting membrane potential in nerve cells is determined primarily by resting channels selective for K^+, Cl^-, and Na^+. In general, the membrane potential will be closest to the Nernst potential of the ion or ions with the greatest membrane permeability. The permeability for an ion species is proportional to the number of open channels permeable to that ion.

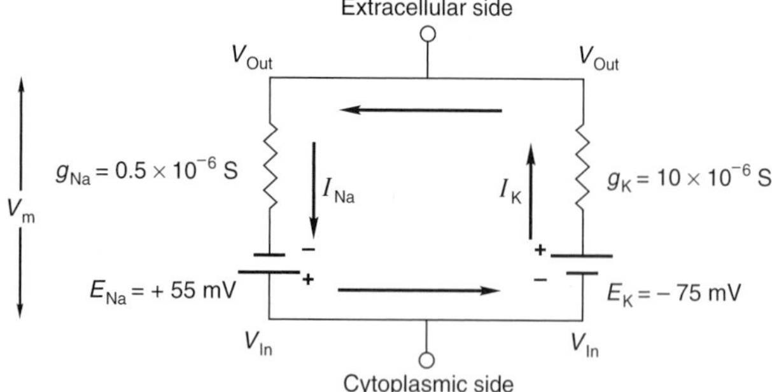

Figure 8–10 This electrical equivalent circuit omits the Cl^- pathway for simplicity in calculating the resting membrane potential.

At rest, the membrane potential is close to the Nernst potential for K^+, the ion to which the membrane is most permeable. However, the membrane is also somewhat permeable to Na^+, and an influx of Na^+ drives the membrane potential slightly positive to the K^+ Nernst potential. At this potential the electrical and chemical driving forces acting on K^+ are no longer in balance, so K^+ diffuses out of the cell. These two passive fluxes are each balanced by active fluxes driven by the sodium–potassium pump.

Chloride is actively pumped out of some, but not all, cells. When it is not, it is passively distributed so as to be at equilibrium. Under most physiological conditions the bulk concentrations of Na^+, K^+, and Cl^- inside and outside the cell are constant. The changes in membrane potential that occur during signaling (action potentials, synaptic potentials, and receptor potentials) are caused by the substantial changes in the relative membrane permeabilities to these three ions, not by changes in the bulk concentrations of ions, which change negligibly. The changes in permeability, caused by the opening of gated ion channels, cause changes in membrane potential by altering the net charge separation across the membrane.

Postscript: An Equation for the Resting Membrane Potential Can Be Derived from an Equivalent Circuit

We illustrate with a simple example how the equivalent circuit model of the neuron may be used to analyze neuronal properties quantitatively. Specifically, we use the equivalent circuit model of the resting membrane to calculate the resting potential. To simplify the calculation, we initially ignore Cl^- channels and begin with just two types of passive channels, K^+ and Na^+, as illustrated in Figure 8–10. Moreover, we ignore the electrogenic influence of the sodium–potassium pump because it is small. Because we only consider steady-state conditions, where V_m is not changing, we can also ignore membrane capacitance. (Membrane capacitance and its delaying effect on changes in V_m are discussed in Chapter 9.) Because there are more open resting channels for K^+ than for Na^+, the membrane conductance for current flow carried by K^+ is much greater than that for Na^+. In our equivalent circuit (Figure 8–10) g_K is 20 times higher than

g_{Na} (10×10^{-6} S compared with 0.5×10^{-6} S). Given these values and the values of E_K and E_{Na}, we can calculate the membrane potential V_m as follows.

Because V_m is constant in the resting state, the net current must be zero; otherwise the separation of positive and negative charges across the membrane would change, causing V_m to change. Therefore I_{Na} is equal and opposite to I_K[3]:

$$I_{Na} = -I_K$$

or

$$I_{Na} + I_K = 0. \quad (1)$$

We can easily calculate I_{Na} and I_K in two steps. First, we add up the separate potential differences across the Na^+ and K^+ branches of the circuit. As one goes from inside to outside across the Na^+ branch, the total potential difference is the sum of the potential differences across E_{Na} and across g_{Na}[4]:

$$V_m = E_{Na} + I_{Na}/g_{Na}.$$

Similarly, for the K^+ conductance branch

$$V_m = E_K + I_K/g_K.$$

Next, we rearrange and solve for I:

$$I_{Na} = g_{Na} \times (V_m - E_{Na}) \quad (2a)$$

$$I_K = g_K \times (V_m - E_K). \quad (2b)$$

As these equations illustrate, the ionic current through each conductance branch is equal to the conductance of that branch multiplied by the difference between membrane potential and the Nernst potential for the given ion. This difference defines the net *electrochemical driving force* on the ion. For example, the conductance for the K^+ branch is proportional to the number of open K^+ channels, and the driving force is equal to the difference between V_m and E_K. If V_m is more positive than E_K (–75 mV), the driving force is positive (outward); if V_m is more negative than E_K, the driving force is negative (inward).

If we now substitute Equations 2a and 2b for I_{Na} and I_K in Equation 1, we obtain the following expression:

$$g_{Na} \times (V_m - E_{Na}) + g_K \times (V_m - E_K) = 0.$$

Multiplying through we see that

$$(V_m \times g_{Na} - E_{Na} \times g_{Na}) + (V_m \times g_K - E_K \times g_K) = 0.$$

This can be rearranged to yield

$$V_m \times (g_{Na} + g_K) = (E_{Na} \times g_{Na}) + (E_K \times g_K).$$

Solving for V_m, we obtain an equation for the resting membrane potential that is expressed in terms of membrane conductances and batteries:

$$V_m = \frac{(E_{Na} \times g_{Na}) + (E_K \times g_K)}{g_{Na} + g_K}. \quad (3)$$

From this equation we can calculate V_m using the values in our equivalent circuit (Figure 8–10):

$$V_m = \frac{(+55 \times 10^{-3}\text{ V}) \times (0.5 \times 10^{-6}\text{ S})}{0.5 \times 10^{-6}\text{ S} + 10 \times 10^{-6}\text{ S}} + \frac{(-75 \times 10^{-3}\text{ V}) \times (10 \times 10^{-6}\text{ S})}{0.5 \times 10^{-6}\text{ S} + 10 \times 10^{-6}\text{ S}}$$

$$= \frac{-722.5 \times 10^{-9}\text{ V} \times \text{S}}{10.5 \times 10^{-6}\text{ S}}$$

$$= -69\text{ mV}.$$

Equation 3 states that V_m will approach the value of the ionic battery that is associated with the greater conductance. This principle can be illustrated with another example as we consider what happens during the action potential. At the peak of the action potential, total membrane g_K is essentially unchanged from its resting value, but g_{Na} increases by as much as 500-fold. This increase in g_{Na} is caused by the opening of voltage-gated Na^+ channels. In our equivalent circuit example (Figure 8–10) a 500-fold increase would change g_{Na} from 0.5×10^{-6} S to 250×10^{-6} S. If we substitute this new value of g_{Na} into Equation 3 and solve for V_m, we obtain +50 mV, a value much closer to E_{Na} than to E_K. V_m is closer to E_{Na} than to E_K at the peak of the action potential because g_{Na} is now 25-fold greater than g_K, so the Na^+ battery becomes much more important than the K^+ battery in determining V_m.

In reality the resting membrane has open conductance channels not only for Na^+ and K^+, but

---

[3]This equality is true only if one makes the simplifying assumption that the sodium–potassium pump is electroneutral.

[4]Because we have defined V_m as $V_{in}-V_{out}$, the following convention must be used for these equations. Outward current (in this case I_K) is positive and inward current (I_{Na}) is negative. Batteries with their positive poles toward the inside of the membrane (e.g., E_{Na}) are given positive values in the equations. The reverse is true for batteries that have their negative poles toward the inside, such as the K^+ battery.

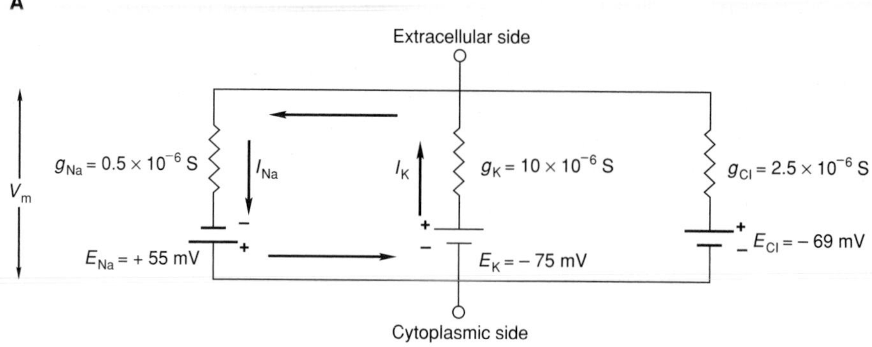

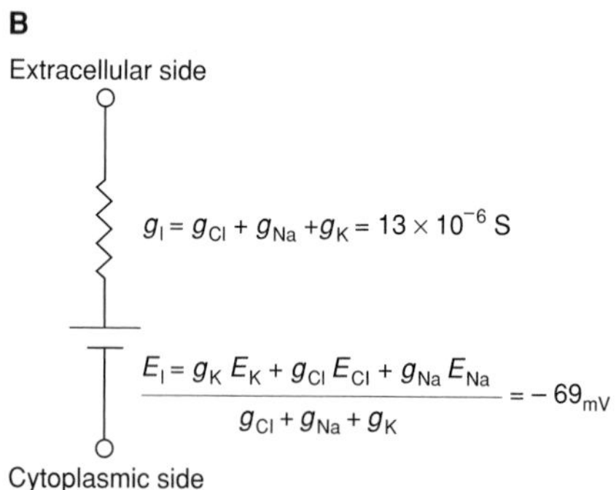

Figure 8–11

A. This electrical equivalent circuit includes the Cl^- pathway. However, no current flows through the Cl^- channels in this example because V_m is at the Cl^- equilibrium (Nernst) potential.

B. In this simplified equivalent circuit the resting membrane conductance $g_l = g_{Cl} + g_{Na} + g_K$, and the electromotive force or battery (E_l) is the resting potential predicted by Equation 8–4.

also for Cl^-. One can derive a more general equation for V_m, following the steps outlined above, from an equivalent circuit that includes a conductance pathway for Cl^- with its associated Nernst battery (Figure 8–11A):

$$V_m = \frac{(E_K \times g_K) + (E_{Na} \times g_{Na}) + (E_{Cl} \times g_{Cl})}{g_K + g_{Na} + g_{Cl}} \qquad (4)$$

This equation is similar to the Goldman equation presented earlier in the chapter (Equation 6). As in the Goldman equation, the contribution to V_m of each ionic battery is weighted in proportion to the conductance (or permeability) of the membrane for that particular ion. In the limit, if the conductance for one ion is much greater than that for the other ions, V_m will approach the value of that ion's Nernst potential.

The contribution of Cl^- ions to the resting potential can now be determined by comparing the value of V_m for the circuit for Na^+ and K^+ only (Figure 8–10) with the value for the circuit for all three ions (Figure 8–11A). For most nerve cells the value of g_{Cl} ranges from one-fourth to one-half of g_K. In addition, E_{Cl} is typically quite close to E_K, but slightly less negative. In our example (Figure 8–11A) Cl^- ions are passively distributed across the membrane, so that E_{Cl} is equal to the value of V_m, which is determined by Na^+ and K^+. Note that if $E_{Cl} = V_m$ (–69 mV in this case), no net current flows through the Cl^- channels. Thus, including g_{Cl} and E_{Cl} from Figure 8–11A in the calculation of V_m does not change the value of V_m from that for Figure 8–10. On the other hand, if Cl^- were not passively distributed but actively pumped out of the cell, then E_{Cl} would be more negative than –69 mV, so adding the Cl^- pathway to the calculation would shift V_m to a slightly more negative value.

The equivalent circuit can be further simplified by lumping the conductance of all the resting channels that contribute to the resting potential into a single conductance g_l and replacing the battery for each conductance channel by a single battery whose value E_l is that predicted by Equation 4 (Figure 8–11B). This simplification will prove useful when we consider gated channels in subsequent chapters.

Selected Readings

Hille, B. 1992. Ionic Channels of Excitable Membranes, 2nd ed. Sunderland, Mass.: Sinauer.

Hodgkin, A. L. 1992. Chance and design: Reminiscences of Science in Peace and War. Cambridge, England: Cambridge University Press.

9

Local Signaling: Passive Electrical Properties of the Neuron

Membrane Resistance Affects the Magnitude of Electrical Signals

Membrane Capacitance Prolongs the Time Course of Electrical Signals

Membrane and Axoplasmic Resistance Affect the Efficiency of Signal Conduction

Passive Membrane Properties and Axon Diameter Affect the Velocity of Action Potential Propagation

While all cells of the body possess a membrane potential, neurons have the unique ability to generate very rapid electrical signals, which can then be conducted long distances along dendrites and axons. In the previous chapter we saw how membrane potentials are generated by the concentration gradients of different ions and the permeability of the cell membranes to those ions. Using the Goldman equation, we were able to calculate the expected membrane potential generated in response to a given set of ionic concentration gradients and membrane permeabilities.

This description does not, however, provide us with any information about *changes* in the membrane potential in response to a stimulus. In fact, the Goldman equation only applies to the steady-state case in which the voltage does not change. When the neuron is signaling—as occurs when the neuron generates receptor potentials, synaptic potentials, or action potentials (see Figure 2–12)—the membrane voltage is constantly changing. What determines the rate of the potential change in response to a stimulus? Will a brief synaptic current always produce a similar potential change, regardless of the size of the postsynaptic cell? What determines whether a given stimulus will be subthreshold or suprathreshold?

To answer these questions we need to consider in detail the properties of the ion channels that generate the active ionic currents that underlie changes in membrane potential. This we do in the next three chapters. Before we can understand how these active currents generate changes in membrane potential, however, we must first consider the passive electrical properties of the neurons, which are constant and do not change during electrical signaling. There are three passive properties: resting membrane resistance, membrane capacitance, and intracellular axial resistance along the length of the axon and dendrites.

These passive properties play a crucial role in electrical signaling because they provide the return pathway by which the active currents flow into or out of the cell. Thus they determine the time course of the synaptic potential generated by the synaptic current. They determine whether a synaptic potential generated in a dendrite will result in a subthreshold or suprathreshold depolarization at the trigger zone on the axon hillock. The passive properties also help to determine the speed at which the action potential is conducted once an action potential is generated.

Membrane Resistance Affects the Magnitude of Electrical Signals

The difference between passive and active properties of neurons was shown in Chapter 8 by considering what happens when we inject hyperpolarizing or depolarizing current pulses into the cell body of a neuron and record the resulting changes in membrane potential (see Box 8–1). When we inject negative charge through our electrode, the charge separation across the membrane increases and the membrane potential becomes more negative, or hyperpolarized. As we inject a larger negative current, we obtain a larger hyperpolarization. In most neurons there is a linear relation between the size of the negative current and the steady-state hyperpolarization (Figure 9–1). The relation between current and voltage defines a resistance, which is called the neuron's *input resistance*, R_{in}.

Injecting a positive charge into the cell produces a depolarization. The neuron still behaves as a simple resistor, but only over a limited voltage range. As the positive current becomes larger, the membrane potential becomes more positive and eventually reaches a point, the threshold, at which the neuron generates an action potential. Above this threshold the neuron no longer behaves as a simple resistor, due to the special properties of its voltage-gated channels. However, much of a neuron's behavior in the subthreshold range of voltages can be explained using simple circuits containing resistors, capacitors, and batteries.

The input resistance of the cell determines the magnitude of the depolarization in response to a steady current. From Ohm's law, the magnitude of the depolarization, ΔV, is given by

$$\Delta V = I \times R.$$

Thus, of two neurons receiving identical synaptic current inputs, the cell with the higher input resistance will experience a larger change in membrane voltage. For an idealized spherical neuron with no processes, the input resistance depends on both the density of the resting ion channels in the membrane (that is, the number of channels per unit area) and the size of the neuron. The larger the neuron, the greater will be its membrane surface area and the lower the input resistance, since there will be more resting channels to conduct ions. To compare the resistance of different-sized neurons, electrophysiologists often use the resistance of a unit area of membrane, termed the *specific membrane resistance*, R_{m}, which is measured in units of ohms times centimeters squared (Ω-cm^2). The specific membrane resistance depends only on the density and conductance of the resting ion channels. To obtain the total input resistance of the cell, we divide the specific membrane resistance by the membrane area of the cell, since a larger membrane area implies more resting channels and, hence, a lower net resistance. For our spherical neuron we obtain:

$$R_{\text{in}} = \frac{R_{\text{m}}}{4\pi a^2}$$

where a is the radius of the neuron. Thus, for a spherical cell the input resistance is inversely proportional to the square of the radius. For the more realistic case of a neuron with extensive dendrites and axons, the input resistance also depends on the membrane resistance of its processes as well as on the intracellular cytoplasmic resistance between the cell body and those processes.

Membrane Capacitance Prolongs the Time Course of Electrical Signals

While the magnitudes of the changes in membrane potential seen in Figure 9–1 resemble the behavior of a resistor, their *time courses* do not.

A1

Outward →
Current
← Inward

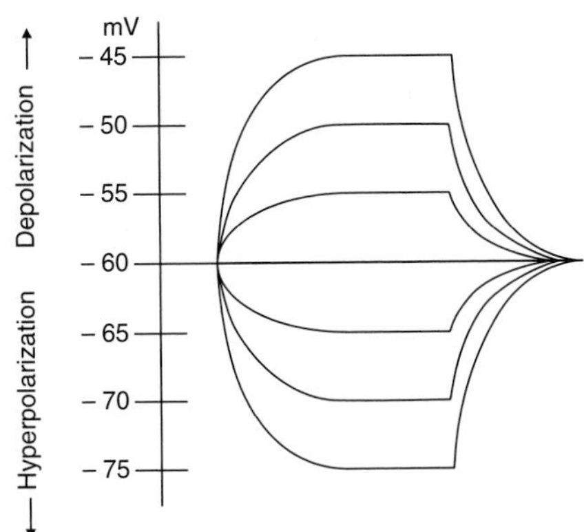

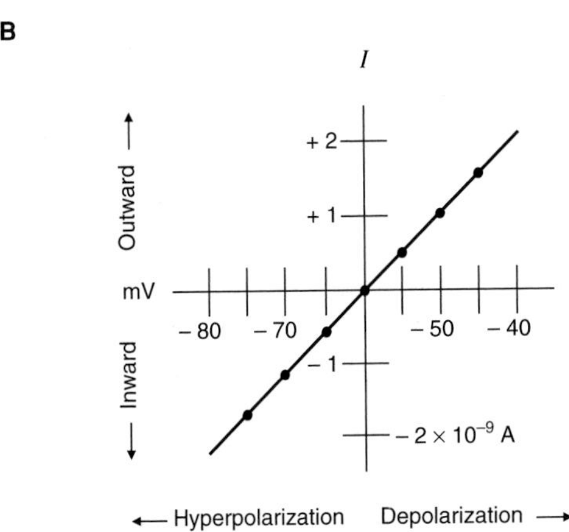

Figure 9–1 Current–voltage relationships. By passing subthreshold, graded, inward- and outward-current pulses into the cell, one can determine the relationship of current (*I*) and voltage (*V*) in the cell. Graded increases in outward- or inward-current pulses (**A1**) produce proportionally graded and symmetrical changes in the membrane potential (**A2**). Note that the potential changes more slowly than the square current pulses. The resulting *I–V* curve (**B**), obtained by plotting the steady-state voltage against the injected current, is linear. The slope of the *I–V* curve defines the input resistance of the neuron. In the example shown here the membrane potential changes by 10 mV for every 1 nA change in current, yielding a resistance of 10 mV/1 nA or 10×10^6 ohms (10 MΩ).

Thus, while a resistor will respond to a step change in current with a similar rapid step change in voltage, the cell of Figure 9–1 shows a voltage response that rises more slowly than the step change in current. As we will see, this property of the membrane is due to its *capacitance*.

To understand how the capacitance slows down the voltage response, recall that the voltage across a capacitor is proportional to the charge stored on the capacitor:

$$V = Q/C$$

where C is the capacitance. Altering the voltage across the capacitor thus requires that charge be either added or removed from the capacitor:

$$\Delta V = \Delta Q/C.$$

The change in charge (ΔQ) is the result of the flow of current across the capacitor (I_c). Since current is simply the flow of charge per unit time ($\Delta Q/\Delta t$), we can thus express the change in voltage across a capacitor as a function of current and the time that the current flows (Δt):

$$\Delta V = \frac{I_c \cdot \Delta t}{C}. \quad (1)$$

Thus we see that the magnitude of the change in voltage across a capacitor in response to a current

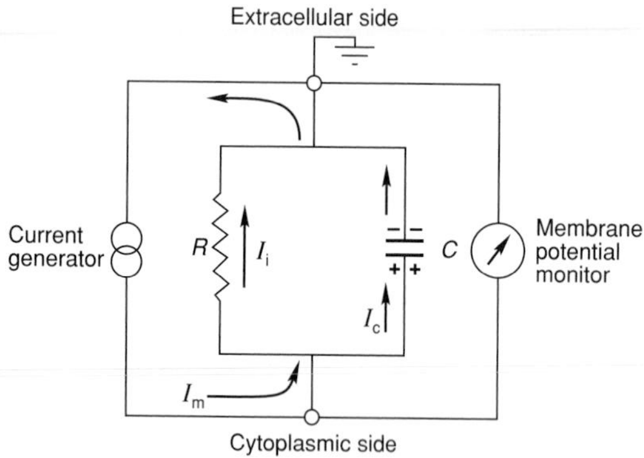

Figure 9–2 A simplified electrical equivalent circuit is used to examine the effects of membrane capacitance (C) on the rate of change of membrane potential in response to current flow. All ion-conducting channels are lumped into a single element (R). Batteries representing the electromotive forces generated by ion diffusion are not included because they affect only the absolute value of membrane potential, not the rate of change. This equivalent circuit represents the experimental setup shown in Box 8–1, in which pairs of electrodes are connected to the current generator and the membrane potential monitor.

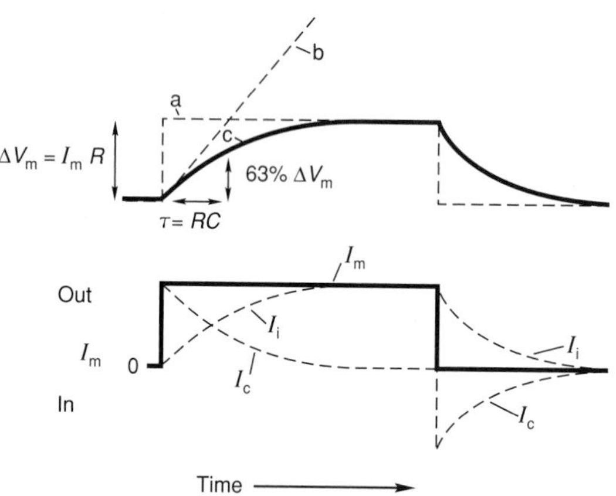

Figure 9–3 The rate of change in the membrane potential is slowed by the membrane capacitance. The **upper plot** shows the response of the membrane potential (ΔV_m) to a rectangular current pulse. The actual shape of the response (**line c**) combines the properties of a purely resistive element (**line a**) and a purely capacitive element (**line b**). The time taken to reach 63% of the final voltage defines the membrane time constant, τ. The time constants of different neurons typically range from 1 to 20 ms. The **lower plots** show the total membrane current (I_m) and its ionic (I_i) and capacitive (I_c) components ($I_m = I_i + I_c$).

pulse depends on the duration of the current. The time is necessary to deposit and remove charge on the plates of the capacitor.

Capacitance is directly proportional to the area of the plates of the capacitor. The larger the area, the more charge can be stored for a given potential difference. The value of the capacitance also depends on the insulating medium and the distance between the two plates of the capacitor. All biological membranes are composed of a similar insulating material and have a similar separation between the two plates (4 nm). The specific capacitance per unit area of all biological membranes, C_m, is approximately 1 μF/cm² of membrane. Thus the total input capacitance of a cell, C_{in} is given by the product of C_m and the area of the cell membrane:

$$C_{in} = C_m (4\pi a^2).$$

From this we can see it takes much more charge and therefore more current, to produce the same change in membrane potential for a larger neuron than for a smaller neuron.

According to Equation 1 the voltage across the membrane will continue to increase with time as long as a current pulse is applied. In reality, however, the voltage levels off after some time (Figure 9–1), because the membrane of a neuron acts as both a resistor (due to its ion-conducting channels) and a capacitor (due to the phospholipid bilayer). In terms of the electronic circuit seen in Chapter 8, we place the resistance and capacitance in parallel, since current crossing the membrane can flow either through ion channels (the resistive pathway) or across the capacitor (Figure 9–2). Thus, the total current crossing the membrane, I_m, is given by the sum of the resistive or ionic current (I_i) and the capacitive current:

$$I_m = I_i + I_c. \quad (2)$$

The capacitance of the membrane has the effect of reducing the rate at which the membrane potential changes in response to a current pulse (Figure 9–3). If the membrane had only resistive properties, a step pulse of outward current passed across it would change the membrane potential instantaneously. On the other hand, if the membrane had only capacitive properties, the membrane potential would change as a linear function of time in response to the same step pulse of current. Because the membrane has *both* capacitive and resistive properties in parallel, the actual change in membrane potential combines features of the two pure responses. Thus the initial slope of V_m as a function of time is the same as that for a purely capacitive element, whereas the final slope and amplitude are the same as those for a purely resistive element.

We can now understand why a step change in current produces the slowly rising voltage wave form seen in Figure 9–3. Since the resistance and capacitance of the membrane are in parallel, the voltage across each element must always be the same and equal to the membrane potential. Let us assume that the membrane potential starts off at 0 mV and that at time $t = 0$ we apply a depolarizing current step from a current generator with magnitude I_m. Initially the voltages across the resistor and capacitor are both equal to 0 mV. Since the ionic current across the resistor is given by Ohm's law ($I_i = V/R$), initially the current flowing through the resistor will be zero and all the current will flow through the membrane capacitance (i.e., $I_c = I_m$). As a result of the large initial capacitive current flow, the potential across the capacitor, and hence the membrane potential, will rapidly become more positive.

As the membrane potential increases, the voltage difference across the membrane begins to drive current across the membrane resistance. Thus after the initial instant, as the voltage across the membrane becomes more positive, more and more current flows through the resistor and less and less flows across the capacitor. As a result, the membrane potential rises more slowly. Eventually, it reaches a value where all the membrane current flows through the resistor ($I_i = I_m$). From Ohm's law this voltage is given by $V = I_m R_m$. At this point the capacitative current is zero, so, from Equation 1, the membrane potential will no longer change. Once the step of current is turned off, the total membrane current, I_m, equals zero so that $I_i = -I_c$. This causes the charge on the capacitor to dissipate by flowing in a loop around the circuit through the resistive pathway and the membrane potential returns to zero.

The rising phase of the potential change can be described by the following equation:

$$\Delta V_m(t) = I_m R(1 - e^{-t/\tau}) \tag{3}$$

where e, which has the value of 2.72, is the base of the system of natural logarithms, and τ is the *membrane time* constant, the product of the resistance and capacitance of the membrane (RC). The time constant can be measured experimentally (Figure 9–3). We return to discuss the importance of the time constant when we consider synaptic integration in Chapter 13.

Membrane and Axoplasmic Resistance Affect the Efficiency of Signal Conduction

So far we have considered the effects of the passive properties of neurons on signaling within the cell body. Because the neuron's soma can be approximated as a simple sphere, we did not have to take into account the effect of distance on the propagation of a signal. However, a subthreshold voltage signal along dendrites and axons decreases in amplitude with distance from its site of initiation. To understand how this attenuation occurs we will again refer to an equivalent circuit, one that shows how the geometry of a neuron influences the distribution of current flow.

Synaptic potentials that originate in dendrites are conducted along the dendrite toward the cell body and the trigger zone. The cytoplasmic core of a dendrite offers significant resistance to the longitudinal flow of current because it has a relatively small cross-sectional area and ions flowing down the dendrite collide with other molecules. The greater the length of the cytoplasmic core, the greater the resistance since the ions experience more collisions the further they travel. Conversely, the larger the diameter of the cytoplasmic core, the lower the resistance will be in a given length due to the greater number of charge carriers at any point.

To represent the incremental increase in resistance along the length of the dendritic core, the dendrite can be thought of as a series of identical cytoplasm-containing membrane cylinders, each with its own measurable membrane resistance and capacitance and axial resistance in the cytoplasmic core. The extracellular fluid has only negligible resistance, due to its large volume, and therefore can be ignored. The equivalent circuit for this simplified model is shown in Figure 9–4.

The *axial resistance* (r_a) of a unit length (1 cm) of the cytoplasmic core is expressed in Ω/cm. Axial resistance depends on both the specific resistivity of the cytoplasm, ρ, measured in Ω · cm, and the

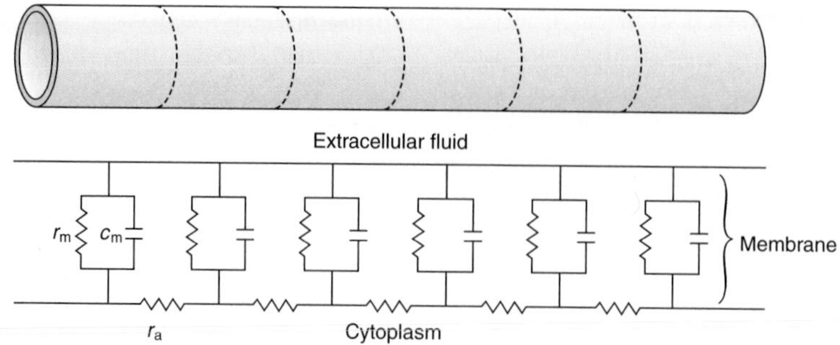

Figure 9–4 A neuronal process, either an axon or dendrite, can be divided into unit lengths, which can be represented in an electrical equivalent circuit. Each unit length of the process is a circuit with its own membrane resistance (r_m) and capacitance (c_m). All the circuits are connected by resistors (r_a), which represent the axial resistance of segments of cytoplasm, and a short circuit, which represents the extracellular fluid.

cross-sectional area of a dendrite with radius a:

$$r_a = \frac{\rho}{\pi a^2}. \qquad (4)$$

The *membrane resistance* (r_m) per unit length of cylinder is expressed in Ω-cm. Membrane resistance depends on both the specific resistance of a unit area of membrane, R_m, measured in $\Omega \cdot cm^2$, and the circumference of the dendrite:

$$r_m = \frac{R_m}{2\pi a}. \qquad (5)$$

For a dendrite of a uniform diameter, r_m is the same for equal lengths of membrane cylinder.

If current is injected into the dendrite at one point, how will the membrane potential change with distance along the dendrite? For simplicity, let us consider the variation of membrane potential with distance after a constant-amplitude current pulse has been on for some time ($t \gg \tau$). Under these conditions the membrane potential will have reached a steady value, so capacitive current is zero. When $I_c = 0$, all of the membrane current is ionic, $I_m = I_i$. The variation of the potential thus depends solely on the relative values of r_m and r_a.

The current that is injected flows out through several pathways across successive membrane cylinders along the length of the process (Figure 9–5). Each of these current pathways is made up of two resistive components in series: the total axial resistance, r_x, and the membrane resistance, r_m, of the unit membrane cylinder. The total axial resistance for each outflow pathway is the cytoplasmic resistance between the site of current injection and the site of the outflow pathway. Since resistors in series add, $r_x = r_a x$, where x is the distance along the dendrite from the site of current injection. The membrane component, r_m, has the same value at each outflow pathway along the cell process.

More current flows across a membrane cylinder near the site of injection than at more distant regions because current always follows the path of least resistance, and the total axial resistance, r_x, increases with distance from the site of injection (Figure 9–5). Because $\Delta V_m = I_m r_m$, the change in membrane potential, $\Delta V_m(x)$, produced by the current across a membrane cylinder becomes smaller as one moves down the dendrite away from the current electrode. This decay with distance has an exponential shape (Figure 9–5), expressed by the following equation:

$$\Delta V_m(x) = \Delta V_0 e^{-x/\lambda} \qquad (6)$$

where λ is the membrane *length constant*, x is the distance from the site of current injection, and V_0 is the change in membrane potential produced by the current flow at the site of the current electrode ($x = 0$). The length constant is the distance along the dendrite from the site of current injection to the site where ΔV_m has decayed to $1/e$, or 37% of its initial value (Figure 9–5), and is determined as follows:

$$\lambda = \sqrt{(r_m/r_a)}. \qquad (7)$$

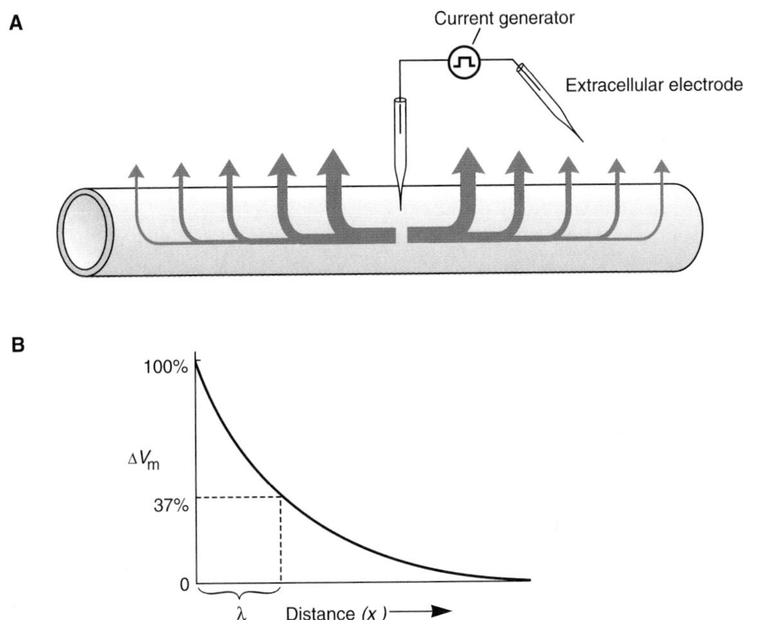

Figure 9–5 Current injected into a neuronal process by a microelectrode follows the path of least resistance to the return electrode in the extracellular fluid (**A**). The thickness of each arrow represents the relative size of the membrane current. Under these conditions the change in V_m decays exponentially with distance along the length of the process (**B**). The distance at which ΔV_m decays to 37% of its initial value (its value at the point of current injection) defines the length constant, λ.

The better the insulation of the membrane (the higher r_m is) and the better the conducting properties of the inner core (the lower r_a is), the greater the length constant of the dendrite. That is, current is able to spread further along the inner conductive core of the dendrite before leaking across the membrane. Since r_m is inversely related to the radius, whereas r_a is inversely related to the square of the radius, the length constant is proportional to the square root of the radius. Thus, larger-diameter axons will have a longer length constant than narrower axons. Typical values of the length constant fall in the range 0.1 to 1.0 mm.

Such passive spread of voltage changes along the neuron is called *electrotonic conduction*. The efficiency of this process, which is measured by the length constant, has two important effects on neuronal function. First, it influences *spatial summation*, the process by which synaptic potentials generated in different regions of the neuron are added together at the trigger zone, the decision-making component of the neuron (see Chapter 13).

A second important feature of electrotonic conduction is its role in the propagation of the action potential. Once the membrane at any point along an axon has been depolarized beyond threshold, an action potential is generated in that region in response to the opening of voltage-gated Na^+ channels (see Chapter 10). This local depolarization then spreads electrotonically along the axon, causing the adjacent region of the membrane to reach the threshold for generating an action potential (Figure 9–6).

The depolarization is thus spread by "local-circuit" current flow resulting from the potential difference between the active and the resting regions of the axon membrane. Once the depolarization of the resting region of the membrane approaches threshold, the voltage-gated Na^+ channels in this region of membrane open up, Na^+ rushes down its electrochemical gradient into the cytoplasm, and the depolarization becomes greater. This increase in depolarization causes more Na^+ channels to open, so that more Na^+ comes in, and so forth.

A

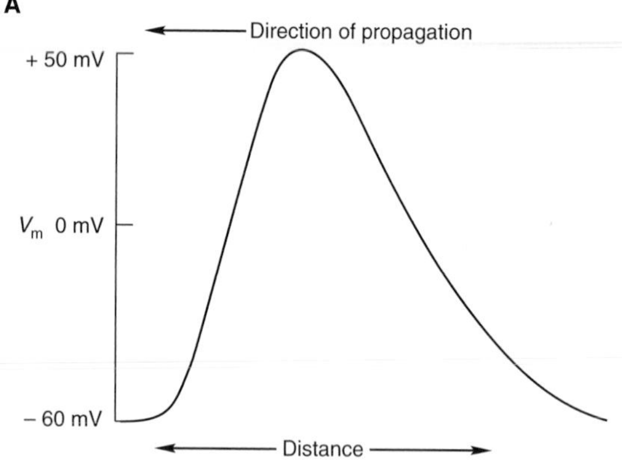

B

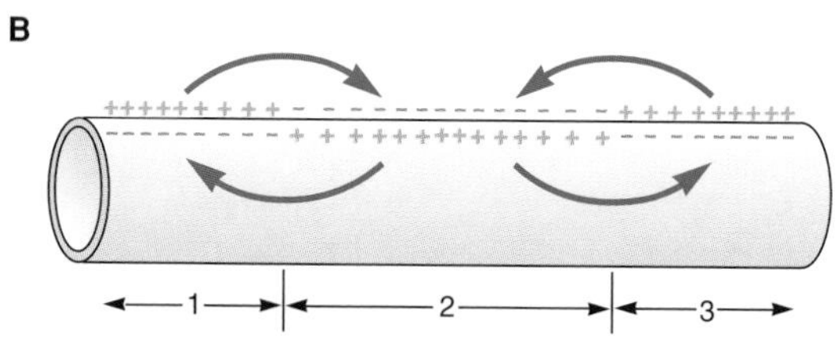

Figure 9–6 Passive conduction of positive (depolarizing) charge along the axon contributes to propagation of the action potential.

A. The wave form of an action potential propagating from right to left.

B. The difference in potential along the length of the axon creates a local-circuit current flow that causes the depolarizing charge to spread passively from the active region (**2**) to the resting region ahead of the action potential (**1**), as well as the area behind the action potential (**3**). Because there is also an increase in g_K in the wake of the action potential (see Chapter 10), the build up of positive charge along the inside of the membrane in area **3** is more than balanced by the local efflux of K^+, allowing this region of membrane to repolarize.

Thus, as the local membrane potential approaches threshold, the depolarization changes from a passive to an active regenerative process. This actively generated depolarization then spreads by passive, local-circuit flow of current to the next region of membrane, and the cycle is repeated.

Passive Membrane Properties and Axon Diameter Affect the Velocity of Action Potential Propagation

The passive spread of depolarization during conduction of the action potential is not instantaneous. In fact, it is a rate-limiting factor in the propagation of the action potential. We can understand this limitation by considering a simplified equivalent circuit of two adjacent membrane segments connected by a segment of axoplasm (Figure 9–7). As described above, an action potential generated in one segment of membrane supplies depolarizing current to the adjacent membrane, causing it to depolarize gradually toward threshold. According to Ohm's law, the larger the axoplasmic resistance, the smaller the current flow around the loop ($I = V/R$), and thus the longer it takes to change the charge on the membrane of the adjacent segment.

Recall that since $\Delta V = \Delta Q/C$, membrane potential will change slowly if the current is small because ΔQ will change slowly. Similarly, the larger the membrane capacitance, the more charge must be deposited on the membrane to change the potential across the membrane, so the current must flow for a longer time to produce a given depolarization. Therefore, the time it takes for depolarization to spread along the axon is determined by both the axial resistance and the capacitance per unit length of the axon (r_a and c_m). The rate of passive spread varies inversely with the product $r_a c_m$. If this product is reduced, the rate of passive spread of a given depolarization will increase and the action potential will propagate faster.

Rapid propagation of the action potential is functionally important, and two distinct mechanisms have evolved to increase it. One adaptive strategy is to increase conduction velocity by *increasing the diameter of the axon core*. Because r_a decreases in proportion to the square of axon diameter, while c_m increases in direct proportion to diameter, the net effect of an increase in diameter is a decrease in $r_a c_m$. This adaptation has been carried to its extreme in the giant axon of the squid, which can be as large as 1 mm in diameter. No larger axons have evolved, presumably because of the opposing need to keep neuronal size small (so that many cells can be packed into a restricted space).

A second mechanism for increasing conduction velocity by reducing $r_a c_m$ is *myelination*, the wrapping of glial cell membranes around an axon (see Chapter 3). This process is functionally equivalent to increasing the thickness of the axonal membrane by as much as 100 times. Because the capacitance of a parallel-plate capacitor such as the membrane is inversely proportional to the thickness of

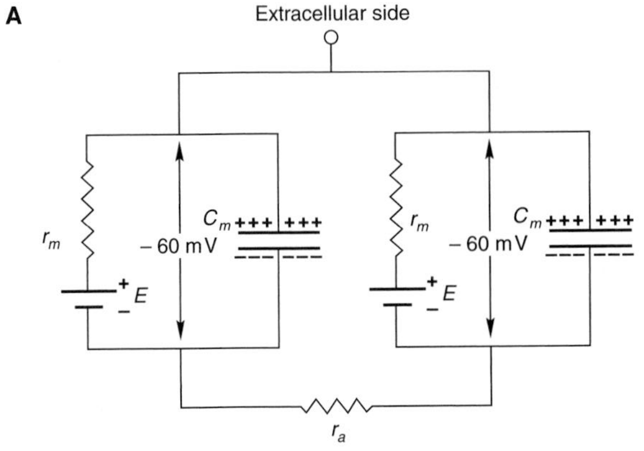

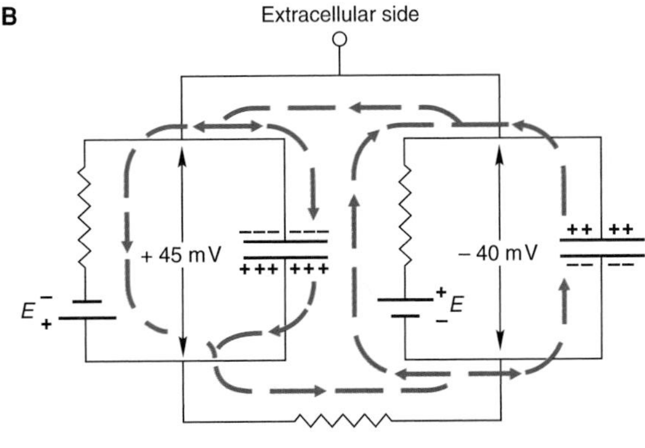

Figure 9–7 The electrical equivalent circuit in part **A** represents two adjacent resting membrane segments of an axon connected by a segment of axoplasm. In part **B** an action potential is spreading from the membrane segment on the left to the segment on the right. **Broken lines** indicate pathways of current flow.

the insulating material, myelination decreases c_m and thus $r_a c_m$. For any given increase in the total diameter of the axon, myelination results in a much larger percentage decrease in $r_a c_m$ than does an increase in the diameter of the axon core. For this reason, conduction in myelinated axons is typically faster than in nonmyelinated axons of the same diameter.

In a neuron with a myelinated axon the action potential is triggered at the nonmyelinated membrane of the axon hillock. The inward current that flows through this region of membrane is then available to discharge the capacitance of the myelinated axon ahead of it. Even though the thickness of myelin makes the capacitance of the axon quite small, the amount of current flowing down the core of the axon from the trigger zone is not enough to discharge the capacitance along the *entire* length of the myelinated axon.

To prevent the action potential from dying out, the myelin sheath is interrupted every 1 to 2 mm by the nodes of Ranvier (see Figure 3–4). The bare patches of axon membrane at the nodes are only about 2 μm in length. Although the area of each nodal membrane is quite small, it contains a relatively high density of voltage-gated Na^+ channels and thus can generate an intense depolarizing inward Na^+ current in response to the passive spread of depolarization from the axon upstream. These regularly distributed nodes thus boost the amplitude of the action potential periodically, preventing it from dying out.

The action potential, which spreads quite rapidly between nodes because of the low capacitance of the myelin sheath, slows down as it crosses the high-capacitance region of each bare node. Consequently, as the action potential moves down the axon, it seems to jump quickly from node to node (Figure 9–8). For this reason, the action potential in a myelinated axon is said to move by *saltatory conduction* (from the Latin *saltare*, "to leap"). Because ionic membrane current flows only at the nodes in myelinated fibers, saltatory conduction is also favorable from a metabolic standpoint. Less energy must be expended by the sodium–potassium pump in restoring the Na^+ and K^+ concentration gradients, which tend to run down as a result of action potential activity.

Several diseases of the nervous system, such as multiple sclerosis and Guillain–Barré syndrome, cause demyelination (see Chapter 3). Because the lack of myelin slows down the conduction of the action potential, these diseases can have devastating effects on behavior. As an action potential goes from a myelinated region to a bare stretch of axon, it encounters a region of relatively high c_m and low r_m. For this unmyelinated segment of membrane to reach the threshold for an action potential, the inward current generated at the node just before this area has to flow for a longer time. In addition,

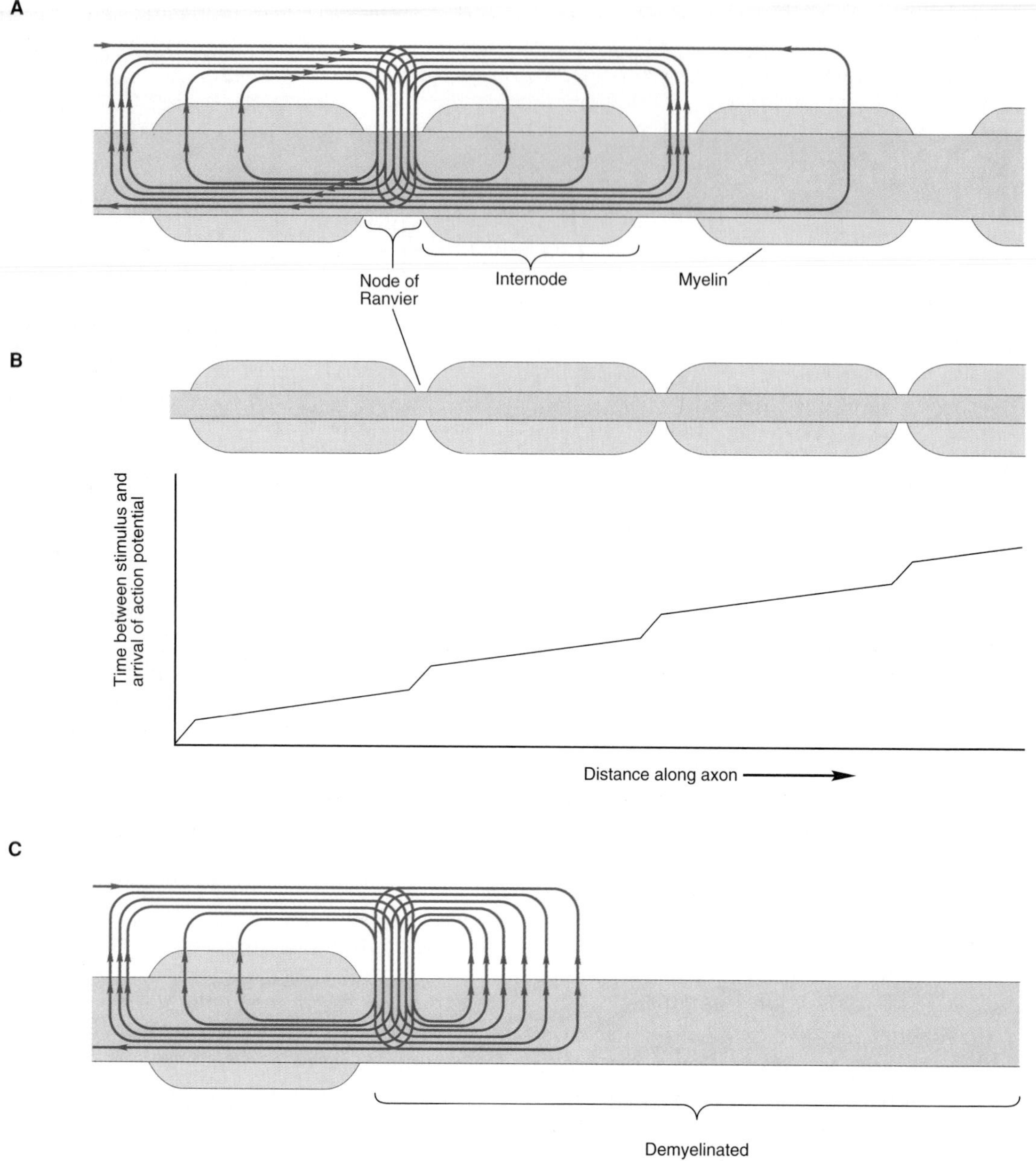

Figure 9–8 Action potentials in myelinated nerves are regenerated at the nodes of Ranvier, interruptions in the myelin sheath that occur at regular intervals.

A. Capacitive and ionic membrane current densities (membrane current per unit area of membrane) are much higher at the nodes of Ranvier than in the myelinated internodal regions of the axon. In this figure membrane current density at any point along the axon is represented by the number of the lines depicting current flow (**arrows**).

B. Because of the higher capacitance of the axon membrane at the unmyelinated nodes, the action potential slows down as it approaches each node and thus appears to skip rapidly from node to node.

C. In regions of the axon that have lost their myelin, because of disease, the spread of the action potential is slowed down or blocked. The local-circuit currents must charge a larger membrane capacitance, and, because of the low r_m, they do not spread effectively down the axon.

this local-circuit current does not spread as far as normal because it is flowing into a segment of axon that, because of its low r_m, has a short length constant (Figure 9–8). These two factors can combine to slow, and in some cases actually block, the conduction of action potentials.

An Overall View

Two competing pressures determine the functional design of neurons. First, to maximize the computing power of the nervous system, neurons must be small so that large numbers of them can fit into the available space. Second, to maximize the ability of the organism to respond to changes in its environment, neurons must conduct signals rapidly. In meeting these two design objectives, evolution has been constrained by the materials from which neurons are made.

Because the nerve cell membrane is very thin and is surrounded by a conducting medium, it has a very high capacitance and thus slows down the conduction of voltage signals. In addition, the currents that change the charge on the membrane capacitance must flow through a relatively poor conductor—a thin column of cytoplasm. The ion channels that give rise to the resting potential also degrade the signaling function of the neuron. They make the cell leaky and, together with the high membrane capacitance, they limit the distance that a signal can travel without being actively amplified.

Several features in the nervous system have evolved to compensate for these constraints. (1) The time constant of a neuron is exploited at the trigger zone, where inputs to the cell are time-averaged over a period of several milliseconds (temporal summation, discussed in Chapter 13). (2) The input zone of a neuron is compact, so that receptor or synaptic potentials are generated fairly close to the trigger zone, thus optimizing spatial summation (also discussed in Chapter 13). (3) As the inputs to the neuron (the receptor potentials or the synaptic potentials) propagate electronically, they become attenuated. These signals must therefore be converted into a pulse code for long-distance signaling. Voltage-gated channels generate these all-or-none action potentials, which are then conducted without attenuation. (4) For pathways in which rapid signaling is particularly important, the conduction velocity of the action potential is enhanced by either myelination or an increase in axon diameter, or both.

Selected Readings

Graubard, K., and Calvin, W. H. 1979. Presynaptic dendrites: Implications of spikeless synaptic transmission and dendritic geometry. In F. O. Schmitt and F. G. Worden (eds.), The Neurosciences; Fourth Study Program. Cambridge, Mass.: MIT Press, pp. 317–331.

Hodgkin, A. L. 1964. The Conduction of the Nervous Impulse. Springfield, Ill.: Thomas, chap. 4.

Hubbard, J. I., Llinás, R., and Quastel, D. M. J. 1969. Electrophysiological Analysis of Synaptic Transmission. Baltimore: Williams & Wilkins, chap. 2, pp. 91-109, 257–264.

Jack, J. J. B., Noble, D., and Tsien, R. W. 1975. Electric Current Flow in Excitable Cells. Oxford: Clarendon Press, chaps. 1–5, 7, pp. 276–277.

Khodorov, B. I. 1974. The Problem of Excitability. New York: Plenum Press, chap. 3.

Moore, J. W., Joyner, R. W., Brill, M. H., Waxman, S. D., and Najar-Joa, M. 1978. Simulations of conduction in uniform myelinated fibers: Relative sensitivity to changes in nodal and internodal parameters. Biophys J. 21:147–160.

Rall, W. 1977. Core conductor theory and cable properties of neurons. In E. R. Kandel (ed.), Handbook of Physiology, Section 1: The Nervous System, Vol. I: Cellular Biology of Neurons, Part 1. Bethesda, Md.: American Physiological Society, pp. 39–97.

10

Propagated Signaling: The Action Potential

The Action Potential Is Generated by the Flow of Ions Through Voltage-Gated Channels

The Action Potential Can Be Reconstructed from the Known Electrical Properties of the Neuron

Variations in the Action Potential in Different Neurons Are Accounted for by Variations on the Basic Theme of the Hodgkin–Huxley Theory

- *The Basic Mechanism of Action Potential Generation Is the Same in All Neurons*
- *The Nervous System Expresses a Rich Variety of Voltage-Gated Ion Channels*
- *Gating of Voltage-Sensitive Channels Can Be Influenced by Intracellular Ion Concentrations*
- *Excitability Properties Vary Among Neurons*
- *Excitability Properties Vary Within Regions of the Neuron*

Voltage-Gated Channels Have Characteristic Molecular Properties

- *Voltage-Gated Sodium Channels Are Sparsely Distributed*
- *Opening of Voltage-Gated Channels Is All-or-None*
- *Charges Are Redistributed Within Voltage-Gated Sodium Channels When the Channels Open*
- *The Voltage-Gated Sodium Channel Selects for Sodium on the Basis of Size, Charge, and Energy of Hydration of the Ion*
- *Potassium, Sodium, and Calcium Channels Stem from a Common Ancestral Gene*

The ability of nerve cells to carry signals over long distances depends on one crucial property of the nerve impulse—its regenerative characteristic. Unlike other electrical signals in the neuron, action potentials do not attenuate as they move away from their site of initiation.

So far in this book we have touched only briefly on the generation of the action potential by voltage-gated Na^+ and K^+ channels. In Chapter 9 we considered how the neuron's ability to generate signals and the speed of conduction are determined by passive electrical properties of the membrane. In this chapter we describe in detail the actions and properties of voltage-gated channels that are critical for generating and propagating the action potential.

The Action Potential Is Generated by the Flow of Ions Through Voltage-Gated Channels

An important early clue about how action potentials are generated came from an experiment performed in 1938 by Kenneth Cole and Howard Curtis. They recorded the activity in the giant axon of the squid and found that during the action potential the conductance of the membrane to ions increases dramatically (Figure 10–1). This discovery suggested that the action potential results from an enhanced movement of ions through the channels of the membrane. It also raised a further question: Which ions are responsible for the action potential?

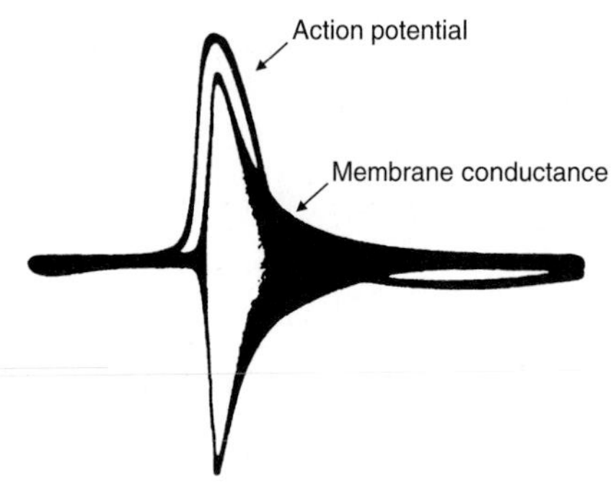

Figure 10–1 A net increase in ionic conductance in the membrane of the axon accompanies the action potential. This historical recording from an experiment conducted in 1938 by K. S. Cole and H. J. Curtis shows the oscilloscope record of an action potential superimposed on a record of the simultaneous membrane conductance.

A decade later Alan Hodgkin and Bernard Katz (Figure 10–2) found that the amplitude of the action potential is reduced when the external Na^+ concentration is lowered. Hodgkin and Katz proposed that the depolarization that initiates an action potential causes the membrane to increase its permeability to Na^+ briefly. This influx dwarfs the primary permeability of the resting cell to K^+ ions. Their data also suggested that the falling phase of the action potential is caused by a later increase in K^+ permeability.

To test this hypothesis it was necessary to vary the membrane potential systematically and then to measure the resulting changes in Na^+ and K^+ membrane conductances that resulted from the opening and closing of voltage-gated Na^+ and K^+ channels. This is difficult to do experimentally because of the strong interdependence of the membrane potential and the Na^+ and K^+ channels. For example, if the membrane is depolarized sufficiently to open some of the voltage-gated Na^+ channels, inward Na^+ current flows through these channels and causes further depolarization.

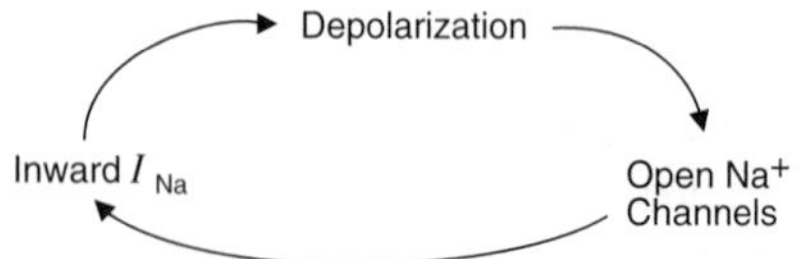

The additional depolarization causes still more Na^+ channels to open and consequently induces more inward Na^+ current. A regenerative cycle is thereby initiated that makes it impossible to achieve a stable membrane potential. This positive feedback cycle eventually drives the membrane potential to the peak of the action potential. A similar technical difficulty hinders the study of the active K^+ channels, which are responsible for the falling phase of the action potential. In 1949 Cole designed an apparatus known as the voltage clamp to overcome these problems. By using the voltage-clamp technique on the squid giant axon in the early 1950s, Hodgkin and Andrew Huxley provided the first complete description of the ionic mechanisms underlying the action potential.

The basic function of the voltage clamp is to interrupt the interaction between the membrane potential and the opening and closing of voltage-gated channels. The amount of current that must be generated by the voltage clamp to keep the membrane potential from changing provides a direct measure of the current flowing across the membrane (Box 10–1). The membrane current can then be analyzed into ionic and capacitive components.

Recall that V_m at any time is proportional to the charge Q_m on the membrane capacitance (C_m). When V_m is not changing, Q is constant and no capacitive current ($\Delta Q/\Delta t$) flows. Capacitive current flows *only* when V_m is changing (see Chapter 9). Therefore, when the membrane potential changes in response to a very rapid step of command potential, capacitive current flows only at the beginning and at the end of the step. This capacitive current is essentially instantaneous; on an oscilloscope record it looks quite different from the later ionic currents. Because the capacitive current has been effectively isolated, the ionic currents that flow through the membrane channels can be analyzed.

From the ionic membrane current and the membrane potential, the voltage- and time-dependence of the changes in membrane conductances caused by the opening and closing of Na^+ and K^+ channels can be calculated. This information provides insights into the properties of these two types of channels.

A.L. Hodgkin

A.F. Huxley

B. Katz

Figure 10–2 Bernard Katz, Alan Hodgkin, and Andrew Huxley developed the ionic hypothesis for signaling within neurons.

Hodgkin and Huxley attended Cambridge University in England. In 1939 they revolutionized thinking in biology by discovering that the action potential in nerve cells does not merely abolish the resting potential, as had generally been believed, but overshoots it by 50 mV. During World War II they both worked on radar for the British government and returned to neurobiological research in 1945. It was then that Hodgkin developed the ionic hypothesis for signaling in nerve cells. A few years later Hodgkin and Huxley, together with Katz, carried out a brilliant series of voltage-clamp experiments in which they showed rigorously that the action potential involves a sequential current flow of sodium moving into the cell and potassium moving out. On completing the analysis of the ionic mechanism of the action potential, Huxley turned to the study of muscle where, together with Hugh Huxley, he developed the *sliding filament hypothesis*, which underlies our current understanding of muscle contraction. Katz was born in Leipzig, Germany, and emigrated to England just before World War II began. After applying the ionic hypothesis to synaptic transmission at the nerve-muscle synapse, he analyzed the mechanism of transmitter release, showing that transmitter substances are released in packets.

The Current Through Voltage-Gated Sodium and Potassium Channels Can Be Isolated

In a typical voltage-clamp experiment we start with the membrane potential clamped at its resting value. If a 10 mV depolarizing potential step is commanded, we observe that an initial, very brief outward current instantaneously discharges the membrane capacitance by the amount required for a 10 mV depolarization. This *capacitive current* (I_c) is followed by a smaller outward ionic current that persists for the duration of this pulse. At the end of the pulse there is a brief inward capacitive current, and the total membrane current returns to zero (Figure 10–4A). The steady ionic current that persists throughout the stimulating current is the current that flows through the resting ion channels of the membrane (Chapter 7) and is called the *leakage current*, I_l. The total conductance of this population of channels is called the *leakage conductance* (g_l). These resting channels, which are always open, are responsible for generating the resting membrane potential (Chapter 8). In a typical neuron most of the resting channels are permeable to K^+ ions; the remaining ones are permeable to Cl^- or Na^+ ions.

If a larger depolarizing step is commanded, the current records become more complicated. The capacitive and leakage currents both increase in amplitude. In addition, an inward current develops shortly after the end of the capacitive current and the start of the leakage current; it reaches a peak within a few milliseconds, declines, and gives way to an outward current. This outward current reaches a plateau that is maintained for the duration of the pulse (Figure 10–4B).

A simple interpretation of these results is that the depolarizing voltage step sequentially turns on active conductance channels for two separate ions: one type of channel for inward current and another for outward current. Because these two oppositely directed currents partially overlap in time, the most difficult task in analyzing voltage-clamp experiments is to determine the separate time courses.

Hodgkin and Huxley achieved this separation by changing ions in the bathing solution. By substituting a larger, impermeant cation (choline•H^+) for Na^+, they eliminated the inward Na^+ current. Subsequently, a simpler technique for separating in-

BOX 10–1 Voltage-Clamp Technique

The voltage-clamp technique, first developed by Kenneth Cole in 1949, was used by Alan Hodgkin and Andrew Huxley in 1952 to study the squid giant axon. When an axon is voltage-clamped, the voltage-gated ion channels still open or close in response to imposed changes in membrane potential, but the voltage clamp prevents the resultant changes in membrane current from influencing the membrane potential. The changes in the membrane's conductance of individual ion species as a function of membrane potential can then be measured.

The voltage clamp consists of a source of current connected to two electrodes, one inside and the other outside the cell. By passing current across the cell membrane, the membrane potential can be stepped rapidly to a predetermined level of depolarization.

These depolarizations open voltage-gated Na^+ and K^+ channels. The resulting movement of Na^+ and K^+ across the membrane would ordinarily change the membrane potential, but the voltage clamp "clamps" the membrane potential at its commanded level. Normally, when Na^+ channels open in response to a moderate depolarizing voltage step, an inward membrane current develops because Na^+ ions flow through these channels driven by electrochemical potential differences. This Na^+ influx depolarizes the membrane by increasing the positive charge on the inside of the membrane and reducing the positive charge on the outside. The voltage clamp intervenes in this process by simultaneously moving positive charges out of the cell into the external solution. By generating an equal and opposite current, the voltage-clamp circuit automatically counteracts the flow of any membrane current that would tend to change the membrane potential from the commanded value (Figure 10–3A). As a result there is no change in the net amount of charge separated by the membrane and, therefore, no significant change in V_m.

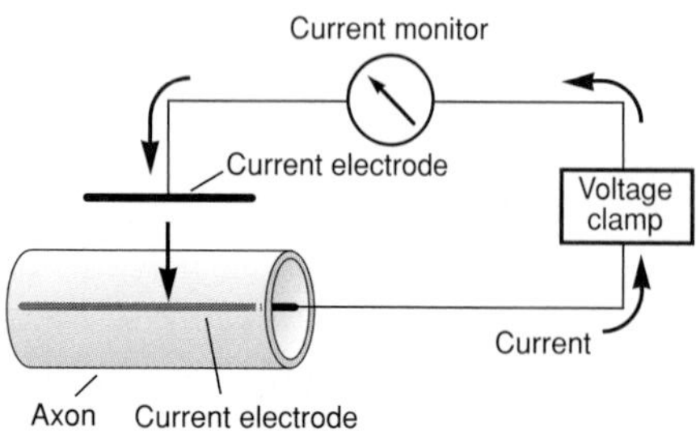

Figure 10–3A

The voltage clamp is a negative feedback system. A negative feedback system is one in which the value of the output of the system (V_m in this case) is "fed back" to the input of the system, where it is compared with a command signal for the desired output. Any difference between the command potential and the output signal activates a "controller" device that automatically reduces the difference. Thus the membrane potential *automatically* follows the command potential exactly (Figure 10–3B).

ward and outward currents was developed, in which the separate voltage-sensitive conductance channels are selectively blocked with drugs. Tetrodotoxin blocks the voltage-gated Na^+ channel and tetraethylammonium blocks the voltage-gated K^+ channel.

When tetraethylammonium is applied to the axon to block the K^+ channels, the total membrane current (I_m) consists of I_c, I_l, and I_{Na}. The leakage conductance, g_l, is constant; it does not vary with V_m or with time. Therefore, I_l, the leakage current, may be readily calculated and subtracted from I_m, leaving I_{Na} and I_c. Because I_c occurs only briefly at the beginning and end of the pulse, it can be eliminated easily by visual inspection, leaving the pure I_{Na}. To measure the current flowing through the voltage-gated Na^+ channels (I_{Na}), various levels of current are injected to change V_m to different levels. By a similar process, I_K may be measured when the Na^+ channels are blocked by tetrodotoxin (Figure 10–4C).

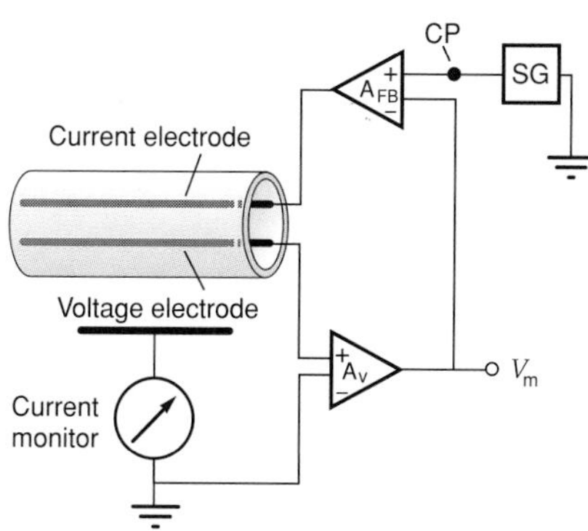

Figure 10–3B Membrane potential is measured by the voltage amplifier (A_v), which is connected to an intracellular electrode (*V*) and to the system ground, which is connected to the bath. The membrane potential signal (V_m) is displayed on an oscilloscope and is also fed into one terminal of the "feedback" amplifier (A_{FB}). This amplifier has two inputs—one for membrane potential and the other for the command potential (**CP**). The command potential, which comes from a signal generator (**SG**), is selected by the experimenter and can be of any desired amplitude and wave form. The feedback amplifier subtracts the membrane potential from the command potential. Any difference between these two signals is amplified several thousand times at the output of the feedback amplifier. The output of this amplifier is connected to a thin wire, the current-passing electrode, which runs the length of the axon. For the measured membrane current–voltage relationship to be meaningful, it is important that the membrane potential be uniform along the entire surface of the membrane. This condition can be maintained because the highly conductive current-passing wire short-circuits the axoplasmic resistance, reducing the axial resistance to zero. The presence of this low-resistance pathway within the axon makes it impossible for a potential difference to exist between different points along the axon core.

For example, assume that an inward Na^+ current through the voltage-gated Na^+ channels causes the membrane potential to become more positive than the command potential. The input to the feedback amplifier is equal to $V_{command} - V_m$. Thus, both the input and the resulting output voltage at the feedback amplifier are negative. This negative output voltage makes the internal current electrode negative, withdrawing net positive charge from the cell through the voltage-clamp circuit. As the current flows around the circuit, an equal amount of net positive charge is deposited into the external solution through the other current electrode.

A refinement of the voltage-clamp technique, the patch-clamp technique, allows the functional properties of *individual* ion channels to be analyzed (see Box 7–2).

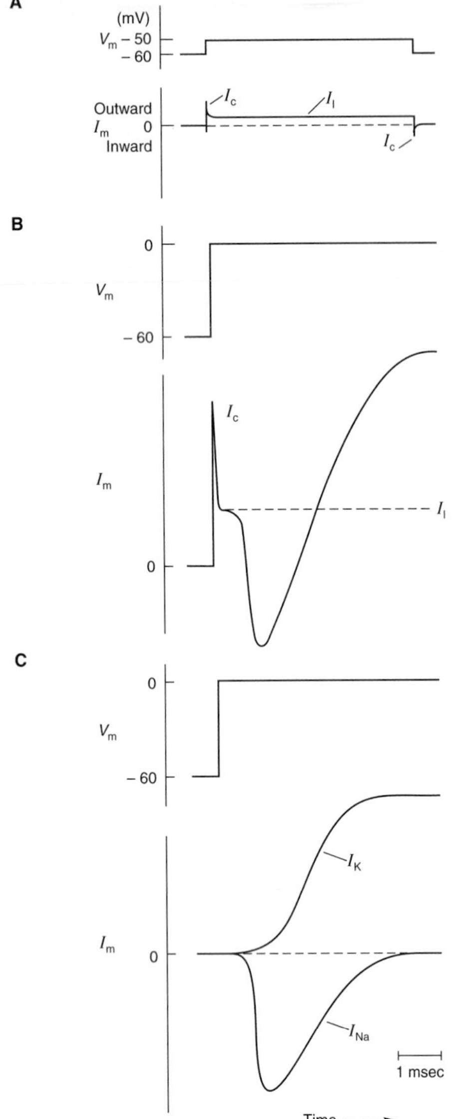

Figure 10–4 This record from a voltage-clamp experiment, performed on the giant axon of the squid, demonstrates the existence of two types of voltage-gated channels.

A. A small depolarization is accompanied by capacitive and leakage currents (I_c and I_l, respectively).

B. A larger depolarizing step results in larger capacitive and leakage currents, plus additional currents caused by the opening of voltage-gated Na^+ and K^+ channels.

C. When the voltage step shown in **B** is repeated in the presence of tetrodotoxin (which blocks the Na^+ current) and again in the presence of tetraethylammonium (which blocks the K^+ current), records of the pure K^+ and Na^+ currents (I_K and I_{Na}, respectively) are obtained by subtraction of I_c and I_l.

Sodium and Potassium Conductances Are Calculated from Their Currents

Once the Na^+ and the K^+ currents have been separated, they can be used to calculate the Na^+ and K^+ conductances (Box 10–2). These in turn can be used to determine the voltage sensitivity and kinetics of opening and closing of the entire population of voltage-gated Na^+ and K^+ channels in the membrane.

Measurements of Na^+ and K^+ conductances at various levels of membrane potential reveal two basic similarities and two fundamental differences between them. The two conductances are alike in that both populations of channels open in response to depolarizing steps of membrane potential, and they both do so more rapidly and to a greater extent for *larger* depolarizations (Figure 10–5). They differ, however, in the rates of onset

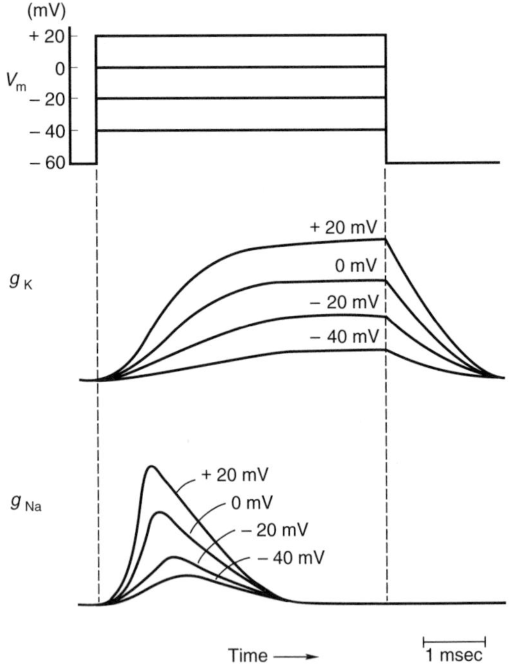

Figure 10–5 Voltage-clamp experiments show that sodium channels (g_{Na}) turn on and off more rapidly than potassium channels (g_K) over a wide range of membrane potentials. The increases and decreases in the Na^+ and K^+ conductances shown here reflect the shifting of thousands of voltage-gated channels between the open and closed states.

BOX 10–2 Calculation of Membrane Conductances from Voltage-Clamp Data

An equivalent circuit of the membrane that includes the membrane capacitance (C_m) and leakage conductance (g_l), as well as g_{Na} and g_K (Figure 10–6), can be used to calculate conductances from voltage-clamp currents. In this context g_l represents the conductance of all of the resting K^+, Na^+, and Cl^- channels (see Chapter 8); g_{Na} and g_K represent the conductances of the voltage-gated Na^+ and K^+ channels. The ionic battery of the resting (leakage) channels, E_l, is equal to the resting potential. The voltage-sensitive Na^+ and K^+ conductances are in series with their appropriate ionic batteries.

The current through each class of voltage-gated channel may be calculated from Ohm's law:

$$I_K = g_K \times (V_m - E_K)$$

$$I_{Na} = g_{Na} \times (V_m - E_{Na})$$

Rearranging and solving for g gives two equations that can be used to compute the conductances for the active Na^+ and K^+ channel populations.

$$g_K = \frac{I_K}{(V_m - E_m)}$$

$$g_{Na} = \frac{I_{Na}}{(V_m - E_{Na})}$$

To solve these equations, one must know V_m, E_K, E_{Na}, I_K, and I_{Na}. The independent variable, V_m, is set by the experimenter. The dependent variables, I_K and I_{Na}, can be measured from the records of voltage-clamp experiments (Figure 10–3C). The remaining variables, E_K and E_{Na}, are constants; they can be determined empirically by finding the values of V_m at which I_K and I_{Na} reverse their polarities. For example, if V_m is stepped to very positive values, I_{Na} assumes a lower value in the inward direction. At E_{Na} it goes to zero, and for values of V_m more positive than E_{Na}, it is directed outward.

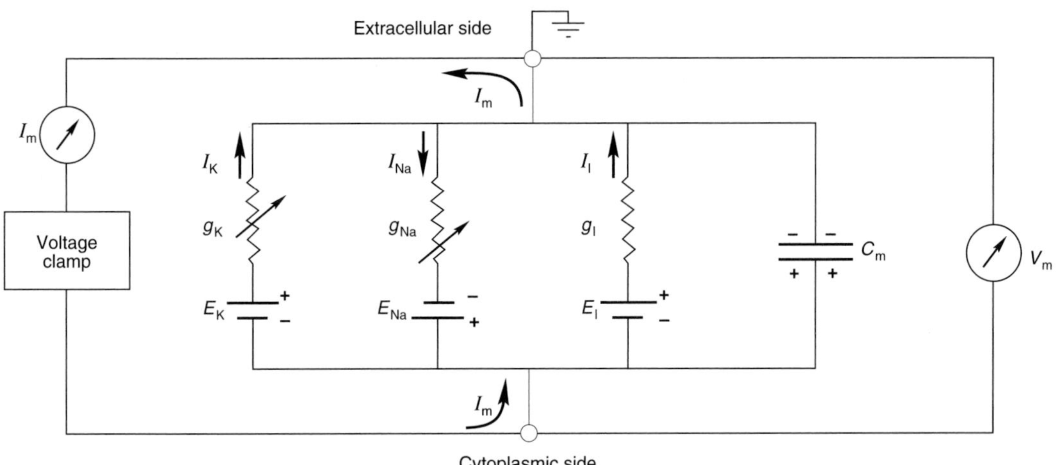

Figure 10–6 Electrical equivalent circuit of a nerve cell under voltage-clamp conditions. The voltage-gated conductance pathways (g_K and g_{Na}) are represented by the symbol for *variable conductance*—a conductor (resistor) with an arrow through it.

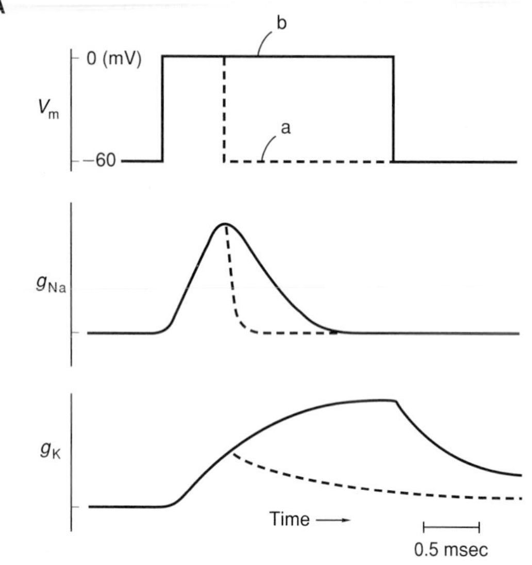

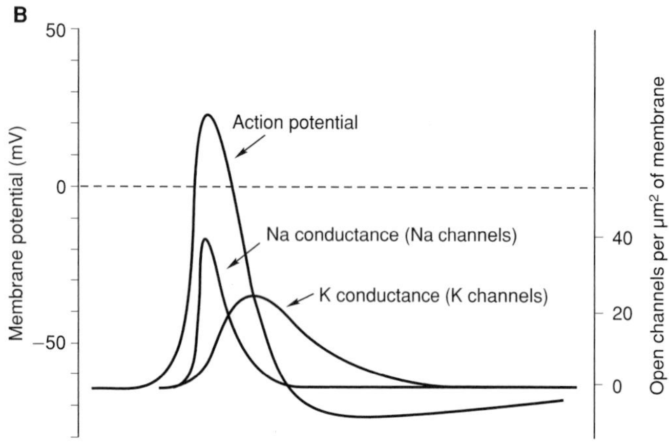

Figure 10–7 The action potential is caused by the sequential opening of voltage-gated Na^+ and K^+ channels.

A. Sodium and potassium channels respond differently to long-term depolarization. The illustration at the **top** shows a brief depolarizing voltage step (line **a**) and a longer-lasting step (line **b**). After the brief depolarization, both g_{Na} and g_K return to their initial values when the cell repolarizes. When depolarization is maintained, however, the Na^+ channels close (inactivate) before the depolarization is terminated, whereas the K^+ channels remain open and g_K increases throughout the depolarization.

B. One of Hodgkin, Huxley, and Katz's great insights was to separate the total conductance change detected by Cole and Curtis (see Figure 10–1) into separate components, attributable to the opening of Na^+ and K^+ channels.

and offset and the response to prolonged depolarization. At all levels of depolarization the Na^+ channels open more rapidly than do the K^+ channels (Figure 10–5). They also close more rapidly when the depolarizing pulse is very brief (Figure 10–7). In addition, when the depolarization is maintained for some time, the Na^+ channels begin to close, or inactivate, leading to a decrease of inward current. In contrast, the K^+ channels remain open as long as the membrane is depolarized (Figure 10–5).

Each Na^+ channel can exist in three different states, which are thought to represent three different conformations of the Na^+ channel protein: resting, activated, and inactivated (see Chapter 7). On depolarization the channel goes from the resting (closed) to the activated (open) state. If depolarization is brief, the channels go directly back to the resting state. If the depolarization is maintained, the channel switches to the inactivated (closed) state. Once the channel is inactivated it cannot be activated (opened) by continued depolarization. Inactivation can be relieved only by repolarizing the membrane, which allows the channel to switch from the inactivated to the resting state.

This switch takes time because channels leave the inactivated state relatively slowly (Figure 10–8). In other words, each Na^+ channel acts as if it has two kinds of gates that must be opened simultaneously for this channel to conduct Na^+ ions: an *activation gate*, which is closed when the membrane is at its resting potential and is rapidly activated by depolarization, and an *inactivation gate*, which is open at the resting potential and closes slowly in response to depolarization. The channel conducts only for a brief period during depolarization when both gates are open. Repolarization reverses the two processes. (The molecular interaction between these two coupled processes is described later in the chapter.) After the channel has returned to the resting state, it is again available to be activated by depolarization.

The Action Potential Can Be Reconstructed from the Known Electrical Properties of the Neuron

After measuring membrane conductance changes generated by depolarizing pulses of various amplitudes and durations, Hodgkin and Huxley fit their data to a set of empirical equations that describe the variations of membrane Na^+ and K^+ conduc-

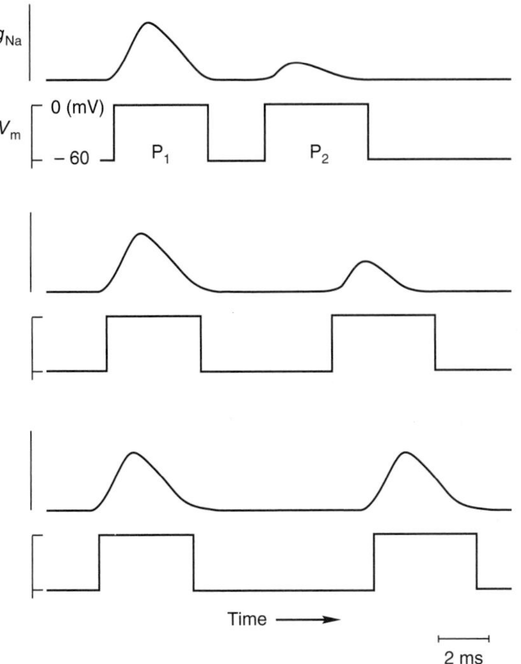

Figure 10–8 Sodium channels remain inactivated for a few milliseconds after the end of an activating pulse. Therefore if the interval between two depolarizing pulses (P_1 and P_2) is brief, the second pulse produces a smaller increase in g_{Na} because many of the Na^+ channels are still inactivated. The longer the interval between pulses, the greater the increase in g_{Na}, because a greater fraction of channels will have returned to the resting state when P_2 begins.

tances as functions of membrane potential and time. Using these equations and measured values for the passive properties of the axon, they computed the shape and the conduction velocity of the propagated action potential. The calculated wave form of the action potential matched the wave form recorded in the unclamped axon almost perfectly. This close agreement indicates that the voltage and time dependence of the active Na^+ and K^+ channels, calculated from the voltage-clamp data, accurately describe the properties of the channels that are essential for generating and propagating the action potential.

According to the Hodgkin–Huxley model, an action potential involves the following sequence of events. A depolarization of the membrane causes Na^+ channels to open rapidly (an increase in g_{Na}), resulting in an inward Na^+ current. This current, by discharging the membrane capacitance, causes further depolarization, thereby opening more Na^+ channels, resulting in increased inward current. This regenerative process causes the action potential. The depolarizing state of the action potential then limits the duration of the action potential in two ways: (1) it gradually inactivates the Na^+ channels, thus reducing g_{Na}, and (2) it opens, with some delay, the voltage-gated K^+ channels, thereby increasing g_K. Consequently, the Na^+ current is followed by an outward K^+ current that tends to repolarize the membrane.

In most nerve cells an action potential is followed by a transient hyperpolarization, the *afterpotential.* This brief increase in membrane potential occurs because the K^+ channels that open during the later phase of the action potential close some time after V_m has returned to its resting value. It takes a few milliseconds for all of the voltage-gated K^+ channels to return to the closed state. During this time, when the efflux of K^+ from the cell is greater than during the resting state, V_m is hyperpolarized slightly with respect to its normal resting value (Figure 10–7B).

The action potential is also followed by a brief period of diminished excitability, or refractoriness, which can be divided into two phases. The *absolute refractory period* comes immediately after the action potential; during this period it is impossible to excite the cell no matter how great a stimulating current is applied. This phase is followed directly by the *relative refractory period,* during which it is possible to trigger an action potential, but only by applying stimuli that are stronger than normal. These periods of refractoriness, which together last just a few milliseconds, are caused by the residual inactivation of Na^+ channels and opening of K^+ channels.

Another feature of the action potential predicted by the Hodgkin–Huxley model is its all-or-none behavior. A fraction of a millivolt can be the difference between a subthreshold depolarizing stimulus and a stimulus that generates a full-blown action potential. This all-or-none phenomenon may seem surprising when one considers that Na^+ conductance increases in a strictly *graded* manner as depolarization is increased (see Figure 10–5). With each increment of depolarization the number of voltage-gated Na^+ channels that switch from the closed to the open state increases, thereby causing a gradual increase in Na^+ influx. Why then is there an abrupt threshold for the generation of an action potential?

Although a small subthreshold depolarization increases the inward I_{Na}, it also increases two *outward* currents, I_K and I_l, by changing the electrochemical driving forces that determine their values. At the same time, the depolarization also causes a slow increase in g_K by gradually increasing the number of open K^+ channels (Figure 10–5). As I_K and I_l increase with depolarization, they tend to resist the depolarizing action of the Na^+ influx. However, the great voltage sensitivity and rapid kinetics involved in Na^+ channel activation ensure that the depolarization will eventually reach a point where the increase in inward I_{Na} exceeds the increase in outward I_K and I_l, and becomes regenerative. That point, the threshold, is the specific value of V_m at which the *net* ionic current ($I_{Na} + I_K + I_l$) just changes from outward to inward, depositing positive charge on the inside of the membrane, capacitance.

Variations in the Action Potential in Different Neurons Are Accounted for by Variations on the Basic Theme of the Hodgkin–Huxley Theory

Hodgkin and Huxley carried out their analysis on the squid, an invertebrate with an unusually large axon. To what degree does their model for the generation of action potentials apply to neurons in vertebrates or to the other components of the neuron—the cell body, dendrites, and presynaptic terminals? From studies designed to test the general applicability of the model, five fundamental conclusions have emerged.

The Basic Mechanism of Action Potential Generation Is the Same in All Neurons

Hodgkin and Huxley proposed that the action potential in the squid axon is caused by an inward membrane current followed by an outward current, and that the currents flow through voltage-gated membrane conductance channels. This mechanism of excitability appears to be universal. Dozens of different types of voltage-gated ion channels have now been described in both invertebrates and vertebrates. Although differences among ion channels do have important consequences for membrane excitability, the basic mechanism producing the all-or-none action potential is the same in all nerve and muscle cells in all species.

The Nervous System Expresses a Rich Variety of Voltage-Gated Ion Channels

The Na^+ and K^+ channels described by Hodgkin and Huxley in the squid axon have been found in almost every type of neuron examined. Nevertheless, many other kinds of channels have also been identified. To take just one example, most neurons contain voltage-gated Ca^{2+} channels that open in response to membrane depolarization. Although the calcium ion has a strong electrochemical gradient driving it into the cell, in most circumstances Ca^{2+} influx contributes only weakly to the rising phase of the action potential, since the Ca^{2+} conductance is much smaller than the Na^+ conductance. Some neurons also have voltage-gated Cl^- channels.

Each type of ion-selective channel has many variants. For example, several types of voltage-gated K^+ channels are found in neurons. They differ from each other in their kinetics of activation, voltage activation range, and sensitivity to various ligands. Four types of K^+ channel variants are particularly common in nervous systems. (1) The slowly activating channel described by Hodgkins and Huxley is called the *delayed rectifier*. (2) A *calcium-activated* K^+ *channel* is also activated by depolarization but its voltage sensitivity is a function of the intracellular Ca^{2+} concentration. Thus, to open this channel requires both a rise in internal Ca^{2+} (mediated by voltage-gated Ca^{2+} channels) and a depolarization. (3) A *fast, transient* K^+ *channel* is activated rapidly by depolarization, almost as rapidly as the Na^+ channel; it also inactivates rapidly, but only if the depolarization is maintained. (4) The *M-type* K^+ *channel* is activated very slowly by depolarization but inactivated by acetylcholine.

In addition, there are at least four types of voltage-gated Ca^{2+} channels and two types of voltage-gated Na^+ channels. Thus a single ion species can cross the membrane through several distinct types of ion channels, each with its own characteristic kinetics and voltage sensitivity.

Gating of Voltage-Sensitive Channels Can Be Influenced by Intracellular Ion Concentrations

In its most basic form a change in membrane potential involves the flow of ionic current through membrane channels, which leads to a change in the net charge stored on the membrane. This process does not result in appreciable changes in the intracellular concentrations of most ion species. However, in some neurons current flow through ion channels does lead to changes in the intracellular concentration of ions, and such changes have important modulatory influences on voltage-gated channels. The ion that most commonly has such a modulatory effect is Ca^{2+}. The concentration of free Ca^{2+} in the cytoplasm of a resting cell is extremely low, about 10^{-7} *M*, which is several orders of magnitude below that for Na^+, Cl^-, or K^+. For this reason the intracellular Ca^{2+} concentration is particularly likely to increase significantly as the result of current flow through Ca^{2+} channels in the membrane.

The increase in Ca^{2+} concentration affects many cellular mechanisms. Even the minute amount of Ca^{2+} that enters the cell during a single action potential may saturate the Ca^{2+} buffering systems of the cell briefly. When this occurs, the transient increase in Ca^{2+} concentration near the inside of the membrane increases the probability that calcium-sensitive K^+ channels will open. A train of action potentials will have an even more significant effect on these channels. Some Ca^{2+} channels are themselves sensitive to levels of intracellular Ca^{2+} and are inactivated when incoming Ca^{2+} binds to their internal surfaces. In other channels the influx of Ca^{2+} activates a calcium-sensitive protein phosphatase, calcineurin, which dephosphorylates the channel, thereby inactivating it.

Thus, in some cells the Ca^{2+} influx during an action potential can have opposing effects—the positive charge that it carries into the cell contributes to the regenerative depolarization, while the increase in Ca^{2+} concentration results in opening more K^+ channels and turning off Ca^{2+} channels. Because of these two effects, outward ionic current increases, inward ionic current decreases, and the cell repolarizes as the net efflux of positive charge increases. Thus, the influx of Ca^{2+} through voltage-gated Ca^{2+} channels is limited by two processes that aid repolarization—an increase in K^+ efflux and a decrease in Ca^{2+} influx.

Excitability Properties Vary Among Neurons

Although the function of each neuron is determined to a great extent by its position in a specific circuit, biophysical properties also play a role. How a neuron responds to synaptic input is determined by the proportions of different types of voltage-gated channels in the cell's integrative and trigger zones. Some cells respond to a constant excitatory input with a decelerating train of action potentials, others respond with an accelerating train, and still others maintain a constant firing frequency. In certain neurons small changes in the strength of synaptic inputs produce a large increase in firing rate, whereas in others the firing rate responds only to large changes in synaptic input. In most neurons a steady hyperpolarizing input makes the cell less responsive to excitatory input by reducing the resting inactivation of the fast, transient voltage-gated K^+ channels. In other neurons such a steady hyperpolarization makes the cell *more* excitable, because it removes the inactivation of a particular class of voltage-gated Ca^{2+} channels.

Only two types of voltage-gated ion channels are required to generate an action potential, but there is great diversity in the molecular properties of these channels in different types of neurons and, therefore, an enormous range of excitability properties among neurons. As a result, different types of neurons respond to the same synaptic input with different temporally patterned spike trains, which in turn results in unique patterns of synaptic output.

Excitability Properties Vary Within Regions of the Neuron

In addition to the variation in the proportions of specific types of ion channels among neurons, the spatial distribution of channel types within individual cells also varies. These regional variations have a direct bearing on function. For example, the dendrites, cell body, axon hillock, and nerve terminals contain a greater variety of channels than does the axon. This may reflect the fact that the input and output zones actively transform the signals they receive, whereas the axon is a simple relay line between the input and output regions.

Voltage-Gated Channels Have Characteristic Molecular Properties

The empirical equations derived by Hodgkin and Huxley successfully describe how the flow of ions through the Na^+ and K^+ channels generates the action potential. But while these equations deal with the process of excitation primarily in terms of changes in membrane conductance and current flow, they tell us little about the molecular structure of the voltage-gated conductance channels and the mechanisms by which they are activated. Fortunately, technical advances such as those described in Chapter 7 now make it possible to examine the structure and function of the voltage-gated Na^+, K^+, and Ca^{2+} channels in detail at the molecular level.

Voltage-Gated Sodium Channels Are Sparsely Distributed

The task of characterizing the distribution of the voltage-gated Na^+ channels has been aided greatly by the availability of several naturally occurring neurotoxins that bind tightly to the channel and therefore can be used as specific probes for labeling channel molecules. These include tetrodotoxin from the puffer fish, saxitoxin from a dinoflagellate that infects shellfish, batrachotoxin from South American poisonous frogs, and the venom from the North African scorpion. For example, studies show that radiolabeled tetrodotoxin molecules bind to a small number of specific sites on the axon membrane. These sites are thought to be the Na^+ channels themselves, because the binding constant and the kinetics of tetrodotoxin binding correspond to the values for tetrodotoxin blockade of Na^+ conductance, as determined by electrophysiological measurement.

The number of voltage-gated Na^+ channels can be estimated by measuring the total amount of tetrodotoxin bound when specific axonal binding sites are saturated. The density of channels is quite low in nonmyelinated axons, ranging in different cell types from 35 to 500 Na^+ channels per square micrometer of axon membrane. Even when the thick channel wall surrounding the pore is included, the channel area taken up by 500 Na^+ channels per square micrometer accounts for less than 1% of the total membrane area.

Despite this small number, however, quite large Na^+ currents can flow during the action potential because the ion flux through each channel is quite high. Patch-clamp recordings demonstrate that a single Na^+ channel can pass up to 10^7 Na^+ ions per second! These studies also showed that the greater the density of Na^+ channels in the membrane of an axon, the greater the velocity at which the axon conducts action potentials. This result is to be expected. When the cell is excited, the greater density of voltage-gated Na^+ channels allows more current to flow through the excited membrane and along the axon core, thus rapidly discharging the capacitance of the unexcited membrane downstream (see Figure 9–6).

Opening of Voltage-Gated Channels Is All-or-None

The current flow through a single channel cannot be measured in ordinary voltage-clamp experiments for two reasons. First, the voltage clamp acts on a large area of membrane in which thousands of channels are opening and closing randomly. Second, the background noise caused by the flow of current through passive membrane channels is much larger than the flow of current through any one channel. Both these problems can be circumvented by electrically isolating a tiny piece of membrane in a patch-clamp electrode (see Box 7–2).

Patch-clamp experiments demonstrate that voltage-gated channels generally have two conductance states, open and closed. Each channel opens in an all-or-none fashion and, when open, permits a pulse of current to flow with a variable duration but constant amplitude. The conductance of a single Na^+ channel in the open state ranges from 8 to 18 pS, depending on the type of channel. The conductances of individual voltage-gated K^+ channels range from about 4 to 20 pS; for single Ca^{2+} channels they range from about 1 to 3 pS.

Charges Are Redistributed Within Voltage-Gated Sodium Channels When the Channels Open

In their classic study of the squid axon, Hodgkin and Huxley suggested that the conductance of Na^+ and K^+ channels may be regulated by a conformational change in an intramembranous gating molecule, which in turn could open the activation gate. They postulated that the gating molecule has a net charge, the *gating charge*, somewhere within its structure. Movement of this net charge to another site in response to a change in membrane potential could change the conformation of the molecule.

Hodgkin and Huxley predicted that when the membrane is depolarized, a positive gating charge would move from near the inner surface to near the outer surface. Such a displacement of positive charge would reduce the net separation of charge across the membrane and hence tend to hyperpolarize the membrane. Therefore, to keep the membrane potential constant in a voltage-clamp experiment, a small extra component of outward capacitive current, called *gating current*, would have to be generated by the voltage clamp. When in the 1970s the membrane current was examined by means of very sensitive techniques, the predicted gating current was found to flow at the beginning and end of a depolarizing voltage clamp step prior to the opening of the Na^+ channels.

Analysis of the gating current provides two critical insights into the properties of the Na^+ channel. (1) *Gating is a multistep process*. Several steps, with different kinetics, must occur before the channel opens in response to depolarization. (2) *Activation and inactivation are coupled processes*. During a short depolarizing pulse, net movement of gating charge within the membrane at the beginning of the pulse is balanced by an opposite movement of gating charge at the end of the pulse. However, if the pulse lasts long enough for Na^+ inactivation to take place, the movement of gating charge back across the membrane at the end of the pulse is delayed. The gating charge is thus temporarily immobilized; only as the Na^+ channels recover from inactivation is the charge free to move back across the membrane. This charge immobilization means that the activation gate cannot close while the channel is in the inactivated state (i.e., while the inactivation gate is closed).

The Voltage-Gated Sodium Channel Selects for Sodium on the Basis of Size, Charge, and Energy of Hydration of the Ion

After the gates of the Na^+ channel have opened, how does this protein channel discriminate between Na^+ and other ions? The channel's selectivity can be assessed by measuring its relative permeability to several types of organic and inorganic cations that differ in size and hydrogen bonding characteristics. The channel behaves as if it contains a filter or recognition site that selects partly on the basis of size, by acting as a molecular sieve, with a pore size of 0.3×0.5 nm (see Figure 7–3). The ease with which ions with good hydrogen bonding characteristics pass through the channel suggests that part of the inner wall of the protein channel is made up of polar or charged amino acid residues that can substitute for water (e.g., glutamic acid, aspartic acid, serine, and threonine). When the pH of the fluid surrounding the cell is lowered, the conductance of the open channel is gradually reduced, consistent with the titration of negatively charged carboxylic acid residues.

The pore of the channel is thought to be made up of four transmembrane domains within the molecule that are quite similar in structure (see below). A glutamic acid residue is situated at equivalent points in two of these domains. (A lysine and an alanine are situated in the other two domains.) The channel is thought to select Na^+ ions by the following mechanism. The negatively charged carboxylic acid groups of the glutamic acid residues, which are located at the outer mouth of the pore, perform the first step in the selection process by attracting cations and repelling anions. Cations that are larger than 0.3×0.5 nm in diameter are too bulky to pass through the pore. Cations smaller than this critical size pass through the pore, but only after losing most of the waters of hydration they normally carry free in solution. The negative carboxylic acid group, as well as the oxygen atoms that line the pore, can substitute for these waters of hydration, but the degree of effectiveness of this substitution varies among ion species. The more effective the substitution, the more readily that ion permeates the Na^+ channel. The lysine and alanine residues also contribute to the selectivity of the channel. When these residues are changed by site-directed mutagenesis, the Na^+ channel can act as a Ca^{2+} channel!

Potassium, Sodium, and Calcium Channels Stem from a Common Ancestral Gene

The fact that a change in two amino acid residues in two of the four membrane-spanning domains that form the Na^+ channel's pore can cause the channel to behave as a Ca^{2+} channel provides a hint that Na^+ and Ca^{2+} channels are closely related. Detailed molecular studies have revealed that K^+, Na^+, and Ca^{2+} channels share several functionally important domains. Indeed, there is now strong evidence from studies of plants, invertebrates, and vertebrates that all cation channels stem from a

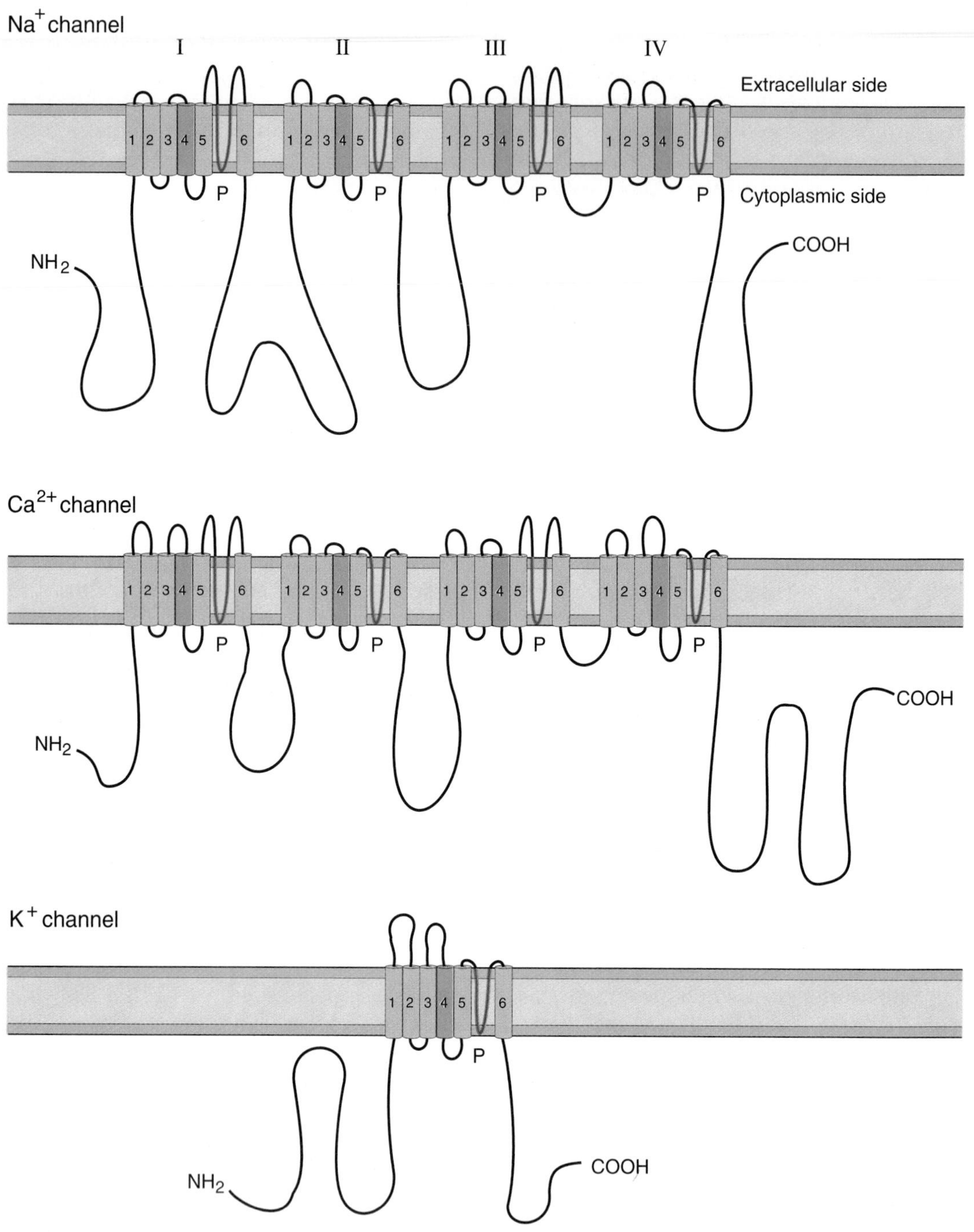

Figure 10–9 The basic sequence of the α subunit, which makes up the core of the voltage-gated Na^+ channel, and the corresponding segments of the voltage-gated Ca^{2+} and K^+ channels, is thought to have the following secondary structures in the membrane. The α subunit of the Na^+ and Ca^{2+} channels consists of a single polypeptide chain with four repetitions of six membrane-spanning α-helical regions. The S4 region, the fourth membrane-spanning α helical region, is thought to be the voltage sensor. A stretch of amino acids, the P region between α-helices 5 and 6, dips into the membrane in the form of two strands. A fourfold repetition of the P region is believed to line the pore. The K^+ channel, by contrast, has only a single copy of the six α-helices and the *P* region. Four such subunits are assembled to form a complete channel. (Adapted from Catterall, 1988; Stevens, 1991.)

common ancestral channel—perhaps a K^+ channel—that can be traced to a single-cell organism living over 1400 million years ago, before the evolution of separate plant and animal kingdoms. The amino acid sequences conserved through evolution have helped us to identify the domains within contemporary cation channels that are critical for function.

Molecular studies of the voltage-dependent cation channels began with the identification of Na^+ channel molecules. Three subunits have been isolated: one large glycoprotein (α) and two smaller polypeptides ($\beta1$) and ($\beta2$). The α subunit apparently is ubiquitous, and insertion of this subunit into an artificial lipid bilayer reconstitutes the function of the purified Na^+ channel as shown by patch-clamp recordings. The α subunit therefore is presumed to form the aqueous pore of the channel. The smaller subunits appear variably in different tissues and different species and are thought to play a regulatory role.

Examination of the amino acid sequence encoded by the cloned Na^+ channel α-subunit gene reveals two fundamental features of the structure of the Na^+ channel. First, the ion-conducting portion comprises four internal repetitions (sequences I–IV), with only slight variations, of an amino acid sequence motif that is approximately 150 amino acids in length. Each of these motifs has been interpreted as having six membrane-spanning hydrophobic domains (S1–S6) and a seventh hydrophobic domain (the P segment) that dips into and out of the membrane (Figure 10–9). The four repeated versions of the motif are thought to be arranged roughly symmetrically, with the walls of the water-filled pore being formed by either one or two of the membrane domains, repeated four times around the circumference of the pore (Figure 10–10).

Second, one of the six putative membrane-spanning regions, the S4 region, is similar in the Na^+ channels of many different species. Such strict conservation suggests that the S4 region has a critical role in Na^+ channel function. Moreover, the S4 region of the Na^+ channel is also similar to corresponding regions of the voltage-gated Ca^{2+} and K^+ channels (Figure 10–9) but is lacking in certain K^+ channels that are not voltage dependent. Because it is present in all three channels that are voltage gated, the S4 region may be the voltage sensor, that part of the protein that transduces a change in membrane potential into a gating transition within the channel that opens the activation gate. This idea is supported by several findings. Although the S4 region is hydrophobic it has a relatively high density of charged amino acid residues along the length of the postulated helical region; every third amino acid along the helix has a net positive charge. The conformation of such a highly charged structure is therefore likely to be quite sensitive to changes in the electric field across the membrane. Reducing the net positive charge in one of the S4 regions of the channel lowers the voltage sensitivity of the entire activation gate.

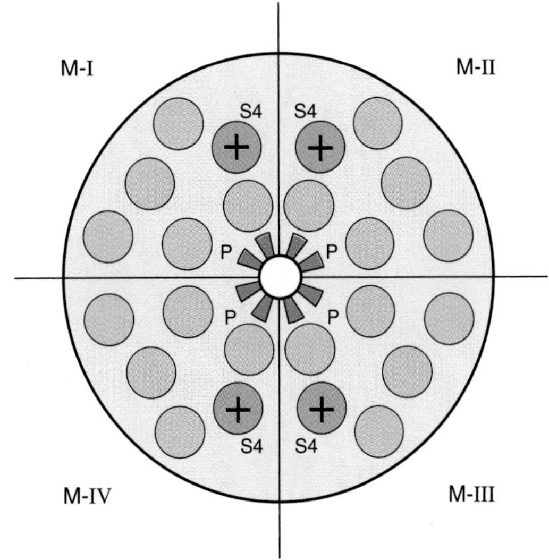

Figure 10–10 The tertiary structure of the voltage-gated Na^+ and Ca^{2+} channels proposed here is based on the secondary structures shown in Figure 10–9. The pore (gray circle) is surrounded by the four internally repeated domains (M-I to M-IV). Each quadrant of the channel includes six cylinders, which represent the six putative membrane-spanning α-helices. The S4 segment is thought to be involved in gating because it is the site of the net charge in the channel. The two central figures in each quadrant represent the pair of strands (the P region) that dip into the membrane to form the wall of the pore. (Adapted from Alsobrook and Stevens, 1988; Stevens, 1991.)

One hypothesis for how the S4 region might control the activation gate of the Na^+ channel

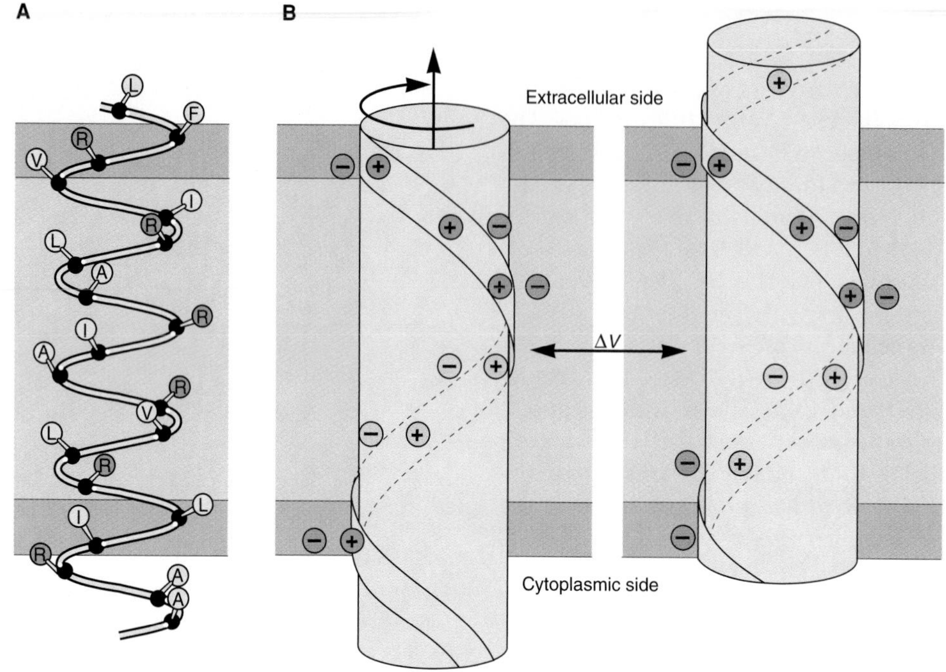

Figure 10–11 Gating of the Na^+ channel is thought to rely on certain specific movements of the S4 region.

A. Ball-and-stick model of the S4 region. The "R" stands for positively charged arginine residues.

B. In the resting state each net positive charge on the region (α-helix) is stabilized by a negative charge on a neighboring portion of the molecule. When the cell is depolarized, the change in electrical field across the membrane allows the positive charges on the S4 region to move toward the extracellular face of the membrane. It is postulated that this movement is translated into a screw-type movement, which stops when each positive charge on S4 has gone through about a 60° turn and is again in register with a stationary negative charge on an adjacent helix. The realignment of positive charges on the helix with negative charges in an adjacent region of the channel stabilizes the channel in a new conformation. (From Catterall, 1988.)

is derived from the analysis of gating current described above. According to this view, the S4 region forms an α-helix (a common secondary structure for proteins). The regularly spaced positive charges on a mobile S4 helix align with immobile, negatively charged residues on adjacent regions of the membrane-spanning portions of the channel protein. When the cell is depolarized, the increase in positivity within the cell causes the positive charges on the S4 region to move outward. This movement results in a screwlike rotation as each positive charge on the S4 region turns and moves closer to the outside of the membrane, to a position where it is stabilized by its electrostatic attraction to the neighboring negative charge next in line (Figure 10–11). This type of internal redistribution of charge could also account for the gating currents recorded in voltage-clamp experiments when the activation gates open or close. It is thought that before the activation gate can open, all four S4 regions in the channel must undergo the type of conformational change illustrated in Figure 10–11.

Cleaving the region of the Na^+ channel that connects the repeating sequences III and IV on the cytoplasmic face of the membrane slows the rate of inactivation of the Na^+ channel without altering channel activation. An antibody directed against the same region of the channel also slows inactivation. Thus, this cytoplasmic segment of the molecule may move into position to block the inner mouth of the pore after the activation gate has opened, thereby causing inactivation.

The genes encoding the major α subunits of several voltage-gated Ca^{2+} channels have also been cloned. Similar to the voltage-gated Na^+ channels, these Ca^{2+} channels are composed of four repeats, each with seven hydrophobic domains that are structurally homologous to the voltage-gated Na^+ channel.

The genes for several subfamilies of K^+ channels have been cloned in both *Drosophila* and the mammalian brain. One subfamily encodes the voltage-dependent A-type K^+ channel. Another subfamily encodes the delayed rectifier K^+ channel described by Hodgkin and Huxley. Unlike the voltage-gated Na^+ and Ca^{2+} channels, the K^+ channel contains only a single repeat. However, this repeat is structurally homologous to a single repeat of the Na^+ and Ca^{2+} channels. A third subfamily encodes a voltage-independent K^+ channel that is thought to contribute to the resting potential. This channel lacks the S4 region.

The diversity of K^+ channel types in mammals is thought to have been generated primarily by gene duplication and mutation. Because the K^+ channel has only one of the four internally repeated motifs found in the Na^+ and Ca^{2+} channels, the functioning K^+ channel is thought to be formed by four similar, perhaps identical, subunits that aggregate around a central pore. The homology among the voltage-gated Na^+, Ca^{2+}, and K^+ channels suggests that all three channels belong to the same gene family and have evolved from a common ancestral structure, presumably a K^+ channel.

An Overall View

The action potential is produced by the movement of ions across the membrane through voltage-gated channels. This movement, which occurs only after the channels are opened, changes the distribution of charges on either side of the membrane. An influx of Na^+, and in some cases Ca^{2+}, depolarizes the membrane; an efflux of K^+ then repolarizes the membrane by restoring the initial charge distribution. A subset of voltage-gated ion channels are opened primarily when the membrane potential is near the action potential threshold, and thus these channels have profound effects on the firing patterns generated by the neuron.

Three major technical advances have led to detailed explanations of the action of the voltage-gated channels. First, the voltage-clamp technique has been extended to allow the analysis of gating currents and patch-clamp recording from single channels. Second, a combined biochemical, genetic, and molecular biological approach has revealed the nucleotide sequence of these channels. Third, a concerted effort involving biophysical, structural, biochemical, and molecular biological approaches is leading to a comprehensive understanding of how these channels function.

Selected Readings

Catterall, W. A. 1988. Structure and function of voltage-sensitive ion channels. Science 242: 50–61.

Hille, B. 1992. Ionic Channels of Excitable Membranes, 2nd ed. Sunderland, Mass.: Sinauer.

Hodgkin, A. 1992. Chance & Design: Reminiscences of Science in Peace and War. Cambridge, England: Cambridge University Press.

Isom, L. L., De Jongh, K. S., and Catterall, W. A. 1994. Auxiliary subunits of voltage-gated ion channels. Neuron 12:1183–1194.

Jan, L. Y., and Jan, Y. N. 1990. How might the diversity of potassium channels be generated? Trends Neurosci. 13:415–419.

Llinás, R. R. 1988. The intrinsic electrophysiological properties of mammalian neurons: Insights into central nervous system function. Science 242:1654–1664.

IV

SIGNALING BETWEEN NERVE CELLS

About the image on the preceding page:

This portrait of Charles Sherrington and John Eccles by the British painter Stan Smith, hangs in the Laboratory of Physiology at the University of Oxford, which also preserves Sherrington's library, equipment, and other memorabilia. Sherrington coined the word *synapse* and initiated the experimental study of spinal reflexes as paradigms of neural integration. Eccles, his last student, was the first to study chemical synaptic transmission in the brain through intracellular microelectrode recording of single nerve cells. He stressed the importance of approaching the brain through nerve cells, its elementary units. (Reproduced with the permission of Stan Smith and Professor Colin Blakemore, University of Oxford.)

Introduction

In our examination of the cell biology of the neuron we have seen that neurons use two classes of channels for signaling *within* the cell: resting channels that generate the resting potential and voltage-gated channels that produce the action potential. We now turn to synaptic transmission, the process for signaling between cells. Here we will learn about two additional classes of channels: gap-junction channels and ligand-gated channels.

During development individual neurons make highly specific synaptic connections with other neurons. One of the key ideas that we shall develop in later sections of this book is that the specificity of synaptic connections between neurons established during development underlies perception, action, emotion, and learning.

On average a neuron makes about 1,000 synaptic connections and receives many more. Thus, the human brain, which contains about 10^{11} neurons, forms about 10^{14} connections! Fortunately, we can achieve a general understanding of synaptic transmission at all of these connections by examining only a few synapses in detail.

The synapse, the site of transmission between two interacting cells, is made up of three elements: the presynaptic terminal, the postsynaptic cell, and a zone of apposition. Based on their zones of apposition, synapses are categorized into two major groups: electrical and chemical. At electrical synapses the synaptic cleft is small, and the current generated by the action potential in the presynaptic neuron flows directly into the postsynaptic cell through specialized bridging channels called *gap-junction* channels. At chemical synapses the synaptic cleft is larger and there are no bridging channels. Here a change in the membrane potential in the presynaptic cell leads to the release of chemical transmitter from the presynaptic terminals. The chemical transmitter diffuses across the synaptic cleft and binds to receptor molecules on the postsynaptic cell membrane, thus opening ion channels through which currents flows.

Transmitter receptor can be classified into two major groups depending on how the receptor controls the ion channel in the postsynaptic cell. One type forms a common structure with the ion channel and thus controls the channel *directly*. The second type acts indirectly on the ion channel by activating a second messenger as well as other molecular reactions within the postsynaptic cell. Within each of these groups there are many types due to different combinations of molecular isoforms.

Both groups of receptors give rise to excitation and inhibition. The sign of the synaptic action depends on the action of the receptor with which the transmitter interacts, not on the properties of the transmitter. The same transmitter can produce different actions in different subtypes of receptors. For example, glutamate is the predominant excitatory chemical transmitter in the brain, but produces different actions by binding to a variety of glutamate receptors. Thus a relatively limited variety of transmitters is able to produce a wide variety of synaptic actions.

Because the structure and molecular architecture of chemical synapses are complex, many diseases may affect chemical synaptic transmission. One particular disorder of the chemical synapse that we consider in this section is myasthenia gravis, a disease that disrupts cholinergic transmission at the synapses between motor neurons and muscle.

11

An Introduction to Synaptic Transmission

Synapses Are Either Electrical or Chemical

Electrical Synapses Provide Instantaneous Signal Transmission

Electrical Transmission Allows for the Rapid and Synchronous Firing of Interconnected Cells

Gap-Junction Channels Connect Communicating Cells at an Electrical Synapse

Chemical Synapses Can Amplify Signals

Chemical Transmitters Bind to Postsynaptic Receptors

Postsynaptic Receptors Gate Ion Channels Either Directly or Indirectly

Nerve cells are the elementary signaling units of the nervous system. In the preceding section of this book we examined the resting potential, electrotonic transmission, impulse generation, and conduction. These mechanisms are essential for the intracellular signals that are propagated *within* the neuron, from the cell body and its dendrites to the axon terminal. In this section we consider signaling *between* neurons, the mechanisms at the synapse where one neuron communicates with another. The mechanisms of synaptic transmission are the basis for many of the processes we consider later in this book: perception, voluntary movement, and learning.

Of all the cells in the body, nerve cells have the special ability to communicate rapidly with one another over great distances and with great precision. The average neuron forms about 1000 synaptic connections and it receives even more, perhaps 10,000 connections. Since the human brain contains at least 10^{11} neurons, this means that about 10^{14} synaptic connections are formed in the brain. Thus, there are more synapses in one human brain than there are stars in our galaxy! Despite this vast number of connections, synaptic transmission throughout the nervous system makes use of only two basic mechanisms: electrical transmission and chemical transmission.

Electrical transmission is not unique to the nervous system; it is also found in the heart and smooth muscle as well as the epithelium of the liver. In the brain electrical transmission is rapid and rather stereotyped. Electrical synapses are used primarily to send simple depolarizing signals; they do not readily allow inhibitory actions or long-lasting changes in effectiveness.

In contrast, synapses that use chemical transmission can mediate either excitatory or inhibitory actions. Chemical synapses are more flexible and tend to produce more complex behaviors than do electrical synapses. Because they are capable of undergoing changes in effectiveness, synapses of this kind have a *plasticity* that is important for memory and other higher functions of the brain. Especially important, chemical synapses can amplify neuronal signals, allowing a small presynaptic nerve terminal to alter the potential of a large postsynaptic cell. Because chemical transmission is central to understanding how the nervous system works—how we perceive, move, feel, learn, and remember—we examine the mechanisms of chemical transmission in detail in Chapters 12, 13, and 14.

Synapses Are Either Electrical or Chemical

At the turn of the century Charles Sherrington introduced the term *synapse* for the specialized zone of contact at which one neuron communicates with another; this site had first been described histologically by Ramón y Cajal. In the 1920s Otto Loewi showed that a chemical compound, acetylcholine, conveys signals from the vagus nerve to the heart. Loewi's discovery provoked considerable debate in the 1930s over how chemical signals could generate electrical activity at nerve–muscle synapses and in the brain. Two schools of thought emerged, one physiological and the other pharmacological. Each championed a single mechanism for all synaptic transmission. The physiologists, led by Sherrington's last student, John Eccles, argued that synaptic transmission is basically electrical: the conduction of the action potential results from the passive flow of current from the presynaptic neuron to the postsynaptic cell. The pharmacologists, led by Henry Dale, argued that transmission is chemical: a chemical substance released by the presynaptic neuron initiates the flow of current in the postsynaptic cell.

Later, when physiological techniques improved in the 1950s and 1960s, it became clear that both kinds of transmission exist. Although most synapses use a chemical transmitter, some operate purely by electrical means. Once the fine structure of synapses was made visible with the electron microscope, chemical and electrical synapses were found to have different morphologies. At chemical synapses there is no cytoplasmic continuity between the cells; instead, the neurons are separated by a small space, the *synaptic cleft*. In contrast, at electrical synapses special ion channels, the gap-junction channels, in the pre- and postsynaptic cell membranes serve as bridges between the cytoplasm of the two cells.

The main functional properties of the two types of synapses are summarized in Table 11–1. Many of the most important differences can be observed by injecting current into the presynaptic cell. In both

Table 11–1 Distinguishing Properties of Electrical and Chemical Synapses

Type of Synapse	*Distance Between Pre- and Postsynaptic Cell Membranes*	*Cytoplasmic Continuity Between Pre- and Post-synaptic Cells*	*Ultra-structural Components*	*Agent of Transmission*	*Synaptic Delay*	*Direction of Transmission*
Electrical	3.5 nm	Yes	Gap-junction channels	Ionic current	Virtually absent	Usually bidirectional
Chemical	20–40 nm	No	Presynaptic vesicles and active zones; postsynaptic receptors	Chemical transmitter	Significant: at least 0.3 ms, usually 1–5 ms or longer	Unidirectional

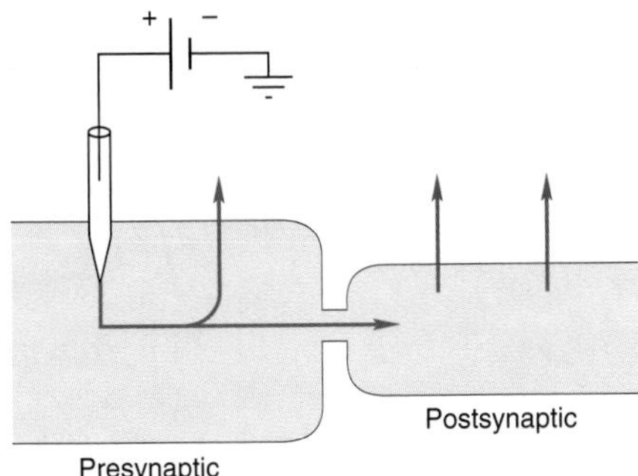

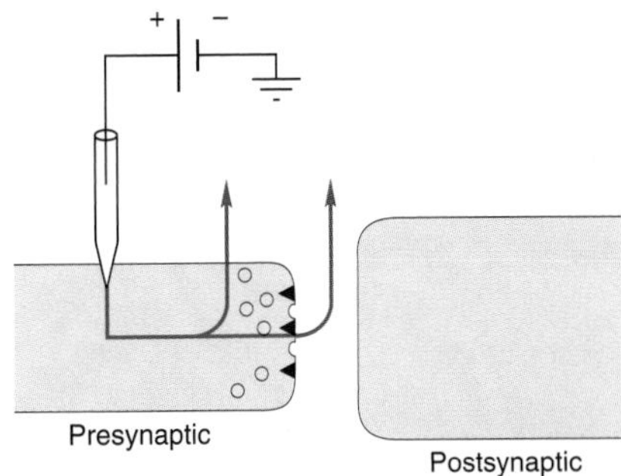

Figure 11–1 Current flows differently at electrical and chemical synapses.

A. At an electrical synapse some of the current injected into a presynaptic cell escapes through resting channels. In addition, some current flows into the postsynaptic cell through specialized ion channels, called gap-junction channels, that connect the cytoplasm of the pre- and postsynaptic cells.

B. At chemical synapses all of the injected current escapes through channels in the presynaptic cell. The resulting depolarization of the cell activates the release of molecules of neurotransmitter packaged in synaptic vesicles (**open circles**), which then bind to receptors on the postsynaptic cell; this action depolarizes or hyperpolarizes the postsynaptic cell.

types of synapses the current flows outward across the presynaptic cell membrane. In so doing, the current deposits a positive charge on the inside of the presynaptic cell membrane, reducing its negative charge and thereby depolarizing the cell (see Chapter 8).

At electrical synapses some current also flows through the channels that connect the pre- and postsynaptic cells. These channels have low resistance and high conductance, and the current flowing through them from the presynaptic neuron deposits a positive charge on the inside of the membrane of the postsynaptic cell, depolarizing it. The current then flows out through postsynaptic resting channels (Figure 11–1). If the depolarization exceeds threshold, voltage-gated channels in the postsynaptic cell open and generate an action potential.

At chemical synapses the outward current injected into a presynaptic cell simply flows out of the resting channels of the presynaptic neuron into the synaptic cleft, seeking the path of lowest resistance. Little or no current crosses the external membrane of the postsynaptic cell, which has a high resistance (Figure 11–1B). Instead, the action potential in the presynaptic neuron of a chemical synapse leads to the release of a chemical transmitter substance; this transmitter diffuses across the synaptic cleft to interact with specific receptors that either depolarize or hyperpolarize the postsynaptic cell.

Electrical Synapses Provide Instantaneous Signal Transmission

Electrical synapses usually occur between a large presynaptic nerve fiber and a small postsynaptic neuron, because a great deal of current is required to depolarize the postsynaptic cell. This current must be generated directly by the voltage-gated ion channels of the presynaptic cell. These channels not only have to depolarize the presynaptic cell and initiate an action potential, but also must generate sufficient ionic current to produce a potential change in the postsynaptic cell. To accomplish this the presynaptic terminals have to be large. At the same time, the postsynaptic cell has to be small, since presynaptic current will produce a larger voltage change across the high resistance of a small cell than across the low resistance of a large cell.

Electrical synaptic transmission was first described at the giant motor synapse of the crayfish, where the presynaptic fiber is in fact much larger than the postsynaptic fiber. An action potential

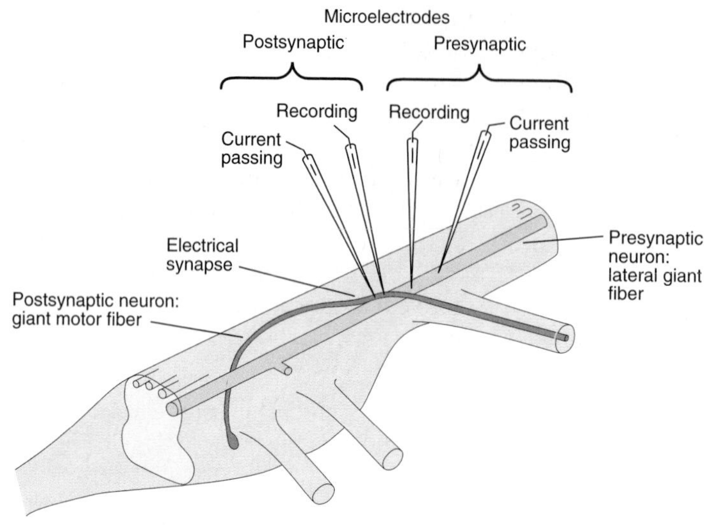

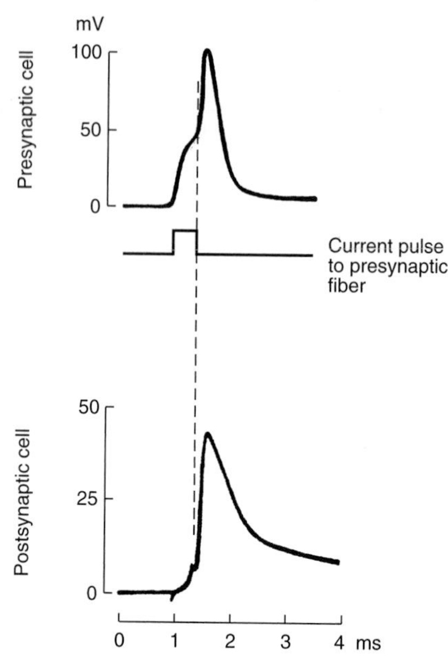

Figure 11–2 Electrical synaptic transmission was first demonstrated at the giant motor synapse in the crayfish.

A. The presynaptic neuron is the lateral giant fiber running down the nerve cord. The postsynaptic neuron is the motor fiber, which projects from the cell body in the ganglion to the periphery. The electrodes for passing current and for recording are placed within both the pre- and postsynaptic cells.

B. Transmission at an electrical synapse is virtually instantaneous—the postsynaptic response follows presynaptic stimulation in a fraction of a millisecond. (See Figure 11–8 for comparison with chemical synapses.) The dotted line shows how the responses of the two cells correspond in time. (Adapted from Furshpan and Potter, 1957 and 1959.)

generated in the presynaptic fiber produces a depolarizing postsynaptic potential that is often large enough to discharge an action potential.

The latency between the presynaptic spike and the postsynaptic potential is remarkably short, in fact almost nonexistent (Figure 11–2). This short latency is incompatible with chemical transmission, a process that requires several steps—the release of a transmitter from the presynaptic neuron, diffusion of the transmitter to the postsynaptic cell, binding to a specific receptor, and subsequent gating of ion channels (all of which are described later in this chapter). Rather, the near-instantaneous transmission at this synapse can only be the result of electric current flowing directly from the presynaptic to the postsynaptic cell.

Further proof for electrical transmission at this synapse came from the finding that a full action potential in the presynaptic cell is not necessary to trigger a response in the postsynaptic cell. Even when a subthreshold depolarizing current is injected into the presynaptic neuron, the injected current flows into the postsynaptic cell and depolarizes it (Figure 11–3).

At most electrical synapses the change in potential of the postsynaptic cell is directly related to the size and shape of the change in potential of the presynaptic cell (Figure 11–3). For example, a presynaptic action potential that has a large hyperpolarizing afterpotential will produce a biphasic (depolarizing–hyperpolarizing) change in potential in the postsynaptic cell. Transmission at electrical synapses is similar to the passive electrotonic propagation of the local subthreshold electrical signals along axons (see Chapter 9) and therefore is often called *electrotonic transmission*. Electrotonic trans-

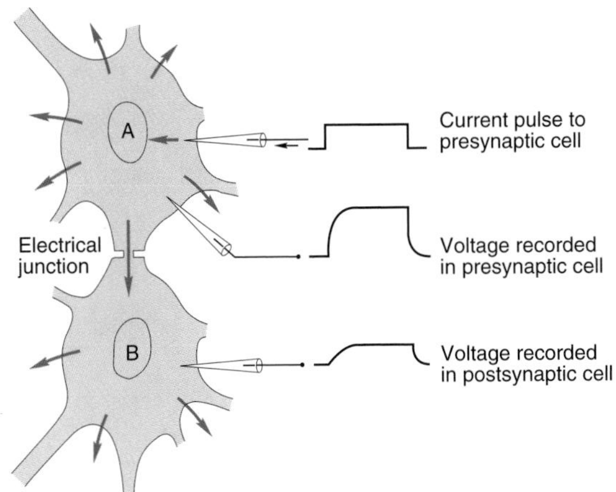

Figure 11–3 Electrical transmission occurs even with subthreshold input. This can be demonstrated by depolarizing the presynaptic cell with a small outward current pulse. Current is passed by one electrode and the membrane potential is recorded with a second electrode. Outward (depolarizing) current is indicated by upward deflections in recordings. A subthreshold depolarizing stimulus causes a passive depolarization in the presynaptic and postsynaptic cells.

mission has been observed even at junctions where the pre- and postsynaptic elements are of similar size. Because of their passive properties, such electrical synapses are often *bidirectional*: they will transmit a depolarization equally well from either the presynaptic to postsynaptic cell or from the postsynaptic to the presynaptic cell.

Electrical Transmission Allows for the Rapid and Synchronous Firing of Interconnected Cells

Why have electrical synapses? As we have seen, transmission across electrical synapses is extremely rapid because it results from the direct flow of current from the presynaptic neuron to the postsynaptic cell. This speed is important for certain responses that mediate escape from danger. For example, the tail-flip response of goldfish is mediated by a specific giant neuron in the brain stem (known as Mauthner's cell), which receives input at electrical synapses with sensory neurons. These electrical synapses bring about the rapid depolarization of Mauthner's cell, which in turn activates the motor neurons of the tail through a chemical synapse, leading to a rapid escape.

Entire groups of neurons can be interconnected by electrical synapses. In such arrangements the synaptic current flowing across the membrane of one cell will also flow into and out of the other electrically coupled cells. In this way several small electrically coupled cells can act as one large cell. Because of the electrical coupling between the cells, the effective resistance of each of the coupled neurons is decreased. From Ohm's law ($\Delta V = \Delta I \times R$) we can see that the lower the resistance (R) of a neuron, the smaller the depolarization (ΔV) produced by an excitatory synaptic current (ΔI). Thus, electrically coupled cells have a high threshold and it is difficult to cause them to fire action potentials. Once this high threshold is surpassed, however, a group of electrically coupled cells tends to fire synchronously.

Behaviors mediated by a group of electrically coupled cells have an important adaptive advantage: once the high threshold is reached the behavior is triggered explosively in an all-or-none manner. For example, when seriously perturbed, the marine snail *Aplysia* releases a massive amount of purple ink that provides a protective screen. This stereotypic all-or-none behavior is mediated by three electrically coupled, high-threshold motor cells that innervate the ink gland. Once the threshold is exceeded, these motor cells fire synchronously (Figure 11–4). In certain fish, rapid eye movements (saccades) are also mediated by the synchronized actions of electrically coupled motor neurons.

Besides providing speed or synchrony in neuronal signaling, electrical synapses may transmit *metabolic signals* between cells. The diameter of the pore in the gap-junction channels of electrical synapses is relatively large, 1.5 nm. Single-channel recordings indicate that gap-junction channels have a large conductance of 100 pS. As a result these channels are nonselective; they readily allow the flow through gap junctions of all inorganic cations and anions. In fact, gap-junction channels are large enough to allow moderate-size organic compounds—such as the second messengers IP_3 (inositol triphosphate), cyclic AMP, and short peptides—to pass from one cell to the next. This form of communication may be particularly important during development.

Figure 11–4 A group of electrically coupled motor neurons firing together is an effective system for producing instantaneous, all-or-none behaviors. Such a system is used by the marine snail *Aplysia* to release a protective cloud of ink when the animal is perturbed.

A. Noxious stimulation of the tail results in release of ink.

B. Sensory neurons from the tail ganglion synapse on three motor neurons that project to the ink gland. The motor neurons are interconnected by means of electrical synapses.

C. A train of stimuli applied to the tail produces a synchronized discharge in all three motor neurons. **1.** When the cells are at rest the stimulus triggers a train of identical action potentials in all three cells. This synchronous activity in the motor neurons results in the release of ink. **2.** When the cells are hyperpolarized, the stimulus cannot trigger action potentials, because the cells are too far from their threshold level. Under these conditions, the inking response is blocked. (Adapted from Carew and Kandel, 1976.)

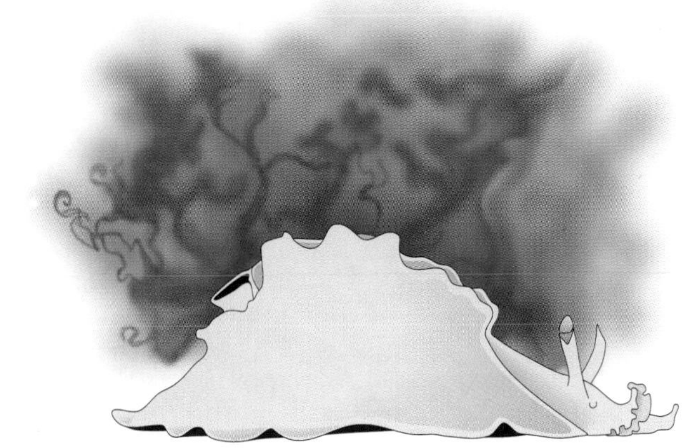

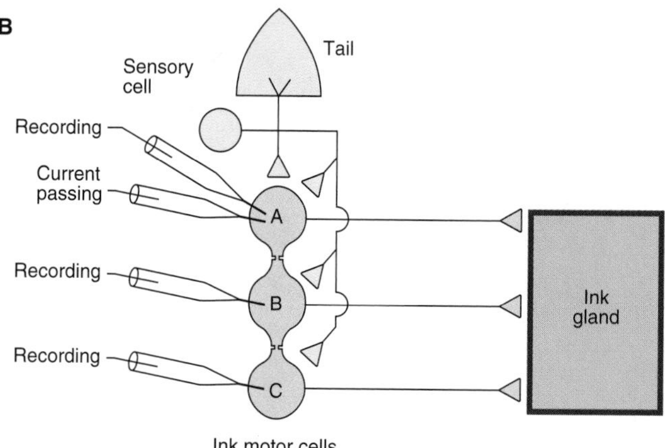

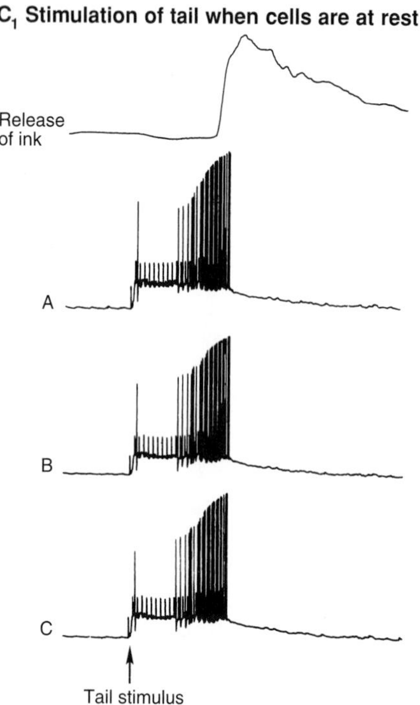

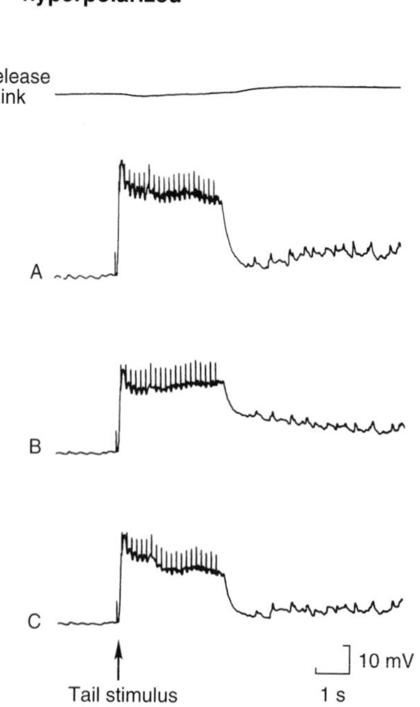

Gap-Junction Channels Connect Communicating Cells at an Electrical Synapse

At electrical synapses the zone of apposition between two neurons, the *gap-junction*, is much narrower (3.5 nm) than the normal space between neurons (20 nm). This gap is bridged by specialized protein structures, the *gap-junction channels*, which conduct the flow of ionic current from the presynaptic to the postsynaptic cell and thus mediate electrical transmission (Figure 11–5). Because of the large-diameter pore of the gap-junction channels, experimental markers such as fluorescent dyes flow readily through the junction.

The conductance of the gap-junction channels can be modulated. For example, most gap-junction channels close in response to lowered cytoplasmic pH or elevated cytoplasmic Ca^{2+}. At some specialized gap junctions the channels are sensitive to voltage, resulting in transmission that is unidirectional: that is only effective from the presynaptic to the postsynaptic cell. The crayfish giant motor synapse is an example of such a rectifying synapse.

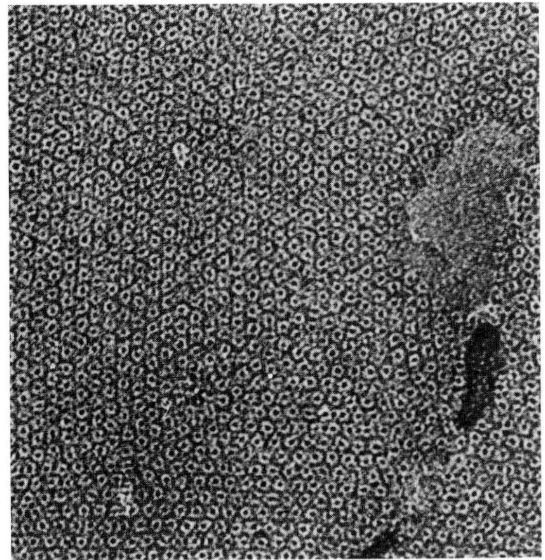

Figure 11–5 At electrical synapses two cells are structurally connected by gap-junction channels. The array of channels shown here was isolated from the membrane of a rat liver. The tissue has been negatively stained, a technique that darkens the area around the channels and in the pores. Each channel appears hexagonal in outline, as explained in Figure 11–6. × 307,800. (Courtesy of N. Gilula.)

Finally, neurotransmitters released from nearby chemical synapses can activate second-messenger–dependent protein kinases that diffuse to the gap junctions and alter the gating of the gap-junction channels by phosphorylating the channel proteins (see Chapter 14).

All gap-junction channels consist of a pair of cylinders (hemi-channels), one in the presynaptic and the other in the postsynaptic cell. The hemi-channels meet in the gap between the two cell membranes. In this way they form a channel, about 1.5 nm in diameter, that connects the cytoplasm of the two cells. Each hemi-channel is called a *connexon*. The connexon is made up of six identical protein subunits, called connexins (Figure 11–6). Each connexin has two recognition functions. First, to form a hemi-channel each connexin must recognize the other five subunits. Second, to form a conductive channel that connects the pre- and postsynaptic cells, each connexin of the hemi-channel must recognize the extracellular domains of the connexin from the opposing hemi-channel.

Connexins from different tissues all belong to one large gene family. Each connexin subunit has four hydrophobic domains thought to span the cell membrane. These membrane-spanning domains in the gap-junction channels of different tissues are quite similar, as are the two extracellular domains thought to be involved in the homophilic recognition of the hemi-channels of the opposing cells (Figure 11–6C). On the other hand, the cytoplasmic regions vary greatly, and this variation may explain why gap junctions in different tissues are sensitive to different modulatory factors.

Given that gap junctions can be modulated by voltage, pH, and Ca^{2+}, how do the channels open and close? One suggestion is that the six connexins in a hemi-channel rotate slightly with respect to each other, much like the elements of the shutter in a camera. The concerted tilting of the connexins by a few angstroms at one end leads to a somewhat larger displacement at the other end, thus exposing the pore (Figure 11–6B). Such conformational changes may be a common mechanism for opening and closing ion channels (see Chapter 7).

Chemical Synapses Can Simplify Signals

Unlike electrical synapses, there is no structural continuity between pre- and postsynaptic neurons

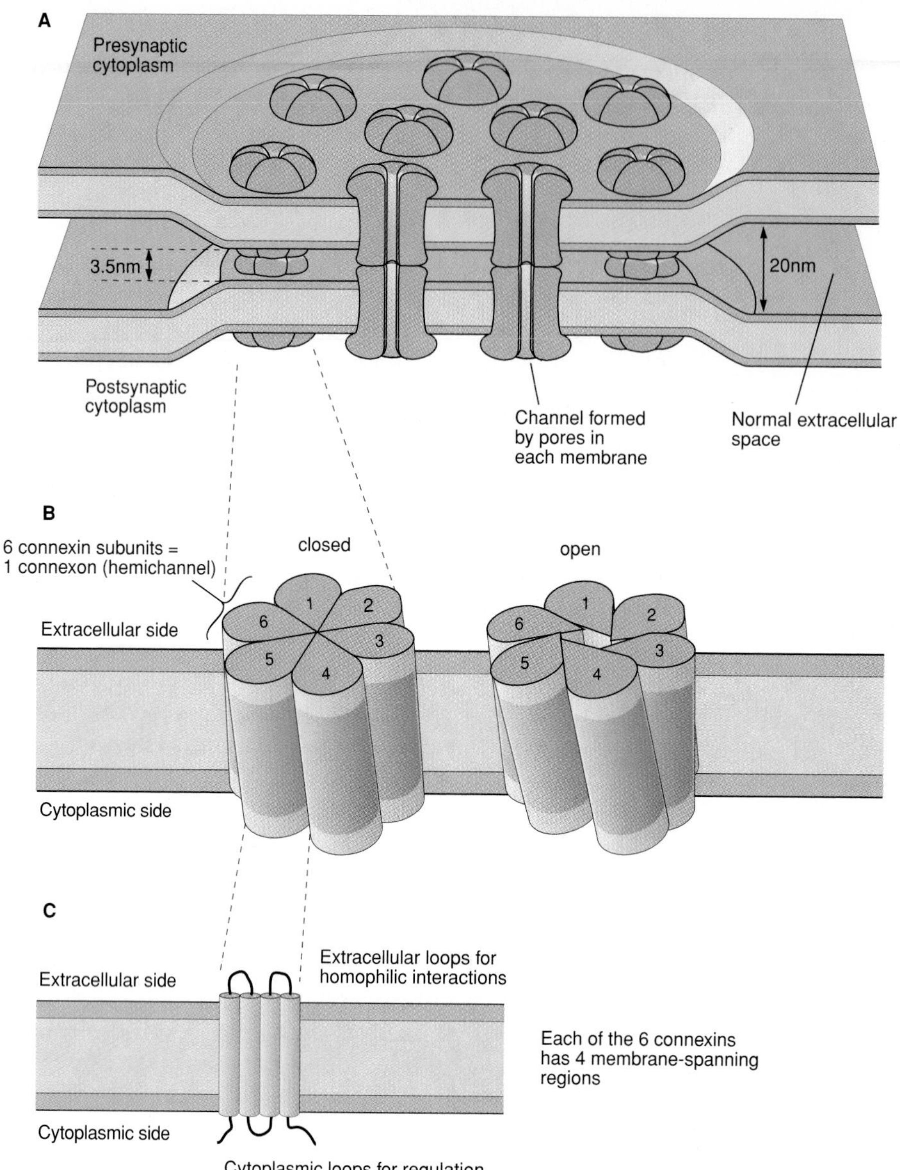

Figure 11–6 A three-dimensional model of the gap-junction channel, based on X-ray diffraction studies.

A. A gap-junction channel is actually a pair of hemi-channels, one in each apposite cell, that match up in the gap of the extracellular space by means homophilic interactions. The channel thus connects the cytoplasm of the two cells and provides a direct means of ion flow between the cells. This bridging of the cells is facilitated by a narrowing of the extracellular space at the gap junction, from the normal 20 nm separation to only 3.5 nm.

B. Each hemi-channel, or connexon, is made up of six protein subunits called connexins. Each connexin is about 7.5 nm long and spans the cell membrane. The connexins are arranged in such a way that a pore is formed in the center of the structure. The resulting connexon, with an overall diameter of about 1.5 to 2 nm, has a characteristic hexagonal outline (see Figure 11–5). The pore is opened when the subunits rotate about 0.9 nm at the cytoplasmic base in a clockwise direction. The dark shading indicates the portion of the connexon embedded in the membrane.

C. A single connexin is thought to have four membrane-spanning regions. The genetic sequences of gap junctions from many different kinds of tissue all show regions with similar amino acid sequences. In particular, four hydrophobic domains with a high degree of similarity among different tissues are presumed to be the regions of the protein structure that traverse the cell membrane. The cytoplasmic loops, however, differ among different channels and are thought to be involved in regulation. In addition, two extracellular regions that are also highly conserved in different tissues are thought to be involved in the homophilic matching of apposite hemi-channels.

at chemical synapses. In fact, the synaptic cleft in chemical synapses is usually slightly wider (20–40 nm), and in some instances substantially wider than the adjacent intercellular space (typically 20 nm). In many neurons with chemical synapses the cell membrane is morphologically specialized at the synapse. Presynaptic terminals contain discrete collections of *synaptic vesicles*, each of which is filled with several thousand molecules of a chemical neurotransmitter (Figure 11–7). The vesicles release their neurotransmitter into the synaptic cleft in response to the Ca^{2+} influx that occurs with each action potential. The transmitter molecules then diffuse into the extracellular space of the synaptic cleft and bind to receptor sites on the postsynaptic cell membrane, causing ion channels to open (or close), thereby altering the membrane conductance and potential of the postsynaptic cell (Figure 11–8). (The mechanisms of transmitter release are discussed in detail in Chapter 15.)

These several steps account for the synaptic delay at chemical synapses, a delay that is often several milliseconds or longer, but can be as short as 0.3 milliseconds. Although chemical transmis-

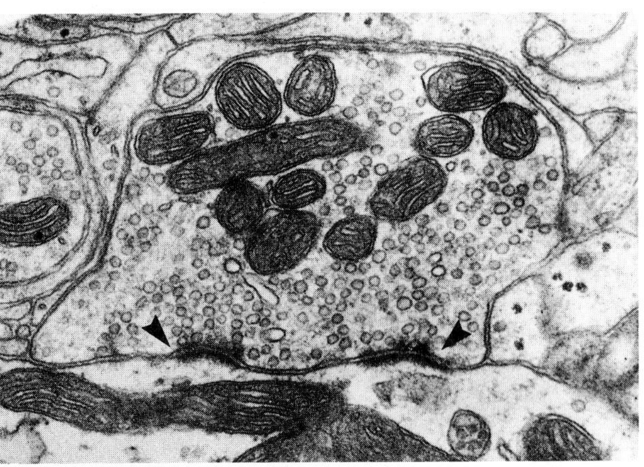

Figure 11–7 At chemical synapses the presynaptic and postsynaptic cell membranes are separated by a space called the synaptic cleft. This electron micrograph shows the fine structure of a presynaptic terminal at a nerve–muscle synapse. The large dark structures are mitochondria. The many round bodies are vesicles that contain the neurotransmitter acetylcholine. The fuzzy dark thickenings along the presynaptic side of the cleft (**arrowheads**) are specialized areas, called active zones, that are thought to be docking sites for vesicles (see Figure 11–8). (Courtesy of J. E. Heuser and T. S. Reese.)

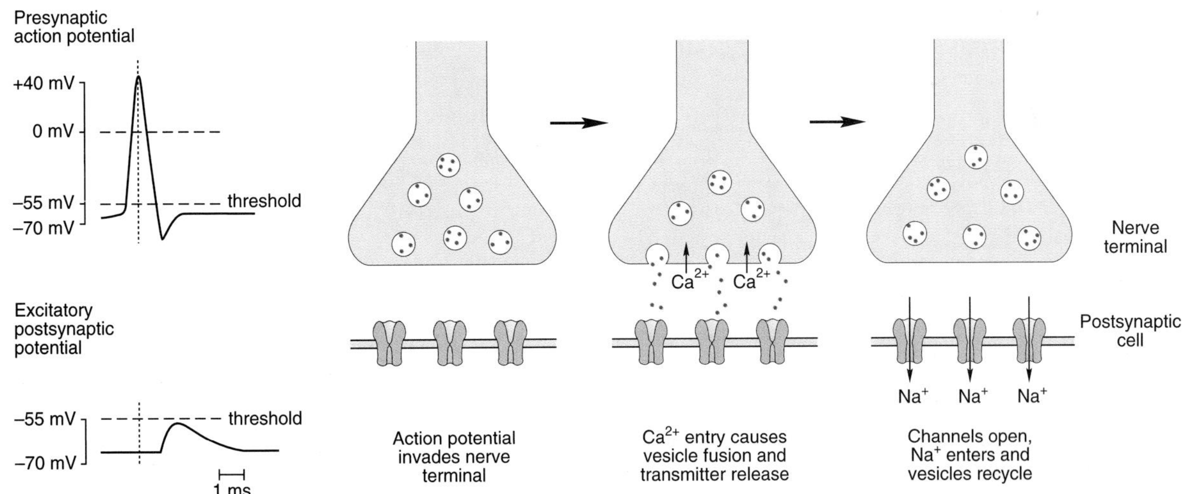

Figure 11–8 Synaptic transmission at chemical synapses involves several steps. An action potential arriving at the terminal of a presynaptic axon causes voltage-gated Ca^{2+} channels at the active zone to open. The influx of Ca^{2+} produces a high concentration of Ca^{2+} near the active zone, which in turn causes vesicles containing neurotransmitter to fuse with the cytoplasmic membrane and release their contents into the synaptic cleft. The released neurotransmitter molecules then diffuse across the synaptic cleft and bind to specific receptors on the postsynaptic membrane. These receptors cause ion channels to open (or close), thereby changing the membrane conductance and depolarizing the cell. The complex process of chemical synaptic transmission is responsible for the delay between action potentials in the pre- and postsynaptic cells, compared with the virtually instantaneous transmission of signals at electrical synapses (see Figure 11–2B).

sion lacks the speed of electrical synapses, it has the important property of *amplification*. With the discharge of just one synaptic vesicle, several thousand molecules of transmitter stored by that vesicle are released. Typically, to open a postsynaptic ion channel the binding of only two molecules of transmitter is required. Consequently, the action of one synaptic vesicle leads to the opening of thousands of ion channels in the postsynaptic cell. In this way a small presynaptic nerve terminal, which generates only a weak electrical current, is nevertheless able to release one or more vesicles, each of which is able to depolarize even a large postsynaptic cell.

Chemical Transmitters Bind to Postsynaptic Receptors

Chemical synaptic transmission can be divided into two steps: a *transmitting* step, in which the presynaptic cell releases a chemical messenger, and a *receptive* step, in which the transmitter binds to receptor molecules on the postsynaptic cell.

The transmitting process of the presynaptic terminals of a chemical synapse resembles the release process of an endocrine gland, and chemical synaptic transmission can be seen as a modified form of hormone secretion. Both endocrine glands and presynaptic terminals release a chemical agent with a signaling function, and both are examples of regulated secretion. Likewise, both endocrine glands and neurons are usually some distance from their target cells. There is one important difference, however. The hormone released by the gland travels through the bloodstream until it interacts with the receptor in the appropriate target cells. A neuron, on the other hand, can usually only communicate with those distant cells that its axon contacts directly.

In neurons, then, communication consists of sending an action potential down its axon. At the axon terminal the electrical signal triggers the focused release of the chemical transmitter, which travels only a small distance to its target. Neuronal signaling, therefore, has two special features: it is fast and it is precisely directed. This highly directed or focused release is accomplished with specialized secretory machinery called *active zones*.

In neurons without active zones the distinction between neuronal and hormonal transmission becomes blurred. For example, the neurons in the autonomic system that innervate smooth muscle are at some distance from their postsynaptic cells and do not have specialized release sites in their terminals; synaptic transmission between these cells is slower and more diffuse.

The distinction is further blurred by the fact that the same neurotransmitter can serve several functions. At one set of terminals the transmitter can be released at an active zone and serve as a conventional transmitter, acting directly on neighboring cells; at another locus it can be released in a less focused way to serve as a modulator, producing a more diffuse action; and at a third locus it can be released into the bloodstream to act as a neurohormone.

Although a variety of chemicals serve as neurotransmitters, including both small molecules and peptides (see Chapter 15), the action of a transmitter in the postsynaptic cell does not depend on the chemical properties of the transmitter but rather on the properties of the receptors, which recognize and bind the transmitter. For example, acetylcholine can excite some postsynaptic cells and inhibit others, and at still other cells it can produce both excitation and inhibition. It is the receptor that determines whether a cholinergic synapse is excitatory or inhibitory, and whether an ion channel will be activated directly by the transmitter or indirectly through a second messenger.

Within a group of closely related animals a given transmitter substance binds to conserved families of receptors and is associated with specific physiological functions. For example, in vertebrates acetylcholine produces synaptic excitation at the neuromuscular junction by acting on a special type of acetylcholine receptor. Similarly, acetylcholine invariably slows the heart in vertebrates by acting on a special type of inhibitory acetylcholine receptor.

The notion of a receptor was introduced in the late nineteenth century by the German biological chemist Paul Ehrlich to explain the selective action of toxins and other pharmacological agents and the exquisite specificity of immunological reactions. In 1900 Ehrlich wrote, "Chemical substances are only able to exercise an action on the tissue elements with which they are able to establish an intimate chemical relationship. . . . [This relationship] must be specific. The [chemical] groups must be adapted to one another . . . as lock and key."

In 1906 the English pharmacologist John Langley postulated that the sensitivity of skeletal muscle to curare and nicotine is due to a "receptive molecule." A theory of receptor function was subsequently developed by Langley's students—in particular, Eliot Smith and Henry Dale—a development that was greatly influenced by concurrent studies of enzyme kinetics and cooperative interactions between small molecules and proteins. As we shall see in the next chapter, Langley's "receptive molecule" has been isolated and characterized as the acetylcholine receptor of the neuromuscular junction.

All receptors for chemical transmitters have two biochemical features in common:

1. They are membrane-spanning proteins. The region exposed to the external environment of the cell recognizes and binds the transmitter from the presynaptic cell.
2. They carry out an effector function within the target cell. The receptors typically influence the opening or closing of ion channels.

Postsynaptic Receptors Gate Ion Channels Either Directly or Indirectly

Chemical neurotransmitter actions fall into two classes, according to whether their gating of the ion channel is direct or indirect. The two classes of transmitter actions are mediated by receptor proteins derived from different gene families.

Receptors that gate ion channels directly, such as those mediating the action of acetylcholine at the neuromuscular junction, are membrane-spanning proteins composed of several peptide subunits. These subunits form a single macromolecule that contains both the recognition element for transmitter and the ion channel. On binding a neurotransmitter, such *ionotropic* receptors undergo a conformational change that opens the channel (Figure 11–9A).

In contrast, receptors that gate ion channels indirectly, like those for norepinephrine or sero-

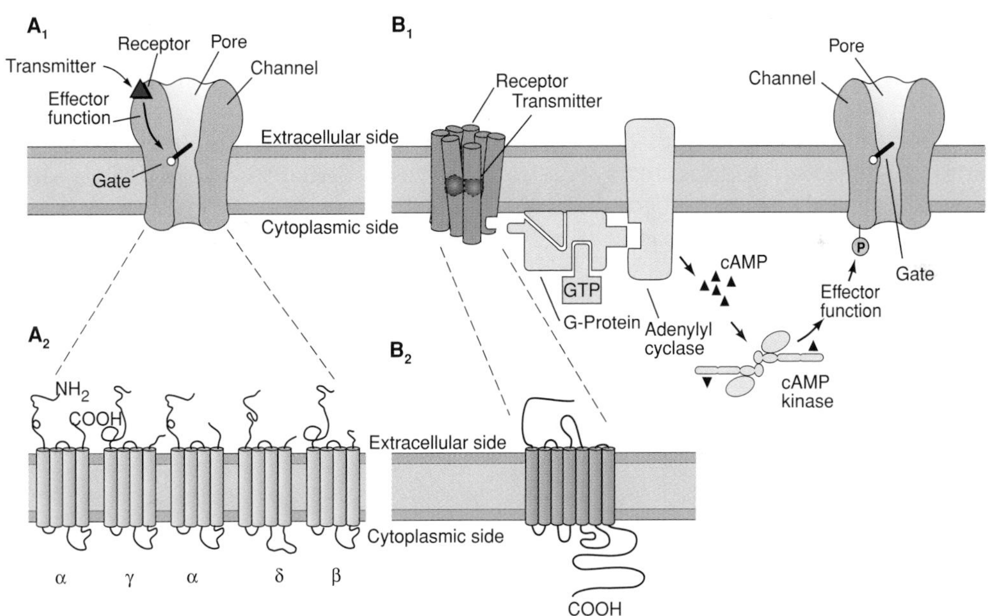

Figure 11–9 Neurotransmitters act either directly or indirectly on ion channels.

A. 1. A transmitter receptor acts directly on an ion channel when the transmitter receptor is structurally part of the channel. **2.** These receptor-channels are usually composed of five subunits, each of which is thought to contain four membrane-spanning α-helical regions.

B. When the neurotransmitter is structurally distinct from the ion channels, the neurotransmitter acts indirectly on the channel. **1.** The receptor activates a GTP-binding protein (G-protein), which in turn activates a second-messenger cascade, that modulates ion channel activity. Here the G-protein stimulates adenylyl cyclase, which converts ATP to cAMP. The cAMP activates the cAMP-dependent protein kinase (cAMP-kinase), which phosphorylates the channel (**P**), leading to a change in function. **2.** The typical receptor of this family of proteins is composed of a single protein with seven membrane-spanning α-helical regions that bind transmitter within the plane of the membrane.

tonin at synapses in the cerebral cortex, are macromolecules that are separate from the ion channels on which they act. Two families of receptors fall into this category.

The best described are the *metabotropic* receptors. These membrane-spanning proteins consist of a single polypeptide chain and couple to any one of a family of GTP-binding proteins (G-proteins). The G-proteins in turn activate enzymes that produce one or another intracellular second messenger, such as cyclic AMP or diacylglycerol. The second messenger then acts on a channel directly or, more commonly, through the action of one of a family of enzymes called serine-threonine protein kinases.

Another type of receptor that indirectly gates ion channels belongs to the tyrosine kinase family. These kinases add a phosphate group to tyrosine residues of substrate proteins, resulting in modulation of the channel (Figure 11–9B). Second messengers and protein kinases are described in greater detail in Chapter 14.

Receptors that directly gate ion channels produce relatively fast synaptic actions lasting only milliseconds. These are commonly found in neural circuits that directly mediate behavior. The two types of receptors that indirectly gate ion channels produce slow synaptic actions lasting seconds and even minutes. These slower actions often serve to modulate behavior by altering the excitability of neurons and the strength of the synaptic connections of the neural circuitry mediating behavior. For example, modulatory synaptic pathways often serve as the crucial reinforcing pathways in the process of learning. Tyrosine kinase receptors not only participate in synaptic transmission in the adult organism, they also are important during development for the initial formation of the synapse.

An Overall View

Signaling between neurons takes place at electrical or chemical synapses. Electrical transmission is mediated by the direct flow of current from the presynaptic to the postsynaptic neuron through channels that connect the cytoplasm of the two cells. These gap-junction channels are permeable to small molecules and some second messengers. Because of this direct mechanism, electrical transmission is the more rapid form of signaling between neurons. Groups of cells with electrical synapses can fire together when their collective threshold is reached. These two properties, speed and synchrony, make electrical synapses suitable for fast, stereotyped behaviors, such as escape and defensive responses.

Chemical synaptic transmission is slower than electrical transmission because the presynaptic neuron must first release a neurotransmitter, which then diffuses across the synaptic cleft and binds to receptors in the postsynaptic cell membrane. It is the receptor, not the transmitter, that determines whether the synaptic response is excitatory or inhibitory. Directly gated chemical transmission is mediated by receptors that are structurally part of an ion channel. Indirectly gated chemical transmission is slower because it involves several additional postsynaptic steps: the receptors couple to enzymes that synthesize second messengers, which then act on structurally separate ion channels.

Although even the fastest chemical synaptic mechanisms are slower than electrical synaptic transmission, chemical synaptic transmission has the advantage that a single action potential releases thousands of neurotransmitter molecules, allowing signals to be amplified from one neuron to another. Perhaps because it is a multistep process, chemical transmission is more modifiable than electrical transmission. As we shall see later in the book, the modifiability of chemical synapses in the brain is an important mechanism of behavioral learning.

Selected Readings

Bennett, M. V. L., Barrio, L. C., Bargiello, T. A., Spray, D. C., Hertzberg, E., and Sáez, J. C. 1991. Gap junctions: New tools, new answers, new questions. Neuron 6:305–320.

Dermietzel, R., and Spray, D. C. 1993. Gap junctions in the brain: Where, what type, how many and why? Trends Neurosci. 16:186–192.

Eccles, J. C. 1976. From electrical to chemical transmission in the central nervous system. The closing address of the Sir Henry Dale Centennial Symposium. Notes Rec. R. Soc. Lond. 30:219–230.

Edelman, G. M., Gall, W. E., and Cowan, W. M. (eds.) 1987. Synaptic Function. New York: Wiley.

Fatt, P. 1954. Biophysics of junctional transmission. Physiol. Rev. 34:674–710.

Furshpan, E. J., and Potter, D. D. 1959. Transmission at the giant motor synapses of the crayfish. J. Physiol. (Lond.) 145:289–325.

Jessell, T. M., and Kandel, E. R. 1993. Synaptic transmission: A bidirectional and self-modifiable form of cell–cell communication. Cell 72/Neuron 10 (Suppl.):1–30.

Ross, E. M. 1989. Signal sorting and amplification through G protein-coupled receptors. Neuron 3:141–152.

Unwin, N. 1993. Neurotransmitter action: Opening of ligand-gated ion channels. Cell 72/Neuron 10 (Suppl.):31–41.

12

Transmission at the Nerve–Muscle Synapse

The Neuromuscular Junction Permits Study of Directly Gated Transmission

Synaptic Excitation at the Nerve–Muscle Synapse Involves Transmitter-Gated Ion Channels

The Ion Channel at the Muscle End-Plate Is Permeable to Both Sodium and Potassium

Patch-Clamp Experiments Reveal the Nature of Current Flow Through Single Ion Channels

Individual Transmitter-Gated Channels Conduct a Unitary Current

Four Factors Determine the End-Plate Current

The Nicotinic Acetylcholine Receptor–Channel Is a Membrane-Spanning Protein

Transmitter-Gated Channels Differ from Voltage-Gated Channels

Postscript: The End-Plate Current Can Be Calculated from an Equivalent Circuit

Chemical synaptic transmission is the predominant form of synaptic communication in the brain. As we saw in the last chapter, chemical synapses have two adaptive advantages: they can amplify signals and their actions are modifiable. All chemical transmission involves the binding of a neurotransmitter from a presynaptic cell to a receptor on the postsynaptic cell, and this action opens or closes ion channels in the postsynaptic cell. Two types of receptors define two broad categories of chemical transmission: one type gates an ion channel directly, the other does so indirectly through second messengers.

We begin our consideration of chemical transmission by describing the directly gated transmission at the synapse between the motor neurons and skeletal muscle fiber. In the next two chapters we discuss chemical synaptic transmission in the central nervous system, which includes synaptic actions mediated by second messengers.

The nerve–muscle synapse is the best understood site of directly gated transmission in vertebrates. It is accessible to direct experimentation because the muscle cell is large enough to accommodate several microelectrodes for electrophysiological measurements. Moreover, unlike synapses in the central nervous system, the region of the synapse where the presynaptic terminal contacts the postsynaptic membrane can be seen with the light microscope in a living cell and the postsynaptic cell is innervated by only one presynaptic axon.

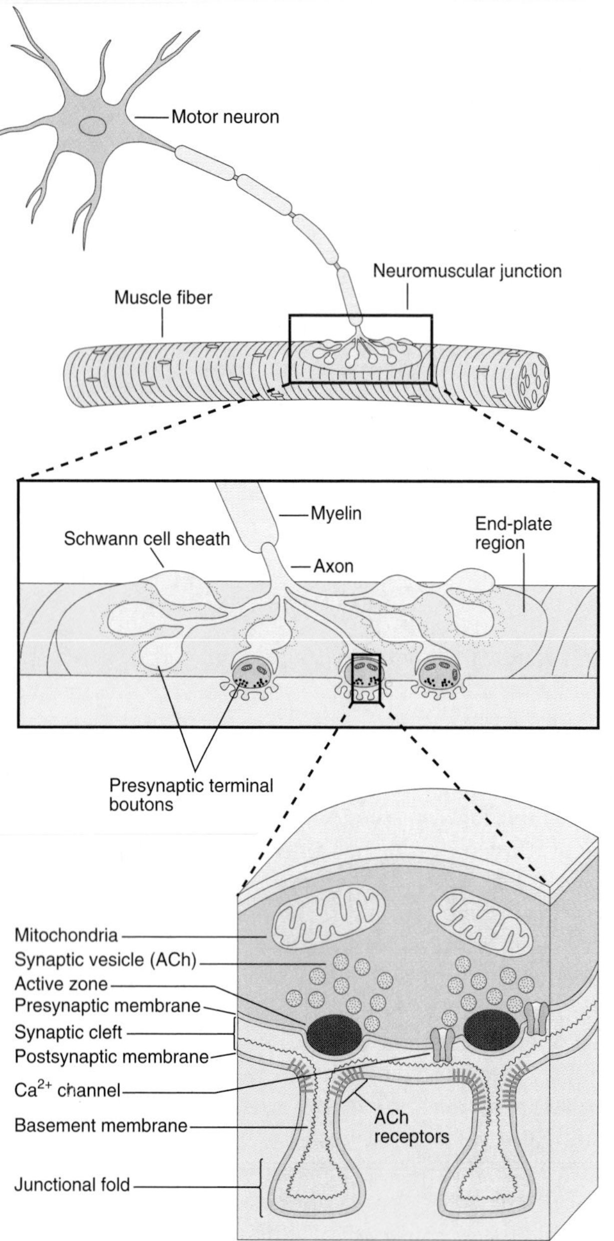

The Neuromuscular Junction Permits Study of Directly Gated Transmission

The anatomy of the nerve–muscle synapse is relatively simple. A single skeletal muscle fiber in vertebrates usually is innervated by only one motor axon. The transmitter released by the axon terminal is acetylcholine (ACh), and the receptor on the muscle membrane is the nicotinic type of ACh receptor.[1]

The motor neuron's axon innervates the *end-plate*, a specialized region of the muscle membrane. As the motor axon approaches the end-plate it loses its myelin sheath and splits into several fine branches. The ends of the fine branches form multiple expansions or varicosities, called *synaptic boutons*, from which transmitter is released. Each bouton lies over *junctional folds*, deep depressions in the surface of the muscle fiber that contain ACh receptors. Junctional folds are lined by the *basement membrane* (or *basal lamina*), a network of connective tissue consisting of collagen and glycoproteins that covers the surface of the entire muscle fiber (Figure 12–1).

Each presynaptic bouton contains all the machinery required to release neurotransmitter. This includes the *synaptic vesicles*, which contain the ACh transmitter, and the *active zone*, a part of the

[1]There are two basic types of receptors to acetylcholine: *nicotinic* and *muscarinic*. Both bind ACh, but the nicotinic receptor is ionotropic and the muscarinic receptor is metabotropic (see Chapter 11). The two receptors can be distinguished further because certain drugs that simulate the actions of ACh—that is, nicotine and muscarine—bind exclusively to one type of ACh receptor or the other. We learn more about muscarinic ACh receptors in Chapter 14.

Figure 12–1 The neuromuscular junction is readily visible with the light microscope. At the muscle the motor axon ramifies into several fine branches approximately 2 μm thick. Each branch forms multiple swellings called *synaptic boutons*, which are covered by a thin layer of Schwann cells. The boutons lie over a specialized region of the muscle fiber membrane, the *end-plate*, and are separated from the muscle membrane by a 50 nm *synaptic cleft*. Each presynaptic bouton contains mitochondria and synaptic vesicles clustered around *active zones* that are the site of release of the ACh transmitter. Immediately under each bouton in the end-plate are several *junctional folds*, sites with a high density of ACh receptors. The muscle fiber is covered by a layer of connective tissue, the *basement membrane* (or basal lamina). Both the presynaptic terminal and the muscle fiber secrete proteins into the basement membrane, including the enzyme acetylcholinesterase, which inactivates the ACh released by the presynaptic terminal by breaking it down into acetate and choline. The basement membrane also acts to organize the synapse by aligning the presynaptic boutons with the postsynaptic junctional folds (see Chapter 4). (Adapted in part from McMahan and Kuffler, 1971.)

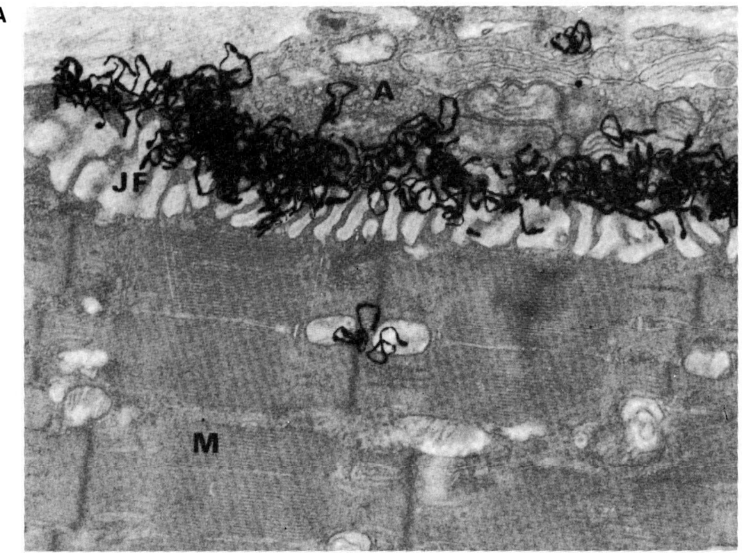

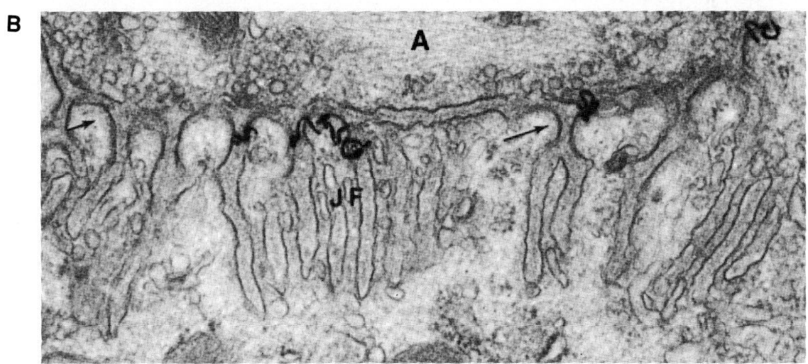

Figure 12–2 Acetylcholine receptors in an end-plate from the sternomastoid muscle of a mouse. The receptors can be labeled with antibodies or the snake venom neurotoxin (α-bungarotoxin), which bind to nicotinic ACh receptors. The toxin has been used extensively because it can be labeled with radioactive iodine or made fluorescent for good visualization. The muscle tissue shown in these electron micrographs was incubated with ^{125}I-labeled α-bungarotoxin until all neurally evoked muscle contractions were blocked, indicating that most ACh receptors were labeled. The labeled receptors appear black. (From Fertuck and Salpeter, 1974.)

A. The label is not uniformly distributed throughout the postsynaptic membrane but is localized in regions of the junctional folds closest to the apposed axon. (**JF**, junctional folds; **A**, axon; **M**, muscle.) × 21,000.

B. The labeled receptors appear as dark outlines along the junctional folds. The receptors are concentrated at the postjunctional membrane nearest the peaks of the folds (**arrows**). × 37,500.

membrane specialized for transmitter release. In addition, each bouton contains voltage-gated Ca^{2+} channels that allow Ca^{2+} to enter the terminal with each action potential; Ca^{2+} triggers fusion of the synaptic vesicles with the active zone, and fusion leads to release of the vesicle's content.

Every active zone is positioned opposite a junctional fold in the postsynaptic cell. At the crest of these folds the receptors for ACh are clustered in a lattice, with a density of about 10,000 receptors per square micrometer (Figures 12–2 and 12–3). The Ca^{2+} channels at the active zone align with the

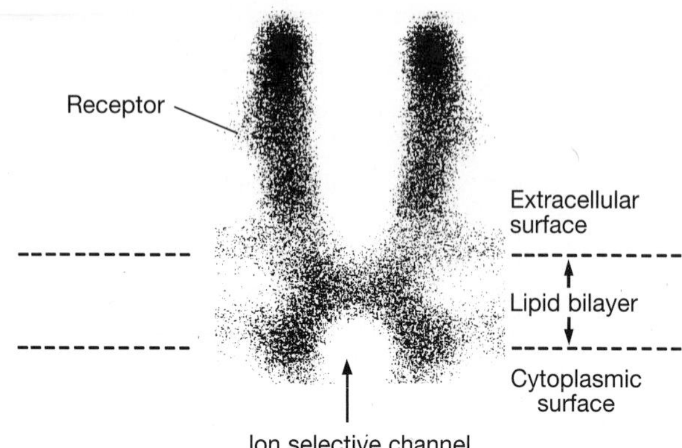

Figure 12–3 Reconstructed electron microscope image of the ACh receptor–channel complex in the fish *Torpedo californica*. The image was obtained by computer processing of negatively stained images of ACh receptors. The resolution is 1.7 nm, fine enough to see overall structures but too coarse to resolve individual atoms. The overall diameter of the receptor and its channel is about 8.5 nm. The pore is wide at the external and internal surfaces of the membrane, but it narrows considerably within the lipid bilayer. The channel extends some distance into the extracellular space. (Adapted from Toyoshima and Unwin, 1988.)

ACh receptor (Figure 12–4). In the region below the crest and extending into the folds, the membrane of the muscle cell is rich in voltage-gated Na^+ channels, which convert the presynaptic action potential into an action potential in the muscle fiber.

The excitatory postsynaptic potential in the muscle cell is called the *end-plate potential*. The amplitude of the end-plate potential is unusually large; a single motor cell produces a synaptic potential of about 70 mV. This change in potential usually is large enough to trigger an action potential in the muscle fiber. In contrast, most neurons in the central nervous system produce synaptic potentials less than 1 mV in amplitude, so that input from many presynaptic neurons is needed to generate an action potential there.

Synaptic Excitation at the Nerve–Muscle Synapse Involves Transmitter-Gated Ion Channels

The synaptic potential at the end-plate was first studied in detail in the 1950s by Paul Fatt and Bernard Katz. They were able to isolate the end-plate potential in intracellular voltage recordings using the drug curare,[2] which reduced the amplitude of the end-plate potential below the threshold for the action potential (Figure 12–5). Fatt and Katz found that the synaptic potential produced in muscle cells by the action of the motor neuron was largest when they placed an intracellular electrode precisely at the end-plate. As the electrode was moved down the muscle fiber away from the end-plate region, the amplitude of the synaptic potential decreased progressively (Figure 12–6). From these findings Fatt and Katz concluded that the synaptic potential is generated by an inward current confined to the end-plate region, which then spreads passively away from the end-plate. The current flow is confined to the end-plate because the ACh-activated channels are localized in that region, opposite the presynaptic terminal from which transmitter is released.

The synaptic potential at the end-plate rises rapidly but decays more slowly. The rapid rise is due to the sudden release of ACh into the synaptic cleft by the action potential in the presynaptic nerve terminal. Once released, ACh diffuses rapidly to the receptors at the end-plate. Not all the released ACh reaches the postsynaptic receptors, however, because two processes act quickly to remove ACh from the cleft: (1) ACh is hydrolyzed by the enzyme *acetylcholinesterase* localized in the basement membrane and (2) ACh diffuses away, out of the synaptic cleft.

The time course and properties of the current that generates the end-plate potential have been studied in voltage-clamp experiments. These studies revealed that the end-plate current rises and decays more rapidly than the resultant depolarizing change in end-plate potential (Figure 12–7). The time course of the end-plate current is thought to reflect the rapid opening and closing of the ion channels activated by ACh. The slower time course of the synaptic potential is partly determined by the passive time constant of the muscle membrane. As we saw in Chapter 9 (Figure 9–3), it takes time for an ionic current to charge or discharge the muscle membrane capacitance (see the Postscript at the end of this chapter).

---

[2]Curare is a mixture of plant toxins used by South American Indians, who apply it to arrowheads to paralyze their quarry. Tubocurarine, the purified active agent, blocks neuromuscular transmission by binding to the ACh receptor and preventing its activation by ACh.

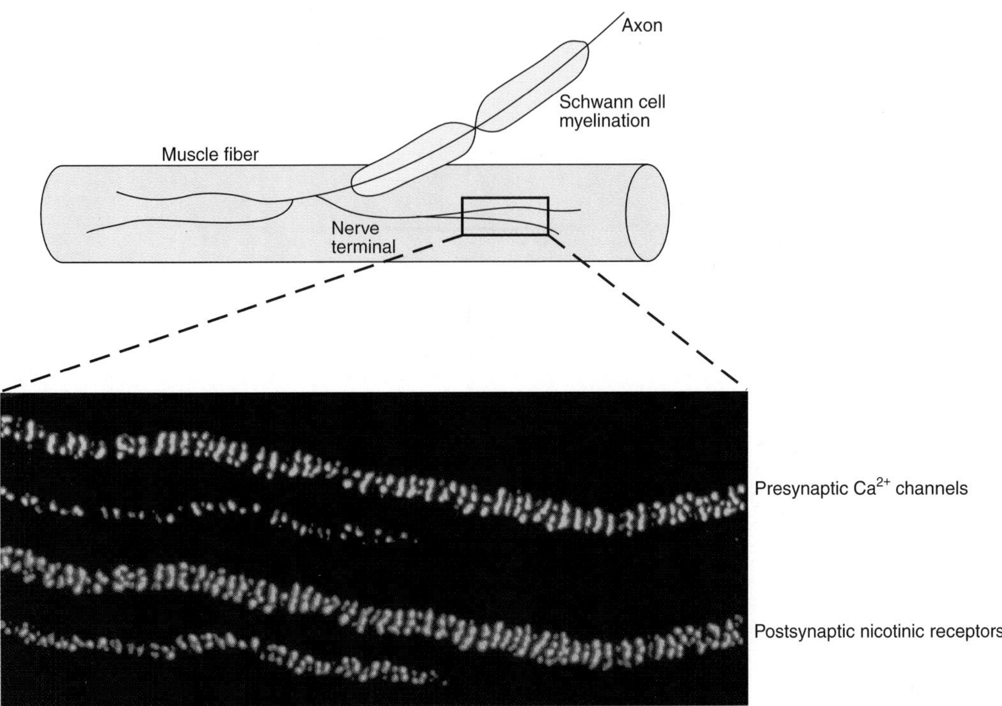

Figure 12–4 Calcium channels are concentrated in regions of the presynaptic nerve terminal opposite the clusters of ACh receptors on the postsynaptic membrane. The fluorescent image shows the presynaptic Ca^{2+} channels in red, after labeling with a Texas red-coupled marine snail toxin that binds to Ca^{2+} channels. Postsynaptic ACh receptors are labeled in green with boron–dipyromethane difluoride-labeled α-bungarotoxin, which binds selectively to ACh receptors. The two images are normally superimposed but have been separated for clarity. The patterns of labeling with both probes are in almost precise register, indicating that the active zone of the presynaptic neuron is in almost perfect alignment with the postsynaptic membrane containing the high concentration of ACh receptors. (From Robitaille et al., 1990.)

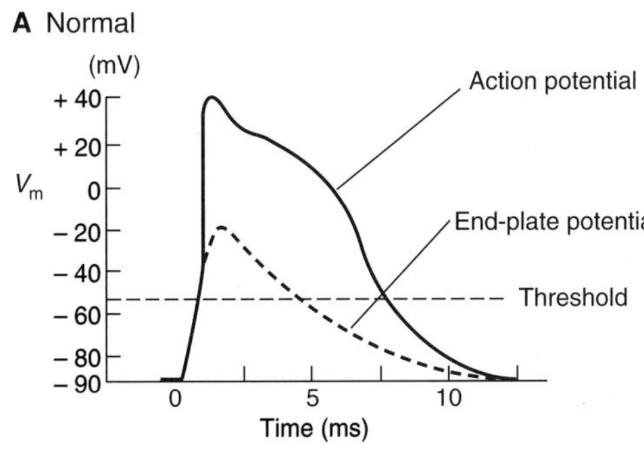

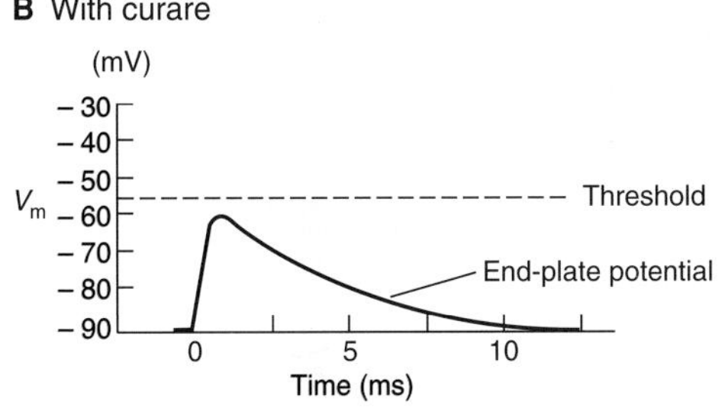

Figure 12–5 The end-plate potential can be isolated pharmacologically for study.

A. In normal circumstances stimulation of the motor axon produces an action potential (**solid line**). The **dashed trace** shows the inferred time course of the end-plate potential that triggers the action potential.

B. The end-plate potential can be isolated in the presence of curare. Curare blocks the binding of ACh to its receptor and so the amplitude of the end-plate potential remains below threshold. This technique is used to study the currents and channels that contribute to the end-plate potential, which are different from those producing an action potential. The values for the resting potential, end-plate potential, and action potential of these intracellular recordings are typical of a vertebrate skeletal muscle.

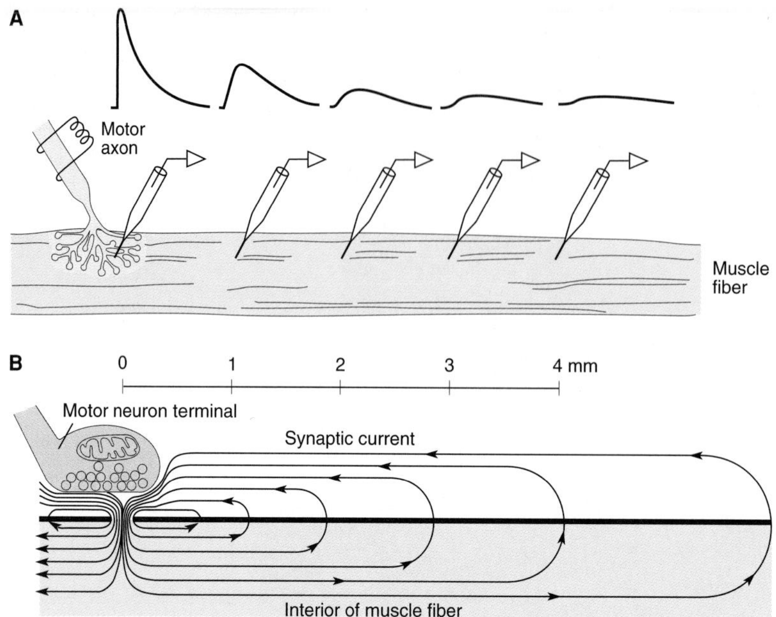

Figure 12–6 The synaptic potential is largest at its site of origin in the end-plate.

A. The amplitude of the synaptic potential decays and the time course of the potential becomes slower with increasing distance from the end-plate region.

B. The decay illustrated in **A** results from the "leakiness" of the muscle fiber membrane. Since current flow must complete a circuit, the inward synaptic current at the end-plate gives rise to a return flow of outward current through resting channels and across the membrane (acting as capacitor). It is this outward flow of current across the capacitor that produces the depolarization. Since current leaks out all along the membrane, the current flow decreases with distance from the end-plate. Thus, unlike the regenerative action potential, the local depolarization produced by the synaptic potential of the membrane decreases with distance. (Adapted from Miles, 1969.)

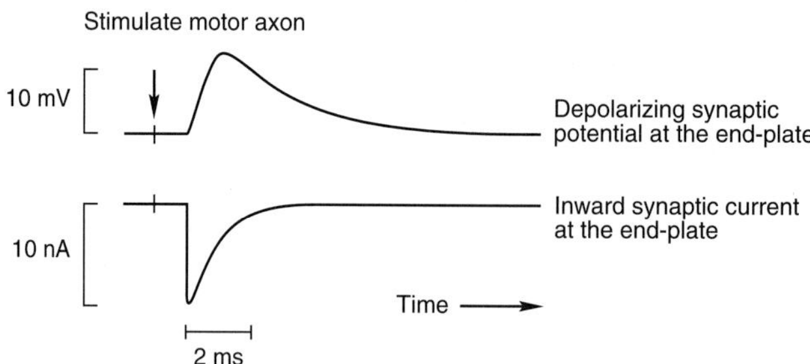

Figure 12–7 The end-plate potential changes slowly, following in time the inward synaptic current, because synaptic current must first alter the charge on the membrane capacitance of the muscle (see Chapter 9). The synaptic current is measured at a constant membrane potential (e.g., –90 mV) using the voltage-clamp technique (see Box 10–1).

The Ion Channel at the Muscle End-Plate Is Permeable to Both Sodium and Potassium

Which ions move through the muscle membrane to produce this synaptic action? An important clue can be obtained by systematically changing the membrane potential and determining the *reversal potential* for the synaptic action. The reversal potential of the end-plate potential is the membrane potential at which the synaptic current has zero amplitude. At the reversal potential there is no net current, because any inward current through the end-plate is balanced by an equal amount of outward current. Does the reversal potential of the end-plate potential coincide with the equilibrium potential for a specific ion species in the muscle fiber, such as 55 mV for Na^+? If so, Na^+ might carry the end-plate current.

To determine the reversal potential we need to examine the synaptic current at different values of membrane potential. At the resting potential of the muscle (–90 mV) the synaptic current flows inward (Figure 12–7). The change in current flowing through transmitter-gated channels at different membrane potentials can be calculated from Ohm's law. According to Ohm's law, the current responsible for the excitatory postsynaptic potential (I_{EPSP}) is given by

$$I_{EPSP} = g_{EPSP} \times (V_m - E_{EPSP}).$$

Here g_{EPSP} represents the conductance of the channels activated by ACh (the synaptic conductance), and the term $V_m - E_{EPSP}$ represents the electrochemical driving force for the ionic current flowing through the channel (where V_m is the membrane potential and E_{EPSP} is the reversal potential for the excitatory postsynaptic potential).

If an influx of Na^+ were solely responsible for the end-plate potential, the reversal potential for the excitatory postsynaptic potential would be the same as the equilibrium potential for Na^+, or +55 mV (Figure 12–8A). Thus, if the membrane potential is experimentally altered from –100 to +55 mV, the end-plate current should diminish progressively because the electrochemical driving force on Na^+ ($V_m - E_{EPSP}$) is reduced. At +55 mV the inward current flow should be abolished, and at potentials more positive than +55 mV the end-plate current should reverse in direction and flow outward.

Instead, as the membrane potential is reduced, the inward current rapidly becomes smaller and is abolished at 0 mV! At values more positive than 0 mV the end-plate current reverses direction and begins to flow outward (Figure 12–8B). This partic-

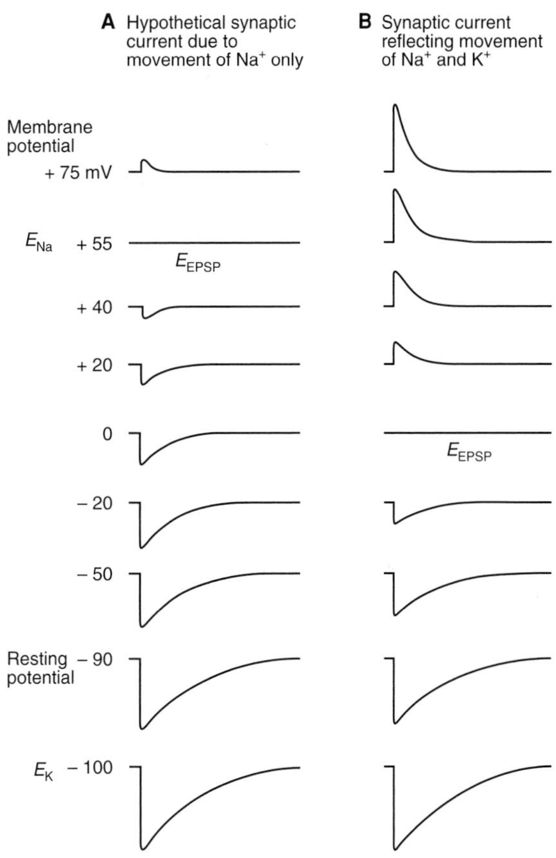

Figure 12–8 The end-plate potential is produced by the simultaneous flow of both Na^+ and K^+.

A. The ionic currents responsible for the end-plate potential can be determined from the reversal potential of the end-plate potential. The membrane is voltage-clamped at different potentials and the nerve is stimulated. When the membrane potential is held at the equilibrium potential for, say, the Na^+ ion, no Na^+ current results. If only Na^+ were responsible for the end-plate current, the reversal potential would occur at +55 mV, the equilibrium potential for Na^+ (E_{Na}).

B. The end-plate current actually reverses at 0 mV because the ion channel is permeable to both Na^+ and K^+, which thus are able to move into and out of the cell simultaneously (see Box 12–1).

ular value of membrane potential is not equal to the equilibrium potential for Na^+, or for that matter any of the major cations or anions. These experiments therefore raise an intriguing question: Could some unidentified ion be responsible for the end-plate potential? In fact, this potential is produced not by a single ion species but by a combination of ions. The synaptic channels at the end-plate are almost equally permeable to both major cations, Na^+ and K^+. Thus, during the end-plate potential Na^+ flows into the cell and K^+ flows out. The reversal potential is at 0 mV because this is a weighted average of the equilibrium potentials for Na^+ and K^+ (see Box 12–1).

Patch-Clamp Experiments Reveal the Nature of Current Flow Through Single Ion Channels

The current underlying an end-plate potential flows through several hundred thousand transmitter-gated channels. This macroscopic current can be analyzed by reducing it to its fundamental unit, the elementary current that flows through a *single* channel. The technique of choice for such a study is the patch-clamp technique (see Chapter 7, Box 7–1).

Individual Transmitter-Gated Channels Conduct a Unitary Current

Studies with patch-clamp recordings reveal that ACh opens individual channels in all-or-none steps. Each time a channel opens it conducts a current of fixed amplitude. At a resting potential of –90 mV the single-channel current is around –2.7 pA, corresponding to a single-channel conductance of 30 pS (Figure 12–9). Although the *amplitude* of the current for a single channel is relatively constant, the *duration* of the opening of the channel is governed by a stochastic (random) process and so varies from one channel opening to the next. However, the *mean open time* (measured from hundreds of individual channel openings) is a fixed property of the end-plate channels (under given experimental conditions) and is around 1 ms. When a single channel opens, about 17,000 Na^+ ions flow into the muscle cell and a somewhat smaller number of K^+ ions flow out.

Changing the membrane potential changes the magnitude of the current through the channels (Figure 12–10). This happens because a change in driving force ($V_m - E_{EPSP}$) has the same effect on single channels that it has on the total current at the end-plate. Recall that Ohm's law applied to synaptic current is

$$I_{EPSP} = g_{EPSP} \times (V_m - E_{EPSP}).$$

The equivalent expression for current flow through a single channel is

$$i_{EPSP} = \gamma \times (V_m - E_{EPSP})$$

where i_{EPSP} is the amplitude of current flow through a single open channel and γ is the conductance of the channel.

Figure 12–9 Individual ACh-activated channels open in an all-or-none fashion, and their current adds up linearly.

A. The patch-clamp technique is used to record from single ACh-activated channels. The patch electrode is filled with salt solution and a low concentration of ACh and then brought into close contact with the surface of the muscle membrane (see Box 7–1).

B. Single-channel currents from a patch of membrane on a frog muscle fiber were recorded in the presence of 100 n*M* ACh at a resting membrane potential of –90 mV. **1.** The opening of a channel results in the flow of inward current, which is recorded as a downward step. Since the patch contained a large number of ACh-activated channels, successive openings probably arise from distinct channels. **2.** A histogram of the amplitudes of these rectangular pulses has a single peak. This distribution indicates that the patch of membrane contains only a single type of active channel and that the size of the elementary current through this channel varies randomly around a mean of 2.69 pA (1 pA = 10^{-12} A). This mean, the *elementary current*, is equivalent to an elementary conductance of about 30 pS. (Courtesy of B. Sakmann.)

C. When the membrane potential is increased to –130 mV, the individual channel currents give rise to all-or-none increments of –3.9 pA, equivalent to 30 pS. Thus, when numerous channels open simultaneously, the individual current pulses add linearly. The record shows one, two, or three channels open at different times. (Courtesy of B. Sakmann.)

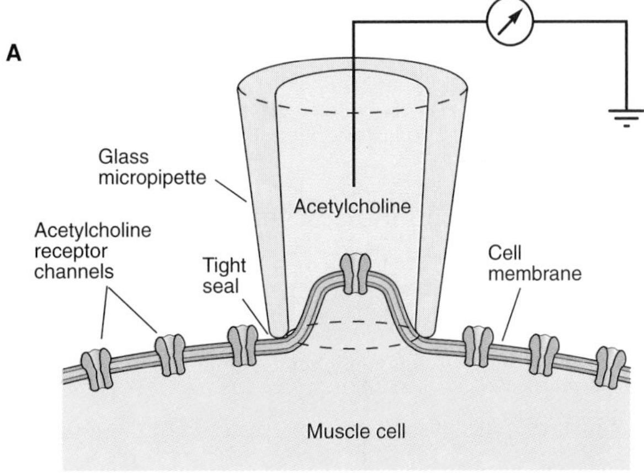
A
Glass micropipette
Acetylcholine
Acetylcholine receptor channels
Tight seal
Cell membrane
Muscle cell

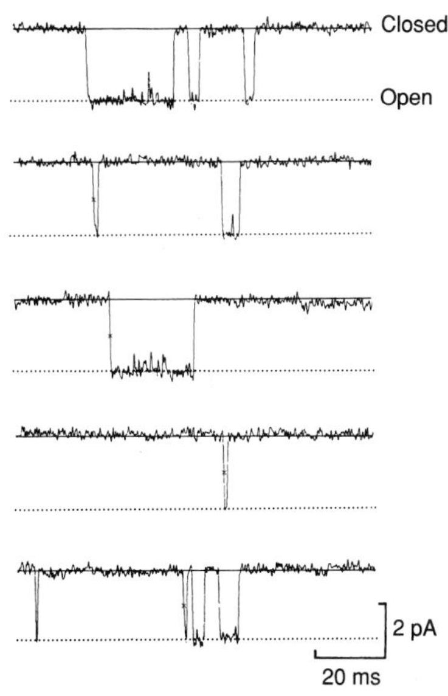
B1 Single-channel currents
Closed
Open
2 pA
20 ms

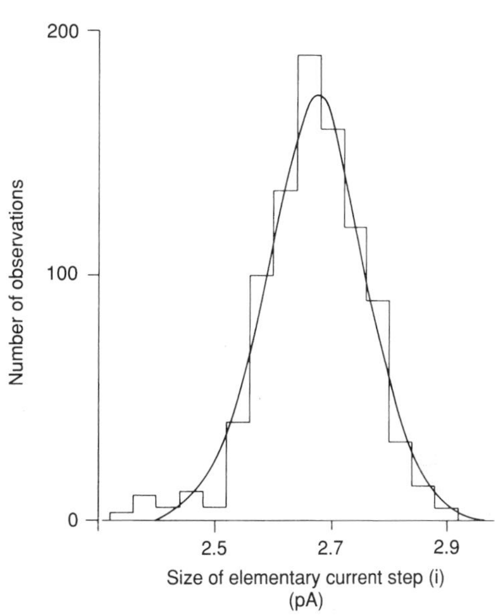
B2 Size of elementary current
200
100
0
Number of observations
2.5
2.7
2.9
Size of elementary current step (i) (pA)

C Total ionic current in a patch of membrane

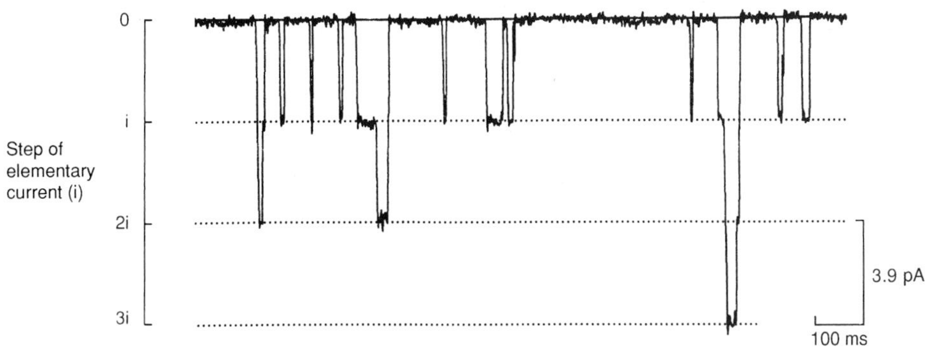
0
i
2i
3i
Step of elementary current (i)
3.9 pA
100 ms

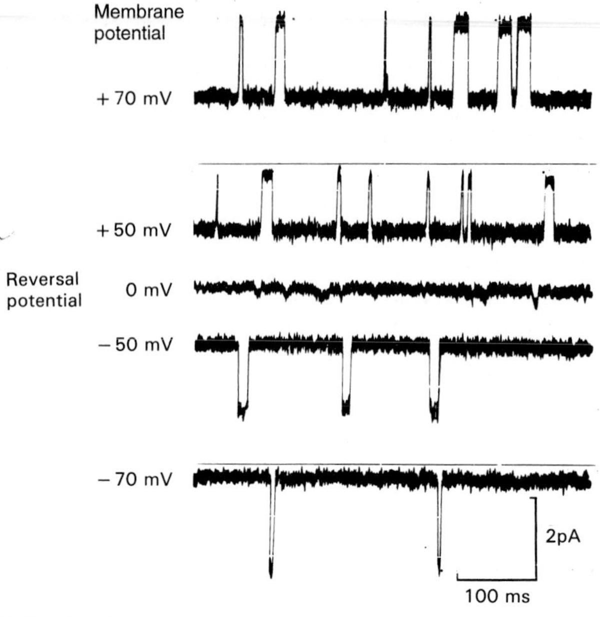

Figure 12–10 A single-channel current has the same reversal potential (0 mV) as does the total end-plate current. The voltage across the patch of membrane where this recording was made was systematically varied during exposure to 2 μ*M* ACh. The current is inward below 0 mV and outward above 0 mV.

BOX 12–1 Reversal Potential of the Excitatory Postsynaptic Potential

The reversal potential of a particular membrane current carried by more than one type of ion, such as the end-plate current through the ACh receptor–channel, is determined by two factors: (1) the relative conductance for the permeant ions (in this case g_{Na} and g_K), and (2) the equilibrium potentials of the ions (E_{Na} and E_K). At the reversal potential for the ACh receptor–channel, inward current carried by Na^+ is balanced by outward current carried by K^+. Thus

$$I_{Na} + I_K = 0. \quad (1)$$

The individual Na^+ and K^+ currents can be obtained from:

$$I_{Na} = g_{Na} \times (V_m - E_{Na}) \quad (2a)$$

$$I_K = g_K \times (V_m - E_K) \quad (2b)$$

Remember that these currents are due to Na^+ and K^+ flowing through a single ACh receptor–channel. Since at the reversal potential $V_m = E_{EPSP}$, and substituting Equations 2a and 2b in Equation 1, we can write

$$g_{Na} \times (E_{EPSP} - E_{Na}) + g_K \times (E_{EPSP} - E_K) = 0 \quad (3)$$

Solving this equation for E_{EPSP} yields

$$E_{EPSP} = \frac{g_{Na} \times E_{Na} + g_K \times E_K}{g_{Na} + g_K}. \quad (4)$$

If we divide the top and bottom of the right side of this equation by g_K, we obtain

$$E_{EPSP} = \frac{\dfrac{g_{Na}}{g_K} \times E_{Na} + E_K}{\dfrac{g_{Na}}{g_K} + 1}. \quad (5)$$

Thus, if $g_{Na} = g_K$, then $E_{EPSP} = (E_{Na} + E_K)/2$.

These equations can also be used to solve for the ratio g_{Na}/g_K if one knows E_{EPSP}, E_K, and E_{Na}. Thus, rearranging Equation 3 yields

$$\frac{g_{Na}}{g_K} = \frac{(E_{EPSP} - E_K)}{(E_{Na} - E_{EPSP})}. \quad (6)$$

At the neuromuscular junction $E_{EPSP} = 0$ mV, $E_K = -100$ mV, and $E_{Na} = +55$ mV. Thus, from Equation 6, g_{Na}/g_K has a value around 1.8, indicating that the conductance of the ACh receptor–channel for Na^+ is slightly higher than for K^+. A comparable approach can be used to analyze the reversal potential and the movement of ions during excitatory and inhibitory synaptic potentials in central neurons (Chapter 13).

Four Factors Determine the End-Plate Current

The summed conductance of all open channels in a large population of ACh channels is the total synaptic conductance, $g_{EPSP} = n \times \gamma$, where n is the average number of channels opened by the ACh transmitter. For an end-plate that contains a large number of ACh channels, the average number of open channels is $n = N \times p_o$, where p_o is the probability that any given ACh channel is open and N is the total number of ACh channels in the end-plate membrane. The total end-plate current is therefore given by:

$$I_{EPSP} = n \times \gamma \times (V_m - E_{EPSP})$$

or

$$I_{EPSP} = N \times p_o \times \gamma \times (V_m - E_{EPSP}).$$

This equation shows that the current for the end-plate potential depends on four factors: (1) the total number of end-plate channels (N); (2) the probability that a channel is open (p_o); (3) the conductance of each open channel (γ); and (4) the driving force that acts on the ions ($V_m - E_{EPSP}$). The probability that a channel is open depends largely on the concentration of the transmitter at the receptor, not on the value of the membrane potential, because the channels are opened by the binding of ACh, not by voltage.

The normal end-plate current is the sum of the opening of more than 200,000 channels. This number is estimated by comparing the end-plate potential of 70 mV with the depolarization of only 0.3 μV caused by the opening of a single channel. The rapid rising phase of the end-plate current is due to the nearly synchronous opening of channels in response to the rapid rise in ACh concentration in the synaptic cleft. The ACh concentration then rapidly falls (in less than a millisecond), owing to the breakdown of ACh and to diffusion. Following the fall in ACh concentration, the ACh-gated channels close in a random manner. Since each closure produces a tiny step decrease in the inward synaptic current, the total synaptic current and the end-plate potential appear to decay smoothly (Figure 12–11).

The relationships between single-channel current, total end-plate current, and end-plate potential are shown in Figure 12–12 for a wide range of membrane potentials.

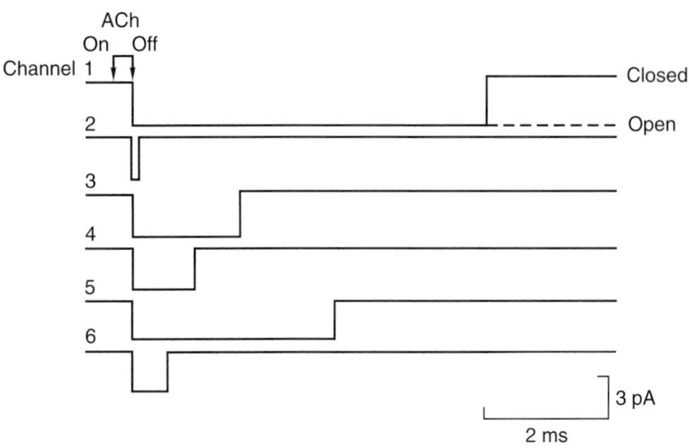

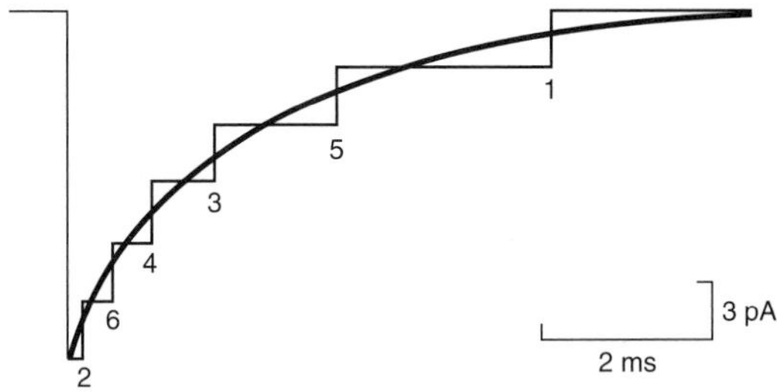

Figure 12–11 Individual ACh-activated channels in a cell open nearly simultaneously but close at different times (**A**). As a result, the total current flowing through all ACh-activated channels at an end-plate appears to decay smoothly. The stepped trace (**B**) reflects the sequential closing of channels at a hypothetical end-plate containing only six channels (the number indicates which channel has closed). In the final period of net current flow only channel 1 is open. In a whole muscle fiber, with thousands of channels, the time course of the ACh-activated current is equivalent to the idealized record but the individual channel closings are not detectable because the total end-plate current (hundreds of nanoamperes) is so much larger than the single-channel current amplitude (2–3 pA). (Adapted from D. Colquhoun, 1981.)

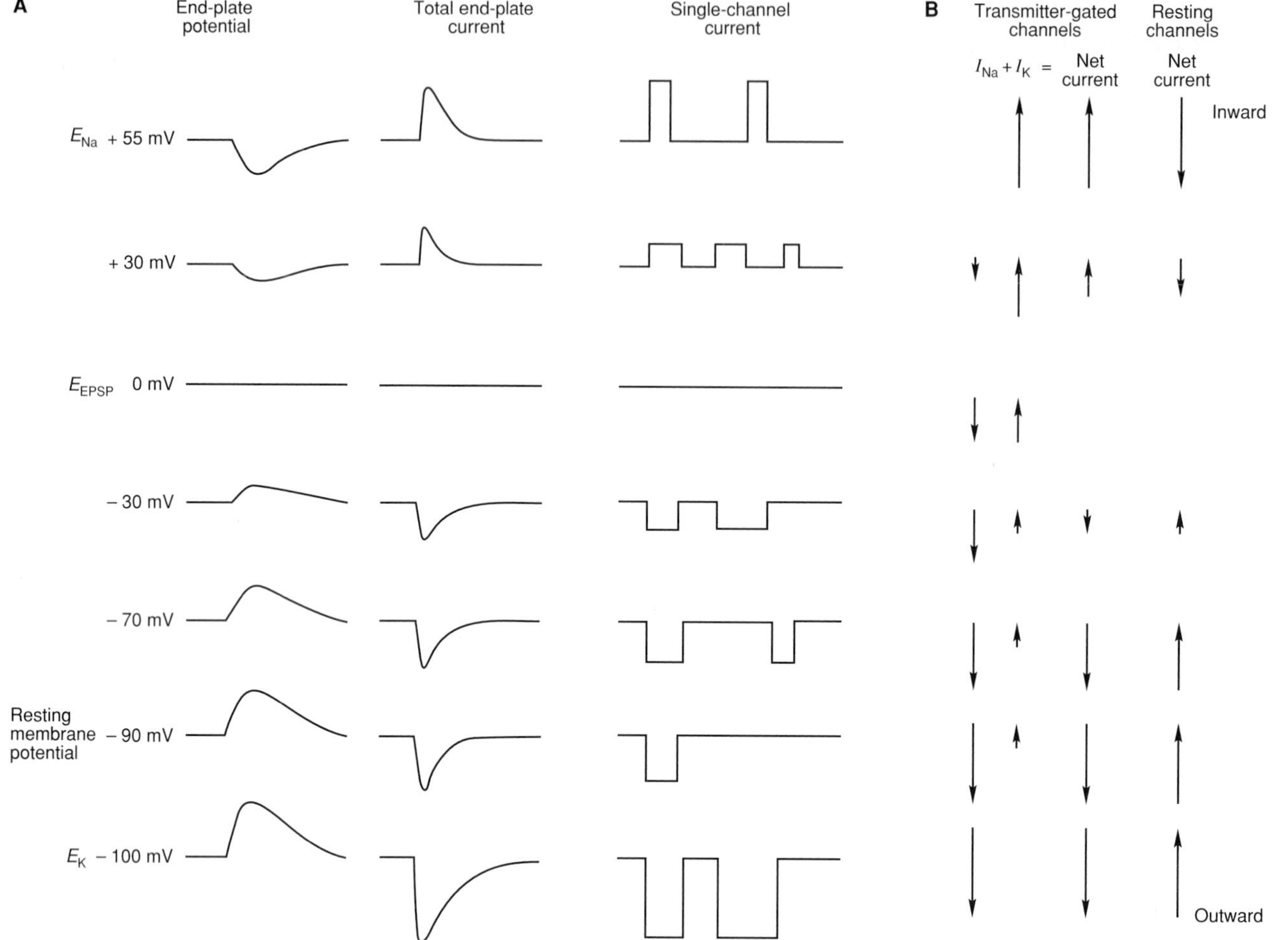

Figure 12–12 The ACh-activated end-plate potential, total end-plate current, and single-channel current are all affected in a similar way by the membrane potential.

A. At the normal muscle resting potential of –90 mV the single-channel currents and total end-plate current (made up of currents from more than 200,0000 channels) are large and inward because of the large inward driving force on current flow through the ACh-gated channels. This large inward current produces a large depolarizing end-plate potential. At more positive levels of membrane potential (increased depolarization) the inward driving force on Na^+ is less and the outward driving force on K^+ is greater. This results in a decrease in the size of the single-channel currents and the magnitude of the end-plate currents, thus reducing the size of the end-plate potential. At the reversal potential (0 mV) the inward Na^+ flux is balanced by the outward K^+ flux, so there is no inward current flow at the end-plate and no change in V_m. Further depolarization to +30 mV inverts the direction of the end-plate current, as there is now a large outward driving force on K^+ and a small inward driving force on Na^+. On either side of the reversal potential the end-plate current drives the membrane potential toward the reversal potential.

B. The direction of Na^+ and K^+ fluxes in individual channels is altered by changing V_m. The algebraic sum of the Na^+ and K^+ fluxes, I_{Na} and I_K, gives the *net current* that flows through the transmitter-gated channels. This net synaptic current is equal in size, and opposite in direction, to that of the net extrasynaptic current flowing in the return pathway of the resting channels and membrane capacitance. (The relative magnitude of a current is represented by the length of the arrow.)

The Nicotinic Acetylcholine Receptor–Channel Is a Membrane-Spanning Protein

A directly gated receptor–channel has two functions: it recognizes and binds the chemical transmitter, and it creates a pore in the membrane through which ions flow. Where in the receptor molecule is the binding site located? How is the pore formed? What are its properties? Insights into these questions have been obtained from molecular studies of the ACh-activated receptor–channel proteins and their genes.

Biochemical and molecular cloning studies indicate that the mature nicotinic ACh receptor is a membrane glycoprotein formed from five subunits: two α units and one each of β, γ, and δ (Figure 12–13). The amino terminus of each of the subunits is exposed on the extracellular surface of the membrane. The amino terminus of the α subunit contains a site that binds ACh with high affinity. One molecule of ACh must bind to each of the two α subunits for the channel to open efficiently (Figure 12–13). The inhibitory snake venom α-bungarotoxin, which is recognized by and binds specifically to the receptor, also binds to the α subunit.

Insight into the structure of the channel pore through which ions pass has come from analysis of the primary sequences of the receptor–channel subunits as well as from biophysical studies. The four subunit types are encoded by distinct but related genes. Sequence comparison of the subunits shows a high degree of similarity among them: half of the amino acid residues are identical or conservatively substituted. This similarity suggests that all subunits have a similar function. Like the voltage-gated channels, with which the ACh receptor–channel shares a common plan (discussed in Chapter 10), all four of the genes for the subunits are thought to be derived from a common ancestral gene.

The distribution of the polar and nonpolar amino acids of the subunits provides important clues as to how the subunits are threaded through

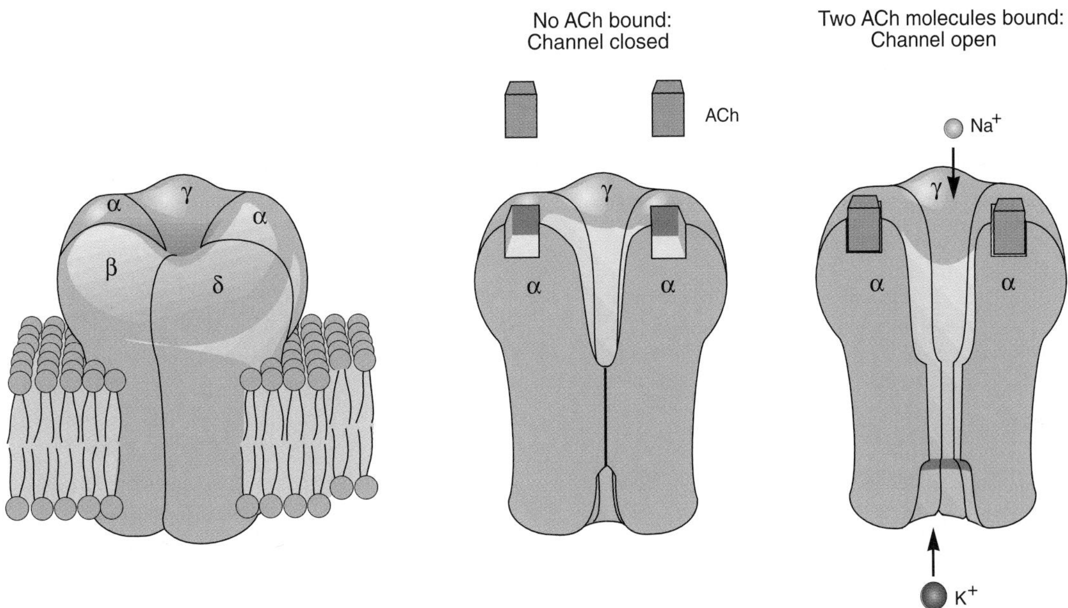

Figure 12–13 This three-dimensional model of the nicotinic ACh-activated ion channel is based on the model of Arthur Karlin and co-workers. The receptor–channel complex consists of several subunits arranged to form a pore. When two molecules of ACh bind to the portions of the α subunits exposed to the membrane surface, the receptor–channel changes conformation, opening the pore in the portion of the receptor embedded in the lipid bilayer (see Figure 12–15A). Both K^+ and Na^+ flow through the open channel down their electrochemical gradients.

the membrane bilayer. Each subunit contains four hydrophobic regions of about 20 amino acids called M1 to M4, each of which is thought to form an α-helix traversing the membrane. Their amino acid sequences suggest that the subunits are symmetrically arranged in such a way that they create a channel spanning the membrane (Figure 12–14).

A A single subunit in the ACh receptor-channel

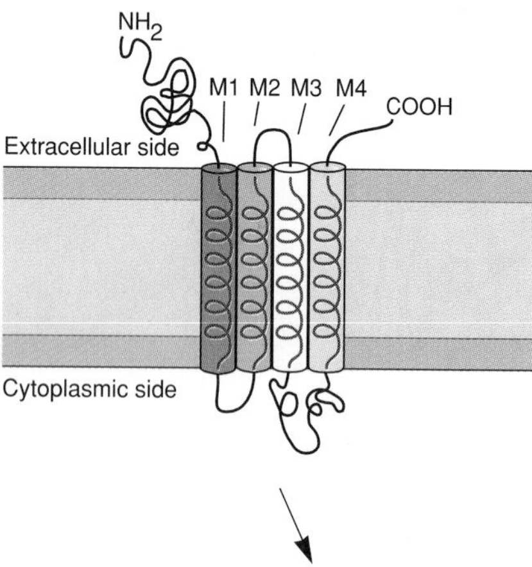

B Hypothetical arrangement of subunits in one channel

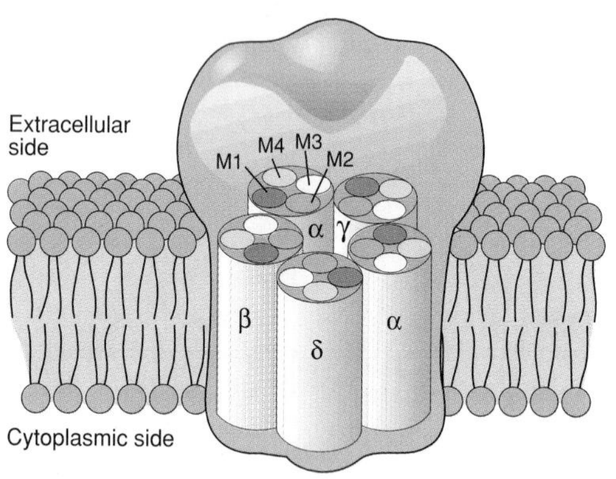

Figure 12–14 A detailed molecular model of the subunits of the transmembrane disposition of the ACh receptor–channel proposed by Shosaku Numa and his colleagues.

A. Each subunit is composed of four membrane α-helices (labeled M1 through M4).

B. The five subunits are arranged such that they form an aqueous channel, with the M2 segment of each subunit facing inside and forming the lining of the channel (see Figure 12–15A).

The lining of the channel pore may be formed by the M2 region and by the segment connecting M2 to M3. For example, chlorpromazine, a photoaffinity-labeled channel blocker, has been found to penetrate the open channel about two-thirds of the way and then plug it by binding to a ring of serine residues on the M2 region within the channel pore. Moreover, the channel's cation selectivity is thought to derive from three rings of negative charge that flank the M2 region (Figure 12–15). Each ring is made up of three or four negative charges contributed by negatively charged amino acids (primarily glutamate) of the different subunits. Evidence from site-directed mutagenesis experiments suggests that replacing the critical glutamate residues with a neutral amino acid decreases the single-channel conductance.

Three-dimensional images of the receptor at a resolution of 1.7 nm reveal a channel made up of the encircling receptor subunits (see Figures 12–4 and 12–15A). The receptor–channel complex is divided into three regions: a large entrance region at the external membrane surface, a narrow transmembrane pore that may determine cation selectivity, and a large exit region at the internal membrane surface. A surprisingly large component of this channel, about 6 nm in length, extends into the extracellular space. At its external surface the channel has a wide mouth about 2.5 nm in diameter. But within the bilayer of the membrane the channel abruptly narrows so markedly that it cannot be resolved with this technique. It is presumably here, where the M2 segments are thought to line the pore, that the selectivity filter lies (Figure 12–15A). Electrophysiological measurements of the channel's permeability to a number of organic cations suggest that the diameter of the pore at the selectivity filter is 0.65 nm. This narrow region is quite short, only about 3 nm in length, corresponding to the lengths of both the M2 segment and the hydrophobic core of the bilayer (Figure 12–15B). As the channel emerges from the inner surface of the membrane, it suddenly widens again.

Transmitter-Gated Channels Differ from Voltage-Gated Channels

At the neuromuscular junction the transmitter-gated channels that produce synaptic potentials

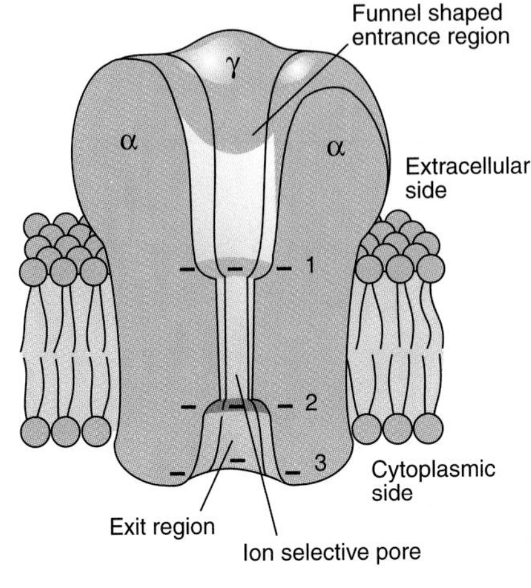

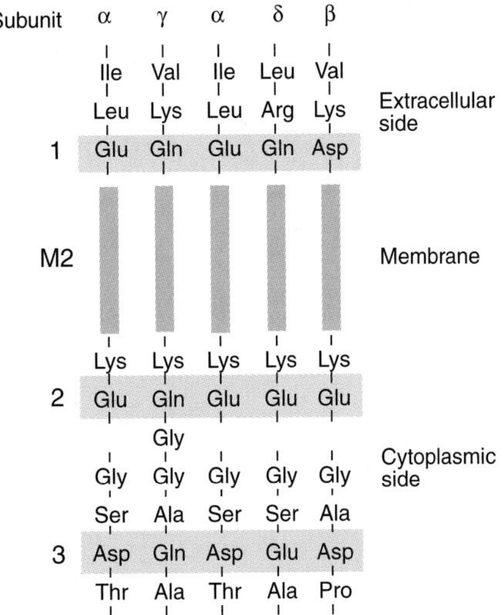

Figure 12–15 A functional model of the nicotinic ACh receptor–channel.

A. This functional model of the ACh receptor–channel is based on experiments by Shosaku Numa, Bert Sakmann, Nigel Unwin, and their colleagues (see Unwin's reconstructed electron microscope image of the channel in Figure 12–4 and Numan's model in Figure 12–14). According to this model negatively charged amino acids on each subunit form three rings of charge around the pore. As a permeant ion traverses the channel it encounters this series of three negatively charged rings. The external (**1**) and internal (**3**) rings may serve as prefilters and divalent blocking sites; the central site (**2**) in the part of the channel embedded in the bilayer may function as a selectivity filter for cations. (Dimensions are not to scale.)

B. Aligned amino acid sequences of the M2 regions and the flanking sequences of each of the five subunits (see Figure 12–14B). The shaded horizantal areas numbered **1**, **2**, and **3** identify the three rings of negative charge (aspartate or glutamate residues) that flank the M2 region and may account for the selectivity of the channel for cations.

differ in two important ways from the voltage-gated channels that generate the action potential. First, two distinct classes of voltage-gated channels, one selective for Na^+ and the other for K^+, are activated sequentially to generate the action potential. In contrast, only one type of transmitter-gated channel, the ACh-activated channel, generates end-plate potentials, and this channel is large enough to allow either Na^+ or K^+ to pass with nearly equal selectivity. Indeed, the channel is so large that it also allows divalent cations, such as Ca^{2+}, and even certain organic cations to pass. Anions are excluded, however, because they are repelled by the negative charges (glutamate) in the pore of the channel.

A second difference between transmitter-gated and voltage-gated channels is that Na^+ flux through voltage-gated channels is regenerative: the increased depolarization of the cell caused by the Na^+ influx opens more voltage-gated Na^+ channels. This regenerative feature is responsible for the all-or-none property of the action potential. In contrast, the number of ACh-activated channels opened during the synaptic potential varies according to the amount of ACh available. The depolarization produced by Na^+ influx through these channels does not lead to the opening of more transmitter-gated channels; it is therefore limited, and by itself cannot produce an action potential. To trigger an action potential, a synaptic potential must recruit neighboring voltage-gated channels (Figure 12–16).

As might be expected from these two differences in physiological properties, there are also pharmacological differences between transmitter-gated and voltage-gated channels. Tetrodotoxin, which blocks

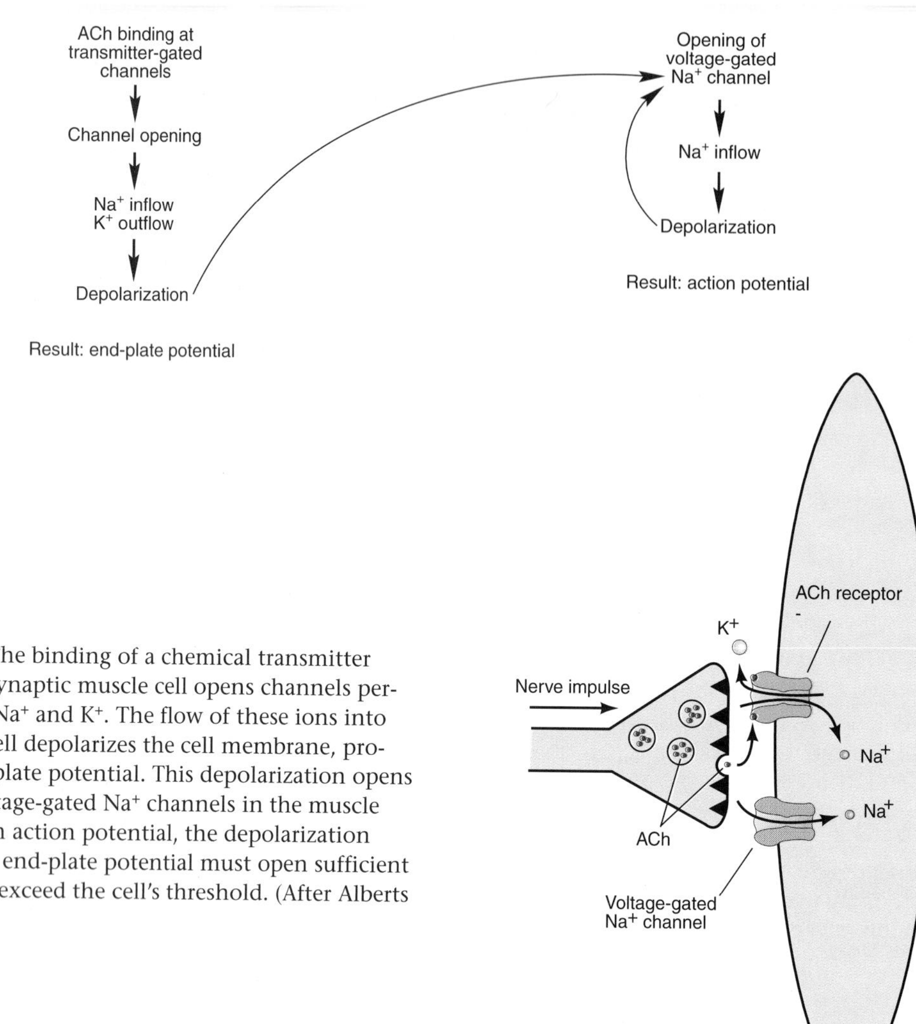

Figure 12–16 The binding of a chemical transmitter (ACh) in a postsynaptic muscle cell opens channels permeable to both Na^+ and K^+. The flow of these ions into and out of the cell depolarizes the cell membrane, producing the end-plate potential. This depolarization opens neighboring voltage-gated Na^+ channels in the muscle cell. To trigger an action potential, the depolarization produced by the end-plate potential must open sufficient Na^+ channels to exceed the cell's threshold. (After Alberts et al., 1989.)

the voltage-gated Na^+ channel, does not block the influx of Na^+ through the nicotinic ACh-activated channels. Similarly, α-bungarotoxin, a snake venom protein that binds to the nicotinic receptors and blocks the action of ACh, does not interfere with voltage-gated Na^+ or K^+ channels.

In Chapter 13 we shall learn about still another type of channel, the NMDA-type (*N*-methyl-D-aspartate) glutamate receptor, which is found in most neurons of the brain. This channel is doubly gated, responding both to voltage *and* to a chemical transmitter.

An Overall View

Motor neurons form synapses with muscle fibers at specialized regions of the muscle membrane called end-plates. An action potential in the presynaptic motor neuron causes ACh to be released from the neuron's terminals. The ACh then diffuses across the synaptic cleft and binds nicotinic ACh receptor–channels in the end-plate. As a result, the channels open to cations (Na^+, K^+, and Ca^{2+}), and a net influx of Na^+ ions produces a depolarizing synaptic potential called the end-plate potential.

Because the number of ACh-activated channels opened is limited by the fixed amount of ACh released onto the postsynaptic cell, these channels by themselves cannot produce a regenerative action potential. But by depolarizing the postsynaptic cell, these transmitter-gated channels activate voltage-dependent Na^+ channels at the end-plate region. As the depolarization of the postsynaptic cell increases, more of the Na^+ channels are activated and the resulting influx of Na^+ generates the current needed to produce an action potential.

The nicotinic ACh-activated channel has been purified, its genes have been cloned, and the amino acid sequence determined. The receptor is a protein integral to the membrane and is composed of five subunits, two of which are identical. Each subunit has four hydrophobic regions that are thought to form membrane-spanning α-helices.

Receptor–channels have two functions: (1) the recognition and binding of neurotransmitter from the presynaptic cell, and (2) the gating of the ion channel. In the ACh-activated receptor the molecular structures underlying these two functions have been identified. The steps that link the binding of ACh to the opening of the ion channel are under investigation. Thus, we should soon be able to answer the question: How does the detailed molecular structure of the ACh receptor account for its various physiological functions?

Acetylcholine is only one of the several neurotransmitters used by the nervous system, and the end-plate potential is one of many examples of postsynaptic transmitter actions. Do other neurotransmitters produce effects similar to those of ACh at the nerve–muscle synapse? Do the same principles also apply to transmitter actions in the central nervous system, or are other mechanisms involved? In the past these questions were difficult to answer because of the small size and great complexity of nerve cells in the central nervous system. However, advances in experimental technique—in particular, patch clamping—have uncovered a wider range of neurotransmitter actions available for study. It is already clear that although many neurotransmitters operate in a way similar to that of ACh at the end-plate, other types of transmitters do not. In the next two chapters we explore some of this rich variety in synaptic transmission in the central and peripheral nervous systems.

Postscript: The End-Plate Current Can Be Calculated from an Equivalent Circuit

The flow of current through a population of ACh-activated end-plate channels can be described simply by Ohm's law. However, to understand fully how the flow of electrical current generates the end-plate potential we also need to consider all the resting channels in the surrounding membrane. Since channels are proteins that span the bilayer of the membrane, we must also take into consideration the capacitive properties of the membrane and the ionic batteries determined by the distribution of Na^+ and K^+ inside and outside the cell.

The dynamic relationship of these various components can be explained using rules that govern the flow of current in passive electrical devices that consist only of resistors, capacitors, and batteries (see Chapter 9). We can represent the end-plate region with an equivalent circuit that has three parallel branches: (1) a branch representing the flow of synaptic current through the transmitter-gated channels; (2) a branch representing the return current flow through resting channels (the nonsynaptic membrane); and (3) a branch representing current flow across the lipid bilayer, which acts as a capacitor (Figure 12–17).

Since the end-plate current is carried by both Na^+ and K^+, we could represent the synaptic branch of the equivalent circuit as two parallel branches, each representing the flow of a different ion species. At the end-plate, however, Na^+ and K^+ flow through the same ion channel. It is therefore more convenient (and correct) to combine the Na^+ and K^+ current pathways into a single conductance, representing the channel gated by ACh. The conductance of this pathway depends on the number of channels opened, which in turn depends on the concentration of transmitter. In the absence of transmitter, no channels are open and the conductance is zero. When a presynaptic action potential causes the release of transmitter, the conductance of this pathway rises to a value of around 5×10^{-6} S (or a resistance of 2×10^5 Ω). This is about five times the conductance of the parallel branch representing the resting or leakage channels (g_l).

The end-plate conductance is in series with a battery (E_{EPSP}), whose value is given by the reversal

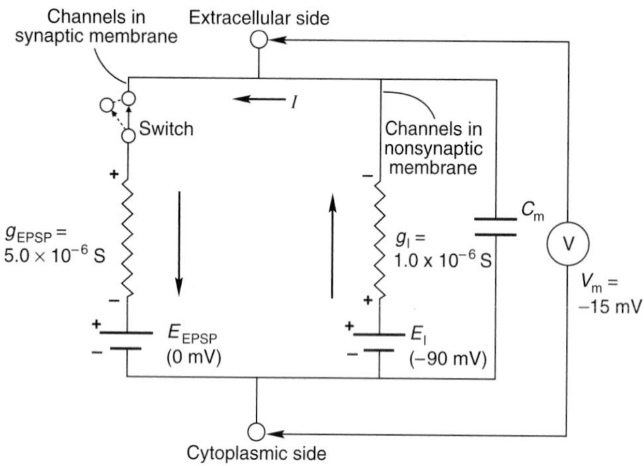

Figure 12–17 The equivalent circuit of the end-plate potential with two parallel current pathways. One pathway consists of a battery representing the synapse, E_{EPSP}, in series with the conductance of the ACh-gated channels, g_{EPSP}. The other pathway consists of the battery representing the resting potential (E_l) in series with the conductance of the resting channels (g_l). In parallel with both of these conductance pathways is the membrane capacitance (C_m). The voltmeter (**V**) measures the potential difference between the inside and the outside of the cell.

When no ACh is present, the gated channels are closed and no current flows through them. This state is depicted as an open electrical circuit in which the synaptic conductance is not connected to the rest of the circuit. The binding of ACh opens the synaptic channel. This event is electrically equivalent to throwing the switch (**S**) that connects the gated conductance pathway (g_{EPSP}) with the resting pathway (g_l). As a result, in the steady state, current flows inward through the gated channels and outward through the resting channels. With the indicated values of conductances and batteries, the membrane will depolarize from –90 mV (its resting potential value) to –15 mV (the peak of the end-plate potential).

potential for synaptic current flow (0 mV) (Figure 12–17). This value is the weighted algebraic sum of the Na^+ and K^+ equilibrium potentials (see Box 12–1).

The current flowing during the excitatory postsynaptic potential (I_{EPSP}) is given by

$$I_{EPSP} = g_{EPSP} \times (V_m - E_{EPSP}).$$

Using this equation and the equivalent circuit of Figure 12–17, we can now analyze the end-plate potential in terms of the flow of ionic current. At the onset of the excitatory synaptic action (the dynamic phase), an inward current (I_{EPSP}) flows through the ACh-activated channels because of the increased conductance to Na^+ and K^+ and the large inward driving force on Na^+ at the initial resting potential (–90 mV). Since current flows in a closed loop, the inward synaptic current must leave the cell as outward current. From the equivalent circuit we see that there are two parallel pathways for outward current flow: a conductance pathway (I_l) representing current flow through the resting (or leakage) channels, and a capacitive pathway (I_c) representing current flow across the membrane capacitance C_m. Thus,

$$I_{EPSP} = -(I_l + I_c).$$

During the earliest phase of the end-plate potential the membrane potential, V_m, is still close to its resting value, E_l. As a result, the outward driving force on current flow through the resting channels ($V_m - E_l$) is small. Therefore, most of the current leaves the cell as capacitive current, and the membrane depolarizes rapidly (phase 2 in Figure 12–18). As the cell depolarizes, the outward driving force on current flow through the resting channels increases, while the inward driving force on synaptic current flow through the ACh-activated channels decreases. Concomitantly, as the concentration of ACh in the synapse falls, the ACh-activated channels begin to close, and eventually the flow of inward current through the gated channels is exactly balanced by outward current flow through the resting channels ($I_{EPSP} = -I_l$). At this point no current flows into or out of the capacitor (i.e., C_m = 0), and since the rate of change of membrane potential is directly proportional to I_c (i.e., $I_c = \Delta V/\Delta t$), the membrane potential will have reached a peak steady-state value ($\Delta V/\Delta t = 0$; phase 3 in Figure 12–18).

As the gated channels close, I_{EPSP} decreases further. Now I_{EPSP} and I_l are no longer in balance and, as the outward current flow I_l becomes larger than the inward synaptic current, I_{EPSP}, the membrane potential starts to repolarize. During most of the declining phase of the synaptic action no current flows through the ACh-activated channels, which are all closed. Instead, current flows only out through the resting channels and in across the capacitor (phase 4 in Figure 12–18).

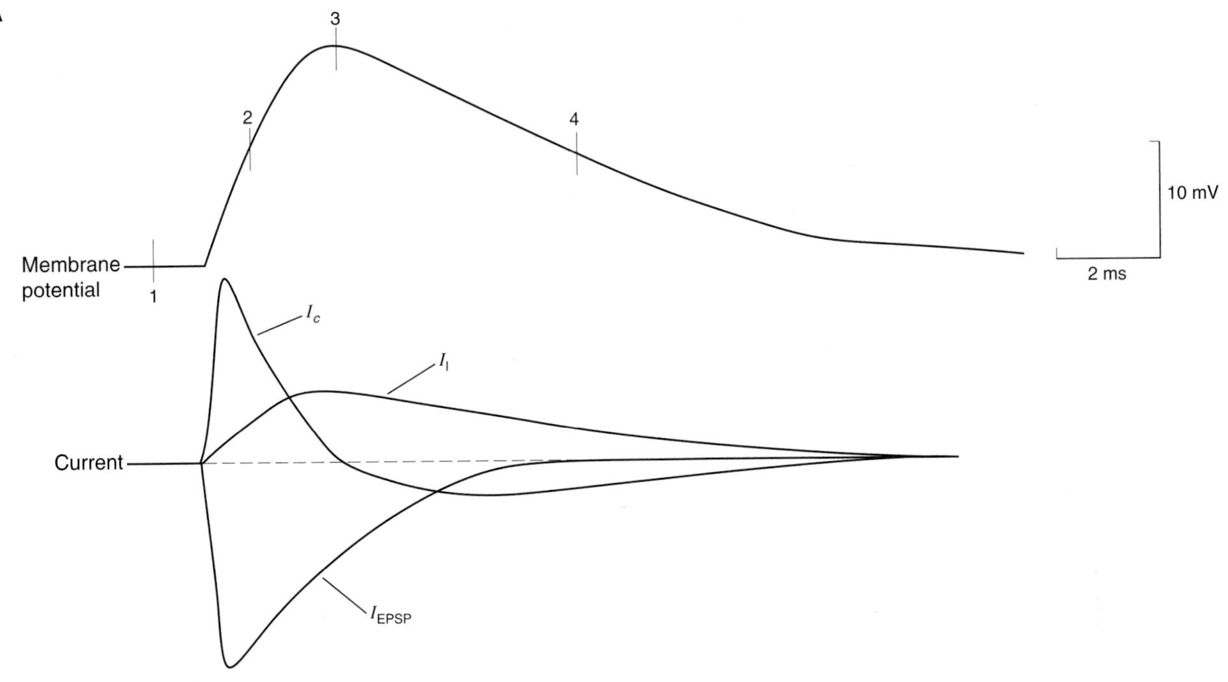

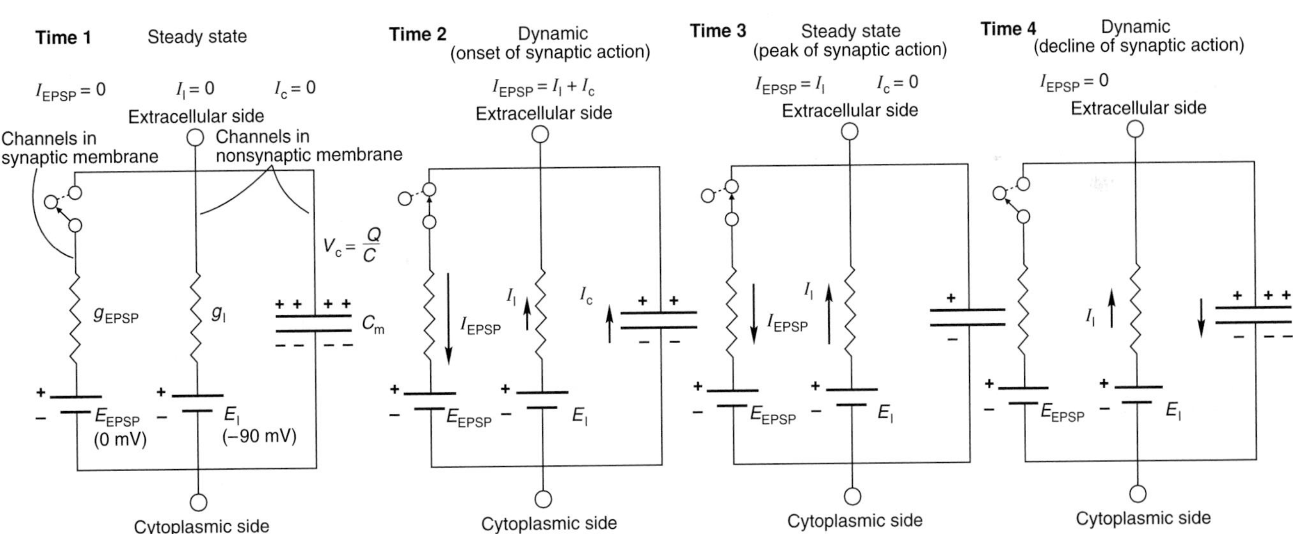

Figure 12–18 Both the ACh-activated synaptic conductance and the passive membrane properties of the muscle cell determine the time course of the end-plate potential.

A. The time course of the end-plate potential (**top trace**) is a summation of the time courses of the components of the current: the currents through the ACh-gated channels (I_{EPSP}), the resting (or leakage) channels (I_l), and the capacitor (I_c). Capacitive current flows only when the membrane potential is changing. In the steady state, such as at the peak of the end-plate potential, the inward flow of ionic current through the ACh-activated channels (I_{EPSP}) is exactly balanced by the outward flow of ionic current across the resting channels (I_l) and there is no flow of capacitive current.

B. Equivalent circuits for the current at times 1, 2, 3, and 4 shown in **A.** (The relative magnitude of a current is represented by the length of the arrow.)

When the end-plate potential is at its peak or steady-state value, $I_c = 0$ and therefore the value V_m can be easily calculated. The inward current flow through the gated channels (I_{EPSP}) must be exactly balanced by outward current flow through the resting channels (I_l):

$$I_{EPSP} + I_l = 0. \qquad (1)$$

The current flowing through the active ACh-gated channels (I_{EPSP}) and through the resting channels (I_l) is given by Ohm's law:

$$I_{EPSP} = g_{EPSP} \times (V_m - E_{EPSP}),$$

and

$$I_l = g_l \times (V_m - E_l).$$

By substituting these two expressions into Equation 1, we obtain

$$g_{EPSP} \times (V_m - E_{EPSP}) + g_l \times (V_m - E_l) = 0.$$

To solve for V_m, we need only expand the two products in the equation and rearrange them so that all terms in voltage (V_m) appear on the left side:

$$(g_{EPSP} \times V_m) + (g_l \times V_m) = (g_{EPSP} \times E_{EPSP}) + (g_l \times E_l).$$

By factoring out V_m on the left side, we finally obtain

$$V_m = \frac{g_{EPSP} \times E_{EPSP} + g_l \times E_l}{g_{EPSP} + g_l} \qquad (2)$$

This equation is similar to that used to calculate the resting and action potentials (Chapter 8). According to Equation 2, the peak voltage of the end-plate potential is a weighted average of the electromotive forces of the two batteries for gated and resting currents. The weighting factors are given by the relative magnitudes of the two conductances. If the gated conductance is much smaller than the resting membrane conductance ($g_{EPSP} \ll g_l$), $g_{EPSP} \times E_{EPSP}$ will be negligible compared with $g_l \times E_l$. Under these conditions V_m will remain close to E_l. This situation occurs when only a very few channels are opened by ACh (because its concentration is low). On the other hand, if g_{EPSP} is much larger than g_l, Equation 2 states that V_m approaches E_{EPSP}, the synaptic reversal potential. This situation occurs when the concentration of ACh is high and a large number of ACh-activated channels are opened. At intermediate ACh concentrations, with a moderate number of ACh-activated channels open, the peak end-plate potential lies somewhere between E_l and E_{EPSP}.

We can now use this equation to calculate the peak end-plate potential for the specific case shown in Figure 12–17, where $g_{EPSP} = 5 \times 10^{-6}$ S, $g_l = 1 \times 10^{-6}$ S, $E_{EPSP} = 0$ mV, and $E_l = -90$ mV. Substituting these values into Equation 2 yields

$$V_m = \frac{(5 \times 10^{-6}\,\text{S}) \times (0\,\text{mV}) + (1 \times 10^{-6}\,\text{S}) \times (-90\,\text{mV})}{(5 \times 10^{-6}\,\text{S}) + (1 \times 10^{-6}\,\text{S})}$$

or

$$V_m = \frac{(1 \times 10^{-6}\,\text{S}) \times (-90\,\text{mV})}{(6 \times 10^{-6}\,\text{S})}$$

$$V_m = -15\,\text{mV}.$$

The peak amplitude of the end-plate potential is then

$$\Delta V_{EPSP} = V_m - E_l$$
$$= -15\,\text{mV} - (-90\,\text{mV})$$
$$= 75\,\text{mV}.$$

As a check for consistency, we can see whether, at the peak of the end-plate potential, the synaptic current is equal and opposite to the nonsynaptic current so that the net membrane current is zero. Thus

$$I_{EPSP} = (5 \times 10^{-6}\,\text{S}) \times (-15\,\text{mV} - 0\,\text{mV})$$
$$= -75 \times 10^{-9}\text{A}$$

and

$$I_l = (1 \times 10^{-6}\,\text{S}) \times [-15\,\text{mV} - (-90\,\text{mV})],$$
$$= 75 \times 10^{-9}\text{A}.$$

Here we see that solving Equation 2 ensures that $I_{EPSP} + I_l = 0$.

Selected Readings

Czajkowski, C., and Karlin, A. 1991. Agonist binding site of *Torpedo* electric tissue nicotinic acetylcholine receptor: A negatively charged region of the δ subunit within 0.9 nm of the α subunit binding site disulfide. J. Biol. Chem. 266:22603–22612.

Fatt, P., and Katz, B. 1951. An analysis of the end-plate potential recorded with an intra-cellular electrode. J. Physiol. (Lond.) 115:320–370.

Heuser, J. E., and Reese, T. S. 1977. Structure of the synapse. In E. R. Kandel (ed.), Handbook of Physiology, Section 1: The Nervous System, Vol. I: Cellular Biology of Neurons, Part 1. Bethesda, Md.: American Physiological Society, pp. 261–294.

Imoto, K., Busch, C., Sakmann, B., Mishina, M., Konno, T., Nakai, J., Bujo, H., Mori, Y., Fukuda, K., and Numa, S. 1988. Rings of negatively charged amino acids determine the acetylcholine receptor channel conductance. Nature 335:645–648.

Jessell, T.M., and Kandel, E.R. 1993. Synaptic transmission: A bidirectional and self-modifiable form of cell-cell communication. Cell 72/Neuron 10 (Suppl.): 1–30.

Karlin, A. 1993. Structure of nicotinic acetylcholine receptors. Curr. Opin. Neurobiol. 3:299–309.

Katz, B., and Miledi, R. 1970. Membrane noise produced by acetylcholine. Nature 226:962–963.

Léna, C. and Changeux, J.-P. 1993. Allosteric modulations of the nicotinic acetylcholine receptor. Trends Neurosci. 16:181–186.

Miller, C. 1989. Genetic manipulation of ion channels: A new approach to structure and mechanism. Neuron 2:1195–1205.

Neher, E. 1992. Nobel lecture. Ion channels for communication between and within cells. Neuron. 8:605–612.

Sakmann, B. Nobel lecture. Elementary steps in synaptic transmission revealed by currents through single ion channels. Neuron. 8:613–629.

13

Synaptic Integration

A Central Neuron Receives Both Excitatory and Inhibitory Signals

Excitatory and Inhibitory Signals Are Integrated into a Single Response by the Cell

Synapses onto a Single Central Neuron Are Grouped According to Function

Excitatory Synaptic Action Is Mediated by Transmitter-Gated Channels Selective for Sodium and Potassium

Opening of Sodium and Potassium Channels Leads to an Inward Depolarizing Current

Glutamate Is a Major Excitatory Transmitter in the Brain

Inhibitory Synaptic Action Is Usually Mediated by Receptor–Channels Selective for Chloride

Opening of Transmitter-Gated Chloride Channels Raises the Threshold for Excitatory Actions

GABA and Glycine Are Inhibitory Transmitters

Excitatory and Inhibitory Synapses Have Distinctive Ultrastructures

Synaptic Receptors for Glutamate, GABA, and Glycine Are Transmembrane Proteins

Transmitter-Gated, Voltage-Gated, and Gap-Junction Ion Channels Share Certain Structural Features

The Signals Produced by Transmitter-Gated and Voltage-Gated Channels Have Features in Common

Like synaptic transmission at the neuromuscular junction, many forms of communication between neurons in the central nervous system involve directly gated ion channels in the cell membrane. Thus many of the principles that govern the workings of synaptic connections in the neuromuscular junction also operate in the central nervous system.

Signaling between central neurons, however, is more complex. First, whereas most muscle fibers are innervated by only one motor neuron, a central nerve cell such as the motor neuron in the spinal cord receives connections from hundreds of neurons. Second, the muscle fiber receives only excitatory input (there are no inhibitory synapses onto vertebrate skeletal muscle). Central neurons, on the other hand, receive both excitatory and inhibitory inputs. Third, all the excitatory connections on muscle fibers are mediated by a single neurotransmitter, acetylcholine, which activates the same kind of receptor–channel, the nicotinic acetylcholine receptor–channel, at all connections. In the central nervous system the inputs to a single cell are mediated by a variety of transmitters and any given transmitter can control different types of ion channels, some of which are directly gated and some indirectly gated by second messengers. As a result, unlike muscle fibers, central neurons must integrate diverse sets of inputs into a coordinated response. Finally, the synapse of a motor neuron at

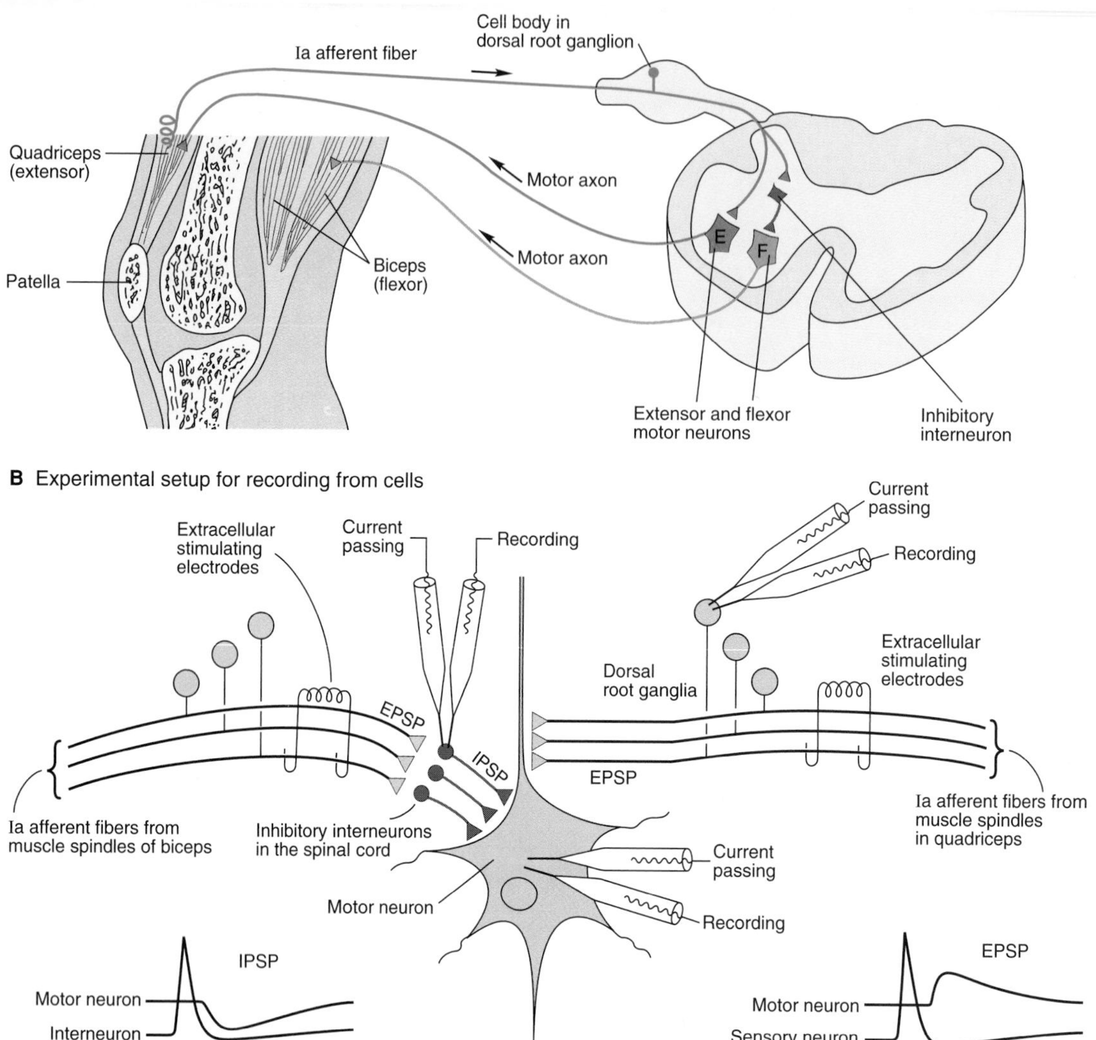

Figure 13–1 The combination of excitatory and inhibitory synaptic connections mediating the stretch reflex of the quadriceps muscle is typical of circuits in the central nervous system.

A. The afferent neurons of the quadriceps muscle make an excitatory connection with the motor neurons innervating this same muscle group (the *extensor* motor neurons). They also make an excitatory connection with an interneuron. This interneuron, in turn, makes an inhibitory connection with the motor neurons innervating the antagonist biceps muscle group (the *flexor* motor neurons).

B. This idealized experimental setup shows the approaches to studying the inhibition and excitation of a motor neuron in the pathway illustrated in **A**. The **right** half of the drawing shows two alternatives for measuring excitatory potentials in the motor neuron: either the whole afferent nerve from the quadriceps can be stimulated electrically with extracellular electrodes, or single axons can be stimulated with intracellular electrodes. The **left** half of the drawing shows the setup for measuring and eliciting inhibitory potentials in the motor neuron; the inhibitory interneurons in the pathway from the biceps are stimulated intracellularly. The type of signal conveyed at each synapse is shown in the idealized electrical recordings at the bottom of the figure. An action potential stimulated in the inhibitory interneuron in the *biceps* pathway causes an inhibitory (hyperpolarizing) postsynaptic potential (IPSP) in the motor neuron. In contrast, an action potential stimulated in the afferent neuron from the *quadriceps* triggers an excitatory (depolarizing) postsynaptic potential (EPSP) in the motor neuron.

a muscle is highly effective—each action potential in a single motor neuron produces a synaptic potential that is invariably suprathreshold and always produces an action potential in the muscle fiber. In contrast, the synaptic connections made by a single presynaptic neuron onto the motor neuron are only modestly effective, and perhaps 50 to 100 excitatory presynaptic neurons must fire together to produce a synaptic potential large enough to trigger an action potential.

The first insight into directly gated synaptic actions in the central nervous system came in the 1950s from John Eccles and his colleagues, who worked on the spinal motor neurons that control the stretch reflex, the simple behavior we considered in Chapters 2 and 3. This work, in turn, was based on studies of the nerve–muscle synapse that we examined in the preceding chapter. The spinal motor neurons have large cell bodies and are useful for examining central synaptic mechanisms because they receive both excitatory and inhibitory connections.

A Central Neuron Receives Both Excitatory and Inhibitory Signals

To analyze the synapses that mediate the stretch reflex, Eccles activated a population of large axons of the sensory cells that innervate the stretch receptor organs in the quadriceps muscle. Nowadays the same experiments can be done by stimulating a single sensory neuron. By passing sufficient current through a microelectrode inserted into the cell body of one of the sensory neurons in the dorsal root ganglion, it is possible to produce an action potential in the cell. This in turn produces a small excitatory postsynaptic potential (EPSP) in the motor neuron innervating the muscle from which the sensory neuron originates (Figure 13–1). The EPSP produced by the one sensory cell depolarizes the motor neuron by less than 1 mV, often only 0.2 to 0.4 mV—far below the threshold required for generating an action potential (a depolarization of 10 mV or more is required to reach threshold).

Stimulating a sensory neuron that innervates the hamstrings, a muscle group antagonistic to the quadriceps, produces a small inhibitory postsynaptic potential (IPSP) in the motor neuron of the quadriceps (Figure 13–1). This hyperpolarizing action is mediated by an inhibitory interneuron, which receives input from the stretch receptor neurons and in turn connects with the motor neurons. The interneurons can also be stimulated intracellularly.

Although a single EPSP in the motor neuron is not nearly large enough to elicit an action potential, the convergence of many excitatory synaptic potentials from many afferent fibers can be integrated by the neuron to initiate an action potential. At the same time, inhibitory synaptic potentials, if strong enough, can counteract the sum of the excitatory actions and prevent the membrane potential from reaching threshold.

Synaptic inhibition, in addition to counteracting synaptic excitation, can exert powerful control over spontaneously active nerve cells. Many cells in the brain are spontaneously active, as are the pacemaker cells of the heart. By suppressing the spontaneous generation of action potentials, synaptic inhibition can shape the pattern of firing in a cell. This function, illustrated in Figure 13–2, is called the sculpturing role of inhibition.

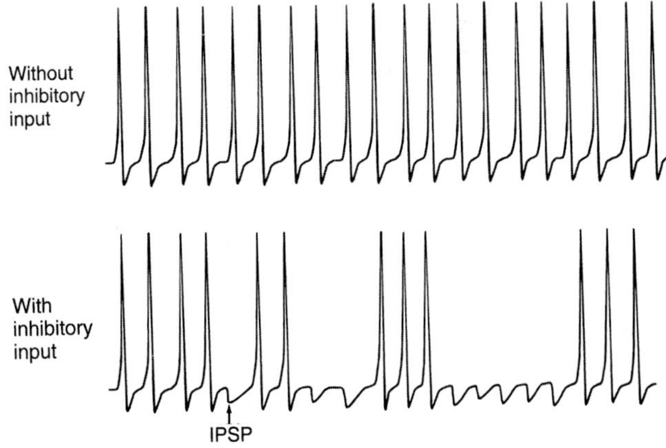

Figure 13–2 Inhibition can shape the firing pattern of a spontaneously active neuron. Without inhibitory input the neuron fires continuously at a fixed interval. With inhibitory input some action potentials are inhibited, resulting in a distinctive pattern of impulses. This effect of inhibition on the firing of a neuron is called *sculpturing*.

Excitatory and Inhibitory Signals Are Integrated into a Single Response by the Cell

Each neuron in the central nervous system, whether in the spinal cord or in the brain, is constantly bombarded by synaptic input from other neurons. For example, there may be as many as 10,000 different presynaptic endings on a single motor neuron. Some are excitatory, others inhibitory; some strong, others weak. Some inputs contact the motor cell on the tips of its apical dendrites, others on proximal dendrites, some on the dendritic shaft, others on dendritic spines. The different inputs can reinforce or cancel one another.

As noted earlier, the synaptic potentials produced by a single presynaptic neuron typically are small and are not capable of exciting a postsynaptic cell sufficiently to reach the threshold for an action potential. The EPSPs produced in a motor neuron by most stretch-sensitive afferent neurons are only 0.2 to 0.4 mV in amplitude. If the EPSPs generated in a single motor neuron were to sum linearly (which they do not), at least 25 afferent neurons would have to fire together to depolarize the trigger zone by the 10 mV required to reach threshold. At the same time the postsynaptic cell is receiving excitatory inputs it may also be receiving inhibitory inputs that tend to prevent the firing of action potentials. The net effect of the inputs at any individual excitatory or inhibitory synapse therefore depends on several factors: the location, size, and shape of the synapse and the proximity and relative strength of other synergistic or antagonistic synapses.

These competing inputs are integrated in the postsynaptic neuron by a process called *neuronal integration*. Neuronal integration, the decision to fire an action potential, reflects at the level of the cell the task that confronts the nervous system as a whole: decision making. A cell at any given moment has two options: to fire or not to fire. Charles Sherrington described the brain's ability to choose between competing alternatives—to select one and suppress the others—as the *integrative action of the nervous system*. He regarded this decision making as the brain's most fundamental capability.

In motor neurons and most interneurons the decision to initiate an action potential is made at the initial segment of the axon, the axon hillock (see Chapter 2). This region of cell membrane has a lower threshold than in the cell body or dendrites because it has a higher density of voltage-dependent Na^+ channels. Thus, for each increment of membrane depolarization, more Na^+ channels open and more inward current flows at the axon hillock than elsewhere in the cell. The depolarization increment required to reach the threshold at the axon hillock (–55 mV) is only 10 mV (from the resting level of –65 mV). In contrast, the membrane of the cell body has to be depolarized by 30 mV before its threshold (–35 mV) is reached. Synaptic excitation will therefore first discharge the region of membrane at the axon hillock. The action potential generated at the axon hillock then brings the membrane of the cell body to threshold and at the same time is propagated along the axon. Thus the membrane potential of the axon hillock serves as the readout for the integrative action of a neuron (Figure 13–3).

Some cortical neurons have one or more additional (booster) trigger zones within the dendritic tree. These dendritic trigger zones amplify weak excitatory input that arrives at remote parts of the dendrite. When a cell has several trigger zones, each one sums the local excitation and inhibition produced by nearby synaptic inputs and, if the net input is above threshold, an action potential may be generated, usually by voltage-dependent Ca^{2+} channels. These action potentials are not conducted along the dendrites in a regenerative manner. Rather, they propagate electrotonically to the cell body and axon hillock, where they are integrated with all other input signals in the cell.

Because neuronal integration depends on the summation of synaptic potentials that spread passively to the trigger zone, it is critically affected by two passive properties of the neuronal membrane discussed in Chapter 9. First, the *time constant* helps to determine the time course of the synaptic potential and thereby affects *temporal summation*, the process by which consecutive synaptic actions at the same site are added together in the postsynaptic cell. Neurons with a large time constant have a greater capability for temporal summation than do neurons with a smaller time constant (Figure 13–4A). As a result, the larger the time constant, the greater is the likelihood that two consecutive inputs from an excitatory presynaptic neuron will summate to bring the cell membrane to its threshold for an action potential (Figure 13–4A).

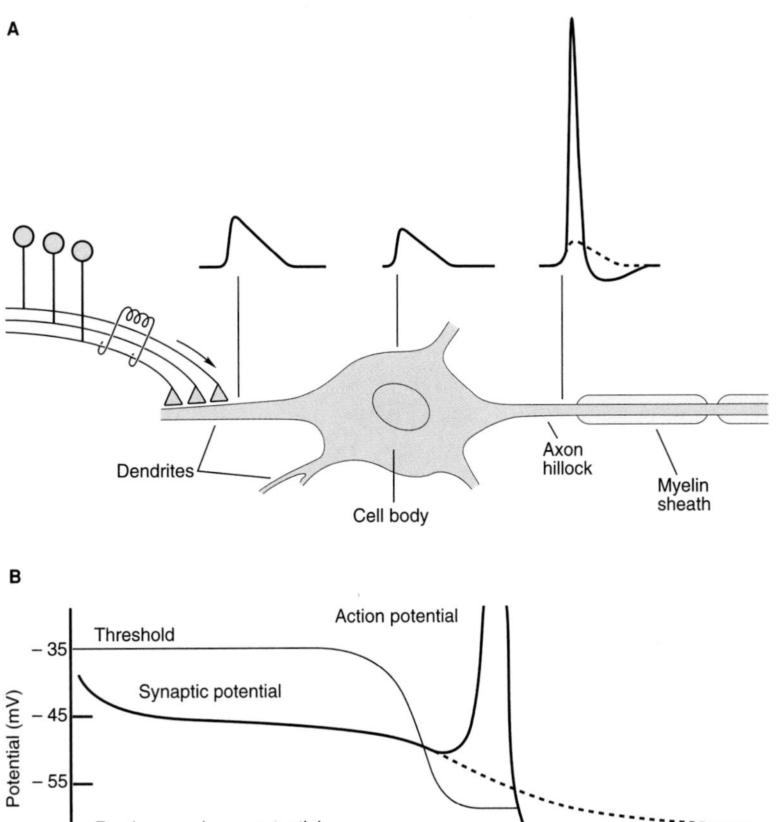

Figure 13–3 Spatial decay of a synaptic potential initiated by an input onto a dendrite. (Adapted from Eckart et al., 1988.)

A. An excitatory synaptic potential originating in the dendrites decreases with distance as it propagates passively in the cell. Nevertheless, an action potential can be initiated at the axon hillock because the density of the Na^+ channels in this region is high and thus the threshold is low.

B. Comparison of the thresholds for initiation of the action potential at different sites in the neuron (corresponding to drawing **A**). An action potential is generated when the amplitude of the synaptic potential crosses the threshold. The dashed line shows the course the decay of the synaptic potential would take if an action potential were not generated.

Second, the *length constant* of the cell determines the degree to which a depolarizing current decreases as it spreads passively. In cells with a larger length constant the signals spread to the trigger zone with minimal decrement; in cells with a small length constant the signals decay rapidly with distance. Since the depolarization produced at one synapse is almost never sufficient to trigger an action potential at the trigger zone, the inputs from many presynaptic neurons acting at different sites on the postsynaptic neuron must be added together. This process is called *spatial summation.* Neurons with a large length constant are more likely to be brought to threshold by two different inputs that contact the neuron at different points than are neurons with a short length constant (Figure 13–4B).

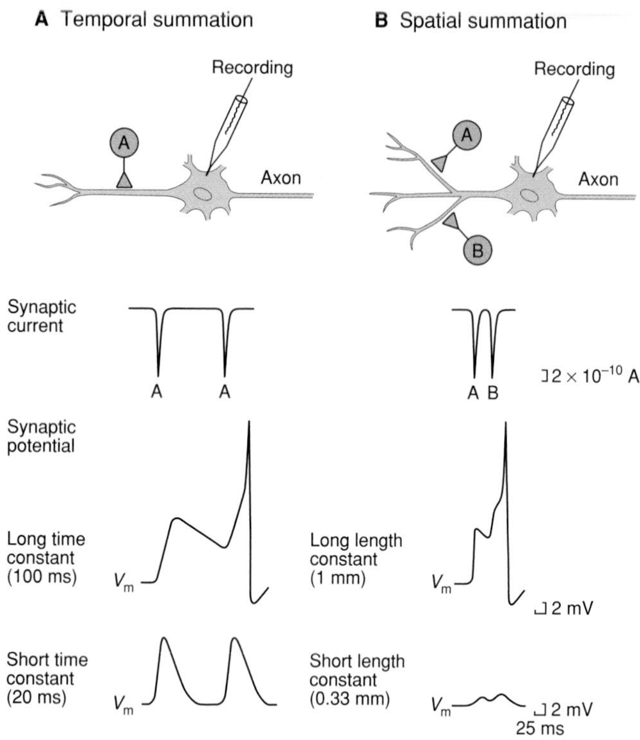

Figure 13–4 Central neurons are able to integrate a variety of synaptic inputs through temporal and spatial summation of synaptic potentials.

A. The time constant of a postsynaptic cell affects the time course of the depolarization caused by consecutive EPSPs produced by a single presynaptic neuron (**A**). The synaptic current generated by the presynaptic neuron is the same for both EPSPs. In the cell with a long time constant the first EPSP does not decay totally by the time the second EPSP is triggered. Therefore the depolarizing effects of both potentials are additive, bringing the membrane potential above the threshold and triggering an action potential. In the cell with a short time constant the first EPSP decays to the resting potential before the second EPSP is triggered. The second EPSP alone does not cause enough depolarization to trigger an action potential.

B. The length constant of a postsynaptic cell affects the passive propagation and overdistance of the EPSPs produced by two presynaptic neurons (**A** and **B**). For illustrative purposes, both synapses are the same distance from the cell's trigger zone and the current produced by each of these synaptic contacts is the same. In the cell with a long length constant the initial segment is only one length constant away from the site of the synaptic contacts. Therefore, the EPSPs produced by each of the two presynaptic neurons will decrease to only 37% of their original amplitude before reaching the trigger zone (see Figure 9–5). This results in enough depolarization to exceed threshold, triggering an action potential. For the cell with a short length constant, the distance between the synapse and the trigger zone in the initial axon segment is equal to three length constants. Therefore, each synaptic potential is barely detectable when it arrives in the postsynaptic cell body, and even the summation of two potentials is not sufficient to trigger an action potential.

Synapses onto a Single Central Neuron Are Grouped According to Function

All three regions of the nerve cell—axon, cell body, and dendrites—can be receptive or transmitting sites for synaptic contact. The most common types of contact, illustrated in Figure 13–5, are axoaxonic, axosomatic, and axodendritic (by convention, the presynaptic element is identified first). Axodendritic synapses can occur at the shaft or spine of the dendrite. Dendrodendritic and somasomatic contacts are also found, but they are rare.

The proximity of a synapse to the trigger zone of the postsynaptic cell is obviously important in determining its effectiveness. Synaptic current generated at an axosomatic site has a stronger signal and therefore a greater influence on the outcome at the trigger zone than does current from the more remote axodendritic contacts (Figure 13–6).

Synapses on Cell Bodies Are Often Inhibitory

The location of inhibitory inputs is also critical for their functional effectiveness. Inhibitory short-circuiting actions, which we discuss later in the chapter, are more significant when they are initiated at the cell body near the initial axon segment. The depolarization produced by an excitatory current from a dendrite must pass through the cell body as it moves toward the initial axon segment. Inhibitory actions at the cell body open Cl^- channels, thus increasing Cl^- conductance and reducing much of the depolarization produced by the spreading excitatory current. (This action is discussed in some detail later in the chapter.) As a result, the influ-

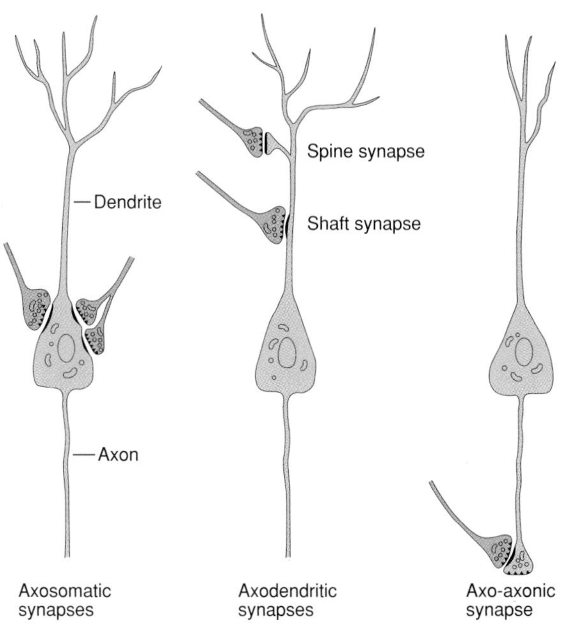

Figure 13–5 Synaptic contact can occur on the cell body, the dendrites, or the axon. The names of various kinds of synapses—axosomatic, axodendritic, and axo-axonic—identify the contacting regions of both the presynaptic and postsynaptic neurons (the presynaptic element is identified first). Axodendritic synapses can occur either on the main *shaft* of a dendritic branch or on a specialized input zone, the *spine*.

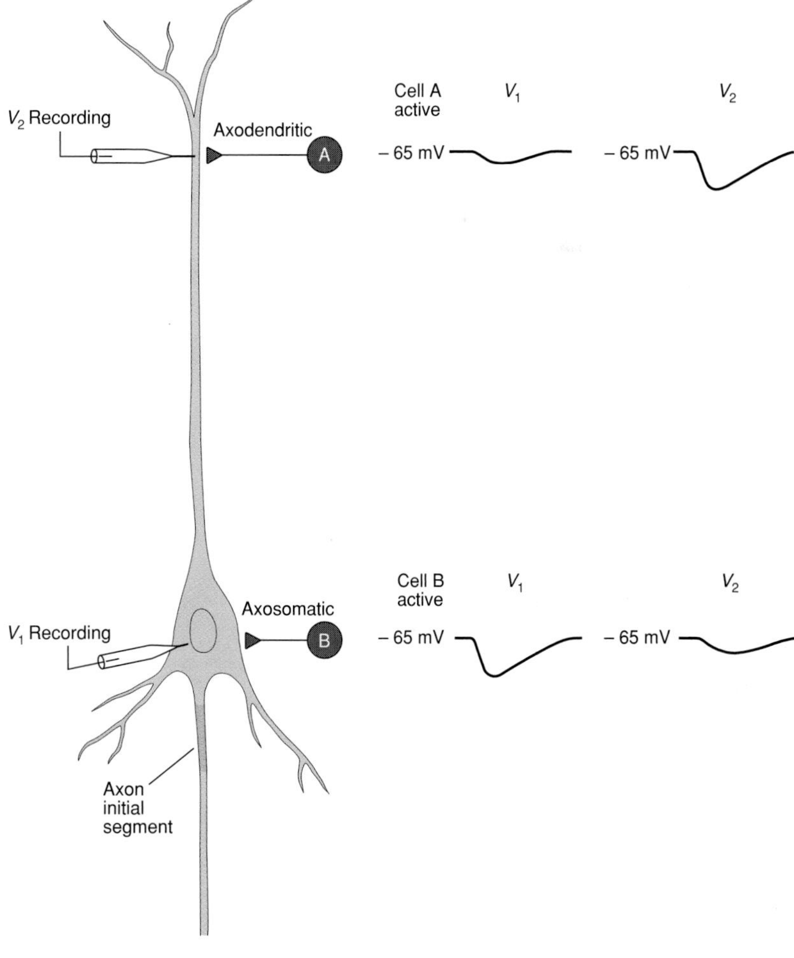

Figure 13–6 The impact of inhibitory synaptic potentials produced by two neurons, A and B, that form synapses at two different sites in the postsynaptic neuron depends on the distance the synaptic potential travels to the cell's trigger zone. In this hypothetical experiment the inputs from axosomatic synapses formed by cell B and the axodendritic synapses formed by cell A are compared by obtaining recordings from both the cell body (V_1) and the dendrite (V_2) of the postsynaptic cell. Stimulating cell B (axosomatic synapse) produces a large IPSP in the postsynaptic cell body. Because the synaptic potential is initiated in the cell body, it will not decay before arriving at the trigger zone in the initial segment of the axon, and thus will have an important influence at the trigger zone. Stimulating cell A postsynaptic (axodendritic synapse) produces only a small IPSP in the cell body. Because the potential is initiated so far from the axon hillock, it decays as it spreads to the cell body and thus has only a minor influence at the trigger zone.

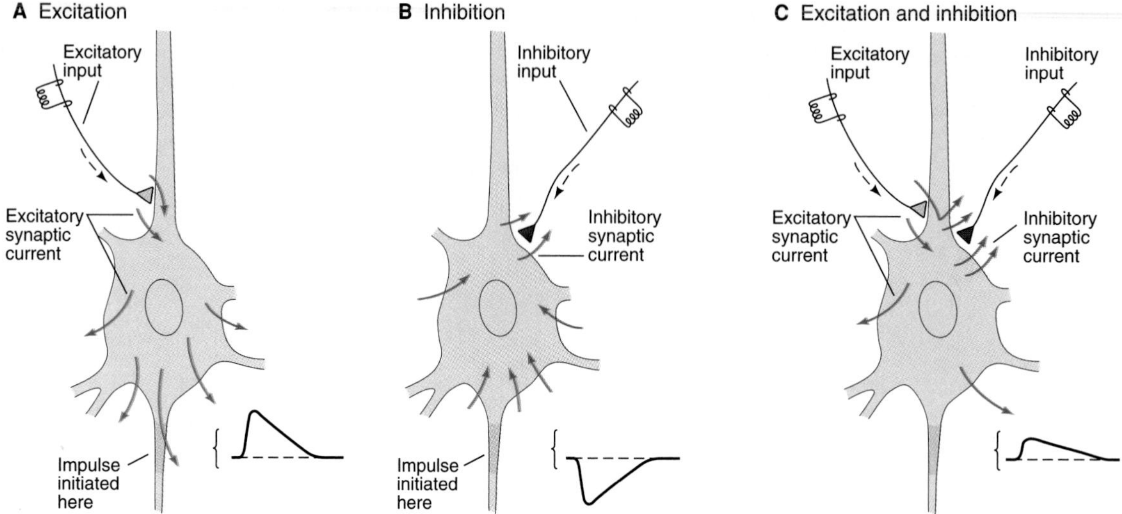

Figure 13–7 Excitatory and inhibitory potentials interact in a single nerve cell. (Adapted from Eckart and Randall, 1989.)

A. An excitatory input at the base of a dendrite causes inward current to flow through cation-selective channels (Na^+ and K^+). This current flows outward at the initial segment and produces a large depolarizing synaptic potential there.

B. An inhibitory input causes an outward (Cl^-) current at the synapse on the cell body and an inward current at other regions of the cell, producing a large hyperpolarization at the initial segment.

C. The stimulation of separate presynaptic pathways gives rise to both excitatory and inhibitory synaptic currents. The channels opened by the inhibitory pathway shunt the excitatory current, thereby reducing the excitatory synaptic potential.

ence of the excitatory current on the potential of the membrane of the trigger zone is strongly curtailed (Figure 13–7). In contrast, inhibitory actions at a remote part of a dendrite are much less effective in shunting excitatory actions or in affecting the more distant trigger zone. Thus, in the brain significant inhibitory input frequently occurs on the cell body of neurons.

Synapses on Dendritic Spines Are Often Excitatory

Central neurons often have as many as 20 to 40 main dendrites that branch into finer dendritic processes (see Figure 3–1B). Each branch has two major sites for synaptic inputs, the main shaft and the spines. The spine is a highly specialized input zone, typically consisting of a thin neck and a more bulbous head. Every spine has at least one synapse on its surface. In certain cortical neurons, such as the pyramidal cells of the CA1 region of the hippocampus, the spine head contains *N*-methyl-D-aspartate (NMDA) receptors (described later in the chapter). Here the postsynaptic density is rich in Ca^{2+}/calmodulin-dependent protein kinase; this kinase can therefore be activated selectively when Ca^{2+} flows through the NMDA-activated channel. Thus each spine represents a distinct biochemical compartment.

Synapses on Axon Terminals Are Often Modulatory

In contrast to axodendritic and axosomatic inputs, most axo-axonic synapses have no direct effect on the trigger zone of the postsynaptic cell. Instead, they affect the activity of the postsynaptic neuron by controlling the amount of transmitter it releases from its terminals onto the next postsynaptic cell (see Chapter 15).

Excitatory Synaptic Action Is Mediated by Transmitter-Gated Channels Selective for Sodium and Potassium

Eccles and his colleagues discovered that the excitatory postsynaptic potential in spinal motor cells results from the opening of transmitter-gated ion channels permeable to both Na^+ and K^+. This ionic mechanism is similar to that produced by acetylcholine (ACh) at the neuromuscular junction described in Chapter 12. As the strength of the extracellular stimulus is increased, more afferent fibers are excited, and the depolarization produced by the excitatory synaptic potential becomes larger. The depolarization eventually becomes large enough to bring the membrane potential of the axon hillock (the integrative component of the motor neuron) to the threshold for generation of an action potential.

Opening of Sodium and Potassium Channels Leads to an Inward Depolarizing Current

As we saw in Chapter 12, the best way to study the movement of ions responsible for the excitatory postsynaptic potential is to measure the reversal potential for the EPSP. Most nerve cells have a resting membrane potential of about –65 mV, considerably lower than that of muscle cells (–90 mV). As the membrane potential of the nerve cell is increased from –65 to –70 mV, the EPSP increases in amplitude, much like the synaptic potential in muscle (Figure 13–8). This occurs because more inward current flows through the synaptic channels as the driving force (V_m–E_{EPSP}) is increased.

As the membrane is progressively depolarized, however, the EPSP diminishes, until it disappears near 0 mV, its reversal potential. At that point the inward Na^+ current that flows through the synaptic channels is reduced because the membrane potential is now closer to E_{Na}, and the outward K^+ current is increased because it is further from E_K. The inward Na^+ current is thus balanced by the outward K^+ current, with the result that no net current flows through the synaptic channels. Additional depolarization (beyond 0 mV) produces a hyperpolarizing EPSP. The outward K^+ current now becomes greater than the inward Na^+ current, resulting in a net outward ionic current because the membrane potential is closer to E_{Na} than to E_K.

Similar results have been obtained in voltage-clamp experiments that have examined synaptic current as a function of different membrane potentials. The current flow is nullified at 0 mV and reverses from inward to outward as the membrane is depolarized further (Figure 13–8D). The synaptic current always tends to drive the membrane potential to the reversal potential, whether from membrane voltages that are either below or above the equilibrium potential. This experiment therefore also illustrates why EPSPs actually excite the motor neuron. As the synaptic current drives the membrane potential from its resting level (–65 mV) toward its reversal potential (0 mV), the membrane potential must pass through threshold (–55 mV).

Glutamate Is a Major Excitatory Transmitter in the Brain

The excitatory transmitter released from the primary afferent neurons is the amino acid L-glutamate, the major excitatory transmitter in the brain and spinal cord. There are several major types of glutamate receptors, and at least some of these have additional subtypes. The glutamate receptors can be divided into two broad categories: those that directly gate ion channels (ionotropic receptors) and those that indirectly gate channels through second messengers (metabotropic receptors).

The ionotropic glutamate receptors can be further classified into two major subtypes, NMDA and non-NMDA, according to the agonists that activate them and antagonists that inhibit them. The NMDA receptor is activated by the amino acid analog NMDA and blocked by the drug 2-amino-5-phosphonovaleric acid (APV). Non-NMDA receptors can be activated by the drugs AMPA, kainate, and quisqualate, and are blocked by CNQX. The metabotropic glutamate receptor can be activated by ACPD (Figure 13–9).

The non-NMDA receptors gate a cation channel with a relatively low conductance (< 20 pS) that is permeable to both Na^+ and K^+ but most non-NMDA receptors are not permeable to Ca^{2+} (Figure 13–9A). At the normal resting potential the non-NMDA receptors generate the predominant component of the EPSP in motor neurons (as well as most other central neurons) in response to stimulation of the primary afferent sensory fibers (see Figure 13–11B).

The NMDA receptor and its channel have four exceptional properties. First, the receptor controls a

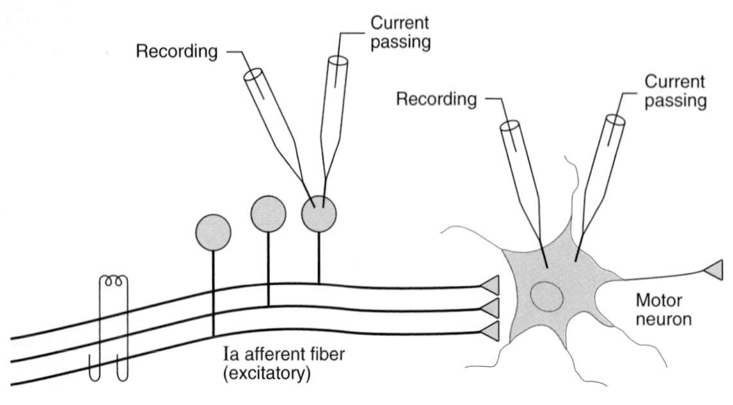
A Experimental setup
Recording
Current passing
Recording
Current passing
Motor neuron
Ia afferent fiber (excitatory)

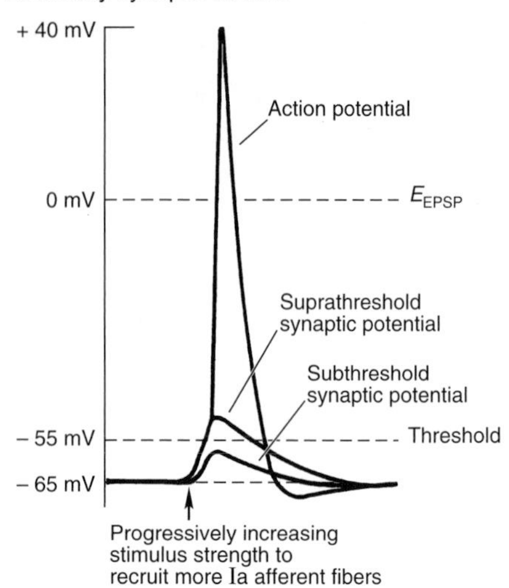
B Excitatory synaptic actions
+ 40 mV
Action potential
0 mV
E_{EPSP}
Suprathreshold synaptic potential
Subthreshold synaptic potential
– 55 mV
Threshold
– 65 mV
Progressively increasing stimulus strength to recruit more Ia afferent fibers

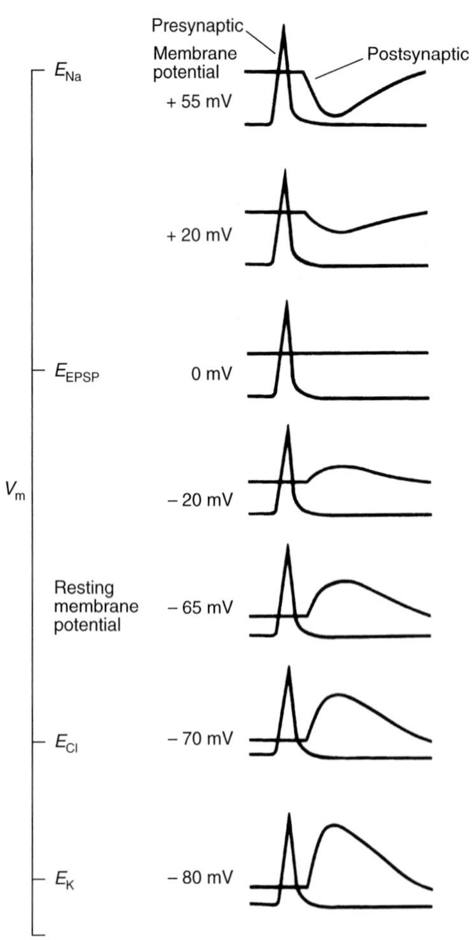
C Reversal potential for synaptic potential
Presynaptic
Membrane potential
Postsynaptic
E_{Na}
+ 55 mV
+ 20 mV
E_{EPSP}
0 mV
V_m
– 20 mV
Resting membrane potential
– 65 mV
E_{Cl}
– 70 mV
E_K
– 80 mV

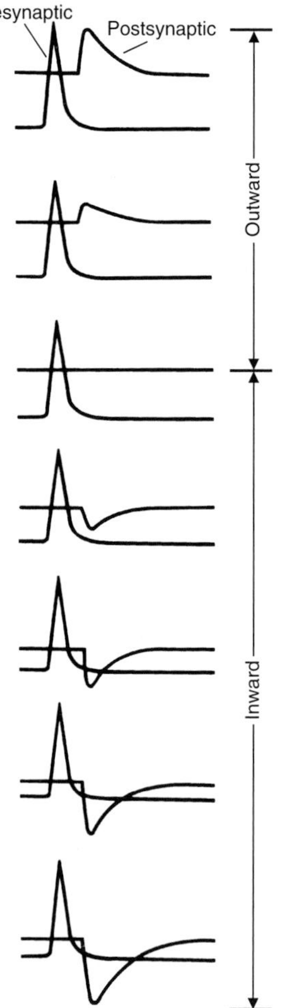
D Reversal potential for synaptic current
Presynaptic
Postsynaptic
Outward
Inward

← *Figure 13–8* Excitatory actions at chemical synapses result from the opening of channels permeable to both Na^+ and K^+. This can be demonstrated by determining the reversal potential for the EPSP.

A. Intracellular electrodes are used to stimulate and record from the neurons. Current is passed in the motor (postsynaptic) neuron either to alter the level of the resting membrane potential prior to presynaptic stimulation (a method of membrane control called *current clamp*) or to keep the membrane potential fixed during the flow of synaptic current (*voltage clamp*).

B. A weak stimulus to the afferent nerve from the quadriceps recruits only a few afferent fibers, resulting in a subthreshold synaptic potential. A strong stimulus recruits more afferent fibers, resulting in a suprathreshold synaptic potential that drives the membrane potential more effectively toward its reversal potential, which is beyond the threshold (–55 mV) for initiating an action potential.

C. The reversal potential for the *synaptic potential* can be determined using a current clamp. When the membrane potential is at its resting value (–65 mV), a presynaptic action potential produces a depolarizing EPSP, which increases when the membrane potential is hyperpolarized to –70 and –80 mV. In contrast, when the membrane potential is depolarized to –20 mV, the EPSP becomes smaller; when the membrane potential reaches the reversal potential (0 mV), the EPSP is nullified. Further depolarization to +20 mV inverts the synaptic potential, causing hyperpolarization. Thus synaptic action, whether hyperpolarizing or depolarizing, always drives the membrane potential toward the reversal potential, E_{EPSP}.

D. The reversal potential for the *synaptic current* can be determined using a voltage clamp. At the resting membrane potential and at more negative clamped potentials (–70 and –80 mV) the synaptic current is large and inward because the electrochemical driving force is inward. This inward current generates the EPSP. When the membrane potential is made less negative (–20 mV), the magnitude of the inward synaptic current decreases; at the reversal potential (0 mV) it becomes zero. When the membrane potential is made more positive than the reversal potential (+20 or +55 mV), the synaptic current is outward.

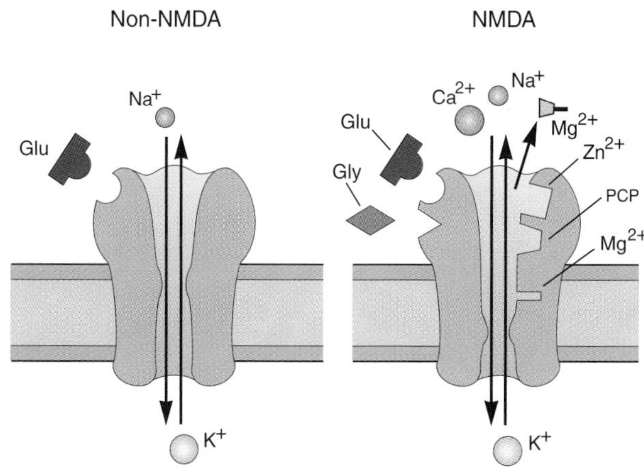

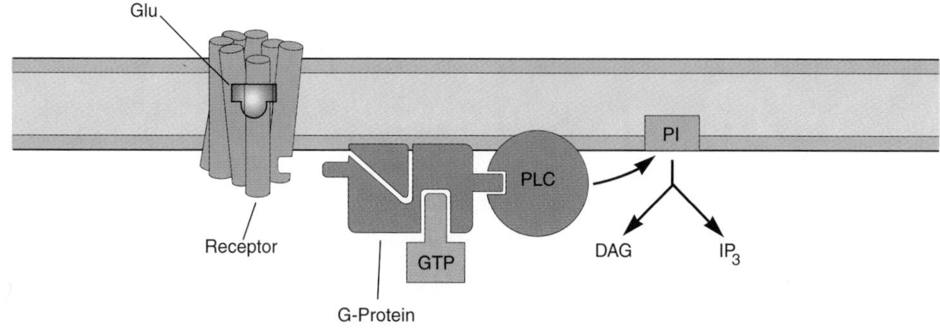

Figure 13–9 Three classes of glutamate receptors regulate excitatory synaptic actions in neurons in the spinal cord and brain.

A. Glutamate directly gates two types of ionotropic receptors. The *N*-methyl-D-aspartate (NMDA) receptor regulates a channel permeable to Ca^{2+}, K^+, and Na^+, and has several binding sites for glycine, zinc, phencyclidine (PCP, "angel dust"), and Mg^{2+}, which regulate the functioning of this channel in different ways (see Figure 13–11). For example, in the resting stage Mg^{2+} blocks the channels. The non-NMDA receptors bind the glutamate agonists kainate, quisqualate, and AMPA, and regulate a channel permeable to Na^+ and K^+.

B. Glutamate also activates a second messenger when it binds to the metabotropic receptor. The binding stimulates the activity of the enzyme phospholipase C (PLC), leading to the formation of two phosphoinositide (PI) second messengers: inositol 1,4,5-triphosphate (IP_3) and diacylglycerol (DAG).

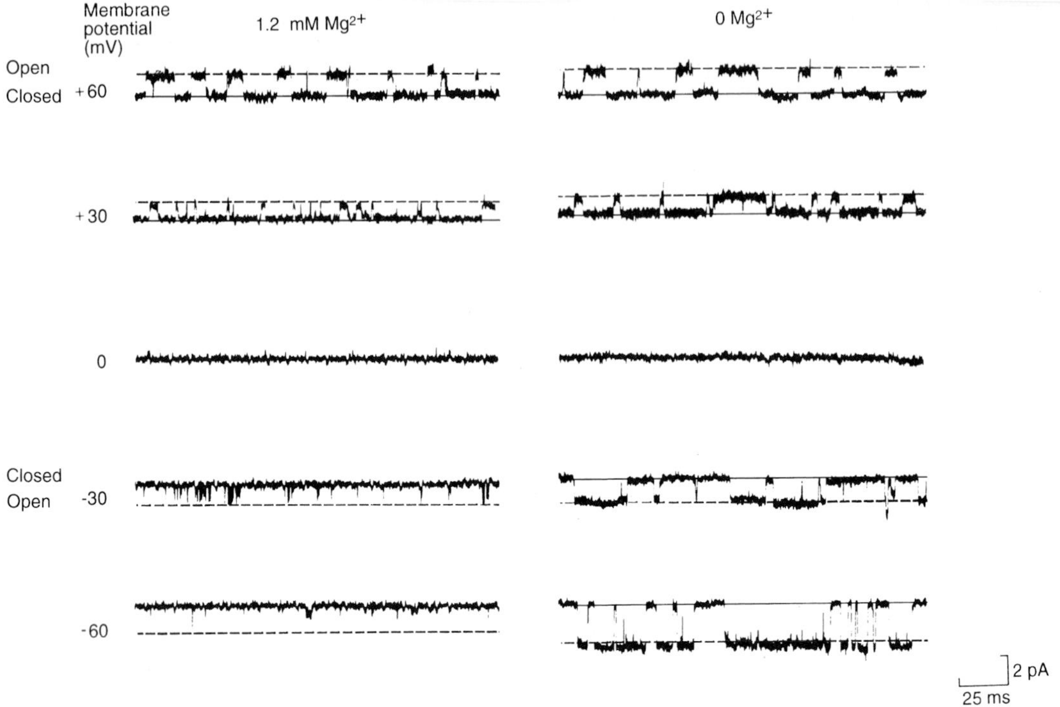

Figure 13–10 Current flow through the NMDA receptor–channel is dependent on voltage. These recordings are from individual channels (from rat hippocampal cells in culture obtained in an outside-out patch configuration, where the extracellular surface of the membrane is exposed to the extracellular bathing medium). When the cell is at rest the NMDA channel is normally blocked by Mg^{2+}. When Mg^{2+} is present in normal concentration (1.2 m*M*) in the bathing solution (recording on left), the NMDA receptor–channel is largely closed at the resting potential and needs substantial depolarization (to +30 mV) before it opens. When Mg^{2+} is removed from the bathing solution (recording on the right), the opening and closing of the channel do not depend on voltage. The channel is open at the resting potential of –60 mV, and the synaptic current reverses near 0 mV, like the total membrane current (see Figure 13–8D). (Courtesy of J. Jen and C. F. Stevens.)

cation channel of high conductance (50 pS) that is permeable to Ca^{2+} as well as to Na^+ and K^+ (Figure 13–9A). Second, Ca^{2+} entry through the NMDA-activated channel is thought to activate calcium-dependent second-messenger cascades. These are important in triggering biochemical changes that contribute to certain forms of long-lasting synaptic modification, which are considered in Chapter 36. Third, the channel only functions in the presence of glycine. Fourth, the channel is unique among transmitter-gated channels thus far characterized because it is gated by both glutamate and voltage. At the resting membrane potential (–65 mV) the mouth of the NMDA receptor channel is plugged by Mg^{2+} from the extracellular space and therefore does not contribute much to the normal synaptic potential. But when the membrane is depolarized in the presence of glutamate, Mg^{2+} is expelled from the channel mouth by electrostatic repulsion, allowing Na^+ and Ca^{2+} to enter. Thus, maximal current flows through the channel when both glutamate is present and the cell is depolarized (Figure 13–10). The NMDA receptor is also inhibited by the hallucinogenic drug phencyclidine (PCP, also known as "angel dust") and by MK801, both of which bind to a site within the open channel pore.

In most cells that have both NMDA and non-NMDA glutamate receptors, the presence of Mg^{2+} in the NMDA-activated channel prevents this channel from contributing importantly to the EPSP

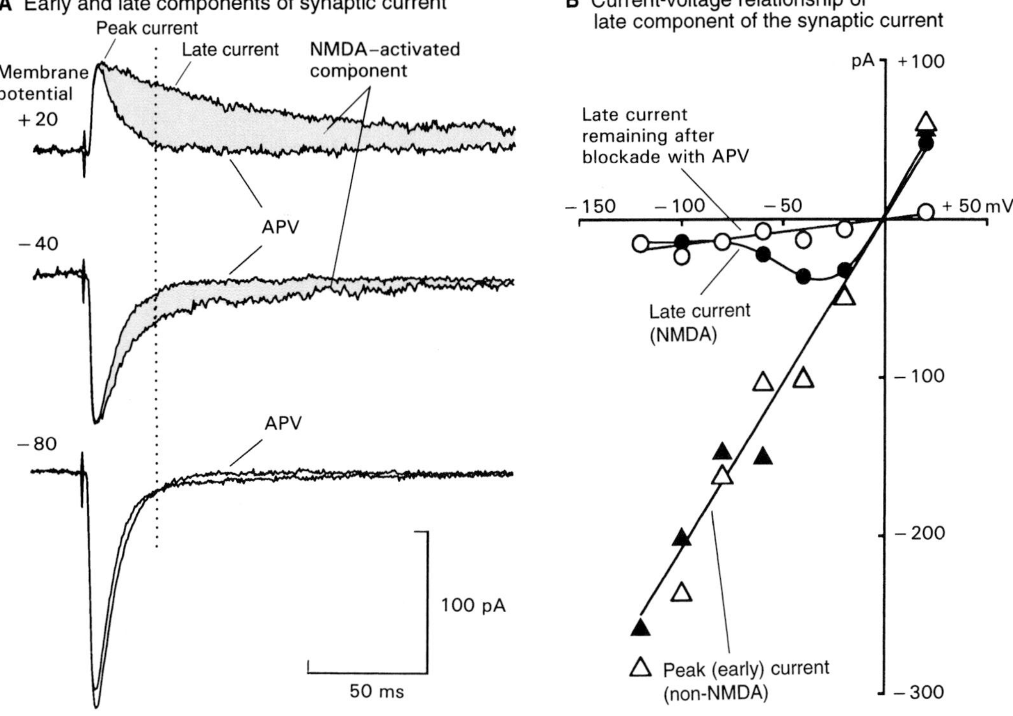

Figure 13–11 The NMDA receptor–channel contributes only a small late component to the normal excitatory synaptic current generated at the resting level of membrane potential. These records are from a cell in the hippocampus. Similar receptor–channels are present in motor neurons. (From Hestrin et al., 1990.)

A. The contribution of the NMDA receptor–channel is revealed by blocking the channel with 2-amino-5-phosphonovalerate) (APV), which selectively binds to the NMDA receptor. The records here show the synaptic current before and during the application of 50 μ*M* APV at three different membrane potentials. At –80 mV there is no current through the NMDA receptor–channel. At –40 mV a small late current is evident, and at +20 mV the contribution of the late current is significantly larger. The shaded areas indicate the size of the NMDA (APV-sensitive) component. The vertical dotted line indicates 25 ms after the peak of current and is used for calculations of the late current in **B**.

B. The APV drug has an effect on both the early and late components of the synaptic current. To measure the early component, the peak values of current–voltage relations before (△) and during (▲) the application of APV were obtained (see **A**). Both components are plotted in relation to the membrane potential. The NMDA component is small at negative potentials ranging from –150 to –80 mV because of Mg^{2+} blockade. This blockade is removed when the cell is depolarized (from –80 to +20 mV). Nevertheless, the peak values are identical, because APV has no effect on the early non-NMDA components of current. The NMDA-induced current is reflected in the late current measured 25 ms after the peak of the excitatory postsynaptic current before (●) and during (○) the application of APV. These curves for the late current diverge at approximately –80 mV, indicating that some of the Mg^{2+} blocking the channel is normally removed at this voltage. As the voltage decreases, more Mg^{2+} is removed.

at the resting membrane potential. Thus the EPSP that is generated at the resting level depends largely on the activation of the non-NMDA receptors. However, as the depolarization of the neuron increases, Mg^{2+} is driven out of the mouth of the NMDA-activated channels. As a result, more NMDA-activated channels are opened, and more current flows through these channels.

The NMDA-activated channels have another characteristic property: they open and close very slowly in response to glutamate. As a result, they contribute a characteristic late phase to the EPSP (Figure 13–11A). The late phase of the NMDA-mediated EPSP is small following a single presynaptic action potential. However, when the presynaptic neuron fires repeatedly, so that the EPSPs summate to depolarize the postsynaptic cell by 20 mV or more, the NMDA receptor expells Mg^{2+} and gives rise to a much larger current. Most significant, this NMDA-activated current is carried, to an important degree, by Ca^{2+}. Thus activation of the NMDA receptor by repetitive stimulation leads to the activation of calcium-dependent enzymes and second-messenger kinases in the postsynaptic cell (see Chapter 14).

Surprisingly, an imbalance in excitatory transmitters like glutamate may, in certain circumstances, contribute to disease. Excessive amounts of glutamate are highly toxic to neurons. Since glutamate is the major excitatory transmitter in the brain, most cells in the brain have receptors that respond to it. In tissue culture even a brief exposure to high concentrations of glutamate kills many neurons, an action called *glutamate excitotoxicity*. Although glutamate excitotoxicity may be due in part to the other types of glutamate receptors, in many cell types it is thought to result predominantly from excessive inflow of Ca^{2+} through NMDA-activated channels. High concentrations of intracellular Ca^{2+} may activate calcium-dependent proteases and may produce free radicals that are toxic to the cell. Glutamate toxicity may contribute to cell damage after stroke, to the cell death that occurs with persistent seizures in status epilepticus, and to degenerative diseases, such as Huntington's chorea. Agents that selectively block the NMDA receptor may protect against the toxic effects of glutamate and are currently being tested clinically.

Inhibitory Synaptic Action Is Usually Mediated by Receptor–Channels Selective for Chloride

The firing of a motor neuron is inhibited by stimulating the 1a afferent pathways from antagonist muscles that oppose the movements of muscles innervated by that motor neuron. The afferents from antagonist muscles produce inhibitory postsynaptic potentials that prevent the membrane potential of the initial segment of the axon from reaching the threshold for spike generation. IPSPs usually hyperpolarize the membrane; they also reduce the amplitude of synaptic potentials produced by excitatory synapses.

Opening of Transmitter-Gated Chloride Channels Raises the Threshold for Excitatory Actions

Inhibitory postsynaptic potentials in spinal motor neurons and most central neurons are generated by the opening of transmitter-gated Cl^- channels. In addition, as we shall see in Chapter 14, other inhibitory synaptic actions are mediated by metabotropic receptors and can involve the opening of K^+ channels. The channels for Cl^- and K^+ are similar in that their reversal potentials (the Nernst potentials for K^+ and Cl^-) are more negative than the resting membrane potential. In a typical neuron, E_{Cl} is about –70 mV (the concentration of Cl^- is high outside the cell and low inside), and E_K is –80 mV, whereas the resting membrane potential is –65 mV. Thus, the electrochemical driving force on both Cl^- and K^+ will be positive at the resting potential. As a result, the opening of Cl^- or K^+ channels leads to an outward hyperpolarizing current flow. Remember, the direction of electrical current flow is always defined in terms of the direction of the movement of positive charge. Thus an outward Cl^- current corresponds to an influx of negatively charged Cl^- ions, while an outward K^+ current is generated by an efflux of positively charged K^+ ions. In both cases there is a net reduction in the total positive charge inside the cell so the membrane hyperpolarizes.

The ionic mechanism of the IPSP in spinal motor neurons can be determined by measuring its reversal potential. This was first accomplished by systematically changing the level of the motor neuron resting membrane potential by passing current through an intracellular electrode while stimulating the inhibitory presynaptic neurons to fire an action potential. These experiments showed that the reversal potential for the IPSP was equal to the Nernst potential for chloride (Figure 13–12).

The resting potential of a central neuron is usually close to E_{Cl} and, in some cells, is equal to E_{Cl}. In these cells synaptic actions that increase Cl^- conductance do not change the postsynaptic membrane potential at all—the cell does not become hyperpolarized. How then does an inhibitory transmitter that opens Cl^- channels prevent a cell from firing?

When Cl^- channels are opened, the Cl^- influx drives the membrane potential toward the reversal

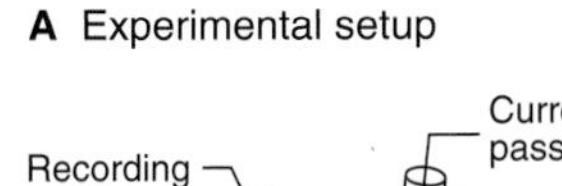

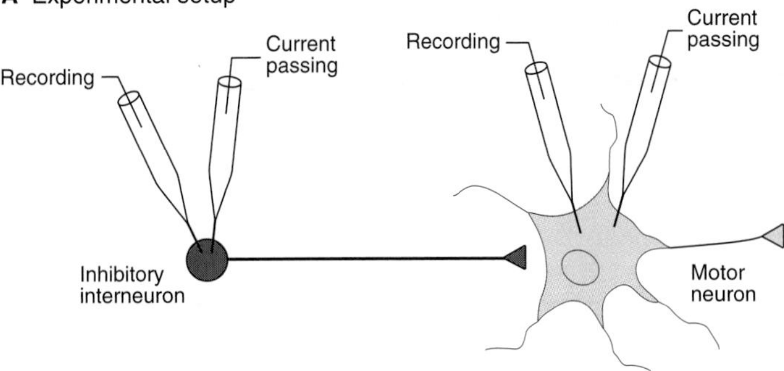

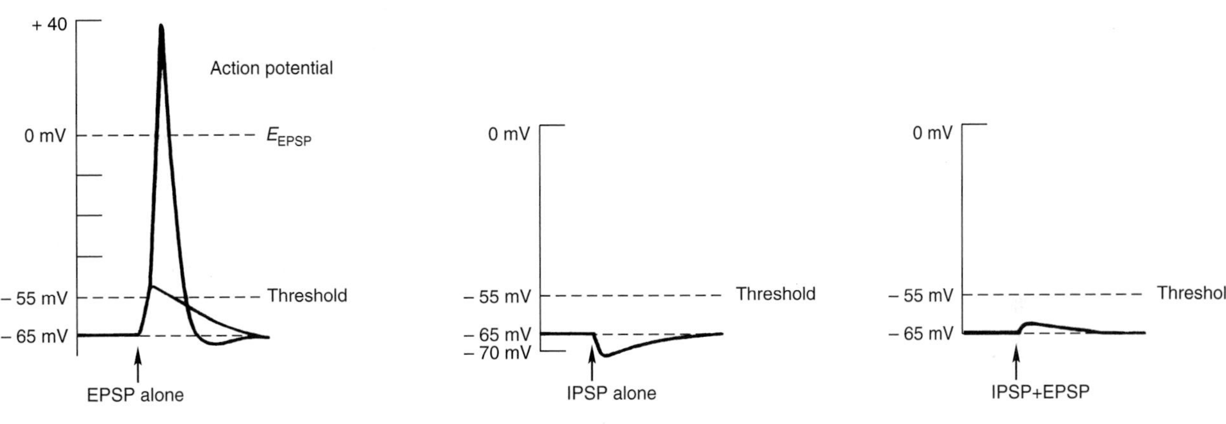

Figure 13–12 Inhibitory actions at chemical synapses result from the opening of ion channels permeable to Cl^-.

A. In this hypothetical experiment two electrodes are placed in the presynaptic interneuron and two in the postsynaptic motor neuron. The current-passing electrode in the presynaptic cell is used to produce an action potential; in the postsynaptic cell it is used to alter the membrane potential systematically (current clamp).

B. Inhibitory actions counteract excitatory actions. **1.** A large EPSP occurring alone moves the membrane potential toward E_{EPSP} and exceeds the threshold for generating an action potential. 2. An IPSP occurring alone moves the membrane potential away from the threshold toward E_{Cl} (–70 mV). **3.** When inhibitory and excitatory potentials occur together, the effectiveness of the EPSP is reduced, preventing it from reaching threshold.

C. The IPSP reverses at the equilibrium potential for Cl^-. At the resting membrane potential (–65 mV) a presynaptic spike produces a hyperpolarizing IPSP, which increases in amplitude as the membrane is artificially depolarized. However, when the membrane potential is hyperpolarized to –70 mV the IPSP is nullified. This reversal potential for the IPSP occurs at E_{Cl}, the Nernst potential for Cl^-. With further hyperpolarization the IPSP is inverted to a depolarizing postsynaptic potential (–80 and –100 mV) because the membrane potential is hyperpolarized in relation to E_{Cl}. Even this depolarizing action has an inhibitory effect, however, because the inhibitory input tends to hold the membrane potential at or below –70 mV, a considerable distance from threshold (–55 mV).

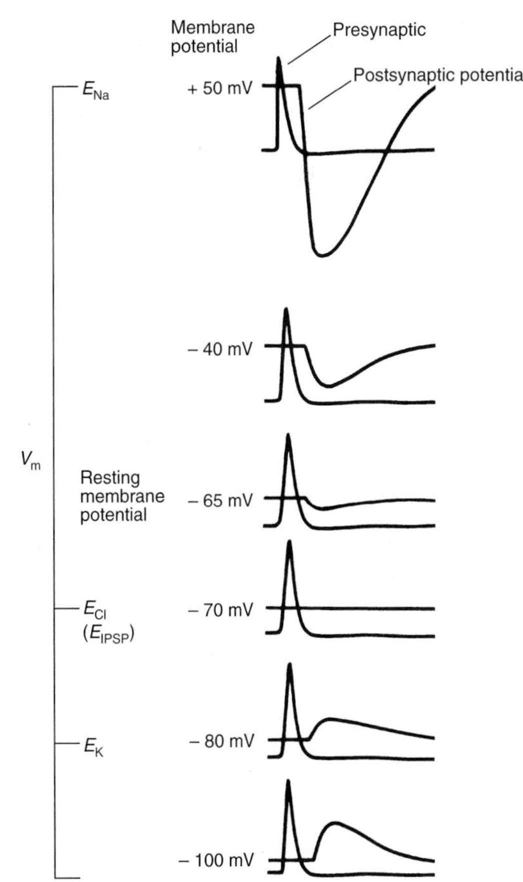

potential for Cl^-, or holds it at E_{Cl} if it already is there. Since this reversal potential is –70 mV and therefore at some distance from the threshold for generating an action potential (–55 mV), the opening of Cl^- channels increases the level of excitatory input needed to drive V_m toward threshold. In addition, the opening of Cl^- channels increases the overall conductance of the membrane of the postsynaptic cell (g_m). Since the amplitude of an excitatory synaptic potential is dependent on g_m ($V_{EPSP} = I_{EPSP}/g_m$), the increased g_m during inhibition will reduce the amplitude of any excitatory potentials that occur during the inhibitory action (see Figure 13–7).

In some cells inhibition is associated with the opening of K^+ channels. The net effect of opening K^+ or Cl^- channels is similar and in each case inhibits the postsynaptic cell in three ways. First, a K^+ or Cl^- mediated IPSP can hyperpolarize the membrane and move the membrane potential further away from threshold (Figure 13–12C). Second, by increasing the cell's permeability to Cl^- (or K^+) channels, the membrane potential is stabilized or clamped near E_{Cl} (or E_K), thereby preventing it from reaching threshold. Finally, opening up Cl^- (or K^+) channels increases the membrane conductance and thereby reduces the amplitude of an EPSP. This result is called the *short-circuiting* or *shunting* action of inhibition.

The opening of Cl^- (or K^+) channels has still another important feature. As with most forms of excitatory synaptic action, the opening of the inhibitory channels is not influenced by membrane voltage—that is, a change in the membrane potential does not alter the number of channels opened by the transmitter.

GABA and Glycine Are Inhibitory Transmitters

Gamma-aminobutyric acid (GABA) is a major inhibitory transmitter in the brain and spinal cord. It acts on two receptors, $GABA_A$ and $GABA_B$. The $GABA_A$ receptor is an ionotropic receptor that gates a Cl^- channel. The $GABA_B$ receptor is metabotropic, activating a second-messenger cascade that activates a K^+ channel. Glycine, a less common inhibitory transmitter, also activates ionotropic receptors linked to Cl^- channels. Glycine is released in the spinal cord by interneurons that inhibit antagonist muscles.

$GABA_A$ and glycine both initiate outward currents of Cl^- through synaptic channels. This inhibitory action can be demonstrated by comparing the reversal potential of single-channel inhibitory currents induced by $GABA_A$ with that of single-channel excitatory currents induced by glutamate. The excitatory current reverses at 0 mV, thereby driving the membrane past threshold. In contrast, the inhibitory current becomes nullified and begins to reverse at values more negative than –60 mV, preventing the membrane from reaching threshold (Figure 13–13).

Excitatory and Inhibitory Synapses Have Distinctive Ultrastructures

As we learned in Chapter 11, the effect of a synaptic potential, whether excitatory or inhibitory, is determined not by the type of transmitter released from the presynaptic neuron but by the type of ion channels gated by the transmitter in the postsynaptic cell. Although most transmitters are recognized by several types of receptor–channels that mediate either excitatory or inhibitory potentials, some act predominantly on receptors that are of one or another sign. For example, in the vertebrate brain, neurons that release glutamate typically act on receptors that produce excitation; neurons that release GABA or glycine act on inhibitory receptors. (An exception is found in the retina, which we discuss in a later chapter, and there are many exceptions in invertebrates.) The synaptic terminals of excitatory and inhibitory neurons can sometimes be distinguished by their morphology.

There are two common morphological types of synaptic connections in the brain, Gray type I and type II (after E. George Gray, who described them). Type I synapses are often glutaminergic and therefore excitatory, whereas type II synapses are often GABAergic and therefore inhibitory. In type I synapses the cleft is slightly widened to approximately 30 nm, the presynaptic active zone is 1 to 2 μm^2 in area, and dense projections, the presumed release sites for the vesicles, are prominent. The synaptic vesicles tend to assume a characteristic round shape when treated with certain electron microscopic fixatives. The dense region on the postsynaptic membrane also is extensive, and amorphous dense basement membrane material appears

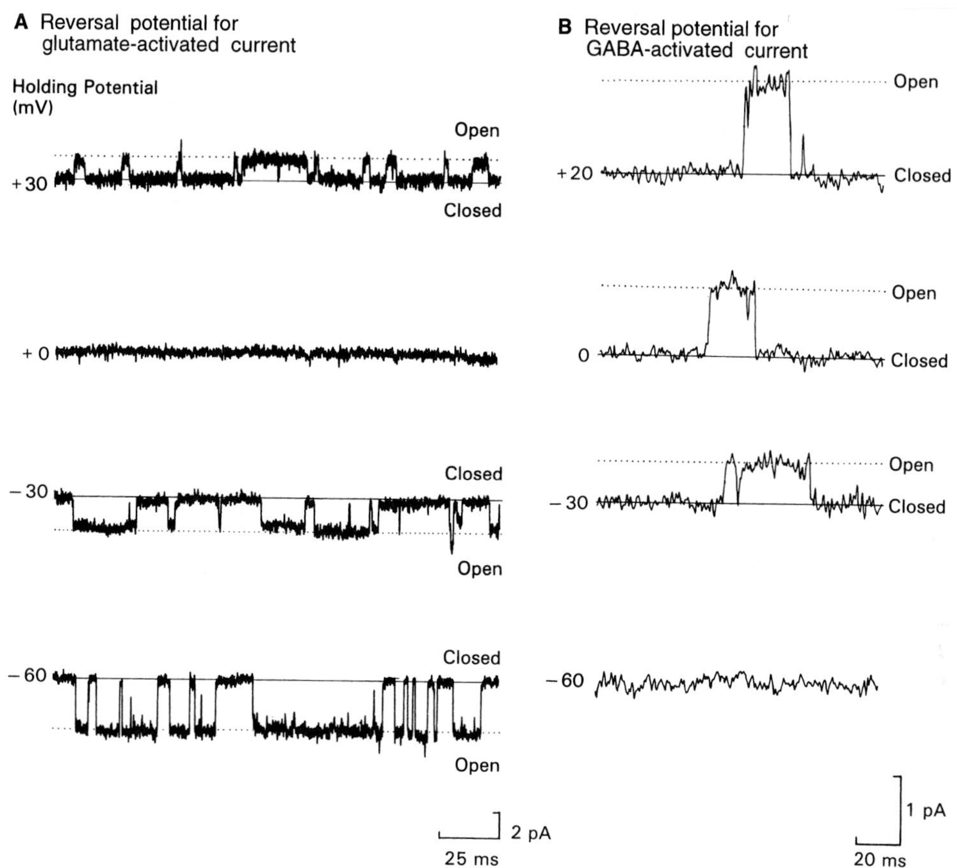

Figure 13–13 Single-channel currents activated by the excitatory transmitter glutamate and those activated by the inhibitory transmitter GABA have different reversal potentials. In the recordings shown here downward deflection indicates pulses of inward current; upward deflection indicates outward current.

A. Excitatory current through a single channel activated by glutamate in a rat hippocampal neuron. As the membrane potential is moved in a depolarizing direction (from –60 to –30 mV) the current pulses become smaller. At 0 mV (the reversal potential for the EPSP) the current pulses are nullified, and at +30 mV they invert and are outward. The reversal potential at 0 mV is the averaged equilibrium potentials for Na^+ and K^+, the two ions responsible for generating this current. (Courtesy of J. Jen and C. F. Stevens.)

B. Elementary-inhibitory current activated by GABA (5 μ*M*) in a rat hippocampal neuron. The current is nullified at approximately –60 mV (the reversal potential for IPSP). At more depolarized levels the current pulses are outward. This reversal potential lies near the equilibrium potential for Cl^-, the only ion contributing to this current. (Courtesy of B. Sakmann.)

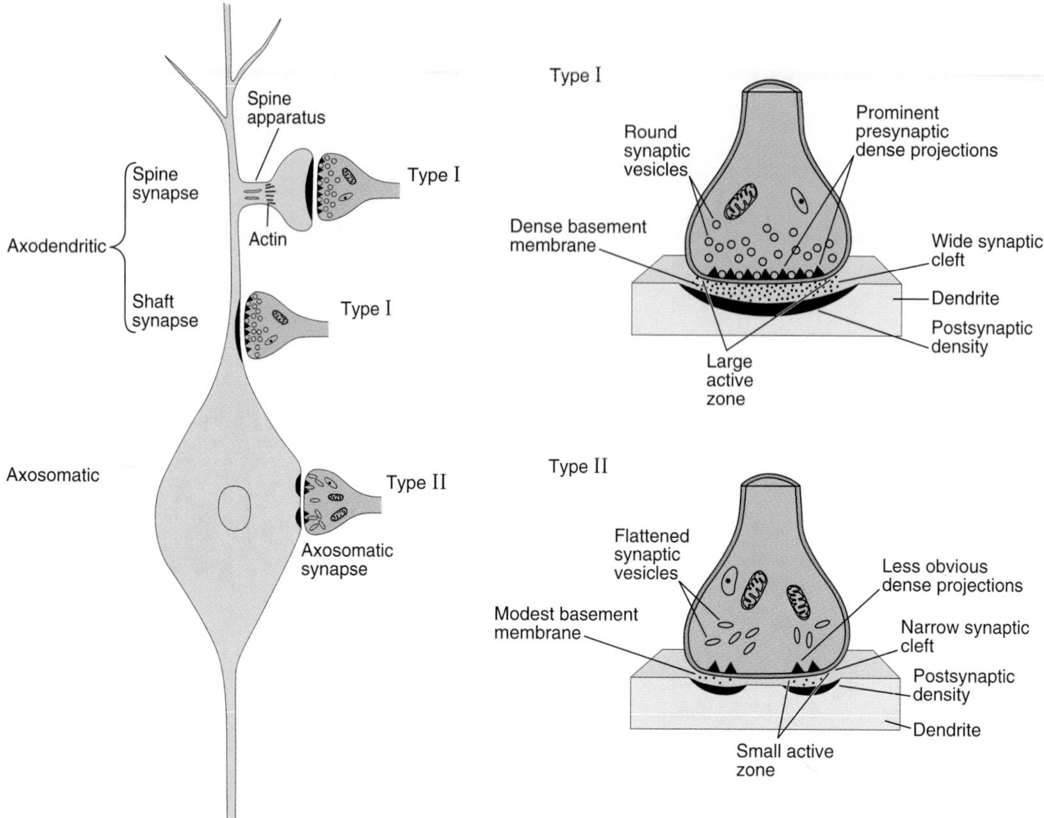

Figure 13–14 The two most common morphologic types of synapses in the central nervous system are Gray type I and type II synapses. Type I is usually excitatory, exemplified by glutaminergic synapses; type II is usually inhibitory, exemplified by GABAergic synapses. Differences include the shape of vesicles, prominence of presynaptic densities, total area of the active zone, width of the synaptic cleft, and presence of a dense basement membrane. Type I synapses contact dendritic shafts, but more frequently they contact dendritic spines. Type II synapses often contact the cell body.

in the synaptic cleft. In type II synapses the cleft is 20 nm across, the active zone is smaller (less than 1 μm^2), the presynaptic membrane specializations and dense projections are less obvious, and there is little or no basement membrane within the cleft. Characteristically, the vesicles of type II synapses tend to be oval or flattened (Figure 13–14).

Although type I synapses are often excitatory and type II inhibitory, these morphological characteristics have proved to be only a first approximation of transmitter biochemistry. As we shall learn in Chapter 15, a much more precise and impressive morphological distinction between transmitter types has been gained through the use of immunocytochemistry.

Synaptic Receptors for Glutamate, GABA, and Glycine Are Transmembrane Proteins

GABA and Glycine Receptors

Like the ACh-activated cation receptor–channel, the $GABA_A$- and glycine-activated channels are each formed from a transmembrane protein, made up of several subunits that contain both a conducting pore that threads through the cell membrane and a transmitter binding site on the outer face of the membrane. (The $GABA_B$ receptor is coupled to a G-protein, a class of receptors we consider in the next chapter.)

The $GABA_A$-activated channel must have at least two other functional domains, since benzodiazepines and barbiturates act at different sites to increase the Cl^- current through this type of channel. The benzodiazepines are antianxiety agents and muscle relaxants that include diazepam (Valium) and chlordiazepoxide (Librium). The barbiturates are a group of hypnotics that includes phenobarbital and secobarbital. The presence of any one of the three ligands—GABA, benzodiazepine, or barbiturate—influences the binding of the other two. For example, a benzodiazepine (or a barbiturate) will bind more strongly to the receptor when $GABA_A$ also is bound. Although all three sites can influence one another, each is distinct. Indeed, analysis of the primary structure of the $GABA_A$ receptor, which by analogy to the Ach receptor is thought to contain five subunits, indicates that the receptor is composed of at least three different types of subunits: α, β, and γ. All of the subunits bind GABA, although the α subunit does so with the greatest affinity. Both the α and β subunits bind barbiturates, but only the α subunit binds benzodiazepines.

This work, and similar studies on a subunit of the glycine receptor, reveals that the molecular structure of the inhibitory anion channels is similar to that of the ACh-activated cation channel. In fact, the genes that encode the $GABA_A$, glycine, and ACh receptor–channels belong to one family. The members of this gene family all have several subunits, and the subunits are structurally similar. Like the nicotinic ACh receptor, the subunits of the $GABA_A$ and glycine receptors each appear to have four membrane-spanning helixes (also labeled M1, M2, M3, and M4), and therefore are similar in transmembrane structure both to each other and to the ACh receptors. In the GABA- and glycine-activated channels, as in the ACh-activated channel, the M2 region is thought to line the channel pore.

Glutamate Receptors

The ionotropic glutamate receptors, by contrast, all belong to a different family. Like the other ionotropic receptors (those activated by ACh, GABA, or glycine), the glutamate receptor–channels have four transmembrane regions. However, their genetic sequence shows only a slight similarity to that of the other ionotropic receptors, even in the M2 region.

Transmitter-Gated, Voltage-Gated, and Gap-Junction Ion Channels Share Certain Structural Features

Although the gene families for the transmitter-gated ion channels we consider here are distinct from those for the voltage-gated and gap-junction channels we considered in Chapters 10 and 11, all of these channels have a similar overall design. Directly gated ion channels can be either cation selective (excitatory) or anion selective (inhibitory). Both types are peptides that have four membrane-spanning segments, one of which, the M2 region, is thought to line the pore. When the M2 segment is flanked by a cluster of acidic amino acids (e.g., aspartate and glutamate), as in the nicotinic ACh receptor, the channel is selective for cations (Figure 13–15A). When the M2 segment has clusters of basic residues (e.g., lysine and arginine), as in the $GABA_A$ and glycine receptors, the channel is selective for anions.

Voltage-gated channels are also selective for ions, specifically Na^+, Ca^{2+}, or K^+. Unlike the transmitter-gated channels, which are formed from several subunits of the peptide, at least some of the voltage-gated channels—those for Na^+ and Ca^{2+}—are formed by a single large subunit. This single subunit has four internal segments with similar molecular motifs (Figure 3–15B). Each repeated segment is analogous to one of the subunits of the multimeric protein that forms the transmitter-gated channel. Thus, each of the four internal motifs is thought to include six membrane-spanning α-helices (labeled S1 – S6). One of these helices, the S4 region, is thought to be the voltage sensor, the region of the channel that transforms changes in membrane potential into the conformational changes in the protein that open and close the channel. Between α-helical segments 5 and 6 a stretch of amino acids called the P region dips into the membrane to line the pore of the voltage-gated ion channel, much as the M2 region lines the pore of the transmitter-gated ion channel.

The gene families for both the voltage-gated and transmitter-gated channels thus represent variations on a common structural plan, a plan shared by the gene family for the gap-junction channels (Figure 13–15C). All three families are membrane-spanning proteins, with five features in common. First, they all share an architectural plan in which the segments that span the membrane are arranged

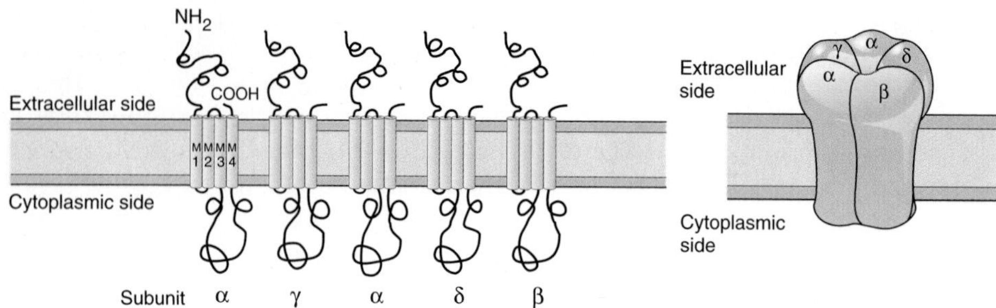

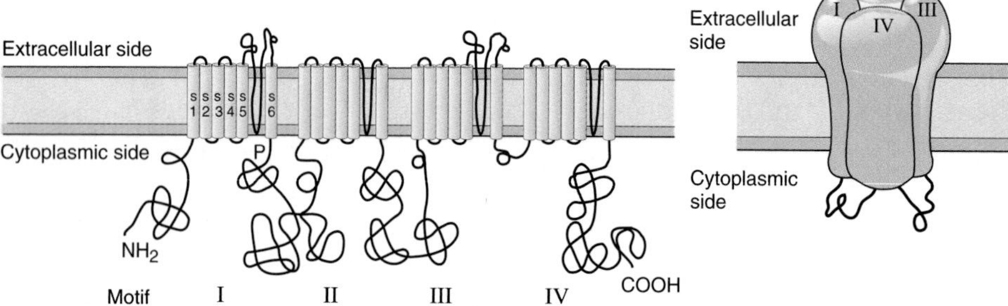

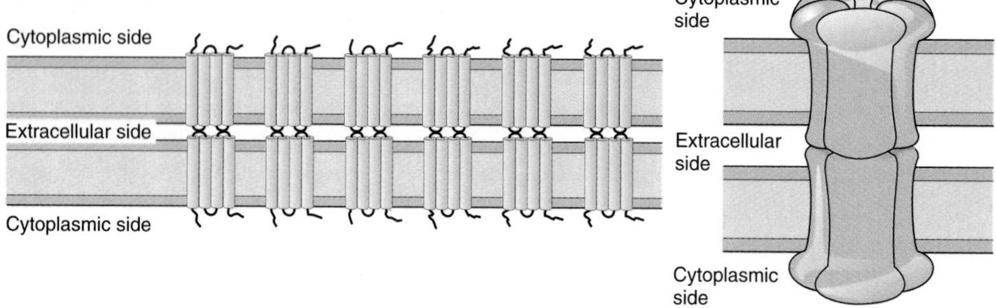

Figure 13–15 All ion channels have similar molecular structures.

A. The nicotinic ACh receptor–channel has five subunits and each subunit consists of four transmembrane regions (M1 to M4). A three-dimensional model of the channel is shown on the **right**.

B. The voltage-gated Na^+ channel is formed from a single (α) polypeptide chain thought to contain four homologous domains (motifs I to IV), each with six α-helical membrane-spanning regions (S1 to S6) and one P region thought to line the pore. Each cylinder represents a single transmembrane α-helix. The figure at the **right** shows a hypothetical model of the channel.

C. The gap-junction channel, found at electrical synapses, is formed from a pair of channels in the pre- and postsynaptic membranes that join in the space between two cells. Each hemichannel is made of six subunits, each with four transmembrane regions. A three-dimensional model of the two apposite hemichannels is illustrated on the right.

around a central axis to form a gated, water-filled pathway for ions. Second, the structural units of the three types of channels are either identical protein subunits, very similar protein subunits, or several similar domains within a single polypeptide chain. Third, the ion selectivity of each type of channel appears to be roughly related to the number of subunits and the diameter of the pore. Of the channels so far described at the molecular level, the most selective channels—the voltage-gated Na^+ and Ca^{2+} channels—have four structural units and the narrowest pore; the least selective channel, the gap-junction channel, has six units and the widest pores. The nicotinic ACh receptor, with five units, has intermediate properties.

A fourth shared feature is that although the channels belong to different gene families and have different numbers of subunits and different-sized central pores, they seem to be similar in overall conformation and may therefore operate by similar molecular mechanisms. For example, in all ion channels the narrower portion of the pore seems to be formed by α-helices and β strands from each of the encircling subunits; the number depends mainly on the size of the pore needed to fit the task. A fifth feature in common is that the change from open to closed states in the gated channels is thought to involve only a slight tilting of subunits, not a radical realignment.

The Signals Produced by Transmitter-Gated and Voltage-Gated Channels Have Features in Common

The various types of membrane-spanning ion channels are not only structurally similar; they also have similar actions and produce electrical signals that have features in common. All channels produce electrical signals that result from the movement of ions down their electrochemical gradients through the channel. The differences in the signals are due to the specific ions involved, the molecular properties of the channels through which the ions move, the state of the channels (open or closed) when the cell membrane is at rest, and the type of stimulus that opens and closes the channel. These stimuli include voltage for the Na^+ and K^+ channels involved in generating the action potential, chemical transmitters for the channels involved in producing synaptic potentials, mechanical pressure for the channels involved in producing the generator potentials of stretch and touch receptors, and light for the photosensitive Ca^{2+} and Na^+ channels in the retina. Thus, by moving through different channels, the same ions can produce different actions. For example, K^+ moves through a resting channel to generate the resting potential, through a voltage-gated channel to repolarize the membrane during the action potential, and through a second-messenger–activated channel to hyperpolarize the membrane in some inhibitory synaptic actions.

In addition, when two species of ions are involved in synaptic signaling, the type of signal generated depends on whether the two species move simultaneously through one type of channel or sequentially through two distinct types of channels. Simultaneous movement of Na^+ and K^+ through the same channel produces synaptic excitation; movement through independent channels in sequence produces an action potential. Finally, most (but not all) of the transmitter-gated channels are unaffected by changes in membrane potential and therefore lack the regenerative link between conductance and voltage that is critical for the explosive all-or-none firing of the action potential. The features of the various signaling potentials are summarized in Table 13–1.

An Overall View

Chemical synaptic transmission in the central nervous system is similar in principle to that in the vertebrate neuromuscular junction, but the two processes also differ in some essential ways. In the central nervous system synaptic transmission can be either excitatory or inhibitory. Excitatory postsynaptic potentials in the central nervous system tend to be less than 1 mV in amplitude, as compared with 70 mV in muscle. However, central neurons receive input from hundreds of presynaptic neurons, whereas a single muscle fiber is innervated by one motor neuron.

The major excitatory transmitter in the brain and spinal cord is glutamate. Several classes of postsynaptic receptors for glutamate have thus far been identified. The non-NMDA (AMPA) receptors are very similar to each other and are thus often classified together. Like the nicotinic ACh receptor, these receptors form channels permeable to both Na^+ and K^+, and they have reversal potentials near

Table 13–1 Features of Different Types of Electrical Signals in Neurons

	Ion Channels		
Signal	*Type*	*Mechanism*	*Signal Properties*
Resting potential	Mostly K^+ and Cl^- channels; some Na^+ channels	Most are leakage channels (some are voltage-gated K^+ channels)	Usually steady, ranging in different cells from –45 to –90 mV
Action potential	Separate Na^+ and K^+ channels	Voltage-gated channels; Na^+ and K^+ channels activated sequentially	All or none; about 100 mV in amplitude, 1–10 ms in duration
Receptor potential	Single class of channels for both Na^+ and K^+	Sensory stimulus	Graded; fast, several ms in duration; several millivolts in amplitude
Synaptic potentials			
Electrical synapse	Gap junctions (permeable to many ions and small organic molecules)	Channels can sometimes be regulated by membrane voltage, pH, or Ca^{2+}	Fast, due to passive spread of current from the presynaptic to the postsynaptic cell
Increased conductance	EPSP depends on a single class of channels for Na^+ and K^+; IPSP depends on separate channels for Cl^- (or K^+)	Channels are gated directly by neurotransmitter	Graded; fast, several ms to seconds in duration; several millivolts in amplitude
Decreased conductance	Closure of K^+, Na^+, or Cl^- channels	Channels are gated indirectly by G-proteins or second messenger	Graded; slow, seconds to minutes in duration; one to several millivolts in amplitude; contributes to the action potential's amplitude and duration

0 mV. Ion flux through these channels contributes to the fast early peak of the EPSP. The second type of receptor, the NMDA receptor, forms a channel permeable to Ca^{2+} in addition to Na^+ and K^+. This receptor–channel is unique among transmitter-gated receptors in that it is also voltage dependent. In the resting state the mouth of this channel is blocked by extracellular Mg^{2+}, which is removed when the membrane is depolarized. Thus, both glutamate and depolarization are needed to open NMDA receptor–channels. Because of the delay in opening, ion flux through this channel contributes to the late component of the EPSP. Calcium influx through NMDA receptor–channels is thought to trigger cellular processes involved in certain types of memory as well as certain cell processes contributing to brain damage. A third class of glutamate receptors, the metabotropic receptors, indirectly couples a channel through a second-messenger pathway.

The thousands of excitatory and inhibitory inputs that converge onto a single central neuron do not simply add together until threshold (–55 mV) is reached. Instead, the temporal and spatial summation of inputs within a single cell depend critically on the passive properties of the cell, specifically its time and length constants. The location of a particular synapse also contributes to its efficacy. Excitatory glutaminergic synapses tend to be located on the dendrites, whereas inhibitory synapses are found primarily on the cell body, where they can effectively override excitatory inputs from the cell's axon and dendrites. The final integration of inputs to the cell is made at the axon hillock, the region of the cell membrane near the initial segment of the axon. This region contains the highest density of Na^+ channels in the cell and thus has the lowest threshold for spike initiation.

The major inhibitory transmitters in the central nervous system are GABA and glycine. GABA can activate two types of receptors: $GABA_A$ receptors form channels permeable to Cl^-, while $GABA_B$ receptors couple to G-proteins and either increase K^+ permeability or block voltage-gated Ca^{2+} channels. Gating of the $GABA_A$ channels permits Cl^- influx into the cell, which hyperpolarizes the membrane. Opening these channels also increases the resting membrane conductance. Thus, opening these channels also short circuits any excitatory current flowing into the cell. Two important classes of drugs, benzodiazepines and barbiturates, both bind to GABA receptors and enhance the Cl^- flux through these channels in response to GABA.

All of the transmitter-gated channels thus far cloned have a certain degree of homology. Like the ACh receptor, both GABA and glycine receptors have multiple subunits, with each subunit containing four membrane-spanning segments. The GABA and glycine receptors are more similar to each other than to the ACh receptor, as might be expected since they conduct anions rather than cations.

Transmitter-gated and voltage-gated channels are genetically coded by separate gene families. Nonetheless, the overall structures of both families of molecules are similar. All classes of channels, including gap junctions, consist of several transmembrane sequences arranged symmetrically around a water-filled pore. The voltage-dependent Na^+ and Ca^{2+} channels thus far cloned are all formed from a single subunit with internal repeats, whereas the transmitter-gated channel molecules have multiple subunits. In the voltage-gated K^+ channel, where the subunit is a small protein, several subunits are thought to be required for the channel to function.

Much of the discussion in this chapter has been based on the idealized model of the neuron considered in Chapters 2 and 3. According to this model, the dendritic arbor is specialized as the receptive pole of the neuron, the axon is the conducting portion, and the axon terminal is the transmitting pole. This model implies that the nervous system is composed of information-receiving and information-transmitting units. Most brain regions are not quite this simple, however. As we shall see in considering the sensory and motor systems, cells in many regions of the brain also transform information in addition to receiving and transmitting it.

Selected Readings

Choi, D. W. 1988. Glutamate neurotoxicity and diseases of the nervous system. Neuron 1:623–634.

Cooper, J. R., Bloom, F. E., and Roth, R. H. 1991. The Biochemical Basis of Neuropharmacology, 6th ed. New York: Oxford University Press.

Eccles, J. C. 1964. The Physiology of Synapses. Berlin: Springer.

Heuser, J. E., and Reese, T. S. 1977. Structure of the synapse. In E. R. Kandel (ed.), Handbook of Physiology, Section 1: The Nervous System, Vol I: Cellular Biology of Neurons, Part 1. Bethesda, Md.: American Physiological Society, pp. 261–294.

Hollmann, M., and Heinemann, S. 1994. Cloned glutamate receptors. Annu. Rev. Neurosci. 17:31–108.

Masu, M., Tanabe, Y., Tsuchida, K., Shigemoto, R., and Nakanishi, S. 1991. Sequence and expression of a metabotropic glutamate receptor. Nature 349: 760–765.

Moriyoshi, K., Masu, M., Ishii, T., Shigemoto, R., Mizuno, N., and Nakanishi, S. 1991. Molecular cloning and characterization of the rat NMDA receptor. Nature 354:31–37.

Nicoll, R. A., Malenka, R. C., and Kauer, J. A. 1990. Functional comparison of neurotransmitter receptor subtypes in mammalian central nervous system. Physiol. Rev. 70:513–565.

Pritchett, D. B., Sontheimer, H., Shivers, B. D., Ymer, S., Kettenmann, H., Schofield, P. R., and Seeburg, P. H. 1989. Importance of a novel $GABA^A$ receptor subunit for benzodiazepine pharmacology. Nature 338: 582–585.

Snyder, S. H. 1984. Drug and neurotransmitter receptors in the brain. Science 224:22–31.

Stevens, C. F. 1987. Molecular neurobiology: Channel families in the brain. Nature 328:198–199.

Unwin, N. 1993. Neurotransmitter action: opening of ligand-gated ion channels. Cell 72/Neuron 10 (suppl.):31–41.

14

Modulation of Synaptic Transmission: Second-Messenger Systems

Second-Messenger Pathways Share a Common Molecular Logic

The Cyclic AMP Pathway Involves a Polar and Diffusible Cytoplasmic Messenger

Some Second Messengers Are Generated Through the Hydrolysis of Phospholipids

The Tyrosine Kinase Pathway Utilizes Both Receptor and Cytoplasmic Kinases

Other Second-Messenger Pathways: Nitric Oxide, Carbon Monoxide, and Cyclic GMP

Second-Messenger Pathways Can Interact with One Another

Second Messengers Often Act Through Protein Phosphorylation to Open or Close Ion Channels

Second Messengers and G-Proteins Can Sometimes Act Directly on Ion Channels

Second Messengers Can Alter the Properties of Transmitter Receptors: Desensitization

Second Messengers Can Endow Synaptic Transmission with Long-Lasting Consequences

Synaptic receptors have two major functions: the recognition of specific transmitters and the activation of effectors. The receptor first recognizes and binds a transmitter in the external environment of the cell; then, as a consequence of binding, the receptor alters the cell's biochemical state.

As described in the preceding chapters, receptors for neurotransmitters so far identified can be divided into two major groups according to how the receptor and effector functions are coupled (Figure 14–1). One group of receptors gate ion channels directly—the receptor and effector functions are carried out by different domains of a single macromolecule. These *ionotropic* receptors include two genetic families. One family includes receptors for acetylcholine (Ach), δ–aminobutyric acid (GABA), glycine, and serotonin. The other family includes two classes of receptors for glutamate, the *N*-methyl-D-asparate (NMDA) and the non-NMDA receptors. The molecular mechanisms underlying the action of the ionotropic receptors have been described in Chapters 12 and 13.

The second group of the receptors gate channels indirectly—the recognition of the transmitter and activation of effectors are carried out by separate molecules. There are two families of this type of receptor. In the first family, the *metabotropic* receptors, the receptor molecule is coupled to its effector molecule by a guanosine nucleotide-binding protein, or G-protein. For this reason, receptors of this family are also called *G-protein-coupled receptors.*

A Direct gating by ionotropic receptor

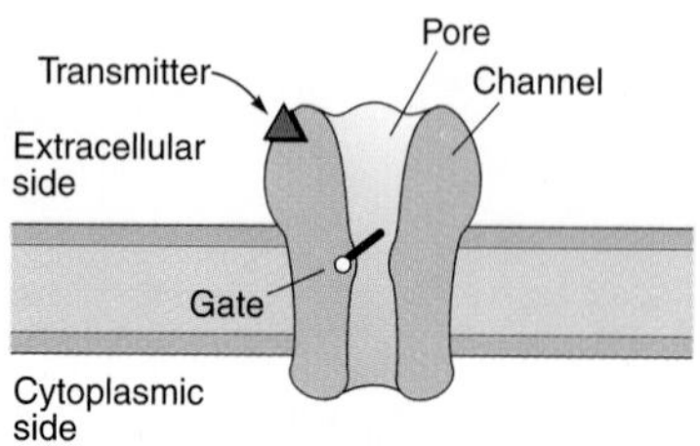

B Indirect gating
1 Metabotropic (G protein-coupled) receptor

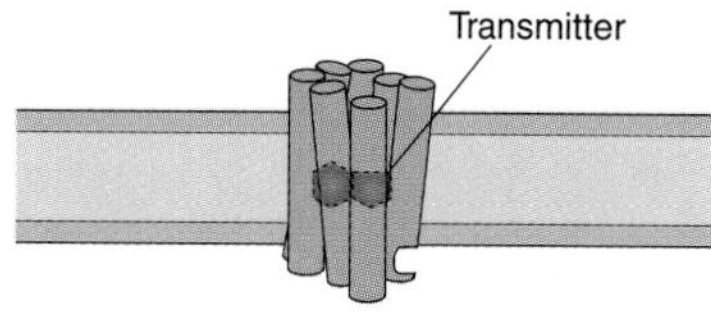

2 Receptor tyrosine kinase

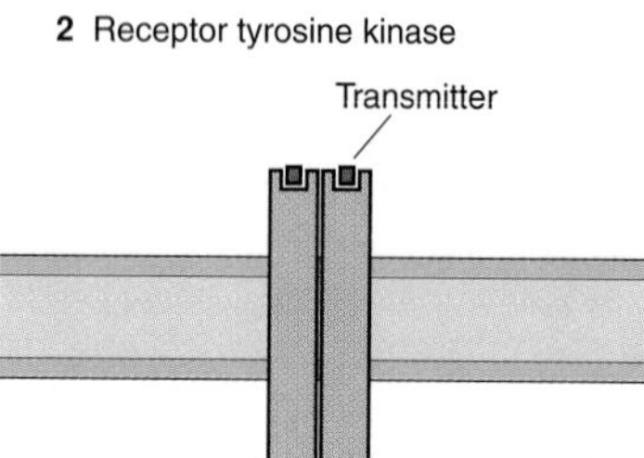

Figure 14–1 All known neurotransmitter receptors can be divided into two groups according to the way in which receptor and effector functions are coupled.

A. Receptors that directly gate ion channels have several subunits. The receptor, located on the extracellular side, and the ion channel, embedded in the cell membrane, represent different domains of the same protein.

B. Receptors that indirectly gate ion channels fall into two families. **1.** Metabotropic receptors activate ion channels and other substrates indirectly by activating a G-protein that engages a second-messenger enzyme. **2.** Receptor tyrosine kinases modulate the activity of ion channels indirectly through a cascade of protein phosphorylation reactions beginning with autophosphorylation of the kinase itself on tyrosine residues.

This family contains the α- and β-adrenergic receptors, the muscarinic ACh receptors, a class of receptors for GABA, glutamate, and serotonin, receptors for neuropeptides, as well as odorant receptors and rhodopsin (the protein that reacts to light initiating visual signals). Activation of the effector component requires the participation of several distinct proteins. Typically the effector is an enzyme that produces a diffusible second messenger. These second messengers in turn trigger a biochemical cascade, either activating specific protein kinases that phosphorylate a variety of the cell's proteins on serine or threonine residues or mobilizing Ca^{2+} ions from intracellular stores and thus initiating the reactions that change the cell's biochemical state. In some instances, however, the G-protein or the second messenger can act directly on an ion channel.

The second family of receptors that act indirectly also includes several different members, the best studied of which is the receptor *tyrosine kinase*. The cytoplasmic domain of the receptor tyrosine kinase is an enzyme that phosphorylates itself and other proteins on tyrosine residues. In addition, the cytoplasmic domain of the receptor recruits proteins to it, including other kinases that are capable of acting on ion channels. These receptors are typically activated by various hormones, growth factors, and neuropeptides. Most tyrosine kinases are not connected directly to a transmembrane component and therefore are classified as *non*receptor tyrosine kinases. Nonetheless, this group of kinases also is (indirectly) responsive to extracellular events that occur, for example, during growth and development.

The structural differences between the subdivisions of receptors are reflected in their functions. Neurobiologists often classify the actions of transmitters on receptors as being fast or slow, a distinction that refers both to speed of onset and to the duration of the postsynaptic effect. Direct gating of ion channels usually is rapid—on the order of milliseconds—because it involves a change in the conformation of only a single macromolecule. In contrast, receptors linked to G-protein or to tyrosine kinases are slower in onset (hundreds of milliseconds to seconds) and longer lasting (seconds to minutes) because they involve a cascade of reactions, each of which takes time.

The number of substances known to act as second messengers in synaptic transmission is far fewer than the number of transmitters. (On the order of 100 substances serve as transmitters, each of which activates a number of different receptors on the cell surface.) Moreover, only a few second messengers have been well characterized. Perhaps the best understood is cyclic adenosine monophosphate (cAMP). The work on cAMP has greatly influenced our thinking about second-messenger

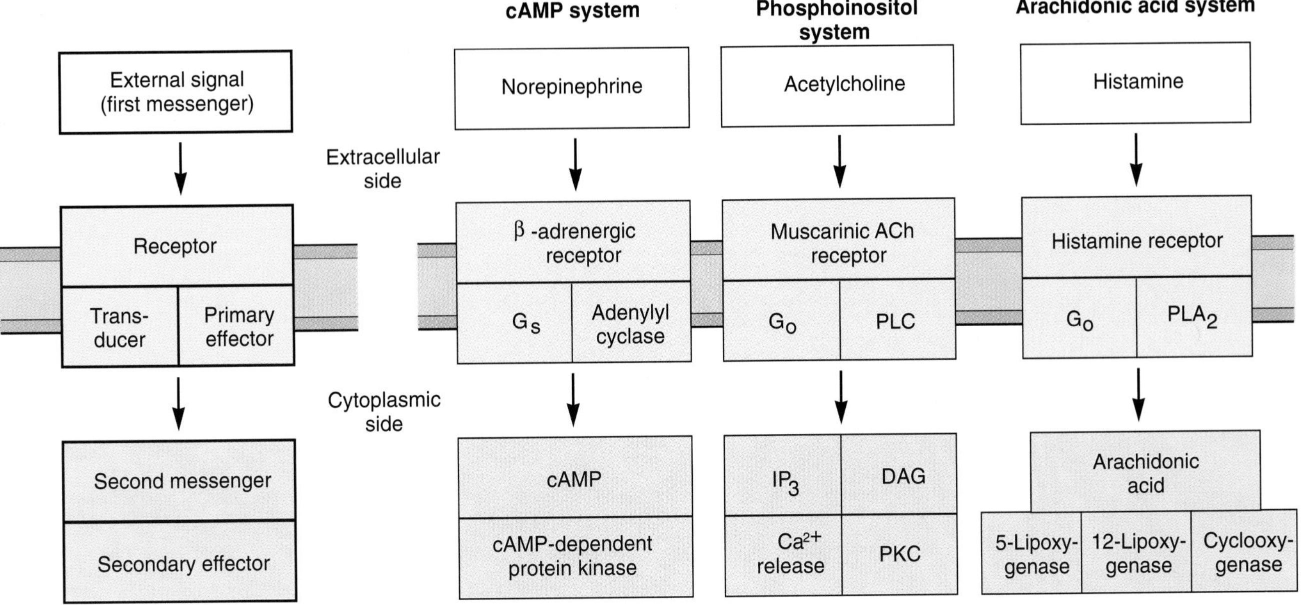

Figure 14–2 General scheme of synaptic second messengers. Only a few key signal transduction pathways have been identified thus far, of which the three illustrated here all follow a common sequence of steps **(left)**. Chemical transmitters arriving at receptor molecules in the plasma membrane activate a closely related family of transducer proteins that activate primary effector enzymes. These enzymes produce a second messenger that activates a secondary effector or acts directly on a target (or regulatory) protein. The first specific pathway shown generates the second-messenger cAMP. This second messenger is produced by adenylyl cyclase when activated by a G-protein, so called because it requires guanosine triphosphate (GTP) to function. The G-protein shown is G_S, because it *stimulates* the cyclase. Some receptors activate G_i, a G-protein that *inhibits* the cyclase. The second pathway, activated by a muscarinic ACh receptor, uses another kind of G-protein (G_o) to activate phospholipase C (PLC). This enzyme yields a pair of second messengers, DAG and IP_3. In turn, IP_3 mobilizes Ca^{2+} from internal stores. DAG activates protein kinase C (PKC). The third major system activates the arachidonic acid cascade through phospholipase A_2 (PLA_2). Three of the major enzymes in this cascade are 5- and 12-lipoxygenase and cyclooxygenase.

mechanisms in general. Another class of second messengers is produced by hydrolysis of the phospholipids in the cell's plasma membrane: inositol polyphosphates and diacylglycerol are liberated by the action of phospholipase C, while arachidonic acid is released by phospholipase A_2. These three second-messenger pathways are examined in detail in this chapter.

Gases that are highly diffusible comprise still another class of second messengers. For example, the enzyme nitric oxide synthetase generates nitric oxide (NO), while heme oxygenase generates carbon monoxide (CO).

Second-Messenger Pathways Share a Common Molecular Logic

Despite their differences, second-messenger pathways share many features (Figure 14–2). As we have seen, the receptors that initiate second-messenger actions belong to a limited number of gene families. Unlike the directly gated receptors, which consist of several subunits that form a channel through the membrane, receptors that are coupled to G-proteins and that generate second messengers consist of a single subunit with seven characteristic membrane-spanning regions (Figure 14–3).

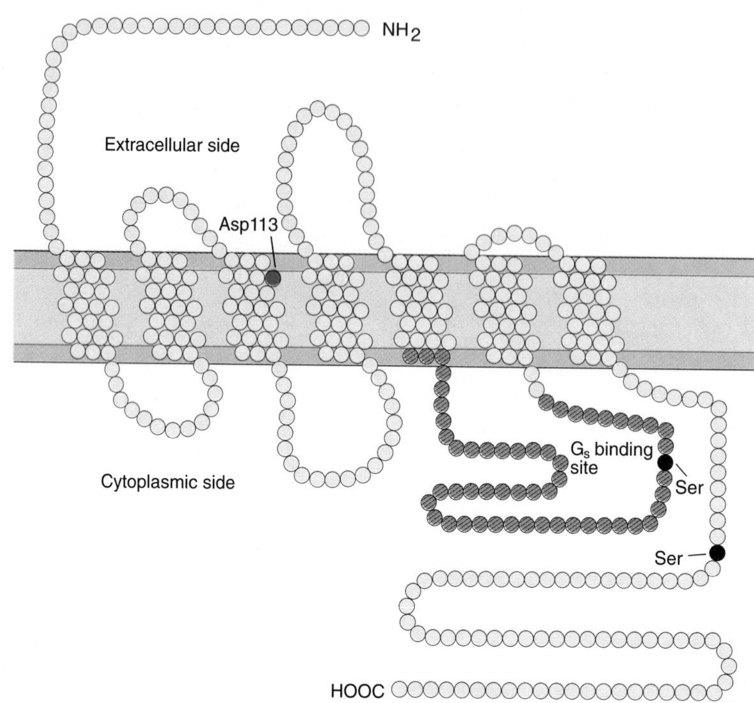

Figure 14–3 A G-protein-coupled receptor contains seven membrane-spanning domains. The structure of the β_2-adrenergic receptor shown here is similar to the structures of the β_1-adrenergic and muscarinic ACh receptors, and to rhodopsin. An important functional feature is that the binding site for the neurotransmitter is just within the lipid bilayer on the extracellular surface of the cell (here, amino acid residue aspartic acid 113). The part of the receptor indicated in **brown** is the part with which G-protein associates. The two serine residues indicated in **black** are sites for phosphorylation. (Adapted from Frielle et al., 1989.)

The binding of transmitter to a receptor with seven membrane-spanning regions activates a *transducing G-protein.* (G-proteins, which couple the receptor to an effector protein, are discussed in detail later.) In the resting state a G-protein binds a molecule of GDP and is inactive. When transmitter is bound, however, the result is a cascade of chemical reactions. The receptor interacts with the G-protein, producing a conformational change that causes GTP to displace GDP, thereby activating the G-protien. The activated G-protein then binds to an effector enzyme: adenylyl cyclase in the cAMP system, phospholipase C in the diacylglycerol–inositol polyphosphate system, and phospholipase A_2 in the arachidonic acid system. In each of these systems the effector enzyme initiates changes in specific proteins within the cell, either by binding to the target (or regulator) protein directly or by activating a protein kinase that phosphorylates the target protein.

A kinase is any enzyme that uses ATP as a donor of phosphoryl groups. The concept of phosphorylation mediated by protein kinases is central to understanding the action of second-messenger pathways. Each of the second-messenger systems—the cyclic AMP, calcium, and diacylglycerol systems, as well as the tyrosine kinase system—all exert a major part of their effect on cells through the actions of protein kinases. Since a single protein kinase can phosphorylate many different target proteins, thereby altering their activities in dramatic ways, protein kinases often lead to the amplification and distribution of signals.

Because of the prevalence of protein phosphorylation, we need ask: What does phosphorylation accomplish? The introduction of the negatively charged phosphoryl group can alter the conformation of a protein to modify the function of a substrate, be it an enzyme, a cytoskeletal protein, a subunit of an ion channel, or a transcriptional activator (a DNA-binding protein that can regulate transcription). For example, by decreasing an enzyme's affinity for its substrates or by changing its location within the cell, phosphorylation can diminish the enzyme's activity. Conversely, activity can be enhanced if phosphorylation increases the enzyme's affinity for its substrates, if it positions the enzyme in a more effective subcellular locale, or if it prevents association of the enzyme with an inhibitor.

In many neurons protein phosphorylation can lead to the closing or opening of ion channels and can thus modulate the signaling properties of those cells. These changes, which depend on transient increases in the concentration of the second messenger in the postsynaptic neuron, usually last from seconds to minutes—significantly longer than the changes in membrane potential produced by receptors that directly gate ion channels. The effects of phosphorylation on the activity of the ion channels are discussed in detail later in the chapter.

The Cyclic AMP Pathway Involves a Polar and Diffusible Cytoplasmic Messenger

The cAMP pathway is the prototype of an intracellular signaling pathway that makes use of a water-soluble second messenger that diffuses within the cytoplasm. This pathway illustrates the typical steps in a neuronal second-messenger pathway (Figures 14–4 and 14–5). The binding of transmitter to a receptor leads to the activation of a stimulatory G-protein called G_S. In its resting, inactive state G_S normally has a molecule of GDP bound to it. On activation, G_S binds a molecule of GTP in exchange for the GDP, thereupon activating the enzyme adenylyl cyclase. This enzyme, an integral membrane protein that spans the plasma membrane many times, in turn catalyzes the conversion of ATP to cAMP. The GTP–G-protein complex and the catalytic subunit of the cyclase together constitute the active form of the enzyme. When associated with the catalytic subunit, G_S also acts as a GTPase, hydrolyzing its bound GTP to GDP. As a result, the G-protein dissociates from the cyclase, stopping the synthesis of cAMP (Figure 14–4). The receptor and the cyclase thus do not interact directly, but are coupled by the transducer protein, G_S. The duration of cAMP synthesis is regulated by the GTPase activity of G_S. After the GTP is hydrolyzed, the G_S is able to bind a new transmitter–receptor complex at the surface of the cell and, thereby, to activate the cyclase again.

G-proteins are not integral components of the membrane, although they are associated with the internal leaflet of the plasma membrane. They consist of three subunits: α, β, and γ. The α subunit is only loosely associated with the membrane and is usually the coupling agent between receptors and primary effector enzymes. There are many types of α subunits, and G-proteins with different α subunits have different actions. For example, the β-adrenergic receptor activates the cyclase by acting on G_S, while other receptors inhibit the cyclase by acting on another G-protein containing a different α subunit. More than a dozen varieties of G-proteins that differ in their α subunit have been identified. Compared with other organs of the body, the brain contains an exceptionally large proportion of these other G-proteins, for example, G_i and G_o. These G-proteins activate guanylyl cyclase, phospholipases A_2 and C, and probably many other signal transduction mechanisms not yet identified (Figure 14–2).

One way the α subunit of G_S can be distinguished from the α subunits of other G-proteins is by its response to bacterial toxins. The α subunit of G_S is permanently activated by an enzymatic reaction catalyzed by the toxin from *Vibrio cholerae*. (The action of the toxin on the G_S of intestinal cells is the initial molecular step in the pathogenesis of cholera.) Other G-proteins are unaffected by this toxin but are inactivated by a similar reaction catalyzed by the toxin from the bacterium *Bordetella pertussis*, which causes whooping cough. Common to these reactions is the formation of a covalent ADP-ribosylated derivative of the α subunit with NAD^+. As we have seen, all of these α subunits bind to similar β and γ subunits, of which fewer isoforms exist. The βγ complex is much more tightly fixed to the membrane than the α subunit. As we shall learn later in this chapter, both the α and the βγ subunits of G-proteins can affect ion channels directly.

Transmitter binding alters conformation of receptor, exposing binding site for G_S protein.

Diffusion in the bilayer leads to association of transmitter-receptor complex with G_S protein, thereby activating it for GTP-GDP exchange

Displacement of GDP by GTP causes the α subunit to dissociate from the G_S complex, exposing a binding site for adenylyl cyclase on the α subunit

α subunit binds to and activates cyclase to produce many molecules of cAMP

Hydrolysis of the GTP by the α subunit returns the subunit to its original conformation, causing it to dissociate from the cyclase (which becomes inactive) and reassociate with a $\beta\gamma$ complex

The activation of cyclase is repeated until the dissociation of transmitter returns the receptor to its original conformation

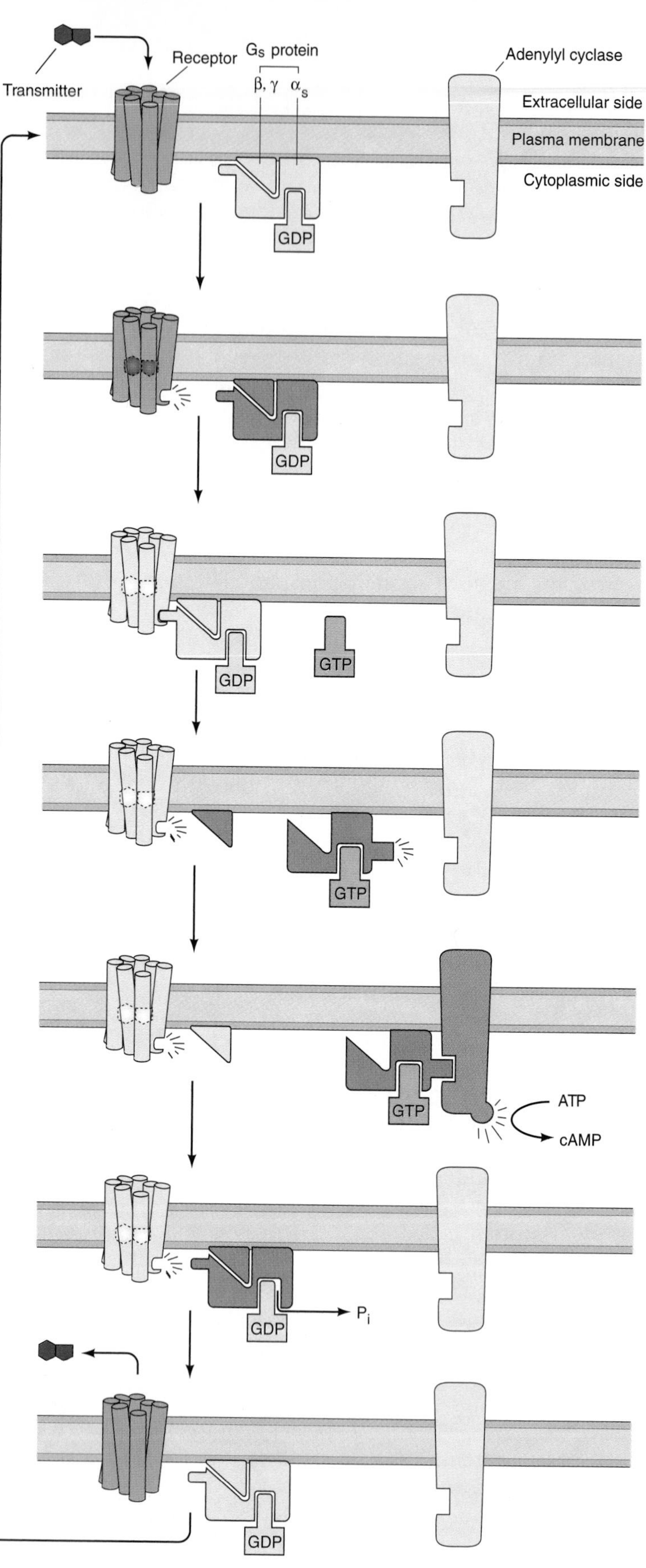

← *Figure 14–4* The cAMP cycle. The binding of a transmitter to the external portion of the receptor allows the stimulatory G-protein (G_S) bearing GDP to bind to an intracellular domain of the receptor. This association causes GTP to exchange with GDP and the α_s subunit to dissociate from the $\beta\gamma$ subunits of the G-protein. The α_s subunit, now bearing GTP, next associates with an intracellular domain of adenylyl cyclase, thereby activating the enzyme to produce many molecules of cAMP from ATP. The α_s subunit, when bound to the cyclase, is a GTPase. The hydrolysis of GTP to GDP and inorganic phosphate (P_i) leads to the dissociation of α_s from the cyclase and its reassociation with the $\beta\gamma$ subunits. The cyclase then stops producing the second messenger. Sometime during this cycle the transmitter dissociates from the receptor. The system returns to an inactive state when the ligand binding site on the receptor is empty, the three subunits of the G-protein have reassociated, and the guanine binding site on the $\beta\gamma$ subunit is occupied by GDP. (Adapted from Alberts et al., 1989.)

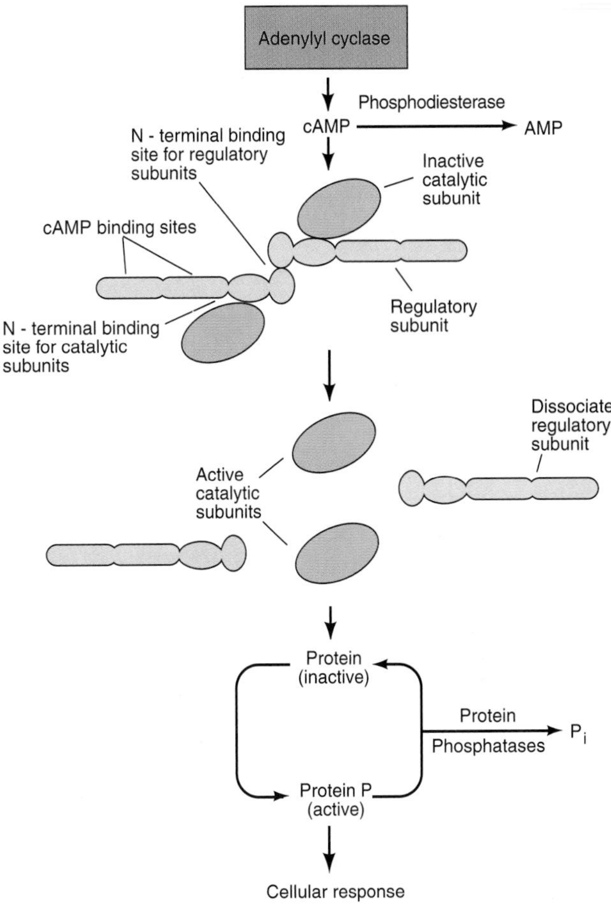

Figure 14–5 The cAMP pathway is typical of neuronal second-messenger pathways. Adenylyl cyclase converts ATP into cAMP. Four cAMP molecules bind to the two regulatory subunits of the cAMP-dependent protein kinase, liberating the two catalytic subunits, which are then free to phosphorylate specific substrate proteins that regulate a cellular response. Two kinds of enzymes regulate this pathway. Phosphodiesterases convert cAMP to AMP (which is inactive), and protein phosphatases remove phosphate groups from the regulator (substrate) proteins.

G-protein molecules outnumber the receptor molecules in a cell. Since a single liganded receptor can activate multiple G-proteins, the G-proteins serve to amplify a small synaptic signal (represented by relatively few chemical transmitter and receptor molecules) into the larger number of activated cyclase complexes needed to catalyze the synthesis of an effective concentration of cAMP within the cell. Further amplification occurs with the protein kinase reaction, the next step in the cAMP cascade (Figure 14–5).

Cyclic AMP activates the cAMP-dependent protein kinase by causing the regulatory subunits (R) to dissociate as a dimer from the catalytic subunits (C), according to the reaction

$$R_2C_2 + 4\ \text{cAMP} \rightleftharpoons R\ (2\ \text{cAMP})_2 + 2C.$$

When combined, the R_2C_2 molecule is inactive. The several isoforms of regulatory subunits all contain three functional domains (Figure 14–5): (1) an N-terminal region responsible for binding to its counterpart, which is always the identical isoform; (2) an N-terminal region responsible for inhibiting the catalytic subunit; and (3) two similar binding sites for cAMP. When bound to these sites, cAMP is not chemically changed; rather, it alters the conformation of the regulatory subunits so that they dissociate from the catalytic subunits. Each catalytic subunit is then free to transfer the γ-phosphoryl group of ATP to the hydroxyl groups of specific serine and threonine residues in protein:

$$\text{Protein} + \text{ATP} \xrightarrow{C} \text{Phosphoprotein} + \text{ADP}.$$

Each catalytic subunit contains a binding site for ATP that is highly conserved among other kinases, as well as a site that recognizes specific sequences of amino acids in substrate proteins. Before describing these sequences further, we need to examine the molecular mechanisms by which the catalytic subunits are regulated. An essential feature of regulation is the association with the regulatory subunits through the N-terminal domain (Figure 14–5). Certain isoforms of the regulatory subunits can be phosphorylated in this domain by the catalytic subunits; this *autophosphorylation* is typical of many protein kinases. Thus the regulatory domain can serve as a substrate for an enzyme's *own* catalytic domain. When the concentration of cAMP falls within the cell, the second messenger dissociates from the free regulatory subunits, which then combine with catalytic subunits again (Figure 14–6A). Phosphorylated regulatory subunits recombine with catalytic subunits at a slower rate than do dephosphorylated regulatory subunits.

The other serine and threonine protein kinases that are discussed in this chapter—the cGMP-dependent, the Ca^{2+}/calmodulin-dependent, and protein kinase C—have regulatory and catalytic domains within the same protein (Figure 16–6B to D). The regulation of cAMP-dependent protein phosphorylation illustrates the molecular principle common to the regulation of almost all protein kinases, whether the regulatory domain is situated on separate subunits or is a part of the same polypeptide chain as the catalytic region. As we saw, some regulatory subunits actually are substrates for phosphorylation. To explain further how the regulatory subunits and the regulatory domains of the other protein kinases operate, it is important to know that a protein can only be a substrate for a kinase if it has a special *phosphorylation sequence* that can be recognized by that kinase. This sequence consists of a specific pattern of amino acids around the serine or threonine residues to be phosphorylated. These phosphorylation sequences are a necessary part of the recognition and binding site on the substrate protein for the catalytic subunit during the phosphorylation reaction itself. One of the suitable sequences for the cAMP-dependent protein kinase is –Arg–Arg–X–Ser–, where X can be any amino acid. Other residues near this sequence also contribute to the affinity of the protein substrate for the kinase.

An important feature of regulatory domains in kinases is that they all contain a sequence similar to the phosphorylation sequence, except that the critical serine (or threonine) is lacking and is replaced by an amino acid residue that does not have a hydroxyl group. Thus, even though this part of the regulatory subunit binds to the catalytic site with high affinity, it cannot be phosphorylated—it is a *pseudosubstrate*. The regulatory domain therefore acts as a competitive inhibitor of the kinase reaction. As already mentioned, some isoforms of regulatory subunits also have a functioning substrate domain. In these subunits the amino acid to be phosphorylated is masked except in the presence of cAMP.

Some Second Messengers Are Generated Through the Hydrolysis of Phospholipids

A similar logic is evident in the way second messengers are generated through the hydrolysis of phospholipids in the inner leaflet of the plasma membrane. Hydrolysis is catalyzed mainly by two specific enzymes, phospholipase C and phospholipase A_2, each of which can be activated by different G-proteins. Phospholipases are named according to the bond that they hydrolyze in the phospholipid.

Before considering the three major second-messenger cascades that stem from these two phospholipases, it is helpful to know the structure of phosphatidyl inositol (PI) and to understand that the fatty acid composition of this phospholipid in the brain is exceptionally uniform. Almost all the phosphoinositides of the brain have the following chemical structure:

Phosphatidyl inositol (PI)

A cAMP-dependent protein kinase

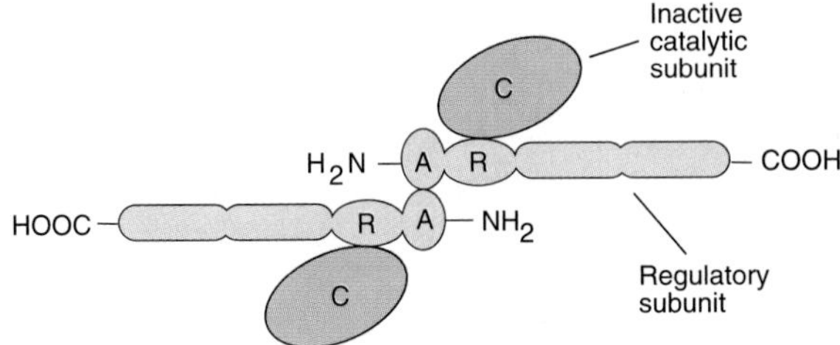

B cGMP-dependent protein kinase

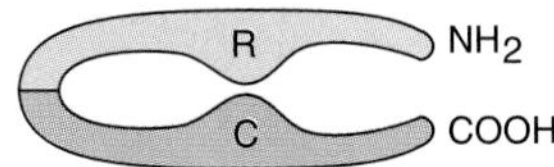

C Protein kinase C

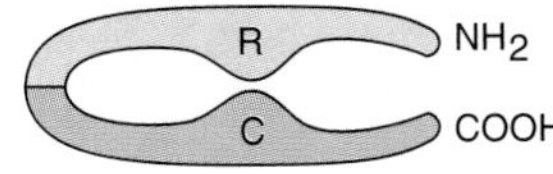

D Ca^{2+}/Calmodulin-dependent protein kinase

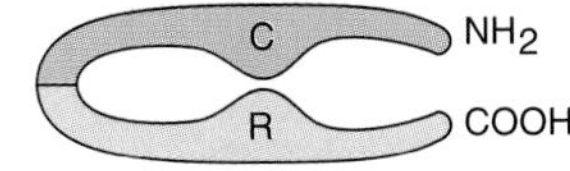

E Tyrosine protein kinase

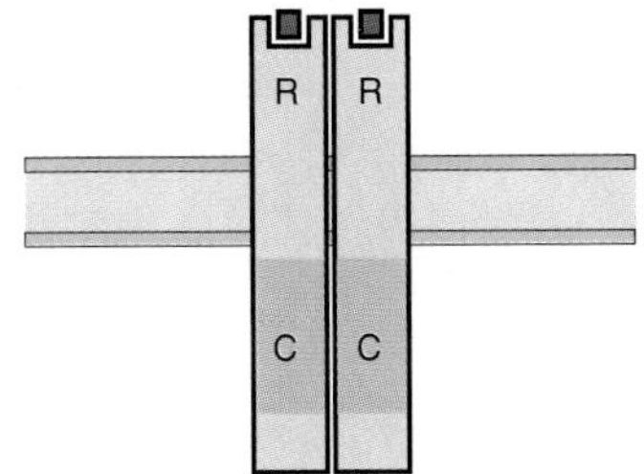

Figure 14–6 All protein kinases are related and are regulated in a similar way. In the absence of an activator, the kinases are enzymatically inactive because their catalytic domains are inhibited. With the serine and threonine protein kinases **(A–D)**, the catalytic domains are actually covered by regulatory domains that have amino acid sequences similar to those required for phosphorylation in substrate proteins. However, unlike the substrate proteins, the serine or threonine residue to which a phosphoryl group would be transferred is absent, and these inhibitory domains are therefore known as *pseudosubstrates*.

A. In the cAMP-dependent protein kinase two identical regulatory subunits associate with each other at site **A** and with the catalytic subunits at site **R**. Each regulatory subunit also contains two binding sites for cAMP. When cAMP is bound, a conformational change in the regulatory domains of the subunits results in their dissociation from the two catalytic subunits. Dissociated catalytic subunits can then phosphorylate substrate proteins.

B–D. In the other major protein kinases the regulatory domains **(R)** and the catalytic domains **(C)** are part of the same polypeptide chain. The cGMP-dependent protein kinase is similar to the cAMP-dependent protein kinase in amino acid sequence and, to a great degree, in the catalytic domain. The Ca^{2+}/calmodulin-dependent protein kinase **(D)**, unlike other kinases, is present in the cell as a complex of several molecules, each with similar biochemical properties. In all these enzymes the binding of second messenger is thought to unfold the molecule, thereby exposing and activating the catalytic region.

E. Activation of receptor-mediated tyrosine kinases is somewhat different. The regulatory domains are extracellular. The binding of a transmitter, hormone, or growth factor to the regulatory domain, as in receptors, causes a conformational change in the molecule through the plasma membrane to activate the intracellular catalytic domain.

The Diacylglycerol–Inositol Triphosphate System. Receptor-activated phospholipase C produces diacylglycerol and inositol 1-phosphate (IP_1) from PI itself, inositol 2-phosphate (IP_2) from phospholipid inositol phosphate (PIP_1), or inositol triphosphate (IP_3) from PIP_2 (Figure 14–7). IP_3, in turn, can be converted to IP_4 and IP_5 by further phosphorylation with specific kinases. Several of these inositol phosphates serve as second messengers. The metabolism of the inositol phosphates is quite complex, and much still remains to be discovered. These compounds are degraded by phosphatases in several reaction sequences to the free alcohol inositol, which is then reincorporated into membrane phospholipids. The final step in all of these degradative pathways is blocked by lithium ions (Li^+), which inhibit the release of free inositol from IP_1.

Protein Kinase C. Diacylglycerol, which is hydrophobic, remains within the membrane, where it activates protein kinase C (Figure 14–7C). The inac-

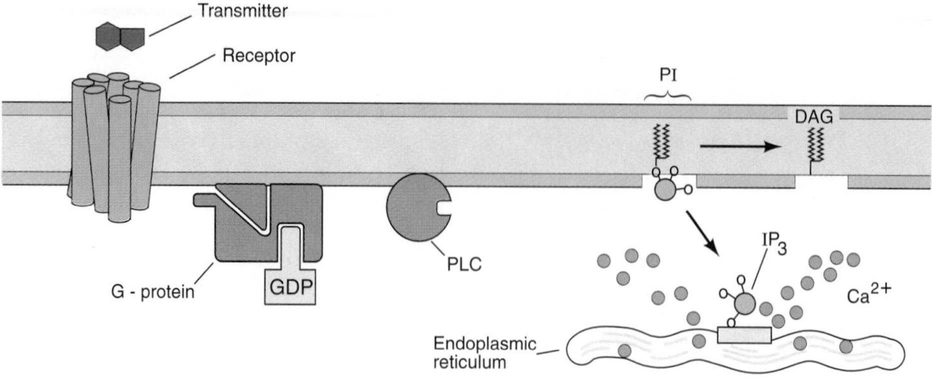

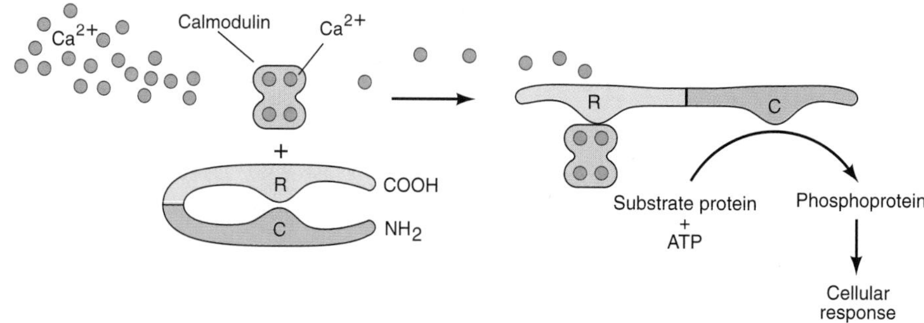

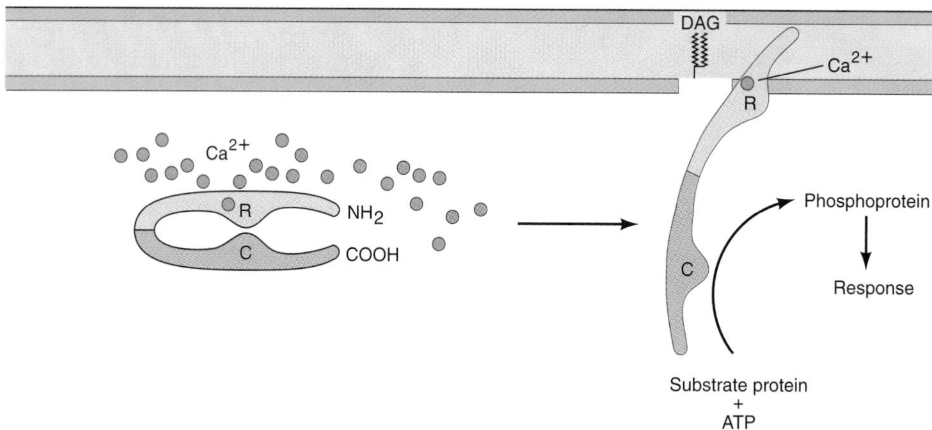

Figure 14–7 Hydrolysis of phospholipids in the cell membrane activates three major second-messenger cascades.

A. In the inositol–lipid pathway the binding of transmitter to a receptor activates a G-protein, which in turn activates phospholipase C (PLC). This phospholipase cleaves the phosphatidyl inositol (PI) into two second messengers, inositol triphosphate (IP_3) and diacylglycerol (DAG). Inositol triphosphate is water soluble and can diffuse into the cytoplasm. There it binds to a receptor on the endoplasmic reticulum to release Ca^{2+} from internal stores.

B. Calcium bound to calmodulin activates the Ca^{2+}/calmodulin-dependent protein kinase.

C. Diacylglycerol the other second messenger produced by the cleavage of PI, remains in the membrane where it activates protein kinase C (**PKC**). Membrane phospholipid is also necessary for this activation. Thus, PKC is active only when translocated from the cytoplasm to the membrane. Some isoforms of PKC do not require Ca^{2+} for activation.

tive form of this kinase is in the cytoplasm. When diacylglycerol is generated, the enzyme is moved to the membrane to form the active complex that can phosphorylate many protein substrates in the cell, both membrane-associated and cytoplasmic. Activation of protein kinase C requires membrane phospholipids in addition to diacylglycerol.

At least nine isoforms of protein kinase C exist, and all have been found in nervous tissue. These different forms are encoded by distinct genes. One gene, however, yields two transcripts, β^{I} and β^{II}, due to alternative RNA splicing. Rather than having different proteins as regulatory and catalytic elements, each enzyme contains regulatory and catalytic domains in a single continuous polypeptide chain (see Figure 14–6). Like the cAMP-dependent protein kinase, protein kinase C can be autophosphorylated.

Two functionally interesting differences have thus far been found among these isoforms. The so-called major forms (α, β^{I}, β^{II}, and γ) all have a Ca^{2+} binding site and are activated by Ca^{2+} ions. The minor forms (e.g., δ, ϵ, and ζ) are molecules that lack the Ca^{2+} binding domain, and therefore their activity is independent of Ca^{2+}. The second interesting difference is that, of the major forms, only protein kinase Cγ is activated by low concentrations of arachidonic acid, a membrane fatty acid, while all the isoforms respond to diacylglycerol or phorbol esters (plant toxins that bind to protein kinase C and promote tumors).

An important aspect of the bifurcating second-messenger pathway that stems from phospholipase C is that the two products of hydrolysis (IP_3 and diacylglycerol) can act independently as well as together. Some transmitter receptors cause the production of IP_3 alone, failing to activate protein kinase C. IP_3 can elevate the intercellular concentration of free Ca^{2+} by causing the release of Ca^{2+} from intracellular membranous stores. This Ca^{2+} then activates a variety of cellular processes. Often, Ca^{2+} acts when it forms a complex with the small protein calmodulin. An important example is the activation of the Ca^{2+}/calmodulin-dependent protein kinase (Figure 14–7B). Although this enzyme in tissue is made up of a complex containing many similar subunits, each subunit contains regulatory and catalytic domains within the same polypeptide chain. Each subunit can be autophosphorylated by an intramolecular reaction at many sites in the enzyme molecule. When Ca^{2+} and calmodulin are absent, the C-terminal regulatory domain behaves as a pseudosubstrate inhibitor of the catalytic portion of this kinase (Figure 14–6). Binding of the Ca^{2+}/calmodulin complex causes conformational changes of the kinase molecule that unfetter the catalytic domain for action (Figure 14–7C).

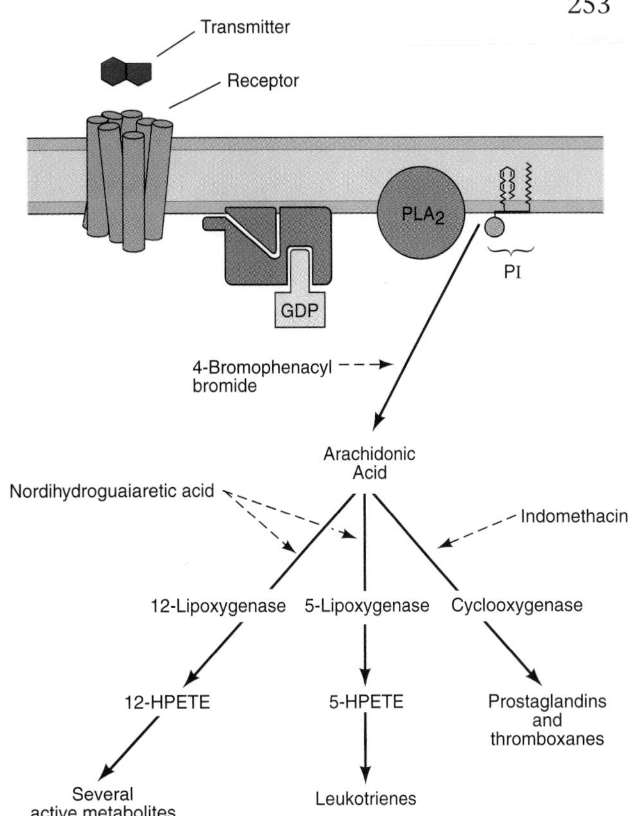

Figure 14–8 Arachidonic acid is released through receptor-mediated activation of a phospholipase. The one shown in the figure is phospholipase A_2 (PLA_2), which hydrolyzes phosphoinositol (PI) in the plasma membrane. This enzyme can be inhibited by alkylation with 4-bromophenacyl bromide. Once released, arachidonic acid is metabolized through several pathways, three of which are shown in the figure. The 12- and 5-lipoxygenase pathways both produce several active metabolites. Lipoxygenases are inhibited by nordihydroguaiaretic acid (NDGA). The cyclooxygenase pathway produces prostaglandins and thromboxanes. This enzyme is inhibited by indomethacin, aspirin, and other nonsteroidal anti-inflammatory drugs.

The Arachidonic Acid Pathway. Receptors that activate phospholipase A_2 cause the release of arachidonic acid from the cell membrane (Figure 14–8).

The arachidonic acid released is rapidly converted to a family of active metabolites, named *eicosanoids* for their 20 (Greek *eicosa*) carbon atoms. Arachidonic acid is metabolized by three types of enzymes: cyclooxygenases (producing prostaglandins and thromboxanes), several lipoxygenases (producing a variety of metabolites to be discussed below), and the cytochrome P450 heme-containing complex, which oxidizes arachidonic acid itself as well as cyclooxygenase and lipoxygenase metabolites.

Most work in nervous tissue has been done with cyclooxygenase and the lipoxygenases. In other tissues, metabolites of arachidonic acid have been extensively characterized because of their potent actions in inflammation, injury, and the control of smooth muscle in blood vessels and lung. Prostaglandins and thromboxanes also are present in the brain, and their synthesis is dramatically increased by nonspecific stimulation such as electroconvulsive shock, trauma, or acute cerebral ischemia (localized loss of blood). The mechanism by which these eicosanoids modulate synaptic transmission and neuronal excitability remains to be discovered.

Lipoxygenases introduce an oxygen molecule into the arachidonic acid molecule, resulting in a hydroperoxyeicosatetraenoic acid (HPETE). Depolarization of brain slices with high concentrations of extracellular K^+ ions, glutamate, or NMDA greatly increases 12-lipoxygenase activity. Also, 12-HPETE and some of its metabolites have been shown to modulate the actions of ion channels at specific synapses.

Arachidonic acid and its metabolites are highly lipid soluble and readily diffuse through membranes. They can be active both within the cell in which they are produced and in neighboring cells, including the presynaptic neuron. Because of this transcellular action, it has been suggested that these substances may act as retrograde synaptic messengers (Figure 14–9).

The Tyrosine Kinase Pathway Utilizes Both Receptor and Cytoplasmic Kinases

Receptor tyrosine kinases bind various peptides such as epidermal growth factor (EGF), fibroblast growth factor (FGF), nerve growth factor (NGF), and insulin. They differ from the typical G-protein-coupled receptor in two ways. First, they span the membrane only once. Second, their cytoplasmic domain contains a protein kinase activity that phosphorylates proteins on tyrosine residues. Like the serine/threonine protein kinases described earlier in this chapter, tyrosine kinases also regulate the function of the neuronal proteins they phosphorylate. The substrates for tyrosine kinase usually belong to a special class of proteins that are thought to be dedicated to producing long-term changes in neuronal function.

In addition to the membrane-spanning domain and the cytoplasmic kinase domain, tyrosine kinase receptors have an extracellular binding domain for a neuropeptide. On binding a ligand, the receptor, originally a pair of monomers, associate to form a dimer. Dimerization causes intracellular kinase to be active. Each monomer now phosphorylates its counterpart on a tyrosine residue. This phosphorylation then further activates the kinase so that it becomes capable of phosphorylating other proteins in the cytoplasm. The ability of tyrosine kinases to recruit cytoplasmic proteins, which then become phosphorylated, is quite dramatic. Within seconds a large number of cytoplasmic proteins become phosphorylated.

Other Second-Messenger Pathways: Nitric Oxide, Carbon Monoxide, and Cyclic GMP

Cyclic GMP also activates a specific protein kinase. Guanylyl cyclase, which is presumably activated through specific receptors by neurotransmitters and hormones, converts GTP to cGMP. Cyclic GMP has been found to act directly on specific ion channels in the outer segment of retinal rod cells, an important regulatory role that will be described in Chapter 22 in detail. The cGMP-dependent kinase differs from cAMP-dependent protein kinases because it is a single polypeptide that contains both regulatory (cGMP-binding) and catalytic domains. As we have seen, these domains are similar to those in other protein kinases with similar function, especially those that are responsible for catalysis. Because of the similarities among the cAMP-dependent, Ca^{2+}/calmodulin-dependent, and tyrosine protein kinases, all of these second-messenger enzymes are believed to be related to an ancestral enzyme.

The function of cGMP-dependent protein phosphorylation is not yet understood, even though it occurs throughout the brain (usually at less than 10% of the level of cAMP-dependent protein phos-

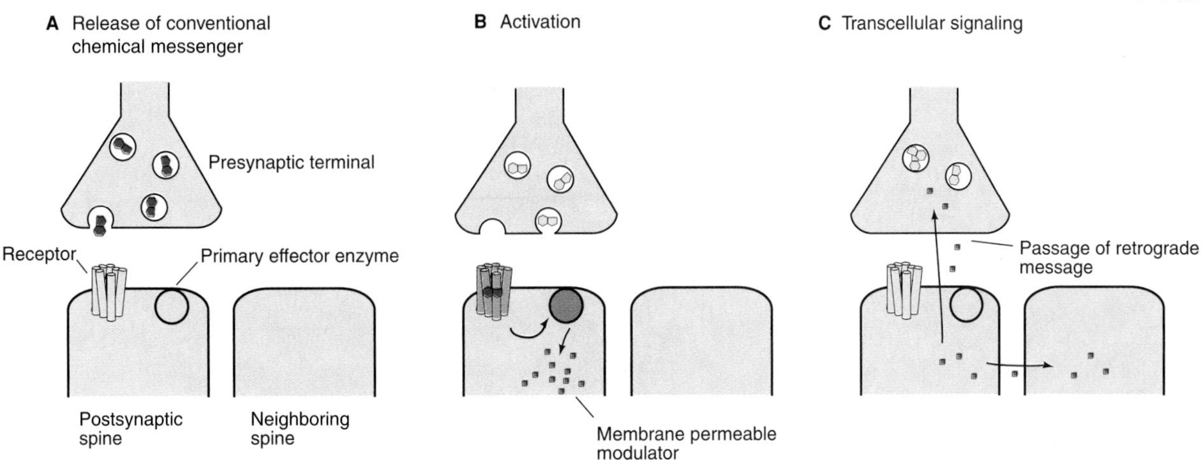

Figure 14–9 Transcellular messengers. Until recently synaptic transmission has been thought to occur only in one direction, from the presynaptic neuron to the postsynaptic target cell, for example, as shown in **A** and **B**. Now there is evidence for transcellular signaling from the post- to the presynaptic neuron as in **C**.

A. A presynaptic terminal releases a neurotransmitter (○), which reacts with the G-protein-coupled receptor in a postsynaptic dendritic spine.

B. This receptor activates an enzyme to produce a membrane-permeable modulator (▪).

C. This modulator can pass through cell membranes easily, and thus can cross over to neighboring postsynaptic spines as well as to presynaptic terminals where it can produce second-messenger-like effects. A transcellular modulator of the presynaptic terminal is called a retrograde messenger. As we shall see in Chapter 36, there is some evidence that this kind of signaling occurs in long-term potentiation in the hippocampus.

phorylation). The greatest amounts of cGMP-dependent protein phosphorylation occur in Purkinje cells of the cerebellum, large neurons with copiously branching dendrites. Synthesis of cGMP is stimulated in neurons and glial cells by the gases nitric acid (NO) and carbon monoxide (CO).

Nitric oxide is produced in neurons by NO synthase in response to glutamate, apparently acting through NMDA receptors and requiring an influx of Ca^{2+} ions. Like other second messengers, NO and CO are not unique to neurons, but operate in other cells of the body. For example, NO is a local hormone released from the endothelial cells of blood vessels, causing relaxation of the smooth muscle of vessel walls and thus allowing the vessels to dilate.

NO and CO, as well as the metabolites of arachidonic acid discussed earlier in this chapter, have three distinctive properties: (1) they pass through membranes readily; (2) they affect nearby cells without acting through a surface receptor; and (3) they are extremely short-lived. Because of these special properties, the term "second messenger" seems an inadequate designation for these substances; a more appropriate name is *transcellular messengers* (see Figure 14–9).

Second-Messenger Pathways Can Interact with One Another

The regulatory effects of the various second-messenger systems—the cAMP, diacylglycerol–IP_3, and arachidonic acid pathways and the tyrosine kinase cascade—are sometimes achieved through interaction between these pathways. Protein phosphorylation, influenced by all three second-messenger systems, has complex regulatory consequences because metabolic transformations consist of many interacting reaction sequences. These reactions are catalyzed by enzymes, many of whose activities are changed by phosphorylation.

Opportunities for interaction (also called *cross-talk*) occur because individual enzymes, channels,

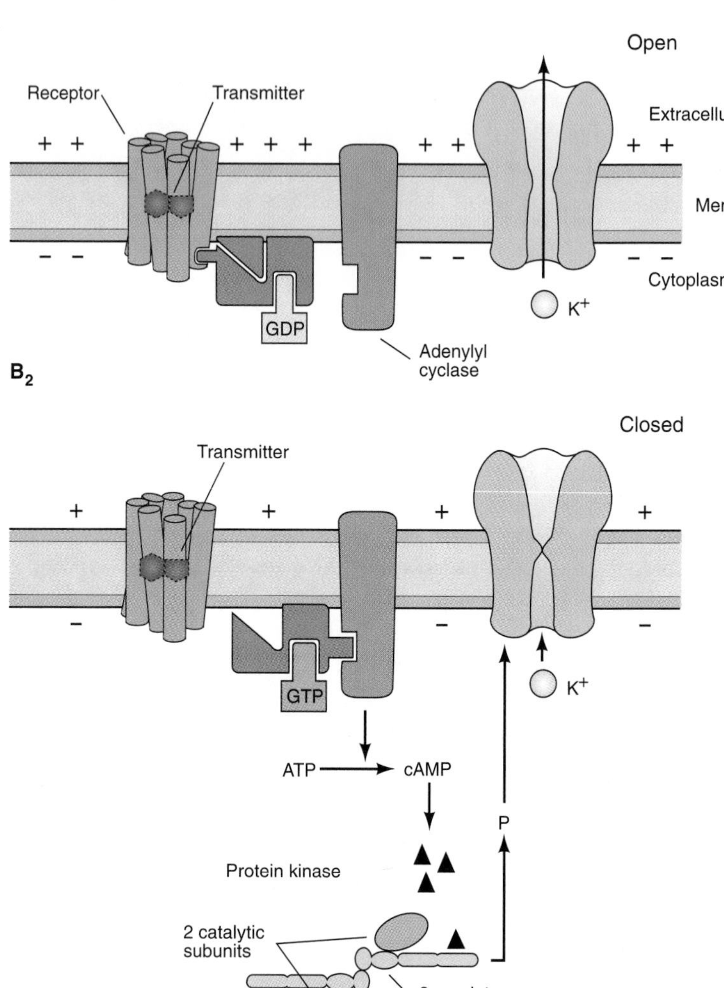

Figure 14–10 Excitatory synaptic action can be initiated by the direct opening of an ion channel by a transmitter or the closing of a channel by a second messenger.

A. A transmitter such as ACh or glutamate acts directly on the channel protein, opening a channel that is permeable to Na^+ and K^+.

B. A transmitter such as serotonin binds to a receptor and causes a coupling protein **(G)** to activate adenylyl cyclase (see Figure 14–4). The cAMP in turn activates a protein kinase, which phosphorylates an unidentified substrate protein (perhaps the channel itself or a regulatory protein that acts on the channel), causing the K^+ channel to close.

or cytoskeletal proteins can be modified at more than one site in the molecule, each by a protein kinase that is dependent on a different second messenger. Specific examples are the β-adrenergic receptor, to be discussed later in this chapter, and synapsin I, discussed in Chapter 15.

Second Messengers Often Act Through Protein Phosphorylation to Open or Close Ion Channels

Second-messenger systems often alter the activity of ion channels by phosphorylating the channel protein. Protein phosphorylation initiated by second-messenger kinases can have two effects. First, it can open channels that are closed at the resting membrane potential. This action appears similar to that produced by the actions of a transmitter on a directly gated ion channel. Second, it can close channels that are open at the resting potential. In certain cells resting K^+ channels are controlled by synaptically activated second-messenger actions (Figure 14–10). Closure of these channels depolarizes the neuron and also increases the excitability of the neuron, overriding the neuron's accommodation—the cell's tendency to raise its threshold—during repetitive firing (Figure 14–11).

One well-characterized example of such a synaptic action is the effect of ACh on the K^+ current in sympathetic ganglion cells. As noted in Chapter 12, there are two basic types of receptors for ACh: nicotinic and muscarinic. The two types of receptors differ from one another biochemically and they serve different functions in the nervous system. The nicotinic ACh receptors in skeletal muscle directly gate an ion channel. Muscarinic ACh receptors are found in sympathetic neurons and certain neurons of the hippocampus and cerebral cortex. They activate an as yet unidentified second-messenger system that closes a voltage-dependent K^+ channel, called the *M-type* (muscarinic) *channel* to distinguish it from other K^+ channels.

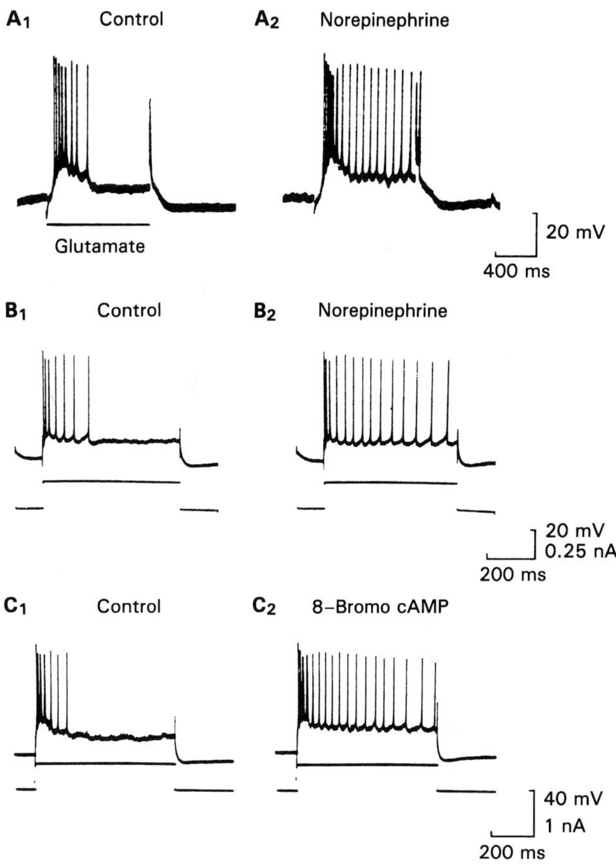

Figure 14–11 Norepinephrine activates cAMP and modulates the excitatory action on hippocampal neurons by closing a K^+ channel. This action prolongs the period of firing by the cell by overcoming the cell's tendency to raise its threshold.

A. The cell's response to glutamate is enhanced in the presence of norepinephrine.

B–C. Likewise, the cell's response to a constant depolarizing current pulse is enhanced by norepinephrine or an analog of cAMP. (Adapted from Madison and Nicoll, 1986, and Nicoll et al., 1987.)

In the sensory neurons of the marine snail *Aplysia*, serotonin produces a slow synaptic excitation by closing a resting K^+ channel called the *S-type K^+ channel*. In this synaptic action serotonin binds to a receptor that activates the cAMP cascade. The cAMP-dependent protein kinase phosphorylates a substrate protein that is either the S-type K^+ channel itself or a regulatory protein that acts on this K^+ channel to close it. In certain cortical neurons of the vertebrate brain, norepinephrine produces a similar slow excitatory synaptic potential. Norepinephrine-releasing terminals in the brain originate in a group of nerve cell bodies in the brain stem called the *locus ceruleus*. Neurons from this nucleus innervate the hippocampus and extend widely over the surface of the cerebral cortex. At the terminals of these cells norepinephrine acts through cAMP to close a K^+ channel that increases excitability and overrides accommodation.

How does closure of K^+ channels that are normally open at the resting potential bring about excitation? As we saw in Chapter 8, the resting membrane is permeable to K^+, Na^+, and Cl^-. The resting potential therefore results from a balance among the permeabilities to these three ions. When the cell is at rest the open K^+ channels hyperpolarize the membrane. Closure of some of these K^+ channels therefore decreases the steady K^+ efflux, leading to a buildup of positive charge in the cell. This drives V_m somewhat closer to E_{Na}, thus driving it closer to the threshold for firing an action potential. In addition, decreasing the number of open K^+ channels reduces the effective membrane conductance, g_m. As a result, any other excitatory input that generates a fast excitatory potential will produce greater depolarization. This is the reverse of the short-circuit effect of synaptic inhibition, which we considered in Chapter 13, where the opening of Cl^- channels increases g_m and thus decreases the effectiveness of excitatory synaptic inputs (see Figure 13–7).

The actions of second messengers are not limited to resting channels. Some voltage-dependent channels are closed by second messengers. For example, in dorsal root ganglion cells norepinephrine is thought to close certain voltage-gated Ca^{2+} channels through protein phosphorylation by protein kinase C.

In addition to being able to close channels that are normally open at the resting potential, second-messenger-mediated synaptic actions differ from directly gated synaptic actions in several other ways. First, second messengers can diffuse intercellularly to affect a distant part of the cell. As a result, K^+ channels closed (or opened) by a transmitter need not be located directly beneath the receptors acted on by the transmitter, but can be at some distance.

Second, the time course of second-messenger-mediated synaptic actions is much slower—in several known instances 10,000 times slower than directly mediated actions. Thus, a peptide similar to luteinizing hormone-releasing hormone, acting as a chemical messenger, closes the channel, producing a long-lasting (10-minute) excitatory synaptic potential. In contrast, an excitatory synaptic potential produced in the same neurons by the nicotinic ACh receptor–channels lasts only 20 ms (Figure 14–12). In *Aplysia* neurons, too, the excitatory synaptic potential produced when serotonin closes the K^+ channels can persist for several minutes. These slow synaptic actions modulate neuronal excitability for a period of minutes (Table 14–1).

Third, with the exception of the NMDA glutamate receptor–channel, directly gated channels are not activated by voltage and therefore are not affected by membrane potential. Consequently they do not contribute to the action potential. In contrast, various K^+, Ca^{2+}, and even Na^+ channels that are modulated by second messengers are voltage sensitive and do contribute to the action potential.

As these examples illustrate, the slow synaptic actions produced by second messengers typically modulate the excitability of neurons, in contrast with fast synaptic actions, which typically mediate synaptic actions between cells. The modulatory actions of second messengers are described further in the next chapter.

Second Messengers and G-Proteins Can Sometimes Act Directly on Ion Channels

Several modulatory actions have been discovered in which a G-protein moves within the membrane and interacts directly with an ion channel, causing it either to open or to close without the intervention of a freely diffusible second messenger or a protein kinase (Figure 14–13). The hyperpolarization produced by muscarinic ACh receptors in the heart is thought to be caused by the direct action of a G-protein that opens a K^+ channel. In addition, several types of K^+ channels in the brain are also modulated directly. In some types of K+ channels the α subunit of the G-protein activates the channel, while in other types the $\beta\gamma$ subunit are responsible.

In certain cells the same ion channel is modulated in two ways in response to a transmitter: directly and relatively rapidly by the binding of a G-protein, and indirectly and more slowly as a result of phosphorylation by the second messenger kinase activated by the same G-protein. This illustrates that G-proteins can act on different effectors, on enzymes that synthesize second messengers as well as on ion channels.

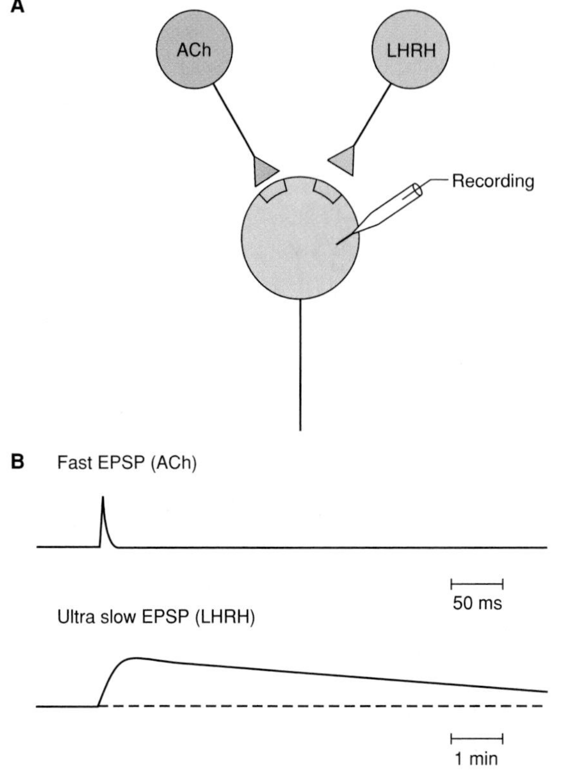

Figure 14–12 Certain neurons in sympathetic ganglia receive convergent excitatory connections from two different sets of neurons. One set of neurons uses ACh as its transmitter, the other uses a peptide, luteinizing hormone-releasing hormone (LHRH). Acetylcholine produces an EPSP through the opening of ion channels permeable to Na^+ and K^+; LHRH produces an EPSP through the closing of K^+ channels. (Adapted from Jan, Jan, and Kuffler, 1979.)

A. The neurons that use ACh make conventional synaptic contact. The neurons that use LHRH make a nondirected contact; the release site of LHRH is some distance from the target cell.

B. In the same neuron the time course of the decreased-conductance EPSP due to the closing of a K^+ channel is much slower than that of an increased-conductance EPSP due to the opening of the channel in the same neuron. The increased-conductance EPSP induced by ACh lasts 20 ms, whereas the decreased-conductance EPSP induced by LHRH is very slow, lasting 10 mins. (Note the different time scales in the two recordings.)

Finally, some channels are modulated directly by the second messengers themselves—by cGMP, cAMP, or metabolites of arachidonic acid—without requiring protein phosphorylation or even the binding of G-proteins. As we shall see later, the cation-selective ion channels of photoreceptors and the depolarizing bipolar cells in the retina are both opened by the direct action of cGMP on the channel. The cation-selective channel activated by the olfactory receptor is directly opened by the action of either cAMP or cGMP. Similarly, the peptide transmitter Phe–Met–Arg–Phe–NH_2 amide produces hyperpolarization and synaptic inhibition in sensory neurons of *Aplysia* by opening the S-type K^+ channel. This action is mediated by the direct action of 12-lipoxygenase metabolites of arachidonic acid.

Table 14–1 Comparison of Synaptic Excitation Produced by the Opening and Closing of Ion Channels

	Ion Channels Involved	*Effect on Total Membrane Conductance*	*Contribution to Action Potential*	*Time Course*	*Second Messenger*	*Nature of Synaptic Action*
EPSP due to opening of ion channels	Cation channel for Na^+ and K^+	Increase	None	Usually fast (milliseconds)	None	Mediating
EPSP due to closing of ion channels	Channel for K^+	Decrease	Modulates current of action potential	Slow (seconds or minutes)	Cyclic AMP (or other second messengers)	Modulating

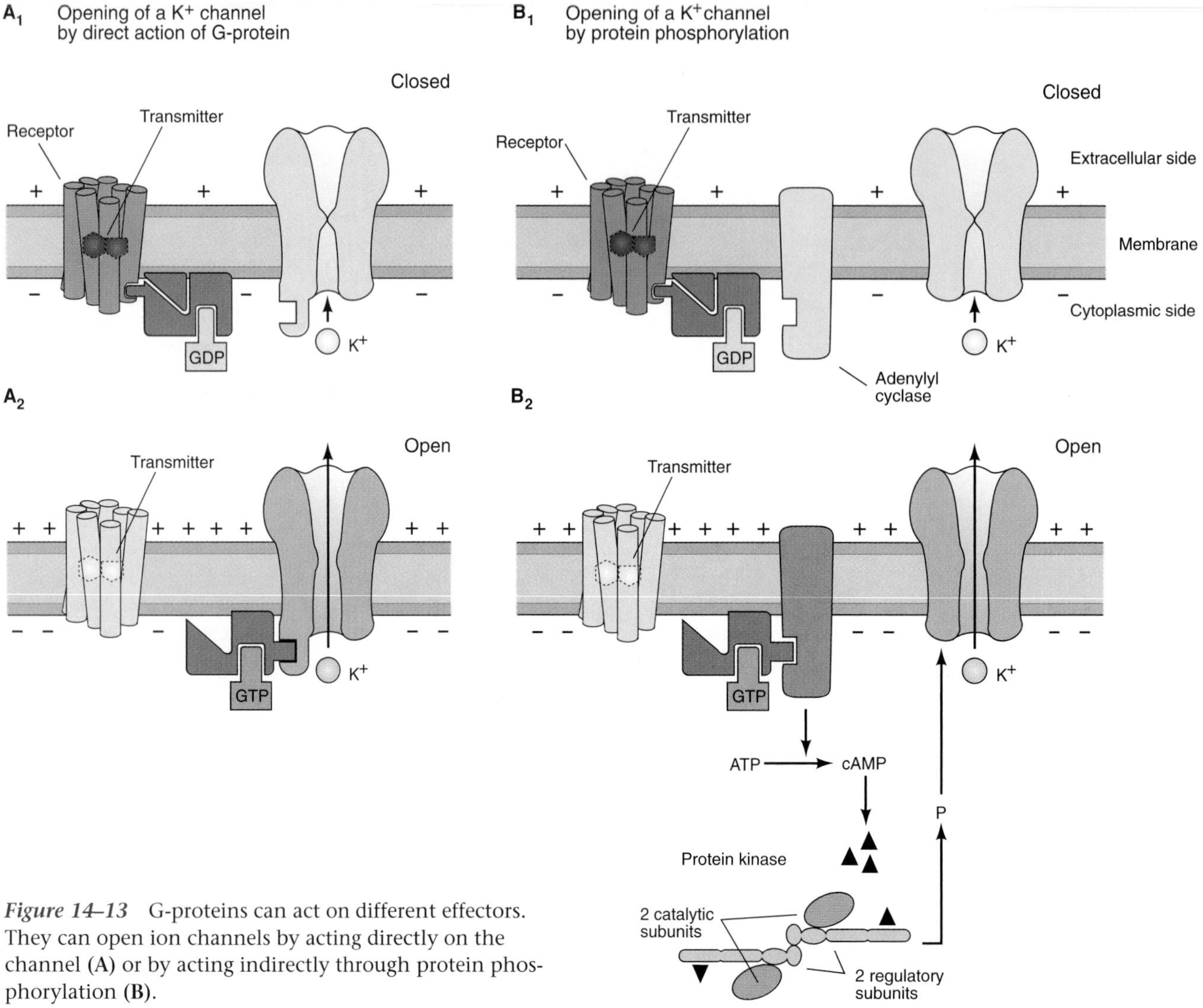

Figure 14–13 G-proteins can act on different effectors. They can open ion channels by acting directly on the channel **(A)** or by acting indirectly through protein phosphorylation **(B)**.

Second Messengers Can Alter the Properties of Transmitter Receptors: Desensitization

Second-messenger systems can act on many target proteins in addition to voltage-sensitive ion channels. A particularly interesting class of target proteins are the receptors for other transmitters, either those that gate ion channels indirectly or those that gate channels directly. Thus a second messenger can regulate the effectiveness of a receptor for another transmitter or the receptor that activates the second messenger system. For example, after prolonged exposure to its own transmitter, a receptor can become refractory to later applications of the same transmitter, a process called *desensitization* (Figure 14–14).

Although many mechanisms produce diminished responsiveness, desensitization has been shown in several instances to result from protein phosphorylation. One example is the β-adrenergic receptor, which is phosphorylated in the cytoplasmic domains of the receptor molecule that interact with G_S by a specific cAMP-independent receptor kinase (β-ARK), as well as by both the cAMP-depen-

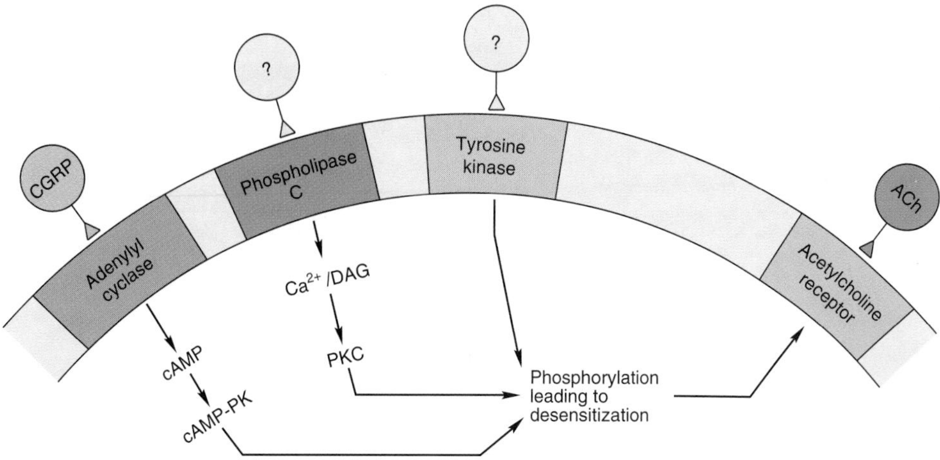

Figure 14–14 A second-messenger system for one receptor can affect other types of receptors and the ion channels they gate directly. Here a peptide, calcitonin gene-related peptide (CGRP), activates the cAMP cascade, which leads to phosphorylation of a different receptor the ACh receptor. Phosphorylation causes the ACh receptor to respond less effectively to ACh, a process called desensitization. Phosphorylation of the ACh receptor can also be produced by protein kinase C and by a tyrosine kinase, for which the initiating factors have not yet been identified. (Adapted from Huganir and Greengard, 1990.)

dent protein kinase and protein kinase C. During phosphorylation of the receptor, 2 to 3 moles of phosphate per mole of receptor is incorporated into the receptor protein, and the degree of desensitization correlates with the extent of phosphorylation. Phosphorylation by the cyclic AMP-dependent kinase or by β-ARK slows the ability of the receptor to activate G_S. But the chief inhibitory effect of β-ARK is to promote the binding of an inhibitory protein to the phosphorylated receptor. This inhibitor is similar to *arrestin*, a protein that regulates the function of rhodopsin in the retina.

Receptors that gate ion channels directly can also be modulated by phosphorylation (Figure 14–14). The cAMP-dependent protein kinase phosphorylates the γ and δ subunits of the nicotinic ACh receptor. Moreover, protein kinase C and tyrosine-specific protein kinases each act on several subunits. In all, three different kinases phosphorylate the ACh receptor at seven different sites, all of which are located in the major cytoplasmic domain of each subunit. Not all of these phosphorylations are known to have functional consequences, but the cAMP-dependent phosphorylation of the γ and δ subunits, as well as tyrosine phosphorylation, does increase the rate at which the receptor desensitizes in response to ACh.

Second Messengers Can Endow Synaptic Transmission with Long-Lasting Consequences

So far we have considered two types of chemically mediated synaptic actions: fast, directly gated synaptic actions lasting milliseconds, and slow, second-messenger-mediated actions involving modifications of ion channels and other substrate proteins and lasting seconds to minutes (Figure 14–15). In a third kind of synaptic action, transmitters, acting through second messengers, phosphorylate transcriptional proteins and thereby alter the cell's gene expression (Figure 14-16 in Box 14–1). Thus, second-messenger kinases not only can modify preexisting proteins, but also can induce the synthesis of new proteins by altering gene expression. This third kind of synaptic action can lead to other changes lasting days or even longer. These long-term changes are likely to be important for

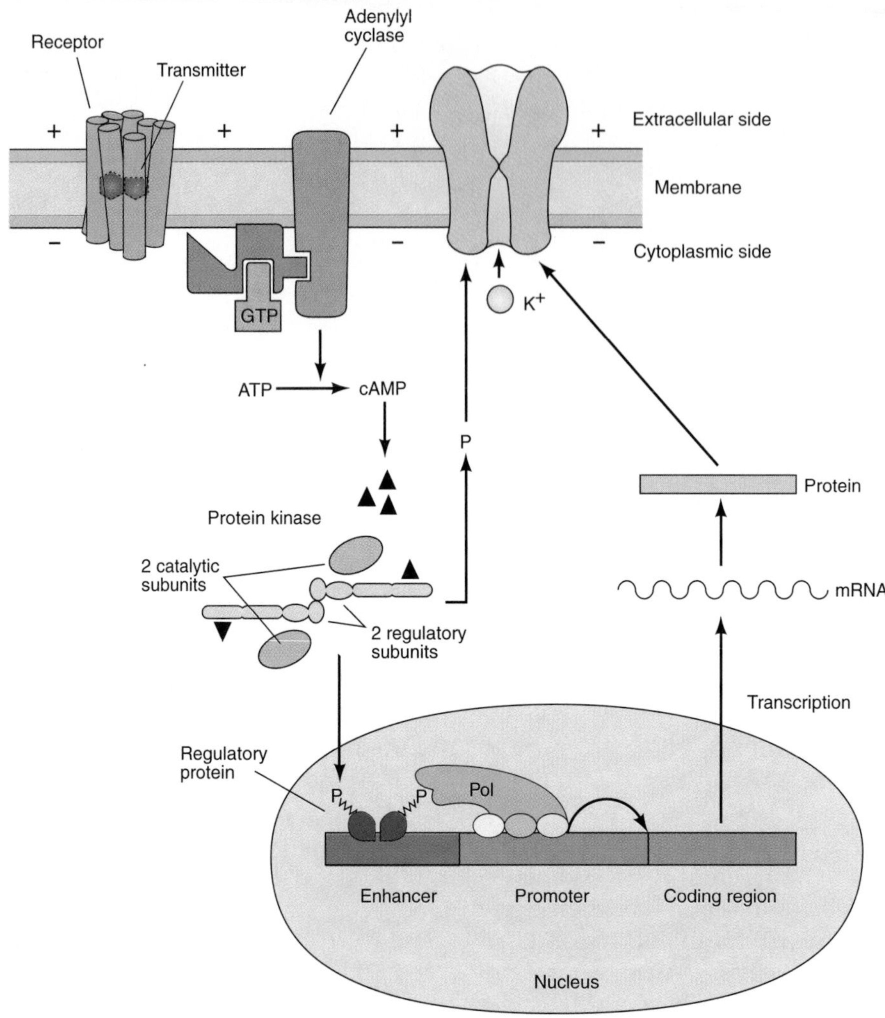

Figure 14–15 A single chemical transmitter can produce synaptic actions with different time courses. In this example a single exposure to the transmitter activates the cAMP second-messenger system, which in turn activates the cAMP-dependent protein kinase that phosphorylates a K^+ channel to produce a synaptic potential, which modifies neuronal excitability for minutes. With repeated activation the transmitter also acts through the cAMP-dependent protein kinase to phosphorylate one or more transcriptional regulatory proteins that activate gene expression. Gene activation produces a protein that modifies the channel itself. The protein produces more enduring closure of the channel and changes in neuronal excitability lasting days or weeks.

neuronal development and for long-term memory.

To illustrate the relative importance of transient modifications of proteins and the more enduring synthesis of new protein by altered gene expression, we now consider how a cholinergic presynaptic neuron can regulate the amount of transmitter substance (norepinephrine) in a postsynaptic target cell. Norepinephrine is a common small-molecule transmitter (see Chapter 15). Because the synthesis of norepinephrine is highly regulated, the amount available for release can keep up with substantial variations in neuronal activity. In certain ganglia the amount of norepinephrine synthesized is regulated through synaptic receptors in response to activity of presynaptic neurons. If the activity in these presynaptic neurons is sufficiently prolonged, the released transmitter will induce relatively long-term changes in the postsynaptic cell. These changes increase the supply of norepinephrine by inducing the synthesis of more *tyrosine hydroxylase,* the first and rate-limiting enzyme in the biochemical pathway for the synthesis of norepinephrine. A similar process of regulation occurs in adrenergic cells of the central nervous system.

In adrenergic neurons immediate short-term mechanisms lead to protein phosphorylation: presynaptic activity releases both ACh and a peptide transmitter, which activates a receptor in the postsynaptic adrenergic cell that stimulates the production of cyclic AMP. Cyclic AMP activates the cyclic AMP-dependent protein kinase, which in turn phosphorylates tyrosine hydroxylase. Normally, the activity of tyrosine hydroxylase is dependent on the concentration of its substrate, tyrosine. The enzyme's activity can be inhibited by norepinephrine and dopamine, the end products of the pathway, which act as *feedback inhibitors*. Phosphorylation frees the hydroxylase from these inhibitory mechanisms. Thus the short-term increase in norepinephrine is rapidly reversible.

Certain circumstances, however, call for a more persistent response. For example, severe or prolonged stress to the animal (such as cold or immobilization) results in intense presynaptic activity and consequently persistent firing of the adrenergic neuron. This puts a greater demand on transmitter synthesis and ultimately results in a long-term increase in norepinephrine, observed within hours in the cell body and somewhat later at nerve endings. The change is maintained for days after the original stress. This persistent increase in the amount of transmitter in the neuron is not the result of phosphorylation of the hydroxylase. Rather, the prolonged activity of the presynaptic neuron activates the cyclic AMP pathway sufficiently so that the kinase phosphorylates not only tyrosine hydroxylase molecules but also a transcriptional regulator—a nuclear protein that can bind to the regulatory region of a gene to alter gene expression. Once phosphorylated, this transactivating protein (called the CRE-binding protein or CREB) binds to a specific DNA regulatory region that is upstream (5′) of the coding region of the hydroxylase gene. Binding of the transcriptional activator to this response element allows RNA polymerase to bind to the promoter of the tyrosine hydroxylase gene and increases the gene's transcription rate (Figures 14–14 and 14–15).

Box 14–1 Regulation of Gene Expression

With the exception of mature lymphocytes, each cell in the human body contains precisely the same complement of genes (thought to number between 100,000 and 300,000) present in every other cell. The reason that cells differ so greatly from one another—why the liver cell is a liver cell and a brain cell a brain cell—is that different combinations of genes are *expressed* in specific cell types. This requires mechanisms for activating and repressing the expression of specific genes. Both repression and activation of genes take place during development, and often both are maintained throughout the life of the differentiated cell. Moreover, in any given cell many genes are not competent for transcription and as a consequence are never transcribed.

In addition to these more or less permanent modifications, the expression of many genes within specific cell types is *regulated*. For example, the rate at which a gene transcribes messenger RNA can be transiently enhanced or reduced by the action of proteins that bind to regulatory regions of the gene. The activity of these regulatory proteins is in turn controlled by receptors located within the cell or on the cell surface. These receptors recognize molecules such as steroid hormones, peptide growth factors, and neurotransmitters.

The template DNA sequence of genes can be divided into two major regions: a *coding region* and a *regulatory region*. For most genes that encode proteins, messenger RNA is transcribed from template DNA in the coding region by the actions of the

(continued)

Box 14–1 Regulation of Gene Expression (continued)

enzyme RNA polymerase II. The regulatory region usually lies upstream of the coding region. Because these DNA regulatory regions are usually near the coding region and linked to it directly, they are called *cis*-regulatory elements. In contrast, the transcription regulatory proteins that bind to these regulatory regions are called *trans*-regulatory elements, because they often are encoded by genes that are not linked to the gene being regulated.

The DNA regulatory region itself consists of two types of control elements. The first (or proximal) type of control element is called the *promoter region*. In many genes this element is an 8-base-pair region rich in adenine and thymine (A and T), called the TATA box ; this box is surrounded by a region rich in guanine and cytosine (G and C). The TATA box is usually located about 30 base pairs upstream from the start site for transcription (Figure 14–16).

The TATA box and adjacent DNA elements in the promoter region are involved in positioning the RNA polymerase in the region where transcription of messenger RNA begins. In eukaryotic organisms RNA polymerase II does not bind directly to the TATA box. Rather, the TATA box is occupied by a complex of other proteins called TATA box proteins. These TATA box-binding proteins are thought to interact directly with RNA polymerase II and direct its binding to a region of DNA adjacent to the site where transcription starts. In addition, other DNA regulatory modules are often near the TATA box; these include the CAAT box and the GC-rich modules that may facilitate the initial binding of the polymerase.

The second (or distal) type of DNA region is called the *enhancer region*. The enhancer elements can be located within a few hundred base pairs from the promoter, or as far as 100 kilobases away. Each individual control element of the enhancer region is generally 7 to 20 base pairs long and functions as a binding site for proteins that control whether the gene will be transcribed by RNA polymerase II.

The enhancer elements that bind cell-specific regulator proteins are called *response elements*. Thus, the *c*yclic AMP *r*esponse *e*lement (CRE) consists of the sequence ACGTCA; this sequence recognizes CREB proteins that are activated by phosphorylation under the control of the cAMP-dependent protein kinase. The *s*erum or *p*horbol ester *r*esponse *e*lement (SRE or PRE) consists of the sequence TGACTCAG. The *g*lucocorticoid *r*esponse *e*lement (GRE) has a longer sequence (TGGTACAAATGTTCT), which recognizes the protein receptor that is activated by binding of glucocorticoid hormones.

Certain enhancer elements bind activator proteins continuously, which permits basal levels of transcription. Other enhancer elements are bound by regulatory proteins only intermittently. This permits the gene to be regulated (induced or repressed) by appropriate transcriptional regulators. Thus, whether the RNA polymerase binds and transcribes a gene and how often it does so in any given period is determined by transcriptional regulators that bind to different segments of the promoter and enhancer regions. In addition to the proteins (the TATA box factors) that bind to the promoter region, regulatory factors need to bind to the enhancer region to bring about the expression that is induced by hormones, stress, and the process of learning. Some of these enhancers are several hundred base pairs away from the TATA box. How can proteins at a distance facilitate the binding of the polymerase to the TATA box? It is thought that the intervening DNA sequence loops out, thereby bringing together proteins bound to the distal enhancer region and the proximal promoter region (Figure 14–16).

Transcriptional factors that bind to these control regions typically have three functional domains: (1) a *DNA-binding domain*, which contains many basic residues that permit the protein to recognize and bind selectively to a specific DNA sequence; (2) an *activator domain*, which is often acidic in nature and permits the protein to contact and activate the basal transcription machinery (the TATA box-binding factors and RNA polymerase II); and (3) one or more *ligand binding* or *phosphorylation domains*, which are required for activating the transcription factors.

The DNA binding domains generally fall into one of three families. (1) *Helix–turn–helix proteins* consist of factors containing at least two α helices.

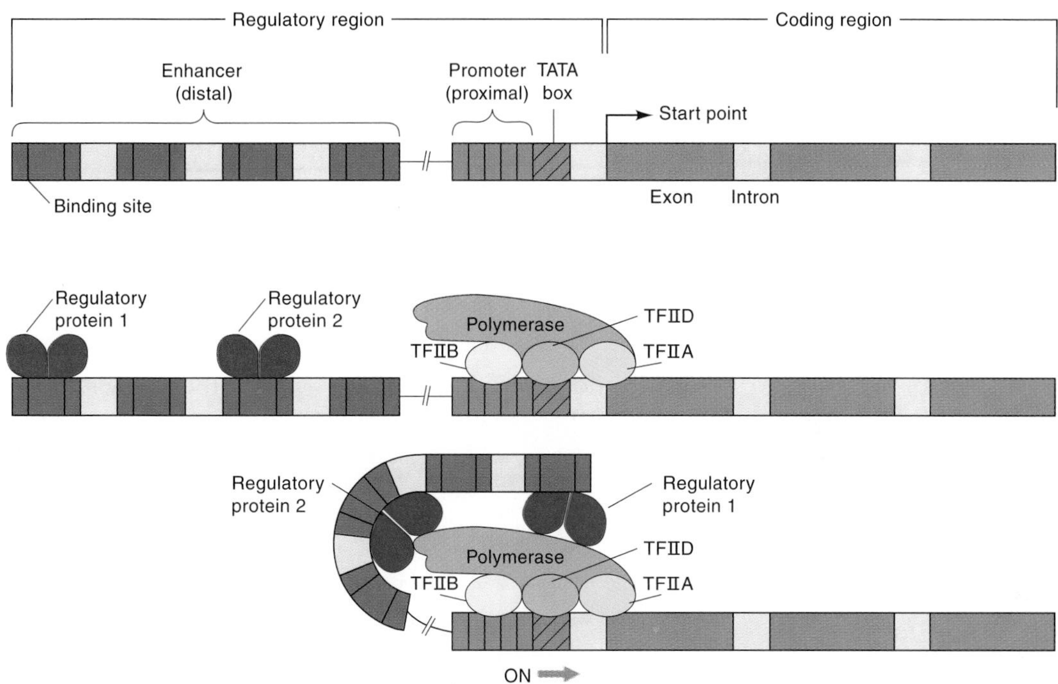

Figure 14–16 Transcriptional control.

A. The typical eukaryotic gene has two regions. The *coding region* is transcribed by RNA polymerase II into a messenger RNA and is then translated into a specific protein. The *regulatory region*, consisting of *enhancer elements* and a *promoter element*, regulates the initiation of transcription of the structural gene.

B. Transcriptional regulatory proteins bind both the promoter and the enhancer regions. **1.** A set of proteins binds to the TATA box, to the promoter, and to the distal enhancer regions. **2.** Proteins that bind to the enhancer region cause looping of the DNA, thereby allowing the regulatory proteins that bind to distal enhancers to contact the polymerase.

One helix occupies the major groove of the DNA and interacts with the regulatory nucleotide backbone. A second helical region is located at an angle across the DNA and interacts less directly with DNA. (2) *Zinc finger* proteins are so called because a stretch of about 23 amino acids containing alternating cystines and histidines forms a finger-like loop whose structure is maintained by the binding of a zinc ion. The proteins interact with DNA through the loop regions. Glucocorticoid, estrogen, vitamin A, progesterone, thyroid, and retinoic acid receptors each contain two zinc fingers. (3) *Amphipathic helical proteins* include two submotifs: helix–loop–helix proteins and leucine zipper proteins.

In fact, each of these three classes of transcriptional regulatory proteins forms dimers, proteins made up of two subunits. These dimers can bind to each other through protein–protein interactions as well as to the DNA by means of protein–DNA interactions. One important consequence of dimer formation is that additional specificity can be achieved by forming heterodimers, such that between related members of one gene family the two subunits are not identical. Heterodimers can bind to distinct DNA sequences and can then regulate the activity of different combinations of target genes. In this way a greater diversity of regulatory interactions can be achieved by a limited number of regulatory proteins.

An Overall View

The molecular actions of the receptor mechanisms considered here generally conform to the speed of the synaptic action they mediate. Directly gated ion channels, which operate most rapidly, are used for physiological processes that need speed. The simple knee-jerk reflex, illustrated in Chapters 2, 3, and 13, is one example. Similar fast processes also include synaptic connections that produce much of the animal's perceptual and motor behavior.

In recent years it has become increasingly clear that neurons also have longer-lasting, regulatory effects in target cells. Indeed, even in muscle contraction, sustained activity requires that the muscle cell's metabolism be regulated. This regulation is achieved by neural receptor mechanisms that are slower in onset and that persist for longer periods.

In the integrating circuits of the brain, neurons make use of both rapid synaptic transmission, through receptors that gate ion channels directly, and long-lasting transmissions by means of second messengers. Synaptic actions that gate ion channels directly and open channels that are closed at the resting potential invariably increase the overall conductance of the postsynaptic membrane. In contrast, actions mediated by second messengers, which can close ion channels that are open at the resting potential, decrease the conductance of the membrane.

Finally, in addition to gating ion channels, second messengers can alter the biochemical state of a nerve cell. For example, second messengers can alter gene expression to initiate persistent changes in function. As we shall see in Chapter 36, synaptically induced activations in gene expression are critical for the storage of long-term memory.

Selected Readings

Bishop, W. R., and Bell, R. M. 1988. Assembly of phospholipids into cellular membranes: Biosynthesis, transmembrane movement and intracellular translocation. Annu. Rev. Cell Biol. 4:579–610.

Casey, P. J., and Gilman, A. G. 1988. G protein involvement in receptor-effector coupling. J. Biol. Chem. 263:2577–2580.

Comb, M., Hyman, S. E., and Goodman, H. M. 1987. Mechanisms of transsynaptic regulation of gene expression. Trends Neurosci. 10:473–478.

Cooper, J. R., Bloom, F. E., and Roth, R. H. 1991. The Biochemical Basis of Neuropharmacology, 6th ed. New York: Oxford University Press.

Dennis, E. A. (ed.) 1991. Phospholipases. Methods in Enzymology, Vol. 197. San Diego: Academic Press.

Edelman, A. M., Blumenthal, D. K., and Krebs, E. G. 1987. Protein serine/threonine kinases. Annu. Rev. Biochem. 56:567–613.

Fantl, W. J., Johnson, D. E., and Williams, L. T. 1993. Signalling by receptor tyrosine kinases. Annu. Rev. Biochem. 62:453–481.

Hanks, S. K., Quinn, A. M., and Hunter, T. 1988. The protein kinase family: Conserved features and deduced phylogeny of the catalytic domains. Science 241:42–52.

Hanson, P. I., and Schulman, H. 1992. Neuronal Ca^{2+}/calmodulin-dependent protein kinases. Annu. Rev. Biochem. 61:559–601.

Huganir, R. L., and Greengard, P. 1990. Regulation of neurotransmitter receptor desensitization by protein phosphorylation. Neuron 5:555–567.

Hunter, T., and Sefton, B. M. (eds.) 1991. Protein Phosphorylation. Part A, Methods in Enzymology, Vol. 200. San Diego: Academic Press.

Hunter, T., and Sefton, B. M. (eds.) 1991. Protein Phosphorylation. Part B, Methods in Enzymology, Vol. 201. San Diego: Academic Press.

Jessell, T. M., and Kandel, E. R. 1993. Synaptic transmission: A bidirectional and self-modifiable form of cell-cell communication. Cell 72/Neuron 10 (Suppl): 1–30.

Lefkowitz, R. J., and Caron, M. G. 1988. Adrenergic receptors. Models for the study of receptors coupled to guanine nucleotide regulatory proteins. J. Biol. Chem. 263:4993–4996.

Majerus, P. W. 1992. Inositol phosphate biochemistry. Annu. Rev. Biochem. 61:225–250.

Murphy, R. C., and Fitzpatrick, F. A. 1990. Arachidonate Related Lipid Mediators. Methods in Enzymology, Vol. 187. San Diego: Academic Press.

Nathanson, N. M., and Harden, T. K. (eds.) 1990. G Proteins and Signal Transduction. Society of General Physiologists Series, Vol. 45. New York: Rockefeller University Press.

Needleman, P., Turk, J., Jakschik, B. A., Morrison, A. R., and Lefkowith, J. B. 1986. Arachidonic acid metabolism. Annu. Rev. Biochem. 55:69–102.

Nestler, E. J., and Greengard, P. 1984. Protein Phosphorylation in the Nervous System. New York: Wiley.

Nicoll, R. A., Malenka, R. C., and Kauer, J. A. 1990. Functional comparison of neurotransmitter receptor subtypes in mammalian central nervous system. Physiol. Rev. 70:513–565.

Nishizuka, Y. 1992. Intracellular signaling by hydrolysis of phospholipids and activation of protein kinase C. Science 258:607–614.

Schwartz, J. H. 1993. Cognitive kinases. Proc. Nat. Acad. Sci. U.S.A. 90:8310-8313.

Siegel, G. J., Agranoff, B. W., Albers, R. W., and Molinoff, P. B. (eds.) 1994. Basic Neurochemistry: Molecular, Cellular, and Medical Aspects, 5th ed. New York: Raven Press.

Sternweis, P. C., and Pang, I.-H. 1990. The G protein-channel connection. Trends Neurosci 13:122–126.

Taylor, S. S., Buechler, J. A., and Yonemoto, W. 1990. cAMP-dependent protein kinase: Framework for a diverse family of regulatory enzymes. Annu. Rev. Biochem. 59:971–1005.

Waite, M. 1987. The Phospholipases. New York: Plenum Press.

Zigmond, R. E., Schwarzchild, M. A., and Rittenhouse, A. R. 1989. Acute regulation of tyrosine hydroxylase by nerve activity and by neurotransmitters via phosphorylation. Annu. Rev. Neurosci. 12:415–461.

15

Transmitter Release

Transmitter Release Is Not Controlled by Sodium Influx or Potassium Efflux

Transmitter Release Is Triggered by Calcium Influx

Transmitter Is Released in Quantal Units

Each Quantum of Transmitter Is Stored in a Synaptic Vesicle

Transmitter Is Discharged from Synaptic Vesicles by Exocytosis at the Active Zone

The Docking of Synaptic Vesicles, Fusion, and Exocytosis Are Controlled by Calcium Influx

Synaptic Vesicles Are Recycled

The Number of Transmitter Vesicles Released by an Action Potential Is Modulated by Calcium Influx

- *Intrinsic Cellular Mechanisms Regulate the Concentration of Free Calcium*
- *Synaptic Connections on Presynaptic Terminals Regulate Intracellular Free Calcium*

Some of the most remarkable activities of the brain, such as learning and memory, are thought to emerge from the elementary properties of chemical synapses. The distinctive feature of these synapses is that the action potentials in the presynaptic terminal lead to the release of a chemical messenger. In the last three chapters we examined how postsynaptic receptors control ion channels that generate the synaptic potential. In this chapter we consider how electrical events in the presynaptic terminal are coupled to the secretion of neurotransmitters. In the next chapter we examine the chemistry of the neurotransmitters themselves.

Transmitter Release Is Not Controlled by Sodium Influx or Potassium Efflux

The action potential results from two sequential steps. First, voltage-gated Na^+ channels open, allowing Na^+ to move into the presynaptic cell. Then voltage-gated K^+ channels open and K^+ moves out of the cell. Is either of these two processes responsible for triggering the release of the transmitter substance?

To explore the contribution of Na^+ and K^+ to transmitter release, Bernard Katz and Ricardo Miledi in the 1960s used the giant synapse of the squid. This synapse is large enough to permit the insertion of two electrodes into the presynaptic terminal (one for stimulating and one for recording) and an electrode into the postsynaptic cell for recording the synaptic potential and for use as an index of transmitter release.

The presynaptic cell typically produces an action potential of 110 mV, which leads to transmitter release and the generation of a large synaptic potential in the postsynaptic cell. When the voltage-gated Na^+ channels are blocked by tetrodotoxin, successive presynaptic action potentials become progressively smaller, owing to the progressive blockade of Na^+ channels, and the postsynaptic potential is reduced accordingly. When the Na^+ channel blockade becomes so profound as to reduce the amplitude of the presynaptic spike below 40 mV, the synaptic potential disappears altogether (Figure 15–1B).

Even though influx of Na^+ into the presynaptic cell appears to be in some way necessary for trans-

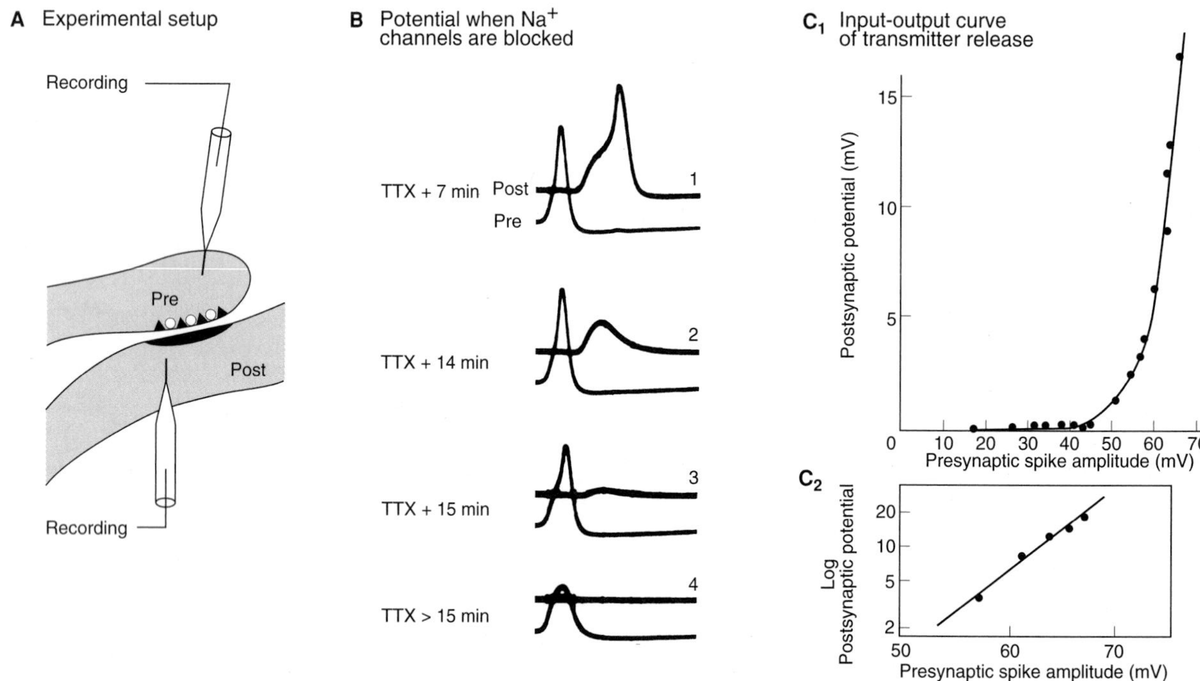

Figure 15–1 The contribution of voltage-gated Na^+ channels to transmitter release is tested by blocking the channels and measuring the amplitude of the presynaptic action potential and the resulting postsynaptic potential. (Adapted from Katz and Miledi, 1967a.)

A. Recording electrodes are inserted in both the pre- and postsynaptic fibers of the giant synapse in the stellate ganglion of a squid.

B. Tetrodotoxin (TTX) is added to the solution bathing the cell to block the voltage-gated Na^+ channels. The amplitudes of both the presynaptic action potential and the postsynaptic potential gradually decrease. After 7 minutes the presynaptic action potential can still produce a suprathreshold synaptic potential that triggers an action potential in the postsynaptic cell (**1**). After 14 and 15 minutes the presynaptic spike gradually becomes smaller and produces smaller synaptic potentials (**2** and **3**). When the presynaptic spike is reduced to 40 mV or less, it fails to produce a synaptic potential (**4**).

C. An input–output curve for transmitter release can be inferred from the amplitude of the synaptic potential. This is obtained when the Na^+ channels for the presynaptic action potential are blocked. **1.** A 40 mV presynaptic spike is required to produce a synaptic potential. Beyond this threshold there is a steep increase in amplitude of the synaptic potential, corresponding to small changes in the amplitude of the presynaptic potential. **2.** The semilogarithmic plot of the data in the input–output curve illustrates that the relationship between the presynaptic spike and the postsynaptic potential is logarithmic. A 10 mV increase in the presynaptic spike produces a 10-fold increase in the synaptic potential.

mitter release, it is not sufficient. While the Na^+ channels are still fully blocked, the presynaptic membrane can be depolarized in steps up to 150 mV above the resting potential by passing current out of the terminal through the second intracellular microelectrode. Beyond a threshold of about 40 mV from the resting potential, progressively greater amounts of transmitter are released (as judged by the appearance and amplitude of the postsynaptic potential). In the range of depolarization at which chemical transmitter is released (40 to 70 mV above the resting level), a 10 mV increase in depolarization produces a 10-fold increase in transmitter release. Thus, the presynaptic terminal is able to release transmitter without the influx of Na^+. The Na^+ influx is only important insofar as it depolarizes the membrane enough to allow transmitter release to occur. Some other ion flux associated with depolarization of the cell must actually cause the release.

The next candidate for this role is potassium. To examine the contribution of K^+ efflux to transmitter release, the voltage-sensitive Na^+ channels are blocked with tetrodotoxin and the K^+ channels are blocked with tetraethylammonium (TEA). A depolarizing current is then passed through the presynaptic terminals: The postsynaptic potentials are of normal size, indicating that transmitter is released normally (Figure 15–2). Indeed, the presynaptic potential is maintained throughout the current pulse because the K^+ current that normally repolarizes the presynaptic membrane is blocked. As a result, transmitter release is sustained (Figure 15–2C). Increases in the presynaptic potential above an upper limit produce no increase in postsynaptic potential (Figure 15–2D). Thus neither Na^+ nor K^+, the two ions responsible for the action potential in the axon, is required for transmitter release.

Transmitter Release Is Triggered by Calcium Influx

Katz and Miledi then turned their attention to Ca^{2+} ions. Earlier, José del Castillo and Katz had found that increasing the extracellular Ca^{2+} enhances transmitter release; lowering the extracellular Ca^{2+} reduces and ultimately blocks synaptic transmission. These findings, however, raised an obvious question. How can external Ca^{2+} influence transmitter release, an intracellular process?

Previous work on the squid giant axon had identified a class of voltage-gated Ca^{2+} channels. As there is normally a very large inward electrochemical driving force on Ca^{2+}, the opening of voltage-gated Ca^{2+} channels will result in a large Ca^{2+} influx. Although these Ca^{2+} channels are sparsely distributed along the axon, Katz and Miledi proposed that they might be much more abundant at the axon terminal and that Ca^{2+} might serve two functions: as a carrier of charge during the action potential (like Na^+ and K^+) and as a special signal conveying information about changes in membrane potential to the intracellular machinery responsible for transmitter release.

Consistent with this idea, when Na^+ and K^+ channels are blocked by tetrodotoxin and tetraethylammonium, graded depolarizations of the terminals activate a graded inward Ca^{2+} current, which in turn results in graded release of transmitter (Figure 15–3). Unlike the voltage-gated Na^+ channels, the Ca^{2+} channels in the squid terminals do not inactivate quickly, but stay open as long as the presynaptic depolarization lasts. This Ca^{2+} current does not occur throughout the length of the axon because of the low density of Ca^{2+} channels. Even in the axon terminal, where the Ca^{2+} channels are more abundant, Ca^{2+} currents are small and are normally masked by Na^+ and K^+ currents, which are 10 to 20 times larger. Nevertheless, in the region of the active zone Ca^{2+} influx is 10 times greater than elsewhere in the terminal. This localization is consistent with the distribution of the intramembranous particles thought to be Ca^{2+} channels (see Figure 15–7 in Box 15–2).

These findings suggested to Katz and Miledi that depolarization of the terminals by the action potential serves to open voltage-dependent Ca^{2+} channels, permitting the Ca^{2+} concentration at the active zone to rise a thousandfold within a few hundred microseconds to micromolar levels. This large and rapid increase is well suited for the synchronous release of transmitter. The characteristic delay of chemical synaptic transmission—the time from the onset of the action potential in the presynaptic terminals to the onset of the postsynaptic potential—is due in large part to the time required for Ca^{2+} channels to open in response to depolarization. Calcium does not begin to flow until the end of the action potential in the presynaptic cell, when the membrane potential begins to return to

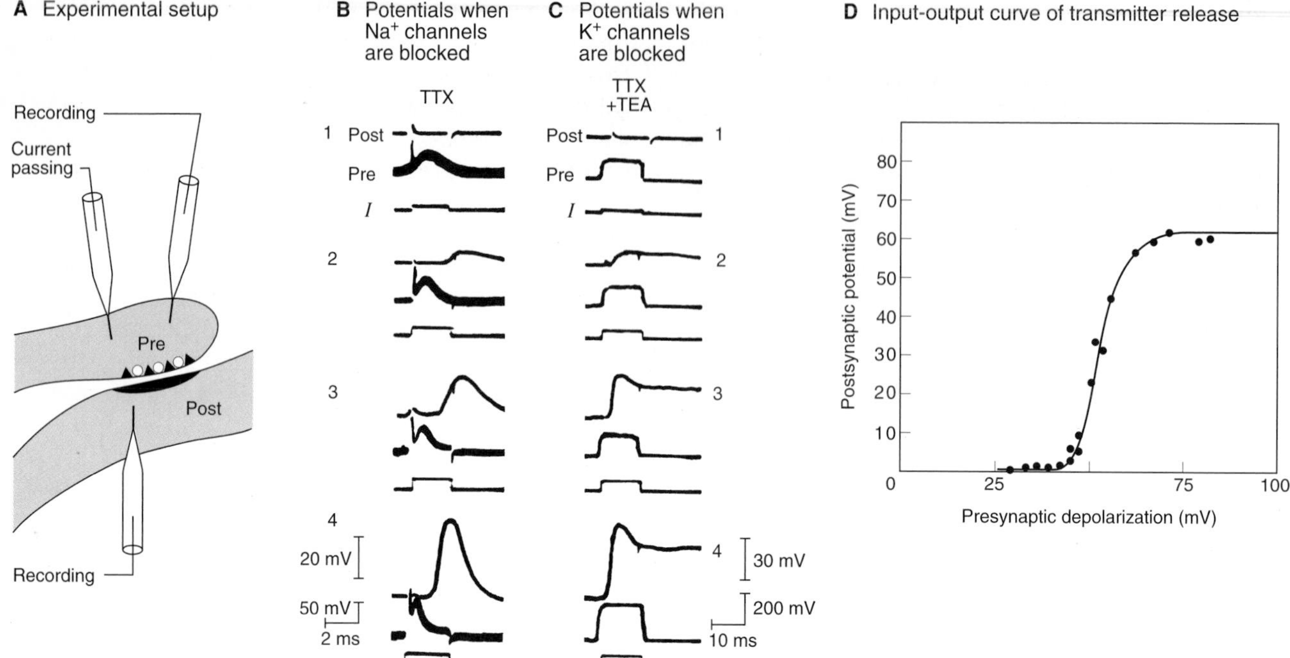

Figure 15–2 Blocking the voltage-sensitive Na^+ channels and K^+ channels in the presynaptic terminals affects the amplitude and duration of the presynaptic action potential and the resulting postsynaptic potential but does not block the release of transmitter. (Adapted from Katz and Miledi, 1967a.)

A. The experimental arrangement is the same as in Figure 15–1, except that a current-passing electrode has been inserted into the presynaptic cell.

B. The voltage-gated Na^+ channels are blocked by adding tetrodotoxin to the cell bathing solution. The three traces represent (from **bottom** to **top**) the current pulse injected into the presynaptic terminal (*I*), the resulting potential in the presynaptic terminal (**Pre**), and the postsynaptic potential generated as a result of transmitter release in the postsynaptic cell (**Post**). Progressively stronger current pulses (**2–4**) are applied to produce correspondingly greater depolarizations of the presynaptic terminal. These presynaptic depolarizations cause postsynaptic potentials even in the absence of Na^+ flux. The greater the presynaptic depolarization, the larger the postsynaptic potential, indicating that membrane potential exerts a direct control over release. The presynaptic depolarizations are not maintained throughout the duration of the depolarizing current pulse because of the delayed activation of the voltage-gated K^+ channel, which causes repolarization.

C. After the voltage-gated Na^+ channels of the action potential have been blocked, tetraethylammonium is injected into the presynaptic terminal to block the voltage-gated K^+ channels as well. The three traces represent current pulse, presynaptic potential, and postsynaptic potential as in **B.** Because K^+ channels are blocked presynaptically, the electrotonic depolarization is maintained throughout the current pulse. Under these conditions, larger presynaptic depolarizations still produce a larger postsynaptic potential (**2–4**). This indicates that neither Na^+ nor K^+ is required for effective transmitter release.

D. By blocking both the Na^+ and K^+ channels it is possible to generate, as a function of different depolarization steps, an input–output curve more complete than that shown in Figure 15–1. In addition to the steep part of the curve, there is now a plateau. Thus, beyond a certain level of presynaptic depolarization, further depolarization does not cause any further release of transmitter. (The initial level of the presynaptic membrane potential was about –70 mV).

the resting level (Figure 15–4). Because the voltage-dependent Ca^{2+} channels are located very close to the transmitter release sites, they can act within 0.2 ms to trigger transmitter release.

As we shall see later in this chapter, the duration of the action potential is an important determinant of the amount of Ca^{2+} that flows into the terminal. If the action potential is prolonged, more Ca^{2+} flows in and therefore more transmitter is released, causing a greater postsynaptic potential.

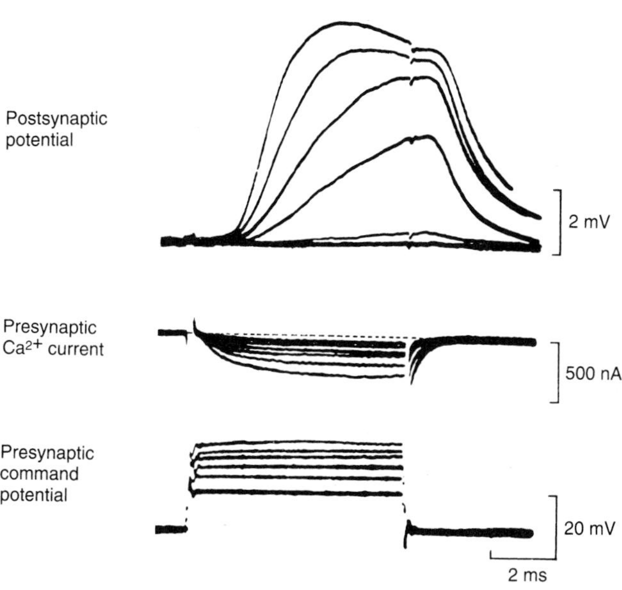

Figure 15–3 A simple experiment demonstrates that transmitter release is proportional to Ca^{2+} influx into the presynaptic terminal. The voltage-sensitive Na^+ and K^+ channels in a squid giant synapse are blocked by tetrodotoxin and tetraethylammonium. The presynaptic terminal is voltage-clamped and the membrane potential is stepped to six different command levels of depolarization. The amount of presynaptic inward Ca^{2+} current that accompanies the depolarization correlates with the amplitude of the resulting postsynaptic potential. This is because the amount of Ca^{2+} current through voltage-gated channels determines the amount of transmitter released, which in turn determines the size of the postsynaptic potential. The notch in the postsynaptic potential trace is an artifact that results from turning off the presynaptic command potential. (Adapted from Llinás and Heuser, 1977.)

In most nerve cells other than the giant axon of squid there are at least three (and probably more) classes of voltage-sensitive Ca^{2+} channels. One class (the *L-type channel*) is characterized by a slow rate of inactivation, so that it remains open during a prolonged depolarization of the membrane. The other two classes (*N-* and *P-type channels*) inactivate more rapidly. In many cells the influx of Ca^{2+} through the N-type and P-type channels contributes most directly to transmitter release.

Transmitter Is Released in Quantal Units

How and where does Ca^{2+} produce its actions? To answer that question we must first consider how transmitter substances are released. Even though the release of synaptic transmitter appears smoothly graded, it is actually quantized. Each *quantum* of transmitter produces a postsynaptic potential of fixed size, called the *unit synaptic potential.* The total postsynaptic potential is made up from an integral number of unit potentials. Synaptic potentials seem smoothly graded in recordings only because each unit potential is extremely small relative to the total potential.

Paul Fatt and Katz discovered the quantal nature of transmission when they recorded from the nerve–muscle synapse of the frog without presynaptic stimulation and observed small spontaneous potentials of about 0.5 to 1.0 mV. Similar results have since been obtained in mammalian muscle and in central neurons. Because the synaptic potentials at vertebrate nerve–muscle synapses are called end-plate potentials, Fatt and Katz called these spontaneous potentials *miniature end-plate potentials.*

The time course of the miniature end-plate potentials and the effects of various drugs on them are indistinguishable from the effects on the end-plate potential evoked by nerve stimulation. Miniature end-plate potentials are only recorded at the site of contact between nerve and muscle. Because acetylcholine (ACh) is the transmitter at the nerve–muscle synapse, the miniature end-plate potentials, like the end-plate potentials, are enhanced and prolonged by prostigmine, a drug that inhibits the hydrolysis of ACh by acetylcholinesterase. Similarly, the miniature end-plate potentials are reduced and finally abolished by agents that block the ACh receptor. In the absence

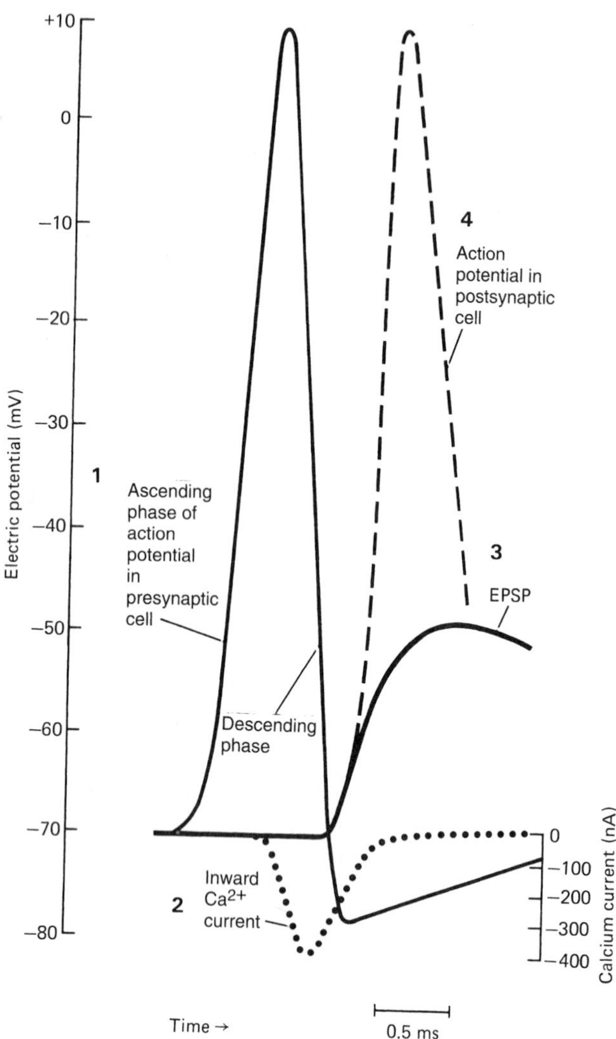

Figure 15–4 Calcium influx in the presynaptic cell affects the time course of synaptic transmission. An action potential in the presynaptic cell (**1**) causes voltage-gated Ca^{2+} channels in the terminal to open and a Ca^{2+} current to flow into the terminal (**2**). (Note that the Ca^{2+} current is turned on during the descending phase of the presynaptic action potential.) The Ca^{2+} influx triggers release of neurotransmitter. The postsynaptic response to the transmitter begins soon afterward (**3**) and, if sufficiently large, will trigger an action potential in the postsynaptic cell (**4**). (Adapted from Llinás, 1982.)

of stimulation the miniature end-plate potentials occur at random intervals; their frequency can be increased by depolarizing the presynaptic terminal. They disappear if the presynaptic nerve degenerates but reappear with reinnervation, indicating that small amounts of ACh are continuously released at the presynaptic nerve terminal.

What could account for the fixed size (0.5 to 1.0 mV) of the spontaneous miniature end-plate potential? Del Castillo and Katz first tested the possibility that it represented a fixed response of a single ACh receptor to one ACh molecule. By applying small amounts of ACh to the frog muscle end-plate, they were able to elicit depolarizing responses much smaller than 0.5 mV. From this it became clear that the miniature end-plate potential must reflect the opening not of one but of many ACh receptor–channels.

Later, Katz and Miledi were able to measure the ion flux through a single synaptic channel—the elementary ionic current—initiated by the interaction of ACh with the receptor (see Chapter 7). From this measurement, Katz and Miledi estimated that the elementary ACh potential produced by the opening of a single ACh receptor channel is only about 0.3 μV, or about 1/2000th of the amplitude of a spontaneous miniature potential. This estimate was later confirmed when the currents through single channels responsive to ACh could be measured directly using patch-clamp techniques (see Box 7–2).

A miniature end-plate potential of 0.5 mV would therefore require summation of the elementary current produced by the opening of about 2000 channels. For a single channel to open, two ACh molecules must bind to the receptor. In addition, some of the ACh released is lost, either by diffusion out of the synaptic cleft or by hydrolysis, and never reaches the receptor molecules. Thus, about 5,000 molecules are needed to produce one miniature end-plate potential. This number has been confirmed by direct chemical measurement of the ACh released per unit synaptic potential.

We can now ask some important questions: Do the spontaneously released quanta correspond to the unit potentials that occur during normal synaptic transmission? How does Ca^{2+} influence the amount of ACh released? Does Ca^{2+} influence the number of ACh molecules packaged into each quantum? Or does it affect the probability that a

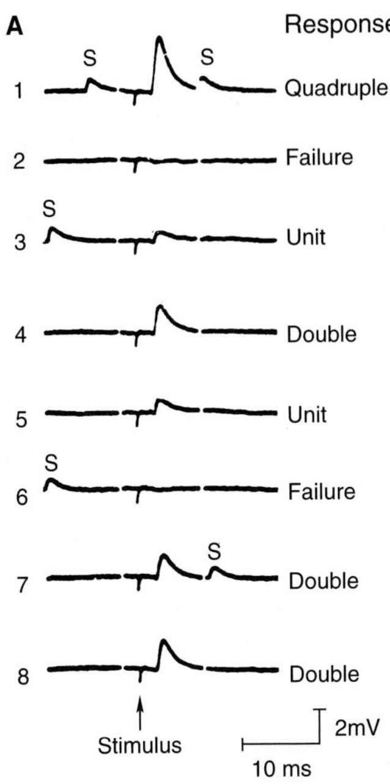

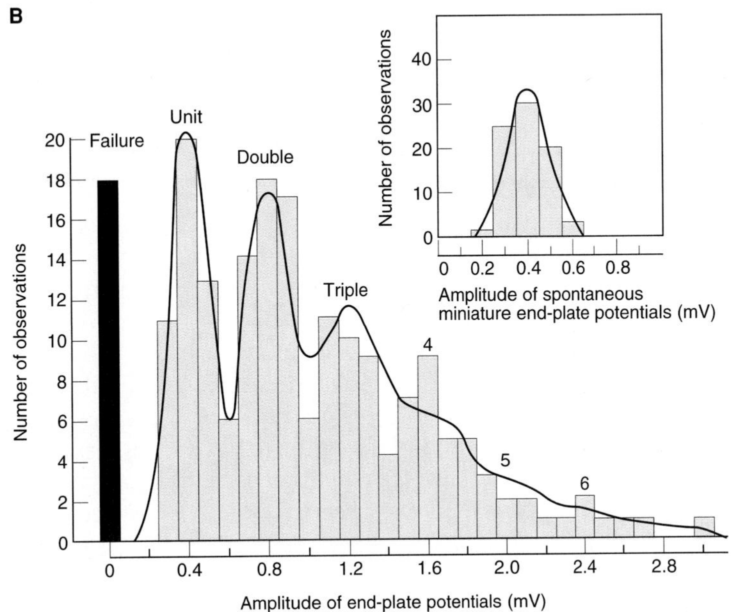

Figure 15–5 Neurotransmitters are released in fixed increments, or quanta. Each quantum of transmitter produces a unit postsynaptic potential of fixed amplitude. The amplitude of the postsynaptic potential is equal to the unit amplitude multiplied by the number of quanta of transmitter.

A. Intracellular recordings show the change in potential when eight consecutive stimuli of the same size are applied to the motor nerve. To reduce transmitter output and to keep the end-plate potentials small, the tissue was bathed in a Ca^{2+}-deficient (and Mg^{2+}-rich) solution. The responses to the stimulus vary. Two impulses produce complete failures, two produce unit potentials, and the others produce responses that are approximately two to four times the amplitude of the unit potential. Note that the spontaneous miniature end-plate potentials (**S**) are the same size as the unit potential. (Adapted from Liley, 1956.)

B. After recording many end-plate potentials, the number of responses at each amplitude was counted and then plotted in the histogram shown here. The distribution of responses falls into a number of peaks. The first peak, at 0 mV, represents failures. The first peak of responses, at 0.4 mV, represents the unit potential, the smallest elicited response. This unit response is the same amplitude as the spontaneous miniature potentials (**inset**). The other peaks in the histogram occur at amplitudes that are integral multiples of the amplitude of the unit potential. The solid line shows a theoretical Gaussian distribution fitted to the data of the histogram. In this Gaussian distribution each peak is slightly spread out, reflecting the fact that the amount of transmitter in each quantum varies randomly about the peak. The distribution of amplitudes of the spontaneous miniature potentials, shown in the inset, also fits a Gaussian curve (**solid line**). (Adapted from Boyd and Martin, 1956.)

quantum will be released? These questions can be answered by experiments in which the external concentration of Ca^{2+} is decreased. When the neuromuscular junction is bathed in a solution low in Ca^{2+}, the evoked end-plate potential (normally 70 mV in amplitude) is reduced markedly, to about 0.5 to 2.5 mV. Moreover, the amplitude of successive evoked end-plate potentials varies from one stimulus to the next, and often cannot be detected at all. However, the minimum response above zero—the unit synaptic potential in response to an action potential—is identical in size and shape to the spontaneous miniature end-plate potentials. All end-plate potentials larger than the unit synaptic potential are integral multiples of the unit potential (Figure 15–5).

The amplitude of the unit synaptic potential does not change when the external Ca^{2+} concentration is increased. However, the number of failures decreases, and the incidence of higher-amplitude responses increases. These observations illustrate that alterations in external Ca^{2+} concentration do not affect the *size* of a quantum (the number of ACh molecules) but rather the *probability* that a quantum will be released (Box 15–1). The greater the Ca^{2+} influx into the terminal, the larger the number of quanta released.

The finding that the amplitude of the end-plate potential increases in a stepwise manner at low levels of ACh release, that the amplitude of each step increase is an integral multiple of the unit potential, and that the unit potential has the same mean amplitude as that of the spontaneous miniature end-plate potentials, led del Castillo and Katz to conclude that the normal end-plate potential is caused by the release of about 150 quanta, each about 0.5 mV in amplitude. In the absence of an action potential only one quantum per second is released at the end-plate. This low probability of release is reflected in the small number of spontaneously released miniature end-plate potentials. The Ca^{2+} that enters the cell with an action potential transiently increases the rate of quantal release by 100,000-fold, bringing about the synchronous release of about 150 quanta in 1 or 2 ms. The amplitude of end-plate potentials actually varies slightly in response to consecutive action potentials, because the precise number of quanta released varies randomly from stimulus to stimulus.

Quantal transmission has been demonstrated at all chemical synapses so far examined, with the possible exception of the retina, which we examine in Chapter 21. However, at most synapses in the central nervous system each action potential releases between 1 and 10 quanta, many fewer than the 150 quanta at the nerve–muscle synapse. The reason for this is that the surface area of a presynaptic motor terminal ending on a muscle fiber is large (about 2,000 to 6,000 μm^2). Distributed along this large presynaptic surface are about 300 active zones. In contrast, a typical excitatory afferent fiber from a dorsal root ganglion cell forms only about four synapses on a motor neuron, each of which is about 2 μm^2 and contains only one active zone release site. Quantal analysis of transmitter release from these afferent neurons indicates that each site releases no more than one quantum of transmitter and alternates between one and none at all. Similar results have been obtained for other central synapses. Thus, variations in the overall response of a central neuron to a single presynaptic neuron result from the all-or-none release of one quantum from each of a few terminals.

Each Quantum of Transmitter Is Stored in a Synaptic Vesicle

How is a quantum of transmitter released by a neuron? The physiological observations indicating transmitter release is quantal coincided with the discovery, through electron micrographs of synapses, that the presynaptic terminal contains an accumulation of small vesicles. The electron micrographs suggested to del Castillo and Katz that the vesicles were storage organelles for the transmitter quanta. They argued that each vesicle stored one quantum of transmitter (amounting to several thousand molecules), and that the vesicle fuses with the inner surface of the presynaptic terminal at specific release sites. Here, the vesicle opens transiently and extrudes its entire contents into the synaptic cleft.

Synaptic vesicles are not uniformly distributed throughout the presynaptic terminal, but rather are clustered at regions where the presynaptic membrane appears thicker and more dense than elsewhere. The specialization appears as dense bodies attached to the internal face of the presynaptic membrane directly above the junctional folds in the muscle (Figure 15–6). At the nerve–muscle synapse these dense bodies are bar-shaped; at central synapses they take the form of discrete pyramidal projections. The vesicles collect in rows between these dense bodies. The region with dense bodies, where transmitter is actually released from small synaptic vesicles, is called the *active zone*. The neuromuscular junction in frogs contains about 300 active zones with about 10^6 vesicles. Here, and at central synapses, the vesicles are typically small and ovoid, with a diameter of about 50 nm.

Transmitter Is Discharged from Synaptic Vesicles by Exocytosis at the Active Zone

The vesicles discharge their contents at the active zone by fusing with the presynaptic plasma membrane, a process called *exocytosis*. But are these the only sites at which exocytotic release of transmitter occurs? This point is difficult to investigate in con-

BOX 15–1 *Calculating the Probability of Transmitter Release*

Transmitter is released in quanta in a random manner. The fate of each quantum of transmitter in response to an action potential has only two possible outcomes—the transmitter is or is not released. This event resembles a binomial or Bernoulli trial (similar to tossing a coin in the air to determine whether it comes up heads or tails). The probability of a quantum being released by an action potential is independent of the probability of other quanta being released by that action potential. Therefore for a population of releasable quanta, each action potential represents a series of independent binomial trials (comparable to tossing a handful of coins to see how many coins come up heads).

In a binomial distribution p stands for the average probability of success and q (or $1–p$) stands for the mean probability of failure. Both the average probability (p) that individual quanta are released and the store (n) from which the quanta are released are assumed to be constant. (Any reduction in the store is assumed to be quickly replenished after each stimulus.) Once n and p are known, the binomial probability law allows one to estimate the mean number of quanta (m, called the *quantal content* or *quantal output*) that are released to make up the end-plate potentials following a series of stimuli where $m = np$.

Calculation of the probability of transmitter release can be illustrated with the following example. A terminal has a releasable store of five quanta ($n = 5$). If we assume that $p = 0.1$, then q (the probability that a quantum is not released from the terminals) is $1 - p$, or 0.9. We can now determine for any given number of stimuli, say 100, the probability that a stimulus will release no quanta (failure), a single quantum, two quanta, three quanta, or any number of quanta (up to n). The probability that none of the five available quanta will be released by a given stimulus is the product of the individual probabilities that each quantum will not be released: $q^5 = (0.9)^5$, or 0.59. We would thus expect to see 59 failures in a hundred stimuli. The probabilities of observing zero, one, two, three, four or five quanta are represented by the successive terms of the binomial expansion:

$$(q + p)^5 = q^5 \text{ (failures)} + 5\,q^4p\text{(1 quantum)} + 10\,q^3p^2\text{(2 quanta)} + 10\,q^2p^3\text{(3 quanta)} + 5\,qp^4\text{(4 quanta)} + p^5\text{(5 quanta)}.$$

Thus, in 100 stimuli the binomial expansion would predict 33 unit responses, 7 double responses, 1 triple response, and 0 quadruple and quintuple responses.

Values for m vary, from about 100 to 300 at the vertebrate nerve–muscle synapse, the squid giant synapse, and *Aplysia* central synapses, to as few as 1 to 4 in the synapses of the sympathetic ganglion and spinal cord of vertebrates. The probability of release p is thought to be high, ranging from 0.7 at the neuromuscular junction in the frog to 0.9 in the crab. Estimates for n range from 1,000 (at the vertebrate nerve–muscle synapse) to 3 (at single terminals of the crayfish).

The parameters n and p are statistical terms; the physical processes represented by them are not yet known. Although the parameter n is usually referred to as the readily releasable (or readily available) store of quanta, it may actually represent the number of release sites in the presynaptic terminals that are loaded with vesicles. The number of release sites is thought to be fixed, but the fraction that is loaded with vesicles is thought to be variable. The parameter p probably represents a compound probability depending on at least two functions: the probability of mobilizing a vesicle into a release site (reloading) after an impulse, and the probability that an action potential discharges a quantum from an active site.

The mean quantal content m is a measure of the amount of transmitter released and is therefore a reflection of certain properties of the presynaptic cell: (1) the size of the presynaptic terminals; (2) the number of terminal branches of a single presynaptic fiber; and (3) possible alterations of quantal release associated with changes in synaptic efficacy. The average quantal size (q) indicates the response of the postsynaptic membrane to a single quantum of transmitter, assuming that each vesicle contains the normal number of transmitter molecules. Quantal size therefore depends on properties of the postsynaptic cell, such as the input resistance (which can be independently estimated) and the sensitivity of the postsynaptic receptor to the transmitter substance, as measured by the receptor's response to the application of a constant amount of transmitter.

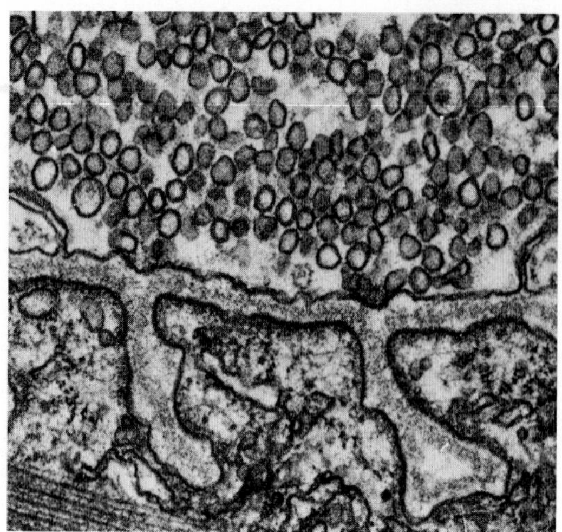

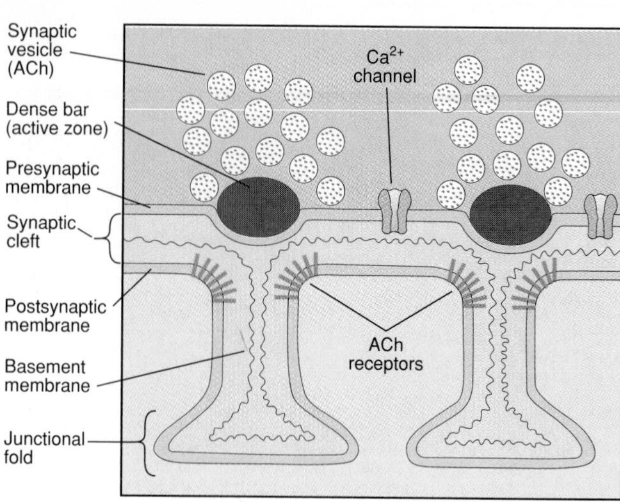

Figure 15–6 The topography of the transmitter release site at the neuromuscular junction is shown in this thin-section micrograph and drawing. Small synaptic vesicles (40–50 nm in diameter) containing ACh molecules are clustered around dense bars in the presynaptic neuron. The region of dense bodies, or active zone, is located just above ACh receptors at the apexes of the junctional folds in the muscle membrane. (Adapted from Kuffler, Nicholls, and Martin, 1984.)

BOX 15–2 Freeze-Fracture Technique

Freeze-fracture reveals the structural details of synaptic membranes. In this technique frozen tissue is broken open under a high vacuum and coated with platinum and carbon. The frozen membranes tend to break at the weakest plane, which is between the two molecular layers of lipids. Two complementary faces of the membrane are thus exposed: The leaflet nearest the cytoplasm (the interior half) is the protoplasmic (**P**) face, while the leaflet that borders the extracellular space is the external (**E**) face.

Freeze-fracture exposes an intramembranous view of a large expanse of the presynaptic area. Hence deformations of the membrane that occur at the active zone, where vesicles are attached, are readily apparent. The panoramic view of the region of active zones that the freeze-fracture technique offers is best appreciated by comparing this figure with the conventional transmission electron microscopic image of the active zone (see Figure 15–7).

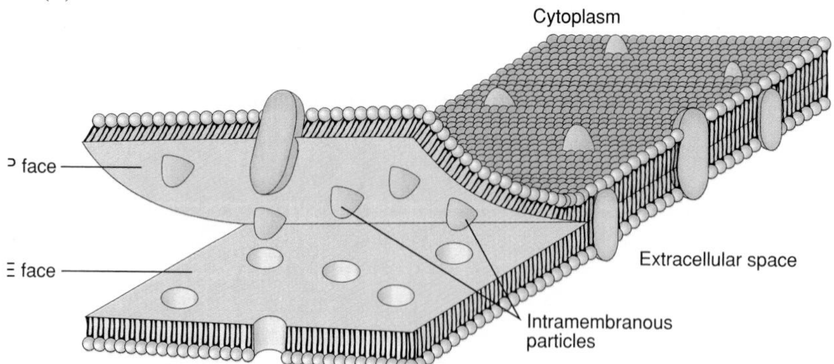

Figure 15–7A The path of membrane cleavage is along the hydrophobic interior of the lipid bilayer, resulting in two complementary fracture faces. The P face contains most of the membrane-spanning proteins (particles), because these are anchored to cytoskeletal structures. The E face shows pits complementary to the membrane-spanning protein particles. (After Fawcett, 1981.)

ventionally fixed tissue sections, because the chance of finding a vesicle in the act of being discharged is extremely small. For example, a thin section through a terminal at the neuromuscular junction of the frog shows only 1/4000th of the total presynaptic membrane.

Because the exocytotic opening of each small vesicle is smaller than the thickness of the ultrathin (50–100 nm) sections required for transmission electron microscopy, investigators began to apply freeze-fracture techniques to this problem in the 1970s (Box 15–2). These techniques produced three important observations. (1) Along both margins of each of the dense bars there are one or two rows of unusually large intramembranous particles (Figure 15–8A). Although the function of these particles is not yet known, they may be voltage-gated Ca^{2+} channels. Their density (about 1,500 per μm^2) is approximately that of the voltage-gated Ca^{2+} channels essential for transmitter release. Moreover, the proximity of the particles to the release site is consistent with the short interval between the onset of the Ca^{2+} current and the release of transmitter. (2) Deformations alongside the rows of intramembranous particles become apparent during synaptic activity (Figure 15–8B). These deformations could represent invaginations of the cell membrane during exocytosis. (3) The deformations do not persist after the transmitter has been released; they seem to be transient distortions that occur only when vesicles are discharged.

To catch vesicles in the act of exocytosis, it is necessary to quick-freeze the tissue with liquid helium. The tissue is rapidly frozen at precisely

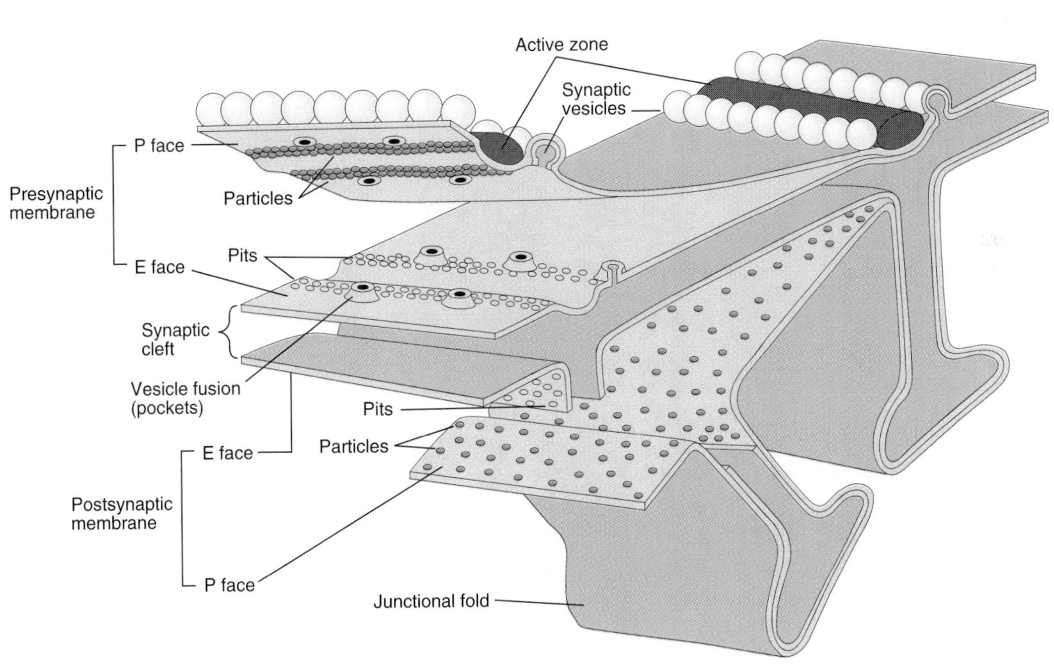

Figure 15–7B This three-dimensional view of pre- and postsynaptic membranes shows active zones with adjacent rows of synaptic vesicles, as well as places where the vesicles are undergoing exocytosis. The split membrane shows the idealized image of these structures in freeze-fracture. The rows of particles on either side of the active zone are intramembranous proteins thought to be Ca^{2+} channels. (Adapted from Kuffler, Nicholls, and Martin, 1984.)

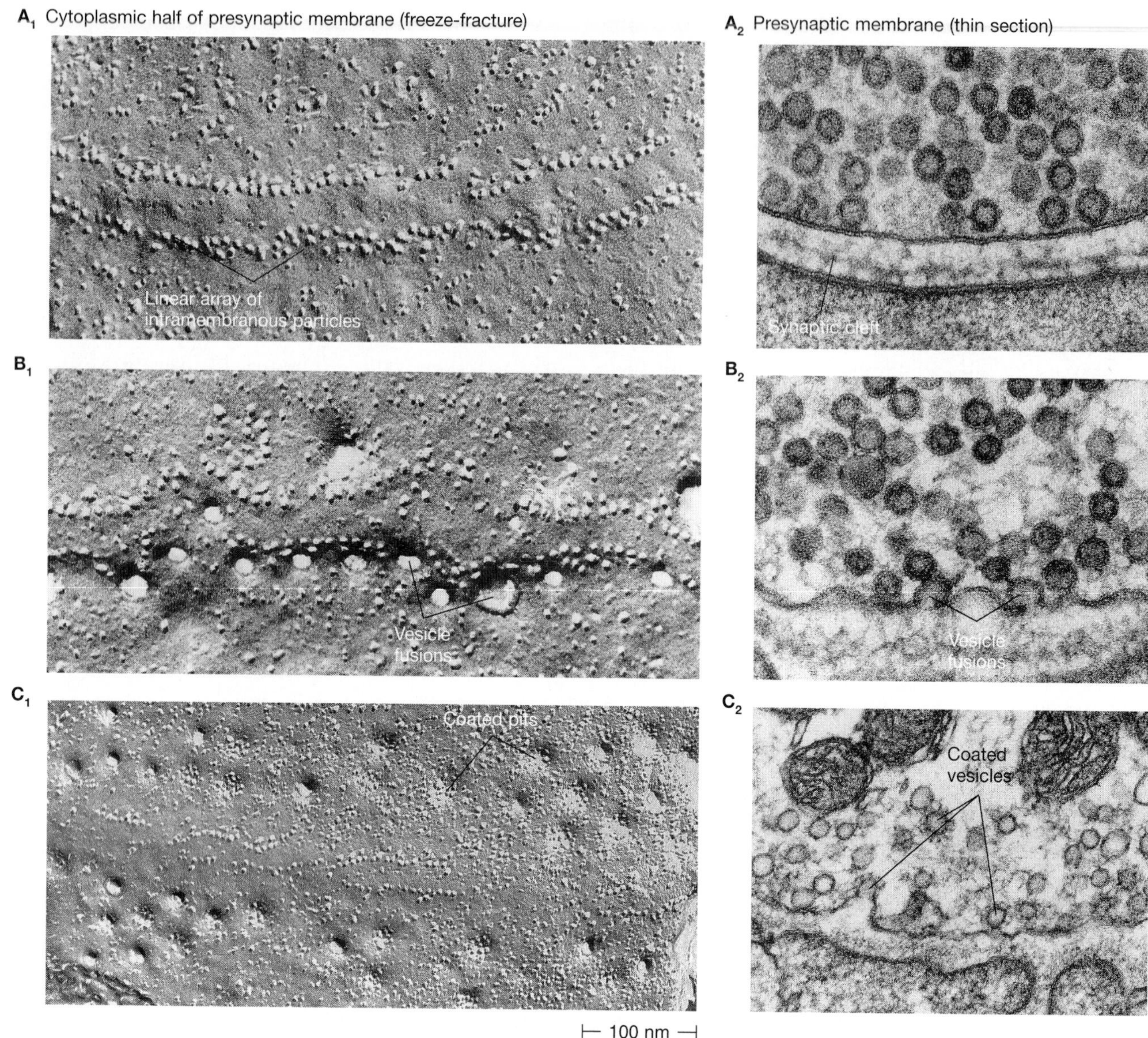

Figure 15–8 The events of exocytosis at the presynaptic terminal are revealed by electron microscopy. The images on the **left** are freeze-fracture electron micrographs of the cytoplasmic half of the presynaptic membrane. Thin-section micrographs of the presynaptic membrane are shown on the **right**. (Adapted from Alberts et al., 1989.)

A. The active zone in the resting state is marked by parallel arrays of intramembranous particles that may be the voltage-gated Ca^{2+} channels essential for transmitter release.

B. 1. Synaptic vesicles begin fusing with the plasma membrane within 5 ms of the stimulus. **2.** Fusion is complete within another 2 ms. Each opening in the plasma membrane represents the fusion of one synaptic vesicle.

C. 1. Membrane retrieval becomes apparent as coated pits form within about 10 sec of fusion of the vesicles with membrane. After another 10 sec the coated pits begin to pinch off by endocytosis to form coated vesicles. **2.** These vesicles include the original membrane proteins of the synaptic vesicle and also contain molecules captured from the external medium. The vesicles are transported to the cell body, where the membrane constituents are degraded or recycled (see Chapter 4).

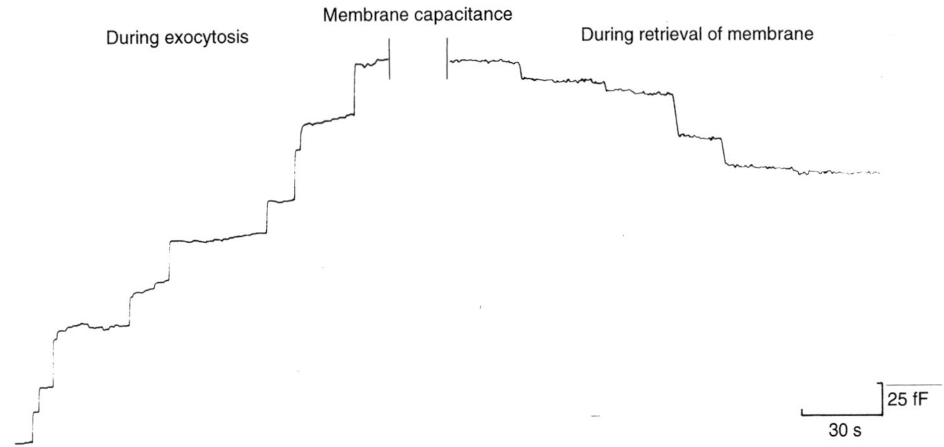

Figure 15–9 Exocytosis of synaptic vesicles and subsequent retrieval of the excess membrane produce changes in the surface area of the membrane that can be detected by electrical measurement of membrane capacitance. The increases in capacitance occur in a stepwise fashion and reflect the fusion of individual synaptic vesicles with the cell membrane. The unequal step increases indicate a variability in the diameter of the vesicles. After transmitter is released from the vesicles, the membrane added by fusion of the synaptic vesicles is excised, retrieved, and transported to the cell body. In this way the cell maintains a constant size. The recording here is from a rat connective tissue cell, a mast cell undergoing massive exocytotic release of its large secretory granules (units are femtofarads, fF, where 1 fF = 0.1 μm^2 membrane area). (Adapted from Fernandez, Neher, and Gomperts, 1984.)

defined intervals after the presynaptic nerve has been stimulated. The neuromuscular junction can thus be frozen just as the action potential invades the terminal and exocytosis occurs. This technique was used with the drug 4-aminopyridine—a TEA-like substance that blocks certain voltage-gated K^+ channels, broadens the action potential, and increases the number of quanta discharged with each nerve impulse—in a quantitative study of the morphological events accompanying exocytosis at the neuromuscular junction of the frog.

This study indicates that one vesicle undergoes exocytosis for each quantum of transmitter released. Statistical analyses of the spatial distribution of discharge sites along the active zones show that individual vesicles fuse with the plasma membrane independently of one another. This result is consistent with physiological studies indicating that quanta of transmitter are released independently. These morphological studies therefore provide independent evidence that the synaptic vesicles store the transmitter and that exocytosis is the mechanism by which transmitter is released into the synaptic cleft.

The fusion of the synaptic vesicles with the plasma membrane during exocytosis increases the surface area of the plasma membrane. In certain favorable circumstances this series of events can be detected in electrical measurements of membrane capacitance. As we saw in Chapter 9, the capacitance of the membrane is proportional to its surface area. In certain cell types, such as mast cells of the rat peritoneum, individual vesicles are large enough so that the increase in capacitance associated with fusion of a single vesicle can be detected (Figure 15–9). Massive release of transmitter accompanied by stepwise increases in capacitance is followed somewhat later by stepwise decreases in capacitance, which presumably reflect the retrieval and recycling of the excess membrane (Figure 15–9).

Exocytosis is not the only mechanism used by neurons to release substances. For example, prostaglandins, lipoxygenase metabolites, and the gases CO and NO (see Chapter 14) are membrane permeable and can traverse the lipid bilayer by diffusion.

These substances may act at synapses either as chemical messengers or as retrograde signals that diffuse from the postsynaptic to the presynaptic neuron to regulate transmitter release. Other substances can be moved out of nerve endings by carrier proteins if their intracellular concentration is sufficiently high. Transporters for glutamate or γ-aminobutyric acid (GABA) normally take up transmitters into the nerve cell from the extracellular space; in certain retinal cells these carriers reverse their direction and release the transmitter into the extracellular space. Still other substances simply leak out of nerve terminals at a low rate: about 90% of the ACh that leaves the presynaptic terminal at the neuromuscular junction can be traced to continuous leakage. Because this leakage is so diffuse and not targeted to receptors at the end-plate region, and because it is continuous and low level rather than synchronous and concentrated, it is functionally ineffective.

The Docking of Synaptic Vesicles, Fusion, and Exocytosis Are Controlled by Calcium Influx

We saw earlier that Ca^{2+} influx affects the probability of transmitter release. What specific actions of Ca^{2+} contribute to transmitter release? Only a small number of vesicles actually are positioned at the active zones at any one time. Most vesicles in the presynaptic terminal are located near the active zone, anchored by actin filaments. Calcium may allow these vesicles to move toward the release site and may aid in the fusion of the vesicles with the plasma membrane.

Proteins have been identified that are thought to (1) restrain the vesicles so as to prevent their accidental mobilization, (2) target the freed vesicles to the active zone, (3) dock the targeted vesicles at the active zone, and (4) allow fusion and exocytosis (Figure 15–10).

We first consider proteins involved in restraint and mobilization. The vesicles outside the active zone represent a reserve storage pool of transmitter. They do not move about freely in the terminal, but rather are restrained or anchored to a network of cytoskeletal filaments. The synapsins are a family of four proteins (Ia, Ib, IIa and IIb) thought to be important for anchoring vesicles to the cytoskeleton prior to mobilization. Of these four, synapsins Ia and Ib are the best studied. These two proteins are substrates for both the cAMP-dependent protein kinase and the Ca^{2+}/calmodulin-dependent kinase. When synapsin I is nonphosphorylated, it is thought to link the synaptic vesicles to actin filaments and other components of the cytoskeleton. When the nerve terminal is depolarized and Ca^{2+} enters, synapsin I is thought to become phosphorylated by the Ca^{2+}/calmodulin-dependent protein kinase. Phosphorylation would free the vesicles from the cytoskeletal constraint, making them available to move into the active zone (Figure 15–11).

The targeting of vesicles to docking sites for release is thought to be carried out by Rab3a and Rab3b, two members of a class of small proteins that bind GTP. These Rab proteins, first discovered in the brain, are related to the superfamily of the oncogene p^{23}*Ras*.

How does docking occur? A general scheme for vesicle docking has recently been discovered that is shared by all cells. As we have seen in Chapter 4, all secretory proteins are synthesized on ribosomes and inserted into the lumen of the endoplasmic reticulum (ER). When these proteins leave the ER they are targeted to the Golgi apparatus in vesicles formed from the membrane of the ER. The vesicles then dock and fuse with the Golgi membrane, discharging their protein into the lumen of the Golgi where the protein is modified. Other vesicles shuttle the secretory protein between the *cis* and the *trans* compartments (the cisternae) of the Golgi apparatus until the protein becomes fully modified and mature. Vesicles containing the mature protein then bud off and are targeted to the cell surface, where the protein is released by exocytosis (Figure 15–12).

By reconstituting the targeting system in a test tube, two essential sets of cytoplasmic proteins have been identified that form the commonly shared components of a universal molecular fusion machine. This molecular assembly is used in all cells to fuse vesicles with their target membranes. One set of proteins is called the *N*-ethyl-malemide *s*ensitive *f*usion protein (NSF); the other group is called *s*oluble *NSF* *a*ttachment *p*roteins (SNAP). These general cytoplasmic proteins intervene between specialized *donor* and *receptor* proteins that are thought to be specific for the particular vesicle and the particular target membrane component.

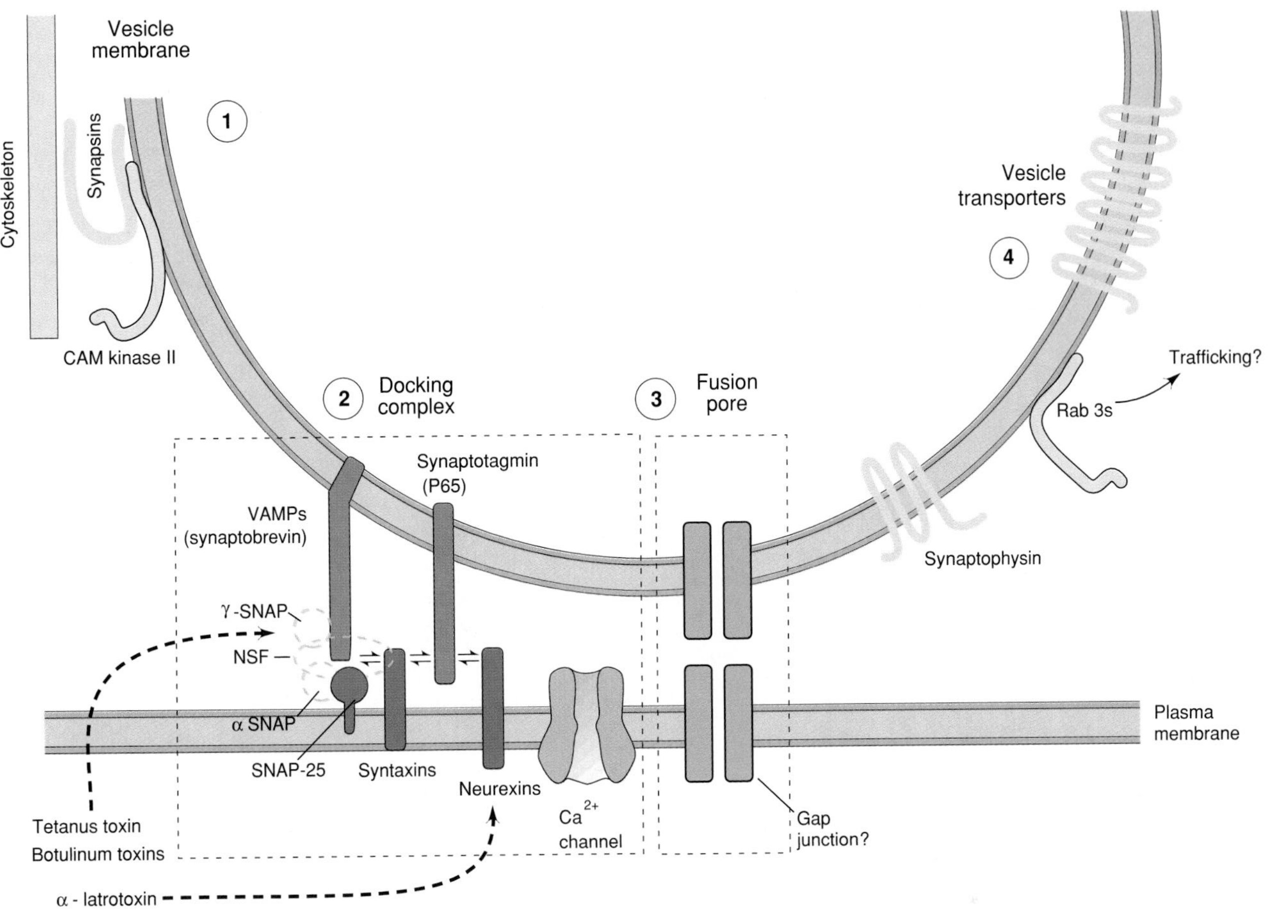

Figure 15–10 This diagram depicts the known synaptic vesicle proteins and some of their postulated receptors and functions. Separate compartments are assumed for storage (where vesicles are tethered to the cytoskeleton) and for docking and release (see Figure 15–11). Also depicted is one current view—that the molecular machinery for docking vesicles is distinct from that for fusion with the cell membrane. Some of these proteins appear to represent the targets for neurotoxins that modify transmitter release. VAMPs (synaptobrevins) are the targets for tetanus and botulinum toxins, two Zn^{2+} dependent metalloproteases, and are cleaved by these enzymes. Alpha-latrotoxin, a spider toxin that generates massive vesicle depletion and transmitter release, binds to the neurexins. **1.** Synapsins are vesicle-associated proteins that are thought to mediate interactions between the synaptic vesicle and the cytoskeletal elements of the nerve terminal. **2.** The docking, fusion, and release of vesicles appears to involve distinct interactions between vesicle proteins and proteins of the nerve terminal plasma membrane: the VAMPs (synaptobrevins) and synaptotagmin (P65) on the vesicle membrane and syntaxins and neurexins on the nerve terminal membrane. Arrows indicate potential interactions suggested on the basis of *in vitro* studies. **3.** The identity of the vesicle and plasma membrane proteins that constitute the fusion pore remains unclear. Synaptophysin may form the vesicle contribution to the fusion pore, while a partially characterized protein, physophilin, may constitute the plasma membrane contribution to the pore protein. Synaptophysin may also have other potential functions. **4.** The Rab proteins appear to be involved in vesicle trafficking within the cell and also in docking of vesicles within the nerve terminal membrane (see Figure 15–11). Vesicle transporters are involved in accumulation of neurotransmitter within the synaptic vesicle. (Modified from Jessell and Kandel, 1993.)

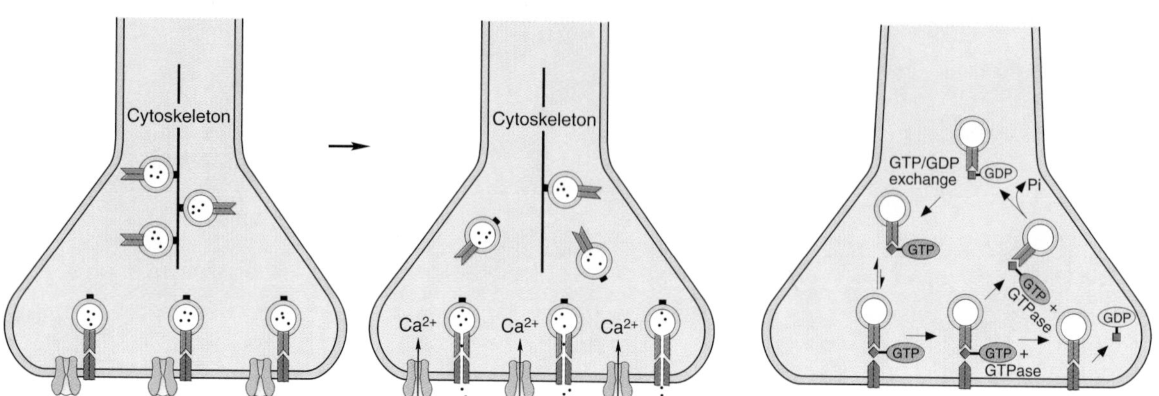

Figure 15–11 The mobilization, docking, and function of synaptic vesicles are controlled by low-molecular-weight GTP-binding proteins and Ca^{2+}.

A. Synaptic vesicles in nerve terminals are present in a sequestered compartment (sometimes called the *storage* compartment) where they are tethered to the cytoskeleton, as well as in a compartment (sometimes called the *releasable* compartment) where vesicles are docked to the fusion pore complex. Entry of Ca^{2+} into the nerve terminal leads to the opening of the fusion pore complex and neurotransmitter release. Calcium entry also mobilizes vesicles from the storage compartment, thus increasing their availability for docking into the presynaptic plasma membrane.

B. The possible role of the low-molecular-weight GTP-binding Rab proteins in vesicle docking to the fusion pore complex. Docking of fusion proteins on the vesicle surface binds Rab proteins (**black diamonds**). In its GTP-bound form the Rab protein prevents vesicle docking with the plasma membrane fusion pore complex. Activation of a GTPase at the docking site leads to the dissociation of the Rab protein, which leads to docking and fusion. (Modified from Jessell and Kandel, 1993.)

The donor protein is probably intrinsic to the vesicle membrane, while the receptor protein is intrinsic to the plasma membrane (Figure 15–12A). In the brain syntaxin, a nerve terminal membrane protein, seems to have the properties of the receptor protein, while VAMPs (vesicle-associated membrane proteins, or synaptobrevins), located in the vesicles, have the properties of the donor proteins. Indeed, VAMPs resemble a viral fusion peptide and thus may also participate in vesicle fusion (Figure 15–12B).

Two additional proteins associated with the membrane of the vesicle are thought to be important for exocytosis: *synaptotagmin* has a membrane-spanning region, while *synaptophysin* has several α-helices that are thought to span the vesicle membrane, and might function as a channel. Synaptotagmin contains a Ca^{2+}-binding domain homologous to the regulatory region of protein kinase C. Since synaptotagmin also binds phospholipids in the presence of Ca^{2+}, it might insert into the bilayer in response to Ca^{2+} influx (see Figure 15–11). It is thought to be an important Ca^{2+} sensor that exerts a brake on docking and vesicle fusion. This inhibitory constraint would have to be relieved for docking or fusion to occur.

Patch-clamp studies of mast cells suggest that fusion of the secretory vesicle to the plasma membrane leads to the temporary formation of an ion channel at the point of contact. These studies support the morphological observation that early in exocytosis there is rapid formation of a *fusion pore*, a narrow cytoplasmic bridge that unites the vesicle membrane with the plasma membrane at the

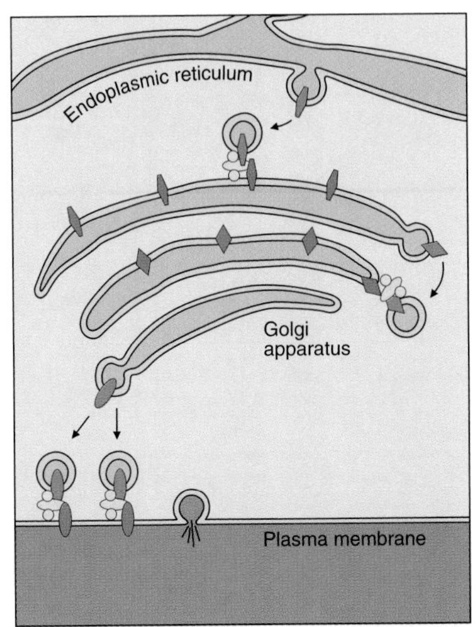

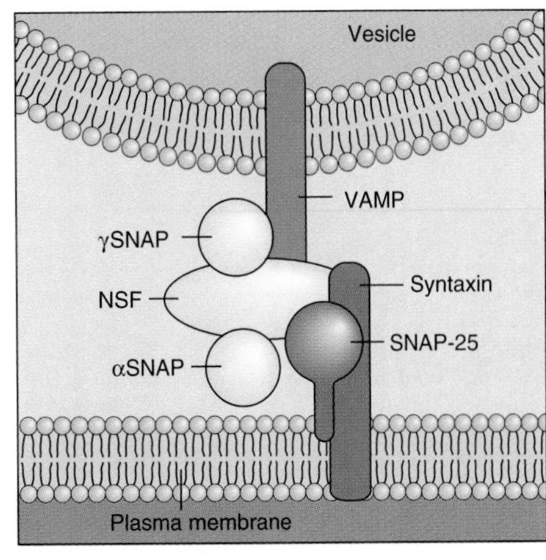

Figure 15–12 The molecular machinery for fusion and exocytosis.

A. In this model two cytoplasmic proteins, NSF and the SNAPs, participate in all stages of trafficking, while the membrane donor and receptor proteins for each step are separate members of a family of related proteins.

B. The docking of synaptic vesicles seems to involve the formation of a complex that includes proteins from the cytoplasm, such as NSF and SNAP, as well as proteins permanently inserted in the vesicle (VAMP) and target membranes (SNAP25 and syntaxin).

active zone and connects the lumen of the vesicle with the extracellular space. This small pore initially has a mean conductance of about 230 pS, similar to that of the gap-junction channels at electrical synapses. The typical diameter of these pores is about 1.5 nm. During exocytosis the pore dilates to 50 nm and, in parallel, the conductance increases dramatically.

How the fusion pore is produced is not known. Since transmitter release is so fast, fusion must occur within a fraction of a millisecond. Therefore, the protein that fuses synaptic vesicles to the plasma membrane is most likely already present in the vesicle before fusion occurs. Much like the gap-junction channels, the fusion pore may consist of two halves—one in the vesicle membrane and the other in the plasma membrane—which join in the course of vesicle docking (Figure 15–12). Calcium influx would then cause the preexisting pore to dilate and allow release of transmitter.

Synaptic Vesicles Are Recycled

As we have seen, when transmitter is released from vesicles by exocytosis, the membrane of the vesicle fuses with the membrane of the synaptic terminal to gain access to the extracellular space. If there were no process to compensate for exocytosis, the surface membrane of a synaptic terminal would enlarge as a result of continued nerve activity, because vesicle membrane would be added continuously to the plasmalemma. This increase does not occur, however, because the vesicle membrane added to the terminal membrane is retrieved rapidly and recycled.

Although the number of vesicles in a nerve terminal decreases transiently during release, the membrane is conserved. The total amount of membrane in vesicles, cisternae, and plasma membrane remains constant, indicating that membrane is retrieved from the surface membrane into the

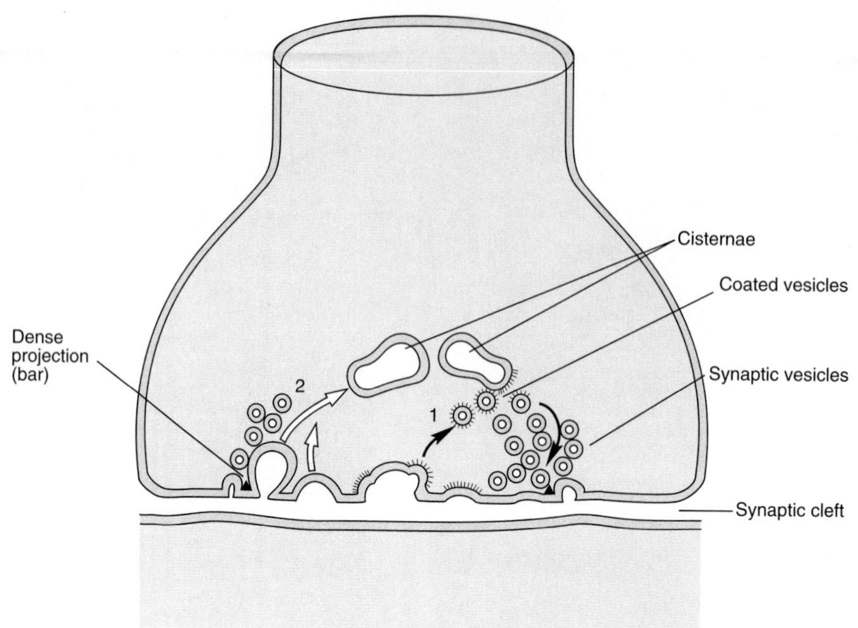

Figure 15–13 Recycling of vesicle membrane at the frog neuromuscular junction may occur in two pathways. In the first and physiologically more important pathway, excess membrane is retrieved by means of coated pits (**1**). These coated pits are selective, concentrating intramembranous particles into small packages. The pits are found throughout the terminal except at the active zones. As the plasma membrane enlarges during exocytosis, more membrane invaginations come to have coated cytoplasmic surfaces. (The path of the coated pits is shown by **solid arrows.**) In the second pathway, excess membrane reenters the terminal by budding from uncoated pits (**2**). These uncoated cisternae are formed primarily at the active zones. Nearly all of the uncoated pits form during the first few seconds after exocytosis. During the usual functioning of the synapse this second pathway may not be used at all. (Adapted from Miller and Heuser, 1984.)

internal organelles. How recycling takes place in nerve endings has not yet been resolved, but the mechanisms suggested by electronmicrograph studies are similar to known mechanisms in epithelial cells (Figure 15–13). According to this explanation, excess membrane from synaptic vesicles that have undergone exocytosis is recycled. The major pathway for recycling membrane at normal physiological rates of stimulation reaches its peak of activity within 30 seconds of exocytosis but persists for more than one minute. In this pathway excess membrane anywhere in the terminal except at the active zone creates a pit coated with clathrin. The clathrin coat forms a regular lattice around the pit, which finally pinches off as a small coated vesicle. The endocytosed vesicles fuse with endosomes that then generate synaptic vesicles.

The Number of Transmitter Vesicles Released by an Action Potential Is Modulated by Calcium Influx

As we saw in Chapter 11, the effectiveness of chemical synapses can be modified for both short and long periods, whereas the action of electrical synapses cannot. This modifiability, or *synaptic plasticity*, is controlled by two types of processes: (1) processes within the neuron, such as changes in the membrane potential and the firing of action potentials, and (2) extrinsic processes, such as the synaptic input from other neurons.

Long-term changes in chemical synaptic action are crucial to development and learning, and we consider these changes in detail later in the book. Here we discuss the short-term changes—modifications of the influx or accumulation of Ca^{2+} within the presynaptic terminal that affect the amount of transmitter released.

Intrinsic Cellular Mechanisms Regulate the Concentration of Free Calcium

Because transmitter release depends strongly on intracellular Ca^{2+} concentration, mechanisms within the presynaptic neuron that affect the concentration of free Ca^{2+} in the presynaptic terminals also affect the amount of transmitter released. In some cells there is a small steady influx of Ca^{2+} through the membrane of the presynaptic terminals, even at the resting membrane potential. This influx occurs through a class of voltage-gated Ca^{2+} channels that inactivate little, if at all.

This steady-state Ca^{2+} influx is enhanced by depolarization and decreased by hyperpolarization. A slight depolarization of the membrane can increase the steady-state influx of Ca^{2+} and enhance the amount of transmitter released by subsequent action potentials. A slight hyperpolarization has the opposite effect (Figure 15–14). By altering the amount of Ca^{2+} influx into the terminal, small changes in the resting membrane potential can make an effective synapse inoperative or a weak synapse highly effective. These changes in membrane potential can be produced experimentally by injecting current. The changes in membrane potential can also be produced by other neurons releasing transmitter at axo-axonic synapses, which regulate ion channels.

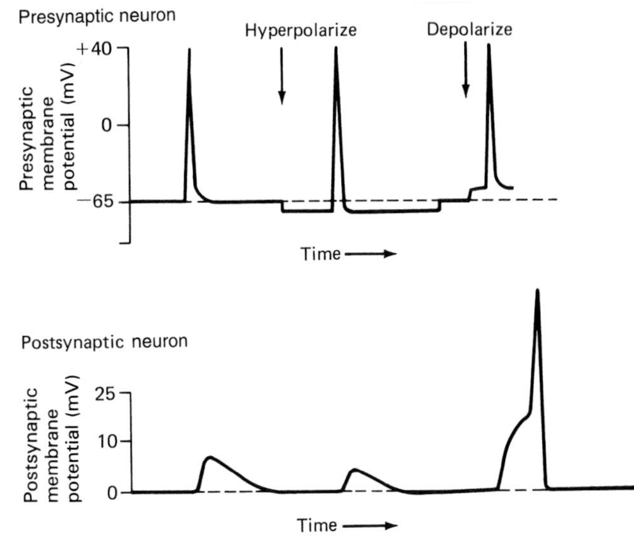

Figure 15–14 Changes in membrane potential of the presynaptic terminal affect the intracellular concentration of Ca^{2+} and thus the amount of transmitter released. When the presynaptic terminal is at the resting potential, an action potential (**top**) produces a postsynaptic potential of a given size (**bottom**). If the presynaptic terminal is hyperpolarized by 10 mV, the steady-state Ca^{2+} influx is decreased, and the same size of action potential now produces a smaller postsynaptic potential. In contrast, if the presynaptic neuron is depolarized by 10 mV, the steady-state Ca^{2+} influx is increased, and the same size of action potential produces a larger postsynaptic potential, which triggers an action potential in the postsynaptic cell.

Synaptic effectiveness can be altered in some nerve cells by intense activity. In these cells a train of high-frequency action potentials is followed by a period during which action potentials produce successively larger postsynaptic potentials. High-frequency stimulation of the presynaptic neuron (which in some cells can generate 500 to 1,000 action potentials per second) is called *tetanic stimulation*. The increase in size of the postsynaptic potentials during tetanic stimulation is called *potentiation*; the increase that persists after tetanic stimulation is called *posttetanic potentiation*. This enhancement usually lasts several minutes, but it can persist for an hour or more (Figure 15–15).

Posttetanic potentiation is thought to result from a transient saturation of the various Ca^{2+} buffering systems in the terminals, primarily the smooth endoplasmic reticulum and mitochondria. The excess, called *residual* Ca^{2+}, builds up after the relatively large influx accompanying a series of action potentials. The resulting increase in the resting concentration of free Ca^{2+} enhances synaptic transmission for many minutes or longer by activating the Ca^{2+}/calmodulin-dependent protein kinase, which stimulates Ca^{2+}-dependent mobilization steps in the terminals. Calcium/calmodulin dependent protein kinase allows more synaptic vesicles to be freed from their cytoskeletal restraint and to be mobilized into and docked at release sites. As a

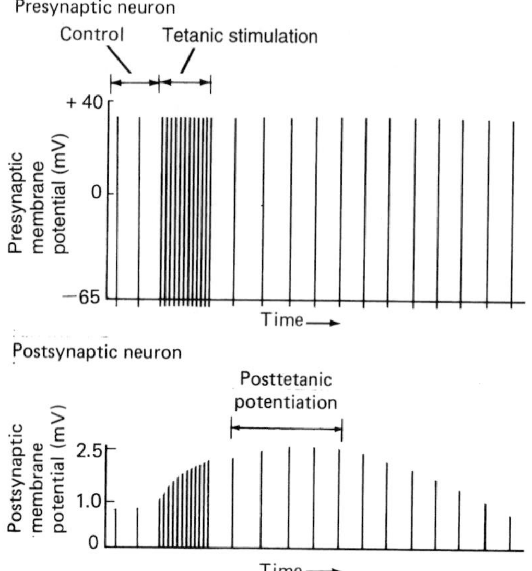

Figure 15–15 After a high rate of stimulation of the presynaptic neuron the amplitude of the postsynaptic potentials gradually increases. This enhancement in the strength of the synapse is a means by which the neuron can store information about its previous activity and thereby remember events. The time scale of this experimental record has been compressed (each presynaptic and postsynaptic potential appears as a simple line indicating its amplitude). To establish a baseline (control), the presynaptic neuron is stimulated at a rate of 1 per second, producing a postsynaptic potential of about 1 mV. The presynaptic neuron is then stimulated for several seconds at a higher rate of 5 per second. During this *tetanic stimulation* the postsynaptic potential increases in size, a phenomenon known as *potentiation*. After several seconds of stimulation the presynaptic neuron is returned to the control rate of firing (1 per second). However, the postsynaptic potentials continue to increase for minutes, and in some cells for several hours. This persistent increase is called *posttetanic potentiation*.

result, each action potential sweeping into the terminals of the presynaptic neuron releases more transmitter than before.

Here then is a simple kind of cellular memory! The presynaptic cell stores information about the history of its activity in the form of the residual Ca^{2+} in its terminals. This storage of biochemical information in the nerve cell, following a brief period of activity, leads to a strengthening of the presynaptic connection that persists for many minutes. In Chapter 36, we see how posttetanic potentiation at certain synapses is followed by an even longer-lasting process (also initiated by Ca^{2+} influx), called *long-term potentiation*, which can last many hours or even days.

Synaptic Connections on Presynaptic Terminals Regulate Intracellular Free Calcium

Neurons form synapses on the axon terminals as well as the cell body and dendrites of other neurons. Whereas axosomatic synaptic actions affect all the axonal branches of the postsynaptic neuron (because they affect the probability that the neuron will fire an action potential), axo-axonic actions selectively control the behavior of individual branches of a neuron. One important action of axo-axonic synapses is control of the Ca^{2+} influx into the terminals, allowing the presynaptic neuron to either depress or enhance transmitter release in the postsynaptic cell.

As we saw in Chapter 13, when one neuron hyperpolarizes the cell body (or dendrites) of another, it decreases the likelihood that the postsynaptic cell will fire. This action is called *postsynaptic inhibition*. In contrast, when a neuron contacts the axon terminal of another cell, it can reduce the amount of transmitter that will be released by the second cell onto a third cell. This action is called *presynaptic inhibition* (Figure 15–16A). Likewise, *presynaptic facilitation* increases the amount of transmitter released by the postsynaptic cell (Figure 15–16B). For reasons that are not well understood, presynaptic modulation often occurs early in the pathway of sensory inflow. For example, presynaptic inhibition is found in the relay nuclei for sensory information concerned with vision, touch, and the body's sense of position (that is, in the retina, spinal cord, and dorsal column nuclei).

The best-analyzed mechanisms of presynaptic inhibition and facilitation are in the neurons of invertebrates and in the mechanoreceptor neurons (in dorsal root ganglia) of vertebrates. Three mechanisms for presynaptic inhibition have been identified in these cells. One is the simultaneous closure of Ca^{2+} channels and opening of voltage-gated K^+ channels, which both decreases the influx of Ca^{2+}

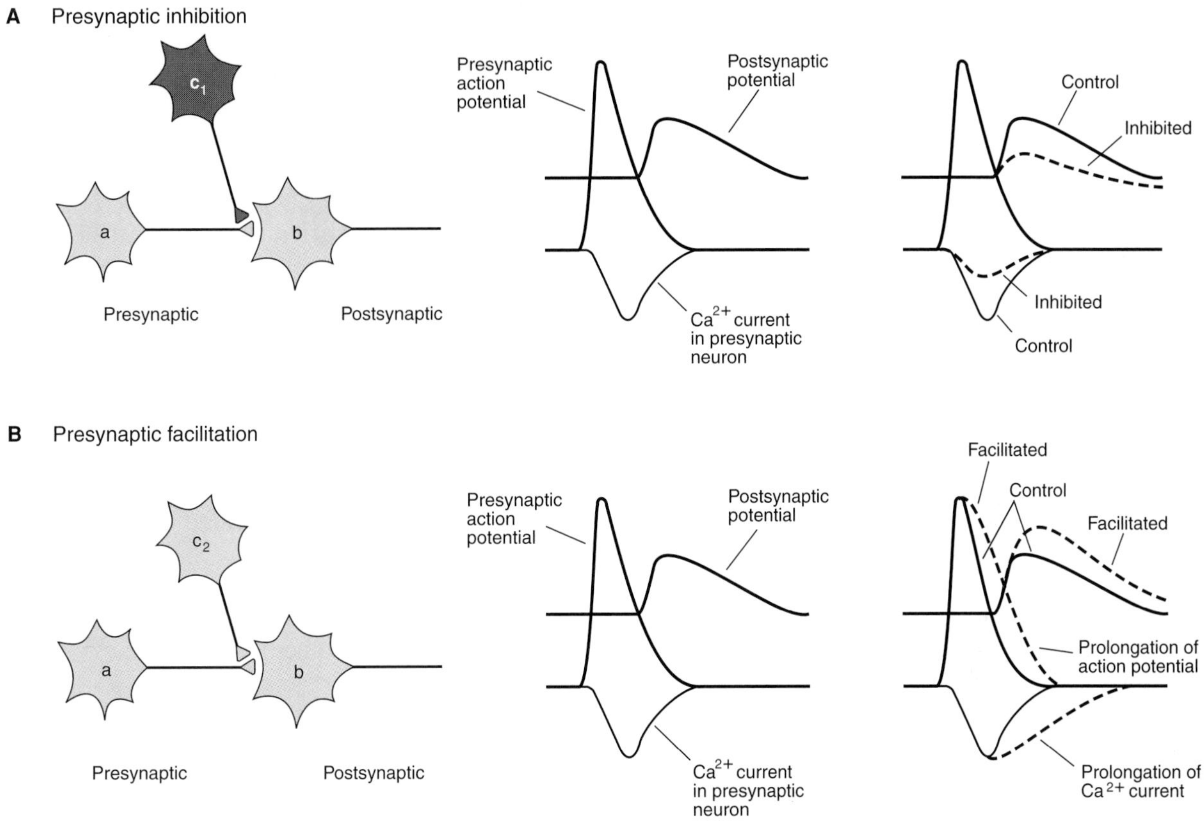

A Presynaptic inhibition

B Presynaptic facilitation

Figure 15–16 Presynaptic contact at the axon terminal of another cell can inhibit or facilitate transmitter release by the postsynaptic cell.

A. An inhibitory neuron ($\mathbf{c}_1$) contacts the terminal of a second presynaptic neuron (**a**). An action potential in cell $\mathbf{c}_1$ depresses the Ca^{2+} current in cell **a**, thereby reducing the amount of transmitter released by cell **a**. As a result, the postsynaptic potential in cell **b** is depressed.

B. A facilitating neuron ($\mathbf{c}_2$) contacts the terminal of a second presynaptic neuron (**a**). An action potential in cell $\mathbf{c}_2$ depresses the K^+ current in cell **a**, thereby increasing the Ca^{2+} current. This action prolongs the action potential in cell **a** and therefore increases the amount of transmitter released by cell **a**. As a result, the amplitude of the postsynaptic potential in cell **b** is enhanced.

and enhances repolarization of the cell. In certain intervertebrate neurons this is due to a specific type of second messenger, lipoxygenase metabolites of arachidonic acid, which we learned about in the previous chapter. The second mechanism is an increased conductance to Cl^-, which decreases (or short-circuits) the amplitude of the action potential in the presynaptic terminal. As a result, less depolarization is produced and fewer Ca^{2+} channels are activated by the action potential. The third mechanism involves direct inhibition of the transmitter release machinery, independent of Ca^{2+} influx. This is thought to work by decreasing the sensitivity for Ca^{2+} of one or more steps involved in the release process.

Conversely, presynaptic facilitation can be caused by an enhanced influx of Ca^{2+}. In mollusks serotonin acts through cAMP-dependent protein phosphorylation to close K^+ channels, thereby broadening the action potential and allowing the Ca^{2+} influx to persist for a longer period. In addition, the cAMP-dependent protein kinase also acts directly on the machinery of exocytosis to enhance release.

Thus, regulation of the free Ca^{2+} concentration in the presynaptic terminal is the basis for a variety of mechanisms that give plasticity to chemical synapses. Although we know a fair amount about short-term changes in synaptic effectiveness—changes that last minutes and hours—we are only beginning to learn about changes that persist days, weeks, and longer. It seems quite likely that these long-term changes require growth and an increase in the number of synapses in addition to alterations in Ca^{2+} influx and enhancement of release from preexisting synapses.

An Overall View

In his book *Ionic Channels of Excitable Membranes*, Bertil Hille summarizes the importance of Ca^{2+} in regulating neuronal function:

> Electricity is used to gate channels and channels are used to make electricity. However, the nervous system is not primarily an electrical device. Most excitable cells ultimately translate their electrical excitation into another form of activity. As a broad generalization, excitable cells translate their electricity into action by Ca^{2+} fluxes modulated by voltage-sensitive Ca^{2+} channels. Calcium ions are intracellular messengers capable of activating many cell functions. Calcium channels . . . serve as the only link to transduce depolarization into all the nonelectrical activities controlled by excitation. Without Ca^{2+} channels our nervous system would have no outputs.

Neither Na^+ influx nor K^+ efflux is required for synaptic release of neurotransmitters. Only Ca^{2+}, which enters the cell through voltage-dependent channels in the presynaptic terminal, is essential. Synaptic delay—the time between the onset of the action potential and the release of transmitter—reflects the time it takes for incoming Ca^{2+} to diffuse to its site of action and trigger the discharge of transmitter quanta from synaptic vesicles.

One quantum of transmitter consists of approximately 5,000 ACh molecules, all contained in a single synaptic vesicle. Larger synaptic potentials evoked under such conditions are integral multiples of the unit potential. Increasing the extracellular Ca^{2+} does not change the size of the spontaneous miniature synaptic potentials or the unit synaptic potentials. Rather, it increases the probability that a synaptic vesicle will discharge its transmitter, so that action potentials evoke fewer failures and higher-amplitude postsynaptic potentials.

Rapid-freezing experiments have shown that the vesicles fuse with the presynaptic plasma membrane in the vicinity of the active zone. Freeze-fracture studies have also revealed rows of large intramembranous particles along the active zone, and these are thought to be Ca^{2+} channels. These highly localized channels may be responsible for the observed rapid increase, as much as a thousandfold, in the Ca^{2+} concentration of the axon terminal during an action potential. One hypothesis about how Ca^{2+} triggers vesicle fusion is that this ion gates a pore that traverses both the vesicle and the plasma membrane and through which the contents of the vesicle are released into the extracellular space.

Calcium also regulates the targeted transport of the small synaptic vesicles to the active zone. These vesicles appear to be bound to the cytoskeleton, and Ca^{2+} is thought to free the vesicles by triggering Ca^{2+}/calmodulin-dependent phosphorylation.

Finally, the amount of transmitter released from a neuron is not fixed, but can be modified by both intrinsic and extrinsic modulatory processes. High-frequency stimulation produces an increase in transmitter release called posttetanic potentiation. This potentiation, which lasts a few minutes, is caused by transient saturation of Ca^{2+} buffering in the terminal following the large Ca^{2+} influx that occurs during the action potential. Tonic depolarization or hyperpolarization of the presynaptic neuron can also modulate release by altering steady-state Ca^{2+} influx. The action of neurotransmitters on receptors in the axon terminal of another neuron can facilitate or inhibit transmitter release by altering the steady-state level of Ca^{2+} influx or the Ca^{2+} influx during the action potential.

What are the molecular mechanisms by which Ca^{2+} affects transmitter release? This is one of the pressing questions in neurobiology today. Calcium may have a direct role in the fusion of the vesicle membrane with the cell membrane, or it may act through one or more calcium-sensitive proteins, such as synaptotagmin. Several molecular candidates have been identified that could account for the three earlier components of release: mobilization, targeting, and docking. The next step in understanding transmitter release is clear. To follow the trail of Ca^{2+} influx we need to move from the membrane channel to the molecular machinery for release. Thus in the next chapter we examine the nature of the transmitter molecules that are used for chemical transmission.

Selected Readings

Almers, W., and Tse, F. W. 1990. Transmitter release from synapses: Does a preassembled fusion pore initiate exocytosis? Neuron 4:813–818.

Barinaga, M. 1993. Secrets of secretion revealed. Science 260:487–489.

Bennett, M. K., and Scheller, R. H. 1994. A molecular description of synaptic vessel membrane trafficking. Annu. Rev. Biochem. 63:63–100.

Breckenridge, L. J., and Almers, W. 1987. Currents through the fusion pore that forms during exocytosis of a secretory vesicle. Nature 328:814–817.

Jessell, T. M., and Kandel, E. R. 1993. Synaptic transmission: A bidirectional and self-modifiable form of cell-cell communication. Cell 72/Neuron 10 (Suppl.): 1–30.

Kamiya, H. and Zucker, R. S. 1994. Residual Ca^{2+} and short-term synaptic plasticity. Nature. 371:603–606.

Kandel, E. R. 1981. Calcium and the control of synaptic strength by learning. Nature 293:697–700.

Katz, B. 1969. The Release of Neural Transmitter Substances. Springfield, Ill.: Thomas.

Kelly, R. B. 1993. Storage and release of neurotransmitters. Cell 72/Neuron 10 (Suppl.):43–53.

Smith, S. J., and Augustine, G. J. 1988. Calcium ions, active zones and synaptic transmitter release. Trends Neurosci. 11:458–464.

Söllner, T., Whiteheart, S. W., Brunner, M., Erdjument-Bromage, H., Geromanos, S., Tempst, P., and Rothman, J. E. 1993. SNAP receptors implicated in vesicle targeting and fusion. Nature 362:318–324.

16

Neurotransmitters

Chemical Messengers Must Fulfill Four Criteria to Be Considered Transmitters

There Are a Small Number of Small-Molecule Transmitter Substances

- *Acetylcholine*
- *Biogenic Amine Transmitters*
- *Amino Acid Transmitters*
- *ATP and Adenosine*
- *Small-Molecule Transmitters Are Actively Taken Up into Vesicles*

There Are Many Neuroactive Peptides

Peptides and Small-Molecule Transmitters Differ in Several Ways

Peptides and Small-Molecule Transmitters Can Coexist and Be Co-released

Removal of Transmitter from the Synaptic Cleft Terminates Synaptic Transmission

Our general scheme of chemical synaptic transmission divides the process into four steps—two presynaptic and two postsynaptic. These steps are (1) synthesis of transmitter substance, (2) storage and release of transmitter, (3) interaction of transmitter with receptor in the postsynaptic membrane, and (4) removal of transmitter from the synaptic cleft. In the previous chapter we considered steps 2 and 3 the release of the transmitter and its interaction with the postsynaptic receptor. We now turn to the initial and final steps of chemical synaptic transmission—the synthesis of the molecules used as transmitters and their removal from the synaptic cleft following synaptic action.

Chemical Messengers Must Fulfill Four Criteria to Be Considered Transmitters

Before we consider the biochemical processes involved in synaptic transmission in detail, it is important to make clear what is meant by a *chemical transmitter*. The concept had become familiar by the early 1930s, after Otto Loewi demonstrated the release of acetylcholine (ACh) from vagus terminals in frog heart and Henry Dale reported his work on cholinergic and adrenergic transmission. The terms *cholinergic* and *adrenergic* were introduced to indicate that a neuron uses ACh or norepinephrine (or epinephrine) as neurotransmitter. Since that time many new substances have been discovered to act as transmitters, and the concept has been continually modified to accommodate new information about the cell biology of neurons and the pharmacology of receptors.

As a first approximation, we can define a transmitter as a substance that is released at a synapse by one neuron and that affects another cell, either a neuron or effector organ, in a specific manner. As with many other operational concepts in biology, the concept of a transmitter is quite clear at the center but can be somewhat fuzzy at the edges. A small number of substances of low molecular weight are generally accepted as transmitters, but many other substances have been accepted as transmitter candidates with varying degrees of consensus. It is often difficult to demonstrate a transmitter function for even those substances commonly assumed to be neurotransmitters. Because of these difficulties, a substance is not accepted as a neurotransmitter unless the following four criteria are met:

1. It is synthesized in the neuron.
2. It is present in the presynaptic terminal and is released in amounts sufficient to exert a defined action on the postsynaptic neuron or effector organ.
3. When administered exogenously (as a drug) in reasonable concentrations, it mimics the action of the endogenously released transmitter exactly (e.g., it activates the same ion channels or second-messenger pathway in the postsynaptic cell).
4. A specific mechanism exists for removing it from its site of action (the synaptic cleft).

Needless to say, it is often difficult to demonstrate *all* of these features experimentally at any one synapse.

The nervous system makes use of two main classes of chemical substances for signaling: small-molecule transmitters and neuroactive peptides, which are short chains of amino acids. The cell-biological distinctions between these two classes are fundamental, so we consider each in turn.

There Are a Small Number of Small-Molecule Transmitter Substances

Nine low-molecular-weight substances are generally accepted as neurotransmitters (Table 16–1). Eight are amines; of these, seven are amino acids or their derivatives. The non-amine is ATP or its degradation products. The amine chemical messengers share many biochemical similarities. All are charged small molecules that are formed in relatively short biosynthetic pathways, and all are synthesized from precursors derived from the major carbohydrate substrates of intermediary metabolism. Like other pathways of intermediary metabolism, synthesis of these neurotransmitters is catalyzed by enzymes that, almost without exception, are cytosolic.

Table 16–1 Small-Molecule Transmitter Substances and Their Key Biosynthetic Enzymes

Transmitter	*Enzyme*
Acetylcholine	Choline acetyltransferase (specific)
Biogenic amines	
Dopamine	Tyrosine hydroxylase (specific)
Norepinephrine	Tyrosine hydroxylase and dopamine β-hydroxylase (specific)
Epinephrine	Tyrosine hydroxylase and dopamine β-hydroxylase (specific)
Serotonin	Tryptophan hydroxylase (specific)
Histamine	Histidine decarboxylase (specificity uncertain)
Amino acids	
γ-Aminobutyric acid	Glutamic acid decarboxylase (probably specific)
Glycine	General metabolism (specific pathway undetermined)
Glutamate	General metabolism (specific pathway undetermined)

In biosynthetic pathways for the amine transmitters, as in any biosynthetic pathway, the overall synthesis of the transmitter is regulated at one enzymatic reaction. The controlling enzyme ordinarily is characteristic of a type of neuron and usually is absent in other types of neurons.

Acetylcholine

The biosynthetic pathway for ACh has only one enzymatic reaction, that catalyzed by choline acetyltransferase (step 1 in the following reaction):

Acetyl CoA + choline

$$\Big\Updownarrow (1)$$

$$CH_3-\overset{\overset{\displaystyle O}{\|}}{C}-O-CH_2-CH_2-\overset{+}{N}-(CH_3)_3 + CoA$$

Acetylcholine

This transferase is the characteristic enzyme in ACh biosynthesis. Nervous tissue cannot synthesize choline, which is derived from the diet and delivered to neurons through the bloodstream. The co-substrate, acetyl coenzyme A, participates in many universal metabolic pathways and is not restricted to cholinergic neurons.

Acetylcholine is the transmitter used by the motor neurons of the spinal cord and, therefore, is released at all vertebrate neuromuscular junctions. In the autonomic nervous system it is the transmitter for all preganglionic neurons and for the parasympathetic postganglionic neurons as well. It is used at many synapses throughout the brain. In particular, cell bodies synthesizing ACh are numerous in the nucleus basalis, which has widespread projections to the cerebral cortex.

Biogenic Amine Transmitters

The term *biogenic amine*, although chemically imprecise, has been used for decades to designate certain neurotransmitters. This group includes the catecholamines and serotonin. Histamine, an imidazole, is often referred to as a biogenic amine, although its biochemistry is remote from that of the catecholamines and the indolamines.

The catecholamine transmitters—dopamine, norepinephrine, and epinephrine—are all synthesized from the amino acid tyrosine in a common biosynthetic pathway containing five enzymes: tyrosine hydroxylase, aromatic amino acid decarboxylase, dopamine β-hydroxylase, pteridine reductase, and phenylethanolamine-*N*-methyltransferase. Catecholamines have a catechol nucleus, a 3,4-dihydroxylated benzene ring.

The first enzyme, tyrosine hydroxylase (**1**), is an oxidase that converts tyrosine to L-dihydroxyphenylalanine (L-DOPA). This enzyme is rate-limiting for the synthesis of both dopamine and norepinephrine. It is present in all cells producing catecholamines and requires a reduced pteridine (Pt-2H) cofactor, which is regenerated from pteridine (Pt) by another enzyme, pteridine reductase (**4**). (This reductase is not specific to neurons.)

Tyrosine + O_2 —(1)→ [HO, HO-substituted benzene ring]—CH_2—C(COOH)(H)—NH_2
Pt (2H) → Pt; Pt —(4)→ Pt (2H)

L-DOPA

L-DOPA is next decarboxylated by a decarboxylase (**2**) to give dopamine and CO_2.

L-DOPA —(2)→ [HO, HO-substituted benzene ring]—CH_2—CH_2—NH_2 + CO_2

Dopamine

The third enzyme in the sequence, dopamine β-hydroxylase (**3**), converts dopamine to norepinephrine.

Dopamine —(3)→ [HO, HO-substituted benzene ring]—CH(OH)—CH_2—NH_2

Norepinephrine

In the central nervous system norepinephrine is used as a transmitter by neurons whose cell bodies are located in the locus ceruleus, a nucleus of the brain stem with many complex modulatory functions. Although these neurons are relatively few in number, they project diffusely throughout the cortex, cerebellum, and spinal cord. In the peripheral nervous system norepinephrine is the transmitter in the postganglionic neurons of the sympathetic nervous system.

In addition to these four catecholaminergic biosynthetic enzymes, a fifth enzyme, phenylethanolamine-*N*-methyltransferase (**5**), methylates norepinephrine to form epinephrine in the adrenal medulla. This reaction requires *S*-adenosylmethionine as methyl donor. Some neurons in the brain also are thought to use epinephrine as a transmitter.

Norepinephrine —(5)→ [HO, HO-substituted benzene ring]—CH_2(OH)—CH_2—NH—CH_3

Epinephrine

Not all cells that release catecholamines express all five of these biosynthetic enzymes, although cells that release epinephrine do. Neurons that use norepinephrine do not express the methyltransferase, and neurons releasing dopamine do not express the transferase or dopamine β-hydroxylase. Thus, the expression of the genes encoding the enzymes that synthesize catecholamines can be regulated independently. There is a high degree of similarity in amino acid sequence and in the nucleic acid sequences encoding three of the biosynthetic enzymes: tyrosine hydroxylase, dopamine β-hydroxylase, and phenylethanolamine-*N*-methyltransferase. The genes for these three enzymes appear to be linked together on the same chromosome and their gene expression can be regulated coordinately.

Serotonin (5-hydroxytryptamine) and the amino acid tryptophan from which it is derived belong to a group of aromatic compounds called *indoles*, with a five-member ring containing nitrogen joined to a benzene ring. Two enzymes synthesize serotonin—tryptophan hydroxylase (**1**), an oxidase similar to tyrosine hydroxylase, and 5-hydroxytryptophan decarboxylase (**2**):

$$\text{Try} \xrightarrow{(1)} \text{5-HTP} \xrightarrow{(2)} \text{Serotonin (HO-indole-}CH_2-CH_2-NH_2\text{)}$$

Serotonin

The controlling region is tryptophan hydroxylase, catalyzed by the first enzyme in the pathway. Interestingly, L-DOPA decarboxylase and 5-hydroxytryptophan decarboxylase seem to be identical. An enzyme with similar activity, L-aromatic amino acid decarboxylase, is present in many nonnervous tissues, but it is not yet certain whether these decarboxylases are identical in structure or whether different molecular forms of the enzyme (specific isozymes) exist in the different tissues.

Cell bodies of serotonergic neurons are found in and around the midline raphe nuclei of the brain stem, which are involved in regulating attention and other complex cognitive functions. The projections of these cells (like those of the noradrenergic cells in the locus ceruleus) are widely distributed throughout the brain and spinal cord.

As we shall learn in later chapters, serotonin, dopamine, and norepinephrine are important in mechanisms underlying some of the most common and important mental and neurological dysfunctions: depression, schizophrenia, drug addiction, and Parkinson's disease. Thus, most antidepressant drugs enhance neurotransmission at serotonergic or adrenergic brain synapses. Moreover, defects in brain dopaminergic neurotransmission are thought to play an important role in schizophrenia. A marked decrease in production of dopamine is characteristic of Parkinson's disease, which is treated by administering L-DOPA. In some cases of Parkinson's disease fetal adrenal medullary tissue has been successfully implanted intraventricularly; the implanted tissue can survive for years and produce substantial amounts of dopamine.

Histamine, like the amino acid histidine from which it is derived, contains a characteristic five-member ring with two nitrogen atoms. It has long been recognized as a local hormone, or autocoid, active in the inflammatory reaction and in the control of vasculature, smooth muscle, and exocrine glands (e.g., secretion of gastric juice of high acidity). Histamine is a transmitter in invertebrates, but binding sites for certain kinds of antihistaminic drugs have been localized to neurons in the vertebrate brain. This transmitter is concentrated in the hypothalamus, one of the centers for regulating the hormonal function. It is synthesized from histidine by decarboxylation. Although not extensively analyzed, the decarboxylase (**1**) catalyzing this step appears to be characteristic of histaminergic neurons:

$$\text{Histidine} \xrightarrow{(1)} \text{Histamine (imidazole-}CH_2-CH_2-NH_2\text{)} + CO_2$$

Histamine

Histamine also is a precursor of two dipeptides that are found in nervous tissue. A synthetase catalyzes the formation of carnosine (β-alanyl histidine) from the amino acid β-alanine and ATP. (Although β-alanine is normally present in tissues, only α-amino acids—amino acids with both carboxyl and amino groups on the α-carbon—can be incorporated into proteins.) The same enzyme forms homocarnosine (β-aminobutyrylhistidine) from α-histidine and γ-aminobutyric acid (GABA). Although the roles of these peptides are not known, carnosine may have a special function in olfactory areas of the brain, where it is highly concentrated.

Amino Acid Transmitters

Acetylcholine and the biogenic amines are not intermediates in general biochemical pathways and, as a rule, are produced only in certain neurons. In contrast, a group of amino acids that function as neurotransmitters are also universal cellular constituents. Glycine and glutamate are 2 of the 20 amino acids common in the proteins of all cells.

Glutamate is a product of the Kreb's cycle of intermediary metabolism, which we do not review here. Glycine, which is probably synthesized from serine, is one of the two transmitters in spinal cord inhibitory interneurons. Its specific biosynthesis in neurons has not been studied, but its biosynthetic pathway in other tissues is well known. GABA is synthesized from glutamate in a reaction catalyzed by glutamic acid decarboxylase (**1**):

$$\underset{\text{Glutamate}}{\begin{array}{c} COOH \\ | \\ CH_2 \\ | \\ CH_2 \\ | \\ H_2N-CH \\ | \\ COOH \end{array}} \xrightarrow{(1)} \underset{\text{GABA}}{\begin{array}{c} COOH \\ | \\ CH_2 \\ | \\ CH_2 \\ | \\ H_2N-CH_2 \end{array}} + CO_2$$

GABA is present at high concentrations throughout the central nervous system and is also detectable in other tissues (especially islet cells of the pancreas and the adrenal gland). An important class of inhibitory interneurons in the spinal cord uses GABA as transmitter. GABA is thought to be the major inhibitory transmitter at many sites in the brain, for example in various inhibitory interneurons and in the granule cells in the olfactory bulb. It is also thought to be released by amacrine cells of the retina, Purkinje cells of the cerebellum, and basket cells of both the cerebellum and the hippocampus.

It might at first seem puzzling that common amino acids can act as transmitters in some neurons but not in others. This phenomenon shows that the presence of a substance, even in substantial amounts, is not in itself sufficient evidence that the substance is used as a transmitter. To illustrate this point, let us consider the following example. GABA is inhibitory at the neuromuscular junction of the lobster (and of other crustacea and insects), and glutamate is excitatory. The concentration of GABA is about 20 times greater in inhibitory cells than in excitatory cells, and this supports the idea that GABA is the inhibitory transmitter at the lobster neuromuscular junction. But the concentration of glutamate, the excitatory transmitter, is the same in both excitatory and inhibitory cells. Glutamate therefore must be compartmentalized within these neurons, that is, *transmitter* glutamate must be kept separate from *metabolic* glutamate. Transmitter glutamate is compartmentalized in synaptic vesicles.

ATP and Adenosine

At some synapses ATP and its degradation products —for example, adenosine—act as chemical messengers. Adenine and guanine and their derivatives are called purines; the evidence for *purinergic* transmission is especially strong for sympathetic neurons that synapse on the vas deferens and muscle fibers of the heart, for nerve plexuses that make connections with smooth muscle in the gut, and for dorsal root ganglion cells that synapse onto some neurons in the dorsal horn of the spinal cord. Some of these nerve endings appear to contain considerably more ATP in vesicles than do others. At some synapses where ATP is released, purines have no effect on postsynaptic targets. These common metabolites can serve as transmitters only if receptors that are sensitive to purines are present. Well-characterized *presynaptic* receptors for adenosine have also been described, at which the purine may act as an autocoid.

Small-Molecule Transmitters Are Actively Taken Up into Vesicles

Although a specific set of biosynthetic enzymes determines which small molecule will be used as a transmitter by a neuron, it does not in itself determine transmitter specificity. Before a substance can be released as a transmitter, other biochemical processes must take place, for example, concentration of the transmitter into the synaptic vesicles.

Catecholamines are carried across the membrane of aminergic vesicles by a specific transporter protein that uses the energy provided by a pH gradient. The pH within the vesicle is 5.5; that of the cytoplasm is 7. This chemiosmotic mechanism is similar to the one that explains oxidative phosphorylation in mitochondria. A proton pump in the vesicle membrane, powered by the hydrolysis of ATP, brings in protons. The influx of H^+ makes the inside of the vesicle more acidic than the cytoplasm and generates an electrochemical gradient, with the inside of the vesicle becoming positive.

How is the energy stored in the proton gradient coupled to transport of the transmitter? One plausible model assumes that only uncharged biogenic amine molecules are transported. (Biogenic amine transmitters exist as charged and uncharged species.) The p*K* of the primary amine group in catecholamines is about 9; therefore, at the neutral pH of cytoplasm only about 1% of the amine exists in uncharged form. In its protonized (cationic) form the biogenic amine binds to the portion of the transporter protein that is on the cytoplasmic surface of the vesicle's membrane.

Once bound, the amine becomes uncharged by the dissociation of a proton. The neutral transmitter molecule is then carried through the membrane into the vesicle. Because of the low pH inside the vesicles, a molecule of uncharged amine coming into the vesicle is protonated and does not readily escape (the proportion of uncharged amine inside is about 70-fold lower than in the cytoplasm). The transmitter remains in the vesicle because of ion trapping and because of the formation of complexes with ATP. Acetylcholine enters small cholinergic vesicles by similar mechanisms. Uptake of small-molecule transmitters occurs in preparations of purified synaptic vesicles, and transporter proteins for amine transmitters have been characterized by molecular cloning.

There Are Many Neuroactive Peptides

With rare exceptions the enzymes that catalyze the synthesis of the low-molecular-weight neurotransmitters are found in the cytoplasm. These enzymes are synthesized on free polysomes in the cell body and are distributed throughout the neuron by slow axoplasmic transport (see Chapter 4). Because these biosynthetic enzymes are distributed throughout the cell, the small-molecule transmitter substances can be formed in all parts of the neuron; most important, these transmitters can be synthesized at the nerve terminals where they are released.

In contrast, the neuroactive peptides are derived from secretory proteins that are formed in the cell body. Like other secretory proteins, neuroactive peptides or their precursors are first processed in the endoplasmic reticulum and then move to the Golgi apparatus for further processing. They leave the Golgi apparatus within secretory granules and are moved to terminals by fast axonal transport.

More than 50 short peptides are pharmacologically active in nerve cells (Table 16–2). These peptides cause inhibition or excitation, or both, when applied to appropriate target neurons. Some of these peptides had been previously identified as hormones, with known targets outside the brain (e.g., angiotensin and gastrin), or as products of neuroendocrine secretion (e.g., oxytocin, vaso-

Table 16–2 Neuroactive Peptides: Mammalian Brain Peptides Categorized According to Tissue Localization

Hypothalamic-releasing hormones	*Invertebrate peptides*	*Heart*
Thyrotropin-releasing hormone	FMRFamide*	Atrial naturetic peptide
Gonadotropin-releasing hormone	Hydra head activator	
Somatostatin	Proctolin	*Others*
Corticotropin-releasing hormone		Angiotensin II
Growth hormone-releasing hormone	*Small cardiac peptides*	Bradykinin
	Myomodulins	Sleep peptide(s)
Neurohypophyseal hormones	Buccalins	Calcitonin
Vasopressin	Egg-laying hormone	Calcitonin gene-related peptide (CGRP)
Oxytocin	Bag cell peptides	Neuropeptide Y
		Neuropeptide Yy
Pituitary peptides	*Gastrointestinal peptides*	Galanin
Adrenocorticotropic hormone	Vasoactive intestinal polypeptide	Substance K (neurokinin)
β-Endorphin	Cholecystokinin	
α-Melanocyte-stimulating hormone	Gastrin	
Prolactin	Substance P	
Luteinizing hormone	Neurotensin	
Growth hormone	Methionine-enkephalin	
Thyrotropin	Leucine-enkephalin	
	Insulin	
	Glucagon	
	Bombesin	
	Secretin	
	Somatostatin	
	Thyrotropin-releasing hormone	
	Motilin	

*Phe–Met–Arg–Phe–NH_2.
Expanded from Krieger (1983).

pressin, somatostatin, luteinizing hormone, and thyrotropin-releasing hormone). These peptides, in addition to being hormones in some tissues, also act as transmitters when released close to the site of intended action.

The study of neuroactive peptides is particularly important because some of them have been implicated in modulating sensibility and emotions. For example, some peptides (substance P and enkephalins) are preferentially located in many of the regions of the brain involved in the perception of pain; others regulate complex responses to stress (γ-melanocyte-stimulating hormone, adrenocorticotropin, and β-endorphin).

Although the diversity of neuroactive peptides is enormous, as a class these chemical messengers share a common cell biology. A striking generality is that neuroactive peptides are grouped in families. The main families are listed in Table 16–3; at least 10 have been identified. Members of each family contain long stretches of similar amino acid residues. How is relatedness between peptides determined? The most direct way is to compare either the actual amino acid sequences of the peptides or the nucleotide base sequences in the genes that encode them.

Divergent and convergent evolution of genes explains how some neuroactive peptides are related. In *divergent evolution* the mRNA, which is the template for several copies of the same peptide, is transcribed from genomic DNA that might have evolved by several duplications of a simpler DNA ancestor. Reduplication followed by divergence could result in the production of related but diversified sets of peptides. In *convergent evolution* independent nucleotide sequences with the potential to code for similar physiologically active peptides might originally have been located at a variety of sites in the ancestral chromosome. During evolution these sequences could have been spliced together and organized in a similar way in all of the genes that encode the protein precursors of a given family.

In most instances several different neuroactive peptides are encoded by a single continuous mRNA, which is translated into one large protein precursor, or *polyprotein* (Figure 16–1). Production from a large precursor can sometimes serve as a mechanism for amplification, since more than one copy of the same peptide can be produced from the one polyprotein. Examples can be found in the opioid peptide family: many distinct peptides with opioid activity all contain the same sequence of amino acids, Tyr–Gly–Gly–Phe. (Opioid peptides arise from three different polyprotein precursors, each of which is the product of a distinct gene.) Another example is the precursor of glucagon, which contains two copies of the hormone. In other instances the biological purposes served are more complicated, since peptides with either related or antagonistic functional capacities can be generated from the same precursor.

Table 16–3 Some Families of Neuroactive Peptides

Opioid	Opiocortins, enkephalins, dynorphin, FMRFamide
Neurohypophyseal	Vasopressin, oxytocin, neurophysins
Tachykinins	Substance P, physalaemin, kassinin, uperolein, eledoisin, bombesin, substance K (neurokinin A)
Secretins	Secretin, glucagon, vasoactive intestinal peptide, gastric inhibitory peptide, growth hormone-releasing factor, peptide histidine isoleucineamide
Insulins	Insulin, insulin-like growth factors I and II
Somatostatins	Somatostatins, pancreatic polypeptide
Gastrins	Gastrin, cholecytokinin

The processing of more than one functional peptide from a single polyprotein is by no means unique to peptide chemical messengers. The mechanism was first described for proteins encoded by small RNA viruses. Several viral polypeptides are produced from the same viral polyprotein, and all contribute to the generation of new virus particles. As with the virus, where the different proteins obviously serve a common biological purpose (formation of new viruses), in many instances a neuronal polypeptide will yield peptides that work together to serve a common behavioral goal.

The processing of precursors to neuroactive peptides takes place within the neuron's major membrane system and in vesicles, as discussed below. Several peptides are produced from a single polyprotein by limited and specific cleavages catalyzed by specific proteases. These enzymes are called serine proteases because they all have a ser-

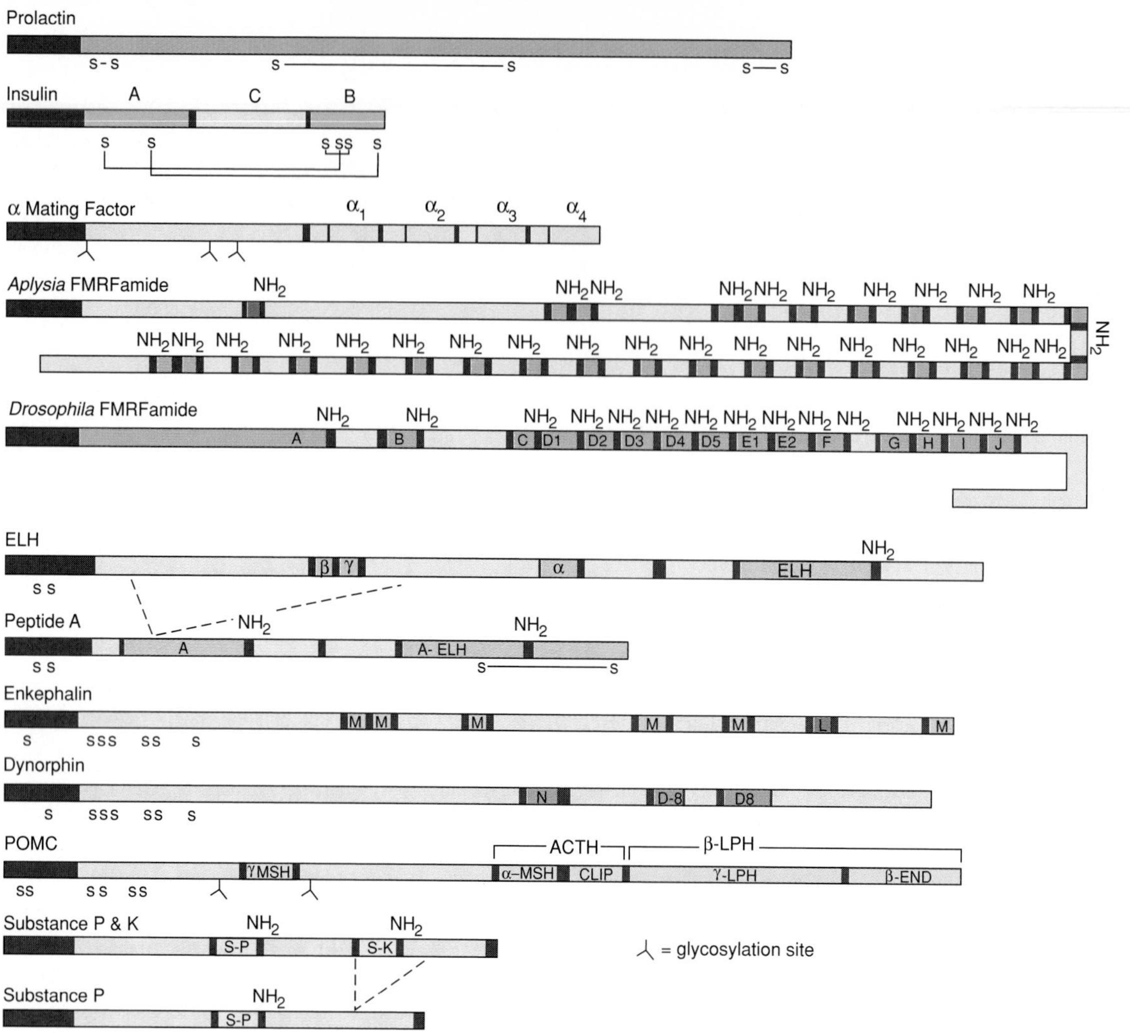

Figure 16–1 Structures of several hormone and neuropeptide precursors. Each pre-prohormone is initiated by a hydrophobic signal sequence (**black bars**). Internal cleavages at basic residues are indicated by the **gray vertical lines** within the sequence. Several of the active peptides are named, and cystine (S) and sugar (⅄) residues are indicated below the schematic. In *prolactin* the mature hormone arises from the removal of the signal sequence and formation of three pairs of disulfide bonds. The *insulin* precursor is cleaved at two internal sites, resulting in the disulfide-linked A and B chains of mature insulin and the C peptide. The α *mating factor* from yeast is processed by endoproteolytic cleavage at dibasic residues, followed by diaminopeptidyl peptidase trimming to generate four copies of the mating factor (α_1–α_4). The *Aplysia FMRFamide* precursor encodes 28 copies of the tetrapeptide (**light purple shading**) and a single copy of the related FMRFamide peptide (**dark purple shading**). The *Drosophila FMRFamide* precursor encodes at least 15 predicted peptides with 10 different structures. The *egg-laying hormone* (ELH) precursor encodes at least four physiologically active peptides, α, β, and γ bag cell peptides as well as ELH. The *peptide A* precursor is quite similar to the ELH precursor; the major differences include a 240-amino-acid deletion encompassing the β and γ bag cell peptides (indicated by **dashed lines**) as well as single base changes that affect the patterns of cleavage, amidation, and disulfide linkage. The family of peptides giving rise to the opioid peptides is illustrated. The *enkephalin* precursor gives rise to six Met (M) and one Leu (L) enkephalin peptides. The *dynorphin* precursor is cleaved to at least three peptides, which are related to Leu enkephalin. The *POMC* precursor is processed differently in different lobes of the pituitary gland, resulting in different peptides. The endoproteolytic cleavage within ACTH and β-lipotropin are cleaved in the intermediate lobe but not the anterior lobe. Alternative RNA splicing generates two prohormones, giving rise to the tachykinins (substance K). One prohormone includes exons encoding both *substance P* and *substance K*; the other skips over this exon, generating a precursor that encodes only substance P. For scale, substance P (S-P) is composed of 11 amino acid residues. (Adapted from Sossin et al., 1989.)

ine residue at the catalytic center whose hydroxyl group participates in the cleavage reaction.

Processing is a critical step in determining which peptides will be released by a peptidergic neuron. Neurons with the same gene encoding a polyprotein may release different neuropeptides because of differences in the way they process the polyprotein precursor. An example is proopiomelanocortin (POMC), one of the three branches of the opioid family. The same mRNA for POMC is found in the anterior and intermediate lobes of the pituitary in the hypothalamus and in several other regions of the brain, as well as in the placenta and the gut, but different peptides are produced and released in each of these tissues.

It is not yet known how differential processing occurs, but current information on the biochemistry of membrane proteins and secretory products (see Chapter 4) suggests two plausible mechanisms. Two neurons might process the same polyprotein differently because each cell contains proteases with different specificities within the lumina of its internal membrane systems and vesicles. Alternatively, the two neurons might contain the same processing proteases, but each cell might glycosylate the common polyprotein at different sites, thereby protecting different regions of the polypeptide from cleavage.

Peptides and Small-Molecule Transmitters Differ in Several Ways

Some peptides satisfy many of the four established criteria for identifying a substance as a neurotransmitter, and several peptides satisfy all of the criteria. Moreover, certain features of the metabolism and action of peptides differ from those of the accepted small-molecule transmitters. Although neuroactive peptides are present in relatively high concentrations in some neurons, they are made only in the cell body because their synthesis requires peptide bond formation on ribosomes, whereas the small-molecule transmitters can be synthesized locally at terminals. Distinguishing between the two classes of chemical messengers by mode of synthesis can present some semantic difficulty because formation of the peptide bond can also be catalyzed by cytosolic enzymes called *synthetases*. Synthesis of peptides from amino acids without the participation of mRNA, however, usually results in short polymers, many of which involve the carboxyl group in the γ position of an amino acid rather than the α position, as in carnosine, homocarnosine, and glutathione as well as other γ-glutamyl peptides.

Furthermore, although the Ca^{2+}-dependent synaptic release of some neuroactive peptide messengers has been demonstrated, the conditions of release of peptides and small-molecule transmitters can be expected to be quite different. Because vesicles can be refilled rapidly with the small-molecule transmitters that are resynthesized at terminals, release can be both rapid and sustained. With peptides, however, once release occurs, a new supply of the peptide must arrive from the cell body before release can again take place.

Peptides and Small-Molecule Transmitters Can Coexist and Be Co-released

Neuroactive peptides, small-molecule transmitters, and other neuroactive molecules can coexist in the same neuron. In mature neurons the combination usually consists of one of the small-molecule transmitters and one or more peptides derived from one kind of polyprotein. For example, ACh and vasoactive intestinal peptide (VIP) can be released together by a presynaptic neuron and work synergistically on the same target cells.

Another example is calcitonin gene-related peptide (CGRP), which is present in some spinal motor neurons together with ACh. We have already considered the action of ACh on the nicotinic ACh receptor of skeletal muscle in earlier chapters. CGRP activates adenylyl cyclase, raising cAMP and cAMP-dependent protein phosphorylation in the muscles (see Chapter 14). Increased protein phosphorylation results in an increase in the force of contraction. Thus, at the neuromuscular junction a small-molecule transmitter (ACh) and a peptide (CGRP) are both released from the same presynaptic neuron. One other example is the co-release of glutamate and dynorphin in the hippocampus, where glutamate is excitatory and the opioid peptide inhibitory. Since nearby postsynaptic cells have receptors for both chemical messengers, this is an example of *co-transmission*.

Neurons that contain peptides processed from a single polyprotein can release several neuroactive peptides with potentially different postsynaptic actions. As described in Chapter 15, the vesicles that release peptides differ from those that release small-molecule transmitters. They are larger and do not require specialization of the presynaptic membrane for their exocytotic release. These peptide-containing vesicles may or may not contain small-molecule transmitter, but all vesicles do contain ATP, and ATP is released with both types of chemical messengers.

The co-release of ATP (which after release can be degraded to adenosine) is an important illustration that coexistence and co-release do not necessarily signify co-transmission. ATP, like many other substances, can be released from neurons but still not be effective if appropriate receptors are not close by —just like the unheard falling of a tree in a forest. On the other hand, some postsynaptic neurons have receptors for both ATP and adenosine or for one but not the other.

Recall that one of the four criteria for judging whether a neuron uses a particular substance as a transmitter is that the substance be present in sufficient amounts. Histochemistry provides an important method for detecting chemical messengers in neurons (Box 16–1).

Removal of Transmitter from the Synaptic Cleft Terminates Synaptic Transmission

Timely removal of transmitters from the synaptic cleft is critical to synaptic transmission. Persistence of transmitter in the cleft following release would prevent new signals from getting through and the synapse would become refractory, mainly because of receptor desensitization due to continued exposure to transmitter. There are three mechanisms by which nervous tissue disposes of soluble or unbound transmitter substances: diffusion, enzymatic degradation, and reuptake. *Diffusion* removes some fraction of *all* chemical messengers; it can be an important means by which the synaptic cleft is cleared of transmitter.

Enzymatic degradation of transmitter substance is used primarily by the cholinergic system where the extracellular enzyme involved is acetylcholinesterase. At the neuromuscular junction the active zones of the presynaptic nerve terminal are located just above the junctional folds of the muscle membrane where the ACh receptors are situated. The receptors are concentrated at the surface of the muscle and do not extend deep into the folds (see Figure 12–1), whereas the esterase is anchored to the basement membrane only within the folds. This anatomical arrangement of the molecules serves two functions. Since any ACh after dissociation from the receptor most likely will be diluted in the relatively large volume within the junctional folds and hydrolyzed to choline and acetate, the transmitter molecules will only be used once. Thus, one function of the esterase is to *punctuate* the synaptic message rapidly. The second function is recapture of choline that otherwise might be lost by diffusion away from the synaptic cleft. Once hydrolyzed by the esterase, the choline is held at a low concentration in the reservoir provided by the junctional folds and is later taken back up into cholinergic nerve endings by a high-affinity uptake mechanism (see below and Chapter 15).

Many other enzymatic pathways degrade transmitter substances within the neuron and in non-neural tissues but are not involved specifically in terminating synaptic transmission. They can be important for controlling the concentration of the transmitter within the neuron or for inactivating transmitters that have diffused from the synaptic cleft. Many are important clinically. They provide sites for drug action and opportunity for diagnosis. Monoamine oxidase inhibitors, for example, which block the degradation of amine transmitters within the cell, are currently used for the control of high blood pressure and for treating depression. Another example is the intracellular enzyme catechol-*O*-methyltransferase, which is important for degrading biogenic amines. It is found in the cytoplasm of most cells, including neurons, but is most prominent in liver and kidney. The concentrations of this enzyme's metabolites in body fluids provide a diagnostic indication of the efficacy of drugs that affect the synthesis or degradation of the biogenic amines in nervous tissue.

Neuroactive peptides are removed more slowly than small-molecule transmitters from the synaptic cleft. The only mechanisms of peptide removal are probably diffusion and proteolysis by extracellular peptidases. The slow removal of peptides contributes to the long duration of their effects and makes their metabolism seem more like that of hormones.

Reuptake of transmitter substance is the most common mechanism for inactivation. High-affinity uptake mechanisms for released transmitter are mediated by transporter molecules in the membranes of nerve terminals and glial cells with binding constants of 25 μ*M* or less. High-affinity uptake mechanisms were first described for norepinephrine, dopamine, and serotonin. Similar uptake mechanisms for amino acid transmitters, glutamate, GABA, glycine, and choline (but not ACh) were found later. Specific neurons each have their own characteristic uptake mechanisms; as an example, noncholinergic neurons do not take up choline with high affinity. Certain powerful psychotropic drugs can block uptake processes (e.g., cocaine blocks the uptake of norepinephrine, and the tricyclic antidepressants block that of serotonin). The application of appropriate drugs to block uptake can prolong and enhance the action of the biogenic amines and GABA.

Transporter molecules have been cloned; most of these are carriers of small-molecule neurotransmitters and appear to belong to a superfamily. These membrane-spanning proteins thread through the cell membrane 12 times. Like most other membrane proteins, these molecules in eukaryotes are glycosylated at several sites in extracellular domains of the protein. The transporters that take up transmitters usually depend on the presence of Na^+, and operate using the energy of ATP. An example is a carrier for GABA (GABA-1) cloned from rat brain. From its inferred amino acid sequence, the GABA-1 transporter has a molecular weight of 6,700 and has four potential sites for glycosylation. Its hydrophobicity plot predicts a secondary structure with 12 membrane-spanning regions. When expressed in frog eggs, the GABA-1 transporter takes up GABA into the oocyte with a binding constant between 3 and 10 μ*M*. Because no other exogenous mRNA is needed in these experiments, it can be assumed that the cloned protein alone serves as the transporter, either as a single molecule or as multiple (identical) subunits. Uptake of GABA is dependent on both Na^+ and Cl^- ions.

An Overall View

The information carried by a neuron is encoded in electrical signals that travel along the axon and into the nerve terminal. At the synapse these signals are carried by one or more chemical messengers across the synaptic cleft. None of these chemical messengers carries unique information, as RNA and DNA do. Indeed, some have several functions within cells as metabolites in other biochemical pathways. For example, amino acids are polymerized into proteins; glutamate and GABA act as substrates in intermediary metabolism; and ATP is the principal means of transferring metabolic energy. To fulfill a signaling function these molecules must bind to membrane receptors. Once the molecules of transmitter are bound, the receptor generates electrical or metabolic signals in the postsynaptic cell. The co-release of several neuroactive substances from a presynaptic neuron onto appropriate postsynaptic receptors, permits an extraordinary diversity of information transfer at one synapse.

Chemical messengers are packaged in vesicles within the presynaptic neuron. Vesicles play different roles in the life cycles of the two major classes of chemical messengers—small-molecule transmitters and neuroactive peptides. Small-molecule transmitters, after synthesis in the cytoplasm, are taken up and concentrated in vesicles that protect them from intracellular degradative enzymes whose job is to maintain a constant level of the transmitter substance in the cytoplasm. Because the small-molecule transmitters are released in relatively large amounts at a fast rate, much of the small-molecule transmitter in the neuron must be synthesized at the terminals. Nerve endings thus contain a high concentration of these synaptic vesicles.

In contrast, the protein precursors of neuroactive peptides are synthesized only in the cell body. There they become packaged in secretory granules and synaptic vesicles that are transported from the cell body to terminals. Thus, unlike the vesicles that contain small-molecule transmitters, these vesicles are not refilled at the terminal.

It seems axiomatic that an understanding of the molecular strategy of chemical transmission begins with the identification of the contents of the

BOX 16–1 *Histochemical Detection of Chemical Messengers Within Neurons*

Powerful histochemical techniques are available for detecting the chemical messengers neurons use, including small-molecule transmitter substances and neuroactive peptides, in sections of nervous tissue. Specific histochemical and autoradiographic methods are used to localize the biogenic amines within neurons in these tissue sections and to show that vesicles contain transmitter.

Catecholamines and serotonin, when reacted with formaldehyde vapor, form fluorescent derivatives. Under properly controlled conditions the reaction can be used to locate transmitters with the fluorescence (light) microscope. Because individual vesicles are too small to be resolved by the light microscope, histofluorescence can only localize transmitters to particular regions of a nerve cell. The position of the vesicles can be inferred by comparing the distribution of fluorescence under the light microscope with the position of vesicles under the electron microscope.

Histochemical analysis can be extended to the ultrastructural level under special conditions: the fixation of nervous tissue can intensify the electron density of vesicles containing biogenic amines. Thus, fixation in the presence of potassium permanganate, chromate, or silver salts brings out the large number of dense-core vesicles that are characteristic of aminergic neurons.

It is also possible to identify neurons in which the gene for a particular transmitter enzyme or peptide precursor is expressed. Many methods for detecting specific mRNAs use the phenomenon of nucleic acid *hybridization*, which works on the principle that two single strands of a nucleic acid polymer will pair if their sequences of bases are complementary. A particularly elegant method is *in situ* hybridization (Figure 16–2), in which the strand of noncoding DNA (the negative or antisense strand or its corresponding RNA) is applied to tissue sections under conditions suitable for hybridizing with endogenous (sense) mRNA. If the probes are labeled with a radioactive substance, autoradiography reveals the locations of neurons that contain the complex formed between the labeled complementary nucleic acid strand and the mRNA. The hybrid can be localized immunocytochemically with greater sensitivity and more precision than with autoradiography, using probes labeled with immunoreactive base analogs.

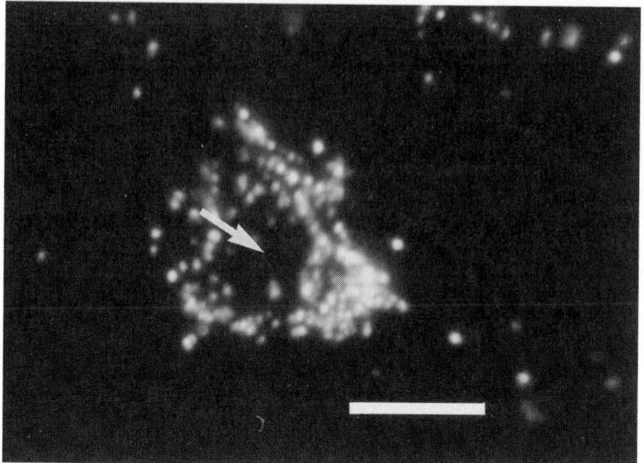

Figure 16–2 *In situ* hybridization of a region of the rat hypothalamus reveals cRNAs encoding pro-opiomelanocortin, which is the precursor of several important hormones. Because the mRNA is translated in the endoplasmic reticulum, the silver grains over neurons have collected predominantly in the cytoplasm. Silver grains are scarcer over the nucleus (**arrow**). The section was photographed in a microscope equipped with polarized light epiluminescence. Bar = 25 μm. (From Fremeau et al., 1989.)

Transmitter substances can also be localized directly to vesicles by electron microscopic autoradiography and by immunocytochemistry (Figure 16–3). Amino acid transmitters and biogenic amines can be successfully located by autoradiography because they have a primary amino group that permits their covalent fixation in place within the neuron; this group becomes crosslinked to proteins by aldehydes, the usual fixatives used in microscopy. For immunohistochemical localization, specific antibodies to the transmitter substance are necessary. Specific antibodies have been raised to serotonin, to histamine, and to many neuroactive peptides. These transmitter-specific antibodies, in turn, can be detected by a second antibody (in a technique called indirect immunofluorescence). As

Figure 16–3 A frozen section labeled with an antibody against histamine shows the distribution of histamine-containing neurons in the abdominal ganglion of *Aplysia*. The immunostained cluster (C) is a large group of small cells that participate in controlling respiration. The cell bodies surrounding this cluster are not immunoreactive. Nearby is the cell body of R2, one of the largest nerve cells in the animal kingdom. The nerve (N) contains some processes that are immunoreactive. Bag cells, which synthesize the precursor of the egg-laying hormone (ELH, see Figure 16–1), lie outside the field of this micrograph and are not immunoreactive. (From Elste et al., 1990.)

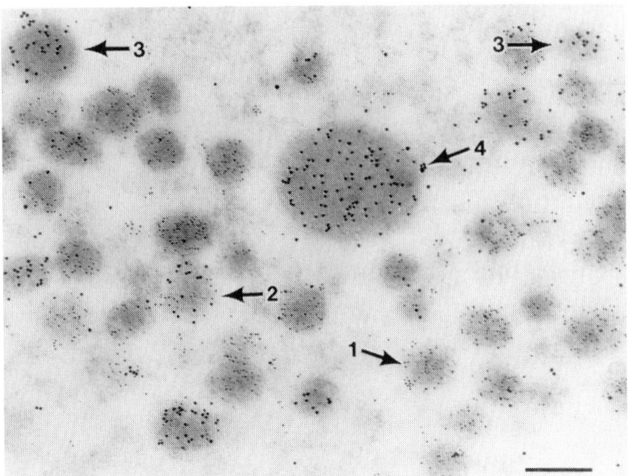

Figure 16–4 An electron micrograph of a section through an *Aplysia* bag cell body, treated with two antibodies against different regions of the prohormone, illustrates the use of immunogold to locate two antigens in a single electron microscopic tissue section. The bag cells, which control reproductive behavior by releasing a group of neuropeptides cleaved from the egg-laying hormone (ELH, see Figure 16–1), contain several kinds of dense-core vesicles. One of these antibodies was raised in rabbits and the other in rats. These antibodies were detected with anti-rabbit or anti-rat immunoglobulins (secondary antibodies) raised in goats. Each secondary antibody was coupled to colloidal gold particles of a distinct size. The four physiologically active peptides cleaved from ELH (see Figure 16–1) are located in different vesicles, numbered 1 through 4. Bar = 240 nm. (From Fisher et al., 1988.)

an example, if the first antibody is rabbit antihistamine, the second antibody can be a goat antibody raised against rabbit immunoglobulins. These antibodies are commercially available labeled with fluorescent dyes (fluorescein, rhodamine, and Texas red, for example). They can be used under the fluorescence microscope to locate antigens to regions of individual neurons—cell bodies, axons, and sometimes terminals.

Ultrastructural localization can be achieved by immunohistochemical techniques, usually involving a peroxidase–antiperoxidase system. Another method is to use antibodies linked to gold particles, which are electron dense (Figure 16–4). Spheres of colloidal gold can be generated with precise diameters in the nanometer range and, because they are electron dense, can be seen in the electron microscope. This technique has the additional useful feature that more than one specific antibody can be used to examine the same tissue section, if each of the antibodies is linked to gold particles of a different size.

synaptic vesicle: only if a molecule can be released does it have the potential of activating a receptor. However, not all molecules released by a neuron are chemical messengers; only those capable of binding to appropriate receptors can serve as transmitters.

While certain aspects of vesicle function vary considerably among different types of neurons, one aspect is shared by most neurons. With the exception of neurons that release transmitters by diffusion or by carrier mechanisms, vesicles release chemical messenger through exocytosis.

Selected Readings

Amara, S. G., and Kuhar, M. J. 1993. Neurotransmitter transporters: Recent progress. Annu. Rev. Neurosci. 16:73–93.

Cooper, J. R., Bloom, F. E., and Roth, R. H. 1991. The Biochemical Basis of Neuropharmacology, 6th ed. New York: Oxford University Press.

Koob, G. F., Sandman, C. A., and Strand, F. L. (eds.). 1990. A Decade of Neuropeptides: Past, Present and Future. Ann. N.Y. Acad. Sci. 579:1–281.

Kupfermann, I. 1991. Functional studies of cotransmission. Physiol. Rev. 71:683–732.

Lee, T. N. H. (ed.) 1991. Molecular Approaches to Drug Abuse Research, Vol. I: Receptor Cloning, Neurotransmitter Expression and Molecular Genetics. Research Monograph 111. Rockville, Md.: National Institute on Drug Abuse.

Siegel, G. J., Agranoff, B. W., Albers, R. W., and Molinoff, P. B. (eds.) 1993. Basic Neurochemistry: Molecular, Cellular, and Medical Aspects, 5th ed. New York: Raven Press.

Sossin, W. S., Fisher, J. M., and Scheller, R. H. 1989. Cellular and molecular biology of neuropeptide processing and packaging. Neuron 2:1407–1417.

Steiner, D. F., Smeekens, S. P., Ohagi, S., and Chan, S. J. 1992. The new enzymology of precursor processing endoproteases. J. Biol. Cham. 267:23435–23438.

Weisskopf, M. G., Zalutsky, R. A., and Nicoll, R. A. 1993. The opioid peptide dynorphin mediates heterosynaptic depression of hippocampal mossy fibre synapses and modulates long-term potentiation. Nature 362:423–427.

17

A Clinical Example: Myasthenia Gravis

Myasthenia Gravis Affects Synaptic Transmission at the Nerve–Muscle Synapse

Physiological Studies Show a Disorder of Neuromuscular Transmission

Immunological Studies Indicate that Myasthenia Is an Autoimmune Disease

Antibodies to the Acetylcholine Receptor Cause the Physiological Abnormality

The Basis of Antibody Binding in Myasthenia Gravis Has Been Defined

Not All Myasthenic Patients Produce Antibodies to the Acetylcholine Receptor

The Molecular Basis of the Autoimmune Reaction Has Been Defined

Myasthenia Gravis Is More Than One Disease

In the preceding chapters we examined the mechanisms of chemical synaptic transmission—the synthesis and release of transmitters by neurons and the activation of postsynaptic receptors by the transmitter. Many human diseases can disrupt chemical transmission between neurons and their target cells. Analysis of these abnormalities in synaptic transmission associated with human disease has provided some critical insights into the mechanisms underlying normal synaptic function. The most prevalent disease that affects synaptic transmission, and the most thoroughly studied, is myasthenia gravis—a disorder of function at the synapse between motor neurons and skeletal muscle.

Myasthenia Gravis Affects Synaptic Transmission at the Nerve–Muscle Synapse

Myasthenia gravis (literally, severe muscle weakness) is an autoimmune disorder in which the individual produces antibodies against the nicotinic acetylcholine (ACh) receptor. These antibodies interfere with synaptic transmission by reducing the number of functional receptors or by impeding the interaction of ACh with its receptors. Because ACh is the sole transmitter at the neuromuscular junction, the skeletal muscle becomes weakened.

The first well-documented example of myasthenia gravis was reported in 1877 by Samuel Wilks. By 1900 neurologists had described the important clinical characteristics of the disease. At that time, however, diseases were still defined primarily in terms of lesions observed by microscopy at post-mortem examination rather than in terms of physiological or etiological factors. Since the brain, spinal cord, peripheral nerves, and muscles all appeared normal in myasthenic patients at autopsy, the disease was considered a functional disorder rather than a structural defect.

Physiological Studies Show a Disorder of Neuromuscular Transmission

Two discoveries in the mid-1930s helped to identify the functional defect in myasthenia as a disturbance of neuromuscular transmission. First, synaptic transmission at the neuromuscular junction is mediated by a chemical transmitter, ACh. Second, the symptoms of myasthenia gravis are reversed by inhibitors of acetylcholinesterase, the enzyme that degrades ACh in the synaptic cleft (Figure 17–1B).

A

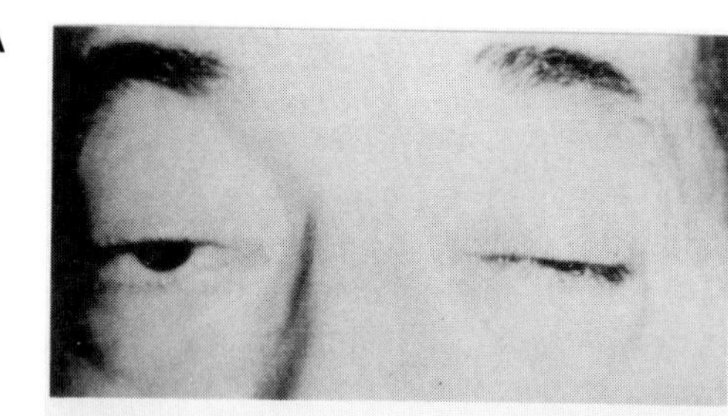

B

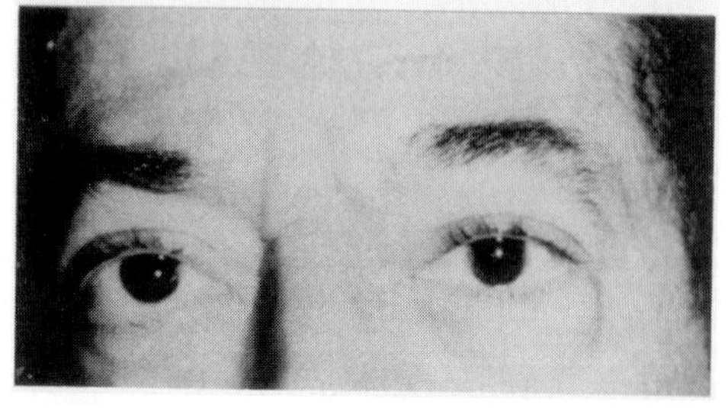

Figure 17–1 Myasthenia gravis typically affects the cranial muscles. (From Rowland, Hoefer, and Aranow, 1960.)

A. Severe drooping of the eyelids, or ptosis, is characteristic of myasthenia gravis. This patient also could not move his eyes to look to either side.

B. One minute after an intravenous injection of 10 mg edrophonium, an inhibitor of acetylcholinesterase, the patient has both eyes open and can move them freely.

In the years between 1945 and 1960, A. McGhee Harvey and his colleagues described in detail the physiological basis of the disorder. When a motor nerve is stimulated electrically, the total electrical activity of a population of muscle fibers (known as the *compound action potential*) can be measured with surface electrodes. In normal human muscle the amplitude of an evoked compound action potential remains constant at stimulation rates of 2 to 5 per second. In myasthenia gravis, however, the amplitude of evoked compound action potentials decreases rapidly. This abnormality resembles the pattern induced in normal muscle by the toxin *d*-tubocurarine (curare), which blocks ACh receptors and inhibits the action of ACh at the neuromuscular junction. The decrease in amplitude of evoked compound action potentials in myasthenic patients can be reversed by administration of neostigmine, which inhibits acetylcholinesterase and thereby increases the duration of action of ACh at the neuromuscular junction (Figure 17–2).

The reduced efficacy of neuromuscular transmission in myasthenia can be assessed by the clinical technique of single-fiber electromyography, which measures the intervals between discharges of different muscle fibers innervated by the same motor neuron. The normal variation in intervals is called *jitter*. The extent of jitter depends on the velocity of conduction in nerve terminals, transmitter release, and activation of the postsynaptic membrane. Although jitter may increase in other neurogenic diseases, it is especially pronounced in myasthenia gravis.

Immunological Studies Indicate that Myasthenia Is an Autoimmune Disease

Soon after the clinical syndrome had been identified, it was recognized that about 15% of adult patients with myasthenia had a benign tumor of the thymus (thymoma). In 1939 Alfred Blalock first reported that the symptoms in myasthenic patients were improved by removal of the thymoma. In the 1950s, on the basis of this finding, Blalock and Harvey found that removing the thymus altogether also resulted in a reduction in symptoms of myasthenia gravis, and this procedure has now become standard therapy. At that time it was not clear why these tumors were associated with myasthenia or why thymectomy was beneficial, because the immunological role of the thymus had not yet

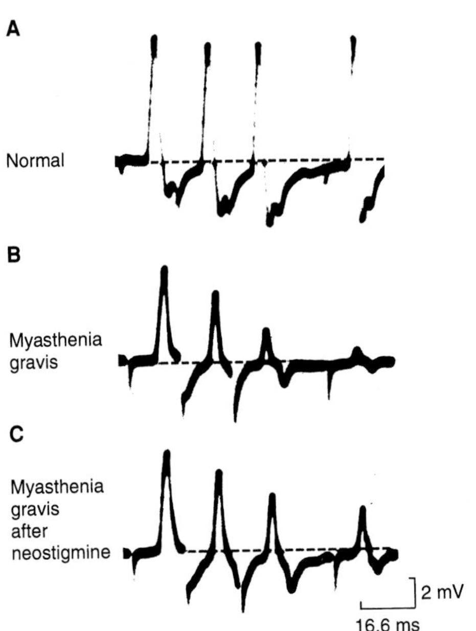

Figure 17–2 Neostigmine increases the duration of action of ACh and thus can compensate for the reduced ACh activity in myasthenia. (From Harvey, Lilienthal, and Talbot, 1941.)

A. In a normal person the amplitude of action potentials evoked by a train of four stimuli at 16.6 ms intervals remains constant.

B. In the myasthenic patient there is a rapid decrease in amplitude.

C. After injection of 2 mg neostigmine into the brachial artery of the myasthenic patient, the decrease in amplitude is partially reversed. (Calibration: 2.0 mV.)

been established. The neurologist Ian Simpson was one of the first to suggest that myasthenia was an immunological disorder, because it frequently occurs in patients with other diseases that are thought to have an autoimmune basis, such as rheumatoid arthritis.

The modern concept of myasthenia emerged with the isolation and characterization of the nicotinic ACh receptor (see Chapter 12). The breakthrough came in 1966 with the isolation of toxins from snake venoms. One of these toxins, α-bungarotoxin, was found to cause paralysis by binding essentially irreversibly to ACh receptors not the end-plates of muscle fibers. In 1971 α-bungarotoxin was used to isolate and purify ACh receptors from the electric organ of the electric eel.

Shortly thereafter radioactive α-bungarotoxin was used to label the ACh receptors in human end-plates. Fewer binding sites were observed in myasthenic muscle than in controls (Figure 17–3). In another study at the same time, ACh receptors purified from the eel electric organ (which is related to the skeletal muscles of higher vertebrates) were injected into rabbits for the purpose of producing antibodies to study the properties of ACh receptors in the eel. Unexpectedly, however, the rabbits' production of antibodies was accompanied by the onset of myasthenia-like symptoms. The characteristic muscle weakness was reversed by cholinesterase inhibitors (neostigmine or edrophonium). As with human myasthenia gravis, the animals were abnormally sensitive to neuromuscular blocking agents, such as curare; moreover, the evoked compound action potentials in the rabbits decreased with repetitive stimulation. All these clinical signs pointed to the rabbits' immune response to the injection as the agent of a myasthenia-like illness in these animals. Experimental autoimmune myasthenia gravis was induced in mice, rats, and monkeys by injecting ACh receptors, suggesting that the structure of ACh receptors is conserved

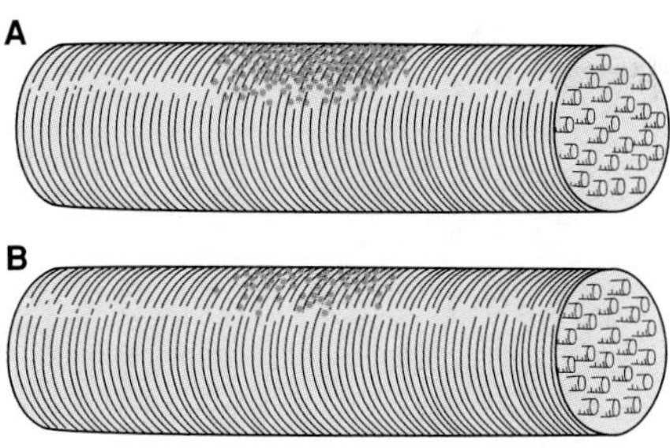

Figure 17–3 Myasthenia gravis reduces the density of ACh receptors in human muscle fibers. ACh receptors are marked with ^{125}I-labeled α-bungarotoxin and detected in autoradiograms (drawn here). (Adapted from Fambrough, Drachman, and Satyamurti, 1973.)

A. In normal fibers there is a dense accumulation of silver grains in the end-plate and a paucity of grains outside this region.

B. In myasthenic fiber the grains are also localized in the end-plate region, but the number per unit area is markedly reduced, indicating a reduced density of functional receptor sites.

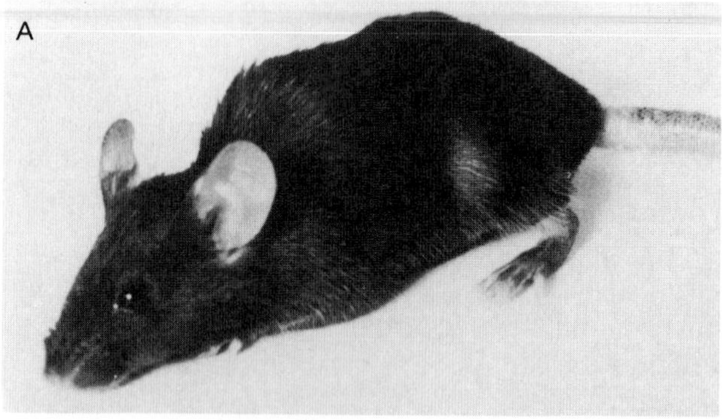

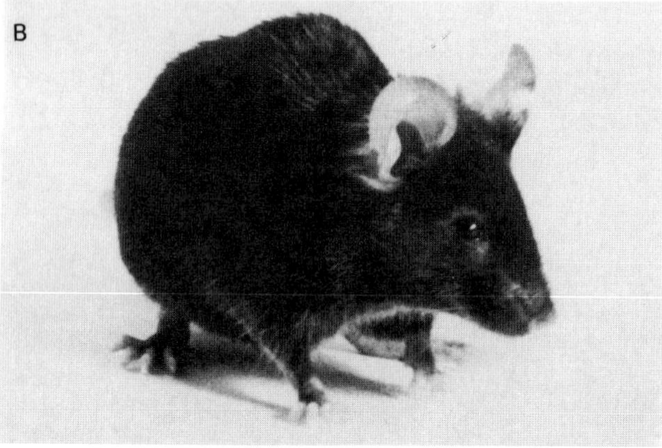

Figure 17–4 Posture of a myasthenic mouse before and after treatment with neostigmine. To produce the syndrome, the mouse received an injection of 15 μg ACh receptors from *Torpedo californica* and another injection 45 days later. (From Berman and Patrick, 1980.)

A. Before treatment the mouse is inactive.

B. Twelve minutes after receiving an intraperitoneal injection of 37.5 μg/kg neostigmine bromide, the mouse is standing.

across a wide range of mammalian species (Figure 17–4).

By 1975 all the essential characteristics of the human disease had been reproduced in experimentally induced myasthenia gravis. These characteristics included reduction in the amplitude of the miniature end-plate potentials, smoothing of the normal convoluted appearance of the postjunctional folds, loss of ACh receptors, and the deposition at postsynaptic sites of antibody and of complement (a serum protein that participates in antibody-mediated cell breakdown).

Once the characteristics of experimental myasthenia gravis had been defined, researchers were able to identify the antibodies directed against ACh receptors in the serum of patients with myasthenia. Cultured B lymphocytes from patients with myasthenia also produced antibodies to ACh receptors. The idea that the human antibodies actually cause the symptoms of myasthenia was supported by other observations as well. For instance, repeated injections with serum from myasthenic patients cause mice to develop the same electrophysiological abnormalities as the human patients by reducing the number of ACh receptors in skeletal muscle.

Further support for the role of antibodies against ACh receptors was provided by the detection of antibodies in infants with neonatal myasthenia. These children of myasthenic mothers find it difficult to swallow and have impaired limb movements. The syndrome lasts from 7 to 10 days and, as the symptoms abate, the level of antibodies declines. Similarly, draining lymph from the thoracic lymph ducts reduces myasthenia symptoms in adults. The symptoms recur if the same lymph fluid is returned to the patient, but not if lymphocytes are replaced. Furthermore, symptoms improve and antibody levels decline when patients undergo *plasmapheresis*, a procedure that entails removing the blood from a patient, separating the cells from plasma, and returning only the cells to the patient (the plasma, which contains the antibodies, is discarded). In another treatment, the antibodies are disabled by intravenous administration of normal immunoglobulin ("IVIG therapy").

Antibodies to the Acetylcholine Receptor Cause the Physiological Abnormality

How do the immunological observations that we have just considered account for the characteristic decrease in the response of myasthenic muscle to repetitive stimulation?

Normally, an action potential in a motor axon releases enough ACh from synaptic vesicles to induce an excitatory end-plate potential with an amplitude of about 70 to 80 mV (see Chapter 12). Since the threshold for generation of an action potential in the end-plate is about –45 mV, the normal end-plate potential is greater than the threshold needed to initiate an action potential. Thus, in normal muscle the difference between the

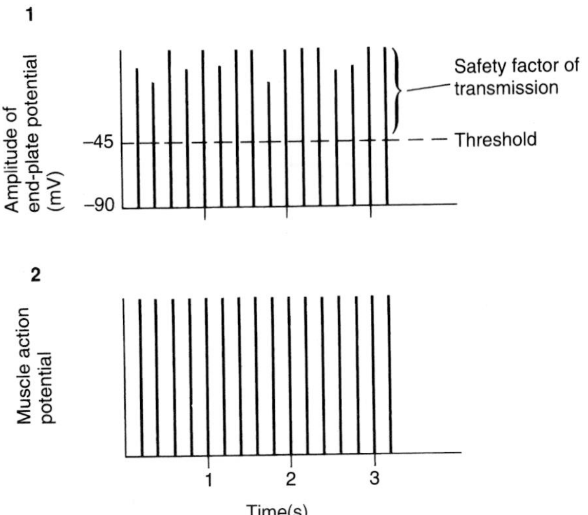

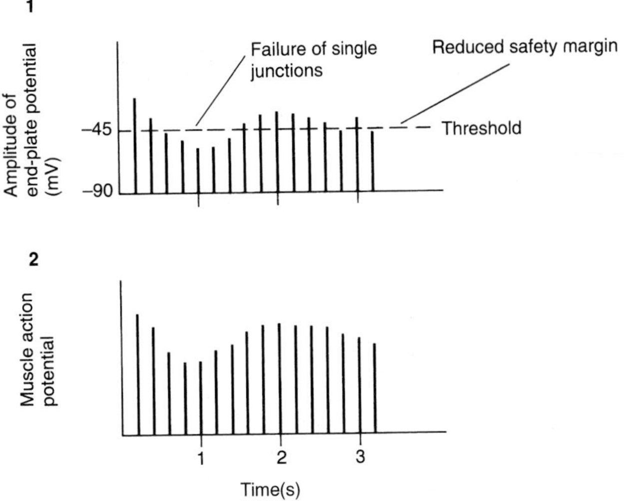

Figure 17–5 Failure of transmission at the neuromuscular junction in myasthenia gravis. (From Lisak and Barchi, 1982.)

A. In the normal neuromuscular junction the amplitude of the end-plate potential is so large that all fluctuations in the efficiency of transmitter release occur well above the threshold for a muscle action potential (**1**). Therefore, the amplitude of a compound muscle action potential during repetitive stimulation is constant and invariant (**2**).

B. In the myasthenic neuromuscular junction, postsynaptic changes reduce the amplitude of the end-plate potential in response to presynaptic release of a given amount of ACh, so that in optimal circumstances the end-plate potential may be just sufficient to produce a muscle action potential. Fluctuations in transmitter release that normally accompany repeated stimulation now cause the end-plate potential to drop below this threshold, leading to conduction failure at that junction (**1**). When the action potential is recorded from the surface of a myasthenic muscle, the amplitude of the compound action potential—a measure of contributions from all fibers in which synaptic transmission is successful—shows a progressive decline and only a small and variable recovery (**2**).

threshold and the actual end-plate potential—the *safety factor*—is quite large (Figure 17–5A). In fact, in many muscles the amount of ACh released during synaptic transmission can be reduced to as little as 25% of normal before it fails to initiate an action potential.

Most of the ACh released into the synaptic cleft by an action potential is rapidly hydrolyzed by acetylcholinesterase. When the number of ACh receptors is reduced, as it is in myasthenia, a molecule of ACh is more likely to be hydrolyzed before it finds a receptor. Moreover, the geometry of the end-plate is also disturbed in myasthenia. The normal infolding is reduced, and the synaptic cleft is enlarged (Figure 17–6). These morphological changes increase the diffusion of ACh away from the synaptic cleft and thus further reduce the probability that ACh will interact with the few remaining functional receptors. As a result, the amplitude of the end-plate potential is reduced to the point where it is barely above threshold (Figure 17–5B).

Thus transmission is readily blocked, even though the vesicles in the presynaptic terminals contain normal amounts of ACh and the processes of transmitter release are intact. Drugs that inhibit acetylcholinesterase partially reverse both the physiological abnormality (the decremental response) and the clinical symptoms (muscle weakness), because the longer the released ACh molecules can remain in the synaptic cleft without being hydrolyzed, the greater the probability they will interact with receptors.

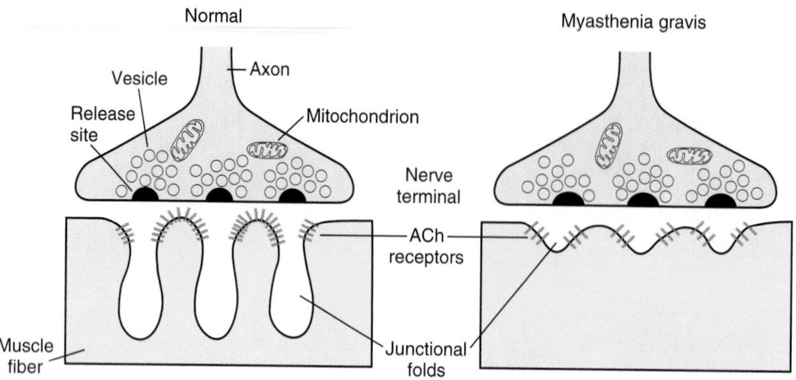

Figure 17–6 In myasthenia morphological changes in the neuromuscular junction reduce the likelihood of synaptic transmission. The myasthenic junction has a normal nerve terminal but the number of ACh receptors is reduced, the junctional folds are sparse and shallow, and the synaptic space is widened. (Adapted from Drachman, 1983.)

The Basis of Antibody Binding in Myasthenia Gravis Has Been Defined

As discussed in Chapter 12, the genes for each of the subunits of mammalian ACh receptor have now been cloned and sequenced, and peptides corresponding to specific domains of ACh receptor subunits have been synthesized. In experimental animals, antibodies that cause myasthenia are usually active against one of two peptide sequences on the native receptor—either the bungarotoxin binding site or an area on the α subunit called the *main immunogenic region.* Circulating antibodies in humans are often directed against the main immunogenic region.

Even though it has been well established that antibodies to the α subunit of ACh receptors play a central role in the pathogenesis of myasthenia—so much so that myasthenia is now the prototype of human autoimmune disease—several questions remain unanswered. What, for example, initiates the production of antibodies to the ACh receptor? One possibility is that persistent viral infection could alter the properties of the surface membrane, rendering it immunogenic, but this has not been shown. Another possibility is that viral or bacterial antigens may share *epitopes*—that is, certain antigenic features—with the ACh receptor. Thus, when a person is infected, the antibodies generated against the foreign organism may also attack the ACh receptor.

How do antibodies cause the symptoms of myasthenia? The antibodies do not occupy the receptor site; this much is known from the test used to detect antireceptor antibodies in human serum. The circulating antibodies react with purified ACh receptors that have been labeled by radioactive α-bungarotoxin. Because the toxin itself occupies and blocks the agonist site, the antibody must react with epitopes elsewhere on the receptor molecule.

One way that the antibodies might work is by hindering the interaction of ACh and the receptor. The loss of receptors is, however, probably due to an increase in turnover and degradation. Myasthenic antibodies are able to bind and crosslink ACh receptors, in this way triggering the breakdown and recycling of the receptors by means of endocytosis (Figure 17–7). In addition, some antibodies to ACh receptors in myasthenic patients bind proteins of the complement cascade, which may result in lysis of the postsynaptic membrane.

Not All Myasthenic Patients Produce Antibodies to the Acetylcholine Receptor

Although the evidence implicating ACh receptor antibodies in myasthenic symptoms is compelling, the antibodies are not found in all myasthenic patients. To make the picture more confusing, there is no consistent relationship between the

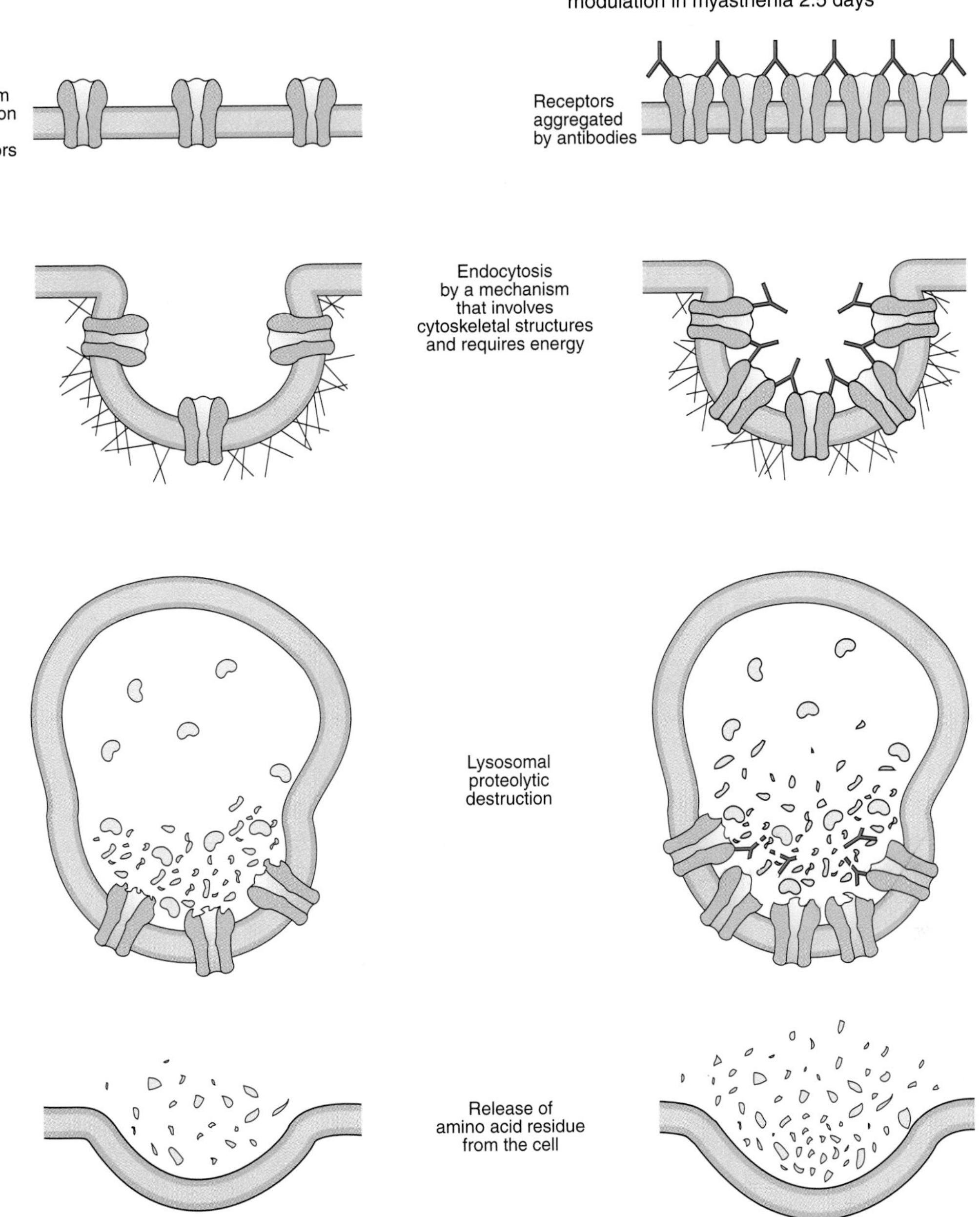

Figure 17–7 The normal rate of turnover of ACh receptors is increased in myasthenia. The degradation of the receptor is schematically illustrated as occurring in consecutive steps. (Adapted from Lindstrom, 1983, and Drachman, 1983.)

A. Normal turnover of randomly spaced ACh receptors takes place every 5 to 7 days.

B. In myasthenia gravis and in experimental myasthenia gravis the crosslinking of ACh receptors by the antibody facilitates the normal endocytosis and digestion of the receptors, which leads to a two- to threefold increase in the turnover rate. The binding of antireceptor antibody activates the complement cascade, which is involved in focal lysis, or breakdown of the postsynaptic membrane. This lysis is probably primarily responsible for the characteristic alterations of postsynaptic membrane morphology observed in myasthenia (see Figure 17–6).

concentration of antibodies directed against ACh receptors and the severity of symptoms.

One explanation of this dissociation is that the antibodies found in the serum of myasthenic patients or in animals with experimentally induced myasthenia gravis are polyclonal; they are produced by different B cells in response to different antigenic determinants, and therefore the serum of each patient contains antibodies with distinct specificities. As a consequence, some people with high levels of antibodies to the receptor but few or no clinical symptoms might have a type of antibody that is limited in its ability to interfere with synaptic transmission or to influence ACh receptor turnover. In contrast, other patients with severe myasthenia might have low levels of antibodies that are more effective in interfering with the function of the receptor and its turnover.

The Molecular Basis of the Autoimmune Reaction Has Been Defined

The autoimmune reaction depends on the interaction of three molecules: (1) the antigen, a peptide segment of the ACh receptor, or a peptide that mimics the receptor, which is antigenic; (2) an antigen-specific T-cell receptor; and (3) class II molecules of the major histocompatibility complex (MHC) that are expressed on the antigen-presenting cell. It is the T cells that become reactive against the ACh receptor. This could come as the result of an infection in which a viral protein happens to include a peptide similar to one in the ACh receptor—a form of "molecular mimicry." Once activated, the T cells would recognize the ACh receptor on myoid cells in the thymus. Antigen-specific T cells have actually been identified in the thymus glands of patients with myasthenia.

Myasthenia Gravis Is More Than One Disease

The modern analysis of myasthenia has lent support to the idea that myasthenia gravis is more than one disease. This had been suspected earlier but was difficult to prove. For instance, it had long been recognized that congenital myasthenia (symptoms present from birth in children whose mothers do not have myasthenia) is often hereditary. Now it seems that patients with congenital myasthenia do not have the expected antibodies to ACh receptors. Therefore, there may be two distinct categories of myasthenia: an acquired autoimmune form in older children and adults (with ACh receptor antibodies) and a nonimmune, heritable, congenital myasthenia (without ACh receptor antibodies). Even among adults, antibodies are much more likely to be found in patients with generalized myasthenia than among those who are affected only in the eye muscles.

Biophysical and immunocytochemical studies of congenital myasthenia also indicate that myasthenia is a heterogeneous syndrome. Some cases seem to be due to an abnormality in the presynaptic terminals, impairing release of ACh from the terminals. Others are apparently due to postsynaptic disorders, including congenital lack of acetylcholinesterase, impaired capacity of ACh receptors to respond to ACh, or abnormally low numbers of ACh receptors. We here consider two illustrative examples: loss of acetylcholinesterase and the slow-channel syndrome.

Lack of acetylcholinesterase presents the following characteristics: There is a decrease in response to repetitive stimulation at 2 Hz, as in myasthenia gravis, but the muscle responds repetitively to a single stimulus, a feature not seen in other conditions. End-plate potentials and miniature end-plate potentials are not small, as in myasthenia gravis, but are markedly prolonged, a feature that could explain the repetitive response to a single stimulus. Cytochemical studies reveal that the enzyme acetylcholinesterase is absent from the postsynaptic membranes. In contrast, the ACh receptors are preserved.

The slow-channel syndrome is characterized by prominent limb weakness but little weakness of cranial muscles (just the reverse of the pattern usually seen in myasthenia gravis, where muscles of the eyes, speech, and swallowing are almost always affected). The end-plate potentials in the slow-channel syndrome are prolonged in a manner similar to that observed with acetylcholinesterase deficiency, and spontaneous miniature end-plate potentials are also prolonged. However, acetylcholinesterase is present and shows normal enzyme kinetics. These features suggest that the opening of the ACh receptor–channel is abnormally prolonged. In addition, the abnormally low amplitude of the miniature end-plate potentials suggests a degeneration of the junctional folds in the muscle membrane and a loss of ACh receptors.

The etiology of the slow-channel syndrome is still uncertain. However, the ACh receptor–channel is known to be slow in newly formed end-plates. It is therefore possible that the syndrome arises when, for whatever reason, the developmental transition from slow to fast channels is prevented. It is also possible that a mutation has altered the ACh receptor in a way that modifies the amount of time the channel spends in the open state. Discriminating between the several different kinds of myasthenia—congenital and adult varieties, autoimmune and those due to other mechanisms, familial and acquired—remains a major challenge.

An Overall View

Myasthenia gravis is a neuromuscular disability caused by a reduced number of ACh receptors at the nerve–muscle synapse. It is improved by drugs that inhibit cholinesterase and thereby prolong the action of the transmitter, acetylcholine. In another neuromuscular disorder, facilitating neuromuscular block, the amount of transmitter released is reduced because of a loss of Ca^{2+} channels.

In principle, these findings suggest a strategy for treating diseases of synaptic function. First, the origin of the disorder in either the presynaptic neuron (a disease of transmitter release) or the postsynaptic neuron (a disease of the receptor) must be determined. Once the cause has been identified, the most effective treatment is likely to be one that corrects the affected step in transmission or eliminates the pathogenic agent. This insight emphasizes the importance of a theoretical understanding of synaptic transmission for analyzing and treating neurological diseases.

The history of work on myasthenia gravis also illustrates that progress in our understanding of neurological diseases often depends on the interplay of clinical and basic research. For example, it was first observed clinically that thymectomy is therapeutic; only later was the physiological evidence of the immunological role of the thymus discovered. Likewise, clinical evidence associating myasthenia with rheumatoid arthritis and other diseases of autoimmunity identified the disease as autoimmune. Similarly, the clinical observation that neostigmine is an effective treatment was the clue that established myasthenia as a disease of neuromuscular transmission, because neostigmine is an inhibitor of acetylcholinesterase.

Selected Readings

Drachman, D. B. 1994. Myasthenia gravis. N. Engl. J. Med. 330:1797–1810.

Engel, A. G. 1988. Congenital myasthenic syndromes. J. Child Neurol. 3:233–246.

Lindstrom, J. 1983. Using monoclonal antibodies to study acetylcholine receptors and myasthenia gravis. Neurosci. Comment. 1:139–156.

Lisak, R. P. (ed.) 1994. Handbook of Myasthenia Gravis. New York: Marcel Dekker.

Lisak, R. P., and Barchi, R. L. 1982. Myasthenia Gravis and Myasthenic Syndromes. Philadelphia: Saunders.

Numa, S. 1989. Molecular structure and function of acetylcholine receptors and sodium channel. In S. Chien (ed.), Molecular Biology in Physiology. New York: Raven Press, pp. 93–118.

Pachner, A. R. 1988. Myasthenia gravis. Immunol. Allerg. Clin. North Am. 8:277–293.

Penn, A. S., Richman, D. P., Ruff, R. L., and Lennon, V. A. (eds.) 1993. Myasthenia gravis and related disorders: Experimental and clinical aspects. Ann. NY Acad. Sci. 681.

Rowland, L. P. 1980. Controversies about the treatment of myasthenia gravis. J. Neurol. Neurosurg. Psychiatry 43:644–659.

Steinman, L., and Mantegazza, R. 1990. Prospects for specific immunotherapy in myasthenia gravis. FASEB J. 4:2726–2731.

Swift, T. R. 1981. Disorders of neuromuscular transmission other than myasthenia gravis. Muscle Nerve 4:334–353.

Toyka, K. V., Drachman, D. B., Pestronk, A., and Kao, I. 1975. Myasthenia gravis: Passive transfer from man to mouse. Science 190:397–399.

Vincent, A., Pinching, A. J., and Newsom-Davis, J. 1977. Circulating anti-acetylcholine receptor antibody in myasthenia gravis treated by plasma exchange. Neurology 27:364.

Walker, M. B. 1934. Treatment of myasthenia gravis with physostigmine. Lancet 1:1200–1201.

Wilks, S. 1883. Lectures on Diseases of the Nervous System Delivered at Guy's Hospital, 2nd ed. Philadelphia: P. Blakiston, Son & Co.

V

COGNITIVE NEURAL SCIENCE

About the image on the preceding page:

When human subjects must change the direction of an intended movement, the time taken to initiate this new movement (reaction time) increases with how much the new direction differs from the original one. It has been proposed that the increased reaction time needed to reprogram movements in a new direction corresponds to a progressive shift in cortical representation—or mental rotation—toward a new direction. This idea was corroborated in the experiment shown here by recording from a population of neurons in the motor cortex of a live monkey trained either to aim straight at a visual target (in one of eight randomized directions) or at 90° counterclockwise to it.

Representations of intended movements are encoded as a "population vector" in directionally selective activity of different populations of neurons in the monkey's cortex. When the monkey aims directly at a target, a particular population becomes active before and during the movement. When the monkey must change its mind and aim its movement at 90° to the original target, the populations of neurons activated during the reaction time (which, as expected, is prolonged) changes gradually. First, the population vector points toward the original direction, then turns progressively counterclockwise until the required 90° rotation has been achieved. The population vectors sampled at successive times are seen as a series of green lines (the orange lines are the projection of the population vectors onto a horizontal plane in the figure; the upward white line represents time). The progressive rotation of the neuronal population signal observed supports the idea that the planned change in direction is mentally represented in the cortex. (From Georgopoulos et al., *Science,* Vol. 243, January 13, 1989)

Introduction

> It has been said that beauty is in the eye of the beholder. As a hypothesis . . . this statement points clearly enough to the central problem of cognition . . . : the world of experience is produced by the man who experiences it. . . . There certainly is a real world of trees and people and cars and even books, and it has a great deal to do with our experience of these objects. However, we have no direct immediate access to the world, nor to any of its properties. . . . Whatever we know about reality has been *mediated* not only by the organs of sense but by complex systems which interpret and reinterpret sensory information. . . .
>
> The term "cognition" refers to all the processes by which the sensory input is transformed, reduced, elaborated, stored, recovered and used.
>
> ULRIC NEISSER, 1967

In the first part of this book we focused our attention on individual nerve cells and how they communicate at synapses to give rise to simple reflex behaviors. We now turn to systems of neurons to examine large interconnected networks of neurons. The transformation of neural signals in these networks gives rise to mental activity—to perception, planned action, and thought. To explain how these networks produce cognitive functions of the brain is clearly one of the ultimate challenges of science.

How is sensory information perceived? How are perceptions assembled into inner representations and formulated into plans for immediate behaviors or into concepts for future actions? How are memories formed, and how are percepts, concepts, and feelings transformed into language?

These questions have been asked at least since the beginning of recorded history. Why then do we need a new field called cognitive neural science? Why not simply continue the independent study of higher neural systems, of psychology, linguistics, information theory, or even philosophy, all of which contribute independently to the study of behavior? As we learned in Chapters 1 and 2, neural science, the modern *science of the brain,* became a more powerful science twenty years ago as a result of the convergence on problems of brain function of several previously independent subdisciplines: neuroanatomy, electrophysiology, cell biology, and developmental biology. Cognitive neural science represents a similar synthesis, an attempt to create a modern *science of the mind* by merging cognitive psychology with neural science.

The aim of cognitive neural science is to examine, in cell biological terms, all the classical, philosophical, and psychological questions about mental functions. This is a bold undertaking. How do we begin to think about perception, ideas, and feelings in biological terms?

Progress so far in understanding the major functional systems of the brain—the sensory, motor, motivational, and attentional systems—has benefited from a reductionist approach, an approach that uses the analytic power of cellular neurobiology to analyze the problems posed by cognitive psychology. This approach is based on the assumption that mental functions emerge from the biological properties of nerve cells and their interconnections. According to this view, the mind is essentially a set of functions carried out by the brain, an information processing organ made powerful by its complexity, by the enormous number, variety, and interactions of its basic units—its nerve cells.

The chapters that follow attempt to outline this cell biological agenda for cognitive neural science. We focus specifically on the major domains of congitive neural science: perception, action, emotion, language, learning, and memory.

18

From Nerve Cells to Cognition

The Major Goal of Cognitive Neural Science Is to Study Internal Representations of Mental Events

Cognitive Neural Science Relies on Five Main Approaches

The Brain Has an Orderly Representation of Personal Space

- *Reliable Cortical Maps of the Body Are the Basis of the Accuracy of Clinical Neurological Examinations*
- *The Cortex Has a Map of the Body for Each Submodality of Sensation*

The Internal Representation of Personal Space Is Modifiable by Experience

- *The Internal Representation of the Human Hand Area Can Be Modified*
- *Modifiability of the Internal Representation May Account for the Phantom Limb Syndrome*

The Internal Representation of Personal Space Can Be Studied at the Cellular Level: Each Central Neuron Has a Specific Receptive Field

- *Inputs to the Somatic Sensory Cortex Are Organized into Columns by Submodality*
- *In Later Stages of Cortical Processing the Submodalities Are Integrated by Cells that Have Complex Feature-Detecting Properties*

Real As Well As Imagined and Remembered Space Is Represented in Posterior Parietal Association Areas

- *Lesions in the Posterior Parietal Association Cortex Produce Defects in Attention*
- *Attention to a Stimulus Affects the Activity of Single Nerve Cells in the Posterior Parietal Cortex*

Considering that the brain is made up of a hundred billion nerve cells, it is remarkable how much we have learned about mental activity by examining the brain one cell at a time. Cellular studies have shown that neurons in different regions of the vertebrate nervous system, and indeed in all nervous systems, are similar. What distinguishes one brain region from another and one brain from the next are the number of neurons and the way they are interconnected.

Cellular studies of the sensory systems of the brain have provided insight into how information is received by various regions of the brain and transferred to others, and how these regions contribute to perception and planned action. For example, cellular analyses of information processing in the visual pathways have shown that the brain does not simply receive pictures of the external world but rather constructs visual images based on a highly selective integration of inputs from discrete parallel pathways. (The processing of visual information in the brain is the subject of Section VI.)

Different modes of interaction with the world—an object seen, a face touched, or a melody heard—are processed in parallel by different sensory systems. The receptors in each system first analyze and deconstruct stimulus information. Each sensory system then abstracts this information and represents it in the brain in specific pathways and brain regions. From moment to moment, this constant flux of information is edited into an apparently seamless continuum of unified perceptions. Thus, the appearance of our perceptions as *direct* and *precise* images of the world is an illusion.

The brain accomplishes these feats because its basic components, the nerve cells, are wired together in precise and orderly ways. Although the connections are precise in each brain, they are not exactly the same in all individuals. Connections between cells can be altered by activity and by learning. We remember events because the structure and function of the connections between nerve cells become modified by experience.

It is the conviction of neural scientists that a cellular approach is necessary for understanding how the brain works. But it is also their conviction that this approach is not sufficient. To understand how people think, behave, feel, act, and relate to one another, it is also essential to understand how events in individual cells lead to cognition. Toward this end the methods of cell biology need to be combined with techniques that relate the activities of interconnected populations of neurons to behavior. For example, imaging methods can be used to study the activities of populations of cells in animals and human beings during normal activities (see Chapter 5). The results of these studies can be explored quantitatively with computational models designed to test the inferences made from these observations. This combination of methods—cell biology, systems neural science, brain imaging, cognitive psychology, behavioral neurology, and computer science—has given rise to a concerted approach called *cognitive neural science*, designed to understand the neural mechanisms that give rise to behavior.

In this chapter we first outline some of the historical trends that led to the emergence of cognitive neural science. We then illustrate how the various methods of cognitive neural science can be combined to study information processing by the brain, using as an example the representation of personal and extrapersonal space, both real and imagined. In the next chapter we consider the role of association cortices in integrating sensory modalities. In subsequent sections of the book we take up in turn the five major subjects of cognitive neural science: perception, action, emotion, language, learning, and memory.

The Major Goal of Cognitive Neural Science Is to Study Internal Representations of Mental Events

Until the nineteenth century the academic study of mental activity was a branch of philosophy, and the chief method for understanding mental activity was introspection. By the middle of the nineteenth century this introspectionist tradition began to give way to empirical studies of the mind, which emerged as the independent discipline of experimental psychology. In its early years experimental psychology was primarily concerned with the study of sensation: the sequence of events by which a stimulus gives rise to a subjective response. By the end of the nineteenth century psychologists turned to analyzing subjective experiences themselves—learning, memory, attention, perception, and voluntary action. The development of simple experimental means for studying learning and memory—first in humans by Hermann Ebbinghaus in 1885, and a few years later in experimental animals by Ivan Pavlov and Edgar Thorndike—extended the quantitative approach of experimental psychology to higher mental processes.

This movement toward an objective psychology culminated in a rigorous empirical tradition called *behaviorism*. Behaviorists, led by J. B. Watson, argued that behavior could be studied with the same rigor as other natural sciences only if students of behavior focused on *observable* aspects of behavior and abandoned speculation about nonobservable mental activity.

Subsequently, experimental psychology came to be dominated by the behaviorists' exclusive concern with measuring observable responses to controlled stimuli. To behaviorists, all processes that intervene between the stimulus input and behavioral output were considered irrelevant to the scientific study of behavior. Thus, the behaviorists largely ignored the constructive brain processes that underlie perception, action, planning, thinking, attention, and complex forms of memory. In fact, during its most influential period in the 1950s many psychologists accepted the behaviorists' most radical position, that observable behavior is *all* there is to mental life. They defined all of behavior in terms of the limited techniques they used for studying it. This emphasis reduced the domain of experimental psychology to a restricted set of problems, and it excluded from study some of the most fascinating features of mental life.

It therefore was not difficult in the 1960s for the founders of cognitive psychology, Frederick Bartlett, Edwin Tolman, George Miller, Noam Chomsky, Ulric Neisser, Herbert Simon, and others, to convince the scientific community of the narrowness of behaviorism. The work of the early cognitive psychologists, as well as the earlier evidence from Gestalt psychology, psychoanalysis, and European neurology, indicates that perception *shapes* behavior and that perception itself is a *constructive* process that depends not only on the information inherent in the stimulus but also on the mental structure of the perceiver. Thus, the task of cognitive psychology is to analyze the brain processes that intervene between stimulus and behavior—precisely the area considered irrelevant by behaviorists. Only in this way, cognitive psychologists argue, can we hope to understand the relationship between a person's actions and what that person sees, remembers, or believes. According to this view any study of mental activity that fails to consider the internal processing and representation of mental events is inadequate to account for all but the simplest forms of behavior. In retrospect, it is now clear that the behavior studied by behaviorists was largely restricted to simple reflex behaviors that do not require conscious mental activity.

In turning to the study of mental operations, cognitive psychologists have been interested in following the flow of sensory information from its transduction by appropriate sensory receptors to its eventual *internal representation* in the brain for use in memory and action. The idea of an internal representation implies that each perceptual or motor act is correlated with a characteristic pattern of activity in a specific set of interconnected cells. The pattern of connections also stores information about the perception and the motor act.

However, this new emphasis on internal representation is not without its own problems. Although behaviorists had approached behavior too narrowly by excluding its internal representation from analysis, they were correct in emphasizing that behavior be studied objectively and that internal representations are not readily accessible to objective analysis. Once psychologists acknowledged that internal representations are important for understanding behavior, they had to come to grips with the stern reality that most of what we conceive of as mental processes are still largely inaccessible to experimental analysis.

Fortunately, the emergence of cognitive psychology in the 1960s and 1970s was paralleled by significant progress in systems neural science and by improvements in brain imaging methods. Together these fields are beginning to make possible the direct study of internal sensory and motor representations in both experimental animals and human subjects. As a result, mental activity no longer needs be approached only indirectly from observations of externally manifested behavior. The goal of the combined efforts of cognitive psychology, systems neurobiology, and brain imaging is to map the elements of cognitive function onto specific neuronal systems.

Cognitive Neural Science Relies on Five Main Approaches

Cognitive neural science has emerged from five main technical and conceptual developments. First, in the 1960s and 1970s techniques were developed for examining the activity of single cells in the brains of intact and active primates, making it possible to study perceptual and motor processes at the cellular level while animals were engaged in typical behaviors. Second, cellular studies led to the finding that even complex cognitive processes, such as attention and decision making, are correlated with the pattern of firing of individual cells in specific brain regions. This has changed the way behavior is studied both in experimental animals and in humans, so that the focus now is not on how a stimulus elicits a response, but on how a subject *arrives* at a response—on information processing.

Third, developments in systems neurobiology and cognitive psychology stimulated a renewed interest in behavioral neurology. Since the neural basis of cognition begins with local operations in the brain, a great deal has been learned from the study of cortical lesions. Many patients with lesions of specific regions of the brain exhibit quite specific cognitive deficits. These studies indicate that within any cognitive system there are many independent information processing modules. The visual system, for example, has specialized pathways for processing information about color, form, and movement.

Fourth, the development of imaging techniques—positron emission tomography (PET), magnetic resonance imaging (MRI), magnetoencephalography, and voltage-sensitive dyes—has made it possible to relate changes in activity in populations of neurons to specific mental processes in the living human brain.

Finally, computer science has made two distinct contributions to cognitive neural science. First, computers have made it possible to model the activity of large populations of neurons and to begin to test specific ideas about complex brain functions. Complex systems of neurons, such as those mediating perception and action, require that we understand not only the cellular properties but also the circuitry of the systems. Analysis of circuits provides us with an understanding of *network properties*, which, although dependent on the properties of individual neurons in the network, need not be identical or even similar to the cellular properties. Computation is helpful in overcoming the problem by allowing us to model large neuronal networks. Second, cognitive processes are in some ways analogous to computer programs in that both are concerned with information processing, transformation, storage, and retrieval. In this sense computer science provides a language that is potentially useful for analyzing cognitive events.

In the remainder of this chapter we consider in detail how cognitive neural science is applied to the study of tactile perception, as the sensation of touch involves three particularly simple examples of internal representation. First, we consider the representation of *personal space* in the form of a cortical map of the tactile sensibilities of the body surface, and how modifications of this neural map following the loss of a body part may account for the phantom limb syndrome. Second, we consider how the representation of personal space is elaborated into a more complex representation of *peripersonal space,* which includes objects within reaching distance, or into *extrapersonal space*, which includes objects beyond arm's length. Finally, we illustrate the further transformation of these actual and immediate representations into *imagined and remembered representations.*

The Brain Has an Orderly Representation of Personal Space

Touch and proprioception are two sensory modalities carried by the somatic sensory system. Touch is the sense of the texture of objects and their movement across the skin. Proprioception is the sense of the static position and of movement of our fingers and limbs. Touch and proprioception enable us to sense the size and shape of objects placed in the hand. The somatic sensory system is also concerned with nociception (the signaling of tissue damage, often perceived as pain) and temperature sense (warmth and cold), but we do not consider these modalities here.

The somatic sensory system contains primary receptor neurons, located in the skin and other body tissues, that transform stimulus energy into neural events. The sense of touch is most discriminating in the fingertips. Touch receptors in the fingers send information on the texture of objects to the brain so we can read braille or play a musical instrument. Receptors in the muscles and joints of the fingers send information on the angles of the finger joints, which is important for recognizing the size and shape of grasped objects.

In the somatic sensory system there are only three synaptic relays between the sensory neurons and the cerebral cortex (Figure 18–1A). The first is made by the central processes of the dorsal root ganglion cells on neurons in the gracile and cuneate nuclei of the medulla. Next, the axons of these medullar neurons cross the midline and ascend to the ventral posterior nucleus of the thalamus. Finally, the thalamic neurons send axons to a region of the anterior parietal lobe termed the *primary somatic sensory cortex*, or S-I (Figure 18–1B). Removal of S-I produces deficits in proprioception

Figure 18–1 Neural architecture of the somatosensory system.

A. General organization of the dorsal column–medial lemniscal system, which mediates tactile sensation and limb proprioception. The main pathway of the system contains three synaptic relays between the periphery and the cerebral cortex. (1) The central processes of the dorsal root ganglion cells form synapses on neurons in the gracile and cuneate nuclei in the lower medulla. (2) The axons of neurons in these nuclei ascend in the medial lemniscus and synapse on neurons in the ventral posterior lateral nucleus of the thalamus. (3) The neurons of the lateral nucleus send axons to the somatic sensory cortex. (Adapted from Carpenter and Sutin, 1983.)

B. The somatic sensory centers in the cortex include the primary (S-I) and secondary (S-II) somatosensory cortices in the anterior parietal lobe and the posterior parietal cortex. S-II is located deep within the lateral sulcus.
1. The relationship of S-I to S-II and to the posterior parietal cortex (Brodmann's areas 5 and 7) is seen best from a lateral perspective of the surface of the cerebral cortex.
2. The primary somatic cortex (S-I) is subdivided into four distinct cytoarchitectonic regions. This sagittal section shows these four regions (Brodmann's areas 3a, 3b, 1, and 2) and their spatial relationship to area 4 of the motor cortex and areas 5 and 7 of the posterior parietal cortex.

A Dorsal column - medial lemniscal system

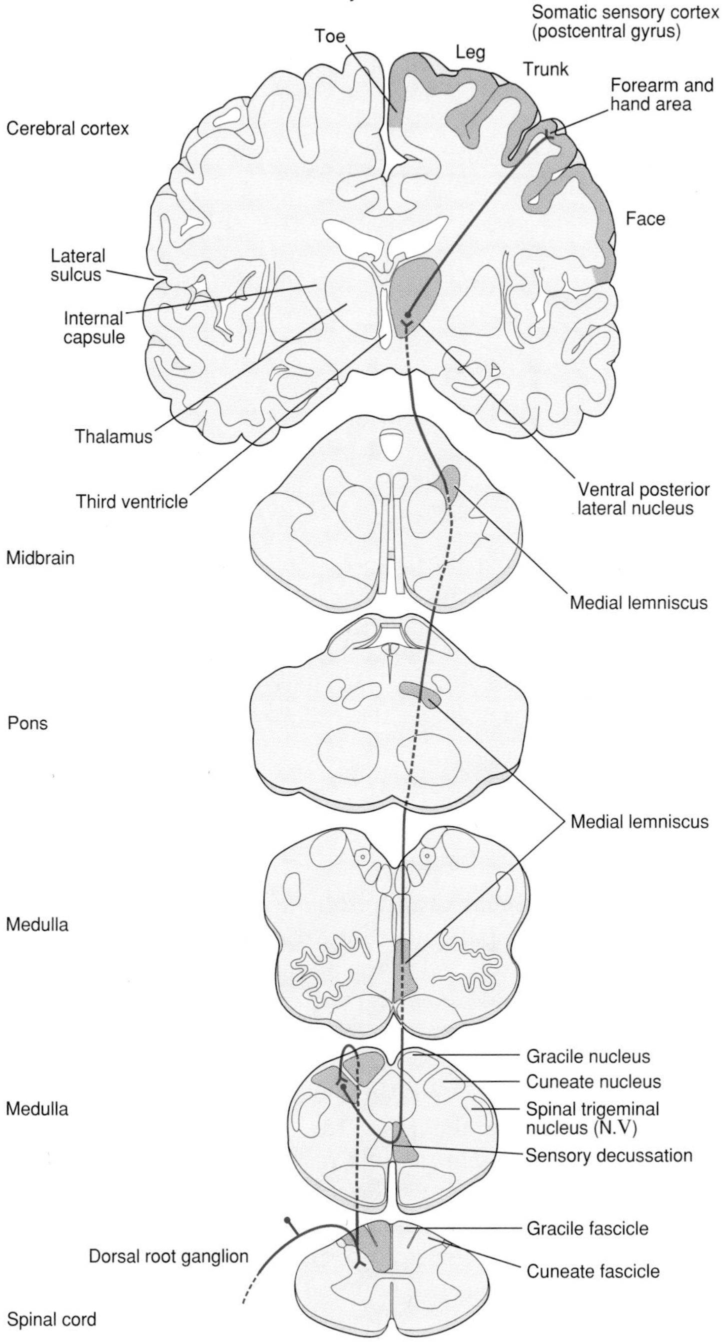

Somatic sensory cortex
(postcentral gyrus)
Toe
Leg
Trunk
Forearm and
hand area
Cerebral cortex
Face
Lateral
sulcus
Internal
capsule
Thalamus
Third ventricle
Ventral posterior
lateral nucleus
Midbrain
Medial lemniscus
Pons
Medial lemniscus
Medulla
Gracile nucleus
Cuneate nucleus
Medulla
Spinal trigeminal
nucleus (N.V)
Sensory decussation
Gracile fascicle
Dorsal root ganglion
Cuneate fascicle
Spinal cord

B The somatosensory cortex

B1
S-I
Posterior
parietal lobe
Central sulcus
S-II
Lateral
sulcus
B2
Postcentral gyrus
Central
sulcus
Postcentral
sulcus
1
2
3b
3a
4
5
7

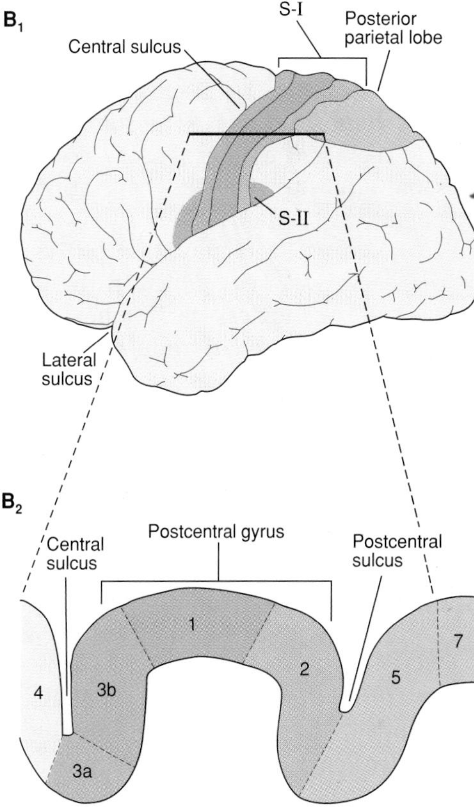

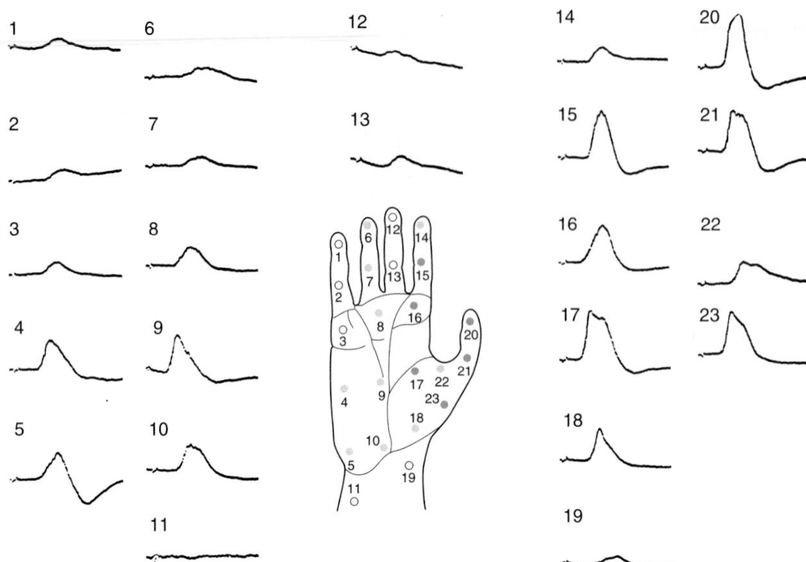

Figure 18–2 The neural representation of specific areas of the body in the somatosensory cortex was first determined from patterns of evoked potentials. The potentials shown here were recorded from one large group of neurons in the left postcentral gyrus of a monkey and evoked by a light tactile stimulus applied to different points on the right palm. The evoked potentials in the area of the digits are strongest when the stimulus is applied to the thumb and forefinger (**points 15, 16, 17, 20, 21** and **23**), and weakest when the stimulus is applied to the middle or the small finger (**points 1, 2, 3, 12** and **13**). (Adapted from Marshall, Woolsey, and Bard, 1941.)

and the ability to discriminate the size, texture, and shape of objects. The somatic sensory system and each of the other sensory systems is organized serially but also includes several components that operate in parallel.

The cortical representation of the somatic sensory system was first studied in the late 1930s using electrophysiological techniques. Wade Marshall found that he could produce an *evoked potential* in the cortex by touching a specific part of the animal's body surface. Evoked potentials are recorded electrical signals that represent the summed activity of thousands of cells and are obtained by using large-tipped metal electrodes or electrolyte-filled glass capillaries. Evoked responses were later used by Marshall, Clinton Woolsey, and Philip Bard to map the representation of the body surface in the postcentral gyrus in monkeys (Figures 18–2 and 18–3).

A similar representation of the body surface was found in the human cortex by the neurosurgeon Wilder Penfield during operations for epilepsy and other brain disorders. Working with locally anesthetized patients, Penfield stimulated the surface of the postcentral gyrus at various points in the area of S-I and asked the patients what they felt. (This procedure was used to ascertain where the epilepsy starts and therefore to avoid unnecessary damage during surgery.) Penfield found that stimulation of points in the postcentral gyrus produced tactile sensations in discrete parts of the opposite side of the body. From these studies Penfield constructed a map of the neural representation of the body in the somatic sensory cortex.

As shown in Figure 18–4, the leg is represented most medially, followed by the trunk, arms, face, and finally, most laterally, the teeth, tongue, and esophagus. Note that in Figure 18–4 each part of the body is represented in proportion to its relative importance in sensory perception. The face is large compared with the back of the head; the index finger is gigantic compared with the big toe. As we shall see later in this chapter, this distortion reflects differences in innervation density in differ-

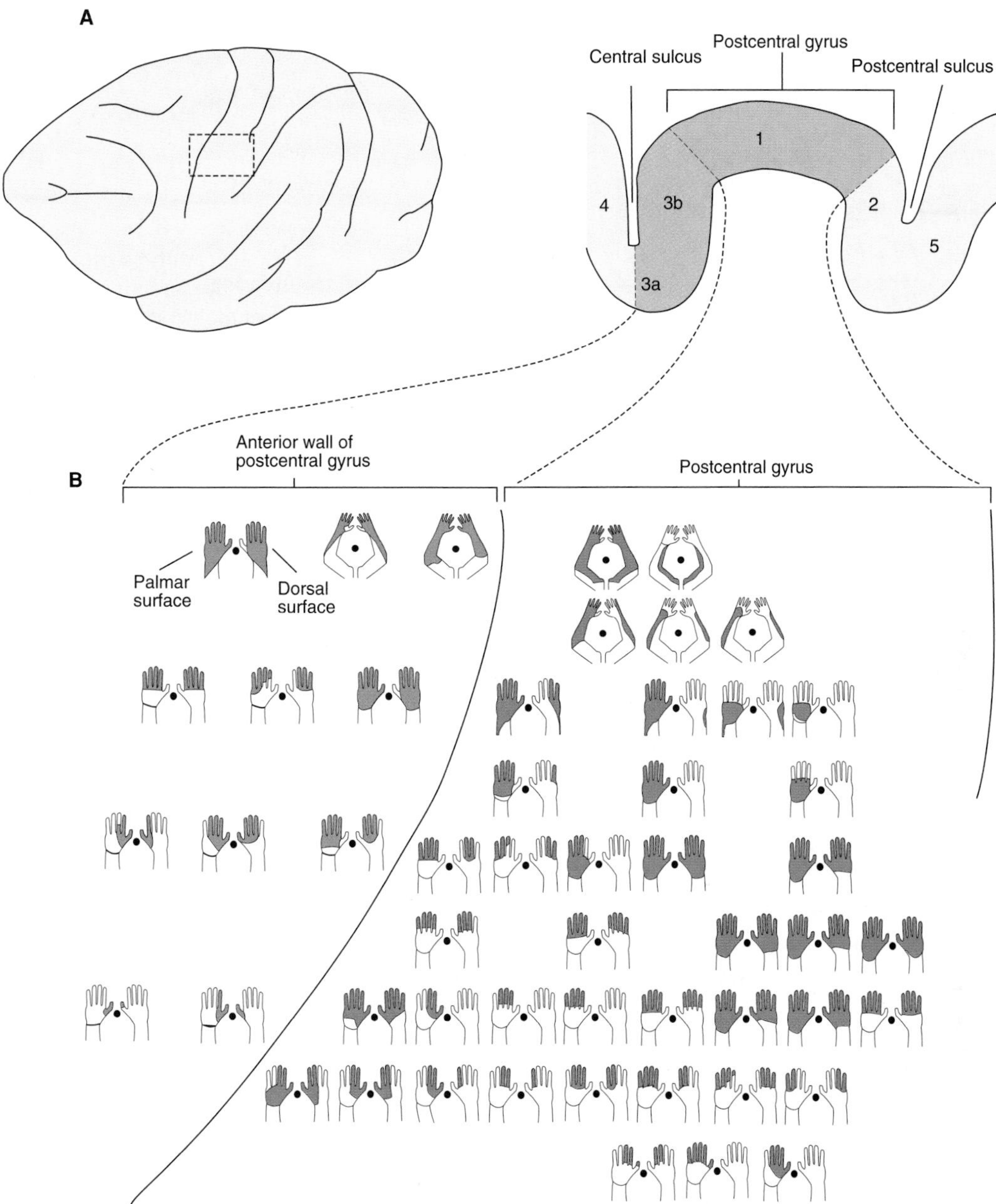

Figure 18–3 An early map of cortical responses to tactile stimulation in monkeys. (Adapted from Marshall, Woolsey, and Bard, 1941.)

A. Recordings were made in the primary somatic sensory cortex (S-I). The lateral view of the brain (**left**) shows the recording site. A sagittal view of S-I (**right**) shows Brodmann's subdivisions.

B. The map shows the sites (**black dots**) in Brodmann's areas 3b and 1 at which responses were evoked by stimulation of the palmar and dorsal surfaces of the right hand. At each site the colored areas of the hand indicate areas of stimulation that evoke a response at that site. The sites on the **left** of the figure are in the anterior wall of the postcentral gyrus, corresponding roughly to areas 3b and 3a in S-I. The sites on the **right** of the figure are on the dorsal surface of the postcentral gyrus, corresponding roughly to area 1 in S-I.

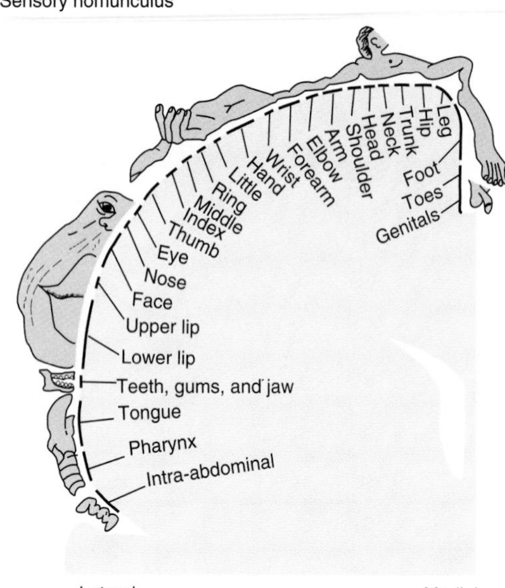

Figure 18–4 Somatic sensory and motor projections from and to the body surface and muscle are arranged in an orderly way in the human cortex. The map illustrated here is for Brodmann's area 1 in the postcentral gyrus of the parietal cortex. Each area within the somatosensory cortex (areas 3a and 3b, and 1 and 2) contains a full representation of the body (see Figure 18–6). Parts of the body that are important for tactile discrimination, such as the tip of the tongue, the fingers, and the hand, have disproportionately large representations reflecting greater degrees of innervation. (Adapted from Penfield and Rasmussen, 1950.)

ent areas of the body. Similar distortions are observed in the body representations of other animals. In rabbits, for example, the face and snout have the largest representation because they are the rabbit's most sensitive means for exploring its environment (Figure 18–5).

Reliable Cortical Maps of the Body Are the Basis of the Accuracy of Clinical Neurological Examinations

The cortical map of the body surface and the parallel motor map (Figure 18–4) explain why neurology is an accurate diagnostic discipline, even though for many decades in the past it relied on only the simplest tools—a wad of cotton, a safety pin, a tuning fork, and a reflex hammer. Disturbances within the somatic sensory system can be located clinically with remarkable accuracy because there is a direct relationship between the anatomical organization of the brain and specific perceptual and motor functions.

A dramatic example of this relationship is the Jacksonian seizure, a characteristic sensory epileptic attack described by the neurologist John Hughlings Jackson. Early in this type of seizure numbness and paresthesia (inappropriate sensations such as burning or prickling) begin in one place and spread throughout the body. For example, numbness might begin at the fingertips, spread to the hand, up the arm, across the shoulder, into the back, and down the leg on the same side. The orderly spread of the seizure—starting laterally in the hand area and propagating toward the midline—is explained by the arrangement of the sensory projections in the brain (Figure 18–4).

The Cortex Has a Map of the Body for Each Submodality of Sensation

The first maps of the body surface were plotted using gross recording and stimulation techniques that explored the functional organization of the surface of the postcentral gyrus, the only portion accessible for experimentation with these techniques. Later, the use of microelectrodes revealed not one but four fairly complete maps in the primary somatic sensory cortex (S-I), in each of the four Brodmann areas: areas 3a, 3b, 1, and 2.

Each of these areas represents different features of touch and proprioceptive information. Sensory information from muscles and joints, important for limb proprioception, projects into area 3a, and information from the skin, important for touch,

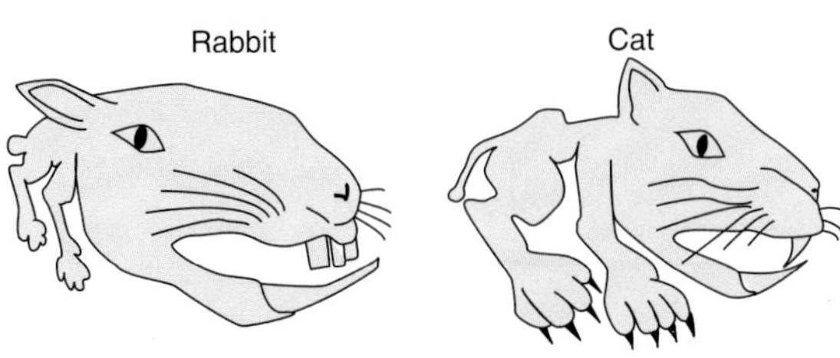

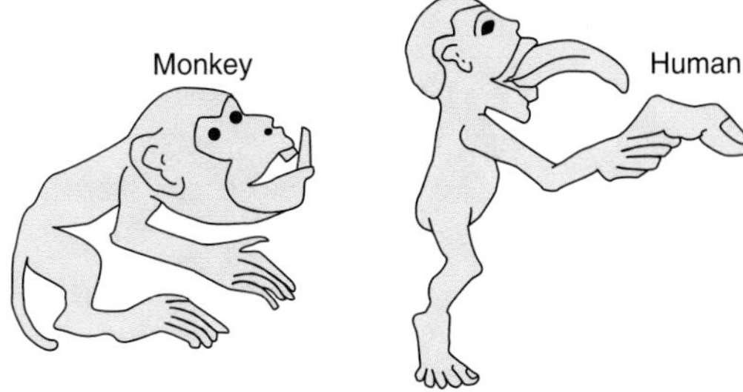

Figure 18–5 Different species rely on different parts of the body for adaptive somatosensory information. These drawings show the relative importance of body regions in the somatic sensibilities of four species, based on studies of evoked potentials in the thalamus and cortex.

into area 3b. Sensory information from the skin is further processed within area 1 and combined with information from the muscles and joints in area 2. This explains why a small discrete lesion in area 1 impairs tactile discrimination, whereas a small lesion in area 2 impairs the ability to recognize the size and shape of a grasped object. Of the four representations of the body in the primary somatic sensory cortex, those in areas 3a and 3b are elementary whereas those in areas 1 and 2 are complex because they represent elaboration of sensory information.

As illustrated in Figure 18–6, the somatosensory maps in Brodmann's areas 3b and 1 lie parallel to one another and correspond in their medial-to-lateral representation of the body surface. This explains why earlier studies, which probed a limited area of the postcentral gyrus and used techniques with poor resolution, concluded that there was only one large representation of the body surface in the cortex.

S-I projects to other areas of the parietal lobe. For example, S-I projects to the secondary somatic sensory areas, where somatic sensory inputs are used for learning new sensory discriminations, and to the posterior parietal cortex, where somatic information is used for shaping body image and for planning movements in extrapersonal space.

The Internal Representation of Personal Space Is Modifiable by Experience

How fixed is the representation of the body in the somatosensory cortex? Recent studies by Michael Merzenich and his colleagues show that the cortical maps differ systematically among individuals in a manner that reflects their use. Two studies have been particularly important. First, Merzenich carried out experiments to determine the relative contributions of genes and experience to variations in somatic sensory maps among normal monkeys. In one experiment monkeys were encouraged to use only their three middle fingers to rotate a disk to obtain food. After several thousand disk rotations the area in the cortex devoted to the middle finger was greatly expanded at the expense of adjacent fingers. Thus, practice strengthens and expands the cortical representation of those fingers that are used (Figure 18–7).

Intense use or disuse brings about dramatic changes in these connections. Several monkeys have been studied 10 years or more after an upper limb had been completely deafferented by severing all sensory nerves. In all of them the adjacent representation from the face had expanded into the hand area, so that stimulation of the face evoked responses in the area traditionally occupied by the hand. A third of the entire map devoted to the body surface, about 1.3 cm^2 of cortex, had switched from connections representing the hands and arms to those representing the face.

What mechanisms underlie these changes? Recent evidence (to be considered in Chapter 25) indicates that the input connections to cortical neurons in the somatic sensory system are formed on the basis of correlated firing. Cells that fire together, wire together! Merzenich and his colleagues tested this idea by surgically connecting the skin surfaces of two adjacent fingers of a mon-

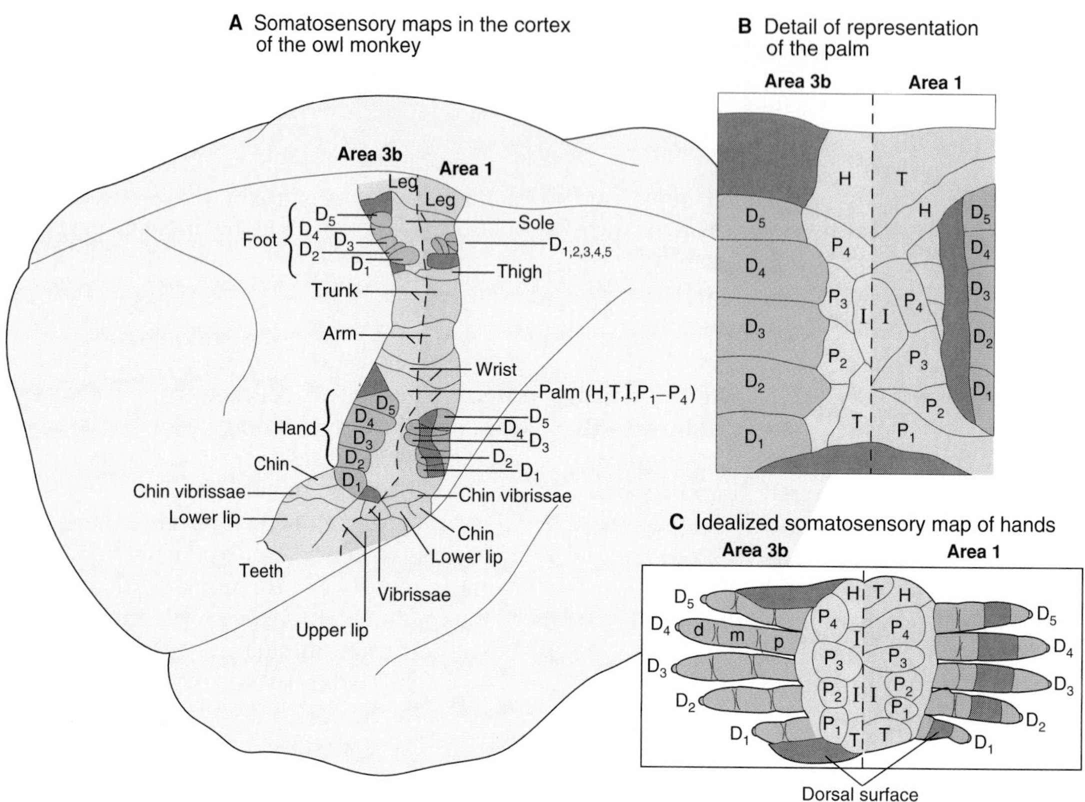

Figure 18–6 Each of the four areas of the primary somatic sensory cortex (Brodmann's areas 3a, 3b, 1, and 2) has its own complete representation of the body surface. (Adapted from Kaas et al., 1983.)

A. Somatosensory maps in areas 3b and 1 are shown in this dorsolateral view of the brain of an owl monkey. The two maps are roughly mirror images. Each digit of the hands and feet is individually represented ($\mathbf{D_1}$ to $\mathbf{D_5}$).

B. A more detailed illustration of the representation of the glabrous pads of the palm in areas 3b and 1. These include the palmar pads (numbered in order, $\mathbf{P_4}$ to $\mathbf{P_1}$), two insular pads (**I**), two hypothenar pads (**H**), and two thenar pads (**T**).

C. An idealized map of the hands based on studies of a large number of monkeys. The representations of the palm and digits reflect the extent of innervation of each palmar area in the cortex. The five digital pads ($\mathbf{D_1}$ to $\mathbf{D_5}$) include **d**istal, **m**iddle, and **p**roximal segments (**d**, **m**, **p**).

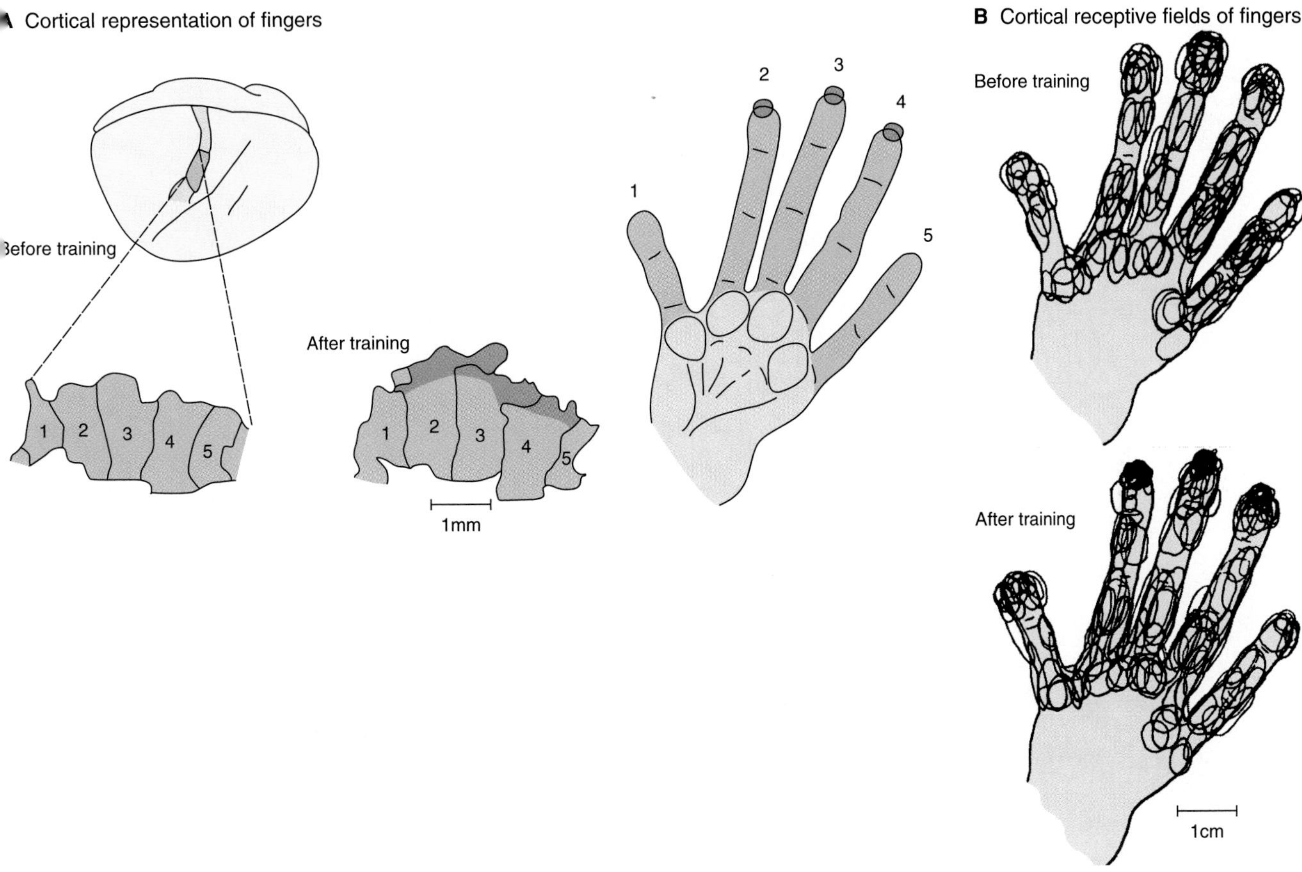

Figure 18–7 Increased use of selected fingers enlarges the cortical representation of the fingers. (Adapted from Jenkins et al., 1990).

A. The regions in cortical area 3b representing the surfaces of the digits of an adult monkey are shown 3 months before training and after training. Training consisted of a task that required repeated use, for 1 hour per day, of the tips of the distal phalanges of digits 2, 3, and occasionally 4. After training there is a substantial enlargement of the cortical representation of the stimulated fingers (**darker color**).

B. A cortical receptive field is the area on the skin where a tactile stimulation either excites or inhibits a cortical cell. All receptive fields on the surfaces of the digits were identified before and after training to determine the extent of the recording sites within area 3b. The number of receptive fields in the distal phalanges of digits 2, 3, and 4 is larger following training.

key. This procedure ensures that the connected fingers are always used together and therefore increases the correlation of inputs from the skin surfaces of the adjacent fingers. Increasing the correlation of activity from adjacent fingers in this way abolishes the normally sharp discontinuity between the areas of cortex that receive inputs from these digits (Figure 18–8). Thus, the normal discontinuity in the representation of adjacent fingers in the cortical map is established not only by a genetical-

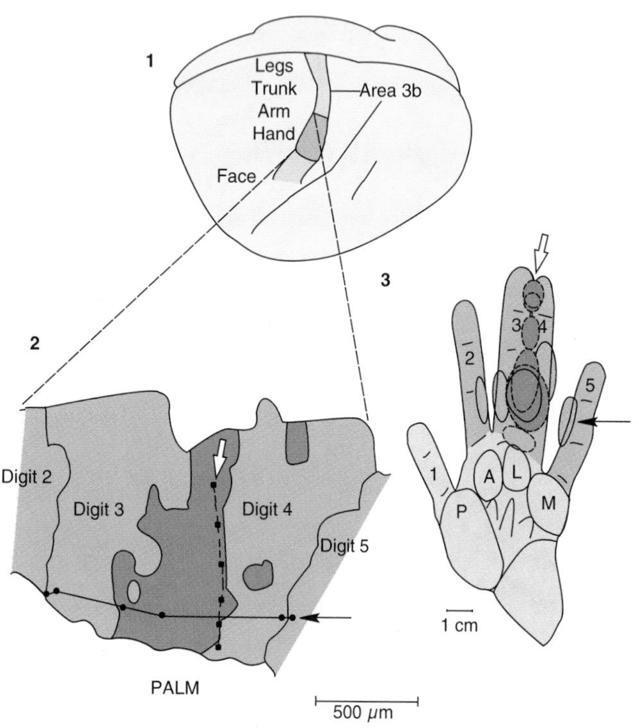

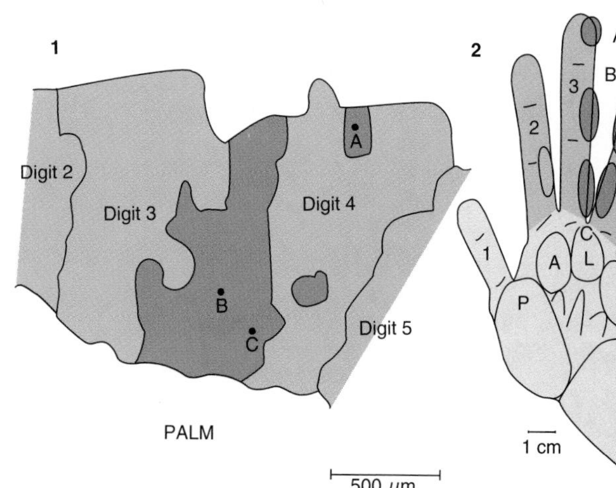

Figure 18–8 Normal discontinuities in the cortical representation of the digits of an adult owl monkey become blurred following surgical fusion of the digits. (Adapted from Clark et al., 1988.)

A. 1. A dorsolateral view of the neocortex of an owl monkey shows the representation of the animal's body in area 3b of the primary somatosensory cortex. 2. This detailed drawing of portions of the representation of the hand shows the areas for digits 3 and 4 and surrounding skin surfaces 5.5 months after surgical fusion of these digits. The areas of representation that changed following digit fusion are indicated by the **darker shading**. Instead of the normal discontinuity between the two digits, there is now a large common area (340–1000 μm in width) representing the parts of the digits that are fused. Threshold stimulation of the surfaces of either one of the two fused digits evoked responses in cortical cells within this zone. In contrast, the representation of the borders of the fused digits with the adjacent free digits (2 and 5) remains sharp. Evoked potentials were obtained in two series of sites corresponding to sequential stimulation of the digits in two axes: a rostral-to-caudal axis (▪) and medial-to-lateral •). 3. The receptive fields for the neurons at the recording sites shown in Figure 2. The **solid** and **white arrows** indicate sequences of stimulation corresponding to the sequence of recording sites shown in Figure 2.

B. Even after the fused digits are separated, the common area of representation remains. Thus, the intermingling of the representation of digits 3 and 4 is achieved centrally, and does not result from peripheral regeneration that spares the site of contact. Evoked potentials were obtained at points A, B, and C in area 3b of the cortex (1) by stimulation of digits 3 and 4 at discrete sites (2).

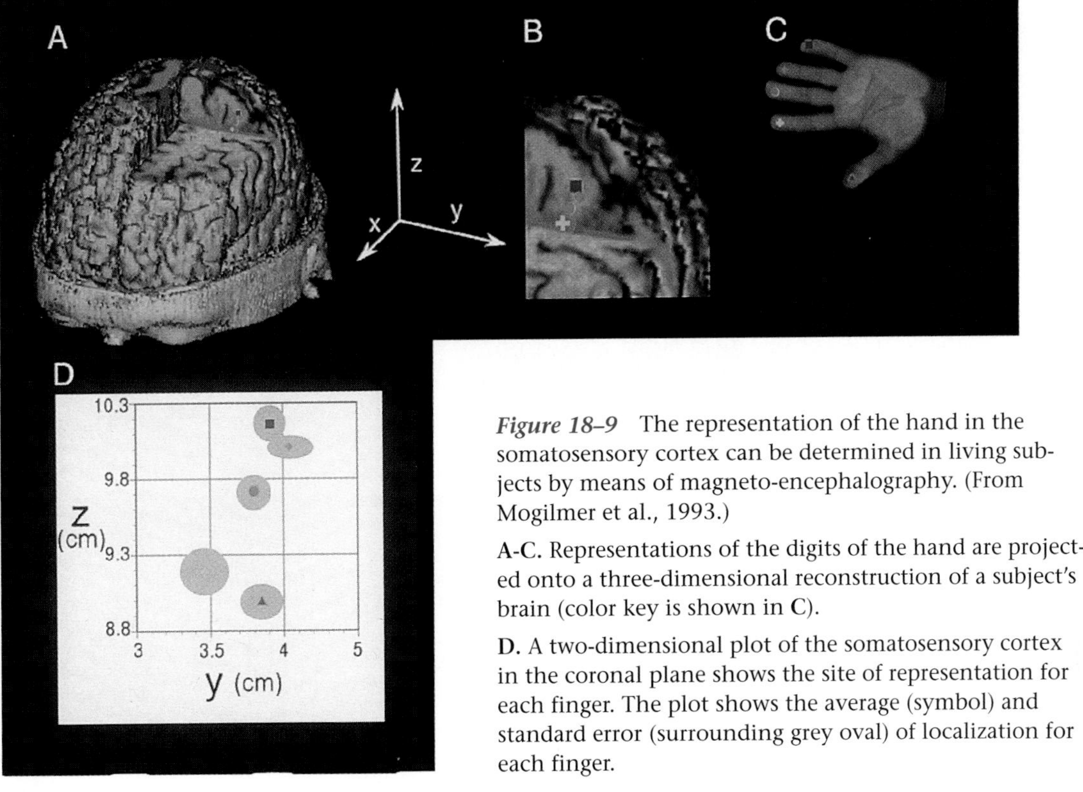

Figure 18–9 The representation of the hand in the somatosensory cortex can be determined in living subjects by means of magneto-encephalography. (From Mogilmer et al., 1993.)

A-C. Representations of the digits of the hand are projected onto a three-dimensional reconstruction of a subject's brain (color key is shown in C).

D. A two-dimensional plot of the somatosensory cortex in the coronal plane shows the site of representation for each finger. The plot shows the average (symbol) and standard error (surrounding grey oval) of localization for each finger.

ly programmed demarcation in the pattern of connections, but also by temporal correlations in patterns of input, that is, through learning.

The Internal Representation of the Human Hand Area Can Be Modified

Does this reorganization also occur in the human brain? Studies using magneto-encephalography have allowed functional maps of the hand in normal subjects to be defined with a precision of millimeters. This imaging technique has been used to compare the hand area in the cortex of normal adult humans with that in patients with a webbed-finger syndrome, called *syndactyly*, where the fingers are fused together from birth. Patients with this syndrome do not have individual fingers; their hand is much like a fist. Activity in one part of the hand is always correlated with activity in all other parts of the hand. The representation of the syndactylic hand in the cortex is considerably smaller than that of a normal person, and within this shrunken representation the fingers are not organized somatotopically (Figures 18–9 and 18–10).

When the fingers of one patient were surgically separated, each of the newly separate fingers became individually represented in the cortex within weeks (Figures 18–9 and 18–10). This new representation of the hand occupied an area of cortex almost corresponding to the normal representation of the hand and the normal distance between each digit.

Modifiability of the Internal Representation May Account for the Phantom Limb Syndrome

Another aspect of cortical representation has recently emerged from studying patients with amputated limbs. Many of these patients continue to sense the missing limb and even feel pain from it, a phenomenon known as phantom limb syndrome. In the past, phantom limb sensations have been attributed to impulses entering the spinal cord from the scar of nervous tissue in the stump. Removing the scar or cutting the sensory nerves just above it may bring relief from the pain. However, recent experiments on patients who have lost a hand suggest another explanation: afferent path-

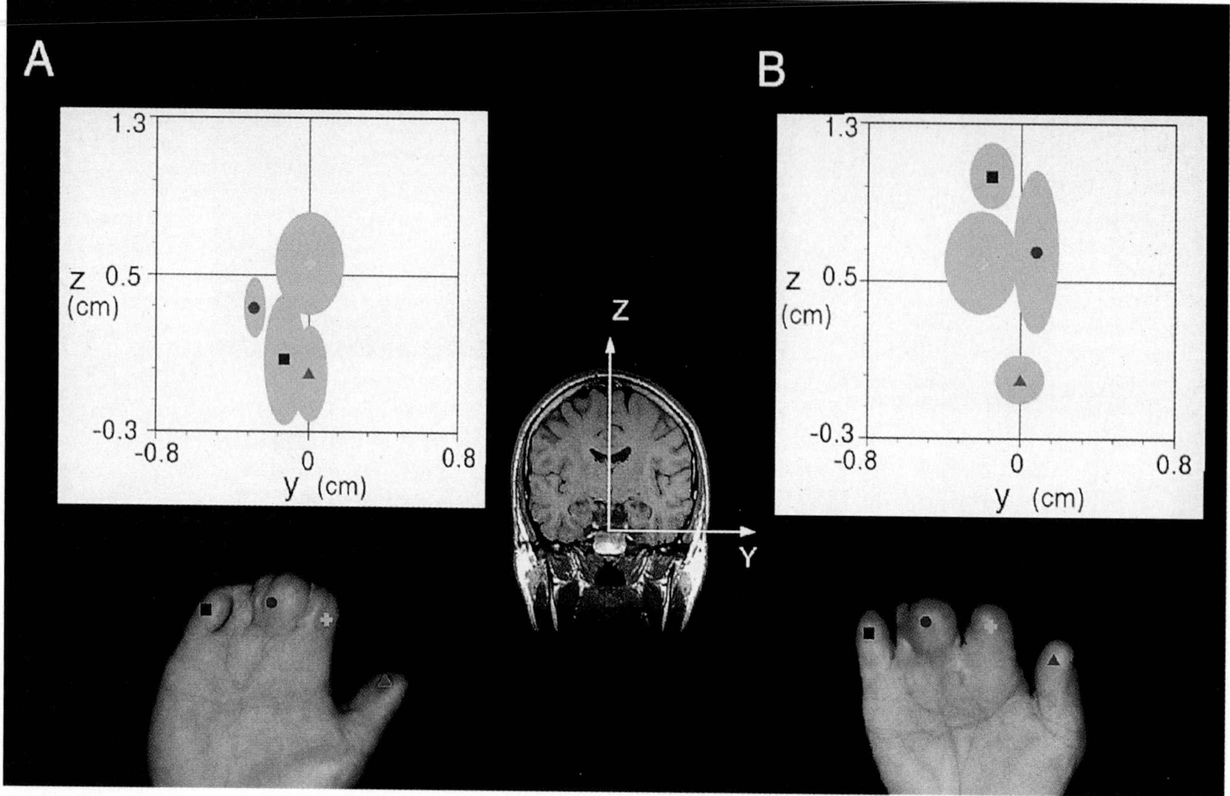

Figure 18–10 The representation of the hand in the somatosensory cortex changes following surgical correction of syndactyly of digits 2 to 5. (From Mogilmer et al., 1993.)

A. A preoperative map shows that the cortical representation of the thumb, index, middle, and little fingers is abnormal and lacks any somatotopic organization. For example, the distance between sites of representation of the thumb and little finger is significantly smaller than normal (see Figure 18–9D).

B. Twenty-six days after surgical separation of the digits the organization of the hand area is somatotopic, and the distance between the sites of representation of the thumb and little finger has increased to 1.06 cm.

ways that normally occupy areas surrounding the hand representation (areas representing the face, shoulder, and upper torso) expand into the area formerly occupied by afferents from the hand, just as they do in monkeys with deafferented limbs (Figure 18–11A). This has been called *remapping of referred sensations*.

In all patients examined at least one remapping of the representation of the lost hand has occurred, and sometimes even two. Touching an area on the face or the upper arm produces a tactile sensation associated with the lost hand. This occurs because the hand area in the cortex lies next to and comes under the control of the areas that innervate the face and the upper arm (see Figure 18–4). Thus, the referred sensations are to be expected if the sensory inputs from the face and upper arm formerly invade the cortical territory formerly occupied by inputs from the amputated hand.

In each of these patients there is a precise, direct correspondence between a point on the face or upper arm and an individual digit. Also, there is a topography: adjacent points on the face or upper arm map onto adjacent points of the phantom hand.

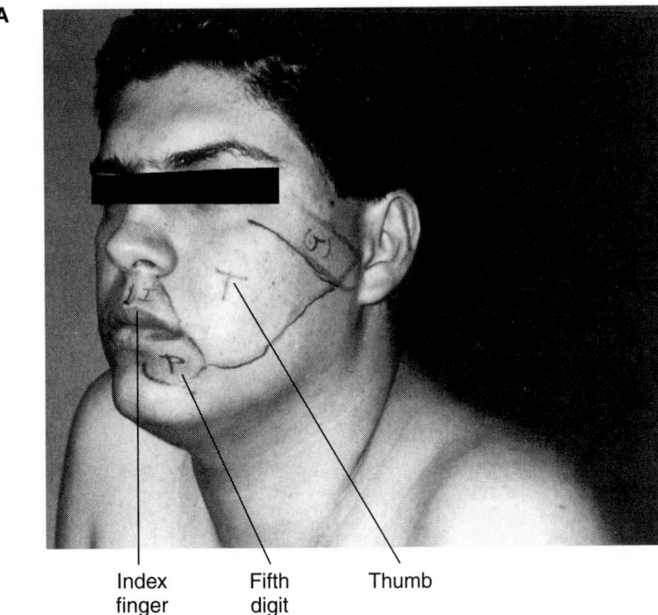

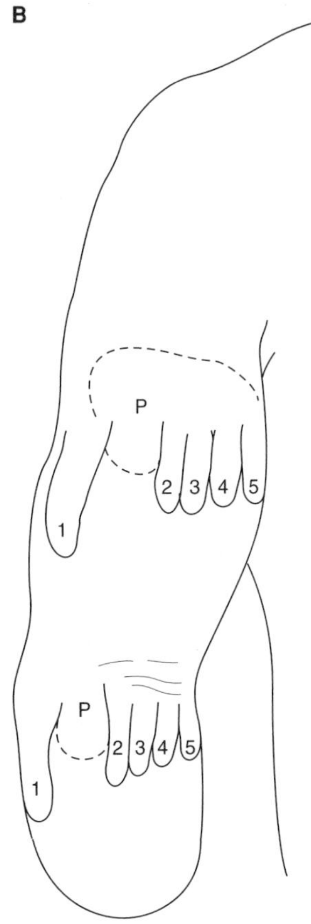

Figure 18–11 Phantom limb sensations can be evoked by touching the face. (From Ramachandran, 1993.)

A. This photograph of the left side of the face of a patient with an amputated hand shows the sites on the face where stimulation (brushing the face with a Q-tip) elicited precisely localized referred sensations in the phantom digits. Regions that evoke referred sensations are called *reference fields*. The region labeled **T** always evoked sensations in the phantom thumb; likewise, the regions labeled **P**, **I**, and **B** consistently evoked sensations in the phantom "pinkie," index finger, and ball of the thumb respectively. This patient was tested 4 weeks after amputation.

B. Referred sensations are localized to two distinct spatial maps—one close to the line of amputation and a second one 6 cm above the elbow crease. The maps are almost identical except for the absence of fingertips in the upper map. Stimulating the skin region between these two maps did not elicit sensations in the phantom limb.

The Internal Representation of Personal Space Can Be Studied at the Cellular Level: Each Central Neuron Has a Specific Receptive Field

How are sensory and motor representations reflected at the cellular level? In the 1950s Vernon Mountcastle and his colleagues began to examine the somatic sensory system at the cellular level. They found that cortical neurons in the somatic sensory system are mostly silent, with little or no spontaneous activity, and that each cell responds only to stimulation of a specific area of the skin. As we shall learn in greater detail in Chapter 20, the region on the sensory sheet that excites a cell is called the *receptive field*.

Each point on the skin is monitored by a population of cells rather than a single cell because receptive fields overlap. When a point on the skin is touched, the population of cortical neurons whose receptive fields include that site are excited. Stimulation of another point on the skin activates another population of cortical neurons (see Chapter 20). Thus, we perceive that a particular point on

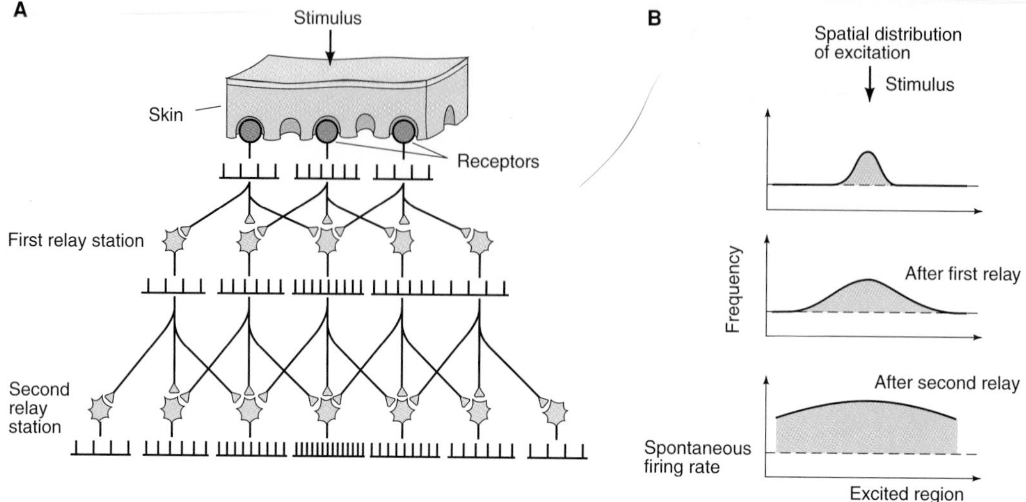

Figure 18–12 Internal representation in the brain of a stimulus on the skin. (Adapted from Dudel, 1983.)

A. Excitatory synaptic connections among three receptors and the interneurons at the next two relays. The relative rate of discharge of each cell during stimulation is shown.

B. The discharge zone (the *excitatory region*) becomes larger at each successive population of relay cells in response to a stimulus in the excitatory region of the receptive field.

the skin is being stimulated because a specific population of neurons in the brain is activated (Figure 18–12). Conversely, as Penfield illustrated, when a point on the cortex is stimulated electrically, we experience tactile sensations on a specific part of the skin.

Inputs to the Somatic Sensory Cortex Are Organized into Columns by Submodality

Most nerve cells in S-I mediate only one modality: touch, pressure, temperature, or pain. Neurons mediating touch are responsive to stimuli applied to the superficial skin but not to deep pressure. Some of these neurons are responsive to movement of hairs, while others respond to a steady indentation of skin. Throughout the somatosensory system cells responding to one submodality tend to be grouped together.

The first example of such grouping was discovered by Mountcastle while studying the distribution of the input to the somatic sensory cortex from various receptors in skin and deep tissue. Because the cortex consists of six major cellular layers, he first looked for a correlation between receptor type and cell layer. He found none. Instead, he discovered that in all six layers within a column or slab of cortex running from the cortical surface to the white matter the neurons respond to a single class of receptors. Some columns of cells are activated by superficial receptors in skin, some by deep receptors, others by movement of hairs, and still others by subcutaneous receptors. All neurons in a column receive inputs from the same local area of skin. Neurons lying within a column therefore constitute an elementary functional module of the cortex (Figure 18–13). We shall see in later chapters that columnar organization is a basic structural principle of the cerebral cortex.

Although each of the four areas of the primary somatic sensory cortex (3a, 3b, 1, and 2) receives input from all areas of the body surface, one modality tends to dominate in each area. In area 3a the dominant input is from muscle stretch receptors; in area 3b it is input from superficial skin receptors; in area 2, deep pressure receptors; and in area 1, rapidly adapting skin receptors.

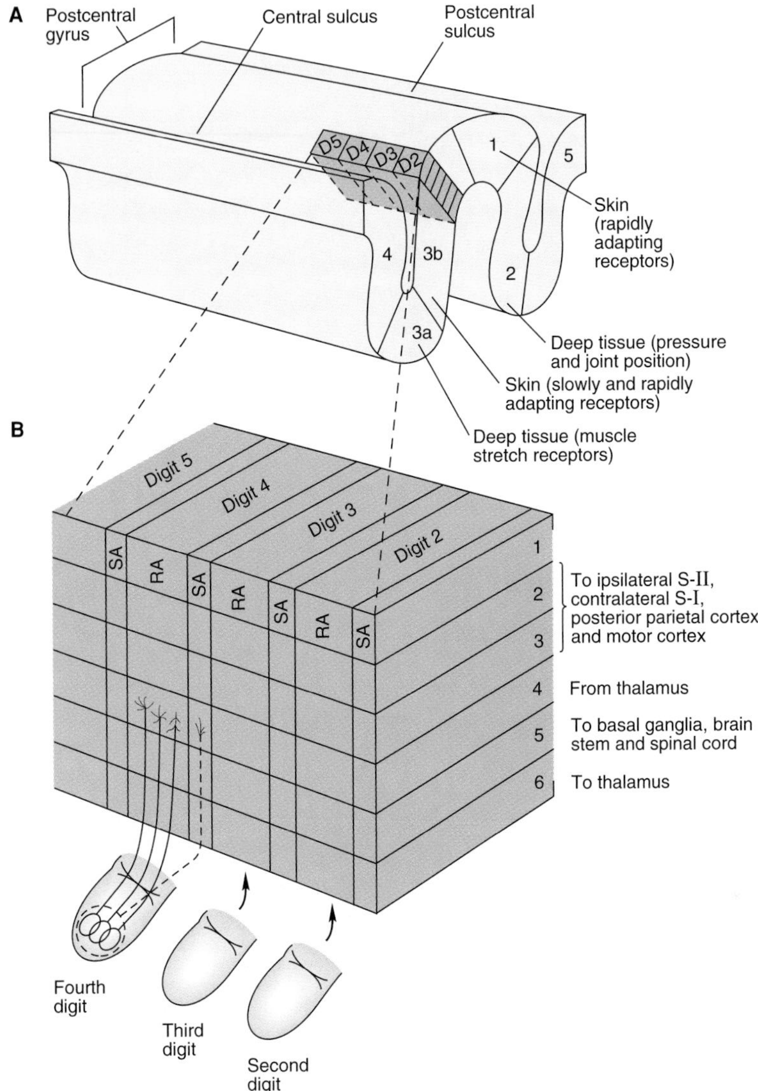

Figure 18–13 Somatic sensory input to the cortex is organized by submodality. (Adapted from Kaas et al., 1979.)

A. Each region of the somatic sensory cortex—Brodmann's areas 3a, 3b, 1, and 2—receives inputs primarily from one type of receptor. Within each region, inputs from specific parts of the body are organized in columns of neurons that run from the surface to the white matter. Thus each part of the body is represented in four distinct neural maps of the body, each comprising inputs from specific types of receptors. The drawing here shows the columnar arrangement of inputs from the skin of digits 2, 3, 4, and 5 to Brodmann's area 3b.

B. Brodmann's area 3b receives inputs from two types of cutaneous receptors of tactile stimuli—rapidly adapting receptors (RA) and slowly adapting receptors (SA). Rapidly adapting receptors cease firing when the stimulus is prolonged; slowly adapting receptors continue firing as long as the stimulus is present. The inputs from each type of receptor in each digit are organized into separate columns.

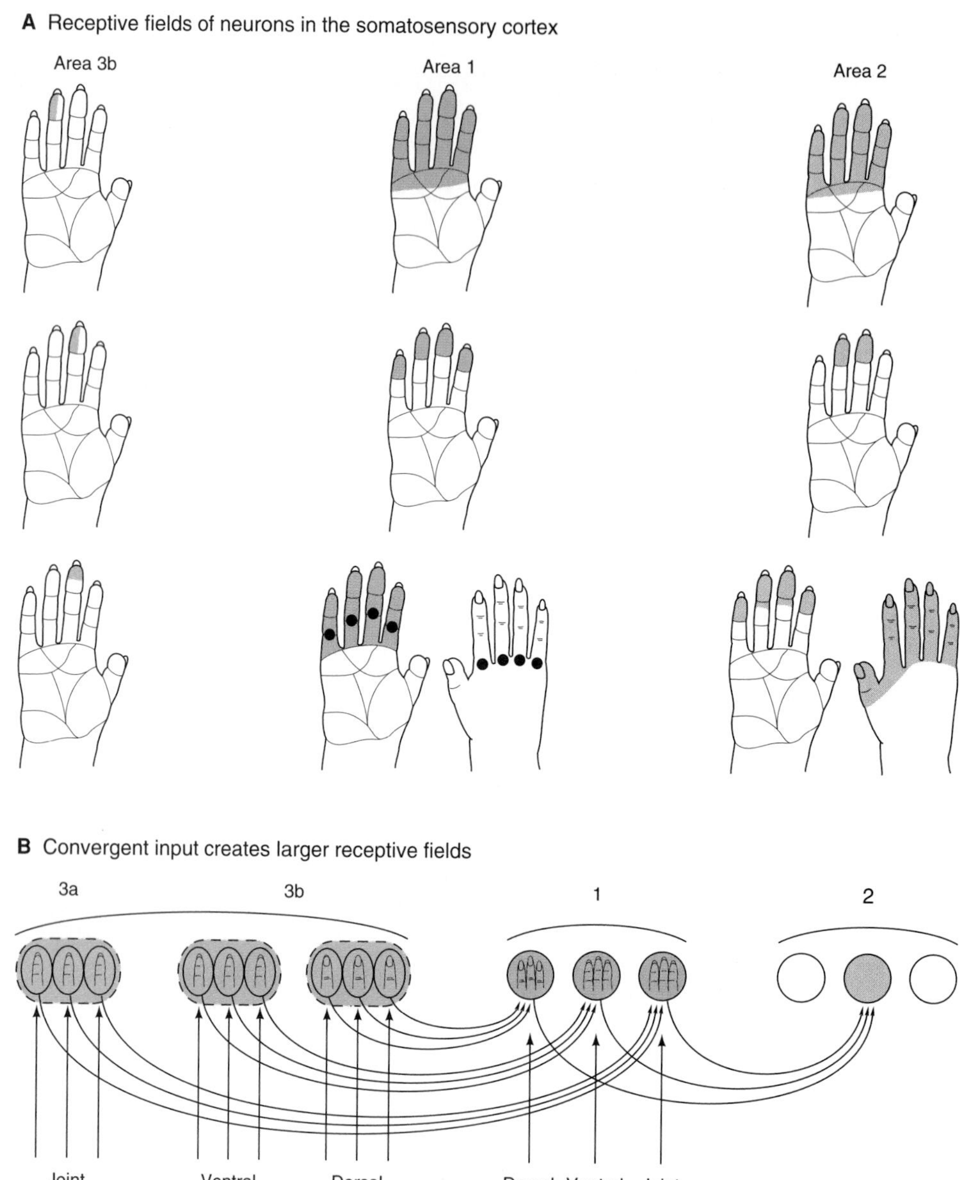

Figure 18–14 Neurons that participate in later stages of cortical processing of somatosensory information (Brodmann's areas 1 and 2) have larger receptive fields.

A. The receptive fields of cells in areas 1 and 2 are compared with those of cells in area 3b, which receives input from the thalamus. The maps are based on recordings made during a single electrode penetration close to the central sulcus. (Adapted from Gardner, 1988.)

B. The larger receptive fields of neurons in areas 1 and 2 are the result of convergent inputs from several neurons with contiguous receptive fields. (Adapted from Iwamura et al., 1985.)

In Later Stages of Cortical Processing the Submodalities Are Integrated by Cells that Have Complex Feature-Detecting Properties

To integrate information about personal space (the body surface) and peripersonal space (the space of objects within reaching distance) the brain must combine information from many different mechanoreceptors sensitive to superficial touch, deep pressure, and the position of the fingers and hand. How is this integration accomplished? At least three factors are involved: (1) the submodalities converge on common cells; (2) the response properties of neurons at successive levels of sensory processing become more complex; and (3) the size of the receptive field becomes larger at each level of processing.

Neurons sensitive to the direction and orientation of the stimulus are not found in Brodmann's areas 3a and 3b, the first two stages of cortical processing, but they are encountered in area 1 and even more extensively in area 2, the areas concerned with the tactile perception of three-dimensional shape (*stereognosis*) and with discrimination of the direction of movement of objects on the skin. Thus, these complex stimulus properties arise from more elementary properties. Similarly, the convergent projections from areas 3a and 3b permit neurons in areas 1 and 2 to respond to other complex features, such as edge orientation. As we shall see below, this convergence provides the necessary tactile clues for skilled movement of the fingers.

The receptive fields of neurons become progressively larger with each successive stage of information processing. For example, neurons in areas 3a and 3b, the sites of initial input into S-I, have quite small receptive fields that usually encompass one or two phalanges on a finger. In contrast, neurons in area 1 and 2, which receive inputs from areas 3a and 3b, have receptive fields that include several fingers. Thus, the receptive fields and response properties of neurons in areas 1 and 2 reflect convergent input from different regions of the hand and fingers, areas that are separately represented in areas 3a and 3b (Figure 18–14). Inputs for the finger areas are commonly adjacent to one another, and the cells respond most effectively when adjacent fingers are stimulated, as when the hand is used to hold and manipulate an object. These complex cells in areas 1 and 2 become active during movements of the hand around an object, and seem to have a role in stereognosis.

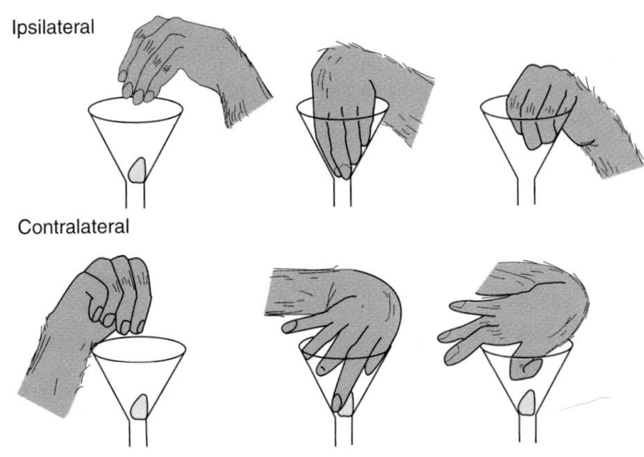

Figure 18–15 A monkey's finger coordination is disrupted following the injection of muscimol, which inhibits synaptic transmission in the somatic sensory cortex. The left hand (ipsilateral) is able to pick up an apple piece from a funnel. Two hours following the injection of muscimol into Brodmann's area 2 on the left side, the finger coordination of the right hand (contralateral) is severely disorganized. (Adapted from Hikosaka et al., 1985).

This increase in the complexity of neuronal response is important not only for perception but also for executing skilled movements. Indeed, somatosensory inputs from the entire body surface are sent from area 2 to the primary motor cortex. Moreover, experimentally induced inhibition of neural activity in area 2 leads to an inability to assume functional postures of the hand or coordinate the fingers for picking up small objects (Figure 18–15).

Real As Well As Imagined and Remembered Space Is Represented in Posterior Parietal Association Areas

We have so far considered the function of the primary and secondary somatic sensory areas in the anterior parietal cortex. As we have seen, areas 3b and 3a are the cortical receiving areas for a variety of somatosensory information from the thalamus, while areas 1 and 2 merge the inputs from areas 3b and 3a, providing for the recognition of three-dimensional shapes of objects and the execution of skilled movements.

The primary and secondary somatic sensory areas of the anterior parietal cortex project to the other major subdivision of the parietal lobe: the *posterior parietal cortex* (Brodmann's areas 5 and 7). The posterior parietal areas also receive input from the visual and auditory systems and are concerned with integrating somatic sensory input with these other sensory modalities so as to give rise to perceptions and to focus attention in extrapersonal space. Lesions in these areas do not produce simple sensory deficits such as blindness, deafness, or loss of tactile sensibility. The deficits are rather more complex, including defects in spatial perception, visuomotor integration, and directed attention.

In Chapter 1 we learned that damage to the dominant (usually left) posterior parietal lobe can produce *aphasia* (a disorder of language, which we return to in Chapter 34). In addition, these lesions produce *agnosia* (an inability to perceive objects through otherwise normally functioning sensory channels). An agnosia, seen after damage to the left posterior parietal cortex, is *astereognosia*, an inability to recognize the form of objects by touch even though there is no pronounced impairment of somatosensory pathways.

Lesions of the nondominant (usually right) posterior parietal lobe do not cause disturbances of language. Instead, patients with right-sided lesions show some of the most remarkable cognitive defects that can be seen in a neurological patient. They often show an agnosia which is manifest as a striking deficit in body image and in the perception of spatial relations. The patients lose conscious awareness of the spatial aspects of all sensory input from the left side of the body as well as of external space. Although their somatic sensations are intact, these patients ignore half of the body (*neglect syndrome*), failing to dress, undress, and wash the affected side. Patients may even deny or disown an affected arm or leg, going so far as to say: "Who put this arm in bed with me?" They may also deny the existence of any paralysis in this limb and may attempt to leave the hospital prematurely since they feel there is nothing wrong with them.

Patients with neglect syndrome not only ignore half of their body, they also ignore half of the outside world. For example, these patients sometimes exhibit a severe disturbance in their ability to copy drawn figures. Thus the patient may draw a flower with petals only on the right side of the plant. When asked to copy a clock, the patient ignores the digits on the left and tries to put all 12 digits on the right, or draws them down the side running off the clock face (Figure 18–16).

A particularly dramatic example of visual neglect is evidenced in the paintings of a German artist who suffered a stroke that affected his right posterior parietal cortex. Asked to paint self-portraits at different times during the course of his recovery, the pictures showed a profound neglect of the left side of the face in portraits done at 2 and 3.5 months. There was partial recovery at 6 months (Figure 18–17), but neglect persisted, albeit to a minor degree, even after the patient's other symptoms had essentially recovered at 9 months. Some aspects of the neglect result from loss of awareness of the contralateral body half. This loss of awareness includes not only the loss of abstract perception but also the loss of representation of personal space. There is also a memory loss, which explains why these patients are not usually aware of their deficits.

Another fascinating instance of unilateral visual neglect occurred in a group of patients in Milan, Italy. As the patients were sitting in the hospital's examining room they were asked to imagine that they were standing in the city's main square, the Piazza del Duomo, facing the famous cathedral of Milan, and to describe from memory the key buildings around the square. These patients, all of whom had injury to the *right* posterior parietal lobe, identified all the buildings on the right side of the square (ipsilateral to the lesion) but could not recall the buildings on the left, even though these buildings were thoroughly familiar to them.

The patients were next asked to imagine that they were standing on the opposite side of the square, on the steps of the cathedral, so that right and left were reversed. In this imagined position the patients were now able to name the buildings they previously had been unable to identify, but were unable to name the buildings they had previously listed (Figure 18–18). The fact that the patients could describe what they previously neglected, and neglected what they had previously described, suggests that they retained in memory full knowledge of the city square. They had not lost stored memo-

Figure 18–16 The three drawings on the **right** were made from the models on the **left** by patients with unilateral visual neglect. (From Kolb and Whishaw, 1990.)

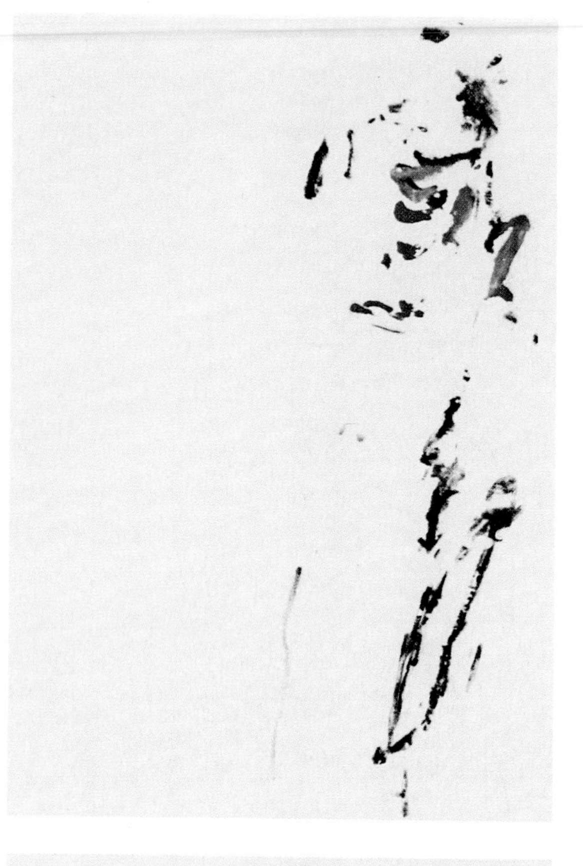

← *Figure 18–17* Self-portraits by an artist following damage to his right posterior parietal cortex. The portraits were drawn 2 months after a stroke (**upper left**), at 3.5 months (**upper right**), at 6 months (**lower left**), and at 9 months (**lower right**), by which time the artist had largely recovered. The early portraits show severe neglect of the side of face opposite the lesion. (From Jung, 1984.)

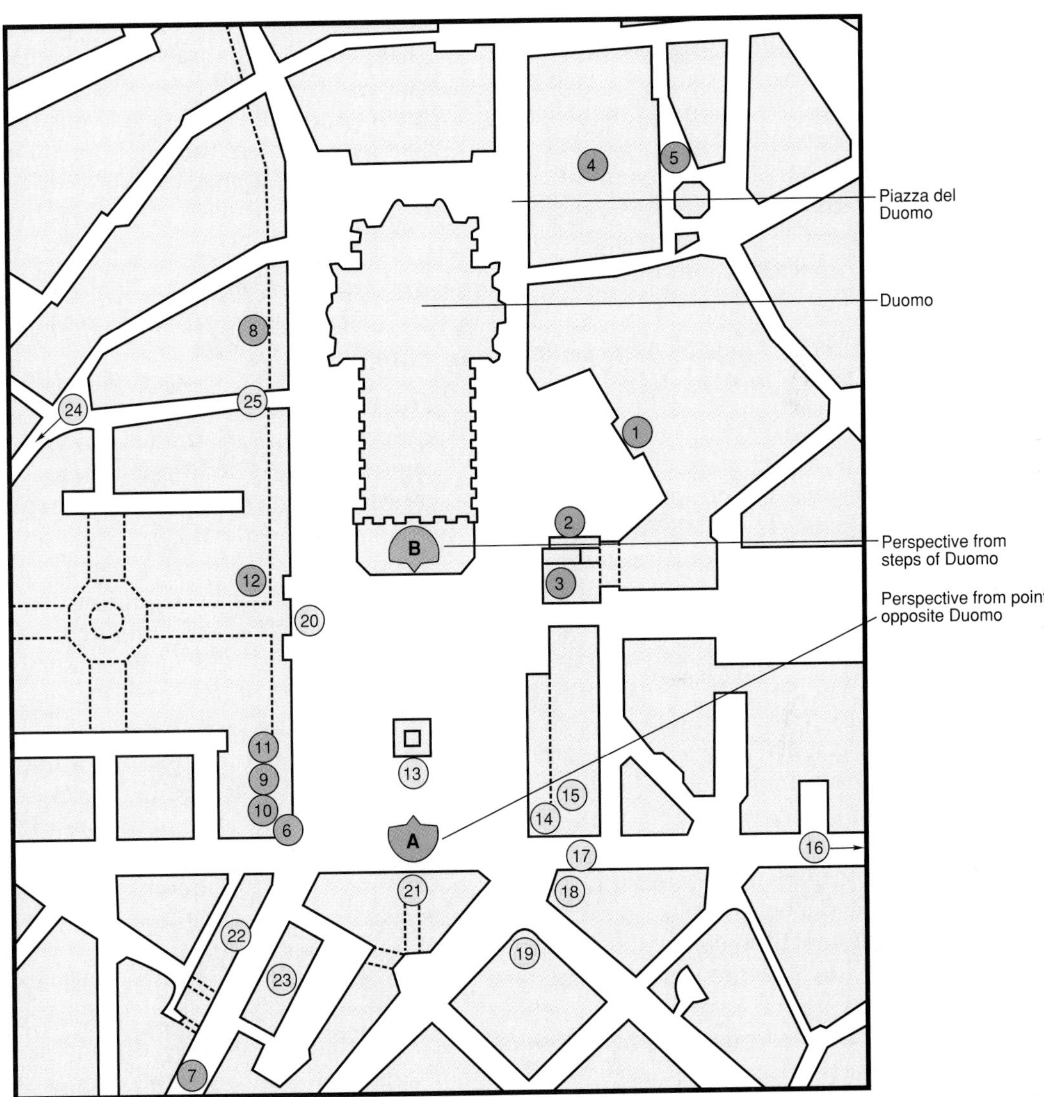

Figure 18–18 Example of a recall deficit limited to the extrapersonal space opposite the side of the lesion. The figure is a map of the Piazza del Duomo in Milan. The numbers represent different landmark buildings. Patients with lesions of the right posterior parietal cortex were asked to recall from memory landmarks bordering the square from two perspectives. The numbered red circles represent landmark buildings recalled from perspective A, the point opposite the Duomo; the numbered dark green circles represent landmark buildings recalled from perspective B, on the steps of the Duomo. (Based on Bisiach and Luzzatti, 1978.)

ries but lacked *access* to memories associated with the side of the body opposite the lesion, no matter which way they imagined themselves facing. Thus, spatial memory is referenced to the body. Memories for each half of the visual field are accessed through the contralateral hemisphere.

Study of these disorders of body image produced by lesions of the posterior parietal lobe yield three insights. First, patients who neglect the left side of the body after damage to the right parietal lobe also neglect visual stimuli in the left visual field. Second, patients with neglect syndrome lack access to memories of earlier perceptions on the neglected side. Thus, these patients neglect not only real external objects but also objects in memory. Finally, the lesions not only lead to disorders in perceptive–spatial relationships, they also commonly lead to a disturbance in directed attention.

Lesions in the Posterior Parietal Association Cortex Produce Defects in Attention

A critical feature of perception is a shifting of attention from one object in the visual field to another. PET imaging of normal subjects has shown that when subjects shift their attention in the visual field both the superior (posterior) parietal cortex and the frontal cortex are activated. Each area is concerned with different aspects of attention. The parietal region is active when attention is shifted on the basis of sensory clues, independent of whether an overt motor act is executed. The frontal region is active only when the selective stimuli lead to an overt motor response.

Observations of patients with lesions of the parietal or frontal cortices support this view. When patients with parietal lobe lesions are presented with simultaneous stimuli in both the left and right visual fields, they are unable to detect the sensory stimulus on the side contralateral to the lesion. This defect, known as extinction, seems to reflect a deficit in shifting attention to-ward contralateral space—the space opposite the lesion—when attention is engaged elsewhere. Although a deficit in focused spatial attention does not by itself explain the wide range of observed phenomena that contribute to neglect, it seems to be an important contributor to the clinical syndrome.

As is the case with neglect, extinction is much more frequent for stimuli on the right side. Why is this so? PET data suggest that the asymmetry arises because the right hemisphere is able to direct attention to both the left and the right visual fields, while the left hemisphere can only control attention to the right visual field. Attention to the left visual field leads to activity in the right superior parietal cortex. Attention to the right visual field yields bilateral activation—one region in the left superior parietal cortex and distinct regions in the right superior parietal cortex. Thus, there are two distinct representations in the right superior parietal lobule for directing attention separately into the left or right visual field, while there is only one representation in the left superior parietal lobule for directing attention mainly to the right visual field.

Attention to a Stimulus Affects the Activity of Single Nerve Cells in the Posterior Parietal Cortex

The role of the posterior parietal cortex in attentiveness has been studied on the cellular level. Studies of cells that respond to visually guided movements reveal that the responses of cells in the posterior parietal cortex to a series of identical stimuli are not uniform but vary remarkably depending on the subject's attention. In one study the response of cells to a spot of light was examined in awake primates; cells in the posterior parietal cortex responded more intensely when the animal attended to the spot than when it ignored it (Figure 18–19). Not only does the firing of the posterior parietal cells increase in response to an object of interest, but it also attenuates in response to objects that are being ignored. These results are consistent with the notion that the parietal cortex is responsible for attention to the manipulation of objects in space.

These studies suggest that when a subject attends visually to an object, cells in the posterior parietal cortex that respond to the object begin to discharge powerfully. As the subject moves its eyes toward the object to examine it further, cells in other areas of the brain involved in visuomotor coordination (the superior colliculus and primary visual cortex) also discharge more briskly. Thus, attention selectively enhances the activity of cells in a number of regions of the cortex responsible for processing visual information related to movement.

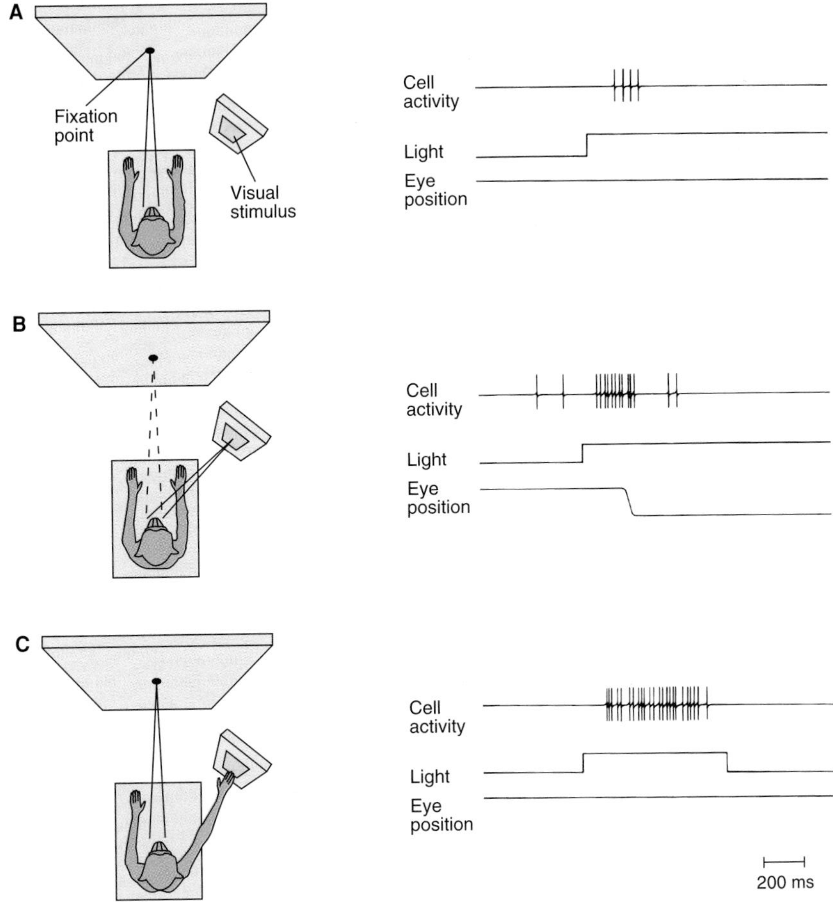

Figure 18–19 Neurons in the posterior parietal cortex of a monkey respond more effectively to a stimulus when the animal is attentive to the stimulus. (From Wurtz and Goldberg, 1989.)

A. A spot of light elicits only a few action potentials in a cell when the animal's gaze is fixed away from the stimulus.

B. The same cell's activity is enhanced when the animal takes visual notice of the stimulus through saccadic eye movement.

C. The cell's activity is further enhanced when the monkey touches the spot, but without moving its eyes.

An Overall View

To come to grips with the biological substrates of cognition it is necessary to move beyond the individual neuron and to consider how information is processed in neural networks. This requires not only the methods and approaches of cellular and systems neuroscience but also the methods and insights of cognitive psychology.

Studies of the sense of touch and its cortical representation in the anterior regions of the parietal lobe provide elementary examples of the internal representation of the body surface and of peripersonal space. This representation is not fixed, but can be modified by experience. Analysis of such modifications indicates that attentiveness is a factor in integrating the internal representation of the body with vision and movement, a process that incorporates the representation of peripersonal space with representations of extrapersonal space. Here, body space becomes related to visual space, whether actual, imagined, or remembered. Thus, a simple representation of self is gradually enlarged into one that includes the external world, both real and imagined—the world in which the self functions.

Selected Readings

Andersen, R. A. 1987. Inferior parietal lobule function in spatial perception and visuomotor integration. In F. Plum (ed.), Handbook of Physiology, Section 1: The Nervous System, Vol. V: Higher Functions of the Brain, Part 2. Bethesda, Md.: American Physiological Society, pp. 483–518.

Beaumont, J. G. 1983. Introduction to Neuropsychology. New York: Guilford Press.

Bisiach, E., and Luzzatti, C. 1978. Unilateral neglect of representational space. Cortex 14:129–133.

Bisach, E. 1993. Mental representation in unilateral neglect and related disorders: The twentieth Sir Frederic Bartlett lecture. Q. J. Exp. Psychol. 46A:435–461.

Corbetta, M., Miezin, F. M., Shulman, G. L., and Petersen, S. E. 1993. A PET study of visuospatial attention. J. Neurosci. 13:1202–1226.

Darian-Smith, I. 1982. Touch in primates. Annu. Rev. Psychol. 33:155–194.

Gardner, E. P., Hamalainen, H. A., Palmer, C. I., and Warren, S. 1989. Touching the outside world: Representation of motion and direction within primary somatosensory cortex. In J. S. Lund (ed.), Sensory Processing in the Mammalian Brain: Neural Substrates and Experimental Strategies. New York: Oxford University Press, pp. 49–66.

Gazzaniga, M. S. (ed.) 1994. The Cognitive Neurosciences. Cambridge, Mass.: M.I.T. Press.

Hyvärinen, J., and Poranen, A. 1978. Movement-sensitive and direction and orientation-selective cutaneous receptive fields in the hand area of the post-central gyrus in monkeys. J. Physiol. (Lond.) 283:523–537.

Kaas, J. H., Nelson, R. J., Sur, M., Lin, C.-S., and Merzenich, M. M. 1979. Multiple representations of the body within the primary somatosensory cortex of primates. Science 204:521–523.

Kaas, J. H., Nelson, R. J., Sur, M., and Merzenich, M. M. 1981. Organization of somatosensory cortex in primates. In F. O. Schmitt, F. G. Worden, G. Adelman, and S. G. Dennis (eds.), The Organization of the Cerebral Cortex: Proceedings of a Neurosciences Research Program Colloquium. Cambridge, Mass.: MIT Press, pp. 237–261.

Kolb, B., and Whishaw, I. Q. 1990. Fundamentals of Human Neuropsychology, 3rd ed. New York: Freeman.

McCarthy, R. A., and Warrington, E. K. 1990. Cognitive Neuropsychology: A Clinical Introduction. San Diego: Academic Press.

Mesulam, M.-M. 1985. Principles of Behavioral Neurology. Philadelphia: Davis.

Mountcastle, V. B. 1984. Central nervous mechanisms in mechanoreceptive sensibility. In I. Darian-Smith (ed.), Handbook of Physiology, Section 1: The Nervous System, Vol. III: Sensory Processes, Part 2. Bethesda, Md.: American Physiological Society, pp. 789–878.

Neisser, U. 1967. Cognitive Psychology. New York, Appleton-Century Crofts.

Pandya, D. N., and Seltzer, B. 1982. Association areas of the cerebral cortex. Trends Neurosci. 5:386–390.

Pons, T. P., Garraghty, P. E., Friedman, D. P., and Mishkin, M. 1987. Physiological evidence for serial processing in somatosensory cortex. Science 237:417–420.

Ramachandran, V. S. 1993. Behavioral and magnetoencephalographic correlates of plasticity in the adult human brain. Proc. Natl. Acad. Sci. USA. 90:10413–10420.

Steinmetz, M.A., Connor, G.E., Constantinidis, C., and McLaughlin, J.R. 1994. Covert attention suppresses neuronal responses in area 7A of the posterior parietal cortex., J. Neurophysiol. 72:1020–1023.

Vallbo, Å. B., Olsson, K. Å., Westberg, K.-G., and Clark, F. J. 1984. Microstimulation of single tactile afferents from the human hand: Sensory attributes related to unit type and properties of receptive fields. Brain 107:727–749.

19

Cognition and the Cortex

The Three Association Areas Are Involved in Different Cognitive Functions

The Frontal Association Areas Are Involved in Motor Strategies and Motor Planning

- *Lesions of the Principal Sulcus in Monkeys Interfere with Specific Motor Tasks*
- *Lesions of the Inferior Prefrontal Convexity Interfere with Appropriate Motor Responses*

The Parietal Association Areas Are Involved in Higher Sensory Functions and Language

The Temporal Association Areas Are Involved in Memory and Emotional Behavior

- *The Two Hemispheres Have Different Cognitive Capabilities*
- *The Two Hemispheres Are Differentially Specialized*
- *The Two Hemispheres Are Anatomically Asymmetrical*
- *Why Are the Hemispheres Specialized?*

Cognitive Functions Can Now Be Simulated by Artificial Neural Networks that Employ Parallel Distributed Processing

As we have seen in the last chapter, even highly complex cognitive functions are localized to specific areas of the brain. We shall now learn that localization of cognitive function does not imply that a specific function is mediated *exclusively* by one region of the brain. Rather, it means that certain areas are more concerned with one kind of function than with others. In fact, most functions require the integrated action of neurons in different regions.

Some of the most compelling evidence for the nature and localization of cognitive functions has come from studies of the *association areas* of the cerebral cortex. In considering the function of the posterior parietal cortex in the last chapter, we saw that association areas are concerned with the integration of somatic information with other sensory modalities and with the planning of movement. In this chapter we focus primarily, but not exclusively, on the functions of the three major association cortices: the prefrontal, the parietal–temporal–occipital, and the limbic association areas.

Because the association cortices produce few or no obvious motor or sensory effects when electrically stimulated, they were at one time called silent areas. They were thought to have two main functions: to integrate the activity of the various primary sensory cortices, and to link sensory cortices to motor cortices. Accordingly, the association cortices were thought to represent the exclusive sites for the highest cognitive functions: perception and

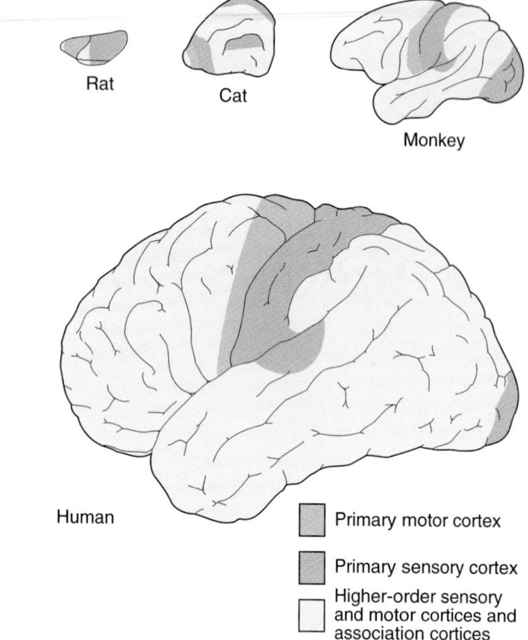

Figure 19–1 Drawings (approximately to scale) of the cerebral hemispheres of four mammals. Note the increase in both size and relative amount of higher-order sensory, motor, and association cortices.

thought. This idea seemed to be supported by the observed increase in the relative extent of the association cortices throughout phylogeny, reaching a maximum in humans (Figure 19–1). Modern evidence, however, indicates that cognitive functions are not restricted to association cortices but involve a variety of cortical areas. Indeed, as our understanding of sensory and motor processes has increased, some areas that were thought to be association cortex instead have proved to be higher-order processing centers for sensory or motor information.

Much of what we know about the function of association areas has come from the study of patients with two types of lesions—those with damage to the cortex, as a result of trauma, cerebrovascular disease, or tumors, and those who have undergone brain surgery for a neurological or behavioral disorder. Evidence from the second group of patients has been particularly instructive because the surgical lesions are relatively well defined. In some instances insight obtained from clinical studies has been amplified by experiments on animals, in which it is possible to make local lesions and to obtain detailed behavioral and electrophysiological information. Recently, brain imaging techniques have provided additional information on the function of these cortices in normal humans (see Chapter 5).

We begin by considering the structure and function of the three association areas.

The Three Association Areas Are Involved in Different Cognitive Functions

Table 19–1 summarizes the major primary and higher-order sensory and motor areas and the three major association areas of the cerebral cortex. Each primary sensory area of cortex is adjacent to higher-order sensory processing centers, to which it connects. For example, the primary visual cortex, in Brodmann's area 17, is adjacent to area 18a, a higher-order visual cortex (Figure 19–2). The higher-order areas are concerned with more detailed processing of sensation and project to one or more of the three association areas.

Each association area appears to be specialized in function, although all three association areas participate in more than one cognitive function, including voluntary movement, sensory perception, emotional behavior, memory, and language. The prefrontal cortex is concerned with the planning and execution of complex motor actions; the parietal–temporal–occipital area with integration of sensory functions and with language; and the limbic area with memory and with emotional and motivational aspects of behavior. We begin our survey of the three association cortices by briefly considering their intracortical connections.

The *prefrontal association cortex* is one of two regions in the frontal lobe that lies anterior to the primary motor area (Figure 19–2). The other region is the *premotor cortex*, a higher-order motor area that includes the supplementary motor area as well as premotor area. The premotor area is important in the initiation of action. The prefrontal area, as we shall see later, is important for the planning of action.

The prefrontal and premotor areas receive input from various higher-order sensory cortices. Higher-order sensory regions that are closely connected with primary sensory areas project to the premotor cortex, which in turn projects to the primary motor cortex. Higher-order sensory regions that are

Table 19–1 Major Sensory, Motor, and Association Areas of the Cerebral Cortex

Functional Designation	*Lobe*	*Brodmann's Area*
Primary sensory cortex		
Somatic sensory	Parietal	1,2,3
Visual	Occipital	17
Auditory	Temporal	41,42
Higher-order sensory cortex		
Somatic sensory II	Parietal	2 (opercular portion)
Visual II	Occipital	18
Visual III, IIIa, IV, V	Occipital, temporal	19 and area rostral to 19
Visual Inferotemporal area	Temporal	21,20
Posterior parietal cortex (somatic sensation, vision)	Parietal	5 (somatic) 7 (visual)
Auditory	Temporal	22
Primary motor cortex	Frontal	4
Higher-order motor cortex (premotor, including supplementary motor area)	Frontal	6,8
Association cortex		
Parietal–temporal–occipital (polymodal sensory, language)	Parietal, temporal, and occipital	39,40, and portions of 19, 21,22,37
Prefrontal (cognitive behavior and motor planning)	Frontal	Area rostral to 6
Limbic (emotion and memory)	Temporal, parietal, and frontal	23,24,38,28,11

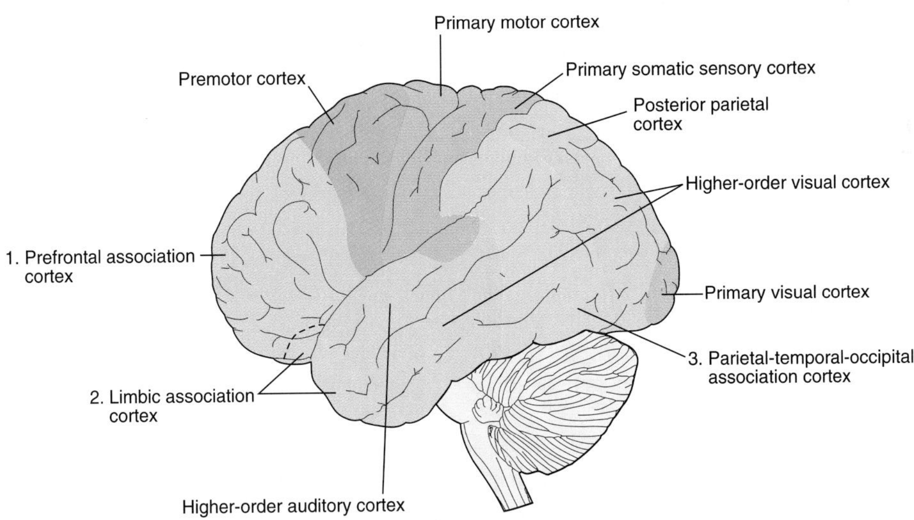

Figure 19–2 The lateral surface of the human brain shows the regions of the primary sensory and motor cortices, the higher-order motor and sensory cortices, and the three association cortices.

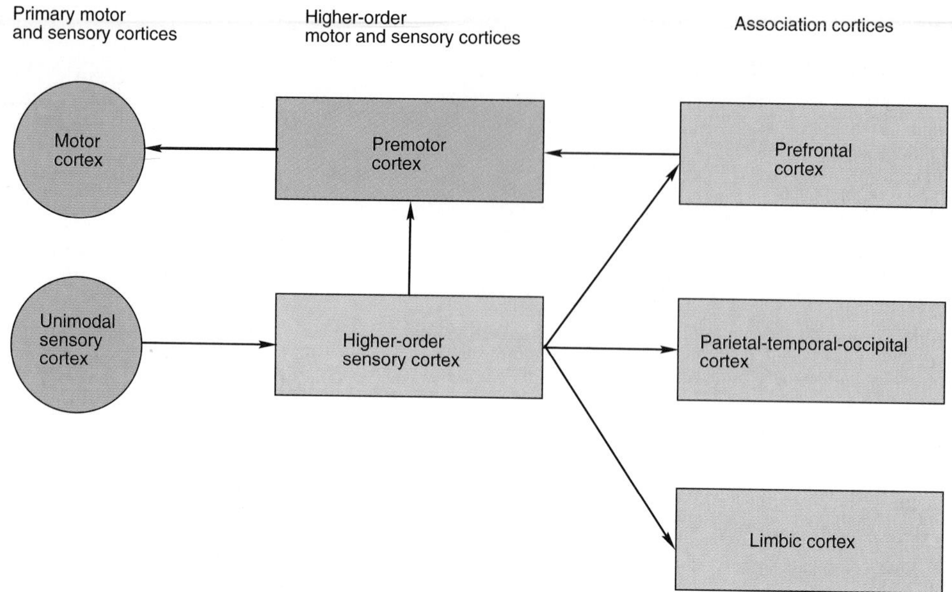

Figure 19–3 The intercortical connections of primary motor and sensory cortices, higher-order motor and sensory cortices, and association cortices are shown here in the form of a flow chart. The same general pattern is repeated for each of the main primary sensory cortices (for vision, touch, and hearing). For simplicity, a number of pathways, such as those interconnecting the three association cortices, have been omitted.

less directly connected to primary sensory areas project to the prefrontal association cortex, which in turn projects to the premotor cortex (Figure 19–3). These patterns of connection permit sensory information to influence the execution of movement (by way of successive projections to the premotor and then to the motor cortex). At the same time, sensory information can influence the planning of movement by way of successive projections to the prefrontal cortex, the premotor cortex, and then the motor cortex.

The *parietal–temporal–occipital association cortex* consists of several functional areas that are intercalated among higher-order somatic, visual, and auditory areas and that receive projections from them. The parietal–temporal–occipital association cortex is therefore thought to provide linkages that are important in the processing of sensory information for perception and language.

The *limbic association* cortex is located in the medial and ventral surfaces of the frontal lobe, the medial surface of the parietal lobe, and the anterior tip of the temporal lobe (called the *temporal pole*). The limbic association cortex includes the orbitofrontal cortex, the cingulate region, and the parahippocampal area. It receives projections from higher-order sensory areas and sends projections to other cortical regions, including the prefrontal cortex. This pathway allows emotions to affect motor planning.

Even though the association areas are concerned with higher mental functions, their organization is similar in principle to that of the primary sensory and motor cortices. Specifically, the intercortical connections between regions of association cortex of the parietal lobe and the frontal lobe terminate in distinct, vertically oriented columns. These columns are 200 to 500 μm wide and extend across all layers of cortex. Thus, the columnar organization we encountered in Chapter 18 is not unique to the sensory cortices, but is a general feature of all neocortex.

The Frontal Association Areas Are Involved in Motor Strategies and Motor Planning

Perhaps the most important association function of the prefrontal lobes is to weigh the consequences of future actions and to plan accordingly. To select the appropriate motor responses from the many available choices, the frontal lobes must integrate internal and external sensory information.

Functional and anatomical studies of monkeys suggest that the prefrontal association cortex can be divided into two main subregions: the *prefrontal association cortex* proper, located on the dorsolateral surface of the frontal lobes, and the *orbitofrontal cortex*, located on the medial and ventral surface of the brain. These two regions are very large in nonhuman primates and humans (Figure 19–4). The orbitofrontal cortex is part of the limbic association cortex and connects directly to limbic structures, such as the amygdala. We first consider the prefrontal cortex, which in the monkey has three subdivisions: principal sulcus, superior prefrontal convexity, and inferior prefrontal convexity (Figure 19–5).

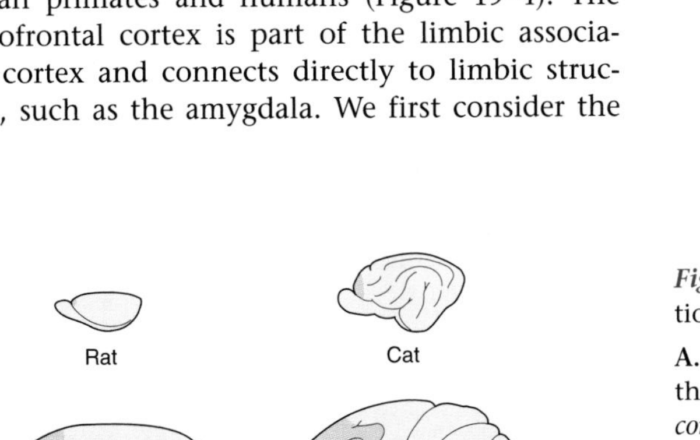

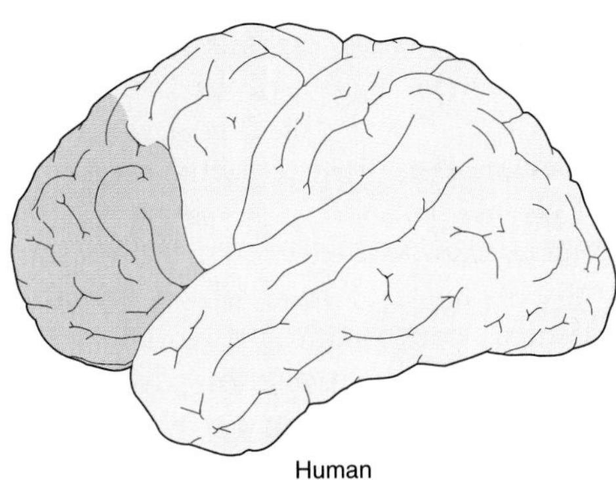

Figure 19–4 Proportion of the brain taken up by the frontal association cortex (**blue**) in five species. (Drawings not to scale.)

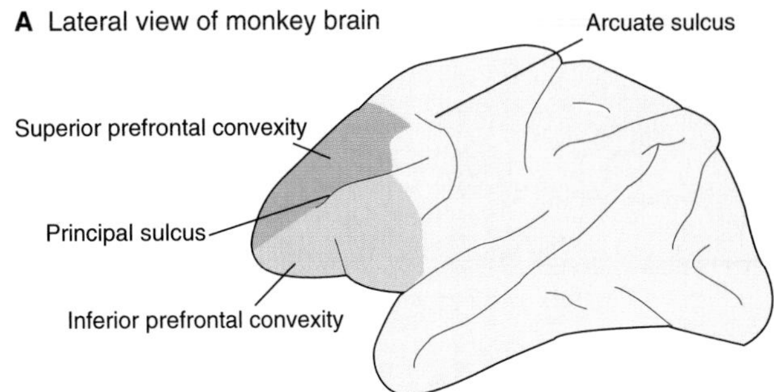

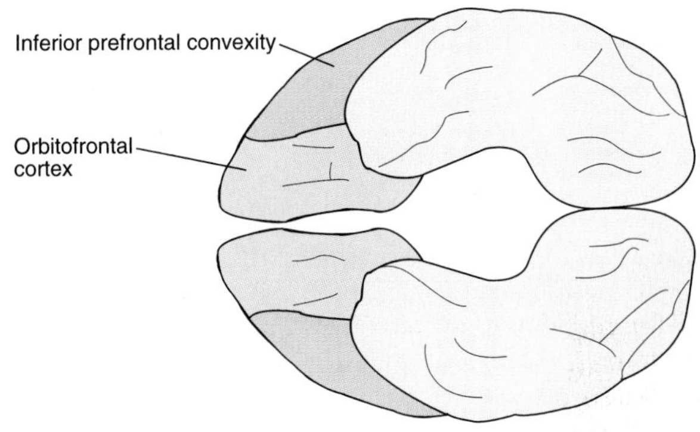

Figure 19–5 Basic subdivisions of the frontal association cortex of the monkey. (Data from Rosenkilde, 1979.)

A. The lateral view illustrates the dorsoventral surface of the prefrontal lobe, the region of the *prefrontal association cortex*.

B. A ventral view illustrates the orbitofrontal cortex, a subdivision of the *limbic association cortex*.

Lesions of the Principal Sulcus in Monkeys Interfere with Specific Motor Tasks

The cortex at the principal sulcus is concerned with the strategic planning for higher motor actions, including cognitive tasks. Relatively small lesions in chimpanzees and monkeys around the principal sulcus impair the ability to perform a task involving a *delayed spatial response*. For example, in

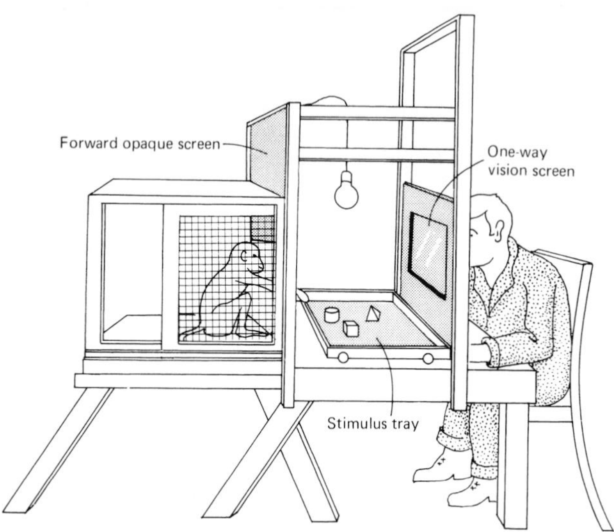

Figure 19–6 The Wisconsin General Testing Apparatus is used to test monkeys in a variety of discrimination and learning problems. (Adapted from Harlow, 1958.)

one experiment a hungry animal was shown a piece of food and, while the animal watched, the food was placed randomly under one or the other of two identical opaque containers, one on the left, one on the right. After a delay of 5 seconds or longer, the monkey was permitted to select one of the containers. Normal animals quickly learned to select the container covering the food, whereas animals with damage to the principal sulcus did poorly on this task. The lesioned animals performed well only if there was no delay after the experimenter covered the food. These experiments suggest that the prefrontal region might be involved in a form of short-term memory, and that the frontal association areas are needed for the execution of complex motor tasks when the essential cues are not present in the environment at the time of responding but must be recalled by a short-term memory process (Figure 19–6).

These lesions do not produce a generalized deficit involving all types of short-term memory. Rather, the deficit is specific for *working memory,* a temporary storing of information used to guide a future action. This deficit is highly specific to delayed spatial tasks. Animals with lesions in the principal sulcus have no difficulty with discrimination tasks involving no delay or with tasks for which spatial cues are not important. For example, lesioned animals can be trained so that when shown several objects, they correctly select the one object that was shown on the previous trial. This task requires a response after a delay but does not have any spatial aspects, such as responding to the left or right cue.

The idea that the principle sulcus of the prefrontal association cortex is involved in delayed response tasks is also supported by cellular studies. In one type of study monkeys are trained to fix their gaze on a central spot while a cue is briefly presented someplace in the visual field. After a delay of 1 to 6 seconds the animal must turn its eyes to where the cue had been. Thus, for 1 to 6 seconds the animal must keep in working memory the information about the position of the cue. Many neurons in the prefrontal region increase their rate of firing when the cue is first presented, but they continue to fire throughout the delay period even when the cue is no longer there (Figure 19–7, upper traces). Furthermore, during the delay many individual neurons respond only to stimuli at a particular location in the visual field—usually a location in the contralateral hemifield. When the animal occasionally makes a mistake and turns its eyes to a position other than where the cue had been, neurons that normally would have fired during the delay period fail to fire (Figure 19–7, lower trace). This pattern of cellular response suggests that the prefrontal region contains a complete map of the contralateral visual field that can be used for the purpose of working memory. Thus, small lesions of the principal sulcus interfere with the capacity of monkeys to remember the position of objects in specific regions of the visual field contralateral to the side of the lesion. These highly focal effects are described as blind spots (or scotomas) of visual memory.

Anatomical studies suggest that the dorsolateral prefrontal association cortex, which contains the principal sulcus concerned with working memory, works closely with the posterior parietal association cortex, which is concerned with extrapersonal space and spatial memory (see Chapter 18). The two regions are highly interconnected and both project to many common cortical and subcortical structures. Like prefrontal neurons, certain posterior parietal neurons are active during a period when

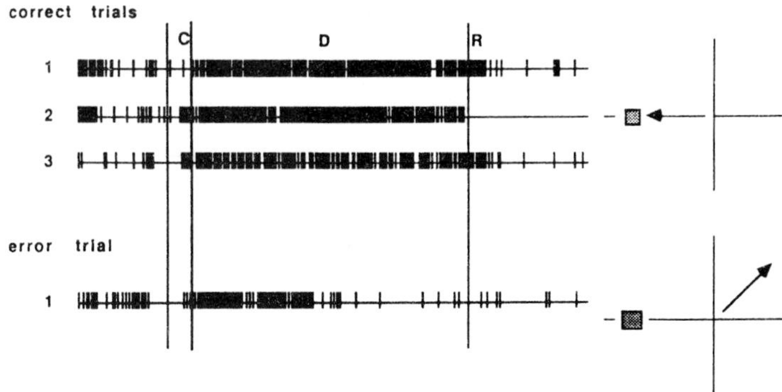

Figure 19–7 Firing of a neuron in the principal sulcus of a monkey during three trials (**upper traces**) of correct performance of an oculomotor delayed-response task and during a trial (**lower trace**) in which the animal made an error (i.e, did not turn its eyes to the cue). In this experiment the neuron was located in the right hemisphere and responded best when the cue was presented directly to the left (indicated by coordinates in the figure). The animal fixated at the center of the visual field, and the visual cue was presented for 0.5 seconds (indicated by the letter **C**). For the three trials in which the animal responded correctly during the subsequent 6-second delay period (**D**), the animal continued to stare at the center of the visual field, and although the cue was no longer present the neuron began to fire at a high rate (**upper trace 1**). At the end of the delay period a signal was given for the animal to turn its eyes to where the cue had been (response, **R**, direction indicated by arrow). On the trial in which the animal incorrectly turned its eyes to the right (**lower trace**) the neuron began to fire but then almost completely stopped firing after 3 seconds. Some neurons failed to show any increase in firing during the delay when the animal made an incorrect response. (Adapted from Funahashi, Bruce, and Goldman-Rakic, 1989.)

the animal has to remember the position of a visual target to which eye movement will be made. Thus, the activity of these neurons corresponds to the plan to make an eye movement with a specific direction and amplitude.

The prefrontal area of primates and other animals is particularly rich in dopaminergic terminals. This dopaminergic innervation is thought to be an important modulatory input for the storage of working memory. In fact, the delayed response is disrupted when dopamine is depleted from the principal sulcus by means of a localized injection of 6-hydroxydopamine, a drug that selectively destroys terminals that use dopamine as a transmitter. Moreover, disturbances of this dopaminergic system contribute to the symptoms of schizophrenia, a disorder of thought.

Imaging studies of the brains of schizophrenic patients support this idea. The frontal lobe tends to be smaller in schizophrenics than in normal individuals. Furthermore, whereas blood flow to prefrontal areas increases in normal subjects challenged by a task that engages prefrontal functions, such as with the Wisconsin card sort test, the increase is less in people with schizophrenia.

Lesions of the Inferior Prefrontal Convexity Interfere with Appropriate Motor Responses

In contrast to lesions of the principal sulcus, lesions of the cortex in the area of the inferior prefrontal convexity affect the ability of an animal to perform any type of delayed response, whether or not it has a spatial element. These lesions appear to interfere with tasks that require the animal to inhibit certain motor responses at appropriate times. Lesions of the arcuate concavity, adjacent to the principal sulcus, do not disturb delayed responses but diminish the animal's ability to choose among different motor responses to particular sensory cues. For example, lesioned animals have difficulty learning a task in which they must move left when an auditory cue comes from above the cage or move right when the cue comes from below the cage.

The Parietal Association Areas Are Involved in Higher Sensory Functions and Language

The anterior parietal lobe contains the primary somatic sensory cortex, whereas the more posterior region contains higher-order sensory areas and an association area (the posterior parietal association area) with extensive convergence from several sensory systems. As we have seen in Chapter 18, lesions in the posterior parietal cortex (Brodmann's areas 5 and 7) of animals and humans produce subtle deficits in the learning of tasks requiring spatial knowledge of surroundings as well as knowledge of the body in space.

As discussed in detail in Chapter 36, patients with damage to the parietal lobes often show striking deficits, including abnormalities in body image and in perception of spatial relations. Damage to the dominant parietal lobe tends to produce *aphasia* (a disorder of language, see Chapter 34) and *agnosia* (an inability to perceive objects although the sensory pathways are otherwise normally functioning).

The Temporal Association Areas Are Involved in Memory and Emotional Behavior

The limbic association cortex consists of several major subareas located in different lobes: the orbitofrontal cortex, the cingulate gyrus, and portions of the temporal lobe. We will consider in Chapter 32 the role of the orbitofrontal cortex and the cingulate gyrus in emotion. Here we focus on the temporal lobe and its role in memory.

In monkeys, lesions of the inferior temporal region, a higher-order visual region, result in deficits in the speed of learning of visual tasks. The deficits, which are not due to blindness, are most pronounced when the visual task is complex. For example, lesions in this area interfere with the ability of an animal to improve performance progressively (to develop a learning set) when a long series of related visual problems is presented. In addition to interfering with the *acquisition* of a learned visual task, these lesions interfere with the *retention* or *memory* of visual tasks. Similarly, damage to the superior temporal cortex of animals does not produce deafness, but impairs learning of auditory patterns (such as discriminating dah–dit–dah, from dit–dah–dit).

As we saw in Chapters 1 and 18, major insights into the functions of the human temporal lobes have come from the work of the neurosurgeon Wilder Penfield. Penfield stimulated various points on the temporal lobe electrically in awake patients before he removed diseased epileptic tissue. As expected, stimulation of the primary auditory areas produced crude auditory sensations. In contrast, stimulation of the superior temporal gyrus produced altered *perception* of sounds and auditory illusions and hallucinations, which seemed remarkably real, almost as if the patients were reexperiencing a past event.

Many of the patients studied by Penfield and others had a portion of their temporal lobe resected as a treatment for epilepsy. The lesion did not include Wernicke's speech area, but did typically include portions of the hippocampus. In the few patients in whom both the left and the right temporal lobes were removed, there was a profound and irreversible impairment of the capacity to form certain types of long-term memories (see Chapter 35).

Patients with unilateral damage of a temporal lobe also experienced some loss of memory, although the deficit was mild compared with that following bilateral lesions. Furthermore, the degree of impairment depended on which side of the brain had the lesion and on the type of material to be memorized. Patients with left temporal lesions had difficulty remembering verbal material, such as a list of nouns. Patients with a right-sided lesion retained normal verbal memory but their ability to remember patterns of sensory input was impaired. When presented with a series of pictures of human faces, some of which were repeated, patients in whom the right temporal lobe was removed had difficulty remembering whether they had previously seen a given face. These patients had no difficulty with series of geometric figures (squares and circles) but they did have difficulty with a series of irregular patterns.

Indeed, geometric patterns can easily be named and then stored as verbal data (as "square," "triangle,"), but faces and complex patterns cannot be readily encoded verbally. Studies of brain-damaged patients show that lesions of several different regions of the left hemisphere impair processing of verbal material, while right-hemisphere lesions interfere with the processing of nonverbal information.

Stimulation (or removal) of the temporal portion of the limbic association cortex affects emotions. For example, stimulation of the anterior and medial temporal cortex produces fear.

The Two Hemispheres Have Different Cognitive Capabilities

Perhaps the most dramatic evidence for the localization of function to one hemisphere rather than the other comes from research on epileptic patients in whom the major fiber pathways connecting the two hemispheres (the corpus callosum and anterior commissure) have been cut to prevent the spread of epileptic activity from one side of the brain to the other. Studies of these patients show that each hemisphere is capable of functioning independently when the two are isolated from each other. Although the right hemisphere is generally mute and cannot communicate its experience verbally, it can do many of the things that the verbal hemisphere is capable of doing. Such basic processes as sensory analysis, memory, learning, and calculation can be performed by either hemisphere. The ability of the right hemisphere alone is limited, however, when the task involves complex reasoning or analysis.

Intuitively, it seems obvious that the corpus callosum and other commissures integrate the functions of the two hemispheres. Yet it is difficult to tell from casual observation what, if any, impact the severance of the corpus callosum and anterior commissure have on behavior. Indeed, early investigators failed to find any deficiencies. By 1940 Warren McCulloch concluded with irony that the only certain role of the corpus callosum was "to aid in the transmission of epileptic seizures from one to the other side of the body." In 1950 Lashley facetiously reiterated his feeling that the purpose of the corpus callosum "must be mainly mechanical . . . to keep the hemispheres from sagging."

The functional role of the hemispheric commissures first became apparent in the mid-1960s in split-brain studies of animals by Roger Sperry. In addition to sectioning the corpus callosum, Sperry limited visual input to one hemisphere by cutting the optic chiasm, thereby destroying the crossed visual fibers (i.e., those fibers from the temporal portion of the retina, Figure 19–8). The *split-brain animals* were trained in complex visual discriminations using one eye. Unlike normal animals, when tested with the untrained eye, the split-brain animals behaved as if they were completely naive. The effects of the training experience were limited to the hemisphere receiving the visual input. This indicated that the corpus callosum conveyed visual information from one hemisphere to the other during training, and also that it provided the pathway by which information stored in one hemisphere can be used by the other hemisphere.

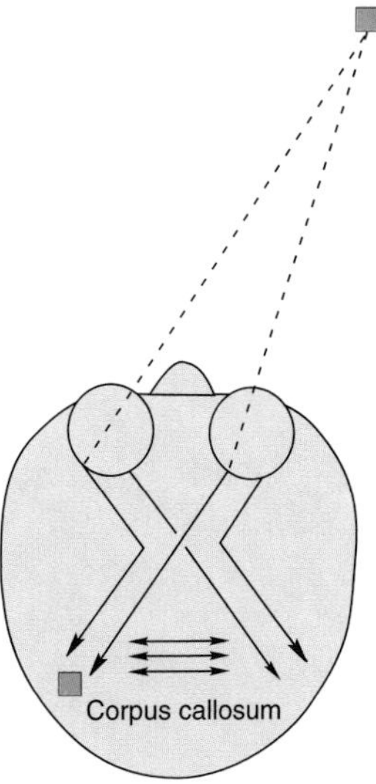

Figure 19–8 An image in the right visual field stimulates the left temporal retina and right nasal retina. Because projections from the nasal retina project contralaterally, whereas those from the temporal retina project ipsilaterally (as shown here in a superior view of the brain), the information projects to the left hemisphere, although it can secondarily reach the right hemisphere if the corpus callosum is intact. (Adapted from Sperry, 1968.)

Sperry next turned to the function of the corpus callosum in humans by studying a group of epileptic patients whose corpus callosum had been sectioned to treat their epilepsy. By studying these patients, Sperry and his colleagues confirmed in humans the findings from earlier studies on animals, but also demonstrated that under experimental conditions these patients were severely limited in the ability to perform certain types of tasks, even when the tasks requried the operation of only a single hemisphere.

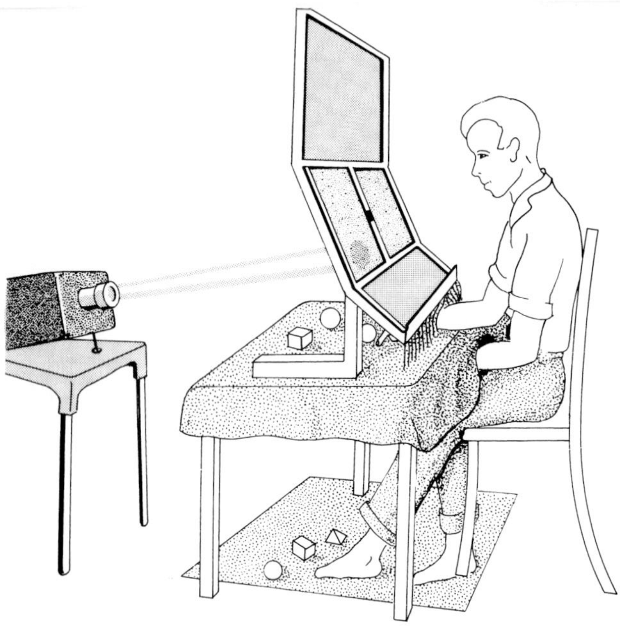

Figure 19–9 In this experimental setup a commissurotomized subject's gaze is fixed between the two screens. Words or images of objects can be briefly flashed on the translucent screens in either the left or right visual field of the subject. The person can identify the stimuli either verbally or nonverbally by touching or pointing to objects hidden behind the screen. (Adapted from Sperry, 1968.)

One reason these patients do moderately well in real-life situations, despite the absence of direct interhemispheric communication, is that ordinarily both hemispheres work to some degree independently and obtain similar information, a paralleling of function that allows for effective integration. For example, as we shall see in Chapter 21, each hemisphere receives a complete representation of the visual field (Figure 19–8). But the experimental situation can be arranged so that these cross-cues are eliminated. One simple way to do this is to present visual stimuli briefly in either the right or left visual field. Such visual stimuli project only to the opposite hemisphere, for in the absence of callosal fibers the visual information is unable to gain access to the ipsilateral hemisphere.

Even a simple experiment using this technique immediately reveals a deficit when the callosum has been sectioned. When a subject was presented with an apple in the right visual field and questioned about what he saw, he said—not surprisingly—"apple." If the apple was presented to the left visual field, however, the patient denied seeing anything, or if prompted to give an answer, he guessed or confabulated. This is not because the right hemisphere was blind or unable to remember a simple stimulus, such as an apple. The patient could readily identify the object if he could point to it. Moreover, he could pick out the apple from several other objects under a cover using only tactile clues (Figure 19–9). What the patient could not do was *name* the visual stimulus that was limited to the right hemisphere. This suggests that, although the right hemisphere cannot talk, it can perceive, learn, remember, and issue commands for motor tasks.

We speak after all, as Broca first illustrated, with our left hemisphere. Nevertheless, the right hemisphere may be capable of a primitive understanding of language. For example, some words projected to the right hemisphere can be read and understood. If the letters **D-O-G** are flashed in the left visual field (the right hemisphere), the patient selects a model of a dog with his left hand. More complicated verbal inputs to the right hemisphere, such as commands, are comprehended poorly. Although the right hemisphere appears almost totally incapable of language *output*, it does seem able to process simple linguistic *inputs*.

The right hemisphere is not, however, merely a copy of the left hemisphere without verbal capacity. On certain perceptual tasks the right hemisphere actually performs better than the left. For example, in a task that calls for arranging pieces of colored wooden blocks into a coherent pattern, commissurotomized patients perform better with the left hand than with the right. Thus, as indicated earlier, the nonspeech (right) hemisphere is superior on spatial–perceptual problems.

In commissurotomized patients the two hemispheres are sufficiently independent that they can even interfere with each other's function. This can be seen when a patient is shown a visual pattern and asked to reproduce that pattern by arranging wooden blocks, each of which has a portion of the total pattern. When the task is performed with the nondominant left hand (the hand ipsilateral to the left hemisphere), the dominant right hand sometimes attempts to interfere with the placement of a correct block, thereby impeding the successful solution of the problem. In addition, the dominant left hemisphere sometimes comments on the performance of the nondominant hemisphere, frequently exhibiting a false sense of confidence on problems for which it cannot know the solution, since the information was projected exclusively to the nondominant hemisphere.

In sum, studies of commissurotomized patients suggest that these individuals function with two independent minds, the left aware and verbal, the right functioning largely automatically. In these patients either hemisphere is capable of directing behavior. Which hemisphere gains control seems to depend on which is best suited for the task to be performed. This is seen clearly in experiments with chimeric figures—for example, a face in which the right half is male and the left half is female (Figure 19–10). Commissurotomized patients say that the face is that of a man, but if asked to select the face from a series of whole faces, they point to a woman. Presumably either hemisphere is capable of directing the pointing; nevertheless, the more competent right hemisphere (which receives information from the left visual field) is in control of this task. When the task requires a verbal answer, of which the right hemisphere is incapable, the left hemisphere controls the task.

Figure 19–10 Hybrid (chimeric) figures, like this face, are used in experiments with commissurotomized patients to clarify the circumstances in which each hemisphere exerts dominant control. After fixating on the dot in the center of the figure, the patient is asked to describe verbally what she or he sees or to point to a face that matches the one seen.

The Two Hemispheres Are Differentially Specialized

Each isolated hemisphere has its own strengths and weaknesses with regard to a given task. Certain tasks are best performed by analysis, by breaking down the problem into logical elements. This type of task is well suited to verbal encoding. Other tasks may be best performed not by sequential analysis but by some type of simultaneous processing of the whole input. For example, we ordinarily recognize a face not by adding up a set of features—a mustache, glasses, a small nose, and so on—but rather by some process by which all these elements are immediately perceived. A face simply looks familiar or not familiar. If we had to verbalize all the features we recognize, we would find it difficult and time consuming.

It is sometimes said that our brain consists of a left hemisphere that excels in intellectual, rational, verbal, and analytical thinking and a right hemisphere that excels in sensory discrimination and in emotional, nonverbal, and intuitive thinking. However, in the normal brain, with extensive commissural interconnections, the interaction of the two hemispheres is such that we cannot dissociate clearly their specialized functions. In fact, there is now evidence that the capacity of one hemisphere to perform a particular task may deteriorate when it is disconnected from the other hemisphere after commissurotomy. For example, a patient has been described who could perform a tactile task (discriminating the detailed shapes of wire figures by touch) with either hand before split-brain surgery. After the surgery, the subject could not perform the task with either hand. The loss of this ability suggests that interaction between the hemispheres is needed for this task, even though other evidence indicates that it may be mediated primarily by the right hemisphere. Thus, despite the dramatic differences in the capacities of the isolated hemispheres, when they are interconnected they seem to aid one another in a variety of tasks, both verbal and nonverbal.

Brain-damaged patients have provided much of the evidence that the two hemispheres differ in their capabilities, although these differences can now be demonstrated in normal subjects without brain damage. One procedure of great clinical importance for demonstrating hemispheric differences is the *Wada test*, which was developed to determine the dominant hemisphere for speech functions so that inadvertent damage to speech centers would be avoided during neurosurgery. In this test the patient is instructed to count aloud or speak continuously while sodium amytal, a fast-acting barbiturate, is injected into the left or right internal carotid artery. The drug has preferential access to the hemisphere on the same side as it is injected and briefly causes dysfunction of that hemisphere. If it is the hemisphere dominant for speech that is affected, the patient stops speaking and does not respond to a command to continue.

The relationship between handedness and hemispheric specialization was one of the first problems explored with the sodium amytal test. Do left-handed individuals have left-hemisphere speech, as do right-handed people, or do they have right-hemisphere speech? As expected, the test confirms that almost all right-handed people (96%) have left-hemisphere speech. Surprisingly, however, so do the majority of left-handed people. Nevertheless, a significant number (15%) of left-handed people have right-hemisphere speech. Furthermore, in some left-handed people speech is controlled by *both* the right and left hemispheres (Table 19–2). In these patients injections of sodium amytal on either side do not suppress the function of speech.

The sodium amytal test yields another unexpected result: Unilateral injection of the drug affects not only speech but also mood. Some studies indicate that the effect on mood is related to the side of injection: injections on the left tend to produce a brief depression, and injections on the right, euphoria. These effects have been seen at doses lower than those needed to block speech. These results suggest that functions related to mood may also be localized to some degree in one or the other hemisphere. This idea is consistent with the clinical observation that some patients with damage to the left hemisphere are exceptionally upset about their symptoms. In contrast, some patients with damage to the right hemisphere are unexpectedly indifferent to their disability.

Support for this view of brain cortical function has now come from other methods. In one test subjects are briefly shown either figurative images (e.g., a face) or words in either the right or the left visual hemifield, and then asked to identify what they have seen. Because of the crossing of the visual pathways, visual information restricted to one hemifield is conveyed first to the contralateral hemisphere and then, through the corpus callosum, to the ipsilateral hemisphere (see Figure 19–9). Right-handed subjects typically respond with a higher percentage of recognition when the words are presented to the right visual field and the figurative images are presented to the left visual field. Left-handed subjects show greater variability with regard to the visual field superior for the task.

Similar results have been obtained by presenting different sounds to both ears at the same time and determining which ear is better at recognizing the sound. In right-handed subjects the left ear tends to be better with nonverbal sounds (e.g., recognition of music), whereas the right ear is better with words. The results of this test suggest that the crossed auditory pathways are more important for perception than the uncrossed pathways.

Table 19–2 Relationship Between Handedness and Hemispheric Specialization for Language

	Hemisphere for Language (%)		
Handedness	*Left*	*Right*	*Both*
Left- or mixed-handed	70	15	15
Right-handed	96	4	0

Data from Rasmussen and Milner, (1977).

The Two Hemispheres Are Anatomically Asymmetrical

Until 1968 it was widely believed that there was no gross anatomical asymmetry in the human brain. In that year, however, postmortem measurement of the dimensions of 100 human brains with a camera and ruler showed clear-cut bilateral differences in the planum temporale, a region on the upper surface of the temporal lobe that includes the classical speech area of Wernicke (Figure 19–11). The left planum was larger in 65% of the brains; the

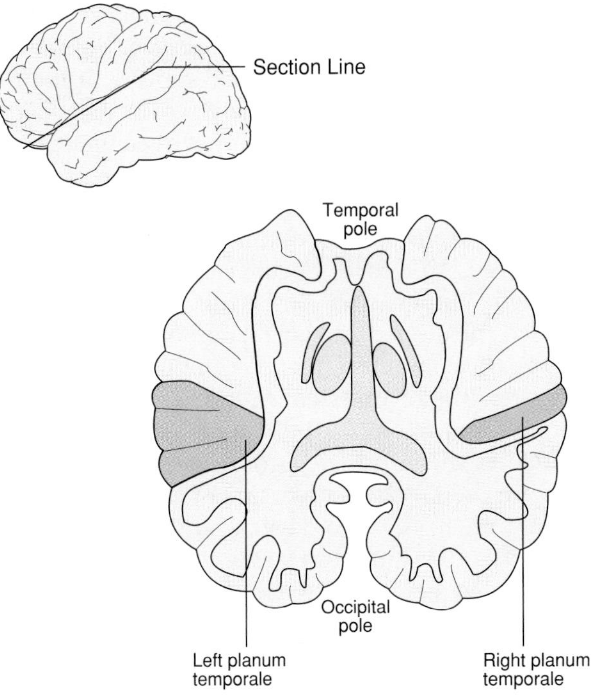

Figure 19–11 In the majority of human brains the planum temporale is larger in the left hemisphere than in the right. (Adapted from Geschwind and Levitsky, 1968.)

right planum was larger in only 11% of the brains; and in 24% of the brains the left and right sides were approximately equal in size. Later work with a variety of techniques, including computerized tomography, confirmed these results and established that similar asymmetries are present even in the human fetus. These observations suggest that an inherent anatomical asymmetry may initially favor the left hemisphere for the development of language functions, a point we return to later.

Why Are the Hemispheres Specialized?

The question why hemispheric specialization exists involves two major issues. First, how does it develop within the life span of the individual? Second, what functional advantages, if any, does it confer? We consider each of these questions in turn.

Studies of children who have sustained damage to either the left or right hemisphere suggest that left dominance is present when language is first expressed. Nevertheless, in contrast to adults, children who sustain damage to the left hemisphere—even substantial damage—usually recover the ability to speak in later life because the right hemisphere can perform language functions if the left is nonfunctional.

If either hemisphere can attain linguistic competence in the developing individual, why is the left hemisphere dominant in most people? It is likely that, at least in part, language ability develops in the left hemisphere because of an inherent anatomical asymmetry in the human brain, which is present in the human fetus and may initially favor the left hemisphere for language functions. Once a region begins to be specialized for a particular function, such as language during early development when neural circuitry is being formed, it is possible that the preexisting functional activity concerned with that function fosters the further development of that area rather than other areas.

Functions requiring extensive connections among several regions of the cortex may become lateralized because of evolutionary design efficiency. The number of fibers within the corpus callosum, which provides connections between the hemispheres, is far smaller than the number of intracortical fibers within a hemisphere. For the corpus callosum to have the richness of connections of a single hemisphere, it would have to be as big as the brain itself. Thus, the concentration of a highly complex function, such as language, in one hemisphere may be computationally advantageous given the limited size of the cranial cavity. Although on theoretical grounds lateralization of function should provide greater efficiency, as yet there is no conclusive evidence establishing this point.

Cognitive Functions Can Now Be Simulated by Artificial Neural Networks that Employ Parallel Distributed Processing

Ideas about how complex neural networks mediate cognitive functions may now be tested rigorously by modeling these functions. Artificial intelligence is an attempt to examine cognitive processes by means of computer programs and other devices that process information. At first, artificial intelligence was not concerned with actual neural mechanisms, but rather with the executive operations and symbolic manipulations that result from cognition. By analogy to computers, artificial intelligence attempted to describe or discover possible *programs* the brain executes. This approach has not been as useful as first expected, in part because it is unlikely that the brain uses a linear series of steps, like a computer program to execute operations. More recently, artificial intelligence has attempted to model cognitive processes by simulating the actual mechanisms that brains use.

Some models are networks in which computations are distributed throughout richly interconnected elements of the network. Hence, these models are called parallel-distributed processing (PDP) models. In PDP models each element is influenced in positive or negative ways by other elements of the network. The effect of the activity of one element on another is the product of its output level and its connection strength. An element sums up the effects of its various inputs and produces an output that is a linear or nonlinear function of the inputs. The elements of PDP models can be considered to be analogous to neurons, but they can also represent higher-order elements such as words, percepts, or ideas.

One of the most powerful and interesting PDP models is the *layered network*. This consists of a set of input elements connected to a set of output elements. This type of network is capable of performing a great variety of computations, particularly if the input layer does not directly connect to the output layer but rather connects through one or more intermediate (so-called "hidden") layers (Figure 19–12A).

The performance of the network depends on the strength and pattern of the connections between the elements. The power of PDP is that the connection strengths required to perform a given calculation are determined through successive applications of algorithms that calculate the appropriate strength for each connection. A popular and powerful technique for modifying the strengths of the connections is *back-propagation*, so named because it involves a series of calculations that start at the last (output) layer of the network and proceed back through successive layers until the first (input) layer is reached.

In simplified terms, back-propagation involves the following steps. A computational problem is defined. For example, one output element may be asked to output a value of 1 if the number of input elements is even, while another may be asked to output a value equal to the total number of input elements. The network is usually set up with random connection strengths between elements of different layers. Different sets of inputs are provided and each time the computer compares the resulting output of each output unit with the output desired. An error is calculated by the computer for each output unit based on the difference between the desired and actual output.

Next, the computer slightly modifies the connection strengths of units connecting to that output unit in proportion to the extent that they contributed to the error. The error terms of the output units are then used as a measure of the error signal of the units in the prior layer, and the connection strengths of the inputs to this penultimate layer are adjusted in proportion to these errors. Over many iterations of training, this self-correcting network typically improves in performance until it solves or comes close to solving the problem. Back-propagation has been used to design networks that can perform a variety of sensory tasks, for example, classifying visual stimuli such as letters presented through a two-dimensional array of input elements (see Figure 19–12A2). These networks can also perform motor computations, such as calculating the correct joint angles required to move an artificial limb to a particular point in space. Back-propagation is one of many techniques used to optimize connections so that networks can solve interesting problems. One recently developed technique simulates the evolutionary forces that may have determined the connections during phylogenesis. These *genetic algorithms* appear to be capable of finding optimum sets of connections, even when other algorithms fail.

A second type of network, based on the mathematics of chaos and physical theory, is the *recurrent network* (Figure 19–12B). In these networks there is feedback between the output and input elements. Therefore, unlike layered networks, recurrent networks sometimes have no stable state; once presented with an input they continue to cycle interminably from one state to another. Algorithms have been developed, however, that allow connection strengths to be set, such that the networks will assume one of several steady states. The network then has the property that if only a portion or a distorted replica of one of the stable states is pre-

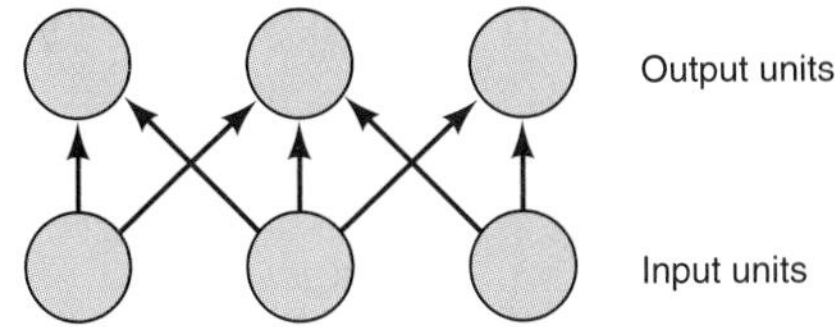

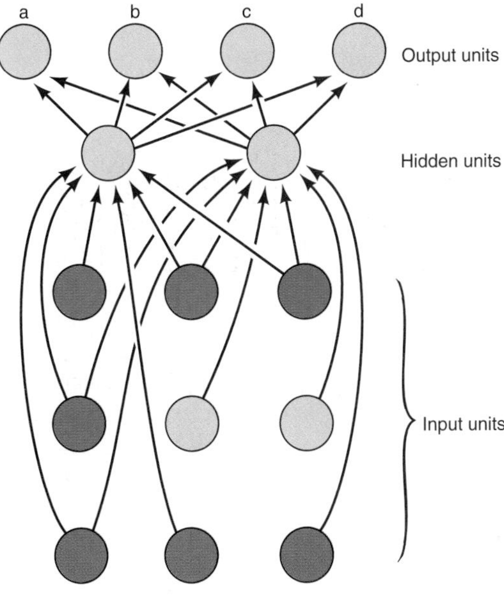

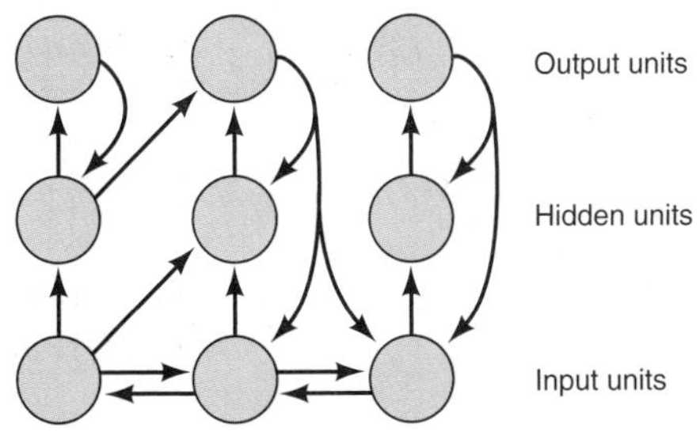

Figure 19–12 Networks capable of parallel distributed processing can simulate certain cognitive functions.

A. In layered networks the input units are connected to the output units either directly (**A1**) or indirectly (**A2**). There are no "back" connections from a later layer to an earlier layer in the network. Each connection strength can be different. For a given trial, the input units are each provided with a level of activation (a value from 0 to 1). Activation of each output unit is then determined by the sum of the activity of the input units. The degree of activation provided by an input unit is a function of the product of the level of activation of the input unit and the magnitude of the connection strength to the output unit. **1.** A layered network consists at a minimum of an input and output layer. **2.** A layered network can also contain a layer of "hidden units." The input layer in this example is a two-dimensional array, such as a retina. It has been stimulated with the letter *c*. The output layer has four choices: *a*, *b*, *c*, or *d*. By training of the network, the connection strength between units can be adjusted in such a way that presentation of each letter primarily activates the corresponding output unit. For the sake of simplicity in these examples, not all connections typically present in connectionist models have been indicated.

B. In a recurrent network some units provide feedback (indicated by **curved arrows**) to earlier units in the circuit.

sented, the network assumes the complete stable state. Thus, these networks form associative memories in which a part of a stimulus evokes a larger related memory.

The PDP models are not likely to be replicas of actual processing in the nervous system. Rather, their interest lies in their ability to illustrate the types of operations that can be performed by networks of interconnected units. The main similarity between these circuits and real brains is the extensive parallel processing that occurs in both. A second similarity is that the operations of parallel-distributed networks, like the brain, are not dependent on any single element, but are a function of the ensemble of elements. They exhibit "graceful degradation," which means that removal of a small number of elements in the circuit results in only minimal computational dysfunction. Furthermore, the PDP models can generalize, so that the correct output can be calculated from an incomplete or distorted input. Current thinking is that some areas of cortex may be specialized for executive (or linear) computations, while in other areas something like PDP may occur, in which the action of specific neurons is not critical to the functioning of the area. That is, both localized and nonlocalized processes may occur in cortical functioning.

An Overall View

An analysis of cognitive processes indicates that even the most complex functions of the brain are localized. Localization has great clinical importance and explains why certain syndromes are characteristic of disease in specific regions of the brain. Nevertheless, the question whether function is a localized or an ensemble property of the nervous system appears to be a dialectic issue, perhaps formally similar to the question in quantum mechanics whether light is a particle or a wave. The answer depends on the interests of the experimenter and on the particular experimental approach. No part of the nervous system functions in the same way alone as it does in concert with other parts. When a part of the brain is removed in a lesion study, the behavior of the animal afterward is more a reflection of the adjusted capacities of the remaining brain than of the part of the brain that was removed. It is unlikely, therefore, that any cognitive behavioral function—especially functions such as thought, perception, and language—will be understood by focusing on one region of the brain without considering the relationship of that region to other regions.

Furthermore, just because a lesion does not disrupt a particular task does not mean that area of brain is not involved in that task. The brain can reorganize—sometimes very quickly, sometimes more slowly—so that other areas take over the function. In addition (and perhaps more important), it is possible that in one functional system some individual units or small groups of units, and perhaps even large ensembles of units, can be removed without dramatically altering the function of the system because of the parallel organization of brain circuits. This is a general property of parallel computers in which the computations are in effect distributed throughout the network.

Finally, modern imaging techniques have revealed that multiple areas of cortex are typically activated even by the performance of simple tasks. Nevertheless, the current approach to the nervous system, reducing its activities into anatomically discrete units, gives us the essential clues we need to examine the contribution of individual parts to the functioning of the whole.

Selected Readings

Andersen, R.A. 1987. Inferior parietal lobule function in spatial perception and visuomotor integration. In F. Plum (ed.), Handbook of Physiology, Section 1: The Nervous System, Vol. V: Higher Functions of the Brain, Part 2. Bethesda, Md.: American Physiological Society, pp. 483–518.

Andreasen, N. C. 1988. Brain imaging: Applications in psychiatry. Science 239:1381–1388.

Bates, E. 1992. Language development. Curr. Opin. Neurobiol. 2:180–185.

Churchland, P. S., and Sejnowski, T. J. 1992. The Computational Brain. Cambridge: MIT Press.

Damasio, A.R., and Damasio, H. 1994. Human Brain Anatomy: A Computerized Imaging Atlas. Oxford: Oxford University Press.

Frith, V. 1989. Antisense. Oxford: Basil Blackwell.

Funahashi, S., Bruce, C. J., and Goldman-Rakic, P. S. 1989. Mnemonic coding of visual space in the monkey's dorsolateral prefrontal cortex. J. Neurophysiol. 61:331–349.

Fuster, J. M. 1989. The Prefrontal Cortex: Anatomy, Physiology, and Neuropsychology of the Frontal Lobe, 2nd ed. New York: Raven Press.

Geschwind, N. 1979. Specializations of the human brain. Sci. Am. 241(3):180–199.

Goldberg, D. E. 1989. Genetic Algorithms in Search, Optimization, and Machine Learning. Reading, Mass.: Addison-Wesley.

Goldman-Rakic, P. S. 1987. Circuitry of primate prefrontal cortex and regulation of behavior by representational memory. In F. Plum (ed.), Handbook of Physiology, Section 1: The Nervous System, Vol. V: Higher Functions of the Brain, Part 1. Bethesda, Md.: American Physiological Society, pp. 373–417.

Hardyck, C., and Petrinovich, L. F. 1977. Left-handedness. Psychol. Bull. 84:385-404.

Kolb, B., and Whishaw, I. Q. 1990. Fundamentals of Human Neuropsychology, 3rd ed. New York: Freeman.

Milner, B. 1974. Hemispheric specialization: Scope and limits. In F. O. Schmitt and F. G. Worden (eds.), The Neurosciences: Third Study Program. Cambridge, Mass.: MIT Press, pp. 75–89.

Pandya, D. N., and Seltzer, B. 1982. Association areas of the cerebral cortex. Trends Neurosci. 5:386–390.

Passingham, R. E. 1993. The Frontal Lobes and Voluntary Action. Oxford: Oxford University Press.

Rumelhart, D. E., and McClelland, J. L. (eds.) 1986. Parallel Distributed Processing: Explorations in the Microstructure of Cognition, Vol. 1: Foundations. Cambridge, Mass.: MIT Press.

Wilson, F. A. W., O'Scalaidhe, S. P., and Goldman-Rakic, P. S. 1993. Dissociation of object and spatial processing domains in primate prefrontal cortex. Science 260:1955–1958.

VI

PERCEPTION

I l e f t the w e l l - h ou s e

e a g er to l(ea)r n . Ever y th ing

h a d a n a m e , and e a ch

n a m e g a v e b i r th to a

n e w th ou gh t . A s w e

r e t u r n ed to the h ou s e ,

ever y o b j e c t I t ou ch ed

s e e m ed to q u i v er with l i f e .

About the image on the preceding page:

". . . My teacher placed my hand under the spout. As the cool stream gushed over one hand, she spelled into the other the word water. . . . I knew then that 'w-a-t-e-r' meant the wonderful cool something that was flowing over my hand." Deaf and blind, the seven-year-old Helen Keller learned the meaning of words as symbols through an unusual sensory modality, touch. In *The Story of My Life* (1906), written as an undergraduate at Radcliffe College, she described how she began to understand language. A sentence from her autobiography is shown here in braille, a system of printing devised for reading words through touch. (Braille setting courtesy of Michael Helmers, The Associated Blind, New York.)

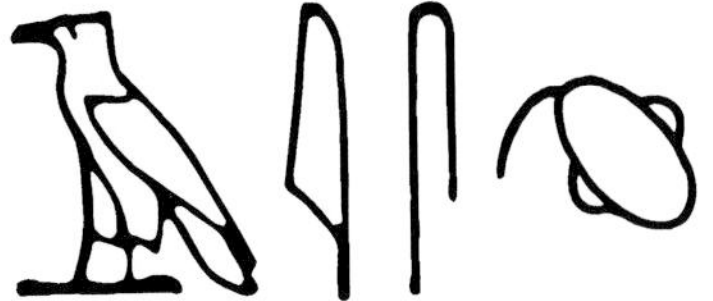

Introduction

Studies of artificial intelligence have shown that the human brain recognizes objects in ways no current computer can even begin to approach. Merely to see—to look onto the world and recognize a face or enjoy a landscape—entails an amazing computational achievement. Indeed, all our perceptions—seeing, hearing, smelling, and touching—are analytical triumphs. The brain accomplishes these computational feats because its many components—its nerve cells—are wired together in very precise ways. Equally remarkable is that the precision of this wiring is not immutable. The connections between cells can be altered by learning. We remember events because the structure and function of these connections can be modified by experience.

We construct our knowledge of the world with sight, sound, touch, pain, smell, taste, and the sensation of body movements. How do our perceptions arise from the neural activity initiated by sensory stimuli? Perception begins in receptor cells that are sensitive to one or another kind of stimulus. Most sensory inputs are perceived as a sensation identified with a specific stimulus. For example, the feeling of cold is associated with snow. Specific neurons in the sensory system, both peripheral receptors and central cells, encode critical attributes of sensations: the location of the stimulus and its properties. Other attributes are encoded by the pattern of activity in different populations of cells.

Sensory pathways connect the receptor at the periphery with the spinal cord, brain stem, thalamus, and cerebral cortex. We feel a tactile stimulus on the hand when a population of touch receptors produce a discharge of action potentials in the afferent fibers of the hand. These cause certain populations of cells to discharge in the dorsal column nuclei in the thalamus and in several connected areas of the cortex. Initially, sensory information is processed serially in discrete pathways. In each sensory system these serial pathways operate in parallel with each other. This parallel processing of sensory information by different components of one sensory system, and by all sensory systems together, is essential to the way the brain forms our perceptions of the external world.

What emerges from the studies of parallel processing in somatic and visual sensations is our first glimpse into the neural mechanisms of perception. We have learned that contrary to the intuitive analysis of our personal experience, perceptions are not precise and direct copies of the world around us. Sensation is an abstraction, not a replication of the real world. The brain does not simply record the external world like a three-dimensional photograph. Rather, the brain constructs an internal representation of external physical events after first analyzing them into component parts. In scanning the visual field the brain simultaneously but separately analyzes the form of objects, their movement, and their color, all before putting together an image according to the brain's own rules. How this reconstitution occurs—the *binding problem*—is one of the most pressing questions in cognitive neural science.

In this section of the book we will not treat extensively all aspects of sensory processing. Rather, we attempt to outline the principles essential for understanding how perception occurs in the brain, and in doing so we focus on the visual system because it is the best understood sensory system. There is now a fairly complete understanding of how photons of light of different wave lengths are transduced by the photoreceptors in the retina into electrical activity, and how the retina processes this activity along parallel visual pathways. Since the central pathways are fairly well delineated, we now have some insight into how this visual information is processed at the cellular level in several parallel pathways. What is still not known is how these components are brought together and bound into a unified image. Clearly one of the major goals of cognitive neural science is to determine how the information that reaches the cortex by means of these parallel pathways interacts to construct a unified conscious perception. Indeed, one of the hopes driving this work is that solution of the binding problem will yield our first insights into selective attention and consciousness.

20

The Sensory Systems

Sensory Information Underlies Motor Control and Arousal As Well As Perception

Modality, Intensity, Duration, and Location Are the Major Attributes of Sensation

All Sensory Systems Have a Common Plan

- *Sensory Receptors and Sensory Neurons in the Central Nervous System Have a Receptive Field*
- *Sensory Systems Have Parallel Pathways Organized in a Hierarchical Fashion*
- *Sensory Pathways to the Brain Convey Neural Maps of the Receptive Sheet*

Stimulus Information Is Encoded at the Entrance to the Nervous System

- *Stimulus Energy is Converted Into Electrochemical Energy*
- *Stimulus Features Are Represented in the Discharge Patterns of Neural Activity*
- *Information About the Strength of the Stimulus Is Conveyed by Frequency and Population Codes*
- *Information About the Duration of the Stimulus Is Encoded in the Discharge Patterns of Rapidly and Slowly Adapting Receptors*
- *Modality Is Encoded by a Labeled Line Code*
- *Discrimination Between Stimuli Is Sharpened by Lateral Inhibition*

Different Modalities Place Specific Demands on the Common Neural Architecture of Sensory Systems

Sensation has historically been the starting point for the scientific study of mental processes. The modern origins of this field date to the early nineteenth century, when the French philosopher Auguste Comte defined a new philosophy concerned with applying the empirical methods of natural science to the study of human behavior. Comte was influenced by the British empiricists who maintained that all knowledge comes through sensory experience—what can be seen, heard, felt, tasted, or smelled. One of these, John Locke, proposed that at birth the human mind is as blank as a "white Paper void of all Characters without any Ideas," a *tabula rasa*, on which experience leaves its marks.

The empiricist view ultimately led to the emergence of psychology as a distinct academic discipline. Separated from philosophy, psychology developed as a science concerned with the experimental study of mental processes, emphasizing in its early years sensation as the key to the mind. The founders of experimental psychology were particularly concerned with questions about the sequence of events by which a stimulus leads to subjective experience.

They soon found that although sensory reception differs for each of the senses, three steps were common to all senses: (1) a physical stimulus, (2) a set of events by which the stimulus is transduced into a message of nerve impulses, and (3) a response to the message, often as a perception or inner representation of the sensations. This sequence lent itself to two modes of analysis, giving rise to the fields of *psychophysics* and *sensory physiology*. Psychophysics focused on the relationship between the physical characteristics of a stimulus and the attributes of its perception. Sensory physiology examined the neural consequences of a physical stimulus—how the stimulus is transduced by sensory receptors and processed in the brain. Much of the current excitement in the neurobiology of perception comes from the recent merging of these two approaches in experiments on human subjects, using noninvasive probes of brain function such as positron emission tomography (PET) and magnetic resonance imaging (MRI).

The early findings in psychophysics and sensory physiology quickly revealed a weakness in the empiricist argument: the newborn mind is not blank, nor is our perceptual world formed from an accumulation of passive encounters with the physical properties of objects. In fact, our perceptions differ qualitatively from the physical properties of stimuli. This is because the nervous system extracts only certain information from a stimulus and neglects other information, and then interprets this information in the context of earlier experience.

We *receive* electromagnetic waves of different frequencies but we *perceive* colors: red, green, orange, blue, or yellow. We receive pressure waves but we hear words and music. We come in contact with a myriad of chemical compounds dissolved in air or water but we experience smells and tastes.

Colors, sounds, smells, and tastes are mental constructions created in the brain by sensory processing. They do not exist, as such, outside of the brain. Therefore, we can answer the traditional question raised by philosophers: Does a falling tree in the forest make a sound if no one is near enough to hear it? We can say with certainty that while the fall causes pressure waves in the air, it does not create a sound. Sound occurs only when pressure waves from the falling tree reach and are perceived by a living being.

Thus, our perceptions are not direct records of the world around us but are constructed internally according to innate rules and constraints imposed by the capabilities of the nervous system. The philosopher Immanuel Kant referred to these inherent constraints as *preknowledge.* The mind, Kant argued, is not a passive receiver of sense impressions, but rather is constructed to conform with ideal or objective preexisting categories, such as space, time, and causality, that exist independent of physical stimulation from outside the body. Knowledge, according to Kant, is based not only on sensory information but also on the preknowledge that organizes sensory experience.

In subsequent chapters we shall see that the philosophical dialectic between Kant's idealism and experimental pragmatism continues to influence studies of perception. As we shall see in Chapter 22, Kant's idea of preknowledge influenced the emergence of *Gestalt psychology*, which holds that aspects of perception reflect an inborn capacity of the brain to organize simple sensations in characteristic ways. Similarly, pragmatic experimentation influenced the emergence of *behaviorist psychology*, which focuses on observable behavior, on the individual's motor response to a stimulus.

All sensory systems—not only in humans, but throughout animal evolution—rely on the same basic principles of information processing and organization. The basic mechanisms of perception seem therefore to have been remarkably conserved during evolution. Since our understanding of perception is most complete for vision, we examine in detail in the next four chapters the principles of information processing in the visual system. By understanding the principles of visual processing we shall be well served for understanding other aspects of perception.

In this chapter we introduce the study of perception by tracing the experimental path followed by the early students of sensation. We first consider the psychophysical studies that allow us to evaluate the several functions of sensory information in behavior. It is important, however, to have some idea of the neural organization of the other senses. Thus, we also examine the organizational features common to all sensory systems. Specifically, we consider how stimuli are transduced by sensory receptors and encoded into neural signals that allow us to interpret the physical world in which we live.

In considering perception in these chapters we use the term *internal sensory representation.* By that we simply mean, here as in Chapter 18, the pattern of activity in a specific set of interconnected cells.

Sensory Information Underlies Motor Control and Arousal As Well As Perception

Sensory systems receive information from the environment through specialized cells at the periphery of the body and transmit this information to the central nervous system. There the information is used for four main functions: perception, control of movement, regulation of the functions of internal organs, and maintenance of arousal. While we tend to think of sensation as a conscious experience, not all sensory information reaches consciousness. When we withdraw a hand after touching a hot surface, the sensory information drives the motor response automatically before we are consciously aware that the surface is hot.

In addition to stimulation from the external world, we receive sensory information from within the body: from blood vessels, the viscera, and the actions of skeletal muscles on joints. This information is used to regulate temperature, blood pressure, heart rate, respiratory rate, and reflex movements. Regulation of these essential body functions usually is not available to consciousness. The brain also uses sensory information for maintaining a heightened level of arousal. A portion of the brain stem, called the *reticular formation*, is essential for the sensory regulation of arousal.

Modality, Intensity, Duration, and Location Are the Major Attributes of Sensation

Despite their diversity, all sensations share certain basic properties. Different forms of energy—light, mechanical, thermal, chemical—are transformed by the nervous system into different qualities of sensations or *sensory modalities*. Five major sensory modalities have been recognized since ancient times: vision, hearing, touch, taste, and smell.[1] Each modality has many *submodalities* that constitute elementary senses or building blocks from which more complex sensations are formed. For example, a multitude of flavors are formed by combinations of the four basic submodalities of taste: sweet, sour, salty, and bitter.

In 1826 Johannes Müller advanced his "laws of specific sense energies." He proposed that modality is a property of the sensory nerve fiber, that each nerve fiber is activated by a specific type of stimulus. This idea has since been conclusively demonstrated. For example, nerve fibers activated by noxious stimuli, which evoke pain, are not activated by mechanical stimuli, which evoke touch. Different fibers also make unique connections within the nervous system. For example, nerve fibers activated by noxious stimuli and those activated by mechanical stimuli make distinctly different connections in the spinal cord and brain. Thus the many sensory modalities are mediated by different sensory nerve fibers, which have different sites of termination in the brain.

The *intensity* or amount of sensation depends on the strength of the stimulus. The lowest stimulus intensity a subject can detect is termed the *sensory threshold*. Sensory thresholds are not invariant; they can be influenced by experience, fatigue, or the context in which the stimulus is presented. Modification of sensory thresholds by contextual cues is particularly intriguing. The threshold for pain, for example, is often raised during competitive sports or in childbirth. Sensory thresholds can also be lowered, as when some stimulus is anticipated.

The sensitivity of the sensory system to differences in stimulus strength depends on the difference in strength of two stimuli. For example, we easily perceive that 1 kg is different from 2 kg but have difficulty distinguishing between 50 and 51 kg. Yet both sets of weights differ by 1 kg! This phenomenon was examined in 1834 by Weber, who proposed a quantitative and linear relationship between stimulus strength and discrimination, now known as Weber's law:

$$\Delta S = K \times S$$

where ΔS is the *just noticeable difference* (JND) representing the minimal perceptible difference in strength between a reference stimulus S and a second stimulus, and K is a constant. Thus, the difference in magnitude between two stimuli must increase in proportion to the strength of the reference stimulus for a difference to be detected.

[1] Touch is a short-hand term for the body, or somatic, sensations. In addition to touch, these include pain, thermal sensation, and limb proprioception (limb position sense).

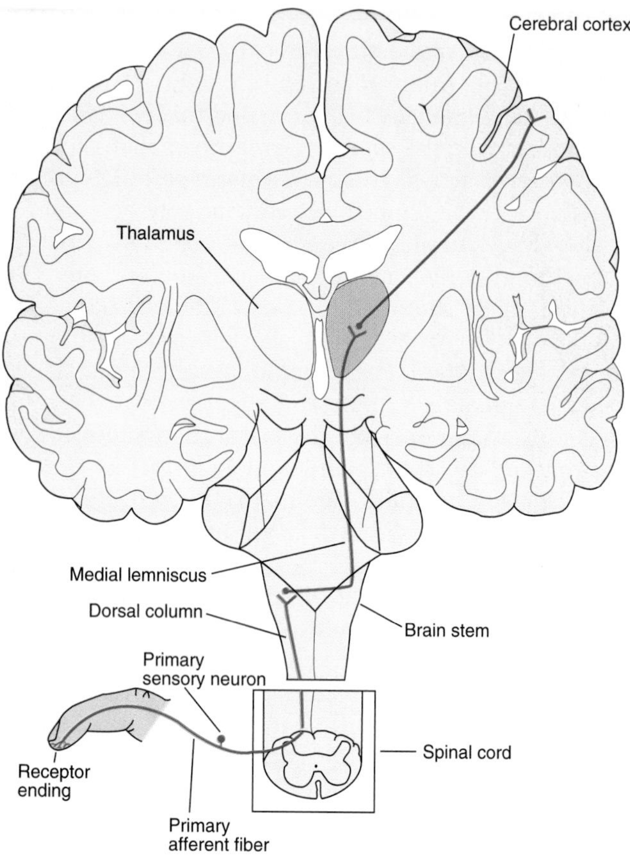

Figure 20–1 The general organization of sensory systems is demonstrated by the sensory pathway for touch. Three neurons link the peripheral receptor sheet (the skin) with the cerebral cortex. The first neuron (primary sensory neuron) is a receptor that transduces stimulus energy into neural signals. Its axons project from the receptor to the brain stem. The second neuron, whose axon is located in a brain stem tract called the medial lemniscus, carries sensory information to the thalamus, the brain's major sensory relay station. From the thalamus, the third neuron in the circuit transmits sensory information to the cerebral cortex.

The *duration* of a sensation is a function of both the duration and the strength of the stimulus. Typically, if a stimulus persists for a long time the intensity of the sensation diminishes. This decrease is called *adaptation*. The strength of a stimulus may diminish over time and eventually fall below the sensory threshold, when the sensation is lost. Naturally, a weak stimulus fades away faster than a strong stimulus.

Most sensations are perceived as having a *specific source* in space, either on the body or outside. The ability to locate the source of stimulation depends on the ability to distinguish between closely spaced stimuli. This ability can be quantified by determining the minimum detectable distance between two stimuli, a measurement that Weber called the *two-point threshold*. The two-point threshold may be as small as 1 mm on the fingertips.

All Sensory Systems Have a Common Plan

The common properties of sensations—modality (or quality), intensity, duration, and location in space—correspond to common attributes of the physical stimuli that activate the different sensory systems. This may be one reason why all the sensory systems have a similar anatomical organization. The organization of the somatic sensory system (Figure 20–1) is a bit simpler than that of the other sensory systems and is therefore a good starting point in our exploration of the neural mechanisms of perception.

In each sensory system the initial contact with the external world occurs through specialized cells called *sensory receptors*. Each receptor is sensitive primarily to one form of physical energy (Table 20–1). Nevertheless, all stimulus energies are transformed into electrochemical energy, so that all sensory systems share a common signaling medium.

The process by which stimulus energy is translated into a neural discharge consists of two stages. The first is *stimulus transduction*, the conversion of stimulus energy into a local depolarization or hyperpolarization of the receptor cell membrane. The second stage is *neural encoding*, whereby the local neural signal produced by the stimulus evokes a discharge of action potentials that represent stimulus information, such as intensity or duration.

Information about the stimulus may be conveyed by the activity of a single neuron or a population of neurons. For example, in a single neuron the code may consist of the average impulse activity of the receptor as well as the time intervals between impulses. At the same time, the distribution of neural activity in the receptor population is itself information used by the brain. For example, as the stimulus becomes stronger, the size of the responding receptor population also increases. The

Table 20–1 Mammalian Sensory Systems

Modality	*Stimulus*	*Receptor Type*	*Specific Receptor*
Vision	Light	Photoreceptor	Rods, cones
Hearing	Air-pressure waves	Mechanoreceptor	Hair cells (cochlear)
Balance	Head motion	Mechanoreceptor	Hair cells (semicircular canals)
Touch	Mechanical, thermal, noxious (chemical)	Mechanoreceptor, thermoreceptor, nociceptor, chemoreceptor	Dorsal root ganglion neurons
Taste	Chemical	Chemoreceptor	Taste buds
Smell	Chemical	Chemoreceptor	Olfactory sensory neurons

neural coding of sensory information is described in more detail later in this chapter.

In the somatic sensory and olfactory systems the sensory receptor is a neuron, called a *primary sensory neuron*. The terminal portion of the neuron transduces stimulus energy and the axon conveys action potentials to the central nervous system. In the gustatory, visual, auditory, and balance (or vestibular) systems the receptors are separate epithelial cells. These nonneural receptor cells communicate with a primary sensory neuron by a mechanism similar to synaptic transmission. Thus the functions of stimulus transduction and neural encoding are combined in a single receptor cell in the somatic sensory and olfactory systems, but in the visual, auditory, gustatory, and vestibular systems they are carried out by separate cells. The morphology and organization of receptors for different modalities are shown in Figure 20–2.

Animals have five major types of sensory receptors: chemoreceptors, mechanoreceptors, thermoreceptors, photoreceptors, and nociceptors (see Table 20–1). Some animals have additional types of specialized receptors, such as electroreceptors in fish and infrared receptors in snakes. The particular type of stimulus energy to which a receptor is sensitive is known (in a term coined by Sherrington) as the *adequate stimulus*. Moreover, each receptor is sensitive or *tuned* to a narrow range of stimulation as well. For example, individual auditory receptors are not sensitive to all frequencies of sound but only to a small part of the spectrum.

Thus, in physiological experiments we can quantify this limited sensitivity by generating a *tuning curve*. For an auditory receptor, for example, the tuning curve describes the frequency sensitivity of the receptor. The curve illustrates the minimum stimulus intensity at which the receptor is activated (i.e., threshold) and by how much (Figure 20–3). The stimulus amplitude must be increased to excite the receptor at higher or lower frequencies. In Figure 20–3 for example, the auditory receptor is most sensitive to a 2 kHz stimulus, but will respond to a 1 kHz stimulus if the amplitude is increased 30 times.

Sensory Receptors and Sensory Neurons in the Central Nervous System Have a Receptive Field

Each sensory receptor and primary sensory neuron can be activated only by stimuli that impinge on a circumscribed area of the receptive sheet in which

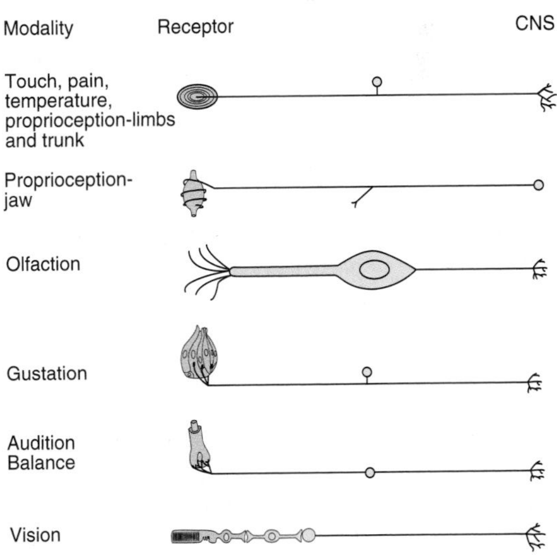

Figure 20–2 Sensory receptors in each sensory system have a characteristic organization, morphology, and size. (Adapted from Martin, 1989.)

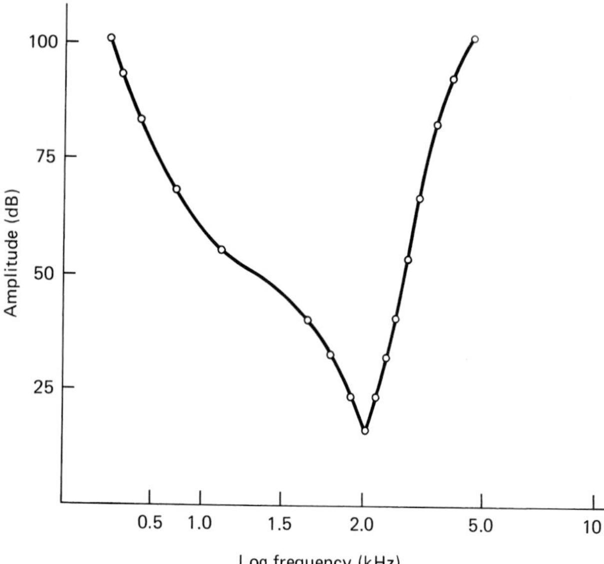

Figure 20–3 Each sensory receptor responds primarily to one type of stimulus energy and to a narrow range of intensities of that energy. This range can be identified in physiological experiments by constructing a *tuning curve* for individual receptors. The tuning curve shown here is for an auditory receptor that is most sensitive to sound at 2.0 kHz. The further the stimulus is from 2.0 kHz, the greater the amplitude of the stimulus must be to excite the receptor.

the receptor is located. This area, the *receptive field,* has a discrete location. For example, the receptive field of a mechanoreceptor (for touch) is the portion of the skin innervated by the peripheral nerve branch of the primary sensory neuron that gives rise to the receptor terminals (Figure 20–4). The receptive field, through which the external environment communicates with the brain, is the most important property of a primary sensory neuron. The size of the field is an important factor in determining the spatial resolution of a sensory system: greater resolution is made possible by cells with smaller receptor fields.

The central nerve branch of primary sensory neurons converges onto second-order neurons in the central nervous system, which in turn converge on higher-order neurons. Sensory information is transmitted from one neuron to the next in *relay nuclei.* The thalamus (Figure 20–1) is the relay for virtually all sensory pathways to the cerebral cortex; separate relay nuclei in the thalamus are specific to each sensory system and project to specific primary sensory areas of the cerebral cortex. The only exception is the olfactory system, which transmits sensory information from the first relay nucleus directly to the primitive cortex of the medial temporal lobe.

Relay nuclei contain *projection* (or *relay*) neurons that send their axons to the next relay nucleus in the sensory pathway. In addition to activating relay cells, sensory fibers also activate interneurons, both excitatory and inhibitory, which contribute to the processing of sensory information. The firing patterns of most projection neurons differ from those of the primary sensory fibers, indicating that the relay nuclei have a role in *transforming* sensory information as it travels through the brain.

Despite the fact that each relay neuron has multiple inputs, the synaptic action of some sensory fibers is so effective that activity in one of these fibers can discharge a relay cell. At such synaptic connections information can be transmitted with high fidelity. Thus, neurons in sensory relay nuclei also have a receptive field, one that encompasses all the receptive fields of the presynaptic sensory receptors (Figure 20–4). This explains why neurons in the sensory areas of the cortex also have receptive fields. We shall see in Chapter 23 that the receptive fields of second-order and higher-order sensory neurons in the visual system are larger and their properties are more complex than those of receptors.

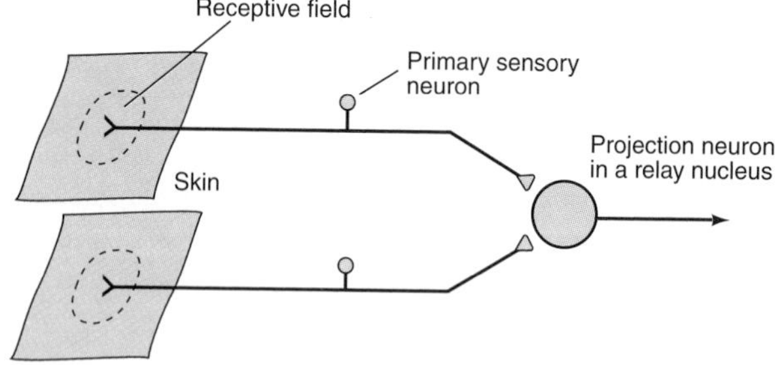

Figure 20–4 Sensory neurons in the central nervous system have a receptive field. In this example two primary sensory neurons each have a discrete receptive field on the skin. These neurons converge onto a projection neuron in the central nervous system. The receptive field of the projection neuron is the sum of the two fields of the primary sensory neurons.

Sensory Systems Have Parallel Pathways Organized in a Hierarchical Fashion

Most sensory systems consist of several distinct afferent pathways into the brain, each conveying a particular submodality. Each of these pathways has a serial or *hierarchical* organization: Receptors project on first-order neurons in the central nervous system, which in turn project on second- and higher-order neurons (see Figure 20–1). As a result, each *component* of a pathway has a distinct function in the processing of sensory information from the periphery to higher brain centers. For example, primary sensory fibers, which represent the lowest level in the hierarchy, provide the sensory input to and define the sensory properties of neurons at the next higher level.

While individual pathways in a system are organized serially, the several pathways of a system are organized *in parallel*. For example, the visual system has separate and parallel pathways for information about the form, color, or movement of an object. The four distinct submodalities of somatic sensation—touch, limb position sense, pain, and temperature sense—are carried by two parallel paths (see Chapter 18). Parallel pathways converge in the cerebral cortex, often in higher-order sensory cortical regions. This is where information from the various submodalities is combined for perceptions. Form and color are combined so that we can distinguish objects. Similarly, touch is linked with position sense, so that we can recognize the shape of grasped objects.

Sensory Pathways to the Brain Convey Neural Maps of the Receptive Sheet

Except for the chemical senses of taste and smell, sensory pathways from the periphery to the central nervous system are anatomically organized to preserve the spatial relations of receptors in the periphery. Nerve fibers of primary sensory neurons from neighboring sites in the receptive sheet make connections with adjacent second-order neurons, and so on through higher levels of processing in the central nervous system. As we saw in Chapter 18, in the somatic sensory system this preservation of the spatial relationships of receptor inputs is called *somatotopy* (see Figures 18–4 and 18–5), in the visual system *retinotopy*, and in the auditory system *tonotopy*. Spatially organized inputs in the central nervous system are often referred to as *neural maps*.

Stimulus Information Is Encoded at the Entrance to the Nervous System

Although each sensory system responds to a different type of stimulus and presents the brain with unique information, all sensory systems use similar mechanisms to process stimulus information. Each system must perform three tasks. First, the stimulus energy, such as mechanical or electromagnetic energy, must be converted into electrochemical neural signals (*stimulus transduction*). Second, key attributes of the stimulus must be represented in the signals of the primary sensory neuron (*neural encoding*). Third, sensory information must be fine-tuned to achieve maximal discriminative capacity by a mechanism called *lateral inhibition*.

Stimulus Energy is Converted Into Electrochemical Energy

How is the energy of the stimulus transduced into a neural signal? Sensory receptors have a specialized receptive portion of the membrane that contains the molecular machinery for transduction. The stimulus produces a local change in the membrane potential of the primary sensory neuron that spreads electrotonically but is restricted to the receptive membrane. This *receptor potential* results from a depolarizing potential that is produced by the opening of channels for Na^+, K^+, and Ca^{2+} the same process that initiates the excitatory synaptic potential (see Chapter 12). In the mechanoreceptor, for example, mechanical stimulation deforms the membrane of the receptor, changing its physical characteristics. As a result, additional cation-selective channels open and more Na^+ and K^+ ions flow through the membrane. The precise mechanism by which deformation of the membrane triggers the opening of channels is not known, but is thought to involve physical interactions between the membrane protein that forms the channel and the structural components of the membrane (Figure 20–5A).

Stimulus transduction in other systems is similar to that of the somatic sensory system. The impact of the stimulus on the receptor activates ion chan-

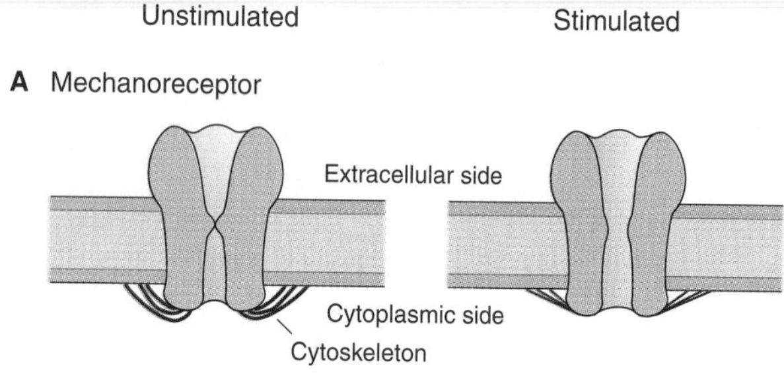

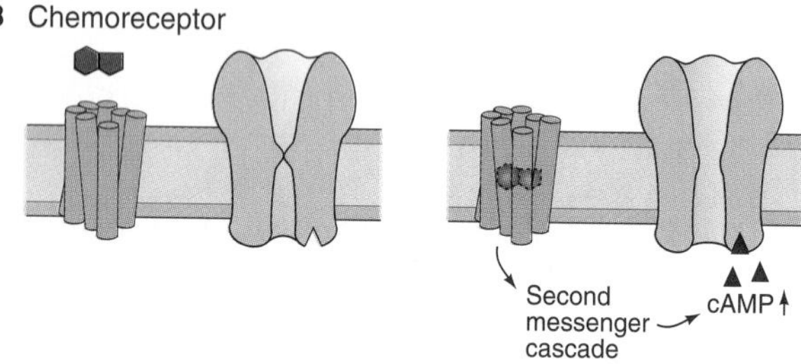

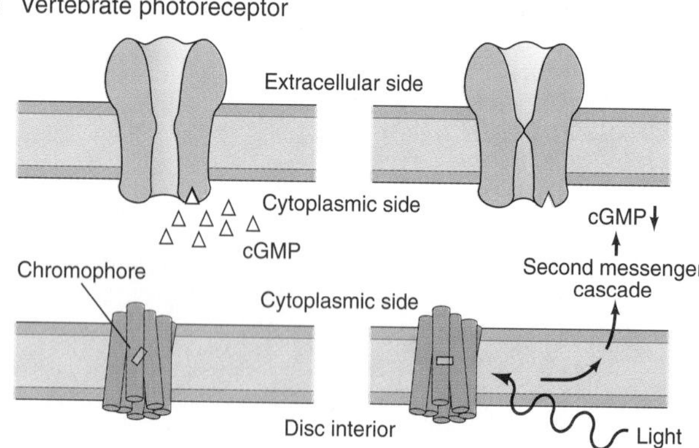

Figure 20–5 Different stimulus energies are all transduced into neural signals by distinct molecular mechanisms. (Adapted from Shepherd, 1983.)

A. Mechanical energy deforms the membrane of the receptor cell and thus affects the configuration of ion channels in the membrane. Few channels are open in the unstimulated membrane, whereas mechanical stimulation causes more channels to open. An influx of Na^+ and K^+ then causes the receptor terminal to depolarize locally, producing the receptor potential.

B. Chemoreceptors bind ambient chemicals and this action opens ion channels. In olfactory receptors and certain gustatory receptors a second messenger mediates channel opening.

C. Photoreceptors absorb light, and the interaction between the photons and a chromophore on intracellular membranes produces a three-dimensional change in the chromophore. A second-messenger system mediates the resulting change in membrane permeability.

nels, leading to a change in current flow across the membrane (Figure 20–5B). In olfactory receptors and certain taste receptors the interaction of the ligand and the receptor activates second messengers that open the channels. The use of a second messenger permits amplification of the stimulus signal (see Chapter 14). Transduction of light in the retina (Figure 20–5C) is accomplished similarly except that in the absence of a stimulus, in the dark, there is a continuous flow of an inward current in photoreceptors. The channels close in response to light, thereby reducing the inward current. In vision, as in smell, stimulation activates a second-messenger system that regulates ion channel functions.

Stimulus Features Are Represented in the Discharge Patterns of Neural Activity

Because the receptor potential spreads passively in the sensory receptor, the response of a receptor to a stimulus is a purely local event. Unless the activity in the receptor is *transmitted* to the central nervous system, the information about the stimulus will be lost. In the somatic sensory and olfactory systems, stimulus transduction and the transmission of information are both performed in specialized regions of one cell, the primary sensory neuron (see Figure 20–2). The potential elicited in the receptor is a graded response proportional to the strength of the stimulus (Figure 20–6). The receptor potential is not propagated along the axon like an action potential. In fact, its amplitude decreases as a function of distance from the site of increased membrane conductance, just like a synaptic potential.

When the receptor potential reaches the cell's trigger zone, action potentials are generated if the amplitude of the receptor potential is above the cell's threshold for firing (Figure 20–6; see Chapter 12). The action potentials are then transmitted over the length of the axon to a central neuron. In the visual, vestibular, auditory, and gustatory systems stimulus transduction and neural encoding are performed by separate receptor cells and primary sensory neurons that communicate through a chemical synapse. In this arrangement the amount of transmitter released by the receptor (or, for photoreceptors, the reduction in transmitter release) is proportional to the strength of the stimulus.

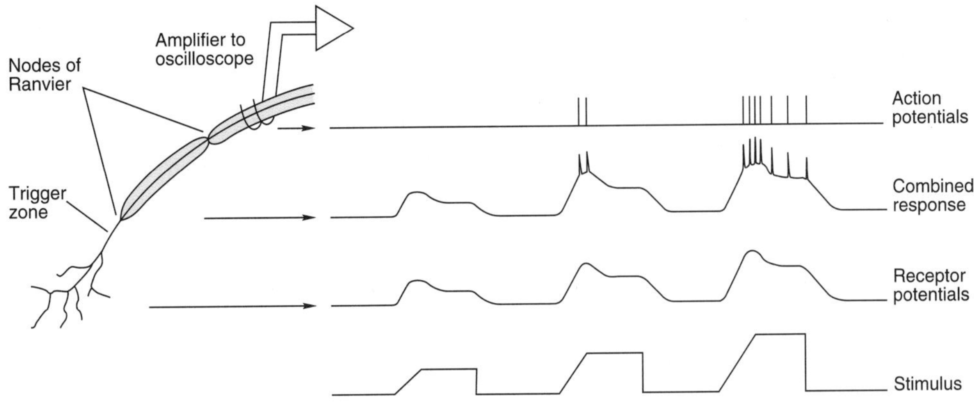

Figure 20–6 Different parts of a primary sensory neuron sensitive to skin stretch have different physiological characteristics. Stimulation of the receptive membrane of the axon terminal produces a local change in membrane potential called the receptor potential. The receptor potential is a graded response; that is, the number of ion channels opened by a stimulus is proportional to the stretch of the stimulus. The receptor potentials travel passively to the cell's trigger zone at the first node of Ranvier, where action potentials are produced. Because the trigger zone is close to the receptive membrane, the potential across the trigger zone membrane reflects the sum of the receptor potential and action potentials. Further from the receptive membrane and trigger zone only action potentials are recorded.

Action potentials are the same everywhere in the nervous system. The action potentials generated by mechanoreceptors are identical to those generated by photoreceptors. An action potential evoked by a weak stimulus is identical to that evoked by a strong stimulus. How can this stereotyped signal convey diverse information at different times and in different sensory systems? In fact, individual action potentials do not themselves carry information. Instead, the neural code for the intensity and duration of the stimulus is represented by the discharge *patterns* of action potentials. The pattern may be that of the action potentials in a single neuron or the activity in an entire population of neurons.

Earlier we noted that the sensory pathways, except for taste and smell, are anatomically arranged to preserve spatially organized maps of the receptive surfaces. These neural maps are the key to our capacity to localize stimuli precisely. The location of a visual or tactile stimulus therefore is encoded by the distribution of active neurons in the entire population of topographically organized sensory neurons. Stimulus localization is one of the four basic stimulus attributes. How then are other stimulus attributes—strength, duration, and modality—encoded? How do we discriminate between stimuli? We now address these questions.

Information About the Strength of the Stimulus Is Conveyed by Frequency and Population Codes

The discharge frequency of a primary sensory neuron, that is, the number of action potentials per unit of time, increases with the strength of the stimulus. This property of sensory fibers underlies the neural code for stimulus strength, called the *frequency code*. Stronger stimuli produce receptor potentials with faster rates of rise and greater amplitude, and these potentials in turn evoke trains of action potentials with higher frequencies (Figure 20–7). The relationship between stimulus strength and discharge frequency underlies the relationship between the strength of a stimulus and its perceived intensity. This can be demonstrated by comparing the response of primary sensory neurons in anesthetized experimental animals to a mechanical stimulus with perception of touch in human subjects. First, the activity in an individ-

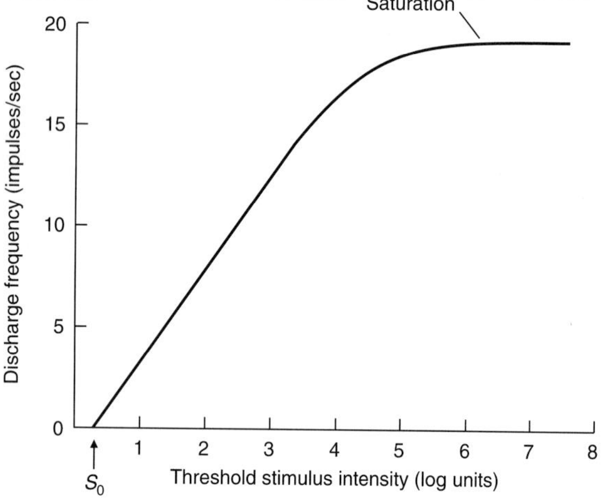

Figure 20–7 The frequency of discharge of a sensory neuron is a function of the stimulus intensity. (The plot is based on recordings from the myelinated portion of an axon a few millimeters from the terminal.) Sensory fibers begin generating action potentials when the stimulus amplitude reaches S_0 (the absolute physiological threshold). At lesser amplitudes only passively propagated receptor potentials are generated. The absolute physiological threshold is important because stimulus information reaches the central nervous system only when action potentials are generated.

ual slowly adapting mechanoreceptive neuron in an experimental animal is recorded while a stimulus is applied within the receptive field of the neuron. As the stimulus intensity increases—in this case the amount of indentation of the skin produced by a blunt probe—so too does the response of the sensory neuron (Figure 20–8A). (The stimuli used in the experiment are never strong enough to saturate the response of the receptor; compare Figure 20–7). The same kind of stimulus is also applied to the skin of human subjects, who describe their subjective estimate of the intensity of touch, and the perceived intensity is plotted against stimulus intensity (Figure 20–8B). Both the neural response and the perceived intensity increase in proportion with the strength of the stimulus. This simple example demonstrates the importance of the frequency code for stimulus intensity.

However, the range of stimulus strengths to which a primary sensory neuron can respond is limited by the transductive capabilities of the cell (e.g., the number of channels) and the conductive properties of the axon (e.g., the refractory period). While the response of a primary sensory fiber has an upper limit beyond which the discharge rate cannot increase further (Figure 20–7), the maximum stimulus strength that one neuron can encode is much lower than the maximal strength that the entire sensory system can register or that an organism can sense. This is because a stimulus activates more than one receptor; the stronger the stimulus, the greater the number of receptors that are activated. Thus, in addition to the discharge frequency of individual sensory neurons, the size of the population of responding neurons also provides a neural code, called a *population code*, for the strength of the stimulus. Frequency and population codes are also used by motor systems, where the size of the population of active motor neurons and their frequency of firing determine the strength of muscle contraction.

Information About the Duration of the Stimulus Is Encoded in the Discharge Patterns of Rapidly and Slowly Adapting Receptors

All sensory receptors adapt to constant stimulation—the receptor potential invariably decreases in amplitude in response to a persistent stimulus. The adaptation of receptors is thought to be an important component of the overall process of perceptual adaptation. A receptor can adapt rapidly or slowly. Rapidly adapting receptors respond transiently to stimulation, and only at the onset and offset of the stimulus. Slowly adapting receptors, on the other hand, fire throughout the application of a stimulus. The duration of a stimulus can therefore be measured in two ways: by the discharge of rapidly adapting receptors, which defines the onset and termination of the stimulus, or by the response of slowly adapting receptors (Figure 20–9).

Adaptation may result from characteristic response properties of the membrane of the receptor or primary sensory neuron, such as inactivation of Na^+ and Ca^{2+} channels or activation of a calcium-dependent K^+ channel. It may also depend on the nonneural accessory cells that surround the recep-

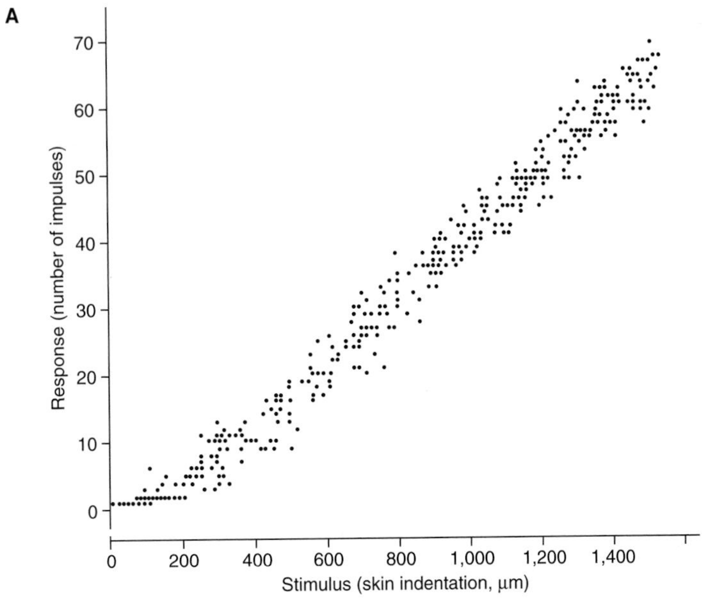

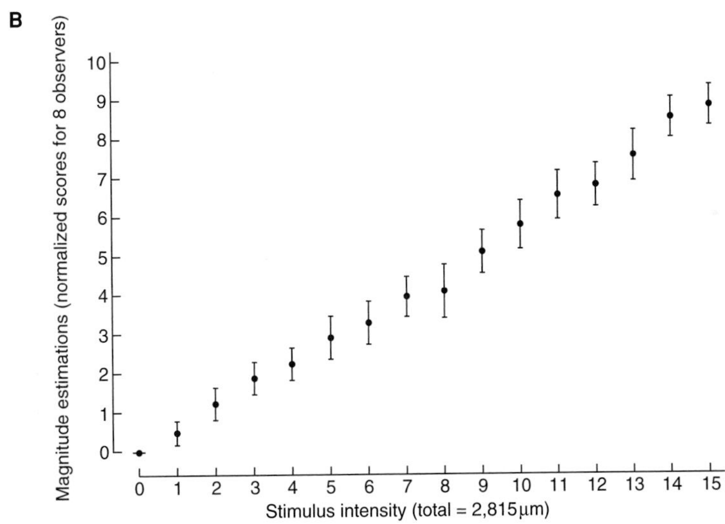

Figure 20–8 The frequency code underlies perception of stimulus strength (Adapted from Mountcastle, 1974).

A. The response of a slowly adapting mechanoreceptor to a step indentation of the skin was recorded in an anesthetized monkey. The relationship between the strength of the stimulus and the number of action potentials produced by the stimulus is plotted.

B. Estimates by human subjects of the strength of the same type of stimulus used in the neurophysiological experiments in **A** are plotted against actual stimulus intensities. A comparison of these psychophysical findings with the neurophysiological findings in part A shows the importance of the frequency code in processing information about stimulus intensity. (The bar at each data point represents six times the standard error.).

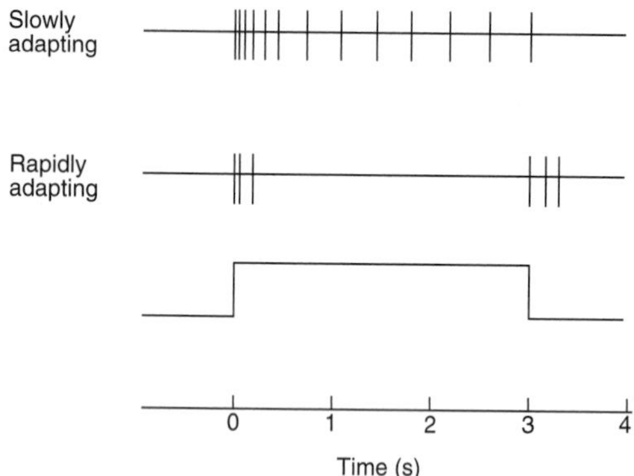

Figure 20–9 Slowly adapting mechanoreceptors continue firing during a steady stimulus, whereas rapidly adapting mechanoreceptors respond only at the beginning and ending of the stimulus.

tive ending of the primary sensory neuron or receptor cell.

Receptor adaptation acts as a signal filter, eliminating neural signals related to the steady or slow components of a stimulus, thereby focusing the neuron's responses on abrupt changes in stimulus intensity. This aspect of encoding, whereby certain features of the stimulus are selectively detected and accentuated, is called *feature extraction.*

Modality Is Encoded by a Labeled Line Code

As noted at the beginning of this chapter, specificity of function in neural pathways was first proposed by Müller in 1826. We now know that most sensory receptors are maximally sensitive to one type of stimulus energy, a property called *receptor specificity*. Receptor specificity defines the modality of the entire sensory pathway from the receptor to the cortex and underlies the most important coding mechanism for stimulus modality, the *labeled line code*. Whether a person feels "pain" or "touch" depends on which receptors are excited by the stimulus and on the connections the receptor makes in the central nervous system. Thus, excitation of a particular receptor by a stimulus always elicits the same sensation. For example, photoreceptors are most sensitive to light but a blow to the head can evoke the sense of a light flash. This is because a powerful mechanical stimulus can excite photoreceptors, which in turn is sensed as the light flash. Similarly, patients who suffer deafness because of inner ear damage can be made to "hear" tones of different frequencies by means of electrical stimulation of the auditory nerve.

Discrimination Between Stimuli Is Sharpened by Lateral Inhibition

The ability of an animal to discriminate one stimulus from another is facilitated by neural mechanisms that enhance the contrast of stimulus features. *Lateral inhibition,* first discovered in the visual system of the horseshoe crab *Limulus,* is a common mechanism in the sensory systems of many species. Lateral inhibition works because the receptive fields of sensory relay neurons have both excita-tory and inhibitory components (Figure 20–10B). The sensory relay neurons receive convergent excitatory input from many presynaptic neurons. For example, primary sensory neurons carrying information from the central portion of the receptive field of a relay neuron discharge the neuron more effectively than those at the periphery of the receptive field. This results in a gradient of excitation within the excitatory field.

To reinforce this gradient, receptor neurons also contact inhibitory interneurons (Figure 20–10B). The receptor neurons that innervate the center of the field are less vulnerable to the effects of the inhibition than those at the periphery of the field because the strength of the excitatory connection is stronger. The combined actions of the excitatory gradient and inhibitory field have the effect of reducing the likelihood that stimuli in the periphery of the field will activate the relay neuron. This focuses the size of the receptive field and enhances the acuity of the sensory system.

We can see the adaptive significance of lateral inhibition by comparing the effect of a stimulus on

A

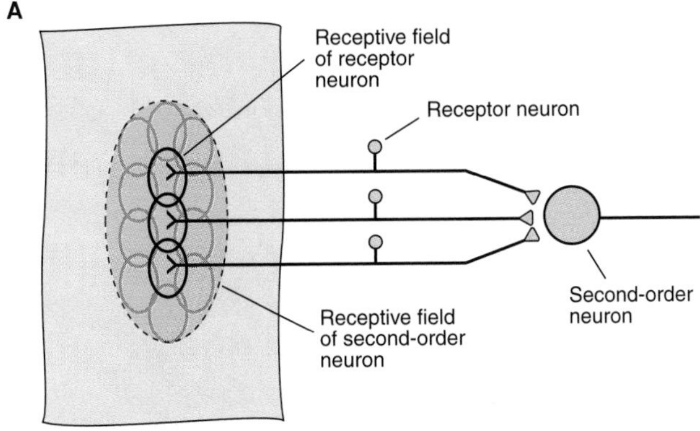

B

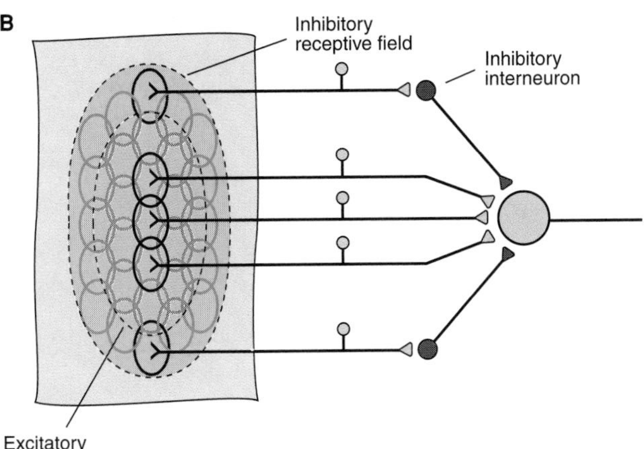

Figure 20–10 The receptive field of a higher-order neuron in a sensory pathway has a characteristic pattern of excitation and inhibition that increases spatial resolution.

A. Many peripheral receptors converge onto a single second-order sensory neuron in the central nervous system. As a consequence, the receptive field for the central neuron is made up of the receptive fields of all the presynaptic cells.

B. The receptive field of a central sensory neuron typically has a central excitatory receptive field surrounded by an inhibitory region.

a population of sensory relay neurons with and without lateral inhibition (Figure 20–11). The central terminals of individual receptor neurons diverge to contact many neurons in a central relay nucleus. The effect of this divergence is that a discrete stimulus activates a large population of central neurons. If this process were not limited, the population of successive relay nuclei would grow very large (Figure 20–11A). The effect of such an increase in the size of a population of active neurons would be a loss of discriminative capacity. The neurons in the central relay nucleus would respond the same way to different stimuli. Lateral inhibition restricts the tendency for divergent excitation to increase the size of the postsynaptic neural population (Figure 20–11B).

Lateral inhibition is a mechanism by which the most active neurons limit the activity of adjacent, less active neurons. Another name for lateral inhibition is *feedback inhibition* (Figure 20–12). Two other types of inhibition occur in sensory systems: feed-forward and distal inhibition. *Feed-forward* (or *reciprocal*) *inhibition* is the means by which activity in one group of neurons can inhibit another group (Figure 20–12). Feed-forward inhibition permits what Sherrington called a singleness of action, a winner-take-all strategy, which ensures that only one of two or more competing responses is expressed. Both feedback and feedforward inhibition create a central zone of intense activity surrounded by a ring of lesser activity in the inputs to higher-order neurons. By enhancing or amplifying the contrast between highly active cells and their neighbors, these circuits contribute to *selective perception*, in which we attend to one stimulus and not another.

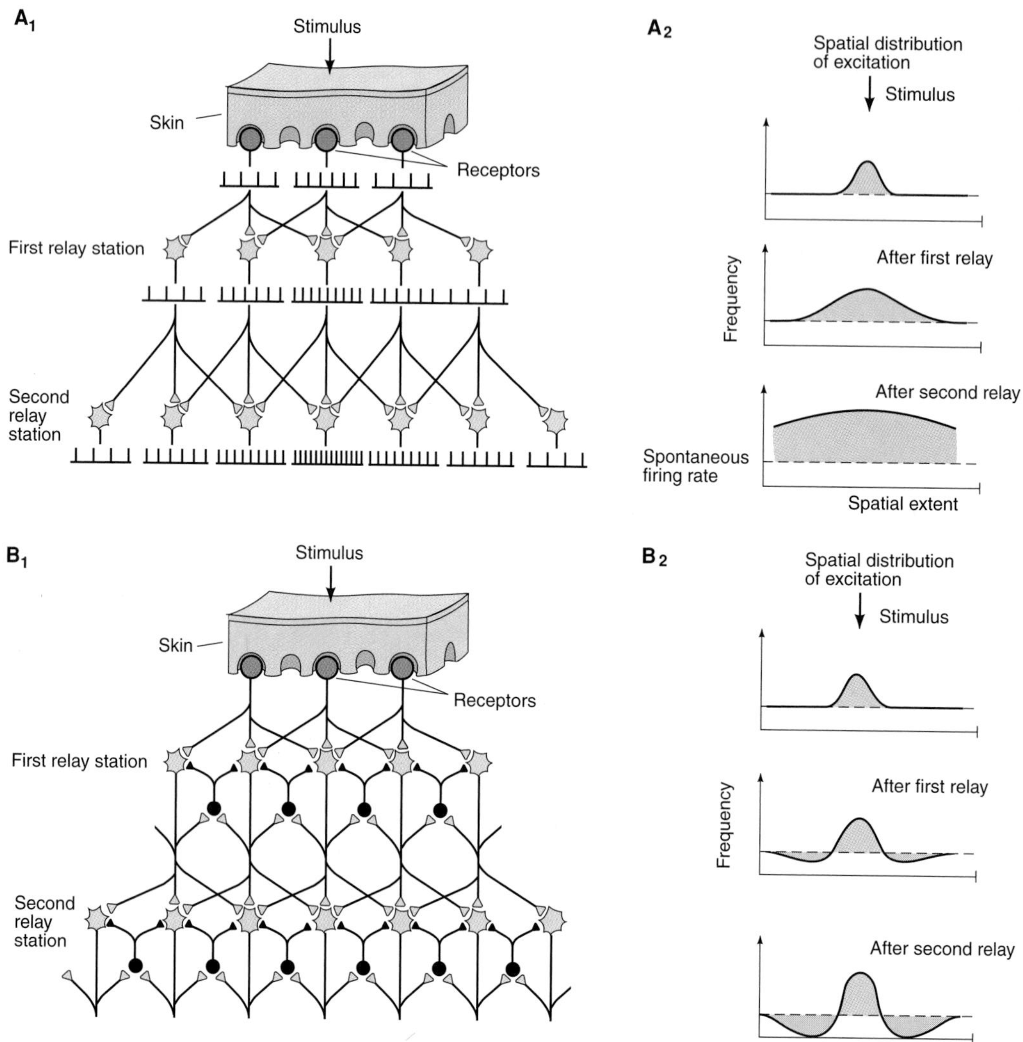

Figure 20–11 The function of lateral inhibition in sensory pathways is to enhance stimulus contrast. (Adapted from Dudel, 1983.)

A. 1. Diagram of the excitatory synaptic connections from three receptors onto projection neurons at the next two relays. The relative rate of discharge of each cell during stimulation is shown. **2.** The plots show the distribution of excitation in a population of neurons at the site activated by the stimulus. The *y* axis is the frequency of discharge, the *x* axis the spatial distribution of the population of cells. In the absence of lateral inhibition, due to the lack of inhibitory interneurons, the excitatory region (pink) becomes progressively larger at each of the relays in response to a stimulus.

B. 1. In the presence of lateral inhibition, due to the addition of inhibitory interneurons (**black**), a narrow excitatory region is maintained through successive relays. **2.** On either side of the excitatory region the discharge rate is driven below the resting level by feedback inhibition (gray). The effect of restricting excitation in this manner is to enhance contrast.

Feed-forward and feedback inhibitory pathways are local circuits only; they involve only the cells within one relay nucleus. *Distal inhibition* is the mechanism by which neurons from higher centers, such as the motor cortex and the brain stem, can inhibit and thereby control the flow of information into relay nuclei (Figure 20–12). Distal inhibition illustrates still another principle of organization in the sensory system: Higher areas of the brain are able to control the incoming information from the periphery.

Different Modalities Place Specific Demands on the Common Neural Architecture of Sensory Systems

In the next four chapters we examine the principles underlying sensory coding in the visual system. We focus on this system because our understanding of the processing of information for visual perception is advanced and because, for humans, vision is the most important sensory modality. As a result, a significant component of cognitive neural science is concerned with vision. To place the studies of vision in a broader context we briefly describe some of the defining features of other sensory systems. Although all of the sensory systems use the same coding mechanisms and have a similar anatomical organization, each must also cope with the special needs that the different modalities present.

The Somatosensory System Mediates Several Different Sensations

The somatic sensory system is distinctive for two reasons. First, the receptors for somatic sensation are distributed throughout the body, whereas those for other sensory systems are restricted to small specialized organs in the ears, eyes, nose, or mouth. Second, the somatic sensory system processes different kinds of stimuli, many of which have a different energy. The sensations or submodalities that the somatic sensory system mediates, often called the *body senses*, are diverse.

1. *Touch* is elicited by mechanical stimulation of the body surface.
2. *Limb proprioception* is elicited by mechanical displacements of the muscles and joints.

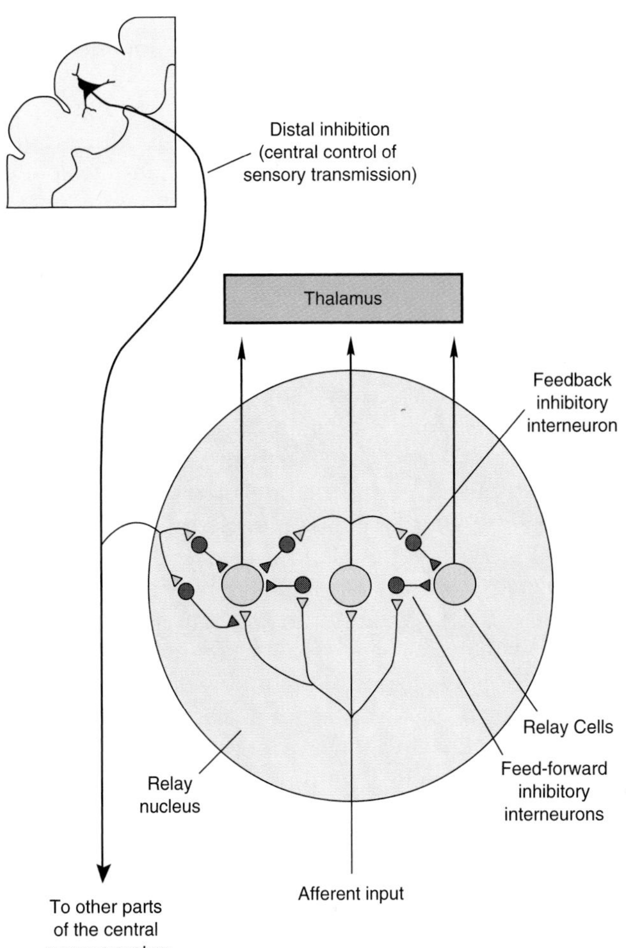

Figure 20–12 Three types of inhibition occur in sensory relay nuclei: feedback, feedforward, and distal. Each form is mediated by inhibitory interneurons located within the nucleus. The inhibitory interneurons that mediate *feedback inhibition,* of which lateral inhibition is an example, are activated by collateral axons from projection neurons within the nucleus. Inhibitory interneurons driving *feedforward inhibition* are activated by afferent information from the next lower-order station. Inhibitory interneurons for *distal inhibition* are activated by signals from higher-order portions of the sensory pathway or from nonsensory structures, like the motor cortex.

3. *Pain* is elicited by noxious (tissue damaging) chemical, mechanical, or thermal stimuli.
4. *Thermal sensations* are elicited by cool and warm stimuli.

These sensations are combined to produce more complex sensory experiences. For example, touch and proprioception of the hand are combined to produce the sense of the three-dimensional shape of a grasped object, a sensation termed *stereognosia.* This sensory capacity enables us to choose the correct coin from a pocket of change without looking.

The mechanisms we examined earlier in this chapter are clearly evident in the pathways for pain sensation. Specialized sensory receptors, termed *nociceptors,* are selectively activated by noxious stimuli. The information they transmit to the central nervous system is processed like any of the other sensory modalities. Lateral inhibition enhances contrast and enables better localization of the noxious stimulus on the body. Distal inhibition, which we saw, is a mechanism for higher brain centers to control the inflow of sensory information. For pain, distal inhibition confers important adaptive advantages. Under special behavioral conditions, such as during childbirth or military combat, pain is disadvantageous because the sensory experience makes it difficult to function effectively. Serotonin and the neuropeptide enkephalin are two key neurotransmitters in the pathways mediating the central inhibition of pain.

Pain is important because it signals information of a protective nature. In humans pain is further important because it can evoke strong emotions., For example, we may become concerned that a pain will lead to suffering or signal a serious health problem. Not surprisingly, the pathways that signal the discriminative aspects of pain, such as the location and intensity of a noxious stimulus, are distinct from those that signal the affective aspects of pain. Our response to a noxious stimulus reflects the actions of both systems. Indeed, analgesic drugs like morphine diminish both the discriminative and affective aspects of pain. In contrast, marijuana diminishes only the affective aspect.

Population Codes May Play an Important Role for Coding Stimulus Quality in Taste and Smell

The sense of smell (olfaction) is carried by receptors that lie deep within the nasal cavity, confined to a patch of specialized tissue, the olfactory epithelium. The small unmyelinated axons of the olfactory receptor neurons terminate on projection neurons and interneurons in the *olfactory bulb,* the first relay in the olfactory system. The sense of smell is unique among the sensory systems in that olfactory pathways project from the olfactory bulb first to phylogenetically older portions of the cerebral cortex before reaching the thalamus and eventually the neocortex. In fact, the olfactory bulb is itself a primitive form of cortex. This may not be surprising since chemical senses are phylogenetically the oldest senses.

The olfactory system has the unique challenge of anticipating the myriad chemical configurations that can come in contact with the olfactory epithelium. How does it do so? This question is beginning to be answered using various cell biological, molecular, and genetic approaches. Genes encoding olfactory receptors have been cloned and shown to consist of a large family of G-protein coupled receptors. In rats and other mammals this family may contain up to 1,000 genes. Nevertheless, there are not nearly enough olfactory receptors to encompass the entire spectrum of possible odorants.

Moreover, anatomical studies have shown that cells of an individual receptor subtype are not clustered in a small discrete area in the olfactory epithelium but are widely scattered. This may help to explain why in the olfactory system, unlike the somatosensory and visual systems, the spatial relationships of the receptors in the receptive sheet (the olfactory epithelium) are not maintained in the sensory pathways to the brain.

Taste receptor cells transduce soluble chemical stimuli into electrical signals that can be transmitted to the brain. They are clustered in sensory organs called *taste buds,* located in various parts of the oral cavity. These chemoreceptors are selective for one of four basic qualities: bitter, salty, sour, and sweet. More complex tastes result from activation of combinations of these receptors. As with somatic, visual, and auditory information, information about taste is relayed through the thalamus before reaching the cortex.

In both the smell and taste systems the transduction of sensory stimuli involves the activation of membrane receptors that trigger intracellular second messengers (see Figure 20–5). For example, the structure of the olfactory receptors and of other components of signaling pathways in the olfactory epithelium, including G-proteins and adenylyl cyclases, suggests that the processing of olfactory sensory signals is similar to that of visual transduction in photoreceptors. The ion channels of olfactory and gustatory receptor cells are similar to those of other sensory neurons. In addition, many of the same general rules that we have encountered in the other senses—including population coding, analysis of contrast, and parallel processing—seem also to apply to smell and taste.

The Auditory System Locates Sound in Space

The auditory system, composed of the ear and the auditory pathways of the brain, enables us to translate air pressure waves into sound and to locate sound sources. The effectiveness of this system is particularly striking when we consider that the energy in a sound wave, even a loud one, is exceedingly small and that most sounds are composed of many different frequencies in the midst of a noisy environment. This remarkable signal analysis is accomplished by the sophisticated mechanoelectric transduction system of the inner ear working in conjunction with neural systems within the brain that compare signals from the two ears. Owing to the mechanical design of the ear and the specificity of wiring in the brain, humans are capable of detecting sounds ranging from 20 to 20,000 Hz, over a millionfold range of intensities, with a spatial resolution as fine as one degree of arc.

In the brain, information from the two ears is combined in the brain stem. The auditory pathways separate information about the *timing* and the *intensity* of signals, the two binaural cues for sound localization. Sound frequency is also represented in the central auditory pathways. Individual neurons are tuned to a limited range of frequencies, or bandwidth (see Figure 20–3). This information about the timing, intensity, and frequency of sound ascends in parallel to separate processing areas in the auditory cortex that analyze location, loudness, and pitch, just as in the visual cortex form, color, and motion are processed in separate areas.

An Overall View

The sensory systems accomplish the extraordinary task of keeping the brain constantly informed about the external world. We are beginning to understand at the molecular level the mechanisms for transducing the energy of the stimulus into electrochemical signals, the language of the nervous system. Knowledge obtained from the study of different systems has provided important insights into these common neural mechanisms. By simply modulating the patterns of action potentials, whose wave forms are stereotypic, a remarkable diversity of stimulus attributes are encoded.

Stimulus encoding is, however, only one of many functions of the sensory systems. As we will see in later chapters, sensory systems are also essential for maintaining arousal, forming our body image, and controlling movements.

Our present understanding of the myriad functions of the sensory systems is based on studies of isolated components of individual sensory systems under highly controlled conditions. Understanding how the various sensory systems interact "on the fly" in dealing with a constant stream of information represents the greatest challenge for a neurobiological analysis of perception. The real-time processing of sensory information, the basic operation of each sensory system, is beyond the capacity of even the most sophisticated computers.

Selected Readings

Buck, L., and Axel, R. 1991. A novel multigene family may encode odorant receptors: A molecular basis for odorant recognition. Cell 65:175–187.

Darian-Smith, I. 1982. Touch in primates. Annu. Rev. Psychol. 33:155–194.

Fields, H. L. 1987. Pain. New York: McGraw-Hill.

Finger, T. E., and Silver, W. L. (eds.) 1987. Neurobiology of Taste and Smell. New York: Wiley.

Hudspeth, A. J. 1989. How the ear's works work. Nature 341:397–404.

Kaas, J. H., Nelson, R. J., Sur, M., Lin, C.-S., and Merzenich, M. M. 1979. Multiple representations of the body within the primary somatosensory cortex of primates. Science 204:521–523.

Kinnamon, S. C., and Cummings, T. A. 1992. Chemosensory transduction mechanisms in taste. Annu. Rev. Physiol. 54:715–731.

Margolskee, R. F. 1993. The biochemistry and molecular biology of taste transduction. Curr. Opin. Neurobiol. 3:526–531.

McMahon, S. B., Lewin, G. R., and Wall, P. D. 1993. Central hyperexcitability triggered by noxious inputs. Curr. Opin. Neurobiol. 3:602–610.

Miller, G. A. 1962. Psychology: The Science of Mental Life. New York: Harper & Row.

Mountcastle, V. B. 1975. The view from within: Pathways to the study of perception. Johns Hopkins Med. J. 136:109–131.

Mountcastle, V. B. 1980. Sensory receptors and neural encoding: Introduction to sensory processes. In V. B. Mountcastle (ed.), Medical Physiology, 14th ed., Vol. 1. St. Louis: Mosby, pp. 327–347.

Norgren, R. 1984. Central neural mechanisms of taste. In I. Darian-Smith (ed.), Handbook of Physiology, Section 1: The Nervous System, Vol. III. Sensory Processes, Part 2. Bethesda, Md.: American Physiological Society, pp. 1087–1128.

Pickles, J. O., and Corey, D. P. 1992. Mechanoelectrical transduction by hair cells. Trends Neurosci. 15:254–259.

Shepherd, G. M. 1994. "Chemical senses," In Neurobiology, 3rd ed. New York: Oxford University Press.

Shepherd, G. M. 1994. Discrimination of molecular signals by the olfactory receptor neuron. Neuron. 13:771–790.

Stevens, S. S. 1975. Psychophysics: Introduction to Its Perceptual, Neural, and Social Prospects. New York: Wiley.

Von Békésy, G. 1960. Experiments in Hearing. E. G. Wever (ed. and trans.) New York: McGraw–Hill.

Willis, W. D., Jr. 1985. The Pain System: The Neural Basis of Nociceptive Transmission in the Mammalian Nervous System. Basel: Karger.

21

Construction of the Visual Image

Visual Perception Is a Creative Process

Three Parallel Pathways Process Information for Depth and Form, Motion, and Color

Psychological Evidence Indicates that the Separate Pathways Carry Different Visual Information

Clinical Evidence Is Consistent with Parallel Processing of Visual Information

Attention Focuses Visual Perception by Facilitating Coordination Between Separate Visual Pathways

The Analysis of Visual Attention May Provide Important Clues About Conscious Awareness

Most of our impressions of the world and our memory of it are based on sight. Yet the mechanisms that underlie vision are not at all obvious either to the perceiver or even to the student of perception. How do we see form? How do we perceive the movement of objects in space? How do we perceive color?

Studies in artificial intelligence and analyses of pattern recognition by computers have made us realize that the brain recognizes form and movement using strategies that no existing computer begins to approach. Simply to look out into the world and recognize a face or a landscape is an amazing feat, more difficult than the information processing required to solve logic problems or play chess.

How is this feat accomplished? This question, which we address in this and the subsequent three chapters, appears all the more puzzling when we learn that the neural information related to form, motion, and color is carried not by one single hierarchical pathway or processing stream, but by at least three (and possibly more) parallel and interacting processing pathways in the brain.

The existence of parallel processing pathways in turn raises a second problem we consider here—the *binding problem*. How is the information carried by three separate pathways brought together into a single image? In addressing the binding problem we focus on one of the central questions of cognition: How does the brain construct a perceived world from sensory information, and how does it bring it into consciousness?

Visual Perception Is a Creative Process

Until recently, visual perception was often compared with the operation of a camera. Like the lens of a camera, the lens of the eye focuses an inverted image onto the retina. This analogy breaks down rapidly, however, because it misses what vision really does, which is to create a three-dimensional perception of the world that is different from the two-dimensional images projected onto the retina. It also fails to capture an important feature of our visual system—that we are able to recognize an object as the same even though the actual image on the retina varies widely under different light conditions.

As we move about, or as the ambient illumination changes, the size, shape, and brightness of images on the retina also change. Yet under most conditions we do not perceive the object itself to be changing. When people walk toward you, you perceive them as coming closer; you do not perceive them as growing larger, even though the image on the retina does enlarge. As we move from a brightly lit garden into a dim room, the intensity of light reaching the retina may vary a thousandfold. Yet in the dim light of a room, as in the bright light of the sun, we see a white shirt as white and a red tie as red. Our ability to perceive an object's size or color as constant illustrates clearly what is so remarkable about the visual system. It does not record images passively, like a camera. Instead, it transforms transient light stimuli on the retina into mental constructs of a stable three-dimensional world.

The degree to which visual perception is transformational and therefore creative has been fully appreciated only recently. Earlier psychophysical thinking was greatly influenced by the British empiricist philosophers of the seventeenth and eighteenth centuries, notably John Locke and George Berkeley, who thought of perception as assembling elementary sensations in an additive way, component by component. The modern cognitive view that perception is an active and creative process, involving more than just the intake of sensory information, was first articulated in the early twentieth century by the German psychologists Max Wertheimer, Kurt Koffka, and Wolfgang Köhler, who founded the school of *Gestalt psychology*.

The German word *Gestalt* means form. The central idea of the Gestalt psychologists is that the process of perception actively shapes, from the details of a stimulus, the complete form that emerges in consciousness. A perceived image is not the sum of its perceptual elements as the empiricist philosophers believed. Rather, the elements are selectively organized by our brain so as to create a form that is more than the sum of its parts. The Gestalt psychologists liked to compare the perception of visual form with the perception of a melody. What we recognize in a melody is not the sequence of particular notes but their *interrelationship*. A melody played in different keys will still be recognized as the same melody because the relationship among the notes remains the same. Likewise, we are able to recognize different images under a variety of visual conditions, including differences in illumination, because the relationships between the components of the image are maintained.

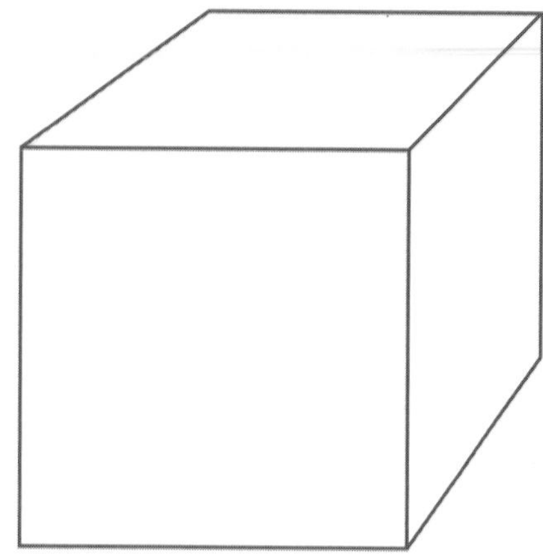

Figure 21–1 This drawing is *two*-dimensional, yet we perceive the object as being *three*-dimensional because we have considerable experience with three-dimensional boxes of similar shape.

The Gestalt psychologists argued that the brain actively builds up complete percepts from details in the visual image by searching for and combining parts that correspond most satisfactorily to objects in the real world. The brain does this by making certain assumptions about what will be seen in the world, assumptions that seem to derive in part from experience and in part from the built-in neural wiring for vision. Thus, the simple two-dimensional drawing of Figure 21–1 appears to us to be a three-dimensional cube because we have had extensive experience with real boxes.

The visual system organizes these perceptual tasks by following certain innate laws that govern the pattern, shape, color, distance, and movement of objects in the visual field. The Gestalt psychologists illustrated these *laws of perception* with various examples of visual patterning. Consider the array of dots in Figure 21–2A. The dots in the figure are equally spaced, yet the brain organizes them alternatively into either rows or columns. This tendency to perceive one pattern rather than another can be enhanced according to the laws of *similarity* and *proximity*. Thus, when some of the dots in one direction have a similar appearance there is a tendency to see the whole pattern in that direction (Figure 21–2B). Likewise, if some of the dots in one direction are brought nearer together (proximity), the brain is more disposed to see a pattern in that direction (Figure 21–2C).

This process of perceptual organization is continuous and dynamic, as is evident in the well-known alternation of figures on a background, first illustrated by the psychologist Edgar Rubin (Figure 21–3). The image can be seen as two black profiles

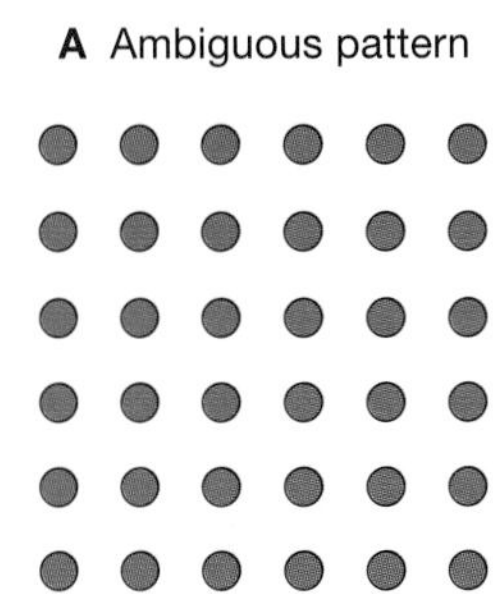

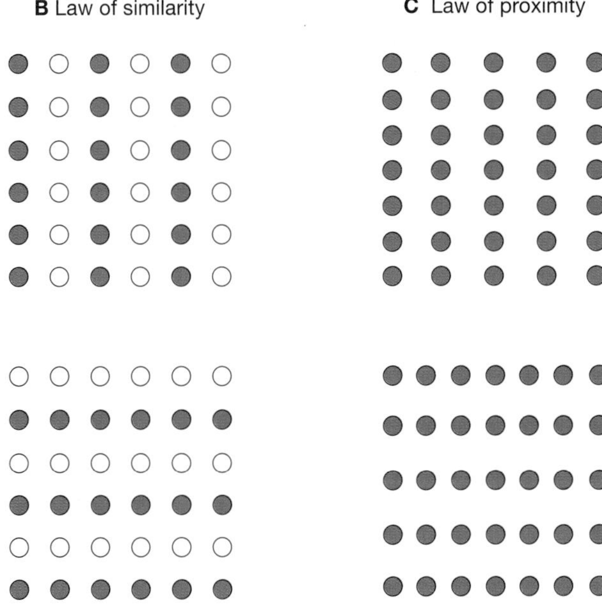

Figure 21–2 The array of identical dots in **A** can be seen alternatively as a pattern of rows or columns. The arrays in **B** and **C** are not ambiguous because of additional cues. In **B** similarities in the appearance of some dots create a strong pattern of columns or rows, while in **C** the spatial arrangement alone determines whether we see a vertical or horizontal pattern. (From Gleitman, 1981.)

Figure 21–3 In this famous image by the Danish psychologist Edgar Rubin we sometimes see a pair of faces, sometimes a vase. The perceptual decision on what is *figure* (object) and what is *ground* (background) resembles the radar engineer's distinction between signal and noise. By focusing on one signal, we cause the other to recede into the background.

A

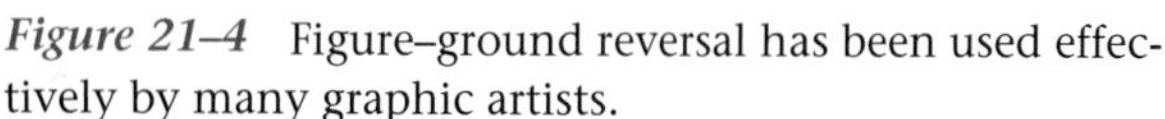

B

Figure 21–4 Figure–ground reversal has been used effectively by many graphic artists.

A. In Victor Vasarely's fluid rendering of two lovers embracing, either the black or white form can be seen as figure or ground. (From Gleitman, 1981.)

B. In this repeating pattern of fishes and birds by Maurits Escher, the same outline is shared by the two different figures. Normally, contours serve only to outline an object against its background. (From Escher, 1971.)

Figure 21–5 A highly schematic drawing, typical of children's drawings, nevertheless presents clearly recognizable objects, judged simply from their outlines.

against a white background or as a white vase against a black background, but it is almost impossible to see both images simultaneously. This *winner-take-all* strategy has been extensively exploited by artists (Figure 21–4). Maurits Escher writes: "Our eyes are accustomed to fixing on specific objects. The moment this happens everything *around* is reduced to background. . . . The human eye and mind cannot be busy with two things at the same moment, so there must be a quick and continual jumping from one side to the other." The figure–ground dichotomy thus illustrates one principle of visual perception: only part of the image is selected as the focus of attention, while the rest becomes submerged into the background. (We shall encounter the winner-take-all strategy again in the singleness of action in the motor system.)

The part of an image that is the central focus of attention is its boundaries and edges. As a result, we recognize objects or a scene with clarity even when they are presented to us as simple line drawings without shading or color (Figure 21–5).

Illusions, which are "misreadings" of visual information by the brain, illustrate how the brain applies certain assumptions about the visual world to the sensory information it receives. Certain organizational mechanisms of vision—selection, distortion, filling in of omissions—are powerfully demonstrated by illusions. In the classic Müller–Lyer illusion two lines of equal length look unequal (Figure 21–6). As is characteristic of many illusions, knowing that the lines are equal does not prevent us from being fooled by this illusion. We consistently see the line with inwardly directed barbs as smaller than the line with outwardly directed barbs. We perceive them to be unequal because of the logic built into the brain, and because our experience teaches us to use shape as an indicator of size.

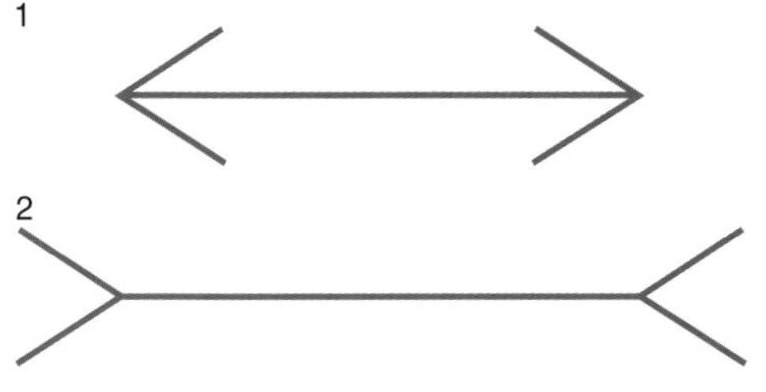

Figure 21–6 Perceived length can differ from measured length, as illustrated by the classic Müller–Lyer illusion. The two horizontal lines are identical in length, but **line 1** appears shorter than **line 2**.

The creative *filling in* process of vision is nicely illustrated by a famous triangle illusion (Figure 21–7). The outlines of the white and the black triangles emerge from illusory contours supplied by the brain—contours that do not actually exist on the page!

In addition to the built-in logic and experience of the brain, the spatial relationship of objects to each other also helps us to interpret an image. That is, we judge the size of an object by comparing it with its immediate surroundings. For example, we do not judge the sizes of two people at different distances, by comparing them with each other, but by comparing each person with objects immediately around him (Figure 21–8). In this comparison we also rely on our familiarity with objects in the visual field.

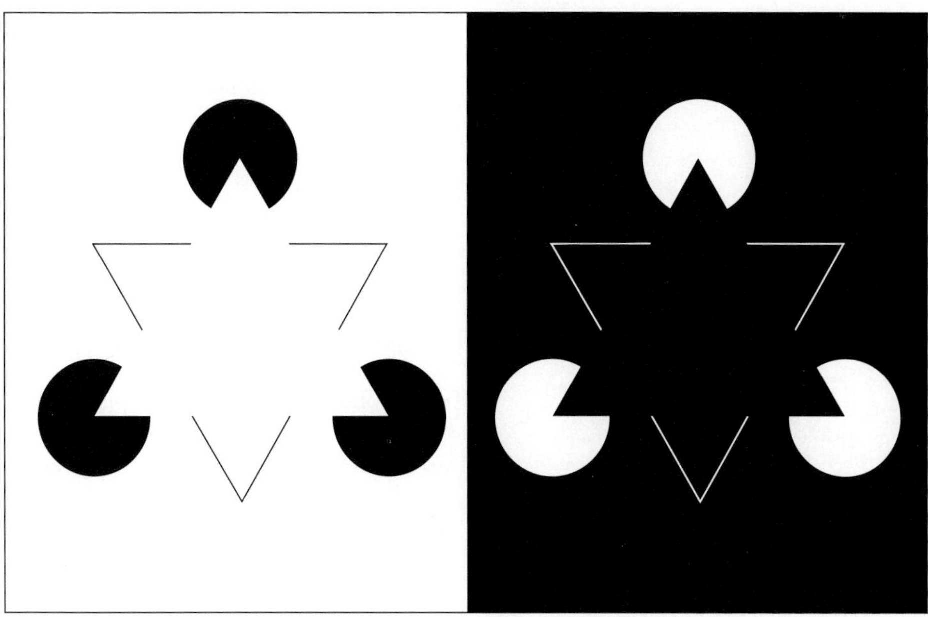

Figure 21–7 The Kanizsa triangle. The contours of the triangles are subjective and do not exist in the drawings.

Figure 21–8 To judge size, we automatically compare objects in the visual field. In the picture on the **left** the woman in the foreground is 9 ft from the camera and the second one is 27 ft away. Both appear to be the same size. The photograph on the **right** was taken with the woman in the foreground only; the photograph of the second woman was pasted into the picture. In the doctored photo the second woman seems small, not far away, because the corridor and tiles around her are not proportional, as they are in the photo at the left. To convince yourself that she is the same size as in the photo at the left, you will probably need to measure her. (From Brown and Herrnstein, 1975.)

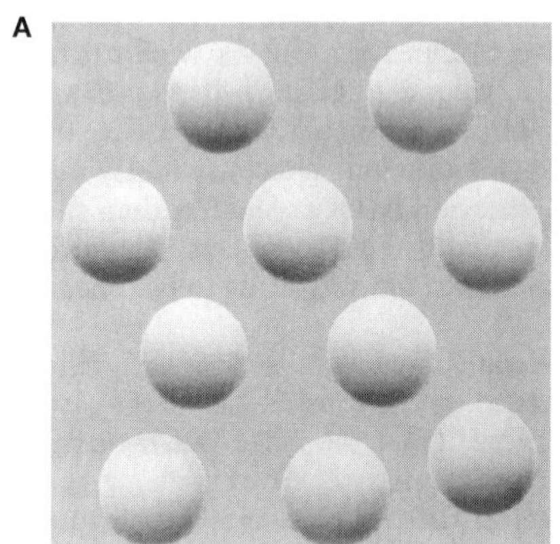

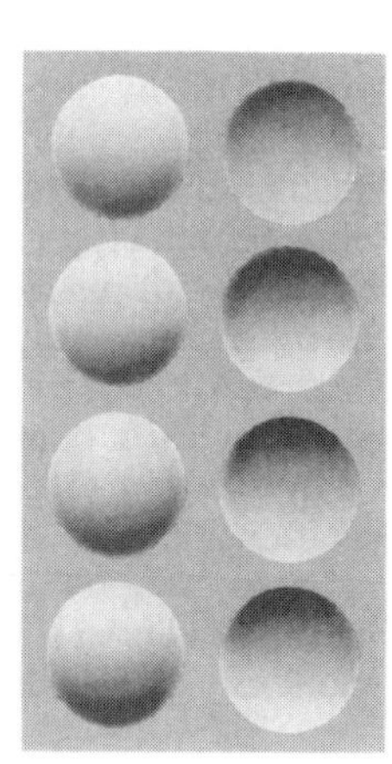

Figure 21–9 Spheres or cavities? The decision depends on where you assume the light source is.

A. You can reverse the apparent curvature of these objects by imagining a shifting in the light source from above to below.

B. In this array, once you see a convex shape in the spheres of one column, the spheres in the other column will appear concave. It is almost impossible to see only convex or concave shapes in both columns together. (Adapted from Ramachandran, 1987.)

As we have seen, some strategies used by the visual system seem to reflect inferences built into the brain (Figure 21–1). For example, the shadows on a sphere lit from above appear to have a convex shape like the outside of a bowl, whereas the shadows on a sphere lit from below appear to have a concave shape like the inside of a bowl (Figure 21–9). When we do not know the direction of the light source, the curvature is ambiguous. Under these conditions, the way we interpret the curvature of one object determines how we see all the other objects in a scene (Figure 21–9). In interpreting the curvature of several objects together the brain seems to infer that the entire visual image is illuminated by only one source of light. We make this inference probably because we evolved in an environment with one source of light, the sun, and our brain assumes that the source of light is always above rather than below. This notion is consistent with the Gestaltist's contention that our reading of shape involves interpretation of sensory information.

As a result of the influence of Gestalt theorists, most psychologists studying vision no longer ask the empiricist's question, What are the building blocks of this perception? Rather, they—and we—are interested in the question, What produces this perception in the brain? This question provides common ground for the meeting between contemporary psychology and neurobiology in studies of vision. In this chapter we first consider the anatomical pathways and psychophysical evidence for the processing of form, motion, and color in three parallel pathways that extend from the retina to the lateral geniculate nucleus and from there to the cortex. Then we examine how visual attention may bring these parallel transformations together into a single conscious image. In the next chapter (Chapter 22) we examine how the retina first processes visual scenes. In the following two chapters we focus in detail on the processing of form, motion, and color.

Three Parallel Pathways Process Information for Depth and Form, Motion, and Color

Before we examine the physiology of the visual system, we need briefly to consider some of the anatomical features of the visual pathways, which will be discussed in detail in the following chapters. The axons of ganglion cells of the retina form the optic nerve, which projects in an orderly fashion to the lateral geniculate nucleus in the thalamus. The lateral geniculate in turn projects to the ipsilateral primary visual cortex, or V1, in Brodmann's area 17 (also called the striate cortex). As a result of this orderly projection, the striate cortex contains a complete map of the retina.

Beyond the striate cortex lie the extrastriate areas, a set of higher-order visual areas also containing representations of the retina (Figure 21–10). There are about 32 representations of the retina in the extrastriate areas (Figure 21–11). These 32 areas concerned with vision occupy more than half of the total surface area of the cortex! Some of these representations are complete, others are only partial. They also differ in the precision of their retinotopic organization. What is the function of these cortical regions? These regions differ in the selectivity of cells for different features of stimuli. For example, area V5 (MT) is primarily concerned with motion in the visual field, while V4 is much more concerned with color and with the orientation of edges. Thus, the visual system (like the somatic sensory system) has several distinct representations of its receptive sheet, the reti-

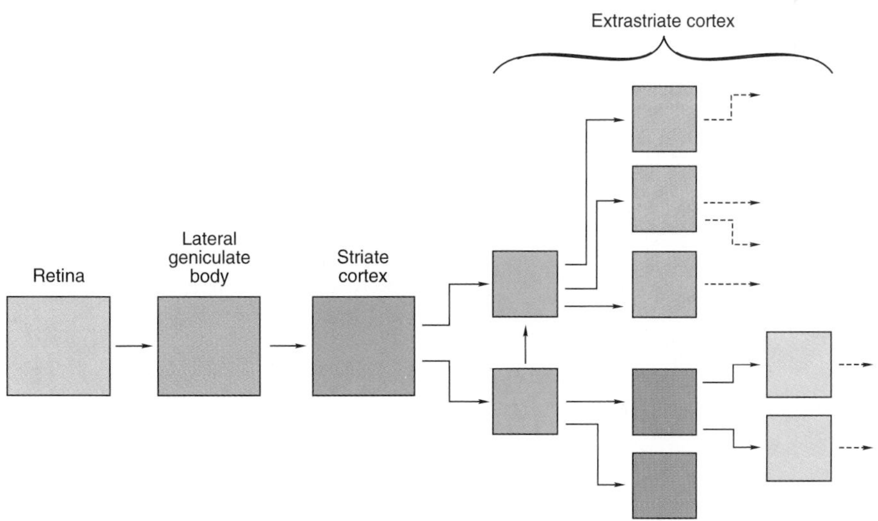

Figure 21–10 The parallel pathways of the visual system. Each structure, shown as a box, consists of millions of cells, aggregated into sheets. Each receives inputs from one or more structures at lower levels in the path and each sends its output to several structures at higher levels. The path has been traced only for four or five stages beyond the primary visual cortex. (From Hubel, 1988.)

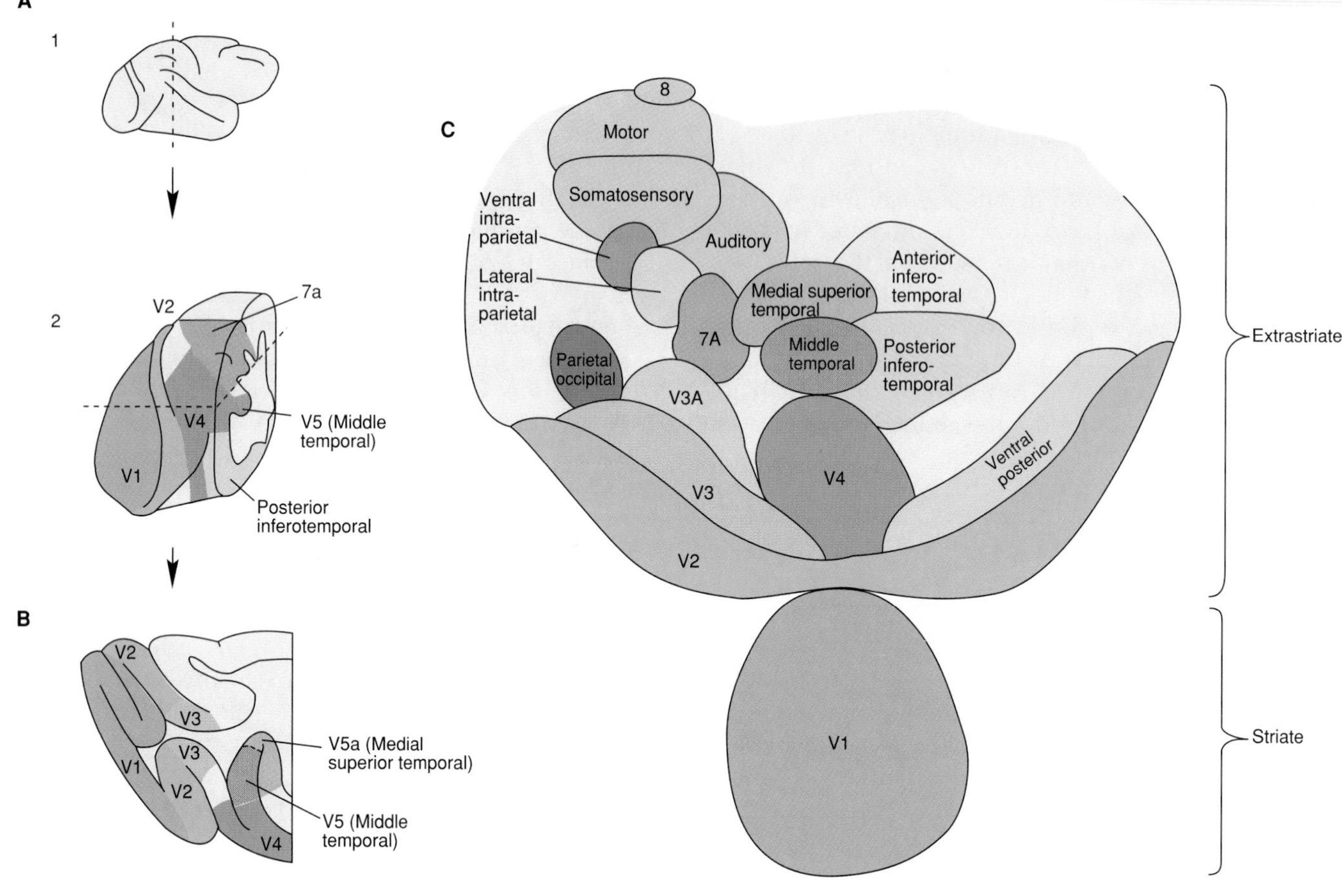

Figure 21–11 Areas of the human cortex dedicated to vision.

A. 1. This lateral view of the right hemisphere shows exposed portions of the striate cortex (V1) and extrastriate cortex (V2 and V4). (The **vertical line** indicates the location of the coronal section in **2**.) **2.** An expanded view of the occipital lobe (anterolateral angle) shows areas V1, V2, V4, and V5. Area V5 lies in the buried MT area at the rostral border of the occipital lobe.

B. This horizontal section through the occipital lobe at the location shown in **A2** illustrates the approximate locations of known visual processing areas at this level.

C. In this drawing the cortical surface of the right hemisphere of the monkey is unfolded and flattened to show the visual areas. Fourteen areas are indicated, but more may be present. The abbreviations for the various visual areas (V1, V2, V3, V3a, V4, V5) were originally based on the belief, no longer thought to be correct, that visual processing was strictly serial. In addition, some terms such as 7a derive from old architectural maps of the cerebral cortex.

na, and each is involved in processing different aspects of visual information.

How does information about different aspects of the visual image reach these cortical areas? As we shall learn in the next chapter, the segregation of visual information begins in the retina, which contains two sizes of ganglion cells—large cells (called magnocellular or type M) and small cells (called parvocellular or type P). Each of these cells carries slightly different information, which it conveys to different layers in the lateral geniculate nucleus of the thalamus and then on to different regions of the cortex. The axons of the M cells project to the magnocellular layers of the lateral geniculate

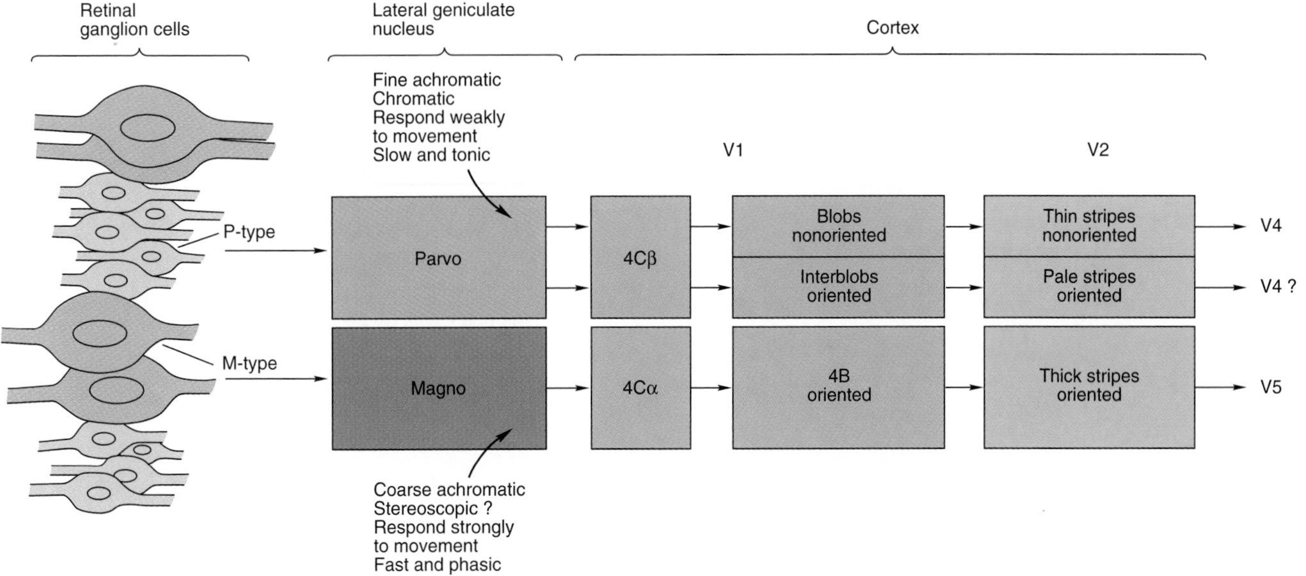

Figure 21–12 Three major parallel pathways for visual perception arise from the lateral geniculate nucleus. Each pathway is concerned with one type of visual information (see Figure 21–14). (Adapted from Livingstone and Hubel, 1984.)

nucleus, whereas the axons of the P cells project to the parvocellular layers. These two sets of layers give rise to three major pathways—two from the parvocellular layers and one from the magnocellular layer of the lateral geniculate nucleus. The three pathways extend from the lateral geniculate nucleus to V1, the primary visual cortex, and from there to V2 and several other extrastriate cortices (Figure 21–12).

Both V1 and V2 have important subdivisions that stain intensely for the mitrochondrial enzyme cytochrome oxidase. In V1 the heavily staining regions represent the *blobs*, which form a repeating polka dot-like pattern of peg-like structures about 0.2 mm in diameter separated by pale regions called the *interblob region*. In V2 the heavily staining regions form two types of dark *stripes*—thick and thin—separated by pale *interstripe* regions (Figure 21–13).

The first pathway that originates in neurons intercalated between the parvocellular layers of the lateral geniculate nucleus—the *parvocellular-blob pathway*—is concerned with the perception of color (Figures 21–12 and 21–14). This pathway synapses in the blobs of the superficial layers of V1. From V1 the pathway goes to V2, and from there it projects to area V4, an area having many color-responsive cells. It eventually terminates within the inferior temporal cortex, an area concerned with the perception of color and form.

The second pathway that originates in the parvocellular pathway—the *parvocellular-interblob pathway*—is concerned with the perception of shapes (Figures 21–12 and 21–14). This pathway terminates in the deeper layers of V1, and also receives a small contribution from the magnocellular pathway. Like the pathway concerned with color, this pathway eventually projects to the inferior temporal cortex. This system is sensitive to the outline and orientation of images, important elements in the perception of form. The neurons in this system are capable of high resolution, which is probably important for seeing stationary objects in detail. Thus, this system is concerned with *what* is seen. It

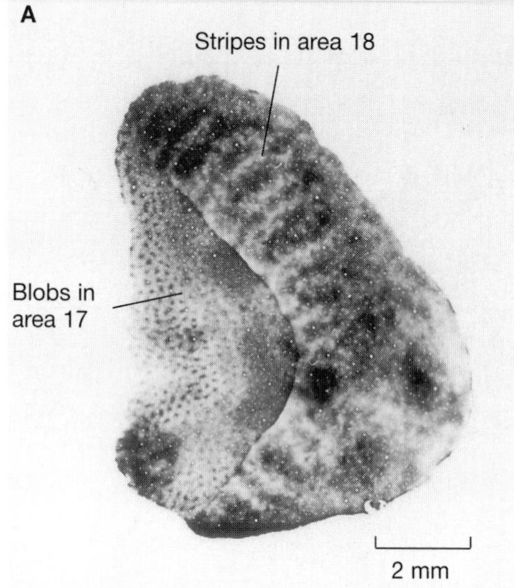

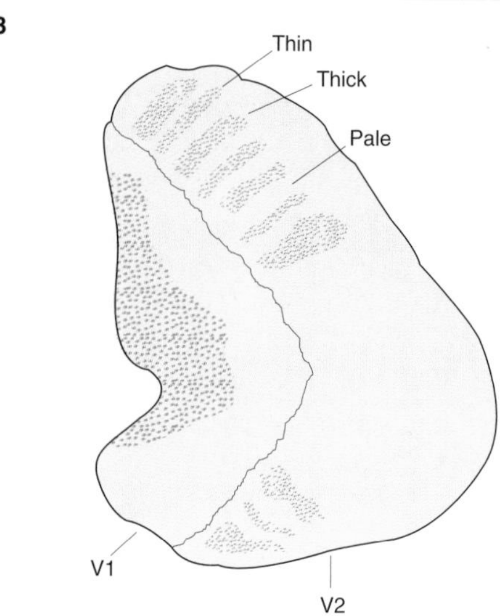

Figure 21–13 Mitochondrial stain reveals the blob regions of area 17 and the thin and thick stripes of area 18.

A. Blobs in area 17 of the cortex are revealed by cytochrome oxidase stain in this section from the occipital lobe of a squirrel monkey. The stain also reveals the thick and thin stripes separated by pale stripes in area 18. (Courtesy of M. Livingstone.)

B. Parts of the staining pattern shown in **A** are highlighted in this drawing. (Adapted from Dowling, 1992.)

is also important for depth perception (and to some degree color). Lesions in the inferior temporal cortex produce deficits related to the recognition of complex objects and the recognition of faces.

The third pathway, the *magnocellular-thick stripes pathway* is specialized for detecting motion and spatial relationships and contributes to depth perception (Figures 21–12 and 21–14). It extends from the magnocellular layers of the lateral geniculate nucleus, through the interblob regions of V1, to the thick stripes of V2 and then to MT (V5), the area concerned with depth and motion. MT projects to other areas in the parietal cortex concerned with visuospatial function. This pathway is concerned with *where* objects are rather than what they are. The neurons in this system are relatively insensitive to color and poor at analyzing stationary objects. As we shall see later, lesions in this pathway result in a selective deficit in motion perception and in eye movements directed toward moving targets.

The three specialized pathways interact at many levels (Figure 21–14).

Psychological Evidence Indicates That the Separate Pathways Carry Different Visual Information

Can one experimentally relate distinct aspects of what is normally perceived as a unified whole to the specific pathways that run from the retina to the temporal or parietal cortex? How separable is movement from form, and either of these from color?

The contribution of the parvocellular system can be examined in partial isolation by reducing the input from the magnocellular system. This can be achieved by using *equiluminant stimuli*, images that vary in color but not in their degree of brightness. A border between two equiluminant colors has color contrast but no brightness contrast. In a black-and-white photograph two equiluminant colors appear to be the same shade of gray. In theory, the magnocellular system is largely color blind. It therefore relies only on brightness cues and would not be able to distinguish borders between equiluminant red and green. Thus, equiluminant stimuli would

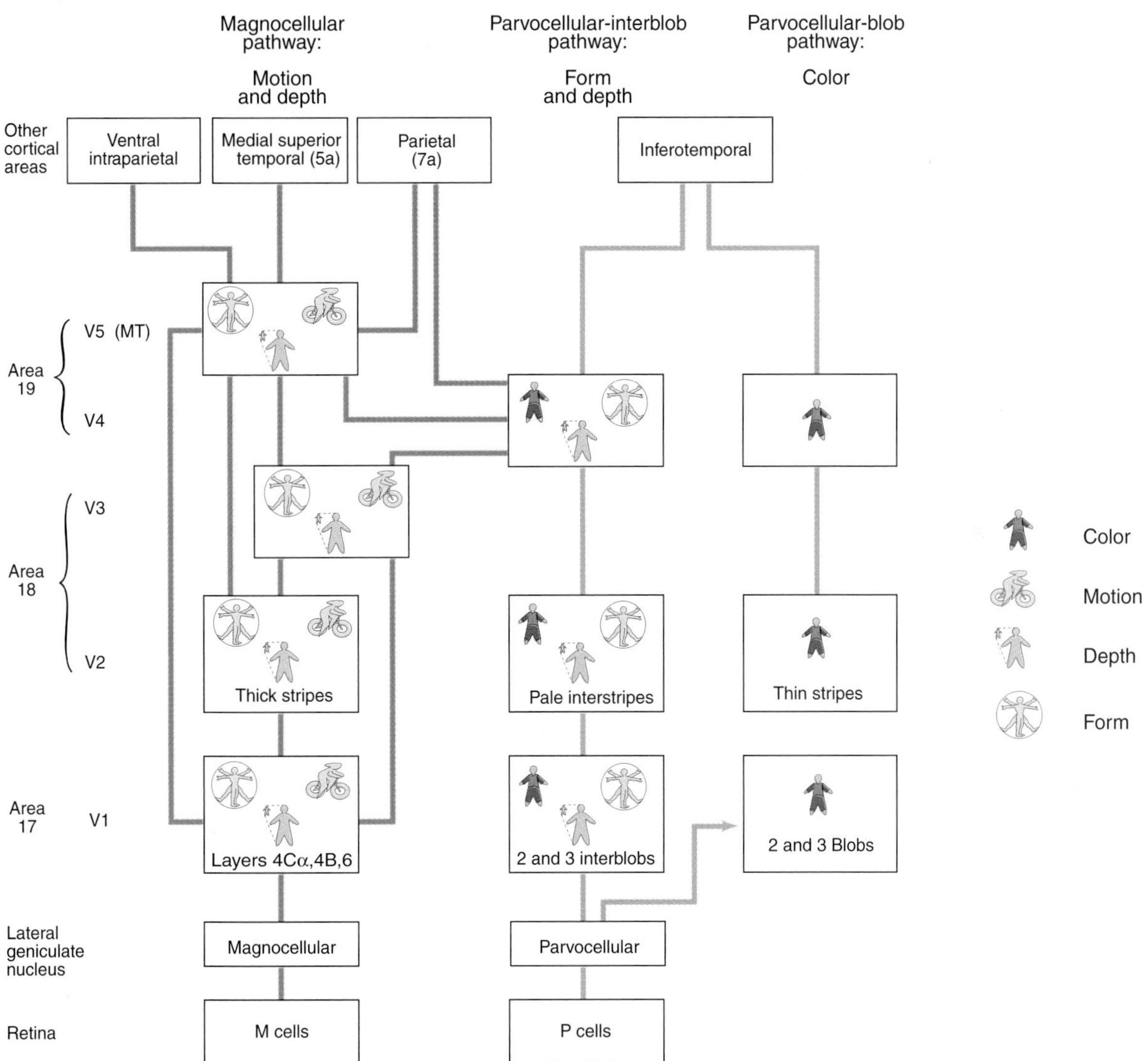

Figure 21–14 Each of the three major parallel visual pathways in the visual system carries primarily one kind of visual information. The organization shown here is based on studies of the macaque monkey. (Adapted from DeYoe and Van Essen, 1988.)

reduce the magnocellular system's contribution to perception. The color-sensitive cells of the parvocellular–blob system, however, should distinguish between red and green at any relative degree of brightness.

Studies of human responses to equiluminant stimuli suggest that the perception of motion is substantially compromised at equiluminance. As a result, it is thought that information on motion may be processed in good part independently of color information—presumably by the magnocellular system and largely independent of the parvocellular system. Perspective, the relative sizes of objects, depth perception, figure–ground relations, and visual illusions also disappear at equiluminance and therefore seem also to be mediated, to an important degree, by the magnocellular system.

Why would all of these relationships, and *these* relationships in particular, be mediated by one system? To separate figure from ground we seem to organize the visual field into *coherent* components, each representing a cluster of values of depth, brightness, and texture. In addition, when an object moves, a cluster of these elements will have a specific direction and velocity of motion, so that a moving cluster can also be a cue for distinguishing between objects. Depth, brightness, and texture can be analyzed efficiently at low resolution, whereas form requires high-resolution vision. Thus, the ability to discriminate figure and ground, to link parts of a scene, and to perceive spatial relationships may all be mediated by a low-resolution system that organizes the whole image by defining its components. In contrast, the color system is less concerned with movement.

Although there is a consensus that the visual system uses parallel processing, researchers disagree on how neatly the various functions are parceled out among the three pathways (even on exactly how many—two or three—key pathways there are). One disagreement revolves around the question of whether equiluminant stimuli fail to affect *only* the magnocellular pathway. Another focuses on the extent of the interaction between the pathways. A number of researchers contend that, however much one pathway may specialize in processing a particular component of the visual image, other pathways probably contribute to processing those components as well because of the extensive crosstalk between pathways.

Clinical Evidence Is Consistent with Parallel Processing of Visual Information

The idea that different aspects of visual perception may be handled in separate areas of the brain actually dates to the end of the nineteenth century, when Sigmund Freud concluded that the inability of certain patients to recognize specific features of the visual world was due not to a sensory deficit but to cortical defects that affect the ability to combine components of visual impressions into a meaningful pattern. These defects, which Freud called *agnosias* (loss of knowledge), can be quite specific depending on the area of the cortex damaged (Table 21–1). For example, a patient may sustain a selective defect for the perception of depth as a result of a specific lesion in the visual cortex. One such patient with a *depth agnosia* had an "inability to appreciate depth or thickness of objects seen. . . . The most corpulent individual might be a moving cardboard figure; everything is perfectly flat." Similarly, a *movement agnosia* can occur after bilateral damage to the MT or MST areas of the cortex and is manifested by an inability to perceive movement, without the loss of any other perceptual capabilities (Box 21–1).

Still other patients lose color vision (*achromatopsia*) because of localized damage to the region of the temporal cortex that in humans contains the homolog of V4 (see Chapter 24). These patients, nonetheless, have reasonably good perception of form. This color processing area in the brain can be identified in normal living human subjects using positron emission tomography (PET) scanning. In addition to movement agnosia and achromatopsia, there is an agnosia for form, which can be selective for either inanimate or animate objects.

Indeed, there is striking evidence for a discrete pathway for face recognition from studies of patients who, following a stroke, suffer specific impairment of face recognition, or *prosopagnosia*. These patients with lesions in the interior temporal cortex can identify a face as a face, its parts, and even specific emotions expressed on the face, but they are unable to identify a particular face as belonging to a specific person. Patients with prosopagnosia often cannot recognize their close relatives, and may not even recognize their *own* faces in the mirror, even though they recognize that they are looking at a face. It is not the identity

Table 21–1 The Visual Agnosias

Type	*Deficit*	*Most Probable Site of Lesion*
Agnosia for form and pattern		
Object agnosia	Naming, understanding of purpose, recognition of real objects	Areas 18, 20, 21 on left and corpus callosum
Agnosia for drawings	Recognition of drawn objects	Areas 18, 20, 21 on right
Prosopagnosia	Recognition of faces	Areas 20, 21 bilaterally
Agnosia for color		
Color agnosia	Association of colors with objects	Area 18 on right
Color anomia	Naming colors	Speech zones or connections from areas 18, 37
Achromatopsia	Distinguishing hues	Areas 18, 37
Agnosia for depth and movement		
Visual spatial agnosia	Stereoscopic vision	Areas 18, 37 on right
Movement agnosia	Discerning movement of object	Bilateral medial-temporal area (junction of occipital and temporal cortex)

Source: Kolb and Whishaw (1980).

BOX 21–1 *Movement Agnosia: A Case*

A patient with intact visual fields lost all perception of motion and could not distinguish between stationary and moving objects. The visual disorder complained of by the patient was a loss of movement vision in all three dimensions. She had difficulty, for example, in pouring tea or coffee into a cup because the fluid appeared to be frozen, like a glacier. In addition, she could not stop pouring at the right time since she was unable to perceive the movement in the cup (or a pot) when the fluid rose. Furthermore, the patient complained of difficulties in following a dialogue because she could not see the movements of the face and, especially, the mouth of the speaker. In a room where more than two other people were walking she felt very insecure and unwell, and usually left the room immediately, because "people were suddenly here or there but I have not seen them moving." The patient experienced the same problem but to an even more marked extent in crowded streets or places, which she therefore avoided as much as possible. She could not cross the street because of her inability to judge the speed of a car, but she could identify the car itself without difficulty. "When I'm looking at the car first, it seems far away. But then, when I want to cross the road, suddenly the car is very near." She gradually learned to 'estimate' the distance of moving vehicles by means of the sound becoming louder. (From Zihl et al., 1983.)

of people that has been lost, but the *connection* between a particular face and a particular identity. To recognize even a close friend, patients must rely on the friend's voice or other nonvisual clues (Box 21–2). In the purest form of prosopagnosia only recognition of faces is impaired; recognition of other objects is not affected.

These visual agnosias rarely occur in a pure form. A combination of deficits is not surprising because lesions of the brain due to vascular accidents or tumors are not normally restricted to functionally discrete regions. In animal experiments, on the other hand, a single region can be surgically removed without damage to adjacent areas. Although the clinical evidence for the anatomical basis of visual agnosias is not always precise, it is nevertheless consistent with the experiments showing that vision is mediated by interconnected parallel pathways.

BOX 21–2 Prosopagnosia: A Case

He was of above average intelligence and his general level of awareness was extremely keen. His memory was remarkable. . . . His span of digit retention was 8 forward and 6 backwards. There was no hesitation in his speech and he could obey complex orders. He read smoothly and there was no trouble in understanding and later describing what he had read. . . . He promptly recognized, named, and demonstrated the use of a wide variety of test objects. . . . The significance of line drawings was immediately apparent to him, and he could accurately describe the content of various pictures he was shown.

The patient analysed his difficulty in identifying faces with considerable insight. "I can see the eyes, nose, and mouth quite clearly but they just don't add up. They all seem chalked in, like on a blackboard. I have to tell by the clothes or voice whether it is a man or woman, as the faces are all neutral, a dirty grey colour. The hair may help a lot, or if there is a moustache." "All the men appear unshaved." "I cannot recognize people in photographs, not even myself. At the club I saw someone strange staring at me and asked the steward who it was. You'll laugh at me. I'd been looking at myself in a mirror." "I later went to London and visited several cinemas and theatres. I couldn't make head or tail of the plots. I never knew who was who." "I bought some copies of *Men Only* and *London Opinion*. I couldn't enjoy the usual pictures. I could work out what was what by accessory details, but it's no fun that way. You've got to take it in at a glance."

He had some difficulty with animal faces also. "They are easier to recognize than human faces, because the rest of an animal varies so much more than does the rest of a man. In real life I could tell a cat from a dog, or a horse from a cow, but it would be more difficult from pictures, especially if only the face is shown."

He had no difficulty in revisualization. "I can shut my eyes and can well remember what my wife looked like or the kids." (From C. A. Pallis, *Journal of Neurology, Neurosurgery and Psychiatry*, Vol. 18, pp. 218–224, 1955.)

Attention Focuses Visual Perception by Facilitating Coordination Between Separate Visual Pathways

The evidence for parallel processing of visual information raises a deep question: How is information about color, motion, depth, and form, all of which are carried at least to some degree by separate neuronal pathways, organized into cohesive perceptions? When we see a square purple box, we combine into one perception the sensations of color (purple), form (square), and solidity (box). We can equally well combine purple with a round box, a hat, or a coat. Clearly, the possible combinations are so great that the existence of distinct feature-detecting cells, each responsive to only one set of combinations, is improbable.

Instead, as we have seen in this chapter, visual images typically are built up from the inputs of parallel pathways that process different features—movement, solidity, form, and color. To express the specific combination of properties in the visual field at any given moment, therefore, independent groups of cells with different functions must be temporarily brought into *association*. That is, there must be a mechanism whereby the brain momentarily associates the information being processed independently in different cortical regions. This mechanism, as yet unspecified, is called *the binding mechanism*.

Ann Treisman and her colleagues and Bella Julesz have independently shown in psychophysical studies that formation of these associations requires *attention*. They began by trying to understand one of the problems addressed by the early Gestalt psychologists: How is attention focused on *one* object in the visual field? What features of the object make that object stand out from the background? Treisman and Julesz found that distinctive boundaries are created from elementary properties: brightness, color, and orientation of line. Consider, for example, Figure 21–15A. Here a rectangle composed of small +'s within a field of large L's creates distinctive boundaries between the two. The rectangle easily stands out against its background.

In contrast, a second square composed of T's can be found only after carefully scanning the figure because the T's are only subtly different from the background of L's. Treisman and Julesz have found that when the boundaries are made up of elements that are clearly different, the boundaries pop out almost automatically within 50 ms.

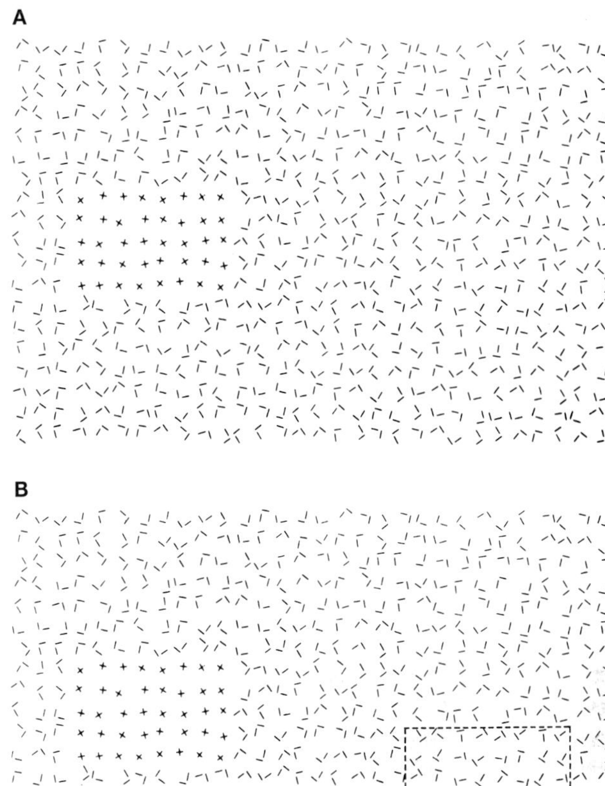

Figure 21–15 Some perceptions are produced by pre-attentive scanning; others require focal attention. In **A** the small subarea composed of +'s is easy to pick out from the surrounding area by simply looking at the figure. The figure also includes a subarea composed of T's. Can you find it? To do so, you must focus on each region of the figure. The area of T's is outlined in **B**. (From Bergen and Julesz, 1983.)

On the basis of these observations, Treisman and Julesz suggest that there are two distinct processes in visual perception. A *pre-attentive process* acts as a rapid scanning system and is concerned only with the detection of the major outline of an object.

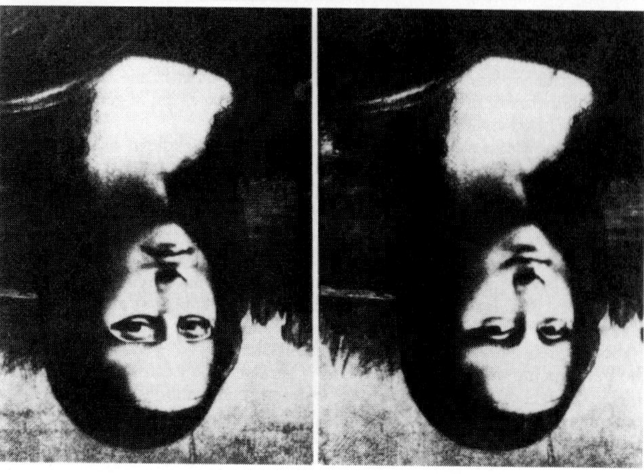

Figure 21–16 These pictures appear similar at first glance. When the pictures are seen upright, the true detail in the two faces is revealed. (From Julesz, 1986, after an idea of Thompson, 1980.)

This process rapidly scans the object's overall texture or features and encodes the useful elementary properties of the scene. At this point, variation in a simple property may be discerned as a border or contour, but complex differences in combinations of properties are not detected. A later *attentive process* directs attention to specific more subtle features of an object, selecting and highlighting combinations of features that are segregated in the separate feature maps. It is this attentive process that intrigued the Gestaltists. As we have seen, it represents a winner-take-all strategy, whereby certain features of one object are emphasized and attended to while other features and other objects are ignored.

The pre-attentive and attentive processes can be illustrated by examining the two pictures in Figure 21–16. At first glance (pre-attentive processing) each face seems to be the same. However, when we turn the pictures right side up and examine the individual features (attentive processing), we recognize, to our surprise, that one face is distorted. Treisman has proposed that different properties are encoded in different *feature maps* in different brain regions. To solve the binding problem, Treisman has postulated that there may be a *master map* or *saliency map* that codes only for key aspects of the image (Figure 21–17). The master map receives input from all feature maps but retains only those features that distinguish the object of attention from its surround. Once these salient features have been represented in the master map, the detailed information associated with each feature can be retrieved by referring back to the individual feature maps. In this way the master map can combine details from the feature maps that are essential for recognition. Recognition thus occurs when the salient features in different feature maps are associated or bound together.

How does this binding occur? How is attention achieved in the visual system? Treisman speaks metaphorically of the *spotlight of attention*. What might be the switch for this spotlight? What turns it on?

As may be appreciated from the evidence presented in this chapter, the neuronal mechanisms of focused attention and conscious awareness are now emerging as one of the great unresolved problems in perception and indeed in all of neurobiology. Francis Crick and Christoff Koch suggest that visual attention may be mediated by one or more subcortical structures such as the pulvinar, claustrum, and superior colliculus. In addition, as we saw in Chapter 18, some cortical areas, specifically the posterior parietal cortex, are particularly important for some types of focused attention. These structures may represent Treisman's saliency map for different types of focused attention and bursts of action potentials in these structures may modulate the activity of the appropriate cells in the different feature maps.

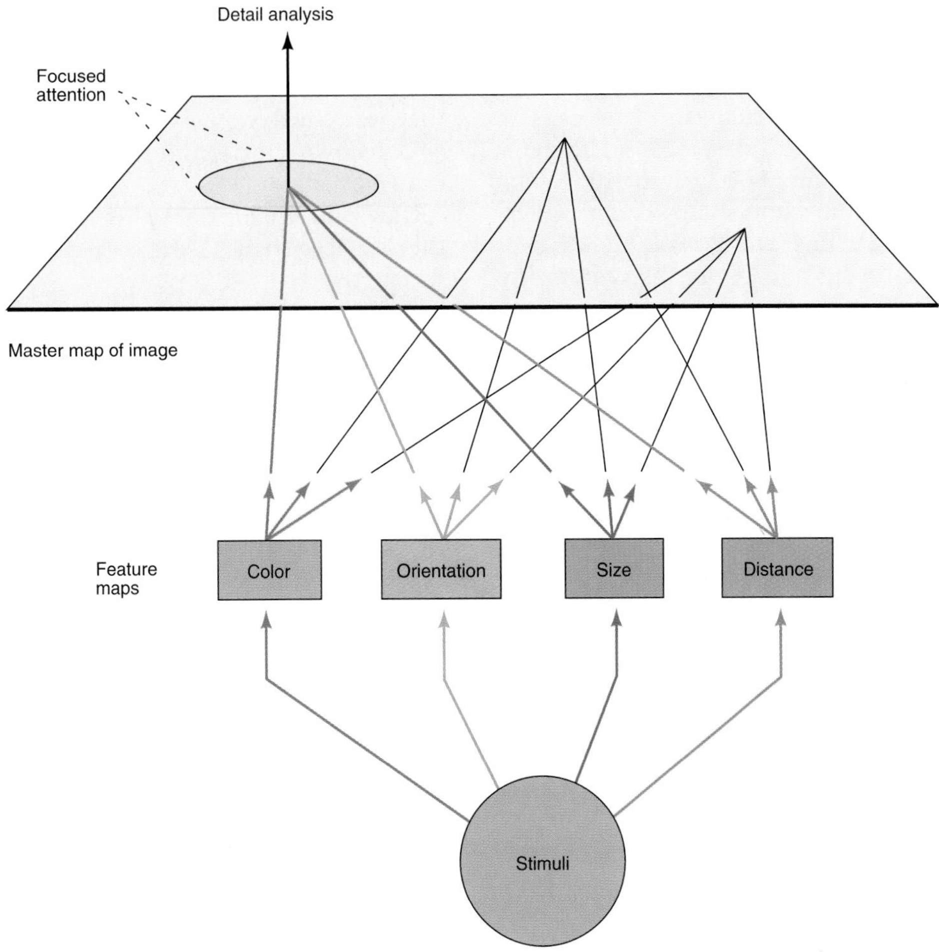

Figure 21–17 A hypothetical model of the stages in visual perception based on experiments by Ann Treisman. Some basic and useful properties of a scene (color, orientation, size, distance) are encoded in separate parallel pathways, each of which generates a *feature map*. Selected features from these maps are then integrated into a *master map*, which is a representation of those features of the image that distinguish it from the background. Focused attention, perhaps by recruiting the posterior parietal lobe, then draws on the information from the feature and master maps to analyze more carefully the features associated in a small region of the image. (From Treisman, 1988.)

The Analysis of Visual Attention May Provide Important Clues About Conscious Awareness

The problem posed by selective attention was first defined in 1890 by William James in his *Principles of Psychology*:

> Millions of items . . . are presented to my senses which never properly enter my experience. Why? Because they have no *interest* for me. *My experience is what I agree to attend to*. . . . Everyone knows what attention is. It is the taking possession by the mind, in clear and vivid form, of one out of what seem several simultaneously possible objects or trains of thought. Focalization, concentration of consciousness are of its essence. It implies withdrawal from some things in order to deal effectively with others.

Much of the sensory information received by the peripheral receptors in our body must eventually be filtered out and eliminated within the brain, much as we disregard the ground of an image when we focus on the figure. Although the visual system contains extensive parallel pathways for simultaneously processing different streams of information, the amount of this information that reaches the highest centers of processing in the brain is limited by the mechanism of selective or focused attention. As we saw in considering the figure–ground dichotomy, selective attention both filters out some features and sharpens our perception of others. In this winner-take-all strategy, some stimuli stand out in consciousness while others recede into dim awareness.

It is attractive to think that exploration of visual attention will lead us to define the neural mechanisms of a specific instance of consciousness. Despite its central importance for a neurobiological understanding of mental processes, the problem of consciousness has so far eluded reductionist approaches. But as this and later chapters illustrate, biological insights into any component of consciousness are likely to give us at least a glimmer of understanding of some of the most complex components: volition, intention, and self-awareness. If consciousness in its various forms is the product of a generalized set of neural mechanisms, then the study of visual attention could put us on the path to a new level of self-understanding. We shall therefore consider cellular studies of visual attention in Chapter 23.

An Overall View

David Marr began his important book on the computational tasks of vision with the question: "What does it mean, to see?" His answer is that vision is the process of discovering from images *what* is present in the visual world, and *where* it is.

Mortimer Mishkin and his colleagues first pointed out that these two tasks, identifying *what* and *where*, are carried by distinct anatomical pathways. In addition, identifying *what* involves two subsystems: one system conveys information about form and a second system reports color. Both terminate in the inferior temporal cortex, in a complex of areas identified by Mishkin as being important for the recognition of form. Location of the object in space is the task, in large part, of a third system that terminates in the posterior parietal cortex, in a complex of areas identified by Mishkin as important for spatial organization.

Only recently has it become clear that visual processing involves parallel pathways rather than one serial pathway. But this important discovery has posed a new problem for the study of visual perception. Integration of information in a serial pathway is achieved *progressively*, from one area to the next. In a system of parallel pathways, each with its own function, integration can be achieved only *interactively*.

How and where does this interaction occur in the visual system? Semir Zeki has put this issue succinctly:

> At first glance, the problem of integration may seem quite simple. Logically it demands nothing more than that all the signals from the specialized visual areas be brought together, to 'report' the results of their operations to a single master cortical area. This master area would then synthesize the information coming from all these diverse sources and provide us with the final image, or so one might think. But the brain has its own logic . . . . If all the visual areas report to a single master cortical area, who or what does that single area report to? Put more visually, who is "looking" at the visual image provided by that master area? The problem is not unique to the visual image or the visual cortex. Who, for example, listens to the music provided by a master auditory area, or senses the odour provided by the master olfactory cortex? It is in fact pointless pursuing this grand design. For here one comes across an important anatomical fact, which may be less grand but perhaps more illuminating in the end: *there is no single cortical area to which all other cortical areas report exclusively, either in the visual or in any other system. In sum, the cortex must be using a different strategy for generating the integrated visual image.*

There are in fact extensive interactions between the major visual pathways at almost all cortical levels, as well as reciprocal connections from higher to lower levels both within and between pathways. Crosstalk is not deferred until late stages of sensory processing. Perceptual integration is therefore likely to be a multistage process, not a single grand synthesis. Finally, other brain centers that make connections in the visual system and which are known to affect visual attention, such as the prefrontal cortex, the claustrum, and the pulvinar, may allow attention mechanisms to bind the visual process.

In this chapter we have focused on the role of vision in gathering information about our environment. Obviously, vision is also important in guiding body movement. It is likely that much visual processing, particularly in the magnocellular pathway concerned with motion and spatial relationships, is essential for the control of our own movement. Simply moving about in the world requires complex analyses of visual stimuli. We return to the visual guidance of movement later in this book when considering the motor system and guidance of action.

Selected Readings

Crick, F. 1993. The Astonishing Hypothesis: The Scientific Search for the Soul. New York: Charles Scribner & Sons.

Crick, F., and Koch, C. 1990. Towards a neurobiological theory of consciousness. Semin. in Neurosci. 2:263–275.

Dowling, J. E. 1992. Neurons and Networks: An Introduction to Neuroscience. Cambridge, Mass.: Belknap Press.

Farah, M. J. 1990. Visual Agnosia—Disorders of Object Recognition and What They Tell Us About Normal Vision. Cambridge, Mass: MIT Press.

Gregory, R. L. 1966. Eye and Brain. The Psychology of Seeing, 2nd ed. New York: McGraw-Hill.

Hendry, S. H. C., and Yoshioka, T. 1994. A neurochemically distinct third channel in the Macaque dorsal lateral geniculate nucleus. Science. 264:575–577.

Hochberg, J. E. 1968. Perception, 2nd ed. Englewood Cliffs, N.J.: Prentice-Hall.

Horgan, J. 1994. Can science explain conscience? Sci. Amer. 271(1):88–94.

Hubel, D. H. 1988. Eye, Brain, and Vision. New York: Scientific American Library.

Kanizsa, G. 1979. Organization of Vision: Essays on Gestalt Perception. Westport, Conn.: Greenwood.

Livingstone, M. S. 1988. Art, illusion and the visual system. Sci. Am. 258(1):78–85.

Marr, D. 1982. Vision: A Computational Investigation into the Human Representation and Processing of Visual Information. San Francisco: Freeman.

Riddoch, G. 1917. Dissociation of visual perceptions due to occipital injuries with especial reference to appreciation of movement. Brain 40:15–57.

Rock, I., and Palmer, S. 1990. The legacy of Gestalt psychology. Sci. Am. 263(6):84–90.

Teuber, M. L. 1974. Sources of ambiguity in the prints of Maurits C. Escher. Sci. Am. 231(1):90–104.

Treisman, A. 1986. Features and objects in visual processing. Sci. Am. 255(5):114B–125.

Van Essen, D. C. and Gallant, J. L. 1994. Neural mechanisms of form and motion processing in the primate visual system. Neuron, in press.

Zeki, S. 1993. A Vision of the Brain. London: Blackwell.

Zihl, J., Von Cramon, D., and Mai, N. 1983. Selective disturbance of movement vision after brain damage. Brain 106:313–340.

22

Visual Processing by the Retina

The Retina Contains the Eye's Receptive Sheet

There Are Two Types of Photoreceptors: Rods and Cones

Light Is Absorbed by Visual Pigments in the Photoreceptors

Phototransduction Results from a Cascade of Biochemical Events in the Photoreceptors

Light Activates Pigment Molecules in the Photoreceptors

Activated Pigment Molecules Affect the Cytoplasmic Concentration of Cyclic GMP

Cyclic GMP Gates Specialized Ion Channels that Close to Hyperpolarize the Photoreceptor

Photoreceptors Slowly Adapt to Changes in Light Intensity

The Output of the Retina Is Conveyed by the Ganglion Cells

The Receptive Field of the Ganglion Cell Has a Center and an Antagonistic Surround

Ganglion Cells Are Specialized for the Detection of Contrasts and Rapid Changes in the Visual Image

Specialized Ganglion Cells Process Different Aspects of the Visual Image

Signals from Photoreceptors to Ganglion Cells Are Relayed Through a Network of Interneurons

Cone Signals Are Conveyed to Ganglion Cells Through Direct or Lateral Pathways

The Receptive Fields of Bipolar Cells Have a Center–Surround Organization

Different Classes of Bipolar Cells Have Excitatory Connections with Corresponding Classes of Ganglion Cells

Visual perception begins in the retina and occurs in two stages. Light entering the cornea is projected onto the back of the eye, where it is converted into an electrical signal by a specialized sensory organ, the retina. These signals are then sent through the optic nerve to higher centers in the brain for further processing necessary for perception. In this chapter we analyze the neural processing of visual signals in the retina. The next two chapters are devoted to considering in detail how processing in higher centers gives rise to the perception of form, motion, and color.

The retina bears careful examination for several reasons. First, it is useful for understanding sensory transduction in general because photoreceptors, specialized retinal neurons that convert light into electrical signals, are perhaps the best understood of all sensory cells. Second, unlike other sensory structures, such as the cochlea and somatic receptors in the skin, the retina is not a peripheral organ but part of the central nervous system. The retina develops directly from the neural ectoderm, the specialized part of the ectoderm that gives rise to the brain (Chapter 6), and the synaptic organization of the retina is similar to that of other central neural structures.

At the same time, the retina is relatively simple compared with other brain regions. It contains only five major classes of neurons, linked in an intricate pattern of connections but with an orderly, layered anatomical arrangement. This combination of physiological diversity and relatively simple structural organization makes the retina useful for understanding how information is processed by complex neural circuits in the brain.

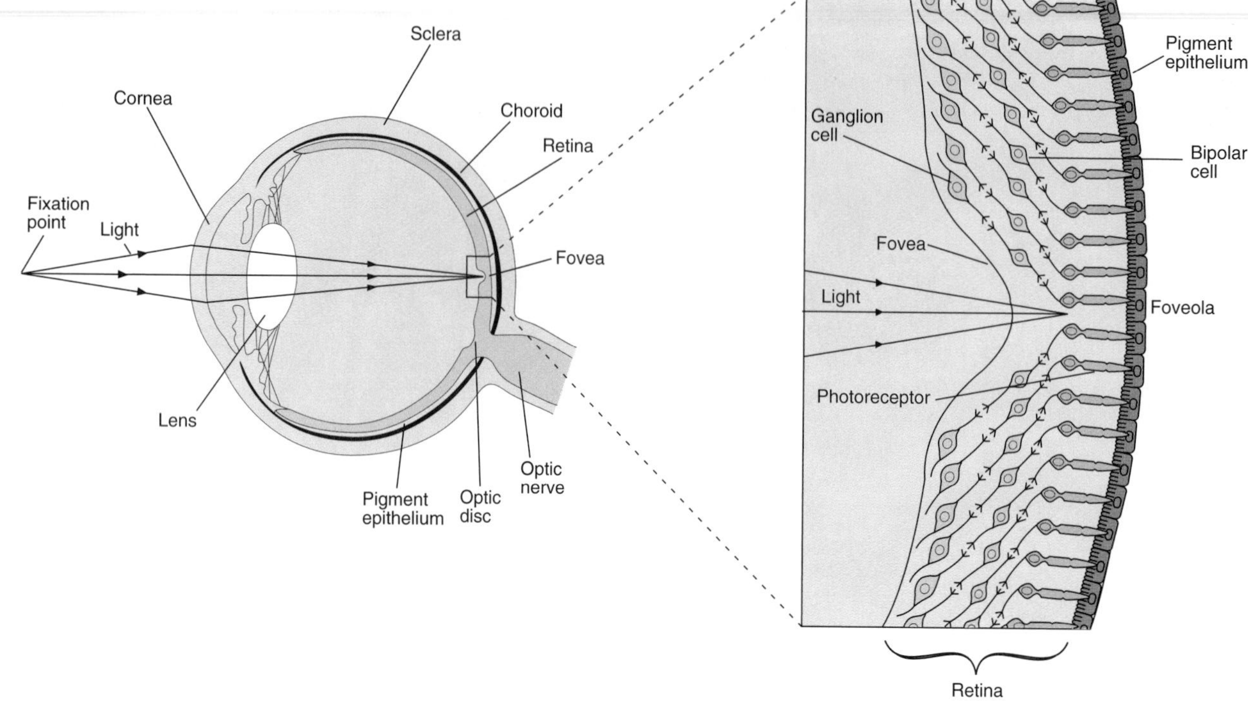

Figure 22–1 Photoreceptors are located in the retina. The location of the retina within the eye is shown at the **left**. Detail of the retina at the fovea is shown on the **right** (the diagram has been simplified by eliminating lateral connections mediated by horizontal and amacrine cells; see Figure 22–6). In most of the retina light must pass through layers of nerve cells and their processes before it reaches the photoreceptors. In the center of the fovea, or foveola, these proximal neurons are shifted to the side so that light has a direct pathway to the photoreceptors. As a result, the visual image received at the foveola is the least distorted.

For these reasons we describe neural processing in the retina in considerable detail. This chapter is divided into two parts. In the first part we describe how photoreceptors transduce light into an electrical signal. In the second we consider how these signals are shaped by other retinal neurons before being sent to the brain, and how synaptic connections among the retinal neurons are organized to accomplish this processing. Before discussing phototransduction, however, we review the organization of the retina and the basic physiological properties of the photoreceptor cells.

The Retina Contains the Eye's Receptive Sheet

The eye is designed to focus the visual image on the retina with minimal optical distortion. Light is focused by the cornea and the lens, then traverses the vitreous humor that fills the eye cavity before being absorbed by the photoreceptor cells (Figure 22–1). The retina lies in front of the *pigment epithelium* that lines the back of the eye. Cells in the pigment epithelium are filled with the black pigment *melanin*, which absorbs any light not captured by the retina. This prevents light from being reflected off the back of the eye to the retina again (which would degrade the visual image).

As shown in Figure 22–1, the photoreceptors directly contact the pigment epithelium, whereas the other retinal cells are closer to the lens. One remarkable consequence of this arrangement is that light must travel through layers of other retinal neurons before striking the photoreceptors. To allow light to reach the photoreceptors without being absorbed or greatly scattered (which would distort the visual image), the proximal neural layers of the retina are unmyelinated and therefore relatively transparent. Moreover, in one region of the retina, the *fovea*, the cell bodies of the proximal retinal neurons are shifted to the side, enabling the photoreceptors there to receive the visual image in its least distorted form (Figure 22–1). This shifting is most pronounced at the center of the fovea, the *foveola*. Humans therefore constantly move their eyes so that scenes of interest are projected onto the fovea. The retina also contains a region called the optic disc, where the optic nerve fibers leave the retina. This region has no photoreceptors and therefore creates a *blind spot* in the visual field (see Figure 23–2). The projection of the visual field onto the two retinas is described in Chapter 23.

There Are Two Types of Photoreceptors: Rods and Cones

The human retina contains two types of photoreceptors, rods and cones. Cones are responsible for day vision; people who lose functioning in the cones are legally blind. Rods mediate night vision: they function in the dim light that is present at dusk or at night, when most stimuli are too weak to excite the cones. Rods are exquisitely sensitive to light. In fact, in psychophysical experiments with humans performed over 50 years ago, Selig Hecht showed that rods can be excited by a single photon. A total loss of rods produces only night blindness.

Cones perform better than rods in all visual tasks except the detection of dim stimuli. Cone-mediated vision is of higher acuity than rod-mediated vision and provides better resolution of rapid changes in the visual image (i.e., better *temporal resolution*). Cones also mediate color vision. The rod system is more sensitive than the cone system, but it is achromatic. These differences in performance are due partly to properties of the rods and cones themselves and partly to the connections they make with other neurons in the retina (the rod and cone systems). The most important factors that contribute to these differences are summarized in Table 22–1 and discussed next.

Rods detect dim lights. Rods contain more photosensitive visual pigment than cones, enabling them to capture more light. Even more important, rods amplify light signals more than cones do. A single photon can evoke a detectable electrical response in a rod; in contrast, hundreds of photons must be absorbed by a cone to evoke a similar response. In addition, the rod system is *convergent*: many rods synapse on the same target interneuron, which is known as the bipolar cell (see below). The signals from these rods are pooled in the bipolar

Table 22–1 Differences Between Rods and Cones and Between Their Neural Systems

Rods	*Cones*
High sensitivity, specialized for night vision More photopigment, capture more light High amplification, single photon detection	Lower sensitivity, specialized for day vision Less photopigment Lower amplification
Low temporal resolution: slow response, long integration time	High temporal resolution: fast response, short integration time
More sensitive to scattered light	Most sensitive to direct axial rays
Rod system	*Cone system*
Low acuity: highly convergent retinal pathways, not present in central fovea	High acuity: less convergent retinal pathways, concentrated in fovea
Achromatic: one type of rod pigment	Chromatic: three types of cones, each with a distinct pigment that is most sensitive to a different part of the visible spectrum

cell and reinforce one another, strengthening the response evoked by light in the cell and increasing the ability of the brain to detect dim lights. In contrast, fewer cones converge on each bipolar cell.

Cones mediate color vision. There are three types of cones, each containing a visual pigment that is sensitive to a different part of the spectrum (see below). As we shall see in Chapter 24, the brain obtains information about color by comparing the responses of the three types of cones. In contrast, rods contain only one type of pigment, so that all rods respond in the same way to different wavelengths. Rod vision is therefore achromatic.

Cones provide greater spatial resolution. Although rods outnumber cones by roughly 20 to 1, the cone system has better spatial resolution for two reasons. First, cones are concentrated in the fovea, where the visual image is least distorted. Second, because many rods converge onto a single bipolar cell, the rod system is less able to transmit spatial variations in the visual image because differences in the responses of neighboring rods are averaged out in the interneuron. Since few cones converge on bipolar cells, cones provide better spatial resolution. In fact, in the foveola there is no convergence at all: each bipolar cell receives input from a single cone.

Cones provide better temporal resolution. Unlike most neurons, rods and cones do not fire action potentials. Instead, they respond to light with graded changes in membrane potential. Rods respond slowly, so that the effects of all the photons absorbed during a 100 ms interval are summed together. This helps rods detect small amounts of light, but prevents them from resolving light that is flickering faster than about 12 Hz. The response of cones is much faster, so that they can detect flicker up to 55 Hz. Thus, cones also provide greater temporal resolution of the visual image.

Light Is Absorbed by Visual Pigments in the Photoreceptors

Rods and cones have similar functional regions: (1) the *outer segment*, a region specialized for phototransduction, located at the outer or distal surface of the retina; (2) the *inner segment*, a region containing the cell's nucleus and most of its biosynthetic machinery, located more proximally within the retina; and (3) a synaptic terminal that makes contact with the photoreceptor's target cells. The outer segment is connected to the inner segment by a thin stalk or cilium (Figure 22–2).

The outer segments of rods and cones are filled with light-absorbing *visual pigments*. Each pigment molecule comprises a small light-absorbing molecule attached to a large membrane-spanning protein. Rods and cones can contain a remarkably large number of these membrane proteins (as many as 10^8 in each cell), because they have evolved an elaborate system of stacked membranous *discs* in their outer segments that dramatically increase the surface area of the membrane in these cells (Figure 22–2B). These discs develop as a series of invaginations of the cell's plasma membrane, ultimately arranging themselves like a roll of pennies in a bank wrapper. In cones the discs are continuous with the plasma membrane, while in rods they pinch off from the plasma membrane to become intracellular organelles.

Like most other mature neurons, photoreceptors do not divide, but their outer segments are constantly renewed. New discs are formed at a rapid rate; in rods about three discs are synthesized every hour. Old discs are discarded at the tips of photoreceptors and removed by the phagocytotic activity of the pigment epithelial cells.

Phototransduction Results from a Cascade of Biochemical Events in the Photoreceptors

The absorption of light by visual pigments in rods and cones triggers a cascade of events that leads to a change in ionic fluxes across the plasma membrane of these cells and a consequent change in membrane potential. A key molecule in the cascade is the cyclic nucleotide *cyclic guanosine 3′,5′-monophosphate* (cGMP), which in rods acts as a second messenger carrying information through the cytoplasm connecting the freely floating discs, where light is absorbed, to the cell's plasma membrane, where ionic fluxes are altered. (In cones, since the discs are continuous with the plasma membrane, a cytoplasmic messenger is not necessary; nonetheless, cGMP is used in these cells in the same way as in rods.) Cyclic GMP controls ionic fluxes by opening a specialized species of ion channels, the *cGMP-gated* ion channels, which allow an inward current carried largely by Na^+ ions to flow into the cell.

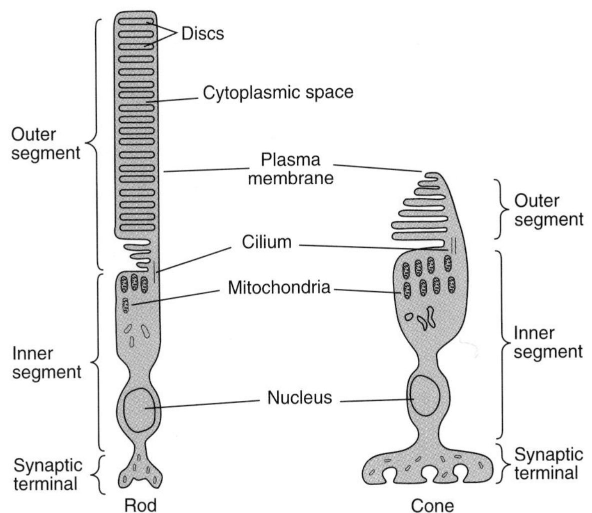

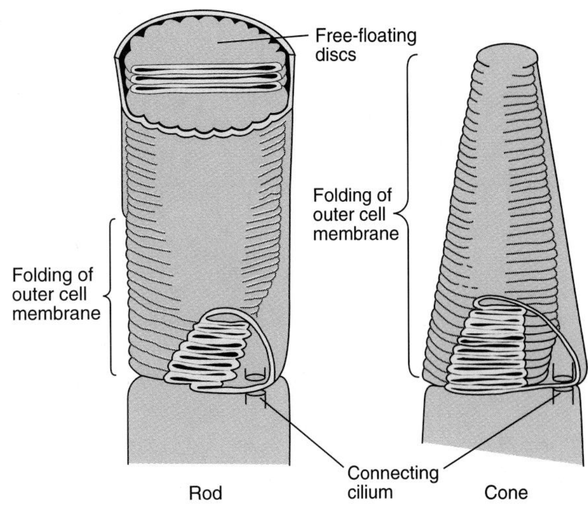

Figure 22–2 The two types of photoreceptors, rods and cones, have similar structures. (Adapted from O'Brien, 1982, and Young, 1970.)

A. Both rod and cone cells have inner and outer segments connected by a cilium. The inner segment contains the cell's nucleus and most of its biosynthetic machinery. The outer segment contains the light-transducing apparatus. The conical shape of the outer segment in cones makes these cells most sensitive to direct axial rays.

B. The outer segment consists of a stack of membranous discs, which contain the light-absorbing photopigments. In both types of cells these discs are formed by infolding of the plasma membrane. In rods, however, the folds pinch off from the membrane so that the discs are free-floating within the outer segment.

In the dark the concentration of cGMP is relatively high, opening the cGMP-gated channels and allowing the inward current they carry to maintain the cell in a relatively depolarized state. Phototransduction then occurs in three stages: (1) light activates visual pigments; (2) these activated molecules stimulate cGMP phosphodiesterase, an enzyme that *reduces* the concentration of cGMP in the cytoplasm; (3) the reduction in cGMP concentration closes the cGMP-gated channels, thus hyperpolarizing the photoreceptor (Figure 22–3). We now examine these events step by step.

Light Activates Pigment Molecules in the Photoreceptors

Rhodopsin, the visual pigment in rod cells, has two parts. The protein portion, *opsin,* is embedded in the disc membrane and does not by itself absorb light. The light-absorbing portion, *retinal,* is a derivative of vitamin A. Retinal can assume several different isomeric conformations, two of which are important in different phases of the visual cycle. In its nonactivated form rhodopsin contains the 11-*cis* isomer of retinal, which fits snugly into a binding site in the opsin molecule (Figure 22–4).

Activation of rhodopsin starts with the absorption of light, which causes retinal to change from the 11-*cis* to the all-*trans* configuration (Figure 22–4B). *This reaction is the only light-dependent step in vision.* As a result of this conformational change, retinal no longer fits into the binding site in opsin. The opsin, therefore, undergoes a conformational change to a semistable conformation called *metarhodopsin II,* which triggers the second step of phototransduction.

Metarhodopsin II is unstable and splits within minutes, yielding opsin and all-*trans*-retinal. The all-*trans*-retinal is then transported from rods to pigment epithelial cells, where it is reduced to all-

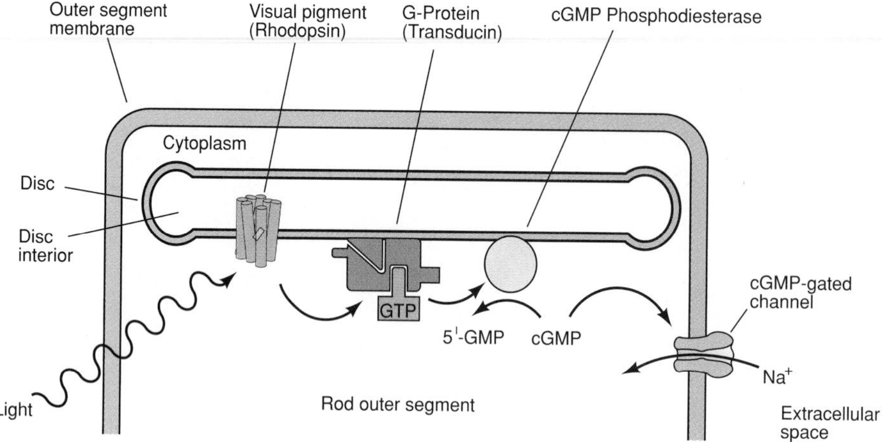

Figure 22–3 Phototransduction involves the closing of cation channels in the outer segment of the photoreceptor membrane, illustrated here in a rod cell. In the absence of light these cation channels are kept open by intracellular cGMP and conduct an inward current, carried largely by Na^+. When light strikes the photoreceptor these channels are closed by a three-step process. (1) Light is absorbed by and activates pigment molecules (rhodopsin in rods) in the disc membrane (the **blue** rectangle in the rhodopsin molecule represents the light-absorbing compound, retinal). (2) The activated pigment stimulates a G-protein (transducin in rods), which in turn activates cGMP phosphodiesterase. This enzyme catalyzes the breakdown of cGMP to 5′-GMP. (3) As the cGMP concentration is lowered, the cGMP-gated channels close, thereby reducing the inward current and causing the photoreceptor to hyperpolarize.

trans-retinol (vitamin A), the precursor in the synthesis of 11-*cis*-retinal, which is transported back to rods. All-*trans*-retinol is thus a crucial compound in the visual system, and because it cannot be synthesized by humans, it must form a regular part of the diet. Deficiencies of vitamin A can lead to night blindness and, if untreated, to a deterioration of receptor outer segments and eventually total blindness.

In cones, as in rods, the visual pigments are composed of two parts: a protein called *cone opsin* and a light-absorbing portion that, as in rods, appears to be 11-*cis*-retinal. Each of the three types of cone cells in the retina of primates contains a different pigment optimized for absorption of light in a different part of the visible spectrum. These three pigments contain different isoforms of cone opsin, each of which interacts with 11-*cis*-retinal in a distinct way, causing it to be most sensitive to a particular part of the visible spectrum. The existence of three types of cones with different absorption characteristics is what underlies trivariant color vision in humans (see Chapter 24).

Activated Pigment Molecules Affect the Cytoplasmic Concentration of Cyclic GMP

The activation of pigment molecules by light leads to a reduction in the cytoplasmic concentration of the second messenger cGMP. The concentration of cGMP is controlled by two enzymes. It is synthesized from GTP by guanylyl cyclase and it is broken down to 5′-GMP by cGMP phosphodiesterase, a protein peripherally associated with the disc membrane (Figure 22–3). The concentration of cGMP is affected by light because cGMP phosphodiesterase is itself controlled by the visual pigments. In darkness cGMP phosphodiesterase is only weakly active, and the concentration of cGMP is therefore relatively high. Activation of pigment molecules by light leads to the activation of the phosphodiesterase, which breaks down cGMP and lowers its concentration.

Photoactivation of a single rhodopsin molecule can lead to the hydrolysis of more than 10^5 molecules of cGMP in one second. This amplification is achieved through the regulatory protein *transducin*.

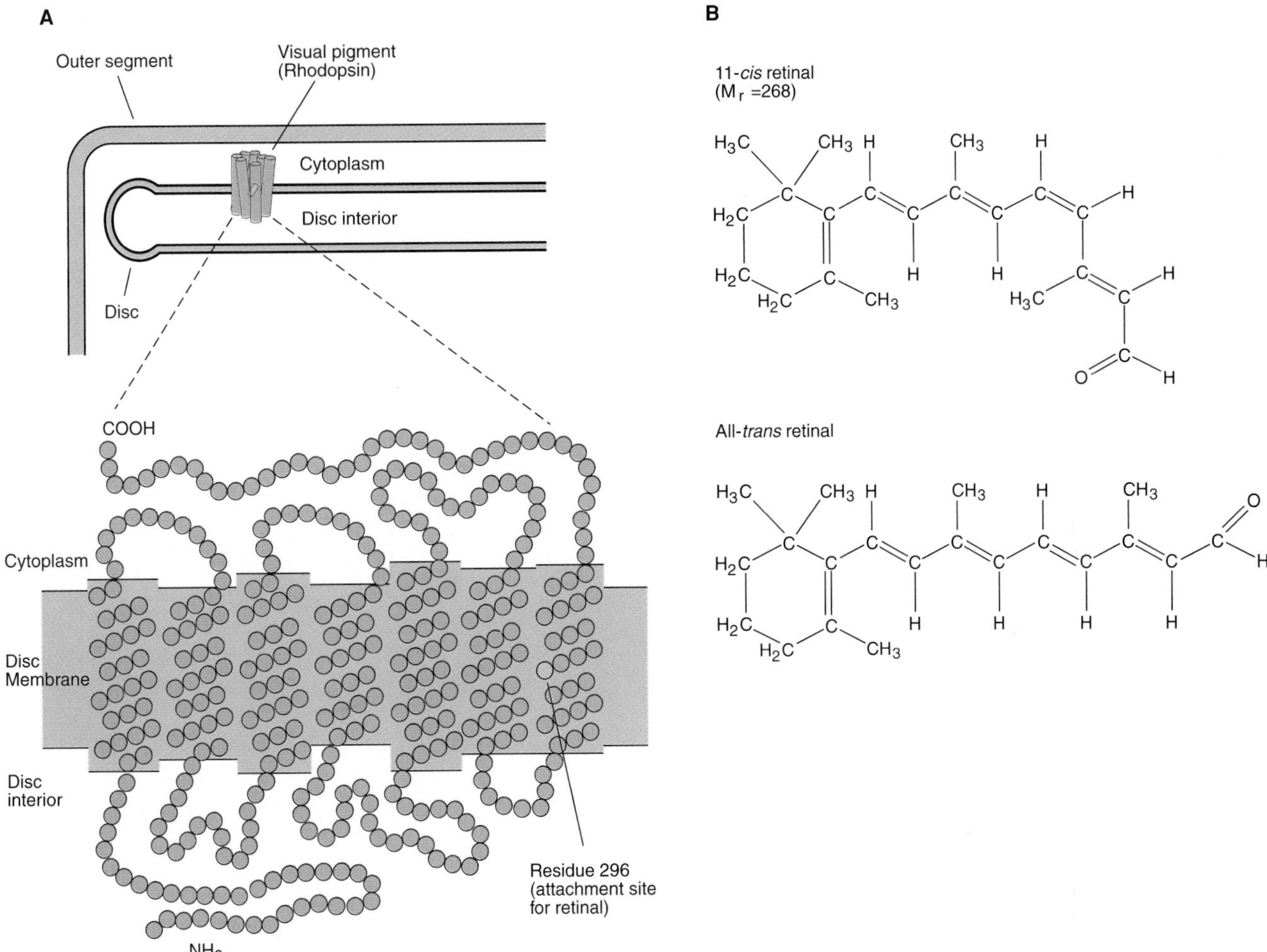

Figure 22–4 Rhodopsin, the visual pigment in rod cells, is located in the disc membrane. Rhodopsin is the covalent complex of a large protein, *opsin*, and a small light-absorbing compound, *retinal* (the small blue rectangle in the figure). The absorption of light by retinal causes a change in the three-dimensional structure of rhodopsin.

A. Opsin has 348 amino acids and a molecular weight of about 40,000. It loops back and forth seven times across the membrane of the rod disc. Retinal is covalently attached to the side chain of residue 296 in the protein's seventh membrane-spanning region. (Adapted from Nathans and Hogness, 1984.)

B. In its nonactivated form rhodopsin contains the 11-*cis* isomer of retinal. The first event in visual transduction is the absorption of light by 11-*cis* retinal. This causes a rotation around the 11-*cis* double bond. As retinal returns to its more stable all-*trans* configuration, it brings about a conformational change in the opsin portion of rhodopsin, which triggers the other events of visual transduction.

One rhodopsin molecule can diffuse within the disc membrane and activate hundreds of transducin molecules, each of which stimulates a phosphodiesterase molecule. Each phosphodiesterase molecule in turn is capable of hydrolyzing more than 10^3 molecules of cGMP per second.

The biochemical cascade initiated by the photoactivation of rhodopsin resembles the cascades triggered by the binding of many hormones and neurotransmitters to their receptors. Indeed, the rod and cone opsins show a high degree of structural similarity with the family of hormone and transmitter receptors that couple to G-proteins (see Chapter 14). Moreover, transducin is a member of the G-protein family. Like other G-proteins, the activation of transducin involves a characteristic interaction with guanine nucleotides (see Figure 14–3). Inactive transducin binds a molecule of GDP tightly; on interaction with activated rhodopsin in the disc membrane, however, transducin exchanges GDP for GTP and itself becomes active. Transducin becomes inactivated because it also has GTPase activity, which breaks down the bound GTP molecule into GDP (see Figure 14–4).

Two mechanisms terminate the light response. As described, transducin inactivates itself by breaking down bound GTP. Also, once activated, rhodopsin becomes a target for phosphorylation by a specific protein kinase, *opsin kinase*; the phosphorylated rhodopsin then interacts with a specific regulatory protein called *arrestin*, leading to its rapid inactivation.

Cyclic GMP Gates Specialized Ion Channels that Close to Hyperpolarize the Photoreceptor

The light-evoked decrease in cGMP results in the closure of cGMP-gated channels in the photoreceptor. Cyclic GMP gates these channels by binding directly to the cytoplasmic face of the channel; a channel becomes activated by the cooperative binding of at least three molecules of cGMP. The cGMP-gated channel of photoreceptors was the first known example of an ion channel regulated by a cyclic nucleotide acting directly on the channel rather than through a protein kinase. Similar channels are also present in some retinal bipolar cells (see below).

When open in darkness the cGMP-gated channels conduct an inward current, carried largely by Na^+ ions, that tends to depolarize the photoreceptor. The closing of these channels during illumination reduces this current and therefore makes the cell hyperpolarize (Box 22–1).

Photoreceptors Slowly Adapt to Changes in Light Intensity

During prolonged illumination several changes in the functional properties of photoreceptors occur that are important in mediating *light adaptation*. This phenomenon occurs, for example, when one steps from a dark environment into bright daylight. At first the light is blinding, but over a period of several seconds the eyes adapt. Adaptation involves many changes in the retina and eye (such as contraction of the pupil to reduce the amount of light reaching the retina), but the two most important changes occur in the cone photoreceptors.

The first change is the slow recovery of the cone membrane potential. A very bright light closes all cGMP-gated channels, hyperpolarizing the cones to –70 mV, the equilibrium potential for K^+. In this state the cones cannot respond to further increases in light intensity. If this illumination is maintained, however, the cones slowly depolarize to a membrane potential between –70 and –40 mV (the resting potential) and are once again capable of hyperpolarizing in response to further increases in light intensity—the bright light is no longer blinding.

The second change is the desensitization of the cone. During prolonged illumination by a background light, the smallest increment in light intensity capable of evoking a detectable change in the cone membrane potential increases in proportion to the background intensity, in accordance with Weber's law (Chapter 20).

These changes in the responses of cones to light are due to a slow decrease in Ca^{2+} concentration that occurs in the cones during prolonged background illumination, a decrease that affects the function of several proteins in the phototransduction pathway (Box 22–2).

The Output of the Retina Is Conveyed by the Ganglion Cells

We now turn to the second topic of this chapter: How does the retina modify and process the signals evoked by light in photoreceptors before sending them to higher centers? The output of the retina is

BOX 22–1 The Dark Current

In darkness the level of cGMP in a photoreceptor is high, and two currents predominate in the cell: an inward current that flows through the cGMP-gated channels, which are confined to the photoreceptor outer segment, and an outward K^+ current that flows through nongated K^+-selective channels, which are like those of other neurons and are confined to the inner segment. In darkness, therefore, current circulates through the photoreceptor, entering through the cGMP-gated channels in the outer segment and flowing out through the nongated K^+ channels. The photoreceptor is able to maintain steady intracellular concentrations of Na^+ and K^+ in the face of these large fluxes because its inner segment has a high density of sodium–potassium pumps, which pump Na^+ out and K^+ in (Figure 22–5A).

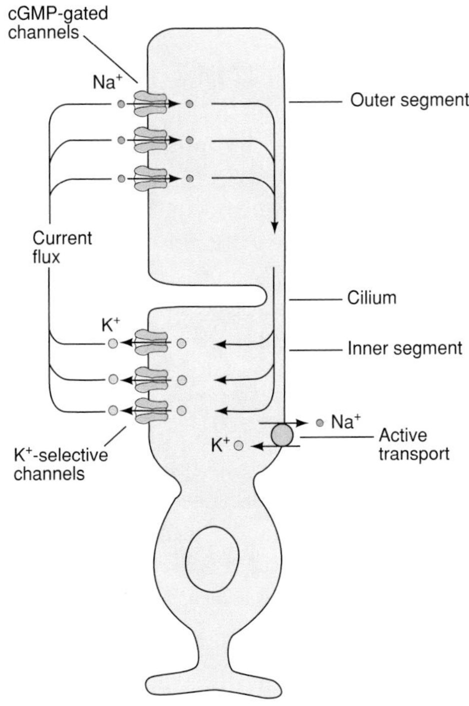

Figure 22–5A An inward current (the *dark current*), carried largely by Na^+ ions, flows into a photoreceptor through cGMP-gated channels in the outer segment of the membrane. The current flows out of the cell, carried largely by K^+ ions, through nongated channels in the inner segment membrane. Sodium–potassium pumps in the inner segment maintain the cell's Na^+ and K^+ concentrations at steady levels.

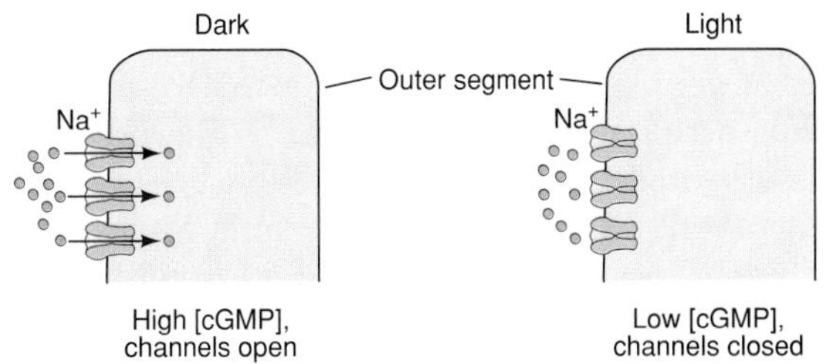

Figure 22–5B A reduction in the cytoplasmic concentration of cGMP closes the cGMP-gated channels.

The outward current carried by the K^+ channels tends to hyperpolarize the photoreceptor toward the equilibrium potential for K^+ (around –70 mV). The steady inward current, called the dark current, tends to depolarize the photoreceptor. As a result, in darkness the photoreceptor's membrane potential is around –40 mV, significantly more depolarized than that of most neurons. When light reduces the level of cGMP, closing cGMP-gated channels, the inward current that flows through these channels is reduced, and the cell becomes hyperpolarized (Figures 22–5B and C).

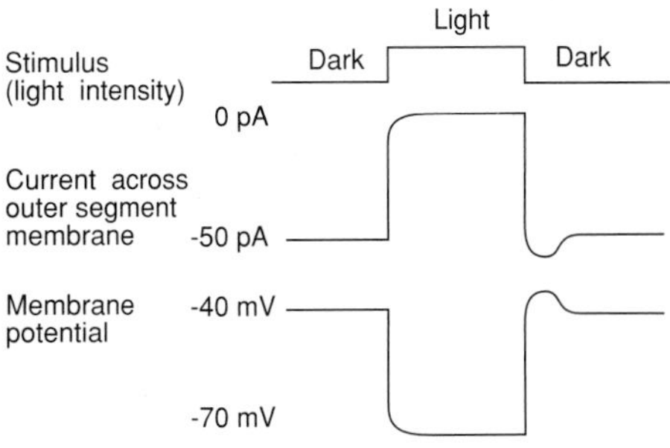

Figure 22–5C A dark current of 50 pA is suppressed by a bright light, hyperpolarizing the cell to –70 mV, the equilibrium potential for K^+. A light of intermediate intensity would hyperpolarize the cell to potentials between –40 and –70 mV.

conveyed by the *ganglion cells*. Unlike photoreceptors, which respond to light with graded changes in membrane potential, ganglion cells transmit information as trains of action potentials. Their axons form the optic nerve, which projects to the lateral geniculate nucleus of the thalamus and to the superior colliculus as well as to the pretectum (see Chapter 23).

Between the photoreceptors and the ganglion cells are three classes of interneurons: bipolar, horizontal, and amacrine cells (Figure 22–6). These cells do not simply transmit signals from the photoreceptors to the ganglion cells; they also combine signals from several photoreceptors in such a way that the electrical responses evoked in ganglion cells depend critically on the precise spatial and temporal patterns of the light that stimulates the retina. In this section we discuss the types of processing of the visual image that occur in the retina by examining how ganglion cells respond to different patterns of light. In the final section of this chapter we discuss how the synaptic connections between the photoreceptors, interneurons, and ganglion cells are organized to carry out the processing of the visual image.

The Receptive Field of the Ganglion Cell Has a Center and an Antagonistic Surround

Single ganglion cells are never silent, even in the dark, but their spontaneous activity is modulated by the input from retinal interneurons. The inputs

BOX 22–2 Calcium and Light Adaptation

Calcium modulates the function of several proteins of the phototransduction pathway. The recovery of the cone membrane potential and the desensitization of the cone that underlie light adaptation are mediated by a slow decrease in Ca^{2+} concentration in the cone outer segment during prolonged illumination.

In darkness Ca^{2+} constantly flows into the outer segment of the cone through the cGMP-gated channels. (Calcium accounts for about one-seventh of the current that flows through these channels.) The Ca^{2+} that enters is extruded by a specialized Ca^{2+} carrier in the outer segment membrane, and this process maintains a constant Ca^{2+} concentration in the outer segment. In contrast, during prolonged illumination the cGMP-gated channels close, thus reducing the influx of Ca^{2+}; this process causes a slow decrease in the intracellular Ca^{2+} concentration because the extrusion of Ca^{2+} continues.

This slow decrease in Ca^{2+} concentration under illumination allows the cone membrane potential to recover from its initial hyperpolarizing response to bright illumination because Ca^{2+} inhibits guanylyl cyclase, the enzyme that synthesizes cGMP from GTP. In darkness, when the Ca^{2+} level is relatively high, guanylyl cyclase is therefore maintained in a partially inhibited state. The slow decrease in Ca^{2+} concentration during illumination relieves the inhibitory effect of Ca^{2+} on guanylyl cyclase. As a result, more cGMP is synthesized, and the concentration of cGMP slowly increases. This results in the reopening of cGMP-gated channels and, consequently, slow depolarization of the cone.

The slow decrease in Ca^{2+} concentration also causes the desensitization of the cone during light adaptation, at least partly through effects on the visual pigments and the cGMP-gated channels. Lowering Ca^{2+} is believed to speed up the inactivation of the visual pigments, so that the effectiveness of a given light flash in activating cGMP phosphodiesterase is reduced. A lower concentration of Ca^{2+} also decreases the sensitivity of the cGMP-gated channels to changes in cGMP. Because of these effects of Ca^{2+}, a larger light stimulus is required to close the same number of cGMP-gated channels. Although these effects contribute to the desensitization, it is not known whether they account for it entirely.

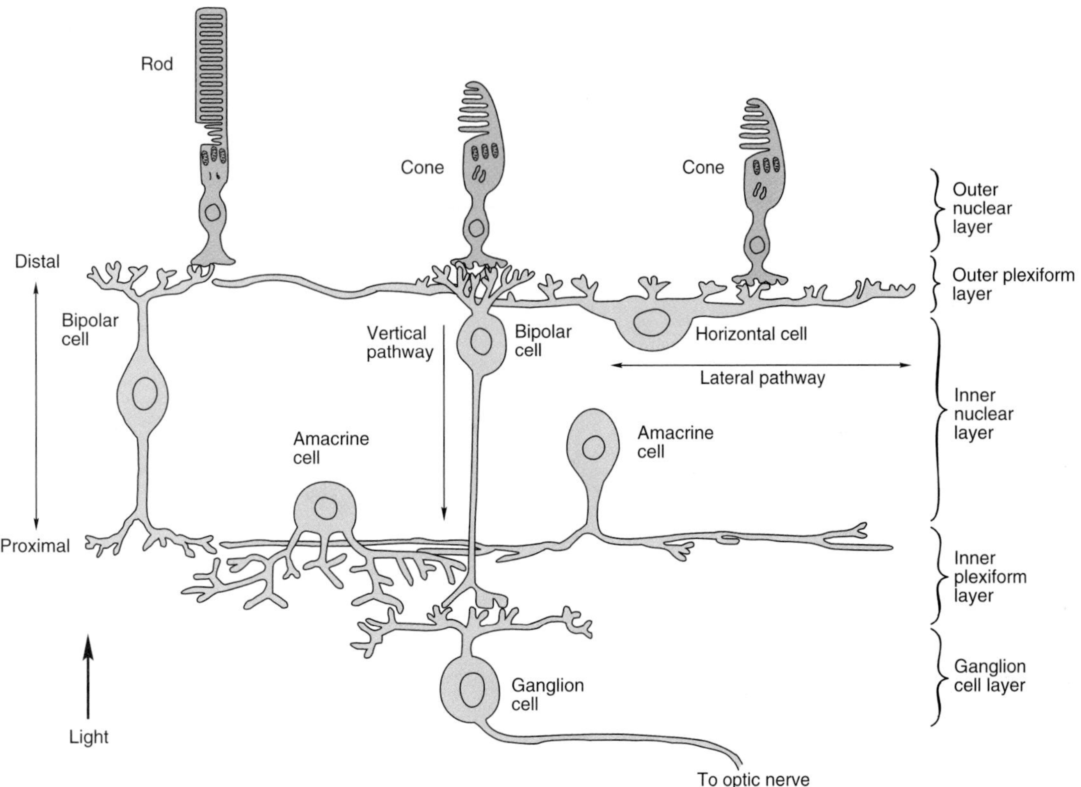

Figure 22–6 The three nuclear layers of the retina contain three major functional classes of neurons: photoreceptors (rods and cones) in the *outer nuclear layer*, interneurons (bipolar cells, horizontal cells, amacrine cells) in the *inner nuclear layer*, and ganglion cells in the *ganglion cell layer*. Photoreceptors, bipolar cells, and horizontal cells make synaptic connections with each other in the *outer plexiform layer*. The bipolar, amacrine, and ganglion cells make contact in the *inner plexiform layer*. Bipolar cells bridge the two plexiform layers, and information flows vertically from photoreceptors to bipolar cells to ganglion cells. Information also flows laterally, mediated by horizontal cells in the outer plexiform layer and amacrine cells in the inner plexiform layer. (Adapted from Dowling, 1979.)

to each ganglion cell always originate from the same photoreceptors in a circumscribed area of the retina, the *receptive field* for that cell. In effect, the ganglion cell's receptive field is the area of retina that the ganglion cell monitors. The receptive fields of ganglion cells have two important features.

First, ganglion cell receptive fields are circular. When small spots of light are used to probe the retina for the properties of ganglion cell receptive fields, the receptive fields prove to be roughly circular and to vary in size across the retina. In the foveal region of the primate retina, where visual acuity is greatest, the receptive fields are small, with centers that are only a few minutes of arc (60 minutes = 1 degree). At the periphery of the retina, where acuity is low, the fields are larger, with centers of 3° to 5° (1° on the retina is equal to about 0.25 mm).

Second, in most ganglion cells the receptive field is divided into two parts: a circular zone at the center, called the *receptive field center*, and the remaining area of the field, called the *surround*. Ganglion cells respond optimally to differential illumination of the receptive field center and surround.

Two classes of ganglion cells can be distinguished by their response to a small spot of light applied to the center of their receptive field. *On-*

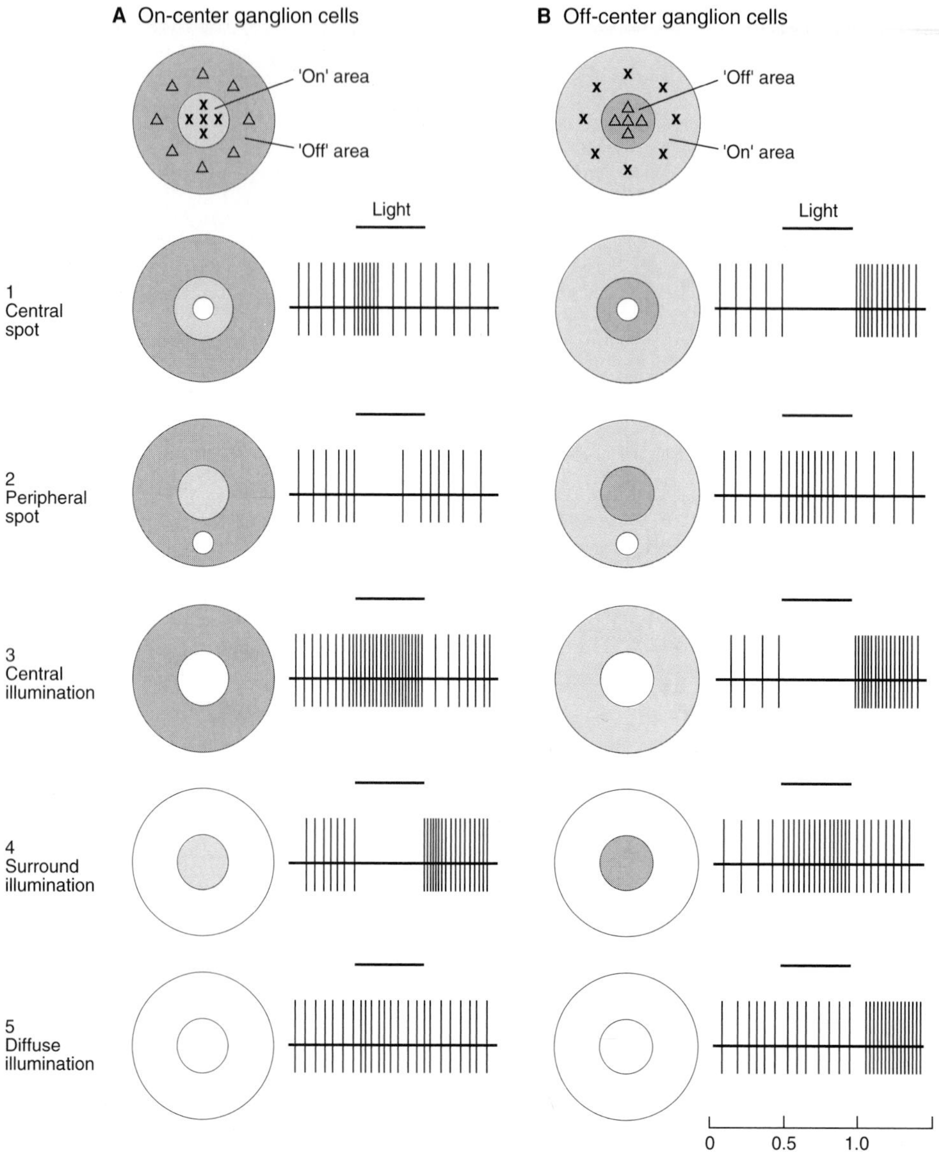

Figure 22–7 Retinal ganglion cells respond optimally to contrast in their receptive fields. Ganglion cells have circular receptive fields, divided into a center area and a surround. *On-center* cells are excited when stimulated by light in the center and inhibited when stimulated in the surround; *off-center* cells have the opposite responses. The figure shows the responses of both types of cells to five different light stimuli (the **white** portion of the receptive field represents the stimulated area). The pattern of action potentials fired by the ganglion cell in response to each stimulus is shown in extracellular recordings. Duration of illumination is indicated by a bar above each record. (Adapted from Kuffler, 1953.)

A. On-center cells respond best when the entire central part of the receptive field is stimulated (**3**). These cells also respond well, but less vigorously, when only a portion of the central field is stimulated by a spot of light (**1**). Illumination of the surround with a spot of light (**2**) or ring of light (**4**) reduces or suppresses the cell firing, which resumes more vigorously for a short period after the light is turned off. Diffuse illumination of the entire receptive field (**5**) elicits only a relatively weak discharge because the center and surround oppose each other's effects.

B. The spontaneous firing of off-center cells is suppressed when the central area of the receptive field is illuminated (**1**,**3**) but accelerates for a short period after the stimulus is turned off. Light shone onto the surround of the receptive field excites the cell (**2**,**4**).

center ganglion cells are excited when light is directed to the center of their receptive field (i.e., they are excited when the spot of light on the center is turned *on*). Light applied to the surround inhibits the cell; the most effective inhibitory stimulus is a ring of light on the entire surround. *Off-center ganglion cells* are inhibited by light applied to the center of their receptive field. Their firing rate increases for a short period after the light is removed (i.e., they are excited when the spot of light on the center is turned *off*). Light excites an off-center ganglion cell when it is directed to the surround of the receptive field. In both types of cells the response evoked by a ring of light on the entire surround cancels almost completely the response evoked by light directed to the center. For this reason, diffuse illumination of the entire receptive field evokes only a small response (Figure 22–7).

Both types of ganglion cells are present in roughly equal numbers and every photoreceptor sends output to both types. Thus, ganglion cells provide two *parallel pathways* for the processing of visual information.

Ganglion Cells Are Specialized for the Detection of Contrasts and Rapid Changes in the Visual Image

Why do ganglion cells have a center–surround receptive field organization, and why are there parallel on-center and off-center pathways?

As we have just seen, because of the center–surround structure of their receptive fields ganglion cells respond only weakly to uniform illumination; they respond best when the light intensities in the center and surround are quite different. They therefore report principally on *contrast* in visual input, rather than its absolute intensity.

Most of the useful information in a visual scene is contained in the pattern of contrasts. The absolute amount of light reflected by objects is relatively uninformative because it is largely determined by the intensity of the light source; doubling the ambient light intensity doubles the amount of light reflected by objects, but does not alter contrasts between them. By having a center–surround receptive field organization, ganglion cells are therefore specialized for detecting useful information in the visual scene.

As we shall see in Chapters 23 and 24, perception of the brightness and color of objects also relies principally on information about contrast rather than the absolute amount of light and can therefore be influenced by the contrast between the object and its surround. For example, the same gray ring looks much lighter against a black background than against a white one (Figure 22–8).

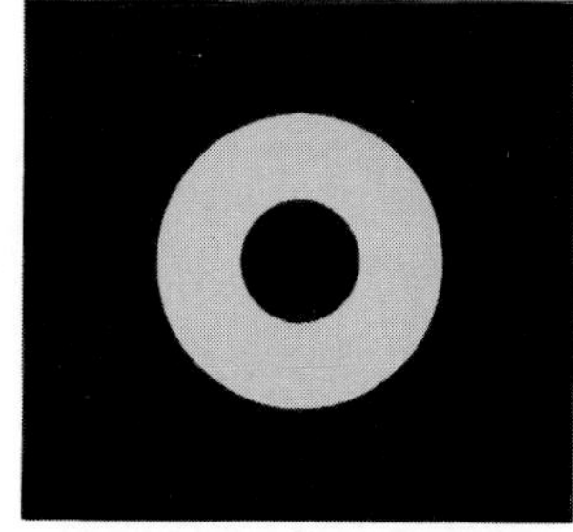
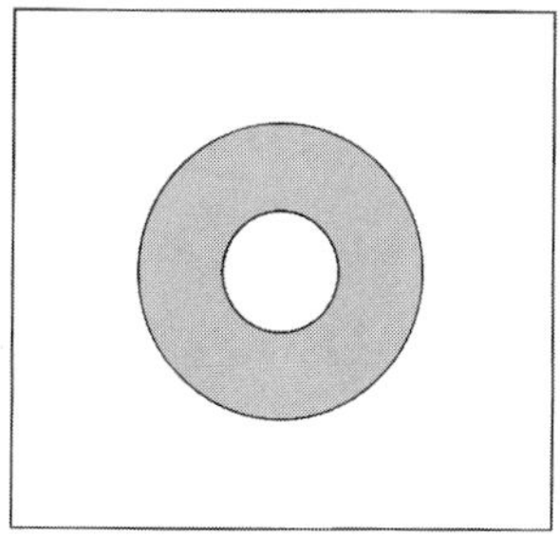

Figure 22–8 The appearance of an object depends principally on the contrast between the object and its background, not on the intensity of the light source. The two gray rings in the figures are identical in hue, but they appear to have different brightnesses because the different backgrounds produce different contrasts.

Parallel on-center and off-center pathways enhance the performance of the visual system because the properties of each type of ganglion cell are best suited to signaling either a rapid increase or decrease in illumination. On-center ganglion cells, which have a low rate of firing under dim illumination, can signal rapid *increases* in light intensity in their receptive field center by rapidly increasing their firing rate. These same cells cannot signal a rapid decrease in light intensity because they are firing at a low rate in the resting state. In contrast, off-center ganglion cells, which have a low discharge rate in the light, can signal rapid *decreases* in illumination in their receptive field center by rapidly increasing their firing rate.

Specialized Ganglion Cells Process Different Aspects of the Visual Image

In addition to contrast and rapid changes in illumination, the visual system also analyzes several other aspects of the visual image, such as color, form, and movement. As we mentioned in the preceding chapter and will discuss again in the following chapter, in the visual cortex these features are processed by parallel pathways. This parallel processing begins in the retina with parallel networks of ganglion cells.

Each region of the retina has several functionally distinct subsets of ganglion cells that serve the same photoreceptors in parallel. Most ganglion cells in the primate retina fall into two classes, M (for *magni* or large) and P (for *parvi* or small). Within each class there are both on-center and off-center cells.

M cells have large receptive fields (reflected in their large dendritic arbors) and show a relatively transient response to sustained illumination. They respond to large objects and, as we shall see in Chapter 23, they appear to be concerned with the analysis of the gross features of a stimulus and its movement. The smaller P cells are more numerous, have small receptive fields, are for the most part wavelength selective, and are involved in the perception of form and color. It is thought that P cells are responsible for the analysis of fine detail in the visual image, although some M cells may also be involved in this function.

Signals from Photoreceptors to Ganglion Cells Are Relayed Through a Network of Interneurons

How do the relatively simple signals provided by photoreceptors give rise to the complex responses of the ganglion cells? Although the circuitry connecting these cells appears complicated, on close examination it is relatively simple. Each type of interneuron in the retina (horizontal, bipolar, and amacrine cells) plays a specific role in shaping photoreceptor signals as they are transmitted through the retina. To simplify understanding of the role of retinal interneurons, we focus on the bipolar cells as they represent the most direct pathway between receptors and ganglion cells. We further restrict our attention to the circuitry for cones, the circuitry that mediates vision in normal daylight.

Cone Signals Are Conveyed to Ganglion Cells Through Direct or Lateral Pathways

Visual information is transferred from cones to ganglion cells along two types of pathways in the retina. Cones in the *center* of a ganglion cell's receptive field make direct synaptic contact with bipolar cells that in turn directly contact the ganglion cells; these connections are known as *direct* or *vertical pathways*. Signals from cones in the *surround* of the ganglion cell's receptive field are also conveyed to the ganglion cell through bipolar cells but only indirectly by means of horizontal and amacrine cells; these indirect connections are called *lateral pathways*. Horizontal cells, which have large dendritic trees, transfer information from distant cones to bipolar cells. (Horizontal cells are also electrically coupled to each other by gap junctions and thus are able to respond to inputs from even more distant cones that contact neighboring horizontal cells.) Some types of amacrine cells transfer information from distant bipolar cells to ganglion cells (see Figure 22–6).

Most synaptic contacts in the retina are grouped in two plexiform (network-like) layers. The *outer plexiform layer* contains the processes of receptor, bipolar, and horizontal cells, while the *inner plexiform layer* contains the processes of bipolar, amacrine, and ganglion cells (see Figure 22–6). Thus the bipolar cells bridge the two plexiform layers by having processes in both, an arrangement that reflects the key role of these cells in the retina.

We have seen that photoreceptors respond to light with graded changes in membrane potential rather than by firing action potentials. The same is true of horizontal and bipolar cells. These cells lack voltage-gated Na^+ channels capable of generating action potentials and, instead, transmit signals passively (Chapter 9). Because these cells are small and have short processes, the signals spread to their synaptic terminals without significant reduction. (Passive signal spread in cells with short processes occurs in many different parts of the brain.) In contrast, the axons of ganglion cells project considerable distances to their targets in the brain and transfer information in the form of trains of action potentials. Many types of amacrine cells also fire action potentials.

The Receptive Fields of Bipolar Cells Have a Center–Surround Organization

Like ganglion cells, the bipolar cells have receptive fields with an antagonistic center–surround organization and the cells are either on-center or off-center. When cones in the center of the receptive field are active, on-center bipolar cells depolarize, whereas off-center bipolar cells hyperpolarize. When cones in the surround are active, the response of the bipolar cell is opposite that evoked by illumination of the center (Box 22–3).

BOX 22–3 The Center–Surround Receptive Field of Bipolar Cells

Each cone cell synapses on both on-center and off-center bipolar cells. Cone cells release a single neurotransmitter, glutamate, which inhibits (hyperpolarizes) on-center bipolar cells and excites (depolarizes) off-center cells. Recall though that in the dark the cone is depolarized (around –40 mV), so that voltage-gated Ca^{2+} channels in its synaptic terminals are open, allowing Ca^{2+} to enter the terminals and trigger the release of glutamate. This constant release of glutamate in the dark maintains the on-center bipolar cells in a hyperpolarized state. When the cone is illuminated, however, it becomes hyperpolarized, and this change causes the closure of the voltage-gated Ca^{2+} channels, which reduces the Ca^{2+} influx and therefore reduces the amount of glutamate the cell releases; as a result, the on-center bipolar cells depolarize. Conversely, off-center bipolar cells are maintained in a depolarized state in the dark by the cone cells, and hyperpolarize when glutamate release is reduced by light (Figure 22–9).

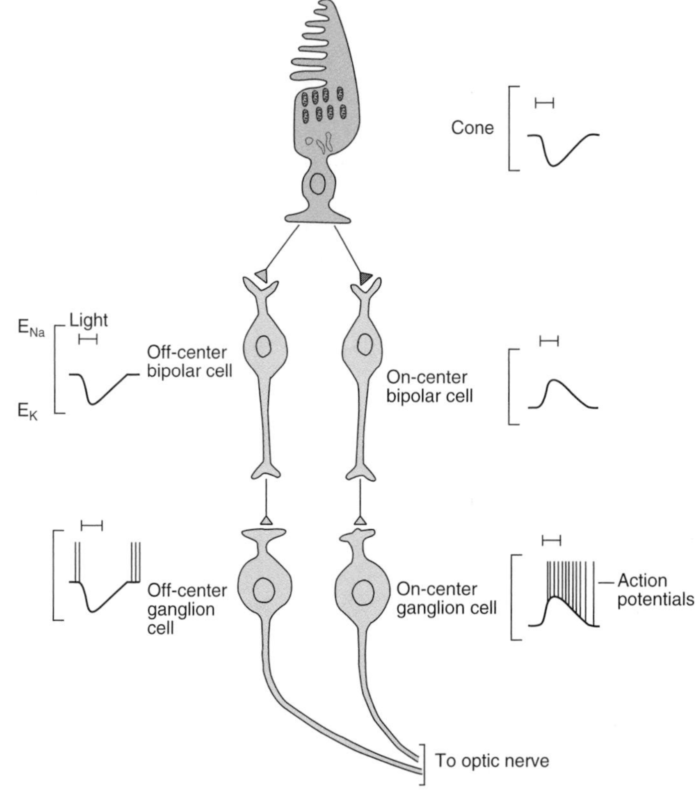

Figure 22–9 Visual information is segregated into on-center and off-center pathways at the bipolar cells. In the illustration a single cone photoreceptor synapses on two bipolar cells, each of which makes an excitatory connection with a ganglion cell of the same type. When the cone is hyperpolarized by light, the on-center bipolar cell is excited and the off-center bipolar cell is inhibited. These opposite and simultaneous actions are initiated by the transmitter glutamate. In the dark the cone releases large amounts of transmitter because it is depolarized. Light, by hyperpolarizing the cone, causes a reduction in transmitter release. The same transmitter has different actions because the two types of bipolar cells have different postsynaptic receptors that gate different types of ion channels. The responses of the ganglion cells are largely determined by the inputs from the bipolar cells. The on-center bipolar cell, which becomes depolarized by illumination of its receptive field center, depolarizes the on-center ganglion cells; the off-center cells show the opposite response.

(continued)

BOX 22–3 The Center–Surround Receptive Field of Bipolar Cells (continued)

Glutamate produces different responses in the two classes of bipolar cells by gating different cation channels. In off-center bipolar cells it opens one type of cation channel that carries an inward (depolarizing) Na^+ current into the cells. In on-center bipolar cells glutamate *closes* another type of cation channel that also carries an inward Na^+ current, and which, like the cGMP-gated channels in photoreceptors, is gated by intracellular cGMP. In the absence of transmitter this type of channel is kept open by a high intracellular concentration of cGMP in the on-center bipolar cell. Glutamate appears to cause the closure of these channels in precisely the same way that light causes the closure of cGMP-gated channels in photoreceptors—by activating a specific glutamate receptor that, like rhodopsin, activates a G-protein, which in turn activates cGMP phosphodiesterase and lowers the cytoplasmic concentration of cGMP in the on-center bipolar cell.

Horizontal cells mediate the antagonistic inputs from cones in the surround of the bipolar cell's receptive field. Horizontal cells do not make direct synaptic contact with the bipolar cells. Instead, they synapse onto cones in the *center* of the bipolar cell's receptive field. When the surround is illuminated, the horizontal cells *depolarize* the cones in the center, the opposite effect of light absorption by these cones (Figure 22–10). Whether this mechanism alone accounts for the antagonism between center and surround in bipolar cells is not yet known.

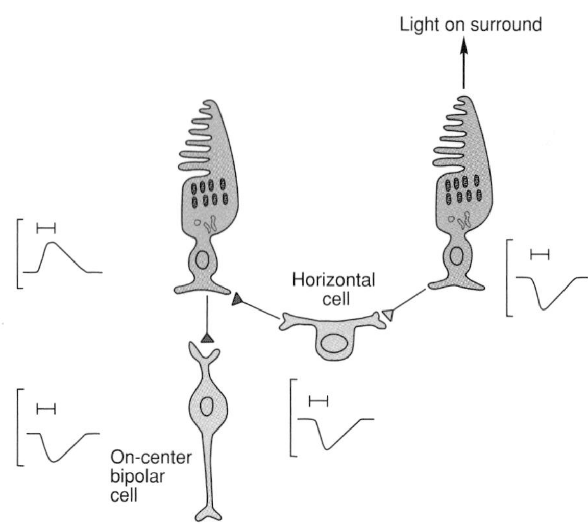

Figure 22–10 The response of a bipolar cell to the stimulation of photoreceptors in the center of its receptive field is antagonized by stimulation of photoreceptors in the surround of its receptive field. This antagonistic interaction between neighboring retinal areas is mediated by the action of *horizontal cells*. Center–surround antagonism is illustrated here for an on-center bipolar cell. In the dark, horizontal cells release an inhibitory transmitter that maintains postsynaptic cones in the receptive field center in a slightly hyperpolarized state. Illumination of a cone in the bipolar cell's *surround* hyperpolarizes that cone, which in turn hyperpolarizes the postsynaptic horizontal cell. (In the dark the cone in the surround is maintained in a depolarized state and thus excites those horizontal cells.) This hyperpolarization of the horizontal cell reduces the amount of inhibitory transmitter released by the horizontal cell onto postsynaptic cones in the receptive field center, and as a result these cones become depolarized (the opposite effect of light absorption by these cones). This in turn allows the on-center bipolar cell to become hyperpolarized, the opposite effect of illumination in the receptive field center.

Different Classes of Bipolar Cells Have Excitatory Connections with Corresponding Classes of Ganglion Cells

The receptive field properties of a ganglion cell largely reflect those of the bipolar cells connected to it, because each type of bipolar cell (on-center or off-center) makes *excitatory* synaptic connections with the corresponding type of ganglion cell. When on-center bipolar cells are depolarized by light, they depolarize on-center ganglion cells (see Figure 22–9 in Box 22–3).

Although the responses of ganglion cells are largely determined by these direct inputs from bipolar cells, they are also shaped by amacrine cells (see Figure 22–6). There are more than 20 morphologically distinct types of amacrine cells that use at least eight different neurotransmitters. Some amacrine cells function like horizontal cells: they mediate antagonistic inputs from bipolar cells in the ganglion cell's surround. Others have been implicated in shaping the complex receptive field properties of specific classes of ganglion cells, such as the M-type ganglion cells that process orientation information (see Chapter 23).

An Overall View

The absorption of light and its transduction into electrical signals are carried out by the photoreceptors. Visual information is then transferred from the receptors to the ganglion cells via the bipolar cells. The ganglion cells in turn project to the brain: their axons form the optic nerve. Two types of interneurons (horizontal cells and amacrine cells) provide lateral inputs to bipolar cells and ganglion cells.

The cyclic nucleotide cGMP plays a central role in phototransduction. Absorption of light by the photosensitive visual pigments in the photoreceptor triggers a second-messenger cascade. The activated pigment molecules stimulate a G-protein, transducin, which in turn activates a phosphodiesterase that catalyzes the hydrolysis of cGMP. Light absorption therefore causes a reduction in the cytoplasmic concentration of cGMP. In darkness cGMP opens specialized ion channels that carry a depolarizing current into the cell, so that the reduction in the level of cGMP makes the photoreceptor hyperpolarize.

As visual information is transferred from photoreceptors to ganglion cells, it is segregated into parallel on-center and off-center pathways. An on-center ganglion cell is excited when light stimulates the center of its receptive field and inhibited when light stimulates its surround. An off-center ganglion cell has the opposite responses; it is inhibited when light stimulates its center and excited by light on its surround. These transformations of the visual signal assist higher centers in detecting weak contrasts and rapid changes in light intensity. In addition, ganglion cells are specialized for processing different aspects of the visual image like movement, fine spatial detail, or color.

The pattern of synaptic connections in the retina explains how the various responses of ganglion cells arise. Bipolar cells, like ganglion cells, fall into two classes, on-center and off-center. The transmitter released by cones excites bipolar cells of one class and inhibits the others. Each cone makes contact with both types of bipolar cells. Cones in the receptive field center of a ganglion cell synapse onto bipolar cells that make direct contact with the ganglion cell. Inputs from cones in the receptive field surround are relayed along lateral pathways by horizontal and amacrine cells.

Selected Readings

De Vries, S. H., and Baylor, D. A. 1993. Synaptic circuitry of the retina and olfactory bulb. Cell 72/Neuron 10(Suppl);139–149.

Dowling, J. E. 1987. The Retina: An Approachable Part of the Brain. Cambridge, Mass.: Belknap Press.

Hurley, J. B. 1994. Termination of photoreceptor responses. Curr. Opinion Neurobiol. 4(4):481–487.

Lagnado, L., and Baylor, D. 1992. Signal flow in visual transduction. Neuron 8:955–1002.

Nakanishi, S. 1995. Second-order neurones and receptor mechanisms in visual- and olfactory-information processing. Trends Neurosci. 18(8):359–364.

Shiller, P. H. 1992. The ON and OFF channels of the visual system. Trends Neurosci. 15(3):86–92.

Stryer, L. 1987. The molecules of visual excitation. Sci. Am. 257(1):32–40.

23

Perception of Form and Motion

The Retinal Image Is an Inversion of the Visual Field

The Retina Projects to the Lateral Geniculate Nucleus

Information About Contrast from the Retina Is Not Altered Significantly in the Lateral Geniculate Nucleus

Form Is Analyzed in the Primary Visual Cortex by Cells Whose Receptive Fields Have Linear Features

Simple and Complex Cells Deconstruct Light Contrasts into Short Line Segments

Some Simple and Complex Cells Are End-Stopped

Progressive Convergence in the Primary Visual Cortex May Lead to the Extraction of the Linear Features of an Image

The Primary Visual Cortex Is Organized into Columns and Layers

Cells with the Same Response Properties Are Organized in Columns

Horizontal Connections Interconnect Columns with One Another

Beyond the Primary Visual Cortex: The Representation of Faces and Other Complex Forms Occurs in the Inferotemporal Cortex

Motion in the Visual Field Is Analyzed by a Special Neural System

Motion Is Represented in the Middle Temporal and the Medial Superior Temporal Areas

Perception of Motion Direction Can Be Influenced by Microstimulation of Cells Within the Middle Temporal Area

Visual Attention Can Now Be Studied on the Cellular Level

In the previous two chapters we began to consider the three visual pathways concerned with the perception of form, movement, and color. In this chapter we review the evidence that form is analyzed by the parvocellular–interblob system and motion by the magnocellular system. The parvocellular–blob system, which is concerned with the assessment of color, is considered in Chapter 24. To gain insight into the cellular mechanisms at work in the processing of the different types of visual information, we consider the response properties of neurons in each of these pathways.

We first consider the cellular mechanisms that give rise to the perception of form. This component of perception is mediated in part by a serial pathway—a processing stream—that carries information about contrast and contours. This new pathway extends from the retina to the lateral geniculate nucleus of the thalamus and from there to striate and extrastriate cortices. Next we focus on the analysis of motion.

In examining the cellular mechanisms underlying the perception of form and motion, we shall see that the architecture of the cortex—specifically its modular organization—plays a critical role in the segregation, analysis, and processing of information for vision.

The Retinal Image Is an Inversion of the Visual Field

Each retina is divided into quadrants with reference to an imaginary vertical line down the middle that delineates the *nasal hemiretina* and the *temporal hemiretina*. Each half of the retina is divided into a *dorsal* and a *ventral* quadrant.

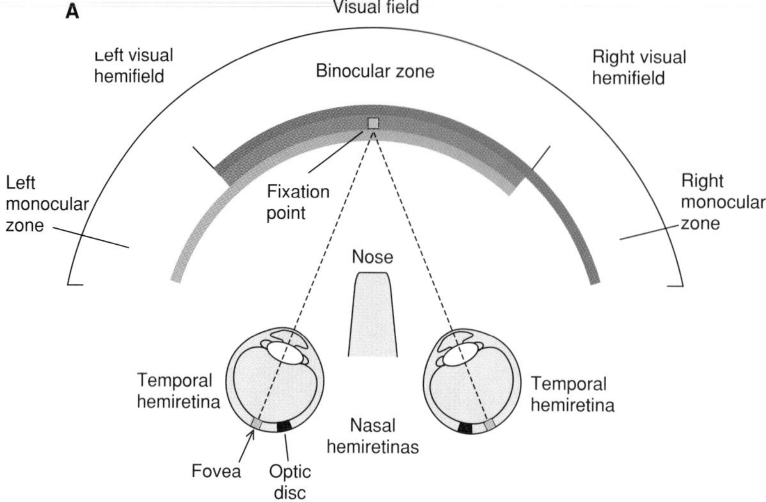

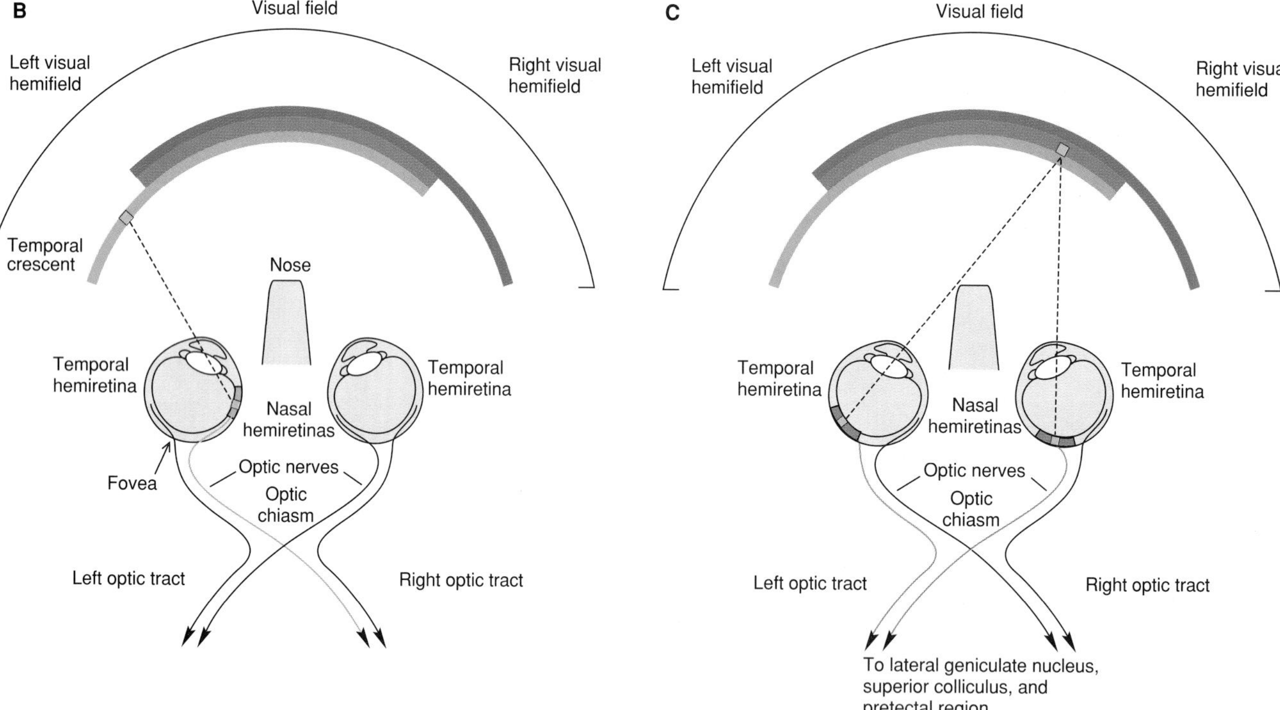

Figure 23–1 The visual field has both binocular and monocular zones.

A. Light from the binocular strikes both eyes. Light in the center of the field impinges on the fovea, a region lateral to the center of the retina with the highest acuity. The hemiretinas are defined with respect to the fovea. Light from the binocular zones activates photoreceptors in the temporal hemiretina and part of the nasal hemiretina of each eye. The optic disc, the region at the true center if the retina where the ganglion cell axons leave the retina, is free of photoreceptors and therefore creates a gap or blind spot (see Figure 23–2).

B. Light from a monocular zone (the temporal crescent) strikes only one eye. It falls on the ipsilateral nasal hemiretina and does not project upon the contralateral retina because it is blocked by the nose.

C. Each optic *tract* carries a complete representation of one-half of the visual field. Fibers from the nasal hemiretina of each eye cross to the opposite side at the optic chiasm, whereas fibers from the temporal hemiretina do not cross. In the illustration, light from the right half of the binocular zone falls on the left temporal hemiretina and right nasal hemiretina. Axons from these two sites thus contain a complete representation of the right hemifield of vision (see Figure 23–6).

Figure 23–2 Find the blind spot in the left eye by shutting the right eye and fixating on the upper cross with the left eye. Hold the book about 1.5 ft from the eye and move it back and forth slightly to find the location where the circle on the left disappears. At that point, the image of the circle has fallen within the blind spot. Likewise, by fixing the left eye on the lower cross, the gap in the black line falls on the blind spot and the black line is seen as continuous. (Adapted from Hurvitch, 1981.)

The visual field is the view seen by the two eyes without movement of the head. Imagine that the foveas of both eyes are fixed on a single point in space. It is then possible to define a *left half* and a *right half* of the visual field. The *left hemifield*, or left half of the visual field, projects on the nasal hemiretina of the left eye and on the temporal hemiretina of the right eye. The *right hemifield* projects on the nasal hemiretina of the right eye and on the temporal hemiretina of the left eye (Figure 23–1).

Light originating in the central region of the visual field enters *both* eyes; this area is called the *binocular zone*. In either half of the visual field there is also a *monocular zone*: light from the lateral or temporal portion of the hemifield projects only onto the medial or nasal hemiretina of the eye on the same side, because the nose blocks this light from reaching the eye on the opposite side (Figure 23–1B). This monocular zone is known as the temporal crescent.

The optic disc, the region of the retina from which the ganglion cell axons exit, contains no photoreceptors and therefore is insensitive to light. Since the disc is located closer to the midline than the fovea in both eyes (Figure 23–1A), light coming from any single point in the binocular zone never strikes both optic discs simultaneously, so that we are normally unaware of this blind spot. The blind spot can be demonstrated by looking at Figure 23–2 with the left eye. This exercise simulates what blind people experience—not blackness, but simply nothing. It also reveals why damage to large regions of the peripheral retina can go unnoticed. In these instances no large dark zone appears in the periphery; it is usually by accidents, such as bumping into an unnoticed object, or by clinical testing of the visual fields, that blind spots are noticed.

The correspondence between regions of the visual field and the *retinal image* may seem complicated at first. One complication is that the lens of the

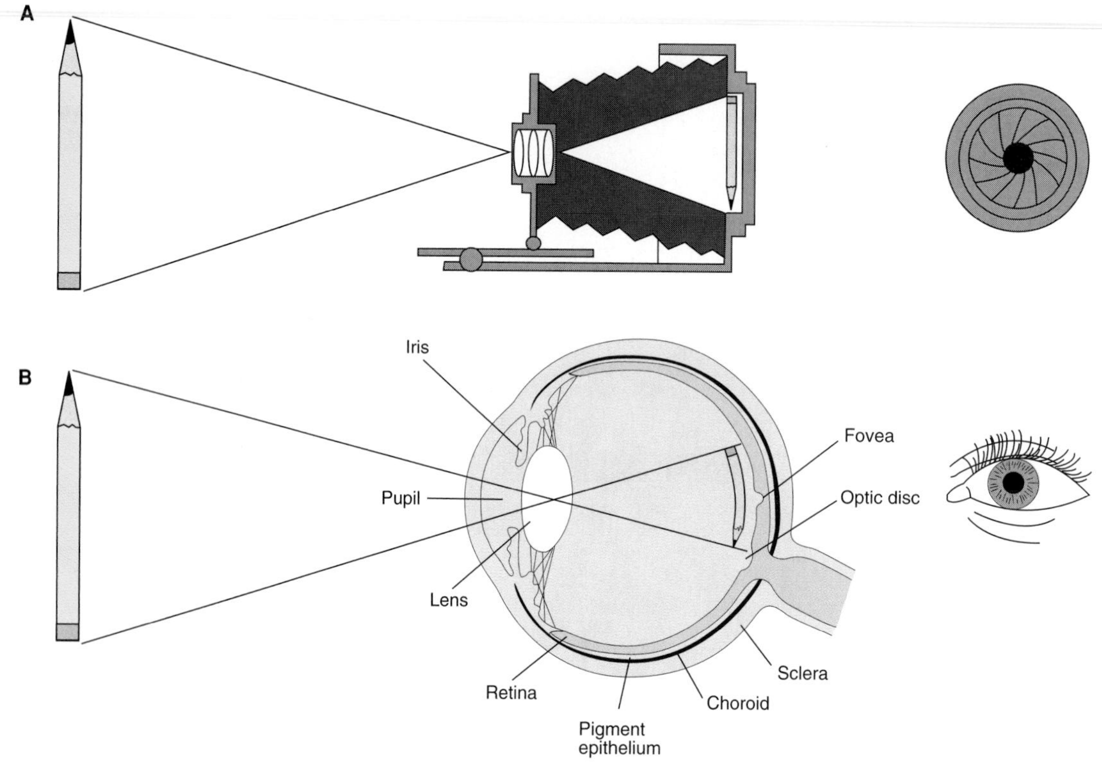

Figure 23–3 The lens of the eye projects an inverted image on the retina in the same way as the lens of a camera inverts an image on the photographic film. (Adapted from Groves and Schlesinger, 1979.)

eye inverts the visual image on the retina (Figure 23–3). The superior half of the visual field is projected onto the inferior (or ventral) half of the retina, and the inferior half of the visual field is projected onto the superior (or dorsal) half of the retina. (The brain of course adjusts this inversion.) Thus, damage to the *inferior half of the retina* of one eye causes a monocular deficit in the *superior half of the visual field*. Another complication is that the binocular portion of each visual hemifield stimulates different halves of each retina (temporal in one retina, nasal in the other; see Figure 23–1C).

The Retina Projects to the Lateral Geniculate Nucleus

The axons of all retinal ganglion cells stream toward the optic disc, where they become myelinated and together form the *optic nerve*. The optic nerves from each eye join at the *optic chiasm*. There the fibers from the nasal hemiretina cross to the opposite side of the brain; the axons of ganglion cells in the temporal hemiretina do not cross. The resulting separation of fibers forms the left and right *optic tracts* (Figure 23–1C). The left optic tract contains axons from the left half of each retina—the temporal hemiretina of the left eye and the nasal hemiretina of the right eye—so the left optic tract carries a complete representation of the *right* hemifield of vision (Figure 23–1C). Likewise, the right optic tract conveys a representation of the entire *left* visual hemifield.

The left and right optic tracts project to three subcortical targets (Figure 23–4). Of the three, only the lateral geniculate nucleus processes visual information for perception. (The pretectal area of the midbrain uses inputs from the retina to produce pupillary reflexes, while the superior colliculus uses its input to generate eye movements.) The right optic tract, containing fibers from the right

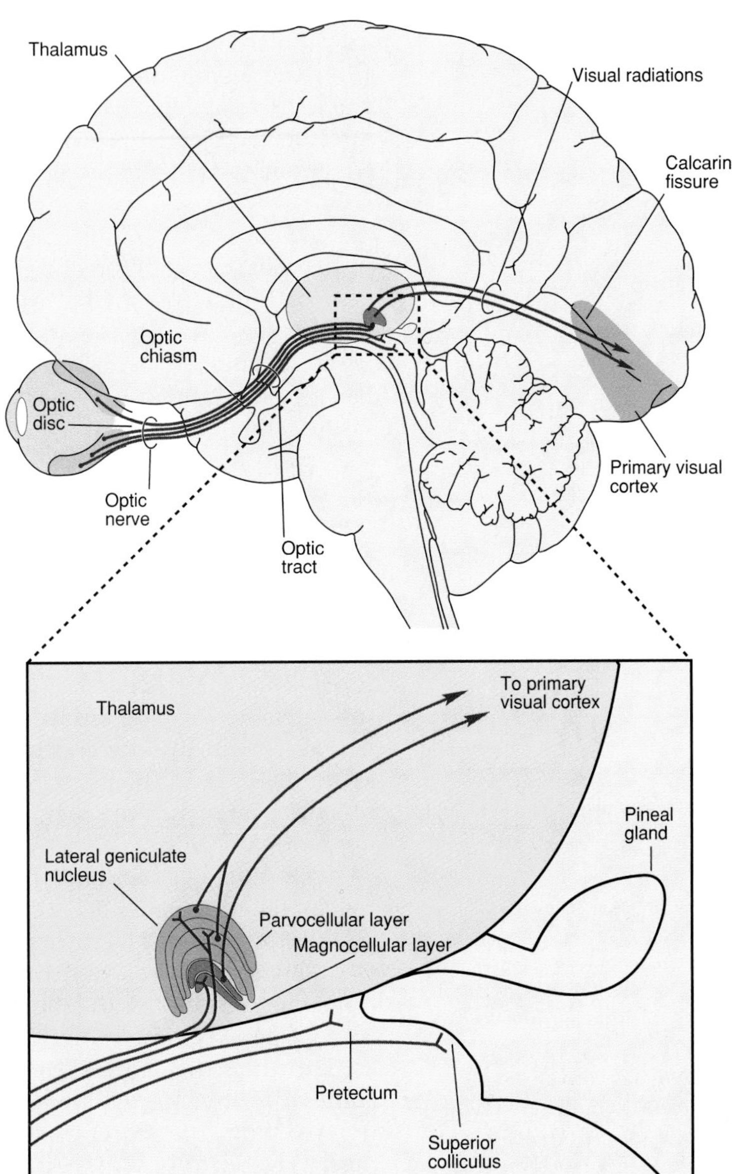

Figure 23–4 A simplified diagram of the projections from the retina to the visual areas of the thalamus (lateral geniculate nucleus) and midbrain (pretectum and superior colliculus) and to the primary visual cortex. Only the lateral geniculate nucleus and visual cortex process visual information for perception. The retinal projection to the pretectal area is important for reflexes of the pupil, while the projection to the superior colliculus mediates eye movements that are guided by visual information.

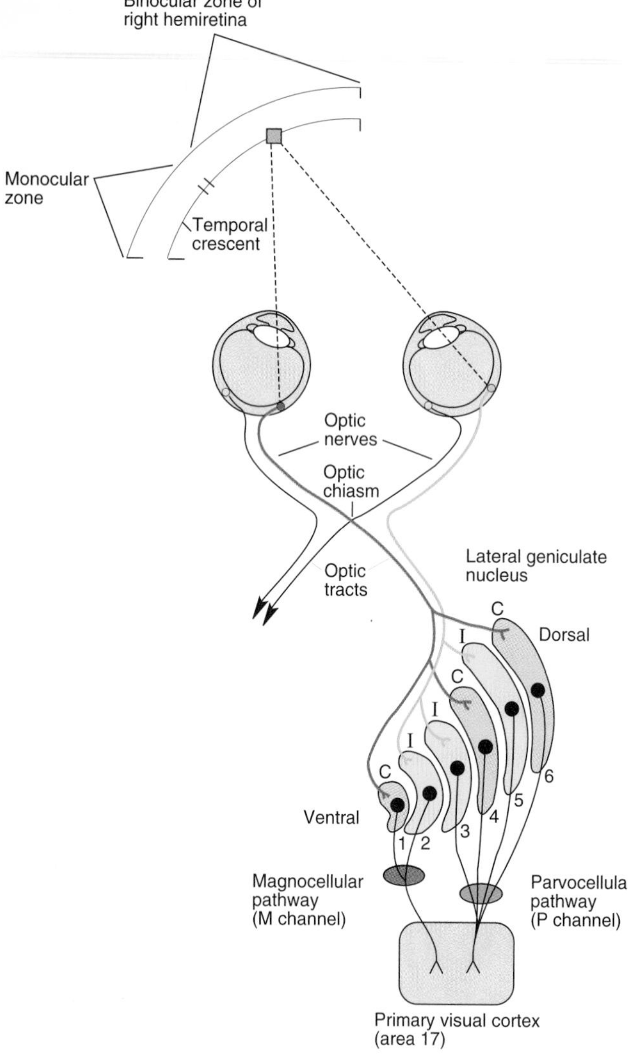

Figure 23–5 The lateral geniculate nucleus is the principal subcortical site for processing information for vision. Inputs from the right hemiretina of each eye project to the right lateral geniculate nucleus and convey a complete representation of the left visual hemifield. Similarly, fibers from the left hemiretina of each eye project to the left lateral geniculate nucleus. The temporal crescent is not represented in contralateral inputs (see Figure 23–1B). The lateral geniculate nucleus comprises six layers of cell bodies. In each nucleus (left and right) the fibers from the contralateral (C) nasal hemiretina terminate in layers 1 and 4; fibers from the ipsilateral (I) temporal hemiretina terminate in layers 2, 3, and 5. Layers 1 and 2 are the magnocellular layers; layers 3 through 6 are the parvocellular layers. Neurons in all layers project to the primary visual cortex and each layer contains a complete neural representation of the contralateral hemifield. In this representation, or neural map, the inputs are organized to preserve the spatial arrangement of the receptors on the retina.

hemiretina of each eye (the nasal hemiretina of the left eye and the temporal hemiretina of the right eye), projects to the right lateral geniculate nucleus. Similarly, the left optic tract, containing fibers from the left hemiretina of each eye, projects to the left lateral geniculate nucleus (Figure 23–5). As a result, each lateral geniculate nucleus receives all the inputs from the contralateral half of the visual field.

The axons of the retinal ganglion cells terminate in a topographically organized way in the lateral geniculate nucleus, so that each lateral geniculate nucleus has a *retinotopic* representation of the contralateral visual field. As a result, the lateral geniculate nucleus contains a *neural map* of the retina. Like the neural map of the body in the somatosensory pathways (see Chapter 20), the neural map of the retina in the visual pathways is not a point-for-point representation of the retina. The fovea, the area of the retina with greatest acuity, has proportionately a much larger representation than does the periphery of the retina. This is because more photoreceptors per unit area are packed into the fovea. The fovea therefore also has more inputs in the lateral geniculate nucleus. About half of the neural mass in the lateral geniculate nucleus (and, as we shall see, in the primary visual cortex) represents the fovea and the region just around it (Figure 23–6).

In the human brain and that of other primates the lateral geniculate nucleus contains six layers of cell bodies separated by intervening layers of axons and dendrites. The layers are numbered 1 to 6, from the ventral to the dorsal end (see Figure 23–5). The two most ventral layers of the nucleus are known as the *magnocellular layers* because they contain relatively large cells; their main retinal input is from the M cells in the retina. The four dorsal layers are known as *parvocellular layers* and receive input from the P cells in the retina.

Each layer contains a representation of the contralateral visual hemifield but receives input from one eye only. Layers 1, 4, and 6 each receive input from the nasal hemiretina in the contralateral eye. Layers 2, 3, and 5 each receive input from the temporal hemiretina in the ipsilateral eye; in these layers only 90% of the hemifield is represented because light from the temporal crescent of the contralateral visual hemifield does not reach the ipsilateral eye (see Figure 23–1B). The layers of the nucleus are stacked one on top of the other,

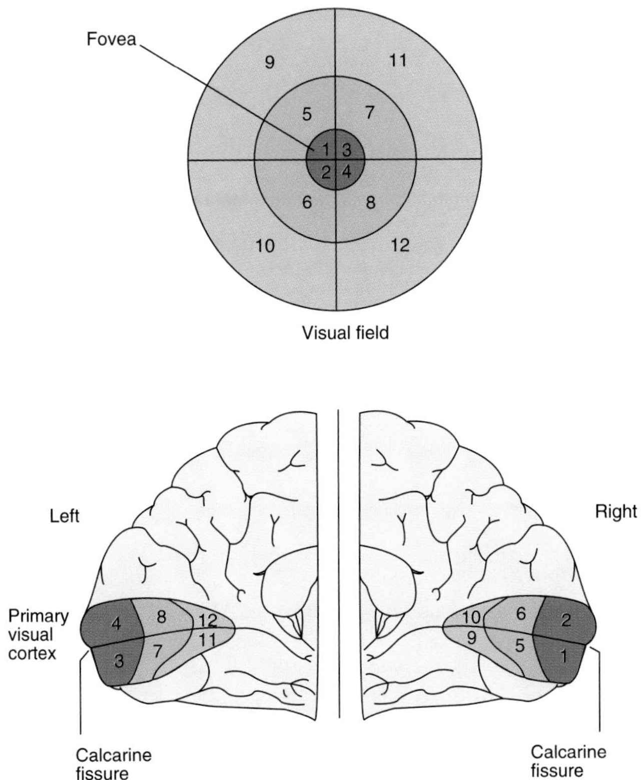

Figure 23–6 In humans the primary visual cortex is located at the posterior pole of the cerebral hemisphere. Each half of the visual field is represented in the contralateral hemisphere. Areas in the primary visual cortex are devoted to specific parts of the visual field, indicated in the drawing by the corresponding numbers. The upper fields are mapped below the calcarine fissure, and the lower fields above it. The striking aspect of this map is that about half of the neural mass is devoted to representation of the fovea and the region just around it. This area has the greatest visual acuity.

and the six maps of the contralateral hemifield are in precise vertical register. If an electrode were to penetrate the nucleus at a right angle to the layers, it would detect activity responding to a single point in the visual field.

Information About Contrast from the Retina Is Not Altered Significantly in the Lateral Geniculate Nucleus

Although we know a great deal about the cell types and circuitry of the lateral geniculate nucleus and about the receptive field properties of different cell types, the actual function of the nucleus is not yet clear. Only 10 to 20% of the presynaptic connections onto geniculate relay cells come from the retina! Most connections are from other brain regions, and many of these, particularly those from the reticular formation in the brain stem and from the cortex, are feedback connections. This input to the lateral geniculate nucleus may control the flow of information from the retina to the cortex.

As we saw in Chapter 22, retinal ganglion cells have concentric receptive fields, with an antagonistic center–surround organization that allows the ganglion cells to detect the contrast in light intensity between their receptive field center and their surround (see Figure 22–7). The receptive fields of

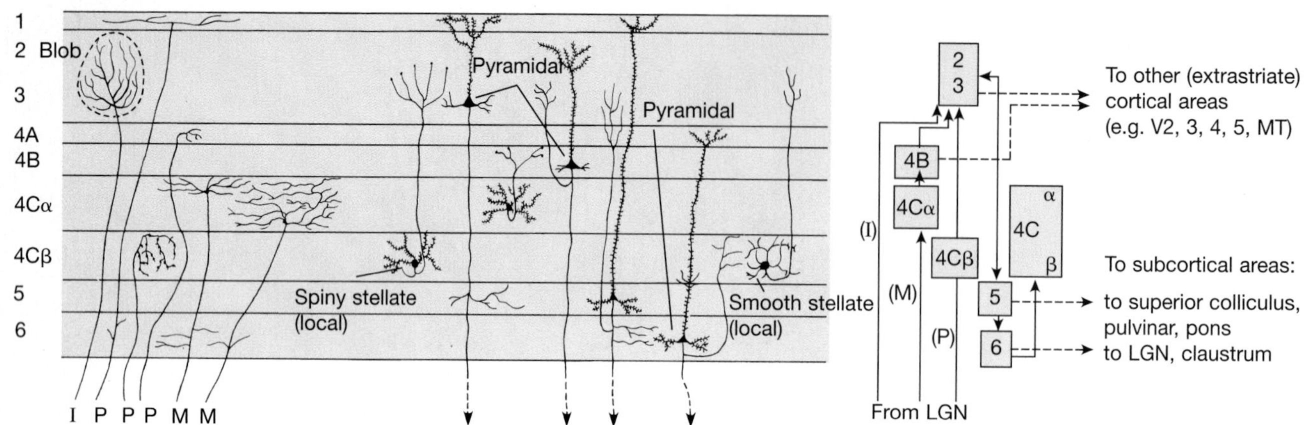

Figure 23–7 The primary visual cortex has distinct anatomical layers, each with characteristic input and output connections.

A. Most afferent fibers from the lateral geniculate nucleus terminate in layer 4C. Axons of P cells (in the parvocellular layers of the geniculate nucleus) terminate primarily in layer 4Cβ, with minor inputs to 4A and 1. Axons from M cells (in the magnocellular layer of the geniculate nucleus) terminate primarily in layer 4Cα. Collaterals of both types of cells also terminate in layer 6. Cells of the interlaminar regions of the geniculate nucleus terminate in layers 2 and 3.

B. Several types of resident neurons make up the primary visual cortex. Spiny stellate and pyramidal cells, both of which have spiny dendrites, are excitatory. Smooth stellate cells are inhibitory. Pyramidal cells project out of the cortex, whereas both types of stellate cells are local interneurons.

C. The local circuitry within the primary visual cortex is complex. Afferents from M and P cells in the lateral geniculate nucleus end on spiny stellate cells in layer 4C, and these cells project axons to layer 4B and the upper layers 2 and 3. Cells from the interlaminar zones (I) in the lateral geniculate nucleus project directly to layers 2 and 3. From there, pyramidal cells project axon collaterals to layer 5 pyramidal cells, whose axon collaterals project both to layer 6 pyramidal cells and back to cells in layers 2 and 3. Axon collaterals of layer 6 pyramidal cells then make a loop back to layer 4C onto smooth stellate cells. Each layer, except for 4C, has different outputs. The cells in layers 2, 3, and 4B project to higher visual cortical areas. Cells in layer 5 project to the superior colliculus, the pons, and the pulvinar. Cells in layer 6 project back to the lateral geniculate nucleus and the claustrum. (Adapted from Lund, 1988.)

neurons in the lateral geniculate nucleus are similar to those found in the retina: small concentric fields about 1° in diameter. Like the retinal ganglion cells, the cells are either on- or off-center. They, too, respond best to small spots of light within their receptive field center; diffuse illumination of the whole receptive field produces only weak responses. The receptive properties of cells in the lateral geniculate nucleus are similar to those of the retina, presumably because each geniculate neuron receives its main retinal input from only a few ganglion cells, and because little of the incoming information is transformed before it is relayed to higher processing centers.

As in the retina, the on- and off-center cells in the lateral geniculate nucleus form separate pathways. Each of these pathways includes separate pathways for magnocellular and parvocellular cells. As we saw in Chapter 21, the parvocellular and the magnocellular layers in the lateral geniculate nucleus give rise to three major parallel pathways that extend to the primary visual cortex and from there to higher-order regions of the cortex concerned with the perception of form, color, and movement. The *magnocellular pathway* is concerned with the location and movement of the visual image—with *where* the object is. The *parvocellular–blob* and *parvocellular–interblob* pathways are concerned more with color and form in vision—with *what* the object is. As we have seen, neuroscientists often refer to these pathways as the *where* and the *what* pathways.

Figure 23–8 David Hubel and Torsten Wiesel together have provided key insights into how the visual system operates. In examining the visual system from the retina through the cortex they discovered that individual nerve cells in the visual pathways are highly specialized, each responding to certain stimuli while ignoring others. At each stage in the visual pathways they found different classes of cells with progressively more complex properties for processing information about movement, directional orientation, contours, and other components of visual images. The discovery that individual nerve cells could convey such complex and sophisticated information was quite unexpected.

Wiesel and Hubel have paralleled their investigations of the normal visual system with studies on the immature and developing brain in the early phases of life. They found that if an eye in an early period of life is deprived of light, or if the brain receives inappropriate information from an eye, critical physiological changes occur in cortical cells, resulting in specific behavioral defects. The demonstration that a critical period in early development is decisive for the formation of appropriate connections between nerve cells has broad implications for understanding human development

Form Is Analyzed in the Primary Visual Cortex by Cells Whose Receptive Fields Have Linear Features

Like the lateral geniculate nucleus, the primary visual cortex in each cerebral hemisphere receives information exclusively from the contralateral half of the visual field. In humans the primary visual cortex is about 2 mm thick and, as with all areas of neocortex, it consists of six layers of cells (Figure 23–7).[1] These layers each have a distinct appearance in the light microscope because of differences in cell and fiber density. Layer 4, the principal site of inputs from the lateral geniculate nucleus, is subdivided into four sublayers (4A, 4B, 4Cα, and 4Cβ). The upper half of layer 4C (sublayer 4Cα) receives the axons of most M cells in the lateral geniculate nucleus. These cells give rise to the magnocellular pathway concerned with the detection of movement. The lower half (sublayer 4Cβ) is the principal target for the axons of one group of P cells that form the parvocellular–interblob pathway. Axons from a second group of P cells terminate in layers 2 and 3, where they innervate repeating patches of cells called *blobs* (see Figure 21-13). This second parvocellular–blob pathway is concerned with color (see Chapter 24).

The primary visual cortex contains two basic classes of cells. *Pyramidal cells* are large with long spiny dendrites; they are excitatory projection neurons that use glutamate as their transmitter and send axons to other cortical regions or to other regions of the brain. *Nonpyramidal cells* are small and stellate in shape. Some of the stellate cells (spiny stellates) have many dendritic spines; these cells are also excitatory and use glutamate as their transmitter. Other stellate cells (smooth stellates) have no dendritic spines; these cells are inhibitory and use γ-aminobutyric acid (GABA) as their transmitter. Both types of stellate cells are local interneurons whose axons do not project out of the primary visual cortex (Figure 23–7B).

Simple and Complex Cells Deconstruct Light Contrasts into Short Line Segments

How are the receptive fields of visual neurons transformed in the cerebral cortex? This question

[1]The primary visual cortex (Brodmann's area 17 or visual area 1, abbreviated V1) is also called the striate cortex, because it contains a prominent stripe of white matter consisting of myelinated axons from the thalamus and other areas of the cortex.

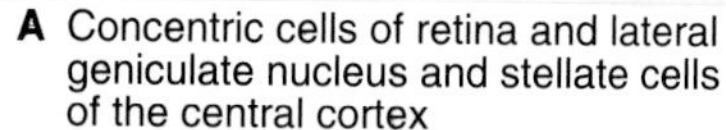

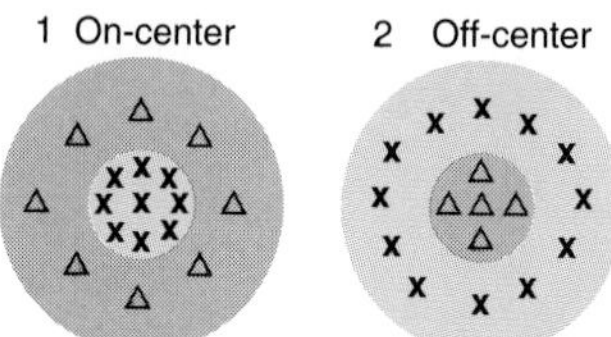

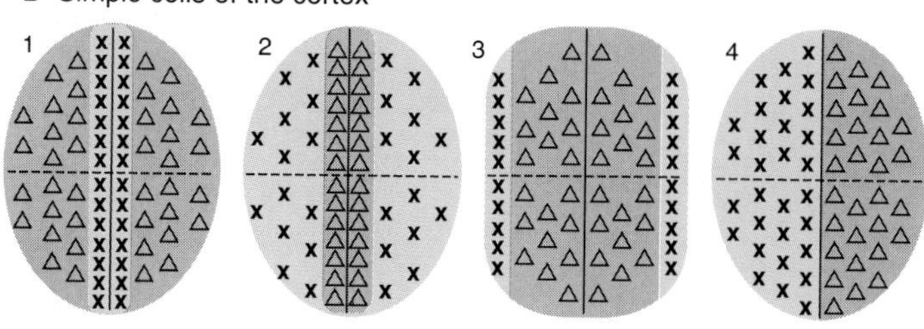

Figure 23–9 The receptive fields of simple cells in the primary visual cortex differ significantly from those of neurons in the retina and lateral geniculate nucleus. (Adapted from Hubel and Wiesel, 1962.)

A. Neurons in the retina and lateral geniculate nucleus have circular receptive fields with either on-center or off-center zones of excitation. ×, excitatory; Δ, inhibitory.

B. Neurons in the primary visual cortex have rectilinear receptive fields with discrete excitatory and inhibitory zones. Two types of cells, simple and complex, respond differently to linear images. A simple cell responds to a bar of light with an axis of orientation that corresponds to the axis of the cell's receptive field. Several different types of receptive fields of simple cells are illustrated here. For simplicity, only those with a vertical axis of orientation (12–6 o'clock) are shown. In fact, all axes of orientation—vertical, horizontal, and various oblique axes—are represented in the receptive fields of neurons in the primary visual cortex. Some cells have a single central excitatory area flanked by inhibitory regions (**1**), others have a reverse arrangement (**2**, **3**), and still others have a single excitatory region in apposition to an inhibitory one (**4**).

was first addressed in the early 1960s by David Hubel and Torsten Wiesel (Figure 23–8) in the two pathways concerned with form and motion. The output of the lateral geniculate nucleus ends on spiny stellate cells in sublayer 4Cβ of the primary visual cortex. These stellate cells have circular receptive fields similar to those of the cells in the lateral geniculate nucleus. However, the circuitry of the primary visual cortex (see Figure 23–7C) transforms the input, so that most cells above and below layer 4C respond selectively to stimuli that are substantially more complex than those that excite the stellate cells in sublayer 4Cβ.

Surprisingly, Hubel and Wiesel found that small spots of light, which are so effective in stimulating cells in the retina and lateral geniculate nucleus and the spiny stellate cells of the cortex, are substantially less effective stimuli for almost all other cells of the cortex. Cells in all regions except the blobs in the superficial layers respond best to stimuli that have *linear* properties, stimuli such as lines and bars.

Based on their responses to linear stimuli, Hubel and Wiesel classified these cortical cells into two major groups, simple and complex. The simple cells are thought to receive their input from the spiny stellate cells and to converge on the complex cells.

Simple cells are pyramidal cells located near the input layer 4Cβ. Their receptive fields are larger than those of the spiny stellate cells that receive the input from the lateral geniculate nucleus. Moreover, they are elongated and rectilinear. For example, the field of a simple cell may have a rectangular *on* (excitatory) zone with its long axis running from 12 o'clock to 6 o'clock, flanked on each side by *off* (inhibitory) zones, or it may have a similarly oriented central rectangular *off* zone flanked by excitatory regions (Figure 23–9B). Thus, to be optimally effective the stimulus must match the orientation of the cell's receptive field exactly.

In addition to having a specific orientation, the linear stimulus must also occupy a specific position within the receptive field to be optimally effective. That is, the stimulus must lie only within the excitatory region of the receptive field and not extend into the inhibitory region. If the stimulus is not well positioned in the receptive field and does not have the correct axis of orientation, it will be less effective because it both misses portions of the central excitatory zone and activates the flanking inhibitory zones (Figure 23–10).

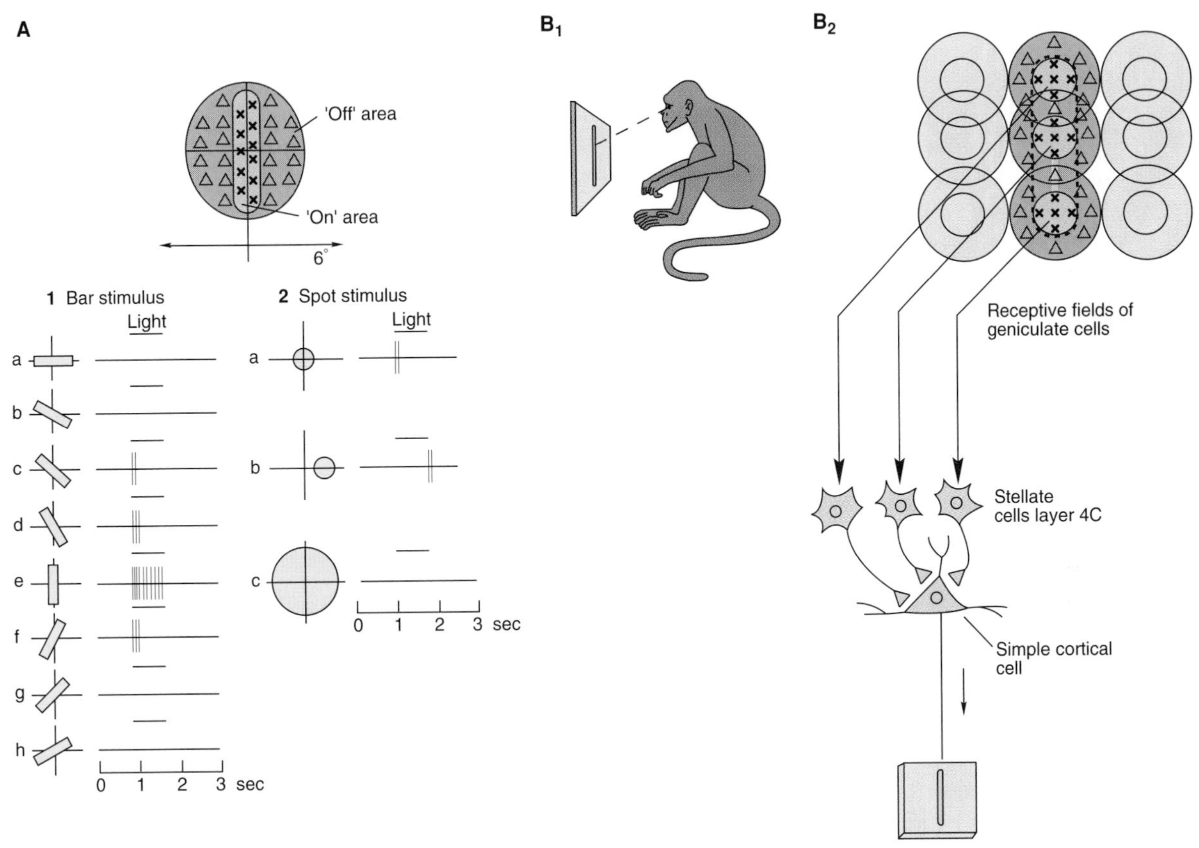

Figure 23–10 A simple cell in the primary visual cortex responds maximally to a bar of light that is parallel to the axis of the cell's receptive field.

A. In this example the cell's receptive field has a narrow rectangular excitatory area (×) in the center flanked by symmetrical inhibitory areas (Δ). The patterns of action potentials fired by the cell in response to two types of stimuli are shown. The bar above each record indicates the period of illumination. **1.** The cell's response to a bar of light is strongest when the orientation of the bar is parallel to the axis of the excitatory zone in the cell's receptive field (record **e**). Other orientations of the bar of light (oblique or horizontal) are less effective. **2.** Spots of light consistently elicit weak responses or no response. A small spot in the excitatory center of the field elicits only a weak excitatory response (**a**). A small spot in the inhibitory area elicits a weak inhibitory response (**b**). Diffuse light produces no response (**c**). (Adapted from Hubel and Wiesel, 1959.)

B. The exact process by which the circular receptive fields of geniculate cells are mapped to the rectilinear fields of simple cells in the visual cortex is not known. **1.** In the arrangement used by Hubel and Wiesel to study simple cells, a subject faces a target screen on which bars of light are projected. **2.** The rectilinear receptive field of a simple cell may be generated by convergent excitatory connections from three (or more) cells in the lateral geniculate nucleus. According to this idea the converging cells have similar center–surround organizations and are excited together when light falls along a straight line in the retina and strikes the receptive fields of all the cells. The convergence of these cells on a single cell would thus create an elongated excitatory region in the receptive field of the simple cell, indicated by the **dashed outline**. (Adapted from Hubel and Wiesel, 1962.)

Because of these features, each simple cell is tuned to detect stimuli with orientations in a range of about 10°. As a stimulus is rotated by 180° through all possible axes of orientation, it will sequentially activate about 20 different populations of cells, each with a slightly different axis of orientation: vertical, oblique, or horizontal.

Different cortical cells that receive input from the same point on the retina have similar receptive fields but different axes of orientation. In this way, all axes of orientation are represented for all points on the retina.

How are the rectilinear receptive fields of simple cells in the primary visual cortex generated? It is thought that they are created by the topographic arrangement of the input neurons during development, when a set of stellate cells with similar center–surround responses are aligned in sublayer 4Cβ in such a way that they represent slightly different retinal positions. Together, these inputs represent a minute rectangular window on the visual field with a particular axis of orientation (Figure 23–10B).

Complex cells are also pyramidal cells, but at a greater distance from sublayer 4Cβ than are simple cells. Most of them lie in layers 2, 3, 5, and 6. They, too, have rectilinear receptive fields with specific axes of orientation. However, the fields are larger than those of simple cells and have no clearly defined on or off zones, so the precise position of the stimulus within the field is less crucial. Movement in a particular direction across the receptive field is an effective stimulus for certain complex cells (Figure 23–11A). Although some complex cells receive direct connections from the input cells—the stellate cells of layer 4Cβ—it is thought that others receive significant input from those simple cells whose receptive fields have the same axis of orientation but cover slightly different positions in the retina (Figure 23–11B).

What is the function of simple and complex cells? As we have seen, these cells do not respond to small spots of light, but respond best to linear stimuli with particular orientations. They may therefore be important for analyzing the edges of a visual image—its form.

The combined actions of simple and complex cells may explain why we can maintain an accurate perception of an edge or the outlines of a form even when our head or our eyes are moving. Consider the outlines of a dark square on a light background in front of you. A vertical edge (or line) of the square excites a population of simple cells and a population of complex ones, each with the same vertical axis of orientation. If you now move your eye, or the square is moved against the background, a new population of simple cells will be excited, since these cells are sensitive to the exact position of the line in the receptive field. If the movement is small, however, the same population of complex cells will be excited in both positions, because complex cells have large receptive fields without clearly delineated excitatory regions and are responsive to movement within the receptive field. This response of complex cells to orientation over a range of retinal positions may be an elementary neuronal mechanism for *positional invariance,* the ability to recognize the same feature anywhere in the visual field.

Simple and complex cells in the primary visual cortex are thus sensitive to the *outline* of an object, to its contours, boundaries, and contrasts, but not to the interior or background of objects. These monotonous interior or background surfaces contain no visual information! David Hubel describes these unexpected features of perception in the following way:

> Many people, including myself, still have trouble accepting the idea that the interior of a form . . . does not itself excite cells in our brain . . . , that our awareness of the interior as black or white . . . depends only on cells' sensitivity to the borders. The intellectual argument is that perception of an evenly lit interior depends on the activation of cells having fields at the borders and on the absence of activation of cells whose fields are within the borders, since such activation would indicate that the interior is not evenly lit. So our perception of the interior as black, white, gray or green has nothing to do with cells whose fields are in the interior—hard as that may be to swallow. . . . What happens at the borders is the only information you need to know: the interior is boring. (Hubel, 1988, p. 87)

It is the information carried by edges that allows us to readily recognize objects in a picture even when the objects are sketched only in rough outline (see Figure 21–5).

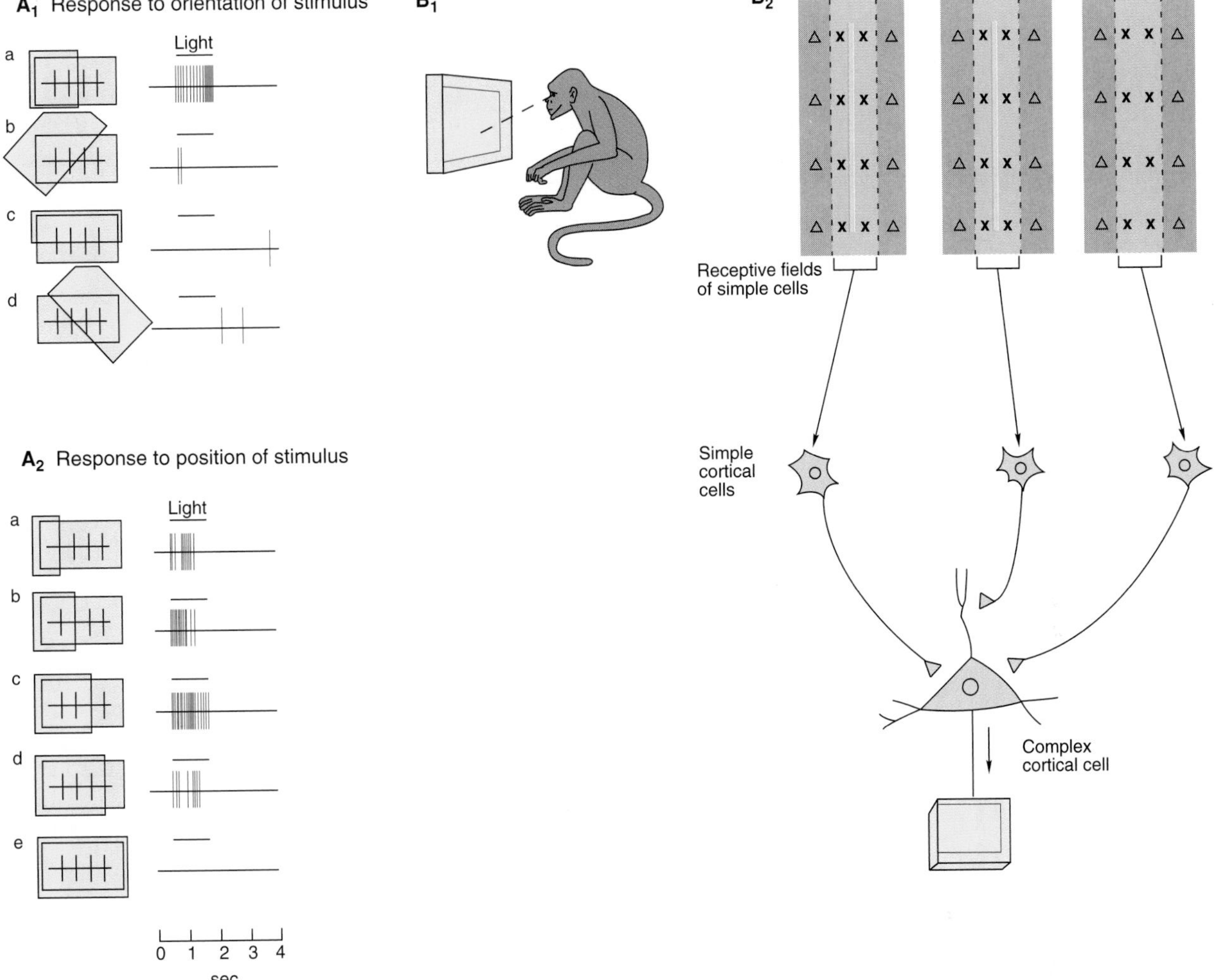

Figure 23–11 The receptive fields of complex cells in the primary visual cortex do not have clearly defined excitatory or inhibitory zones. The receptive field of an individual complex cell can be characterized by testing the cell's responses to variations in the orientation or position of a bar of light. Complex cells respond to specific orientations of the image in the visual field, but the exact position of the stimulus within the receptive field is not critical and the movement often enhances the response. In the example shown here, the cell responds best to a light stimulus with a vertical edge. The light is turned on for the period indicated by the bar above each record of the pattern of action potentials fired by the cell in response to the light. (Adapted from Hubel and Wiesel, 1962.)

A1. The rectangular receptive field is first mapped onto the tangent screen. An edge of light is then projected onto the screen so as to cross the receptive field of the cell in various orientations (see **B**). The cell responds optimally to a certain orientation of the stimulus. A vertical edge with light coming from the left of the receptive field produces a strong excitatory response (**a**). Orientations other than vertical are less effective in the cell (**b–d**) **2.** The position of the excitatory stimulus within the receptive field does not affect the response of the cell. The vertical edge evokes a strong excitatory response wherever it is placed in the field (**a–d**). Illumination of the entire field, with no edges, does not excite the cell (**e**).

B. According to Hubel and Wiesel, the properties of the receptive fields of complex cells may be explained by the pattern of inputs. They suggest that each complex cell receives convergent excitatory input from several simple cortical cells, each of which has a receptive field with the same organization: a central rectilinear excitation zone (×) and two flanking inhibitory regions (Δ). In this way the receptive field of the complex cell is built up from the individual fields of the presynaptic cells.

Some Simple and Complex Cells Are End-Stopped

The response of most simple or complex cells is enhanced when the bar of light within the excitatory zone of the receptive field is lengthened, although extending the bar beyond the excitatory zone has no effect (Figure 23–12A). However, some simple and complex cells are *end-stopped*. The discharge frequency of these cells increases as the bar of light in their receptive field lengthens, but only up to a point. Exceeding a certain limit in either direction attenuates and ultimately abolishes the cell's response (Figure 23–12A).

A

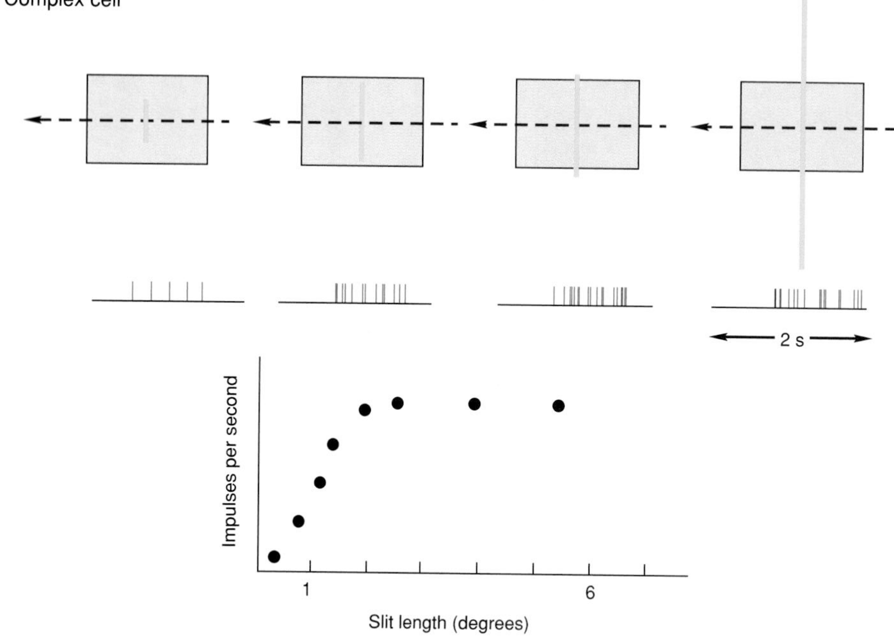

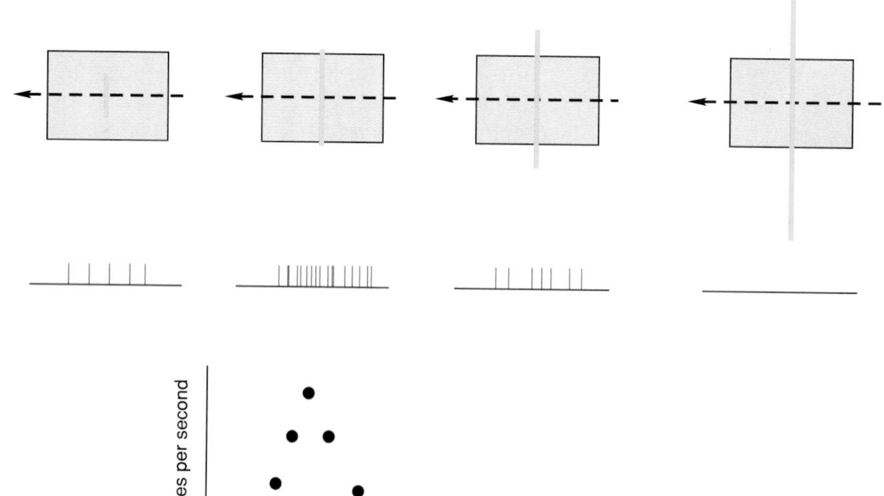

This limitation can be in one direction (*single end-stopped* cells) or in both directions (*double end-stopped* cells). Thus, the excitatory component of the receptive field of these cells has a central activating region and an inhibitory region at one or both ends. An end-stopped cell can be thought of as receiving convergent input from three cells with similar properties but slightly staggered receptive fields, the central one being excitatory and the two flanking regions being inhibitory (Figure 23–12B).

What is the optimal stimulus for end-stopped cells? For a single end-stopped cell the optimal

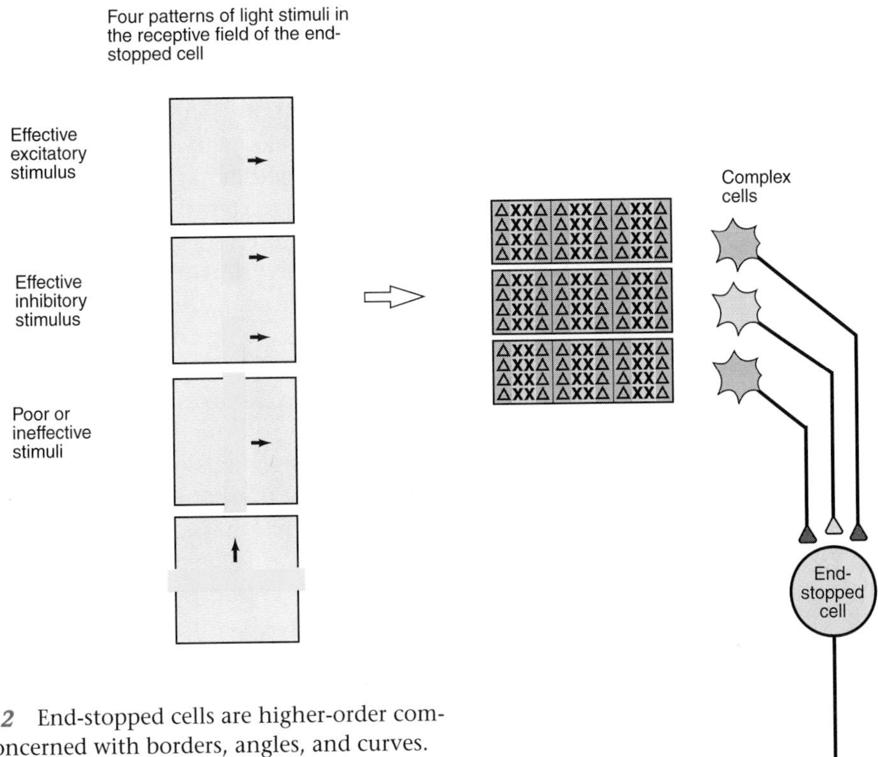

Figure 23–12 End-stopped cells are higher-order complex cells concerned with borders, angles, and curves. (Adapted from Hubel and Wiesel, 1965.)

A. Cells in the visual cortex respond selectively to a specific range of lengths of a vertical slit of light. **1.** The response of this complex cell increases as the length of the slit is increased to about 2°, after which there is no change. **2.** The response of this end-stopped cell improves as the slit increases to 2° but then declines, so that a line 6° or larger gives no response.

B. Hubel and Wiesel's scheme for generating the receptive field of double end–stopped cells. Three complex cells with similar receptive field properties, lined up one below the other, converge on cell which therefore becomes stopped. Two of the complex cells are inhibitory; the central one is excitatory. **Left:** An end-stopped cell responds differently to different patterns of light stimuli in its receptive field. The arrows indicate direction of movement of the stimulus.

stimulus is one that extends along its axis of orientation and then stops abruptly. For a double end-stopped cell, a stimulus consisting of a short line or a gradually curved border is particularly effective (Figure 23–12).

What, then, is the function of end-stopping? End-stopped cells measure the length of a line and therefore can signal the borders of an image: the corners, curvatures, or sudden breaks in a line. For these measurements both orientation and length are important.

Progressive Convergence in the Primary Visual Cortex May Lead to the Extraction of the Linear Features of an Image

The elaborate convergent connections of cells in the primary visual cortex are thought to be the initial steps in perception. In its simplest form this scheme suggests that each complex cell surveys the activity of a group of simple cells. Each simple cell surveys the activity of a group of stellate cells in layer 4Cβ, which in turn survey the activity of a group of geniculate cells, which themselves survey the activity of a group of retinal ganglion cells. The ganglion cells survey the activity of bipolar cells, which survey a group of receptors. *At each level every cell has a greater capacity for abstraction than do the cells at the lower levels.*

At the lowest level, the level of the retinal ganglion and the geniculate cells, neurons respond primarily to contrasts in light within small circular receptive fields. At the next level, the primary visual cortex, the simple and complex cells respond to relatively precise line segments and boundaries within rectilinear receptive fields that comprise the fields of several lower-order neurons. Thus, at each level of the afferent system the stimulus requirements necessary to activate a cell are different. In the retina and lateral geniculate, only the relative position of the stimulus within a receptive field is important. In simple cortical cells the location of the stimulus must be as precise as its axis of orientation. In complex cells, whose receptive fields are larger, the axis of orientation is still important but not the precise position of the stimulus.

Moreover, different populations of simple and complex cells receive input from one of two populations of cells in the lateral geniculate nucleus. The magnocellular pathway is concerned with movement and the coarse outlines of the stimulus, while the parvocellular pathway is concerned more with texture and pattern. Both pathways could contribute to the initial two-dimensional approximation of stimulus shape, which the theoretical biologist David Marr called the *primal sketch.*

The Primary Visual Cortex Is Organized into Columns and Layers

Cells with the Same Response Properties Are Organized in Columns

As we have seen, cells in the primary visual cortex are selectively responsive to stimuli with a specific retinal position and axis of orientation. Some cells, for example, respond only to horizontally oriented stimuli. Since cells with similar properties tend to be grouped together, and the cerebral cortex is organized into six layers, it is tempting to think that cells with different axes of orientation might be distributed in different layers of the cerebral cortex. In fact, however, the neurons of the primary visual cortex receiving input from the parvocellular–interblob and magnocellular pathways are organized into *columns* running from the pial surface to the white matter. Neurons within a column respond to a single axis of orientation. (As we saw in Chapter 18, the somatosensory cortex is also organized in functional columns.)

Each *orientation column* is about 30 to 100 μm wide, and each contains stellate cells in layer 4C with concentric receptive fields. Just above and below layer 4C are simple cells. The fields of all cells in one column are centered on almost the same retinal positions, although some cells have larger fields than others. Each orientation column also contains many complex cells. The properties of these complex cells can thus most easily be explained by postulating that each complex cell receives direct connections from the simple cells in the column (see Figure 23–11B). Thus, the local interconnection of cells in columns permits a new level of abstraction of visual information from the properties of circular receptive fields: the generation of receptive fields with axes of orientation. This is the first step in the deconstruction of the visual world into line segments of various orientations, an essential prerequisite for the visual analysis of form.

As we shall see later, in the extrastriate cortices cells selective for complex forms or to motion are also organized in functional columns.

How are columns of cells with *different* axes of orientation organized in relation to one another? Detailed electrophysiological mapping of adjacent columns in area 17 first revealed a systematic shift in axis of orientation (of about 10°) from one column to the next. This anatomy was next delineated using 2-deoxyglucose, a glucose analog that can be radiolabeled and injected into the bloodstream. Cells that are metabolically active take up the label. When lines of a given orientation are presented to a subject, an orderly array of active and inactive stripes of cells is revealed when the cortex is cut and a section is overlaid with X-ray film to make an autoradiograph (Figure 23–13).

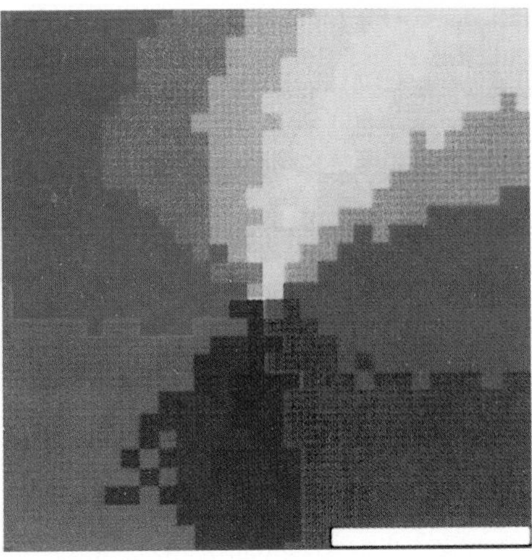

Figure 23–14 Optical imaging shows that the orientation columns are spatially arranged in a pinwheel at the center of which all columns of cells come together. Each orientation is represented only once in each pinwheel. Such pinwheel foci occur at regular distances from each other. Scale is 300 μm. (After Bonhoeffer and Grinvald, 1991.)

Figure 23–13 These orientation columns in the visual cortex of a monkey were made visible by an injection of 2-deoxyglucose. This glucose analog can be labeled with a radioactive isotope and injected into the bloodstream. Cells that are metabolically active take up 2-deoxyglucose avidly. The cells so labeled can then be detected by placing an X-ray film over a section of cortex. The monkey was binocularly stimulated with vertical lines. Bright areas indicate neurons responding to the stimulus. (From Hubel, Wiesel, and Stryker, 1978.)

More recently, orientation columns have been visualized on the surface of the living cortex. A highly sensitive camera can detect the pattern of active and inactive orientation columns during presentation of a bar of light with a specific axis of orientation using a voltage-sensitive dye or inherent differences in the light scattering of active and inactive cells. These studies have revealed that the orientation columns are spatially organized in a *radial* rather than linear fashion (Figure 23–14). The different axes of orientation radiate like the spoke of a wheel from an *orientation center,* and each orientation axis is represented only once in each pinwheel.

The orientation columns are regularly interrupted by the blob regions of cells (Figure 23–15). As we see in the next chapter, the cells in the blob regions are not sensitive to axis of orientation and are concerned primarily with color.

In addition to a system of columns devoted to axis of orientation and a system of blobs concerned with color, the striate cortex has a third component: a system of *ocular dominance columns*

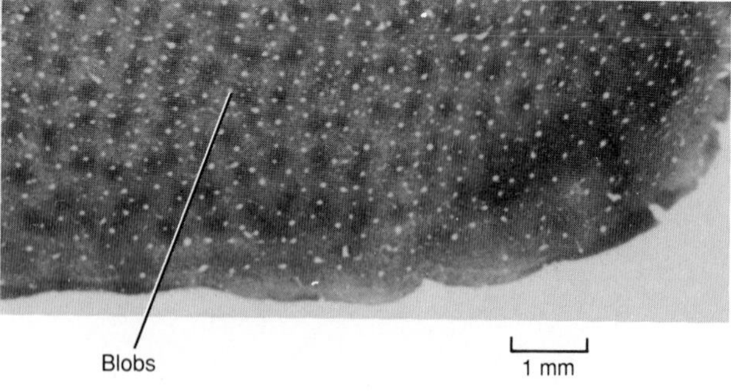

Figure 23–15 Blobs are visible as rows of dark patches in this autoradiograph of the superficial layers of the visual cortex (area 17) of a macaque monkey. The blobs have been enriched in cytochrome oxidase, an energy-producing enzyme. The heightened enzymatic activity in the blobs is thought to represent heightened neural activity. The cortex was sectioned tangentially. (Courtesy of D. Ts'o, C. Gilbert, and T. Wiesel.)

devoted to binocular vision, which is important for the perception of depth (stereopsis). Each ocular dominance column receives input from one or the other eye, and columns for the left and right eye alternate regularly (Figure 23–16A). In Chapter 25 we shall learn how the monocular fields of fibers from the lateral geniculate nucleus are transformed into binocular receptive fields of cells.

A complete sequence of right and left ocular dominance columns, a set of orientation columns (representing 360°), and a set of blobs occupy a region of about 1 mm^2 called a *hypercolumn*. Hypercolumns are repeated regularly and precisely over the surface of the primary visual cortex. A single hypercolumn represents the neural machinery necessary to analyze one point on the retina. In effect, each hypercolumn is an elementary computational module. The relationship between the known components of a hypercolumn—the orientation columns, the ocular dominance columns, and the

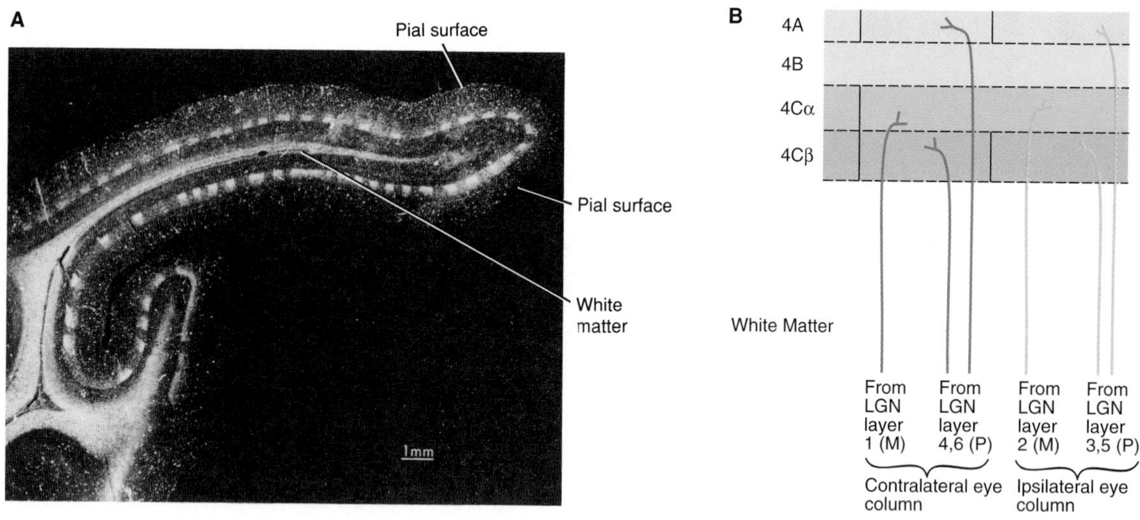

Figure 23–16 Ocular dominance columns in the primary visual cortex.

A. This autoradiograph of the primary visual cortex of an adult monkey shows the ocular dominance columns as alternating white and dark (labeled and unlabeled) patches in layer 4 of the cortex, below the pial surface (see Figure 23–7). One eye of the monkey was injected with a cell label, which over the course of 2 weeks was transported to the lateral geniculate nucleus and then across synapses to the geniculocortical relay cells, whose axons terminate in layer 4 of the visual cortex. Areas of layer 4 that receive input from the injected eye are heavily labeled and appear white; the alternating unlabeled patches receive input from the uninjected eye. In all, some 56 columns can be counted in layer 4C. The underlying white matter appears white because it contains the labeled axons of geniculate cells. (From Hubel and Wiesel, 1979.)

B. The inputs to two ocular dominance columns are shown in this schematic representation. (The sources in the lateral geniculate nucleus, LGN, are shown in Figure 23–5.)

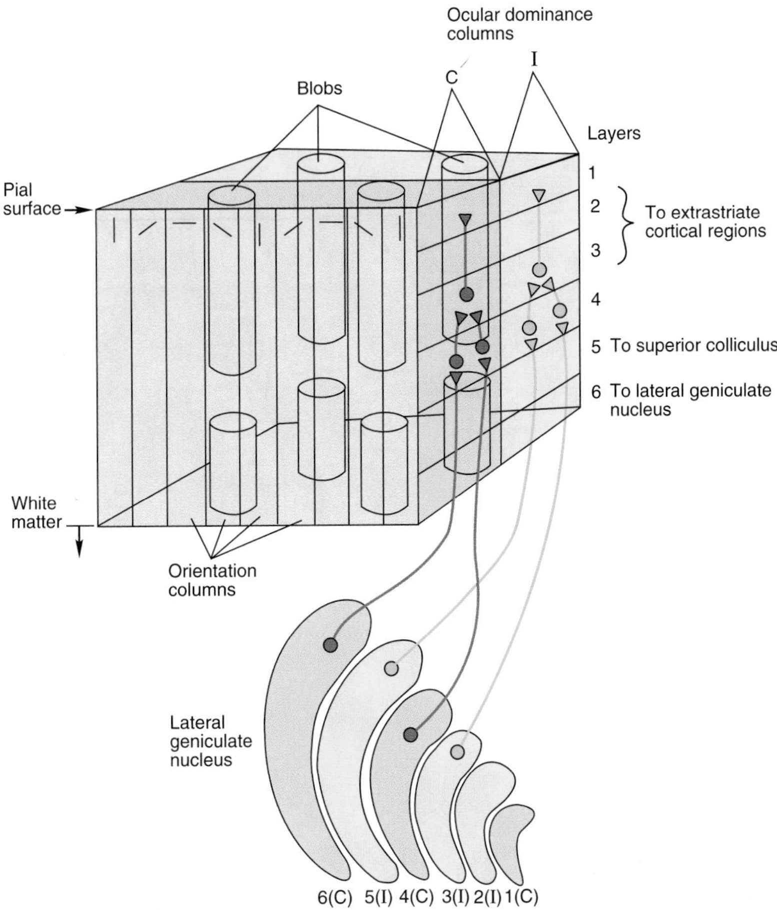

Figure 23–17 An array of different functional columns of cells in the visual cortex constitute the neural machinery necessary to analyze a discrete region of the visual field. Each array, or *hypercolumn*, contains a complete set of orientation columns, representing 360°, a set of left and right ocular dominance columns, and several blobs (regions of the cortex that process color information). The entire visual field is represented in the visual cortex by a regular pattern of hypercolumns. This modular organization is characteristic of the cerebral cortex.

blobs—is shown in Figure 23–17. This repeating organization illustrates nicely the characteristic modular organization of the cerebral cortex. The hypercolumns also have a layered organization corresponding to the layers of the primary visual cortex.

Horizontal Connections Interconnect Columns with One Another

The regularly spaced orientation columns in the primary visual cortex communicate with one another by means of the horizontal connections between cells within a single layer. As a result, cell pairs separated by about 1 mm in superficial layers of cortex—the distance that typically separates orientation columns with the same orientation in contiguous hypercolumns—can fire simultaneously in response to stimuli with a specific orientation and direction of movement (Figure 23–18B). Also, color-selective cells in one blob region are linked to cells with similar responses in other blobs. These connections integrate information over many millimeters of cortex.

Thus, area 17, the primary visual cortex, is organized functionally in two intersecting dimensions: functional columns are formed by local connections between the different cortical layers, while functional columns with the same response properties are connected across columns.

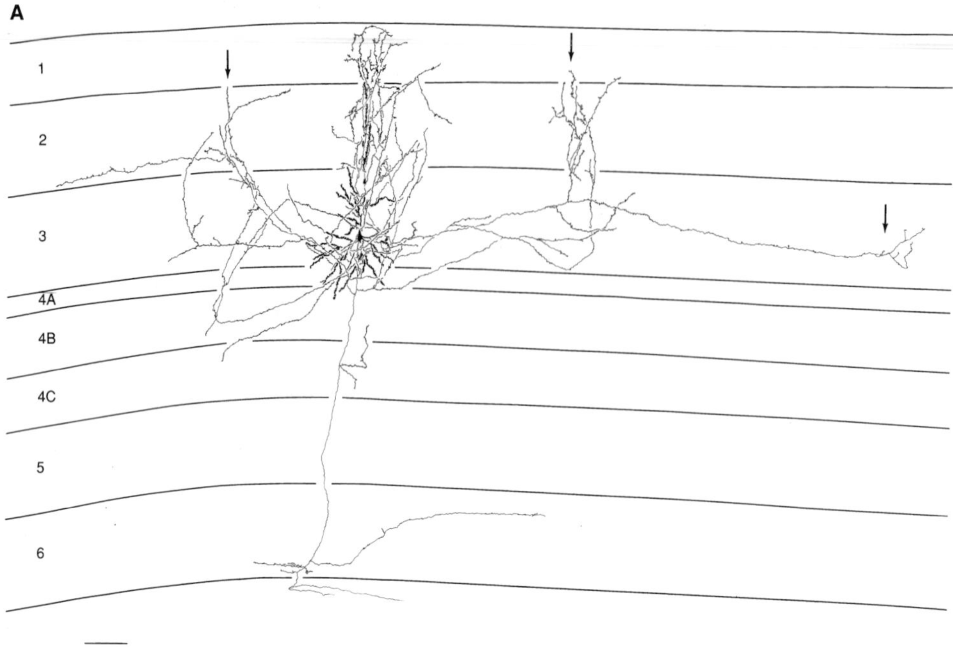

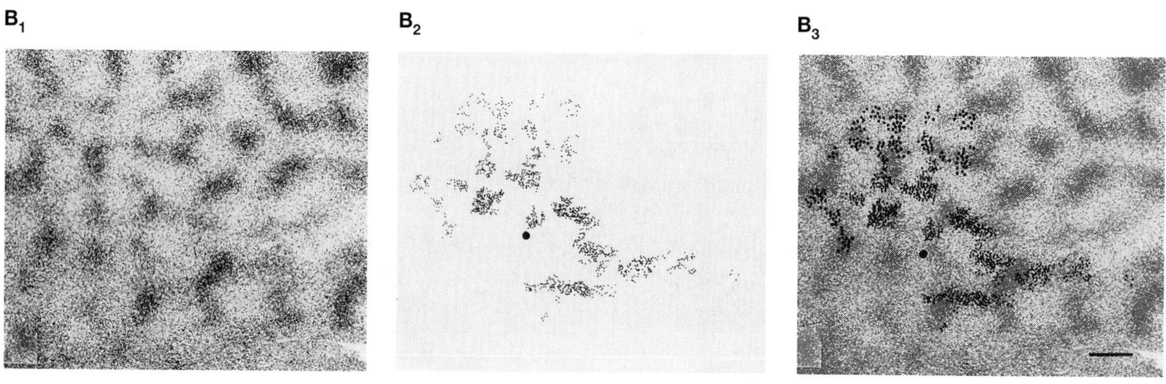

Figure 23–18 Columns of cells in the visual cortex with similar function are linked through horizontal connections.

A. A camera lucida reconstruction of a pyramidal cell injected with horseradish peroxidase in layers 2 and 3 in a monkey. Several axon collaterals branch off the descending axon near the dendritic tree and in three other clusters (**arrows**). The clustered collaterals project vertically into several layers at regular intervals, consistent with the sequence of functional columns of cells. (From McGuire et al., 1990.)

B. Cells in different columns of the visual cortex that respond to stimuli with the same axis of orientation are interconnected. **1.** A section of cortex labeled with 2-deoxyglucose shows a pattern of stripes representing activity of cells in the columns that respond to a stimulus with a particular orientation. **2.** Microbeads injected into the same recording site are taken up by the terminals of neurons and transported to the cell bodies. Reconstruction of the distribution of bead-labeled cells in the same region is visualized in the same section. **3.** Superimposition of **1** and **2**. The clusters of bead-labeled cells lie directly over the 2-deoxyglucose-labeled areas, showing that groups of cells in different columns with the same axis of orientation are connected. (From Gilbert and Wiesel, 1989.)

Beyond the Primary Visual Cortex: The Representation of Faces and Other Complex Forms Occurs in the Inferotemporal Cortex

When we perceive an object or a person, we see a fully formed image. Moreover, we consistently recognize an almost infinite variety of shapes. How is this accomplished? Clearly, the analysis of the visual image does not end in the primary visual cortex with the simple and complex cells. In fact, the primary visual cortex represents only the input stage in the cortex. In the cortex there are at least 32 other representations of the retina, most of which are in the extrastriate regions outside the primary visual cortex. Which of these is related to the perception of complex forms, objects, and people? How are these complex forms represented?

As we have seen in this chapter and Chapter 21, clinical studies of humans support the finding from experimental studies in monkeys that the processing of information for the perception of form is carried out by the parvocellular–interblob system. This system extends from the parvocellular layers of the lateral geniculate nucleus to the interblob regions of the striate cortex (V1). From there the information is conveyed to V2 and then to V4, which is retinotopically organized and includes neurons sensitive to both form and color. From V4 the pathway projects to the *inferior temporal cortex* (inferotemporal cortex). It is this region that is most concerned with the *what* of the visual image.

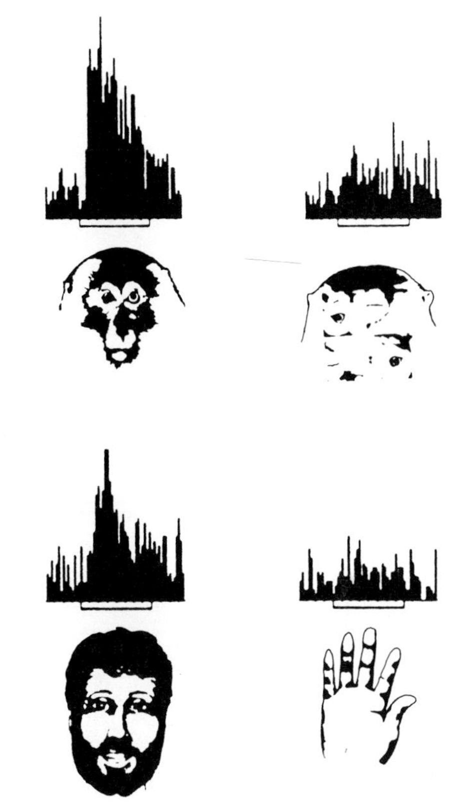

Figure 23–19 Response of a cell in the inferior temporal cortex of a monkey to different images. This cell responds preferentially to faces. (From Gross et al., 1981.)

The response properties of cells in the inferotemporal cortex are those we might expect for an area involved in a later stage of form recognition. For example, the receptive field of almost every cell includes the foveal region, where fine discriminations are made. Unlike the striate cortex and most extrastriate visual areas, the cells in the inferotemporal cortex are not organized retinotopically. The receptive fields for most of these cells are large (on average 25° × 25°), and occasionally may include the entire visual field (both visual hemifields).

About 10% of these cells are selective for specific images, such as hands or faces (Figure 23–19). For cells that respond to a hand, all orientations of the hand elicit similar responses, but the fingers are a particularly critical stimulus feature; these cells do not respond when the spaces between the fingers are filled in. Some neurons selective for faces respond best to a frontal view, others to a side view. Still other neurons respond preferentially to facial expressions. Lesions of this region of the cortex in humans lead to prosopagnosia, a specific deficit in face recognition (see chapter 21). These observations indicate that the inferior temporal cortex is involved in face recognition. In support of this idea, surgical removal of the inferior temporal cortex in monkeys impairs visual recognition of shapes and patterns without disturbing other basic functions of visual perception, such as acuity or recognition of color and movement.

Motion in the Visual Field Is Analyzed by a Special Neural System

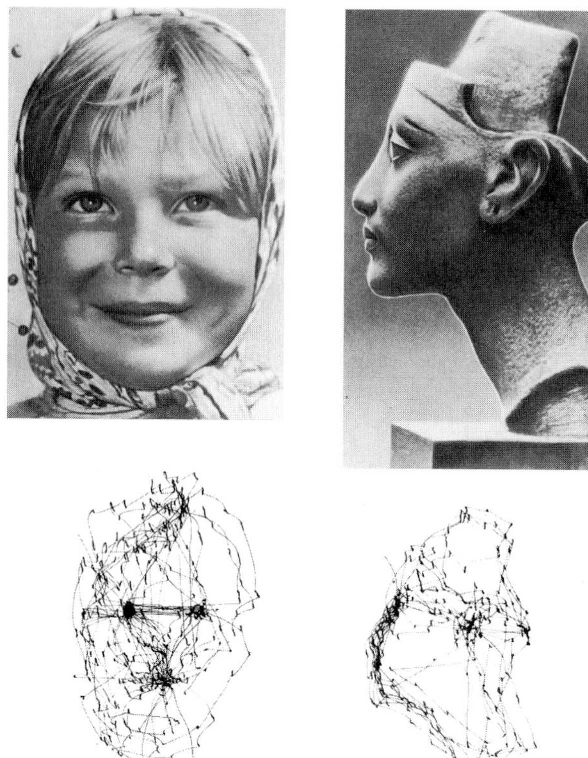

Figure 23–20 When we look at an object our eyes constantly scan areas of the object. The scanning consists of saccades, imperceptible rapid eye movements between points of interest in the image. During these scanning movements the entire image moves around the retina, yet we perceive it as stationary. Below each photograph is a record of the eye movements made during a 2-minute examination of the photograph. **Left**: photograph of "Girl from the Volga"; **right**: head of the Egyptian queen Nefertiti, about 1350 BC. (From Yarbus, 1967.)

Because we are constantly moving, we need accurate information about the motion of objects relative to our own motion. Even when we or the objects that interest us do not move, the images cast on the retina move because our eyes and head are never entirely still (Figure 23–20). The visual system has two ways of detecting motion: one based on the motion of the image, and the other based on the movement of the head and eyes (Figure 23–21). Here we shall consider motion of the image in the visual field.

Detection of motion is so important to adaptive behavior in most animals that only humans and other evolved primates can respond to objects that do *not* move. Many vertebrates (such as frogs and deer) cannot see objects unless they move. In humans this limitation persists at the peripheral part of the retina. We cannot detect motion at the

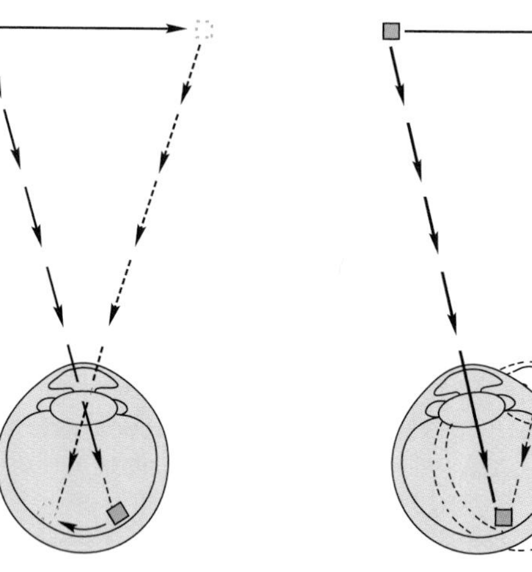

Figure 23–21 Movement in the visual field can be analyzed in two ways by the brain.

A. When the eyes are held still, the image of a moving object traverses the retina. Information about movement is relayed through sequential firing of receptors in the retina.

B. When the eyes follow an object, the image of the moving object falls in one place on the retina and the brain infers the object's movement by measuring the movement of the eyes or the head.

extreme periphery of the visual field; instead, a moving object in the periphery triggers an unconscious reflex that causes the eyes to rotate, thereby bringing the moving object into the central visual field.

Motion in the visual field could be detected by comparing the position of images perceived at different times. Most cells in the visual system respond only to a stimulus that projects on a particular position on the retina and can resolve events separated in time by a few tenths of milliseconds. In principle, therefore, the visual system should be able to extract information about motion by comparing the present location of an object on the retina with its previous location.

Evidence for the existence of a separate afferent pathway for motion detection has come from psychophysical observations of *apparent motion*, an illusion of motion such as occurs when lights separated in space are turned on and off at appropriate intervals (Figure 23–22), giving the illusion of motion even though the lights have not changed position. Movies are a dramatic example of the illusion of apparent motion. The impression of movement is actually caused by the rapid succession of still pictures on the screen. This illusion shows that the image of an object does not have to sweep continuously across the retina for movement to be perceived, and suggests that information about position and motion is carried by separate pathways.

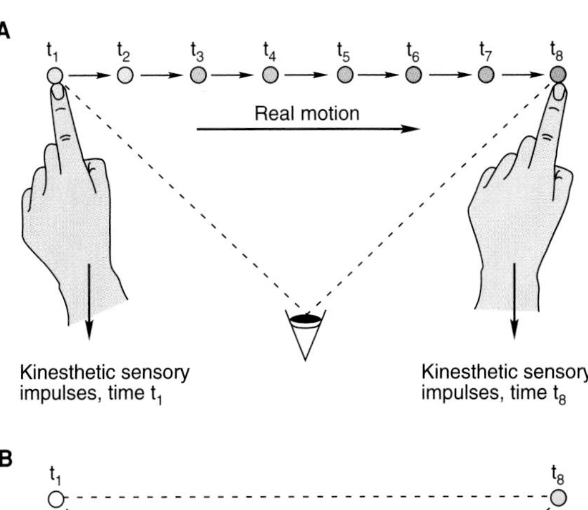

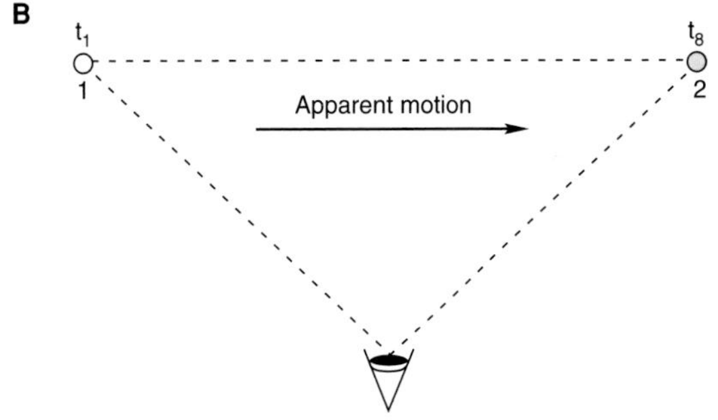

Figure 23–22 Apparent motion is as "real" as actual motion.

A. Actual motion of an object causes photoreceptors along the trajectory of the image in the retina to fire in sequence (see Figure 23–21A). The real motion of the object can be verified, as illustrated here, by touching the moving object.

B. Apparent motion may actually be more convincing than actual movement in the visual field, and is the basis for motion pictures. When two lights at positions **1** and **2** are turned on and off at suitable intervals, we perceive a single light moving between the two points. Since in fact a single light does not sweep across the retina, this perceptual illusion cannot be explained as the sequential firing of photoreceptors along a trajectory on the retina, and therefore constitutes evidence for the existence of a special visual system for the detection of motion. (From Hochberg, 1968.)

Motion Is Represented in the Middle Temporal and the Medial Superior Temporal Areas

Information on motion originates in the large, type M retinal ganglion cells (see Figure 21–14). These cells have no special sensitivity to motion per se, but respond best to circular targets in the visual field whose contrast varies over time. The signals generated by the type M cells are then conveyed through the magnocellular layers of the lateral geniculate nucleus to various layers of visual cortex V1. In V1 the signals are processed by simple and complex neurons that respond selectively to direction of motion; each cell responds to the direction of motion perpendicular to its axis of orientation. These type M signals are further elaborated in the midddle temporal area (MT or V5) and the medial superior temporal area (MST or V5a) and then relayed to the visual motor area of the parietal lobe, where the firing pattern of neurons reflects both the speed and direction of motion of objects in the visual field.

Information about motion is used for three different behavioral purposes: perception of moving objects, maintenance of pursuit eye movements, and guidance of bodily movement through the environment. We will here focus on the perception of moving objects.

When two or more objects are moving simultaneously in a limited region of the visual field, we need to distinguish between the motion of different parts of a single object and the motion of different objects. How does the brain make this distinction? This question has been addressed in human psychophysical experiments using moving patterns of parallel lines or gratings. Two different gratings are first moved independently (at right angles to their axis of orientation) and then moved together, one superimposed on the other but each moving at a right angle to its axis of orientation (Figure 23–23). A subject is asked whether he perceives the superimposed patterns as two independent gratings sliding past each other or as a single coherent *plaid* moving in one direction. When the gratings are appropriately superimposed, the observer in fact sees a single coherent plaid moving in one direction (Figure 21–23C).

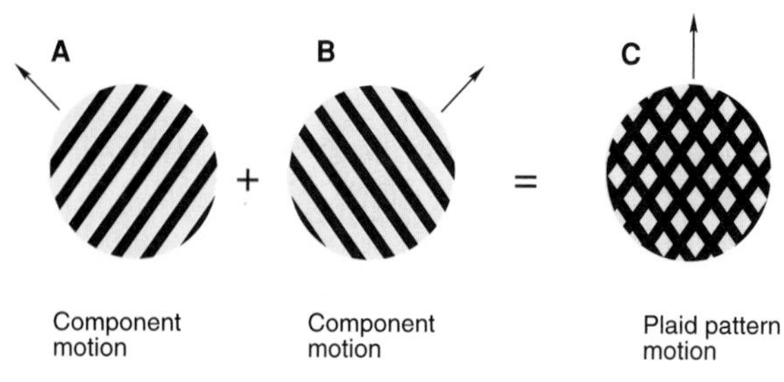

Figure 23–23 If grating **A** or **B** is moved in the direction indicated by the arrow, we perceive the pattern moving in that direction. However, if the two gratings are superimposed while in motion, we perceive a single plaid pattern moving *upward* rather than two gratings moving independently in different directions. The brain seems to compute the loci of possible motion for the two gratings separately and then determines the single point where the loci intersect. This point determines the direction and velocity of the coherent moving plaid. In the visual system the motions of the component gratings are separately encoded by neurons at the first stage of motion processing in the cortex. The signals from these neurons are integrated by higher-order neurons at the second motion processing stage so that the perceived direction and speed of the plaid differ from the detected motion of either component grating. (Based on Stoner, Albright, and Ramachandran, 1990.)

These psychophysical findings led to the idea that information about motion is analyzed in two stages. The first is concerned with motion in one direction, that is, information about one component of motion of a multisided object. Each neuron active in this stage would signal movement perpendicular to the axis of orientation of its receptive fields. The second stage is concerned with establishing the motion of complex patterns. Neurons active in stage two would integrate the different directions of motion detected by neurons active in stage one processing.

The two-stage hypothesis was tested in neurophysiological experiments on monkeys in which the activity of nerve cells in the cortex was recorded while the monkey was exposed to the sequence of stimuli illustrated in Figure 23–23. Neurons in V1 as well as the majority of neurons in MT respond well only to the motion of one component of the crosshatching or plaid, i.e., to one set of parellel lines moving in a direction perpendicular to the axis of orientation of the receptive field of the cell (Figure 23–23a,b). They do not respond to the direction of motion of the plaid (Figure 23–23c). These neurons are therefore called *component direction-selective* neurons. A second, small population of neurons in MT (about 20%) responds to the direction of motion of the plaid presumably because they integrate inputs from the component direction-sensitive cells. These neurons are therefore called *pattern direction-selective* neurons.

Thus, as suggested by the two-stage hypothesis, the *global motion* of an object is computed by pattern-selective neurons in MT based on the inputs of the component direction-selective neurons in V1 and MT. This global motion is consistent with our perception of motion.

Is the activity of motion-selective cells in MT causally related to the perception of motion? Smooth-pursuit eye movements, which allow humans and other primates to keep a moving target on the fovea, have been used in experiments to measure the ability of monkeys to perceive motion. MT has a retinotopic map that conveys information about speed and direction of motion in the contralateral visual hemifield. Small chemical lesions within different MT regions containing inputs from different regions of the retina prevent a monkey from estimating correctly the speed of a moving target in the region of the visual field monitored by the damaged area. The lesions do not affect eye pursuit of targets in other regions of the visual field, nor do they affect eye movements to stationary targets. Thus, MT is required for the analysis of motion in the visual field.

Perception of Motion Direction Can Be Influenced by Microstimulation of Cells Within the Middle Temporal Area

Like other cortical areas, the neurons in MT are organized functionally in columns. Within a single column, neurons are excited by motion in one direction but not by motion in other directions. The preferred direction of motion varies systematically from column to column, such that MT contains a complete representation of motion in all directions at each point in the visual field.

If cells in MT are directly involved in the analysis of motion, the firing patterns of neurons in these motion-selective columns presumably participate in forming the perceptual judgments about motion. How well does the firing pattern of these neurons actually correlate with behavior? To address this question, the activity of motion-selective neurons in MT was recorded in monkeys while the animals carried out a task designed to monitor the direction of motion in a random dot display. In this display some dots moved coherently while the remainder moved randomly. The strength of the motion signal could be varied by varying the proportion of dots that moved coherently. At zero correlation the motion of all dots was random, and at 100% correlation all dots moved in one direction. The firing of neurons correlated so well with the performance of the monkey that the directional information encoded by the neurons of a single column in MT was considered sufficient to account for the subject's judgment.

If this inference is correct, modifying the firing rates of the neurons that constitute a single column should alter the monkey's perception of motion. To test this idea, monkeys were trained to report the direction of motion in the random dot display while the experimenters stimulated clusters of directionally sensitive neurons in a single column with currents designed to increase the discharge rate of neurons. Stimulation altered the animal's judgment, biasing it toward the particular direction of motion encoded by the neurons that were stimulated. Thus, the firing of a relatively small population of motion-sensitive neurons in MT, perhaps about 200 cells, directly contributes to perception! The relationship between functioning of the magnocellular pathway, the cells in the columns of MT, and the perception of motion is impressive.

Visual Attention Can Now Be Studied on the Cellular Level

The neuronal mechanisms of attention and awareness are now emerging as one of the unresolved problems in perception. Cellular studies of visual attention have shown that attention involves either enhanced firing of cells that respond to the object of interest or attenuated firing of cells that respond to objects that are being ignored. These studies have demonstrated the importance of the posterior parietal cortex, a region first known from clinical studies to be involved in attention to visual form.

As we saw in Chapter 18, the response of cells in the posterior parietal cortex to a spot of light was tested in alert primates under two conditions: when the animal looked elsewhere and did not attend to the spot, and when the animal was forced to fix its gaze on the spot by making saccadic eye movements to the spot of light. The response of these cells was compared to that of cells in area V1 and in the superior colliculus (a region of the midbrain involved in the control of eye movements). When the animal used sacadic eye movements to attend to the spot, cells in all three brain regions responded more intensely. In V1 and the superior colliculus however, the enhancement did not result from selective attention per se, but from changes in the general level of arousal and from the neural mechanisms involved in the initiation of eye movement. Indeed, the enhancement occurred only when eye movements were required to focus attention. In contrast, in the posterior parietal cortex the enhancement was independent of eye movements or any other behavior of the animal in relation to the stimulus.

Based on these studies, Robert Wurtz and his colleagues have proposed that when a person attends visually to an object, cells in the posterior parietal cortex that respond to the object begin to discharge powerfully. As the person moves his eyes toward the object to examine it further, cells in the superior colliculus and V1 also discharge more briskly.

An Overall View

Information important for vision flows from the retina to the lateral geniculate nucleus. Cells in these structures have small circular receptive fields. From the lateral geniculate nucleus, visual information flows to the primary visual cortex, where it is transformed in at least three ways. (1) The visual field is deconstructed into short line segments by cells in the parvocellular–interblob and magnocellular pathways. These cells are sensitive to specific orientations of bars of light. This is an early step in the process thought to be necessary for discrimination of form and movement. (2) Information about color is processed through regions called blobs, which are not sensitive to orientation. (3) The inputs from the two eyes are combined through the ocular dominance columns; this is thought to be the initial step in the transformations of neuronal signals necessary for fusion and depth perception.

The cells in each of these visual pathways are selectively responsive to different features in the visual field. In light of the research on form and motion selectivity, it is attractive to think that the selectivities of particular classes of neurons are related to specific aspects of visual perception. Thus, orientation-selective neurons seem to provide information for the perception of shape and form. These cells seem to be important for perceiving *what*. Neurons responsive to components of motion seem to tell us *where*.

The success of such extensive parallel processing in the visual system depends on remarkably specific connections. Separate regions in the retina project to the lateral geniculate nucleus in the thalamus in such a way that a complete visual field for each eye is represented in the nucleus. Each geniculate axon terminates in the primary visual cortex, principally in layer 4C. Cells in this layer communicate with cells in other layers to form a rich circuitry for transforming visual information in the primary visual cortex.

In addition to being organized in six layers, cells in the visual cortex are also arranged functionally into columnar systems: orientation-specific columns, ocular dominance columns, and blob regions. Neurons within each columnar system that have similar response properties are linked by horizontal connections. Thus information flows both vertically and laterally within the visual cortex in a highly systematic manner. The columnar units seem to function as elementary computational modules—they receive varied inputs, transform them, and send their output to a number of different regions of the brain.

Although we have advanced steadily in our understanding of perception, we still have a long way to go before we understand perception in its entirety.

Selected Readings

Albright, T. D. 1993. Cortical processing of visual motion. In F. A. Miles and J. Wallman, eds., Visual Motion and Its Role in the Stabilization of Gaze. Amsterdam: Elsevier.

Bishop, P. O., and Pettigrew, J. D. 1986. Neural mechanisms of binocular vision. Vision Res. 26:1587–1600.

Gilbert, C. D., Hirsch, J. A., and Wiesel, T. N. 1990. Lateral interactions in the visual cortex. Cold Spring Harbor Symp. Quant. Biol. 55:663–677.

Hendry, S. H. C., and Yoshioka, T. 1994. A neurochemically distinct third channel in the Macaque dorsal lateral geniculate nucleus. Science. 264:575–577.

Hubel, D. H. 1988. Eye, Brain, and Vision. New York: Scientific American Library.

Hubel, D. H., and Wiesel, T. N. 1979. Brain mechanisms of vision. Sci. Am. 241(3):150–162.

Lam, D. M.-K., and Gilbert, C. D. (eds.) 1989. Neural Mechanisms of Visual Perception. Proceedings of the Retina Research Foundation Symposia, Vol. 2. The Woodlands, Tex.: Portfolio Publishing.

Marr, D. 1982. Vision: A Computational Investigation into the Human Representation and Processing of Visual Information. San Francisco: Freeman.

Meigan, W. H., and Maunsell, J. H. R. 1993. How parallel are the primary visual pathways? Annu. Rev. Neurosci. 16:369–402.

Moran, J., and Desimone, R. 1985. Selective attention gates visual processing in extrastriate cortex. Science. 229:782–784.

Newsome, W. T., Shadlen, W. N., Zohary, E., Britton, K. H., and Movshon, J. A. 1994. Visual motion: Linking neuronal activity to psychophysical performance. In: M. S., Gazzaniga ed., The Cognitive Neurosciences. Cambridge, Mass.: M. I. T. Press.

Pallis, C. A. 1955. Impaired identification of faces and places with agnosia for colours. J. Neurol. Neurosurg. Psychiat. 18:218–224.

Rock, I. 1984. Perception. New York: Sci Am. Library.

Salzman, C. D., Britten, K. H., and Newsome, W. T. 1990. Cortical microstimulation influences perceptual judgments of motion detection. Nature 346:174–177.

Stoner, G. R., Albright, T. D., and Ramachandran, V. S. 1990. Transparency and coherence in human motion perception. Nature 344:153–155.

Stryker, M. P. 1992. Elements of visual perception. Nature 360:301–302.

Wurtz, R. H., Goldberg, M. E., and Robinson, D. L. 1982. Brain mechanisms of visual attention. Sci. Am. 246(6):124–135.

Zeki, S. 1993. A Vision of the Brain. London: Blackwell.

24

Color

Three Separate Cone Systems Respond to Different Parts of the Visible Spectrum

Color Discrimination Requires at Least Two Types of Photoreceptors with Different Spectral Sensitivities

Color Opponency, Simultaneous Color Contrast, and Color Constancy Are Key Features of Color Vision

Color Is Coded in the Retina and the Lateral Geniculate Nucleus by Color-Opponent Cells

Color Information Is Processed in the Cortex by Double-Opponent Cells in the Blob Zones

Cortical Area V4 Contains a Neural Substrate for Color Constancy

Color Blindness Can Be Caused by Genetic Defects in Photoreceptors or by Retinal Disease

Color greatly enriches our visual experience. But in addition to its esthetic value, color is important for detecting objects and patterns that would otherwise not be seen. In natural scenes the gradients of light energy are often small. To distinguish an object from its background, it is often helpful also to exploit the differences in wavelength between the light reflected from the object and the light reflected from the background. In a black and white reproduction so many nuances of contrasting shapes evident in color are lost (Figure 24–1). Thus, color perception serves to enhance contrast. This faculty has evolved from simple brightness perception, which we considered in Chapters 22 and 23. The superiority of color perception over brightness perception is evident in the fact that we can discriminate only 500 gradations of brightness but we can discriminate more than 7 million gradations of color.

Color is a property of objects. However, the wavelength composition of the light reflected from the object is determined not only by its color, but also by the wavelength composition of the ambient light, which can vary considerably over time and from place to place. Color vision compensates for this variation so that the object appears to be the same color. A lemon, for example, seems yellow whether seen in sunlight (which is whitish), under the light of a tungsten filament bulb (which is reddish), or by fluorescent (bluish) light. This property of color vision is known as *color constancy*. Color constancy is not entirely foolproof, however, as anyone can testify who has bought paint or clothing in artificial light and later was startled to see it appear a different shade in daylight.

Figure 24–1 A black and white reproduction of Claude Monet's *Bend in the Epte River, near Giverny* loses not only the esthetic power of the original but also much of its information content. (Reprinted with permission of the Philadelphia Museum of Art: The William L. Elkins Collection. From an idea in Gleitman, 1981.)

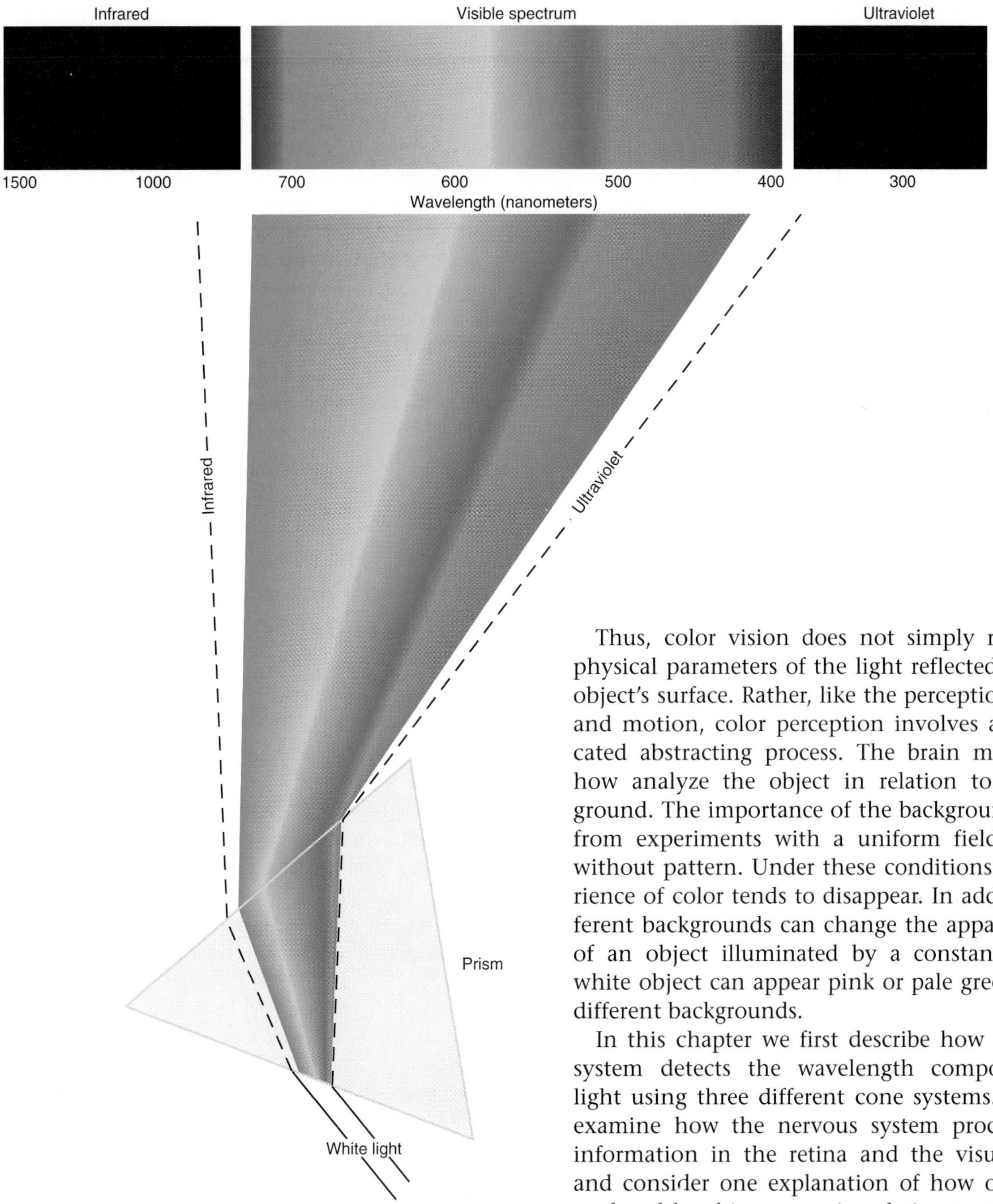

Figure 24–2 Visible light represents a small part of the enormous range of electromagnetic radiation, which varies in wavelength from 10^{-14} m (cosmic rays) to several meters (radio waves). Light containing all the wavelengths in the visible spectrum is called white light. When passed through a prism at a particular angle, white light yields the visible spectrum shown here. (Adapted from Gleitman, 1981.)

Thus, color vision does not simply record the physical parameters of the light reflected from the object's surface. Rather, like the perception of form and motion, color perception involves a sophisticated abstracting process. The brain must somehow analyze the object in relation to its background. The importance of the background is clear from experiments with a uniform field of color without pattern. Under these conditions the experience of color tends to disappear. In addition, different backgrounds can change the apparent color of an object illuminated by a constant light. A white object can appear pink or pale green against different backgrounds.

In this chapter we first describe how the visual system detects the wavelength composition of light using three different cone systems. We then examine how the nervous system processes this information in the retina and the visual cortex, and consider one explanation of how objects are analyzed by this process in relation to their background.

Three Separate Cone Systems Respond to Different Parts of the Visible Spectrum

The human eye is sensitive to wavelengths from 400 to 700 nm. Throughout this range the color of monochromatic light changes gradually from blue, through green, to red (Figure 24–2). People with

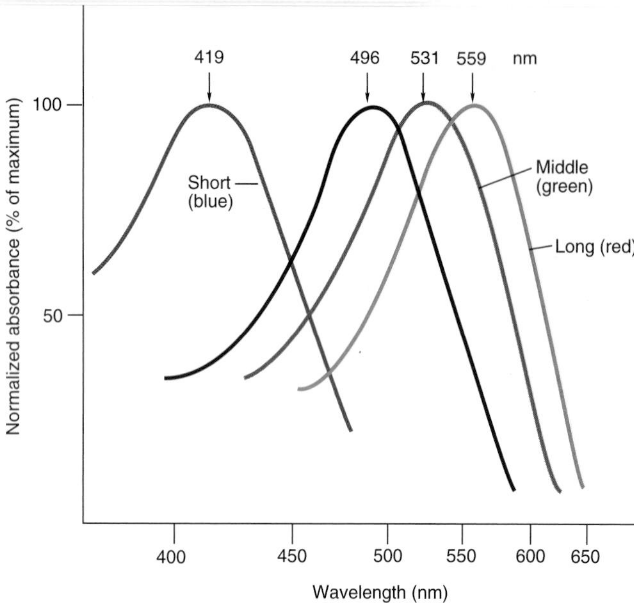

Figure 24–3 In humans three types of cone cells respond preferentially, but not exclusively, to short, middle, or long wavelengths as illustrated here by the absorbance spectra of the photopigments of each type of cell. The short-wavelength cone type contributes to the perception of blue, the middle-wavelength type to green, and the long-wavelength type to red. The **black curve** shows the absorbance spectrum of the rod photopigment rhodopsin. (Adapted from Dartnall, Bowmaker, and Mollon, 1983.)

normal color vision can readily match the color of any spectral composition of light by combining appropriate proportions of three primary colors—blue, green, and red. This property of color vision, called *trivariancy*, is the basis for color television, which uses three types of phosphors to reconstruct the full spectrum of perceived colors. Trivariancy results from three types of light-absorbing cone photoreceptors, each with a different visual pigment (see Chapter 22). These pigments have different but overlapping absorption spectra. As we shall see, the visual system extracts information on color by comparing the responses of the three classes of cones.

The idea that color vision is mediated by three classes of photoreceptors, each responsive to a primary color, was proposed at the beginning of the nineteenth century by Thomas Young. Trivariancy was confirmed in the 1960s through direct measurement of the absorption spectra of visual pigments in the cones of the human retina. A single cone contains only one of three pigments. One pigment is sensitive primarily to *short* wavelengths in the visible spectrum and makes a strong contribution to the perception of *blue*; this pigment is referred to as B. Another is selective for *middle* wavelengths and makes a strong contribution to the perception of *green*; this one is known as G. The third pigment, R, is responsive to *longer* wavelengths and makes a strong contribution to the perception of *red*. Recent measurements show that B pigments absorb most strongly at 419 nm, G pigments at 531 nm, and R pigments at 559 nm (Figure 24–3).

Color Discrimination Requires at Least Two Types of Photoreceptors with Different Spectral Sensitivities

The cone pigments, like rhodopsin in rods, are composed of a protein called opsin and the light-sensitive compound 11-*cis*-retinal. The three cone pigments each contain a different opsin (Box 24–1).

When an individual cone absorbs a photon, its electrical response is always the same, whatever the wavelength of the photon. Cones have this property because light absorption triggers the isomerization of the light-absorbing portion of the pigment, retinal, from the 11-*cis* to the all-*trans* form (see Chapter 22). Thus, individual cones do not transmit information about the wavelength of a light stimulus. This is not to say that cones do not respond preferentially to particular wavelengths; they do. But what varies with wavelength is the *probability* that a photon will be absorbed, not the electrical response itself. Thus, although individual cones respond preferentially to a particular color, the nervous system cannot determine from the response of, say, just the R cones whether the eye is being illuminated with red light or with much more intense blue light, or with a combination of lights of different wavelengths. It is for this reason that people with only a single type of cone are unable to experience color.

The visual experience of people with only one type of cone is similar to normal vision in dim light, which relies completely on rod cells. In dim light vision is achromatic, as we have all experi-

Box 24–1 The Cone Pigments

The three cone pigments each contain different opsins. All three opsins are transmembrane proteins with seven membrane-spanning regions. This basic structure indicates that the cone pigments belong to the family of genes that also encodes rhodopsin, bacteriorhodopsins, and the invertebrate photopigments (Figure 24–4), as well as a variety of transmitter receptors that also interact with G-proteins (see Chapter 14).

Genes for the three types of cone opsin are quite similar to each other and to the rhodopsin gene, suggesting that all four evolved from a common precursor by duplication and divergence. A comparison of amino acid sequences suggests that the blue cone pigment gene arose first from the rod gene. This short-wavelength gene then seems to have given rise to a single long-wavelength gene, an arrangement still found in contemporary New World monkeys, which have only two color pigments. The long-wavelength gene is thought to have duplicated and diverged to give rise to the red and green pigment genes only recently—about 30 million years ago, when Old World monkeys (which have all three pigments) separated from New World monkeys (which have only two). The red and green pigment gene products are indeed closely related, with 90% identity in their amino acid sequences (Figure 24–4).

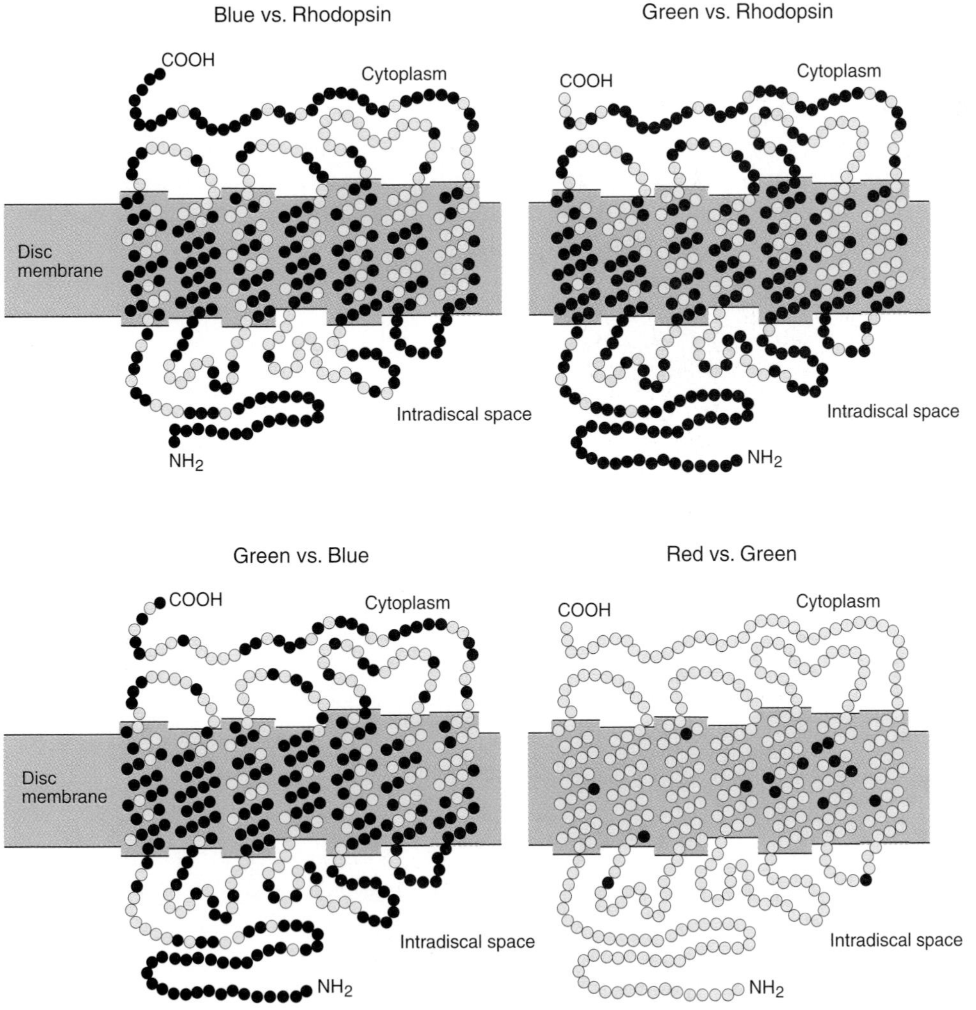

Figure 24-4 Pairwise comparisons of the amino acid sequences of the three photopigments of cones (red, green, blue) with the photopigment of rods (rhodopsin). Each **dark circle** indicates a difference of amino acids. (Adapted from Nathans et al., 1986a.)

enced at twilight (see Box 24–2), because there is only one type of rod with one visual pigment (rhodopsin). Nevertheless, we can still distinguish an object if its *brightness* is stronger than that of the background—we just do not normally see it as colored. When an object in dim light does appear colored (as when a full moon looks yellow), it is because its brightness is sufficient to excite cones.

Color vision therefore requires at least two sets of photoreceptors with different spectral sensitivities. A two-receptor, or *divariant*, system would convey two values of brightness for each object. By comparing the two brightnesses, the brain would be able to distinguish colors. For example, if an object reflected primarily light of a long wavelength, the response in the longer-wavelength cone system would be stronger than the response in the other system, and higher processing centers would interpret the object as being red or yellow. If the object reflected primarily shorter wavelengths, it would evoke a stronger response in the short-wavelength system and the object would be seen as blue. If the object reflected long and short wavelengths equally, it would be perceived as white, gray, or black, depending on the brightness of the background.

Many color combinations of object and background are nevertheless invisible to a divariant system. An object that reflects light at both ends of the spectrum, and which appears against a background that reflects light in the midspectrum, would be invisible—because both the object and the background would produce the same response in both types of photoreceptors. These ambiguities are greatly reduced by a three-receptor or trivariant system.

Despite its discriminatory potential, however, color vision is not useful for discriminating fine detail. It has two limitations. First, since color vision depends on a comparison of the outputs of different cones, it fails when color varies so rapidly across the surface of an object that neighboring cones are stimulated by light emanating from different colored parts of the object. Second, in the fovea, where visual acuity is greatest, color vision is *divariant* only. Short-wavelength (B) cones are absent from the fovea, presumably because optical images in this spectral region are blurred. This blurring is due to the refractive properties of the lens: when the lens focuses longer-wavelength light on the retina, the short-wavelength light is focused in front of the retina, a phenomenon known as *chromatic aberration*. For these reasons, color vision is not used for fine spatial discrimination.

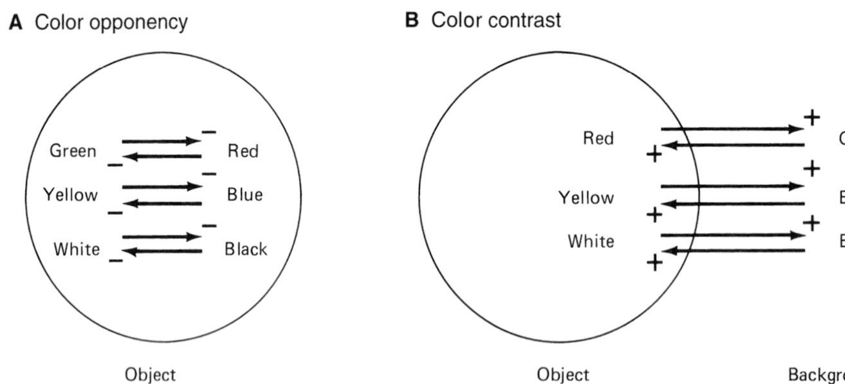

Figure 24–5 Color opponency and simultaneous color contrast are characteristic properties of human color vision.

A. Certain colors cannot be combined in vision when they emanate from the *same* point in space. Red and green, yellow and blue, and white and black cancel each other in these circumstances. We cannot experience a reddish green, for example.

B. Opponent colors enhance each other when they emanate from *different* but adjacent points in space, such as an object and its background. For example, a green object stands out more on a red than a blue background.

Color Opponency, Simultaneous Color Contrast, and Color Constancy Are Key Features of Color Vision

The theory of trivariant color vision attributes color perception to the activity of three primary cone classes. This theory explains a large variety of data on color perception. For example, the combination of green and red light is seen as yellow, and the combination of all three (what we perceive individually as blue, green, and red) is seen as white. However, trivariancy alone fails to explain at least three important aspects of color perception: color opponency, simultaneous color contrast, and color constancy.

Color opponency refers to the fact that certain colors are never perceived in combination. For example, we cannot perceive reddish green or bluish yellow colors, even though we can readily see reddish blue (magenta), reddish yellow (orange), greenish yellow (chartreuse), or bluish green (cyan). Red and green lights can be mixed so that all traces of the original redness or greenness are lost and a pure yellow is seen; yellow and blue can be mixed to produce white without any trace of the original blue or yellow.

This perceptual cancelation of colors led Ewald Hering in 1877 to propose the *opponent process* theory. According to this theory there are six primary color qualities, which are processed in three mutually antagonistic (or opponent) pairs: red–green, yellow–blue, and white–black. A modern formulation of Hering's model is that these three pairs are analyzed in three pairs of *color-opponent* neural channels. In one pair the sensations of red and green are opposed: one channel is excited by red and inhibited by green while the other is inhibited by red and excited by green. When red and green inputs are precisely balanced both channels are silent. A second pair of channels opposes the sensations of yellow and blue, and a third opposes white and black (Figure 24–5A). As we shall see later in this chapter, the outputs from the three types of cones are indeed combined in opponent fashion, starting in the retina and continuing on in the lateral geniculate nucleus, and then in the cortex.

Simultaneous color contrast occurs when opponent colors emanate from neighboring points in space, such as an object and its background (Figure 24–5B). For example, a green object stands out more against a red than a blue background. A more dramatic example of this phenomenon is that a gray object seen against a background of red has a green tinge; in a background of green it has a red tinge (Figure 24–6). In all these situations cone mechanisms appear to facilitate rather than cancel one another (Figure 24–5B). As we shall see later, so-called double-opponent cells in the visual cortex have properties that may in part explain simultaneous color contrast.

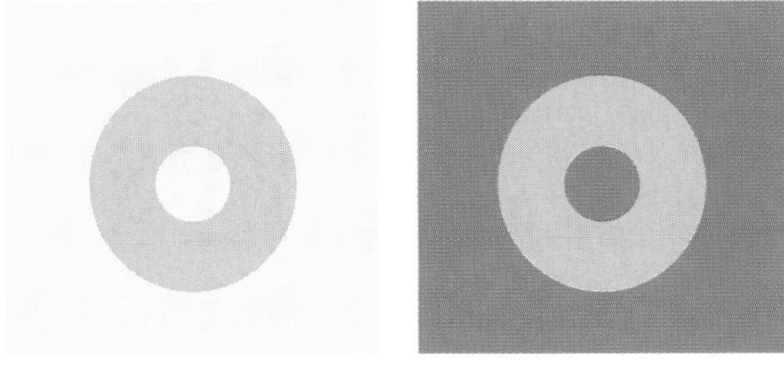

Figure 24–6 A gray object in a yellow background appears to have a purple tinge; in a background of purple the same gray object has a yellow tinge. (Adapted from Gleitman, 1981.)

Color constancy is perhaps the most important phenomenon that a theory of color vision needs to explain. As described at the beginning of this chapter, we perceive the color of an object as relatively constant even though there are enormous changes in the spectral composition of ambient light. We return later to what is known about the constancy of color perception.

Color Is Coded in the Retina and the Lateral Geniculate Nucleus by Color-Opponent Cells

In primates information about color and brightness is conveyed in separate pathways at the earliest stages of processing—in the retina and lateral geniculate nucleus. Retinal ganglion cells and the principal cells of the lateral geniculate nucleus fall into two major functional classes: the concentric broad-band cells and the color-opponent cells.

Concentric broad-band cells convey information about brightness and were described in detail in Chapters 22 and 23. They have receptive fields with a center–surround organization. A spot of white light on the center of the receptive field excites (or inhibits) the cell, whereas light applied to the surround elicits the opposite response. Diffuse light is a poor stimulus for these cells (see Figure 22–7). Although the broad-band cells receive inputs from the G and R cones (but not B cones), the inputs from G and R cones are summed in the center and also independently in the surround (Figure 24–7A). The broad-band cells therefore respond to a contrast in brightness within their receptive field and do not contribute to the perception of color.

In most *color-opponent cells*, on the other hand, the inputs of the R and G cones are antagonistic. In the retina and lateral geniculate nucleus the inputs of R and G cones are segregated in the receptive fields of most color-opponent cells: the center receives inputs from one type of cone (say, R) while the larger antagonistic surround receives inputs from the other type of cone (G). These cells are called *single-opponent cells*, to distinguish them from double-opponent cells in the visual cortex (see below). More specifically, they are called *concentric single-opponent* cells to denote the center–surround organization of their receptive fields (Figure 24–7A). Like broad-band cells, these cells do not receive inputs from B cones.

Concentric single-opponent cells respond strongly to large spots of monochromatic light of the appropriate wavelength. The R–center/G–surround cells (R cone inputs in the center, G cone inputs in the surround) respond best to red light (Figure 24–7B), whereas the G–center/R–surround cells respond best to green light. These cells can transmit information about achromatic brightness contrast as well. Because the G and R cones in the center and the surround absorb white light to similar extents, concentric single-opponent cells behave like broad-band cells in response to achromatic stimuli (Figure 24–7B).

Finally, information from B cones is transmitted by a distinct class of single-opponent cells, the *coextensive single-opponent* cells. The receptive fields of these cells do not have separate zones of excitation and inhibition; instead, inputs from B cones oppose the combined inputs of R and G cones throughout the receptive field (Figure 24–7A).

The different types of retinal ganglion cells involved in color processing can be further classified as M (magnocelluar) or P (parvocellular) cells, depending on whether the cell axon terminates in a magnocellular or parvocellular layer of the lateral geniculate nucleus (see chapters 22 and 23). The broad-band ganglion cells can be either M type or P type, while single-opponent cells are exclusively

Figure 24–7 Retinal ganglion cells and cells of the lateral geniculate nucleus can be classified according to the way in which inputs from the three types of cones are summed in their center and their surround.

A. In *concentric broad-band* cells, G and R cones act together in each zone of the receptive field but their actions in the two zones are opposite. *Concentric single-opponent cells* are the most common type of color-coding cell among retinal ganglion cells and lateral geniculate neurons. One type of cone (R or G) predominates in the center of the receptive field and has the opposite action of the type of cone that predominates in the surround. *Coextensive single-opponent* cells are less common; they have an undifferentiated receptive field in which the action of the B cones is opposed to the combined actions of the G and R cones.

B. Single-opponent cells transmit information about both color and achromatic brightness contrast because the center and surround are activated by different cone mechanisms. The recordings shown here illustrate the sensitivity of a single-opponent cell to different stimuli. The center of the receptive field of this cell is activated by the R cones and the surround by the G cones. The cell is excited by small spots of white light (**1**) but is unresponsive to diffuse white light (**4**) because the inputs from the center and the surround cancel each other (R and G cones absorb white light to similar extents). Small spots of red or green light in the center (**2**, **3**) elicit a slightly weaker response than does a centered white spot. Diffuse colored light reveals the color selectivity of these cells. Diffuse red light excites the cell (**5**), while diffuse green light does exactly the opposite (**6**). However, a centered red spot on a green background has little effect because the inputs from the center and the surround cancel each other (**7**). The duration of the light stimulus is indicated by the bar below each recording.

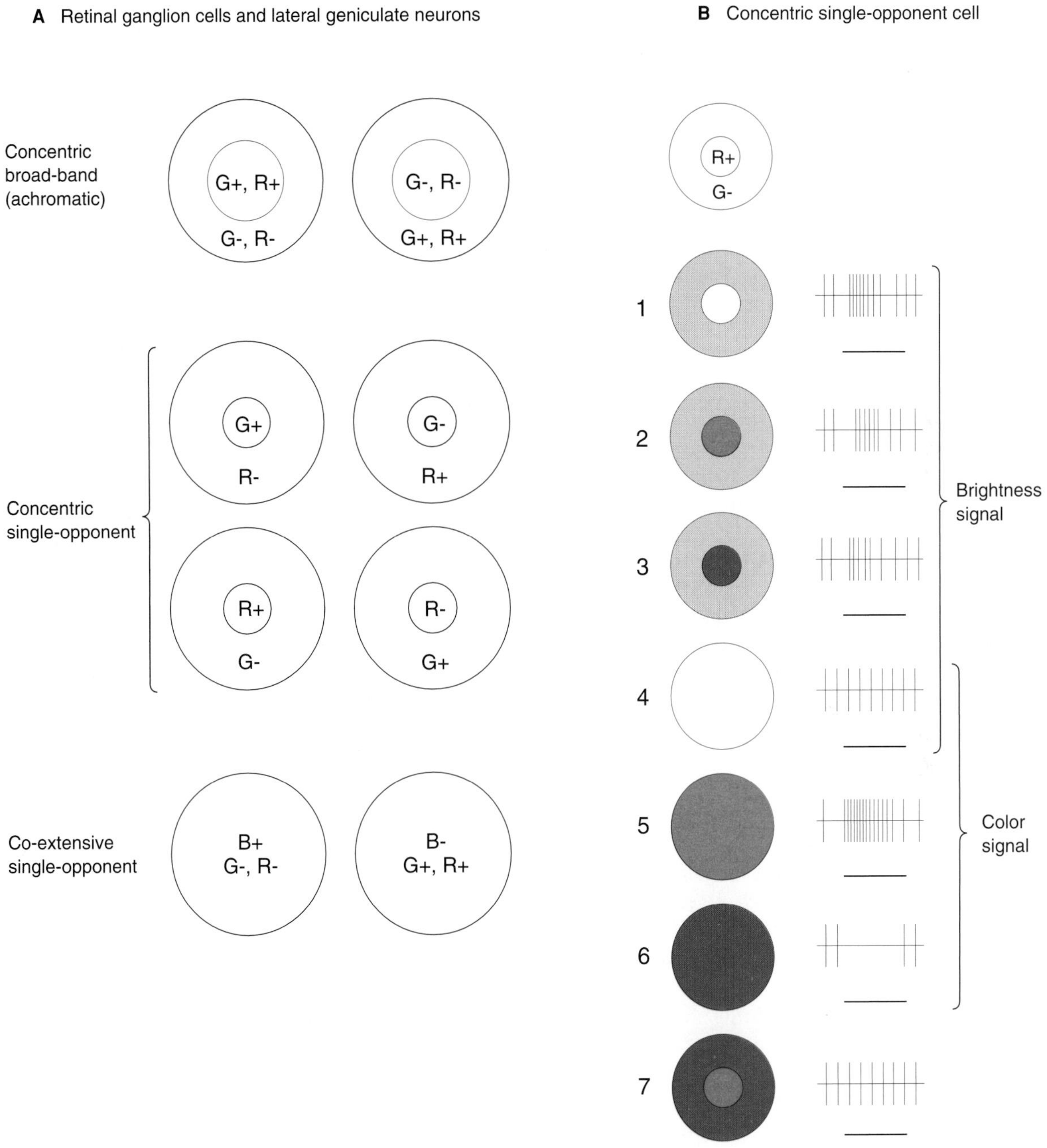

A Retinal ganglion cells and lateral geniculate neurons
Concentric broad-band (achromatic)
G+, R+
G-, R-
G-, R-
G+, R+
Concentric single-opponent
G+
R-
G-
R+
R+
G-
R-
G+
Co-extensive single-opponent
B+
G-, R-
B-
G+, R+
B Concentric single-opponent cell
R+
G-
1
2
3
4
5
6
7
Brightness signal
Color signal

P type. Thus, the parvocellular layers of the lateral geniculate nucleus relay all color information to the cortex in addition to some information about achromatic contrast. The magnocellular layers are involved only in achromatic vision.

Color Information Is Processed in the Cortex by Double-Opponent Cells in the Blob Zones

Inputs from the single-opponent cells of the lateral geniculate nucleus converge in area 17 (or V1) of the cortex on so-called *concentric double-opponent cells*. The receptive fields of these cells also have an antagonistic center-surround organization, but the inputs of different types of cones are not segregated in the receptive field as they are for single-opponent cells in the retina and lateral geniculate nucleus. Instead, all parts of the receptive field receive input from more than one type of cone but the inputs of each type vary depending on their location. For example, in some double-opponent cells the R cones excite in the center and inhibit in the surround. In such cells the G cones have the opposite action: they inhibit in the center and excite in the surround (Figure 24–8A). These cells respond best to a red spot in the center against a green background. They do not respond well to white light, whatever the size or the intensity of the stimulus, because the inputs of R and G cones (which absorb white light to similar extents) cancel each other at *all* points in the receptive field. These cells are therefore more selective for chromatic stimuli than are concentric single-opponent cells (Figure 24–8B).

There are three other classes of double-opponent cells: those that respond best to a green spot in a red background, and those that respond to a blue spot in a yellow background or vice versa (Figure 24–8). Although these cells respond to some other contrasts, they respond most strongly to these specific contrasts.

Double-opponent cells are heavily concentrated in the blob zones in area V1 of the cortex (see Figure 21–13). As we saw in Chapter 21, the *parvocellular–blob system* is a discrete pathway dedicated to processing color information. Cells in the blobs have circular receptive fields and therefore, are not selective for orientation of a stimulus. In contrast, most cells in the large interblob areas are selective for orientation.

From V1 the parvocellular–blob system projects to the thin stripes in V2 and from there to V4, an area with color-selective cells.

Cortical Area V4 Contains a Neural Substrate for Color Constancy

Since sets of different cone inputs oppose each other throughout the receptive fields of the double-opponent cells, the actions of these cells may provide a neural basis for the psychological phenomenon of color opponency. They may also underlie the phenomenon of simultaneous color contrast. For example, a double-opponent cell that is excited by red and inhibited by green in its center will give the same response to either a red light in its center or a green light in the surround. This may explain why a gray object seen against a background of green has a red tinge (see Figure 24–6).

A neural substrate for the psychophysical phenomenon of color constancy has been identified by Semir Zeki in area V4 (none has yet been identified in areas V1 or V2). That area contains many neurons whose firing correlates with the apparent color of the object that is being viewed (as perceived by a normal observer), not the particular wavelength composition of the light reflected from the object. Each of these cells responds best to a particular color (e.g., green). Changes in the wavelength composition of the illuminating light do not greatly alter the cell's firing rate. Thus the cells report on color, not wavelength.

The output from these cells may be used directly by higher centers in the brain to form our perceptions of the color of objects. Macaque monkeys with experimental lesions of V4 lose color constancy, even though they are still capable of discriminating wavelengths. A dramatic example of this kind of loss is the case of a painter who, following a mild concussion, lost color perception entirely even though his ability to discriminate wavelengths was totally unaffected. The perceptual and emotional consequences of this loss have been vividly documented (Box 24–2).

Somehow, then, cells in V4 extract information about the color of objects from the information supplied by cells in V1 and V2, which are concerned with wavelength discrimination. The way in which the visual system accomplishes this is not known. However, Edwin Land, the inventor of the

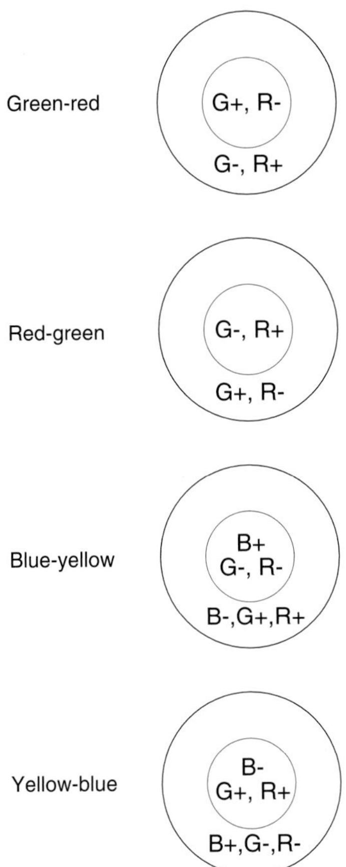

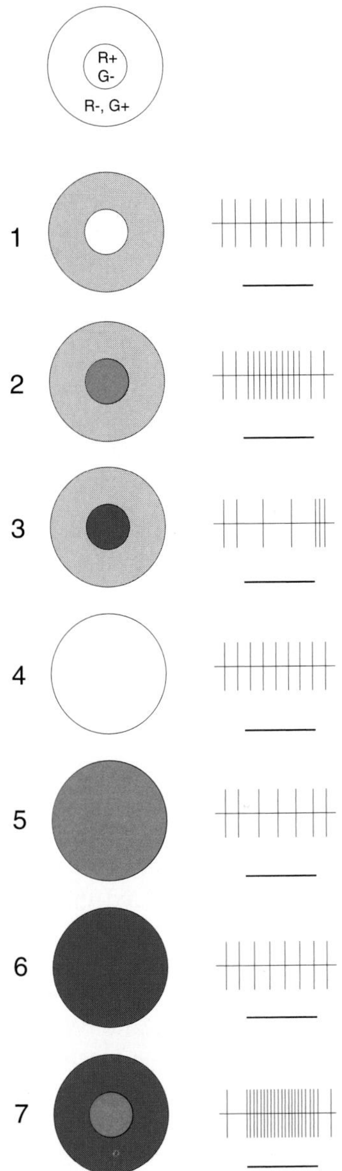

Figure 24–8 Concentric double-opponent cells in the cortex are highly sensitive to color contrast.

A. There are four types of double-opponent cells: two respond preferentially to red–green contrasts, the other two to yellow–blue contrasts. For example, a green–red contrast cell (top) is activated by G cones in its center and inhibited by G cones in its surround. R cones have the opposite effect; they activate the cell in its surround and inhibit it in its center. The yellow–blue and blue–yellow contrast cells oppose the input of B cones to the summed inputs of R and G cones.

B. Recordings from a red–green contrast cell show the cell's sensitivity to different stimuli. Large spots of any color (**4–6**) have little effect because of center–surround antagonism. Small spots of white light (**1**) have little effect because R and G cones in the center absorb white light to similar extents, so that their inputs cancel out. The cell responds strongly to a small spot of red light in its center (**2**) or an annulus of green light in its surround (not shown). The cell is inhibited by a small spot of green light in its center (**3**) or an annulus of red light in its surround (not shown). The cell responds most vigorously to simultaneous illumination of the center with red light and the surround with green light (**7**).

Polaroid camera, developed a quantitative method, the *retinex* (*retin*a and cort*ex*) method, for predicting the perceived colors of objects from the responses of the three types of cones. The predicted colors accurately match our perceptions; moreover, the method correctly predicts that perceived colors remain roughly constant as lighting conditions change and that the perceived color of an object can change if its background is changed in an appropriate way.

Land's method predicts colors in three steps. First, the brightness of each object in a scene is measured for each type of cone in turn, as if the scene were viewed with just that cone type. (This corresponds roughly to measuring the brightness of objects illuminated with monochromatic red, green, or blue light.) For each cone type, the brightness value obtained for each object in the scene is then normalized to the brightest object in the scene. In this way three numbers (one for each cone type) are assigned to each object. These numbers can then be used to predict the color of all objects in the scene using a formula devised by Land.

Does the cortex use this method to detect colors? Probably not in its simplest form. Indeed, the inputs from the different types of cones converge at a very early stage of processing in the retina, so that measurements by a single type of cone are not available to the cortex. Land has shown, however, that his method works equally well using the responses of the three classes of double-opponent cells as measures of the "red–green" brightness, "blue–yellow" brightness, and "white–black" brightness of objects. In this method each object is assigned a relative value of "red–green" brightness, "blue–yellow" brightness, and "black–white" brightness, and these three values are used to predict the perceived color of the object. It is certainly plausible that the cortex uses the outputs of the different classes of double-opponent cells to determine colors in exactly the way that Land has suggested, although this has not been demonstrated.

Color Blindness Can Be Caused by Genetic Defects in Photoreceptors or by Retinal Disease

Certain acquired forms of color blindness, like the case described in Box 24–2, result from cortical lesions (see Table 23–1). Color vision can also be impaired by disease in the retina. But the most common forms of color blindness, red or green blindness, are caused by genetic mutations of the X chromosome. In the general population about 1% of men are red blind and 2% are green blind. These forms of blindness result from a defect not in the neuronal circuitry mediating color vision, but in the red or green cone pigments themselves. Recent molecular studies by Jeremy Nathans and his colleagues have shown that the red and green pigment genes are indeed located on the X chromosome and are next to each other (Figure 24–9).

Because of the great similarity and close linkage of the red and green pigment genes, recombination can lead to the loss of one gene or the generation of a hybrid gene (the pattern of gene rearrangement that occurs with red–green blindness). Recombination can also lead to the duplication of a gene, so that many people with normal color vision actually have two or even three copies of the gene for the green pigment (Figure 24–9).

Mutations in the B (blue) pigment gene also occur, but very rarely. This form of color blindness is not sex-linked; the gene encoding the blue pigment has now been localized to the seventh chromosome. The rhodopsin gene is on the third chromosome.

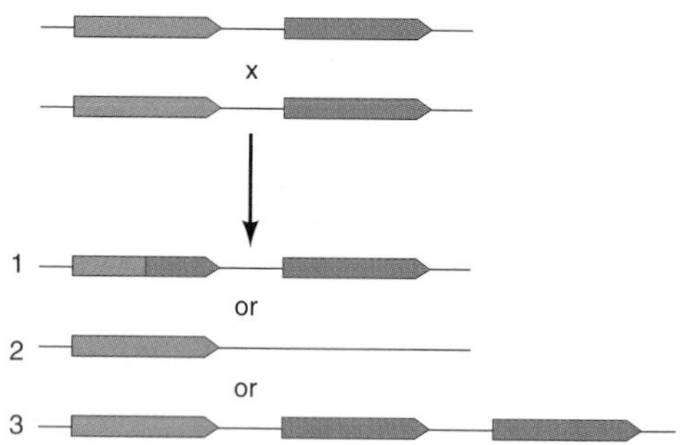

Figure 24–9 The arrangement of red and green pigment genes on the X chromosome may explain variations in these genes observed in both normal and color-blind individuals. Because they are next to each other on a normal X chromosome, recombination between these genes can lead to the generation of a hybrid gene (**1**) or the loss of a gene (**2**), the patterns observed in color-blind men. It can also lead to the duplication of a gene (**3**), a pattern observed in some people with normal color vision. (Adapted from Stryer, 1988.)

Box 24–2 A Life Without Color Vision: The Case of the Colorblind Painter

Mr. I. had seen normally all his life, had been born with a full complement of cones, or color receptors. . . . He had become colorblind, after sixty-five years of seeing colors normally. And he did not just confuse some colors or see them as gray, as is usually the case with the congenitally colorblind. He had become totally colorblind—as if "viewing a black and white television screen." All this came on suddenly when he had an accident. The suddenness of the event was incompatible with any of the slow deteriorations that can befall the retinal cone cells, and suggested, instead, a mishap at a higher level, in those parts of the brain specialized in perceiving color.

[Some time after the car accident] he decided to go to work again. It seemed to him as if he were driving in a fog, even though he knew it to be a bright and sunny morning. Everything seemed misty, bleached, grayish, indistinct. His bewilderment and fear now became a feeling of horror. . . .

Mr. I arrived at his studio with relief, expecting that the horrible mist would be gone, that everything would be clear again. But as soon as he entered, he found his entire studio, which was hung with brilliantly colored paintings . . . now utterly gray and void of color. His canvases, the abstract color paintings which he was known for, all were grayish or black and white, unintelligible. Now to horror there was added despair: even his art was without meaning, and he could no longer imagine how to go on. . . .

Mr. I. could hardly bear the changed appearances of people ("like animated gray statues") any more than he could bear his own changed appearance in the mirror; he shunned social intercourse and found sexual intercourse impossible. He saw people's flesh, his wife's flesh, his own flesh, as an abhorrent gray; "flesh-colored" now appeared "rat-colored" to him. This was so even when he closed his eyes, for his preternaturally vivid ("eidetic") visual imagery was preserved but now without color, and forced on him images, forced him to "see" but see internally with the wrongness of his achromatopsia. He found foods disgusting in their grayish, dead appearance and had to close his eyes to eat. But this did not help very much, for the mental image of a tomato was as black as its appearance. . . .

Thus reds were seen (or not seen) as black. Yellows and blues, in contrast, were almost white. Further, there was an excessive tonal contrast, with loss of delicate tonal gradations. . . . Objects stood out, if they stood out at all, with inordinate contrast and clarity, like silhouettes. But if the contrast were normal, or low, they might disappear from sight altogether. . . .

His despair of conveying what the world looked like, and the uselessness of the usual black-and-white analogies, finally drove him, some weeks later, to create an entire "gray room," a gray universe, in his studio, in which tables, chairs, and an elaborate dinner ready for serving were all painted in a range of grays. The effect of this, in three dimensions and in a different tonal scale from the "black and white" we are all accustomed to, was indeed macabre, and wholly unlike that of a black-and-white photograph. As Mr. I. pointed out, "we accept drawings, films, television—small, flat images in black and white you can look at, or away from, when you want. It is only an image, it is not supposed to be real. But imagine black and white all around you, 360 degrees, all solid and three-dimensional, and there all the time—a total black and white world. . . . You can't imagine it: the only way I can express it is to make a complete gray room, with everything in it gray—and you yourselves would have to be painted gray, so you'd be part of the world, not just observing it." It was, he once said, like living in a world "molded in lead." . . .

Music, curiously, was impaired for him too, because he had previously (like the Russian composer, Scriabin, and others) had an extremely intense synesthesia, so that different tones had immediately been translated into color, and he experienced all music simultaneously as a rich tumult of inner colors. With the loss of his ability to generate colors, he lost this ability as well—his internal "color-organ" was out of action, and now he heard music with no visual accompaniment; this, for him, was music with its essential chromatic counterpart missing, music now radically impoverished.

When we asked Mr. I. to examine and paint a copy of a colored spectrum . . . he could see only black and white and varying shades of gray, and painted it as such. Intriguingly, his perception of the spectrum

(continued)

Box 24–2 A Life Without Color Vision: The Case of the Colorblind Painter (continued)

bore no resemblance to that of the retinally color–blind (which has a single peak of luminosity in the green around 500 nanometers) but did resemble that of people with normal ("photopic") vision, whose perception of luminosity reaches a peak in the yellow–green (around 560 nanometers). This showed that his cone mechanisms and discrimination of wavelengths were intact, and only color "perception" (or "construction") was deficient. . . .

Testing up to this point—other forms of visual testing, and a general neurological examination, were entirely negative—had shown an isolated but total achromatopsia or colorblindness. . . .

Efforts had . . . been made to delineate the brain damage in Mr. I.'s case (by the use of special scan techniques: CAT scan, NMR scan), and to measure the physiological reactions of visual cortex (with evoked potential tests), but these tests were all negative. With more sophisticated brain imaging we might well be able to identify the minute brain areas affected; but Mr. I. was getting tired of "all those tests," and for the present it seemed best to return to perceptual testing, but in a more elaborate form.

"Higher" forms of color perception have engaged the interest of Edwin Land in this country and S. Zeki in England, who have both devised a number of experimental and clinical tests. These use complex, subtly juxtaposed blocks of different colors, with a vague resemblance to some paintings of Mondrian (and hence sometimes called "Mondrians"). The colored shapes are projected on a screen through filters that can quickly be changed. In January 1987, with the patient, we met with Professor Zeki, and performed more elaborate testing. A "Mondrian" of great complexity was used as a test object, and this was projected with white light and with extremely narrow-range gel filters allowing the passage of only red, green, and blue light. . . .

Mr. I., it was evident, could distinguish most of the geometric shapes, though only as consisting of differing shades of gray, and he instantly ranked them on a one-to-four gray scale, although he could not distinguish some color boundaries (for example, between red and green, which both appeared to him, in white light, as "black"). With rapid, random switching of the filters, the gray-scale value of all the shapes dramatically changed, some shades previously indistinguishable now becoming very different, and all shades (except actual black) changed, either grossly or subtly, with the wavelength of the illuminating beam. (Thus a green area would be seen by him as "white" in green [medium-wavelength] light, but as "black" in white or red [long-wavelength] light.)

All Mr. I.'s responses were consistent and immediate. . . . Such a response was utterly unlike that which would be made by someone with retinal color blindness—i.e., an absence of receptors sensitive to wavelengths in the eye. Mr. I., it was clear, could discriminate wavelengths—but he could not go on from this to "translate" the discriminated wavelengths into color, could not generate the cerebral or mental construct of color. . . .

This showed us with great clarity how his ability to discriminate different wavelengths was preserved, while his color perception was obliterated, how there was a clear dissociation of the two. Such a dissociation could not occur unless there were separate processes for wavelength discrimination and color construction. Thus, Mr. I.'s situation only becomes intelligible with a theory of multistage processing such as Land's or Zeki's; and such a theory can only be grounded, finally and elegantly, in such a patient.

Postscript (October 1987)

It is almost two years since Mr. I. lost his color vision. The intense sorrow that was so characteristic at first, as he sat for hours before his (to him) black lawn, desperately trying to perceive or imagine it as green, has disappeared, as has the revulsion (he no longer sees his wife, or himself, as having "rat-colored" flesh).

There has, we think, been in his case a real "forgetting" of color—a forgetting at once psychological and physiological, at once strategic and structural. Perhaps this has to occur in someone who is no longer able to imagine or remember, or in any physiologically based way generate, a lost mode of perception. . . .

(continued)

Box 24–2 *A Life Without Color Vision: The Case of the Colorblind Painter (continued)*

In the past few months Mr. I has been changing his habits and behavior—"becoming a night-person," in his own words. . . . He drives, at random, to Boston, Baltimore, or small towns and villages, arriving at dusk, and then wandering about the streets for half the night, occasionally talking to a fellow walker, occasionally going into little diners: "Everything in diners is different at night, at least if it has windows. The darkness comes into the place, and no amount of light can change it. They are transformed into night places. I love the nighttime," Mr. I. says. . . .

Richard Gregory, speaking of those who have never had color vision (owing to absence of cones, or normal cone function, in their eyes) said, "They live in a scotopic world, in a world of bright moonlight," and this now seems to be the only world that Mr. I. can bear. Our world—our "photopic" world, dazzlingly bright and colored—must appear discordant and painful to an achromatope (whether he has been born colorblind, like Gregory's subjects, or become colorblind, like Mr. I.); given this, along with an enhanced, compensatory sensitivity to the nocturnal and scotopic, it is not surprising, it is perhaps inevitable, that achromatopes should be drawn to the only world in which they feel at ease and at home—and that they should, like the loris and the potto, the big-eyed primates that only emerge and hunt at night, turn wholly, or as much as they can, to becoming night creatures in a night world.

(Abridged with permission from the *New York Review of Books*, O. Sacks, and R. Wasserman, "The Case of the Colorblind Painter," November 19, 1987, pp. 25–34.)

An Overall View

Individual cones do not transmit information about the wavelength of light. The wavelength affects only the probability that a photon will be absorbed, not the properties of the electrical response of the cone. To detect color, the brain compares the responses of three types of cone cells, each most sensitive to a different part of the visible spectrum.

As information is transmitted to the brain, inputs from the three classes of cones are combined in a variety of ways. Many retinal ganglion cells, as well as cells in the lateral geniculate nucleus and the cortex, are excited by one type of cone and inhibited by another. These opponent interactions of the three cone systems underlie the phenomena of color opponency and simultaneous color contrast. However, the brain computes color perception of an object by comparing not only the responses of the cones stimulated by the object, but also the responses of all cones throughout the retina. In this way the brain is able to compensate for changes in the spectral characteristics of ambient light, so that an object's color appears roughly the same whatever the composition of the light—a phenomenon known as color constancy.

Color information is processed separately in the brain, in the parvocellular–blob system. Cells in the retina and lateral geniculate nucleus, as well as in areas V1 and V2 of the cortex, respond selectively to wavelength. In area V4, however, some cells respond best to a particular color; that is, they respond preferentially to the color of objects regardless of the precise wavelength composition of the illuminant. Monkeys with lesions in area V4 lose color constancy although they are still able to discriminate wavelengths. A selective loss of color constancy has also been observed in some humans with specific cortical lesions; such individuals report a complete loss of the subjective experience of color.

Two challenges for the future are to determine how wavelength-based information is translated into specific color information in the cortex, and to understand how this information is subsequently brought together with information about form and movement to generate a unified perception of objects.

Selected Readings

Albers, J. 1991. Interaction of Color. New Haven, Conn.: Yale University Press.

Boynton, R. M. 1979. Human Color Vision. New York: Holt, Rinehart and Winston.

Daw, N. W. 1984. The psychology and physiology of colour vision. Trends Neurosci. 7:330–335.

Gouras, P. 1984. Color vision. In N. N. Osborne and G. J. Chader (eds.), Progress in Retinal Research, Vol. 3. Oxford: Pergamon Press, pp. 227–261.

Hurvich, L. M. 1972. Color vision deficiencies. In D. Jameson and L. M. Hurvich (eds.), Handbook of Sensory Physiology, Vol. 7, Part 4. Visual Psychophysics. Berlin: Springer, pp. 582–624.

Land, E. H. 1977. The retinex theory of color vision. Sci. Am. 237(6):108–128.

Livingstone, M., and Hubel, D. 1988. Segregation of form, color, movement, and depth: Anatomy, physiology and perception. Science 240:740–749.

Nathans, J., Merbs, S. L., Sung, C.-H., Weitz, C. J., and Wang, Y. 1992. Molecular genetics of visual human pigments. Annu. Rev. Genet. 26:403–424.

Zeki, S. 1993. A Vision of the Brain. London: Blackwell.

25

Sensory Experience and the Formation of Visual Circuits

The Development of Visual Perception Requires Sensory Experience

Early Sensory Deprivation Alters Perceptual Development

Early Sensory Deprivation Alters the Development of Neural Circuits

The Development of Ocular Dominance Columns Serves as a Model for Understanding the Fine Tuning of Visual Circuitry by Activity

Cooperation and Competition Are Important for Segregating Afferent Inputs into the Ocular Dominance Columns

Cooperation Requires Synchronous Activity

Different Regions of the Brain Have Different Critical Periods of Development

There Is an Early Critical Period in the Development of Social Competence

Visual perception—the brain's synthesis of a coherent image from discrete signals carrying form, motion, and color information—is mediated by precisely connected nerve cells in several parallel pathways. How is this pattern of interconnections established?

Until about 30 years ago most neurobiologists believed that the connections in the brain were programmed by a set of molecular events, independent of activity or experience, whereby the appropriate pre- and postsynaptic elements at every synapse recognized each other through molecular clues. It is now clear that molecular recognition indeed is critical for development, but only for the first two steps of a three-step program for synapse formation. Molecular recognition is important for the neuron's initial selection of a specific pathway for the outgrowth of its axon and for the axon's subsequent selection of a particular target region within the nervous system (see Chapter 6). For example, in the visual system molecular cues guide the growing axons of the retinal ganglion cells from the retina into the optic nerve and then from the optic nerve to its target in the lateral geniculate nucleus.

Once the axon reaches its target, however, the retinal ganglion cell axon is matched to a specific set of postsynaptic target neurons in the lateral geniculate nucleus through activity-dependent mechanisms. These activity-dependent mechanisms are responsible for the final topographic organization of inputs that are required for the mature function of this sensory relay nucleus (see Chapter 23). Indeed, as we shall see later in the book, synaptic connections can be modified throughout adult life by neural activity induced by experience. This suggests that there is likely to be a continuity in the neural mechanisms of development and learning.

In Chapter 6 we examined the early development of the nervous system at the molecular level. In this chapter we look at the later activity- and experience-dependent stage of prenatal and postnatal development. We focus on vision because studies of the effects of experience on visual perception are particularly insightful and have been instructive in furthering our understanding of how experience shapes the neural circuitry of the brain in general.

The Development of Visual Perception Requires Sensory Experience

At critical stages of prenatal and early postnatal development the integrative action of the brain—in fact, the details of its wiring—depends on specific interactions between the organism and its environment. The influence of the environment on the brain, and therefore on behavior, changes with age. Abnormal environmental experiences usually have more profound effects during early stages of postnatal development than in adulthood.

Early Sensory Deprivation Alters Perceptual Development

One dramatic example of the dependence of the brain on environmental experience was discovered in studies of people born with cataracts. Cataracts are opacities of the lens that interfere with the optics of the eye but not with the nervous system. Today, congenital cataracts are typically removed in infancy. Formerly, however, removal of congenital cataracts was usually delayed until much later in life. A study in 1932 of several patients who had congenital binocular cataracts removed between the ages of 10 and 20 found that this delay had resulted in permanent impairment of the ability to perceive form, although color vision was normal. Even long after the cataracts were removed, these patients had difficulty recognizing shapes and patterns.

The idea that visual experience is required for the development of normal vision also drew support from studies of newborn monkeys raised in the dark for the first 3 to 6 months of their lives. When these monkeys were later introduced to a normal visual world, they could not discriminate even simple shapes. It took weeks or even months of training to teach them to distinguish a circle from a square, whereas normal monkeys learn such discrimination in days.

Early Sensory Deprivation Alters the Development of Neural Circuits

A next important step toward understanding the development of perception was made by Hubel and Wiesel studying the response characteristics of cells in the visual cortex of newborn kittens and monkeys. The perception of three-dimensional depth for objects less than 100 feet away is achieved by comparing the disparity in the retinal images of the two eyes (stereoscopic vision). The first opportunity for binocular interaction (convergence of input from the two eyes on a common target cell) occurs in the cortical cells above and below layer 4C (Figure 25–1). Thus, in the visual cortex of monkeys, most cells above and below layer 4C respond to an appropriate stimulus presented to either eye; only a small proportion of cells, those in layer 4C, respond exclusively to the left or the right eye (Figure 25–2A and B).

This convergence of inputs from the two eyes on single cells in the visual cortex was used to study the effect of visual experience on the development of cortical connections. If a monkey is raised from birth to 6 months of age with one eyelid sutured shut, it will permanently lose useful vision in that eye. When the sutures are removed and the eye is exposed to light, the retinal ganglion cells in the deprived eye and cells in the lateral geniculate nucleus receiving the projections from that eye respond well to visual stimuli and have essentially normal receptive fields. In the visual cortex, however, most cells no longer respond to the deprived eye. The few cortical cells that can still be activated by the deprived eye are not sufficient for visual perception (Figure 25–2C). Comparable deprivation in an adult has no effect on the responses of cortical cells to stimulation of the deprived eye or on perception. Yet as little as one week of deprivation during the first 6 months of life will lead to a nearly complete loss of cortical responsiveness to stimulation of a deprived eye and a loss of vision in that eye.

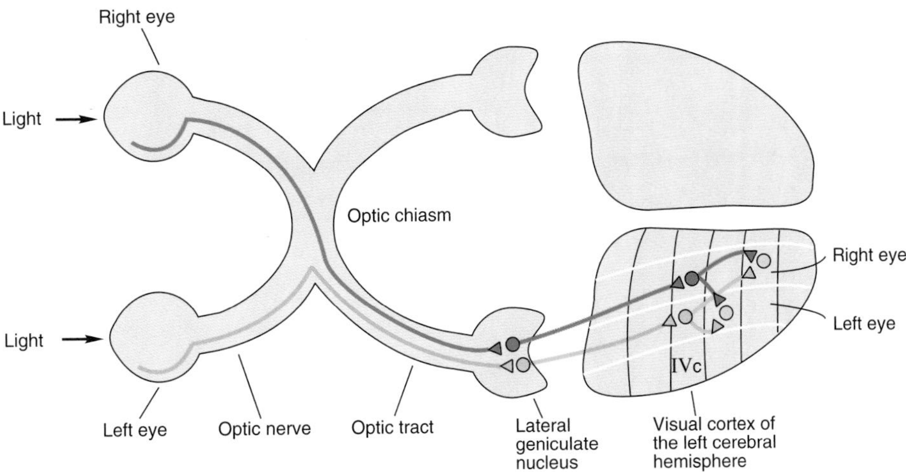

Figure 25–1 The input from the two eyes is segregated up to the level of the visual cortex. The retinal ganglion cells of each eye project to separate layers of the lateral geniculate nucleus. The axons of cells in the lateral geniculate nucleus form synaptic connections with neurons in layer 4C of area 17, the primary visual cortex. Neurons in layer 4C are organized in alternating sets of ocular dominance columns; each set receives input from only one eye (see Figure 23–15). The axons of the cells in layer 4C project to adjacent columns as well as to the upper and lower layers in the same column (see Figure 23–17). As a result of these connections, most cells in the upper and lower layers of the visual cortex receive information from both eyes.

The Development of Ocular Dominance Columns Serves as a Model for Understanding the Fine Tuning of Visual Circuitry by Activity

Are these permanent changes in the response properties of cortical cells accompanied by structural changes in the ocular dominance columns? As we first saw in Chapter 23, the segregation of pathways from the two eyes to the cortex is the anatomical basis of the ocular dominance columns in the primary visual cortex. The inputs from each eye terminate in alternating columns of stellate cells in layer 4C, which project to cells in higher and lower layers of the same column, as well as to cells in adjacent columns. These connections form the binocular receptive fields of the cells in cortical layers above and below 4C. The projections from the input layer 4C are thus essential for the processing of convergent input from the two eyes by the cortical layers above and below.

To examine whether visual deprivation alters the architecture of the ocular dominance columns in the cerebral cortex, newborn monkeys were deprived of input from one eye. Labeled amino acids were then injected into one or the other eye, and the transport of the label to the cortex was observed using autoradiography. This procedure thus labels the synaptic terminals of axons from the lateral geniculate nucleus in layer 4C of the striate cortex. After closure of one eye, the columns receiving input from the normal eye were greatly widened at the expense of those receiving input from the deprived eye (Figure 25–3B and C).

Here, then, is direct evidence that sensory deprivation early in life can alter the structure of the cerebral cortex. It is possible that rearing of an animal in the dark also leads to a deterioration or distortion of specific connections in the brain.

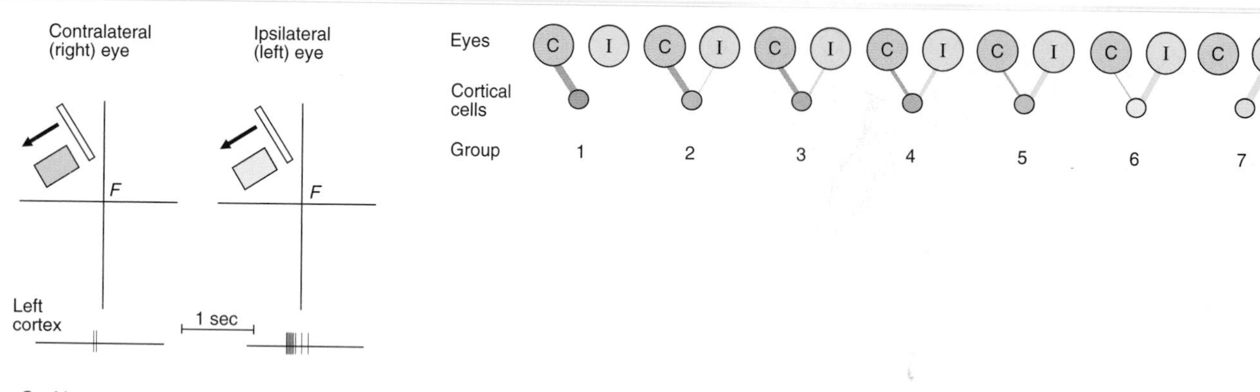

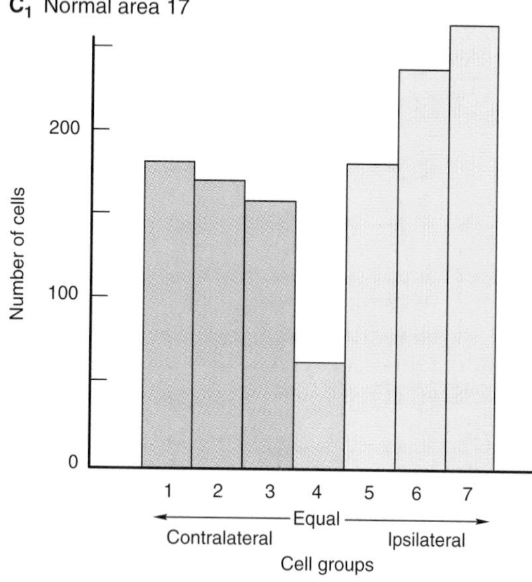

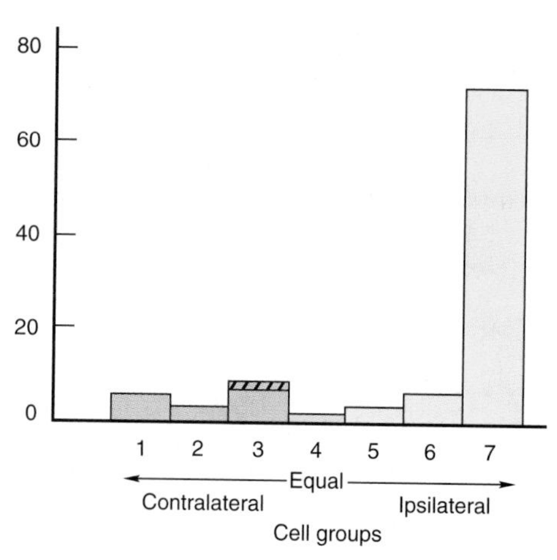

Figure 25–2 Inputs from the two eyes first converge in area 17 of the monkey's visual cortex. (Adapted from Hubel and Wiesel, 1977.)

A. The two retinal images, for the right and left eyes, are drawn separately for clarity. The center of the visual field falls on the fovea (**F**), the region of the retina with greatest acuity. The figure shows a diagonal bar of light moving leftward across the two eyes in the path of the receptive fields (**colored rectangles**) of two cells, each conveying input from one eye. The receptive fields of the two cells are similar in orientation, position, shape, and size, and respond to the same form of stimulus. The inputs from these cells converge on a single neuron in area 17 of the cortex. Action potential recordings (**bottom**) of the cortical cell show that it responds more effectively to input from the ipsilateral eye than from the contralateral eye.

B. On the basis of recordings from cells in area 17, Hubel and Wiesel classified the responses of individual neurons into seven ocular dominance groups. Cells receiving input only from the contralateral eye (C) fall into group 1, while cells that receive input only from the ipsilateral eye (I) fall into group 7. Other cells receive inputs from both eyes but one eye may influence the cell much more than the other (groups 2 and 6), or the differences may be slight (groups 3 and 5). Some cells respond equally to input from both eyes (group 4). According to these criteria, the cell shown in part A would fall into group 6.

C. Histograms of the responsiveness of cells in area 17 to stimulation of one eye or the other. **1.** A histogram of the responses of 1,256 cells in area 17 in the left hemisphere of normal adult and juvenile monkeys. The cells in layer 4 that normally receive only monocular input were excluded. Most cells responded to input from both eyes. **2.** A histogram of the responses of cells in the left hemisphere of a monkey in which the contralateral (right) eye was closed from the age of 2 weeks to 18 months and then reopened. Most of the cells responded only to stimulation of the ipsilateral eye.

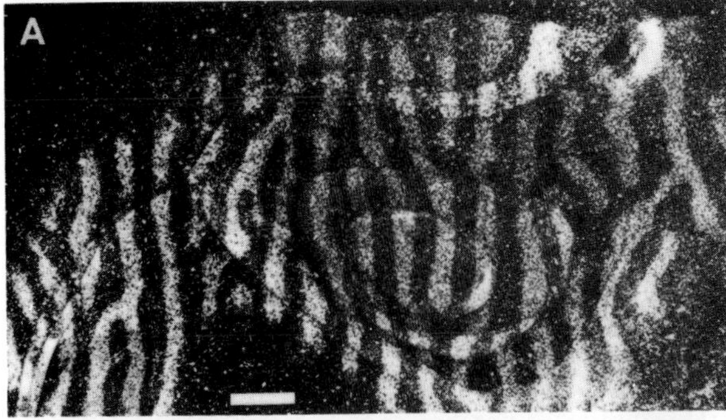

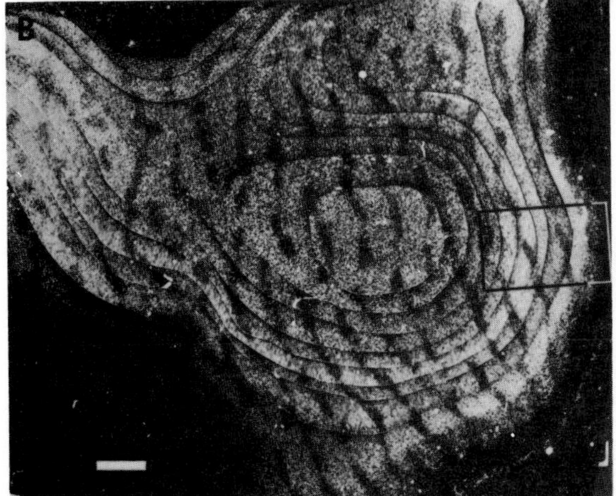

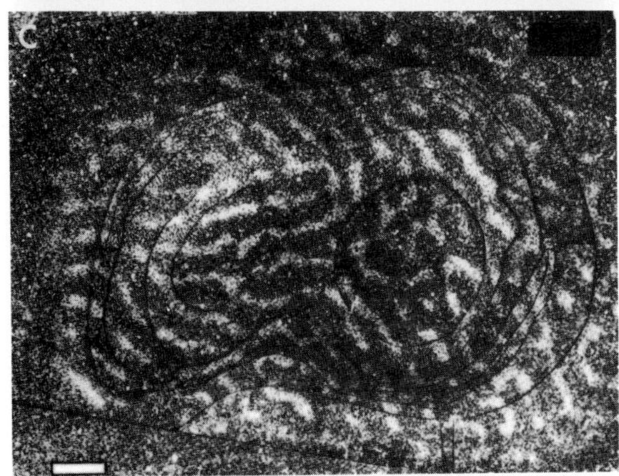

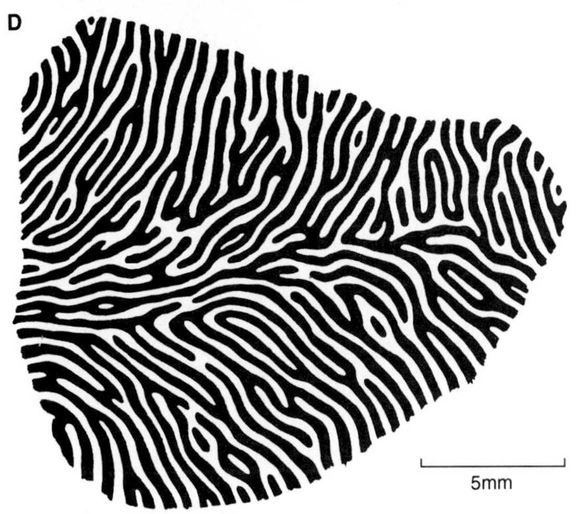

Figure 25–3 Visual deprivation of one eye during the critical period of development of the ocular dominance columns reduces the size of the columns for that eye.

A. The right eye of a normal adult monkey was injected with a radiolabeled amino acid. This autoradiograph, taken 10 days later, shows a tangential section through the dome-shaped area 17 of the right hemisphere. Here, the radioactivity forms white stripes in layer 4C of the visual cortex, indicating the areas innervated by afferents from the lateral geniculate nucleus that carry input from the injected eye. The alternating dark stripes correspond to regions innervated by afferents from the uninjected eye. The section cuts through layer 5, which is seen as the dark oval central area. Scale bars in micrographs **A**, **B**, and **C** = 1 mm.

B. A comparable section through the visual cortex of an 18-month-old monkey whose right eye had been surgically closed at 2 weeks of age. The label was injected into the left eye. The white stripes of label correspond to the terminals of afferent axons carrying signals from the open (left) eye; the narrower dark stripes correspond to inputs for the closed (right) eye. The plane of section cuts across layer 6, which is seen as the central oval shape.

C. A section comparable to that in Figure B from an 18-month-old animal whose right eye had been shut at 2 weeks. In this case, however, the label was injected into the eye that had been closed, giving rise in the cortex to narrow white stripes and expanded dark ones.

D. Complete reconstruction of ocular dominance columns in area 17 of the right hemisphere of a normal animal, showing the intricate organization of the complete map. (A–C adapted from Hubel, Weisel, and LeVay, 1977; D From S. LeVay, 1981.)

Cooperation and Competition Are Important for Segregating Afferent Inputs into the Ocular Dominance Columns

How does monocular deprivation change the dimensions of ocular dominance columns in layer 4C of the visual cortex? Are the columns already present at birth, so that sensory deprivation alters an established columnar structure? Or does deprivation interfere with *establishment* of the columns? Ocular dominance columns in monkeys do not mature until 6 weeks after birth. It is only then that the afferent fibers from the lateral geniculate nucleus become completely segregated. In cats, segregation occurs even later (Figure 25–4).

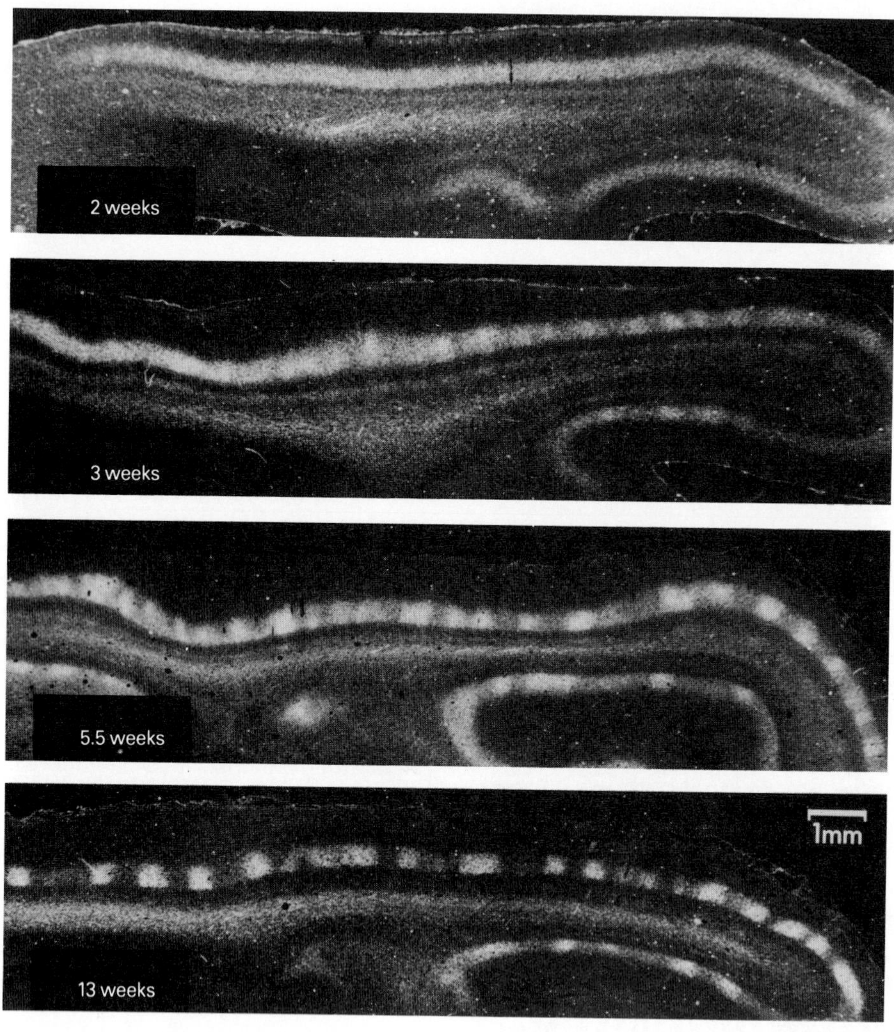

Figure 25–4 Autoradiographs of four stages in the development of the visual cortex illustrate postnatal development of the ocular dominance columns in a cat. (The autoradiographs are horizontal sections through columns ipsilateral to an eye that was injected with a radiolabeled amino acid. The geniculocortical afferents serving the injected eye are labeled by transneuronal transport.) At about 2 weeks (15 days) after birth, the afferents have spread uniformly along layer 4, completely intermingled with the (unlabeled) afferents serving the contralateral eye. At 3 and 5.5 weeks the emerging columns are visible, but only as modest fluctuations in labeling density. At 13 weeks the borders of the labeled bands become more sharply defined as the afferents segregate—the anatomical basis for ocular dominance columns. (Adapted from LeVay, Stryker, and Shatz, 1978.)

When the afferent fibers from the lateral geniculate nucleus first reach layer 4C during development, they overlap extensively. With further development they become segregated into ocular dominance columns, through retraction processes to be described later. Monocular deprivation interferes with this normal pattern of development. Since the columns have not yet been fully established at birth, reducing the input from one eye by suturing the eyelid shut places the axon terminals of that eye at a selective disadvantage, and more of the terminals from the deprived eye retract than in normal circumstances. This explains the widening of the columns for the nondeprived eye (see Figure 25–3B and C).

The development of ocular dominance columns can be followed at the level of individual neurons. Changes in the morphology of the afferent terminals of individual neurons of the lateral geniculate nucleus can be observed during development by labeling the terminals at different development stages. This technique shows that, at first, each afferent fiber from the geniculate nucleus branches extensively over areas covering several future ocular dominance columns for each eye (Figure 25–5A). As each afferent neuron matures, its axon terminal in the cortex loses some of its branches by retraction while expanding and strengthening others, so that finally the neuron connects almost

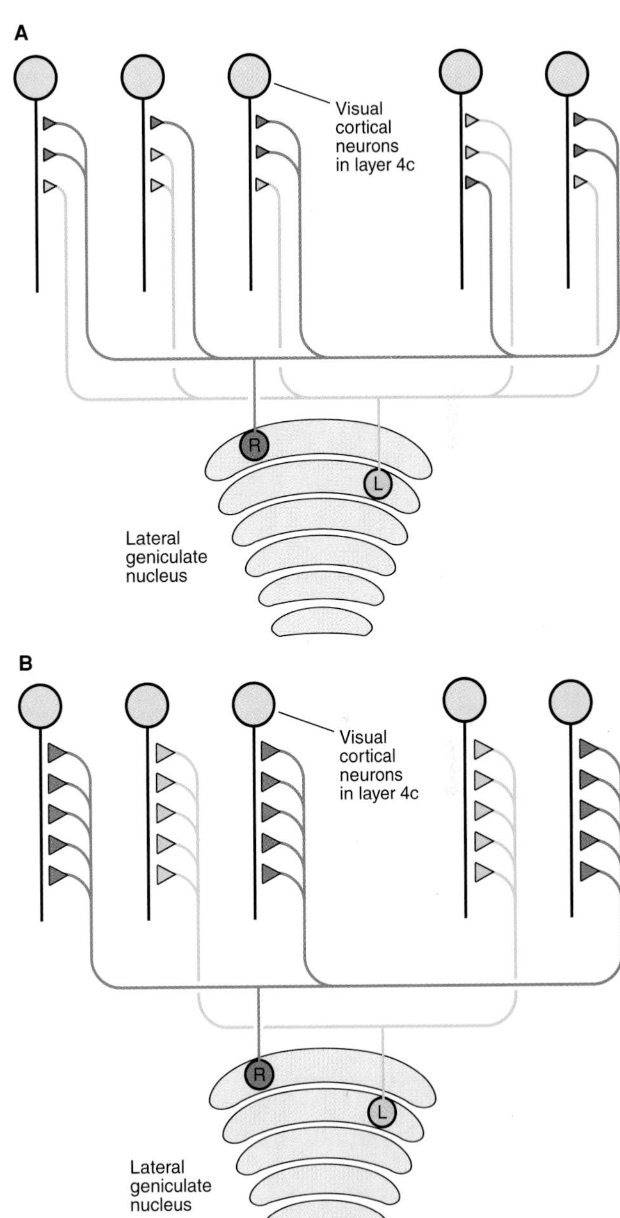

Figure 25–5 Competition between the afferent fibers of the two eyes is a factor in the establishment of ocular dominance columns. (Based on unpublished experiments of Gilbert, Wiesel, and Katz.)

A. Early in development the afferent axons from cells in the lateral geniculate nucleus converge on neurons in layer 4C of the visual cortex. Initially, because of genetic or possibly random developmental processes, the afferent axons serving one eye tend to predominate in alternating clusters of cells in layer 4C (illustrated here as single cells). In the illustration the predominance of one eye is indicated by two synaptic endings rather than one.

B. As a result of the intrinsic bias illustrated in A, the more numerous terminals carrying input from one eye edge out the competing terminals that carry inputs from the other eye. The resulting dominance of inputs from one or the other eye gives rise to the ocular dominance columns (see Figures 25–3 and 25–4).

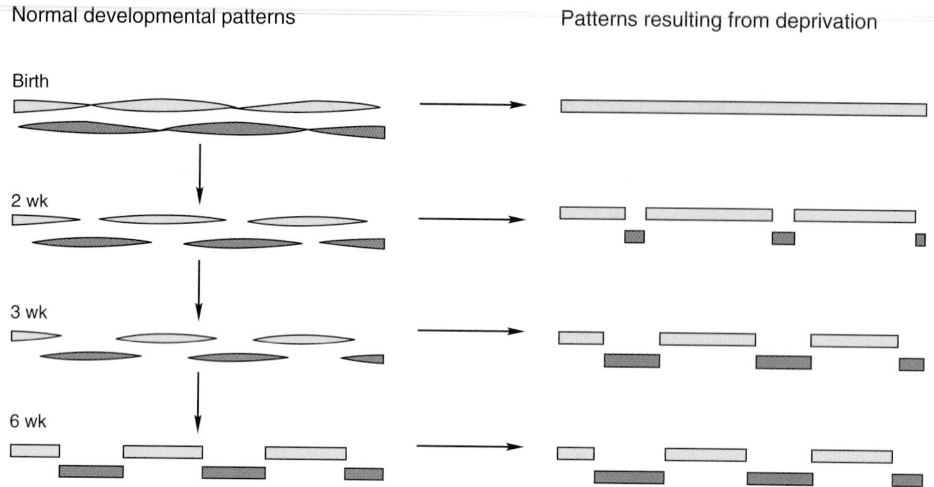

Figure 25–6 A comparison of normal development and the effects of eye closure on columns in layer 4C. The yellow shapes represent the terminals of geniculate afferents in layer 4C from one eye; the blue shapes represent the terminals from the other eye. The lengths of the shapes represent the density of the terminals at each point along layer 4C. For clarity, the columns are shown here as one above the other, whereas in reality they are side by side in the cortex. During normal development (**left**) layer 4C is gradually divided into an alternating group of fibers from each eye. The consequences of depriving one eye (**right**) depend on when closure occurs. Closure at birth leads to complete dominance by the open eye (**yellow**) because little segregation has occurred at this point. Closure at 2, 3, and 6 weeks has a progressively weaker effect on the ocular dominance columns, since the columns become more segregated with time. (Adapted from Hubel, Wiesel, and LeVay, 1977.)

exclusively to cells that receive input only from one eye (Figure 25–5B). This process of synapse retraction (or pruning) is widespread in the nervous system (see Chapter 6). The segregation of afferent inputs into ocular dominance columns can be accounted for by the schema illustrated in Figure 25–6.

Why do some terminals retract during normal development while others survive and grow stronger? The reason may be that at birth there are small, developmentally determined differences in the proportion of terminals from each eye contacting common target cells. Neighboring axons from the same eye tend to fire synchronously and thereby *cooperate* to depolarize and excite a target cell. As we shall see below, this cooperative action on common target cells strengthens the cooperating axons at the expense of competing axons. Thus, if afferents from one eye are initially more numerous in one region of cortex, they are likely to have an advantage because the cooperation between them allows them to make additional connections onto cells in the region and even to grow further branches and spread to adjacent cells. At the same time, the afferent fibers from the two eyes *compete* with each other in the formation of strong connections on common target cells, with the result that strong connections from one eye tend to inhibit the growth of the axons from the opposite eye. Together, cooperation and competition permit two populations of afferent fibers to share a common space without overlapping.

During the early stages of this critical period of segregation, if an animal is deprived of the use of one eye, the axon terminals of the lateral geniculate cells from the closed eye are at a competitive disadvantage and therefore retract to an abnormal extent. At the same time, the terminals from the normal eye continue to occupy areas they normally would have relinquished. If the animal is

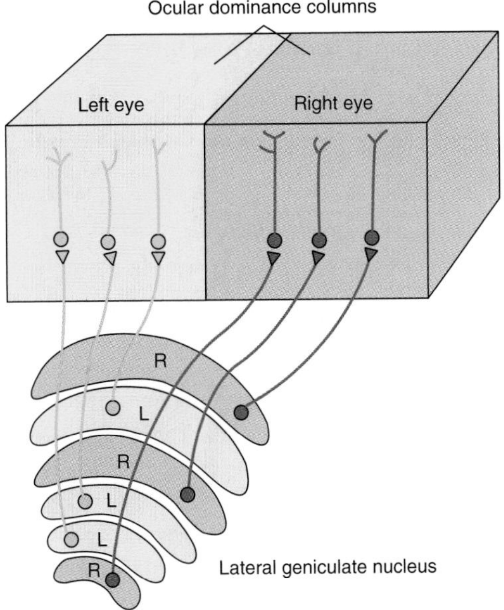

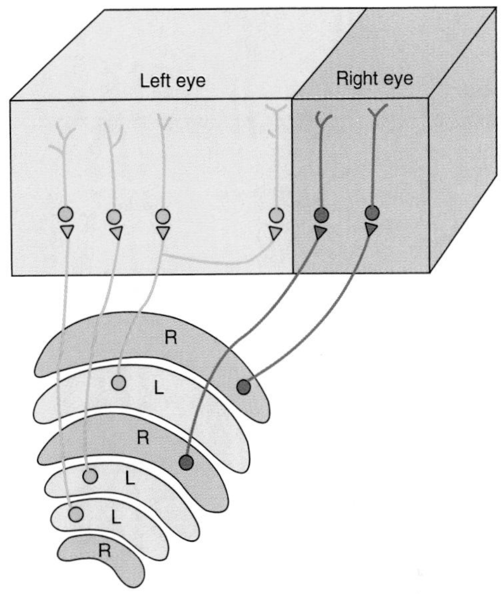

Figure 25–7 Deprivation of vision in one eye during the later phase of development of the ocular dominance columns affects the balance between the columns.

A. Under normal conditions the alternating ocular dominance columns for each eye are equal in size.

B. Without vision in the right eye the afferent terminals carrying input from the nonfunctional (right) eye retract due to low levels of activity. At the same time the comparatively more robust activity of the geniculate cells carrying input from the left eye causes these cells to establish new connections with cortical cells, replacing the lost afferent terminals for the right eye. As a result, the columns devoted to the functional eye become larger than those devoted to the nonfunctional eye.

deprived of the use of one eye during a *later* stage of this critical period, when the ocular dominance columns are almost fully segregated, a second mechanism comes into play. The axons serving the open eye actually sprout and grow collateral branches that extend into areas they had earlier vacated (Figure 25–7).

If the development of ocular dominance columns depends on competition between afferent fibers, it should be possible to induce the formation of columns where columns normally are *not* present by establishing competition between two sets of axons. This possibility has been examined in developing frogs, where the axons of the retinal ganglion cells in one eye project only to the opposite side of the brain. There they terminate in the optic tectum in an orderly way that forms a map of the visual world; there is no competition from a second retina, and indeed this neural map has no columnar organization.

To establish a potential source of competition, a third eye was transplanted into a region of the head near one of the normal eyes. The retinal ganglion cells of the transplanted eye sent out axons that terminated in the contralateral optic tectum. This projection was then mapped by injecting a radioactive tracer into the transplanted eye. Axons from the transplanted and normal eyes were found to terminate in a regular pattern of alternating columns (Figure 25–8). Thus, in the frog optic tectum, as in the cerebral cortex, columnar organization results when two sets of afferent fibers are forced to compete for the same population of target neurons.

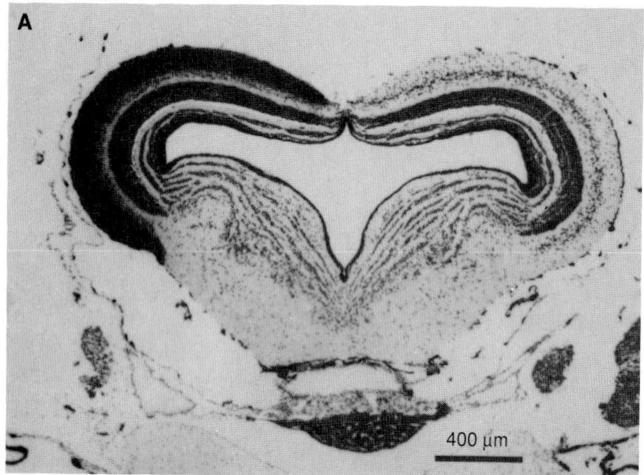

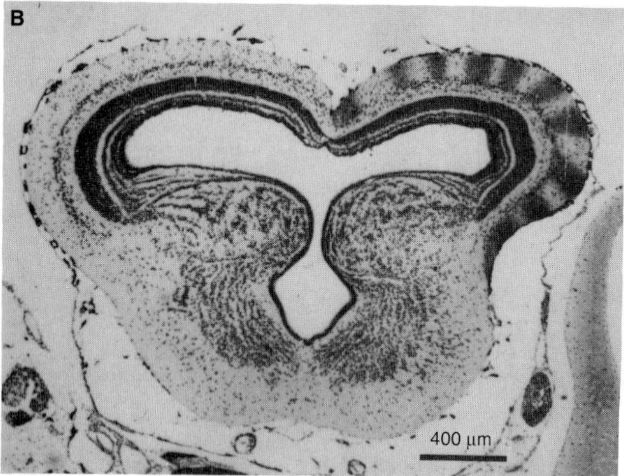

Figure 25–8 Ocular dominance columns can be experimentally induced in a frog by the transplantation of a third eye. These autoradiographs show a coronal section through the frog's midbrain. (Adapted from Constantine-Paton, 1981.)

A. The left eye of a normal frog was injected with a radiolabeled amino acid 3 days before the transplant. The entire superficial neuropil of the right optic lobe (**left**) is filled with silver grains, indicating the region occupied by synaptic terminals from the labeled (contralateral) eye.

B. A third eye was transplanted in the animal near the normal right eye. Some time later the normal right eye was injected with a radiolabeled amino acid. The left optic lobe (**right**) receives inputs from both the labeled eye and the supernumerary eye. The retinotectal synaptic zone of the contralateral eye, normally continuous, here is divided into alternating bands of terminal endings from each eye.

Cooperation Requires Synchronous Activity

What factors lead to cooperation between adjacent fibers from the same eye and competition between afferent fibers from the two eyes? The critical factor regulating both competition and cooperation may be neural activity.

Ocular dominance columns do not form in kittens if all impulse activity in the retinal ganglion cells is blocked by injecting tetrodotoxin (which selectively blocks the voltage-sensitive Na^+ channel) into each eye prior to the critical period of development. Because activity is blocked in the retina but not along the optic nerve, which contains the ganglion cell axons, both optic nerves can be stimulated using implanted electrodes. When the two nerves are stimulated *synchronously*, ocular dominance columns do not form. In contrast, when the optic nerves are stimulated *asynchronously*, ocular dominance columns begin to take shape. Thus, although the formation and maintenance of normal binocular vision require synchronous electrical activity among the optic nerve fibers from one eye, they also require asynchronous activity *between* the two eyes.

A similar mechanism appears to be at work in the lateral geniculate nucleus. As we have seen in Chapter 23, inputs from each eye in the lateral geniculate nucleus are segregated in alternating layers, much as they are in the ocular dominance columns. Unlike the ocular dominance columns, this segregation is completed before birth. However, the segregation of the projections from each eye can be disrupted irreversibly by applying tetrodotoxin to the optic chiasm of a fetus, thus suppressing the generation of action potentials during the time axons still overlap extensively in the lateral geniculate nucleus.

There is thus an important difference in the way activity affects the segregation of afferents in the visual cortex and in the lateral geniculate nucleus. In the cortex the segregation occurs *after* birth, and the activity required for the formation of these connections is driven by visual experience. In the lateral geniculate nucleus segregation occurs *before* birth, while the infant is still *in utero*, so that the activity essential for segregating optic nerve fibers cannot be driven by visual experience. What then drives the retinal afferents?

It has been discovered that the retinal fibers of the optic nerve are spontaneously active *in utero*, independent of any visual information. Moreover, neighboring cells in the fetal retina tend to be active together, firing in synchronous bursts that last a few seconds separated by silent periods lasting 1 to 2 minutes. Thus, spontaneous activity *in utero* may have an important instructive function in development, not only in the visual system but also generally. The spontaneous firing of a group of fibers and the resulting synchronous excitation of its target seem to strengthen those synapses whose presynaptic fibers are active together (cooperation) and to weaken those synapses whose presynaptic fibers are inactive or out of synchrony (competition).

These events in the lateral geniculate nucleus and cerebral cortex are consistent with an important idea first proposed by the psychologist Donald Hebb: coincident activity in the pre- and postsynaptic elements of a synapse leads to the strengthening of that synapse. Hebb's idea has been incorporated into many neuronal models of competition and cooperation, and we shall encounter them again in Chapter 36 in connection with cellular mechanisms of learning and memory. Indeed, based on the assumptions of a Hebbian coincidence mechanism, a mathematical model of activity-dependent competition between the two eyes simulates quite accurately the segregation of ocular dominance columns during development (Figure 25–9).

A clue as to how this sort of cooperation might work has come from studies of the three-eyed frog discussed above, which have implicated specific postsynaptic receptors in this process. As in the mammalian visual system, the formation of ocular dominance columns in the three-eyed frog appears to require synchronous activity among neighboring neurons. The feature of retinal activity that appears critical is that neighboring fibers fire together. It appears as if cells that fire together become wired together!

How does this occur? The optic nerve fibers are glutaminergic and terminate on target cells in the tectum that have both *N*-methyl-D-aspartate (NMDA) and non-NMDA receptors (see see Figure 13–9). The temporal summation of synaptic excitation seems to be designed to ensure a level of synaptic depolarization in the common target cell that is sufficient to activate the NMDA-type glutamate receptors. The ion channel of this receptor is normally blocked by Mg^{2+} (see Figure 13–10). Depolarizing the cell removes the Mg^{2+} block and allows Ca^{2+} to flow into the postsynaptic cell to activate calcium-dependent second-messenger systems, which are thought to be essential for stabilizing the active synapses. Consistent with this idea, the segregation of columns in the frog brain can be blocked by exposing the tectum to aminophosphonovaleric acid (APV), a selective antagonist of the NMDA receptor (see Chapter 13). In contrast, exposure of the tectum to NMDA, the agonist of this receptor, leads to a sharpening of the columnar organization.

These considerations raise one final question: What are the molecular mechanisms by which synchronous activity strengthens cooperating synapses and weakens the synapses of competing fibers? That is, how do the calcium-dependent second-messenger systems stabilize the cooperating synapses?

We do not yet know the answer to this question. One possibility is that the postsynaptic target cells in layer 4C release a neurotrophic or growth factor. This factor may be in short supply, so that the intermingled afferent fibers have to spread over the entire target structure as they compete for the factor. Afferent fibers that fire together and connect to the same target cell could increase the secretion of the growth factor from the postsynaptic cell. Axons from different eyes, or even axons from the same eye that originate from cells that are not neighbors, would not increase secretion of the growth factor because they do not fire in synchrony.

According to this view, then, the growth factor is only released when synchronously firing axons depolarize the postsynaptic cell sufficiently to activate the NMDA receptor–channels and allow Ca^{2+} inflow, which activates calcium-dependent second-messenger systems that increase the release of the growth factor. Equally important, the growth factor is taken up only by those presynaptic terminals that are active in exciting the postsynaptic cell (by firing synchronously with other axons), perhaps because these active postsynaptic cells are in the process of endocytic membrane retrieval when the postsynaptic cell is active (see Chapter 15). As a result, afferent fibers that are not able to take up the trophic factor, that is, those competing fibers from the other eye that do not fire in phase with the predominate inputs, are at a competitive disadvantage and are likely to be eliminated (Figure 25–10).

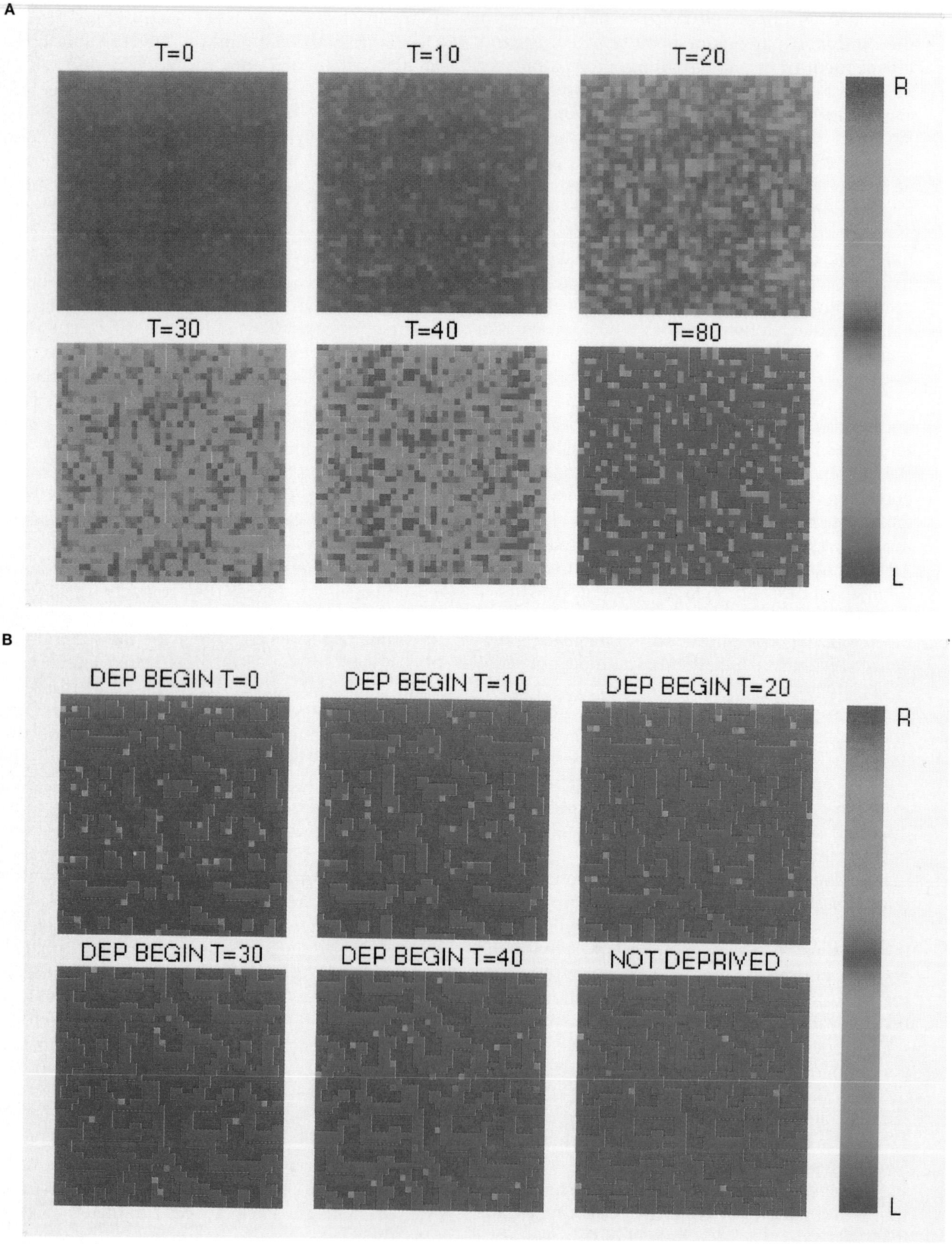

A

T=0
T=10
T=20
R
T=30
T=40
T=80
L

B

DEP BEGIN T=0
DEP BEGIN T=10
DEP BEGIN T=20
R
DEP BEGIN T=30
DEP BEGIN T=40
NOT DEPRIVED
L

← *Figure 25–9* A computer simulation of the development of ocular dominance columns. (From Miller et al., 1989.)

A. These results of the simulation illustrate different stages in the normal development of ocular dominance columns. Each square represents a single cortical cell. Red and green signify complete dominance of inputs from one or the other eye. Shades of blue indicate degrees of convergent input from both eyes. With time (T), in arbitrary units roughly comparable to days, this model generates a progressive segregation of inputs from each eye, until at $T = 80$ there is almost no overlap. The resulting pattern is similar to that observed experimentally in Figure 25–3A.

B. These results show the effects on segregation of the columns when the use of one eye has been lost at different starting times ranging from 0 to 40. When segregation is started early ($T = 0$), the input of the functioning eye expands significantly during the initial period. As the onset of monocular deprivation is delayed ($T = 10, 20, 30, 40$), the expansion of the input of the intact eye becomes more limited. In these simulation results the critical period lasts from T = 0 to T = 20. After T = 30 monocular deprivation has little effect, as can be seen in a comparison with the normal case in A.

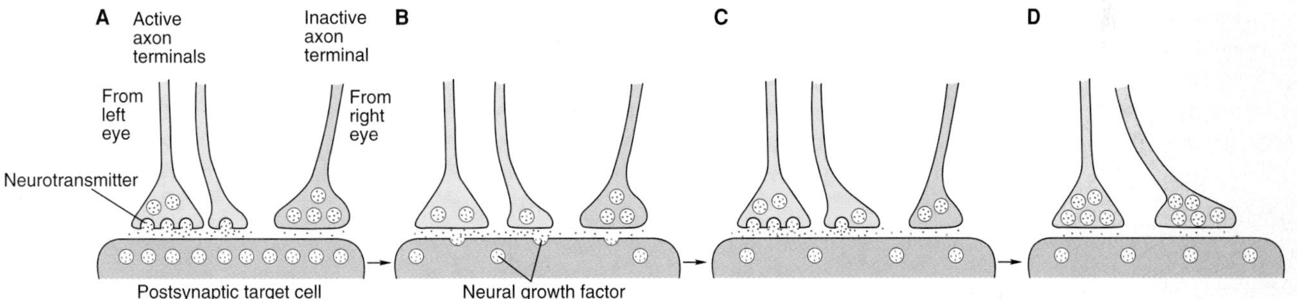

Figure 25–10 A possible mechanism for cooperation and competition among neurons during fine tuning of synaptic connections in the visual cortex.

According to this model, stabilization of a synapse depends on growth factors released from the postsynaptic (cortical) cell that stimulate the growth of active presynaptic terminals (in this case a cell conveying inputs from the left eye). Synchronous release of the growth factor is triggered by the depolarization produced in the cortical neurons by the synchronous action potentials produced in neighboring afferent neurons. These postsynaptic potentials are sufficiently large to depolarize and activate the NMDA receptors only when there is synchronous activity among the presynaptic terminals. In the visual system, cells carrying information from neighboring regions in the retina tend to fire in synchrony. In the absence of activity there is only a low level of spontaneous release of the growth factor.

According to the model, the presynaptic terminal can take up the growth factor only when it is active, and it does so when it has just released neurotransmitter by exocytosis and is in the process of endocytic retrieval of the vesicle membrane (see Chapter 15). Active presynaptic terminals that take up the factor grow additional terminals that reinforce the strength of the contact with the cortical cell. The more frequently the postsynaptic cortical cell is stimulated, the more its internal store of the factor is depleted and the lower is the rate of spontaneous release. Thus, inactive presynaptic terminals competing with active ones fail to obtain adequate amounts of growth factor and consequently shrink and eventually withdraw (**B–D**).

When two axon terminals are active at different times, they compete with one another for the limited amount of growth factor that the postsynaptic cell contains. A large terminal takes up more growth factor, which in turn makes it larger still. The outcome of such competition may thus depend on slight differences in the initial size of the synaptic terminals (**A**). (Modified from Alberts et al., 1989.)

Different Regions of the Brain Have Different Critical Periods of Development

It is becoming clear that, as with other aspects of behavior, both the ability to perceive form and the binocular vision necessary for depth perception develop in stages after birth. Each stage culminates in one or more developmental decisions, many of which are irreversible. At each stage appropriate sensory experiences are necessary to validate, shape, and update normal developmental processes. Consequently, the effects of sensory deprivation are most severe during a critical period early in postnatal life.

In the past it has been difficult to relate the development of behavior to the development of the nervous system. However, research on ocular dominance columns is providing an important bridge between the two. For example, developmental studies have shown that neurons in the visual cortex mature and become sensitive to binocular disparity toward the end of the fourth postnatal month. Psychophysical studies have shown that stereopsis develops at the same time. Thus, stereoscopic vision seems to parallel the maturation of the ocular dominance columns.

Critical periods of development generally do not have *sharp* time boundaries. Different layers within one region of the brain may have different critical periods of development, so that even after the critical period for one layer has passed, rearrangement of the layer may still be possible because the entire region has not yet fully developed. As an example, 8 weeks after birth layer 4C in the visual cortex of the monkey is no longer affected by monocular deprivation, whereas the upper and lower layers continue to be susceptible for almost the entire first year.

It is thought that each cortical area and each layer within each area has its own timetable, and thus its own critical period, for the segregation of connections. Experiences that interfere with the development of a primary sensory region of the brain, such as visual deprivation, might produce their behavioral consequences early in postnatal development, whereas other experiences, such as social deprivation, which we consider next, might act on association cortices and perhaps exert their actions later. Such differential maturation of regions of the brain would explain two well-known features of cognitive and behavioral development: (1) certain capabilities, such as those for language, music, or mathematics, usually must be developed well before puberty if they are to develop at all; and (2) traumatic insults at particular stages of postnatal life affect particular aspects of perceptual or character development.

There Is an Early Critical Period in the Development of Social Competence

Just as there are critical periods in the development of the brain during which critical sensory experiences are necessary to the normal development of sensory abilities, likewise the normal development of behavior also depends on the occurrence of particular social activities at specific stages of development. During these critical periods the infant must undergo certain social interactions if development of behavior is to proceed satisfactorily.

A particularly well-studied example of a critical period in the acquisition of a normal social behavior is *imprinting*, a form of learning in birds, which was examined in detail by the Austrian ethologist Konrad Lorenz. Just after birth, birds become attached to a prominent moving object in their environment, typically the mother. Imprinting is important for the protection of the hatchling; it is acquired rapidly and, once acquired, the attachment generally persists. However, imprinting can be acquired only during a critical period (which in some species lasts just a few hours) early in postnatal development. Imprinting therefore illustrates the close relationship between programmed development and learning.

The clearest way to show that certain social or perceptual experiences are important for development is to deprive an infant of these stimuli and to examine the consequences on later perceptual or social competence. There are reliable histories of children who suffered complete abandonment in the wild and who later returned to human society. There is also anecdotal evidence on the fate of newborn infants left in isolation during the major part of each day, being fed but not otherwise cared for. Not surprisingly, severely isolated children become socially maladjusted, usually in an irreversible way. These abandoned children are often mute and incapable of learning language.

The first compelling evidence that early social interaction with other humans is essential for normal development came from studies in the 1940s.

The development of infants raised in a foundling home for abandoned children was compared with the development of infants raised in a nursing home attached to a women's prison. Both institutions were clean, and both provided adequate food and medical care. The babies in the nursing home were all cared for by their mothers, who, because they were in prison and away from the rest of their families, tended to shower affection on their infants in the limited time allotted to them each day. In contrast, infants in the foundling home were cared for by nurses, each of whom was responsible for seven babies. As a result, children in the foundling home had much less contact with other humans than those in the prison's nursing home.

The two institutions also differed in another respect. In the nursing home the cribs were open, so that the infants could readily watch the activity in the ward; they could see other babies play and observe their mothers and the staff go about their business. In the foundling home the bars of the cribs were covered by sheets that prevented the infants from seeing outside. This dramatically reduced the infants' environment. In effect, the babies in the foundling home lived under conditions of sensory and social deprivation.

A group of newborn infants at each of the two institutions was followed throughout their early years. At the end of the first 4 months, the infants in the foundling home scored better than those in the nursing home on several developmental tests. This suggested that genetic factors did not favor the infants in the prison nursing home. However, 8 months later, at the end of the first year, the motor and intellectual performance of the children in the foundling home had fallen far below that of children in the nursing home, and many had developed a syndrome called *hospitalism* (now often called *anaclitic depression*). These children were withdrawn, showed little curiosity or gaiety, and were prone to infection.

By their second and third years, children in the nursing home were similar to children raised in normal families at home: they walked well and talked actively. In contrast, the development of the children in the foundling home was delayed. Only 2 of 26 children in the foundling home were able to walk and speak, and even those two could only say a few words. Normal children at this age are agile, have a vocabulary of hundreds of words, and speak in sentences. Thus, severe social and sensory deprivation in early childhood can have catastrophic consequences for later development. In contrast, isolation later in life (although often unpleasant) is ordinarily much better tolerated.

This work was carried one important step further in the 1960s, when psychologists studied monkeys reared in isolation. They found that newborn monkeys isolated for 6 months to 1 year were physically healthy but behaviorally devastated. These monkeys crouched in a corner of their cages and rocked back and forth like autistic children. They did not interact with other monkeys, nor did they fight, play, or show any sexual interest. Thus, a 6-month period of social isolation during the first one and a half years of life produces persistent and serious disturbances in behavior. By comparison, isolation of an older animal for a comparable period was found to be innocuous.

An Overall View

The precise neural connections within the sensory areas of the brain are achieved by two different mechanisms. First, as we have seen in Chapter 6, various molecular cues guide axons from specific regions of the periphery to particular, yet broadly defined, target regions. Once this initial alignment is accomplished a second set of processes takes over.

This second set of mechanisms, based on cooperation of the axons from the same eye and competition of the outgrowing axons from the opposite eye, matches each axon to its specific target neuron and thereby introduces a point-to-point order in the map of the target region. In the primary visual cortex cooperation between afferent fibers from the same eye and competition between afferents from the two eyes result in alternating ocular dominance columns. In this precise matching, cooperation among afferent fibers from local regions of the retina of the same eye enhances the ability of these afferents to predominate where they have common connections to target cells, thereby helping to segregate the axons from the two eyes. At the same time, competition between fibers from the two eyes also separates the axons, because the weaker of the inputs from the two retinas onto a common target will decline until eventually the axon terminal withdraws, eliminating the overlap and leading to the almost complete segregation of the terminals. These two mechanisms—cooperation and competition—interact to establish a precise topographical map.

During a critical period in postnatal development cooperation and competition are regulated by activity in the afferent fibers. During this period the segregation of afferent fibers and the establishment of ocular dominance columns can be dramatically affected by experimentally changing the balance of input activity from the two eyes. After the critical period existing connections become stable and much less susceptible to modification.

Studies of the development of the ocular dominance columns allow us to understand how other, more complex, sensory experiences early in development may change the circuitry and structure of the growing brain. These studies also suggest that the use of drugs such as narcotics and alcohol during pregnancy can have profound effects on the brain wiring of the infant by interfering with activity-dependent development of neural connections.

Studies of sensory development provide a striking example of how genetic factors and experience interact in the maturation of the brain and how environmental deprivation can dramatically alter developmental processes. In addition to providing insights into the mechanisms governing development, these studies have obvious clinical relevance. For example, the clinical treatment of strabismus, a misalignment of the visual axes of the two eyes, has changed because of studies of its effects on the development of visual perception. Children with strabismus initially have good vision in each eye. However, because these children cannot fuse the images in the two eyes, they often tend to favor one eye. Ophthalmologists used to delay correcting strabismus in children until they had reached 8 or 9 years of age, long after the critical period for development of visual processing centers in the brain. As a result, these children often lost useful vision in the neglected eye. Because of the studies of the effects of strabismus on the development of vision, ophthalmologists now surgically correct the strabismus very early, when normal binocular vision can still be restored.

Selected Readings

Constantine-Paton, M., Cline, H. T., and Debski, E. 1990. Patterned activity, synaptic convergence, and the NMDA receptor in developing visual pathways. Annu. Rev. Neurosci. 13:129–154.

Frith, U. 1989. Autism: Explaining the Enigma. Oxford: Basil Blackwell.

Harlow, H. F. 1958. The nature of love. Am. Psychol. 13:673–685.

Hebb, D. O. 1949. The Organization of Behavior: A Neuropsychological Theory. New York: Wiley.

Held, R. 1989. Perception and its neuronal mechanisms. Cognition 33:139–154.

Hubel, D. H. 1988. Eye, Brain, and Vision. New York: Scientific American Library.

Hubel, D. H., and Wiesel, T. N. 1977. Ferrier Lecture: Functional architecture of macaque monkey visual cortex. Proc. R. Soc. Lond. [Biol.] 198:1–59.

Knudsen, E. I. 1984. The role of auditory experience in the development and maintenance of sound localization. Trends Neurosci. 7:326–330.

Leiderman, P. H. 1981. Human mother–infant social bonding: Is there a sensitive phase? In K. Immelmann, G. W. Barlow, L. Petrinovich, and M. Main (eds.), Behavioral Development: The Bielefeld Interdisciplinary Project. Cambridge, England: Cambridge University Press, pp. 454–468.

Meister, M., Wong, R. O. L., Baylor, D. A., and Shatz, C. J. 1991. Synchronous bursts of action potentials in ganglion cells of the developing mammalian retina. Science 252:939–943.

Miller, K. D., Keller, J. B., and Stryker, M. P. 1989. Ocular dominance column development: Analysis and simulation. Science 245:605–615.

Moreau, E. 1913. Histoire de la guerison d'un aveugle-ne. Ann. Ocul. 149:81–118.

Rakic, P. 1981. Development of visual centers in the primate brain depends on binocular competition before birth. Science 214:928–931.

Riesen, A. H. 1958. Plasticity of behavior: Psychological aspects. In H. F. Harlow and C. N. Woolsey (eds.), Biological and Biochemical Bases of Behavior. Madison: University of Wisconsin Press, pp. 425–450.

Shatz, C. J. 1990. Impulse activity and the patterning of connections during CNS development. Neuron 5:745–756.

Shatz, C. J., and Stryker, M. P. 1988. Prenatal tetrodotoxin infusion blocks segregation of retinogeniculate afferents. Science 242:87–89.

VII

ACTION

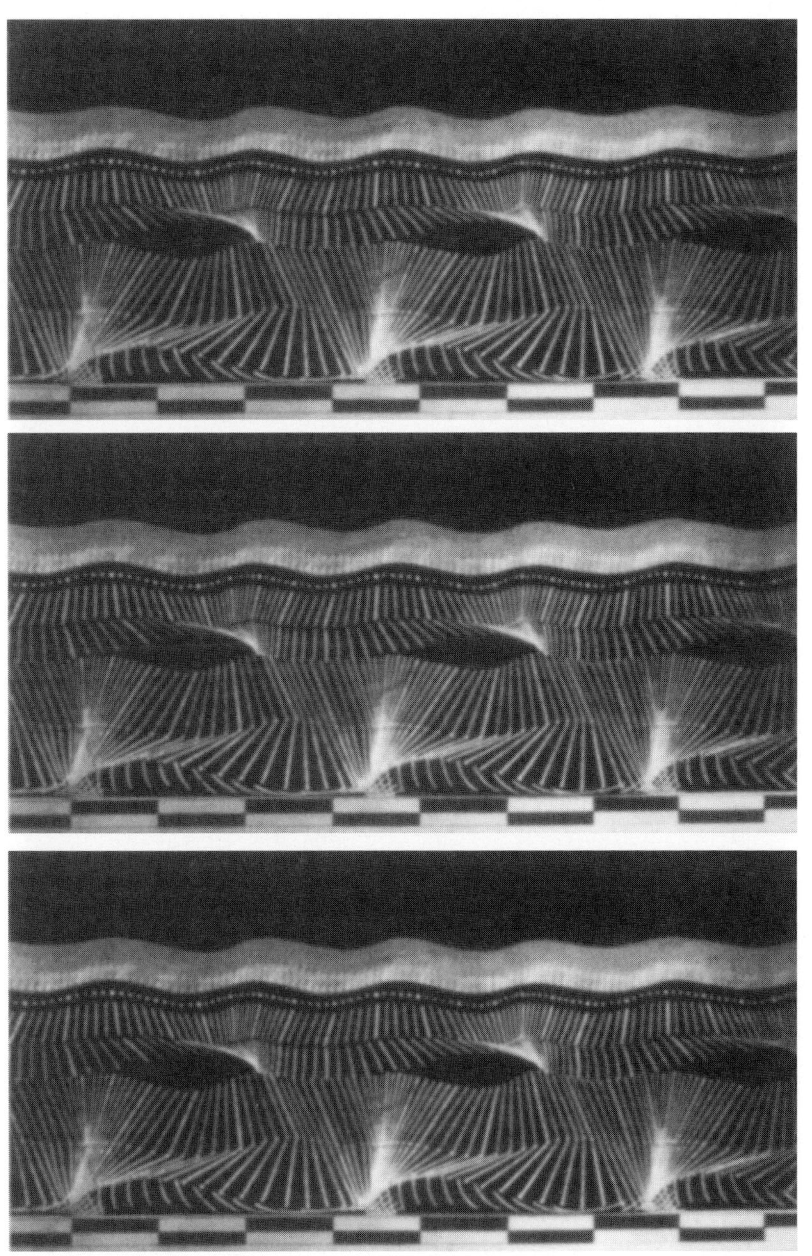

About the image on the preceding page:

The French physiologist Etienne-Jules Marey perfected several instruments for charting changes in the parts of the body during movement. Many of these instruments were based on the rotating smoked drum, or kymograph, introduced by the German physiologist Carl Ludwig in 1847, for recording change in arterial blood pressure. This apparatus operates with a stylus connected to a mercury manometer. Marey's outstanding contribution was the use of the camera to photograph multiple exposures on a single glass plate for analyzing animal locomotion. In 1883, realizing that the camera collected too much detail, he simplified the images by dressing a walking soldier in black and placing white lines and buttons at points on the body that would provide the essential information about the progressive positions of the body in time and space. Marey's images were influential not only in the study of animal movement but also in the development of modernism in art. Perhaps the clearest example of his impact on art is the work of Marcel Duchamp, whose "Nude Descending a Staircase" (1912) shows the artist's interest in the nature of movement. (Marey photograph in the Collège de France, reproduced in *Picturing Time, The Work of Etienne-Jules Marey (1830–1904)*, M. Braun, University of Chicago Press, 1992.)

Introduction

> To move things is all that mankind can do; . . . for such the sole executant is muscle, whether in whispering a syllable or in felling a forest.
>
> CHARLES SHERRINGTON, 1924

One of the main purposes of the internal representation of the outside world created by the sensory systems is to guide movement. For example, the visual system provides us with information on the location and motion of objects around us. Of course, the sensory and motor systems use a common medium, neural signals, to communicate within the nervous system. The sensory systems, however, are the entry points to the nervous system—they transform physical energy into neural signals—while the motor systems use neural signals to translate plans for action into contractile force in the muscles that produce movements. In the words of Charles Sherrington, the motor neurons are the final common path of the nervous system.

Purposeful action in complex organisms requires fine coordination of the simultaneous activity of numerous motor pathways. One of the key tasks of the motor systems is to select one appropriate response at any given moment and focus the complex machinery of movement on that one action. We humans, in particular, exercise a remarkable range of purposeful behaviors that reflects not only our highly evolved cognitive abilities but also a high degree of plastic control of movement.

Sherrington pointed out that the integrative action of the nervous system that makes purposeful behavior possible depends on a hierarchy of motor controls within a single system. He correctly recognized that spinal reflexes are the simplest example of purposeful action, and that these spinal mechanisms are critical for the execution of all movement. Thus, in the hierarchy of structures that control motor function, the spinal cord is the most basic and fundamental—it is the final decision point for voluntary as well as reflex actions. When two or more reflexes use the same motor neurons, the "singleness of action" necessary for effective behavior is achieved because only one of a family of competing reflexes is expressed while others are inhibited.

The next level of the motor hierarchy is the brain stem, and the highest level is the cerebral cortex. Each higher level contains several anatomically distinct areas that project in parallel to the spinal cord. Like the sensory systems, most of these motor areas are organized in a somatotopic fashion—movements of adjacent body parts are controlled by neighboring parts of each area of the brain. Two associated structures, the cerebellum and the basal ganglia, are not directly involved in producing movement. Rather, they modulate the actions of the corticospinal and the brain stem systems.

Movements can be divided into three broad overlapping classes that can be distinguished on the basis of their complexity and the degree to which we exert voluntary control over the movements. Reflex responses, such as the knee jerk, are the simplest motor behaviors and are least affected by voluntary controls. Rhythmic motor patterns, such as walking and running, combine features of voluntary and reflex acts. Voluntary movements represent the most complex. These movements are purposeful, goal-directed, and largely learned.

Some functions of the motor systems and their disturbance by disease are now understood at the level of the biochemistry of specific transmitter systems. The discovery that a transmitter deficiency in neurons of the basal ganglia was responsible for Parkinson's disease first suggested the possibility that other neurological and psychiatric disorders also result from altered functioning of chemical transmitter systems—malfunctions of synthesis, transport, release, and interaction with the postsynaptic receptor. We now know that the symptoms of Huntington's disease are due to a mutation that can lead to premature death of nerve cells. With the identification of the genes and proteins important for motor function, we may soon understand the molecular mechanisms of the integrative action of the nervous system.

26

An Introduction to Movement

The Psychophysics of Movement

The Motor Systems Generate Three Types of Movement

The Spinal Cord, Brain Stem, and Motor Cortex Represent Three Levels of Motor Control

The Spinal Cord Contains the Cell Bodies of Motor Neurons

The Brain Stem Modulates Motor Neurons and Interneurons in the Spinal Cord Through Two Systems

Medial Descending Pathways Control Axial and Proximal Muscles

Lateral Pathways Control Distal Muscles

The Motor Cortex Acts on Spinal Motor Neurons Directly via the Corticospinal Tract and Indirectly Through Brain Stem Pathways

The Corticospinal Tract Is the Largest Descending Fiber Tract from the Brain

Cortical Control of Movement Is Achieved Only Late in Phylogeny

The Motor Areas of the Cerebral Cortex Are Organized Somatotopically

The Cerebellum and Basal Ganglia Control the Cortical and Brain Stem Motor Systems

Sensory systems provide an internal representation of the outside world. A major function of this representation is to guide the movements that make up our behavioral repertory. These movements are controlled by the motor systems of the brain and spinal cord, which allow us to maintain balance and posture, to move our body, limbs, and eyes, and to communicate through speech and gesture. In contrast to sensory systems, which transform physical energy into neural signals, the motor systems translate neural signals into contractile force in muscles in order to produce movements.

Just as our perceptual skills are a reflection of the capabilities of the sensory systems to detect, analyze, and estimate the significance of physical stimuli, our agility and dexterity are reflections of the capabilities of the motor systems to plan, coordinate, and execute movements. The accomplished pirouette of a ballet dancer, the powered backhand of a tennis player, the fingering technique of a pianist, and the coordinated eye movements of a reader all require a remarkable degree of motor skill that no robot approaches. Yet, once trained, the motor systems execute the motor programs for these skills with ease, and for the most part automatically.

In this chapter we introduce the study of movement by examining how both the flow of sensory information and the hierarchy of motor controls govern different classes of movement. In later chapters we examine the individual components of the motor systems and the pathways through which they act on motor neurons and muscles to produce movements. In particular we shall see how the motor systems of the brain and spinal cord cooperate to control different classes of movements.

The Psychophysics of Movement

The task of the motor systems in controlling movement is the reverse of the task of sensory systems in perception. Whereas the end-product of sensory processing is to generate an internal representation of the outside world or the state of the body, motor processing *begins* with an internal representation—an image of the desired result of movement. In writing, for example, the form of a letter is independent of the particular machinery of the body used to write it. Whether you write the letter A small (using only your fingers) or large (using your fingers and wrist), whether you write it with your preferred hand, your opposite hand, your foot, or even with the pen in your mouth, the general form of the letter in all instances is the same (Figure 26–1). This principle, that the same result can be accomplished in different ways by the motor systems, is referred to as *motor equivalence*, a term first used by the psychologist Donald Hebb.

As we have seen, sensory processing involves analyzing physical events in the external world into elementary properties and assembling these sensory elements into coherent percepts. Motor processing, on the other hand, involves constructing behaviors, or motor acts, by assembling and coordinating elementary motor components. To understand how perception is formed, we investigate the ways in which sensory stimuli are encoded by neural structures, in the study of sensory psychophysics (Chapter 20). Likewise, to understand how movements are generated, we need to investigate the ways in which the elementary features of movements are coded in the nervous system.

In sensory psychophysics we look at the organization of perception in terms of four elementary attributes of a stimulus: its quality (or modality), intensity, location, and duration. In motor psychophysics we look at analogous attributes in considering the organization of action.

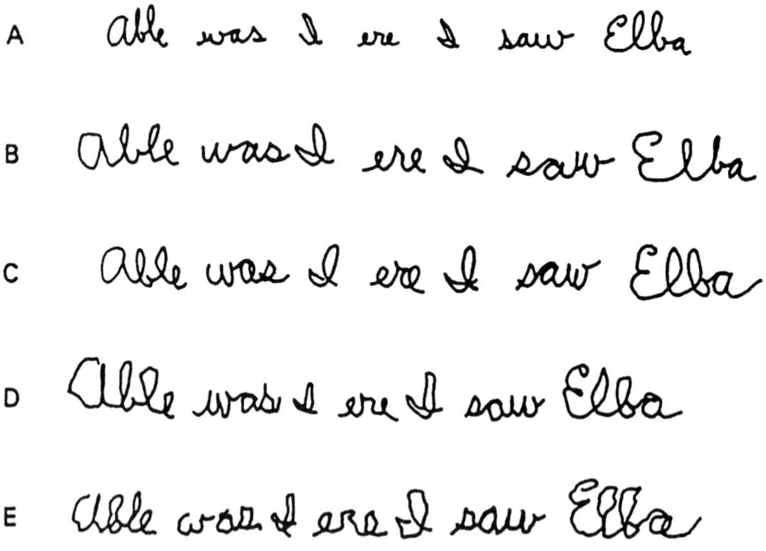

A Able was I ere I saw Elba

B Able was I ere I saw Elba

C Able was I ere I saw Elba

D Able was I ere I saw Elba

E Able was I ere I saw Elba

Figure 26–1 Writing can be performed using different parts of the body: with the right (dominant) hand (**A**); with the right arm but with the wrist immobilized (**B**); with the left hand (**C**); with the pen gripped between the teeth (**D**); and with the pen attached to the foot (**E**). The ability of different motor sets to achieve the same behavior is called motor equivalence. (From Raibert, 1977.)

Intensity of muscle contraction is signaled in a manner analogous to signaling the intensity of sensory stimuli—both by the rate of firing individual motor neurons and the recruitment of distinct populations of motor neurons (Chapter 27). Location and duration of sensory stimuli also have their counterparts in motor psychophysics, in the accuracy of a movement (how close the movement comes to a target location in space) and in the speed of movement. The relationship between accuracy and speed is perhaps the best studied aspect of motor psychophysics. In general, fast movements are less accurate than slow ones. This "speed–accuracy trade-off" was first described almost 100 years ago by the psychologist Robert Woodworth. He found that when subjects had their eyes open and moved a pen at different speeds to hit a target, the movement errors were larger with faster movements. However, when subjects moved the pen with their eyes closed, errors were large but did not change appreciably with changes in speed. Thus the tendency for faster movements to be less accurate results in part from the use of visual information to make corrections during movements. In faster movements there is less time to process visual input about errors, and therefore the errors are larger.

The Motor Systems Generate Three Types of Movement

Just as there are distinct modalities of sensation, there are three distinct categories of movement: reflexes, rhythmic motor patterns, and voluntary movements. Each of these categories of movement in turn depends on a combination of two basic modes of muscular control. First, with *phasic control* muscles are activated transiently to make discrete movements, such as reaching for a cup, pushing a button, or throwing a ball. In repetitive movements such as walking, muscles are physically activated in a rhythmic fashion, a topic we consider in Chapter 28. Second, with *tonic control* muscles are activated in steady contractions to stabilize joints, as in maintaining a certain posture or holding a pen while writing.

Movements can be divided into three broad, overlapping classes: reflex responses, rhythmic motor patterns, and voluntary movements. These classes differ in their complexity and degree of voluntary control.

Reflex responses, such as the knee jerk, the withdrawal of a hand from a hot object, or swallowing, are the simplest motor behaviors and are least affected by voluntary control. Reflexes are rapid, stereotyped involuntary responses; as a rule they are controlled in a graded way by the eliciting stimulus.

Rhythmic motor patterns, such as walking, running, and chewing, combine features of voluntary and reflex acts. Typically only the initiation and termination of the sequence are voluntary. Once initiated, the sequence of relatively stereotyped, repetitive movements may continue almost automatically in reflex-like fashion.

Voluntary movements, such as combing one's hair, driving a car, or playing the piano, are the most complex. These movements are characterized by two features: they are purposeful (goal directed) and they are largely learned (their execution improves greatly with practice). The more these skilled movements are mastered with practice, the less they require conscious direction. Once we have learned to drive a car we no longer think through the steps of shifting gears or stepping on the brakes.

In each of these classes of movement—in fact, in all behavior—muscles relax and contract. Most movements occur at joints where two or more bones form a lubricated contact point with low friction. Since muscles can only *pull* on their bony attachments when they contract or relax (they cannot push), separate sets of muscles are required at the opposite sides of the joint, which serves as a fulcrum. Each movement at a joint thus brings into play two opposing sets of muscles. *Agonists*, the prime movers, are counterbalanced by the *antagonists* that move the limb in an opposing direction and help brake movement produced by the agonists.

Beyond simply controlling the contraction of individual muscles, the motor systems have three other tasks. First, they must convey accurately timed commands to *many* groups of muscles, since even a simple motor act involves movement at many different joints (reaching forward with the hand involves the wrist, elbow, and shoulder). Second, motor systems must take into account the distribution of body mass and must plan postural

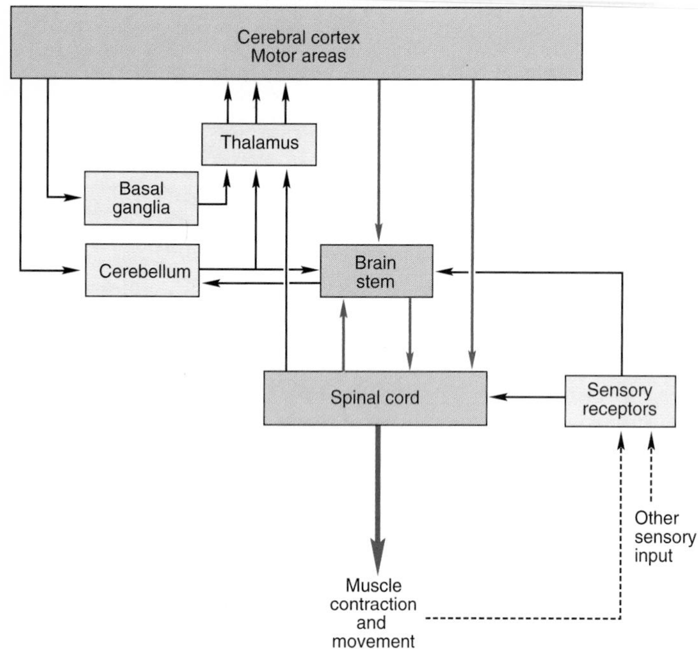

Figure 26–2 The motor systems are organized hierarchically (on three levels) and in parallel. The motor areas of the cerebral cortex can influence the spinal cord both directly and through descending systems in the brain stem. All three levels of the motor systems receive sensory inputs and are also under the influence of two independent subcortical systems: the basal ganglia and the cerebellum. Both the basal ganglia and cerebellum act on the cerebral cortex through relay nuclei in the thalamus.

adjustments appropriate for particular movements. For example, when we are standing, the leg muscles must contract before we raise an arm; otherwise the arm movement would shift our center of gravity, causing us to fall. Finally, the motor systems must take into account the *motor plant*: the mechanical properties of the muscles, bones, and joints. With each movement the motor systems must adjust their commands to compensate for the inertia of the limbs and the mechanical arrangement of the muscles, bones, and joints prior to the movement.

The Spinal Cord, Brain Stem, and Motor Cortex Represent Three Levels of Motor Control

To integrate these three features into voluntary and reflex acts, motor systems rely on two important and interrelated organizational features. First, motor systems have available to them a continuous flow of sensory information about events in the environment, the position and orientation of the body and limbs, and the degree of contraction of the muscles. Motor systems use this information to select an appropriate response and to make adjustments in ongoing movement.

Second, the motor systems are organized into three control levels: the spinal cord, the descending systems of the brain stem, and the motor areas of the cerebral cortex (Figure 26–2). Each level is provided with the sensory information that is relevant for the functions it controls. Integration is achieved by feed-back, feed-forward, and adaptive mechanisms distributed among the three levels of motor control. Thus the different levels of the motor systems are organized both *hierarchically* and in *parallel*, much as the sensory systems are organized.

The hierarchical organization of motor control permits lower levels to generate reflexes—complex patterns of muscle activation that do not require intervention of higher levels. As a result, higher centers are free to give general commands without having to specify all the details of the motor action. At the same time, the parallel organization of the motor systems permits higher centers to adjust the operation of spinal circuits (see below). In addition, parallel organization of descending systems allows certain functions to be controlled relatively independently.

Much as parallel processing of sensory information allows the brain to control the flow of discrete types of sensory information, so parallel organization in the motor systems allows the brain to control the pathways that produce discrete types of movement. For example, maintaining balance during a reaching movement is controlled through pathways different from those responsible for moving the arm itself.

The *spinal cord* is the lowest level of the hierarchy and contains neuronal circuits that mediate a rich variety of automatic and stereotyped motor patterns and reflexes. These responses can function even when the cord is disconnected from the brain. At the beginning of this century Sherrington demonstrated that virtually all reflexes and locomotor movements involve the integrated activation and inhibition of different muscle groups, and that these are coordinated by local spinal circuits. For example, both the reflexive withdrawal from a noxious stimulus and the alternating activity in flexors and extensors during locomotion are organized by networks of spinal interneurons. Moreover, simple descending commands can produce complex effects through these interneurons. Indeed, the same networks of interneurons that organize reflex behavior are also involved in voluntary movements.

All of these control signals ultimately converge on the motor neurons that innervate skeletal muscles. To stress the importance of this convergence, Sherrington called the motor neurons the *final common path*.

The next level of the motor hierarchy, the *brain stem*, contains two parallel neuronal systems, medial and lateral, whose axons project to and regulate the networks of spinal cord interneurons and motor neurons necessary for behavioral acts. The *medial systems* of the brain stem play an important role in the control of posture by integrating visual and vestibular information with somatosensory inputs. The *lateral systems* control distal muscles of the limbs and are thus important for discrete goal-directed movements, especially of the arm and hand. In addition, specific brain stem nuclei control movements of the eyes and head.

The highest level of motor control consists of three areas of cerebral cortex: the *primary motor cortex*, the *lateral premotor area* (or premotor cortex), and the *supplementary motor area*. Each area projects directly to the spinal cord through the corticospinal tract, as well as indirectly through brain stem motor systems. In addition, the premotor and supplementary motor areas, which are important for coordinating and planning complex sequences of movement, receive information from the posterior parietal and prefrontal association cortices, (see Chapter 19) and project to the primary motor cortex.

Each component of the motor system contains somatotopic maps of the head, body, and limbs. This organization is retained from level to level. Thus, regions of primary motor cortex that control the arm receive input from arm-control areas in the premotor cortex and, in turn, make connections with arm-control areas of the descending brain stem pathways.

The Spinal Cord Contains the Cell Bodies of Motor Neurons

As we have previously seen (Chapter 5), the spinal cord contains a central region of gray matter, made up mostly of the cell bodies of the constituent neurons, and a surrounding region of white matter, consisting of axons. The cell bodies of the motor neurons are located in the ventral part of the gray matter, the *ventral horns* (Figure 26–3). The motor neurons that innervate individual muscles are clustered in motor nuclei (also called *motor neuron pools*) arranged in longitudinal columns extending one to four spinal segments. The spatial organization of the different motor nuclei conforms to two important anatomical and functional rules: a proximal–distal rule and a flexor–extensor rule.

According to the *proximal–distal rule* the motor neurons innervating the most proximal (axial) muscles are located most medially, while those innervating more distal muscles are located progressively more laterally (Figure 26–3). According to the *flexor–extensor rule* the motor neurons that innervate extensor muscles lie ventral to those innervating flexor muscles (Figure 26–3).

These anatomical relationships correspond to an important functional distinction: axial (postural) muscles and proximal muscles (especially the extensor muscles of the legs) are predominantly used to maintain equilibrium and posture. In contrast, distal muscles (especially those of the upper extremity) are used for fine manipulatory activities. The medial and lateral motor neurons are controlled by separate populations of interneurons and descending pathways.

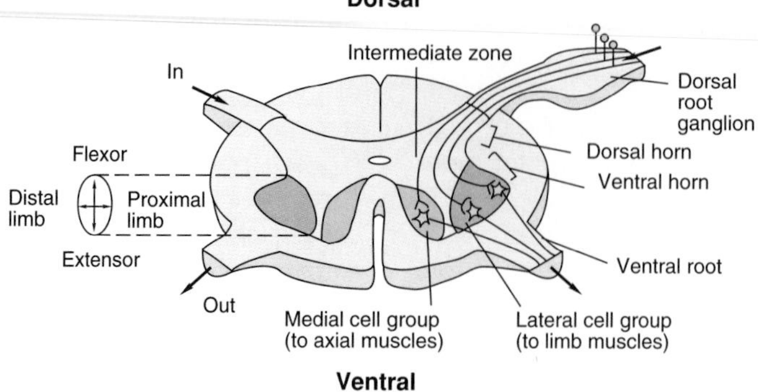

Figure 26–3 The motor nuclei of the spinal cord are grouped functionally in distinct medial and lateral positions. The medial group contains the motor neurons innervating axial muscles of the neck and back. Within the lateral group, the most medial motor neurons innervate proximal muscles, while the most lateral innervate distal muscles. Ventral motor neurons innervate extensors; dorsal ones innervate flexors.

The Brain Stem Modulates Motor Neurons and Interneurons in the Spinal Cord Through Two Systems

Many groups of neurons in the brain stem project to the spinal gray matter. These projections can be classified into two main systems of pathways on the basis of their location and distribution in the spinal cord. The *medial pathways*, concerned with the control of posture, terminate in the ventromedial part of the spinal gray matter and thus influence motor neurons that innervate axial and proximal muscles. The *lateral pathways* are concerned with goal-directed movements, especially of the arms and hands. These pathways terminate in the dorsolateral part of the spinal gray matter and influence motor neurons that control distal muscles of the limbs.

Medial Descending Pathways Control Axial and Proximal Muscles

The medial system of descending pathways has three major components: the vestibulospinal, reticulospinal, and tectospinal tracts. These pathways descend in the ipsilateral ventral columns of the spinal cord and terminate predominantly on interneurons in the intermediate zone and on medial motor neurons, which innervate axial muscles (Figure 26–4A). The medial pathways are important in maintaining balance and posture, both of which rely on proximal and axial muscles. Individual axons in this system terminate over a wide area, thus distributing control to a variety of different motor nuclei that are functionally related.

The *vestibulospinal tracts* originate in the vestibular nuclei and carry information for the reflex control of balance and posture from the vestibular labyrinth.

The *reticulospinal tracts* originate from several nuclei in the reticular formation of the pons and medulla and are important for the maintenance of posture. These systems integrate information from a variety of inputs, notably the vestibular nuclei and cerebral cortex, and make both excitatory and inhibitory connections with spinal interneurons and motor neurons.

The *tectospinal tract* is important in coordinating head and eye movements. It originates in the superior colliculus of the midbrain and is controlled from the cerebral cortex.

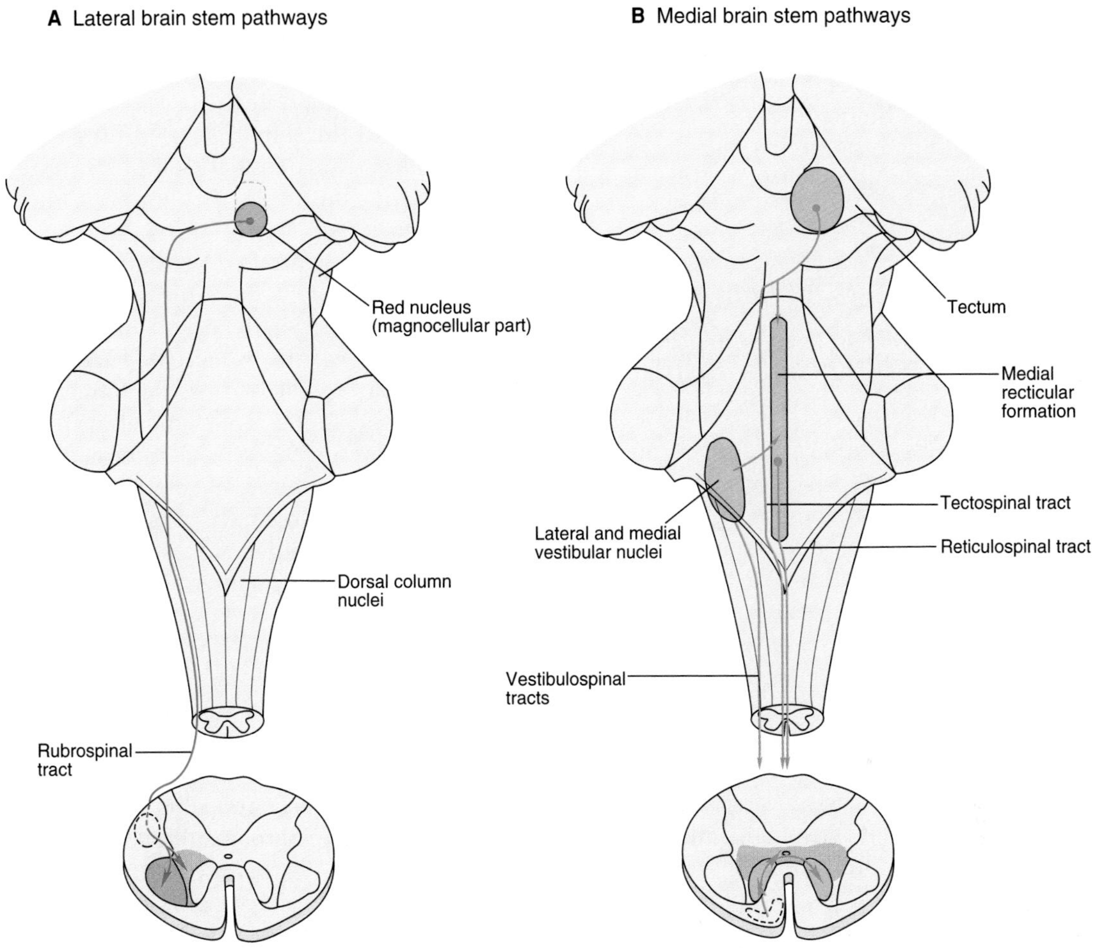

Figure 26–4 Two groups of descending pathways from the brain stem control different groups of motor neurons and different groups of muscles.

A. The *medial pathways* include the reticulospinal, the vestibulospinal, and the tectospinal tracts. These tracts descend in the ventral column and terminate in the spinal gray matter.

B. The main *lateral pathway* is the rubrospinal tract, which originates in the caudal, magnocellular portion of the red nucleus. The rubrospinal tract descends in the contralateral dorsolateral column and terminates in the spinal gray matter.

Lateral Pathways Control Distal Muscles

The main lateral descending pathway from the brain stem is the *rubrospinal tract*, which originates in the magnocellular portion of the red nucleus in the midbrain (Figure 26–4B). Rubrospinal fibers descend through the medulla to the dorsal part of the lateral column of the spinal cord and terminate among the dorsolateral group of motor neurons innervating the distal limb muscles. These distal muscles are used in fine movements, such as reaching and manipulating objects with the fingers and hand.

The Motor Cortex Acts on Spinal Motor Neurons Directly via the Corticospinal Tract and Indirectly Through Brain Stem Pathways

The ability to organize complex motor acts and to execute fine movements with precision depends on control signals from the motor areas in the cerebral cortex. Cortical motor commands descend in two tracts. The *corticobulbar fibers* control the cranial motor nerve nuclei, and thus the facial muscles, while the *corticospinal fibers* control the motor neurons innervating the trunk and limb muscles. Corticospinal axons act directly on the motor neurons and local interneurons. The motor cortex also influences motor activity indirectly through the descending brain stem pathways, notably through corticoreticulospinal, corticorubrospinal, and corticobulbar projections.

The Corticospinal Tract Is the Largest Descending Fiber Tract from the Brain

The corticospinal tract is a massive bundle of fibers containing about one million axons. About half of these originate in the primary motor cortex (Brodmann's area 4), located in the precentral gyrus of the frontal lobe. Electrical stimulation of the primary motor cortex evokes movements of different contralateral muscle groups. Another third of the corticospinal fibers originate in the premotor area (area 6), a larger zone rostral to area 4. The remaining axons originate in areas 3, 2, and 1 in the somatic sensory cortex and regulate the transmission of sensory input to different parts of the brain.

The corticospinal fibers course through the internal capsule to the ventral portion of the midbrain. As they descend through the pons they separate into small bundles that run between the pontine nuclei. The fibers regroup in the medulla to form the *medullary pyramid*, a conspicuous landmark on the ventral surface of the medulla where it forms the pyramidal tract.

At the junction of the medulla and the spinal cord about three-quarters of the corticospinal fibers cross the midline in the *pyramidal decussation*. The crossed fibers descend in the dorsal part of the lateral columns of the spinal cord, forming the *lateral corticospinal tracts*. The uncrossed fibers descend in the ventral columns as the *ventral corticospinal tracts* (Figure 26–5).

The lateral and ventral corticospinal tracts terminate in approximately the same regions of spinal gray matter as do the lateral and medial descending brain stem systems, respectively (compare Figure 26–4). The lateral corticospinal tract projects primarily to the lateral motor nuclei of the ventral horn and to interneurons in the intermediate zone. The ventral corticospinal tract projects bilaterally to the medial cell column, which contains the motor neurons that innervate axial muscles, and to adjoining portions of the intermediate zones.

Cortical Control of Movement Is Achieved Only Late in Phylogeny

Phylogenetically, the corticospinal and corticobulbar pathways first appear in mammals, where they initially serve to control and adjust the flow of sensory information to spinal interneurons and neurons.

Higher mammals have distinct sensory and motor representations of the body in the cortex and have additional corticospinal terminations within the intermediate zone of the spinal cord. During evolution the number of corticospinal fibers to more ventral regions of the spinal cord gradually increases, so that corticospinal neurons make direct connections to motor neurons in the lateral motor nuclei that control distal limb muscles and, later, also to neurons in medial motor nuclei.

More complex control is seen in primates. The number of corticospinal axon terminals ending on spinal motor neurons increases progressively from prosimians to monkeys, anthropoid apes, and finally to humans. In more primitive primates direct connections are present only in the most dorsolateral cell groups innervating the most distal muscles. In monkeys the entire lateral group of

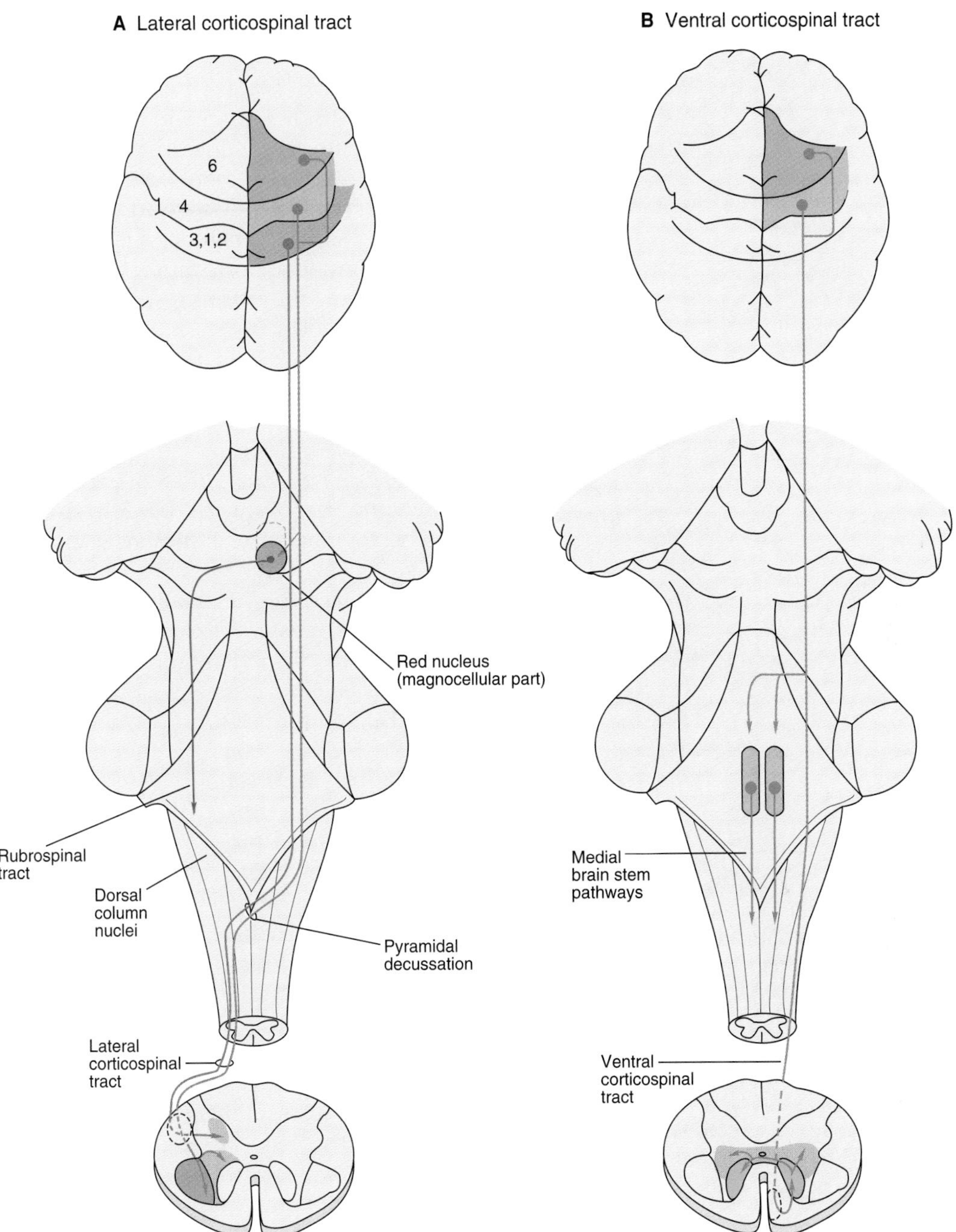

Figure 26–5 The cortex directly controls motor neurons in the spinal cord through two descending pathways.

A. The crossed *lateral corticospinal tract* originates in two motor areas (Brodmann's areas 4 and 6) and three sensory areas (3, 2, and 1). The tract crosses at the pyramidal decussation, descends in the dorsolateral column, and terminates in the spinal gray matter. Corticorubral neurons are located mainly in area 6. The corticospinal neurons in the sensory cortex terminate primarily in the medial portion of the dorsal horn. Collateral fibers project to dorsal column nuclei.

B. The uncrossed *ventral corticospinal tract* originates principally in Brodmann's area 6 and in zones in area 4 controlling the neck and trunk. Terminations are bilateral and collaterals project to the medial pathways from the brain stem.

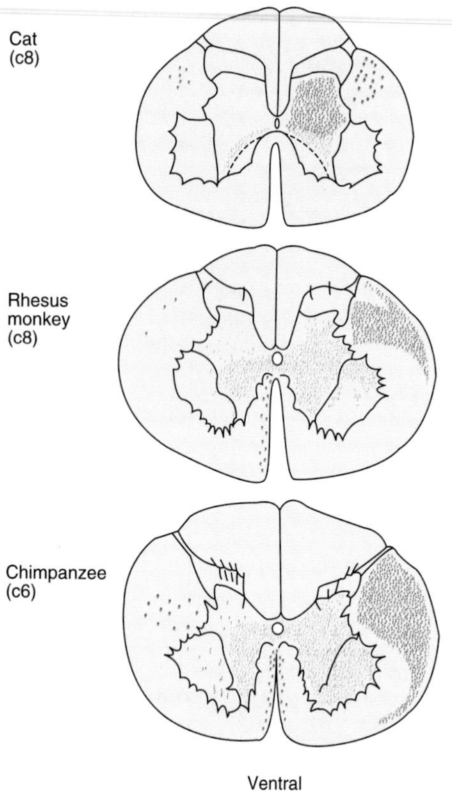

Figure 26–6 Motor cortical neurons in different species have different patterns of termination. In cats the corticospinal fibers terminate principally on neurons in the ventral parts of the dorsal horn and in the spinal intermediate zone. In lower primates, such as the rhesus monkey, most terminations remain in the intermediate zone but a small number also reach the motor neurons. In more highly evolved primates, such as chimpanzees and humans, there are extensive terminations throughout the contralateral intermediate zone and both medial and lateral motor neuron groups. A substantial ipsilateral fiber tract is also present and terminates primarily on proximal muscles important for postural control. (Adapted from Kuypers, 1985.)

motor nuclei receives corticospinal input; in higher apes and humans, the medial motor nuclei also receive dense corticospinal terminations. In most carnivores corticospinal fibers terminate exclusively in the dorsal horn and dorsolateral parts of the intermediate zone and do not make any direct connections with motor neurons (Figure 26–6).

The Motor Areas of the Cerebral Cortex Are Organized Somatotopically

In 1870 Gustav Fritsch and Eduard Hitzig provided the first direct evidence that distinct areas of the brain control movements on the contralateral side of the body. They discovered that electrical stimulation of different parts of the cortex of dogs produces contractions of different contralateral muscles. These observations were soon extended to monkeys by David Ferrier and to apes by Sherrington, who discovered that in primates motor effects are elicited most readily from the precentral gyrus. This region corresponds to Brodmann's area 4 and is now called the primary motor cortex.

The discovery that different areas of cerebral cortex control movements of different parts of the body had immediate clinical relevance. It explained why damage to different areas of the contralateral frontal lobe results in weakness of the face, arm, or leg. It also enabled clinicians to understand the mechanism of focal motor seizures. For example, the Jacksonian seizure, described by the neurologist Hughlings Jackson, typically begins with a series of clonic (abrupt, intense, and repetitive) involuntary contractions of muscles on one side, commonly the finger flexors. The contractions gradually spread proximally to the wrist, then to the elbow, shoulder, trunk, and other muscles. The abrupt, intense muscle jerks that occur during the seizure resemble those elicited by electrical stimulation of the cortex. Jackson correctly surmised that the sequential activation of different muscle groups during the seizure results from the progressive spread of abnormal motor activity from a site in the cortex controlling distal extremity muscles to the site controlling more proximal ones. Frequently, these focal seizures are triggered by tumors, scars, or other abnormalities in nearby areas of the brain.

Later work in patients and in monkeys showed that the primary motor cortex contains a *motor map* of the body (see Chapter 1). The head is represented close to the lateral sulcus; above it are representations of the arms, trunk, and legs (Figure 26–7A). As with the sensory maps, not all body parts are represented equally in the motor map. The parts of the body used in tasks requiring precision and fine control, such as the face and hands, have a proportionately larger representation in the motor map (Figure 26–7B).

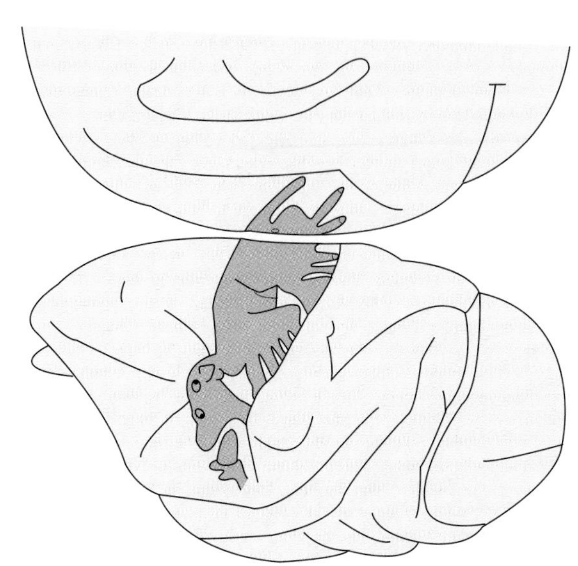

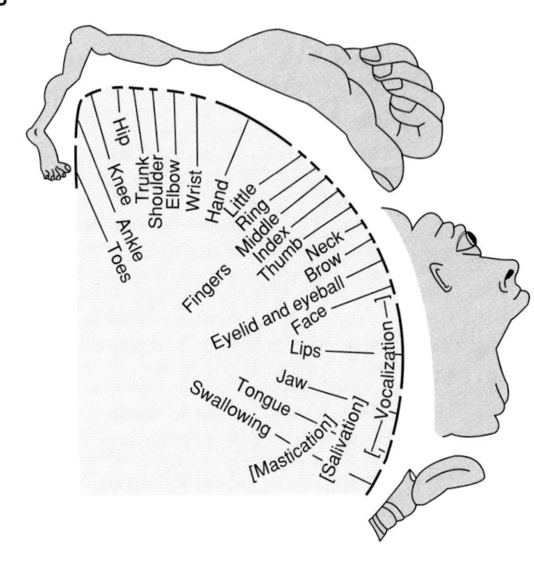

Figure 26–7 Comparison of the somatotopic representations in the primary motor cortex of monkeys (**A**) and humans (**B**). This sequence of representation is similar, with the ankles being medial and the face, mouth, and muscles of mastication lateral. But the human motor cortex has a much larger representation of the face and digits. (**A** from Woolsey, 1958; **B** adapted from Penfield and Rasmussen, 1950.)

The Cerebellum and Basal Ganglia Control the Cortical and Brain Stem Motor Systems

In addition to the three hierarchical levels—spinal cord, brain stem, and corte—two other parts of the brain also regulate motor function: the cerebellum and basal ganglia. The cerebellum improves the accuracy of movement by comparing descending motor commands with information about the resulting motor action. The cerebellum makes these corrections by acting on the brain stem and on the cortical motor areas that project directly to the spinal cord, monitoring both their activity and the sensory feedback signals they receive from the periphery.

The basal ganglia receive inputs from all cortical areas and project principally to areas of frontal cortex that are concerned with motor planning. Diseases of the basal ganglia produce a range of motor abnormalities, including loss of spontaneous movements, abnormal involuntary movements, and disturbances in posture.

The actions of the cerebellum and basal ganglia are described in detail in connection with voluntary movement (Chapter 29).

An Overall View

All behavior involves movement and depends on the contraction of muscles controlled by the motor systems. These systems are hierarchically organized, from the spinal circuits that control automatic reflexes to higher centers in the brain stem and motor cortex. These three components—spinal cord, brain stem, and cortex—also function in parallel, so that any one of them can control movement somewhat independently of the other two.

Different components of the motor systems carry out distinct but interrelated functions. Thus, while the spinal cord and brain stem mediate reflexive and simple, automated postural and locomotor movements, the cortical motor areas initiate and control more complex voluntary movements. The prefrontal motor cortex and basal ganglia are thought to be involved in planning movement and in coordinating movement of body parts. The cerebellum is responsible for coordinating precisely timed activity by integrating motor output with ongoing sensory feedback.

The corticospinal pathway (and its corticobulbar counterpart that controls muscles of the head) is the most direct and powerful route by which the cerebral cortex can control the motor neurons that innervate all of our muscles. The direct projection from primary motor cortex to distal motor neurons is the only route by which the brain can individually control the muscles of our hands and fingers. Therefore, injury to these fibers results in the permanent loss of all the skilled movements we use in manipulation of small objects and in tactile exploration. All other muscle groups are controlled both directly, by corticospinal fibers, and indirectly, by the influence of the motor cortex on the various descending brain stem systems. This redundancy allows for significant recovery of function in cases of injury.

Selected Readings

Alexander, G. E., and DeLong, M. R. 1986. Organization of supraspinal motor systems. In A. K. Asbury, G. M. McKhann, and W. I. McDonald (eds.), Diseases of the Nervous System, Vol. I: Clinical Neurobiology. Philadelphia: Saunders, pp. 352–369.

Evarts, E. V., Wise, S. P., and Bousfield, D. (eds.) 1985. The Motor System in Neurobiology. New York: Elsevier.

Gallistel, C. R. 1980. The Organization of Action: A New Synthesis. Hillsdale, N.J.: Erlbaum.

Kuypers, H. G. J. M. 1985. The anatomical and functional organization of the motor system. In M. Swash, and C. Kennard (eds.), Scientific Basis of Clinical Neurology. New York: Churchill Livingstone, pp. 3–18.

Marsden, C. D., Rothwell, J. C., and Day, B. L. 1984. The use of peripheral feedback in the control of movement. Trends Neurosci. 7:253–257.

Miles, F. A., and Evarts, E. V. 1979. Concepts of motor organization. Annu. Rev. Psychol. 30:327–362.

Rosenbaum, D. A. 1990. Human Motor Control. San Diego: Academic Press.

Rothwell, J. C. 1987. Control of Human Voluntary Movement. Rockville, Md.: Aspen.

Schmidt, R. A. 1988. Motor Control and Learning: A Behavioral Emphasis, 2nd ed. Champaign, Ill.: Human Kinetics Publishers.

Sherrington, C. 1947. The Integrative Action of the Nervous System, 2nd ed. New Haven: Yale University Press.

27

Muscles and Muscle Receptors

A Motor Unit Consists of a Single Motor Neuron and the Muscle Fibers It Innervates

The Nervous System Grades the Force of Muscle Contraction in Two Ways

- *Increases in Firing Rate of Motor Units Produce Increasing Force Output*
- *Motor Units Are Recruited in a Fixed Order from Weakest to Strongest*

Muscles Contain Specialized Receptors that Sense Different Features of the State of the Muscle

- *Muscle Spindles Respond to Stretch of Specialized Muscle Fibers*
- *Golgi Tendon Organs Are Sensitive to Changes in Tension*
- *Functional Differences Between Spindles and Tendon Organs Derive from Their Different Anatomical Arrangements Within Muscle*
- *The Primary and Secondary Afferent Endings Respond Differently to Phasic Changes in Length*

The Central Nervous System Controls the Sensitivity of the Muscle Spindles Through the Gamma Motor Neurons

Much of the elaborate information processed by the brain is used to control the contractile force in our skeletal muscles. It is the controlled contraction of muscle that allows us to move our limbs, maintain posture, and perform a variety of tasks with great precision. The force produced in contracting muscle and the resulting change in length of the muscle are dependent on three factors: the initial length, the velocity of length change, and the external loads acting to oppose movement. Thus, the central nervous system needs information about the lengths of the muscles and the forces they generate. This *proprioceptive information* is monitored by two types of receptors, the muscle spindles and Golgi tendon organs.

Muscle spindles are of special interest because their activity is modulated in turn by the brain. Muscle spindles therefore provide a particularly well-studied example of how the central nervous system can control the inflow of sensory information. Information from muscle spindles and tendon organs reaches all levels of the nervous system. At the cerebral cortex it is used for the perception of limb position and for controlling voluntary movements. At lower levels it is used to control reflexes, the simplest of motor behaviors. (Reflexes are discussed in detail in Chapter 28.)

A Motor Unit Consists of a Single Motor Neuron and the Muscle Fibers It Innervates

In mammals each skeletal muscle fiber is innervated by only one motor neuron. A single motor neuron, however, innervates several muscle fibers, and

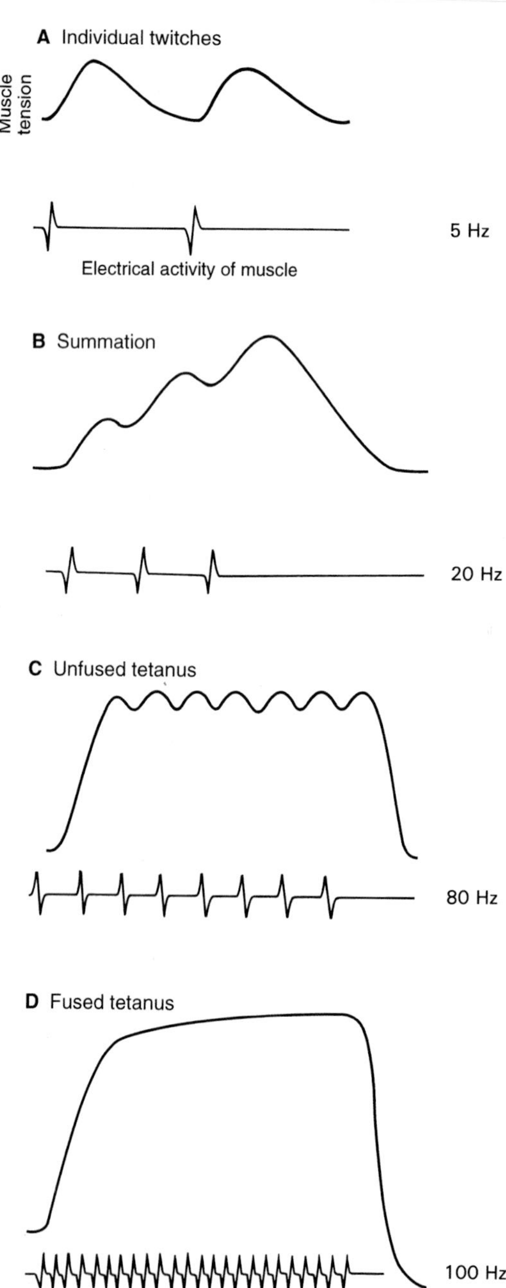

Figure 27–1 Active muscle tension varies with the rate of firing of the motor neurons.

A. Successive twitches evoked at 5 Hz.

B. Summation of successive twitches.

C. Unfused tetanus, in which individual twitches can be distinguished.

D. Fused tetanus, in which individual twitches are no longer identifiable.

all the fibers innervated by one motor neuron contract in response to an action potential in the neuron. Together, a single motor neuron and all the muscle fibers it innervates constitute the smallest functional unit within the motor systems. In 1925 Sherrington introduced the term *motor unit* to describe this basic unit of motor function that underlies all motor behavior.

The number of muscle fibers innervated by one motor neuron is called the *innervation ratio*. This ratio varies considerably among muscles but is roughly proportional to the size of the muscle. A low innervation ratio allows for finer grading of the muscle's total force by the nervous system, much as small receptive fields allow greater spatial resolution in the somatic sensory and visual systems. In human extraocular muscles, which are very small, the ratio is about 10; in hand muscles, which are somewhat larger, it is about 100; in the still larger gastrocnemius muscle of the knee the ratio is very high—about 2,000 fibers are innervated by a single motor neuron.

The nervous system can grade the force of a muscle contraction in two ways. First, it can vary the numbers of motor units activated—the more that are activated, the higher the force the whole muscle will produce. This mechanism is called *recruitment*. Second, it can vary the rate of action potentials in a motor neuron—the higher the rate of firing in the motor neuron, the higher the force the motor unit will produce. This mechanism is called *rate modulation*. Each of these mechanisms is discussed in detail in this chapter.

The Nervous System Grades the Force of Muscle Contraction in Two Ways

Increases in Firing Rate of Motor Units Produce Increasing Force Output

As noted earlier, one mechanism the nervous system can use to vary force in muscles is to modulate the rate of firing of motor neurons. The action potential in a single nerve or muscle lasts only 1 to 3 ms, whereas the contraction and relaxation time for a muscle *twitch*, a brief contraction, takes 10 to 100 ms. Thus, an increase in the frequency of firing allows the forces of successive twitches to summate (Figure 27–1B). If the muscle is not allowed to relax between successive action potentials, the forces produced by each twitch add until a plateau of force, or *tetanus*, is reached.

Two forms of tetanus can be distinguished based on the degree of summation of force. When the frequency of action potentials is relatively low, successive stimuli activate the muscle *after* the peak force of each twitch, so that individual twitches can still be detected (Figure 27–1C and D). This form, called an *unfused tetanus*, produces a characteristic ripple in the contractile force of the isolated muscle. As the rate of firing of the motor neuron increases, the force produced in the muscle increases progressively to a steady maximum value. This is called a *fused tetanus*, because individual twitches can no longer be distinguished (Figure 27–1D).

Usually, firing rates stay within a narrow range, however. For steady voluntary muscle contractions, the lowest firing rate of motor neurons is about 8 Hz; frequencies are rarely higher than 25 Hz, even during strong contractions (Figure 27–2). Although the low firing rates produce an unfused tetanus in individual muscle fibers, the total muscle contracts smoothly because not all motor units are activated at the same time. Thus, individual twitches in a motor unit do not occur at the same time and the peaks and troughs of the individual twitches are averaged out in the net muscle force. Higher rates of firing, which produce a fused tetanus in individual motor units, occur only transiently during the early phase of rapid contraction.

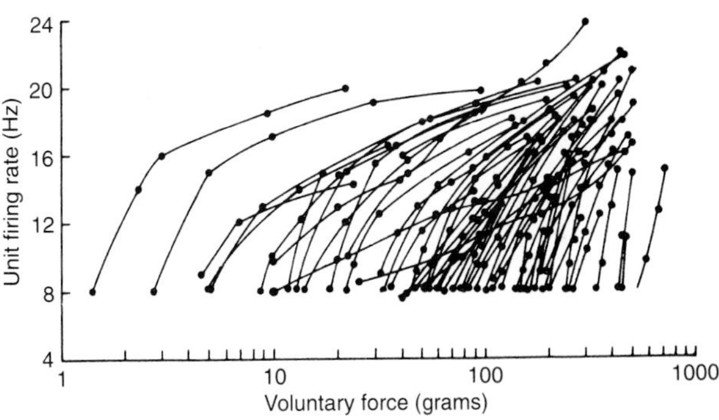

Figure 27–2 The number of motor units recruited and their firing frequency increase as the force of a steady voluntary movement increases. These recordings of motor units in the extensor muscle of the fingers show that the motor units fire at about 8 Hz when they are recruited. Their firing rate rises progressively as the subject produces more force, but the rate rarely exceeds 25 Hz. (Adapted from Monster and Chan, 1977.)

Motor Units Are Recruited in a Fixed Order from Weakest to Strongest

Since all the muscle fibers belonging to a motor unit have similar physiological and biochemical properties, motor units can be classified according to these properties. Three types of motor units have been distinguished according to the time the fibers take to achieve peak force during a twitch and the rate at which they fatigue.

In the first type, called *fast fatigable*, the fibers contract and relax rapidly but fatigue rapidly when stimulated repeatedly. These fibers generate the greatest force during a twitch or tetanic contraction (Figure 27–3). In the second type, termed *slow fatigue-resistant*, the fibers have a much longer contraction time and are highly resistant to fatigue. These fibers can generate only 1 to 10% of the force of the fast fatigable fibers. In the third type, called *fast fatigue-resistant*, the physiological properties of the fibers are intermediate between those of the other two. Their contraction time is only slightly slower than fast fatigable fibers but they are almost as resistant to fatigue as are the slow fatigue-resistant fibers and are able to produce about twice as much force as the latter.

All three types of motor units are found in most muscles, but in different proportions. Moreover, in individual muscles the muscle fibers belonging to one motor unit are widely distributed within the muscle (Figure 27–4).

The three types of motor units are also distinguished by the size of the cell body of the motor neuron. In slow fatigue-resistant motor units the motor neurons have small cell bodies. These units generate the least force and are the most resistant to fatigue (Figure 27–5). Motor neurons in the fast fatigue-resistant units have larger cell bodies. Neurons in the fast fatigable units have the largest cell bodies.

When a motor neuron pool is activated synaptically, the initial weak inputs activate the cells with the lowest threshold for synaptic activation (i.e., those with the smallest cell bodies). As the synaptic input increases in strength, motor neurons with larger cell bodies are recruited in turn according to

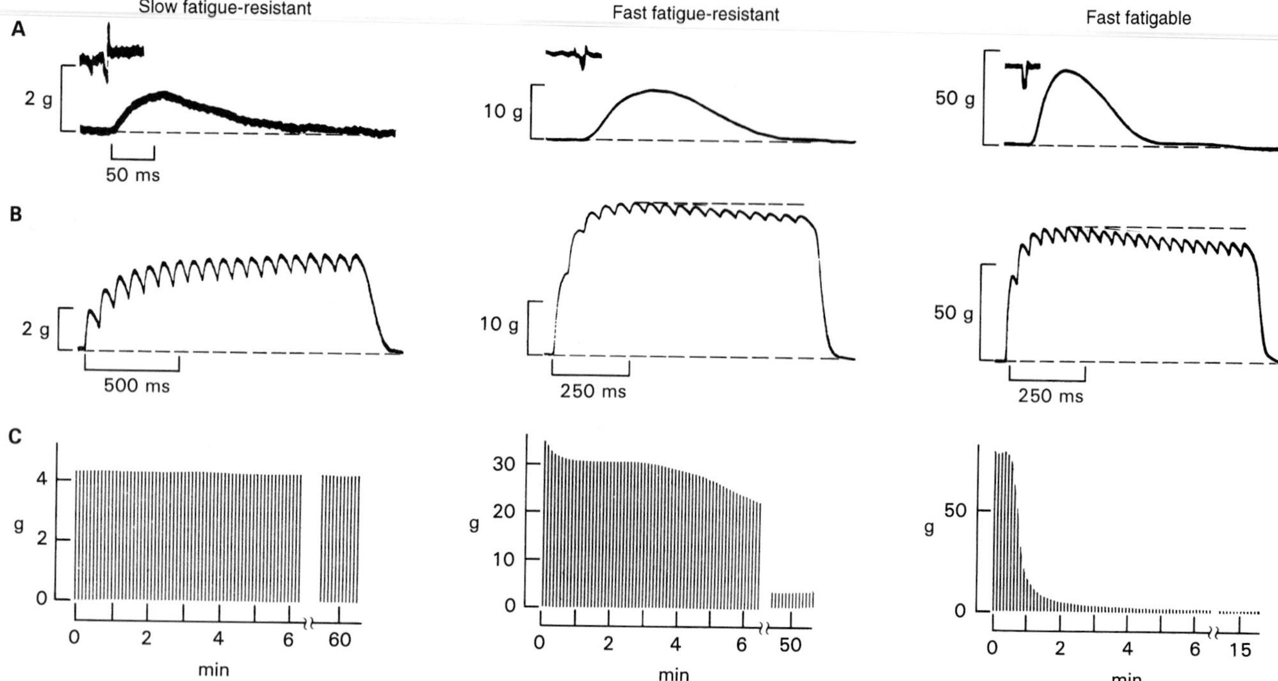

Figure 27–3 Three types of motor units are distinguished by the tetanic force during a twitch and by fatigability of the muscle: slow fatigue-resistant, fast fatigue-resistant, and fast fatigable. The activity records in this illustration are the result of stimulating motor neurons intracellularly. (From Burke et al., 1974.)

A. The muscle twitches characteristic of the three motor units.

B. The tetanic tensions produced by a train of 12 Hz stimuli. Note the markedly greater twitch and tetanic forces produced by fast fatigue-resistant and fast fatigable units relative to the slow fatigue-resistant units.

C. The muscle is activated by repeated (tetanic) stimuli lasting 330 ms and repeated every second. Each vertical line represent the force produced by each tetanus, recorded at slow speed. In the slow fatigue-resistant unit the force remains constant throughout an hour of repeated stimulation, whereas in the fast fatigable unit the force drops abruptly after only a minute. The fast fatigue-resistant unit shows substantial resistance to fatigue and the force declines slowly over many minutes; some residual force remains after 50 minutes.

Figure 27–4 A single motor neuron innervates many muscle fibers, and these fibers are typically distributed quite widely throughout the muscle. In this example of a soleus muscle the **pink area** is the approximate lengthwise territory of one motor unit as projected onto the surface of the muscle. The cross-sectional area occupied by the motor unit is shown at the **right** (at the level indicated by the **dashed line**), where each **dot** represents one fiber in the unit. The location of individual muscle fibers making up the motor unit was determined by stimulation of a single motor neuron for a prolonged period. This caused all the muscle fibers connected to that motor neuron to contract and deplete their stores of glycogen. The fibers were then identified histochemically with a stain selective for glycogen. (Adapted from Burke et al., 1974.)

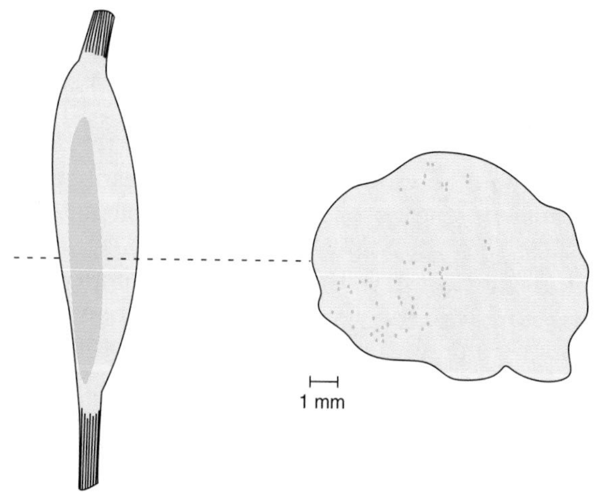

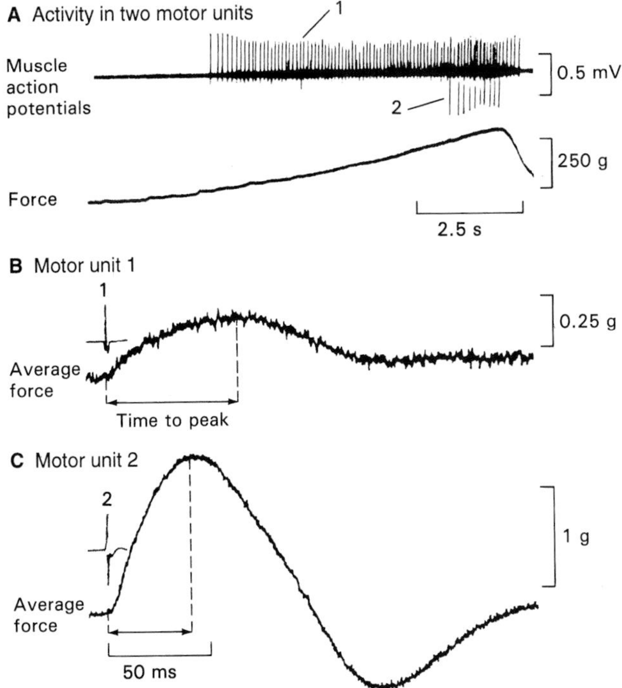

Figure 27–5 Motor units producing a small force are recruited before motor units producing a large force. Two motor units in the first dorsal interosseus muscle of a human subject were recorded through a single electrode (**A**). The two units together produce a slowly rising force. Spike-triggered averaging shows that the unit recruited first produces a lower force and has a longer contraction time (**B**) than does the second unit (**C**). (Adapted from Desmedt and Godaux, 1977.)

size. This fixed order of recruitment, called the *size principle,* occurs during both reflexive and voluntary contraction.

This organization of recruitment has two important functional consequences. First, ordered recruitment simplifies the task of modulating muscle force. Higher centers need only determine how much synaptic drive to deliver to the motor neuron pool as a whole and thus do not have to manage the selection of specific combinations of motor units to produce the needed amount of force.

The second functional consequence is that the slow motor units are the most heavily used, since they are recruited in all contractions. The fast motor units are less frequently used, since they are recruited only when greater forces are required. For example, half the motor units in the hind limb of the cat are slow and are used for standing and walking, actions that require only about 20% of the total force of the muscle. The remaining units, which are fast fatigable, generate the greatest forces but are used only occasionally for such strenuous activities as running and jumping. Although half of the motor units are active only rarely, the metabolic cost of this reserve capacity is relatively low, since fast fatigable units depend primarily on anaerobic metabolism.

Orderly recruitment and rate modulation are not mutually exclusive. When motor tasks require slowly increasing force, motor units are recruited one at a time in an orderly fashion and their rate is modulated as the force increases.

Muscles Contain Specialized Receptors that Sense Different Features of the State of the Muscle

Muscles contain two types of sensory receptors that provide the central nervous system with information on changes in muscle length and the forces these changes generate. *Muscle spindles* are located within the fleshy portions of muscles, in parallel with the skeletal muscle fibers (Figure 27–6A). They are innervated by two types of myelinated afferent fibers: *group Ia* (large diameter) and *group II* (small diameter). *Golgi tendon organs* are located at the junction between muscle fibers and tendon; they are therefore connected in series to a group of skeletal muscle fibers. Golgi tendon organs are innervated only by *group Ib* afferent fibers, which are also myelinated and have slightly smaller diameters than group Ia fibers. Both receptors are distributed extensively throughout most skeletal muscles.

Muscle Spindles Respond to Stretch of Specialized Muscle Fibers

Muscle spindles are elongated structures with a slightly swollen center and tapered ends, giving them a fusiform or spindle-like shape. Spindles range in length from 4 to 10 mm. They have three main components: a group of specialized (intrafusal) muscle fibers, sensory terminals in the intrafusal muscle fibers, and motor terminals that regulate the sensitivity of the spindle. The central part of the spindle is enclosed by a connective tissue capsule filled with a gelatinous fluid that facilitates sliding of the intrafusal muscle fibers within it (Figure 27–6; see also Figure 3–3).

The specialized muscle fibers of the spindle are called *intrafusal* to distinguish them from ordinary skeletal muscle fibers, the *extrafusal* fibers. Intrafusal fibers are smaller than extrafusal fibers and do not contribute significant force to muscle contraction. Instead, changes in their lengths are sensed by the sensory terminals. Their central regions have few myofibrils and are essentially noncontractile; only the polar regions actively contract.

Intrafusal muscle fibers can be classified as nuclear chain or nuclear bag fibers. The *nuclear chain fiber* is short and slender; its nuclei lie in a single file within the fiber. The *nuclear bag fiber* is thicker in diameter; its nuclei are clustered in the central region of the fiber, which thus appears slightly swollen. Physiological studies have further distinguished two types of nuclear bag fibers, dynamic and static. A typical mammalian muscle spindle contains one of each type of nuclear bag fibers and a variable number of nuclear chain fibers, usually about five.

The myelinated sensory axons enter the muscle spindle capsule in its central part and terminate on or near the central portions of the intrafusal fibers, where they typically spiral around individual fibers (Figure 27–6B). When the intrafusal fibers are stretched, referred to as *loading* the spindle, the sensory endings increase their firing rate. This happens because stretching of the spindle lengthens the central region of the intrafusal fibers around which the afferent endings are entwined. The elongation of the afferent endings activates stretch-sensitive ion channels; opening of the channels alters the ionic current into the cell, thus depolarizing the cell membrane and generating action potentials. When the stretch is released, referred to as *unloading,* the intrafusal fibers slacken and the firing rate in the afferent endings decreases.

The sensory endings in muscle spindles are classified as primary or secondary. There is usually just one primary ending in each spindle, consisting of all the branches of a single group Ia afferent axon. The group Ia afferent terminates on all three types of intrafusal fibers. There is also usually just one secondary ending in a spindle, consisting of the terminations of a single group II afferent. The group II fibers terminate only on chain fibers or static bag fibers. The primary and secondary endings have different signaling characteristics: primary endings are much more sensitive to the rate of change of length than are secondary endings.

Golgi Tendon Organs Are Sensitive to Changes in Tension

Golgi tendon organs are slender encapsulated structures about 1 mm long and 0.1 mm in diameter. They are typically located at the junction of muscle and tendon, where collagen fibers arising from the tendon attach to the ends of groups of extrafusal muscle fibers (Figure 27–6C). Collagen bundles within the capsule of the tendon organ divide into fine fascicles that form a braided structure.

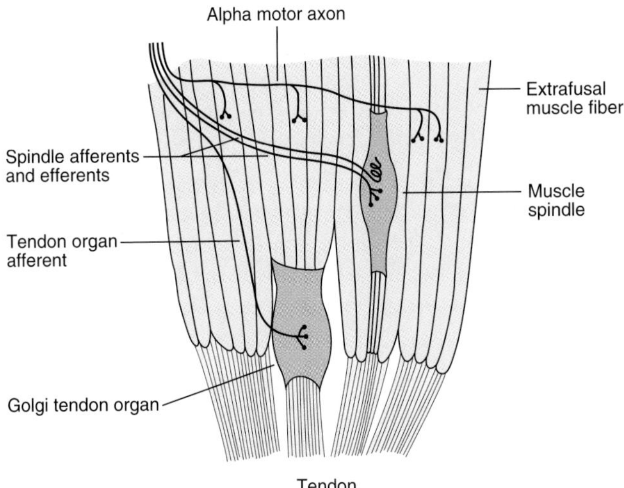

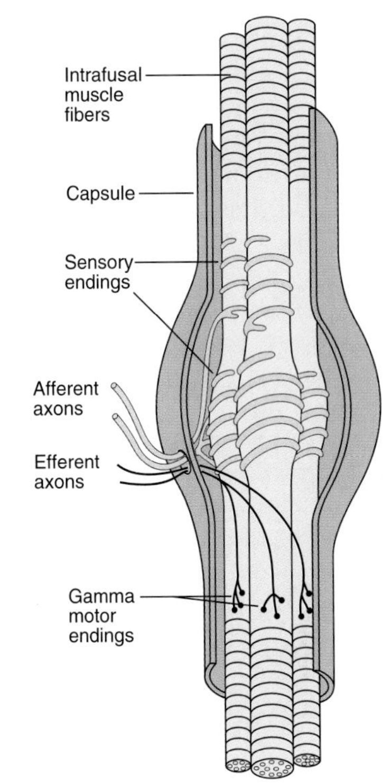

Figure 27–6 Muscle spindles and Golgi tendon organs are encapsulated structures found in skeletal muscle.

A. The main skeletal muscle fibers, or extrafusal fibers, are innervated by large-diameter alpha motor axons. The muscle spindle is arranged in parallel with extrafusal fibers; it is innervated by both afferent and efferent (gamma motor neuron) fibers. The Golgi tendon organ is found at the junction between a group of extrafusal fibers and the tendon; it is therefore in series with extrafusal fibers. Each tendon organ is innervated by a single afferent axon. (Adapted from Houk et al., 1980.)

B. The muscle spindle contains three functional components: intrafusal fibers, sensory endings, and motor axons. The intrafusal fibers are specialized muscle fibers that run through the muscle spindle. The sensory endings spiral around the noncontractile central regions of the intrafusal fibers and are responsive to stretch of these fibers. Gamma motor neurons innervate the contractile polar regions of the intrafusal fibers. Contraction at both ends of the intrafusal fiber pulls on the central region, causing the stretch-sensitive sensory endings to fire. (Adapted from Hulliger, 1984.)

C. The Golgi tendon organ is attached to the muscle fibers by collagen fibers. A single Ib afferent axon enters the capsule and branches into many unmyelinated endings that wrap around and between the collagen fibers. When the tendon organ is stretched (usually because of contraction of the muscle), the afferent axon is compressed by the collagen fibers (see inset at lower right) and its rate of firing increases. (Adapted from Schmidt, 1983; inset adapted from Swett and Schoultz, 1975.)

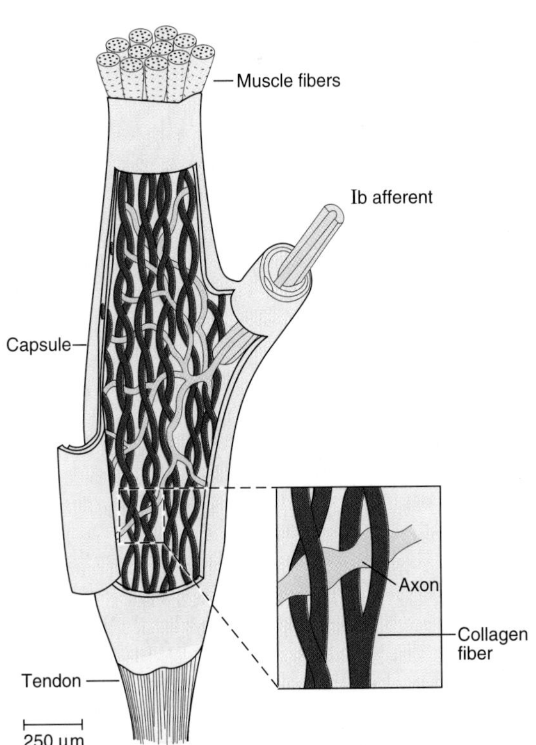

Each tendon organ is innervated by a single group Ib axon that loses its myelination after it enters the capsule and branches into many fine endings, which intertwine among the braided collagen fascicles. Stretching of the tendon organ straightens the collagen bundles. This compresses and elongates the nerve endings and causes them to fire. Because the free nerve endings intertwine among the collagen fiber bundles, even very small stretches of the tendon organ can deform the nerve endings. The firing rate of tendon organs is thus very sensitive to changes in tension of the muscle. Tendon organs stretch most easily when the muscle tension increases due to contraction (see below).

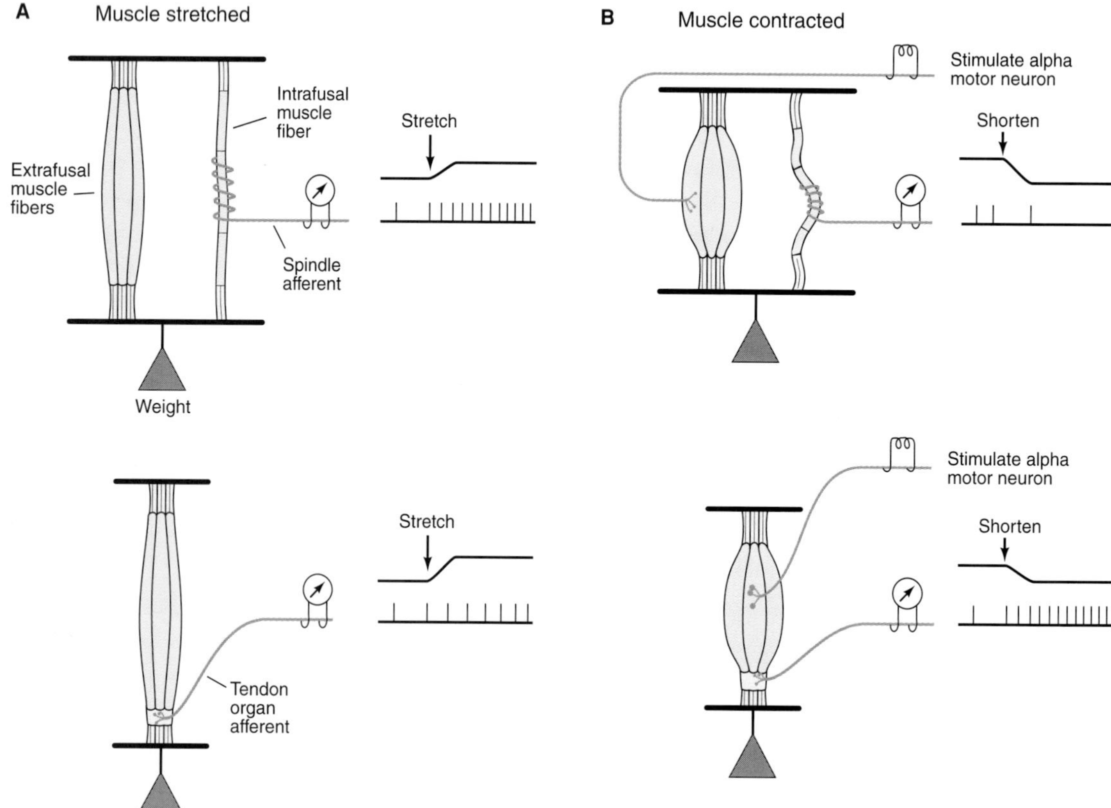

Figure 27–7 The muscle spindle and Golgi tendon organ have different responses to muscle stretch and muscle contraction. Both afferents discharge when the muscle is stretched (**A**), the Golgi tendon organ much less than the spindle. However, when the muscle contracts (**B**), the spindle is unloaded and therefore goes silent, whereas the tendon organ firing rate increases.

A. Stretching of the muscle elongates the intrafusal fibers, stretching the sensory endings in the spindle and leading to increased firing. In tendon organs, however, the collagen fibers from the tendon are stiffer than the muscle fibers with which they are in series. Therefore, most of the stretch is taken up by the more compliant muscle fibers and little direct mechanical deformation of the tendon organ takes place.

B. When the muscle contracts, the muscle fibers themselves pull directly on the collagen fibers and transmit the stretch to the tendon organ more effectively. As a result, tendon organs always respond more robustly to contraction than to stretch of the muscle. Spindles, in contrast, decrease their firing rate when the muscle contracts because, as the extrafusal fibers shorten with contraction, the parallel intrafusal fibers are unloaded. (Adapted from Patton, 1965.)

Functional Differences Between Spindles and Tendon Organs Derive from Their Different Anatomical Arrangements Within Muscle

Spindles and tendon organs provide complementary information about the mechanical state of the muscle, its length, and its degree of tension. The information on muscle length from the spindles is used by the brain to determine the relative positions of the limb segments, because the length of a muscle varies with the angle of the joint on which it acts. The information on muscle tension from Golgi tendon organs is useful for a variety of motor acts, such as maintaining a steady grip on an object or compensating for the effects of fatigue (when a steady neural drive to the muscle would produce decreasing levels of tension).

The two types of muscle receptors provide different information because each is activated by different events in the muscle. When muscle is stretched, spindle afferents rapidly increase their rate of discharge, while tendon organs show only a slight and inconsistent increase. On the other hand, when the muscle contracts after its motor nerve is electrically stimulated, the firing rate of the tendon organ increases markedly, whereas the firing rate of the spindle decreases or ceases altogether.

This difference in response results from the different anatomical relationships of the two types of receptors to the extrafusal muscle fibers. The spindles are arranged *in parallel* with extrafusal fibers, whereas the Golgi tendon organs are arranged *in series* (Figure 27–7).

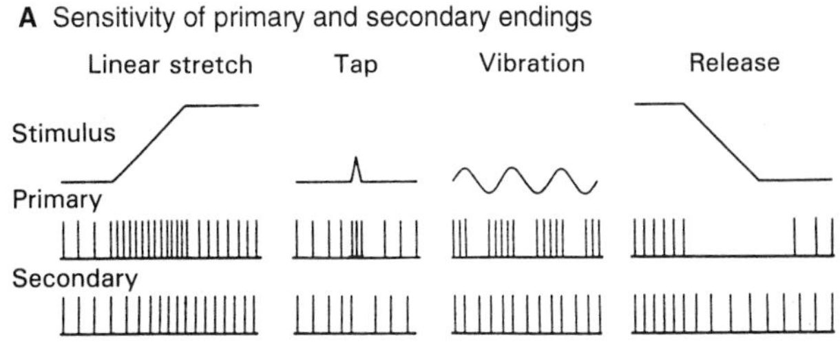

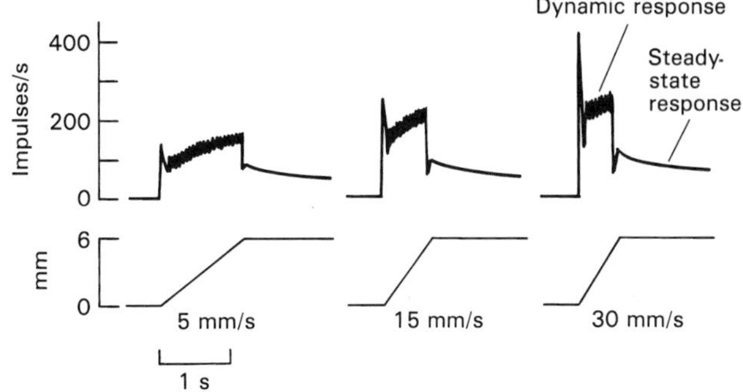

Figure 27–8 Primary and secondary endings in muscle spindles have different firing properties.

A. When a muscle is stretched, or when a stretched muscle is released and allowed to shorten, both endings reach a steady-state firing rate that reflects the new muscle length. In addition, there is a burst of firing in the primary endings during the dynamic phase of stretch and a transient silence during release. Primary endings are therefore very sensitive to transient or changing stimuli, such as taps or vibration of the muscle. (Adapted from Matthews, 1964.)

B. The primary ending is highly sensitive to the velocity of stretch. Its firing rate during the dynamic phase of a fast stretch reaches higher levels than that of secondary endings. Primary endings are particularly sensitive to very small stretches; this is reflected in the transient increase in the firing rate at the very beginning of the stretch. (Adapted from Matthews, 1972.)

The Primary and Secondary Afferent Endings Respond Differently to Phasic Changes in Length

When muscle is stretched or is released from stretch, its change in length has two phases. During the *dynamic phase* the length is actually changing, while during the static or *steady-state phase* the new muscle length is stabilized. During these two phases, primary and secondary sensory endings behave differently. When a muscle lengthens, both endings increase their firing to a higher steady-state rate. During the dynamic phase of the stretch, however, the primary ending fires at a much higher rate than during the later steady-state phase. The firing of the secondary ending increases only gradually and is not much higher during the dynamic phase than during the steady-state phase (Figure 27–8A).

Because of their high degree of sensitivity during the dynamic phase, primary endings are very sensitive to small changes in the length of the muscle. In addition, the increase in firing rate in primary endings during the dynamic phase of stretch reflects the rate of change in muscle length—higher rates occur during faster stretches (Figure 27–8B). This property of the primary endings is referred to as *velocity sensitivity*. Velocity sensitivity is also seen when the muscle shortens: as the muscle shortens rapidly, primary endings stop firing, and then resume at a lower rate when shortening stops. Transient stimuli, such as brief taps or vibration of the muscle, produce bursts of firing in primary endings because of the high degree of sensitivity of these endings to changes in muscle length (the dynamic phase). In contrast, such short-lasting stimuli have virtually no effect on secondary endings because the changes in muscle length occur too quickly to alter the steady-state discharge of these endings (Figure 27–8A). For this reason secondary endings provide information only about the static length of muscles.

Since primary endings encode not only the length of a muscle but also how fast the muscle changes its length, they provide information about the speed of movements as well as the static positions of joints.

The Central Nervous System Controls Sensitivity of the Muscle Spindles Through the Gamma Motor Neurons

The parallel arrangement of intrafusal and extrafusal fibers raises an interesting problem. Because intrafusal fibers tend to slacken when a muscle shortens, the spindles should stop firing when the muscle shortens. Were this to occur, however, the spindles would fail to transmit information about changes in length at the very time that information is most critical—when the contracting muscle is shortening. To ensure that information on changes in muscle length is transmitted during contraction, the central nervous system regulates the loading of the intrafusal fibers through *gamma motor neurons*.

In amphibians and other lower vertebrates the intrafusal fibers are only innervated by collaterals of the motor axons that innervate the extrafusal fibers. In mammals, however, the motor innervation of intrafusal fibers is generally separate from that of extrafusal fibers; the intrafusal fibers are innervated by the small gamma motor neurons, while the extrafusal fibers are innervated by larger *alpha motor neurons*. The gamma motor neurons that innervate muscle spindles are often referred to as the *fusimotor system*, while the alpha motor neurons that innervate extrafusal fibers are referred to as the *skeletomotor system*.

When isolated alpha motor neurons alone are electrically stimulated, the afferents pause in firing during the resulting contraction because they are relaxed (Figure 27–9). When, however, a gamma motor neuron is stimulated at the same time as the alpha motor neuron, the afferents do not pause. Electrical stimulation of the motor cortex and other higher centers in anesthetized animals typically leads to simultaneous activation of alpha and gamma motor neurons, a pattern called *alpha–gamma coactivation*.

How does activity of gamma motor neurons modulate the activity of spindle afferents? The gamma neurons innervate the polar regions of the intrafusal fibers, where the contractile elements are located (see Figure 27–6B). Activation of a gamma motor neuron causes contraction and shortening of the polar regions, which in turn stretches the central region from both ends. This action increases the firing rate of the sensory endings and also makes the afferent endings more sensitive to stretch of the intrafusal fibers.

In slow voluntary movements the Ia afferent fibers often increase their rate of discharge even though the muscle shortens during the movement (Figure 27–10). This increased firing could only occur if the gamma motor neurons were activated in synchrony with the alpha motor neurons. If there were no gamma activation, the firing of the spindle would decrease or pause during a contraction, because of slackening of the intrafusal fibers.

The maintenance of spindle sensitivity by alpha–gamma coactivation is useful to the nervous system, as shown in Figure 27–10. The trajectory of movement of a finger during flexion shows small deviations in velocity due to variation in the rate of contraction of the flexor muscle—at times the muscle shortens quickly, at other times more slowly. The firing of the Ia afferent mirrors the irregularities in the trajectory. When the velocity of flexion increases transiently, the rate of discharge of the Ia axons decreases because the muscle shortens more rapidly, thus exerting less tension on the intrafusal fibers. When the velocity decreases, the discharge of the Ia increases because the muscle

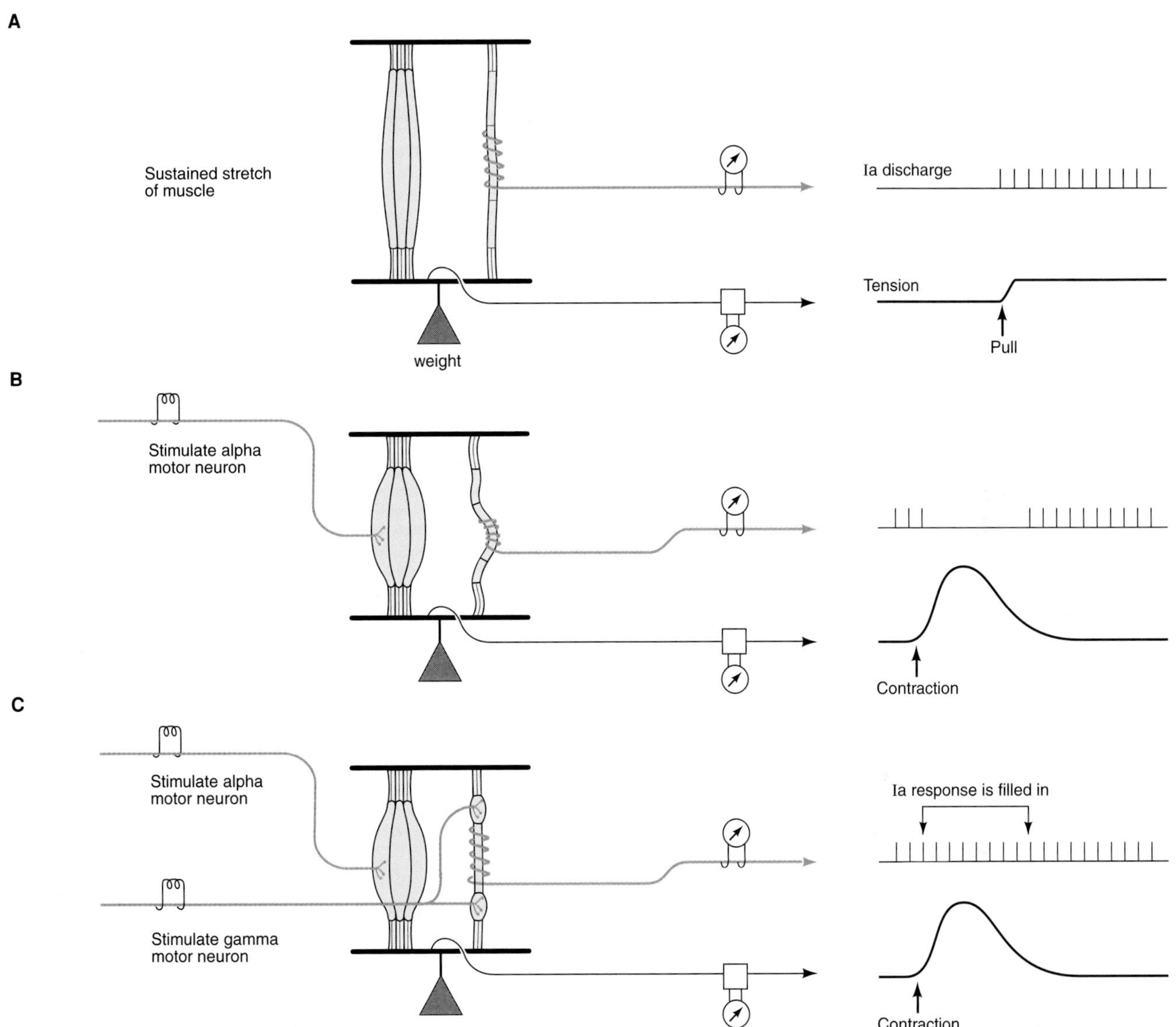

Figure 27–9 The muscle spindles do not slacken during active contractions because the tension of the intrafusal fibers is maintained by gamma motor neurons. (Adapted from Hunt and Kuffler, 1951.)

A. Under normal conditions sustained tension elicits steady firing of the afferent (Ia) fibers in the muscle spindle.

B. If the alpha motor neuron alone is stimulated electrically, the afferent fibers stop firing when the muscle contracts. The firing stops because the spindle is relaxed by the contraction.

C. If both the alpha and gamma motor neurons are stimulated together, the spindle does not slacken during the contraction and the pause in firing of the afferent is "filled in."

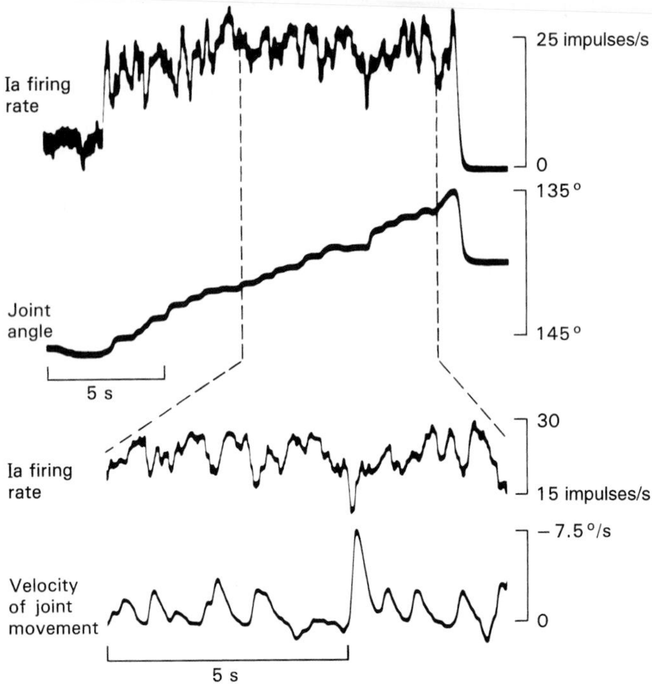

Figure 27–10 Gamma motor neurons are coactivated during voluntary movements. Recordings from a spindle afferent during slow flexion of a finger show that the spindle's firing mirrors irregularities in the movement velocity (**bottom**). These changes in spindle firing can be used by the nervous system to compensate for the irregularities and thus smooth the movement. The ability of the spindle to signal these irregularities depends on alpha–gamma coactivation. If the gamma motor neurons were not active, the spindle would slacken as the muscle shortened, as shown in Figure 27–9. (Adapted from Vallbo, 1981.)

shortens more slowly, thereby increasing relative tension on the intrafusal fibers.

Thus, in movement the Ia afferent's discharge rate is very sensitive to variations in the rate of change of muscle length. This sensitivity illustrates the functional importance of a property of the primary afferent endings that we discussed earlier: primary endings are most sensitive to small changes in length. This information can be used by the nervous system to compensate for irregularities in the trajectory of a movement and to detect fatigue of local groups of muscle fibers.

An Overall View

The muscle spindle is a remarkable sensory organ whose elegant design and operation have intrigued physiologists for over a century. It contains specialized elements that sense muscle length and the velocity of changes in length. In conjunction with the Golgi tendon organ, which senses muscle tension, it provides the central nervous system with continuous information about the mechanical state of the muscle. Innervation of the muscle spindle by an independent system of gamma motor neurons allows the central nervous system to adjust the sensitivity of the spindle and thus to fine-tune the information it receives. In this sense the muscle spindle is like the eye, in which motor innervation of the lens and extraocular muscles allows the nervous system to control how light impinges on the retina.

The finer control conferred by this motor innervation has a cost: the added complexity of perceptual processing by higher centers. Since the firing rate of muscle spindles depends on both muscle length and the level of gamma activation of the intrafusal fibers, the nervous system, in interpreting the signals from muscle spindles, must also monitor and take into account the fusimotor drive. This illustrates the close relationship between sensory and motor processing. For virtually all higher-order perceptual processes, the brain must correlate sensory input with motor output to assess the body's interaction with its environment accurately.

Selected Readings

Bizzi, E., and Abend, W. 1983. Posture control and trajectory formation in single- and multi-joint arm movements. In J. E. Desmedt (ed.), Motor Control Mechanisms in Health and Disease. Advances in Neurology, Vol. 39. New York: Raven Press, pp. 31–45.

Boyd, I. A. 1980. The isolated mammalian muscle spindle. Trends Neurosci. 3:258–265.

Desmedt, J. E. 1985. Patterns of motor commands during various types of voluntary movement in man. In E. V. Evarts , S. P. Wise, and D. Bousfield (eds.), The Motor System in Neurobiology. New York: Elsevier, pp. 133–139.

Eckert, E. 1988. Muscle and movement. In, Animal Physiology: Mechanisms and Adaptations, 3rd ed. New York: Freeman.

Freund, H.-J. 1983. Motor unit and muscle activity in voluntary motor control. Physiol. Rev. 63:387–436.

Hasan, Z., and Stuart, D. G. 1984. Mammalian muscle receptors. In R. A. Davidoff (ed.), Handbook of the Spinal Cord. Vols. 2 and 3: Anatomy and Physiology. New York: Marcel Dekker, pp. 559–607.

Houk, J. C., and Rymer, W. Z. 1981. Neural control of muscle length and tension. In V. B. Brooks (ed.), Handbook of Physiology, Section 1: The Nervous System, Vol. II: Motor Control, Part 1. Bethesda, Md.: American Physiological Society, pp. 257–323.

Hulliger, M. 1984. The mammalian muscle spindle and its central control. Rev. Physiol. Biochem. Pharmacol. 101:1–110.

Huxley, A. F. 1974. Review lecture: Muscular contraction. J. Physiol. (Lond.) 243:1–43.

Matthews, P. B. C. 1981. Evolving views on the internal operation and functional role of the muscle spindle. J. Physiol. (Lond.) 320:1–30.

Partridge, L. D., and Benton, L. A. 1981. Muscle, the motor. In V. B. Brooks (ed.), Handbook of Physiology, Section 1: The Nervous System, Vol. II: Motor Control, Part 1. Bethesda, Md.: American Physiological Society, pp. 43–106.

Polit, A., and Bizzi, E. 1978. Processes controlling arm movements in monkeys. Science 201:1235–1237.

Prochazka, A., and Hulliger, M. 1983. Muscle afferent function and its significance for motor control mechanisms during voluntary movements in cat, monkey, and man. In J. E. Desmedt (ed.), Advances in Neurology, Vol. 39: Motor Control Mechanisms in Health and Disease. New York: Raven Press, pp. 93–132.

Stuart, D. G., and Enoka, R. M. 1983. Motoneurons, motor units and the size principle. In R. N. Rosenburg (ed.), Clinical Neurosciences, Vol. 5. Neurobiology. New York: Churchill Livingstone, pp. 417–517.

28

Spinal Reflexes

The Stretch Reflex Is a Simple Model of a Stereotyped Reflex

The Neural Circuit for the Stretch Reflex Is Monosynaptic

Stretch Reflexes Regulate Muscle Tone

Most Spinal Reflexes Are Mediated by Polysynaptic Circuits That Allow the Reflex to Be Modified

Muscle Action Around a Joint Is Coordinated by Inhibitory Interneurons

Group Ia Inhibitory Interneurons Inhibit Antagonist Muscles

Group Ib Inhibitory Interneurons Inhibit the Homonymous Muscle

Complex Reflexes That Serve Protective and Postural Functions Are Initiated by Stimulation of the Skin

Flexion Reflexes Involve Movement of Entire Limbs

Certain Reflexes Consist of Rhythmic Movements

The Main Features of Walking Movements Are Controlled by the Spinal Cord

Spinal Circuits Generate Rhythmic Locomotor Pattern

Goal-Directed Locomotion Requires Intact Supraspinal Systems

Every movement requires the coordinated action of many muscles. Even a relatively simple act—reaching to pick up a pen—requires contraction of dozens of muscles to move the hand to the pen and to grasp it, and contraction of many more muscles in the trunk and legs to maintain stability of the body. The process of linking the contractions of independent muscles so that they act together is called *motor coordination*.

The most elementary form of motor coordination is the reflex. A reflex action is a relatively stereotyped response to a specific sensory stimulus. Two features of the stimulus are particularly important in shaping the reflex response. First, the *locus* of the stimulus determines which muscles will contract to produce the reflex response. Second, the *strength* of the stimulus determines the amplitude of the response—reflexes typically are graded in intensity.

The neural circuitry responsible for a spinal reflex is entirely contained within the spinal cord, and receives sensory information from muscles, joints, and skin directly. Spinal reflexes have an essential role in all voluntary action. Since these elementary circuits are recruited by higher brain centers to generate more complex behaviors, an appreciation of how they are organized is essential for understanding more complex motor sequences. Spinal reflexes also are valuable for clinical diagnosis. They can be used to assess the integrity of both afferent and motor connections as well as the general excitability of the spinal cord.

In this chapter we begin by examining the neural circuitry of one particularly important and well-studied spinal reflex: the stretch reflex. This reflex consists of contractions of a muscle that occur when that muscle is lengthened. Such contractions were once thought to result from intrinsic properties of muscles. But at the turn of the century Sherrington discovered that stretch reflexes could be abolished by cutting either the dorsal or the ventral roots of the spinal cord, thereby establishing that these reflexes require both sensory input from the muscle to the spinal cord and a motor path to the muscle.

The stretch reflex depends only on the monosynaptic connections between primary afferent fibers from muscle spindles and motor neurons innervating the same muscle. Other spinal reflexes, such as those produced by cutaneous stimuli (e.g., pinching the skin), have a more complex circuitry: one or more interneurons are interposed between the primary afferent fibers and the motor neurons. We shall also consider the circuitry of these more complex spinal reflexes and examine their role in posture, locomotion, and other complex motor acts.

The Stretch Reflex Is a Simple Model of a Stereotyped Reflex

To study the central circuits that mediate stretch reflexes, Sherrington cut the brain stem in animals at the level of the midbrain. In this *decerebrate preparation* the cerebrum is disconnected from the spinal cord, blocking sensations of pain as well as normal modulation of reflexes by higher brain centers. In decerebrate animals many spinal reflexes are heightened and become more stereotyped, making it is easier to examine the factors controlling their expression. Decerebrate animals also show a dramatic increase in extensor muscle tone, sometimes sufficient to support the animal in the standing position.

When Sherrington attempted to passively flex the rigidly extended hind limb of a decerebrate cat, he encountered increased contraction of the extensor muscles being stretched. He called this the *stretch* (or *myotatic*) reflex and found that it has two components: (1) a brisk but short-lasting *phasic* contraction, which is triggered by the change in muscle length; and (2) a weaker but longer-lasting *tonic* contraction, determined by the static stretch of the muscles at the new longer length. He also discovered that stretching one set of muscles caused the antagonist (opposing) muscles to relax. He concluded that stretch causes excitation of the motor neurons to the stretched muscles (and their synergists) and inhibition of motor neurons to the antagonist muscles and called this dual action *reciprocal innervation*.

Stretch reflexes are weaker and considerably less stereotyped in intact animals than in decerebrate animals. In intact animals facilitation and inhibition of reflex circuits are balanced; descending pathways from the cerebral cortex and other higher centers of the brain are continuously modulating the strength of stretch reflexes to meet the changing requirements of ongoing action. In decerebrate animals, brain stem pathways powerfully facilitate the reflex circuits involved in stretch reflexes.

The Neural Circuit for the Stretch Reflex Is Monosynaptic

Group Ia afferent fibers from muscle spindles enter the spinal cord through the dorsal roots. Within the dorsal horn they separate into numerous branches. Some branches make direct excitatory connections with the motor neurons that innervate the same muscle from which the Ia afferent arises (called the *homonymous* muscle). Thus, the stretch reflex is monosynaptic; it is produced by a two-neuron circuit consisting of a single set of synaptic connections between the Ia afferent and the alpha motor neurons (Figure 28–1A). The neurons that mediate this reflex are described in Chapter 3.

Branches of the Ia afferent excite not only the motor neurons innervating the homonymous muscle, but also those innervating *synergist* muscles (muscles that control the same joint and have a similar mechanical action). Each Ia afferent makes excitatory connections with *all* motor neurons innervating the homonymous muscle and with many of the motor neurons innervating synergist muscles (up to 60% for some synergists). Still other branches make excitatory connections with interneurons that inhibit antagonist motor neurons. (The interneurons are described later in the chapter.) This set of connections accounts for the reciprocal innervation described by Sherrington.

How does this spinal circuitry produce the stretch reflex? A simple stretch reflex in flexor

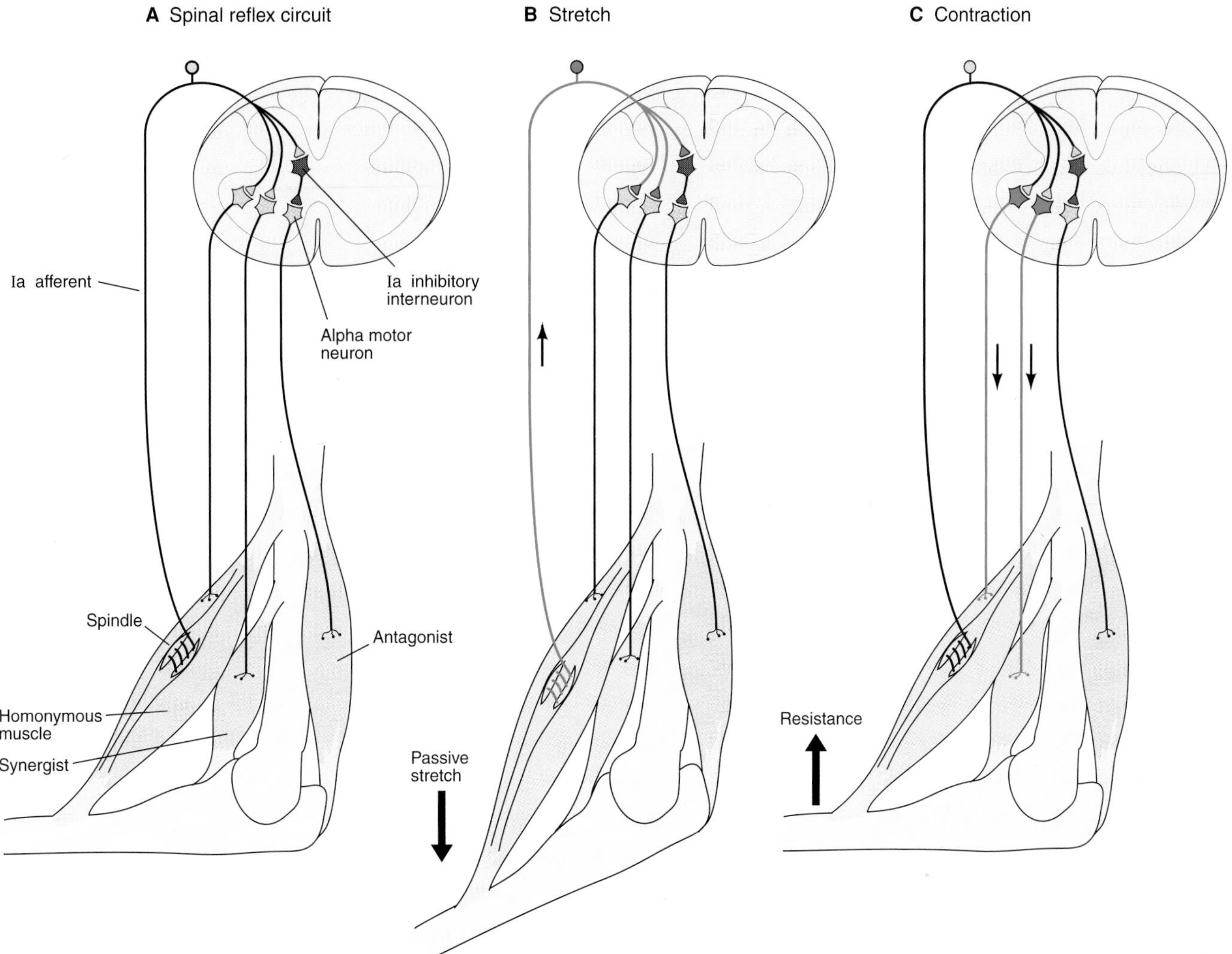

Figure 28–1 Excitation of muscle spindles initiates the stretch reflex.

A. Ia afferent fibers make monosynaptic excitatory connections to two sets of alpha motor neurons: those innervating the same (homonymous) muscle from which the Ia fibers arise and those innervating synergist muscles. They also inhibit motor neurons to antagonist muscles through an inhibitory interneuron.

B. When a muscle is stretched, the Ia afferents increase their firing rate.

C. The increase in firing rate leads to contraction of the same muscle and its synergists and relaxation of the antagonist. The reflex therefore tends to counteract the stretch, enhancing the springlike properties of the muscles.

muscles is illustrated in Figure 28–1. A brisk passive extension of the limb lengthens the flexor muscles, causing an increase in the discharge rate of Ia fibers arising from these muscles. The discharge of the Ia afferents excites the motor neurons to both homonymous and synergist muscles, causing contractions that oppose the lengthening. Because the discharge of the Ia afferent also inhibits antagonist motor neurons (through the Ia inhibitory neuron), the antagonist muscles tend to relax. This action assists the resistance of the flexors to the stretch. Thus, the circuitry of the stretch reflex acts as a *negative feedback* loop to resist changes in muscle length.

Testing the strength of stretch reflexes, by trapping the muscle or its tendon with a reflex hammer, is useful in clinical diagnosis. Absent or weak (hypoactive) stretch reflexes often indicate a disorder of one or more components of the reflex circuit (the sensory or motor axons, the cell bodies of motor neurons, or the muscle itself). Hypoactive stretch reflexes can also result from lesions of the central nervous system, because the excitability of motor neurons is dependent on both excitatory and inhibitory inputs from higher motor centers. Hyperactive stretch reflexes, in contrast, always result from central lesions that lead to increased excitatory input to motor neurons. They are often associated with disorders of tone, such as spasticity and rigidity.

Stretch Reflexes Regulate Muscle Tone

Muscle tone is the force with which a muscle resists being lengthened. It is assessed clinically by passively extending and flexing the patient's limbs and feeling the resistance offered by the muscles. One component of muscle tone derives from a muscle's intrinsic stiffness—muscle behaves like a spring. In addition to this intrinsic stiffness, however, there is a neural component of muscle tone. As we have seen, the stretch reflex also acts to resist lengthening of the muscle. Thus, stretch reflexes enhance the spring-like quality of muscles. Through its excitatory and inhibitory influences on the stretch reflex pathway, the central nervous system is able to adjust muscle tone appropriate to a given behavioral task.

Normal muscle tone serves several important functions. First, it assists in maintaining posture. For example, as we sway back and forth while standing, muscles resist being stretched, preventing the sway from becoming too wide. Second, normal tone allows muscles to store energy, like springs, and to release it later, a feature that is particularly important in walking and running. As the foot presses against the ground, leg muscles stretch and store mechanical energy. When the leg pushes off, some of this energy is released and assists the active contraction of muscles; less contraction of muscles is required to propel the leg forward, making the locomotion more efficient. Finally, the spring-like qualities of muscles help to smooth movements: if muscles acted simply like the motors that control a robot's limbs, movements would be jerky, with sudden starts and stops. The elasticity of muscle smoothes out these jerks: like a spring, the muscle achieves an equilibrium length more gradually.

Most Spinal Reflexes Are Mediated by Polysynaptic Circuits That Allow the Reflex to Be Modified

Spinal reflexes provide the nervous system with elementary and automatic motor patterns that can be activated either by sensory stimuli or by descending signals from the brain stem and cerebral cortex. The stretch reflex, as we have seen, is mediated by a simple monosynaptic circuit. Most reflex pathways, however, are *polysynaptic*—one or more interneurons are interposed between sensory and motor neurons. These interneurons typically receive convergent input from more than one source. This feature allows signals from higher centers in the brain as well as other afferent inputs to modify the expression of the reflex. In the absence of supraspinal intervention, reflexes are stereotyped and automatic.

Most spinal reflexes are mediated by complex circuits that coordinate the actions of groups of muscles, sometimes spanning several joints. Thus, spinal reflexes are governed by three main levels of control: (1) control of individual muscles, (2) coordination of muscle action around a single joint, and (3) coordination of muscles at several joints. So far, we have focused on the control of individual muscles—the stretch reflex. In the next section we examine how two types of inhibitory interneurons coordinate the actions of different muscles around one joint. Following that, we consider the coordination of muscles at multiple joints by examining several reflexes initiated by stimulation of the skin.

Muscle Action Around a Joint Is Coordinated by Inhibitory Interneurons

We have previously noted that group Ia fibers from muscle spindles make connections not only with the motor neurons innervating the homonymous muscle, but also with motor neurons innervating synergist muscles and with interneurons that inhibit motor neurons innervating the antagonist muscles (see Figure 28–1). These divergent connections of spindle afferents establish strong neural

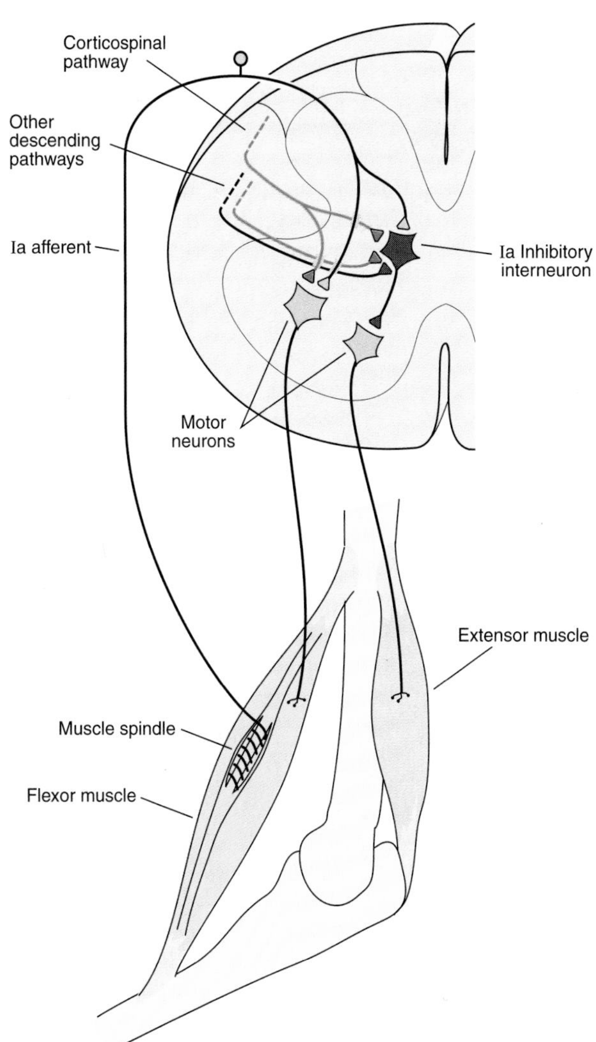

Figure 28–2 Higher motor centers are able to coordinate opposing muscles at a joint through a single command to a Ia inhibitory interneuron. This inhibitory interneuron mediates reciprocal innervation in stretch reflex circuits. (Only a few of the many inputs are shown in this highly simplified diagram.) In addition, the interneuron receives inputs from corticospinal descending axons, so that a descending signal to activate one set of muscles automatically leads to relaxation of the antagonists. Other descending pathways make excitatory and inhibitory connections to this interneuron. When the balance of inputs is shifted to greater inhibition, reciprocal inhibition is decreased, and co-contraction of opposing muscles occurs.

links between muscles acting around a joint, so that the muscles do not act independently of each other. The group of muscles around a joint that are linked together by this system of reflex pathways is called the *myotatic unit*. Inputs from Golgi tendon organs and the secondary endings of muscle spindles influence the myotatic unit through interneurons.

Most joints are not simple hinges, but rather allow movement in two or three planes. The muscles that surround these joints act in different combinations and with different relative forces depending on the particular direction of movement. When a stretch stimulus is applied to such a joint, the divergent connections of spindle afferents on synergist muscles and inhibitory interneurons ensure that the appropriate combination of synergists is activated and the appropriate combination of antagonist muscles inhibited. Thus, the myotatic unit regulates the tone, or stiffness, of the whole joint.

Group Ia Inhibitory Interneurons Inhibit Antagonist Muscles

The actions of opposing muscles in the stretch reflex are coordinated by *group Ia inhibitory interneurons*, which mediate reciprocal inhibition through excitatory inputs from the muscle spindle afferents (see Figure 28–1). This mode of coordination is useful not just in stretch reflexes, but also in voluntary movements. Relaxation of the antagonist muscle during movement enhances speed and efficiency because the muscles that act as prime movers are not working against the contraction of opposing muscles. Descending axons from the motor cortex, which make direct excitatory connections to spinal motor neurons, also send collaterals to Ia inhibitory interneurons. In this way higher motor centers need not send separate commands to opposing muscles (Figure 28–2).

Reciprocal inhibition of opposing muscles is not the only function of the Ia inhibitory interneuron. Sometimes it is advantageous to contract both the prime mover and the antagonist muscles simultaneously. This *co-contraction* has the effect of stiffening the joint and is most useful when precision and joint stabilization are critical. Because the Ia inhibitory interneuron receives both excitatory and inhibitory connections from all of the major descending motor pathways, brain centers can

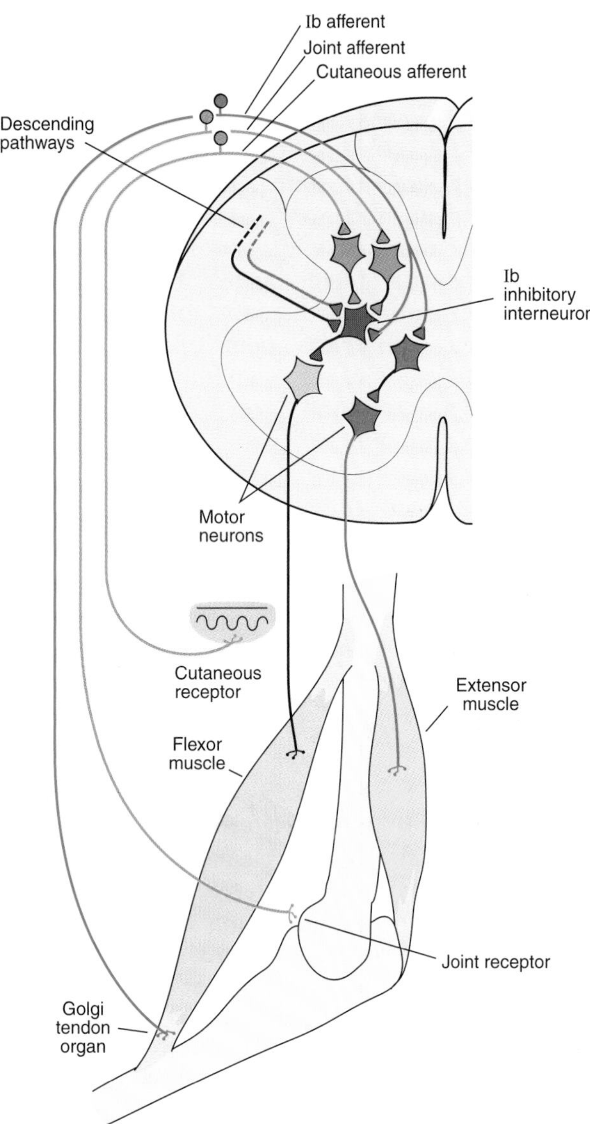

Figure 28–3 Afferent fibers from Golgi tendon organs provide a negative feedback system for regulating tension of the homonymous and synergist motor neurons through the Ib inhibitory interneuron. They also excite antagonist motor neurons through an excitatory interneuron. Thus, the motor effect of stimulating tendon organs is opposite that of stimulating muscle spindles. Ib inhibitory interneurons receive convergent input from receptors in the joints and skin and from descending pathways, and thus mediate control of movements in which integration of different sensory modalities is important, as in touch.

change the balance of excitatory and inhibitory inputs onto this interneuron and thus control the relative amount of joint stiffness appropriate for a motor act.

Group Ib Inhibitory Interneurons Inhibit the Homonymous Muscle

Homonymous motor neurons are influenced by a second type of inhibitory interneuron, the *Ib inhibitory interneuron,* which receives inputs from the Golgi tendon organs (Figure 28–3). The input from these muscle receptors provides a negative feedback mechanism for regulating muscle tension, parallel to the negative feedback from muscle spindles that regulates muscle length. Increasing muscle tension is signaled by an increase in activity of the Golgi tendon organs. This increased activity excites the Ib inhibitory interneuron, which in turn inhibits the homonymous motor neurons, thereby decreasing muscle tension.

The influence of the Ib interneuron on motor neuron excitability depends on combined inputs from many sources, both central and peripheral. In addition to input from tendon organs, this interneuron receives convergent input from Ia afferents from muscle spindles, low-threshold cutaneous afferents, and joint afferents, as well as both excitatory and inhibitory input from various descending pathways (Figure 28–3). These connections have important functions. For example, they provide a spinal mechanism for the fine control of exploratory movements, such as active touch. When the hand first contacts an object, muscle force is strongly inhibited by combined activation of tendon organs and cutaneous afferents, allowing immediate reduction in muscle force to soften the contact. Descending pathways can modulate the strength of this inhibitory effect, enhancing it if the object is fragile or decreasing the inhibition if more forceful contact is desired.

Complex Reflexes That Serve Protective and Postural Functions Are Initiated by Stimulation of the Skin

Tactile stimulation of many areas of skin causes reflex contraction of specific muscles, usually those under the area of stimulation. For example, stroking the abdomen (usually done from lateral to medial) causes reflex contraction of abdominal muscles, often visible as a deviation of the umbilicus toward the side of stimulation. Stroking the upper abdomen causes contraction of upper abdominal muscles, whereas stimulation of the lower abdomen causes contraction of lower abdominal muscles. This fixed spatial relationship between the locus of the stimulus and the particular muscles that contract is called *local sign*.

The reflex effect of cutaneous stimulation also depends on the quality of the stimulus. For example, stroking the plantar surface of the foot (usually from heel to toe) normally causes flexion of the toes and, in some cases, of the whole foot (referred to as a *plantar reflex*). In contrast, light pressure on the plantar surface leads to a generalized extensor response in the whole leg (referred to as *extensor thrust*). Finally, a painful stimulus such as a pinch or pin prick applied to the same area produces contraction of all flexor muscles of the limb, resulting in flexion withdrawal of the limb from the stimulus.

Flexion Reflexes Involve Movement of Entire Limbs

Flexion withdrawal from a noxious stimulus is a protective reflex involving coordinated muscle contractions at multiple joints through polysynaptic reflex pathways (Figure 28–4). Flexion reflexes, like stretch reflexes, involve reciprocal innervation: flexor muscles of the stimulated limb are contracted at the same time that extensor muscles of the limb are inhibited. Along with flexion of the stimulated limb, the reflex produces an opposite effect in the contralateral limb: extensor muscles are excited and flexor muscles are inhibited. This *crossed extension reflex* enhances postural support during withdrawal from a painful stimulus.

Although the flexion reflex is a relatively stereotyped response to a variety of painful stimuli, both the size and strength of muscle contraction reflect stimulus intensity. Touching a warm stove may produce moderately fast withdrawal only at the wrist and elbow, while touching a hot stove invariably leads to a forceful contraction at all joints and to rapid withdrawal of the entire limb. Moreover, the flexion reflex always outlasts the stimulus, and its duration usually increases with stimulus intensity. Thus like most reflexes, flexion reflexes are not simply repetitions of the same stereotyped movement pattern, but are modulated in different ways according to the properties of the stimulus.

The spinal circuits responsible for flexion withdrawal and crossed extension do more than mediate protective reflexes—they also serve to coordinate limb movements in voluntary movements. Interneurons in these pathways receive convergent inputs from different types of afferent fibers, not just pain fibers, and also from descending pathways.

Why is there so much multisensory convergence onto the same interneurons? This convergence would seem to mix together inputs from many different sources, resulting in the loss of specificity of sensory information. However, these same interneurons mediate descending commands for voluntary movements. This convergence makes sense, therefore, because voluntary movements lead to excitation not just of muscle receptors, but also of cutaneous and joint receptors, and these combined sensory inputs are used to regulate movements.

Certain Reflexes Consist of Rhythmic Movements

Some spinal reflexes consist of repeating rhythmic movements. For example, the *scratch reflex*, most often observed in furry animals, removes an annoying stimulus such as a flea. Like the flexion withdrawal reflex, it is a complex behavior with an obvious purpose. The reflex begins with movement of the hind limb into a position close to the location of the stimulus; rhythmic scratching occurs only after the limb has been brought to the correct starting position. *Crossed extension reflexes* maintain the standing posture of the animal. The end result is that the animal is able to execute a well-coordinated series of scratches appropriately directed to an irritated area of skin.

Sherrington studied the neural mechanism of the scratch reflex in dogs. He found that the scratch reflex was fully manifested in animals in which the spinal cord had been severed at the cervical level, thus demonstrating that a sustained rhythmic alternation of movements does not depend on supraspinal inputs.

Like flexion reflexes, many of the properties of the reflex are dependent on the intensity and duration of the stimulus. With increasing intensity of stimulation the latency of the reflex decreases, while the strength of muscle contractions and the number of repetitions increase. A salient feature of this reflex, however, is that the *rhythm* of the reflex

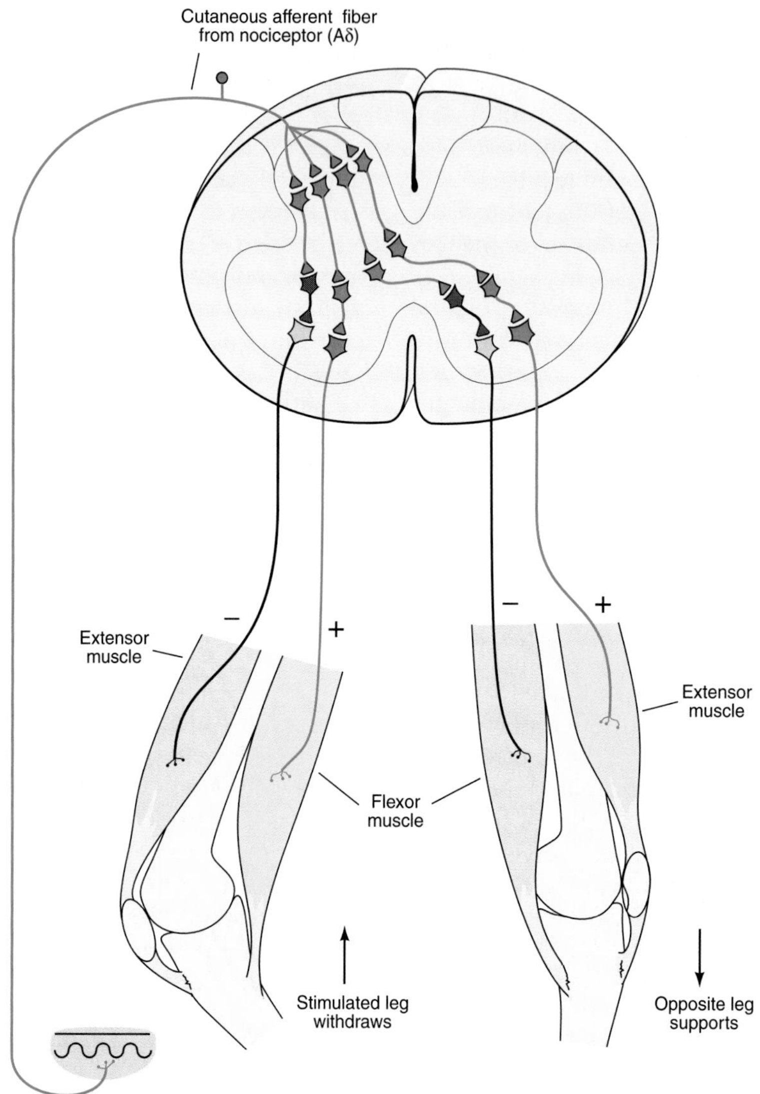

Figure 28–4 The flexion withdrawal reflex produces flexion of the stimulated limb and extension of the opposite limb. Stimulation of cutaneous afferents, such as a fiber from a pain receptor, produces excitation of the ipsilateral flexor muscles and inhibition of the ipsilateral extensor muscles, while producing the opposite response in the contralateral limb (the crossed extension reflex). The cutaneous input is distributed over many spinal segments, so that the full reflex involves contraction of muscles at all joints of both limbs. The illustration shows the pathways for one spinal segment only. (Adapted from Schmidt, 1983.)

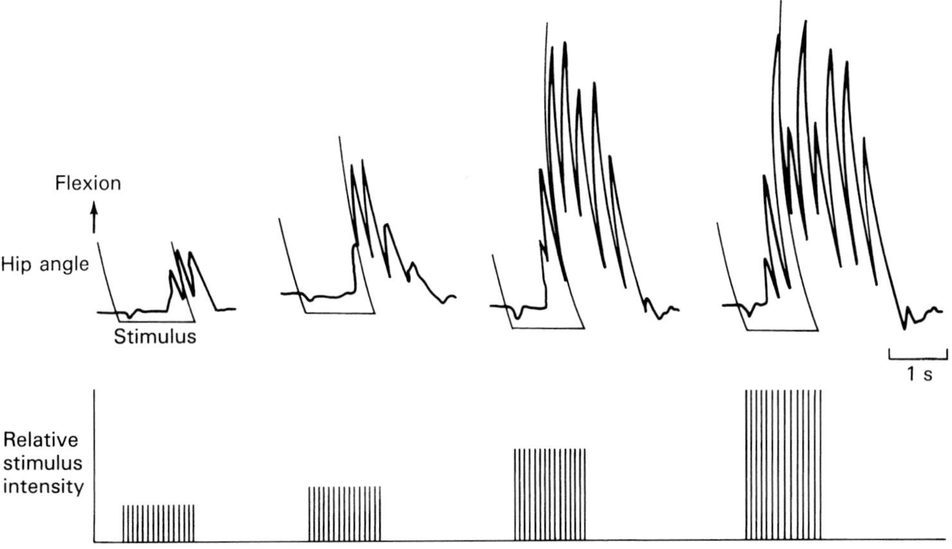

Figure 28–5 The rhythmic movements of the scratch reflex are generated by the spinal cord without a rhythmic stimulus. Early in the century Sherrington studied this reflex in dogs in which the spinal cord was severed at the cervical level to isolate the spinal circuit of the reflex. These records from his experiments show the effect of increasing stimulus strength on the scratch reflex. The tracings show movement of the hind limb in response to a mild electrical stimulus. As the intensity of the stimulus is increased, the amplitude of movement and duration of the reflex increase but the rhythm remains constant. (Adapted from Sherrington, 1947.)

is stereotyped and automatic, being independent of the features of the stimulus (Figure 28–5).

Once Sherrington demonstrated spinal control of the scratch reflex, the possibility remained that such movements might require rhythmic input from muscle receptors in the moving limb. To exclude this input, Sherrington cut the dorsal roots carrying sensory input from the hind limb. Even after this, however, cutaneous stimulation in spinal animals elicited rhythmic scratching movements, although the movements were less well coordinated and not as effective in reaching the site of the stimulus. This result showed that the circuits responsible for alternation of flexion and extension do not require input either from supraspinal centers or from the periphery.

How are these circuits organized? One model of such a central neural circuit is called the *half-center model*. The simplest form of this model proposes that interneurons controlling flexor and extensor motor neurons have reciprocal inhibitory connections. The model assumes that the duration of reciprocal inhibition is limited by some intrinsic factor, for example, *fatigue* of the inhibitory synapses, which causes the strength of synaptic transmission to diminish with time (Figure 28–6). Other possible intrinsic processes, now considered more plausible, are *accommodation*, in which a neuron responds to constant excitatory input with a declining rate of output, and *postinhibitory rebound*, in which the threshold for excitation of a neuron is transiently lower following inhibition.

The Main Features of Walking Movements Are Controlled by the Spinal Cord

Spinal Circuits Generate a Rhythmic Locomotor Pattern

Although locomotion is a voluntary action, once initiated it does not require conscious direction under normal conditions. When we walk we barely notice the alternating stepping movements of our legs that propel our body forward, and we superimpose on these movements all manner of purposeful actions, such as carrying a suitcase, throwing a ball, speaking to a friend, and even reading a book.

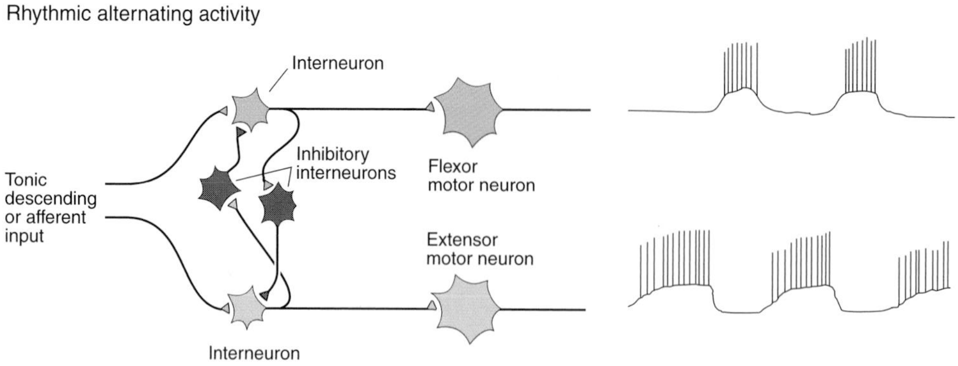

Rhythmic alternating activity

Interneuron

Inhibitory interneurons

Flexor motor neuron

Tonic descending or afferent input

Extensor motor neuron

Interneuron

Figure 28–6 The half-center model of rhythmic alternating activity in the flexor and extensor motor neurons. Flexor and extensor motor neurons, together with their associated interneurons, each constitute a half-center. The two half-centers are assumed to inhibit each other reciprocally, so that when one is active, the other is inactive. Tonic input causes both interneurons to fire, but because of random fluctuations in excitability or other inputs, one half-center initially dominates and inhibits the other. For rhythmic activity to be generated, some mechanism is needed to switch the activity from one half-center to the other. One possible switching mechanism would be an intrinsic limit to the duration of inhibition by the active half-center, which would have the effect of lessening the inhibition of the other half-center, enabling it to become active. (Adapted from Pearson, 1976.)

To a large extent, the automaticity, and hence the ease, of normal locomotion can be attributed to local spinal circuits that coordinate the contraction of the several muscle groups needed to generate rhythmic stepping. Such spinal circuits act as *central pattern generators*, similar to those that generate rhythmic movements like the scratch reflex.

The neural mechanisms of locomotion have been studied in cats in which the spinal cord is severed at the lower thoracic level, thus preventing descending signals from reaching the part of the cord that controls the hind limbs (Figure 28–7A). These *spinal cats* are able to walk on a moving treadmill with a near-normal stepping pattern, although they do require external support for balance.

The overall stepping pattern consists of a rhythmic alternation in the contractions of flexor and extensor muscles. The step cycle during locomotion has two phases: the *swing phase* (when the foot is off the ground and flexing forward) and the *stance phase* (when the foot is planted and the leg is extending relative to the body). The swing phase is generally controlled by contraction of flexor muscles, the stance phase by contraction of extensors. However, the pattern of muscle activation in

Figure 28–7 Rhythmic motor patterns in mammals are generated by local spinal circuits that are activated by descending signals from the midbrain. (Adapted from Pearson, 1976.)

A. Transection of the spinal cord at the level of **b′–b** isolates the hindlimb segments of the cord. The hind limbs are still able to walk on a treadmill after recovery from surgery. Transection at the level of **a′–a** isolates the spinal cord and lower midbrain of the cat from the cerebral hemispheres. Locomotion can be produced in the animal by electrical stimulation of specific regions of the midbrain called *locomotor regions*.

B. Locomotion of a cat (transected at **b′–b**) on a motorized treadmill. Reciprocal bursts of electrical activity can be recorded from flexors during the swing phase of walking and from extensors during the stance phase.

C. Locomotion of a cat (transected at **a′–a**) on a free-moving treadmill. The mesencephalic locomotor region is stimulated with increasing intensity. **Bottom traces** show the movements of the left and right hind limbs at different stimulus intensities. As the stimulus intensity increases, the gait becomes faster. As the cat progresses from trotting to galloping, the hind limbs shift from alternating to simultaneous flexion and extension.

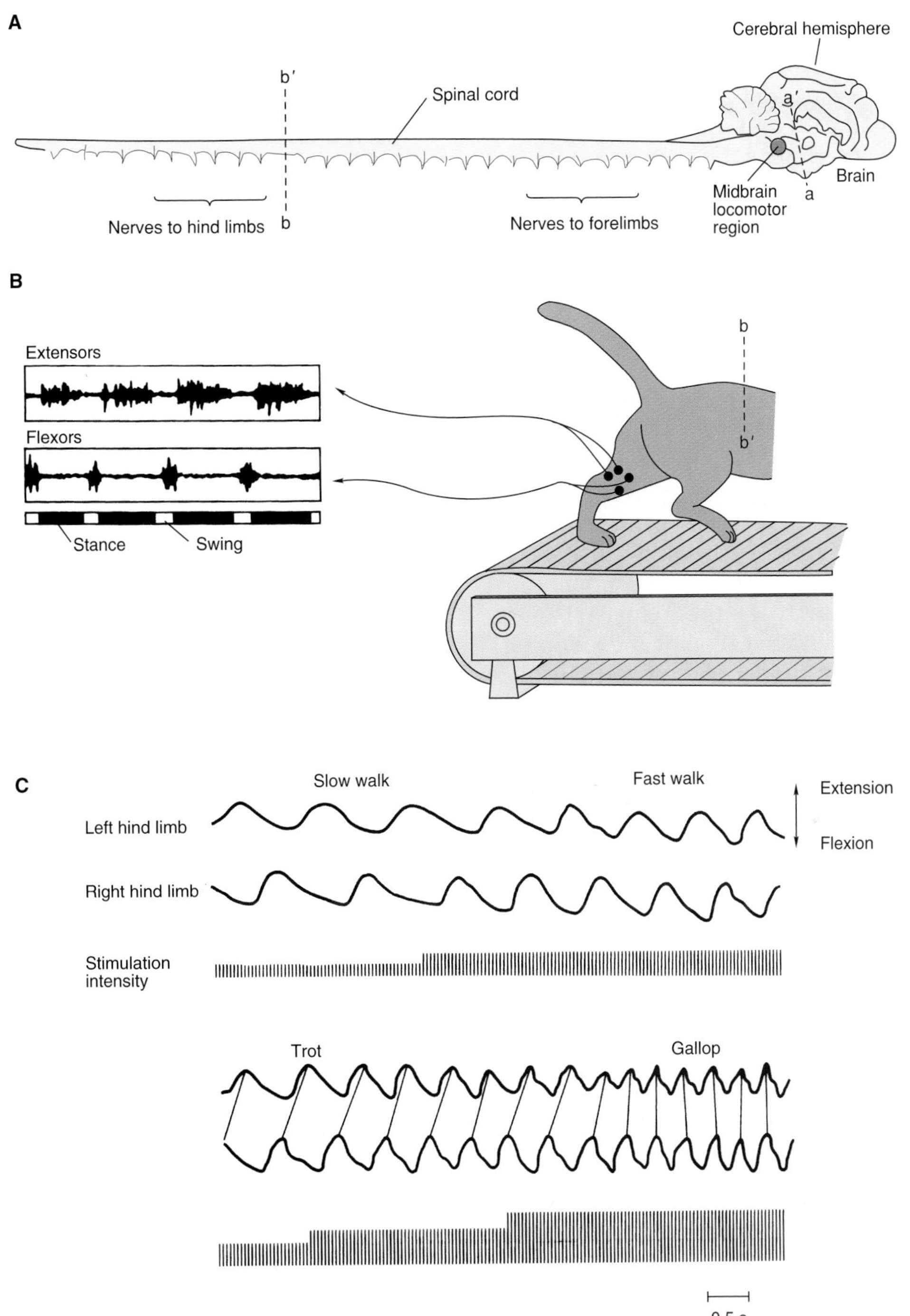
A
Cerebral hemisphere
b′
Spinal cord
a′
Brain
Midbrain
locomotor
region
a
Nerves to hind limbs
b
Nerves to forelimbs
B
b
Extensors
b′
Flexors
Stance
Swing
C
Slow walk
Fast walk
Extension
Left hind limb
Flexion
Right hind limb
Stimulation
intensity
Trot
Gallop
0.5 s

the limbs of the spinal cat is not simply stereotyped flexion and extension, but rather consists of a differentially timed and spatially distributed *synergy* of muscle contractions, similar in form to that of normal cats. A similar pattern is evident in cats whose spinal cords are transected at 1 to 2 weeks of age. Thus, spinal pattern generators are innate, built into the architecture of the spinal circuitry.

If one hind limb is prevented from moving on the treadmill, the other continues stepping normally—the restrained limb is frozen in midcycle, while the other limb continues to cycle rhythmically. Thus, each limb has its own pattern generator, each of which can act independently of the other generators. In normal locomotion the pattern generators for each limb are coupled to one another. When a cat walks on a treadmill, for example, the movements of the left and right hind limbs are exactly out of phase with each other, so that while one limb is in the stance phase the other is in the swing phase. Increasing the treadmill speed dramatically shifts the temporal coupling between the limbs as the animal changes from walking to trotting to galloping. In galloping, the hind limbs are in phase with each other; that is, they flex and extend together (Figure 28–7C).

Goal-Directed Locomotion Requires Intact Supraspinal Systems

Normal locomotion requires multiple levels of neural control. To support the body against gravity and to propel it forward, the nervous system coordinates muscle contractions at many joints through local spinal circuits. These circuits are activated by tonic signals from areas in the midbrain that control locomotion (Figure 28–7). Other descending systems, including the reticulospinal, rubrospinal, and corticospinal pathways (see Chapter 26), are phasically active during locomotion and appear to be important for modulating the strength of muscle contractions.

At the same time, the nervous system must maintain the balance of the moving body, and it must adapt the locomotor pattern to the environment and to the overall behavioral goals. The spinal cat can produce relatively normal stepping patterns but it cannot maintain its balance. Adequate balance depends on other descending signals from the brain stem, especially from the vestibular system.

Finally, although adaptation of the locomotor pattern is partially accomplished by spinal reflex pathways and brain stem areas, successful goal-directed locomotion requires the participation of other cortical and subcortical structures, including the motor cortex, basal ganglia, and cerebellum. The role of these higher motor centers in directing complex movements is the subject of the next chapter.

An Overall View

The stretch reflex presents us with a model of sensory–motor integration. Although relatively simple—it is the only known monosynaptic reflex in the mammalian nervous system—it embodies many of the principles seen in more complex motor circuits. Because the participating afferent and efferent axons have large diameters and are among the most rapidly conducting neurons in the nervous system, the stretch reflex pathway is adapted for speed of operation. The economy of the neural circuit for the stretch reflex allows muscle tone to be regulated quickly and efficiently without direct intervention by higher centers. Descending control signals adjust the strength or sensitivity of the reflex, adapting them to the requirements of specific motor acts.

In the control of movement by higher centers of the brain, most descending axons influence motor neurons indirectly, through interneurons. Interneurons in spinal circuits are not merely relay neurons that carry messages from sensory neurons to higher centers and from higher centers to motor neurons. Rather, they constitute a powerful neural machinery with the capacity to link muscles together into functional units.

We can see this organization in the spinal reflexes, which are really elementary components of more complex behaviors. Reflex circuits provide higher centers of the brain with a set of elementary patterns of motor coordination, from relatively simple combinations, like reciprocal innervation at a single joint, to more complex spatial patterns of movement, such as flexion reflexes, and temporal patterns, as in the scratch reflex. Although they are relatively stereotyped, these elementary patterns are nevertheless remarkably adaptable to new conditions, as when the dog's scratch reflex adjusts to changes in the animal's posture. Reflexes are thus not entirely stereotyped but rather are adapted to the initial position of the body segments and the external loads acting to oppose movement. Because this information reaches the lower levels directly, higher centers can activate these reflex circuits to produce voluntary movements and need not be concerned with the details of shaping the movement patterns to concurrent circumstances.

Locomotion illustrates how the spinal circuits responsible for simple reflexes simplify the control of an important and frequent voluntary movement. The spinal circuits that generate the locomotor synergy allow higher centers to control complex movement patterns with relatively simple descending signals.

Our understanding of the neural mechanisms involved in locomotion has come almost exclusively from experiments on animals. How relevant is this information to human voluntary movement? Indirect evidence indicates most likely it is of great importance. Human locomotion differs from most animal locomotion in that it is bipedal, placing greater demands on descending systems that control posture during walking. Therefore the spinal networks that contribute to human locomotion are probably more dependent on supraspinal centers. Thus, humans with complete transections of the spinal cord are incapable of rhythmic stepping like that observed in spinal cats. However, developmental studies indicate that human infants are born with innate reflex circuits capable of rhythmic pattern generation. Newborn infants exhibit rhythmic stepping when placed on a moving treadmill, and there is evidence that this reflex pattern is a forerunner of the mature locomotor synergy.

A large body of experimental work on such different types of locomotion as swimming, flying, and walking, in both vertebrates and invertebrates, indicates that all forms of locomotion rely on the same general principles of neuronal organization —intrinsic oscillatory networks are activated and modulated by afferent input and by higher motor centers in the brain stem and cortex. Through evolution similar strategies have developed in a variety of species, and there is no reason to believe that humans are fundamentally different.

Selected Readings

Baldissera, F., Hultborn, H., and Illert, M. 1981. Integration in spinal neuronal systems. In V. B. Brooks (ed.), Handbook of Physiology, Section 1: The Nervous System, Vol. II: Motor Control, Part 1. Bethesda, Md.: American Physiological Society, pp. 509–595.

Boyd, I. A. 1980. The isolated mammalian muscle spindle. Trends Neurosci. 3:258–265.

Grillner, S., and Wallén, P. 1985. Central pattern generators for movement, with special reference to vertebrates. Annu. Rev. Neurosci. 8:233–261.

Grillner, S., Wallén, P., and Viana di Prisco, G. 1990. Cellular network underlying locomotion as revealed in a lower vertebrate model: Transmitters, membrane properties, circuitry, and simulation. Cold Spring Harbor Symp. Quant. Biol. 55:779–789.

Grillner, S. and Matsushima, T. 1991. The neural network underlying locomotion in Lamprey—synaptic and cellular mechanisms. Neuron. 7:1–15.

Hasan, Z., and Stuart, D. G. 1984. Mammalian muscle receptors. In R. A. Davidoff (ed.), Handbook of the Spinal Cord, Vols. 2 and 3: Anatomy and Physiology. New York: Marcel Dekker, pp. 559–607.

Houk, J. C., and Rymer, W. Z. 1981. Neural control of muscle length and tension. In V. B. Brooks (ed.), Handbook of Physiology, Section 1: The Nervous System, Vol. II: Motor Control, Part 1. Bethesda, Md.: American Physiological Society, pp. 257–323.

Hulliger, M. 1984. The mammalian muscle spindle and its central control. Rev. Physiol. Biochem. Pharmacol. 101:1–110.

Lundberg, A. 1975. Control of spinal mechanisms from the brain. In D. B. Tower (ed.), The Nervous System, Vol. 1: The Basic Neurosciences. New York: Raven Press, pp. 253–265.

Matthews, P. B. C. 1981. Evolving views on the internal operation and functional role of the muscle spindle. J. Physiol. (Lond.) 320:1–30.

Pearson, K. 1976. The control of walking. Sci. Am. 235(6):72–86.

Prochazka, A., and Hulliger, M. 1983. Muscle afferent function and its significance for motor control mechanisms during voluntary movements in cat, monkey, and man. In J. E. Desmedt (ed.), Advances in Neurology, Vol. 39: Motor Control Mechanisms in Health and Disease. New York: Raven Press, pp. 93–132.

Sherrington, C. S. 1947. Integrative Action of the Nervous System, 2nd ed. New Haven, Conn.: Yale University Press.

29

Voluntary Movement

Neurons in the Primary Motor Cortex Encode the Force and Direction of Voluntary Movements

Neurons in the Primary Motor Cortex Alter Their Activity in Anticipation of Movements

Movement Direction Is Encoded by Populations of Neurons, Not by Single Cells

Neurons in the Motor Cortex Are Informed of the Consequences of Movements

Premotor Cortical Areas Prepare the Motor Systems for Movement

The Supplementary Motor Area Programs Motor Sequences and Coordinates Bilateral Movements

The Premotor Cortex Controls the Proximal Movements that Project the Arm to Targets

The Cerebellum Regulates Movement Indirectly

The Cerebellum Has Three Functional Divisions

The Cerebellar Cortex Is Divided into Three Distinct Layers

The Purkinje Cells Provide the Output of the Cerebellar Cortex and Receive Excitatory Input from Two Fiber Systems

The Cerebellum Participates in Motor Learning

Cerebellar Diseases Lead to Disorders of Coordination

The Basal Ganglia Integrate Information from Diverse Areas of Cortex

The Basal Ganglia Consist of Five Subcortical Nuclei

The Basal Ganglia Receive Input from the Cortex and Project Back to the Cortex by Way of the Thalamus

Loss of Dopaminergic Cells in the Substantia Nigra Leads to Parkinson's Disease

The Motor Portion of the Basal Ganglia Is Somatotopically Organized and Involved in Higher-Order Aspects of Movement

The impressive perceptual, cognitive, and emotional capacities of the human brain would be of little value if the brain could not use them to organize behavior. As we shall see, pathways from the somatic sensory cortex to the motor areas of the cortex provide the means by which the brain integrates sensory and motor information into purposeful action.

In the previous chapter we saw how spinal and brain stem reflexes provide elementary movement patterns that can be used by the brain to organize more complex behaviors, much like the subroutines of a computer program. This frees the higher centers of the brain from the need to manage all the details of the actual movements.

However, the movements organized by the brain stem and spinal cord would have a stereotyped and therefore nonadaptable quality if higher centers could not modify them to achieve specific aims. Higher motor centers are responsible for two fundamental characteristics of voluntary movement. First, voluntary movement is *planned* according to the goals of the movement. Second, it is *shaped* to specific features of the environment.

Those areas of the somatic sensory cortex that process information about the external environment (Chapter 18) and those that translate motivations into plans for action (see Chapter 5) are able to influence behavior primarily through their connections with the two main motor areas in the frontal lobes of the cerebral cortex: the primary motor area (area 4) and the premotor area (area 6). Both contain somatotopic maps of the body, so that stimulating a local site produces contractions of the muscles controlling a particular body part. Projection neurons from these areas form the major part of the corticospinal pathway, a massive tract of fibers that passes from the cortex to the spinal cord. Thus, these areas exert a direct effect on motor neurons. The same areas also influence motor output indirectly, through their connections to brain stem nuclei that in turn project to the spinal cord.

In addition to this main line for cerebral control of movement, two subcortical structures—the cerebellum and the basal ganglia—play a critical role in shaping and refining movement. Neither of these structures sends fibers directly to the spinal cord. Rather, they exert their influence on movement principally through connections with the motor areas of the cortex. These side loops compare current sensory information and completed movements and are crucial in planning movements.

In this chapter we first consider the specific roles of the primary motor area and the premotor areas in planning, initiating, and guiding voluntary movements. Then we discuss how the cerebellum and basal ganglia refine movements in progress. Our understanding of the control of voluntary movement is far from complete, and anatomical and physiological data can provide only part of the picture. Many important insights into the functions of different motor structures have come from observing the effects of different lesions. Therefore, we also present examples of the disorders of movement that result from damage to specific structures.

Neurons in the Primary Motor Cortex Encode the Force and Direction of Voluntary Movements

Neurons in the Primary Motor Cortex Alter Their Activity in Anticipation of Movements

We have seen in Chapter 26 how the motor areas of the cerebral cortex are somatotopically organized, so that specific cortical motor areas control specific contralateral body parts. However, the ablation and stimulation studies that led to this understanding provided few clues as to how these areas initiate and control movement. This question could only be addressed by examining how neuronal activity changes during the performance of a motor task itself.

In the 1960s, this issue began to be investigated by recording from single neurons in the primary motor cortex of awake monkeys trained to perform various simple tasks. The activity of neurons was initially examined in the wrist area of the primary motor cortex during wrist flexion and extension. Different populations of neurons were found to be active during flexion and extension and their activity typically changed *before* the contraction of the relevant muscles (Figure 29–1). This provided direct evidence that the primary motor cortex actually participates in the initiation or triggering of movement.

Single-neuron recordings during behavior also make it possible to find out what feature of movement is controlled by cell discharge. The discharge frequency of corticospinal tract neurons encodes the amount of force used to move the limb rather than the change in the position of the limb. Thus, the firing rate of a neuron that becomes active during wrist flexion increases when a load that resists flexion is added. When the load is changed to assist flexion and oppose extension, wrist flexion can be achieved solely by relaxation of the antagonist (extensor) muscles. Now the neuron no longer fires before or during the movement (Figure 29–1B).

Movement Direction Is Encoded by Populations of Neurons, Not by Single Cells

The observation that flexion and extension of the wrist or elbow are associated with the firing of different populations of cortical neurons fits well with the idea of a muscle map in the primary motor cortex. However, since individual neurons in the primary motor cortex can influence several muscles at once, the question arises, how can the direction of a multijoint arm movement, which requires the contraction of many different muscles, be encoded by cortical neurons?

This question was addressed by studying how neuronal activity varied when monkeys moved a handle to one of several targets arranged in different directions around a central starting position. The result showed that the activity of individual neurons did indeed vary with the direction of the movement. They fired most briskly for movements in a preferred direction and were silent during

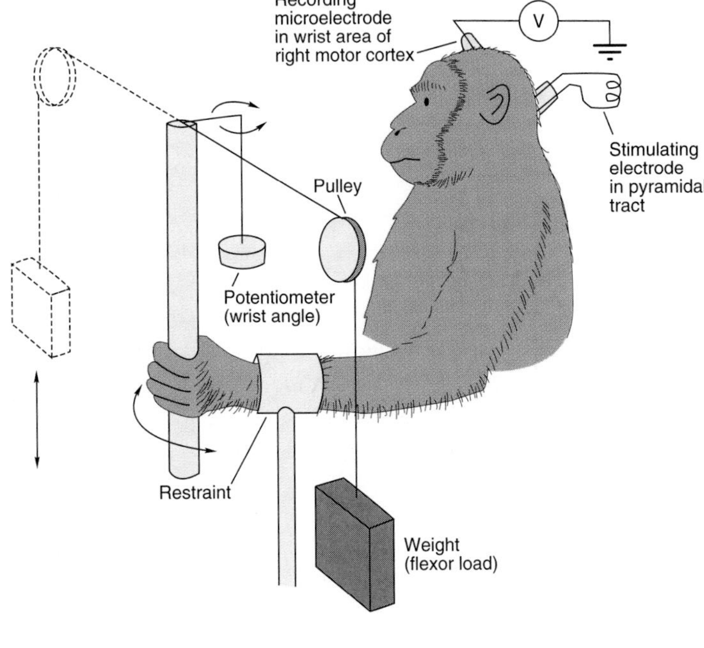

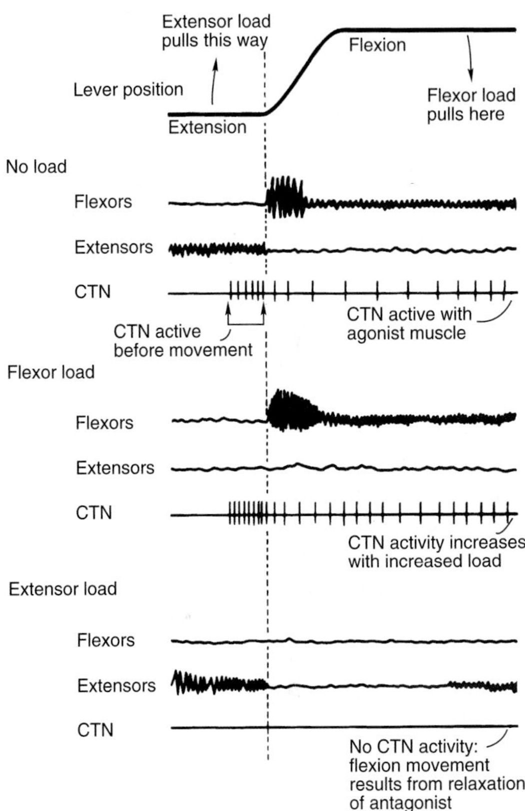

Figure 29–1 The firing rate of neurons in the motor cortex controls the force exerted by the innervated muscle. (Adapted from Evarts, 1968.)

A. Setup for recording specific corticospinal tract neurons in the motor cortex of an awake monkey. The apparatus permits the animal alternately to flex and extend its wrist.

B. A corticospinal tract neuron (CTN) that controls flexion of the wrist starts firing before movement. Electromyograms of flexor and extensor muscles, together with the discharge records of the neuron, are shown under different load conditions. The absence of neuronal activity during *extension* of the wrist indicates that the neuron codes for force rather than displacement.

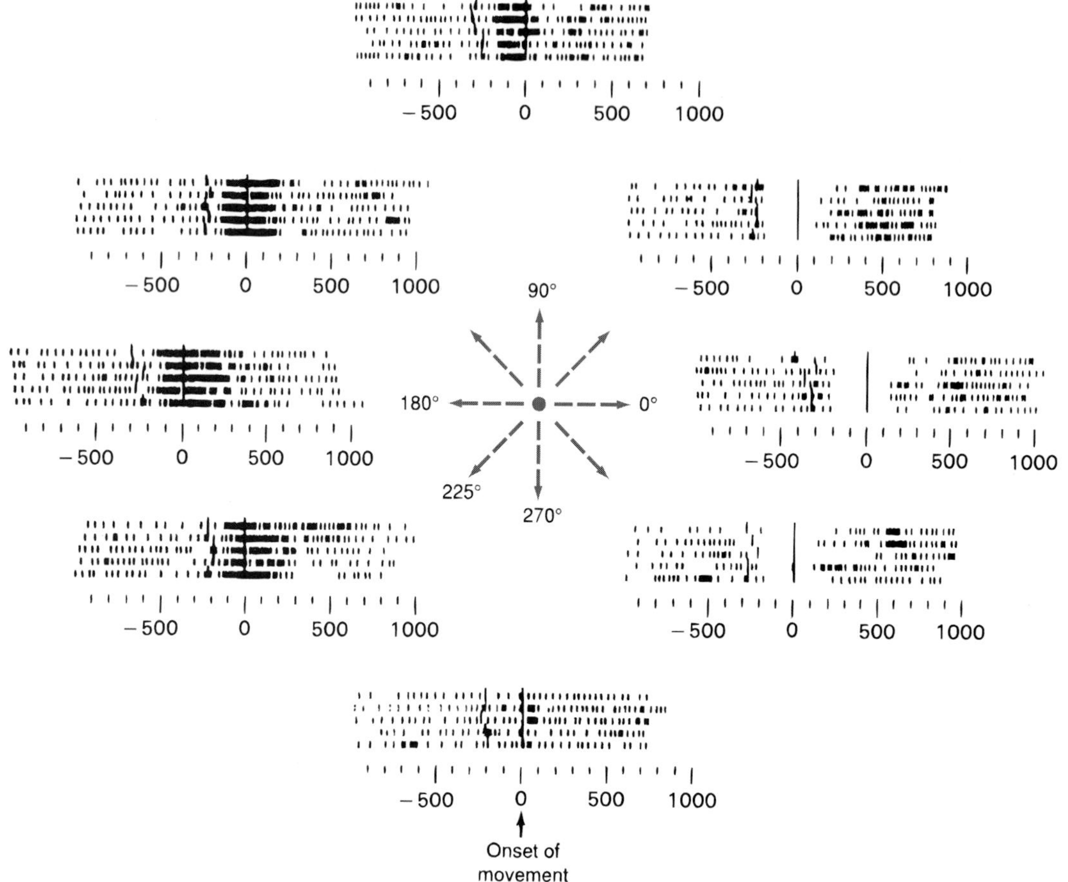

Figure 29–2 Individual cortical neurons are broadly tuned to the direction of movement. Raster plots show the firing pattern of a single neuron during movement in eight directions. A monkey was trained to move a handle to eight different locations, represented by light-emitting diodes, arranged radially in one plane around a central starting position. Each row of ticks in each raster plot represents activity in a single trial. The rows are aligned at zero time (the onset of movement). The center diagram shows the directions of the eight movements. It can be seen that the cell fires at relatively high rates during movements made in a range of directions from 90° to 225°. (From Georgopoulos et al., 1982.)

movements in the opposite direction. Like the primary somatosensory and visual cortices, neurons in the motor cortex are arranged in functional columns—each column governs a particular muscle. Moreover, within a column of cortex the preferred directions of neurons were quite similar. The *directional tuning* of all recorded neurons was, however, surprisingly broad. Thus, individual neurons participate predominantly in movements in a preferred direction, but also (to lesser degrees) in movements in a range of directions (Figure 29–2).

How can movement direction be coded precisely by neurons that are so broadly tuned? Perhaps movement in a particular direction is not determined by the action of single neurons but by the net action of a population of neurons. The contribution of one neuron to movement in a particular direction can be represented as a vector whose length depends on the degree of activity during movements in that particular direction. The contributions of individual cells can then be added vectorally to produce a *population vector*, and the direc-

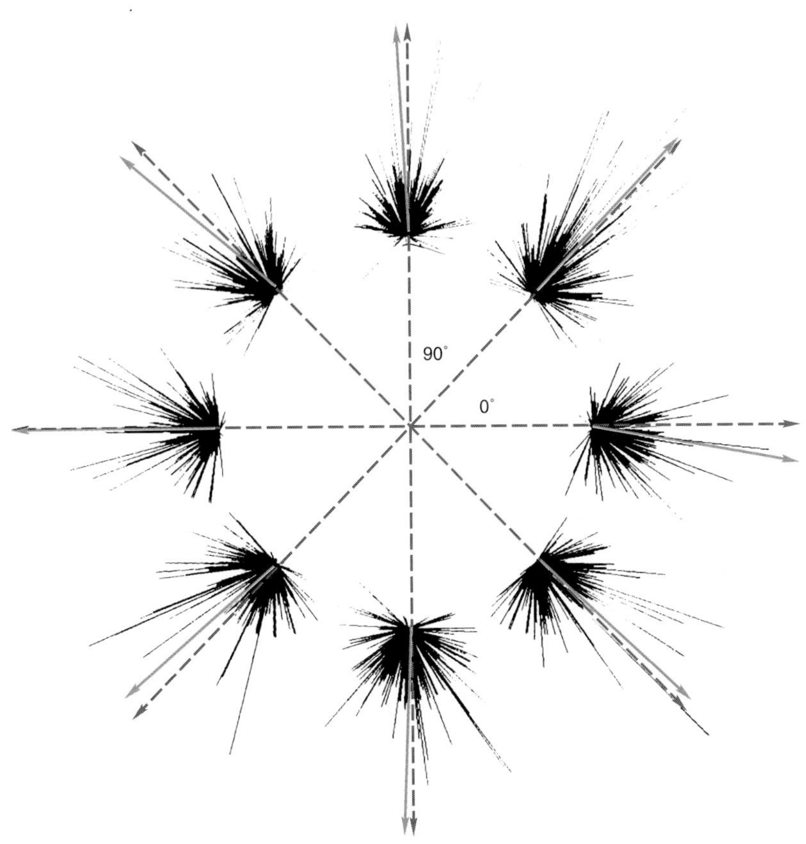

Figure 29–3 During movement in any given direction, cortical neurons with different preferred directions are active, but the direction of the *population vector* closely matches that of the direction of movement. Each cluster represents the activity of one population. The directions of the population vectors (**dashed arrows**) closely match the direction of the targets (**brown arrows**).

tion of the population vector will determine the direction of the hand movement.

To test this idea and determine the relationship between the direction of the population vector and the ensuing movement, the activity of individual neurons in monkeys reaching toward targets in different directions was measured. The directions of the computed population vectors were found to match closely the directions of movement (Figure 29–3). This shows that although the specific features of a reaching movement are determined by the combined activity of a population of neurons, they cannot be predicted from the discharge patterns of any one neuron. This situation contrasts with that of motion detection in the visual cortex, where small populations of cells seem to be critically important for perception (see Chapter 23).

Neurons in the Motor Cortex Are Informed of the Consequences of Movements

Neurons in the primary motor cortex are kept informed about the position and speed of movement because they receive sensory input. Like neurons in the somatic sensory cortex, neurons in the motor cortex have receptive fields in the periphery. Some respond to tactile stimuli, others to movements of the hands, and still others to stretch of individual muscles or rotation of joints.

What is the relationship between the location of these receptive fields on the body and the muscle groups controlled by local sectors of motor cortex? Some neurons in the motor cortex receive proprioceptive input from the muscles they control, while others receive input from regions of skin that are

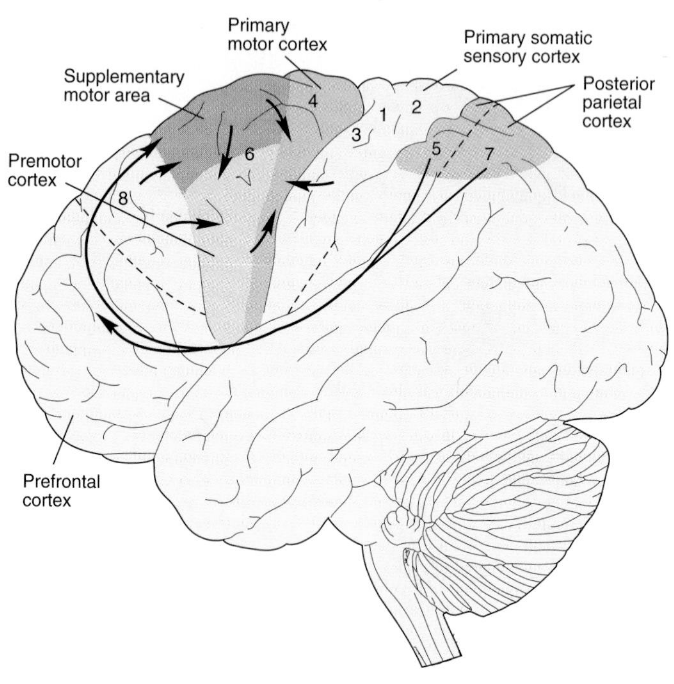

Figure 29–4 The motor areas are connected to each other and to the posterior parietal cortex by corticocortical connections.

likely to be stimulated during contraction of the muscles. This sensory input is transmitted to the motor cortex from the somatic sensory cortex and by direct pathways from the thalamus. Such transcortical circuits may convey afferent information from muscles and control contraction of muscles by a long loop pathway through the motor cortex.

Premotor Cortical Areas Prepare the Motor Systems for Movement

The premotor areas of the cerebral cortex, directly rostral to the primary motor area in the frontal lobe, play a major role in the planning of goal-directed movements. There are two principal premotor areas: the *supplementary motor area* and the *premotor cortex* (Figure 29–4). Both areas are organized somatotopically. While the size of the primary motor cortex in proportion to body weight remains roughly constant across primate phylogeny, the premotor areas in humans are six times larger than in macaque monkeys. Neurons in the premotor areas project to the primary motor cortex, as well as to subcortical structures and to the spinal cord.

Electrical stimulation of the premotor areas typically evokes coordinated contractions of muscles at more than one joint and, in the case of the supplementary motor area, on both sides of the body as well. These movements are more complex and require larger stimulus currents than those produced by stimulation of the primary motor cortex.

While lesions of the primary motor cortex cause weakness, lesions of premotor areas impair the ability to develop an appropriate strategy for movement. For example, when monkeys with lesions of these areas are presented with food behind a transparent shield with an opening to the side of the food, they do not reach through the opening but instead aim directly for the food, bumping their hands into the shield.

These symptoms in monkeys are similar to *apraxias* that occur in humans with lesions of the supplementary motor or posterior parietal association cortices. Patients with apraxia show neither weakness nor sensory loss and are able to make simple movements accurately, but they are unable to perform complex acts requiring sequences of muscle contractions or a planned strategy such as combing their hair or brushing their teeth.

The Supplementary Motor Area Programs Motor Sequences and Coordinates Bilateral Movements

The supplementary motor area plays an important role in programming complex sequences of movement. Movements elicited by stimulating the supplementary motor area require more intense and longer-lasting trains of pulses than do movements evoked from the primary motor cortex. These movements include such complex patterns as orienting the body or opening or closing the hand. Many of the movements are bilateral. Movements involving proximal muscles are mediated through direct projections from the supplementary motor area to the spinal cord. Those involving distal muscles appear to be mediated indirectly by the supplementary area, through connections to the motor cortex, since these movements are abolished by lesions of the motor cortex.

How the different areas of cortex contribute to movement in awake humans was demonstrated in the 1980s in a particularly striking way. Knowing that variations in neuronal activity are accompanied by corresponding changes in local cerebral blood flow, investigators measured blood flow over different areas of the brain while subjects performed various tasks. During simple tasks, blood flow increased dramatically within the contralateral hand areas of both the primary motor and somatic sensory cortices, but did not increase significantly over the premotor areas (Figure 29–5A). During a complex sequence of movements involving all of the fingers, the increase in cerebral blood flow extended to the supplementary motor area (Figure 29–5B). However, when subjects were told

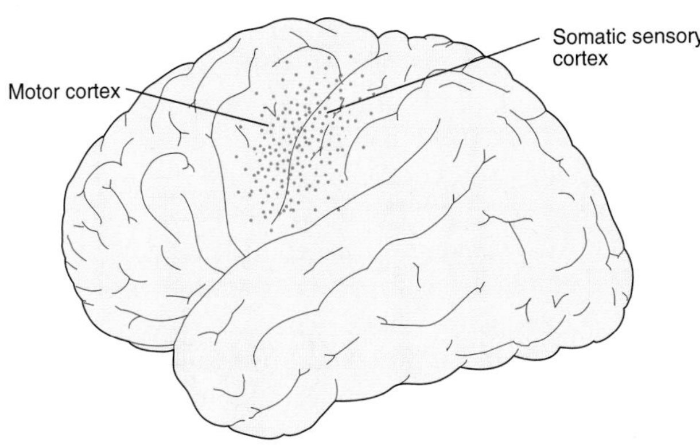

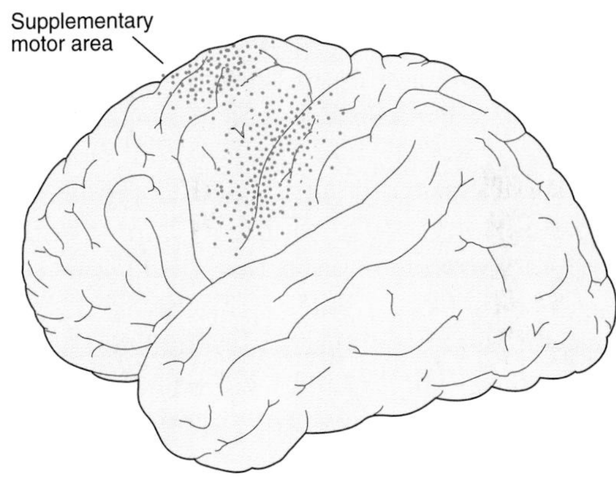

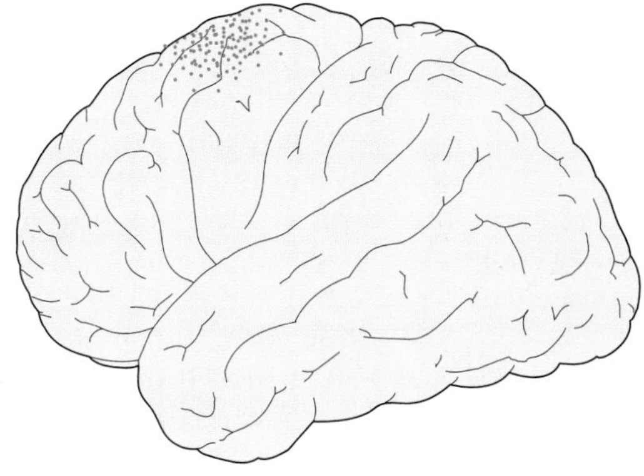

Figure 29–5 Local increases in cerebral blood flow during a simple task indicate which areas of motor cortex participate in the behavior. Blood flow is measured by intravenously injecting radioactive xenon dissolved in a saline solution and measuring the radioactivity over different parts of cortex using arrays of detectors placed over the scalp. Since local tissue perfusion varies with neural activity, the measured radioactivity provides a good index of regional activity in the surface of the brain. (Adapted from Roland et al., 1980.)

A. When a finger is pressed against a spring, blood flow increases in the hand areas of the primary motor and sensory cortices. The increase in the motor area is related to the execution of the response, and the increase in the sensory area reflects the activation of peripheral receptors.

B. During a complex sequence of finger movements, the increase in blood flow extends to the supplementary motor area.

C. During mental rehearsal of the sequence illustrated in **B**, blood flow increases only in the supplementary motor area.

to rehearse the sequence of finger movements mentally but not to perform the sequence, blood flow increased *only* in the supplementary motor area (Figure 29–5C).

This demonstrates that the activity of neurons in the supplemental motor area is linked to the mental activity needed to plan such movements, which involves the same area of cortex whether the movement is executed or simply rehearsed mentally. It is not linked directly to the execution of the movements.

The Premotor Cortex Controls the Proximal Movements that Project the Arm to Targets

Although the premotor cortex is still poorly understood, some preliminary insights into its functions have emerged from studies correlating anatomy, single-cell recording, and behavior. The premotor cortex receives its principal input from the posterior parietal cortex and, like the supplementary cortex, projects to the primary motor cortex. It also sends abundant projections to regions of the brain stem that contribute to the medial descending systems (notably the reticulospinal system) and to the region of the spinal cord that controls proximal and axial muscles. These connections led to the idea that the premotor cortex plays a primary role in the control of proximal and axial muscles that are important in the initial phases of orienting the body and arm to a target.

The prominent inputs to the premotor areas from the posterior parietal cortex also suggest that the premotor areas might participate in the control of movement through visual or somatosensory cues. This idea is supported by recordings from monkeys trained to touch a target when a trigger signal (a visual or auditory cue) was presented to them. Some neurons fired in precise association with the trigger signal, as though they might participate in initiating action. Others fired when the animal was instructed on which target to touch, as though they participated in preparing action (Figure 29–6). Such neurons are termed *set related* to indicate that their activity is associated with the animal's preparation to act and to suggest a role for the premotor cortex in the preparatory process itself.

In the preceding paragraphs we have seen how the premotor and motor cortices play critical roles in triggering the commands and in specifying the spatial details of the movement. Two subcortical areas, however, play important roles in ensuring the accuracy and coordination of movement. These are the *cerebellum* and the *basal ganglia,* whose structure and function we next review.

The Cerebellum Regulates Movement Indirectly

The cerebellum regulates movement and posture *indirectly* by adjusting the output of the major descending motor systems of the brain. Lesions of the cerebellum do not cause paralysis; instead, they disrupt coordination of limb and eye movements, impair balance, and decrease muscle tone. The signs of cerebellar damage thus differ dramatically from those of damage to the motor cortex, which reduce the strength and speed of movement and diminish the ability to contract individual muscles.

How does the cerebellum adjust the output of the motor systems? The most attractive idea is that it acts as a *comparator* that compensates for errors in movement by comparing intended movement with actual performance. First, through comparisons of internal and external feedback signals, the cerebellum is able to correct ongoing movements when they deviate from their intended course. Second, the cerebellum is able to modify central motor programs so that subsequent movements can fulfill their goal with fewer errors. In part, these adaptations appear to depend on the capacity of certain classes of inputs to modify cerebellar circuits for long periods. Thus, the activity of cerebellar neurons is changed by experience and plays an important role in the learning of motor tasks.

The cerebellum occupies most of the posterior cranial fossa. It is composed of an outer mantle of gray matter (the *cerebellar cortex*), internal white matter, and three pairs of deep nuclei, which are the major output structures of the cerebellum: the *fastigial*, the *interposed* (itself composed of two nuclei), and the *dentate* nuclei (Figure 29–7).

The cerebellum receives input from the periphery and from all levels of the central nervous system. Afferent pathways synapse on neurons in both the deep nuclei and the cerebellar cortex. Although the outflow from most regions of the cerebellar cortex projects first to the deep nuclei, the phylogenetically oldest part of the cerebellar cortex (the flocculonodular lobe, described below) projects directly to the vestibular nuclei in the

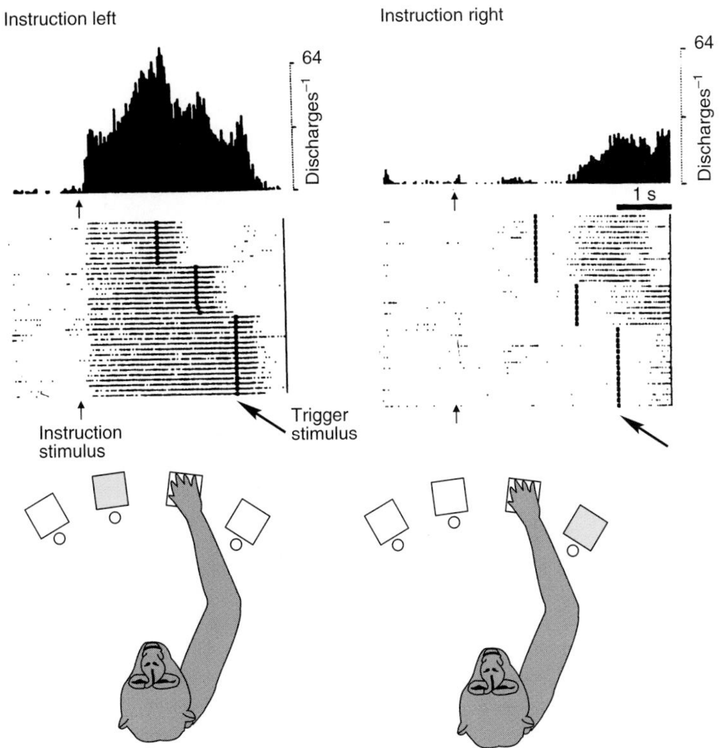

Figure 29–6 Certain neurons in the premotor area are active only during the preparation for movement, not during movement itself.

In the experiment illustrated here a monkey is trained to depress one of four panels arranged within reach of the animal. In each trial an *instruction stimulus* (illumination of one of the four panels) tells the animal which panel it will have to depress when a *trigger stimulus* (illumination of a light-emitting diode next to the panel) is presented.

Records of neuronal activity are for a single neuron in the lateral premotor area. Each line is one trial, and successive trials are aligned on the onset of the instruction signal. Each dot on each line represents a spike in the recorded neuron. The time between the instruction and trigger signals varied randomly among three values. The records of cell activity are grouped according to the time of the trigger signal to show that the discharge of the neuron begins with the instruction signal and lasts until the animal's response after the trigger signal.

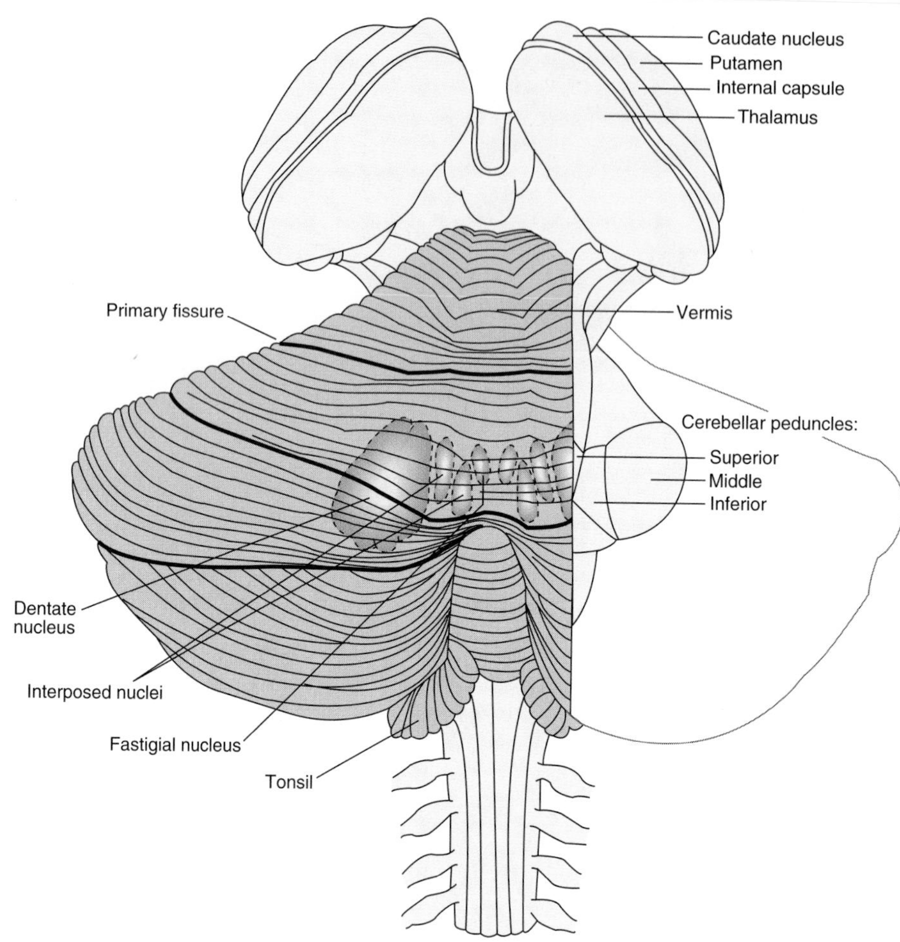

Figure 29–7 The gross anatomy of the cerebellum is shown in this dorsal view. Part of the right hemisphere has been cut out to show the underlying cerebellar peduncles. (Adapted from Nieuwenhuys, Voogd, and van Huijzen, 1988.)

brain stem, which are analogous in function to the deep cerebellar nuclei. The deep cerebellar nuclei and vestibular nuclei transmit the entire output of the cerebellum, primarily to the motor regions of the cerebral cortex and the brain stem. The input and output connections of the cerebellum run through three symmetrical pairs of tracts that connect to the brain stem.

A striking feature of the cerebellar surface is the many parallel transverse convolutions called *folia* (Latin for "leaves") that run from one side to the other. Two deep transverse fissures divide the cerebellum into three major lobes: the *anterior lobe*, the *posterior lobe*, and the small *flocculonodular lobe*. Two longitudinal furrows, most prominent on the undersurface of the cerebellum's posterior lobe, identify the region of the *vermis* (Latin for "worm"), a thin longitudinal strip in the midline, which separates the left and right cerebellar hemispheres (Figure 29–8). The part of the hemisphere closest to the vermis is called the *intermediate part*; the remainder of the hemisphere is the *lateral part*.

The vermis and the two parts of the hemisphere are connected to different deep nuclei and to dif-

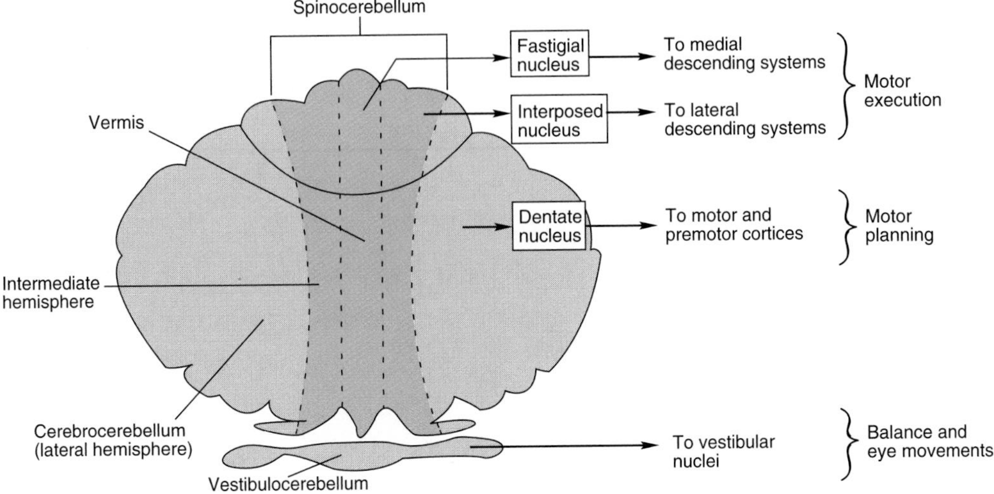

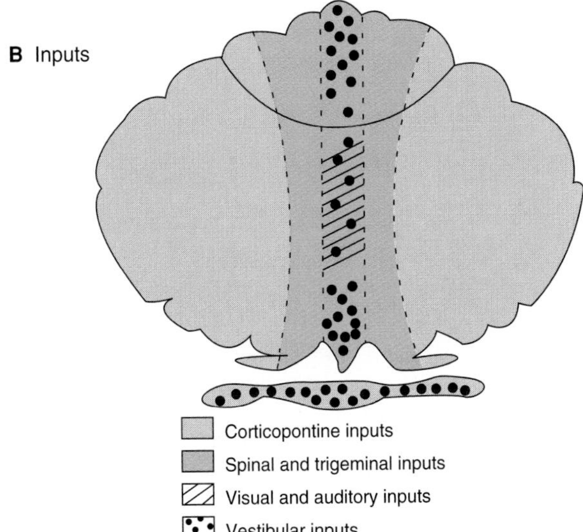

Figure 29–8 The cerebellum has three functional components—the spinocerebellum, cerebrocerebellum, and vestibulocerebellum—and each has different outputs (**A**) and inputs (**B**).

ferent components of the descending systems and constitute functionally distinct regions of the cerebellum (Figure 29–8). The vermis, through direct projections to the vestibular nuclei and through indirect projections to other medial brain stem and cortical descending systems, controls axial and proximal limb muscles. The intermediate zone of each hemisphere projects to the cortical and brain stem regions that give rise to the lateral descending systems, which control distal limb muscles. The lateral zone of each hemisphere projects to the *dentate nucleus*, which connects primarily with motor and premotor regions of the cerebral cortex, the regions involved in the planning of voluntary movements.

The Cerebellum Has Three Functional Divisions

The cerebellum is organized into three functional divisions, each with distinct anatomical connections to the brain and spinal cord: the vestibulocerebellum, the spinocerebellum, and the cerebrocerebellum. These three regions correspond roughly

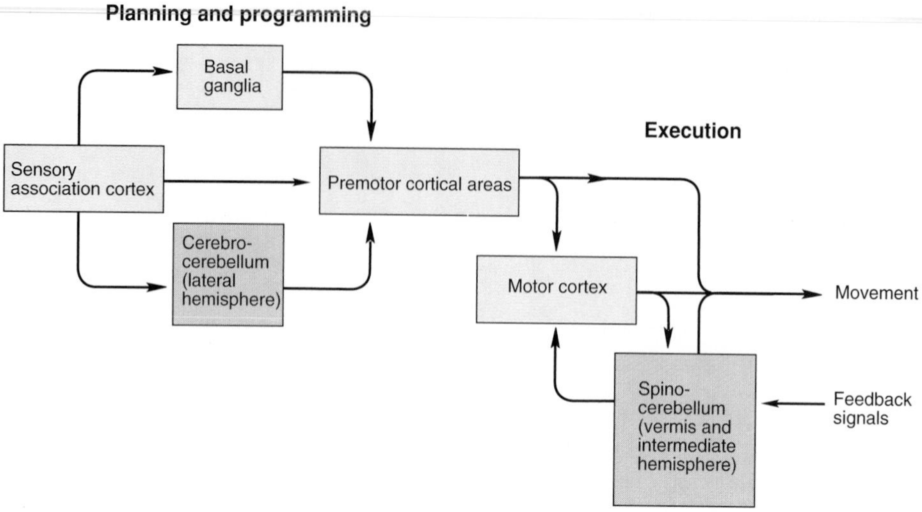

Planning and programming

Basal ganglia

Sensory association cortex

Premotor cortical areas

Cerebro-cerebellum (lateral hemisphere)

Execution

Motor cortex

Movement

Spino-cerebellum (vermis and intermediate hemisphere)

Feedback signals

Figure 29–9 This flow diagram shows the likely role of the cerebellum in the initiation of voluntary movement. Here the cerebrocerebellum participates in the planning and programming of movement, while the spinocerebellum participates in the execution. The thalamic relay for basal ganglia, cerebellar, and somatic sensory input is omitted for simplicity. (Adapted from Allen and Tsukahara, 1974.)

to anatomical subdivisions that have evolved successively in phylogeny. The vestibulocerebellum, which is coextensive with the flocculonodular lobe, appeared first in evolution. The spinocerebellum and cerebrocerebellum constitute the body of the cerebellum. Each region receives its main inputs from a different source and sends its outputs to a different part of the brain (Figure 29–8). Lesions of each region give rise to characteristic clinical syndromes.

The *vestibulocerebellum* receives its input from the vestibular nuclei in the medulla and projects directly back to them, hence its name. Through its afferent and efferent connections with the vestibular nuclei, the human vestibulocerebellum governs eye movements and body equilibrium during stance and gait.

The *spinocerebellum* extends rostrocaudally through the central part of both the anterior and posterior lobes and includes the vermis at the midline and the intermediate part of the hemispheres. These two regions receive sensory information from the periphery. The spinocerebellum is so named because a major source of its input arises in the spinal cord. Through the deep nuclei the spinocerebellum controls the medial and lateral components of the descending motor systems and thus plays a major role in controlling the ongoing execution of limb movement.

The *cerebrocerebellum* is the lateral part of the cerebellar hemisphere. Its inputs originate exclusively in the pontine nuclei, neurons in the base of the pons that are segregated into many small groups by the fascicles of corticospinal and corticobulbar axons and that relay information from the cerebral cortex. Its output is conveyed by the dentate nucleus to the thalamus and from there to the motor and premotor cortices. Because of its connections to these cortical regions, the cerebrocerebellum is thought to play a special role in the planning and initiation of movement.

Current views of the roles of the intermediate and lateral cerebellum are consistent with a model first proposed in the 1970s (Figure 29–9). According to this model, the lateral cerebellum and the basal ganglia process information originating in the parietal–temporal–occipital association cortex, which is concerned with integrating sensory information for purposeful action (see Chapter 19). This processing is critical for planning movement and preparing the motor systems to act and thus shapes the commands for movement issued by the lateral cerebellum and a basal ganglia to the premotor and motor cortical areas (and to subcortical

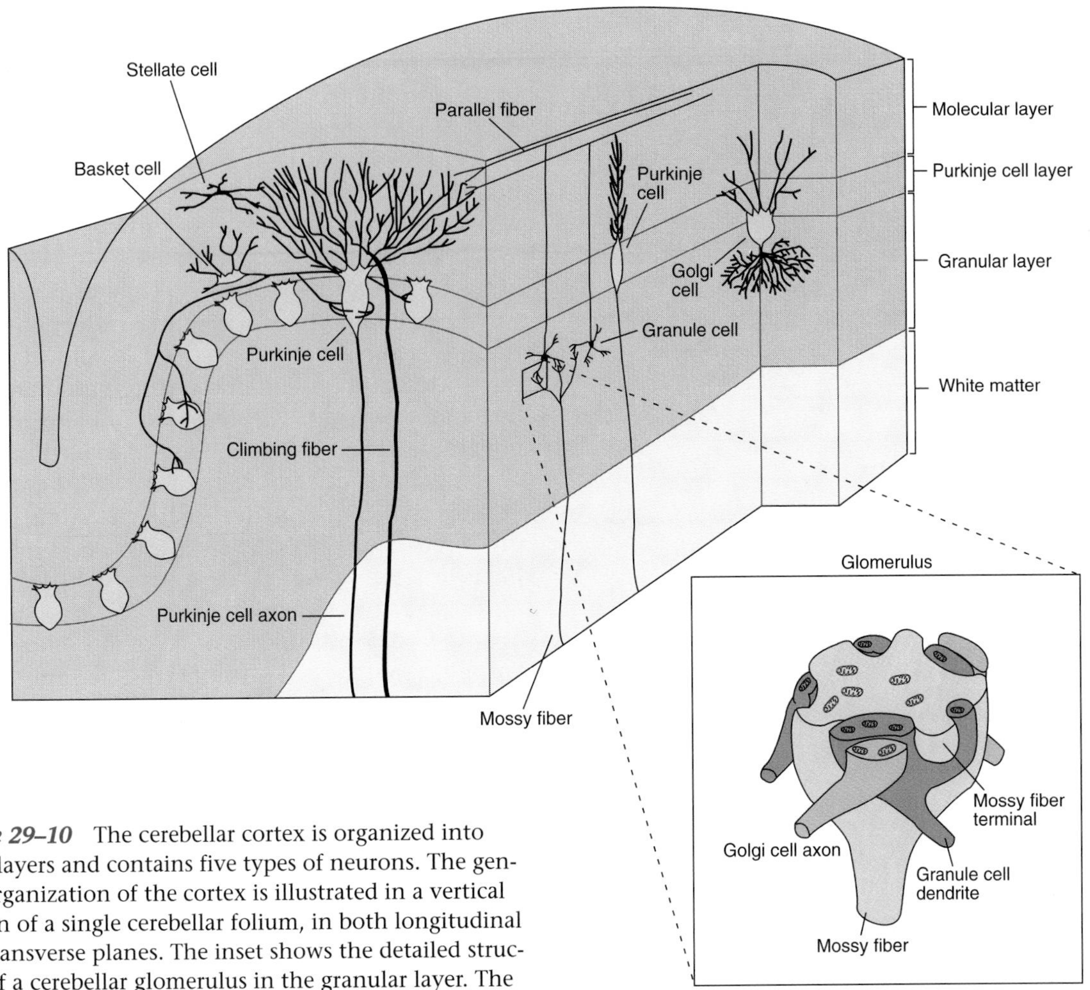

Figure 29–10 The cerebellar cortex is organized into three layers and contains five types of neurons. The general organization of the cortex is illustrated in a vertical section of a single cerebellar folium, in both longitudinal and transverse planes. The inset shows the detailed structure of a cerebellar glomerulus in the granular layer. The glomerulus is where the bulbous terminal of mossy fibers makes synaptic contact with Golgi and granule cell axons.

centers not shown in Figure 29–9). These motor areas execute movement and, at the same time, signal the spinocerebellum of the ongoing commands. In turn, the spinocerebellum monitors actual output and corrects for errors that have occurred or compensates for impending errors in the commands for movement.

The Cerebellar Cortex Is Divided into Three Distinct Layers

The cerebellar cortex is organized in three layers: the molecular, the Purkinje cell, and the Golgi granular cell layers (Figure 29–10).

The outermost *molecular layer* is composed primarily of the axons of granule cells, which run parallel to the long axis of the folium. It also contains scattered stellate and basket cells, which function as interneurons, as well as the dendrites of the underlying Purkinje neurons.

The *Purkinje cell layer* lies beneath the molecular layer. It contains the large Purkinje cell bodies, which are arranged side by side in a single layer. The extensive dendritic trees of the Purkinje cells extend up into the molecular layer in a single plane perpendicular to the main axis of the folium. The Purkinje neurons, which convey all the output of the cerebellar cortex, send their axons down into the underlying white matter.

The innermost *granular layer* contains a vast number of densely packed small neurons, mostly small granule cells. Their number, about 10^{10},

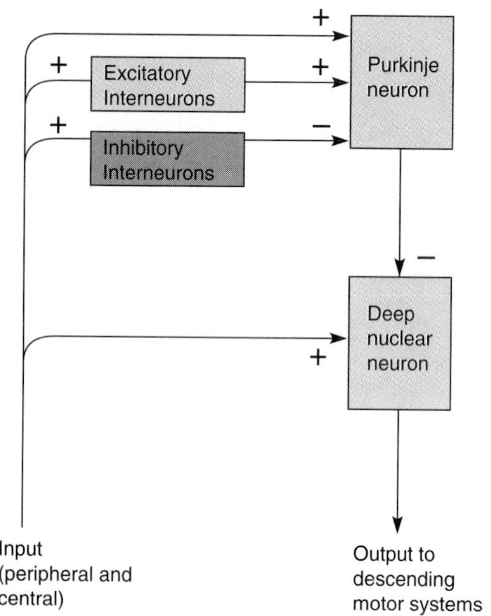

Figure 29–11 The excitatory input to the cerebellar nuclei can be modulated by the Purkinje cells in the cerebellar cortex.

exceeds the total in the cerebral cortex! A few larger Golgi cells are found at the outer border. Cells in the granular layer form complex synaptic contacts with the bulbous terminals of afferent fibers in structures called *cerebellar glomeruli* (Figure 29–10).

The Purkinje Cells Provide the Output of the Cerebellar Cortex and Receive Excitatory Input from Two Fiber Systems

Information flowing through the cerebellum acts initially on the deep nuclei, which together with the vestibular nuclei transmit all output from the cerebellum. The activity of the Purkinje cells, the only output neurons of the cerebellar cortex, is determined by two excitatory afferent inputs: mossy fibers and climbing fibers. Mossy and climbing fibers arise from different sources, terminate in different ways in the cerebellum, and have different functional roles. Both send collateral axon branches to the deep cerebellar nuclei. These collateral pathways form the principal excitatory circuit of the cerebellum (Figure 29–11). The strength of this direct circuit is continuously modulated by an indirect inhibitory circuit through the cerebellar cortex. The same afferent inputs project directly and indirectly onto the Purkinje cells in the cerebellar cortex, which form inhibitory connections with cells in the deep cerebellar nuclei.

The mossy fibers constitute the major afferent input to the cerebellar cortex. They project from a variety of brain stem nuclei and from neurons in the spinal cord that give rise to the spinocerebellar tract. Mossy fibers synapse on the granule cells of the granular layer, whose axons project into the molecular layer. In the molecular layer the axons of the granule cell bifurcate into *parallel fibers* that extend several millimeters along the long axis of the cerebellar folia (see Figure 29–10). The parallel fibers intersect and make contact with the dendrites of a row of Purkinje cells, all of which are oriented perpendicular to the parallel fibers. Each Purkinje cell receives converging input from approximately 200,000 parallel fibers from granule cells and each granule cell collects input from many mossy fibers.

The *climbing fibers*, the other excitatory input to the cerebellar cortex, originate in a single site in the medulla, the inferior olivary nucleus, and receive input from both the cerebral cortex and the spinal cord. Climbing fibers are so named because of the morphology of their terminations on the Purkinje neurons. Their axons enter the cortex and wrap around the soma and dendrites of Purkinje neurons, where they make numerous synaptic contacts, primarily on the proximal portions of the dendrites. Each climbing fiber contacts only 1 to 10 Purkinje neurons, and each Purkinje neuron receives synaptic input from only a single climbing fiber.

The synaptic connection made by climbing fibers on Purkinje neurons is one of the most powerful in the nervous system. A single action potential in a climbing fiber elicits very large excitatory postsynaptic potentials in both the soma and dendrites of the Purkinje cells. Each large presynaptic potential is followed by a high-frequency burst of smaller action potentials.

Climbing fiber input to the Purkinje cells is important in modulating the effect of mossy fibers on Purkinje cells, and it occurs in two distinct ways. Climbing fibers can transiently enhance the effect of mossy fiber inputs to Purkinje cells through mechanisms that are not fully understood, or they can produce a long-lasting depression of the efficacy of selected mossy fiber inputs, through *heterosynaptic* action (a change in activity in one pathway due to activity in another pathway).

The Cerebellum Participates in Motor Learning

Several lines of evidence support the idea that cerebellar circuits are modified by experience and that these changes are important for motor learning. Much of this work has focused on the vestibulo-ocular reflex, which maintains the orientation of the eyes on a fixed target when the head is rotated. In this reflex, motion of the head in one direction is sensed by the vestibular labyrinth, which initiates eye movements in the opposite direction to maintain the image on the retina. When humans and experimental animals wear prism goggles that reverse the left and right visual fields, the vestibulo-ocular reflex is maladaptive at first. This is because the resulting eye movement accentuates motion of the visual field on the retina. With time, however, the direction of the reflex gradually becomes reversed. This learning can be prevented by lesions of the vestibulocerebellum.

The role of the mossy and climbing fibers in motor learning has been examined by observing the activity of these fibers in monkeys trained to move a handle, using the flexors or extensors of the wrist. In one set of experiments the handle was connected to a motor to introduce loads that unexpectedly displaced the handle. When the load was kept constant, the trained animal moved the lever rapidly and smoothly to the required position from trial to trial. However, when the load was unexpectedly increased in several trials, the animal's response was inaccurate for many trials, until gradually it learned to counteract the additional load with the correct force (Figure 29–12A).

When the load was constant and predictable, each movement was accompanied by stereotyped fluctuations in simple spikes from mossy fiber input, with an occasional interspersed complex spike from the climbing fiber input. When the load was changed suddenly, the frequency of complex spikes increased dramatically, and this increase was accompanied by a gradual decrease in the frequency of simple spikes. As the animals adjusted to the

A Neural activity during motor adaptation

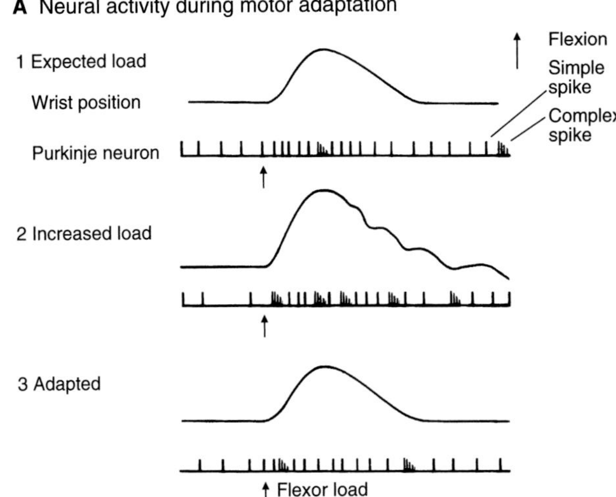

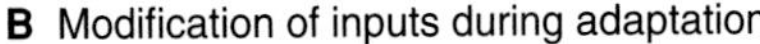

B Modification of inputs during adaptation

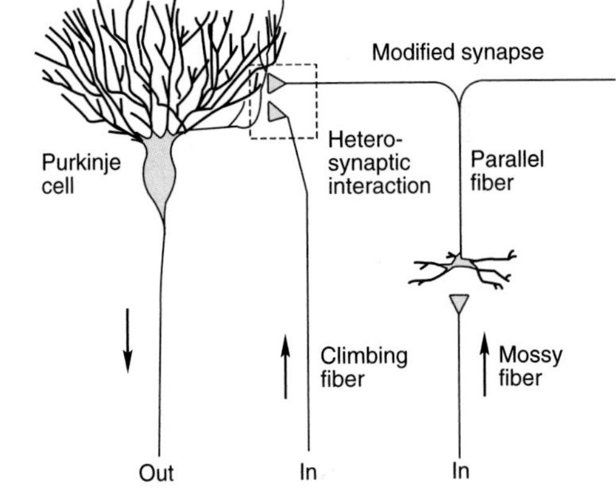

Figure 29–12 Cerebellar circuits are modified during learning.

A. Simple and complex spike activity in a Purkinje neuron change as a monkey learns to adapt to an unexpected increase in load on wrist flexion. **1.** The response to a constant load (control trial) contains only occasional complex spikes. **2.** When the load is unexpectedly increased, the neuron fires numerous complex spikes. **3.** After several trials with the additional load, the frequency of complex spikes returns to the control level, while the frequency of simple spikes remains below the baseline. (Adapted from Gilbert and Thach, 1977.)

B. The two major inputs to the cerebellum, the mossy fibers and climbing fibers, converge on the Purkinje cells. During the learning of a motor task the climbing fibers depress the actions of the parallel fibers on the Purkinje cells. According to this view, the climbing fibers instruct or modulate the action of the mossy fibers. (Adapted from Ito, 1984.)

unexpected load over several trials, the firing of the complex spikes gradually returned to control levels. However, the frequency of simple spikes remained below baseline, as if firing of the mossy fibers had been modified to adjust to the new load.

These findings demonstrate that the activity of the climbing fibers is modulated during motor learning, and suggest that this modulation might in turn, by heterosynaptic inhibition, reduce the strength of the mossy fiber input to the Purkinje neurons (Figure 29–12B). The reduction of mossy fiber input following the perturbing stimulus would in turn lead to a decrease in firing of the Purkinje cells and to an increase in the output (due to disinhibition) of the neurons in the deep nuclei (see Figure 29–11).

But do these cellular changes actually mediate changes in behavior? Pharmacological inactivation of the cerebellar cortex prevents this behavioral adaptation from taking place. The results of cellular studies of the role of the cerebellum in motor tasks therefore support James Albus's idea that mismatches between intended movement and actual movement are corrected in the cerebellum through heterosynaptic inhibition of the mossy fibers by the climbing fibers.

Cerebellar Diseases Lead to Disorders of Coordination

Disorders of the cerebellum result in distinctive symptoms and signs first described in the 1920s and 1930s by Gordon Holmes. In his studies of patients who sustained gunshot wounds to the cerebellum in the first world war, Holmes described a variety of abnormalities in the coordination of voluntary movements, referred to as *ataxia*. These abnormalities include several distinctive defects: a delay in initiating responses with the affected limb, errors in the range and force of movement (or *dysmetria*, i.e., errors in the metrics of movement), and errors in the rate and regularity of movements.

The Basal Ganglia Integrate Information from Diverse Areas of Cortex

Like the cerebellum, the basal ganglia do not have direct output connections to the spinal cord and thus play an indirect role in the control of movement. Unlike the cerebellum, the basal ganglia do not receive precisely organized information directly from sensory receptors. Rather, their inputs arise principally from diverse regions of the cerebral cortex. Their output is directed back to the prefrontal, premotor, and motor cortices through the thalamus.

Diseases affecting the basal ganglia, such as Parkinson's disease, produce disorders of motor function that include involuntary movements, alterations in muscle tone, and a generalized poverty of movement. Research into the treatment of these conditions has led to important insights into the organization of neurotransmitter systems in the brain.

The Basal Ganglia Consist of Five Subcortical Nuclei

The basal ganglia consist of five extensively interconnected nuclei: the caudate nucleus, putamen, globus pallidus, subthalamic nucleus, and substantia nigra (Figure 29–13). The *caudate nucleus* and *putamen* develop from the same structure in the forebrain; as a result, they are composed throughout of identical cell types and are fused anteriorly. The two nuclei are the input components of the basal ganglia and together are called the *neostriatum* (or *striatum*).

The *globus pallidus* (or *pallidum*) is derived from the diencephalon and is divided into internal and external segments (Figure 29–13). The *subthalamic nucleus* lies below the thalamus at its junction with the midbrain. The *substantia nigra* lies in the midbrain and has two zones. A ventral pale zone, the *pars reticulata* (or reticulate zone), resembles the globus pallidus cytologically. A dorsal pigmented zone, the *pars compacta* (or compact zone), comprises dopaminergic neurons whose cell bodies contain neuromelanin. This dark pigment, a polymer derived from dopamine, gives the *substantia nigra* its name (Latin, "black substance"), because in humans this part of the brain appears black in cut sections.

Because the internal segment of the globus pallidus and the substantia nigra are strikingly similar in cytology, connectivity, and function, these two nuclei can be considered as a single structure arbitrarily divided by the internal capsule, much like the caudate and putamen. The globus pallidus and reticulate zone of the substantia nigra constitute the major output nuclei of the basal ganglia.

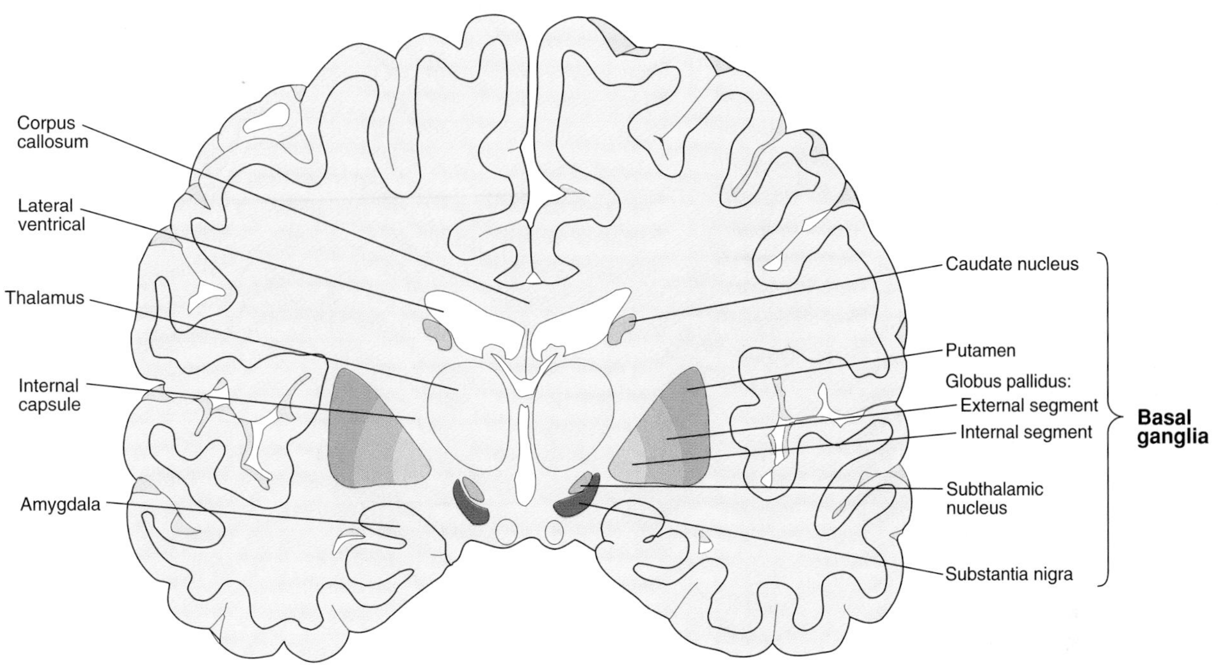

Figure 29–13 A coronal section of the cerebral hemispheres shows the basal ganglia in relation to surrounding structures. (Adapted from Nieuwenhuys, Voogd, and van Huijzen, 1988.)

The Basal Ganglia Receive Input from the Cortex and Project Back to the Cortex by Way of the Thalamus

Almost all afferent connections to the basal ganglia terminate in the neostriatum. The neostriatum receives input from two major sources outside the basal ganglia: the cerebral cortex and the intralaminar nuclei of the thalamus (Figure 29–14A). Each area of the cortex projects to a different nucleus in the neostriatum, and each nucleus has a specific behavioral function. For example, the putamen is primarily concerned with motor control, the caudate is involved in the control of eye movements and with certain cognitive functions, and the ventral striatum is related to other cortical areas that mediate the effects of emotion on behavior. The subthalamic nucleus also receives input from the cerebral cortex (see below).

The connections between the basal ganglia are extensive but orderly. Cells in different parts of the caudate and putamen project to specific parts of the globus pallidus and substantia nigra (Figure 29–14B). Thus, because the pathways are all topographically organized, specific parts of the cortex act through the neostriatum on specific parts of the globus pallidus and substantia nigra.

The subthalamic nucleus receives the output of the external segment of the globus pallidus and has topographically organized projections to both segments of the globus pallidus and to the reticulate zone of the substantia nigra. It also receives topographically organized inputs directly from the primary motor and premotor cortices, providing the motor cortex another means for controlling the output of the basal ganglia. Finally, the neostriatum receives an important dopaminergic pro-

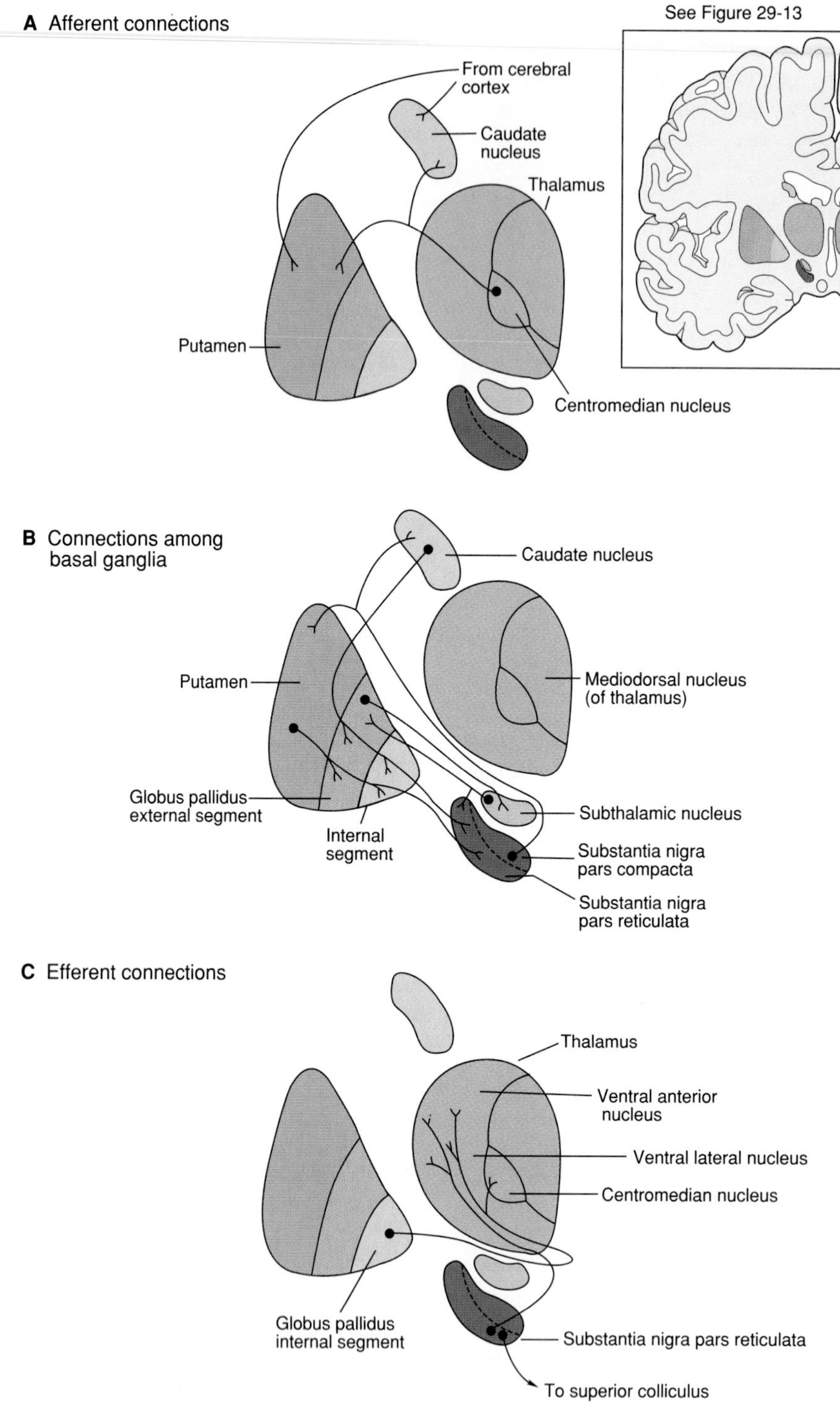

Figure 29–14 Major anatomical connections of the basal ganglia.

A. Almost all afferent input to the basal ganglia is received in the caudate nucleus and putamen. Inputs to the subthalamic nucleus from the cortex are omitted.

B. All of the nuclei of the basal ganglia are interconnected by topographically organized projections.

C. The thalamus is the principal target of the output from the basal ganglia.

jection from the compact zone of the substantia nigra. Loss of this projection leads to Parkinson's disease (see next section). Other neurotransmitters, including γ-aminobutyric acid (GABA), glutamate, and substance P, play an important role in modulating the transformation of information through the various nuclei of the basal ganglia. (See Chapter 16 for a discussion of transmitters.)

The major output pathways of the basal ganglia arise from the internal segment of the globus pallidus and the reticulate zone of the substantia nigra. These pathways terminate in several nuclei in the thalamus, which in turn project to several areas of cortex: the prefrontal cortex, the premotor cortex, the supplementary motor area, and the primary motor cortex. Through these projections the basal ganglia influence other descending systems, such as the corticospinal and corticobulbar systems. In addition to influencing movements of the body and limbs, the basal ganglia also influence eye movements by means of an additional projection to the superior colliculus.

Loss of Dopaminergic Cells in the Substantia Nigra Leads to Parkinson's Disease

Parkinson's disease, the most common disorder affecting the basal ganglia, is named for an English physician who first described it in 1857. More than 500,000 Americans are afflicted with this slowly progressive disease which typically appears in patients in their late 50s. Patients with Parkinson's disease have difficulty initiating movements. Their movements are slowed, and they lack the ability to generate spontaneous movements. Because of increased tone, their muscles become rigid and they show a characteristic resting tremor, most prominently in their hands and fingers. Untreated, these signs gradually worsen, eventually rendering patients completely unable to move voluntarily.

In the 1950s there was a major breakthrough in our understanding of this disease when investigators discovered that the neurotransmitter dopamine is severely depleted in the brains of patients with Parkinson's disease. This depletion results from the degeneration of dopaminergic neurons in the substantia nigra that project to the striatum, where they are critical for controlling the processing of information by the basal ganglia. It soon became clear that patients might be helped if the amount of dopamine in the brain could be restored to normal. Because dopamine does not cross the blood–brain barrier, these patients were given L-Dopa, an immediate precursor of dopamine. L-Dopa had dramatic effects and significantly ameliorates the disorder. The administration of L-Dopa increases the synthesis of dopamine by neurons in the substantia nigra that have not yet degenerated, leading to an increase in the amount of dopamine available at synapses in the striatum.

The importance of these discoveries cannot be overstated. Not only did they lead to a treatment that significantly reduced the suffering of thousands of patients, they also demonstrated that diseases of the brain can target specific neurons and specific molecular processes (see also Chapter 17, on myasthenia). Moreover, they showed that administration of a neurotransmitter and its precursor can be useful therapeutically, an approach now widely used in the treatment of psychiatric disorders. It should be noted, however, that L-Dopa does not cure Parkinson's disease. In fact, because neurons in the substantia nigra continue to degenerate over time, L-Dopa administration has a diminishing effect. Therefore, the search for the root causes of the disease, as well as more permanent treatments, actively continues. One promising approach to treatment, involving the implantation of fetal dopamine cells in the brain, is currently undergoing clinical trials.

The Motor Portion of the Basal Ganglia Is Somatotopically Organized and Involved in Higher-Order Aspects of Movement

The extensive linkage between the basal ganglia and the cerebral cortex in both directions is believed to include several functionally and anatomically distinct circuits responsible for regulating limb movements, eye movements, and more complex cognitive behaviors. The best understood circuit arises in the *corticostriate pathway* from those portions of the cerebral cortex most closely related to the control of movement (supplementary motor area, premotor cortex, motor cortex, somatosensory cortex, and the superior parietal lobule). This pathway provides dense, topographically organized inputs to the motor portion of the putamen. The output of this pathway, termed the *motor circuit* of the basal ganglia, is directed primarily back to the

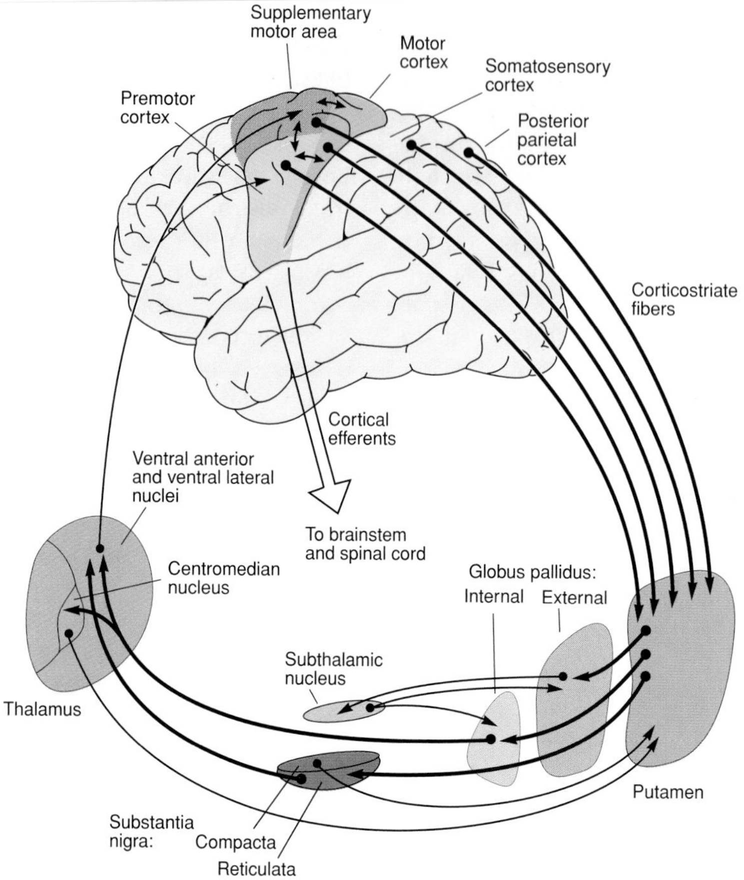

Figure 29–15 The motor circuit of the basal ganglia is one of several functionally distinct feedback circuits between the basal ganglia and the cerebral cortex.

supplementary motor area and premotor cortex (Figure 29–15). These two areas are reciprocally connected with each other and with the motor cortex, and all three of these sites have direct descending projections to brain stem motor centers and the spinal cord. Thus, these three descending pathways are the means by which the basal ganglia influence posture and limb movement.

The activity of some neurons in the basal ganglia resembles that of cells in the motor areas of the cortex. These cells respond to discrete passive and active movements of individual parts of the body, usually movements of a single joint, in specific directions. Despite this resemblance, the activity of neurons in the putamen differs from that of neurons in the motor areas of the cortex in several interesting ways.

First, during visually guided tracking tasks the movement-selective cells in the basal ganglia fire later than cells in the cortical motor areas. Second, neurons in the putamen are more likely to be selective for the direction of limb movement than for the activation of specific muscles. These findings indicate that the basal ganglia do not play a significant role in the initiation of stimulus-triggered movements and do not specify directly the muscular forces necessary for the execution of movement.

What then is the role of the basal ganglia in motor control? Why, for example, do lesions of the basal ganglia and the motor cortex or cerebellum result in distinctly different motor disturbances? One possibility is that the basal ganglia are mainly involved in the processing of information needed for planning and triggering self-initiated movements and organizing associated postural adjustments. This idea is consistent with the difficulty shown by patients with Parkinson's disease in initiating voluntary movements in the absence of triggering stimuli (akinesia). It is also consistent with the breakdown of postural mechanisms that occur in this and other basal ganglia diseases.

It has been speculated that the basal ganglia selectively facilitate some movements and suppress others, analogous to the "inhibitory surround" characteristic of receptive fields in the sensory systems. This idea is attractive because it can explain many of the diverse symptoms, such as involuntary movement and tremors, that appear in diseases of the basal ganglia.

An Overall View

The primary motor cortex is the nodal point for mediating the influence of higher sensory systems on movement. It is essential to organizing movements in which somatosensory input plays a controlling role, such as the manipulation of objects. Motor areas of the cortex also mediate the influence of vision on movement, as when we reach for an object or modify our walking to avoid obstacles.

The basal ganglia and the cerebellum are the major constituents of two important subcortical loops of the motor system. Both receive substantial projections from the cerebral cortex and both project back to the cortex via the thalamus. There are three differences, however, in the cortical connections of the basal ganglia and those of the cerebellum. First, the basal ganglia receive inputs from the entire cerebral cortex; in contrast, the cerebellum receives input only from that part of the cortex that is directly related to sensorimotor functions. It also receives highly specific information from the periphery. Second, the output of the cerebellum is directed back to the premotor and motor cortex, whereas the output of the basal ganglia is directed not only to the premotor and motor cortex but also to the prefrontal association cortex. Finally, the cerebellum receives somatic sensory information directly from the spinal cord and has major afferent and efferent connections with many brain stem nuclei that are directly connected with the spinal cord. In contrast, the basal ganglia have relatively few connections to the brain stem and no direct connections at all to the spinal cord.

These differences suggest that the cerebellum directly regulates the execution of movement, whereas the basal ganglia are involved in higher-order, cognitive aspects of motor control, namely the planning and execution of complex motor strategies. In addition, because of their extensive connections with association cortex and limbic structures, the basal ganglia, unlike the cerebellum, are involved in many functions other than motor control.

Selected Readings

Adams, R. D., and Victor, M. 1993. Principles of Neurology, 5th ed. New York: McGraw-Hill.

Albin, R. L., Young, A. B., and Penney, J. B. 1989. The functional anatomy of basal ganglia disorders. Trends Neurosci. 12:366–375.

Alexander, G. E., and Crutcher, M. D. 1990. Functional architecture of basal ganglia circuits: Neural substrates of parallel processing. Trends Neurosci. 13:266–271.

Alexander, G. E., DeLong, M. R., and Strick, P. L. 1986. Parallel organization of functionally segregated circuits linking basal ganglia and cortex. Annu. Rev. Neurosci. 9:357–381.

Asanuma, C., Thach, W. T., and Jones, E. G. 1983. Anatomical evidence for segregated focal groupings of efferent cells and their terminal ramifications in the cerebellothalamic pathway of the monkey. Brain Res. Rev. 5:267–297.

Brooks, V. B., and Thach, W. T. 1981. Cerebellar control of posture and movement. In V. B. Brooks (ed.), Handbook of Physiology, Section 1: The Nervous System, Vol. II: Motor Control, Part 2. Bethesda, Md.: American Physiological Society, pp. 877–946.

DeLong, M. R. 1990. Primate models of movement disorders of basal ganglia origin. Trends Neurosci. 13:281–285.

DeLong, M. R., and Georgopoulos, A. P. 1981. Motor functions of the basal ganglia. In V. B. Brooks (eds.), Handbook of Physiology, Section 1: The Nervous System, Vol. II: Motor Control, Part 2. Bethesda, Md.: American Physiological Society, pp. 1017–1061.

DiFiglia, M. 1990. Excitotoxic injury of the neostriatum: A model for Huntington's disease. Trends Neurosci. 13:286–289.

Georgopoulos, A. P. 1986. On reaching. Annu. Rev. Neurosci. 9:147–170.

Gilman, S. 1985. The cerebellum: Its role in posture and movement. In M. Swah and C. Kennard (eds.), Scientific Basis of Clinical Neurology. New York: Churchill Livingstone, pp. 36–55.

Glickstein, M., and Yeo, C. 1990. The cerebellum and motor learning. J. Cogn. Neurosci. 2:69–80.

Hepp-Reymond, M.-C. 1988. Functional organization of motor cortex and its participation in voluntary movements. In H. D. Steklis and J. Irwin (eds.), Comparative Primate Biology, Vol. 4: Neurosciences. New York: Liss, pp. 501–624.

Holmes, G. 1939. The cerebellum of man. Brain 62:1–30.

Humphrey, D. R. 1979. On the cortical control of visually directed reaching: Contributions by nonprecentral motor areas. In R. E. Talbott and D. R. Humphrey (eds.), Posture and Movement. New York: Raven Press, pp. 51–112.

Ito, M. 1984. The Cerebellum and Neural Control. New York: Raven Press.

Keele, S. W., and Ivry, R. 1990. Does the cerebellum provide a common computation for diverse tasks? A timing hypothesis. Ann. N.Y. Acad. Sci. 608:179–211.

Llinás, R. R. 1981. Electrophysiology of the cerebellar networks. In V. B. Brooks (ed.), Handbook of Physiology, Section 1: The Nervous System, Vol. II: Motor Control, Part 2. Bethesda, Md.: American Physiological Society, pp. 831–876.

Martin, J. B. 1984. Huntington's disease: New approaches to an old problem. Neurology 34:1059–1072.

Passingham, R. E. 1993. The Prefrontal Cortex. Oxford: Oxford University Press.

Thach, W. T., Kane, S. A., Mink, J. W., and Goodkin, H. P. 1992. Cerebellar output: Multiple maps and motor modes of control in movement coordination. In, R. Llinás and C. Sotelo (eds.), The Cerebellum Revisited. New York: Springer-Verlag.

Wise, S. P. 1985. The primate premotor cortex: Past, present, and preparatory. Annu. Rev. Neurosci. 8:1–19.

Wise, S. P., and Strick, P. L. 1984. Anatomical and physiological organization of the non-primary motor cortex. Trends Neurosci. 7:442–446.

Yurek, D. M., and Sladek, J. R., Jr. 1990. Dopamine cell replacement: Parkinson's disease. Annu. Rev. Neurosci 13:415–440.

VIII

GENES, EMOTIONS, AND INSTINCTS

About the image on the preceding page:
Shakespeare's Hamlet embodies the full range of human emotions. Hamlet's words, "How weary, stale, flat, and unprofitable seem to me all the uses of this world!" express one of the most salient symptoms of major depressive illness. (Courtesy of Rank Film Distributors.)

Introduction

A comprehensive study of behavior must address the question: What do genes contribute to behavior? Of course behavior itself is not inherited—what is inherited is DNA. Genes encode proteins that are important for the development, maintenance, and regulation of the neural circuits that underlie behavior. Behavior gradually emerges as a result of the impact of environmental factors on this developing neural circuitry. The environment begins to exert its influence *in utero*, and becomes of prime importance after birth. Thus, to understand behavior fully, we must understand both innate (genetic and developmental) and environmental determinants.

Traditionally, behavior has been divided into two categories: instinctive and learned. Instinctive behavior is most closely dependent on genetic endowment. The consideration of instinct is therefore a good starting point for examining the genetic determinants of behavior.

During the period from 1920 to 1950, European zoologists such as Konrad Lorenz and Nikolaas Tinbergen laid the groundwork for the comparative study of behavior with particular emphasis on its mechanisms, ontogenesis, and evolution. This approach, known as *ethology*, advanced the study of instincts in two ways. First, whereas scientists had previously only speculated about the role of instinct and behavior, ethologists systematically observed and experimented on inborn behavior. Second, they limited the biological studies of instincts to stereotyped sequences of observable movements that are inborn. Ethological study of many species led to a partial reconciliation of the old concept of instinct with the behaviorists' insistence of explaining behavior only in terms of observed action.

Among instinctive behaviors we first consider sexual behavior. We ask: What factors determine female and male brain anatomy, brain function, and behavior? Next we examine emotional behavior. Like cognitive functions, emotion is controlled by distinct neuronal circuits within the brain. Since we experience emotion consciously, there are also important cognitive elements in emotions, elements that most likely involve the cerebral cortex. At the same time, emotions are accompanied by autonomic endocrine and skeletal motor responses that depend on subcortical parts of the nervous system, especially the nuclei of the amygdala. These peripheral responses serve to prepare the body for action and, in addition, communicate our emotional states to other people.

Finally we consider motivation. The neural systems in the brain that mediate positively motivated behaviors have now been well delineated and involve dopaminergic pathways that connect the midbrain to the limbic system and cortex. Addictive drugs such as nicotine, alcohol, opiates, and cocaine are thought to produce their actions by co-opting the same neural pathways that mediate positively motivated behaviors essential for survival.

30

Genes and Behavior

Is There a Genetic Component to Human Behavior?

What Components of Behavior Are Inherited?

- *Single Genes Can Be Critically Important for Some Simple Behavioral Traits*
- *Single Genes Can Affect Normal Behavior: Color Vision*
- *Mutation in A Single Gene Can Produce Motor and Cognitive Defects*
- *Most Complex Behavioral Traits Are Polygenic*
- *Complex Disorders of Human Behavior Such As Schizophrenia and Bipolar Affective Disorder Are Thought to Be Polygenic*

How Do Genes Organize Behavior?

- *Courtship Behavior Is Genetically Determined in Sexual Differentiation of the Nervous System*
- *Biological Rhythms Are Governed by Oscillations in Gene Expression*

All behavior is shaped by the interaction of genes with the environment. Even the most stereotypic behavior is influenced by the environment, while highly plastic behavior, such as language, is constrained by innate factors. In this chapter we consider only the innate determinants of behavior. In later chapters we examine how environmental factors can modify behavior through learning.

In considering the innate factors that control behavior, we need first to identify the components of behavior that are heritable. Clearly, behavior itself is not inherited; what is inherited is DNA, which comprises the genes. The genes carry information on the sequences of all the proteins that the organism can express, and they transmit this information through replication. Thus each gene provides succeeding generations with copies of itself, and in each cell in which the gene is expressed it directs the manufacture of specific proteins that determine the structure, function, and other biological characteristics of the cell.

The DNA of genes expressed in the brain encodes proteins that are important for the development, maintenance, and regulation of the neural circuits that underlie behavior. The interplay of the development of neural circuits and behavior has been studied in many organisms, from humans to fruit flies (Box 30–1). These studies have attempted to answer three major questions: Is there a genetic component of human behavior? What components of behavior are heritable? How do genes organize behavior?

Is There a Genetic Component to Human Behavior?

Because of ethical and legal considerations in human studies, the issue of whether some components of human behavior are inherited is controversial. Controlled studies of heritable factors in human behavior are difficult to devise because it is not possible (or desirable) to control an individual's environment for experimental purposes, except in some very limited situations. Thus, behavioral studies of identical twins provide important information not otherwise available. Identical twins share all of each others genes and are therefore as alike genetically as is possible for two individuals. Similarities between identical twins that have been separated early in life and raised in different households, as occasionally happens, will thus be more attributable to genes than to environment. (The extent of environmental difference is limited in most cases since twins are generally raised in culturally and economically similar households.) To be meaningful, comparisons of twins must be based on traits that are uniform in the population,

BOX 30–1 Genetic Systems for Studying Behavior: Their Pluses and Minuses

Extensive efforts are being made to understand the connections between genes and behavior in four organisms: the nematode *Caenorhabditis elegans,* the fruit fly *Drosophila melanogaster,* the laboratory mouse *Mus musculus,* and ourselves, *Homo sapiens.* Each has unique behavioral characteristics and each offers a different set of advantages and disadvantages for pursuing the study of genes and behavior.

Caenorhabditis elegans: This little organism has wormed its way into the hearts of many biologists because of the ease of inducing new mutations. Study of its genetics was inaugurated in the mid-1960s by Sidney Brenner. Now, thirty years later, it is clear that the nematode is a formidable genetic organism that has proven to be a source of new mutants. It is possible to screen through more putative mutants, and thus recover rarer and more different types of mutations, than in any other metazoan, and its practitioners perform at a level of genetic virtuosity equalled only by those who work with yeast and bacteria.

The principal focus of work in the nematode during these years has been development, including that of the nervous system. In the course of this work a great many new genes have been identified—genes which we now know are also present in flies and mammals—and their functions and interactions worked out, including homologs of cellular oncogenes that have now been assigned roles in normal development. Some of these mutants also affect the animal's behavior.

These studies have been significantly aided by the small number and ease of identification of the animal's cells, and a complete knowledge of their anatomical connections. At the same time, the animal's small size and very simple repertory of behavior make electrophysiological studies extremely difficult. Nor is there any significant anatomical homology with distantly related organism, such as mammals.

Drosophila: The fruit fly has a deeper and wider background of genetic information than any other organism. As a consequence of 90 years of intensive study, and the development of a wide variety of techiques, it is possible to manipulate the fly genome in a large number of ways, including the mutation of genes, the cloning of mutated genes, and the introduction of transgenes that can be controlled at will. Since *Drosophila's* behavior is relatively sophisticated, the fly has proved a rich source of mutants affecting such behaviors as courtship, circadian rhythms, and learning. These mutant studies have identified genes for ion channels and enzymes for second-messenger systems, as well as developmental genes affecting the nervous system and thus behavior. As with *C. elegans,* these genes have been found in distant relatives such as ourselves.

The fly has some of the same drawbacks as the

there must be a clustering of similarities for several characteristics, and twins must be compared with a control group of individuals that are matched in age, sex, and socioeconomic status.

Such studies have shown that identical twins share a remarkable number of behavioral traits. These include tastes and preferences such as religious and vocational interests, which we normally consider to be distinguishing features of individuality; (Figure 30–1). These findings argue strongly that human behavior has a significant hereditary component. Alone, the findings do not refute the existence of environmental influences, but instead support the idea that behavior is shaped by genetic as well as environmental factors.

Although studies of identical twins are intriguing, they do not tell us anything about how many or which genes are important, let alone how genes affect behavior. For these issues we must turn to animal studies where the environment can be controlled, thus allowing genetic differences to be recognized and studied.

worm: it is not anatomically homologous to vertebrates, its behaviors are not as sophisticated or as plastic as those of mammals, and its central nervous system is not yet amenable to electrophysiology. In some respects it occupies a place midway between the simplicity of the worm and the complexity of mammals.

Mus: The laboratory mouse has a genetic history as long, if not as deep, as the fruit fly. In fact, some of the first mouse mutants to be studied were those affecting behavior—the so-called "waltzing" strains that were popular household pets at the turn of the century and subsequently became laboratory subjects. These were spontaneous, as are most of the mouse mutants. Recently, techniques of gene replacement have permitted the genetic engineering of new mouse mutants in cloned, defined genes. The result is a new stream of "knock-out" mutants designed to test the function of cloned genes. The production of mutants by this technique is relatively slow and must be reiterated each time a new allele is created. It does, however, add significantly to the existing experimental attractiveness of the mouse: its mammalian heritage and its neuroanatomical, physiological, and genetic closeness to humans. Similarly, the advent of improved gene-mapping techniques, such as those used in humans, opens up new possibilities for mapping genes governing natural polymorphisms in behavior.

The principal genetic drawback is the persisting difficulty of using mutant screens to identify unknown genes based on their functional effects. Behavioral studies have lagged far behind those of the mouse's cousin, the white rat, on whom the preponderance of behavioral research, especially as regards learning and memory, has been lavished. But this is likely to be only a temporary drawback.

Homo sapiens: Although almost none of the experimental advantages found in the worm, the fly, or the mouse are available in studies of humans, we nonetheless persist in our investigation of the influence of genes on human behavior out of a desire to understand ourselves. The study of human genes has made a quantum leap in the past decade due to the introduction of DNA markers to facilitate gene mapping. Sequencing of the human genome should also contribute substantially in the future, but for now the principal lever we have is the ability to map genetic polymorphisms in behavior to an extent unimaginable in the past. Once mapped, the genes can then be cloned and identified.

The prospects of gaining insights into the genetic factors influencing the attributes that are so important to us, like personality and intelligence, are exciting and at the same time daunting. They raise some of the most difficult ethical questions yet to be faced in the advance of molecular biology.

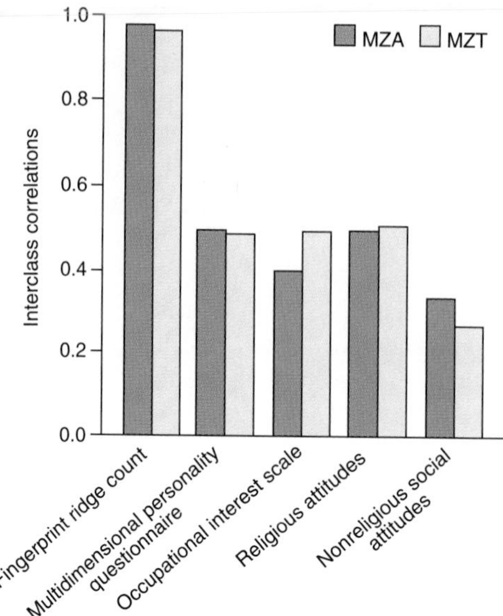

Figure 30–1 Correlations among monozygotic twins reared together (MZT) and those reared apart (MZA) for physiological characteristics, personality traits, interests, and attitudes. Fingerprint ridge count, which is not expected to be subject to significant environmental influence, is virtually identical between MZA and MZT subjects and within each class. Other characteristics, expected to be more subject to environmental influences, are not so highly correlated within the MZA and MZT classes. The multidimensional personality scale and religious attitudes also show virtual identity between MZA and MZT; these results suggest a significant, though not necessarily predominant, genetic influence on those traits. In contrast, the occupational interest scale and nonreligious social attitudes are different between MZA and MZT. (Based on Bouchard et al., 1990.)

What Components of Behavior Are Inherited?

Darwin first postulated that variations in behavior may be due in part to natural selection. If this is so, then some elements of behavioral variation in any population will necessarily have a genetic basis. Some portion of this variation in turn should show up as clearly heritable differences. Since it is difficult to carry out genetic experiments on natural populations, the question of the genetic basis of behavioral variations has been studied in inbred populations of domestic or laboratory animals. In simpler experimental animals we can examine how genes influence behavior. The revolution in molecular genetics has made it possible to look at simple instances in which the relationship between genes and behavior is unambiguous.

Clearly, genes do not code for behavior in a direct way; a single gene cannot code for a single behavior. Behavior is generated by neural circuits that involve many nerve cells, each of which expresses specific genes that direct the production of specific proteins. Thus a wide variety of proteins, structural and catalytic, are required for the development and function of a neural circuit (see Chapter 4). The fact that many genes are involved does not mean, however, that individual genes are not essential for the expression of a behavior. The importance of specific genes to behavior can best be demonstrated in simple animals, such as the fruit fly, in which mutations in single genes can be more easily studied. Mutations of single genes in *Drosophila* can produce abnormalities in learned behavior (see Chapter 36) as well as in innate behaviors such as courtship and locomotion.

Single Genes Can Be Critically Important for Some Simple Behavioral Traits

Although most behavioral traits are polygenic, a few traits in simple animals are influenced in important ways by the actions of single genes. One

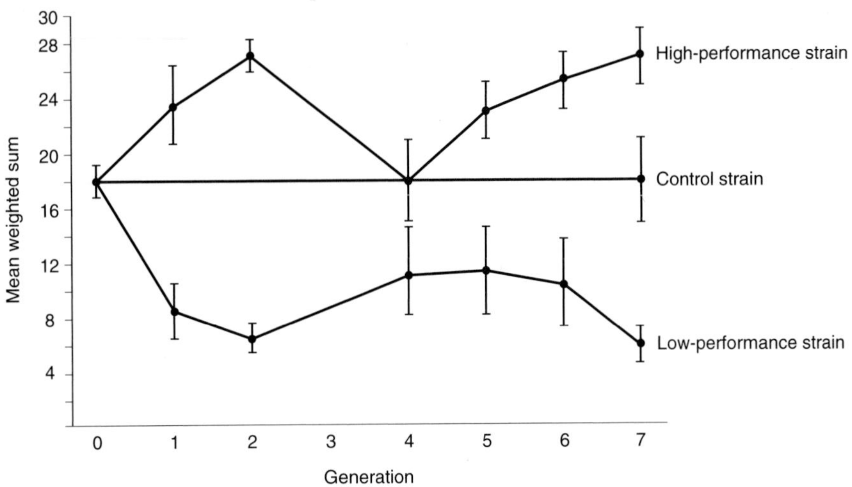

Figure 30–2 The genetic selection of a behavioral trait. Seven generations of the blowfly, *Phormia regina*, were selected for high or low responses to behavioral sensitization of the central excitatory state. The plot shows the mean weighted responses of each selected generation to stimulation of the sensory hairs (see text). (From Tully and Hirsch, 1982.)

of these exceptions is illustrated in a selection experiment on the blowfly, *Phormia regina*, selecting for differences in behavioral sensitization of its central excitatory state. A blowfly extends its proboscis when its sensory hairs are stimulated with a solution containing sucrose, but not with water alone. After application of sucrose to one hair, however, it extends its proboscis in response to water stimulation to a different hair. Selection for high response and low (or no) response in this experiment led to the generation of two pure-breeding lines (Figure 30–2). Matings of progeny to each other revealed that in this laboratory population there was one major genetic locus influencing this difference in trait.

This finding is unusual in its genetic simplicity. It contrasts not only with findings on behavior and learning in mammals, to be discussed later, but also with most of the 70-year history of behavior genetics. Its importance lies in the demonstration that significant behavioral variation in a population can be governed by individually segregating genes, as Darwin would have predicted if he had known about genes. It does not, however, imply that an entire behavior is produced or specified by the action of a single gene. But the fact that some behavioral variations *can* be attributed to single genes means that genetic variations in behavior can, in principle, be identified and studied. Indeed, examples of single gene defects affecting behavior are also found in humans.

Single Genes Can Affect Normal Behavior: Color Vision

The analysis of human genes that control behavior is necessarily limited both by the lack of ability to manipulate those that are known and by the ethical inhibition against such experimentation. However, during the course of the last 100 years, classical genetics has identified several human genes that affect aspects of behavior. In recent years cloning has advanced this analysis by specifying the molecules that affect behavior.

Humans have 46 chromosomes: 22 pairs of autosomes and one pair of sex chromosomes, either two X's in females or one X and one Y in males. Sons thus always inherit their X chromosome from their mothers. Thus, each individual receives 22 autosomes and one sex chromosome from each of its parents. As a result, at each genetic locus each individual has two copies or alleles of

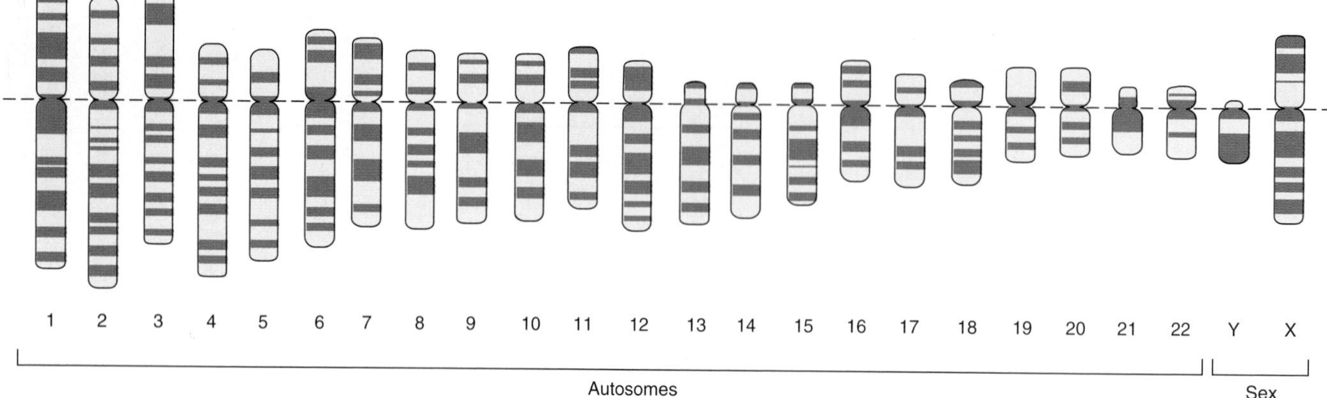

Figure 30–3 This map of normal human chromosomes at metaphase illustrates the distinctive morphology of each chromosome. (Adapted from Watson, Tooze, and Kurtz, 1983.)

each gene, one from the father and one from the mother.

There are morphological distinctions between autosomal pairs and between the X and Y chromosomes, consisting of differences in size, position of centromere, and visible banding patterns, revealed with appropriate histological stains (Figure 30–3). Within a pair of autosomes the chromosomes are morphologically indistinguishable, although genetically they have many sequence differences corresponding to the different alleles at each locus. A *dominant mutation* is one in which the sequence change exerts its effect even if only one of the two homologous chromosomes of a pair is mutant. Dominant mutations are usually the result of inappropriate expression of the gene or the production of an abnormal protein. A *recessive mutation* requires that both homologs be mutant for the phenotype to be manifest. Recessive mutations are usually the result of nonexpression of the gene or the production of little or no active protein. An X-linked gene is present in only one copy in males, so a recessive mutation on a male's X is expressed.

Color vision is one of the few cases in which a normal behavioral variation has been traced to the molecular level. Molecular cloning techniques have been used to identify and clone the genes encoding the proteins for red, green, and blue pigment used in color vision (see Chapter 24). Red–green color blindness is an example of a defect due to mutations in an X-linked gene. The genes for red and green pigments are arrayed in tandem on the X chromosome and are very similar, differing in only about 5% of amino acid residues. This tandem arrangement may explain the origin of many cases of red–green color blindness, in which the pigment genes contain deletions that sometimes appear as hybrids between a red and a green pigment gene. This rearrangement apparently originated in an unequal recombination between X chromosomes in a female, alterations that became heritable, showing up as a *hemizygous* condition in her sons. (Genes on a male's X are termed hemizygous since they only exist in one copy.) Unequal recombination occurs when two homologous chromosomes pair up slightly out of register. It is especially likely to occur when similar genes are arrayed in tandem; during the normal process of reciprocal exchange of chromosome arms, one chromosome ends up missing a small sequence while the other gets two tandem copies.

Even among individuals with normal color vision there are subtle variations in color perception. This is attributable to a sequence variation or *polymorphism* in the red pigment gene: serine is present at amino acid residue 180 in 62% of the normal male population and alanine at that position in the remaining 38%. The difference is revealed in psychophysical tests in which subjects are asked to match the intensity of a mixture of red and green lights. The intensity of red light a subject needs to match as standard depends on the amino acid at position 180. Because of the presence of two X chromosomes, females fall into three groups: homozygotes for Ser^{180}, homozygotes for Ala^{180}, and heterozygotes who display an intermediate phenotype. Thus, a subtle behavioral polymorphism in color perception in humans can be explained by a single gene and is the consequence of a subtle change in protein sequence.

Mutation in A Single Gene Can Produce Motor and Cognitive Defects

One of the first human traits to be traced to a single gene inherited in a Mendelian fashion was a degenerative disease of the nervous system. In 1872 George Huntington, a physician living on Long Island in New York, described a disease that he, his father, and his grandfather had observed in several generations of their patients. The disease was characterized by four features: heritability, chorea (incessant, rapid, jerky movements), dementia, and death 15 or 20 years after onset. This disease, now called Huntington's disease, affects men and women with equal frequency, about 5 per 100,000 population.

In most patients the onset of the disease occurs in the fourth to fifth decade of life. Thus, the disease strikes after most individuals have married and had children. Each child of an affected parent has a 50% chance of inheriting the disease. One of the tragic aspects of the disease in the past was that no test was available to make the diagnosis before the symptoms become apparent. As a result, the children of a patient lived for decades in the fear that they, too, may have inherited the gene for the disease.

Huntington's disease involves the loss of cholinergic and (γ-aminobutyric acid) GABA-ergic neurons in the corpus striatum. The death of these nerve cells is thought to cause the chorea. The impaired cognitive functions and eventual dementia may be due either to a concomitant loss of cortical neurons or to the disruption of normal activity in the cognitive portions of the basal ganglia, namely, the dorsolateral prefrontal and lateral orbitofrontal circuits (see Chapter 19). It is now possible, using imaging techniques, to demonstrate selective loss of neurons in the caudate nucleus of a living patient with Huntington's disease.

The genetic transmission of Huntington's disease became evident when it was discovered that practically all patients with this disease on the East Coast of the United States were descendants of two ancestors born in Suffolk, England, who emigrated to Salem, Massachusetts in 1630. In all likelihood, several of the apparently deranged women in Salem who were executed as witches were actually exhibiting symptoms of the disease. The familial pattern is impressive; traced through 12 generations (over 300 years), the disease has been expressed in each generation.

Huntington's disease is inherited as a highly penetrant, autosomal dominant disorder. (Penetrance is the frequency with which the phenotype appears in those individuals with the mutation.) Using a technique to map genetic polymorphisms (Box 30–2), the gene responsible for Huntington's disease has been identified on chromosome 4 (Figure 30–4). Its product is a large protein, Huntington, of

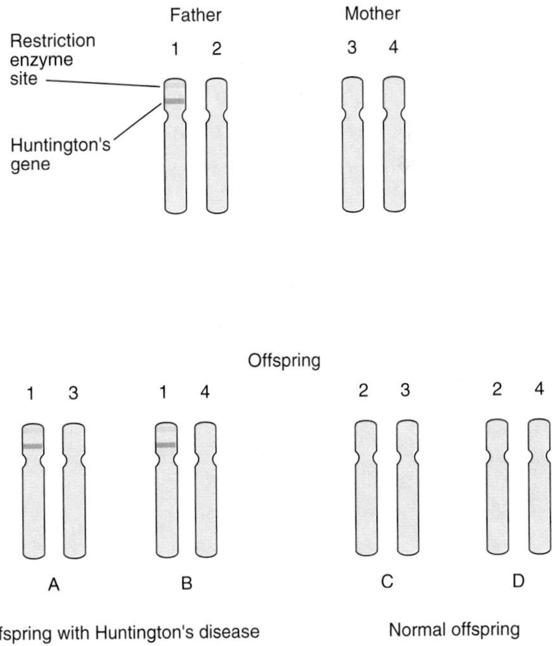

Figure 30–4 Inheritance of the gene responsible for Huntington's disease can be traced by following the inheritance of restriction fragment length polymorphisms for chromosome 4.

unknown function. The basis for the dominant phenotype appears to be a trinucleotide repeat that alters the protein-coding region of the gene. Such repeats are presumably due to random errors in recombination and are not unusual. Cloning of the Huntington's disease gene has had two important results. First, it allows the disease to be diagnosed before symptoms develop or even prenatally. Second, it will eventually lead to determination of how the gene's normal role in the nervous system and a mutation of this gene cause the disease.

Most Complex Behavioral Traits Are Polygenic

Different breeds of dogs provide intriguing examples of vast behavioral differences within a species. Such differences arise, along with equally vast differences in morphology, from the artificial selection imposed by human domestication of dogs. Cross-breeding experiments with various dog breeds have been conducted to test for the mode of inheritance of specific behavioral traits. These traits are heritable but never consist of single or even a few genes. Instead, they are polygenic. Thus, traits such as pointing, gun-shyness, and leash training have major hereditary components but are not controlled by one or a few genes. The same results have been obtained in experiments on the inheritance of behavioral differences within a breed.

By selecting for particular traits, such as learning ability in rats, separate substrains or inbred lines can be established that breed true for the behavior. Cross-breeding experiments between these lines and analysis of the behavioral phenotypes in subsequent generations reveal how many genes govern a particular trait. Consider avoidance learning in rats. When two lines of rats are selected for avoidance conditioning and bred, and their progeny analyzed, the patterns of inheritance are as complex and multigenic as dog traits. Similar findings have been obtained for the honeybee and for nearly all such studies of selection and cross-breeding.

BOX 30–2 Genetic Polymorphisms

If two genes are located very near one another, they are likely to be inherited together. Thus, if abnormality of one gene produces a disease and a nearby gene encodes a readily recognized phenotypic trait (such as hair or eye color) or a readily detectable gene product (such as protein present in the blood), it is likely that people who express the marker will also express the disease—even though the marker may have nothing to do with the disease. These markers or *genetic polymorphisms* vary in the normal population. The traits encoded by these genes, such as eye color, are the expression of the particular genetic locus; both the phenotypic trait and the DNA sequence at the locus vary in the normal population.

In the past, genetic markers were based on variations in the coding regions of DNA expressed as gene products, such as blood groups, enzymes, and antigens of the histocompatibility complex. However, gene products are encoded by only 20% of the total human genome; 80% of the genome contains noncoding regions. Fortunately, it is now possible to saturate the human genome with markers based on variations in DNA sequences throughout the whole genome (including noncoding as well as coding sequences). This broad coverage has made it much easier to trace the inheritance of a disease to a specific region of a particular chromosome.

These new markers, called restriction fragment length polymorphisms (RFLPs), are the result of differences in DNA sequence that produce or eliminate a cutting site for a particular restriction enzyme (an enzyme that cuts DNA only at a specific nucleotide sequence). A restriction enzyme produces DNA fragments of different lengths from the two alleles on the paired chromosomes. These so-called restriction fragments can be separated by electrophoresis in agarose gels and distinguished by specific DNA probes (Figure 30–5A).

When such a polymorphic region of the DNA is closely linked to a particular gene, inheritance of the gene can be traced by following the inheritance of a particular pattern of restriction fragments. The method can be applied to the analysis of polymorphisms in any population of subjects (Figure 30–5B).

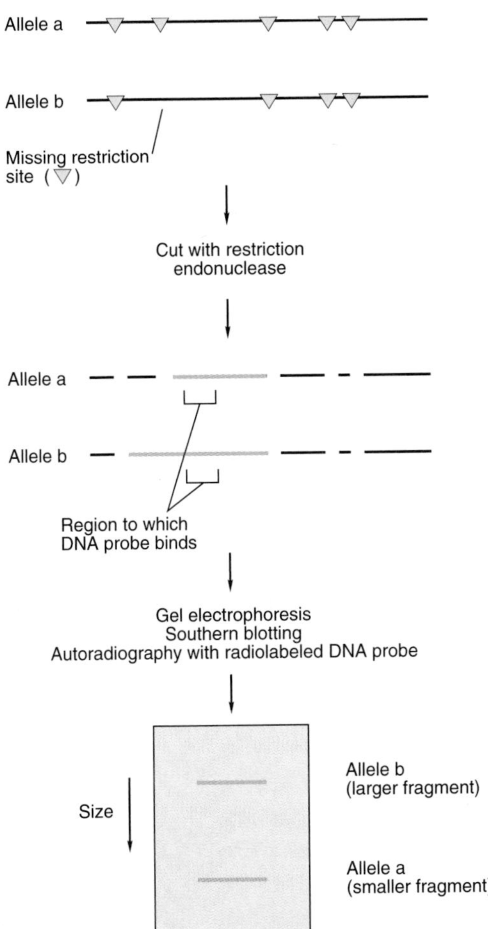

Figure 30–5A The presence of a restriction fragment length polymorphism is detectable by analyzing DNA cleaved by restriction endonucleases, enzymes that cut at specific nucleotide sequences (**yellow triangles**). In this example chromosome *b* is missing a restriction site that is present on chromosome *a*. As a result, cutting chromosome *b* with a restriction enzyme produces a larger than normal DNA fragment. The DNA from both chromosomes is then separated according to size by gel electrophoresis and transferred to nylon filters (a procedure called *Southern blotting*). A radiolabeled DNA probe against this region can then be used to reveal the polymorphism. Because the *b* fragment is larger, it is distinguishable from the *a* fragment.

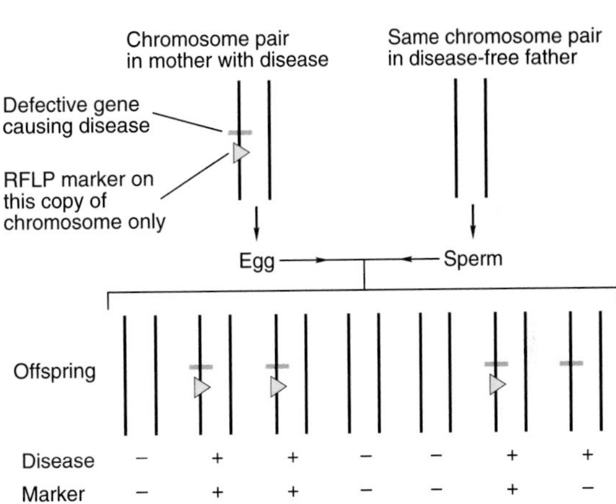

Figure 30–5B Genetic linkage analysis detects the co-inheritance of a mutated gene responsible for a human disease and a nearby restriction fragment length polymorphism (RFLP) marker. In this example the gene responsible for the disease is co-inherited with the RFLP marker in 75% of offspring. (The figure shows three co-inheriting offspring.) Thus, the gene responsible for the disease is located close to the RFLP marker on this chromosome. (Adapted from Alberts et al., 1989.)

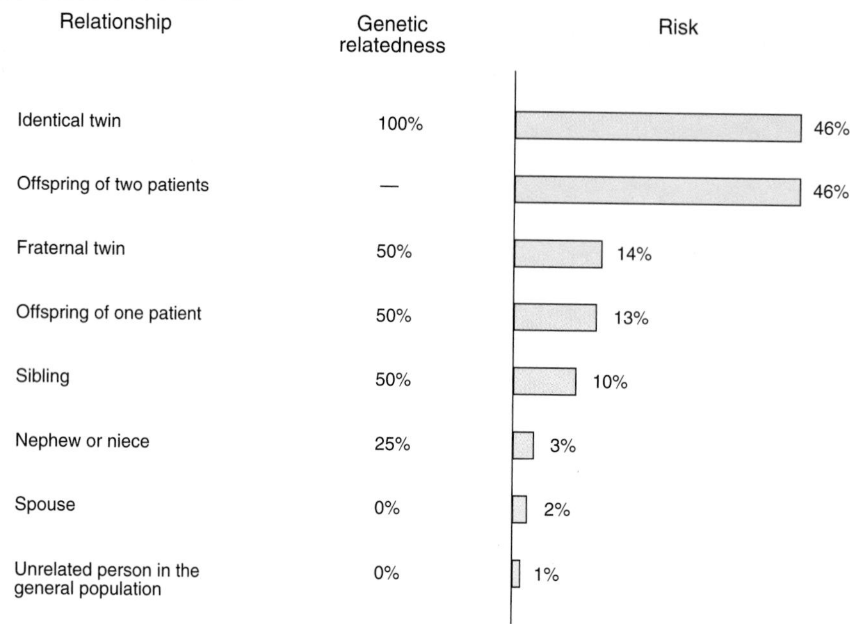

Figure 30–6 Lifetime risks of developing schizophrenia are largely a function of how closely an individual is genetically related to a schizophrenic and not a function of how much the environment is shared. The observed risks, however, are much more compatible with a multifactorial polygenic theory of transmission than with a simple Mendelian model or one involving a single major locus. (From Groves and Rebec, 1992.)

Complex Disorders of Human Behavior Such As Schizophrenia and Bipolar Affective Disorder Are Thought to Be Polygenic

Two of the most striking forms of mental disorder in humans show strong genetic components. These are schizophrenia, a disorder consisting of delusions, auditory hallucinations, and disjointedness of thought, and bipolar affective disorder (manic–depression), consisting of intermittent episodes of extreme mood elevation and overactivity alternating with loss of energy and motivation. Both disorders are present in higher percentages among relatives of affected individuals than in the general population, and both show a significantly higher incidence among identical twins. The likelihood that both twins will be affected is 30 to 50% in identical (monozygotic) but only 10% in fraternal (dizygotic) twins. The higher rate for identical twins holds for twins reared apart as well as those reared together. The 10% fraternal incidence is similar to that of nontwin siblings, but is still much higher than the frequency in the general population (Figure 30–6).

At the outset, any approach to mapping the genes for these syndromes faces major problems due to the lack of a simple, single gene modality in their inheritance patterns. This could have several possible explanations, any one of which would confound mapping attempts. The syndrome could be the product of several genes segregating in the population, so that affected individuals must inherit most or all. The putative genes might affect the individual predisposition, so that the mere fact of having a given genotype only confers a certain probability of developing the syndrome. Each syndrome could be more than one disorder, each with specific but unrelated genetic etiologies.

Despite the daunting nature of the situation, attempts have been made to map genes for schizophrenia and bipolar affective disorder using RFLPs (see Box 30–2). The results thus far have been equivocal, and apparent successes in both disorders have not held up, underscoring the challenge presented by genetic components that do not follow simple modes of inheritance. Schizophrenia probably re-sults from the concerted actions of many genes, each of which makes only a small contribution. This is consistent with evidence from other areas of behavioral genetics, which suggests that the normal range of a given behavioral variation usually reflects the combined actions of many genes, each with only a small effect on the behavior.

Moreover, the consistent finding that only 30 to 50% of monozygotic twins of schizophrenics have schizophrenia indicates that nongenetic factors also are important. Thus, most forms of schizophrenia differ from Huntington's disease, the genetic disease of the nervous system we considered earlier, in which there is transmission of a *dominant* gene. The penetrance of Huntington's disease approaches 100%, and there are only minor nongenetic influences in its expression. Also, the transmission pattern of schizophrenia differs from that of simple *recessive* mutations, by which neither parent may have the phenotype but one in four children will have the disease. For diseases that show classical dominant or recessive Mendelian inheritance, such as Huntington's disease and phenylketonuria, relatively routine studies of pedigrees are sufficient to pinpoint the mode of transmission.

How Do Genes Organize Behavior?

Genes that vary in natural populations, such as those governing pointing in dogs or the central excitatory state in blowflies, provide the substrate for selection, as originally recognized by Darwin. But are these polymorphisms the only genes relevant to behavior? The answer to this question depends on how one defines genetic variation. In principle any gene can vary to some extent—DNA sequence varies even in the most conserved genes. The issue is whether random variation in DNA sequence is sufficient to produce a detectable difference in the phenotypes of a natural population.

To identify critical genes it is necessary to induce and isolate new mutations that affect particular aspects of behavior. This approach, begun in the late 1960s by Seymour Benzer and Sydney Brenner, allows the individual genetic components of behavior to be recognized, studied, and understood. The strategy is much the same as that used in the study of development—obtain mutants with a particular phenotype, track down the genes, identify their products, and analyze their time, place, and mode of action. This strategy has yielded important insights into the genetic makeup of behaviors as diverse as visual tracking, courtship, learning, and circadian rhythms. Moreover, it has shown us that genes affecting processes as diverse as development of specific neuronal cell types, intracellular second-messenger systems, and transcription in the adult nervous system are capable of altering behavior when mutated.

Manipulation of gene expression in intact animals has proven to be one of the most adaptable techniques applicable to a variety of biological problems, including those involving the nervous system. Initially developed as a means of studying and identifying control regions for genes, it was later employed as a way to interfere with normal function by forcing expression of genes in inappropriate places or at inappropriate times. Techniques for injecting DNA into a fly or mouse so that it will be incorporated permanently into the genome were developed in the early 1980s (Box 30–3).

Genetic modification of the mouse embryo can be achieved by adding novel genetic material to somatic tissues, either by introducing genetically marked cells to the embryo or by directly introducing the genetic material into the cells of interest in the intact embryo. In addition, the germline can be modified by injection of DNA into the one-cell zygote, by retroviral infection of the early embryo, or by introduction of pluripotent embryonic stem cells in which specific genes have been mutated into blastocysts (Box 30–4).

Box 30–3 *Manipulating Genes in Transgenic Flies and Mice*

Genes can be manipulated in mice by injecting DNA into the nucleus of a newly fertilized egg (Figure 30–7A). In some of the injected eggs it is incorporated into a random site on one of the chromosomes and, since the embryo is at the one-cell stage, the incorporated gene is replicated and ends up in all (or nearly all) of the animal's cells, including the germline.

Gene incorporation is most easily detected by co-injecting the marker gene for pigment production into an egg obtained from an albino strain. Mice with patches of pigmented fur represent successful injections. Confirmation of the transgene's presence is made by testing a sample of DNA from the injected individuals once they are grown.

In flies a similar approach is used, the main difference being that the DNA need not be injected directly into a nucleus, since the vector used, called a *P element*, is capable of being incorporated into germ cell nuclei at the time the first cells form in the embryo (Figure 30–7B).

These techniques have been used to alter development and function in the nervous system. This is accomplished in a way using promoters that are expressed ubiquitously, such as the inducible heat-shock promoter *hsp*70 in *Drosophila*. The transgene can then be controlled by raising the temperature, which activates transcription from the *hsp*70 promoter. Examples of this in *Drosophila* include transgenic expression of specific inhibitory peptides for the cAMP-dependent protein kinase and the calcium/calmodulin-dependent protein kinase II. More specific patterns of expression in brain cells can be obtained using promoter and enhancer sequences from cell-type specific genes.

Figure 30–7 Standard procedures for generating transgenic mice and flies. Here the gene injected into the mouse causes a change in coat color, while the gene injected into the fly causes a change in eye color. In some transgenic animals of both species the DNA is inserted at different chromosomal sites in different cells (see illustration at bottom). (From Alberts et al., 1994.)

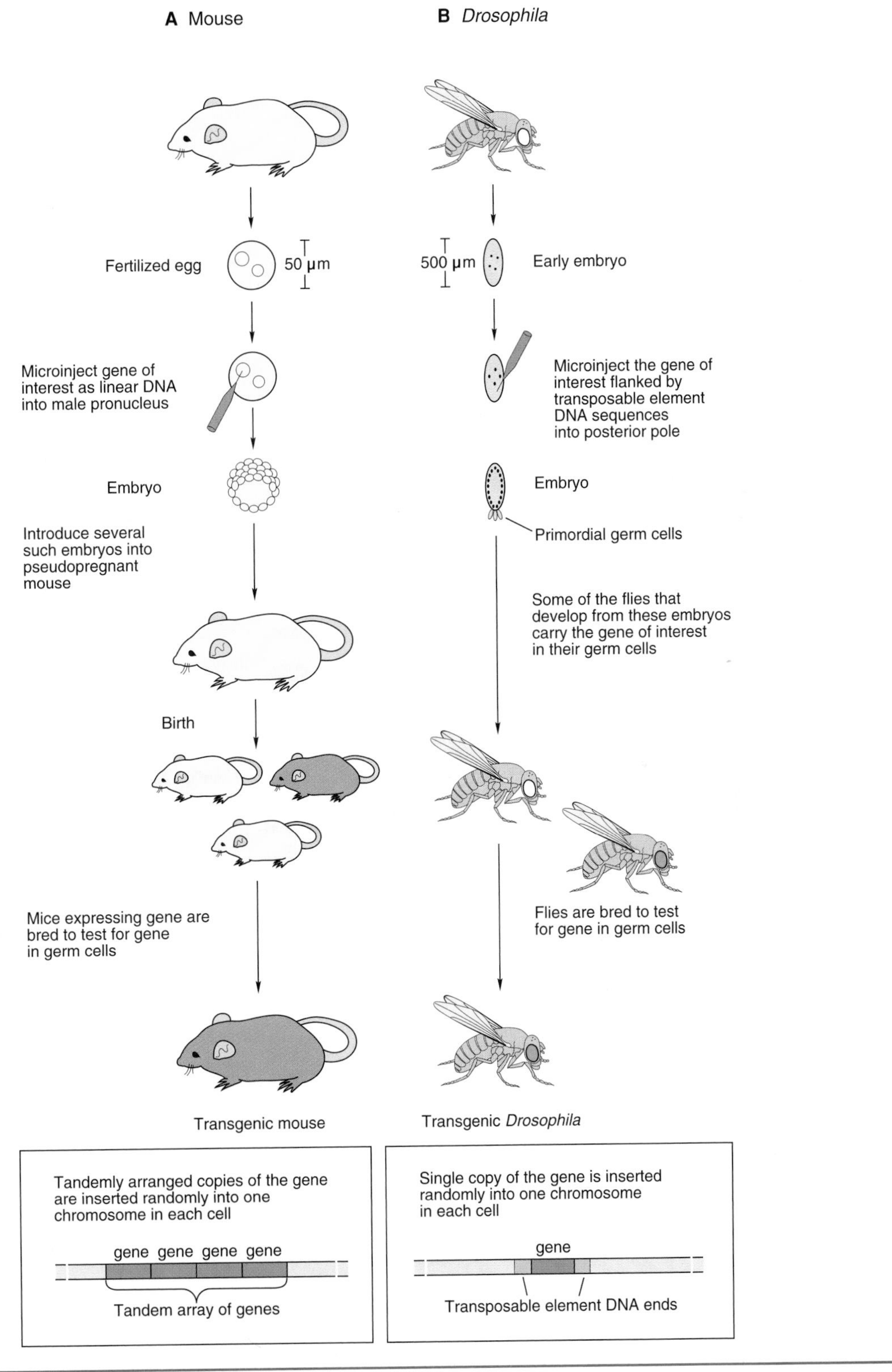
A Mouse
B Drosophila
Fertilized egg
50 μm
500 μm
Early embryo
Microinject gene of interest as linear DNA into male pronucleus
Microinject the gene of interest flanked by transposable element DNA sequences into posterior pole
Embryo
Embryo
Primordial germ cells
Introduce several such embryos into pseudopregnant mouse
Some of the flies that develop from these embryos carry the gene of interest in their germ cells
Birth
Mice expressing gene are bred to test for gene in germ cells
Flies are bred to test for gene in germ cells
Transgenic mouse
Transgenic Drosophila
Tandemly arranged copies of the gene are inserted randomly into one chromosome in each cell
gene gene gene gene
Tandem array of genes
Single copy of the gene is inserted randomly into one chromosome in each cell
gene
Transposable element DNA ends

Box 30–4 Making Targeted Mutations in Mice

Generating new *Drosophila* mutants has been the bread and butter of genetic studies for the past 80 years and has driven much of the progress in understanding the role of genes in development and behavior. Successful mutagenesis in mammals, however, has faced many obstacles. Not the least of these are the expense and labor of breeding enough progeny to screen for relatively rare new mutations. This difficulty was compounded for many years by the lack of an effective mutagen. Moreover, linkage analyses of newly induced mutations are difficult to carry out, as are the tracking and handling of recessive lethal mutations.

Recent advances in molecular manipulation of genes have mitigated some of these problems by allowing for direct *in situ* replacement on the chromosome of a known, normal gene with a mutant version. The process of generating a strain of mutant mice relies on two separate manipulations: the replacement of a gene on a chromosome by homologous recombination in a special cell line known as embryonic stem cells (Figure 30–8), and the subsequent incorporation of this modified cell line into the germ cell population of the embryo (Figure 30–9).

The strategy assumes that one already knows and has cloned the gene of interest. The amino acid sequence is modified so as to mutate or otherwise disrupt the normal gene, and a selectable marker, usually a drug resistance gene, is then introduced into this fragment. This altered gene is then transfected into embryonic stem cells in culture, and clones of cells that incorporate the altered gene are isolated. To identify a clone in which the new gene has been integrated into the homologous (normal) site, rather than some other random site, DNA samples of each clone are tested. Insertion at the homologous site will occur 100 to 1,000 times less often than at some other site.

When a suitable clone has been obtained, cells from it are injected into a mouse embryo at the blastocyst stage (3 to 4 days after fertilization), when the embryo consists of approximately 100 cells. These embryos are then reintroduced into a female that has been hormonally prepared for implantation and allowed to come to term. Embryonic stem cells in the mouse have the capability of participating in all aspects of development, including the germline. Thus, injected cells can become germ cells and pass on the altered gene.

Since incorporated stem cells generally mix into other tissues besides the germline, their presence can be tested when the injected embryo is born. Initially, this can be done simply by using a stem cell line from a mouse strain with a fur color different from that of the strain used to obtain the embryo. The mixed (*chimeric*) offspring appear to have a patchy colored coat. These progeny are then test-mated to determine if any stem cells have become germ cells. If so, their progeny will carry the altered gene on one of their chromosomes, detectable by analyzing DNA samples from each. When the heterozygous individuals are mated together, one-fourth of the progeny will be homozygous mutant. This technique has been used to generate mutations in various genes crucial to development or function in the nervous system.

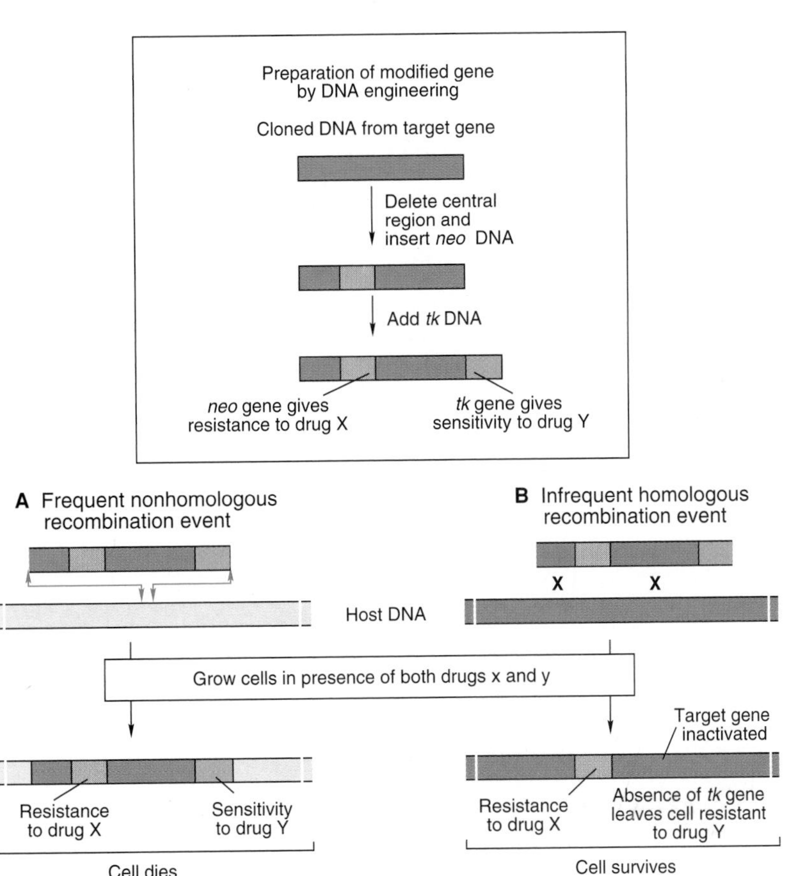

Figure 30–8 Experimentally controlled homologous recombination is the first step in creating mutant mice. Cloned DNA from the mouse gene to be mutated is modified by genetic engineering so that it contains a *bacterial gene*, called *neo*. Integration of *neo* into a mouse chromosome makes the mouse cells *resistant* to a drug that otherwise kills them (drug *X*). A *viral gene*, called *tk*, is also added, attached to one end of the mouse DNA. Integration of *tk* into a mouse chromosome makes the cells *sensitive* to a different drug (drug *Y*). Most insertions occur at random sites in the mouse chromosome, and these nearly always include both ends of the engineered DNA fragment, as shown in **A**. By selecting for those rare mouse cells that grow in the presence of both drugs, colonies of cells are obtained in which homologous recombination has incorporated the center of the engineered DNA fragment without the ends. Most of these cells will turn out to carry the targeted gene replacement shown in **B**. (Adapted from Alberts et al., 1994.)

(continued)

Box 30–4 Making Targeted Mutations in Mice (continued)

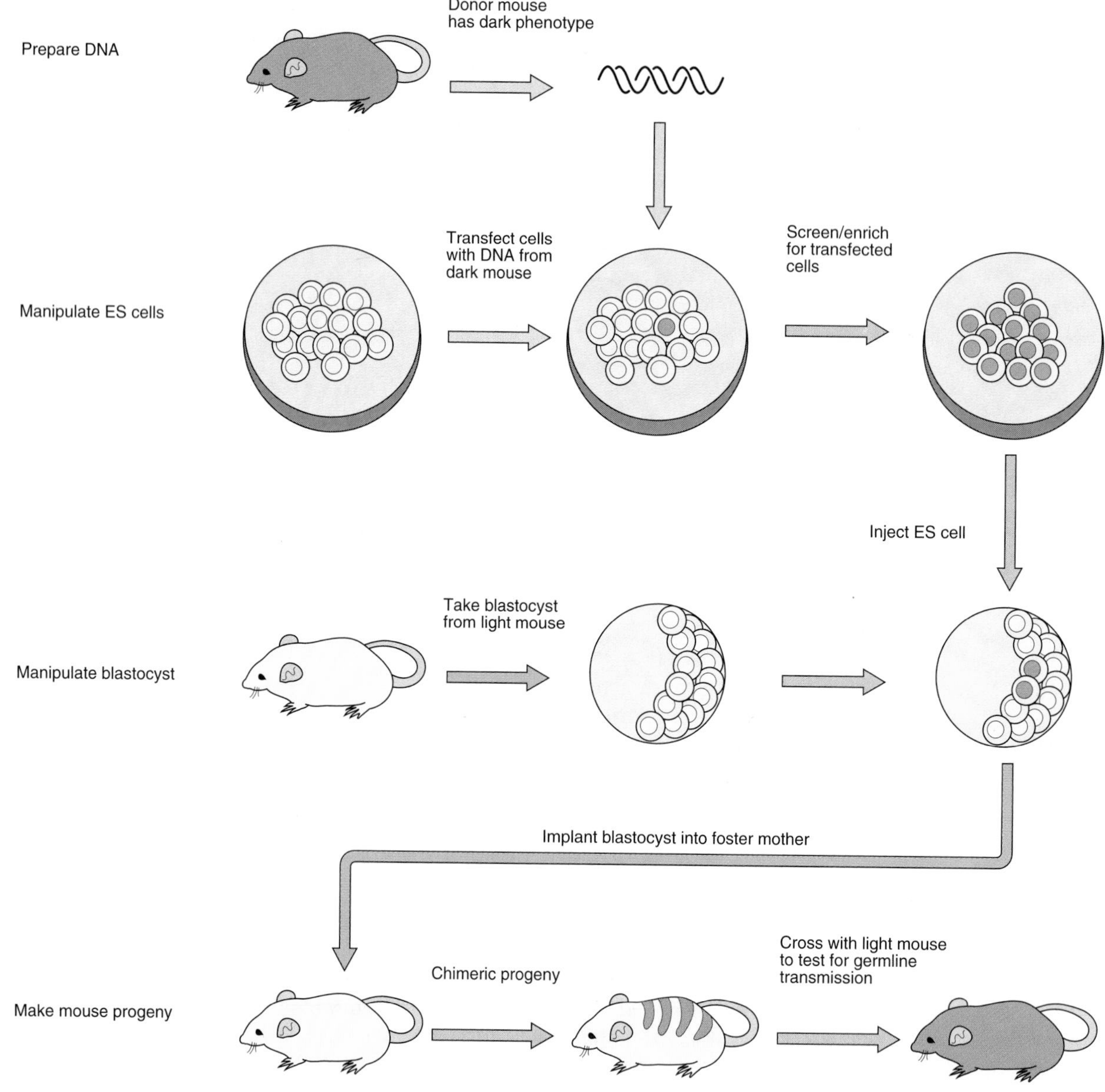

Figure 30–9 Transgenic mice can be generated from embryonic stem cells derived from mouse blastocysts. Embryonic stem (ES) cells are transfected with altered DNA. ES cells that have integrated a transgene for a particular trait can be selected by using a donor that carries an additional sequence, such as a drug resistance gene (see Figure 30–8). An alternative is to assay the transfected ES cells for successful integration of the donor DNA using PCR technology. Once a population of ES cells is obtained in which a high proportion carry the marker, the ES cells are then injected into a recipient blastocyst. This blastocyst is implanted into a foster mother to generate a chimeric mouse. Some of the tissues of the chimeric mice will be derived from the cells of the recipient blastocyst; other tissues will be derived from the injected ES cells. To determine whether ES cells have contributed to the germline, the chimeric mouse is crossed with a mouse that lacks the donor trait. Any progeny that have the trait must be derived from germ cells that have descended from the injected ES cells. By this means, an entire mouse is generated from an original ES cell. (Adapted from Lewin, 1994.)

Courtship Behavior Is Genetically Determined in Sexual Differentiation of the Nervous System

Courtship behavior is an example of what has traditionally been known as innate or instinctive behavior—a programmed series of complex routines that the animal is capable of performing without prior experience. The pioneering ethologist Konrad Lorenz called such routines *fixed action patterns*, although they are rarely fixed in the strict sense. Because such behaviors follow a highly stereotyped pattern among individuals, we consider that they are genetically programmed and so offer interesting cases for the study of how genes organize behavior. At present *Drosophila* is the only organism exhibiting complex but stereotypic behavior that is also amenable to genetic analysis in depth. The nematode *Caenorhabditis elegans* is similarly suited to sophisticated genetic analysis, but does not appear to have as complex a behavioral repertory as the fruit fly.

In the presence of virgin female flies a male will perform a set of stereotypic actions. He orients toward her, taps her abdomen with his foreleg, follows her if she walks away, and then extends one wing and vibrates it to produce the species-specific *love song*. Depending on the female's level of interest at this point, he may go back and repeat these actions, or he may go on to extend his proboscis and lick the female's genitals, attempt to mount her, and copulate.

Since courtship is a sex-specific behavior, parts of the nervous system must naturally develop in a sex-specific manner to subserve the behaviors. Portions of the central nervous system that must develop along the path of male sex differentiation to produce male behavior have been defined. By creating genetically mosaic flies, in which some cells develop as male and others as female, it is possible to show that the early stages of courtship—orienting, tapping, and wing extension—require male tissue only in the dorsal brain and only on one side (Figure 30–10). Later steps in the sequence require male tissue in parts of the ventral brain and (for a proper love song) in the thoracic ganglion.

These results indicate that genes acting in the sex determination pathway cause differential development of certain cells in these crucial portions of the brain. The cells in turn organize the nervous system's behavioral output.

What are these genes and what do they do? One of the best clues to this question is the mutation called *fruitless*. Males that are mutant for *fruitless* display several courtship defects. They court females normally through the first several stages, but never attempt copulation. Moreover, they court other males. Mutant females, in contrast, are behaviorally normal.

A single gene defect that affects only males—that alters both attempted copulation and partner discrimination—is likely to have an important role in the cascade of genes that determines the sex of male flies, acting hierarchically on more than one set of final targets of differentiation. Further support for this kind of role comes from the resemblance of the gene to a ubiquitous male sex determination gene known as *doublesex*. One of the targets of *fruitless* has already been identified—a male-specific muscle in the abdomen that is missing in *fruitless* mutants. Since the differentiation of this muscle is under the control of its motor neuron, the motor neuron is the likely target of *fruitless* action. The *fruitless* gene gives us one perspective on the genetic organization of a complex, programmed behavior. It indicates that genes in a developmental pathway can influence selected aspects of a behavior.

Biological Rhythms Are Governed by Oscillations in Gene Expression

As we shall see further in Chapter 33, many aspects of our physiology and behavior fluctuate in regular rhythmic cycles. Most of these follow a daily (circadian) period, others follow shorter-term (ultradian) periods. Circadian rhythms, which affect everything from locomotor activity in the fly to mood in humans, play a major role in the biology of motivation. Mutants altering biological rhythms have been isolated in several organisms, but it is from studies of the *period* (or *per*) gene in *Drosophila* that we have gained the first real insight into how living things keep time.

The *per* gene seems to be devoted almost exclusively to rhythms. Even when the gene is completely eliminated, the only observable defects are in the organism's rhythms. Mutations in this gene affect circadian rhythms in locomotor activity and eclosion (emergence of the adult from metamorphosis), as well as shorter-term rhythms. The

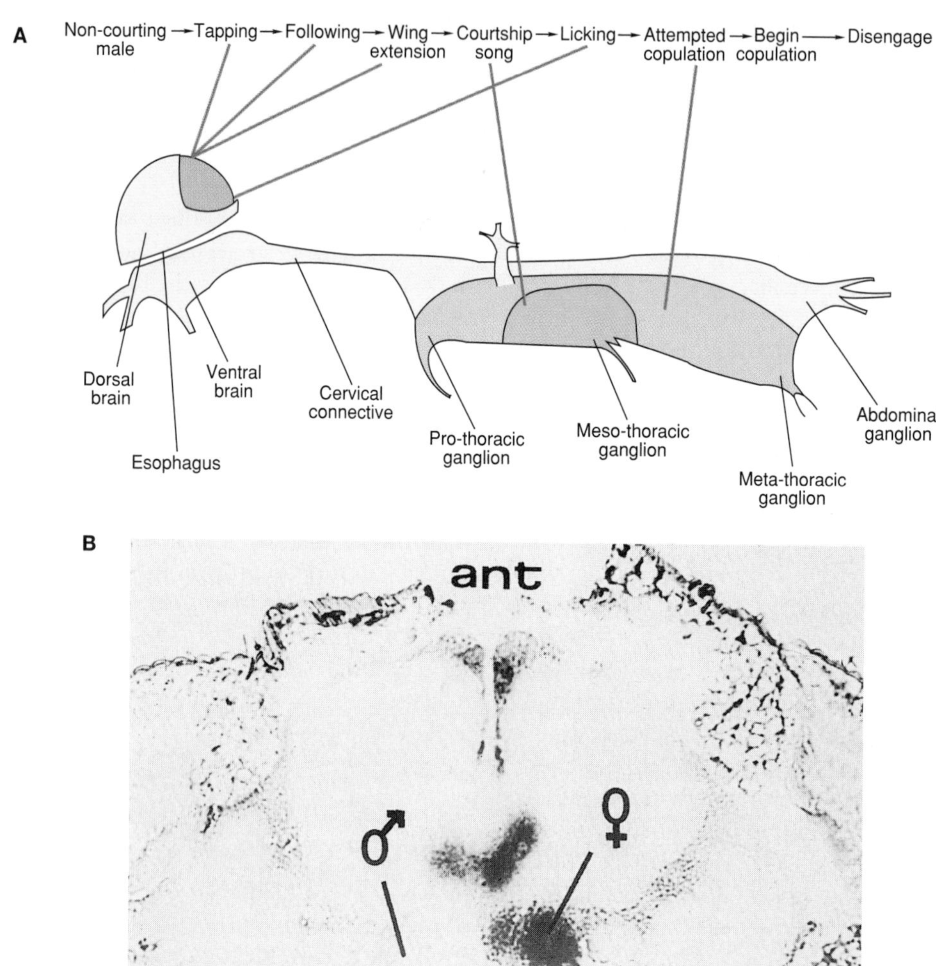

Figure 30–10 Sex-specific development in the brain of *Drosophila*. (From Hall, 1985.)

A. This schematic drawing shows the regions of the brain and thoracic ganglia of *Drosophila melanogaster* in which sex-specific differentiation in the neural pathway is required for courtship behavior.

B. Sex-specific sites in the neural pathway for male courtship behavior are identified by creating genetically mosaic flies with both male and female cells. In this horizontal section through the protocerebrum of a male-behaving fly, the female cortex is darkly stained and the male cortex is unstained.

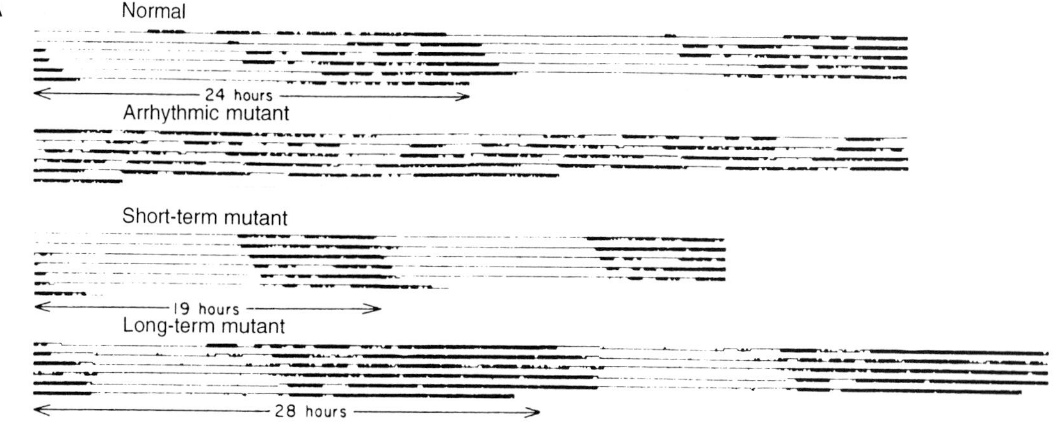

Figure 30–11 A single gene, *period*, governs the circadian rhythms of specific behaviors in *Drosophila*. (From Konopka and Benzer, 1971.)

A. Locomotor rhythms on four *Drosophila* strains: normal, arrhythmic, short-day, and long-day. Flies were exposed to a cycle of 12 hours of light and 12 hours of darkness, and activity was then monitored under infrared light. **Heavy lines** indicate activity.

B. Normal adult flies emerge from metamorphosis (eclosion) in cyclic fashion, even in constant darkness. The plots show the number of flies (in each of four strains) emerging per hour over a four-day period of constant darkness.

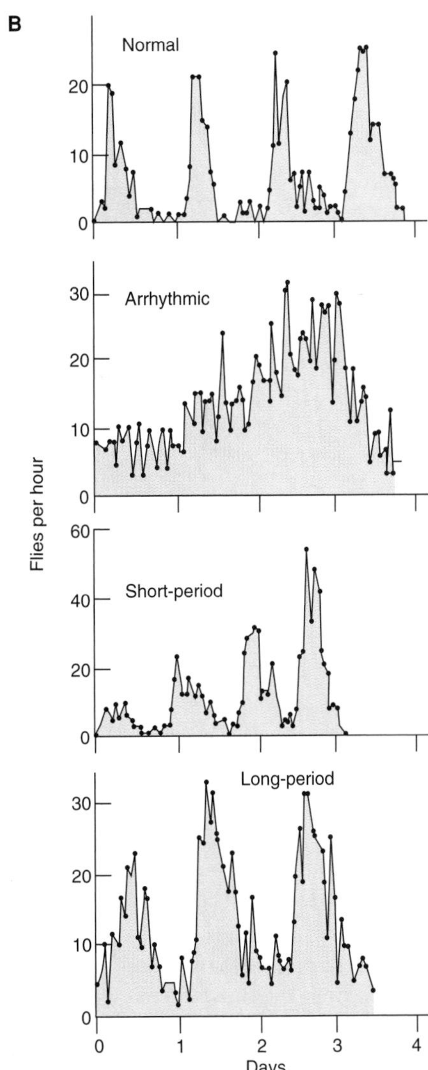

effects of the *per* gene on several rhythms, as well as the similarity in the way each rhythm is altered in mutants, conclusively demonstrate the role of the gene in time keeping (Figure 30–11). That is, arrhythmic mutants show no discernible rhythms for any behavior. A long-day allele produces an extended cycle, 28 hours instead of 24 hours, for all rhythmic behaviors. Similarly, two short-day alleles shorten the cycle (to 19 hours in one case and 16 hours in the other).

Surprisingly, all of these mutants exert an analogous effect on an ultradian rhythm: the song cycle during courtship. This is not a song cycle in the Schubertian sense, of course, but rather a regular oscillation in the interval between songs produced by vibration of the wing. In normal males this cycle repeats every 55 seconds. In long-day mutants it lasts up to 83 seconds, and in short-day mutants, 40 seconds (Figure 30–12).

How does the *per* gene keep time? The answer to this question, at least for circadian rhythms, has emerged from genetic and molecular studies of the gene and the mutants, and it has added another case to the growing number in which transcriptional control governs the output of the mature nervous system (see Chapter 14).

The product of the *per* gene is a nuclear protein, although it does not resemble any known transcription factors. The synthesis and accumulation of both the mRNA and the protein follow a circadian cycle (Figure 30–13). The clue to the role of the protein in regulating transcription came from the finding that the timing of transcription is sped up in short-day mutants and is slowed down in long-day mutants. That is, the protein feeds back to control transcription of its own gene. In arrhythmic mutants that produce a nonfunctional protein, the transcription of *per* is left on at all times. This suggests that the protein feeds back negatively on

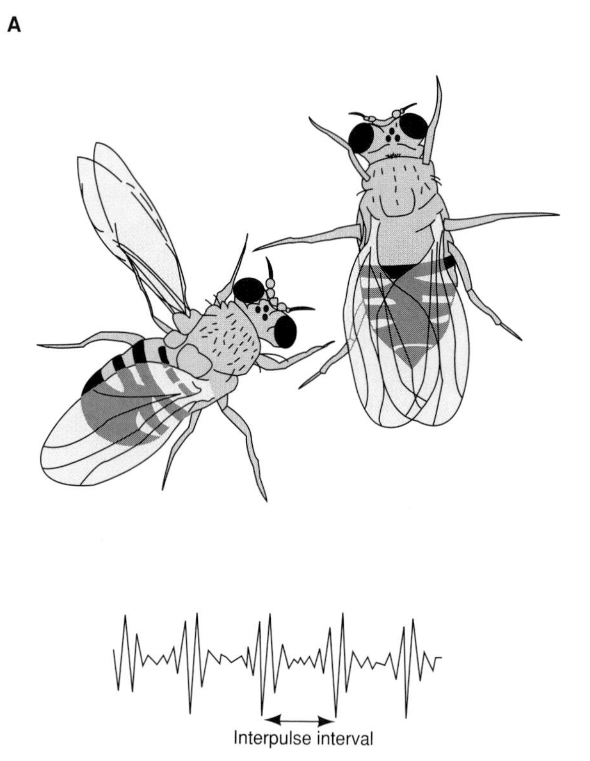

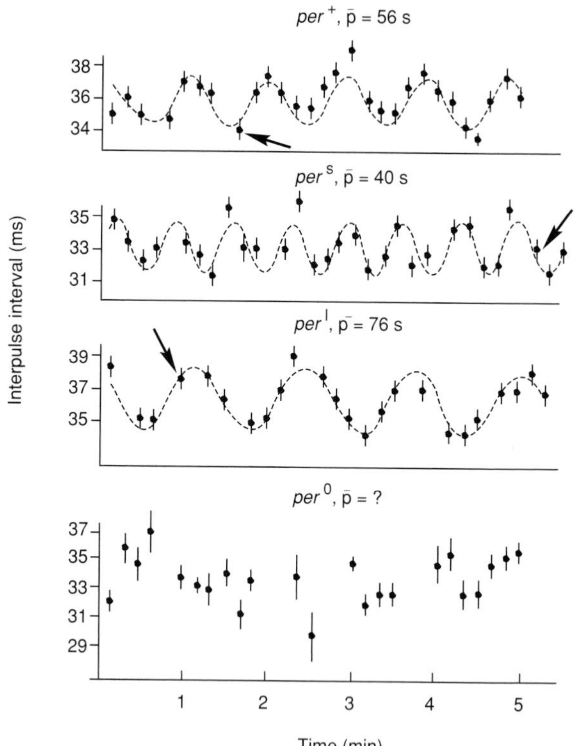

Figure 30–12 The love song of a male fly. (From Hall, 1985.)

A. During courtship, the male *Drosophila melanogaster* extends and vibrates his wing, producing a series of pulses.

B. Courtship song rhythms are affected by mutations of the *period* gene. In the plots shown here each point represents the mean interpulse interval for 10 seconds of song. There are gaps in the data because the flies routinely interrupt their courtship. Nonetheless, when the males resume singing, they tend to remain in phase with the sinusoidal oscillation (**arrows**). Dashed lines represent the results of regression analysis. Period lengths ($\bar{p}$) are calculated for each set of recordings. Recordings are shown for four *D. melanogaster* strains: *per*$^{+}$, normal; *per*^{s}, short-day; *per*^{l}, long-day; *per*^{o}, arrhythmic.

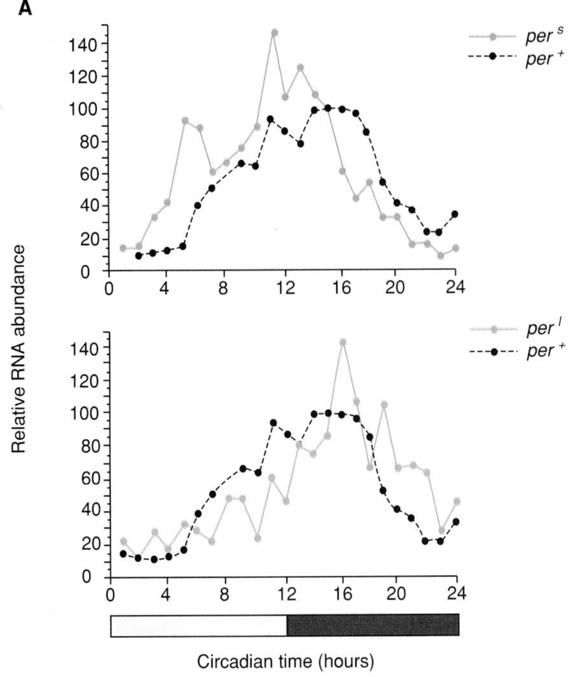

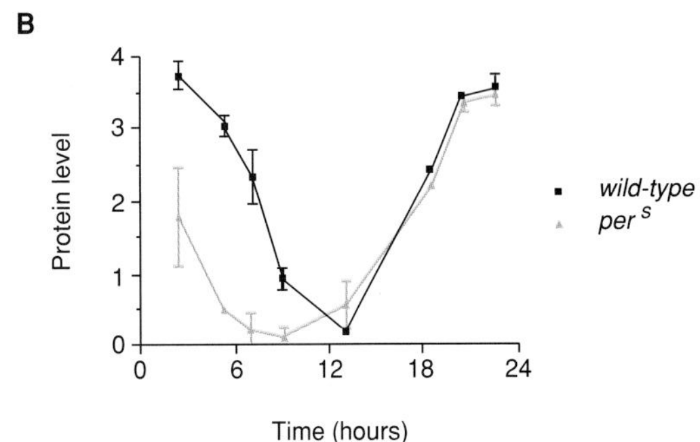

Figure 30–13 Cycling of RNA and protein products of the *period* gene in *Drosophila melanogaster*.

A. The daily peak in RNA comes sooner in the short-day mutant (*per*^{s}) and later in the long-day mutant (*per*^{l}). (From Hardin et al., 1990).

B. The protein level fluctuates in the same manner for wild-type and short-day (*per*^{s}) mutants. Protein levels in long-day mutants (not shown) are very low. (From Zerr et al., 1990.)

its own expression. When it is present in the nucleus, transcription is turned off; when it is absent, transcription is turned on.

Why, then, does *per* not simply short-circuit itself and turn off as soon as it turns on? The answer lies in a built-in delay at the level of translation. The mRNA is not translated as quickly as it is transcribed and not translocated into the nucleus as quickly as it is translated. As a result, there is time for the transcript to accumulate before a significant amount of the protein is made. When protein levels reach a high enough level, they feed back and turn off *per* transcription, reducing the available mRNA to negligible levels. When the mRNA falls so low, translation ceases and protein levels fall. This in turn allows transcription to pick up once again, and so on. The trick is to build in the necessary time delay of translation and translocation and the requisite half-lives for protein and mRNA. Time keeping thus becomes a matter of oscillation between the gene's inherent activation and negative feedback on itself.

The universality of this mechanism has been borne out by the presence of *per*-like genes in circadian pacemaker neurons in other organisms, where electrophysiological recordings from the central nervous system are possible. Retinal pacemaker neurons in the mollusk *Aplysia* oscillate in their firing during the 24-hour cycle, and these cells contain a similar protein that oscillates in its expression. Cells in the suprachiasmatic nucleus, the site of circadian time keeping in mammals (see Chapter 33), also synthesize a *per*-like protein. Even in the fungus *Neurospora*, a gene called *frequency* can mutate to produce long-day, short-day, and arrhythmic phenotypes; here, too, transcription oscillates

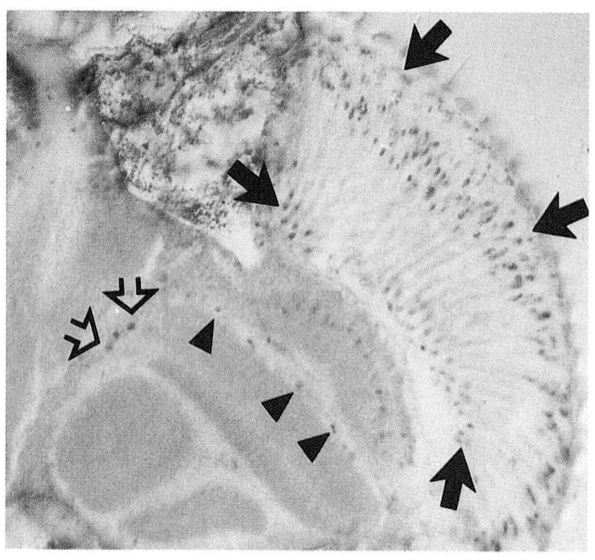

Figure 30–14 This horizontal section of the head of a normal *Drosophila* fly shows the sites of expression of the protein product of the *period* gene (stained with an antibody to the gene product). **Open arrows** label lateral neurons, **filled arrows** label photoreceptors, and **arrowheads** label glia. (From Zerr et al., 1990.)

over a 24-hour period. Thus it appears that the time-keeping mechanism identified for the *per* gene in *Drosophila* may be ubiquitous in the biological world, as are circadian rhythms. Recently, genetic screens in mice have also identified a locus that affects circadian rhythms.

Although the *per* gene is expressed very widely in the nervous system and other tissues of the fly, its role in setting circadian rhythmicity appears to be restricted to a set of neurons in the lateral portion of the brain, proximal to the optic lobes (Figure 30–14). When in experiments the normal *per* gene is selectively expressed in some cells of an otherwise mutant fly, the key spot of expression is this set of lateral cells. These cells apparently exert their effect through a signal that can diffuse freely, since arrhythmic mutants can be rescued by transplantation of a large piece of normal brain tissue into the abdomen. Thus, the oscillation of transcription controlled by the *per* gene may in turn control the secretion of a molecule, which acts as the output of the clock.

Although the *per* gene seems to have a very specific action, its effects are nonetheless widespread, since rhythms appear in many processes. Is there any rule to how genes influence behavior? At present the answer appears to be no. Each example discussed here represents a distinct strategy by which an individual gene contributes to the conglomeration of events we call behavior.

An Overall View

Behavior emerges gradually as the brain develops. At first the development of the brain is largely under the control of genetic and developmental programs. Influences from the environment begin to exert their effect *in utero* and become of prime importance after birth. Knowledge of both innate (genetic and developmental) and environmental determinants is needed to understand behavior fully. This information is also essential for developing rational therapeutic strategies for treating psychiatric disorders.

Many aspects of behavior are under genetic control. Evidence for this can be seen in the striking similarities revealed in studies of human twins and in the ability to select and breed for particular behavioral traits in domestic and laboratory animals. Such breeding experiments generally indicate that behavioral traits have a complex multigenic etiology. In rare instances, however, the source of natural variation in behavior has been traceable to one predominant genetic factor, as in the case of sensitization of the central excitatory state in *Phormia*. Sometimes a single gene is the sole source of variation, as in human red color vision.

Mutant studies in *Drosophila* have revealed a variety of genetic contributions to the building of behavior. The *fruitless* gene contributes to programmed courtship behavior by its regulation of sex-specific development in the male, and the *per* gene controls circadian rhythms by its own periodic expression.

Selected Readings

Bouchard, T. J., Lykken, D. T., McGue, M., Segal. N. L., and Tellegen, A. 1990. Sources of human psychological differences: The Minnesota twin study of twins reared apart. Science 250:223–228.

Bouchard, T. J., Jr. 1994. Genes, environment, and personality. Science 264:1700–1701

Greenspan, R. J. 1995. Understanding the genetic construction of behavior. Sci Am. 272(4):72–78.

Groves, P. M., and Rebec, G. V. 1992. Introduction to Biological Psychology, 4th edition. Dubuque, Iowa: W. C. Brown.

Hall, J. C. 1994. The mating of a fly. Science. 264:1702–1714.

Hall, J. C. 1990. Genetics of circadian rhythms. Annu. Rev. Genet. 24:659–697.

Nathans, J., Merbs, S., Sung, C.-H., Weitz, C. J., and Wang, Y. 1991. Molecular genetics of human vision pigments. Annu. Rev. Genet. 26:403–424.

Plomin, R., Owen, M. J., and McGuffin, P. 1994. The genetic basis of complex human behaviors. Science 264:1733–1739

Rossant, J. (1990) Manipulating the mouse genome: Implications for neurobiology. Neuron 4:323–334.

Takahashi, J. S., Pinto, L. H., and Vitaterna, M. H. 1994. Forward and reverse genetic approaches to behavior in the mouse. Science 264:1724–1733.

Tully, T., and Hirsch, J. 1982. Behavior–genetic analysis of *Phormia regina*. I. Isolation of pure-breeding lines for high and low levels of the central excitatory state (CES) from an unselected population. Behav. Genet. 12:395–415.

Vitaterna, M. H., King, D. P., Chang, A. M., Kornhauser, J. M., Lowrey, P. L., McDonald, J. D., Dove, W. F., Pinto, L. H., Turek, F. W., and Takahashi, J. S. 1994. Mutagenesis and mapping of a mouse gene, *clock*, essential for circadence behavior. Science. 264:719–725.

Watson, J. D., Gilman, M., Witkowski, J., and Zoller, M. 1992. Recombinant DNA, 2nd ed. New York: Scientific American Books.

Wimer, R. E., and Wimer, C. C. 1985. Animal behavior genetics: A search for the biological foundations of behavior. Annu. Rev. Psychol. 36:171–218.

31

Sex and the Brain

A Single Gene Switches the Initial Development of the Gonad from Female to Male

Gonadal Hormones from Both the Mother and the Male Fetus Regulate Continued Development

Perinatal Hormones Impose a Permanent Sex-Specific Blueprint on the Developing Nervous System

- *Fetal Exposure to Male Hormones Causes Pseudohermaphroditism in Genetic Females*
- *Steroid Hormones Influence Perinatal Development Only During Critical Periods*

The Brain Can Be Masculinized Not Only by Male Hormones But Also by Many Other Compounds

- *Female Fetuses Are Protected from Masculinization by Binding of Estrogen*
- *Receptors in the Cell Nucleus Mediate the Effects of Gonadal Steroid Hormones*

Sexually Differentiated Brains Have Different Physiological Properties and Behavioral Tendencies

- *Perinatal Hormones Also Determine the Degree to Which Sex-Linked Behaviors Are Expressed by Normal Males and Females*
- *Sexual Differentiation Is Reflected in the Structure of Certain Neurons*
- *Cellular Mechanisms Involved in the Development of Sex Differences in the Brain Can Be Studied In Vitro*

A Wide Range of Behavior Is Influenced by Sex Differences in the Organization of the Brain

- *Cognitive Development in Monkeys Is Sexually Dimorphic*
- *Human Cerebral Asymmetry Is Sexually Dimorphic*

Reproductive behavior, like all forms of behavior, is a consequence of the developmental plasticity of the brain. The range of potential behaviors is genetically determined, but the particular behaviors expressed are shaped by interactions between developing cells and the environment. As we saw in Chapter 26, these interactions occur during specific *critical periods*. Critical periods are characteristic of the sequential nature of most developmental programs; at each stage of development a choice is made between a limited set of alterations. Once the choice is made, it is nearly impossible to reverse the result.

Although males and females differ in many ways, the developmental program underlying all aspects of sexual determination is the same in both sexes. A single gene determines the type of gonad. The gonad in turn influences the hormonal environment of the developing fetus or infant. Specific tissues develop along sexually dimorphic lines in response to the combination of sex hormones to which they are exposed. (The term *dimorphism* refers to the existence of two distinct forms within a species; *sexual dimorphism* refers to any characteristics that differ in males and females.) Developing target tissues are responsive to hormones only during certain critical phases of differentiation.

To understand this basic template of sexual differentiation we examine the following questions in sequence. Which gene determines gonad type? What and when do the developing gonads secrete? Which are the target tissues for sexual differentiation? When are the critical periods for these events? Finally, we examine the biological and behavioral consequences of sexually differentiated neuronal populations. As we shall see, these extend well beyond the domain of reproductive behavior.

A Single Gene Switches the Initial Development of the Gonad from Female to Male

The chromosomal sex of an individual is established at conception when the sperm contributes either an X or a Y chromosome. The chromosome determines whether the embryonic gonad differentiates into an ovary or a testis. Subsequent steps in sexual differentiation result from the action of hormones. If a Y chromosome is present, testes develop and their hormonal secretions result in the development of a phenotypic male. If only X chromosomes are present, ovaries develop and the female phenotype results. If only one sex chromosome is present (invariably an X), or if a gonad is absent, the individual also develops as a female. Thus, in human development the ovary is the normal (default) pathway. The function of the Y chromosome is to switch the developmental program of the precursor cells in undifferentiated gonads from the pathway for follicle cell development (characteristic of the ovary) to that for Sertoli cells (characteristic of the testis).

The existence of X and Y sex chromosomes was first demonstrated in humans in 1923, yet for nearly 40 years thereafter it was assumed incorrectly that sex in mammals was decided by the *number* of X chromosomes, as it is in *Drosophila*. The study of abnormal sex chromosome combinations, occurring in such syndromes as Klinefelter's (XXY) and Turner's (XO), revealed that mammalian embryos carrying a Y chromosome develop as males regardless of the number of X chromosomes. Then the critical sex-determining region was narrowed down to the short arm of the Y chromosome. All sexually dimorphic characteristics, including the development of the brain, depend on the presence or absence of one or more genes on the short arm of chromosome Y. Thus, this segment of the mammalian Y chromosome is a binary switch for sexual dimorphism. Although sexual differentiation requires many genes that act in conjunction with the Y chromosome, the product of a gene on the Y chromosome—the gene encoding the *testis-determining factor* (*TDF*)—determines the form of the gonad.

Detailed exploration of the short arm of the Y chromosome led to the discovery of a candidate gene for TDF. An equivalent gene is present on the Y chromosome of all mammals so far examined and encodes a transcript that is specifically expressed in the testes. The gene, called *sex-determining region of Y* (*SRY*), encodes a protein that has a DNA binding domain, suggesting that the protein is a transcription activator. The equivalent gene in female mice that are XY is deleted. The *SRY* gene is similar to a gene in yeast that encodes a transcriptional activator important in determining mating type. The mouse homolog of the *SRY* gene (*sry*) has been introduced into transgenic mice (see Box 30–3). Some XX mice receiving the *sry* gene develop as phenotypically male even though they lack all the other genes on the Y chromosome. Thus, the *sry* gene alone can induce male development in mice.

The *SRY* gene determines whether the undifferentiated gonad becomes an ovary or a testis. Subsequent events in the sexual differentiation of the male are determined by factors secreted by cells in the fetal testis. In contrast, the female phenotype can develop in the absence of any gonadal tissue. This was shown by removing gonadal tissue from early fetal rabbits, all of which later developed as females (with oviducts, uterus, cervix, and vagina) whether or not they were XX or XY.

Cells in the fetal testes secrete two major hormones: testosterone and müllerian duct-inhibiting substance (MIS). Testosterone, a steroid secreted by Leydig cells, masculinizes the sex organs, mammary gland rudiments, and nervous system. MIS, a glycoprotein of the transforming growth factor β family secreted by Sertoli cells, causes the resorption of the tissue that would otherwise become the oviducts, uterus, cervix, and vagina. The absence of these two hormones, or of receptors for them, results in the development of a female.

Gonadal Hormones from Both the Mother and the Male Fetus Regulate Continued Development

Although *TDF* genes alone determine whether the undifferentiated gonad will become a testis, later stages in the development of sexual phenotype result from the combined actions of the hormones of the fetal testes and those of the mother. This interaction was first appreciated and extended to the nervous system by the analysis of two syndromes that arise from spontaneously occurring hormonal deficiencies during early development.

The first, *Turner's syndrome*, is a congenital condition in which only one X (and no Y) chromosome is present. In this disorder functional gonadal tissue does not form. Fetal ovaries bud, then atrophy. Wolffian ducts decay; müllerian ducts develop. In the estrogen-dominated environment furnished by the mother and the placenta, a female genital tract forms. Patients with Turner's syndrome are usually regarded as female by their families before adolescence. Some are diagnosed at birth by such accompanying signs as webbing of the neck and impairment of hearing. Many cases are discovered only much later, when these children fail to show the signs of female puberty. If treated with ovarian hormones during adolescence, they respond as normal females. Moreover, the gender identity and sexual behavior of these patients do not differ significantly from those of normal females.

The second genetic anomaly, called *androgen insensitivity syndrome*, is seen in patients who are incapable of responding to androgens. This condition occurs in XY individuals with the *TDF* gene. These individuals develop testes, which secrete both testosterone and MIS during fetal development. Because these genetic males cannot respond to the androgens they produce, they are indistinguishable from phenotypic females in external appearance (Figure 31–1). However, they do respond to MIS and so their müllerian ducts degenerate. Although these individuals develop as women, they lack both a uterus and oviducts.

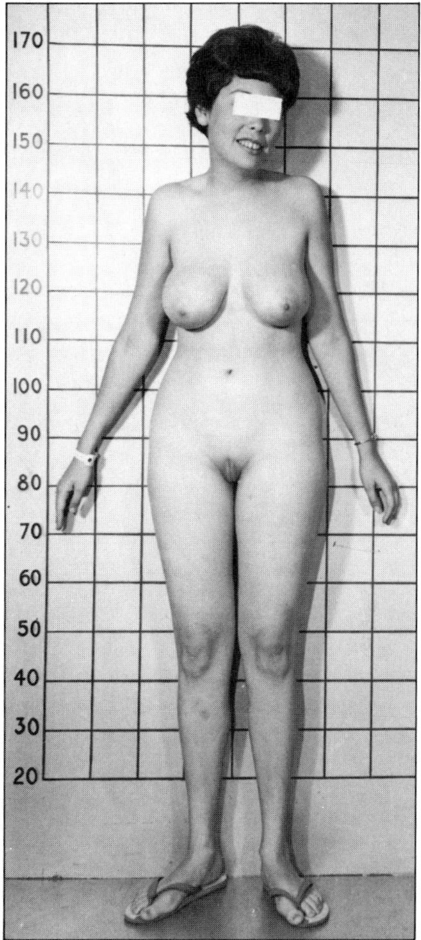

Figure 31–1 This adult patient with androgen insensitivity syndrome is a genetic XY male. The female phenotype developed under the influence of the normal secretion of estrogens by the testes, with no exogenous hormonal treatment. (From Money and Ehrhardt, 1972.)

Studies on mutant mice and rats with androgen insensitivity syndrome demonstrate that the deficiency is caused by the absence of androgen receptors; estrogen receptors are unaffected. As a result, even with normal production of androgens by the testes, the target cells of male sexual differentiation are unable to respond to the hormonal signal. Thus, despite the XY karyotype, the presence of testes, and the absence of ovaries, these patients develop female secondary sex characteristics during adolescence in response to estrogens produced by both their adrenals and testes.

The implication of these clinical states is that the female form and gender identity can develop in the absence of hormonal influences from the fetal gonads. Moreover, in special circumstances individuals who are genetically male can and do develop feminine body characteristics. To integrate these clinical phenomena with experimental observations on the sexual differentiation of the nervous system *in utero*, we must first consider the differential actions of gonadal hormones on the developing and mature nervous system.

Vertebrate reproductive behavior, especially courtship behavior between prospective mates, is richly varied and often distinctive to a species.

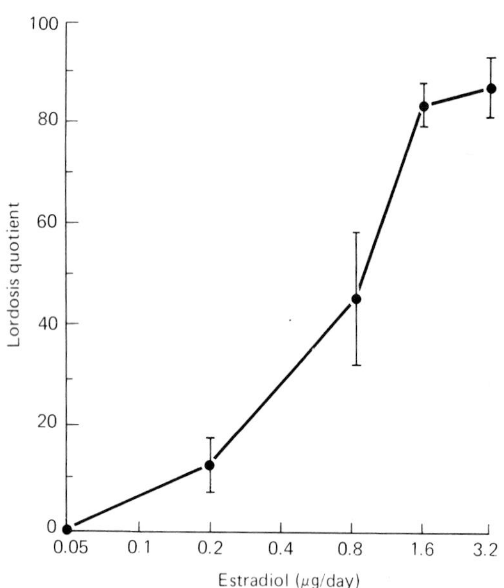

Figure 31–2 Estrogen induces sexual receptivity in ovariectomized female rats in a dose-dependent manner. After eight daily subcutaneous injections of estradiol benzoate, receptivity was measured by the lordosis quotient:

$$\frac{\text{Number of lordotic responses}}{\text{Number of mounts}} \times 100$$

(Data from Bermant and Davidson, 1974.)

Whatever the behavioral ritual, it is virtually always sexually dimorphic. Lordosis and mounting are two sexually dimorphic behavioral markers of neural differentiation frequently studied in rodents. In lordosis, the female rodent elevates her rump and assumes a concave posture of the back. The male mounts from the rear clasping her flanks with his forelimbs.

The secretions of the mature gonads (the ratio of steroids) are also sexually dimorphic. Testes secrete principally androgens; ovaries, estrogens. Is it possible that a special blend of testicular hormones produces male behavior and ovarian hormones female behavior equally well in adult males and females? If so, sex-related patterns of behavior could be explained completely by the type of hormones present in the adult. This line of reasoning was an accepted theory for many years. However, we now know that sexually dimorphic behavior patterns in many vertebrate species, including our own, depend on qualitatively different actions of gonadal hormones at two different stages in the life cycle.

In the adult, circulating steroid sex hormones primarily activate sexual responses. Thus, administration of estrogen increases sexual receptivity in female rats in a dose-dependent manner (Figure 31–2). In ovariectomized females, however, the behavioral stimulation is short-lived. Receptivity declines as the hormone is metabolized. Thus, the actions of gonadal hormones on the mature nervous system are inductive and transitory.

In the developing nervous system, steroid hormones create a gender-specific blueprint, which in adulthood leads to the expression of specific sexual behaviors in response to hormonal stimulation. As the clinical and genetic evidence suggests, the nervous system of a developing fetus is essentially undifferentiated and bipotential. Both male and female genotypes are compatible with either brain phenotype. To a remarkable degree, the sexual phenotype of the brain is determined by exposure to specific steroid hormones during an early critical period. Thus, the actions of gonadal hormones on the developing nervous system are *structural* and *permanent*.

Perinatal Hormones Impose A Permanent Sex-Specific Blueprint on the Developing Nervous System

Although the mature gonads of both sexes are capable of synthesizing both androgens and estrogens, the steroid products of the testis and ovary differ in their ratio, timing, and some synthetic pathways. In considering the development of gender identity, it is helpful to distinguish experimentally between homotypical and heterotypical steroid sex hormones. Homotypical hormones are the prevalent hormones of a given sex administered to an individual of the same sex—for example, estrogens to a female. Heterotypical hormones are the dominant gonadal hormones of one sex given to the other.

The same terms are also often applied to sex-specific behaviors: homotypical behavior patterns are those appropriate to the reference sex and heterotypical behavior patterns are those appropriate to the opposite sex. This behavioral dimorphism is not all-or-none. In the limited repertories of some species, some behaviors that serve a gender-specific reproductive purpose may also serve other purposes. Either sex may exhibit both male and female copulatory patterns in environmental or social circumstances that have little to do with reproduction. For instance, monkeys of both sexes use the female sexual posture as a submissive gesture in social contexts where matters of relative rank or dominance are at issue.

Fetal Exposure to Male Hormones Causes Pseudohermaphroditism in Genetic Females

During gestation the fetuses of both sexes are exposed to the high levels of circulating estrogens in the maternal blood. Since estrogen is homotypical for female fetuses and heterotypical for male fetuses, the following question prompted a now-classic experiment: What would happen if this normal relationship were reversed, bringing female fetuses under heterotypical hormone influences? Injecting high doses of testosterone into guinea pig mothers has two consequences for genetically female offspring. First, they are born as pseudohermaphrodites: their external genitalia are indistinguishable from those of normal males, but, because they are not also exposed to MIS, the müllerian duct derivatives are also present internally.

The second and more intriguing effect is that the adult sexual behavior of hermaphroditic females is also altered. When subsequently treated with estrogen and progesterone as adults, these XX guinea pigs show some elements of homotypical sexual behaviors—lordosis, for example—but their capacity for this behavior is significantly less than that of control females. On the other hand, they display much more mounting behavior than normal females. When treated with testosterone as adults, these hermaphroditic females display a degree of heterotypical mounting comparable to that of normal males, and the female pattern of lordosis is suppressed. These observations permit a distinction in the effects of hormones present early in development and the effects of the same hormones circulating in the adult. The former are confined to a short critical period and might not become evident until adulthood.

Steroid Hormones Influence Perinatal Development Only During Critical Periods

As we have seen, the sexual differentiation of the nervous system can occur along two pathways. The emergence of the male pattern, however, requires the influence of androgens. If androgens are needed for the development of a normal male fetus, where in the uterine environment do they come from? Apparently the developing testes supply them. Most experiments on this question have been carried out in the rat, which has a 21-day gestation period. These studies show that the testes begin to synthesize androgens as early as the thirteenth day of fetal development, and continue to secrete androgen until the tenth day after birth.

This explanation can be checked by removing the testes in newborn males. Even though castration on the day of birth deprives the male rat of testicular androgens for little more than one-half the period that these hormones are normally present, this deprivation has profound effects on the sexual development of genotypic male rats. Rats castrated between 1 and 5 days after birth develop behavioral characteristics of genetic females. If they are injected with estrogen and progesterone as adults, they display lordotic behavior when mounted by normal males. In contrast, males castrated later in development, after 10 or more days of age, show little tendency to display lordosis under comparable conditions.

Another source of steroids that might influence the developing fetuses of both sexes is the placenta, which acts as an endocrine gland thought to be essential for the maintenance of pregnancy. The role of placental secretions in sexual differentiation is not well understood. Maternal cholesterol is metabolized by the placenta into progesterone. To form other steroid hormones, placentally derived progesterone is transported to the fetal adrenal glands and liver and then returned to the placenta for transformation into estrogens and a group of androgens, including testosterone.

There is additional evidence that perinatal exposure to male hormones affects later sexual behavior by influencing the developing central nervous system rather than the peripheral sexual apparatus. Both males and females secrete two gonad-stimulating hormones (gonadotropins) from the anterior pituitary: *luteinizing hormone* (LH) and *follicle-stimu-*

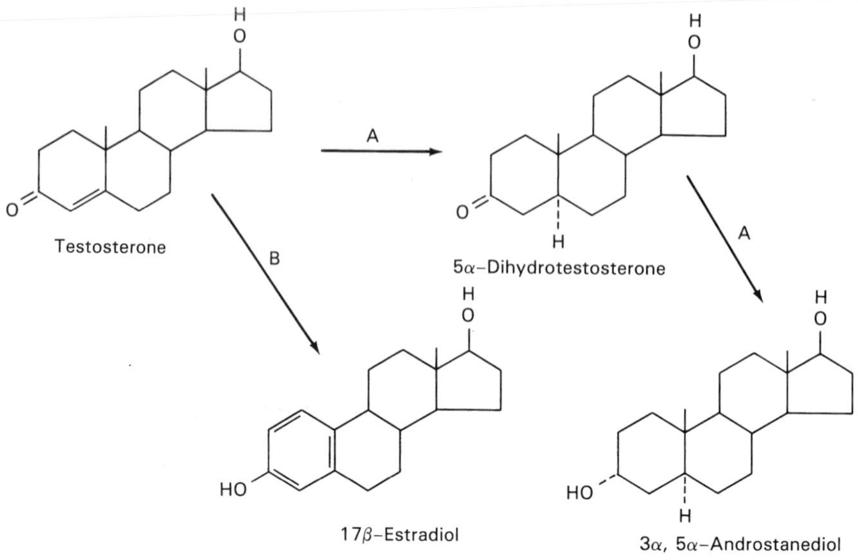

Figure 31–3 There are two metabolic pathways in neurons for testosterone. An androgen-type reduction pathway (**A**) requires the enzyme 5α-reductase. The aromatization route (**B**) is a specific type of chemical conversion by which testosterone is transformed into the female sex hormone 17β-estradiol.

lating hormone (FSH). In males these hormones are secreted at a steady level. In females, surges of the hormones underlie the cyclical activities of the reproductive tract. Cyclical secretion of gonadotropins by the pituitary does not depend directly on the genetic sex of the animal, but rather on the absence of androgen during the perinatal period. In normal circumstances androgen prevents cyclical secretion of gonadotropins in the male.

The Brain Can Be Masculinized Not Only by Male Hormones But Also by Many Other Compounds

The critical developmental period for sexual differentiation corresponds to a time in which the brain is sensitive to a broad spectrum of steroids, many of which are not normally present in the body. Experimental masculinization of the brain can be induced by exposure to such functionally diverse hormones as testosterone, androstenedione, estradiol, and diethylstilbestrol (DES), and even drugs, such as barbiturates, and pesticides, such as dichlorodiphenyl trichloroethane (DDT).

The principal active hormone that determines the normal male brain pattern in newborn rats is estradiol, one of the female sex hormones. Even though the hormone that reaches the brain is testosterone, much of it is converted there to estradiol by enzymes in the cells that are the targets of sexual differentiation. When administered *in vitro,* estradiol has been found to be more effective than testosterone in androgenization.

Female Fetuses Are Protected from Masculinization by Binding of Estrogen

Why are high levels of maternal estrogen not sufficient to masculinize normal female fetuses *in utero*? To understand the process by which the brain is sexually organized during the critical period, it is essential to understand what happens to testosterone when it reaches developing (as well as mature) neurons. As shown in Figure 31–3, there are two tissue-specific metabolic pathways: testosterone can be reduced to 5α-dihydrotestosterone or converted to the female sex hormone, 17β-estradiol. The androgen-type reduction occurs preferentially in the cells of the pituitary gland and the brain stem, whereas the other type of conversion occurs mostly in the neurons of the hypothalamus and limbic system. Conversion into estradiol is also the major metabolic route by which behaviorally relevant neural circuits are permanently modified during the critical period. Inhibitors of this chemical conversion inhibit the sexual differentiation of males and block the facilitation of normal adult sexual behavior induced by estradiol. Masculinization therefore involves estrogen receptors as well as androgen receptors.

For normal development to proceed in females the same estrogen receptors in the same target neurons as in the male must remain *un*occupied during the critical period. Yet maternal blood is rich in estrogens from the gonads and placenta. What protects normal female fetuses from masculinization *in utero* by circulating estrogens? Normal rat fetuses of both sexes are protected from maternal estrogen by an estrogen-binding protein called α-fetoprotein. This protein is synthesized by the fetal liver and is present in blood and cerebrospinal fluid. Unlike estrogen, testosterone is not bound by α-fetoprotein; this is why testosterone has free access to steroid-sensitive neurons in males during the critical period. Once taken up by a neuron, testosterone is converted to estradiol.

Receptors in the Cell Nucleus Mediate the Effects of Gonadal Steroid Hormones

Unlike receptors for neurotransmitters (see Chapters 13 and 14), some gonadal steroid receptors are not situated in the neuronal plasma membrane. Instead they are located in the cell nucleus where they act as transcriptional regulators (Figure 31–4). Steroids can cross the cell membrane and bind to the nuclear receptor. On binding with the hormone the receptor undergoes a conformational

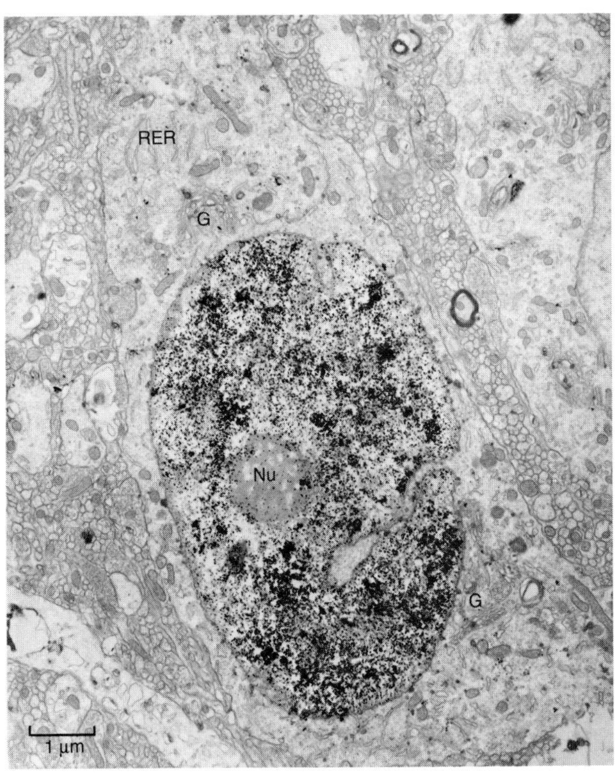

Figure 31–4 Gonadal steroid receptors are concentrated in the nucleus of the cell. This electron micrograph shows estrogen receptors in a cell from the hypothalamic ventrolateral nucleus of the guinea pig. This structure is equivalent to a region that has been identified in the rat as essential for the expression of female reproductive behavior. The prominent nucleolus (**Nu**), large Golgi apparatus (**G**), and extensive rough endoplasmic reticulum (**RER**) are characteristic of peptide-synthesizing cells. (Courtesy of Ann-Judith Silverman, Lydia Don Carlos, and Joan I. Morrell.)

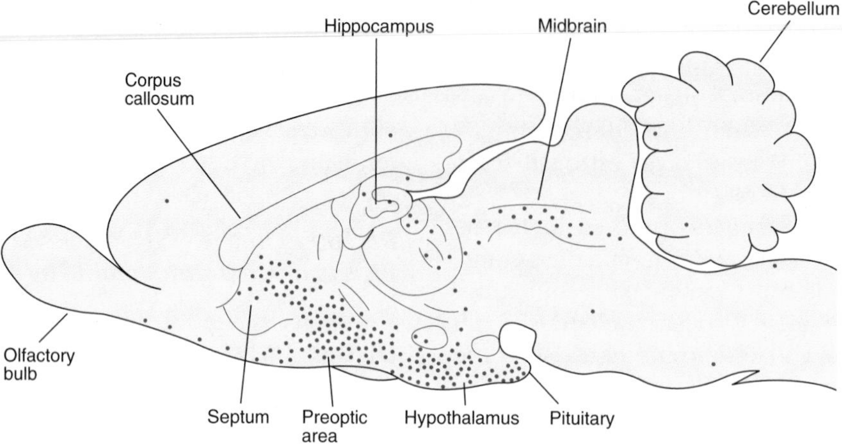

Figure 31–5 The regional distribution of estradiol-sensitive neurons in the brain of an albino rat is shown in a sagittal section just adjacent to the midline. Within the diencephalon the greatest concentration of receptor sites (**dots**) is in the preoptic–suprachiasmatic area and the arcuate–ventromedial area. These are the areas responsible for controlling the release of luteinizing hormone by the pituitary. In more lateral sections (not shown) the amygdala and orbitofrontal cortex also contain a high concentration of receptor sites. (Adapted from McEwen, 1976.)

change, which enables it to bind to specific DNA recognition elements on genes capable of being activated (or repressed) by steroid hormones. Hormone–receptor complexes bind with high affinity only to specific regions of the DNA, called *hormone-responsive elements*. In this manner, gonadal hormones can activate or inhibit transcription of specific genes, resulting in functional changes in the target cells.

In both sexes the same populations of neurons contain receptors for androgens, estrogens, and progesterone. Cells that express steroid receptors are found in the preoptic area, hypothalamus, amygdala, midbrain, and spinal cord. They are also found in cortical areas: the frontal, prefrontal, and cingulate areas of the primate cerebral cortex. As we shall see later, these cortical receptors may be important for the differentiation of nonreproductive but sexually dimorphic behavioral capacities.

Sexually Differentiated Brains Have Different Physiological Properties and Behavioral Tendencies

How does a brain that mediates male behaviors differ from one that mediates female behaviors? A brain exposed to androgens during the critical period secretes LH from the pituitary at a constant rate. In contrast, the secretory pattern generated by the female brain cycles with a periodicity of days. One explanation for this difference involves neurons in the preoptic area of the hypothalamus that project to the neurosecretory cells that produce LH-releasing hormone (LHRH). This peptide regulates the release of both LH and FSH from the anterior pituitary. Thus, LHRH is also known as gonadotropin-releasing hormone (GnRH). LHRH-producing cells do not express any steroid receptors; however, the preoptic cells from which LHRH cells receive synaptic input do contain estrogen receptors. In normal females the estrogen secreted by growing ovarian follicles activates the preoptic estrogen-sensitive cells, which in turn prompts a surge in the production of LHRH in the postsynaptic neurosecretory cells (Figure 31–5). In the androgenized brain, these preoptic cells do not respond to hormonal activation; even direct electrical stimulation fails to alter release of LH from the pituitary.

Second, there is a significant sex difference in the effect of estrogen on the regulation of the number of progesterone receptors in neurons of the hypothalamic ventromedial (HVM) nucleus, the principal site of hormonal activation of lordosis. Significantly increased numbers of progesterone receptors have been found in the HVM of female rats in response to injections of estradiol. In contrast, the same treatment has no effect on progesterone receptors in the male HVM.

Third, the mature animal with an androgenized brain exhibits male mounting behavior when androgens are administered systemically or implanted directly into the anterior hypothalamus. The same behavior cannot be activated by hormones in males that were castrated during the critical developmental period. However, castrated males can be remasculinized, at least in terms of this behavior, by administering replacement androgens within the critical period. A dramatic example of this effect in rats is shown in Figure 31–6 (broken line).

A fourth and distinctive property of mature androgenized brains is that they show little behavioral response to estrogens. In normal males there is an active suppression of the lordotic response, which in mature female brains may be elicited by estrogen. The display of adult lordosis in response to priming with estrogen is not under genetic controls; it is either established or inhibited by the hormonal environment. This has been demonstrated in male rats in the castration-replacement paradigm shown in Figure 31–6 (solid line): early androgen replacement during the critical period effectively suppresses adult heterotypical behavior in response to estrogen.

Finally, events during the critical period result in strong sex differences in the nonreproductive behavioral repertoires of prepubertal juveniles. Because prepuberty is relatively anhormonal, these sex-specific behavior patterns do not depend on the contemporaneous presence of steroid hormones. Genetically female rhesus monkeys exposed to androgen during the critical period show more rough-and-tumble play, more aggressive encounters with normal males, and less maternal-imitative behaviors than normal females. Also, animals exposed to androgen during the critical period spend more time playing with others who were similarly exposed, regardless of their genetic sex.

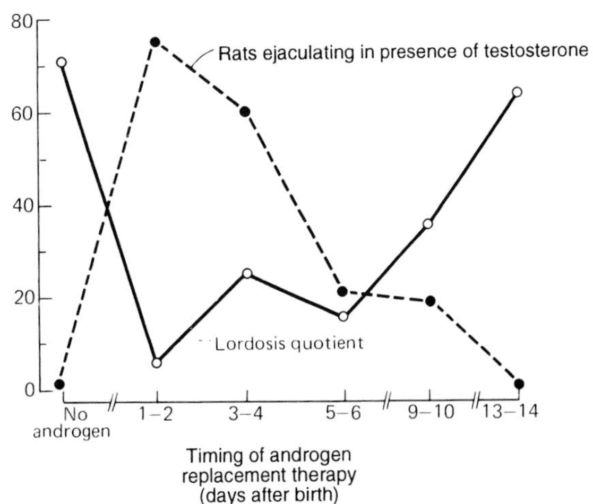

Figure 31–6 The adult sexual behavior of neonatally castrated male rats depends on the age at which testosterone replacement therapy is given. In the experiment presented here, therapy was administered to six groups at different ages. When the androgens were replaced within two days of birth, the castrated males were capable at maturity of homotypical ejaculatory responses in the presence of testosterone (**broken line**). However, as the interval between castration and replacement increases, the remasculinizing effect of early androgen replacement therapy declines and heterotypical behavior increases. Heterotypical behavior was measured in terms of a lordosis quotient: the percentage of mounts by a stud male that elicited lordosis in the castrated males (**solid line**). (Adapted from Beach, Noble, and Orndoff, 1969.)

Perinatal Hormones Also Determine the Degree to Which Sex-Linked Behaviors Are Expressed by Normal Males and Females

Events during the perinatal critical period for sexual differentiation of the nervous system do not result in complete masculinization or feminization. Intermediate degrees are both possible and normal. In humans there is considerable variability in the amounts of testosterone and estrogen to which a developing normal fetus is exposed. Do these perinatal variations affect the degree to which sex-linked behaviors are expressed later in adulthood? Evidence for this possibility has come primarily from studies on rodents, which produce large litters.

Positioning in the uterus is sexually random, but it has sexual consequences. Female mice that develop between two male fetuses have a higher concentration of testosterone in both blood and amniotic fluid than do females that develop between one male and one female or between two other females. The three types of females—defined by intrauterine position as next to two males, one male, or zero males (2M, 1M, 0M)—differ in many characteristics after they are born, including activity, aggressiveness, and acceptability as mating partners to males. Although 2M females reproduce nor-

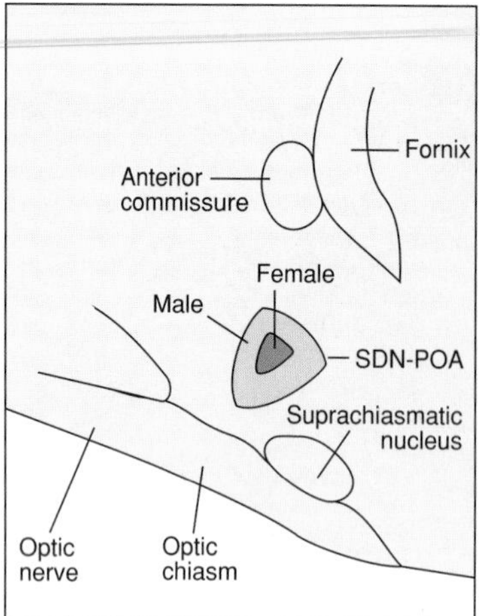

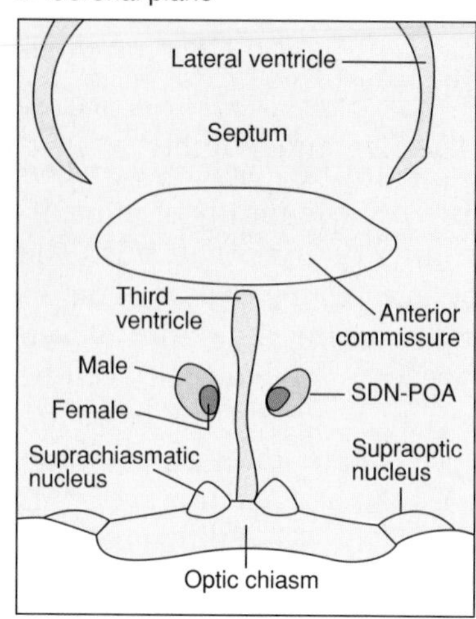

Figure 31–7 The sexually dimorphic nucleus of the preoptic area (SDN-POA) is considerably smaller in female rats than in male rats. (Adapted from Gorski et al., 1978.)

mally, they display erratic estrus cycles, begin to mate later, and cease to bear young earlier than females that develop between two females.

Intrauterine position also has an important effect on certain male characteristics in rodents. The size and weight of the testes of males that develop between two other males are greater than those of males not surrounded by male siblings. The seminal vesicles of males surrounded by two male siblings are more sensitive to testosterone. The dosage of testosterone required to induce aggression in neonatally castrated adult males that developed between two male siblings is lower than that required for similarly castrated adults that developed between two females.

All of the behavioral variations displayed by 2M, 1M, and 0M offspring fall within the normal range of masculine or feminine behaviors. Indeed, together these subgroups define the normal range of behavioral expression in the whole population. The ways in which the 2M, 1M, and 0M offspring differ from each other are the same as those by which the two sexes differ from each other. The degree to which a normal male or female displays a sexually differentiated behavior may be determined by perinatal hormonal mechanisms similar to those that differentiate the two sexes from one another. Individual differences in any behavior with sexually dimorphic components might be due, at least in part, to hormonal exposure during the critical period.

Sexual Differentiation Is Reflected in the Structure of Certain Neurons

Is there a morphological basis for the androgen-dependent process in the developing central nervous system that underlies sex-specific behavioral organization? The size of the cell nucleus in cells in the preoptic area and ventromedial hypothalamus differs in males and females, as does the size of the neuronal processes and synaptic terminals of cells in the arcuate nucleus and the density of dendritic fields in the preoptic area. A sexually dimorphic synaptic organization of afferents to the preoptic area of the hypothalamus has been found. The number of synapses on dendritic spines relative to the number of shafts might represent different patterns of connections in the two sexes. Thus, perinatal exposure to androgen can alter the pattern of connections in the brain as well as the brain's responsiveness.

In mammals the size and number of neurons in a small part of the medial preoptic nucleus in the hypothalamic forebrain region are greater in males than in females. This region is now called the *sexually dimorphic nucleus of the preoptic area* (Figure 31–7). Irreversible sexual differentiation occurs during the perinatal period; in adults the difference is not dependent on the continued presence of gonadal hormones. The size and number of neurons in the sexually dimorphic nucleus increase in the male rat around the time of birth and this continues after birth for 10 days. Although the function of this nucleus is not known, transplantation of the entire preoptic area from newborn males to female littermates enhances both homotypical and heterotypical adult sexual behaviors. Sexually dimorphic areas have also been identified in the preoptic areas of gerbils, ferrets, guinea pigs, hamsters, mice, hyenas, and humans. The volume and number of cells occupied by the sexually dimorphic nucleus decline with age in both males and females (Figure 31–8).

The functions of the morphological sex differences in the superior cervical ganglion, the amygdala, the dorsal hippocampus, and the orbital frontal cortex are unknown, but one example of a sex difference in the spinal cord can be correlated directly with a sexually dimorphic behavior. Testicular androgen released during the critical period produces penile reflexes in the male rat. These behavioral reflexes are dependent on androgen. A discrete cluster of androgen-concentrating motor neurons in the sacral portion of the spinal cord of male rats innervates two striated muscles (the levator ani and bulbocavernosus) that move the penis. In female rats the same muscles are absent or vestigial, as is the corresponding spinal motor nucleus. However, a single, properly timed neonatal injection of testosterone can masculinize the spinal cord of female rats by preventing the death of neurons whose survival depends on the presence of androgens.

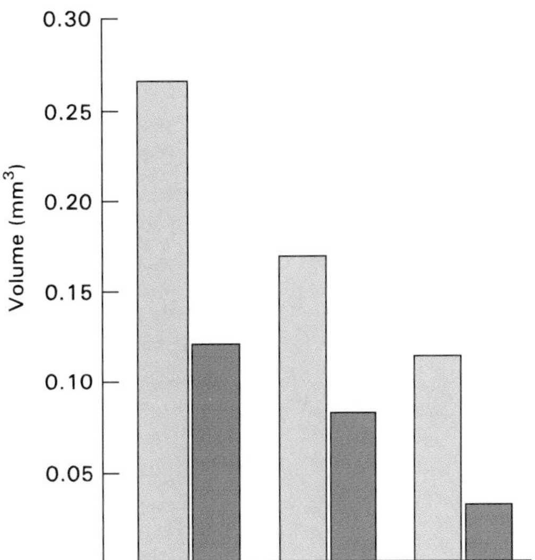

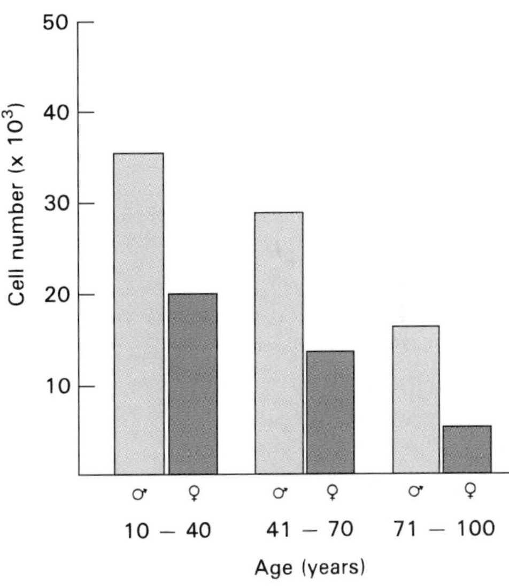

Figure 31–8 In humans the number of cells and the volume of tissue occupied by the sexually dimorphic nucleus decline steadily with age, but at each stage of life the structure is larger in men than in women. (Adapted from Swaab and Fliers, 1985.)

Cellular Mechanisms Involved in the Development of Sex Differences in the Brain Can Be Studied In Vitro

The different neuronal organizations imposed on the brains of males and females by sex steroids may result from alterations in the growth rate of axons and dendrites of select steroid-sensitive cells. During the critical period, steroid sex hormones might bias the rate of axonal differentiation in different regional populations and thereby affect neural circuitry. Since postsynaptic space is limited, the resulting competition for postsynaptic sites between populations of axons of different origins could produce sex differences in neural circuitry.

Morphological sex differences may reflect the growth-promoting effects of gonadal steroids on specific populations of neurons. Slices of the hypothalamus from newborn mouse brains maintained in culture exhibit a marked increase in the outgrowth of neurites (new axons and dendrites) as well as extensive new branching of existing processes after exposure to an androgenizing agent (Figure 31–9). The stimulation of neuritic growth by androgen is dose dependent.

A Wide Range of Behavior Is Influenced by Sex Differences in the Organization of the Brain

We have so far focused on the dimorphic reproductive behaviors of males and females as the primary behavioral markers of the sexual identity of the brain. Several recent lines of evidence, however, suggest that the repertory of behaviors influenced by the perinatal hormonal environment may extend beyond reproductive behaviors.

Cognitive Development in Monkeys Is Sexually Dimorphic

The development of the frontal lobes in rhesus monkeys is sexually dimorphic. In both infant and adult male monkeys, lesions of the orbital prefrontal cortex result in impaired performance in tests involving spatial discrimination and delayed responses. In contrast, identical lesions in infant females do not induce similar deficits until the animal has reached an age of 15–18 months. Thus, the effects of orbital prefrontal lesions are age dependent, and the age at which this part of the cortex becomes involved in spatial learning differs between the sexes. The earlier participation of the masculine frontal cortex in the learning of tasks that assess spatial discrimination may also be related to the later superiority of mature male monkeys in these learning tasks. Prenatal treatment of developing female monkey fetuses with androgens eliminates this sex difference in adults.

The frontal cortex of the monkey is sexually dimorphic in its rate of development. This may be related to the presence of steroid-sensitive neurons in the frontal cortex of infant rats, which decline in number by puberty. If differences exist in the rate of cortical maturation in nonhuman primates, it is plausible that similar sex differences might occur in the brains of humans.

Human Cerebral Asymmetry Is Sexually Dimorphic

The specialization of cognitive functions in the left and right cerebral hemispheres of the mature human brain is described in Chapters 1 and 19. In brief, in most right-handed individuals the left hemisphere is specialized for language and related serial processing of information, and the right hemisphere is specialized for nonverbal processes, including three-dimensional visualization, mental rotation, face recognition, and understanding the meaning of facial expressions. Several lines of evidence suggest that the patterns of cerebral asymmetry may differ in the two sexes. As with the development of prefrontal cortical functions in the monkey, there is evidence that the two sexes differ in the rate of maturation of cognitive functions in the two hemispheres.

The relative participation of the two hemispheres can be assessed in a test of spatial processing. Children of various ages are given 10 seconds to manipulate (out of view) two objects differing in shape, using only the index and middle fingers of one hand. The children then try to identify these objects from pictures. Since tactile shape discrimination by adults depends mainly on the right hemisphere, the test is made as dependent as possible on the right hemisphere, using objects with meaningless shapes, not readily labeled. As early as the age of 6, boys perform in a manner consistent with right-hemisphere specialization (left-hand superiority). Girls show no evidence of bilateral representation (no clear hand superiority) until the age of 13, suggesting that boys develop a greater hemispheric specialization at an earlier age. Therefore, during an extended period of development a sex difference may exist in the hemispheric allocation of cognitive functions.

If the right hemisphere in girls is not specialized for a particular cognitive function, then it may retain greater plasticity for a longer period than in boys. Clinical impressions are consistent with this idea. Language functions appear to transfer more readily to the right hemisphere in females than in males after damage to the left hemisphere in childhood. The extended plasticity of the young female

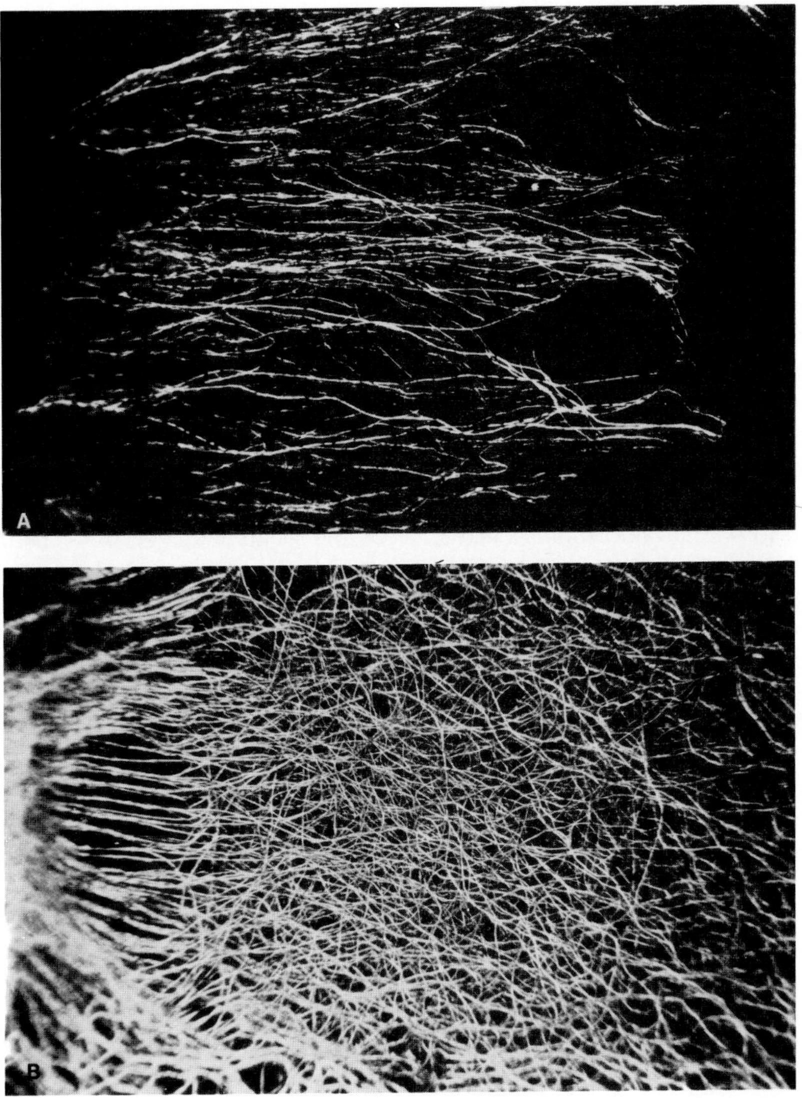

Figure 31–9 Gonadal steroids stimulate the growth of neurites out of slices of the hypothalamus from a newborn mouse. (From Toran-Allerand, 1978.)

A. The control culture shows silver-impregnated neurites coursing outward in hairlike wisps from the margin of the explant.

B. In a slice treated with estradiol (10 ng/mL) neuritic growth is much denser and extends further than that of the control.

brain also suggests that females may have a lower incidence of developmental disorders associated with left-hemisphere dysfunction. Developmental aphasia and infantile autism are more frequent in males, and language deficits are prominent symptoms in all of these syndromes.

In addition, the degree of cerebral cognitive asymmetry differs in adult males and females. In male neurological patients there is a strong association between the side of the brain that has been injured and the type of cognitive deficits observed: verbal functions are disordered by left-hemisphere lesions, and nonverbal functions by right-hemisphere lesions. In female neurological patients this association is much weaker, suggesting that the adult female brain is functionally less asymmetrical than the male brain.

These sex differences in the susceptibility of the developing human brain to early damage and in the cerebral asymmetry of the mature brain have not yet been related to perinatal hormonal events. However, there is increasing evidence for sex differences in cognitive styles that are not simply attributable to differences in social experiences during early development but may reflect fundamental differences in the brain. These differences are not reflected in overall intellectual functions, such as those measured in tests of I.Q.; rather, they reflect differences in specific intellectual skills. Thus, on average women perform better than men on tests of verbal fluency, on tests of perceptual speed (such as the time required to recognize a face), in arithmetic calculations, and in the precision of performing manual tasks. By contrast, men do better on tests that explore spatial relationships, mathematical reasoning, and target-directed motor skills.

An Overall View

It is often tempting to view any intrinsic biological process, such as the perinatal differentiation of the nervous system, as a fixed and permanent constraint on behavior. Nevertheless, most behaviors remain flexible and open to modification, although strongly shaped by neural organization. Humans have relatively few fixed-action patterns. As an example, even though we inherit a finely tuned regulatory system for food consumption and body weight, certain experiences can override the homeostat for body weight and produce obesity in otherwise normal people—a person may develop a passionate interest in fine food and wines or get a job as a restaurant critic. Likewise, although the brain regulates aggressive behavior, individuals can, without apparent neuropathology, become pacifists or terrorists for ideological reasons.

Similarly, there is ample social evidence that the neural organization of reproductive behaviors, while importantly influenced by hormonal events during a critical prenatal period, does not exert an immutable influence over adult sexual behavior or even over an individual's sexual orientation. Within the life of the individual, religious, social, or psychological motives can prompt biologically similar persons to diverge widely in their sexual activities.

Selected Readings

Arnold, A. P., Bottjer, S. W., Nordeen, E. J., Nordeen, K. W., and Sengelaub, D. R. 1987. Hormones and critical periods in behavioral and neural development. In J. P. Rauschecker and P. Marler (eds.), Imprinting and Cortical Plasticity: Comparative Aspects of Sensitive Periods. New York: Wiley, pp. 55–97.

Beato, M. 1989. Gene regulation by steroid hormones. Cell 56:335–344.

Blaustein, J. D., and Olster, D. H. 1989. Gonadal steroid hormone receptors and social behaviors. In J. Balthazart (ed.), Advances in Comparative and Environmental Physiology, Vol. 3: Molecular and Cellular Basis of Social Behavior in Vertebrates. Berlin: Springer, pp. 31–104.

Breedlove, S. M. 1986. Cellular analyses of hormone influence on motoneuronal development and function. J. Neurobiol. 17:157–176.

Byne, W. 1994. The biological evidence challenged. Sci. Am. 270(5):50–55.

Eicher, E. M., and Washburn, L. L. 1986. Genetic control of primary sex determination in mice. Annu. Rev. Genet. 20:327–360.

Evans, R. M. 1988. The steroid and thyroid hormone receptor superfamily. Science 240:889–895.

Gilbert, S. F. 1991. Sex determination. In, Developmental Biology, 3rd ed. Sunderland, Mass.: Sinauer.

Gorski, R. A. 1988. Sexual differentiation of the brain: Mechanisms and implications for neuroscience. In S. S. Easter, Jr., K. F. Barald, and B. M. Carlson (eds.), From Message to Mind. Sunderland, Mass.: Sinauer, pp. 256–271.

Hines, M. 1982. Prenatal gonadal hormones and sex differences in human behavior. Psychol. Bull. 92:56–80.

Kelley, D. B. 1988. Sexually dimorphic behaviors. Annu. Rev. Neurosci. 11:225–251.

Kimura, D. 1992. Sex differences in the brain. Sci. Am. 267(3):118–125.

Knobil, E., and Neill, J. D. (eds.) 1988. The Physiology of Reproduction, 2nd ed. 2 vols. New York: Raven Press.

Koopman, P., Gubbay, J., Vivian, N., Goodfellow, P., and Lovell-Badge, R. 1991. Male development of chromasomally female mice transgenic for *Sry*. Nature. 351:171–121.

LeVay, S. 1993. The Sexual Brain. Cambridge, Mass.: MIT Press.

LeVay, S. and Hamer, D. H. 1994. Evidence for a biological influence in male homosexuality. Sci. Am. 270(5):44–49.

McEwen, B. S., Luine, V. N., and Fischette, C. T. 1988. Developmental actions of hormones: From receptors to function. In S. S. Easter, Jr., K. F. Barald, and B. M. Carlson (eds.), From Message to Mind: Directions in Developmental Neurobiology. Sunderland, Mass.: Sinauer, pp. 272–287.

McLaren, A. 1990. What makes a man a man? Nature 346:216–217.

Pfaff, D. W. and McEwen, B. S. 1983. Actions of estrogens and progestins on nerve cells. Science 219:808–814.

Toran-Allerand, C. D. 1984. On the genesis of sexual differentiation of the central nervous system: Morphogenetic consequences of steroidal exposure and possible role of α-fetoprotein. In G. J. De Vries, J. P. C. De Bruin, H. B. M. Uylings, and M. A. Corner (eds.), Sex Differences in the Brain: The Relation Between

32

Emotional States

A Theory of Emotion Must Explain the Relationship of Cognitive and Physiological States

The Hypothalamus Is a Critical Subcortical Structure in the Regulation of Emotion

The Autonomic Nervous System Participates in Emotional States

The Hypothalamus Plays a Major Role in Controlling the Output of the Autonomic Nervous System

The Hypothalamus Controls the Endocrine System

Manifestations of Emotional States Can Be Selectively Elicited by Stimulating the Hypothalamus

The Search for Cortical and Subcortical Representation of Emotions Has Led to the Amygdala

The Amygdala Is the Part of the Limbic System Most Specifically Concerned with Emotion

The Basolateral Complex Receives Most Sensory Inputs to the Amygdala

The Central Nucleus of the Amygdala Projects to Cortical Areas Concerned with the Representation of Emotion

Pleasure, elation, euphoria, ecstasy, sadness, despondency, depression, fear, anxiety, anger, hostility, and calm—these and other emotions contribute to the richness of our personal life and imbue our actions with passion and character. Although emotions vary and involve many bodily processes, as yet there is no precise scientific definition of the term *emotion.* In everyday language we use the term to refer to our feelings and moods and the way these are expressed both in our overt behavior and in the responses of our bodies. Emotion, like perception and action, is controlled by distinct neuronal circuits within the brain. Many drugs that affect the mind—ranging from addictive street drugs to therapeutic agents—exert their actions by affecting these circuits.

What parts of the brain are involved in emotion? Since we experience emotions consciously, there clearly is an important cognitive element in emotions, an element that most likely involves the cerebral cortex. At the same time, emotions are accompanied by autonomic, endocrine, and skeletomotor responses that depend on subcortical parts of the nervous system: the amygdala, the hypothalamus, and the brain stem. These peripheral responses serve to prepare the body for action and, in addition, communicate emotional states to other people. When frightened, for example, we experience increased heart rate and respiration, dryness of the mouth, tense muscles, and sweaty palms, all of which are regulated in the brain at levels below the cortex. To understand an emotion such as fear, therefore, we need to understand not only how the emotion is represented in the cortex but also how subcortical structures regulate the autonomic and endocrine activities that mediate the peripheral components of emotion.

In this chapter we examine how emotion is represented in the brain. Since emotional experience is integrated with endocrine and autonomic function, we also consider how the the amygdala, the hypothalamus, and the brain stem organize appropriate autonomic and endocrine functions.

A Theory of Emotion Must Explain the Relationship of Cognitive and Physiological States

The development and expression of an emotion have traditionally been thought to consist of several components. There is first the recognition of an important event—seeing your house on fire. This in turn produces a conscious emotional experience in the cerebral cortex—fear—that mediates outgoing signals to peripheral structures including the heart, blood vessels, adrenal glands, and sweat glands. According to this view, the burning house is first recognized as a potential danger, and this cognitive act initiates reflexive autonomic responses in the body.

At the turn of the century the American philosopher William James and the Danish psychologist Karl Lange proposed an alternative hypothesis—that the conscious experience we call emotion occurs *after* the cortex receives signals about changes in our physiological state. Thus, James wrote: "We feel sorry because we cry, angry because we strike, afraid because we tremble and not that we cry, strike or tremble because we are sorry, angry or fearful as the case may be." According to the James–Lange theory, emotions are preceded by certain physiological changes—an increase or decrease in blood pressure, heart rate, and muscular tension. Thus, when you see a fire, you feel afraid because your cortex has received signals about your racing heart, knocking knees, and sweaty palms.

According to this view, emotions are cognitive responses to information coming from the periphery, experienced in a manner akin to the way we perceive thought. Our everyday experience confirms that information from the body contributes to emotional experience. Furthermore, patients in whom the spinal cord has been accidentally severed experience a significant reduction in the intensity of emotion. This reduction is correlated with the spinal level at which the transection occurs: the higher the transection, the more reduced the emotional response.

However, other experiments reveal that this theory explains only one aspect of emotional behavior. For example, one may continue to be emotionally aroused even after the physiological correlates are no longer present. A person can sustain emotion, thoughts, and actions long after a threat has abated. Were physiological feedback the only controlling factor, the emotions should not outlast the physiological change. Conversely, some emotions have a very rapid onset, much more rapid than changes in bodily states. Thus, there may be more to emotions than the interpretation of information from the periphery.

Whereas the James–Lange theory focuses on the initiating role of peripheral signals in emotional experience, Walter B. Cannon and Philip Bard formulated a theory of emotions in which subcortical structures have a key role in mediating emotions. Cannon and Bard observed integrated emotional responses in cats in whom the cerebral cortex had been removed. But such responses were lost when the hypothalamus was removed. These findings led Cannon and Bard to suggest that subcortical structures, the hypothalamus and thalamus, have a double function: they provide the coordinated motor commands that regulate the peripheral signs of emotion and they provide the cortex with the information required for the cognitive perception of emotions.

In recent years emotional behavior has increasingly been viewed as an outcome of the *interaction* of peripheral and central factors. One important contribution to this view has come from Antonio Damasio based on studies of patients with damage to either the amygdala or the prefrontal cortex. Another contribution has come from Stanley Schachter, who in the 1960s elaborated on the James–Lange theory by emphasizing that the cortex actually constructs emotion—much like it does vision—out of often ambiguous signals it receives from the periphery.

According to the James–Lange hypothesis, emotional experience is the direct consequence of information arriving from the periphery in the cerebral cortex. Instead of this simple relation, Schachter proposed an active role for the cortex in transforming peripheral signals. Specifically, he suggested that the cortex creates a cognitive response to peripheral information that is consistent with the individual's expectations and social context. In one study Schachter injected volunteers with epinephrine; some subjects were informed of the side effects (e.g., pounding heart), others were not. All of the subjects were then exposed to either annoying or amusing conditions. When the emotional responses of the subjects were assessed, the subjects who had been warned about the side effects of epinephrine exhibited less anger or euphoria. Schachter interpreted this experiment as indicating that the informed subjects attributed their arousal to the drug, whereas the other group perceived their arousal as an emotional response, as either strong anger or euphoria depending on the conditions.

In the James–Lange–Schachter–Damasio theory the experience of emotion is essentially a story that the brain concocts to explain the bodily reactions. This theory called by Damasio the *somatic marker hypothesis*, is appealing because it accounts for the fact that the same autonomic responses can accompany different emotions (Table 32–1).

Table 32–1 Comparison of Measures Typically Used to Index Fear in Animals and Those in the DSM-III Used to Index Generalized Anxiety in People[a]

Measures of Fear in Animal Models	*DSM-III Criteria—Generalized Anxiety*
Increased heart rate	Heart pounding
Decreased salivation	Dry mouth
Stomach ulcers	Upset stomach
Respiration change	Increased respiration
Scanning and vigilance	Scanning and vigilance
Increased startle	Jumpiness, easy startle
Urination	Frequent urination
Defecation	Diarrhea
Grooming	Fidgeting
Freezing	Apprehensive expectation—something bad is going to happen

[a]*Diagnostic and Statistical Manual of Mental Disorders*, 3rd ed. From Davis (1992).

The Hypothalamus Is a Critical Subcortical Structure in the Regulation of Emotion

How is the peripheral component of emotional states regulated? How is it coordinated with the conscious perception of emotion? The regulation of the peripheral component was first addressed in 1878 by the French physiologist Claude Bernard, who pointed out that the internal environment (*milieux interieur*) of the body is highly regulated so as to be maintained in a stable, narrow range. This *constancy* is the result of homeostatic mechanisms that limit the variability of body state. Cannon, who introduced the term *homeostasis* into behavioral biology in 1929, found together with Bard that the key neuronal mechanisms for maintaining homeostasis are located in the hypothalamus and its two effector systems: the autonomic system and the endocrine system.

Although the hypothalamus constitutes less than 1% of the total volume of the human brain, it contains many of the neuronal circuits that regulate the vital functions that vary with emotional states: temperature, heart rate, blood pressure, and water and food intake. As we shall see later, the hypothalamus also controls the pituitary gland and thereby regulates the endocrine system. Deviations from homeostasis elicit hypothalamic mechanisms that contribute toward rectifying the imbalance. At the same time, however, potentially life-threatening events that do not directly involve homeostatic imbalance also engage hypothalamic mechanisms. As a result, the same autonomic and endocrine mechanisms that maintain homeostasis may also disrupt it, albeit transiently. Thus, a rapid increase in blood pressure may permit you to run faster, and, as long as it is not maintained too long, the increased blood pressure need not be unhealthy.

Since an important feature of all theories of emotion is the involvement of the autonomic nervous system and endocrine systems, we need to examine each of these in turn.

The Autonomic Nervous System Participates in Emotional States

The physiological changes that accompany emotional states typically involve sweating, dryness of the mouth, tenseness in the stomach, rapid respiration, fast heartbeat, and muscle tensions. Most of

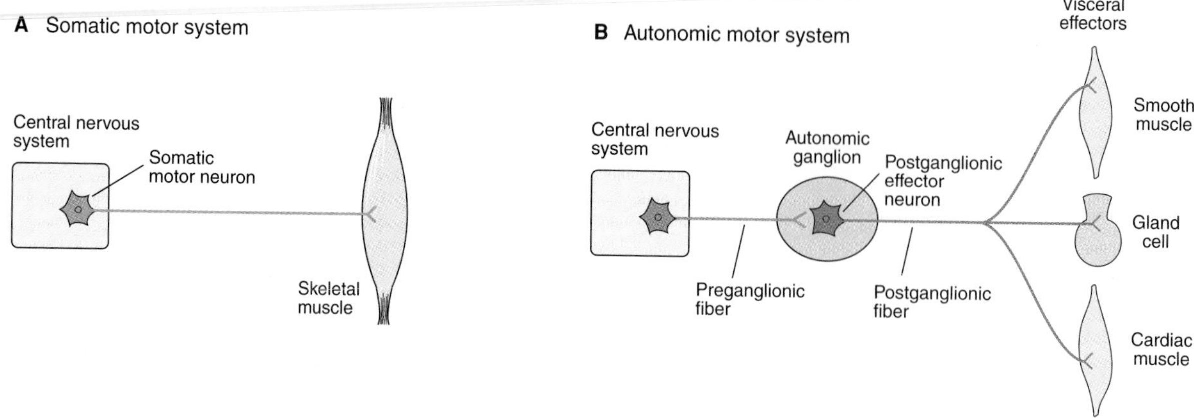

Figure 32–1 Comparison of the anatomical organization of the somatic and autonomic motor systems.

A. All somatic motor neurons are located within the central nervous system, either within cranial motor nuclei or in the ventral horn of the spinal cord. The efferent pathway to skeletal muscle is monosynaptic—the central motor neurons project directly to skeletal muscle.

B. All autonomic motor neurons are located in discrete ganglia that lie outside the central nervous system. These autonomic motor neurons (also called postganglionic neurons) are activated by columns of preganglionic neurons within brain stem nuclei and the spinal cord.

these are mediated by the autonomic nervous system.

The autonomic nervous system is primarily an effector system. In its functions—controlling smooth muscle, heart muscle, and exocrine glands—it is distinct from the somatic motor system, which controls skeletal muscle. The principles underlying the anatomical organization of both the somatic and the autonomic motor systems are similar (Figure 32–1), and the two systems function in parallel to adjust the body to environmental changes.

Nevertheless, the two systems differ in an important way. Most movements initiated by the somatic motor system are under voluntary control, whereas most autonomic adjustments are reflexive. Indeed, the term *autonomic* means self-regulating. Moreover, autonomic adjustments are not normally accessible to consciousness. For this reason the autonomic nervous system is often called the *involuntary motor system*, in contrast to the *voluntary (somatic) motor system*. However, the differences in voluntary control between the two motor systems are relative and not absolute. Some somatic movements, particularly reflex responses, are involuntary. Similarly, some autonomic adjustments (e.g., those involved in regulating blood pressure) can be brought partly under voluntary control with practice.

The autonomic nervous system has three major divisions: sympathetic, parasympathetic, and enteric. We focus here on only the first two divisions (Figure 32–2). Cannon proposed that these two divisions have different functions in regulating emotional behavior and homeostasis: the sympathetic division governs the *fight-or-flight* reaction, whereas the parasympathetic system is responsible for *rest and digest*. In an emergency the body needs to respond to a sudden change in its external or internal environment, be it combat, athletic competition, severe change in temperature, or blood loss. To respond rapidly to the external environment, as in fear during combat, the hypothalamus and the sympathetic nervous system activate an increased sympathetic outflow to the heart and to other viscera, the peripheral vasculature, and sweat glands, as well as to piloerector and ocular muscles.

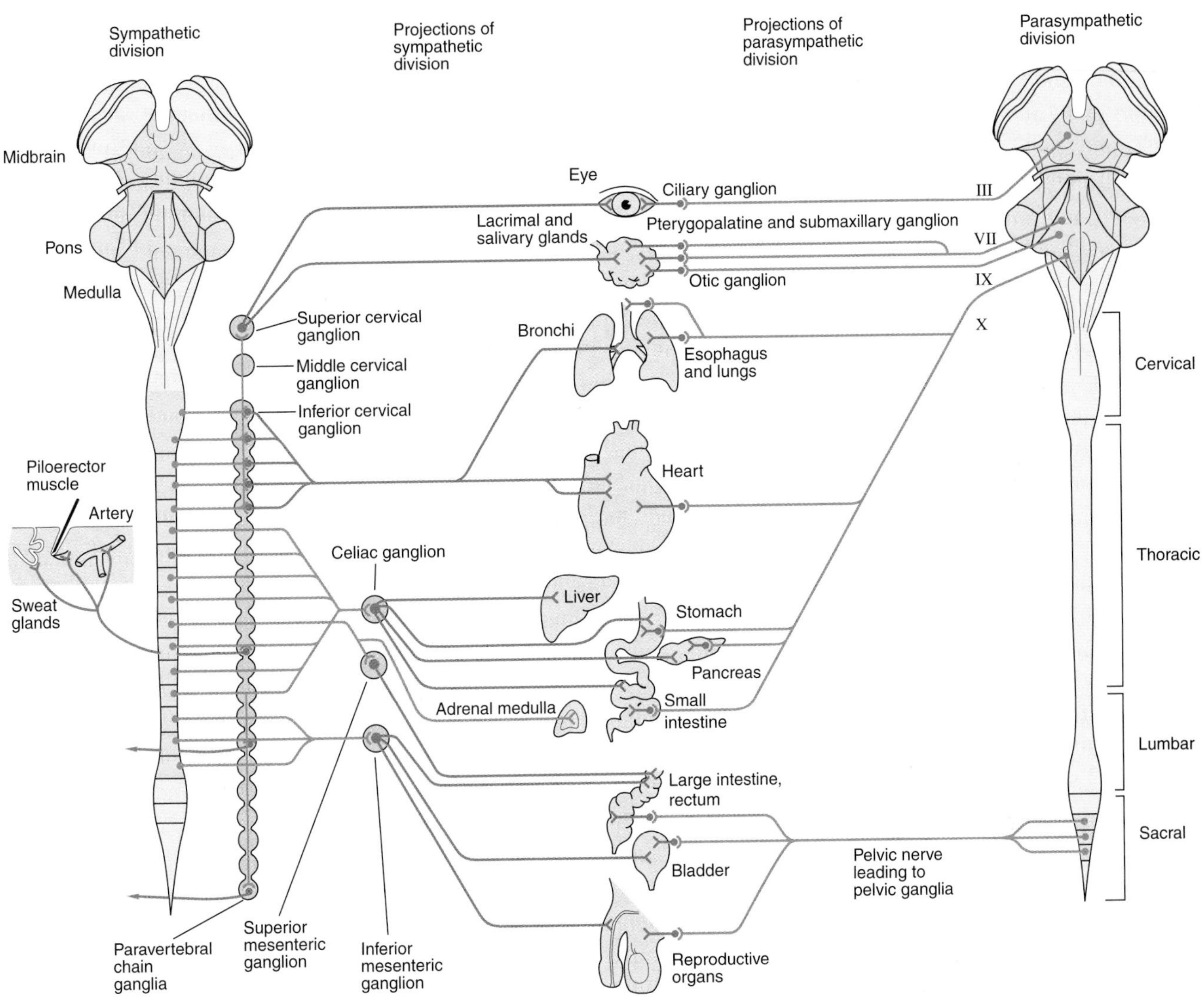

Figure 32–2 The autonomic nervous system has sympathetic and parasympathetic divisions. Preganglionic neurons of the sympathetic division extend from the first thoracic spinal segment to lower lumbar segments. Parasympathetic preganglionic neurons are located within the brain stem and in segments S2 to S4 of the spinal cord. This figure also illustrates the coordinated innervation of selected targets by these two divisions of the autonomic nervous system and the innervation of targets in the skin by sympathetic postganglionic neurons in paravertebral chain ganglia (innervation arises at all levels but only two projections are shown for simplicity).

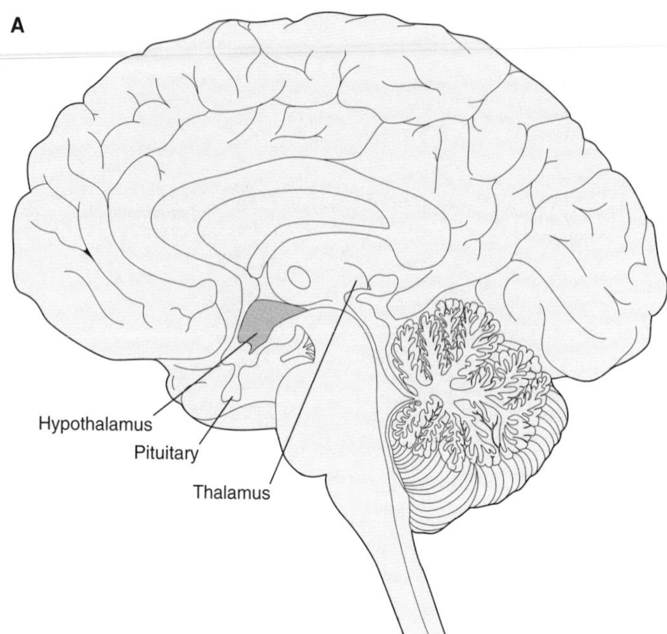

Figure 32–3 The location and structure of the hypothalamus. (Adapted from Nieuwenhuys, Voogd, and van Huijzen, 1981.)

A. Medial view of the brain showing the relationship of the hypothalamus to the pituitary and thalamus.

The increased cardiac output, altered body temperature and blood glucose, and pupillary constriction permit rapid responses to potentially disturbing external conditions. In contrast, the parasympathetic system maintains basal heart rate, respiration, and metabolism under normal conditions.

An animal whose sympathetic nervous system is experimentally eliminated can survive as long as it is sheltered, kept warm, and not exposed to stress. Such an animal cannot , however, carry out strenuous work or fend for itself. It does not mobilize on demand blood sugar from the liver, and it does not react to cold with the normal vasoconstriction or elevation of body hair.

The autonomic nervous system is not recruited only for emergency or restorative purposes. Many sympathetic and parasympathetic pathways are tonically active and operate in conjunction with the somatic motor system to regulate normal behavior and to maintain a stable internal environment in the face of changing external conditions.

The Hypothalamus Plays a Major Role in Controlling the Output of the Autonomic Nervous System

The output of the autonomic nervous system is influenced by many regions of the brain, in particular the cerebral cortex, the amygdala, and parts of the reticular formation. Most of these brain regions produce their actions on the autonomic nervous system through the hypothalamus (Figure 32–3), which in turn integrates the information from these structures into a coherent response.

The hypothalamus acts on the autonomic nervous system by modulating visceral reflex circuitry that is basically organized at the level of the brain stem. It does so in two ways. First, it projects to three important regions in the brain stem and spinal cord.

1. The hypothalamus projects to the *nucleus of the solitary tract*, the major recipient of sensory inflow from the viscera. This nucleus in turn acts on the nucleus of the vagus nerve and other parasympathetic neurons in the brainstem con-

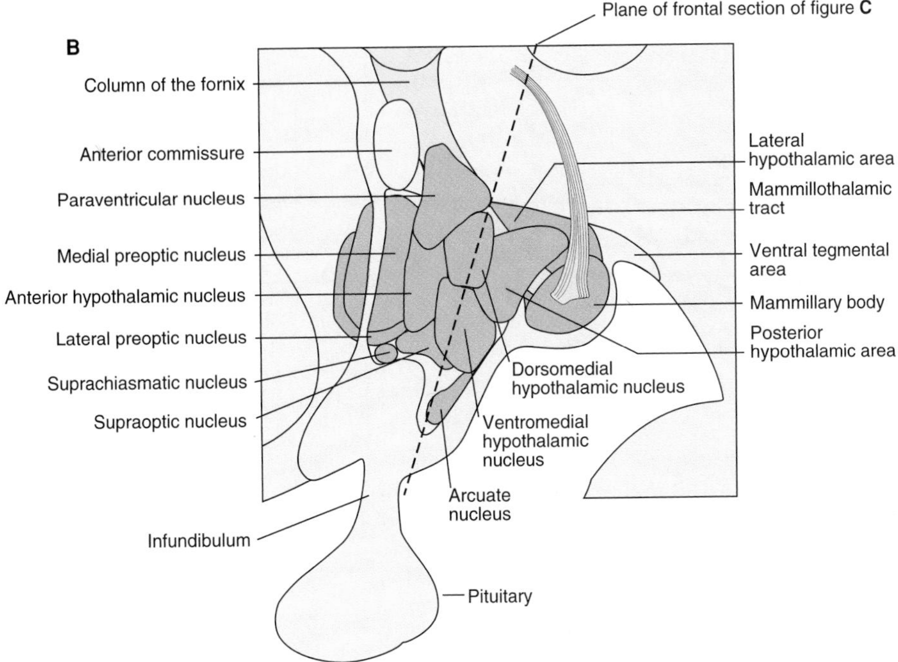

B. Medial view showing positions of the main hypothalamic nuclei. Some nuclei are visible only in the frontal view in **C**.

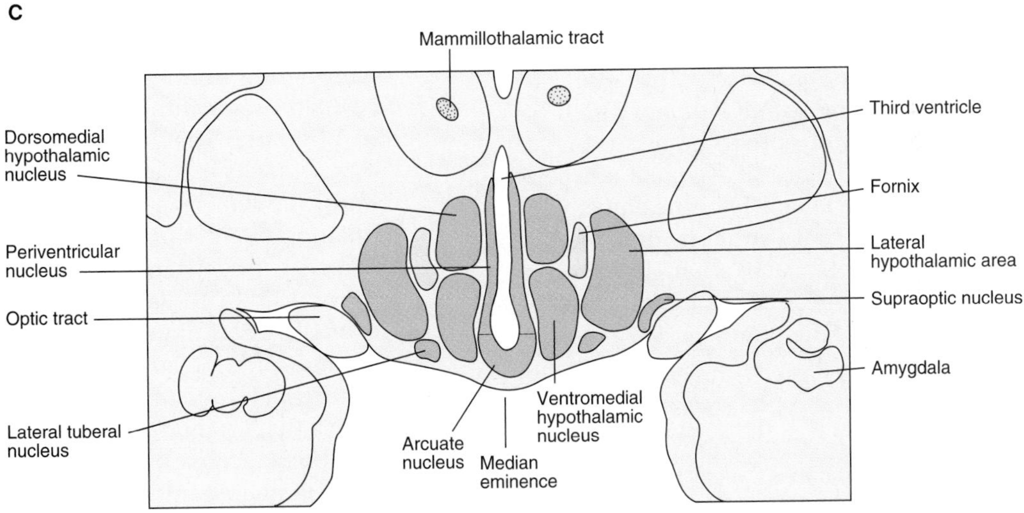

C. Frontal view of the hypothalamus (section along plane shown in **B**).

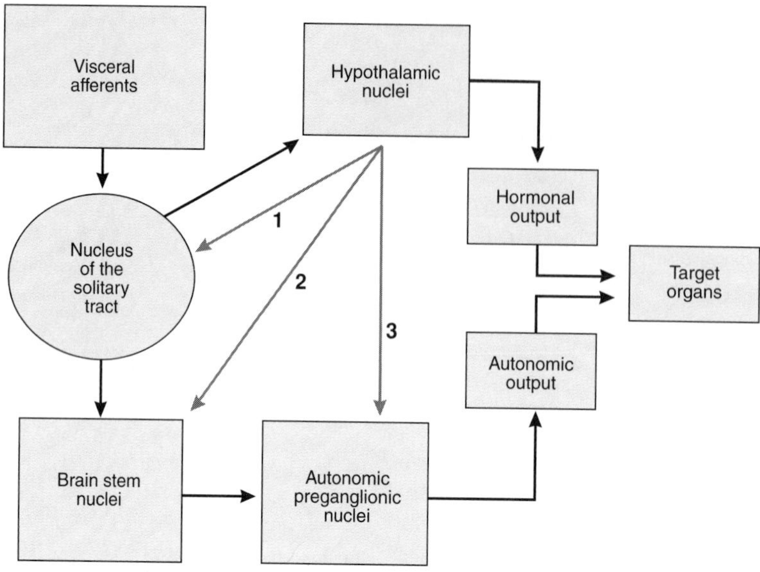

Figure 32–4 Afferent visceral information can be processed either for immediate reflex responses or as a part of a more elaborate central autonomic circuit that affects hormonal functions and overt behavior. (Adapted from Loewy and Spyer, 1990.)

trol temperature, heart rate, blood pressure, and respiration (Figure 32–4).

2. The hypothalamus projects to the brain stem regions in the *rostral ventral medulla*, which control the preganglionic output important for sympathetic function. Thus, as we shall see below, stimulation of the lateral hypothalamus, which connects to the rostral ventral medulla, leads to general sympathetic activation—piloerection, increase in blood pressure and heart rate, sweating, and dilation of the pupils.
3. The hypothalamus projects directly to the autonomic outflow of the spinal cord.

Second, the hypothalamus acts on the endocrine system to release hormones that influence autonomic function (Figure 32–4).

The Hypothalamus Controls the Endocrine System

Control of the endocrine system is accomplished *directly*, by secretion of neuroendocrine products into the general circulation from the posterior section of the pituitary gland, and *indirectly*, by secretion of regulating hormones into the local portal plexus, which drains into the blood vessels of the anterior pituitary. The hypothalamic regulating hormones, which can be either releasing or inhibiting, control the synthesis and release into the general circulation of anterior pituitary hormones.

Our current understanding of the endocrine function of the hypothalamus is based on the analysis of the direct and indirect types of control by Ernst and Berta Scharrer and Geoffrey Harris. The Scharrers developed the concept of *neurosecretion*, the idea that certain neurons function as neuroendocrine transducers capable of directly converting electrical neural information into hormonal information (Figure 32–5). Harris recognized the importance of the blood supply that connects the pituitary to the hypothalamus (the pituitary–hypophyseal–portal system) and showed that this indirect vascular link carries hormonal information from the hypothalamus to the pituitary (Figure 32–6). These two ideas form the basis of modern neuroendocrinology.

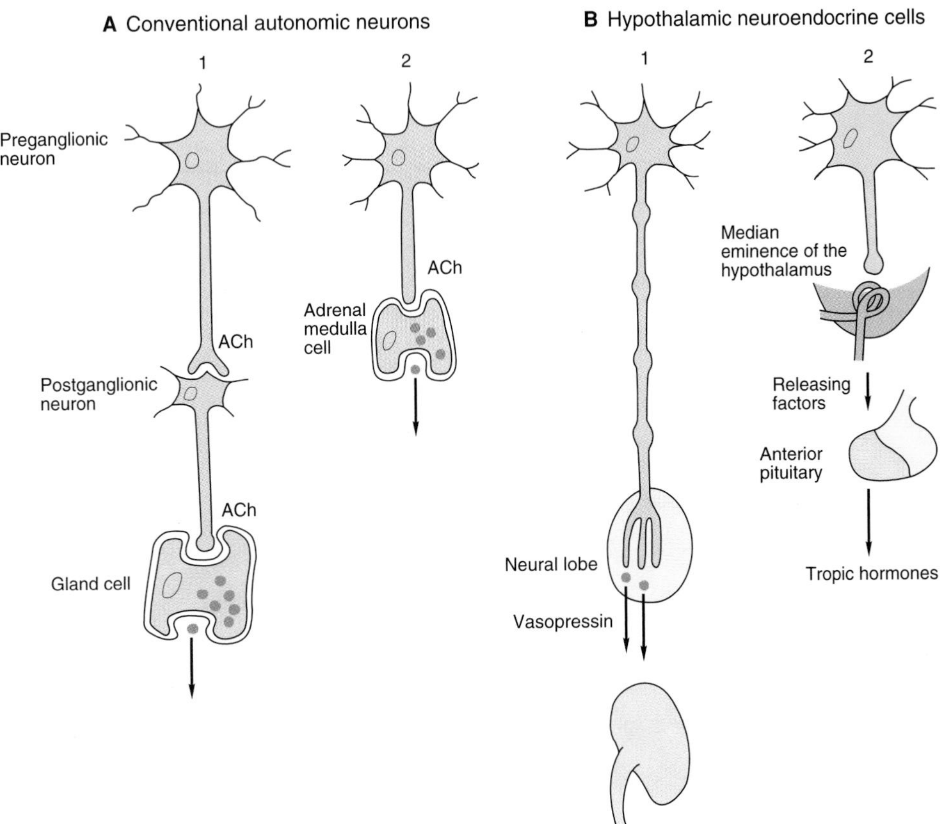

Figure 32–5 Neuroendocrine cells release their own secretory products into the bloodstream. (Adapted from Reichlin, 1978.)

A. Conventional autonomic neurons stimulate secretion from exocrine glands through the synaptic action of acetylcholine (ACh). **1.** Exocrine glands are innervated by postganglionic neurons. **2.** The adrenal medulla is innervated directly by sympathetic preganglionic neurons.

B. In the pituitary gland (hypophysis) neuroendocrine cells synthesize the secretory product or precursor peptide in the cell body and transport it by axoplasmic flow to release sites. **1.** In the *neurohypophyseal system* the seceretory product (a hormone, either vasopressin or oxytocin) is transported to terminals in the posterior lobe of the pituitary (neurohypophysis). The hormones are released into the general circulation when the nuerons are activated.
2. In the *adenohypophyseal system* the secretory product (a releasing or regulating factor) is transported to terminals in the median eminence (and, in some species, the pituitary stalk). Activity in these neurons causes the secretion of hormone releasing factors into the hypophyseal–portal circulation, and these factors in turn cause or inhibit the release of hormones from the anterior lobe of the pituitary (adenohypophysis).

Each type of endocrine control—direct and indirect—is mediated by a distinct class of neuroendocrine neurons. In both classes of neurons the neurohormone or precursor peptide is synthesized in the cell body and packaged in neurosecretory vesicles that are transported down the axon to the terminal, where they are stored and released by secretion when the neuron is stimulated. Some of the larger (magnocellular) peptidergic neurons release the hormone oxytocin and others release vasopressin into the general circulation by way of the posterior pituitary. The smaller (parvocellular) peptidergic neurons secrete peptides into the hypophyseal portal system. These peptides can stimulate or inhibit secretions from the anterior pituitary gland (Figure 32–7).

In fact, most hormones of the anterior pituitary are controlled indirectly, through their vascular link, by peptide neurohormones synthesized by parvocellular neurons. These releasing hormones are thyrotropin-releasing hormone (Figure 32–7), gonadotropin-releasing hormone, somatostatin (Figure 32–7), corticotropin-releasing hormone (CRH), also called corticotropin-releasing factor (CRF), and growth hormone-releasing hormone (GRH).

Release of most of the hormones of the anterior pituitary is regulated by *antagonistic* enhancing and inhibiting substances. For example, the release of growth hormone is stimulated by growth hormone-releasing hormone and inhibited by somatostatin. The known hypothalamic releasing and inhibiting hormones and the anterior pituitary hormones they affect are listed in Table 32–2.

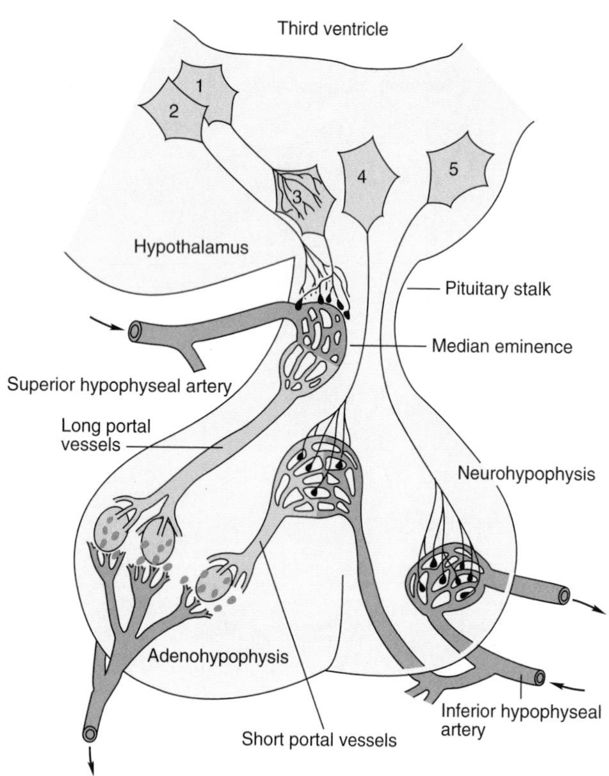

Figure 32–6 Various functional elements participate in the control of the pituitary by the hypothalamus. Peptidergic neurons (**5**) release oxytocin or vasopressin into the general circulation via the posterior pituitary. Two general types of neurons are involved in regulation of the anterior pituitary. Peptidergic neurons (**3**,**4**) form the releasing hormones that enter the hypophyseal–portal vessels. The second type of neuron is the link between the rest of the brain and the peptidergic neuron. These neurons, some of which are monoaminergic, are believed to end on the cell body of the peptidergic neuron in a conventional manner (**1**), or to end on the axon terminal of the peptidergic neuron (**2**) by means of axo-axonic synapses. (Adapted from Reichlin, 1978; Gay, 1972.)

Table 32–2 Hypothalamic Secretions that Release or Inhibit the Release of Anterior Pituitary Hormones

Hypothalamic Substance	*Target Anterior Pituitary Hormone*
Releasing	
Thyrotropin-releasing hormone (TRH)	Thyrotropin, prolactin
Corticotropin-releasing hormone (CRH)	Adrenocorticotropin, β-lipotropin
Gonadotropin-releasing hormone (GnRH)	Luteinizing hormone (LH), follicle-stimulating hormone (FSH)
Growth hormone-releasing hormone (GHRH or GRH)	Growth hormone (GH)
Prolactin-releasing factor (PRF)	Prolactin
Inhibiting	
Prolactin release-inhibiting hormone (PIH), dopamine	Prolactin
Growth hormone release-inhibiting hormone (GIH or GHRIH; somatostatin)	Growth hormone (GH), thyrotropin
Melanocyte-stimulating hormone release-inhibiting factor (MIF)	Melanocyte-stimulating hormone (MSH)

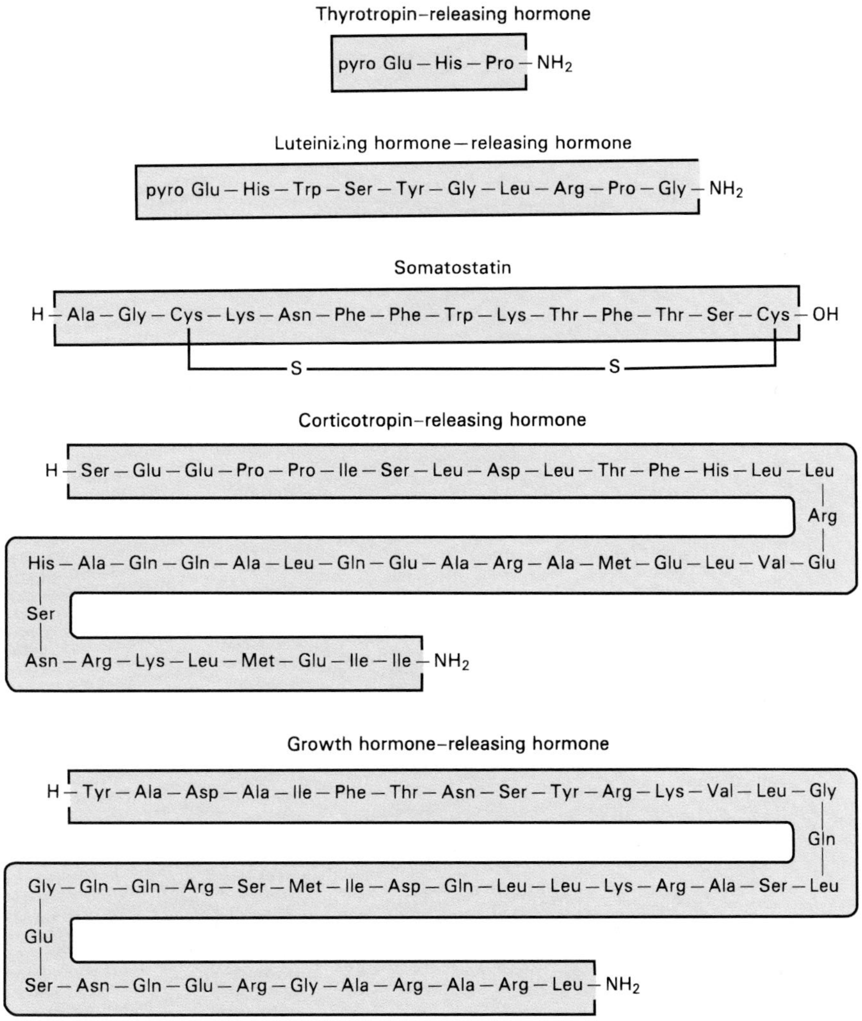

Thyrotropin–releasing hormone

pyro Glu — His — Pro — NH_2

Luteinizing hormone—releasing hormone

pyro Glu — His — Trp — Ser — Tyr — Gly — Leu — Arg — Pro — Gly — NH_2

Somatostatin

H — Ala — Gly — Cys — Lys — Asn — Phe — Phe — Trp — Lys — Thr — Phe — Thr — Ser — Cys — OH
S —— S

Corticotropin–releasing hormone

H — Ser — Glu — Glu — Pro — Pro — Ile — Ser — Leu — Asp — Leu — Thr — Phe — His — Leu — Leu
Arg
His — Ala — Gln — Gln — Ala — Leu — Gln — Glu — Ala — Arg — Ala — Met — Glu — Leu — Val — Glu
Ser
Asn — Arg — Lys — Leu — Met — Glu — Ile — Ile — NH_2

Growth hormone–releasing hormone

H — Tyr — Ala — Asp — Ala — Ile — Phe — Thr — Asn — Ser — Tyr — Arg — Lys — Val — Leu — Gly
Gln
Gly — Gln — Gln — Arg — Ser — Met — Ile — Asp — Gln — Leu — Leu — Lys — Arg — Ala — Ser — Leu
Glu
Ser — Asn — Gln — Glu — Arg — Gly — Ala — Arg — Ala — Arg — Leu — NH_2

Figure 32–7 Structures of some hypothalamic releasing and inhibiting hormones.

There is evidence that at least one inhibiting hormone is not a peptide: prolactin release is inhibited by dopamine. Many releasing hormones affect more than one pituitary hormone.

Manifestations of Emotional States Can Be Selectively Elicited by Stimulating the Hypothalamus

In 1932 Stephen Ranson stimulated different regions of the hypothalamus of anesthetized animals using stereotaxic methods that permit precise and reproducible placement of electrodes in the deep structures of the brain. He evoked a variety of autonomic reactions, including alterations in heart rate, blood pressure, and gastro-intestinal motility, as well as erection of hairs and bladder contraction.

In the 1940s Walter Hess extended Ranson's approach to awake, unanesthetized animals and found that different parts of the hypothalamus produced characteristic constellations of responses that appeared to reflect organized behaviors. For example, electrical stimulation of the lateral hypo-

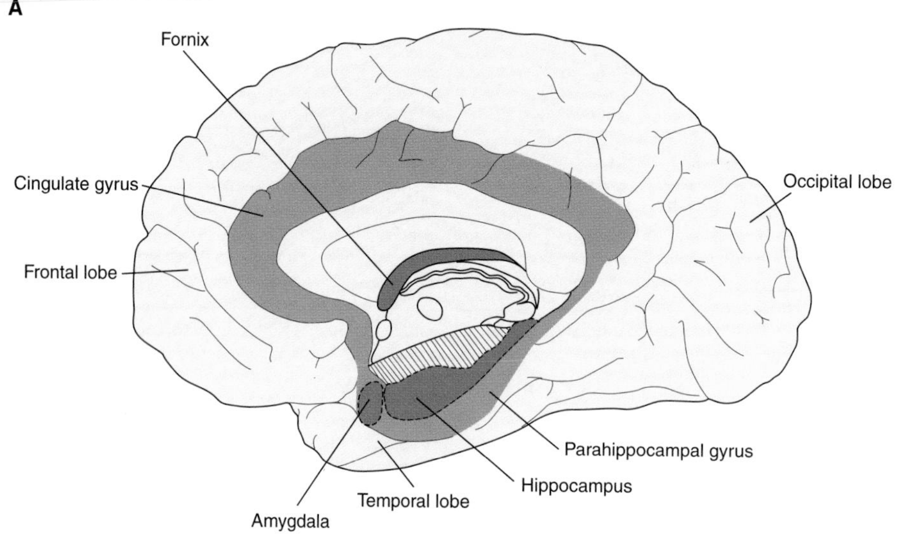

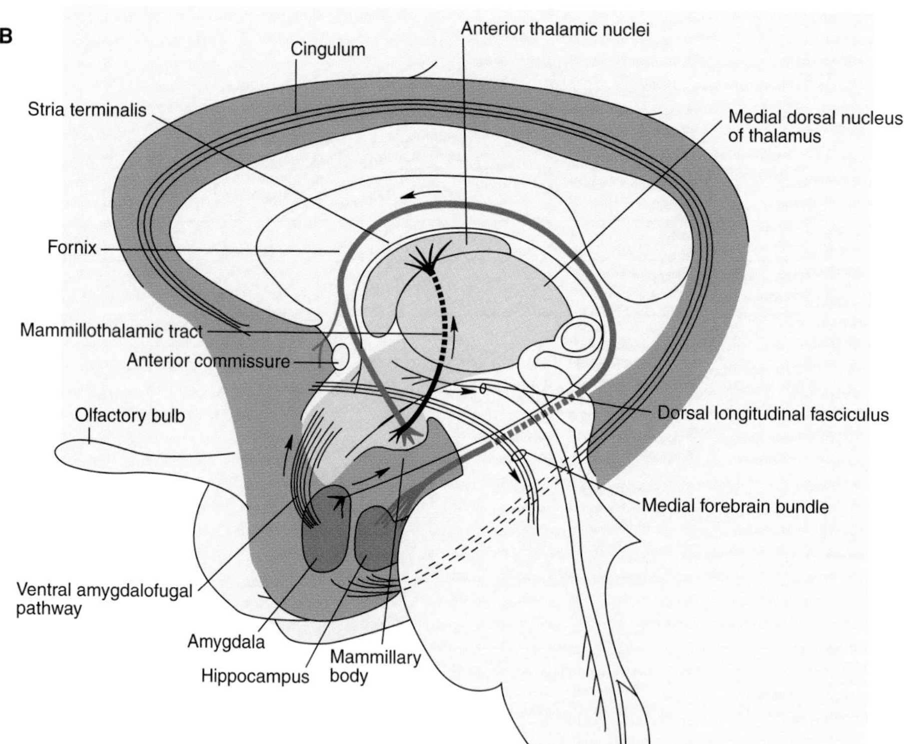

Figure 32–8 The limbic system consists of the limbic lobe and deep-lying structures. (Adapted from Nieuwenhuys, Voogd, and van Huijzen, 1988.)

A. This medial view of the brain shows the limbic lobe, which consists of primitive cortical tissue (**lighter blue area**) that encircles the upper brain stem. The underlying structures (hippocampus and amygdala) are not visible from this medial view.

B. Interconnections of the deep-lying structures included in the limbic system. The predominant direction of neural activity in each tract is indicated by an **arrow**, but activity in the tracts is typically bidirectional.

thalamus in cats elicits autonomic and somatic responses characteristic of anger: increased blood pressure, raising of the body hair, pupillary constriction, arching of the back, and raising of the tail.

These observations provided the first evidence that the hypothalamus is not only a motor command region, for the autonomic nervous system, but also, as envisaged by Canon and Bard, a coordinating center that integrates various inputs into a well-organized set of autonomic and somatic responses. This idea is supported by lesion studies that associate different hypothalamic structures with a wide range of emotional states.

The Search for Cortical and Subcortical Representation of Emotions Has Led to the Amygdala

Emotional stimuli activate sensory pathways that trigger the hypothalamus to modulate heart rate, blood pressure, and respiration. These observations support the James–Lange–Schachter theory. Physiological inputs to the hypothalamus act on the brain stem and autonomic nervous system. In turn, this information projects to the cerebral cortex from the peripheral organs, whose homeostatic state has been disturbed and gives rise to the conscious perception of emotion. Where is the cortical representation of emotion?

In 1937 James Papez proposed that the cortical substratum for emotions is a ring of phylogenetically primitive cortex around the brain stem, a region called the *limbic lobe* by Paul Broca. The limbic lobe comprises (1) the parahippocampal gyrus, which is the anterior and inferior continuation of the cingulate gyrus (Figure 32–8), and (2) the underlying cortex of the hippocampal formation, which is morphologically simpler than the overlying cortex. The hippocampal formation includes the hippocampus proper, the dentate gyrus, and the subiculum.

Papez argued that, since the hypothalamus communicates reciprocally with higher cortical centers, cognition and emotion affect each other reciprocally. Evidence for this idea comes from the clinical observation that patients with encephalitis due to the rabies virus—which characteristically attacks the hippocampus—show profound changes in emotional state, including bouts of terror and rage. Papez proposed that the neocortex influences the hypothalamus through connections of the cingulate gyrus to the hippocampal formation. According to this idea, the hippocampal formation processes information from the cingulate gyrus and projects it to the mammillary bodies of the hypothalamus by way of the fornix (one of the output pathways from the hippocampus, see Figure 32–8). In turn, the hypothalamus provides information to the cingulate gyrus by a pathway from the mammillary bodies to the anterior thalamic nuclei (through the mammillothalamic tract) and from the anterior thalamic nuclei to the cingulate gyrus (Figure 32–9).

A major breakthrough in the search for the cortical representation of emotion occurred in 1939,

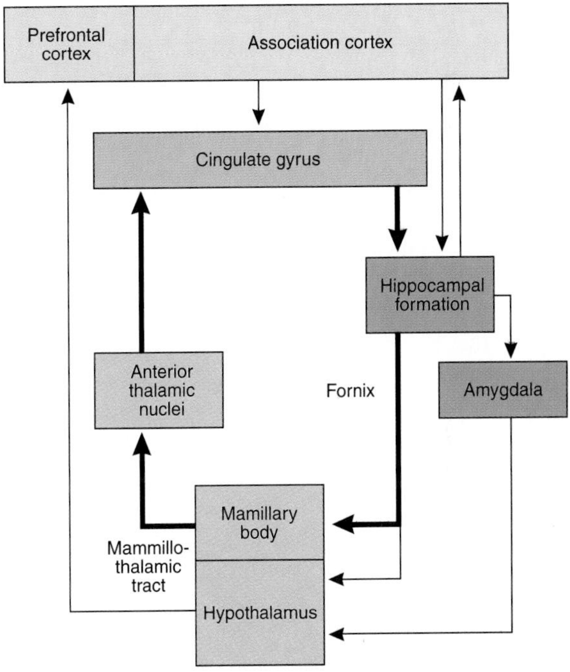

Figure 32–9 A proposed neural circuit for emotion. The circuit originally proposed by James Papez is indicated by **thick lines**; more recently described connections appear in **fine lines**. The more elaborate circuit based on Papez's initial idea includes known projections of the hippocampus through the fornix to hypothalamic regions (mammillary bodies and other hypothalamic areas) and of the hypothalamus to the prefrontal cortex. The ventromedial portion of the prefrontal cortex in particular is important for emotion. A pathway interconnects the amygdala to the hypothalamus and indirectly to the prefrontal cortex. Finally, reciprocal connections between the hippocampal formation and the association cortex are indicated.

when Heinrich Klüver and Paul Bucy found that bilateral removal of the temporal lobe in monkeys (including the amygdala and the hippocampal formation) produces a dramatic behavioral syndrome. The monkeys, formerly quite wild, became tame, showed a flattening of emotions, and became highly oral, putting all kinds of objects, even frightening ones like snakes, into their mouths. They also exhibited an enormous increase in sexual behavior. Finally, the animals showed a compulsive tendency to take note of and react to every visual stimulus, yet failed to recognize familiar objects.

The Amygdala Is the Part of the Limbic System Most Specifically Concerned with Emotion

What structures in the limbic system account for the individual symptoms of the Klüver–Bucy syndrome? Under the influence of Papez's ideas the emotional components of the syndrome were for many years ascribed to the entire limbic system. Only recently has it become clear that only a part of the limbic system plays a role in the Klüver–Bucy syndrome or in emotion generally. The visual deficits in the Klüver–Bucy syndrome are mostly due to damage to the visual association areas of the temporal cortex. The hippocampus, the mammillary bodies, and anterior thalamic nuclei appear to be more closely related to cognitive memory storage. The structure central to the emotional components of the Klüver–Bucy syndrome is the amygdala. In humans, electrical stimulation of the amygdala produces feelings of fear and apprehension. In experimental animals that damage to the amygdala produces oral tendencies, hypersexuality, and tameness. Many of the effects of the amygdala on emotional states are mediated through the hypothalamus and the autonomic nervous system.

Thus lesions and electrical stimulation of the amygdala affect a variety of autonomic responses, emotional behaviors, and feeding. These effects are often similar to those associated with stimulation and lesioning of the lateral or medial regions of the hypothalamus. The key to understanding the role of the amygdala is an appreciation of its input and output structures: the basolateral and central nuclei.

The amygdala is composed of many nuclei that are reciprocally connected to the hypothalamus, hippocampal formation, neocortex, and thalamus. The *basolateral nuclei* of the amygdala receive important afferent information from all sensory modalities. In turn, the *central nucleus* of the amygdala gives rise to two major efferent projections: the stria terminalis and the ventral amygdalofugal pathway (Figure 32–9). The *stria terminalis* innervates the hypothalamus as well as the bed nucleus of the stria terminalis and the nucleus accumbens. The *ventral amygdalofugal pathway* provides input to the brain stem, dorsal medial nucleus of the thalamus, and rostral cingulate gyrus.

The Basolateral Complex Receives Most Sensory Inputs to the Amygdala

The major target of the extensive sensory input to the amygdala is the *basolateral complex* (the basolateral nuclei) (Figure 32–10). Sensory input comes from two sources: (1) the sensory nuclei in the thalamus, and (2) the primary sensory areas of the cortex. For many types of emotion, particularly fear, the projection from the thalamus to the amygdala is especially important, and information mediated by this projection reaches the amygdala more rapidly than does the sensory input from the cortex. Thus, this direct thalamic input may mediate short-latency, primitive emotional responses and prepare the amygdala for the reception of more sophisticated information about the congnitive representation of emotion from higher centers, such as the ventromedial prefrontal cortex. The output of the amygdala, as well as afferent input that is triggered by activity of autonomic effectors, feeds back to cortical structures to give rise to a conscious emotional experience. As a result, the projection from the thalamus to the amygdala may allow primitive sensory representation to activate the amygdala rapidly, an activation that may be important in situations of danger.

There is considerable evidence that the sensory inflow for various learned emotional states, particularly fear and anxiety, comes through the basolateral complex. For example, lesions in the basolateral complex abolish classical conditioning of fear (see Chapter 34). In fear conditioning, an initially neutral stimulus that does not evoke autonomic

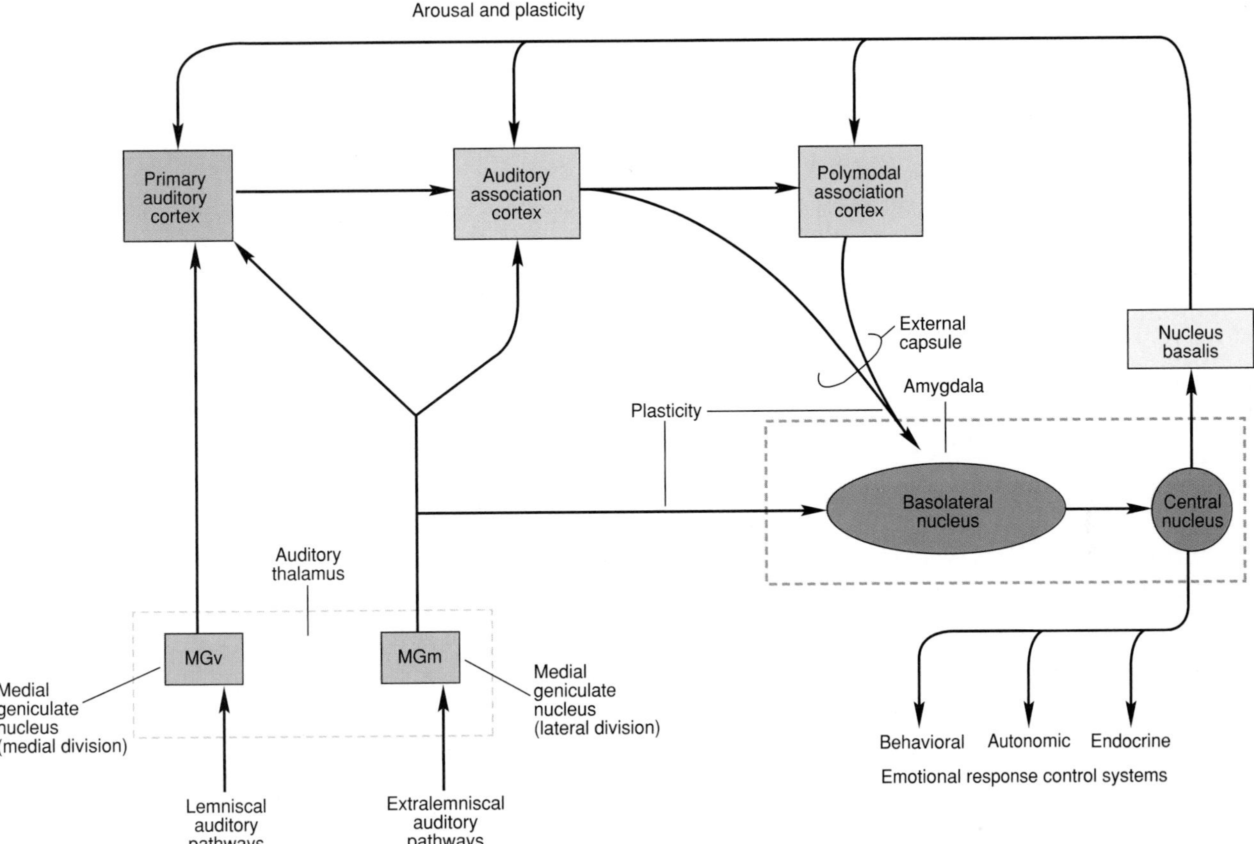

Figure 32–10 Some of the pathways involved in the processing of emotional information. Auditory (and other sensory) information is transmitted to the medial division of the medial geniculate nucleus (MGv) in the auditory thalamus. Other auditory pathways transmit to other parts of the auditory thalamus: the lateral division of the medial geniculate nucleus (MGm) and the posterior intrathalamus nucleus (not shown). The medial division of the medial geniculate nucleus projects only to the primary auditory cortex, whereas the lateral division projects to both primary and association areas of the auditory cortex, as well as to the basolateral nucleus of the amygdala. These pathways from the thalamus to the amygdala have been implicated in emotional learning. The anterior nucleus (not shown) projects widely to cortical areas and to the central nucleus of the amygdala. The central nucleus of the amygdala has extensive connections with brain stem areas involved in the control of emotional responses. It also projects to the nucleus basalis, which projects widely to cortical areas. The pathway from the nucleus basalis to the cortex uses acetylcholine (ACh) as a neurotransmitter. Cholinergic transmission to the cortex from the nucleus basalis has been implicated in cortical arousal and plasticity. (Adapted from LeDoux, 1992.)

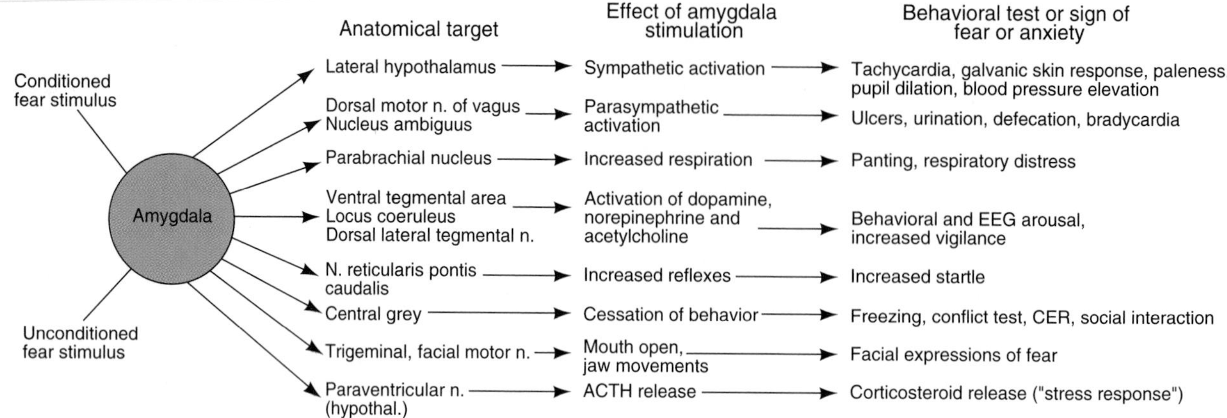

Figure 32–11 The central nucleus of the amygdala has direct connections with a variety of hypothalamic and brain stem target areas that may be involved in different animal tests of fear and anxiety. (EEG, electroencephalographic; N, nerve; ACTH, adrenocorticotropin; CER, conditioned emotional response) (From Davis, 1992.)

responses, such as a tone, is paired with an electric shock, say to the foot, which produces pain, fear, and autonomic responses. After several pairings, the tone itself elicits fearful reactions, such as freezing in place or changes in heart rate or blood pressure (Figure 32–11). This pattern of responses to a once neutral tone resembles those evident in human anxiety states. It is significant that long-lasting alterations of synaptic efficacy (long-term potentiation, see Chapter 36) can be demonstrated in the amygdala following stimulation of thalamic or cortical pathways.

Lesions of the amygdala or local infusion of anxiety-reducing drugs block not only learned but also innate responses to fearful stimuli. For example, these ablations markedly reduce emotionality in wild rats. Thus, the amygdala is not only important for learned fear; it also is critical for innate, unconditioned fear. In fact, lesions of the amygdala also block the memory for emotionally significant sensory stimuli.

In addition to its role in fear or *negative emotional reactions* the amygdala also mediates pleasure or *appetitive emotional reactions*. Thus, when a neural discriminative stimulus such as a tone is paired with a positive reinforcing stimulus such as food, the tone becomes associated with both the rewarding attributes of the food (such as its taste) and the attributes of the food that are not rewarding (such as its visual appearance). Lesions of the basolateral complex leave intact the association between the tone and the nonrewarding aspects (appearance) of the food, but destroy the association of the tone with rewarding attributes (the taste) of the food. This is consistent with the finding in the Klüver–Bucy syndrome that animals with lesions of the amygdala frequently take inedible (nonrewarding) objects into their mouths.

Finally, the amygdala is required for the conditioning of an organism to the environment (or context) in which it lives (*place conditioning*). The survival of an organism depends on behaviors that maximize contact with biologically safe environments and minimize contact with dangerous environments. Many of these dangers are subject to modification through experience. Such a modification can be demonstrated experimentally with *place preference conditioning*, whereby an animal learns to increase its contact with environments in which it has previously encountered positively reinforcing stimuli (sex, food, drink), and minimize contact with environments that are aversive or dangerous. The positive preferences for place can be conditioned to food or sexual partners, but also to the effects of drugs such as stimulants.

Primary rewarding stimuli, ranging from simple rewards such as food to complex ones such as sexual partners, are associated with the constellation of stimuli that make up the distinctive place in which the reward is obtained. As a result, these stimuli later take on positive values and increase the likelihood that the animal will again seek out this place, even in the absence of the primary reward. Place or contextual cues presumably gain positive properties by means of classical conditioning, of which we shall learn more in Chapter 35. There is considerable evidence that the amygdala, particularly its basolateral part, is fundamentally involved in associating place stimuli with reward value.

The Central Nucleus Projects to Cortical Areas Concerned with the Representation of Emotions

From the input zone of the amygdala in the basolateral complex, information flows to the central nucleus of the amygdala, the major output from the amygdala. The central nucleus projects to the lateral hypothalamic and brain stem regions that regulate the autonomic responses to emotionally charged stimuli (Figure 32–11). In addition, the central nucleus projects directly and indirectly (via the bed nucleus of the stria terminalis) to the paraventricular nucleus of the hypothalamus, which may be important in mediating neuroendocrine responses to fearful and stressful stimuli. Thus, electrical stimulation at the central nucleus produces increases in heart rate, blood pressure, and respiration as are seen during fear conditioning. Conversely, lesions of this nucleus block the autonomic changes that occur with fear conditioning.

The central nucleus also projects to cortical association areas, especially the orbitofrontal cortex and the cingulate gyrus. This pathway is important for the conscious perception of emotion. In particular, the central nucleus plays an important role in regulating the state of arousal and the accompanying changes in heart rate. Projections from the central nucleus of the amygdala to the nucleus basalis are thought to mediate these changes in arousal (see Figure 32–10).

Lesions of the orbitofrontal or anterior cingulate cortex result in altered emotional responsiveness. If the lesion is limited to the orbitofrontal cortex, the normal aggressiveness and emotional responsiveness of primates are reduced, and lesioned animals sometimes fail to show rage and anger when they do not receive expected rewards in a training task. Furthermore, electrical stimulation of the orbitofrontal cortex produces many autonomic responses (increases in arterial blood pressure, dilation of the pupils, salivation, and inhibition of gastrointestinal contractions), suggesting that this area is involved in generalized arousal. Finally, lesions that include limbic association cortex also reduce chronic intractable pain, suggesting still another effect of the limbic cortex on emotional behavior.

In 1935 John Fulton and Carlyle Jacobsen reported that removing the frontal cortex (*lobotomy*) had a calming effect in chimpanzees. Within a few months of Fulton and Jacobsen's report, Egas Moniz, a Portuguese neuropsychiatrist, performed the first prefrontal lobotomy. He cut the limbic association connections in humans, thereby isolating the orbitofrontal cortex in an attempt to treat the emotional impairment that often accompanies severe mental illness.

The early results of frontal lobotomy appeared favorable. Many patients seemed to show reduced anxiety. However, results from later, more controlled studies, led to the abandonment of this procedure, in part because lobotomy was associated with a high incidence of complications, including the development of epilepsy and abnormal personality changes, such as a lack of inhibition or a lack of initiative and drive. In addition, the advent of effective psychotherapeutic drugs made unnecessary the need for radical surgical intervention.

An Overall View

The interplay between the amygdala, the hypothalamus, the brain stem, and the autonomic system, on the one hand, and the amygdala and frontal and limbic cortex, on the other hand, results in experiences that we describe as emotional. The behavior of patients in whom the prefrontal cortex has been removed supports this idea. These patients are no longer disturbed by chronic pain. When they do perceive pain and exhibit appropriate autonomic reactions, the pain is no longer associated with a powerful emotional experience.

Thus, noxious or pleasurable stimuli have dual effects. First, these stimuli cause the amygdala to trigger autonomic and endocrine responses that are integrated by the hypothalamus and that alter the internal state, thus preparing the organism for attack, flight, sexual experience, or other adaptive behaviors. These internal reactions are relatively simple to execute and require no conscious control. Once the animal interacts with its external environment, however, a second set of mechanisms involving the cortex come into play. These modulate the animal's behavior much as proprioceptive feedback from an uneven terrain modulates the central program for locomotion.

Selected Readings

Cannon, W. B. 1927. The James–Lange theory of emotions: A critical examination and an alternative theory. Am. J. Psychol. 39:106–124.

Cannon, W. B. 1932. The Wisdom of the Body. New York: Norton.

Damasio, A. R. 1994. Descartes' Error. Emotion, Reason and the Human Brain. New York: Plenum.

Davis, M. 1992. The role of the amygdala in fear and anxiety. Annu. Rev. Neurosci. 15:353–375.

Gallagher, M., and Holland, P. C. 1992. Understanding the function of the central nucleus: Is simple conditioning enough? In J. Aggleton (ed.), The Amygdala: Neurobiological Aspects of Emotion, Memory, and Mental Dysfunction. New York: Wiley–Liss, pp. 307–321.

Guillemin, R. 1978. Control of adenohypophysial functions by peptides of the central nervous system. Harvey Lect. 71:71–131.

Hess, W. R. 1954. Diencephalon: Autonomic and Extrapyramidal Functions. New York: Grune & Stratton.

Hohman, G. W. 1966. Some effects of spinal cord lesions on experienced emotional feelings. Psychophysiology 3:143–156.

LeDoux, J. E. 1992. Brain mechanisms of emotion and emotional learning. Curr. Opin. Neurobiol. 2:191–197.

LeDoux, J. E. 1994. Emotion, memory and the brain. Sci. Am. 270(6):50–57.

Loewy, A. D., and Spyer, K. M. (eds.) 1990. Central Regulation of Autonomic Functions. New York: Oxford University Press.

Papez, J. W. 1937. A proposed mechanism of emotion. Arch. Neurol. Psychiatry 38:725–743.

Ranson, S. W. 1934. The hypothalamus: Its significance for visceral innervation and emotional expression. Trans. Coll. Physicians Phila. [Ser. 4]2:222–242.

Schachter, S. 1964. The interaction of cognitive and physiological determinants of emotional state. In L. Berkowitz (ed.), Advances in Experimental Social Psychology, Vol. I. New York: Academic Press, pp. 49–80.

Silverman, A.-J., and Zimmerman, E. A. 1983. Magnocellular neurosecretory system. Annu. Rev. Neurosci. 6:357–380.

Swanson, L. W., and Sawchenko, P. E. 1983. Hypothalamic integration: Organization of the paraventricular and supraoptic nuclei. Annu. Rev. Neurosci. 6:269–324.

33

Motivation

Motivation Is an Inferred Internal State Postulated to Explain the Variability of Behavioral Responses

Homeostatic Processes Such as Temperature Regulation, Feeding, and Thirst Correspond to Motivational States

Temperature Regulation Involves Integration of Autonomic, Endocrine, and Skeletomotor Responses

Feeding Behavior Is Regulated by a Great Variety of Mechanisms

- *Body Weight Is Regulated Around a Set Point*
- *Dual Controlling Elements Are Involved in the Control of Food Intake*
- *Chemical Stimulation of the Hypothalamus Alters Feeding Behavior*
- *Feeding Behavior Is Regulated by Short-Term and Long-Term Cues*

Thirst Is Regulated by Tissue Osmolality and Vascular Volume

Motivational States Can Be Regulated by Factors Other Than Tissue Needs

- *Ecological Constraints May Shape Motivated Behaviors*
- *Behavior May Anticipate Physiological Events*
- *Pleasure Is an Important but Poorly Understood Factor in Motivating Behavior*

Intracranial Stimulation Can Simulate Motivational States and Reinforce Behavior

The Mesolimbic Dopamine Pathways Important for Reinforcement Are Also Acted on by Some Drugs of Abuse

We have previously considered the cognitive aspects of behavior—what a person knows—their sensations, perceptions, imagery, feelings, and actions. Behavior, however, also reflects what a person needs or wants. We have said little in earlier chapters about *why* people do what they do. That is the domain of motivation, an area of biology concerned with internal needs rather than with the processing of sensory information.

The motivational state of an organism is thought to be controlled in an important way by basic homeostatic regulatory processes essential for survival, such as feeding, respiration, sex, temperature regulation, and self-protection. Motivation varies as a function of deprivation—for example, deprivation of food (hunger), deprivation of water (thirst), or sexual deprivation. Alterations of these motivational states are therefore produced by changes in the internal condition of the animal relative to a given set point of the regulatory process.

As hunger, thirst, and sex imply, several of the most basic issues that surround internal motivational states are related to survival. Activities that enhance immediate survival such as eating or drinking, or those that insure long-term survival such as sexual behavior or caring for offspring, are pleasurable and there is a great natural urge to repeat these behaviors. These internal needs motivate our behavior and steer it in specific directions that lead toward (or away from) specific goals. In addition, motivational states have general effects: they increase our level of arousal and thereby decrease our behavioral threshold and enhance our ability to act. Finally, internal needs require organization of individual behavioral components into a specific, goal-oriented sequence. Attainment of a goal decreases the intensity of the motivational state and increases the behavioral threshold, which leads to a dropping out of one or more components of the behavioral sequence and, finally, to the cessation of the response. A hungry cat is ever alert for the occasional mouse, ready to pounce when it comes into sight. Once satiated, the cat will not pounce again for some time.

Motivational states therefore serve three functions. First, they have a directing function: they steer behavior toward a specific goal. Second, motivational states have an activating function: they increase general alertness and energize the individual to action. Finally, motivational states have an organizing function, combining individual behavioral components into a coherent, goal-oriented behavioral sequence.

These types of motivated behaviors that can be studied in relation to these functions are those related to temperature regulation, hunger, and thirst, which we consider here. We first examine how control systems analysis has made the study of these motivated behaviors amenable to biological experimentation. We next examine how motivated behaviors are regulated by factors other than simple tissue deficits, such as circadian rhythms, ecological constraints, and pleasure. Finally, we discuss systems of the brain concerned with a specific component of motivation: reward or reinforcement. The neural systems that mediate these positively motivated behaviors have now been well delineated. Several addictive drugs such as nicotine, alcohol, opiates, and cocaine are thought to produce their actions by acting on the same neural pathways that mediate positively motivated behaviors essential for survival.

Motivation Is an Inferred Internal State Postulated to Explain the Variability of Behavioral Responses

Specific motivational states, or *drives*, represent urges or impulses to action based on bodily needs. For example, a temperature-regulating drive is said to control behaviors that directly affect body temperature, such as shivering or rubbing the hands together. Other, more complex states, such as curiosity and sexual arousal, do not appear to be based on any well-defined physiological deprivation but are still referred to as drives because, like classic homeostatic drives, they involve *arousal* and *satiation*.

Motivational states, like cognitive processes, are inferred internal states that are postulated to explain the intensity and direction of a variety of behaviors, such as temperature regulation, feeding, thirst, and sex. Behavioral scientists initially posited these internal states because observable stimuli in the external environment are not sufficient to account for the variability of these behaviors. Simple reflexes—for example, the pupillary response—vary largely with the properties of a known stimulus. More complex behaviors, however, cannot be consistently correlated with external stimuli. At certain times food can stimulate vigorous feeding; at other times the same food will produce no feeding or even active rejection. Furthermore, feeding behavior cannot be consistently explained by the objective nutritional requirements of the individual. Thus, a motivational state of hunger was postulated to explain the loose correlation found at different times between the food stimulus and the feeding response.

Homeostatic Processes Such as Temperature Regulation, Feeding, and Thirst Correspond to Motivational States

As with cognitive processes, neurobiologists are now defining the physiological states that correspond to the motivational states that previously were inferred from behavior alone. As more is learned about the physiological correlates of specific behaviors, the need for invoking the concept of motivational state may disappear and be replaced by concepts derived from biology and systems theory. Accordingly, the problem of motivation should be reducible to that of a complex reflex that is regulated by excitatory and inhibitory controls in response to multiple stimuli, some of them internal.

Homeostatic mechanisms can be understood by analogy with the control systems, or *servomechanisms*, that regulate machines. While the specific physiological servomechanisms regulating homeostatic variables have never been directly demon-

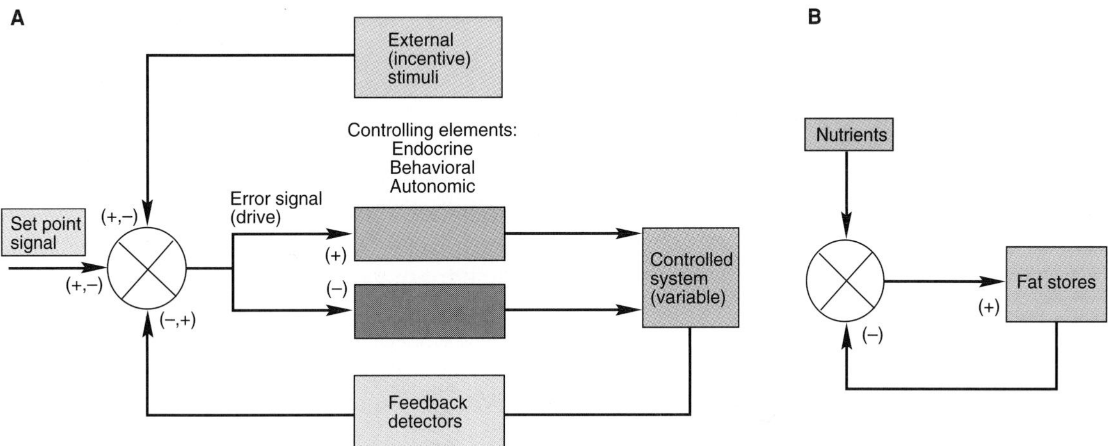

Figure 33–1 Homeostatic processes can be analyzed in terms of control systems.

A. A control system regulates a controlled variable. When a feedback signal indicates the controlled variable is below or above the set point, an error signal is generated. This signal turns on (or facilitates) appropriate behaviors and physiological responses, and turns off (or suppresses) incompatible responses. An error signal also can be generated by external incentive stimuli.

B. A negative feedback system without a set point controls fat stores in cells. (Based on data of Di Girolamo and Rudman, 1968.)

strated, the analogy with servomechanisms not only permits us to organize our thinking about highly complex systems, but also makes it possible to define the problem of physiological control experimentally. This approach has been most successfully applied to temperature regulation, which can be studied as an example of the interaction between external and internal stimuli; it has been less successful when applied to more complex regulatory behaviors, such as feeding, drinking, and sex, where the internal stimuli for hunger, thirst, and sexual behavior are difficult to identify and manipulate experimentally. Nevertheless, the control systems model is probably the best approach to analyzing these poorly understood functions at present.

Control systems maintain a *controlled variable* within a certain range. One way of regulating the controlled variable is to measure it by means of a *feedback detector* and to compare it with a desired value or *set point*. This is accomplished by an *error detector* that generates an *error signal* (i.e., drive) when the value of the controlled variable does not match the set point. The error signal then drives *controlling elements* that adjust the controlled variable in the desired direction. The error signal is not only controlled by internal feedback stimuli but also is affected by external stimuli. All examples of physiological control seem to involve both inhibitory and excitatory effects, which function together to adjust the controlled variable (Figure 33–1).

The control system used to heat a home illustrates these principles. The furnace system is the controlling element, the room temperature is the controlled variable, and the home thermostat is the error detector. The setting on the thermostat is the set point; the output of the thermostat is the error signal that turns the control element on or off.

Animal behavior need not be motivated by internal drives. Stimuli (e.g., the sight or smell of food) that are capable of driving behavior are termed *incentive stimuli*. Incentive stimuli can be innate—sweet tastes—or learned by associating a neutral stimulus with an innate incentive stimulus.

Temperature Regulation Involves Integration of Autonomic, Endocrine, and Skeletomotor Responses

Temperature regulation nicely fits the model of a servocontrol system (or several systems). Normal body temperature is the set point. The integrator and many controlling elements for temperature regulation appear to be located in the hypothalamus. Because temperature regulation requires integrated autonomic, endocrine, and skeletomotor responses, the anatomical connections of the hypothalamus make this structure well suited for this task. The feedback detector appears to collect information about body temperature from two main sources: peripheral temperature receptors situated throughout the body (in the skin, spinal cord, and viscera) and central receptors concentrated in the hypothalamus. Although both anterior and posterior hypothalamic areas are involved in temperature regulation, detectors of temperature, both low and high, are located only in the anterior hypothalamus. The hypothalamic receptors are probably neurons whose firing rate is highly dependent on local temperature, which in turn is determined primarily by the temperature of the blood.

Increases and decreases in body temperature appear to be controlled by separate mechanisms in the hypothalamus. Electrical stimulation of the anterior hypothalamus in unanesthetized animals causes dilation of blood vessels in the skin and a suppression of shivering, responses that result in a decreased body temperature. Electrical stimulation of the posterior hypothalamus produces opposing responses that function to generate or conserve heat (Figure 33–2). As with fear responses evoked by electrical stimulation of the hypothalamus (see Chapter 32), electrophysiologically induced temperature regulation also includes appropriate nonvoluntary responses involving the skeletomotor system. For example, stimulation of the anterior

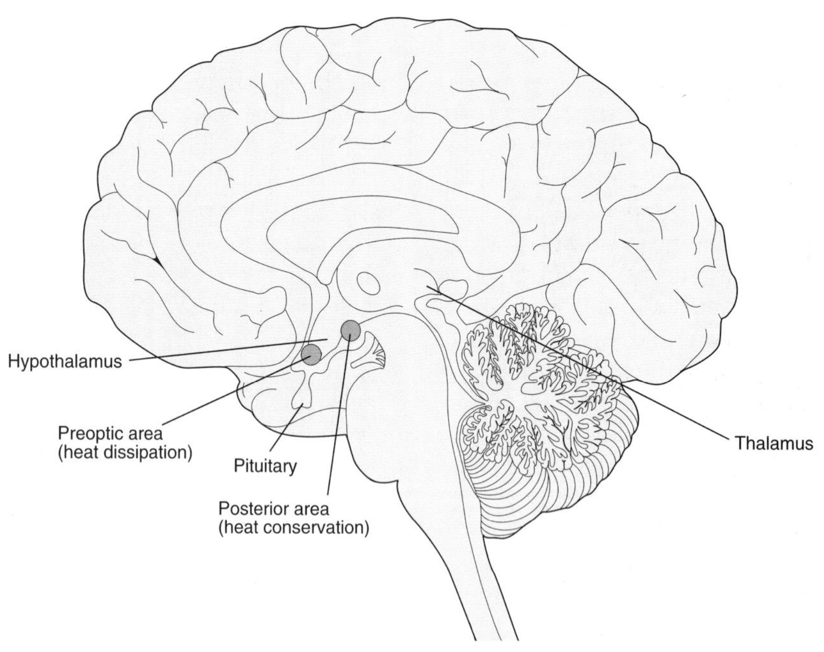

Figure 33–2 This sagittal section of the human brain shows the hypothalamic regions concerned with heat conservation and heat dissipation.

hypothalamus (preoptic area) produces panting, while stimulation of the posterior hypothalamus produces shivering.

The results of ablation experiments corroborate the critical role of the hypothalamus in regulating temperature. Lesions of the anterior hypothalamus cause chronic hyperthermia and eliminate the major responses that normally dissipate excess heat. Lesions in the posterior hypothalamus have relatively little effect if the animal is maintained at room temperature (approximately 22°C). If the animal is exposed to cold, however, it quickly becomes hypothermic because of a failure of the homeostatic mechanisms that generate and conserve heat.

The hypothalamus also controls endocrine responses to temperature challenges. Thus, long-term exposure to cold can enhance an animal's release of thyroxine, which increases body heat by increasing tissue metabolism.

In addition to driving appropriate autonomic, endocrine, and nonvoluntary skeletal responses, the error signal of the temperature control system can also drive voluntary behaviors that minimize the error signal. For example, a rat can be taught to press a button to receive a puff of cool air in a hot environment. When placed in a room at normal temperature, the rat will not press the cool-air button. If the anterior hypothalamus is locally warmed by perfusing warm water through a hollow probe, the rat will run to the cool-air button and press it.

The integration of peripheral and central afferent inputs in the hypothalamus can be demonstrated by heating the environment and concurrently cooling or heating the hypothalamus. When both the environment and hypothalamus are heated, the rat presses the cool-air button faster than when either one is heated alone. Pressing a button for cool air in a hot environment can be suppressed completely by directly cooling the hypothalamus (Figure 33–3).

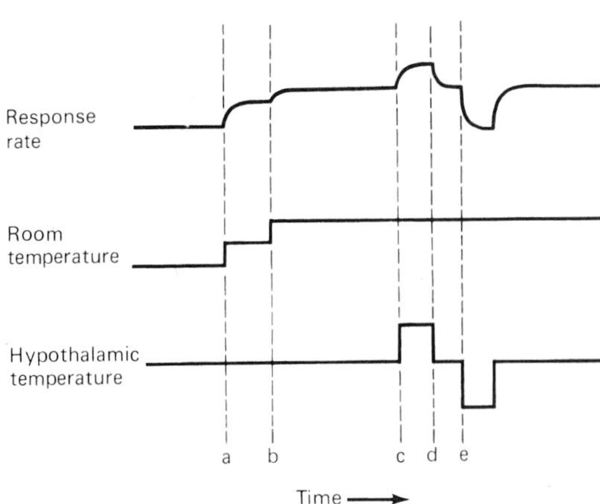

Figure 33–3 Peripheral and central information on temperature is summated in the hypothalamus. Changes in room temperature or local hypothalamic temperature alter the response rate of rats trained to press a button to receive a brief burst of cool air. When the room temperature is increased, thus presumably increasing skin temperature, the response rate increases roughly in proportion to the temperature increase (points **a** and **b**). If the temperature of the hypothalamus is also increased (by perfusing warm water through a hollow probe), the response rate reflects a summation of information on skin temperature and hypothalamic temperature (points **c** and **d**). If the skin temperature remains high enough but the hypothalamus is cooled, the response rate decreases or is suppressed altogether (point **e**). (From data of Corbit, 1973, and Satinoff, 1964.)

Recordings from neurons in the preoptic area and the anterior hypothalamus support the idea that the hypothalamus integrates peripheral and central information relevant to temperature regulation. Neurons in this region, called *warm-sensitive neurons*, increase their firing when the local hypothalamic tissue is warmed. Other neurons, called *cold-sensitive neurons*, respond to local cooling. The warm-sensitive neurons, in addition to responding to local warming of the brain, are usually excited by warming of the skin or spinal cord and are inhibited by cooling of the skin or spinal cord. The cold-sensitive neurons exhibit the opposite behavior. Thus, these neurons could serve to integrate thermal information from the periphery with that from the brain. Furthermore, many temperature-sensitive neurons also respond to nonthermal stimuli, such as osmolarity, glucose, sex steroids, and blood pressure.

In humans the set point of the temperature control system is approximately 98.6°F (37°C). Although it normally varies somewhat diurnally, reaching a minimum during sleep, the set point can be altered by pathological states, most notably by the action of pyrogens, which induce fever. Systemic pyrogens, such as the macrophage product interleukin-1, appear to enter the brain at regions in which the blood–brain barrier is incomplete and act on the preoptic area. This increases the temperature set point: body temperature rises until a new point is reached. When this occurs, an area of the brain known as the *antipyretic area* is activated and limits the magnitude of the fever response.

The antipyretic area includes the septal nuclei, which are located anterior to the preoptic areas, near the anterior commissure. Substantial evidence indicates that the antipyretic area is innervated by neurons that use the peptide vasopressin as a transmitter. Injection of vasopressin into the septal area counteracts fever in a manner similar to that of antipyretic drugs, suggesting that some of the effects of these drugs are mediated by the central release of vasopressin. The antipyretic action of nonsteroidal anti-inflammatory drugs such as aspirin and indomethacin is blocked by injection into the septal nuclei of a vasopressin antagonist. Finally, there is evidence that convulsions brought on by high fevers may in part be evoked by vasopressin that is released in the brain as part of an antipyretic response.

The control of body temperature is a clear example of the integrative function of the hypothalamus in autonomic, endocrine, and drive-state control, and illustrates how the hypothalamus operates directly on the internal environment or provides signals (derived from the internal environment) to control higher neural systems.

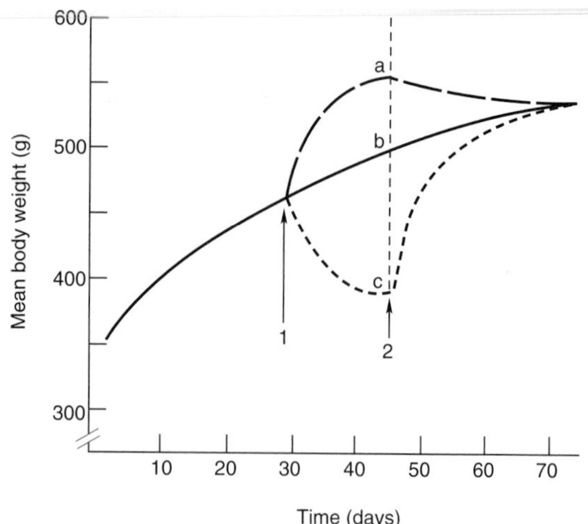

Figure 33–4 Animals tend to adjust their food intake to achieve a normal body weight. The plots show a schematized growth curve for a group of rats. At **arrow 1** one-third of the animals were maintained on their normal diet (**curve b**), one-third were force-fed (**curve a**), and one third were placed on a restricted diet (**curve c**). At **arrow 2** all rats were placed on a normal (ad libitum) diet. The force-fed animals lost weight and the starved animals gained weight until the mean weight of the two groups approached that of the normal growth curve (**b**). (Adapted from Keesey et al., 1976.)

Feeding Behavior Is Regulated by a Great Variety of Mechanisms

Like temperature regulation, feeding behavior may also be approached in terms of a control system, although at every level of analysis the understanding of feeding is less complete.

Body Weight Is Regulated Around a Set Point

One reason for thinking that feeding behavior is subject to a control system is that body weight seems to be regulated by some type of set point. Humans often maintain the same body weight for many years. Since even a small increase or decrease in daily caloric intake could eventually result in a substantial weight change, the body must provide feedback signals that control nutrient intake and metabolism. Control of nutrient intake can be clearly seen in animal studies in which body weight is altered from the set point either by food deprivation or by force-feeding. In both instances animals adjust their subsequent food intake (either up or down) until they regain a weight appropriate for their age (Figure 33–4). Animals are thus said to defend their body weight against perturbations.

Regulation of weight, however, is different from regulation of temperature. Whereas body temperature is remarkably similar from one individual to another, body weight varies greatly from individual to individual. Furthermore, the apparent set point of an individual can vary with stress, palatability of the food, exercise, and many other environmental and genetic factors. One possible explanation is that the set point itself can be changed by a variety of factors. Another possibility is that feeding behavior is regulated by a control system that has no formal set-point mechanism, but that nevertheless functions as if there were set points.

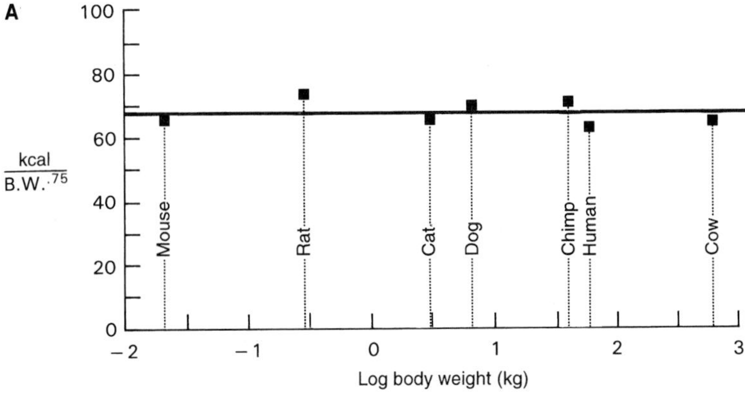

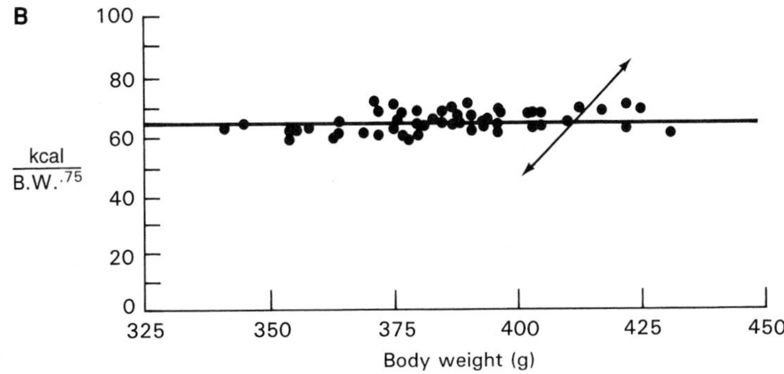

Figure 33–5 Daily energy expenditure is relatively constant across species. (Adapted from Keesey, 1989, and Kleiber, 1947.)

A. The daily energy expenditure of various species of mammals is measured as the ratio of energy expended (kcal) per metabolic mass, defined as body weight (BW) raised to the 0.75 power. This ratio is approximately 70 in all species.

B. Individual male rats of the same strain and age maintain a relatively constant daily energy expenditure at the body weight each animal spontaneously maintains. Rats forced to increase or decrease their weight become, respectively, hypermetabolic (i.e., daily energy expenditure in the graph moves in the direction indicated by the up arrow) or hypometabolic (i.e., daily energy expenditure moves in the direction of the down arrow).

Feedback systems of this type do exist in the body. A negative feedback system for the regulation of fat stores in cells is shown in Figure 33–1B. Apparently, the more fat stored in the cell, the less conversion there is of nutrients to fat. Thus, fat stores may directly or indirectly exert a negative feedback that is proportional to the amount of fat. Because of this feedback mechanism, fat stores tend to be stable in the face of varying intake of nutrients. If nutrient intake is increased over an extended period, however, the system seeks a new set point above the former value. In this system the fat stores cannot increase the negative feedback signal (to meet the demands of higher nutrient input), unless they are first increased somewhat. Automatic physiological feedback systems of this type appear to play an important role in regulating body weight.

Although body weight varies from animal to animal across and within species, there is a remarkable constancy of the daily expenditure of energy, when animals are permitted to eat freely. The expenditure is measured in kilocalories per metabolic mass, defined as body weight (in grams) raised to the 0.75 power. The ratio is approximately 70 in all animals. This constant relationship (sometimes called Kleiber's rule) holds true for freely feeding animals of different species (Figure 33–5) as well as for animals within a species. Animals can be driven way above their normal weight by being

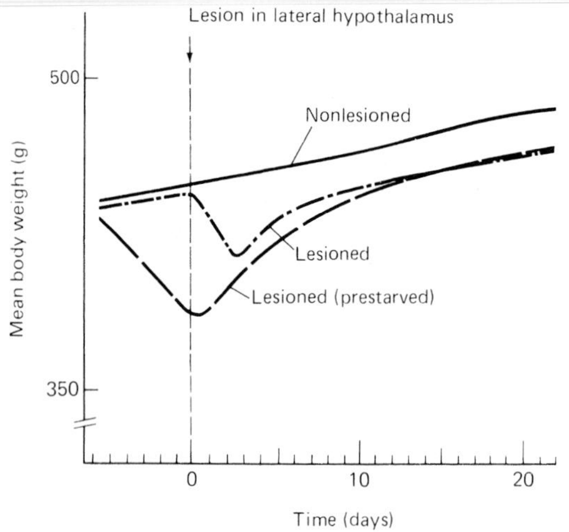

Figure 33–6 The set point for body weight appears to be altered by lateral hypothalamic lesions. Three groups of rats were used in this experiment. The control group was maintained on a normal diet. Animals of the other two groups received small lesions of the lateral hypothalamus on day zero. One of these groups had been starved before the lesion. Following the lesion, all animals were given free access to food. Lesioned animals that had not been starved immediately decreased their food intake and lost body weight before increasing their intake, while those that had been starved immediately gained weight until they reached the level of the other lesioned animals. Note that the body weight of both groups of lesioned rats remained below that of the control group. (Adapted from Keesey et al., 1976.)

force-fed or fed an unusually palatable diet. They can be driven below their normal weight by calorie restriction. In these circumstances the ratio of energy expenditure to metabolic mass gradually changes: underweight animals require fewer calories to maintain their weight, whereas overweight animals require more calories to maintain their weight. For the underweight rat, the ratio falls below what would be expected for the new lower weight, and vice versa for the overweight rat (see arrows in Figure 33–5B). Thus, these self-regulating mechanisms tend to return the animal to a weight at which the ratio of energy expenditure to metabolic mass is close to 70.

An important conclusion from these considerations is that the individual's set point for body weight can be thought of as that weight at which the ratio of energy expenditure to metabolic mass is close to 70. If the organism repeatedly undergoes weight loss (as in human dieting patterns), long-term changes may occur; fewer calories will be needed to maintain a given weight, which has the effect of increasing the set point or else increasing the weight at which the ratio of energy expenditure to metabolic mass is close to 70. Some previously obese people who have maintained a lower weight for years have abnormally low metabolic rates. They maintain their weight loss only by restricting their caloric intake well below that of other people at the same weight.

Dual Controlling Elements Are Involved in the Control of Food Intake

Food intake is thought to be under the control of two regions in the hypothalamus. Bilateral lesions in the ventromedial hypothalamic nuclei (see Figure 33–6) and surrounding regions produce overeating (*hyperphagia*), leading to severe obesity. In contrast, bilateral lesions of the lateral hypothalamus produce severe abstention from food (*aphagia*), which results in death unless the animal is force-fed. Electrical stimulation of these areas produces opposite effects: stimulation of the lateral hypothalamus elicits feeding, whereas stimulation of the medial area suppresses feeding. These observations were originally interpreted to mean that the lateral hypothalamus contains a feeding center, and the medial hypothalamus a satiety center. This conceptually attractive conclusion is probably incorrect, however. Rather, the observed results of lateral or medial hypothalamic lesions on feeding are thought to be due to many different factors, including (1) alteration of sensory information, (2) alteration of set point, (3) alteration of hormonal balance, and (4) effects on fibers of passage. We discuss each of these factors in turn, drawing primarily from experiments done on animals, although one or more of these factors may be seen in humans who have sustained damage to the hypothalamus from vascular disease or a tumor.

Sensory Deficits. Lateral hypothalamic lesions sometimes cause damage to the fibers of the trigeminal system, which innervates the face and mouth. The resulting sensory loss can contribute to the aphagia. In studies on rats, sectioning of either the peripheral or central trigeminal fibers can disturb feeding behavior, since somatosensory input from the face is required for feeding behavior. Large lesions of the lateral hypothalamus also can produce damage in surrounding structures in the basal forebrain, such as the ventral pallidum, and this damage appears to interfere with feeding because it results in an alteration of the sensory attractiveness of foods. The animals develop an active aversion to previously attractive foods, such as sugar.

Damage to the ventromedial nucleus, on the other hand, heightens responsiveness to the aversive or attractive properties of food and other stimuli. On a normal diet animals with ventromedial lesions eat more than nonlesioned animals, but if the food is adulterated with a bitter substance, they eat less than normal animals. This effect is similar to that seen in normal animals that are made obese by force-feeding. It is therefore likely that the heightened responsiveness of animals with ventromedial hypothalamic lesions to the properties of food is, at least in part, a consequence rather than a cause of the obesity. Supporting this interpretation is the observation that some obese humans with no evidence of damage to the region of the ventromedial hypothalamus are also unusually responsive to the taste of food.

Alterations of Set Point. Several experiments indicate that hypothalamic lesions alter the set point for regulating body weight. In some experiments animals were starved to reduce their weight before a relatively small lateral hypothalamic lesion was made. After lesioning, the animals were allowed to resume eating; immediately they ate more than normal and gained weight, whereas the controls (nonstarved) ate less and lost weight (Figure 33–6). Prestarvation apparently brings the weight of these animals below the set point determined by the lateral lesion. Conversely, animals whose weight is increased by force-feeding before the ventromedial hypothalamic lesions are made do not overeat, as they otherwise would without the lesions.

Alteration of Hormonal Balance. Feeding behavior is affected by many hormones, including sex steroids, glucagon, insulin, and growth hormone. Large lesions of the hypothalamus invariably affect many of these hormonal control systems. For example, lesions of the medial hypothalamus result in greatly increased release of insulin when animals eat. This response may explain, at least in part, the hyperphagia and weight gain seen after medial lesions, since a large amount of insulin in the blood can elicit feeding responses and also promotes the conversion of nutrients into fat. Interestingly, the body fat of animals with medial hypothalamic lesions increases even when caloric intake is limited to normal levels.

Effects on Fibers of Passage. Lesions of the lateral hypothalamus have been found to damage dopaminergic fibers coursing from the substantia nigra to the striatum (the nigrostriatal pathway, see Chapter 29). These lesions may also disrupt dopaminergic fibers that emanate from the ventral tegmental area (the mesocorticolimbic projections) and which have extensive connections with structures associated with the limbic system, such as the prefrontal cortex, the nucleus accumbens, septum, olfactory tubercle, amygdala, and the piriform cortex (see Chapter 32). In fact, if the dopaminergic fibers outside of the hypothalamus are sectioned, animals exhibit aphagia and a diminished capacity for arousal similar to that observed after lateral hypothalamic lesions. The aphagia following lateral hypothalamic lesions, however, can be more profound and differs in detail. We discuss the role of these dopaminergic pathways in motivation later in this chapter.

It is possible, of course, that the effects of lateral hypothalamic lesions may be due purely to the interruption of fibers of passage; this idea led investigators to question whether the hypothalamus itself has any role in feeding behavior. This issue was clarified by the use of chemical lesioning of the hypothalamus, a technique whereby the lateral hypothalamus is injected locally with glutamate analogs that destroy neuronal cell bodies without affecting fibers of passage. This technique produces aphagia and certain other aspects of a lateral hypothalamic syndrome. Thus, cell bodies in the hypothalamus as well as fibers passing through are involved in regulation of feeding. Further evidence for a role of the neurons in feeding is that the hypothalamus contains many cells that are activated when food is seen or tasted, particularly when the animal is hungry.

Chemical Stimulation of the Hypothalamus Alters Feeding Behavior

Some of the strongest evidence implicating the hypothalamus in the control of feeding comes from studies in which the hypothalamus is stimulated by a wide spectrum of transmitters. Such stimulation produces profound alterations of feeding behavior. Studies of the paraventricular nucleus and the lateral hypothalamic area clearly illustrate that feeding behavior consists of many different components and that different brain systems are involved in the control of specific aspects of feeding. For example, during feeding, each type of nutrient is regulated differently, and different transmitters appear to be concerned with the regulation of each one. For example, the application of norepinephrine to the paraventricular nucleus greatly stimulates feeding behavior in general, but animals given a choice of carbohydrates, protein, or fat eat more food containing carbohydrates. Application of the peptide galanin selectively increases ingestion of fat, whereas opiates can enhance consumption of protein.

Two peptides have proven to be particularly important in affecting feeding. The first is *neuropeptide Y* (NPY), probably the most powerful stimulant of food intake currently known. Injection of this peptide into the hypothalamus results in voracious eating, particularly of carbohydrates, even in animals that are satiated. Neuropeptide Y and its several receptor subtypes are found in the paraventricular nucleus and in several other regions of the hypothalamus. But the most sensitive site for stimulating feeding is a small region surrounding the fornix (the perifornical area) between the lateral and the anterior hypothalamic nuclei. This is also the region in which amphetamine and dopamine act to suppress feeding. Lesions of this region inhibit feeding. Consistent with the idea that endogenous neuropeptide Y plays an important role in feeding behavior, injections of antisera to the peptide or antisense oligonucleotides directed against its messenger RNA suppress normal food intake in rats. Moreover, the secretion of neuropeptide Y is increased from hypothalamic neurons in food-deprived rats, and this secretion is reduced during the course of eating.

Whereas neuropeptide Y stimulates the ingestion of carbohydrates, a second peptide, *galanin*, also present in high concentrations within the hypothalamus, stimulates fat intake and regulates body weight. Antisense oligonucleotide injections to galanin messenger RNA leads to a dramatic decline in fat ingestion and body weight, suggesting that endogenous galanin contributes to the natural appetite for fat.

Feeding Behavior Is Regulated by Short-Term and Long-Term Cues

Much research has been devoted to analyzing the cues that the organism uses to regulate feeding. Two main sets of cues regulate hunger: *short-term cues* regulate the size of individual meals, and *long-term cues* regulate overall body weight. Short-term cues consist primarily of chemical properties of the food that act in the mouth to stimulate feeding and in the gastrointestinal system and liver to inhibit feeding. Short-term signals for satiety impinge on the hypothalamus through visceral afferent pathways, primarily those to lateral hypothalamic regions. The effectiveness of short-term cues is modulated by some long-term signal reflecting body weight (perhaps related to total fat stores). In this way, body weight is kept reasonably constant over a broad range of activity and diet. Body weight, however, is also maintained relatively constant through self-regulating feedback mechanisms that appropriately adjust metabolic rate when the organism drifts away from its characteristic set point.

Several humoral signals are thought to be important for regulating feeding behavior. The hypothalamus has glucoreceptors that respond to blood glucose levels. This system, however, probably stimulates feeding behavior only in pathological emergency states in which blood glucose falls drastically. (In humans as well as lower animaals, autonomic responses to a relatively modest drop in blood glucose followed by a rise appears to initiate a meal.) Hormones that are released from the gut during a meal may contribute to satiety. The best evidence for this type of hormone is the peptide cholecystokinin. Cholecystokinin is released from the duodenum and upper intestine when amino acids and fatty acids are present in the tract. Moreover, systemic injection of cholecystokinin can inhibit feeding behavior through actions at peripheral receptors. Cholecystokinin slows the rate of emptying of the stomach and enhances the inhibitory effects of gastric distention and perhaps gastric chemical stimuli. Cholecystokinin also appears to be released by neurons of the brain. Injection of cholecystokinin and several other peptides (including neurotensin, calcitonin, and glucagon) into the ventricles also inhibits feeding. Therefore, cholecystokinin released in the brain may also inhibit feeding, independently of its release from the gut.

Cholecystokinin is an example of a hormone or neuromodulator that appears to have independent but functionally related central and peripheral actions. Other examples include luteinizing hormone-releasing hormone (sexual behavior), adrenocorticotropin (stress and avoidance behavior), and angiotensin (responses to hemorrhage). The use of the same transmitter for related central and peripheral functions is a widespread phenomenon. In invertebrates, for example, serotoninergic cells enhance feeding responses by acting directly on muscles that are involved in consuming food, but also promote behavioral arousal by acting on the central motor neurons that innervate these muscles.

Thirst Is Regulated by Tissue Osmolality and Vascular Volume

The hypothalamus regulates water balance by direct physiological action. It also regulates the behavioral aspects of drinking. Unlike feeding, as long as sufficient amounts of water are ingested, the amount of fluid taken in is relatively unimportant. Within broad limits, excess intake is readily eliminated. Nevertheless, a set point or ideal amount of water intake appears to exist, since too much or too little drinking is inefficient. If an animal takes in too little liquid at one time, it must soon interrupt other behaviors and drink again to avoid underhydration. Drinking a large amount at one time results in wasted time spent drinking, as well as in urinating to eliminate the excess fluid.

Drinking is controlled by two main physiological variables: *tissue osmolality* and *vascular (fluid) volume*. These variables appear to be regulated by separate but interrelated mechanisms. Drinking also can be controlled by dryness of the tongue and by abnormally high body temperature, detected at least in part by thermosensitive neurons in the anterior hypothalamus.

Many physiological events can provide feedback signals for water regulation. Osmoreceptors (or sodium-level receptors, probably neurons) in the hypothalamus can respond directly to the osmotic pressure of the extracellular fluid. Changes in vascular volume are detected in the low-pressure side of the circulation—the right atrium and adjacent walls of the great veins. Large changes in volume may also affect arterial baroreceptors in the aortic arch and carotid sinus, and signals from these sources can initiate drinking. Low blood volume (as well as other conditions that decrease body sodium) also results in an increase in renin secreted from the kidney. Renin, a proteolytic enzyme, cleaves plasma angiotensinogen into angiotensin I, which is then hydrolyzed to the highly active octapeptide angiotensin II. The peptide elicits drinking as well as three other physiological actions that compensate for water loss: (1) vasoconstriction, (2) increased release of aldosterone, and (3) increased release of vasopressin.

To affect behavior, bloodborne angiotensin would have to stimulate regions of the brain that permit substances to pass the blood–brain barrier. Angiotensin operates at one of these regions, the subfornical organ, a small neuronal structure that extends into the third ventricle and has fenestrated capillaries that readily permit the passage of bloodborne molecules. The subfornical organ responds to very low concentrations of angiotensin II in the blood. This response is conveyed to the hypothalamus through a neural pathway between the subfornical organ and the preoptic area that uses an angiotensin-like molecule as a transmitter. Thus, the same peptide acts as a hormone and a neurotransmitter to regulate drinking. The preoptic area also receives information from baroreceptors throughout the body. This information is integrated and then conveyed to various brain structures that activate the animal to seek water and to drink. Information from baroreceptors is also sent to structures, such as the paraventricular nucleus, that mediate the release of vasopressin, which in turn regulates water retention.

Signals that terminate drinking are less well understood than those that initiate drinking. It is clear, however, that the termination signal is not always merely the absence of the initiating signal. This principle holds for many examples of physiological and behavioral regulation, including feeding. For example, drinking initiated by a low volume of vascular fluid (e.g., after severe hemorrhage) stops well before the deficit is rectified. This is highly adaptive, since it prevents water intoxication due to excessive dilution of extracellular fluids. It also seems to prevent the overhydration that could result from absorption of fluid in the alimentary system long after the cessation of drinking.

Motivational States Can Be Regulated by Factors Other Than Tissue Needs

We have so far dealt primarily with how physiological deficits initiate appropriate behavioral and physiological responses to minimize or eliminate the deficits. A thorough understanding of motivated behaviors, however, requires knowledge of many factors not related to tissue deficits. For example, sexual responses and curiosity do not appear to be controlled by the lack of specific substances in the body. Even homeostatic responses, such as drinking and feeding, are regulated by innate and learned mechanisms that modulate the effects of feedback signals indicating tissue deficits. In humans, in particular, learned habits and subjective feelings of pleasure can override interoceptive feedback signals. For example, people often choose to go hungry rather than eat food they have learned to avoid. Here we briefly describe three factors that regulate motivated behaviors: the particular ecological requirements of the organism, anticipatory mechanisms, and hedonic (pleasure) factors.

Ecological Constraints May Shape Motivated Behaviors

The details of particular behavior patterns are determined in part by evolutionary selection, which shapes responses so that they are appropriate for the ecology of the particular animal. Cost–benefit analyses, like those done by economists, are helpful in understanding how the environment of a species shapes particular patterns of motivated behaviors. For example, the cost of feeding behavior includes the time and effort to search for and procure food; the benefit consists of nutrient intake that will ultimately support a given level of reproductive success. The spacing and duration of meals can be considered to reflect the operation of brain mechanisms that have evolved to maximize gain and minimize costs. Accordingly, carnivores may eat rapidly not because they have exceptionally powerful feedback signals indicating severe deprivation, but because they have evolved mechanisms that help ensure that their kill will not have to be shared with other animals.

Consideration of ecological factors in motivation need not preclude the consideration of homeostatic mechanisms, since homeostatic mechanisms also have evolved to assist the organism in adapting to its particular environmental conditions.

Behavior May Anticipate Physiological Deficits

Homeostatic regulation often is anticipatory and can be initiated before any physiological deficit occurs. So-called biological clock mechanisms turn physiological behavioral responses on and off before the occurrence of tissue deficit or need. One such common mechanism is a daily rhythm with a free-running period typically close to 24 hours, called a *circadian rhythm* (Latin *circa*, "around," and *dies*, "a day"). In the presence of a repeated 24-hour signal (typically light–dark cycles) this rhythm runs exactly 24 hours. However, circadian rhythms are autogenous—they can continue under constant darkness, although with periods of somewhat more or less than 24 hours.

Circadian rhythms exist for virtually every homeostatic function of the body. Since many of the rhythms are coordinated, the hypothalamus would seem to be the optimal location for a major clock mechanism that would drive them, or at least coordinate independent clock mechanisms located throughout the brain. The results of lesion studies of the suprachiasmatic nucleus of the rat support this suggestion (Figure 33–7). Animals with these lesions lose the 24-hour rhythmicity of corticosteroid release, feeding, drinking, locomotor activity, and several other responses. Furthermore, exposure of animals to light pulses at the appropriate phase of the circadian rhythm leads to an increase in the products of immediate-early genes, such as c-*fos*, in neurons in the suprachiasmatic nucleus. Immediate-early gene products may alter the transcription of later effector genes to regulate the circadian pacemaker.

In primates, including humans, two primary oscillators appear to be linked during normal light–dark cycles but run free with independent cycles under conditions of constant light. One oscillator controls slow-wave sleep, plasma growth hormone, skin temperature, and calcium excretion and appears to be driven by the suprachiasmatic

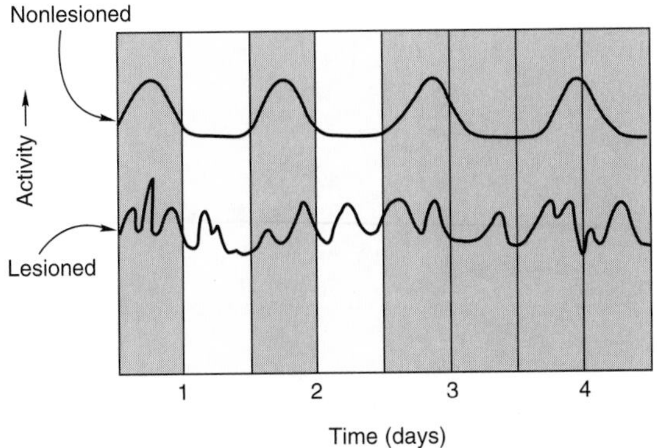

Figure 33–7 Lesions of the suprachiasmatic nucleus affect the daily activity rhythm of the rat. Normal animals exhibit 24-hour rhythms during periods of light and dark (**yellow** and **gray** areas, respectively) and approximately 24-hour rhythms in constant darkness. Animals with lesions of the suprachiasmatic nucleus completely lose the 24-hour rhythm.

nucleus. Direct retinal projections to this nucleus may carry the cyclical signals that are essential for establishing the light–dark cycle. The suprachiasmatic nucleus provides a way to regulate the rhythm of many different systems with a minimal amount of wiring, and it illustrates the advantage of clustering related functions into an anatomically discrete structure. The second oscillator controls rapid eye movement (REM) sleep, plasma corticosteroids, body core temperature, and potassium excretion.

Pleasure Is an Important but Poorly Understood Factor in Motivating Behavior

Pleasure is unquestionably a factor in motivated behaviors of humans. Humans sometimes even deprive themselves to heighten the pleasure obtained when the deprivation is relieved (e.g., skipping lunch to enjoy dinner more) or to satisfy some nonessential need (e.g., dieting to look attractive). Since pleasure is subjective, it is difficult to study in animals, but there are reasons to believe that hedonic factors control motivated behaviors in animals as well. For example, rats eat much more when given a very palatable diet containing a variety of junk foods (chocolate chip cookies, salami) than when they are given a bland but comparably nutritious diet of rat chow. The neural mechanisms underlying pleasure are poorly understood, but it seems reasonable to think that they overlap or even coincide with brain mechanisms (including those in the hypothalamus) that are concerned with reward and the reinforcement of learned behavior.

Intracranial Stimulation Can Simulate Motivational States and Reinforce Behavior

One of the most important discoveries for understanding drives is the finding that intracranial electrical stimulation of the hypothalamus and associated structures can act as reinforcement in operant conditioning of animals (see Chapter 35). In many respects brain stimulation appears to act like ordinary reinforcing stimuli, such as food, but with one important difference. Ordinary stimuli are effective only if the animal is in a particular drive state; for example, food serves as a reinforcing stimulus only in hungry animals. In contrast, brain stimulation seems to work regardless of the drive state of the animal. These considerations led to the idea that brain stimulation acts as a reward because (1) it evokes a drive state and (2) it reinforces responses by activating neural systems that are otherwise activated by ordinary reinforcing stimuli.

The Mesolimbic Dopamine Pathways Important for Reinforcement Are Also Acted on by Some Drugs of Abuse

Animals can not only learn arbitrary responses that allow them to self-administer electrical brain stimulation, they also learn responses that result in the self-administration of several drugs. The self-administered drugs are effective whether they are delivered directly to certain brain sites through a microcannula or systemically by an oral or intravenous route. Drugs that are effective consist of those that induce a sense of craving, including nicotine, opiates such as heroin, psychomotor stimulant drugs such as cocaine and alcohol, and γ-aminobutyric acid (GABA)-enhancing antianxiety drugs such as benzodiazepines. Thus, the effectiveness of the drugs to sustain behavior in animals is highly correlated with the abuse potential of the

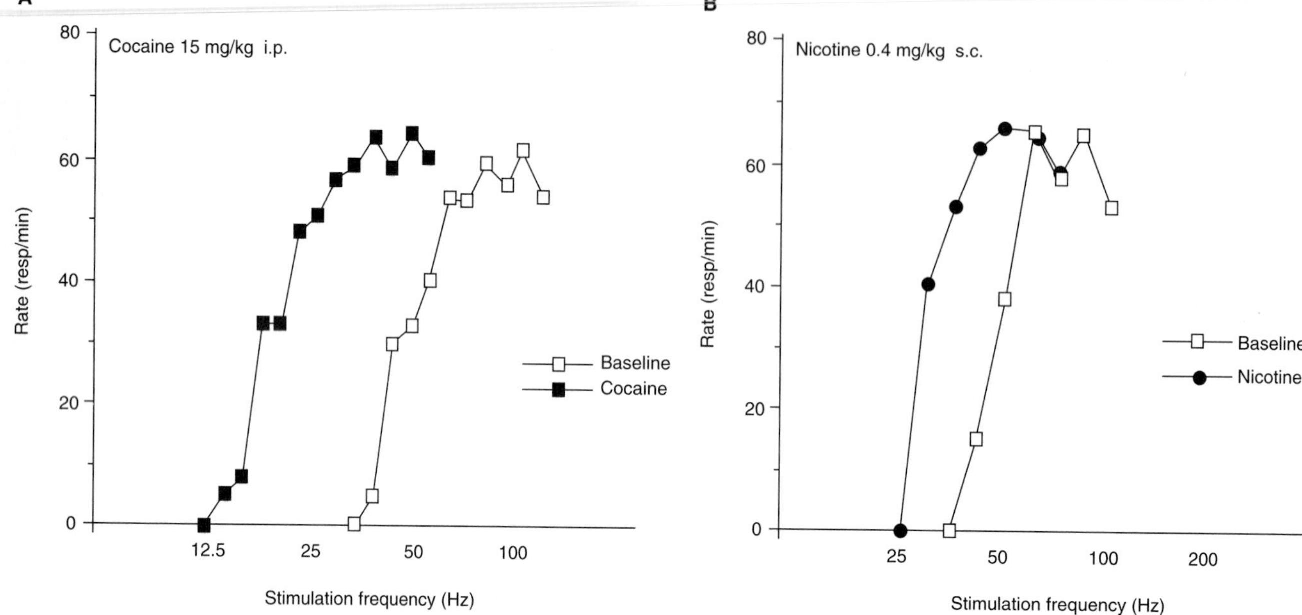

Figure 33–8 Cocaine (**A**) and nicotine (**B**) affect the rate of intracranial self-stimulation. Note that as the pulse frequency of the self-stimulation increases, the rate of responding for the stimulation increases. In the presence of the drugs, animals self-stimulate at stimulation frequencies that were previously ineffective. (Adapted from Wise et al., 1992.)

drug for humans. Animals sometimes prefer injecting the reinforcing drugs to the exclusion of all other behaviors, including feeding and sex.

There are good reasons to believe that drugs of abuse affect the reinforcing systems of the brain. For example, these drugs potentiate the reinforcing effects of electrical brain stimulation. In the presence of these drugs the frequency of shock pulses needed to produce a given level of rewarding stimulation is reduced (Figure 33–8). It is as if the drugs enhance the pleasure produced by a given level of brain stimulation.

Our understanding of addiction and drug abuse is still far from complete. However, one promising focus of the study of addiction is the dopamine pathways that appear to participate in motivating behavior. The anatomy of these neurons is well known. The human brain has relatively few dopaminergic neurons, and these are equally divided between the nigrostriatal pathway (see Chapter 29) and the mesocorticolimbic projections that originate in the ventral tegmental area (VTA). It is primarily the VTA neurons that are involved in motivation.

In animals that are trained to self-reinforce by electrical stimulation, dopaminergic neurons in the ventral tegmental area are indirectely activated, thereby increasing the level of dopamine at synapses of the mesocorticolimbic projection. Psychoactive drugs that are reinforcing also increase the level of dopamine in the neurons of the ventral tegmental area. For example, cocaine raises the level of dopamine in the nucleus accumbens by blocking the dopamine transporter (see Chapter 15), thereby prolonging the time dopamine remains in the synaptic cleft. Nicotine, possibly the most addictive and most widely abused drug, similarly increases the level of dopamine in the mesocorticolimbic pathway. This enhancement of dopamine may therefore serve as a constant reinforcement for cigarette smoking.

While dopamine has an important role in regulating self-stimulation in animals and pleasure in humans, other pathways using other transmitters are also involved. In fact, electrical stimulation of the medial forebrain bundle maintains electrical self-stimulation by activating dopaminergic cells only directly. The most effective stimuli activate a group of nondopaminergic neurons in the medial forebrain bundle that descend into the midbrain and there activate the ascending dopaminergic neurons. (Figure 33–9).

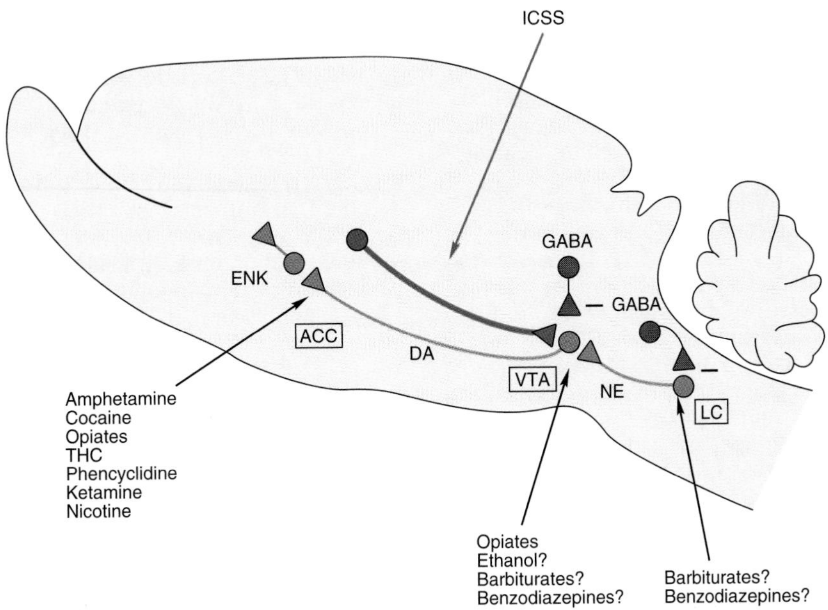

Figure 33–9 The brain-reward circuitry of the rat brain. Intracranial self-stimulation (ICSS) may act directly on descending myelinated fibers (**dark purple line**). Suspected sites of drug actions are indicated. Acc, nucleus accumbens; DA, dopaminergic fibers; ENK, enkephalin and other opioid-containing neurons; GABA, GABA-ergic inhibitory interneurons; LC, locus ceruleus; NE, norepinephrine-containing fibers; VTA, ventral tegmental area. (Adapted from Gardner and Lowinson, 1993.)

An Overall View

The hypothalamus has a major role in regulating various behaviors that contribute to homeostasis, such as feeding, drinking, and temperature control. The hypothalamus influences these behaviors by integrating external stimuli, which act as incentives, and internal stimuli, which report on the state of the animal. Many functions of the hypothalamus can be understood in terms of servocontrol systems.

Although in this and the preceding chapter on emotion, we emphasize the importance of the hypothalamus in a variety of regulatory mechanisms, other structures in the nervous system also contribute to homeostatic regulation. Indeed, when the hypothalamus and structures rostral to it are removed, limited homeostatic regulation of food intake continues. For example, a rat with this type of lesion will eat if food is placed in its mouth and will reject food after ingesting an appropriate amount. These animals, however, will never spontaneously initiate feeding. Initiation of feeding and full integration of the autonomic and endocrine aspects of feeding require the hyothalamus, which plays a central role in regulating the complex behaviors of higher organisms.

Selected Readings

Akabayashi, A., Wahlestedt, C., Alexander, J. T., and Leibowitz, S. F. 1994. Specific inhibition of endogenous neuropeptide Y synthesis in arcuate nucleus by antisense oligonucleotides suppresses feeding behavior and insulin secretion. Mol. Brain Res. 21:55–61

Bligh, J. 1973. Temperature Regulation in Mammals and Other Vertebrates. Amsterdam: North-Holland.

Booth, D. A., Toates, F. M., and Platt, S. V. 1976. Control system for hunger and its implications in animals and man. In D. Novin, W. Wyrwicka, and G. A. Bray (eds.), Hunger: Basic Mechanisms and Clinical Implications. New York: Raven Press, pp. 127–143.

Boulant, J. A. 1981. Hypothalamic mechanisms in thermoregulation. Fed. Proc. 40:2843–2850.

Cromwell, H. C., and Berridge, K. C. 1993. Where does damage lead to enhanced food aversion: The ventral pallidum/substantia innominata or lateral hypothalamus? Brain Res. 624:1–10.

Devane, W. A., Hanuš, L., Bremer, A., Pertwee, R. G., Stevenson, L. A., Griffin, G., Gibson, D., Mandelbaum, A., Eitinger, A., and McChovlam, R. 1992. Isolation and structure of a brain constituent that binds to the cannabinoid receptor. Science 258:1946–1949.

Friedman, M. I., and Stricker, E. M. 1976. The physiological psychology of hunger: A physiological perspective. Psychol. Rev. 83:409–431.

Gardner, E. L., and Lowinson, J. H. 1993. Drug craving and positive/negative hedonic brain substrates activated by addicting drugs. Semin. Neurosci. 5:359–368.

Gleitman, H. 1991. Psychology, 3rd ed. New York: Norton.

Glennon, R. A., Järbe, T. U. C., and Frankenheim, J. (eds.) 1991. Drug Discrimination: Application to Drug Abuse. Rockville, Md.: National Institute of Drug Abuse.

Herkenham, M., Lynn, A. B., Johnson, M. R., Melvin, L. S., de Costa, B. R., and Rice, K. C. 1991. Characterization and localization of cannabinoid receptors in rat brain: A quantitative *in vitro* autoradiographic study. J. Neurosci. 11:563–583.

Keesey, R. E. 1989. Physiological regulation of body weight and the issue of obesity. Med. Clin. N. Am. 73:15–27.

Kissileff, H. R., and Van Itallie, T. B. 1982. Physiology of the control of food intake. Annu. Rev. Nutr. 2:371–418.

Matsuda, L. A., Bonner, T. I., and Lolait, S. J. 1993. Localization of cannabinoid receptor mRNA in rat brain. J. Comp. Neurol. 327:535–550.

Moore-Ede, M. C. 1983. The circadian timing system in mammals: Two pacemakers preside over many secondary oscillators. Fed. Proc. 42:2802–2808.

Rolls, B. J., and Rolls, E. T. 1982. Thirst. Cambridge, England: Cambridge University Press.

Rolls, E. T. 1981. Central nervous mechanisms related to feeding and appetite. Br. Med. Bull. 37:131–134.

Schoener, T. W. 1971. Theory of feeding strategies. Annu. Rev. Ecol. Syst. 2:369–404.

Smith, F. J., and Campfield, L. A. 1993. Meal initiation occurs after experimental induction of transient declines in blood glucose. Am. J. Physiol. (Regul. Integr. Comp. Physiol. 34) 265:R1423–1429.

Stanley, B. G., and Gillard, E. R. 1994. Hypothalamic neuropeptide Y and the regulation of eating behavior and body weight. Curr. Directions Psychol. Sci. 3: 9–15.

Toates, F. 1986. Motivational Systems. Cambridge, England: Cambridge University Press.

Weiss, K. R., Koch, U. T., Koester, J., Rosen, S. C., and Kupfermann, I. 1982. The role of arousal in modulating feeding behavior of *Aplysia*: Neural and behavioral studies. In B. G. Hoebel and D. Novin (eds.), The Neural Basis of Feeding and Reward. Brunswick, Me.: Haer Institute, pp. 25–57.

Wise, R. A., Bauco, P., Carlezon, W. A., Jr., and Trojniar, W. 1992. Self-stimulation and drug reward mechanisms. Ann. N.Y. Acad. Sci. 654:192–198.

IX

LANGUAGE, LEARNING, AND MEMORY

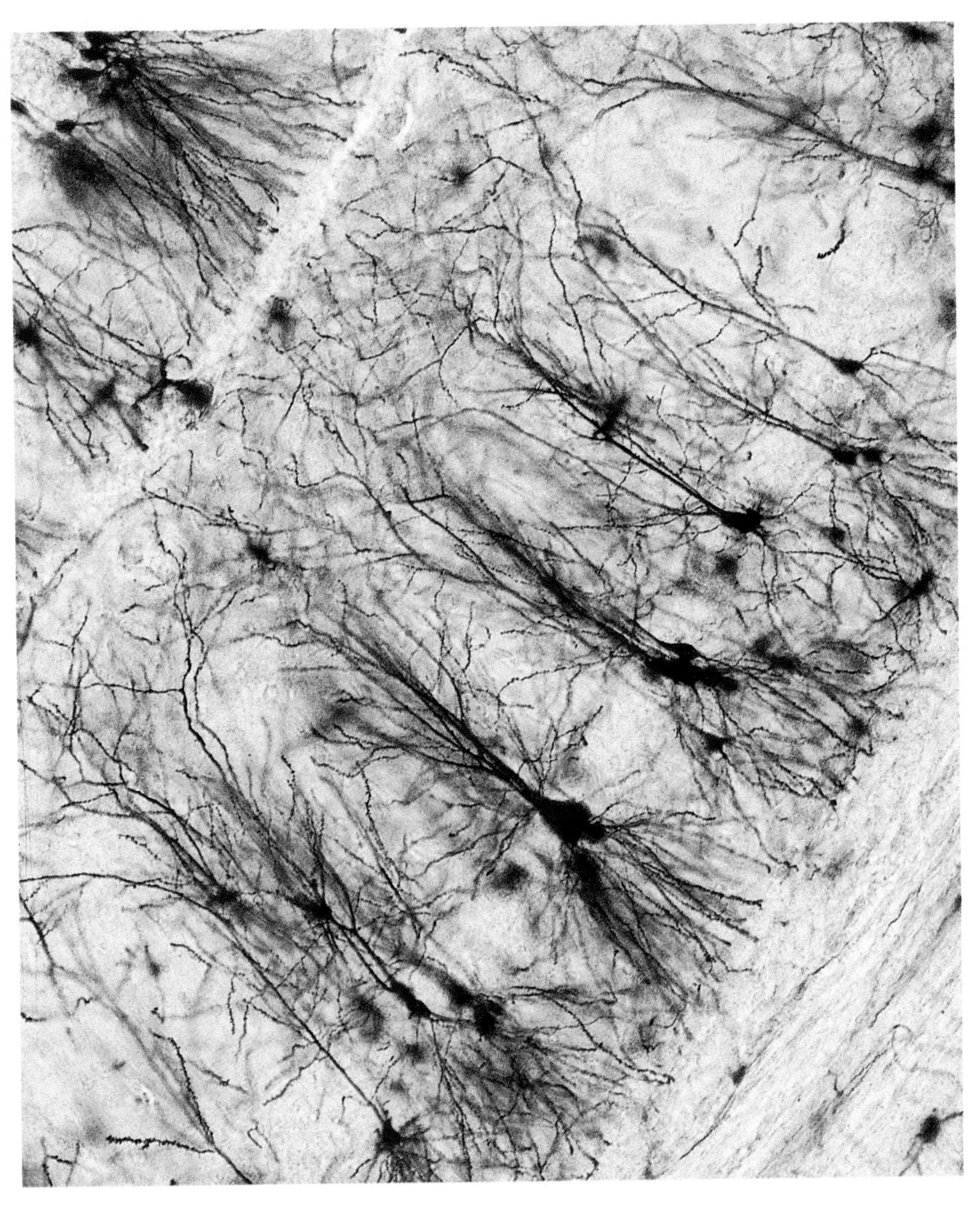

About the image on the preceding page:

Golgi staining reveals the extensive arrays of dendrites from pyramidal cells in the hippocampus. The hippocampus is important for storage of explicit memory, and synapses onto hippocampal pyramidal cells can undergo significant changes in strength, a plasticity that is thought to contribute to memory storage.

Introduction

Modern cognitive neural science is a merger of neurophysiology, anatomy, embryology, cell and molecular biology, and experimental psychology. This merged discipline is providing renewed support for the idea first proposed by Hippocrates more that two thousand years ago that the proper study of the mind begins with the brain. As with more simple behaviors, the higher mental functions—learning, memory, and language—historically have been approached in two complementary ways: through psychological observation and invasive experimental physiology. In the first part of the twentieth century, to avoid untestable concepts and hypotheses, psychology became rigidly concerned with quantifiable behaviors. Extreme behaviorists thought it unnecessary to deal with consciousness, feeling, attention, or even motivation.

Although careful quantitative analysis of stimuli and responses has provided important contributions to the study of behavior, by emphasizing observable actions behaviorists focused on *implicit knowledge*—on the question: What can an organism do and how does it do it? However, as has been emphasized by cognitive psychologists, humans and other higher animals also possess *explicit knowledge* of the surrounding world and past events. Thus, we also need ask: What does the organism know about the world and how does it come to know it? How is that knowledge represented in the brain? Is the representation of explicit knowledge different from that of implicit activity?

The study of learning, memory, and language is now yielding important information about how the processing of explicit and implicit knowledge is distributed in the brain. However, cognitive neural science is only beginning to contribute to the analysis of the richness of the internal representations that cognitive psychologists recognize as intervening between stimulus and response. For example, cognitive neural science has so far not yet directly addressed the subjective sense of individuality, will, and purpose that is common to human experience. Yet these issues are important to us as scientists and as people.

In the past, ascribing a particular behavioral feature to an unobservable mental process essentially excluded the problem from direct study because the complexity of the brain posed a barrier to any kind of biological analysis. However, as the nervous system becomes more accessible to behavior experiments, internal representations of experience can be explored in a controlled manner. Progress in this area encourages us to believe that cognition can now be explored directly and need no longer be merely inferred.

As a first step toward understanding the neural mechanisms of higher mental functions, clinical neurologists (who regarded their patients as "natural experiments") sought to discover where in the brain a particular function is located. This approach began in 1861, when Pierre Paul Broca convinced the scientific establishment, with evidence from the damaged brains of aphasic patients, that speech is controlled by a specific area of the left frontal lobe. This was soon followed by the localization in the brain of the control of voluntary movement and the elucidation of the various primary sensory cortices—for vision, audition, somatic sensation, and taste.

Even though increasing amounts of information about the localization of function became available, psychologists still thought that this information could shed little light on how mental functions are organized in the brain. Nevertheless, since the late 1960s, primarily because of studies on language and memory storage, there has been a general acceptance of the great importance of localizable internal representations for psychological mental functions. Thus, even with complex cognitive skills such as language, the key question no longer is whether study of brain localization is useful for understanding cognitive functions but rather what are the neuronal mechanisms by which these functions are accomplished

34

Language

Language Is Distinctive from All Other Forms of Communication

Animal Models of Human Language Have Been Largely Unsatisfactory

What Is the Origin of Human Language?

Is the Capability for Language Innate or Learned?

Aphasias Are Disorders of Language that Also Interfere with Other Cognitive Functions

The Wernicke–Geschwind Model Is Useful for Distinguishing Damage to the Two Major Language Regions of the Brain

Recent Cognitive and Imaging Studies Have Clarified the Interconnections of the Two Language Regions

Certain Affective Components of Language Are Impaired by Damage to the Right Hemisphere

Some Disorders of Reading and Writing Can Be Localized

Alexia and Agraphia Are Acquired Disorders of Reading and Writing

Dyslexias Are Disorders of Reading that May Be Developmental or Acquired

Comprehension of Graphemes and Ideographs Involves Different Cerebral Pathways

As you are reading these words, you are taking part in one of the wonders of the natural world. For you and I belong to a species with a remarkable ability: we can shape events in each other's brains with exquisite precision. . . . That ability is language. Simply by making noise with our mouths, we can reliably cause precise new combinations of ideas to arise in each other's minds. The ability comes so naturally that we are apt to forget what a miracle it is.

Steven Pinker, *The Language Instinct*

Language is a distinctively human form of communication, a means of conveying complex information from one person to another. The principles underlying language govern a large, varied, and impressive class of human behavior. These principles therefore are fascinating from a purely behavioral point of view.

In a simple conversation you speak on average 180 words a minute. This continuous flow of language not only is effortless, but almost perfect. The words are selected from a mental dictionary (or lexicon) that contains about 60,000–120,000 words. Yet only rarely do you mispronounce or misuse a word; perhaps one word in a million is incorrectly produced. This flawless performance in *generating* language is matched by a remarkable efficiency in *comprehending* language. Although we are largely unaware of it, this use of language requires an extensive knowledge about linguistic structure, including grammar, meaning, and sound forms. Proper use of language also requires the know-how for integrating these verbal and semantic skills with appropriate intonation and expressiveness.

In addition to their behavioral interest, the principles of language are also neurobiologically interesting because the processing of language in the brain occurs in specific and localized structures. Extensive evidence on language processing, both experimental and clinical, has given us the keenest insight into the functional architecture of the dominant hemisphere of the brain. The study of language also represents a striking example of how neurobiology, together with disciplines ranging from anthropology to developmental and clinical neurobiology, might help us understand even the most complex of human behaviors.

Language Is Distinctive from All Other Forms of Communication

Language can be distinguished from other kinds of communication by four features: creativity, form, content, and use.

Creativity. As we saw in earlier chapters, sensory perception does not result simply from summing sensations but from transformational processing of stimulus information by the brain. Speech, too, is a transformational process. When we first learn to talk we do not learn language by memorizing stock sentences, but by understanding the rules for creating meaningful utterances. Each time we speak we literally *create* meaning using grammatical rules, a feature of language called *generativity*. Listening is also creative. We readily understand sentences spoken by others. There seems to be no limit to the number of sentences we can form or the number we can comprehend.

Form. Language makes infinite use of finite means. It is formed from a small set of sounds that signal content when sounded in predictable sequences. Each of the world's languages is based on a fraction of the sounds that humans are capable of making, and not all languages use the same set of sounds. The sounds that make up words are called *phonemes*. These are the smallest unit of sound that will affect a change in meaning, as the difference between the sounds /k/ and /l/ in the words *cog* and *log*. (Phonemes should not be confused with letters, which simply represent sound. One letter can represent several sounds, as the letter *c* does. Two letters together, such as *th* or *ew*, may represent a single phoneme.) The English language uses fewer than 50 phonemes.

The structure, or *grammar*, of a natural language consists of two types of rules: (1) rules for combining phonemes to form words (morphology) and (2) rules for combining words to form phrases and sentences (syntax). To learn a language is to learn this structure.

The smallest units of meaning in a language are called *morphemes*, usually short sequences of phonemes. Morphemes in turn are combined into *words*, which are the elements that correspond most clearly to people, objects, ideas, and actions. The number of words in the English language—several million—is enormous. The typical person, however, uses only a fraction, a few percent of the total. Finally, *sentences* are produced by combinations of words. Sentences correspond to *propositions*, statements about the world. By combining a small number of elementary units in different ways at each level of a hierarchy—phonemes into words and words into sentences—human language achieves almost unlimited possibilities of expression.

The rules of grammar can be codified in a grammar book, but all written grammars are derived by analyzing the way people speak naturally. Native speakers of a language are able to use the rules of grammar (often as early as the age of 3) without recourse to a book and indeed without even being aware that they are using rules of grammar. The rules of language is therefore an outstanding illustration of a general principle that we examine in Chapter 35 when we consider memory. Much of what we learn and remember is *implicit knowledge*, which does not enter our conscious awareness.

Content. Unlike simple sign systems (gestures), in which meaning is tied to highly specific and usually fixed situations, language can shape and communicate *abstractions*, meanings that are independent of the immediate situation. Language also has an emotional dimension that is expressed by such means as gesture, tone of voice (flatness, whining, excitement, positiveness), facial expression, and posture.

Use. Language is fundamentally a means for social communication—whenever we speak or write we have a social purpose. However, language is not merely a neutral medium of exchange of facts and observations about the world. Language organizes our sensory experience and expresses our self-identity—our thoughts, feelings, and expectations.

Each of these features of language can be altered by disease. Form can be affected by disease of the cerebellum, resulting in dysarthria (the inability to talk clearly), or by lesions of the cerebral cortex, resulting in Broca's aphasia. Content is disturbed in Wernicke's aphasia, conduction aphasia, and schizophrenia. Use of language is affected in some psychiatric illnesses, such as manic–depressive illness.

In this chapter we consider the distinctive features of language and examine why animal research has increased our understanding of human language only modestly. In contrast, much has been learned about language from the study of language acquisition in children and of neurological disorders of language. We therefore review the major findings regarding the development of language and then examine the clinical disorders of speech, reading, writing, and gesture. This family of disorders can now be understood with a modern version of the model of language developed by Karl Wernicke in the nineteenth century.

In 1984 Norman Geschwind and Antonio Damasio summarized the progress in our understanding of the biological basis of language:

> As late as the mid 1960s, the standard view regarding cerebral dominance for language stated that [language] had no anatomical correlates, that it did not exist in other species, and that its evolution in humans could not be studied. . . . But the discoveries of the past 15 years have proven that each of these standard views was false and have opened up entirely new avenues of study.

We pursue these avenues here.

Animal Models of Language Have Been Largely Unsatisfactory

The neural analysis of cognitive and other human behavior has often depended on studies of animals. Considerable effort therefore has been expended in developing animal models of language. Animals as simple as crickets and bees show elementary forms of communication. Bird songs are even more elaborate. Patterns of communication between lower animals are often called languages. Nevertheless, the types of communication observed in lower animals cannot be considered *interpersonal* and their form, content, and use are highly stereotyped.

What about our closest relatives, the nonhuman primates? Does ape language have any of the characteristics of human language? Can the behavior of nonhuman primates shed light on human speech? Human communication involves both the meaning of words (semantics) and the rules by which they are used (syntax). Are both present in the communication of nonhuman primates?

In the past few decades opinion on this question has swung back and forth several times. In the 1930s it was generally thought that chimpanzees could learn to speak if they were raised as human children. With this idea in mind, William and Lorna Kellogg raised a chimpanzee, Gua, with their own child. The chimpanzee adopted many human behaviors, understood a few spoken commands, and mastered a few human hand gestures, but never learned to speak. By the early 1960s chimpanzees were thought to lack the intellectual capacity for language. In 1968 Noam Chomsky wrote: "Anyone concerned with the study of human nature and human capacity must somehow come to grips with the fact that all normal humans acquire spoken language whereas acquisition of even the barest rudiments is quite beyond the capacity of an otherwise intelligent ape."

Shortly thereafter it was discovered that the vocal apparatus of chimpanzees is unable to produce the full range of human sounds. The possibility remained, however, that chimpanzees might show a capacity for language if they did not have to produce speech sounds. Allen and Beatrice Gardner circumvented the need for sound production by training a female chimpanzee named Washoe to use signs borrowed from American Sign Language, the language of the American deaf. Within four years, Washoe achieved a vocabulary of 160 words, including signs for objects (bird, hand), attributes (blue, green, different), and modifiers (more, less). Although these results show that a chimpanzee can learn words and use symbols, Washoe's vocabulary of 160 words was much smaller than that of a child of four, whose vocabulary is more than 3,000 words.

To explore whether chimpanzees understand syntactic relationships, David Premack trained a chimpanzee, Sarah, to communicate by means of plastic chips inscribed with different signs. In this training Premack tried to preserve many features that are universal in natural languages. He taught Sarah to interpret commands conveyed by a sequence of chips and to construct her own sentences by arranging the chips in meaningful sequences. Sarah eventually learned the concepts of negation, similarity, and difference, the expression "is the name of," compound sentences, if–then statements, and how to ask questions. Most interesting were experiments in which Premack showed Sarah pairs of objects in which the second object was a transformed version of the first (an apple and an apple cut into pieces, a dry towel and a wet towel). Sarah was then asked to select one of several other objects that would explain the correlation (e.g., a knife or a bowl of water) and insert it between these pairs. She made the appropriate choice about 80% of the time. Sarah appeared to be able to express in symbols her understanding of the causal relationship between physical events.

Thus chimpanzees (and probably gorillas as well) are able to communicate in a rudimentary fashion using symbols. Until recently, however, it was not clear whether they can go beyond that. For example, there is no evidence that chimpanzees can actively manipulate rules that organize words into sentences so that they can recombine the same words to express different ideas. Thus, Washoe can use the words *Washoe, me,* and *banana*, but most students of language think that she cannot distinguish "me give Washoe banana" from "Washoe give me banana." Indeed, until recently, most linguists were impressed by the fact that the artificial language acquired by trained chimpanzees is nongrammatical, imitative, and mechanical—in a word, *noncreative*.

Susan Savage-Rumbaugh has recently reexamined the grammatical capabilities of chimpanzees in an interesting new way. Rather than studying language *production*, as the Gardiners did with Washoe, she examined *comprehension*. Moreover, she used the bonobo or pygmy chimpanzee, a species thought to be particularly talented for language acquisition. She compared the comprehension of spoken English by a bonobo infant named Kanzi and a human infant, and found that the two were comparable in both their lexical and grammatical comprehension up to the age of 2.5 years. At that point, however, the language capabilities of the human infant continued to increase dramatically, whereas Kanzi, having reached a total of 400–500 words, stopped learning new words and stopped making progress in grammatical comprehension.

Thus, the chimpanzee's very earliest language comprehension is comparable to that of a human infant, but language development in the chimpanzee goes only to a certain point and then stops. The limited growth of grammar is paralleled by a lack of growth in vocabulary, a combined failure which may be the result of the inability to generate the later bursts of language acquisition characteristic of human infants.

Thus, despite the fact that Kanzi is somewhat more advanced than Washoe, his overall language capabilities are still very weak compared with the fluent and creative use of language by humans. This, of course, raises the question: Why does the human infant outstrip so greatly all other primates in learning and solving problems at a much higher level? Does this unique capability of human language reflect the retuning of a basic primate brain plan for the specific purpose of language? Or is the language capability of humans due to the development of a completely new neural architecture for language?

Irrespective of the answer to these questions, the studies of chimpanzees have already illustrated that animal models are interesting but have limited usefulness in the study of human language. As a result, students of language now rely primarily on anthropological, developmental, and clinical studies of human language.

What Is the Origin of Human Language?

It is difficult to pinpoint when or how language evolved. Nevertheless, it is clear that some cerebral structures that are prerequisite for language appear to have arisen early in human evolution. This conclusion has come from examination of endocranial casts of human fossils. In most individuals the left hemisphere is dominant for language; the cortical speech area of the temporal lobe (the planum temporale) is larger in the left than in the right hemisphere. Since important gyri and sulci often leave an impression on the skull, the fossil record was searched for the morphological asymmetries associated with speech in modern humans. These asymmetries were found in Neanderthal man (dating back 30,000 to 50,000 years) and in *Homo erectus* (dating back 300,000 to 500,000 years), the predecessor of our own species.

These findings have encouraged the search for a basic primate plan for language. This search has led to the discovery that the left hemisphere in Japanese macaque monkeys is dominant for the recognition of species-specific cries. In addition, asymmetries similar to those of humans are present in the brains of modern-day great apes, such as the chimpanzee. Whether these asymmetries actually evolved for language, for other forms of communication, or for an entirely different function is not known.

Although the anatomical structures that are prerequisites for language may have arisen early (perhaps as long as 500,000 years ago), many linguists believe that language itself emerged late in the prehistoric period of human existence (about 100,000 years ago) and that it probably arose only *once*. According to this view, all human languages are thought to have evolved from a single ancestral language first spoken in Africa. The fact that all human languages have so many features in common is consistent with this view.

But did human language evolve from ape-like communication, as the recent behavioral and anatomical data might suggest? Since human evolution is itself not understood, and since apes, as we have seen, have only rudimentary capabilities for learning human language, answers to this question are speculative only. Two hypotheses about the origin of language have been advanced: gestural and vocal.

Gestural theories propose that language evolved from a system of gestures that became possible when certain apes assumed an erect posture, freeing the hands for social communication. Later, vocal communication may have arisen to free the hands for purposes other than communication. Vocal theories contend that language evolved from an extensive group of instinctive calls that were expressive of emotional states, such as distress, elation, and sexual arousal. About 500,000 years ago changes in the structure of the mouth, jaw, and vocal tract made it possible to control deliberately and reliably the production of different sounds. As a result, sounds could, at least in principle, be used creatively in different combinations. When these ancestors of modern humans dispersed into separate colonies, different sound systems developed because of geographical isolation.

Alternatively, language may have emerged from the coevolution of gesture and vocalization. This possibility might account for the still inexplicable correlation between hand dominance and both verbal and sign language, all of which are localized to the left hemisphere.

Is the Capability for Language Innate or Learned?

Although the acquisition of language clearly involves learning by imitation, studies of the anatomical localization of language and of language development in children provide several indications that a large part of the process is innate. First, both natural and sign language functions are predominantly represented in the left hemisphere of the brain.

Second, the localization of language in the left hemisphere seems to be related to anatomical differences between the two hemispheres. For example, the planum temporale is larger in the left hemisphere in most (67%) right-handed people. Third, this anatomical asymmetry in the planum temporale is present early in development (by the thirty-first week of gestation), suggesting that this asymmetry does not develop in response to experience but is innate.

Fourth, infants at birth are sensitive to distinctions in a broad range of sounds, an ability that is crucial for the comprehension of any human language. Indeed, some of this sensitivity is lost later, when a specific language is acquired. For example, most adult Japanese do not perceive the difference between the sounds /r/ and /l/. Japanese infants can distinguish these sounds, however, and lose this ability only as they mature. Likewise, most native speakers of English do not perceive a difference between the so-called hard /l/ and soft /l/ in Russian. The neural basis of this decline in discrimination of nonnative contrasts is unknown. Some have argued that it may be similar to synaptic retractions underlying the loss of visual acuity in kittens raised in a restricted visual environment (see Chapter 25).

Finally, there are universal regularities in the acquisition of language. Children progress from babbling to one-word speech, to two-word speech with syntax, to complex speech (Table 34–1). Some children move through these stages faster than others, but the average age for each stage is the same in all cultures. Moreover, there is a critical period for the development of language, be it verbal or signed, much as there is for perception. This period lasts from age two to puberty. After puberty language capacity (as measured by the ability to acquire a new language) is reduced dramatically. "Wild" children, who have grown up without human contact, cannot learn to speak after puberty, and even before puberty tend to do poorly at acquiring languages. Among Chinese and Korean children who have immigrated to the United States there is a linear relationship until puberty between the age of arrival and proficiency in English. After puberty English proficiency does not depend on age. People in their thirties do as well as those in their late teens. Although a second language can clearly be learned after puberty, it is not only more difficult to do so but almost impossible to acquire a correct accent.

Presumably this critical period of development corresponds to the maturation of the human brain, although no attempt has yet been made to correlate language acquisition with the maturation of specific areas related to language. During this period children learn the rules of their native language by listening to people around them. Most of these rules, the grammar of the language, are understood, however, by the time the child begins to form sentences.

Clearly a language must be learned. But we have also seen that language is not taught in the ordinary sense of teaching. Noam Chomsky (Figure 34–1) argues that humans have some innate language acquisition device, a neural program that prepares them to learn language. According to Chomsky, an infant learns a language by testing what he or she hears daily against a genetically determined system of rules or *universal grammar*. That is, children have an innate ability to recognize language in the environment and, when exposed to a language, learn it avidly. Because the universal grammar limits the possible characteristics of a natural language, sound patterns that are not consistent with the universal grammar are not recognized and are not learned instinctively.

Table 34–1 Stages of Development in the Acquisition of Language

Average Age	*Language Milestones*	*Motor Milestones*
6 months	Cooing, changes to distinct babbling by introduction of consonants	Sits using hands for support; unilateral reaching
1 year	Beginning of language understanding; one-word utterances	Stands; walks when held by one hand
12–18 months	Words used singly; repertory of 30–50 words (simple nouns, adjectives, and action words), which cannot as yet be joined in phrases but are used one at a time; does not use functors (the, and, can, be) necessary for syntax, but makes good progress in understanding	Grasping and release fully developed; walking; creeps downstairs backward
18–24 months	Two-word (telegraphic) phrases are ordered according to syntactic rules; vocabulary of 50 to several hundred words; understands propositional rules	Runs (and falls); walks stairs with one foot forward
2–5 years	New words every day; three or more words in many combinations; functors begin to appear; many grammatical errors and idiosyncratic expressions; good understanding of language	Jumps with both feet; builds tower of six cubes
3 years	Full sentences; few errors; vocabulary of around 1,000 words	Tiptoes; walks stairs with alternating feet
4 years	Close to adult speech competence	Jumps over rope; hops on one foot; walks on a line

Based on E. H. Lenneberg, 1967.

Thus, linguists and psychologists now believe that certain universal aspects of language acquisition are determined by the innate structure of the human brain. The human brain is prepared to learn and use speech. The particular language spoken and the dialect and accent are determined by the social environment.

Two questions are now being debated by linguists. First, is there a specific cognitive organ unique to human beings that is specialized for language? Second, are the neurally based rules for language specific to language, or are they more general *cognitive* rules? Are the neural mechanisms for representing language completely distinct from all other cognitive processes? Chomsky argues that there is a specific brain organ for language and that specific neural mechanisms exist for acquiring language. But many psychologists disagree. Children are able to understand some of the same abstract rules required for language before they learn to speak. They can, for example, distinguish between causative and noncausative actions.

These problems pose formidable challenges for the neurobiological approach to cognition and language. One avenue of investigation has come from the study of aphasia, disturbances of language caused by vascular damage, trauma, or tumor in specific parts of the brain—usually, but not invariably, regions of the cerebral cortex. Researchers working with aphasic patients are asking two sorts of questions. First, are disorders of language isolated cognitive disorders, or are they related to more general disturbances of cognitive processes? Second, what are the neural structures that underlie the innate "universal grammar"?

Figure 34–1 Noam Chomsky.
In 1957, at the age of 27, Noam Chomsky published *Syntactic Structures*, a book that revolutionized linguistics. Chomsky received his Ph.D. at the University of Pennsylvania, was a junior fellow at Harvard, and then moved to the Massachusetts Institute of Technology, where he is a University Professor. Before Chomsky, linguistics consisted primarily of the anthropological study of various languages and texts. Chomsky made grammar the focus of his approach, formulating explicit rules for generating meaningful sentences. Since the way people learn and use language is at the center of Chomsky's approach, cognitive psychologists have taken great interest in Chomskian linguistics. Chomsky pointed out that language and other cognitive functions could not be explained according to the laws of stimulus and response advocated by B. F. Skinner and other behaviorists. Instead, speech is a creative process: every time we speak we actively organize thoughts into words, rather than repeating stock sentences. Chomsky postulated that the brain must have an organ of language, unique to humans, that can combine a finite set of words into an infinite number of sentences. This capability, he argued, must be innate and not learned, since children speak and understand novel combinations of words they have not previously heard. Children must therefore have built into their brain a *universal grammar*, a plan shared by the grammars of all natural languages.

Aphasias Are Disorders of Language that Also Interfere with Other Cognitive Functions

When language is disturbed by damage to the cerebral cortex, it is not affected in a generalized way, as an overall reduction in language ability. Rather, lesions in different parts of the cerebral cortex disrupt specific language functions. For example, some aphasic patients have difficulty comprehending both speech and writing (Wernicke's aphasia). Others have difficulty expressing thoughts in either written or spoken language (Broca's aphasia).

The major clinical syndromes of aphasia are described in Box 34–1. In practice, the symptoms

BOX 34–1 The Common Aphasias

The major aphasias give rise to characteristic clinical syndromes.

Wernicke's Aphasia

Wernicke's aphasia is characterized by a major deficit in comprehension. The lesion primarily affects Wernicke's area, although it often extends to the superior portions of the temporal lobe (areas 40 and 39) and inferiorly to area 37 (see Figure 34–2). When the lesion is extensive, comprehension of both visual and auditory language input is severely impaired. Speech, however, remains fluent and is normal in rate and melody. Nevertheless, these patients also have some difficulties in the production of language. They may have difficulty finding the right word, or they may use the wrong word or combination of words (*paraphasia*). These patients tend to add syllables to words and additional words to phrases. They may make up new words (*neologisms*). Neologistic or paraphasic distortions most frequently involve key lexical items (nouns, verbs, adjectives, adverbs), especially nouns.

Patients with Wernicke's aphasia fail to convey the ideas they have in mind, an impairment called *empty speech*. Generally, they are unaware of this failure, probably because their language comprehension is impaired. The ability to repeat words and phrases is also impaired because comprehension is severely disturbed. In addition, reading and writing are severely disturbed in patients with Wernicke's aphasia. Language may be excessive (*logorrhea*); this phenomenon has been called *press of speech*. The overabundance of words often conveys little meaning. For example, when asked where he lived, a patient with Wernicke's aphasia replied, "I came there before here and returned there." Except for these symptoms of aphasia, other neurological signs may be absent, but occasionally neural information for the right visual field is defective.

Broca's Aphasia

In Broca's aphasia comprehension is usually in good part preserved, whereas language production is seriously impaired. Patients have damage to the motor association cortex in the frontal lobe, usual-

Table 34–2 Clinical Characteristics of Three Common Aphasias

Type	*Verbal Output*	*Repetition*	*Comprehension*	*Naming*	*Associated Signs*[a]	*Lesions*
Broca's	Nonfluent	Impaired	Normal	Marginally impaired	RHP, apraxia of the left limbs and face	Left posterior inferior frontal
Wernicke's	Fluent	Impaired	Impaired	Impaired	± RHH	Left posterior superior temporal
Conduction	Fluent	Impaired	Normal	Impaired (paraphasic)	± RHS, apraxia of all limbs and face	Left parietal

[a] RHP, right hemiparesis (partial paralysis); RHH, right homonymous hemianopsis (partial blindness); RHS, right hemisensory defect.

ly extending to the posterior portion of the third frontal gyrus (Brodmann's areas 44 and 45), which forms part of the frontal operculum (Broca's area). In severe cases the surrounding premotor and prefrontal regions (areas 6, 8, 9, 10, and 46) are also damaged (see Figure 34–1).

The deficit in language production ranges from almost complete muteness to a slowed, deliberate speech using very simple word forms. Patients express nouns in the singular and verbs in the infinitive or participle, and often eliminate articles, adjectives, and adverbs altogether. For example, instead of saying "I saw some large gray cats," a patient with Broca's aphasia might say "see gray cat."

These omissions are even more dramatic in more complex sentences. Here we can see the second characteristic of this defect: a breakdown in syntax. Consider the sentence: "Ladies and gentlemen, you are now invited into the dining room." A patient with Broca's aphasia may only be able to say "Ladies, men, room." When asked his occupation, a mailman with Broca's aphasia said "Mail . . . Mail . . . M. . . ." In addition to such telegraphic or nongrammatical speech, repetition is always impaired, and naming may be slightly to moderately impaired. Unlike Wernicke's aphasia, however, patients with Broca's aphasia are generally aware of these errors. In fact, patients with Broca's aphasia usually do not overlook important features of their native language. Thus, German-speaking patients will rarely omit the article before the noun, because in German the article carries critical information about case. Chinese patients will usually not omit noun classifiers, and Italians will select from 40 to 45 different inflected forms for the correct verb conjugation!

Although language production is severely disturbed, comprehension of both spoken and written language is less disturbed, because Wernicke's area is not damaged. However, patients with Broca's aphasia have difficulty reading aloud, and writing (like speech) is abnormal. They may also have some difficulty comprehending those aspects of syntax that they have difficulty producing.

Because Broca's area is located near the motor cortex and the underlying internal capsule, a partial paralysis on the right side and a loss of vision almost always accompany this type of aphasia.

Conduction Aphasia

As pointed out in Chapter 1, conduction aphasia was predicted by Wernicke. He proposed that a pathway connected Broca's area with an area in the temporal lobe concerned with the comprehension of language. From this he inferred that a lesion that spared both Broca's and Wernicke's areas could nevertheless disconnect the two. Clinical studies have verified this prediction. Lesions in the arcuate fasciculus, which runs in the white matter and connects Wernicke's and Broca's areas, lead to a conduction aphasia. Damage to the fasciculus occurs with injury of the supramarginal gyrus of the parietal lobe or, less frequently, injury of the posterior and superior aspects of the left temporal lobe (see Figure 34–2). Thus, the lesion is not restricted to white matter but also involves the cortex.

Patients with conduction aphasia are somewhat less fluent in language production than patients with Wernicke's aphasia. They make many paraphasic errors, substituting incorrect words or sounds for correct ones. Although comprehension is good, the ability to repeat is greatly impaired. Other characteristics of conduction aphasia are also consistent with a functional separation of Broca's and Wernicke's areas. Naming is severely impaired. Reading aloud is abnormal, but patients can read silently with good comprehension. Writing may also be disturbed; spelling is poor, with omissions, reversals, and even substitutions of letters. In many patients voluntary movements are impaired to some degree.

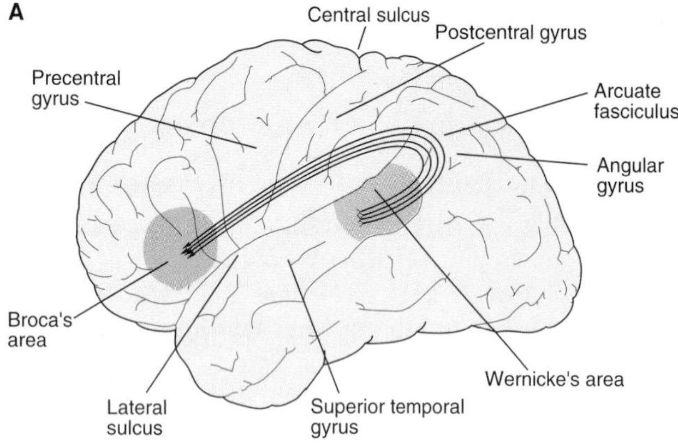

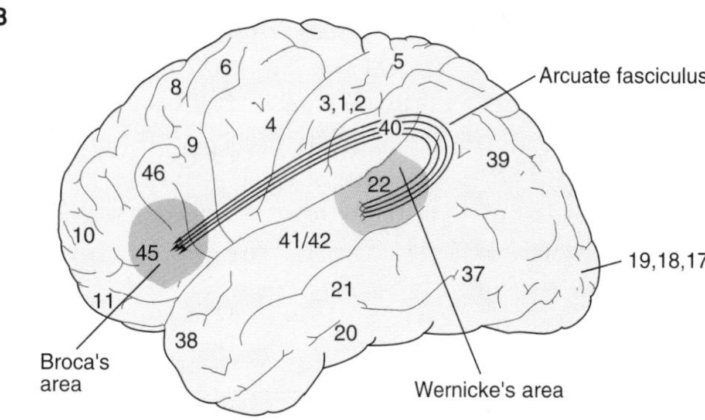

Figure 34–2 Primary language areas of the brain, as seen from the lateral surface of the left hemisphere.

A. Broca's area, the motor–speech area, is adjacent to the region of the motor cortex (precentral gyrus) that controls the movements of articulation, facial expression, articulation, and phonation. Wernicke's area includes the auditory comprehension center and lies in the posterior superior temporal lobe near the primary auditory cortex (superior temporal gyrus). Wernicke's and Broca's areas are joined by a fiber tract called the *arcuate fasciculus*. Broca's and Wernicke's areas are best understood as parts of complex networks rather than independent language centers.

B. The cytoarchitectonic areas (Brodmann's classification). Area 4 is the primary motor cortex; area 41 is the primary auditory cortex; area 22 is Wernicke's area; and area 45 is Broca's area.

of a patient do not always fall exclusively into one category or another, because cortical damage is not always restricted to one functional site.

The aphasias are distinguished from motor disorders of speech, such as *dysarthria*, a disturbance in articulation, and *dysphonia*, a disturbance in vocalization. These are defects in the muscles controlling the vocal apparatus and do not affect language comprehension or the central processes of expression. Patients with cerebellar disorders who are dysarthric, or those with Parkinson's disease who are dysphonic, retain their language ability despite severe speech impairment. In contrast, the hallmark of aphasia is a disturbance in language ability, in either comprehension or production or both, that is not attributable to a mechanical impediment.

Studies of aphasia afford unusual insight into how the brain is organized for language. One of the most impressive insights comes from a study of deaf people who lost their ability for American Sign Language following cerebral damage. Unlike speech, American signing is accomplished with hand gestures rather than by sound and is perceived by visual rather than auditory pathways. Nonetheless, signing, which has the same structural complexities characteristic of spoken languages, is also localized to the left hemisphere. Thus, deaf people can become aphasic for sign language following lesions in the left hemisphere. Lesions in the right hemisphere do not produce these defects. Moreover, the defects in signing following damage to the left hemisphere can be quite specific, involving either sign comprehension and grammar or signing fluency.

These observations illustrate three points. First, the cognitive processing for language occurs in the left hemisphere and is independent of pathways that process the sensory or motor modalities used in language. Second, speech and hearing are not necessary conditions for the emergence of language capabilities in the left hemisphere. Third, spoken language represents only one of a family of cognitive skills mediated by the left hemisphere.

The Wernicke–Geschwind Model Is Useful for Distinguishing Damage to the Two Major Language Regions of the Brain

A useful classification of aphasia was developed by Geschwind and Damasio as an elaboration of the

Wernicke–Geschwind model of language and gesture processing, and we use that scheme here.

This model for language processing can best be illustrated by considering the difference between articulating a word that is heard and one that is read. According to the Wernicke–Geschwind model, hearing a word involves transfer of information from the auditory apparatus to the auditory nerve and the medial geniculate nucleus. The information next flows to the primary auditory cortex (Brodmann's area 41), then to the higher-order auditory cortex (area 42), before it is conveyed to a specific region of the parietal–temporal–occipital association cortex, the *angular gyrus* (area 39), which is thought to process incoming auditory, visual, and tactile information (Figure 34–2). From here the information is projected to Wernicke's area (area 22), which is concerned with the comprehension of the word, and then by means of the arcuate fasciculus to Broca's area (area 45), where this auditory representation is transformed into the grammatical structure of a phrase and where the memory for word articulation is stored. Information about the sound pattern of the phrase is then conveyed to the area of the motor cortex that controls vocalization.

A similar pathway was thought by Wernicke and Geschwind to be involved in reading a word (Figure 34–3). According to their model, visual information about the word is transferred from the retina to the lateral geniculate nucleus, and from there to the primary visual cortex (Brodmann's area 17). The information then travels to a higher-order center (area 18). From there it is conveyed, much like auditory information about a word, first to the angular gyrus of the parietal–temporal–occipital association cortex, and then to Wernicke's area, where it is transformed into a phonetic (auditory) representation of the word. The phonetic pattern is then conveyed to Broca's area by means of the arcuate fasciculus.

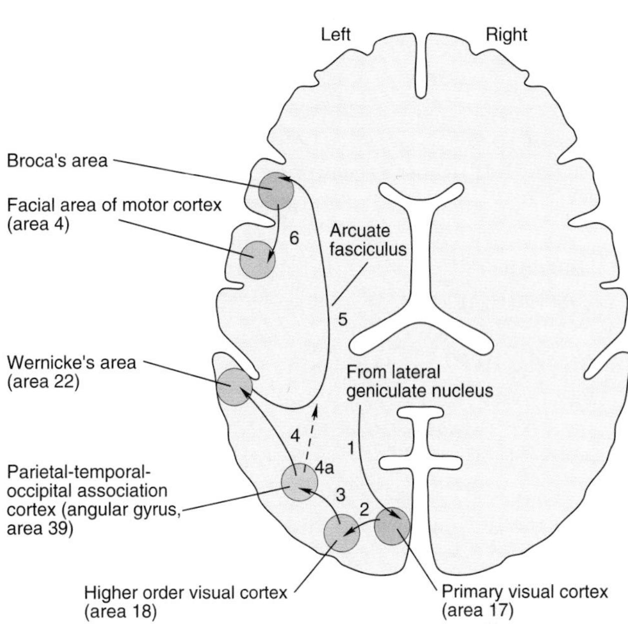

Figure 34–3 The Wernicke–Geschwind model specifies the cortical pathways involved in repeating a printed word. Repeating a printed word begins with input from the retina to the primary visual cortex. Recent evidence suggests that the actual flow of information is similar to the sequence shown here, except that, following step **3**, a component of the arcuate fasciculus (**4a**) conveys information directly from the association cortex to Broca's area, bypassing Wernicke's area. This schematic drawing shows a horizontal section of the human brain at the level of the corpus callosum. (Adapted from Patton, Sundsten, Crill, and Swanson, 1976.)

The original Wernicke–Geschwind model made several interesting predictions that are useful clinically. First, it predicted the outcome of a lesion in Wernicke's area: spoken words reaching the auditory cortex fail to activate Wernicke's area and thus will not be comprehended. If the lesion extended posteriorly and inferiorly beyond Wernicke's area, it would also affect the pathway concerned with the processing of visual input to language, and so, the patient would be incapable of understanding either the spoken or the written word. Second, the model correctly predicted that a lesion in Broca's area would not affect the comprehension of language (either written or spoken) but would cause a major disruption of speech, because the patterns for sounds and for the structure of language would not be passed on to the motor cortex. Third, the model predicts that a lesion in the arcuate fasciculus, by disconnecting Wernicke's area from Broca's, would disrupt the normal flow of language. Specifically, auditory input would not be conveyed to Broca's area, the part of the brain involved with production of language, and the normal feedback from language production to language comprehension would also be disrupted.

Recent Cognitive and Imaging Studies Have Clarified the Interconnections of the Two Language Regions

Even though the Wernicke–Geschwind model continues to be useful clinically, recent neuropsychological and imaging studies comparing the uses of language by normal and aphasic patients indicate that the Wernicke–Geschwind model may be oversimplified in several important ways. First, the emphasis in the Wernicke–Geschwind model on the importance of Broca's and Wernicke's areas for expression and reception (or recognition) was based on lesions that actually affected much larger regions. When lesions are restricted to the limited areas identified by Broca and Wernicke, they usually do not give rise to the full symptoms characteristic of Wernicke's or Broca's aphasia. The characteristic symptoms are usually the result of damage to the surrounding regions as well.

Second, the Wernicke–Geschwind model emphasizes the importance of cortical regions and interconnecting pathways running through subcortical white matter. There now is evidence that subcortical structures, specifically the left thalamus, the left caudate nucleus, and adjacent white matter, also are important for language. For example, the left caudate nucleus is thought to be important for the auditory–motor integration required for linguistic processing because lesions in this nucleus lead to a defect in auditory comprehension.

Third, whereas an auditory input—a spoken word—is indeed projected from the auditory cortex to the angular gyrus and then to Wernicke's area before being conveyed to Broca's area, visual information about a written word is not conveyed to Wernicke's area, but goes from the visual association cortex directly to Broca's area. Words that are read are therefore *not* transformed into an auditory representation. Rather, visual and auditory perceptions of a word are processed independently by modality-specific pathways that have independent access to Broca's area and to the higher-order regions concerned with the meaning and expression of language.

Finally, cognitive studies of language disagree with the Wernicke–Geschwind model more radically than on just the pathway used for processing auditory information. For example, there is good evidence that not all auditory input is processed in the same way. Nonsense sounds—words without meaning—are processed independently from conventional, meaningful words. Thus, it is thought that there are separate pathways for sounds, the *medium* of language, and for meaning, the *content* of language. Similarly, although Broca's area is the common output for both spoken and written words that have meaning, there may be an independent output for nonsense words. Finally, several studies by psycholinguists indicate that patients with both Broca's and Wernicke's aphasia have not only language deficits but also deficits in other cognitive abilities. These difficulties blur the classic distinction between the impairment of reception and production of language.

These and related findings indicate that language involves a larger number of areas and a more complex set of parallel interconnections than just the serial interconnection of Wernicke's area to Broca's area. A more plausible scheme for the neural processing of language is shown in Figure 34–4.

Certain Affective Components of Language Are Impaired by Damage to the Right Hemisphere

We have so far considered some of the cognitive components of language. Human language, and more generally human communication, has important affective components as well. These include the musical elements of stress, pitch, and rhythm; variation in these elements of speech is technically known as *prosody*.

Certain affective components of language rely on specialized processes of the right hemisphere, and disturbances in these components due to damage to the right hemisphere are called *aprosodias*. The neural organization for prosody in the right hemisphere seems to mirror the anatomical organization for the cognitive aspects of language in the left hemisphere. Thus, patients with lesions in the right frontal cortex have a flat tone of voice whether they are happy or sad. Patients with posterior lesions do not comprehend the prosody in other people's speech.

Some Disorders of Reading and Writing Can Be Localized

Reading disorders are either congenital (*dyslexia*) or acquired (*alexia* or *acquired dyslexias*). The alexias are particularly instructive for understanding

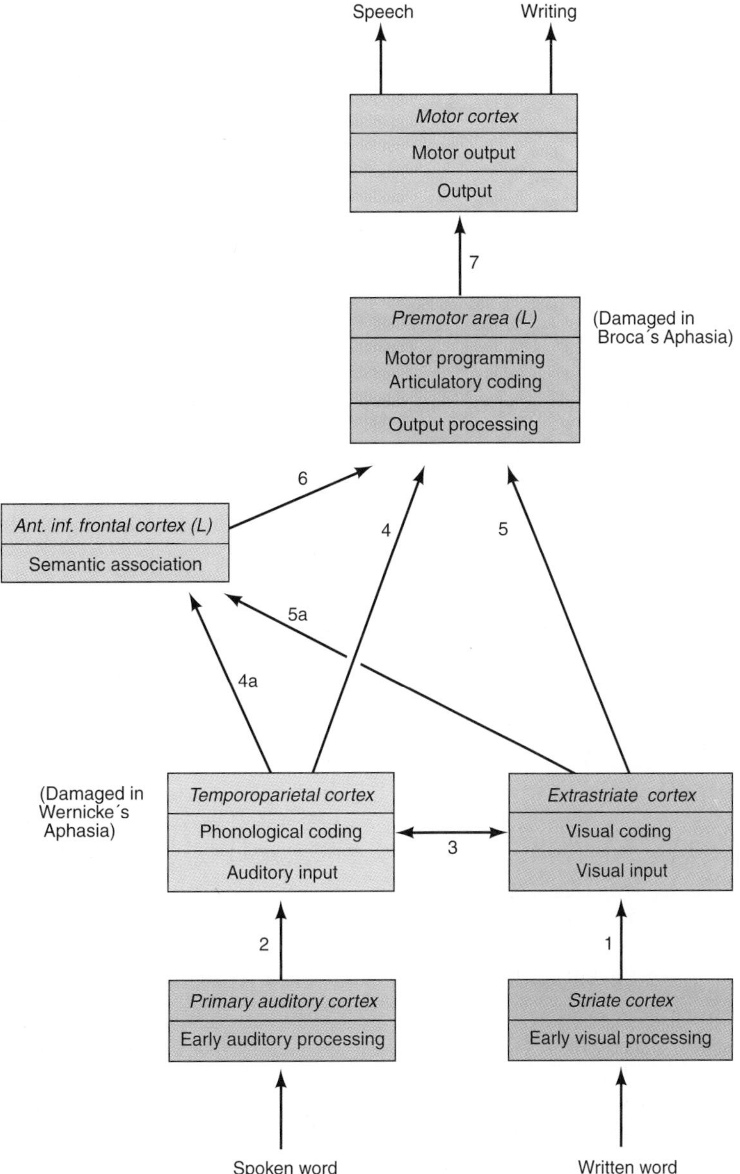

Figure 34–4 Recent models of the neural processing of language are much more complex than the Wernicke–Geschwind model. The model illustrated here represents a fairly simple circuit of what is actually known and shows the relationship between various anatomical structures and steps in the processing of language. Other networks are also plausible. Both visual and auditory inputs as well as spoken and written expression are illustrated. (From Petersen et al., 1988.)

language processing and show that the Wernicke–Geschwind model of language is capable of predicting defects in reading and writing.

Alexia and Agraphia Are Acquired Disorders of Reading and Writing

Alexia (disruption of the ability to read) and agraphia (disruption of the ability to write) are remarkable because they demonstrate that small lesions of the brain in an adult can selectively destroy the ability to read or to write, or both, without interfering with speech or other cognitive functions. This discovery was made in the 1890s by the French neurologist Jules Dejerine, who described a disorder of reading (alexia without agraphia) in one patient with pure word blindness and a disorder of both reading and writing (alexia with agraphia) in another patient.

Pure Word Blindness: Alexia without Agraphia. Dejerine's first patient could speak. An intelligent and highly articulate man, he suddenly observed that he could not read. Even though he could not comprehend written words, he could copy them correctly and could recognize and understand them after writing the individual letters. Moreover, the patient was able to derive meaning from words spelled aloud and was able to spell correctly.

The patient was blind in the right visual field (indicating damage to the left visual cortex) but otherwise had normal visual acuity. Postmortem examination of this and other patients revealed damage to the left occipital (visual) cortex and the the posterior portion of the corpus callosum (the splenium), which carries visual information between the two hemispheres by interconnecting area 18 of the occipital cortex of one hemisphere with that of the other. Although the visual information from the left visual field could still be processed by the right hemisphere, damage to the splenium prevented its transfer to the angular gyrus and to language areas of the left hemisphere.

As might be predicted from the location of the lesion, many patients with damage to Brodmann's area 18—an extrastriate visual cortex—have selective deficits in visual perception. For example, 50% of patients with pure alexia also have either *color agnosia* (they are capable of matching colors but cannot name them) or *achromatopsia* (they cannot perceive color and therefore see objects only as shades of gray).

Surgical disruption of the posterior portion of the corpus callosum results in the loss of reading ability in the left but not the right visual field. In contrast, sectioning of the anterior portion of the corpus callosum (which does not transmit visual information) does not interfere with reading. However, patients in whom the anterior portion of the corpus callosum has been transected cannot write with their left hands (controlled by the right hemisphere), because the right hemisphere no longer has access to the left hemisphere's language centers. Likewise, as we have learned in Chapter 19, these patients cannot name objects held in the left hand, because the somatic sensory information does not reach the language areas in the left hemisphere.

Word Blindness Accompanied by Writing Impairment (Alexia with Agraphia). The second patient described by Dejerine could speak and understand spoken language, but was unable to either read or write. Autopsy of this and later patients revealed that alexia with agraphia is usually associated with lesions of the angular or supramarginal gyrus of the parietal–temporal–occipital association cortex. As we saw in Chapter 19, this association cortex is concerned with the integration of visual, auditory, and tactile information. Once integrated, the information is conveyed to the speech areas of the temporal lobe and then to those of the frontal lobe. Individuals with damage to the association cortex of the angular or the supramarginal gyrus cannot read or write because they cannot connect visual symbols (letters) with the sounds they represent. Similarly, these patients cannot recognize words spelled out loud, nor can they spell. They also are unable to recognize embossed letters by feeling the letters because the angular and supramarginal gyri mediate the transfer of sensory information from the skin surface into language areas of the brain.

Dyslexias Are Disorders of Reading that May Be Developmental or Acquired

Dyslexia can occur as the result of brain injury. This form of dyslexia is considered acquired. A much more prevalent form, however, appears to be developmental and is estimated to affect between

10 and 30% of the population of the United States. This form of dyslexia is grouped with a heterogeneous class of language difficulties that currently are called learning disorders. While it is much easier to classify acquired dyslexia by simply comparing the abilities of a patient before and after sustaining a lesion, the reading ability of individuals with developmental dyslexia is less conveniently assessed, in part because that performance must be normalized against a specific population.

Because of normal variation a variety of social factors, including educational background, some educators do not believe that there is a specific disorder called developmental dyslexia, but think that people who do not read well belong to the lower part of a normal population distribution. Others define developmental dyslexia as an unexpected difficulty in learning to read and spell, unexpected in the sense that there is no obvious cause for the difficulty, such as poor eyesight or deafness, inadequate education, or a low IQ. Although the criteria for the diagnosis of dyslexia change with age and school grade, the current accepted definition is based on a discrepancy between performance on a battery of reading and spelling skill tests and IQ. Thus, the cognitive and intellectual capacities of dyslexic children by definition are normal and may even be superior.

As we mentioned earlier in this chapter, many linguists believe that language emerged quite recently, perhaps late in the prehistoric period a hundred thousand years ago. Writing (and history) obviously developed afterward, no earlier than 6,000 years ago. The history of literacy shows clearly that the ability to read and spell, from an evolutionary point of view, is quite recent. Thus, universal literacy has not been considered necessary until this century. Consequently, the skills required for reading and spelling may not have been selected for, and a wide variety in performance ability might be expected.

Most investigators believe that children with dyslexia seem particularly deficient in phonemic processing—the ability to associate letters with the sounds they represent. However, they can usually understand other signs or symbols of communication, such as traffic signs and words that have a unique visual appearance (e.g., the Coca–Cola trademark). Indeed, American dyslexic children can easily learn to read English when entire words are represented by single characters rather than by a sequence of characters. Other investigators think that defects in visual processing contribute to developmental dyslexia. Similarities between this and other disorders, such as alexia caused by stroke, have led to the suggestion that developmental dyslexia might result from abnormalities in connections between visual and language areas.

Some dyslexic children also exhibit a strong tendency to read words backward (e.g., confusing "saw" for "was") and have particular difficulty distinguishing between letters with the same configuration but in different orientations (e.g., p and q, or b and d). These mistakes occur in both reading and writing. This type of error, together with the disproportionate percentage of left-handers among dyslexics, suggests that dyslexia might involve a deficit in the development of dominance by the left hemisphere. In fact, the normal bilateral discrepancy in the size of the planum temporale is much reduced in dyslexic males. In addition, in the few dyslexic cases that have been studied anatomically, striking cytoarchitectonic abnormalities have been found in the left planum temporale, including an incomplete segregation of cell layers and clusters of misplaced neurons. In contrast, the right hemisphere appeared normal. These observations suggest that migration of neurons to the left cortex during development is slowed in dyslexic patients, a defect that might be attributable to injury of the fetus.

Several findings suggest that one of the problems in dyslexia may be a perceptual defect—an inability to process sensory input with adequate speed. For example, the normally fast rate of conduction in the magnocellular pathway of the visual system is below average in people with dyslexia, whereas that of the parvocellular pathway is normal. Moreover, in a few dyslexic patients examined at autopsy the cells in the magnocellular layers of the lateral geniculate nucleus were smaller relative to cells in the parvocellular layers or to the same cells in control subjects. This defect has psychophysical consequences. Some dyslexic patients have difficulty processing fast, high-contrast, visual stimuli. In fact, when the visual image is deliberately blurred to reduce high-frequency contrast, dyslexic readers can process words in a more normal way. A similar defect is evident in the fast-conducting component of the auditory pathways. Thus, some people with dyslexia may not be able to process normally a variety of fast sensory input, and these defects may interfere with normal language acquisition and processing. All of these neurobiological findings must be considered tentative, however, until there is agreement as to whether the people studied suffer from a single disorder.

Comprehension of Graphemes and Ideographs Involves Different Cerebral Pathways

An interesting disturbance in reading and writing sometimes occurs in Japanese, a language with two distinct systems of writing, one phonetic and one largely ideographic. The phonetic system, *katakana*, uses 71 graphemes (phonetic symbols). *Katakana* words are thus comprehended syllable by syllable but, unlike Western words, are not easily identified at a glance. The ideographic system, *kanji*, uses more than 40,000 ideograms derived from Chinese to represent root words. Affixes, however, are written in *katakana*. In addition, the pronunciation of the *kanji* is often indicated by small *hiragana* symbols (an alternative to *katakana*). The *kanji* system thus contains distinctive phonetic and morphemic references.

Because of these differences, one might expect that certain focal lesions would affect reading or writing in one system but not the other. This is in fact the case. Both systems rely on language centers in the left hemisphere, but each is processed by a different mechanism. Lesions of the angular gyrus of the parietal–temporal–occipital association cortex severely disrupt reading of *katakana* (phonetic writing) but leave comprehension of *kanji* (ideographic writing) largely intact. Such lesions can disrupt reading of *kanji* to some degree, but the disruption entails primarily phonetic processing; patients may be unable to read aloud the *kanji* word but can accurately explain its meaning. In contrast, these patients are unable to understand the same idea expressed in *katakana*.

These observations support the conclusion from brain imaging studies that the angular gyrus of the left hemisphere, concerned with auditory representation, is not involved with the processing of the visual representation of words. Other dissociations between the processing of *katakana* and *kanji* scripts also occur and have provided further insight into the mechanisms of information processing in the production and comprehension of language.

An Overall View

Language is a uniquely human ability. In both its written and spoken forms it represents meaningful interactions between individuals—not just in the present but also across time. The study of language therefore presents problems of common interest to biology and the humanities. Given this special opportunity we may well ask: What can neurobiologists say to psychologists and humanists about the biological process of human cognition?

The first and most important insight is that language abilities can be localized to one of the two cerebral hemispheres. The hemispheric asymmetry that ultimately gave rise to language emerged early in human evolution, perhaps as early as 300,000 years ago, perhaps as much as 200,000 years before the emergence of language. The capability for language may be present at birth, and universal features of language are thought to derive in part from the structure of the cortical regions concerned with language in the left hemisphere.

From a biological standpoint, language is not a single capability but a family of capabilities, two of which—comprehension and expression—are mediated at distinctive sites in the brain. As first suggested by Wernicke, profound aphasia can result by disconnecting these two sites. Success in correlating major components of language with different anatomical regions has led to the development of a simple model of language, the Wernicke–Geschwind model, which can account for a family of language-related disorders. This model, although clinically helpful, is overly simple and no doubt incorrect in detail.

Despite some notable insights, the neurobiological understanding of language is still rudimentary. The Wernicke–Geschwind model, although modified since its introduction, is only a beginning in the localization of cognitive functioning. It has, however, provided an important bridge between the modern analysis of language and its disorders by psycholinguists and the neuroanatomical localization of language function by neural scientists.

Selected Readings

Bates, E. 1992. Language development. Curr. Opin. Neurobiol. 2:180–185.

Bellugi, V., Poizner, H., and Klima, E. S. 1989. Language, modality and the brain. Trends Neurosci. 12:380–388.

Bowler, R. F. (ed). 1993. Dyslexia Research and Its Applications. Ann. Dyslexia 43:5–148.

Caplan, D. 1987. Neurolinguistics and Linguistic Aphasiology: An Introduction. Cambridge, England: Cambridge University Press.

Chomsky, N. 1968. Language and the mind. Psychol. Today 1(9):48–68.

Corballis, M.C. 1991. The Lopsided Ape: Evolution of the Generative Mind. New York: Oxford University Press.

Damasio, A. R., and Geschwind, N. 1984. The neural basis of language. Annu. Rev. Neurosci. 7:127–147.

Galaburda, A. M. 1993. The neurology of developmental dyslexia. Curr.Opin. Neurobiol. 3: 237–242.

Gardner, R. A., and Gardner, B. T. 1969. Teaching sign language to a chimpanzee. Science 165:664–672.

Geschwind, N. 1965. Disconnexion syndromes in animals and man. Brain 88:237–294, 585–644.

Gleitman, L. R., and Gleitman, H. 1991. Language. In H. Gleitman (ed.), Psychology, 3rd ed. New York: Norton.

Hinton, G. E., and Shallice, T. 1991. Lesioning an attractor network: Investigations of acquired dyslexia. Psychol. Rev. 98:74–95.

Kosslyn, S. M., and Koenig, O. 1992. Wet Mind. The New Cognitive Neuroscience. New York: Free Press.

LeMay, M. 1976. Morphological cerebral asymmetries of modern man, fossil man, and nonhuman primate. Ann. N.Y. Acad. Sci. 280:349–366.

McCarthy, R. A., and Warrington, E. K. 1990. Cognitive Neuropsychology. A Clinical Introduction. San Diego: Academic Press, chaps. 6–9.

Miller, G. A. 1981. Language and Speech. San Francisco: W. H. Freeman.

Pennington, B. F. 1991. Diagnosing Learning Disorders: Neuropsychological Framework. New York: Guilford Press.

Petersen, S. E., Fox, P. T., Posner, M. I., Mintun, M., and Raichle, M. E. 1989. Positron emission tomographic studies of the processing of single words. J. Cogn. Neurosci. 1:153–170.

Pinker, S. 1994. The Language Instinct: How the Mind Creates Language. New York: Morrow.

Premack, D. 1976. Intelligence in Ape and Man. Hillsdale, N.J.: Erlbaum.

Shallice, T. 1988. From Neuropsychology to Mental Structure. Cambridge, England: Cambridge University Press.

35

Learning and Memory

Memory Functions Can Be Localized to Specific Regions of the Brain

Memory Is Not Unitary and Can Be Classified as Implicit or Explicit on the Basis of How Information Is Stored and Recalled

Explicit Memory Involves the Medial Temporal Lobe System

Implicit Memory Does Not Require Deliberate Recall and Involves Various Perceptual and Reflex Pathways

Implicit Forms of Learning Can Be Nonassociative or Associative

Classical Conditioning Involves Learning Predictive Relationships Between Stimuli

Operant Conditioning Involves Associating a Specific Behavior with a Subsequent Reinforcing Event

Food-Aversion Conditioning Illustrates How Biological Constraints Influence the Efficacy of Reinforcers

Therapeutic Techniques May Involve Conditioning

The Relationship Between Implicit and Explicit Forms of Memory in Learning

Certain Implicit Forms of Memory Involve the Amygdala and Cerebellum

The Neural Basis of Memory Can Be Summarized in Two Generalizations

Memory Has Stages

Memory Is Often Localized in Different Places Throughout the Nervous System

In Chapter 30 we considered how genetic programs interact with environmental factors to determine behavior. In this and the next chapter we turn our attention to the specific mechanisms by which environmental events shape behavior. The most important way is through learning and memory. Learning is the process by which we and other animals acquire knowledge about the world. Memory is the retention or storage of that knowledge. Even simple animals have the ability to learn from their environment. But clearly this ability reaches its highest form in human beings. Most aspects of human behavior involve some form of learning.

In the study of learning we are interested in several questions. What are the major forms of learning? What types of environmental relationships are learned most easily? What are the stages of memory formation? Here we review the major biological principles of learning, obtained from clinical and behavioral studies of humans and experimental animals. In the next chapter we examine the cellular mechanisms of memory, which have provided us with an important perspective on learning.

Memory Functions Can Be Localized to Specific Regions of the Brain

Until the middle of the twentieth century most students of learning doubted that any memory functions could ever be localized to specific regions of the brain. Indeed, many students of behavior even doubted that memory was a discrete function of the mind, independent of attention, language, and perception. The view that memory storage is widely distributed throughout the brain contrasted sharply with the emerging view on localization for other mental functions.

As we have seen, as early as 1861 Broca had discovered that damage restricted to the posterior portion of the frontal lobe on the left side of the brain (Broca's area) produced a specific language deficit. Following this localization of language, it was not long before neuroscientists turned to the problem of whether memory could also be localized. Is memory a discrete mental capability or is it simply ancillary to other mental processes? Is memory a function of specific systems of the brain? If so, are all memory processes located in one system or in several?

Almost a century passed before these questions could be answered. During much of this period the general opinion was that memory was a general property of the cerebral cortex as a whole. In fact, recent work has shown that memory does involve many regions of the brain. Nevertheless, there are different types of memory and certain regions of the brain are much more important for some types than for others. Furthermore, different types of memory are stored in different neural systems.

The first person to obtain evidence that memory processes are localized in the human brain was Wilder Penfield, a neurosurgeon at the Montreal Neurological Institute. Penfield had been a student of Charles Sherrington, the pioneering neurophysiologist. At the turn of the century Sherrington had mapped the motor representation in the cerebral cortex of anesthetized monkeys by systematically probing the cortex with electrodes and recording the activity of motor nerves. In the 1940s Penfield began to use similar electrical stimulation methods to map motor, sensory, and language functions in the human cortex in patients undergoing neurosurgery for the relief of focal epilepsy. Since the brain itself does not have pain receptors, brain surgery is painless and can be carried out under local anesthesia in fully awake patients. Thus, these patients are able to describe what they experience in response to electrical stimuli applied to different cortical areas. On hearing about these experiments Sherrington, who had always worked with monkeys and cats, told Penfield, "It must be great fun to put a question to the [experimental] preparation and have it answered!"

Figure 35–1 Brenda Milner.
Milner discovered that human memory involves multiple memory systems. Milner received her Ph.D at Cambridge University in England. She then joined the Department of Psychology at McGill University in Montreal, Canada, where she studied with Donald Hebb. She and Hebb collaborated with Wilder Penfield and studied the patients on whom he had operated. In this way she discovered multiple memory systems in the brain by studying patients who have had their hippocampus removed on both sides for epileptic seizure. These patients lacked memory for certain tasks involving conscious knowledge of people, places, and things, but had perfectly good memory for motor skills that were learned unconsciously. These studies provided the initial experimental evidence for the later distinction between implicit and explicit memory.

Penfield explored the cortical surface in more than a thousand patients. Occasionally he found that electrical stimulation produced what he called an *experiential response*, or flashback, in which the patients described a coherent recollection of an earlier experience. These memory-like responses were elicited only from the temporal lobes, never anywhere else. Nevertheless these responses were rare even in the temporal lobes, occurring in only 8% of all attempts at stimulation.

Memory Is Not Unitary and Can Be Classified as Implicit or Explicit on the Basis of How Information Is Stored and Recalled

Penfield's studies were not entirely convincing. All of the patients Penfield studied had epileptic seizure foci in the temporal lobe, and the sites effective in eliciting experiential responses were those near these foci. Thus the experiential responses might have been the result of localized seizure activity. In the early 1950s the role of the temporal lobes in memory became clearer from studies by one of Penfield's collaborators, Brenda Milner (Figure 35–1), of the effects of therapeutic bilateral removal of the hippocampus and neighboring regions of the temporal lobe in a few patients with temporal lobe epilepsy.

The first and best-studied case of the effects on memory of removal of portions of the temporal lobes was a 27-year-old assembly-line worker who had suffered from untreatable seizures for more than 10 years. The patient, known as H.M., was unable to work or lead a normal life. The seizures were greatly reduced following removal of the medial portion of the temporal lobes on both sides of the brain (Figure 35–2).

Immediately after the operation H.M. experienced a devastating memory deficit—he lost the capacity to form new long-term memories. He retained his memories for the events of his life that preceded the surgery, however. He remembered his name, retained a perfectly good use of language, and kept his normally varied vocabulary and his IQ in the range of bright–normal. He remembered the events that preceded his surgery, such as the job he held, and he vividly remembered the events of his childhood. Moreover, H.M.'s short-term memory was perfectly intact.

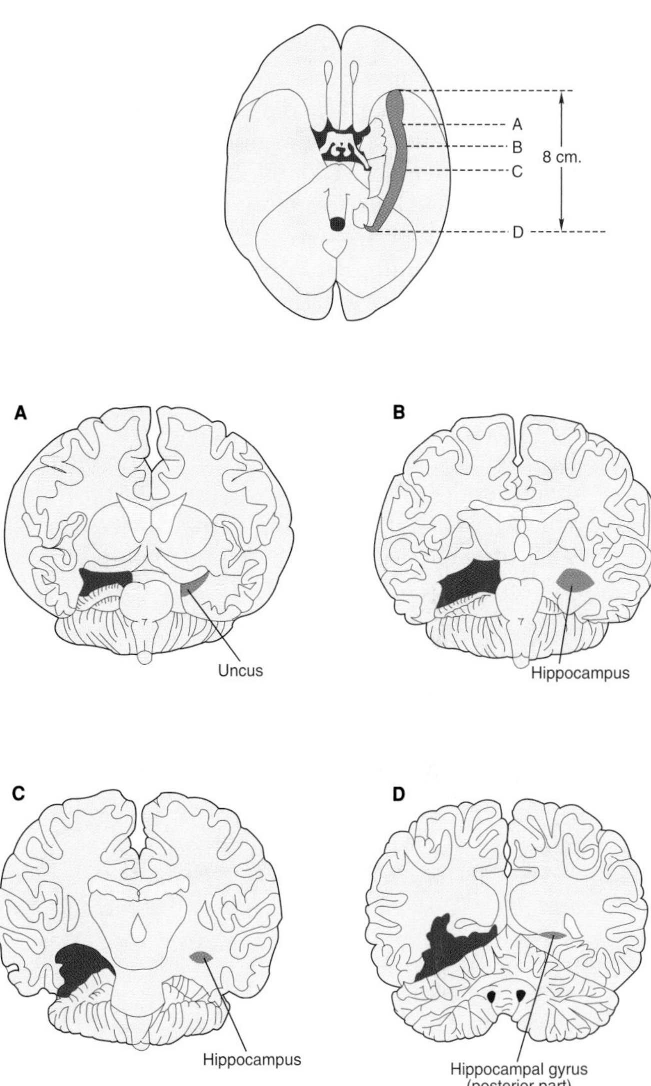

Figure 35–2 Cross sections show the estimated extent of removal of portions of the medial temporal pole, hippocampus, and parts of the hippocampus gyrus. In the patient H. M. Surgery was a bilateral, single-stage procedure, but one side is shown intact here for illustrative purposes.

What H.M. lacked dramatically, however, was the ability to transfer most types of learning from short-term memory, lasting seconds and minutes, to long-term memory, lasting days or longer (Figure 35–3). When he learned a new task he failed to retain the information for more than a minute. Asked to remember the number 584, he could repeat it immediately for many minutes. However, when distracted even briefly he forgot the number. As a result of this difficulty in transferring information from short- to long-term memory, H.M. did not recognize new people he met, even when he met them again and again. He also had profound difficulty with spatial orientation. On moving to a new house, it took him about a year to learn his way around it. This deficit has been described as "forgetting the incidents of daily life as fast as they occur." H.M. is not unique. All patients with extensive bilateral lesions of the temporal lobe show similar memory deficits.

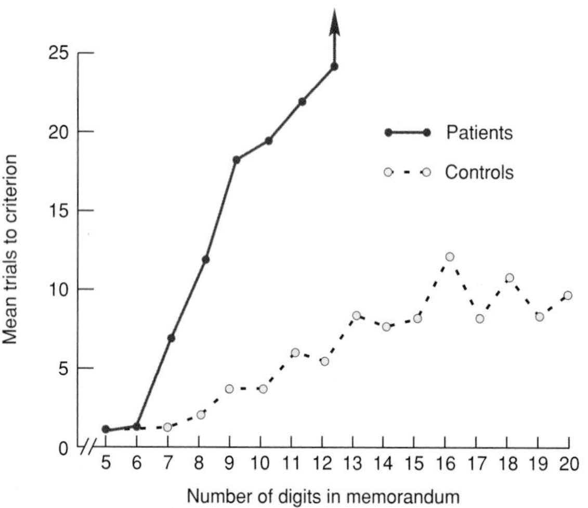

Figure 35–3 Five amnesic patients, including patient H. M., and 20 control subjects were read a sequence of five digits. If the sequence was repeated back correctly, one digit was added to the next sequence. If not, the same sequence was given until it was repeated correctly. Amnesic patients were able to repeat sequences of normal length but required an abnormal number of trials to learn large strings of digits. No amnesic patient recalled more than 12 digits within the testing limit of 25 trials. (From Drachman and Arbit, 1966.)

Milner originally thought that the memory deficit following bilateral temporal lobe lesions was applied equally to all forms of learning and long-term memory. But this proved not to be the case. Even though patients with temporal lobe lesions have profound memory deficits, Milner found that they are able to learn and remember certain types of tasks as well as normal subjects for long periods.

This residual memory capability was first revealed when it was discovered that the patient H.M. could learn new motor skills normally. For example, he learned to draw the outlines of a star while looking at his hand and the star in a mirror (Figure 35–4). Like normal subjects learning this task, H.M. initially made many mistakes, but his performance reached an error-free level after several days of training. Subsequently, other patients were tested and found able to retain various forms of simple reflexive learning, including habituation, sensitization, classical conditioning, and operant conditioning (which we discuss later in this chapter).

Amnestic patients are not limited to learning motor skills, however. They are also able to improve their performance on certain perceptual tasks. For example, they do well with a form of learning called *priming*, in which recognition of words or objects is facilitated by prior exposure to the words or visual clues. Thus, subjects can recall the cued item better than other items for which no cues had been provided. Similarly, when shown the first few letters of a previously studied word, amnestic subjects often correctly select the previously presented word, even though they cannot remember seeing the word before (Figure 35–5).

The tasks that patients with bilateral lesions of the temporal lobe are capable of learning have two things in common. First, they all have an automatic quality. Second, the tasks do not require conscious recall or complex cognitive skills, such as comparison and evaluation. The patient need only respond to a stimulus or cue. He need not deliberately remember anything. Thus, when given a highly complex mechanical puzzle to solve, the patient may learn it as quickly as a normal person, but on questioning will not remember seeing the puzzle or having worked on it previously. When asked why his performance on a task is much better after several days of practice than on the first day, he may respond, "What are you talking about? I've never done this task before."

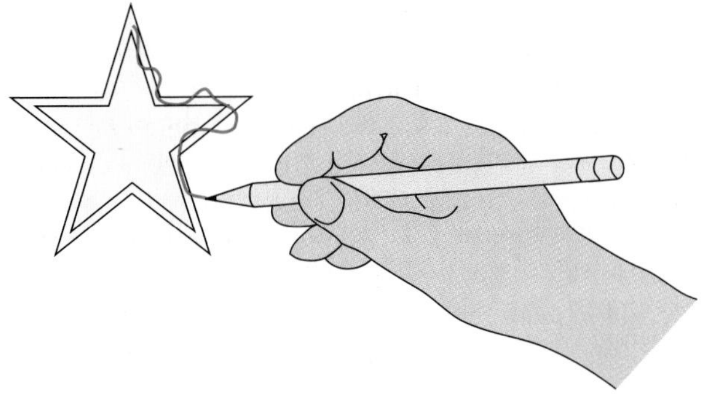

Figure 35–4 H. M. showed definite improvement in any task involving learning skilled movements. In this test he was taught to trace between the two outlines of the shape of a star, while viewing his hand in a mirror. He improved considerably with each fresh test, although he had no idea that he had ever done the task before. (The graph plots the number of times, in each trial, that he strayed outside the outlines as he drew the star.) (From Blakemore, 1977.)

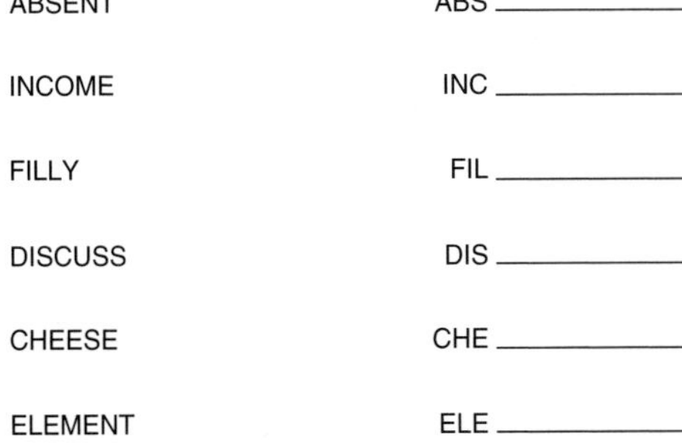

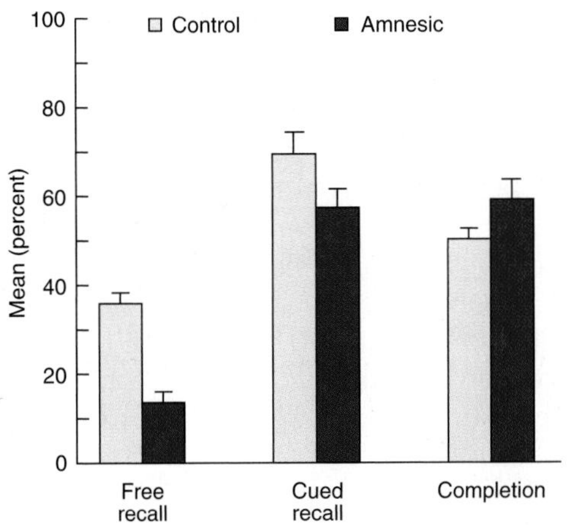

Figure 35–5 Amnesic patients and control subjects saw common words and then were asked to recall the words (*free recall*) or were cued with the first three letters of the words and asked to recall them (*cued recall*). Recall by the amnesic patients was impaired in these two conditions but amnesiacs performed normally when they were given the first three letters of words and instructed simply to form the first word that came to mind (completion). The baseline guessing rate in the word completion condition was 9%. (From Squire, 1987.)

Studies of patients with temporal lobe lesions have revealed two fundamentally different ways of learning, a difference that cognitive psychologists had begun to appreciate from their studies of normal subjects. We learn *what* the world is about—acquiring knowledge of people, places, and things that are available to consciousness—using a form of memory that is commonly called *explicit*. Or we learn *how* to do things—acquiring motor or perceptual skills that are unavailable to consciousness—using *implicit* memory.

Explicit Memory Involves the Medial Temporal Lobe System

Explicit memory encodes information about autobiographical events as well as factual knowledge. Its formation depends on cognitive processes such as evaluation, comparison, and inference. Explicit memories can be recalled by a deliberate act of recollecting. They are sometimes established in a single trial or experience, and often can be concisely expressed in declarative statements, such as "Last summer I visited my grandmother at her country house" (autobiographical event), or "Gold is heavier than water" (factual knowledge).

Evidence from amnestic subjects suggests that autobiographical knowledge (Box 35–1) and factual knowledge may involve different brain mechanisms within the temporal lobe. Thus, following trauma to the temporal lobe, some patients completely lose their capacity to recall any event experienced more than a few minutes previously, including dramatic personal events that occurred before the trauma, such as accidents and deaths in

BOX 35–1 The Transformation of Explicit Memories

How accurate is explicit memory? This question was explored in one series of studies in which the subjects were asked to read stories and then retell them. When the recalled stories were analyzed, they proved to be shorter and more coherent than the original stories. Thus the retold stories reflected reordering, reconstruction, and condensation of the original. The subjects were unaware that they were editing the original stories, and often felt more certain about their own edited version than about the original text. The subjects were not confabulating; they were merely interpreting the original material so it made sense on recall.

Observations such as these lead us to believe that explicit memory for past events is a creative process and reflects a synthesizing or reconstructive process. The information stored as explicit memory is the product of processing by our perceptual apparatus. As we saw in earlier chapters, sensory perception itself is not a faithful record of the external world. It is a transformational process in which incoming information is synthesized according to rules inherent in the brain's afferent pathways. It is also a process by which the individual interprets the external environment not only from the standpoint of a specific point in space, but also from the standpoint of a specific point in his own history. As discussed in Chapter 23 , optical illusions nicely illustrate the difference between perception and objectively measured stimuli.

Likewise, once sensory information is stored, later recall is not a faithful reproduction of the original stored information. Recall involves a process in which representations of past experiences are used as clues that help the brain reconstruct a past event. During this recall, we use a variety of cognitive strategies to generate a consistent and coherent memory, including comparison, inferences, shrewd guesses, and suppositions.

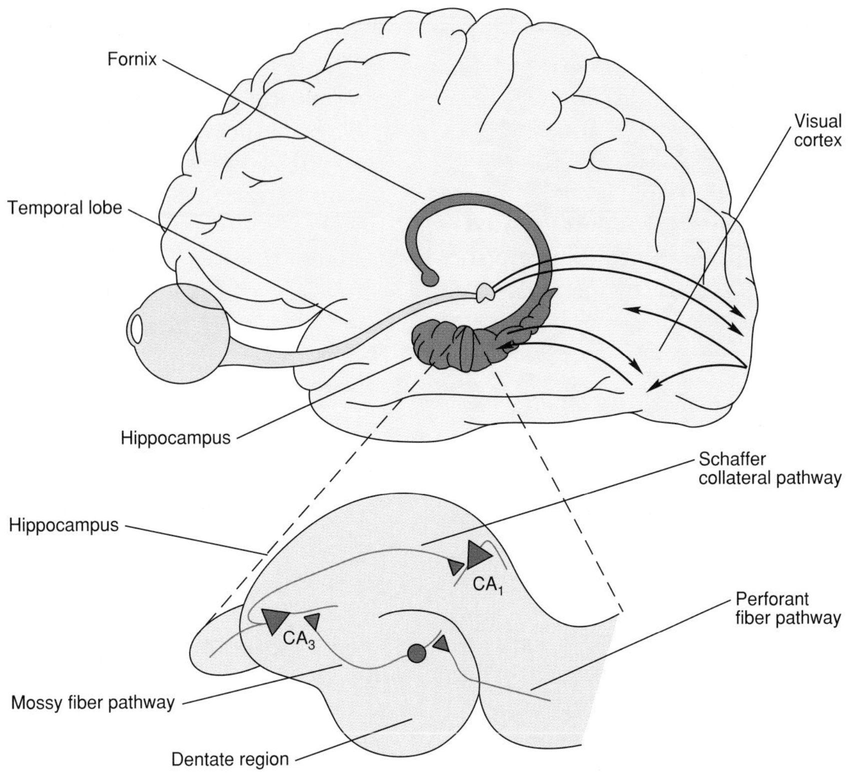

Figure 35–6 The hippocampus may store long-term memory for weeks and gradually transfer it to specific regions of the cerebral cortex. The diagram illustrates this process for visual memory. Neural input travels to the visual cortex and then to the hippocampus, where it is stored for several weeks before it is transferred back to the cortex for long-term memory. The hippocampus (enlargement) has three major synaptic pathways, each capable of long-term potentiaiton (LTP), which is thought to play a role in the storage process.

the family. Remarkably, their ability to recall their vast store of factual information remains intact.

However, lesions of the medial temporal lobe interfere only with the long-term storage of *new* memories. Patients like H. M. retain a reasonably good memory of earlier events. How does this come about?

The hippocampus appears to be only a temporary depository for long-term memory. Ultimately it transfers the learned information to other areas, presumably the cerebral cortex, for permanent storage (Figure 35–6). For example, the sensory information needed for recognizing a face is processed in visual areas of the cerebral cortex concerned with face recognition—the inferotemporal cortex (see Box 21–1 and Chapter 23). Concomitantly, this visual information about faces is also conveyed to the entorhinal cortex, which is the source of the perforant pathway, the major input pathway to the hippocampus. The hippocampus and the related temporal lobe areas are then thought to process this newly learned information for a period of weeks to months—and then transfer the information to the cerebral cortex, perhaps to the area of the visual cortex concerned with face recognition. Alternatively, the hippocampus may not store long-term information at all, but instead may serve as a facilitator that helps store the information initially processed by the inferotemporal cortex.

Thus, the hippocampus may be either a waystation for long-term memory or a facilitation system that is essential for the storage of memories elsewhere in the brain.

Implicit Memory Does Not Require Deliberate Recall and Involves Various Perceptual and Reflex Pathways

Implicit memory has an automatic or reflexive quality, and its formation and recall are not absolutely dependent on awareness or cognitive processes. This type of memory accumulates slowly through repetition over many trials, is expressed primarily by improved performance, and cannot ordinarily be expressed in words. Examples of implicit memory include perceptual and motor skills and the learning of certain types of procedures and rules, such as those of grammar. Verbal learning that depends significantly on repetition, such as rote memorization of the conjugation of verbs of a foreign language, has the characteristics of implicit learning. Such learning can be recalled automatically, without deliberate effort.

The implicit memory for a particular task is thought to be tied to the activity of the particular sensory and motor systems involved in the learning task and retained by storage mechanisms inherent in each of these systems. As a result, implicit memory can be studied in a variety of reflex systems in either vertebrates or invertebrates. Indeed, even simple invertebrates are capable of reflexive learning. Therefore we now turn to a consideration of general forms of implicit learning found in a variety of species.

Implicit Forms of Learning Can Be Nonassociative or Associative

Psychologists often study implicit forms of learning by exposing animals to a controlled sensory experience (Box 35–2). Two major procedures (or paradigms) have emerged from such studies, and these have identified two major classes of implicit learning: nonassociative and associative.

Nonassociative learning results when the animal is exposed once or repeatedly to a single type of

BOX 35–2 Measurement of Learning

Learning can be assessed by repeatedly training a subject in a task and observing the progressive change in performance. This provides a learning curve, an assessment of how performance improves with learning trials and time. However, it is difficult to isolate the variables affecting the learning process using this method. Therefore, an alternative method for assessing learning has been developed.

One group of subjects is given a task, while a second (control) group is given a similar task that lacks a particular variable. The two groups are then tested under identical conditions. The index of learning is the difference in performance between the two groups. For example, one group of rats is placed in a cage containing food in one corner; the control rats are placed in a cage that does not contain food. The next day both groups are placed in a new cage without food, and the researcher measures the time spent by each rat in the corner where the food had been.

This procedure controls for changes that are unrelated to learning, such as developmental changes. It also controls for *performance variables*, such as level of arousal, since the conditions for learning are separated from the performance conditions used to assess the learning.

stimulus. Two forms of nonassociative learning are very common in everyday life: habituation and sensitization. *Habituation* is a decrease in response to a repeated, benign stimulus. For example, most people are startled by the first firecracker on the Fourth of July, but as the day progresses they gradually become accustomed to the noise. *Sensitization* (or pseudoconditioning) is a strengthening of responses to a wide variety of stimuli following an intense or noxious stimulus. For example, a sensitized animal responds more vigorously to a mild tactile stimulus after it has received a painful pinch. Moreover, a sensitizing stimulus can override the effects of habituation. For example, after the startle response to a noise has been reduced by habituation, the response can be restored by delivering a strong pinch. This process is called *dishabituation*. Sensitization and dishabituation are not dependent on the precise timing of the intense stimulus relative to the weaker stimulus; no association between the two stimuli is needed.

Not all examples of nonassociative learning are as simple as habituation or sensitization. Many types of more complex learning have no obvious associational element (although hidden forms of association may be present). These types of learning include imitative learning, a major component of the acquisition of language.

Many types of associative learning have been distinguished based on the experimental procedures used to establish the learning. Two experimental paradigms have been studied extensively. *Classical conditioning* involves learning a relationship between two stimuli, while *operant conditioning* involves learning a relationship between a stimulus and the organism's behavior. (Despite the widespread use of these models in research, not all associative learning readily fits into either paradigm.)

Classical conditioning was introduced into the study of learning at the turn of the century by the Russian physiologist Ivan Pavlov. Since Aristotle, learning has been thought to be achieved through the association of ideas, a concept that was systematically developed by John Locke and the British empiricist philosophers, important forerunners of modern psychology. Pavlov recognized that learning frequently consists of becoming responsive to a stimulus that originally was ineffective in eliciting a response. His brilliant insight was to formulate a testable hypothesis of learning by association. By changing the appearance, timing, or number of stimuli in a tightly controlled set of experiments and observing the changes in selected simple reflexes, Pavlov established a procedure from which reasonable inferences could be made about the relationship of changes in behavior—learning—and the environment.

Pavlov's work marked an important shift in the study of learning from an emphasis on introspection to the objective analysis of stimulus and response. According to Pavlov, what animals and humans learn when they associate ideas can be examined in its most elementary form by studying the association of stimuli.

The essence of classical conditioning is the pairing of two stimuli. The *conditioned stimulus*, such as a light or tone, is chosen because it produces either no overt response or a weak response usually unrelated to the response that eventually will be learned. The *unconditioned stimulus*, such as food or a shock to the leg, is chosen because it normally produces a strong, consistent, overt response (the unconditioned response), such as salivation or withdrawal of the leg. Unconditioned responses are innate; they are produced without learning. When a conditioned stimulus is followed by an unconditioned stimulus, the conditioned stimulus begins to elicit a new or different response called the *conditioned response*. Sometimes the conditioned response resembles the unconditioned response, but the two can also differ.

If the unconditioned stimulus is rewarding (food or water), the conditioning is termed *appetitive*; if the unconditioned stimulus is noxious (an electrical shock), the conditioning is termed *defensive*.

On repeated pairing of the conditioned and unconditioned stimulus, the conditioned stimulus appears to become an anticipatory signal for the unconditioned stimulus. With sufficient experience the animal responds to the conditioned stimulus alone as if it were anticipating the unconditioned stimulus. For example, if a light is followed repeatedly by the presentation of meat, eventually the light itself produces salivation. Thus, classical conditioning is a means by which an animal learns to predict events in the environment.

Pavlov regarded classical conditioning not only as a way to study learning but also as a way to approach the mind—the workings within the brain of higher mental processes. By training animals to respond selectively to stimuli, he hoped to discover which aspects of a stimulus an animal is capable of recognizing and processing. In fact, following Pavlov, psychologists have successfully explored the color sensitivity of animals using different colored lights as conditioned stimuli and observing which colors could be selectively distinguished, using classical conditioning.

The intensity or probability of occurrence of a conditioned response decreases if the conditioned stimulus is repeatedly presented without the unconditioned stimulus. This process is known as *extinction*. If a light that has been paired with food is then repeatedly presented in the absence of food, it gradually ceases to evoke salivation. Extinction is an important adaptive mechanism; it would be maladaptive for an animal to continue to respond to cues in the environment that are no longer significant. The available evidence indicates that extinction is not the same as forgetting, but instead involves learning something new. Moreover, what is learned is not simply that the conditioned stimulus no longer precedes the unconditioned stimulus, but that the conditioned stimulus now signals that the unconditioned stimulus will *not* occur.

Classical Conditioning Involves Learning Predictive Relationships Between Stimuli

For many years psychologists thought that classical conditioning depended only on temporal contiguity, on the conditioned stimulus preceding the unconditioned stimulus by a critical minimum interval. According to this view, each time a conditioned stimulus is followed by a reinforcing or unconditioned stimulus, an internal connection is strengthened between the stimulus and the response or between one stimulus and another, until eventually the bond becomes strong enough to produce conditioning. The only relevant variable determining the strength of conditioning was thought to be the number of pairings of conditioned and unconditioned stimuli.

This theory proved inadequate, however. A substantial body of empirical evidence now indicates that learning cannot be adequately explained only by such simple contiguity. Indeed, depending solely on temporal contiguity would be maladaptive. If animals learned to predict an event simply because it appeared to occur repeatedly with another, they might often take in erroneous information about the true causal relationship between events in the environment.

A striking illustration of the inadequacy of contiguity to explain classical conditioning is called the *blocking phenomenon*, described by Leon Kamin in 1968 in a three-part experiment. First he repeatedly paired a light as a conditioned stimulus with a strong electric shock. He then assessed conditioning by determining to what degree the light as an unconditioned stimulus suppressed ongoing behavior (a reflection of its ability to evoke a strong conditioned fear in the animal). In the second part of the experiment Kamin presented the light simultaneously with a new conditioned stimulus, a tone, so that light–tone was presented as a compound conditioned stimulus and was repeatedly paired with the shock. When, in the third part of the experiment, Kamin presented the tone alone, he found little or no conditioning to the tone. Despite repeated pairing of the light–tone compound stimulus with the shock reinforcer, the tone by itself failed to evoke fear.

These findings were elaborated by Robert Rescorla and Alan Wagner in a theory of classical conditioning that emphasizes the importance of information processing in classical conditioning. According to this theory the amount of conditioning resulting from a trial is dependent on the degree to which the unconditioned stimulus is unexpected or surprising. If the unconditioned stimulus (US) is completely novel and unexpected because it has not been previously paired with a conditioned stimulus (CS), the rate of learning is maximal. But as the US gradually becomes expected because it is predicted by the CS, the rate of learning decreases until the US is fully expected, and the rate of learning becomes zero: no further learning takes place. Thus, in the case of blocking, the tone component of the compound light–tone stimulus is an ineffective CS because the other element of the compound stimulus (the light) successfully and fully signals the occurrence of the US.

This notion has been formalized in simple mathematical terms by Rescorla and Wagner, and predicts several properties of classical conditioning. Classical conditioning develops best when, in addition to contiguity of stimuli, there is also an actual *contingency* between conditioned and unconditioned stimuli. If an animal is presented with a random sequence of conditioned and unconditioned stimuli, some productive CS–US pairings will occur just by chance. Nevertheless, a conditioned response to the CS does not develop. These experiments show that the animal is not just counting the number of CS–US pairings. Rather the animal determines the overall correlation or predictive relationship between the CS and the US.

In fact, a CS that is repeatedly presented so that it specifically does not occur in association with a US will come to predict the *absence* of the US. When that stimulus is later paired with a US, conditioning occurs only very slowly, presumably because the animal must first unlearn the previous predictive property of the stimulus. In some instances CS that have been associated with the absence of the US actually acquire inhibitory properties; that is, they will suppress the response to another stimulus that has been conditioned to the US. Thus, in addition to being paired in time, efficient learning requires that they CS and reinforcer (the US) be positively correlated; the CS must indicate an increased probability that the US will occur.

These considerations suggest why animals and humans acquire classical conditioning so readily. It appears likely that classical conditioning and perhaps all forms of associative learning have evolved to enable animals to distinguish events that reliably and predictably occur together from those that are only randomly associated. In other words, the brain seems to have evolved to detect causal relationships in the environment, as evidenced in correlated or associated events.

All animals that exhibit associative conditioning, from snails to humans, seem to learn by detecting environmental contingencies rather than detecting the simple contiguity of a CS and a US. Why is the faculty of recognizing contingent relationships similar in humans to that in much simpler animals? One good reason is that all animals face common problems of adaptation and survival. Classical conditioning provides a successful solution to this problem, and once a successful biological solution has evolved, it continues to be used.

What environmental conditions might have shaped or maintained a common learning mechanism in a wide variety of species? All animals must be able to recognize prey and avoid predators; they must search out food that is nutritious and avoid food that is poisonous. There are two ways in which an animal arrives at such knowledge. The correct information can be preprogrammed into the animal's nervous system (as we have seen in Chapter 30), or it can be acquired through learning. Genetic and developmental programming may suffice for many of the behaviors of very simple organisms, such as nematode worms and other very simple parasitic invertebrates, but more complex invertebrate animals and all vertebrates must be capable of extensive learning to cope efficiently with varied or novel situations. Complex animals need to establish some minimum order in their world. An effective way to do this is to establish causal or predictive relationships between stimuli or between behavior and stimuli.

Operant Conditioning Involves Associating a Specific Behavior with a Subsequent Reinforcing Event

A second major paradigm of associational learning, discovered by Edward Thorndike and systematically studied by B. F. Skinner and others, is operant conditioning (also called trial-and-error learning). In a typical laboratory example of operant conditioning an investigator places a hungry rat in a test chamber that has a lever protruding from one wall. Because of previous learning as well as innate response tendencies and random activity, the rat occasionally presses the lever. If the rat promptly receives food when it presses the lever, its subsequent rate of pressing the lever will increase above the spontaneous rate. The animal can be described as having learned that, among its many behaviors (e.g., grooming, rearing, and walking), one behavior (lever pressing) is followed by food. With this information the rat is likely to make the appropriate responses whenever it is hungry.

If we think of classical conditioning as the formation of a predictive relationship between two stimuli (the conditioned stimulus and the unconditioned stimulus), operant conditioning can be considered as the formation of a predictive relationship between a response and a stimulus. Unlike classical conditioning, which modifies the properties of specific reflex responses to selected stimuli, operant conditioning modifies the frequency of behaviors (called *operants*) that apparently occur either spontaneously or without a recognizable eliciting stimuli. Thus, operant behaviors are said to be *emitted* rather than elicited, and when the behaviors produce favorable changes in the environment (when they are rewarded or lead to the removal of noxious stimuli), the animal tends to repeat them. In general, behaviors that are rewarded tend to be repeated at the expense of behaviors that are not, whereas behaviors followed by aversive, though not necessarily painful, consequences are usually not repeated. Experimental psychologists agree that this simple idea, called the *law of effect*, governs much voluntary behavior.

Superficially, operant conditioning and classical conditioning seem dissimilar, involving completely different relationships between stimulus and response. Classical conditioning involves the pairing of two stimuli, whereas operant conditioning involves the pairing of a stimulus with a response. However, the laws that govern operant and classical conditioning are quite similar, suggesting that the two forms of learning may be manifestations of the same neural mechanisms.

For example, timing is critical in both forms of conditioning. Typically, the reinforcer or unconditioned stimulus must closely follow the operant response. In operant conditioning, if the reinforcer (reward) is delayed too long, only weak conditioning occurs. There is an optimal interval between response and reinforcement, which varies depending on the specific task and the species. Similarly, in classical conditioning, depending on the task, there is an optimal interval between the CS and the US; learning is generally poor if this interval is too long or if the US precedes the CS.

Predictive relationships are equally important in both types of learning. In classical conditioning the subject learns that a certain stimulus predicts a subsequent event; in operant conditioning the animal learns to predict the consequences of its own behavior.

Food-Aversion Conditioning Illustrates How Biological Constraints Influence the Efficacy of Reinforcers

For many years it was thought that classical conditioning could occur simply by associating any two arbitrarily chosen stimuli or, in the case of operant conditioning, any response and any reinforcer. More recent studies have shown, however, that there are important biological (evolutionary) constraints on learning. As we have seen, animals generally learn to associate stimuli that are relevant to their survival; they do not learn to associate events that are biologically meaningless. These findings illustrate nicely a principle we have encountered in earlier chapters on development of perception and behavior: the brain is not a *tabula rasa*—it cannot be influenced arbitrarily by stimuli—but is limited to detecting and associating only certain stimuli and not others.

For example, not all reinforcers are equally effective with all stimuli. This principle is dramatically illustrated in studies of food aversion (also called bait shyness, as it seems to be the means by which animals in their natural environment learn to avoid bait foods that contain poisons). If a distinctive taste stimulus, such as vanilla, is followed by nausea produced by a poison, an animal quickly develops a strong aversion to the taste of vanilla. Unlike most other forms of conditioning, food aversion develops even when the unconditioned response (the nausea induced by poison) occurs with a very long delay (up to hours) after the conditioned stimulus (the taste of vanilla). This makes biological sense, since the ill effects of naturally occurring toxins are not experienced immediately after swallowing the poison, but only after some delay.

Indeed, for most species, including humans, food-aversion conditioning occurs only when taste stimuli are associated with subsequent illness. Food aversion develops poorly, or not at all, if the taste is followed only by pain. At the same time, visual or auditory stimuli that have been paired with nausea do not produce averse conditioning. Thus, the choice of an appropriate reinforcer depends on the nature of the response to be learned. Evolutionary pressures have predisposed the brains of different species of animals to learn an association between certain stimuli, or between a certain stimulus and a response, much more readily than between others. Within a given species, genetic and experiential factors also can determine the effectiveness of a reinforcer. The results obtained with a particular class of reinforcer vary enormously among species and among individuals within a species, particularly in humans.

Food aversion has been applied in the treatment of chronic alcoholism. The patient is first allowed to smell and taste alcoholic beverages and then given a powerful emetic, such as apomorphine which causes him to feel nauseous and to vomit.

The pairing of alcohol and nausea rapidly results in aversion to the taste of alcohol. Food-aversion learning may be implicated in some common situations. First, it may be a means by which people unintentionally learn to regulate their diets to avoid the unpleasant consequences of inappropriate or non-nutritious food. Second, the malaise associated with certain forms of cancer chemotherapy may induce aversive conditioning to foods in the ordinary diet of the patient. Averse conditioning might account in part for the naseau and depressed appetite of many cancer patients. Specifically, the nausea that follows chemotherapy for cancer can produce aversion to foods that were tasted shortly before the treatment.

Therapeutic Techniques May Involve Conditioning

Various psychotherapeutic procedures involve reeducation of the patient in the context of a trusting relationship with the therapist. Aspects of therapeutic change are likely to involve components of classical and operant conditioning, but the specific contribution that each of these procedures makes to therapy has been delineated in only a few relatively simple instances.

The process of extinction, characteristic of classical conditioning, may underlie the therapeutic changes resulting from a clinical technique known as systematic desensitization (although other interpretations of this method have been offered). Systematic desensitization was introduced into psychiatry by Joseph Wolpe, who used it to decrease neurotic anxiety or phobias evoked by certain definable environmental situations, such as heights, crowds, and public speaking. The patient is first taught a technique of muscular relaxation. Then, over a period of days, the patient is told to imagine a series of progressively more severe anxiety-provoking situations while using the relaxation techniques to inhibit any anxiety that might be elicited. At the end of the series the strongest potentially anxiety-provoking situations can be brought to mind without anxiety. This desensitization, induced in the therapeutic situation, often carries over to real-life situations that the patient encounters.

Biofeedback, another form of operant conditioning that has proved effective clinically, is used to enhance (or suppress) responses of which the patient is unaware. The behavior of interest, such as very slight muscle contractions in a stroke patient, is recorded by an electronic device that provides the patient with an immediate auditory or visual cue signaling that the response has occurred. If the patient desires to increase the frequency or strength of the response, the feedback cue can act as a positive reinforcement.

The Relationship Between Implicit and Explicit Forms of Memory in Learning

Although classical conditioning often leads to implicit forms of memory, even this ostensibly simple form of conditioning may, in some circumstances, lead to explicit memory and involve mediation by cognitive processes. Consider the following experiment. A subject lays his hand, palm down, on an electrified grill. A light (conditioned stimulus) is turned on and he immediately receives an electric shock on one finger. His hand lifts (unconditioned response). After several light–shock conditioning trials, he lifts his hand even when the light alone is presented. The subject has been conditioned; but what exactly has been conditioned?

It appears that the light is triggering a specific pattern of muscle activity that results in lifting of the hand. However, what if the subject now places his hand on the grill upside down, and the light is presented? If a specific pattern of muscle activity has been conditioned, the light should produce a response that moves the hand *into* the grill. Then again, if the subject has acquired the information that the light means "grill shock," his response should be consistent with that information. In fact, the subject will often move his hand away from the grill; that is, he will make an *adaptive* response, even though it involves motor movements antagonistic to the original ones. Therefore, the subject did not simply learn a fixed response to a stimulus, but rather acquired information that the brain could use in shaping an appropriate response in a novel situation.

As this discussion makes clear, many learning experiences have elements of both implicit and explicit learning. Constant repetition, however, can transform explicit memory into the implicit type. For example, learning to drive an automobile at first involves conscious recollection, but eventually driving becomes an automatic and nonconscious motor activity.

Certain Implicit Forms of Memory Involve the Amygdala and Cerebellum

Lesions in several regions of the brain have been found to affect simple, classically conditioned responses; these regions probably represent loci important for implicit types of learning. For example as we have seen in Chapter 32, lesions of the amygdala interfere with conditioned fear responses. Similarly, lesions of the cerebellum interfere with various forms of motor learning (see Chapter 29). The best-studied case is the eyeblink, a protective reflex in rabbits. A conditioned eyeblink reflex can be established by pairing an auditory stimulus with a puff of air to the eye, which causes an eyeblink. The conditioned response can be abolished by a small, precisely placed lesion in the cerebellum. Following this lesion the conditioned auditory stimulus is no longer able to produce an eyeblink, although the unconditioned stimulus (air puff) still evokes the eyeblink. Moreover, during conditioning of the eyeblink in normal rabbits, neurons in this same area of the cerebellum show increases in activity that closely parallel the development of the conditioned response.

The results of these experiments indicate that the cerebellum plays an important role in conditioning the eyeblink as well as other simple forms of classical conditioning of motor responses. Are the neural changes that mediate the learning located in the cerebellum itself, or are they located elsewhere, with the cerebellum playing a crucial role in the performance of the response but not the learning? The bulk of current evidence suggests that neural changes indeed occur in the cerebellum, perhaps in both the cerebellar cortex and the deep cerebellar nuclei.

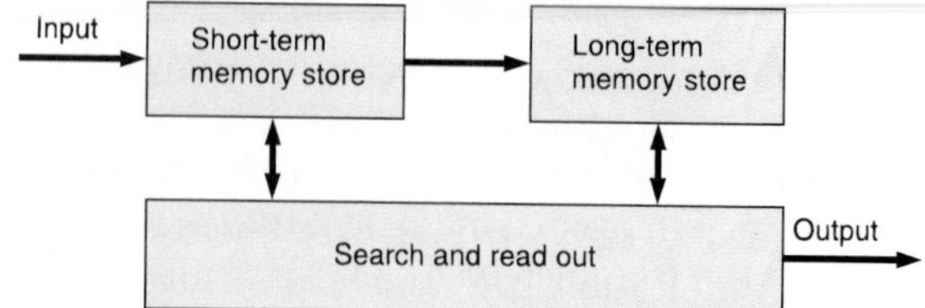

Figure 35–7 A simplified model of memory processes during a typical learning task, such as memorizing a list of nonsense syllables. The three processes indicated have been inferred primarily from the observed time course of the decay of normal memory and from the time course of disruptions of memory resulting from brain trauma.

The Neural Basis of Memory Can Be Summarized in Two Generalizations

Memory Has Stages

It has long been known that a person knocked unconscious can have selective memory loss for events that occurred before the blow (*retrograde amnesia*), as well as for events that occur after the recovery of consciousness (*anterograde amnesia*). This phenomenon has been documented thoroughly in animal studies using such traumatic agents as electroconvulsive shock, physical trauma to the brain, and drugs that depress neuronal activity or inhibit protein synthesis in the brain. Clinical studies, too, provide evidence that brain trauma can produce particularly profound retrograde amnesia for events acquired within a few hours or at most days before the trauma. Thus, recently acquired memories are readily disrupted, whereas older memories remain quite undisturbed. The extent of retrograde amnesia varies from several seconds to several years, depending on the nature and strength of the learning, on the species of animal, and on the nature and severity of the disrupting event.

Studies of retention and disruption of memory have supported a commonly used model of memory storage by stages (Figure 35–7). Input to the brain is processed into a *short-term memory store.* This has very limited capacity (fewer than a dozen items) and, in the absence of rehearsal, persists only for a period of minutes. The information is later transformed by some process into a more permanent *long-term store.* (Some researchers prefer to divide long-term memory into an intermediate form that is relatively sensitive to disruption and a truly long-term form that is relatively insensitive to disruption.) A search-and-retrieval system searches the memory store and makes information available for specific tasks.

According to this model, the retention of memory can suffer because of destruction of the contents of a memory store. Alternatively, memories may be disrupted by interfering with the search-and-retrieval mechanism. This latter conclusion is supported by the observation that trauma sometimes only *temporarily* disrupts memory—considerable memory for past events gradually returns. If stored memory were completely destroyed, it obviously could not be recovered.

Studies of patients undergoing electroconvulsive treatments for depression have confirmed and extended the findings of experiments made on animals. Patients were examined using a memory test that could reliably quantify the degree of memory for relatively recent events (1–2 years old), old events (3–9 years old), and very old events (9–16 years old). The patients were asked to identify by means of voluntary recall the names of television programs broadcast during a single year between 1957 and 1972. The patients were tested before the electroconvulsive therapy and then again afterward (with a different set of television programs). Both before and after electroconvulsive therapy (ECT), recall of the programs was more correct for more recent years. After the ECT, however, the patients showed a significant but transitory loss of memory for more recent programs, while their recall of older programs remained essentially the same as it was before the ECT (Figure 35–8).

One interpretation of these findings is that retrieval of recent memories is easily disrupted until the memories have been converted into a long-term memory form. Once converted, they are relatively stable. With time, however, both the stored information and the capacity to retrieve it gradually diminish, even in the absence of external trauma. Thus, as assessed by its susceptibility to disruption, the memory process undergoes continual change with time.

Several experiments studying the effects of drugs on learning support the idea that memory is time dependent and subject to modification when the memory is first formed. For example, subconvulsant doses of excitant drugs, such as strychnine, can improve the retention of learning of animals even when the drug is administered after the training trials. If the drug is given to the animal soon after training, retention of learning on the following day is greater. The drug has no effect, however, when given several hours after training. Conversely, inhibitors of protein synthesis selectively block long-term memory but not short-term memory when given during the training procedure.

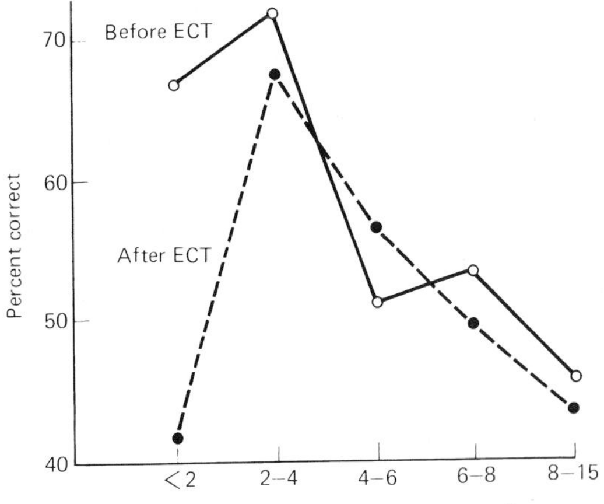

Figure 35–8 Recent memories are more susceptible than older memories to disruption by electroconvulsive shock therapy (ECT). The plot shows the responses of a group of patients who were tested on their ability to recognize the names of television programs that were on the air during a single year between 1957 and 1972. Testing was done before and after the patients received ECT for treatment of depression. After ECT the patients showed a significant (but transitory) loss of memory for recent programs (1–2 years old), but not for older programs. (Adapted from Squire, Slater, and Chace, 1975.)

Memory Is Often Localized in Different Places Throughout the Nervous System

For many different types of learning, memory is not localized to any one brain structure. For example, any one of three visual pathways can sustain conditioning of heart rate responses in pigeons. Although the neuronal mechanisms by which these memories are stored are not known with certainty, it is possible to construct computer models in which functional changes in the specific connections throughout the network represent a simple memory within the model (see Chapter 9). These learning models perform in ways reminiscent of learning in living animals. For example, such models are able to *generalize*; the model can learn even when tested with stimuli somewhat different from the original training stimulus. The information stored in distributed processing models also survives partial damage to the system, as is often the case with an injured brain.

Parallel processing may explain in part why a limited lesion often does not eliminate specific learning, even for a simple implicit task Another important factor that may account for the failure of small lesions to adversely affect learning may reside in the very nature of learning. Although, as we shall see in the next chapter, the synaptic changes that underlie learning are more likely to occur in certain types of neurons, the complex nature of many learning tasks makes it likely that these neurons are distributed throughout the reflex pathway. Therefore, even after a lesion some components of stored information (i.e., some of the synaptic changes) are likely to remain. Furthermore, the brain has the capacity to take even the limited store of remaining information and reconstruct a relatively good reproduction of the original.

Nevertheless, as we have discussed previously, not all regions of the brain participate equally in the process of storing and retrieving memories. The hippocampus and the cerebellum play unique roles in explicit and implicit memories, respectively. It remains to be shown to what degree these structures store the physical representation of memories and to what degree they are necessary for permitting storage or readout of memories stored elsewhere in the brain.

An Overall View

Studies of memory have shown that there are different types of memory processes. Explicit memory includes learning about people, places, and things that can be reported verbally, learning that requires conscious awareness. Implicit memory includes forms of perceptual and motor learning that do not require conscious awareness.

From the extensive literature on the neurobiology of memory, three features seem particularly important: (1) memory has stages, (2) long-term memory is represented in multiple regions throughout the nervous system, and (3) explicit and implicit memories may involve different neuronal circuits. Explicit memory storage requires the temporal lobe system. Implicit memories involve the cerebellum, amygdala, and, for simple forms of learning, the specific sensory and motor systems recruited for the task.

A major task confronting the neurobiology of learning is to determine how alterations in the brain are related to behavioral changes. A second task is to determine the mechanisms underlying synaptic changes associated with memory. To this end, a number of simplified vertebrate and invertebrate animal preparations are being investigated, and some of these studies are reviewed in Chapter 36.

Selected Readings

Bellack, A. S., Hersen, M., and Kazdin, A. E. (eds.) 1990. International Handbook of Behavior Modification and Therapy, 2nd ed. New York: Plenum Press.

Dickinson, A. 1980. Contemporary Animal Learning Theory. Cambridge, England: Cambridge University Press.

Domjan, M., and Burkhard, B. 1986. The Principles of Learning and Behavior, 2nd ed. Monterey, Calif.: Brooks/Cole.

Milner, B. 1985. Memory and the human brain. In: How We Know. Shaffo, M. (ed). San Francisco: Harper and Row.

Rescorla, R. A. 1988. Behavioral studies of Pavlovian conditioning. Annu. Rev. Neurosci. 11:329–352.

Squire, L. R. 1987. Memory and Brain. New York: Oxford University Press.

Steinmetz, J. E., Lavond, D. G., Ivkovich, D., Logan, C. G., and Thompson, R. F. 1992. Disruption of classical eyelid conditioning after cerebellar lesions: Damage to a memory trace system or a simple performance deficit. J. Neurosci. 12:4403–4426.

Tulving, E., and Schacter, D. L. 1990. Priming and human memory systems. Science 247:301–306.

36

Cellular Mechanisms of Learning and Memory

Simple Forms of Implicit Learning Lead to Changes in the Effectiveness of Synaptic Transmission

Habituation Involves Depression of Synaptic Transmission

Sensitization Involves Enhancement of Synaptic Transmission

Long-Term Memory Requires the Synthesis of New Proteins and the Growth of New Synaptic Connections

Classical Conditioning Involves an Associative Enhancement of Presynaptic Facilitation That Is Dependent on Activity

Storage of Explicit Memory in Mammals Involves Long-Term Potentiation in the Hippocampus

Long-Term Potentiation in the CA1 Region Is Associative

Associative Long-Term Potentiation May Be Important for Spatial Memory

Long-Term Potentiation in the CA3 Region Is Nonassociative

Is There a Molecular Alphabet for Learning?

The Somatotopic Map in the Brain Is Modifiable by Experience

Changes in the Somatotopic Map Produced by Learning May Contribute to the Biological Expression of Individuality

Common Cellular Mechanisms May Be Responsible for Changes in the Somatotopic Map

Neuronal Changes Associated with Learning Provide Insights into Psychiatric Disorders

Throughout this book we have emphasized that all behavior is a function of the brain and that malfunctions of the brain are expressed in characteristic disturbances of behavior. All functions of the brain, in turn, are the product of interactions between genetic and developmental processes on the one hand, and learning, on the other. In this final chapter we focus on the mechanisms by which learning alters the structure and function of nerve cells and their connections.

Many aspects of behavior result from the ability to learn from experience. Indeed, we are who we are largely because of what we learn and remember. As we have seen in Chapter 33, we are able to communicate experiences and thereby create cultures that are maintained over generations because we learn language. We also learn dysfunctional behaviors and these can, in the extreme, constitute psychological disorders. Fortunately, what is learned can sometimes be unlearned. Thus, insofar as psychotherapy is successful in treating behavioral disorders, it presumably does so because treatment teaches the patient to acquire new patterns of behavior.

In the last chapter we saw that learning is not a single process but has at least two major forms. Implicit forms of learning are covert and often reflexive and do not require conscious attention, while explicit forms do require conscious awareness. In this chapter we first examine the cellular and molecular mechanisms that underlie simple implicit and explicit forms of learning in both invertebrate and vertebrate experimental animals. We then consider how these mechanisms may contribute to individuality through differences in life experience.

Simple Forms of Implicit Learning Lead to Changes in the Effectiveness of Synaptic Transmission

Most of the progress in the cellular study of implicit forms of learning and memory has come from examining elementary modifications of behavior: habituation, sensitization, and classical conditioning. These elementary modifications have been analyzed in the nervous system of invertebrates and in simple vertebrate behavioral systems, such as the eye-blink response (see Chapter 35). Most of these modifications involve changes in the effectiveness of specific synaptic connections.

Habituation Involves Depression of Synaptic Transmission

As we saw in Chapter 35, habituation is the simplest form of implicit learning. It is a nonassociative form in which an animal learns about the properties of a novel stimulus that is harmless, as that stimulus is repeated. An animal first responds to a new stimulus with a series of orienting reflexes. If the stimulus is neither rewarding nor harmful, the animal learns to suppress its response to the stimulus through repeated encounters with it. This learned suppression of response is *habituation.*

Habituation was first investigated in animals by Ivan Pavlov and Charles Sherrington. While studying posture and locomotion, Sherrington observed that certain reflex forms of behavior, such as the withdrawal of the limb in response to a tactile stimulus (a flexion reflex), habituated with repeated stimulation and only recurred after many seconds of rest. He suggested that the habituation was due to a functional decrease in the synaptic effectiveness of the pathways to the motor neurons that had been repeatedly activated.

This problem was later investigated at the cellular level by Alden Spencer and Richard Thompson. They first carried out a series of behavioral experiments and found close parallels between habituation of the spinal flexion reflex in the cat and habituation of more complex behavioral responses in humans. They thus felt confident that habituation of spinal reflexes is a good model for studying habituation. Next, by recording intracellularly from motor neurons in the spinal cord of cats, they found that habituation does not affect the initial synapse in the spinal cord between the sensory neurons innervating the skin and their central target skills. Rather, habituation leads to a decrease in the strength of the synaptic connection at the next synaptic relay in the chain, that between the interneurons and the motor neurons.

Since the organization of the interneurons in the spinal cord of vertebrates is quite complex, it proved difficult to analyze further the cellular mechanisms of habituation in the flexion reflex. As a result, investigation of habituation required still simpler systems in which the behavioral response could be examined in a series of monosynaptic connections.

This sort of analysis has been carried out in the marine snail *Aplysia californica*, which has a simple nervous system containing only about 20,000 central nerve cells. *Aplysia* has a repertory of defensive reflexes for withdrawing its tail, gill, and siphon, a small fleshy spout above the gill used to expel seawater and waste (Figure 36–1). These reflexes are similar to the leg-flexion reflex studied by Spencer and Thompson. For example, a mild tactile stimulus delivered to the siphon elicits withdrawal of both the siphon and gill; a tactile stimulus to the tail elicits tail withdrawal. With repeated stimulation these reflex withdrawals habituate. As we shall see later, these responses can also be sensitized and classically conditioned.

Gill withdrawal has been studied in detail. In response to a newly encountered stimulus to the siphon, sensory neurons innervating the siphon generate excitatory synaptic potentials in the interneurons and motor cells (Figure 36–1). These synaptic potentials summate both temporally and spatially and cause the motor cells to discharge strongly, leading to a strong reflex withdrawal of the gill. If the stimulus is repeatedly presented, the synaptic potentials produced by the sensory neurons in the interneurons and motor cells become progressively smaller. The synaptic potentials produced by some of the excitatory interneurons in the motor neurons also become weaker, with the net result that the strength of the reflex response is reduced.

The decrease in synaptic transmission in the sensory neurons results from a decrease in the amount of the chemical transmitter (glutamate) released

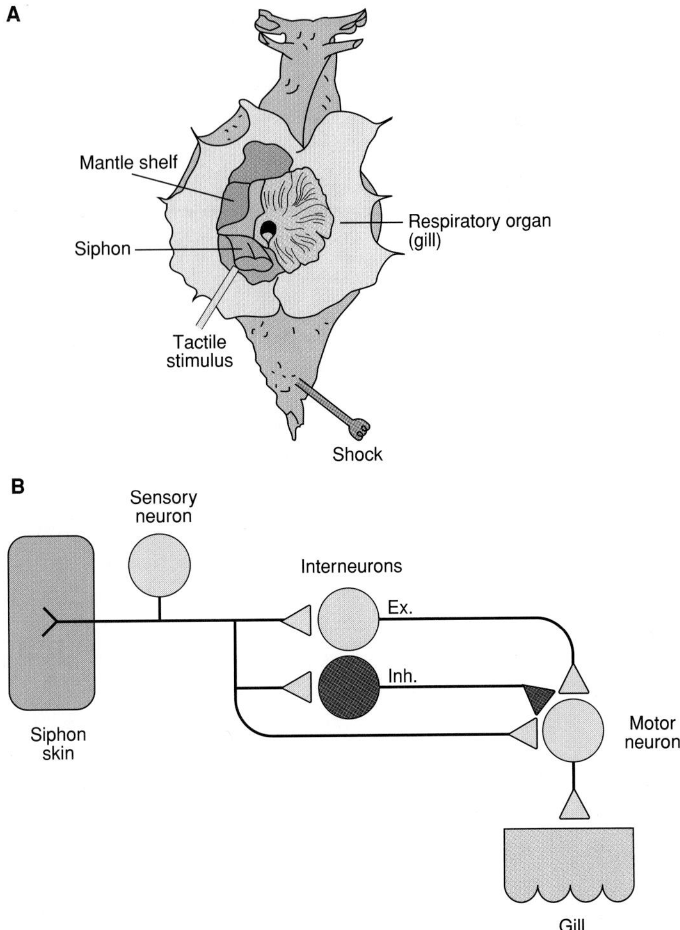

Figure 36–1 The marine snail *Aplysia* has a simple nervous system that makes it an ideal animal model for studying the neural mediation of reflexes at the cellular level. The cellular mechanisms of habituation have been investigated in the animal's gill-withdrawal reflex.

A. A dorsal view of the animal illustrates the respiratory organ, the gill, and the mantle shelf, which ends in a fleshy spout called the siphon.

B. This simplified circuit shows key elements involved in the gill-withdrawal reflex as well as sites involved in habituation. In this circuit about 24 sensory neurons (mechanoreceptors) in the abdominal ganglion innervate the siphon skin. These sensory cells use glutamate as their transmitter and terminate on a cluster of six motor neurons that innervate the gill and on several groups of excitatory (Ex) and inhibitory (Inh) interneurons that synapse on the motor neurons. (For simplicity, only one of each type of neuron is illustrated here.) Repeated stimulation of the siphon leads to a depression of synaptic transmission between the sensory and motor neurons as well as between certain interneurons and the motor cells.

from the presynaptic terminal. How this occurs is not understood. Part of the decrease is thought to be due to a decrease in the ability of transmitter vesicles to be mobilized to the active zone so as to be available for release (see Chapter 15).

This reduction in the effectiveness of the synaptic connections between the sensory neurons and their target cells, the interneurons and motor neu-

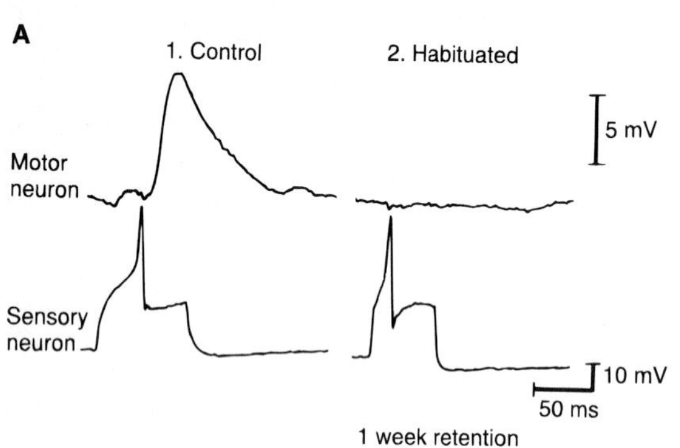

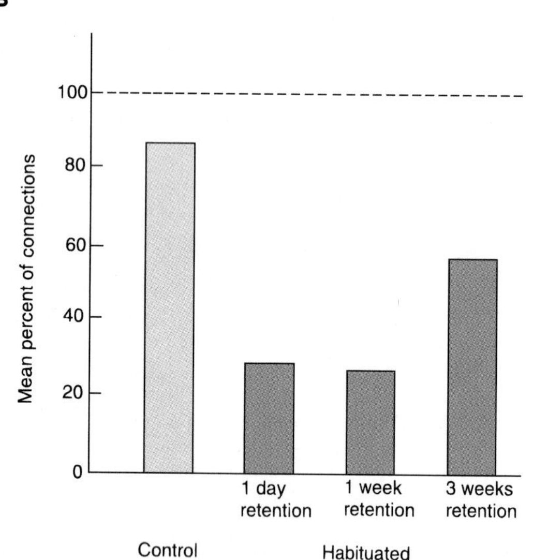

Figure 36–2 Long-term habituation of the gill-withdrawal reflex in *Aplysia* is represented on the cellular level by a dramatic depression of synaptic effectiveness between the sensory and motor neurons. (Adapted from Castellucci, Carew, and Kandel, 1978.)

A. Comparison of the synaptic potentials in a sensory neuron and a motor neuron in a control (untrained) animal and in an animal that has been subjected to long-term habituation. In the habituated animal the synaptic potential in the motor neuron is still undetectable one week after training.

B. The mean percentage of physiologically detectable connections in habituated animals at several points in time after long-term habituation training.

rons, can last many minutes. Similarly, enduring changes occur in the synaptic connections between several interneurons and motor neurons in this circuit. Thus, the storage of even a simple reflexive memory is not restricted to one site but is distributed to several sites within the neural circuit. These distributed *plastic* changes in the functional strength of the synaptic connections represent the short-term memory process for habituation.

Synaptic depression of the connections made by sensory neurons or interneurons, or both, seems to be a fairly common mechanism of habituation. Similarly distributed plastic changes account for habituation of escape responses in crayfish and cockroaches and in the startle reflexes in vertebrates. In each of the instances cellular analyses have shown that memory storage for implicit forms of learning does not depend on specialized "memory neurons" whose only function is to store information. Rather, memory storage results from changes in neurons that are functional components of the normal reflex pathway. It is therefore likely that in the human brain, too, memory can be stored in nerve cells that have a function other than storing information.

What are the limits of this plasticity in neuronal function? How much can the effectiveness of a synapse change and how long can the change last? Can changes in synaptic effectiveness give rise to long-term memory for habituation lasting days, weeks, or years? In *Aplysia* a single training session of 10 stimuli leads to a short-term habituation lasting minutes; four such training sessions spaced over time lead to a long-term memory lasting up to three weeks (Figure 36–2). Whereas 90% of the sensory neurons in control animals make physiologically detectable connections onto an identified motor neuron, in long-term habituated animals the incidence of detectable connections between sensory neurons and motor cells is reduced to 30%. This low incidence persists for one week and does not completely recover until three weeks after habituation training. As we shall see later, this long-term inactivation of synaptic transmission is accompanied by structural changes in the sensory cells.

Not all synapses in *Aplysia* are plastic and adaptable—some synaptic connections in the nervous system do not change their strength, even with repeated activation. However, at synapses involved in learning, such as the connections between the sensory neurons and the motor neurons as well as at some of the interneuronal connections in the withdrawal reflex, a relatively small amount of training can produce large and enduring changes in synaptic strength.

Sensitization Involves Enhancement of Synaptic Transmission

After an animal encounters a harmful stimulus, it typically learns to respond more vigorously to a variety of *other* stimuli, even harmless ones. In particular, its defensive reflexes become sharpened in preparation for withdrawal and escape. This change, called *sensitization*, is a more complex form of nonassociative learning than habituation. Like habituation, sensitization has both a short-term form lasting minutes and a long-term form lasting days and weeks.

Short-term sensitization has been examined at the cellular level in *Aplysia*. After a single noxious stimulus to the head or tail, a number of different synaptic connections in the neural circuit of the gill-withdrawal reflex become modified, including those of the sensory neurons on the motor neurons and interneurons. Thus, a single set of synaptic connections can participate in at least two different forms of learning: they can be depressed by habituation or enhanced by sensitization.

Habituation leads to a *homosynaptic depression*, a decrease in synaptic strength resulting from sustained activity in the stimulated pathway. In contrast, sensitization involves *heterosynaptic facilitation*: the sensitizing stimulus activates a group of interneurons that form synapses on the sensory neurons, including axo-axonal synapses of the sort described in Chapter 15 (Figure 36–3A). These facilitating neurons, some of which are serotonergic, enhance transmitter release from the sensory neurons by increasing the amount of the second messenger cAMP in the sensory neurons.

The sequence of biochemical steps in sensitization of this monosynaptic pathway has been pieced together on the basis of pharmacological and biochemical studies (Figure 36–3B). Serotonin (and the other neurotransmitters released by the facilitating neurons) activate receptors that engage a GTP-binding protein (G_S), which activates the enzyme adenylyl cyclase and increases the concentration of cAMP in the sensory neurons. Cyclic AMP activates the cAMP-dependent protein kinase, which phosphorylates a number of substrate proteins. As discussed in Chapter 14, phosphorylation can lead to an increase or decrease in the activity of a protein by changing its conformation. In sensitization, activation of the cAMP-dependent protein kinase has at least three short-term consequences.

First, the kinase phosphorylates and closes two classes of K^+ channels, thereby reducing two components of the K^+ current that normally repolarizes the action potential. Reduction of these K^+ currents prolongs the action potential and allows the N-type Ca^{2+} channels to be activated for longer periods. More Ca^{2+} is able to enter the terminals, thereby further enhancing transmitter release (Figure 36–3B). Second, the kinase acts to enhance the mobilization of transmitter vesicles and the efficiency of the transmitter release apparatus through a calcium-independent mechanism. Third, the kinase alters an L-type Ca^{2+} channel, whose influx does not directly affect release but increases the mobilization of transmitter vesicles through a calcium-dependent mechanism. In its effects on mobilization and release efficiency, the cAMP-dependent protein kinase works in parallel with protein kinase C, which is also activated by serotonin, presumably through a different receptor (Figure 36–3B).

Long-Term Memory Requires the Synthesis of New Proteins and the Growth of New Synaptic Connections

As with habituation and most other forms of learning, practice makes perfect and prolongs the memory for sensitization. In *Aplysia* a single training trial (or a single application of serotonin) gives rise to short-term sensitization lasting only minutes, whereas four training trials produce long-term sensitization lasting one day, and further repetition produces sensitization that persists for a week or longer. These behavioral studies in *Aplysia* and similar ones in vertebrates suggest that short-term and long-term memory are two points of a graded process. Several findings point to this interpretation. First, both short and long-term sensitization are associated with changes in synaptic strength at the *same* locus: the connections between the sensory and motor neurons (Figure 36–4). Second, in both the long-term and short-term processes the increase in synaptic strength is due to the enhanced release of transmitter. Third, serotonin, a modulatory transmitter that produces the short-term facilitation following a single exposure, pro-

A

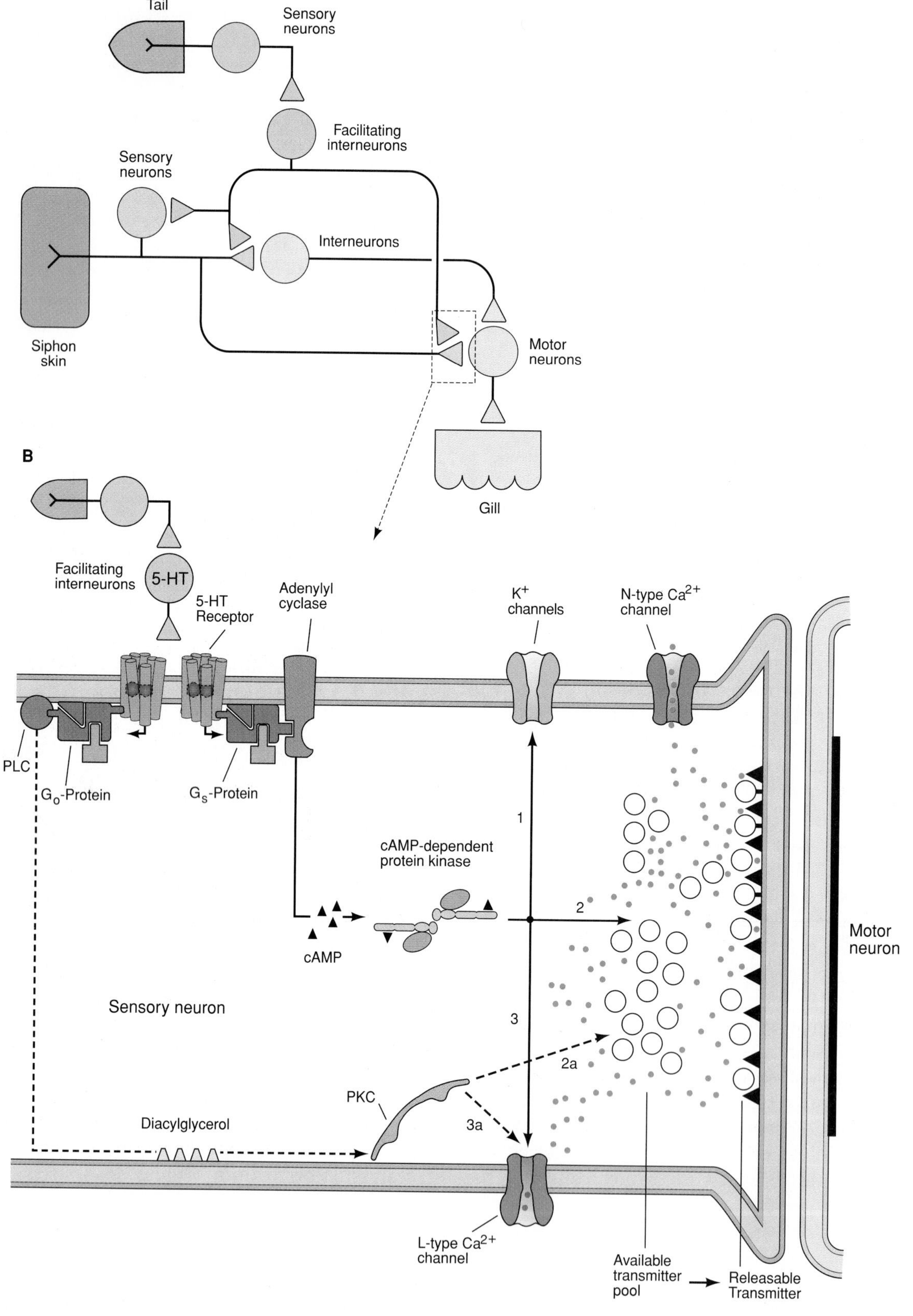

Tail
Sensory neurons
Facilitating interneurons
Sensory neurons
Interneurons
Siphon skin
Motor neurons
Gill
B
Facilitating interneurons
5-HT
5-HT Receptor
Adenylyl cyclase
K$^+$ channels
N-type Ca^{2+} channel
PLC
G$_o$-Protein
G$_s$-Protein
1
cAMP-dependent protein kinase
2
cAMP
Motor neuron
Sensory neuron
3
2a
PKC
3a
Diacylglycerol
L-type Ca^{2+} channel
Available transmitter pool
Releasable Transmitter

← *Figure 36–3* Short-term sensitization of the gill-withdrawal reflex in *Aplysia* involves presynaptic facilitation.

A. Sensitization of the gill is produced by applying a noxious stimulus to another part of the body, such as the tail. Stimuli to the tail activate sensory neurons that excite facilitating interneurons. The facilitating cells, some of which use serotonin as their transmitter form synapses on the terminals of the sensory neurons innervating the siphon skin. There they enhance transmitter release from the sensory neurons by means of presynaptic facilitation.

B. Presynaptic facilitation in the sensory neuron is thought to occur by means of the following biochemical steps. The action of serotonin (5-hydroxytryptamine, 5-HT) and other facilitating transmitters leads to enhanced transmitter release by directly modulating the release process, as well as causing the closure of K^+ channels, which results in broadening of the action potential and a consequent increase in Ca^{2+} influx through a Ca^{2+} channel. Serotonin produces these actions by binding to a receptor that engages a G-protein, which increases the activity of adenylyl cyclase. The adenylyl cyclase converts ATP to cyclic AMP, thereby increasing the level of cyclic AMP in the terminal of the sensory neuron. The cAMP activates the cAMP-dependent protein kinase by attaching to its regulatory subunit, which releases its active catalytic subunit. The catalytic subunit then phosphorylates K^+ channels, thereby changing the conformation of the channel and decreasing the K^+ current (**pathway 1**). This prolongs the action potential, increases the influx of Ca^{2+}, and thus augments transmitter release.

Serotonin also leads to an increase in the availability of transmitter by mobilizing vesicles from a transmitter pool to the releasable pool at the active zone and also directly enhances the efficiency of the machinery of the exocytotic release of transmitter (**pathway 2**). Serotonin also leads to the opening of L-type Ca^{2+} channels (**pathway 3**). The second and third pathways reflect joint action of the cAMP-dependent protein kinase and protein kinase C, a second kinase activated by 5-HT (**pathways 2a and 3a**). Protein kinase C (PKC) is activated by 5-HT through another receptor, which engages another G-protein that activates a phospholipase that in turn stimulates diacylglycerol in the membrane. Diacylglycerol activates protein kinase C.

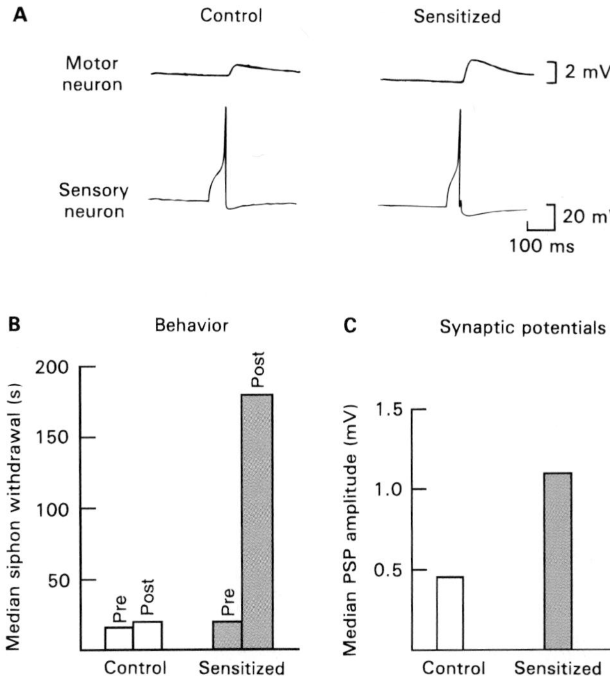

Figure 36–4 Long-term sensitization involves facilitation of the connections between sensory and motor neurons. (Adapted from Frost et al., 1985.)

A. Representative synaptic potentials in a siphon sensory neuron and a gill motor neuron in a control animal and an animal that has received long-term sensitization training by stimulating its tail. The record was obtained one day after the end of training.

B. The median duration of withdrawal of the siphon is used as a measure of the strength of the reflex. A comparison of control and sensitized groups illustrates the effect of sensitization. (Pre = score before training; post = score after training.) The experimental group was tested one day after the end of training.

C. Median values of the amplitudes of postsynaptic potentials (PSP) in an identified gill motor neuron for the control group and for sensitized animals one day after the end of training.

duces long-term facilitation after four or five repeated exposures. Finally, cAMP, an intracellular second messenger involved in the short-term process, also turns on the long-term change.

However, certain clinical conditions such as seizure and head trauma can selectively affect short- or long-term memory in humans. An even clearer behavioral separation between memory processes can be obtained in experimental animals using inhibitors of protein or mRNA synthesis, which selectively block long-term memory without affecting short-term memory. This dependency on macromolecular synthesis, which is also evident in the faciliation at the synapse between the sensory and motor neurons, suggests that genes and proteins not directly involved in short-term facilitation are required for long-term facilitation.

What is the function of these genes and proteins? Molecular studies indicate that with repeated training (or repeated application of serotonin) the cAMP-dependent protein kinase translocates to the nucleus of the sensory neurons to phosphorylate one or more cAMP-dependent transcriptional regulatory proteins such as CREB, the cAMP response element binding protein (Chapter 14). These transcriptional regulators activate genes whose protein products have two long-term consequences.

One consequence is a persistent activation of the cAMP-dependent protein kinase, a heterodimer consisting of two regulatory subunits that inhibit the two catalytic subunits (Chapter 14). With long-term training, the amount of the regulatory subunit in the sensory cells decreases relative to that of the catalytic subunit. This decrease in the regulatory subunit does not occur at the level of transcription but at the level of protein turnover. Indeed, one of the genes induced in long-term memory encodes a protein that is an enzyme in the ubiquitin proteolytic pathway that degrades the regulatory subunits (Figure 36–5). In turn, this decrease in the regulatory subunits causes the kinase to be persistently active, even though the level of cAMP has returned to its basal level. As a result, the short-term phosphorylation of the substrate proteins (Figure 36–3B) can be maintained in the long term. This persistent phosphorylation may explain why long-term facilitation appears to be a graded extension of the short-term process.

A second and more enduring consequence of gene activation is the growth of synaptic connec-

Figure 36–5 Schematic outline of the two major sets of changes in the sensory neurons of the gill-withdrawal reflex that accompany long-term memory for sensitization in *Aplysia*: (1) persistent activity of protein kinase A and (2) structural changes. Serotonin (5-HT), a transmitter released by facilitatory neurons, acts on a sensory neuron to initiate both the short-term facilitation and the long-term facilitation that contribute to the memory processes.

Short-term facilitation (lasting minutes to hours) involves covalent modification of preexisting proteins (**pathways 1 and 2**). Serotonin acts on a transmembrane receptor to activate a GTP-binding protein that stimulates the amplifier, the enzyme adenylyl cyclase, to convert ATP to the second messenger cAMP. In turn, cAMP activates protein kinase A, which phosphorylates and covalently modifies a number of target proteins. These modifications include closing of K^+ channels (**pathway 2**) as well as steps involved in transmitter availability and release (**pathway 1**). The duration of these modifications represents the retention or storage of a component of the short-term memory.

Long-term facilitation (lasting one or more days) involves the synthesis of new proteins. The switch for this inductive mechanism is initiated by the protein kinase A, which translocates to the nucleus where it is thought to phosphorylate one or more transcriptional activators that bind to *c*yclic AMP *r*egulatory *e*lements (CRE) located in the upstream region of cAMP-inducible genes. The transcriptional activators, thought to belong to the protein family of *c*yclic AMP *r*esponse *e*lement-*b*inding (CREB) proteins, activate two classes of effector genes that encode two classes of proteins. Inhibiting protein synthesis during learning blocks the expression of these two classes of proteins. These two sets of proteins have distinct functions. One protein (■), a ubiquitin hydrolase, is a component of a specific protease that leads to downregulation of the regulatory subunit. This results in persistent activity of kinase A, leading to persistent phosphorylation of the substrate proteins of **pathways 1 and 2**. The second set of proteins (▲) is important for the growth of active zones and the development of new synaptic connections.

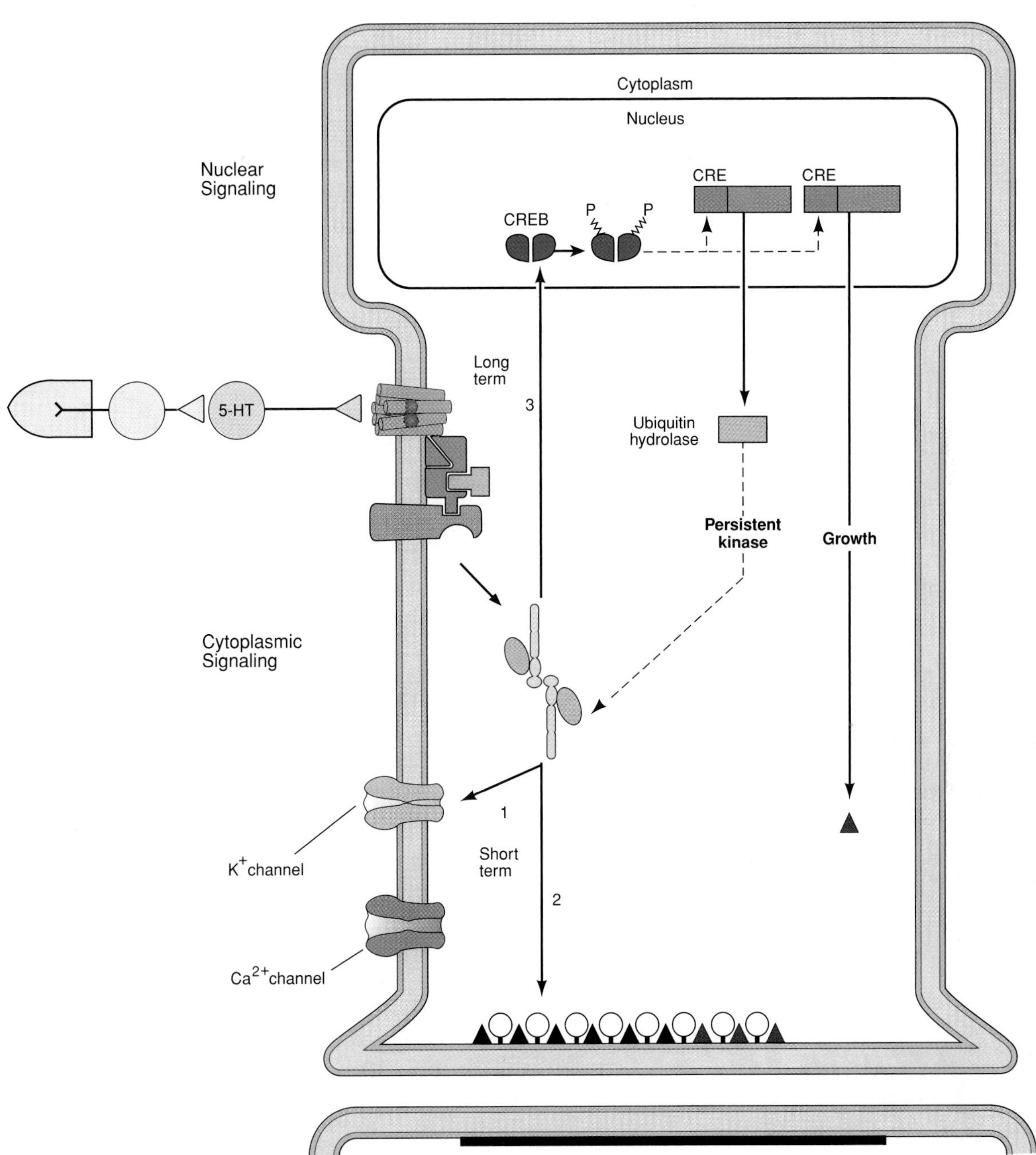
Cytoplasm
Nucleus
Nuclear Signaling
CRE
CRE
CREB
P
P
Long term
3
5-HT
Ubiquitin hydrolase
Persistent kinase
Growth
Cytoplasmic Signaling
1
Short term
2
K+ channel
Ca2+ channel
Motor neuron

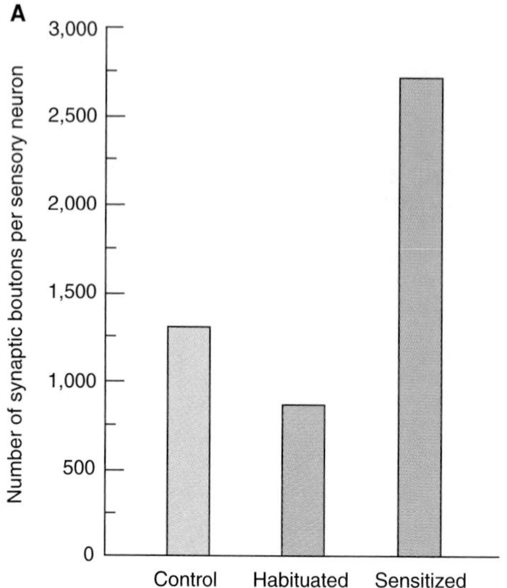

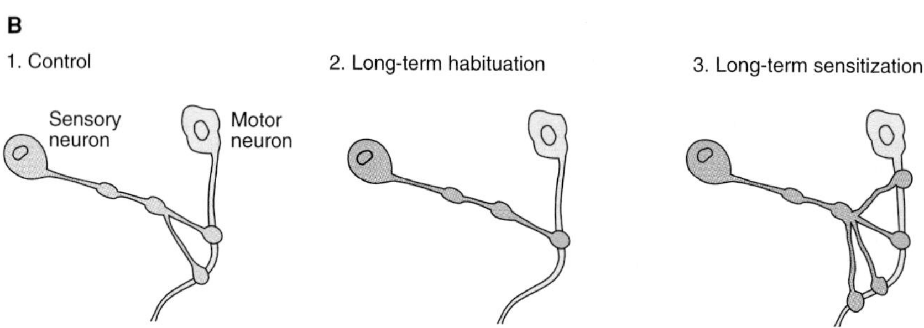

Figure 36–6 Long-term habituation and sensitization involve structural changes in the presynaptic terminals of sensory neurons. (Adapted from Bailey and Chen, 1983.)

A. This histogram compares the number of presynaptic terminals in control animals with those in long-term habituated and sensitized animals. The number is highest in the sensitized animals.

B. Long-term habituation leads to a loss of synapses, and long-term sensitization to an increase.

tions. This change was delineated in the sensory and motor cells involved in the gill-withdrawal reflex by examining the synaptic terminals with the electron microscope. The sensory neurons in sensitized animals had twice as many presynaptic terminals as those in untrained animals (Figure 36–6). Moreover, long-term sensitization increased the number of active zones from 40% of the synaptic terminals in untrained animals to 65% in trained animals. Finally, in the sensitized animals the dendrites of the motor neurons grew to accommodate the additional synaptic input. Such morphological changes seem to be characteristic only of long-term sensitization; they do not occur with short-term sensitization (see Figure 36–5).

In contrast to long-term sensitization, long-term habituation leads to *pruning* of synaptic connections: the inactivation of the functional connections between sensory and motor neurons reduces the number of terminals per neuron by one-third (Figure 36–6) and the proportion of terminals with active zones is reduced from 40% to 10%.

Classical Conditioning Involves an Associative Enhancement of Presynaptic Facilitation That Is Dependent on Activity

Classical conditioning is a more complex form of learning than sensitization. Rather than learning about the properties of one stimulus, the subject learns to associate one type of stimulus with another (see Chapter 35). In classical conditioning an initially weak or ineffective stimulus becomes highly effective in producing a response after it has been paired or associated with a strong unconditioned stimulus. For reflexes that can be modified by both sensitization and classical conditioning, classical conditioning is more effective in enhancing the responsiveness of the reflex, and it lasts longer than sensitization. As we shall see, the cellular mechanism of certain types of classical conditioning is an elaboration of the mechanism for sensitization.

The siphon- and gill-withdrawal reflexes of *Aplysia* can be enhanced by classical conditioning as well as sensitization. These reflexes can be elicited by stimulating, respectively, the siphon and a nearby structure called the mantle shelf. Each of these areas is separately innervated by distinct populations of sensory neurons. Thus, each neural pathway can be conditioned independently by pairing a stimulus to the appropriate area of the body (either the siphon or the mantle shelf) with an unconditioned stimulus (a strong shock to the tail). After such training, the response of the conditioned pathway to stimulation is significantly stronger than that of the unconditioned pathway (Figure 36–7).

Unlike nonassociative learning, timing is critical to associative learning. For classical conditioning to work, the conditioned stimulus must *precede* the unconditioned stimulus, and often it must do so within a critical interval of about 0.5 second. What cellular mechanisms are responsible for this requirement for the temporal pairing of stimuli? In classical conditioning of the gill-withdrawal reflex of *Aplysia*, one important mechanism for the temporal specificity is the convergence of the conditioned and unconditioned stimuli on individual sensory neurons. As we have seen, the unconditioned stimulus to the tail activates facilitating interneurons that have axo-axonic connections with the sensory neurons of the conditioned stimulus. This gives rise to presynaptic facilitation and behavioral sensitization. However, if the two stimuli are timed so that the interneurons are activated by the unconditioned stimulus immediately after the sensory neurons begin to fire in response to the conditioned stimulus, an even greater presynaptic facilitation is produced (Figure 36–7). In contrast, no enhancement of facilitation occurs if the interneurons are activated before the sensory neurons begin firing, that is, if activity in the sensory neurons *follows* the unconditioned stimulus.

This novel property of presynaptic facilitation, whereby the facilitation is amplified if the conditioned stimulus produces action potentials in the sensory neurons just before the unconditioned stimulus arrives, is called *activity dependence*. Thus, one component of the cellular mechanism of classical conditioning in the monosynaptic component of the withdrawal reflex in *Aplysia* is an elaboration of presynaptic facilitation, the mechanism of sensitization in this component of the reflex. A similar enhancement of sensory neurons occurs in the tail of *Aplysia*.

How is activity-dependent presynaptic facilitation achieved? The Ca^{2+} that flows into the cell following an action potential is thought to act through calmodulin to amplify the activation of the adenylyl cyclase by serotonin and other modulatory transmitters (Figure 36–8). Much of the cyclase in the brain is sensitive to Ca^{2+}/calmodulin and generates more cAMP when it is bound to Ca^{2+}/calmodulin than when it is not.

Genetic analyses of learning have also implicated the cAMP system. The fruit fly *Drosophila* can be classically conditioned, and single-gene mutants deficient in learning have been isolated. Three of these mutants, called *dunce*, *rutabaga*, and *amnesiac*, have been studied in detail and show two interesting features. First, all of the mutants that fail to show classical conditioning also fail to show sensitization. Second, all three mutants have a defect in the cAMP cascade. The *dunce* mutant lacks a phosphodiesterase, an enzyme that degrades cAMP. As a result, this fly has abnormally high levels of cAMP that are thought to be out of the range of normal modulation. The *rutabaga* mutant has a defect in the Ca^{2+}/calmodulin-dependent adenylyl cyclase and a low basal level of cAMP. The *amnesiac* mutation lacks a peptide that regulates the activity of the adenylyl cyclase. Finally, blocking the action of the CREB transcription

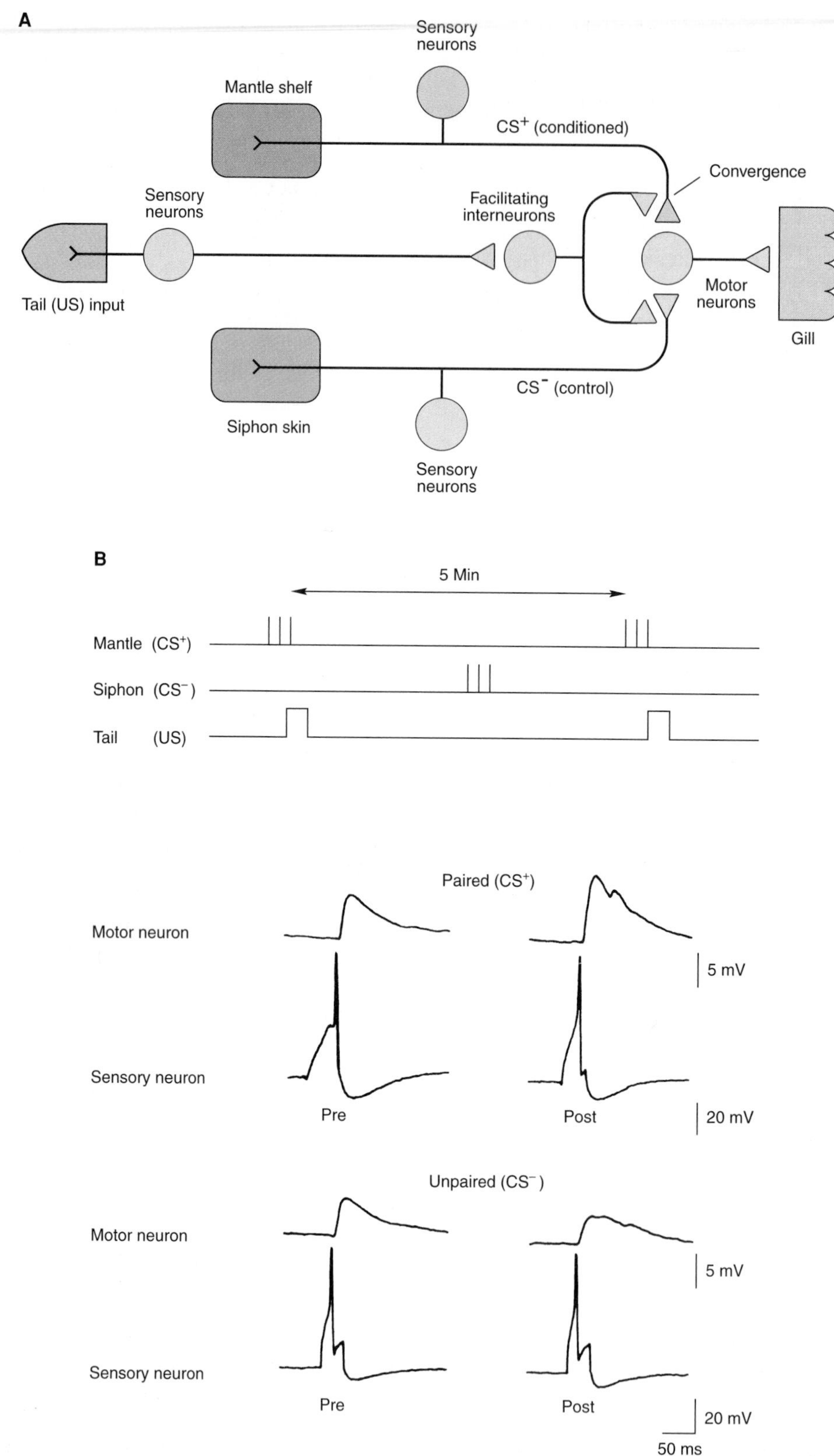
A
Sensory neurons
Mantle shelf
CS+ (conditioned)
Convergence
Sensory neurons
Facilitating interneurons
Tail (US) input
Motor neurons
Gill
Siphon skin
CS− (control)
Sensory neurons
B
5 Min
Mantle (CS+)
Siphon (CS−)
Tail (US)
Paired (CS+)
Motor neuron
5 mV
Sensory neuron
Pre
Post
20 mV
Unpaired (CS−)
Motor neuron
5 mV
Sensory neuron
Pre
Post
20 mV
50 ms

← *Figure 36–7* Classical conditioning of the gill-withdrawal reflex in *Aplysia*. A conditioned stimulus (**CS**) applied to the mantle is paired with an unconditioned stimulus (**US**) to the tail; as a control, a CS applied to the siphon is not paired with the US. (Adapted from Hawkins et. al., 1983).

A. This simplified diagram shows the neural pathways involved. A shock to the tail (**US**) excites facilitatory interneurons that synapse on the presynaptic terminals of sensory neurons innervating the mantle shelf and siphon. This is a mechanism of sensitization. However, when the mantle pathway is activated by a CS just prior to the US, the activity primes the mantle sensory neurons so that they are more responsive to subsequent stimulation from the facilitatory interneurons in the US pathway. This is a mechanism of classical conditioning; it both amplifies the response of the CS pathway and restricts the amplification to that pathway.

B. The activity of individual cells is modified by classical conditioning. Two sensory neurons are each stimulated independently. Stimulation of the mantle sensory neuron (CS^+) is paired with the US (tail shock), while, stimulation of the siphon sensory neuron (CS^-) is not paired with the US. Recordings of the excitatory postsynaptic potentials produced in an identified motor neuron by the two sensory neurons were made before training (Pre) and one hour after training (Post). After training the excitatory postsynaptic potential due to the paired sensory neuron is considerably greater than that due to the unpaired neuron.

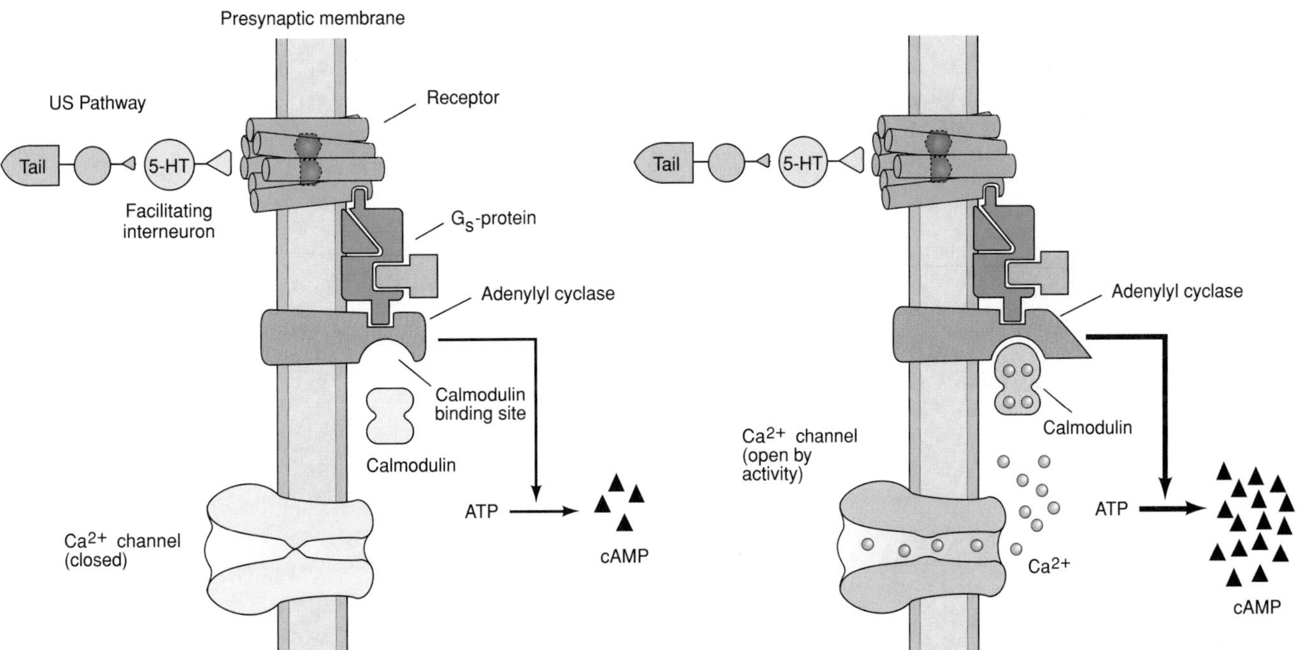

Figure 36–8 A molecular model of the synaptic action underlying classical conditioning. The model is based on the hypothesis that activity in the sensory neurons mediating the conditioned stimulus prior to the presentation of the unconditional stimulus permits an influx of Ca^{2+} that enhances the activity of calcium-dependent adenylyl cyclase.

A. In the unpaired pathway (CS^-) the sensory neuron is not active prior to presentation of the CS, so its Ca^{2+} channels are closed when the unconditioned stimulus (US) is presented. (5-HT, serotonin.)

B. In the paired pathway (CS^+) the sensory neuron is active prior to the CS and thus its Ca^{2+} channels are open when the US is presented. The intracellular Ca^{2+} binds to calmodulin and in turn interacts with adenylyl cyclase. As a result, the adenylyl cyclase undergoes a conformational change that enhances its ability to synthesize cAMP in response to serotonin released in the US pathway. The greater amount of cAMP activates more cAMP-dependent protein kinase and leads to a substantially greater amount of transmitter release than would occur without paired activity.

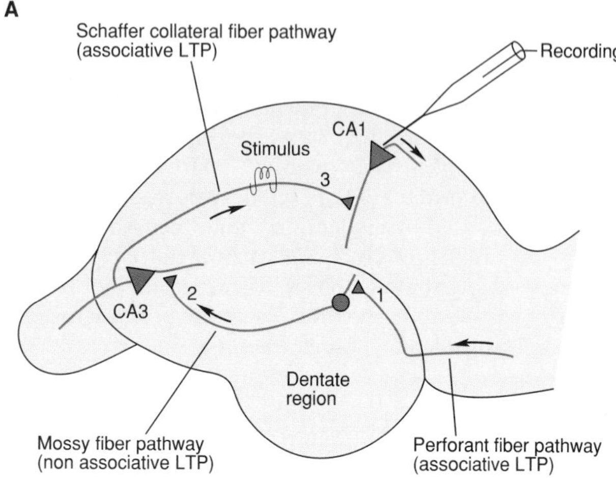

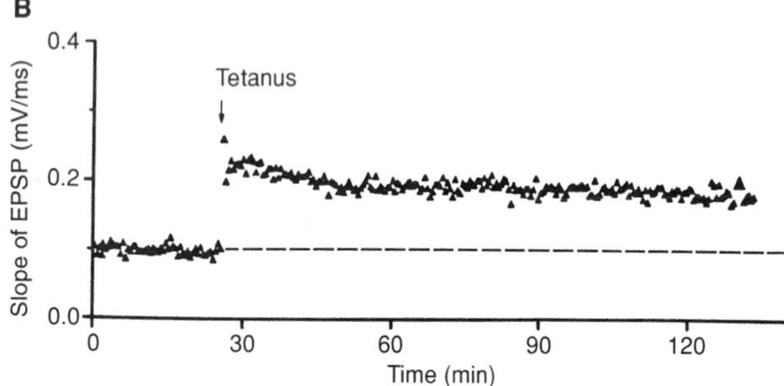

Figure 36–9 Long-term potentiation (LTP) in the hippocampus.

A. There are three major afferent pathways in the hippocampus. (**Arrows** denote the direction of impulse flow.) The *perforant pathway* (**1**) from the entorhinal cortex forms excitatory connections with the granule cells of the dentate gyrus. The granule cells give rise to axons that form the *mossy fiber pathway* (**2**). This pathway connects with the pyramidal cells in area CA3 of the hippocampus. The CA3 cells project to the pyramidal cells in CA1 by means of the *Schaffer collaterals* (**3**).

B. The effect of long-term potentiation in a cell in the CA1 region of the hippocampus is shown in this plot of the slope (rate of rise) of the excitatory postsynaptic potentials in the cell. The slope is a measure of synaptic efficacy. Excitatory postsynaptic potentials (EPSPs) were recorded from outside the cell. A test stimulus was given every 10 seconds. To elicit long-term potentiation two trains of stimuli for 1 second each at 100 Hz tetani and separated by 20 seconds were delivered to the Schaffer collaterals. The resulting LTP lasts several hours. (Adapted from Nicoll et al., 1988.)

factor in *Drosophila* (thereby preventing the expression of cAMP-dependent genes) selectively blocks the protein synthesis-dependent long-term memory in the fly without interfering with learning or short-term memory.

Thus, both cellular studies of *Aplysia* and genetic studies of *Drosophila* indicate that the cAMP cascade is important for certain elementary forms of learning and memory storage. However, the cAMP cascade is not the only second-messenger system important for synaptic plasticity related to implicit forms of learning. In other instances of learning other second messenger cascades are recruited.

That the cellular mechanisms of classical conditioning in *Aplysia* may be an elaboration of those involved in sensitization suggests that, at least in certain instances, more complex forms of learning can be built up from the molecular components of simpler forms. By this means a variety of distinct forms of behavioral modifications could be achieved by a small set of molecular mechanisms.

Storage of Explicit Memory in Mammals Involves Long-Term Potentiation in the Hippocampus

What about explicit forms of learning? Can specific cellular mechanisms for these more complex forms of learning be identified? As we saw in Chapter 35, the hippocampus is important for storage of explicit memory, and there is evidence that neurons in the hippocampus have the plasticity of the sort that would be required for explicit memory.

The hippocampus has three major afferent pathways running from the entorhinal cortex to the CA1 region (Figure 36–9A). The *perforant pathway* runs from the entorhinal cortex to the granule cells in the hilus of the dentate gyrus. The axons of the granule cells form a bundle, the *mossy fiber pathway*, that runs to the pyramidal cells lying in the CA3 region of the hippocampus. Finally, the pyramidal cells in the CA3 region send excitatory collaterals, the *Schaffer collaterals*, to the pyramidal cells in CA1. A brief high-frequency train of stimuli to any one of these pathways increases the excitatory postsynaptic potentials in the hippocampal neurons, an increase that can last for hours and, in the intact animal, for days and even weeks. This facilitation is called *long-term potentiation* (Figure 36–9B). Long-term potentiation is not produced in the same way in all three of these pathways, however, and we discuss the differences next.

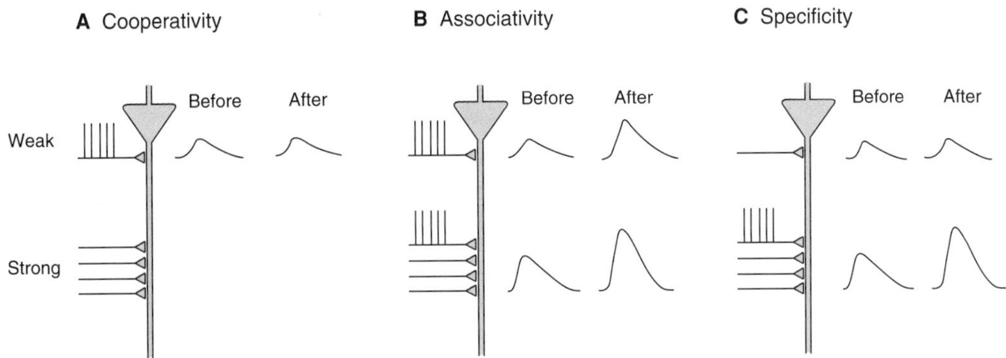

Figure 36–10 Long-term potentiation in area CA1 of the hippocampus shows cooperativity, associativity, and specificity. In the figure a single pyramidal cell receives weak and strong synaptic inputs from two different fascicles of the Schaffer collateral pathway. (Adapted from Nicoll et al., 1988.)

A. Tetanic stimulation of the weak input alone does not cause long-term potentiation in the pathway (compare the potential before and after tetanus).

B. Tetanic stimulation of the strong and weak pathways together causes long-term potentiation in both pathways.

C. Tetanic stimulation of the strong input alone causes long-term potentiation in the strong pathway but not in the weak.

Long-Term Potentiation in the CA1 Region Is Associative

Long-term potentiation (LTP) can be produced in the Schaffer axon collateral pathway, which connects the pyramidal cells of the CA3 region of the hippocampus with those of the CA1 region (Figure 36–9A). To produce LTP it is necessary to use a strong stimulus that activates several afferent fibers together. This *cooperative* activity has associative features similar to those encountered in classical conditioning. When separate weak and strong excitatory inputs arrive at the same region of the dendrites of a pyramidal cell, the weak input will become potentiated only if it is activated in association with the strong one. Finally, LTP is specific to those synapses that are activated by the stimulus. For example, LTP produced by an input to the apical dendrites does not affect an independent input onto the basilar dendrites. These features of LTP in the CA1 region are illustrated in Figure 36–10.

What accounts for these three features? What appears critical for the induction of LTP in the CA1 region of the hippocampus is that the postsynaptic cell be adequately depolarized. LTP can be induced when a weak stimulus train, or even a single test stimulus, not sufficient in itself to produce LTP, is paired repeatedly with a depolarizing current pulse injected in a single postsynaptic cell. Conversely, hyperpolarizing the postsynaptic cell during the tetanus can prevent LTP.

Thus, LTP requires simultaneous firing in both the postsynaptic and presynaptic neurons. This finding provides the first direct evidence for *Hebb's rule*, proposed in 1949 by the psychologist Donald Hebb: "When an axon of cell A . . . excites cell B and repeatedly or persistently takes part in firing it, some growth process or metabolic change takes place in one or both cells so that A's efficiency as one of the cells firing B is increased." As we have seen in Chapter 25, a similar principle seems to be involved in the fine tuning of synaptic connections during the late stages of development.

Why is the simultaneous firing of the pre- and postsynaptic cells important for LTP? The Schaffer collateral axons from the CA3 region of the hippocampus that terminate on the pyramidal cells of the CA1 region use glutamate as their transmitter. Glutamate acts on its target cells in the CA1 region by binding to both *N*-methyl-D-aspartate (NMDA) and non-NMDA receptors (Chapter 13). In normal synaptic transmission the non-NMDA receptors dominate as the NMDA receptor–channels are

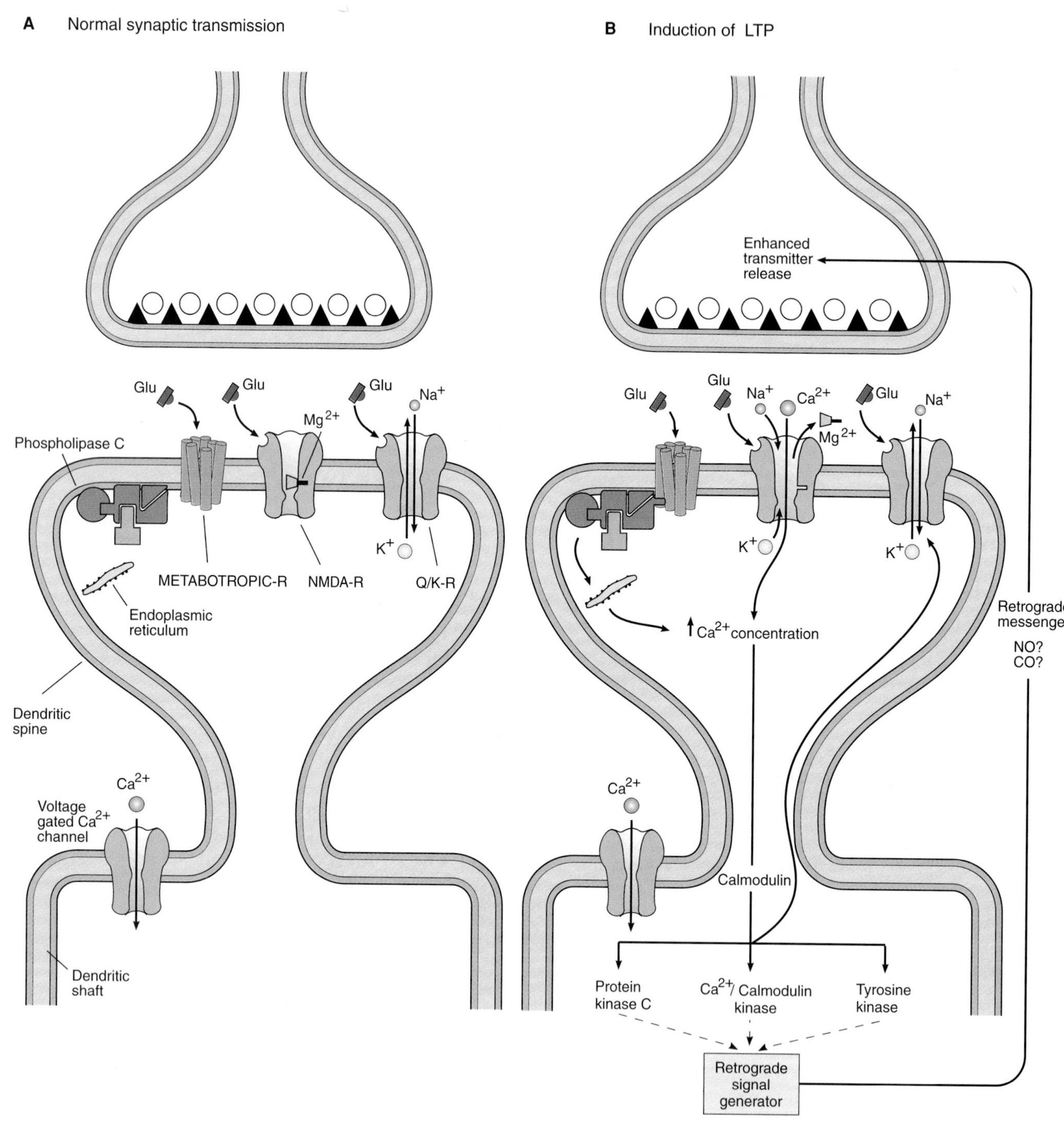

A Normal synaptic transmission
B Induction of LTP
Glu
Glu
Glu
Na+
Mg 2+
Phospholipase C
K+
METABOTROPIC-R
NMDA-R
Q/K-R
Endoplasmic reticulum
Dendritic spine
Ca2+
Voltage gated Ca2+ channel
Dendritic shaft
Enhanced transmitter release
Glu
Glu
Na+
Ca2+
Glu
Na+
Mg 2+
K+
K+
Ca2+ concentration
Retrograde messenger
NO?
CO?
Ca2+
Calmodulin
Protein kinase C
Ca2+/ Calmodulin kinase
Tyrosine kinase
Retrograde signal generator

blocked by Mg^{2+}. They become unblocked, and thus activated, only when the postsynaptic cell is adequately depolarized by strong (cooperative) inputs from many presynaptic neurons. This depolarization causes the positively charged Mg^{2+} ion to pop out of the channel mouth, allowing Na^+ and particularly Ca^{2+} to flow through the channel into the cell. The influx of Ca^{2+} is the signal for the induction of LTP.

Thus, the NMDA receptor–channel is unusual in being a *doubly gated channel.* The channel becomes functional only when glutamate binds to the receptor *and* the membrane is depolarized. This critical depolarization is normally achieved through the activation of many non-NMDA receptors by the firing of many presynaptic neurons (Figure 36–11). Artificially it can be obtained by simply depolarizing the postsynaptic cell.

Figure 36–11 A model for the induction of long-term potentiation. According to this model *N*-methyl-D-aspartate (NMDA) and non-NMDA (quisqualate/kainate) receptor–channels are located near each other in dendritic spines. (Adapted from Gustafsson and Wigström, 1988.)

A. During normal low-frequency synaptic transmission glutamate is released from the presynaptic terminal and acts on both the NMDA and non-NMDA (Q/K) receptors. Sodium and K^+ flow through the non-NMDA receptor–channels but not through the NMDA receptor–channels, owing to Mg^{2+} blockade of these channels at the resting level of membrane potential.

B. When the postsynaptic membrane is depolarized by the actions of the non-NMDA receptor channels, as occurs during a high-frequency tetanus that induces long-term potentiation (LTP), the depolarization relieves the Mg^{2+} blockade of the NMDA channel. This allows Na^+, K^+, and Ca^{2+} to flow through the NMDA channel. The resulting rise in Ca^{2+} in the dendritic spine triggers calcium-dependent kinases (calcium/calmodulin kinase and kinase C) that induce LTP. Once LTP is induced, the postsynaptic cell releases (in ways that are still not understood) a retrograde messenger that is thought to act on kinases in the presynaptic terminal to produce the sustained enhancement of transmitter release that underlies the persistence of LTP.

Calcium influx through the unblocked NMDA receptor–channel is critical for LTP. Blocking Ca^{2+} influx prevents induction of LTP; conversely, injecting Ca^{2+} into the postsynaptic cell initiates the early phase of LTP. In principle, Ca^{2+} could pass through either a voltage-gated Ca^{2+} channel or the NMDA-gated channel. However, the Ca^{2+} influx critical for LTP normally enters through the NMDA receptor–channel, not through the voltage-gated Ca^{2+} channels. Blocking the NMDA receptor–channel blocks LTP.

Receptors for NMDA seem to be clustered on the heads of the spines of dendrites, not on the shafts of the dendrites. (Spines, as we have seen in Chapter 13, are specialized lateral protrusions on the shafts of dendrites that receive excitatory synaptic input.) The activation of non-NMDA receptor–channels depolarizes the spines sufficiently to remove the Mg^{2+} blockade of the NMDA receptor–channels, allowing Ca^{2+} to enter the spines. The spine acts as a functional compartment that restrains the diffusion of Ca^{2+}, so that the synaptic action is restricted to the synapses that are active.

Calcium initiates the persistent enhancement of synaptic transmission by activating two calcium-dependent serine–threonine protein kinases—the Ca^{2+}/calmodulin kinase and protein kinase C—and a tyrosine protein kinase. One of these kinases, an isoform of protein kinase C, then becomes persistently active.

Thus, the induction of LTP in the CA1 region depends on postsynaptic depolarization, Ca^{2+} influx, and Ca^{2+} activation of a second-messenger system. Maintenance of LTP in the CA1 region, however, involves in addition an increase in *presynaptic* transmitter release (Figure 36–12). This finding is based on three lines of evidence. First, LTP is accompanied by an enhancement of glutamate release. Second, LTP involves an increased probability of transmitter release as well as a change in the sensitivity of the non-NMDA glutamate receptors in the postsynaptic cell. Third, the induction of LTP by depolarization of a single postsynaptic cell produces LTP in a small population of surrounding neurons. If the LTP mechanism were strictly postsynaptic, the LTP would be restricted to the cell that was depolarized.

Since the induction of LTP requires a postsynaptic event (activation of NMDA receptors and Ca^{2+} influx) and maintenance of LTP involves a presynaptic event (increase in transmitter release), a mes-

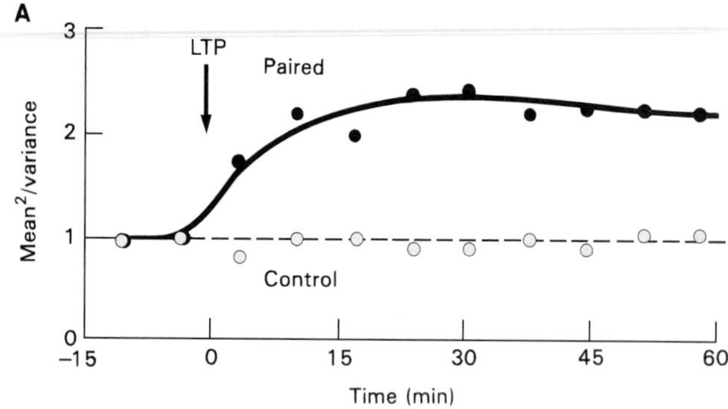

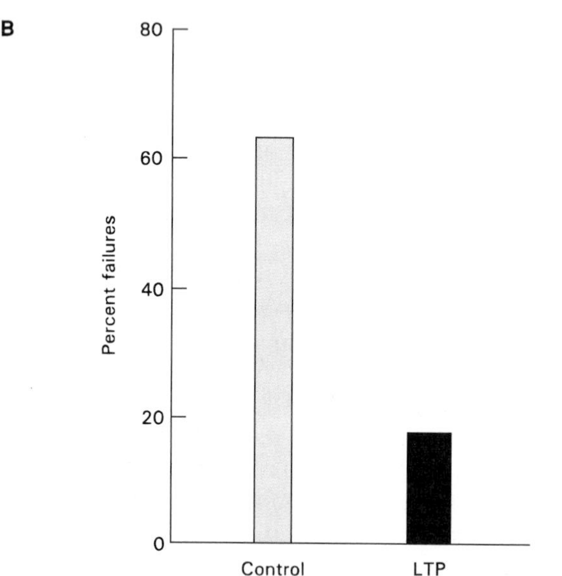

Figure 36–12 Maintenance of long-term potentiation (LTP) in the CA1 region of the hippocampus depends on an increase in presynaptic transmitter release. Quantal analysis of long-term potentiation in area CA1 is based on a coefficient of variation of evoked responses. This analysis assumes that the number of quanta of transmitter released follows a binomial distribution, where the coefficient of variation (mean^2/variance) provides an index of transmitter release from the presynaptic terminal that is independent of quantal size. (From Malinow and Tsien, 1990.)

A. With long-term potentiation there is an increase in the mean^2/variance, indicating an increase in transmitter release. This increase occurs only in the pathway that is paired with depolarization of the postsynaptic cell. It does not occur in a control pathway that is not paired.

B. At normal rates of stimulation the number of failures in transmission is significant. In 60% of cases stimulation of the presynaptic axons leads to no release. Following LTP the percentage of failures decreases to 20%, another indication that LTP is presynaptic.

sage must be sent from the postsynaptic to the presynaptic neurons. There is now evidence that the Ca^{2+}-activated second messenger, or perhaps Ca^{2+} acting directly, causes the release of one or more *retrograde messengers* from the dendritic spines of the active postsynaptic cell. This retrograde factor diffuses to the presynaptic terminals to activate one or more second messengers that act to enhance transmitter release and thereby maintain LTP. The actions of a membrane-permeable retrograde messenger seems to be restricted to recently active presynaptic cells. Indeed, to account for the pathway specificity of LTP, the action of the retrograde messenger must be restricted. Two gases that diffuse readily from cell to cell, nitric oxide and carbon monoxide, have properties that have made them interesting candidates, acting either alone or jointly with other molecules, for the retrograde messenger of LTP.

According to this view, LTP in the CA1 region of the hippocampus uses two associative mechanisms in series: a Hebbian mechanism and activity-dependent presynaptic facilitation. However, LTP differs from the activity-dependent presynaptic facilitation found in *Aplysia* in that the facilitatory substance is released from the postsynaptic target cell by activation of NMDA receptors, rather than from a facilitatory interneuron with diffuse projections, as in *Aplysia*.

What might be the advantage of combining in the hippocampus two associative cellular mechanisms in series (the postsynaptic NMDA receptor and activity-dependent presynaptic facilitation)? One possible advantage is spatial amplification of the signal. The retrograde factor can recruit other nearby presynaptic fibers in addition to those that synapse directly on the active postsynaptic cell.

Associative Long-Term Potentiation May Be Important for Spatial Memory

The finding that LTP occurs in many areas of the brain, including the cerebral cortex and hippocampus, a region known to be important for memory storage, raises the question: Is LTP involved in memory storage? Evidence for this has come from the analysis of a spatial memory task in which a rat has to make its way through a maze in a pool filled with a whitish opaque fluid to find a platform hidden under the fluid. The animal is released at random locations around the pool and must use spatial cues—markings on the walls of the room in which the pool is located—to find the platform. In a simple visual (nonspatial) version of this task, the platform is raised above the water surface or marked with a flag so that it is visible; thus, the rat can navigate to the platform by means of direct sight rather than spatial cues.

When NMDA receptors in the hippocampus are blocked by the injection of an antagonist into the ventricle, the animal can successfully navigate the maze in the simple visual version of the task but cannot find his way to the platform in the spatial version of the task. These experiments suggest that an NMDA receptor mechanism in the hippocampus, perhaps LTP, is involved in spatial learning. Further evidence for this correlation comes from mice with altered genes. Ablation of the gene coding for the calcium/calmodulin-dependent protein kinase or the gene coding for the nonreceptor tyrosine kinase (*fyn*) reduces LTP and also blocks spatial learning in the water maze (Box 36–1 and Figure 36–13).

Long-Term Potentiation in the CA3 Region Is Nonassociative

Although LTP occurs at several synapses in the hippocampus and in many regions of the cerebral cortex, the mechanisms for the induction of LTP are not the same everywhere. Some do not work through the NMDA receptor and do not depend on either Ca^{2+} influx or the activation of calcium/calmodulin-dependent kinases in the postsynaptic cell.

Neurons in the CA3 region of the hippocampus release glutamate as their transmitter, but the synapses they form utilize NMDA receptors only in a minor way. In fact, LTP at these synapses is not blocked by the standard NMDA receptor antagonist. Moreover, this potentiation is not associative—the input need not be paired with another input or with depolarization of the postsynaptic cells.

Blocking Ca^{2+} influx into the postsynaptic cells in the CA3 region does not affect LTP in these cells. Indeed, LTP can be obtained after washing the postsynaptic cell with fluoride, which disrupts various intracellular second-messenger pathways in the postsynaptic cell. Rather, this form of LTP seems to depend on presynaptic Ca^{2+} influx as a result of the tetanus. The Ca^{2+} influx in turn activates a Ca^{2+}/calmodulin dependent (Type I) adenyl cyclase, which increases the level of cAMP and activates the cAMP-dependent protein kinase, much as occurs in the *Aplysia* sensory neurons (Figure 36–14).

Is There a Molecular Alphabet for Learning?

The changes in synaptic efficacy that we have encountered in studies of both implicit and explicit forms of learning raise these surprising reductionist possibilities in a neurobiological approach to learning.

First, synaptic changes can be associative without depending on complex features of the neural *network*. This fact means that the associative activity that contributes to implicit and explicit learning represents a *basic cellular process*. In the two instances we have considered here—activity-dependent enhancement of presynaptic facilitation and associative LTP—the plastic properties of cells seem to derive from the properties of specific proteins, such as the adenylyl cyclase and the NMDA receptor, which are capable of responding to two independent signals.

Second, the finding that the associative forms of synaptic plasticity in *Aplysia* and in the hippocampus are related in certain instances to nonassociative forms suggests that there may be a molecular alphabet for synaptic plasticity—simpler forms of plasticity might represent elements of more complex forms. Of course, these elementary cellular mechanisms are embedded in neural circuitry with considerable additional computational power, which can add substantial complexity to these elementary mechanisms. Finally, there is evidence for a late phase of LTP that parallels long-term memory in requiring gene expression and new protein synthesis. In both the CA1 and the CA3 region this late phase requires cAMP-inducible genes, much as does the long-term memory-related plasticity for implicit learning in *Aplysia* and *Drosophilia*. Thus, even though implicit and explicit forms of learning use different mechanisms for short-term memory storage, both forms of learning seem to share a restricted number of mechanisms for long-term memory storage.

BOX 36–1 *Single Genes, Long-Term Potentiation, and Spatial Learning*

Gene knockout in mice provide a bridge between the actions of single genes and learning. The biological analysis of learning requires the establishment of a causal relation between specific molecules and learning. This relationship, which has been difficult to demonstrate in mammals, can now be studied in mice by generating specific gene mutations through homologous recombination in embryonic stem cells (See Box 30–4).

Gene knockout has made it possible to examine the relationship of two second-messenger kinases—the calcium/calmodulin kinase and the tyrosine kinase—to long-term potentiation (LTP) in the hippocampus and to spatial learning in the intact animal. Ablation of the gene for the alpha subunit of the Ca^{2+}/calmodulin kinase leads to loss of LTP. Animals deficient in this gene also show a deficiency in spatial learning. By contrast, they show no deficit in learning simple visual discrimination tasks which do not involve the hippocampus (Figure 36–13A).

Similarly, ablation of *fyn*, the gene coding for a cytoplasmic tyrosine kinase, results in blunting of LTP and spatial learning (Figure 36–13C). *Fyn* is a member of the *src* family of tyrosine kinases (which also contains the genes *yes* and *src*); knockout of the genes coding for *src* and *yes*, however, has no effect on LTP. *Fyn*$^-$ animals also have a defect in spatial learning. In addition, *fyn*$^-$ animals have abnormally large numbers of granule cells in the dentate gyrus and pyramidal cells in the CA3 region. Thus, a common tyrosine kinase pathway may regulate the growth of neurons in the developing hippocampus and the strength of synaptic plasticity in the mature hippocampus.

This experimental approach represents a beginning in the attempt to bridge molecular mechanisms to cognition. Many more gene ablations will need to be examined before we will have a reasonable understanding of how LTP in the various regions of the hippocampus relates to learning in the intact animal.

Figure 36–13 Effects of gene ablation on long-term potentiation (LTP) in the CA1 region of hippocampus and on spatial learning in the intact mouse.

A. Ablation of the α subunit of calcium/calmodulin-dependent protein kinase blocks LTP and interferes with spatial learning. LTP is measured as an increase in synaptic strength, expressed as a percentage of the average initial excitatory postsynaptic potential (f-EPSP), as a function of time during the experiments for a typical slice from a normal wild-type animal and a mutant animal. The wild-type mice show normal LTP. The mutant mice lack LTP. Testing stimuli were given every 30 seconds. The tetanus (**arrowhead**) was three 20-pulse trains at 100 Hz with an intertrain interval of 11 seconds.

B. All animals were trained to escape from an underwater maze by finding a platform located below the surface of the water. The mean escape latency was then measured. Wild-type and mutant mice (those missing the α subunit of calcium/calmodulin-dependent protein kinase) were given either 3 or 5 days of training. The wild-type mice had lower escape latencies than the mutants. The bars indicate the standard error of the mean.

C. Ablation of the gene *fyn*, which codes for a nonreceptor tyrosine kinase, interferes with spatial learning in the transfer test. After 7 days of training similar to that described in **B**, the platform was removed and the mice were allowed to swim for 60 seconds. The records show the movements of individual mice. The wild-type mouse searches in the quadrant that had contained the platform (**upper left**) indicating that the mouse remembered where the platform had been located. The *fyn*$^-$ mouse swims randomly throughout the pool, indicating that it never had spatial memory for the platform.

A Long-term potentiation

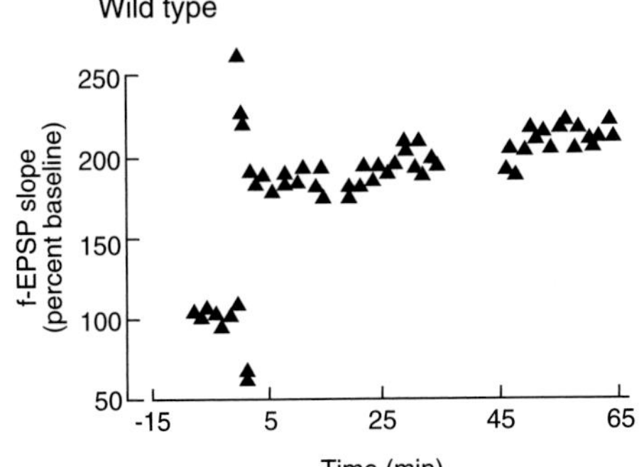

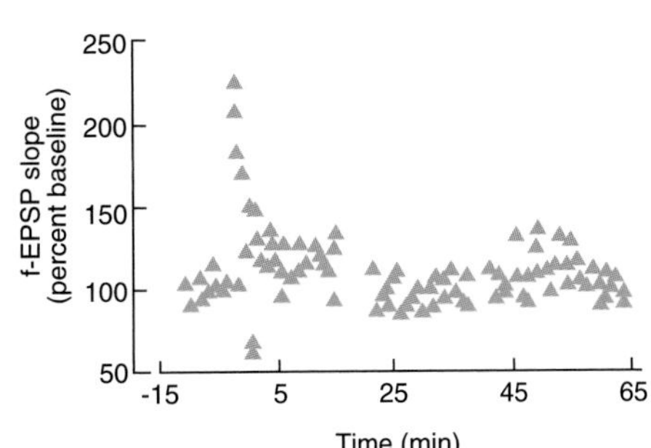

B Spatial Learning

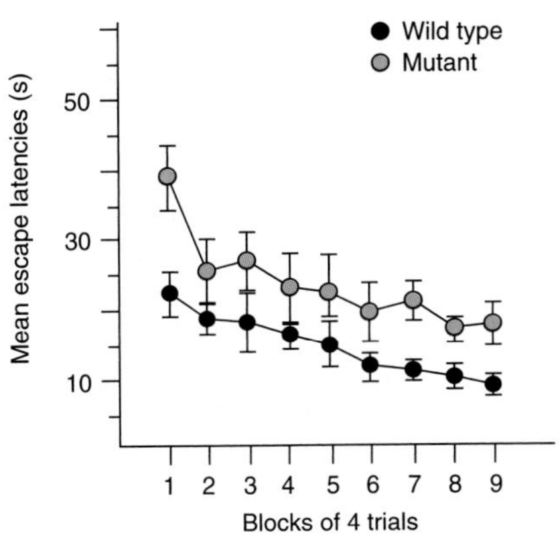

C Transfer test

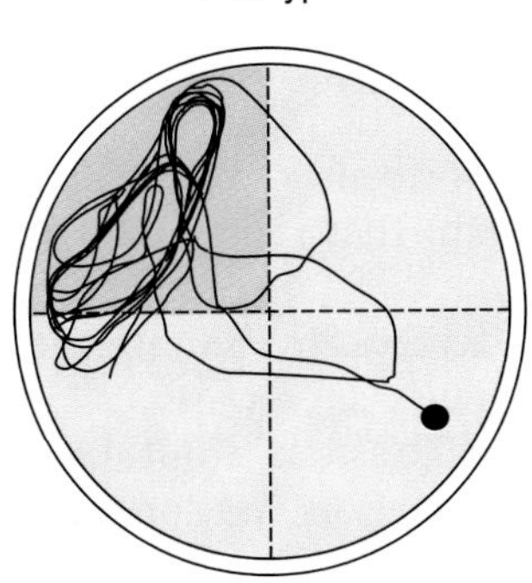

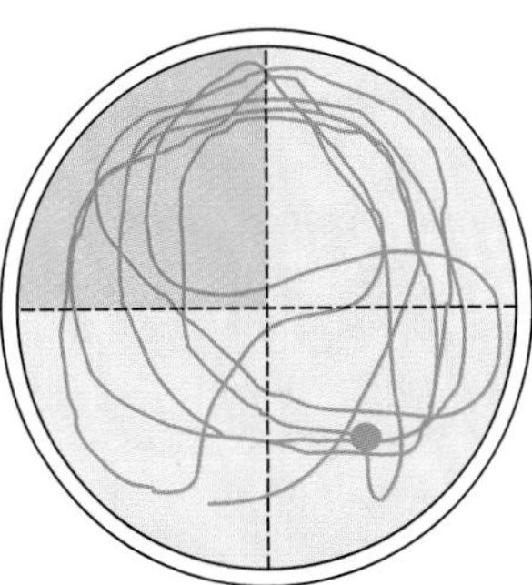

A

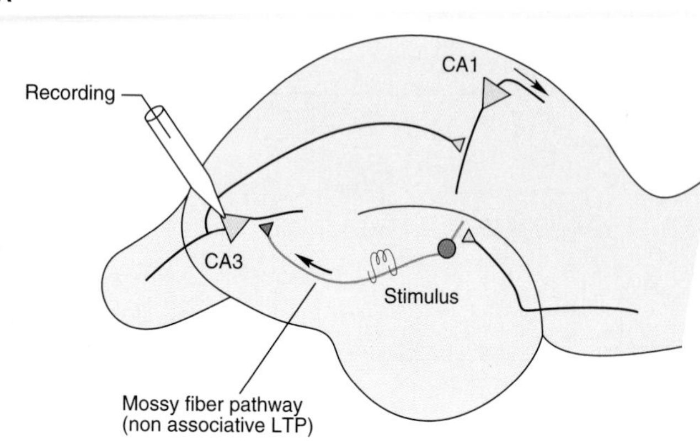

B

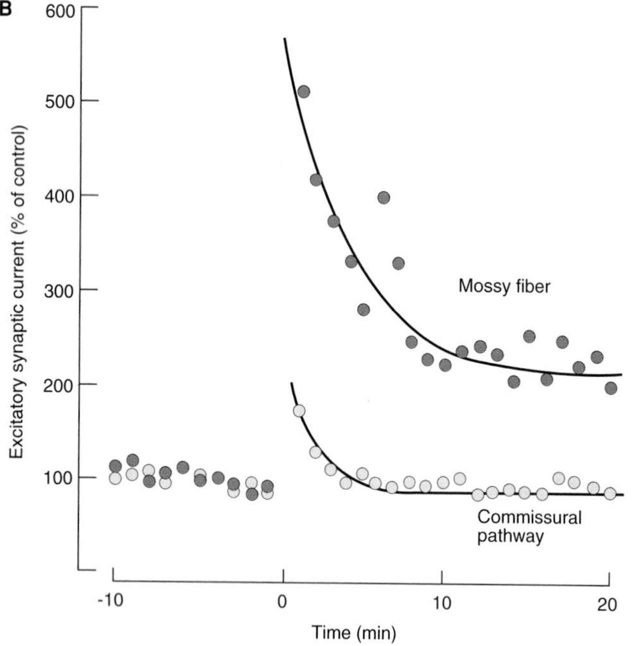

Figure 36–14 Long-term potentiation (LTP) of the mossy fiber synapses on pyramidal cells in the CA3 region of the hippocampus may be induced presynaptically. (Adapted from Zalutsky and Nicoll, 1990.)

A. The arrangement for studying LTP in the CA3 region of the hippocampus.

B. Whole-cell voltage-clamp recording allowed injection of both fluoride and a Ca^{2+} chelator (BAPTA) into the cell body of the CA3 neuron. Together these two drugs are thought to block *all* second-messenger pathways in the postsynaptic cell. Despite this drastic biochemical blockade of the postsynaptic cell, LTP in the mossy fiber pathway is unaffected and is therefore thought to be presynaptically induced. In contrast, these injections do block LTP in the commissural pathway that connects the hippocampus on one side to that on the other. This pathway ends on the NMDA receptor, and here induction of LTP is known to be postsynaptic.

The Somatotopic Map in the Brain Is Modifiable by Experience

Learning can lead to structural alterations in the brain. How common are these changes in determining the functional architecture of the mature brain?

We learned in Chapter 25 that the structure of the ocular dominance columns in area 17 of the cerebral cortex can be altered by experience during an early critical period in development. If one eye is permanently closed during this period, the columns devoted to that eye shrink while those devoted to the seeing eye expand. This modifiability is restricted to a relatively short period just after birth, but it raises an intriguing question: To what degree can altered sensory experience in later life produce changes in the architecture of the brain—in the size of cortical columns, or even in the precise details of the various sensory and motor maps?

The work in *Aplysia* we reviewed earlier indicates that learning produces structural and functional changes in specific nerve cells. In mammals, and especially in humans, in whom each functional component is represented by hundreds of thousands of nerve cells, learning is likely to lead to alterations in many nerve cells and is therefore likely to be reflected in changes in the pattern of interconnections of the various sensory and motor systems involved in a particular learning task. This is indeed what appears to happen. The most detailed evidence has come from studies on the somatic sensory system.

Changes in the Somatotopic Map Produced by Learning May Contribute to the Biological Expression of Individuality

As we have seen in Chapter 18, the primary somatic sensory cortex consists of four areas in the postcentral gyrus (Brodmann's areas 1, 2, 3a, and 3b) that represent separate maps of the body surface (Figure 36–15). These cortical maps differ systematically among individuals, in a manner that reflects their use.

Thus, in adult humans the cortical maps for somatic sensations are dynamic, not static. Functional connections can expand and retract. The cortical maps of an adult are subject to constant modification on the basis of use or activity of the peripheral sensory pathways. Since all of us are brought up in somewhat different environments,

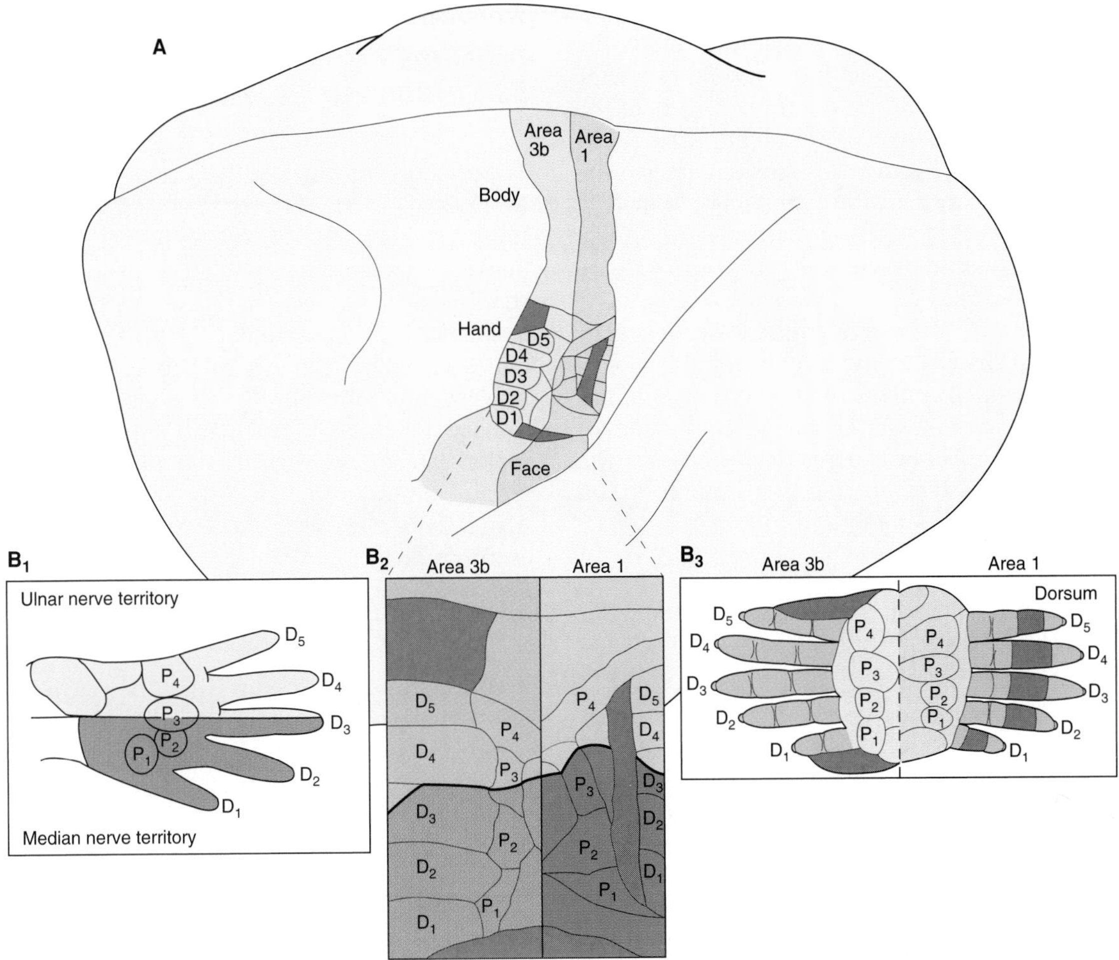

Figure 36–15 The body surface is represented in more than one sensory map in the primary somatic sensory cortex. This example shows the two distinct maps of the hands in the somatic sensory cortex of the owl monkey. (Adapted from Merzenich and Kaas, 1982.)

A. The location of the two representations of the hand areas in the brain is shown in this dorsolateral view.

B.1. The ulnar and median nerves innervate different territories on the ventral surface of the hand. **2.** The areas innervated by the two nerves are represented in adjacent areas of cortex, areas 3b and 1. Cortex devoted to the representation of the ventral surface of the digits is indicated in **color**; that devoted to the dorsal surface is indicated by **gray**. In these cortical maps the five digits (D_1–D_5) and the four palmar pads (P_1–P_4) are arranged in an orderly sequence and their representations have been numbered in order. **3.** The remarkable topographic organization of the cortical map can be appreciated by comparing this diagram of the hands with the map in B_2.

are exposed to different combinations of stimuli, and are likely to exercise our motor skills in different ways, the architecture of each brain is modified in unique ways. This distinctive modification of brain architecture, along with a unique genetic makeup, constitutes the biological basis for individuality.

As we have seen in Chapter 18, two studies have provided evidence consistent with this view. One study found that the somatosensory maps vary considerably between normal animals. However, this study did not separate the effects of different experiences from the consequences of different genetic endowment. Thus, another study was conducted, to see whether activity is important in determining the topographic organization of the somatosensory cortex. Adult monkeys were encouraged to use their middle three fingers at the expense of other fingers to obtain food. After several thousand trials, the area of cortex devoted to the middle finger was greatly expanded. Practice, therefore, may act on existing patterns of connections and strengthen their effectiveness (Figure 36–15 and 36–16).

Reorganization is also evident at lower levels in the brain, specifically at the level of the dorsal column nuclei, which contain the first synapses of the somatic sensory system. Organizational changes are therefore probably a general property of the somatosensory system and occur throughout the somatic afferent pathway.

Common Cellular Mechanisms May Be Responsible for Changes in the Somatotopic Map

What mechanisms underlie the changes in receptive fields of neurons in the somatosensory cortex? As we saw in Chapter 18, recent evidence indicates that the input connections to cortical neurons in the somatic sensory system are formed on the basis of correlated activity, much as cooperative activity shapes the development of ocular dominance columns in the visual system (Chapter 25). These associative mechanisms seem similar to those underlying long-term potentiation and activity-dependent presynaptic facilitation.

To test this idea the skin surfaces of two adjacent fingers on the hand of a monkey were surgically connected to ensure that the connected fingers were always used together. This increases the correlation of inputs from the skin surfaces of the adjacent fingers, abolishing the sharp discontinuity normally evident between the zones in the somatosensory cortex that receive inputs from these digits (see Figure 18–8). Thus, the normal discontinuity in the representation of adjacent fingers in the cortical map appears to be established not only by a genetically programmed demarcation in the pattern of connections, but also through learning, by temporal correlations in patterns of input. Like the development of the ocular dominance columns in the visual cortex, the representation for each finger results from correlated activity, while the normal discontinuity between fingers results from noncorrelated activity.

Neuronal Changes Associated with Learning Provide Insights into Psychiatric Disorders

The demonstration that learning is accompanied by changes in the effectiveness of neural connections has revised our view of the relationship between social and biological processes in the shaping of behavior. There used to be a tendency in medicine and psychiatry to think that biological and social determinants of behavior act on separate levels of the mind. For example, it was customary to classify psychiatric illnesses into two major categories: organic and functional. *Organic* mental illnesses included the dementias, such as Alzheimer's disease, and the toxic psychoses, such as those that follow the chronic use of alcohol. *Functional* mental illnesses included the various depressive syndromes, the schizophrenias, and the neurotic illnesses.

This distinction dated to the nineteenth century, when neuropathologists examined the brains of patients coming to autopsy and found gross and readily demonstrable distortions in the architecture of the brain in some psychiatric diseases but not in others. Diseases that produced anatomical evidence of brain lesions were called organic; those lacking these features were called functional.

The experiments reviewed in this chapter show that this distinction is long outdated. Everyday events—sensory stimulation, deprivation, and learning—can cause an effective weakening of synaptic connections in some circumstances and a strengthening of connections in others. We no longer think that certain diseases (organic diseases) affect mentation by producing biological changes in the brain while other diseases (functional diseases) do not. The basis of contemporary neural science is that all mental processes are biological and any alteration in those processes is organic.

The question has shifted. We now ask: How do the biological processes of the brain give rise to mental events, and how in turn do social factors modulate the biological structure of the brain? In the attempt to understand a particular mental illness, it is more appropriate to ask: To what degree is this biological process determined by genetic and developmental factors? To what degree is it determined by a toxic or infectious agent? To what degree is it environmentally or socially determined? Even those mental disturbances that are

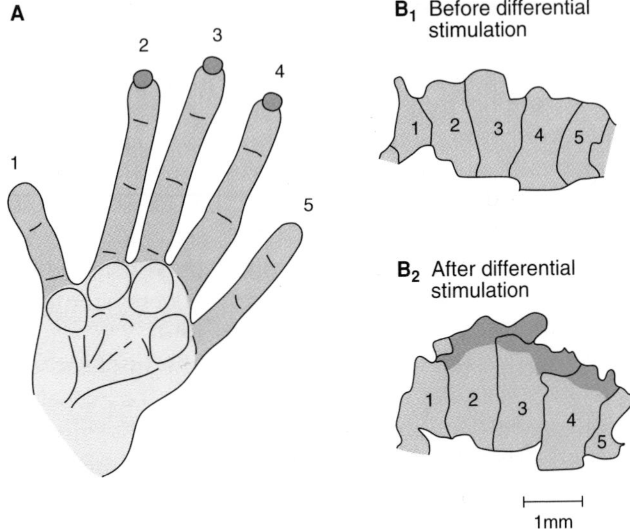

Figure 36–16 Training expands existing afferent inputs in the cortex. (Adapted from Jenkins et al., 1990.)

A.1. A monkey was trained for one hour per day to perform a task that required repeated use of the tips of fingers 2, 3, and occasionally 4 (**dark shading**). **B$_1$.** Representation of the tips of the digits of an adult monkey in cortical area 3b three months before training. **B$_2$.** After a period of repeated stimulation, the portion of area 3b representing the tips of the stimulated fingers is substantially enlarged (**dark shading**).

considered most heavily determined by social factors must have a biological aspect, since it is the activity of the brain that is being modified. Insofar as social intervention works, whether through psychotherapy, counseling, or the support of family or friends, it must work by acting on the brain and, quite likely, on the strength of connections between nerve cells. Moreover, the absence of demonstrable structural changes does not rule out the possibility that important biological changes are nevertheless occurring. They may simply be undetectable with the techniques available to us.

Demonstrating the biological nature of mental functioning requires more sophisticated anatomical methodologies than the light-microscopic histology of nineteenth-century pathologists. To clarify these issues it will be necessary to develop a neuropathology of mental illness that is based on anatomical function as well as anatomical structure. Various new imaging techniques, such as positron emission tomography and magnetic resonance imaging described in Chapter 5, have opened the door to the noninvasive exploration of the human brain on a cell-biological level, the level of resolution that is required to understand the physical mechanisms of mentation and of mental disorders. This approach is now being pursued in the study of schizophrenia and depression.

Since structural changes in mental functions are likely to reflect alterations in gene expression, we should look for altered gene expression in all persistent mental states, normal as well as disturbed. There is now substantial evidence that the susceptibility to major psychotic illnesses—schizophrenia and manic–depressive disorders—is heritable. These illnesses reflect heritable alterations in the nucleotide sequence of DNA, leading to abnormal messenger RNA and abnormal protein. Whereas the genetic data on schizophrenia and depression suggest that these diseases involve alteration in the structure of genes, the cell-biological data on learning and long-term memory reviewed here suggest that neurotic illnesses, acquired by learning, are likely to involve alterations in the *regulation* of gene expression (Figure 36–17).

Development, hormones, stress, and learning are all factors that alter gene expression by modifying the binding of transcriptional activator proteins to each other and to the regulatory regions of genes. It is likely that at least some neurotic illnesses (or components of them) result from reversible defects in gene regulation, which are produced by learning and which may be due to altered binding of specific proteins to certain regulatory regions that control the expression of certain genes.

According to this view, the major mental illnesses—schizophrenia and depression—result primarily from heritable genetic changes in neuronal and synaptic function in a human population carrying one or more, likely several, abnormal alleles. In contrast, neurotic illnesses might result largely from alterations in neuronal and synaptic function produced by environmentally induced alterations in gene expression. It is intriguing to think, then, that insofar as psychotherapy is successful in bringing about substantive changes in behavior, it does so by producing alterations in gene expression.

A corollary to these arguments is that a neurotic illness should involve alterations in neuronal structure and function just as certain psychotic illnesses involve structural (anatomical) changes in the brain. Treatment of neuroses or character disorders by psychotherapeutic intervention should, if successful, also produce structural changes. Thus, we face the intriguing possibility that as brain imaging techniques improve, these techniques might ultimately be useful not only for diagnosing various neurotic illnesses but also for monitoring the progress of psychotherapy.

An Overall View

The studies on synaptic plasticity reviewed here, and those on synapse formation we considered earlier in the book, suggest there are three overlapping stages in the course of synaptic development and its subsequent maintenance. The first stage, synapse formation, occurs primarily in the early stages of development and is under the control of genetic and developmental processes, commonly cell–cell interactions. The second stage, the fine tuning of newly developed synapses, occurs during critical early periods of development and requires an appropriate pattern of activity in neurons, which is usually provided by environmental stimulation. The third stage, the regulation of both the transient and long-term effectiveness of synapses, occurs daily throughout later life and also is determined by experience. An intriguing possibility is that activity-dependent cellular mechanisms involved in associative learning may be similar to the activity-dependent mechanisms at work during critical periods of development.

One of the implications of this view is that the potentialities for all behavior in an individual are created by genetic and developmental mechanisms acting on the brain. Environmental factors and learning bring out specific capabilities by altering the effectiveness (and anatomical connections) of preexisting pathways. It follows from this argument that everything that occurs in the brain—from the most private thoughts to commands for motor acts—are biological processes.

The convergence of neurobiology and neuropsychology that we have emphasized throughout this book is filled with promise. Modern cognitive psychology has shown that the brain stores an internal representation of the world that it experiences, while neurobiology has shown that this representation can be understood in terms of individual nerve cells and their interconnections. From this convergence we have gained a new perspective on perception, learning, and memory.

Although early behaviorist psychology led the way in exploring observable aspects of behavior, advances in modern cognitive psychology indicate that investigations into behavior that fail to consider brain mechanisms cannot adequately account for behavior. The recognition of the importance of internal representations might have been discouraging as recently as 10 years ago, when internal mental processes were essentially inaccessible to experimental analysis. However, as we have seen throughout this book, recent developments in cell and molecular biology have made biological experiments on elementary aspects of internal mental processes feasible. Contrary to the expectations of some, biological analysis is unlikely to diminish our fascination with thinking or to make thinking trivial by reduction when we frame the issues in terms of molecular biology. Rather, cell biology and molecular biology have expanded our vision, allowing us to perceive previously unanticipated

A Alteration in gene structure in inherited psychiatric disease

1 Normal gene

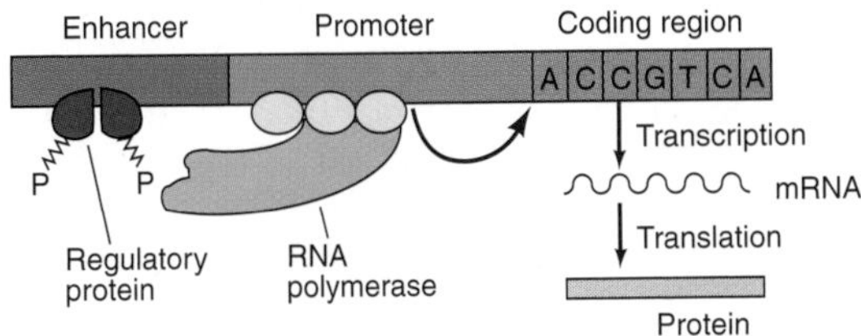

2 Mutation

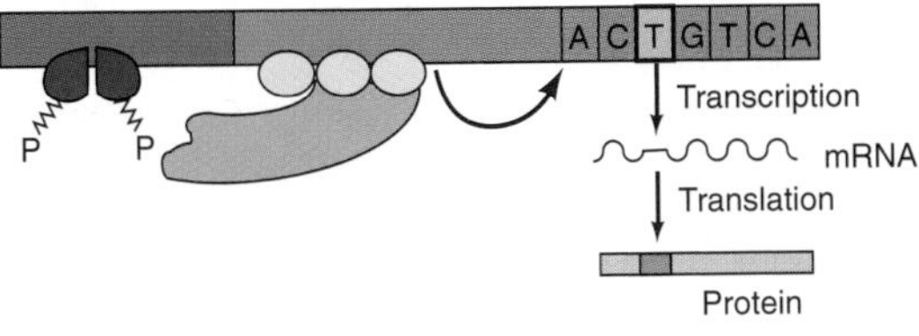

B Alteration in gene regulation in acquired psychiatric disease

1 Gene is not expressed

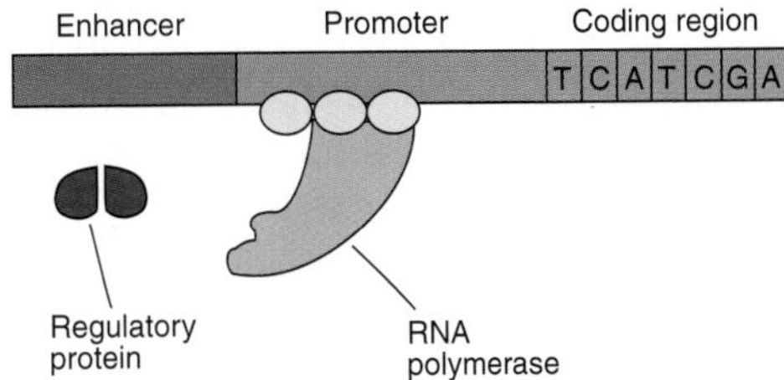

2 Gene is expressed

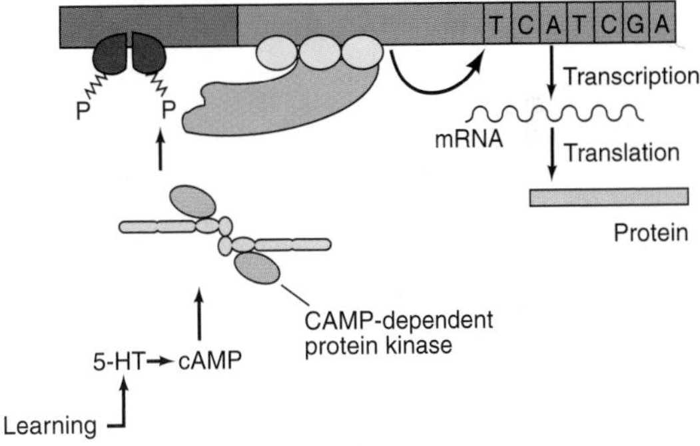

Figure 36–17 Genetic and acquired illnesses both have a genetic component. Genetic illnesses (e.g., schizophrenia) are expressions of altered genes, whereas illnesses acquired as learned behavior (neuroses) involve the modulation of gene expression by environmental stimuli, leading to the transcription of a previously inactive gene. The gene is illustrated as having two segments. A *coding region* is transcribed into a messenger RNA by an RNA polymerase. The mRNA in turn is translated into a specific protein. A *regulatory segment* consists of an *enhancer region* and a *promoter region* (see Chapter 14, Box 14–1). In this example the RNA polymerase can transcribe the gene when the regulatory protein binds to the enhancer region. For binding to occur, the regulatory protein must first be phosphorylated.

A.1. Under normal conditions the phosphorylated regulatory protein binds to the enhancer region, thereby activating the transcription of the gene, leading to the production of the protein. **2.** A mutant form of the coding region of the structural gene, in which a thymidine (T) has been substituted for cytosine (C), leads to transcription of an altered messenger RNA. This in turn produces an abnormal protein, giving rise to the disease state. This alteration in gene structure becomes established in the germline and is heritable.

B.1. If the regulatory protein for a normal gene is not phosphorylated, it cannot bind to the enhancer site, and thus gene transcription cannot be initiated. **2.** In this case a specific experience leads to the activation of serotonin (5-HT) and cAMP, which activate the cAMP-dependent protein kinase. The catalytic unit phosphorylates the regulatory protein, which then can bind to the enhancer segment and thus initiate gene transcription. By this means an abnormal learning experience could lead to the expression of a protein that gives rise to symptoms of a neurotic disorder.

interrelationships between biological and psychological phenomena.

The boundary between cognitive psychology and neural science is arbitrary and always changing. It has been imposed not by the natural contours of the disciplines, but by lack of knowledge. As our knowledge expands, the biological and behavioral disciplines will merge at certain points; it is at these points that our understanding of mentation will rest on more secure ground. As we have tried to illustrate in this book, the merger of biology and cognitive psychology is more than a sharing of methods and concepts. The joining of these two disciplines represents the emerging conviction that scientific descriptions of mentation at several different levels will all eventually contribute to a unified biological understanding of behavior.

Selected Readings

Bekkers, J. M., and Stevens, C. F. 1990. Presynaptic mechanism for long-term potentiation in the hippocampus. Nature 346:724–728.

Bonhoefer, T., Staiger, V., and Aertsen, A. 1989. Synaptic plasticity in rat hippocampal slice cultures: Local "Hebbian" conjunction of pre- and postsynaptic stimulation leads to distributed synaptic enhancement. Proc. Natl. Acad. Sci. USA 86:8113–8117.

Bourtchuladze, R., Frenguelli, B., Blendy, J., Cioffi, D., Schutz, G., and Silva, A. J. 1994. Deficient long-term memory in mice with a targeted mutation of the cAMP responsive element-binding protein. Cell 79:59–68.

Dudai, Y. 1989. The Neurobiology of Memory: Concepts, Findings, Trends. Oxford: Oxford University Press.

Grant, S. G. N., O'Dell, T. J., Karl, K. A., Stein, P. L., Soriano, P., and Kandel, E. R. 1992. Impaired long-term potentiation, spatial learning, and hippocampal development in *fyn* mutant mice. Science 258:1903–1910.

Hawkins, R. D., Kandel, E. R., and Siegelbaum, S. A. 1993. Learning to modulate transmitter release: Themes and variations in synaptic plasticity. Annu. Rev. Neurosci. 16:625–665.

Huang, Y.-Y., Li, X-C, and Kandel, E. R. 1994. cAMP contributes to mossy fiber LTP by initiating both a covalently-mediated early phase and macromolecular synthesis-dependent late phase 79:69–80.

Huang, Y.-Y., and Kandel, E. R. 1994. Recruitment of long-lasting and protein kinase A-dependent long-term potentiation on the CA1 region of hippocampus requires repeated tetanization. Learning & Memory 1:74–82.

Kandel, E. R. 1989. Genes, nerve cells, and the remembrance of things past. J. Neuropsychiatry 1:103–125.

Malinow, R. 1991. Transmission between pairs of hippocampal slice neurons: Quantal levels, oscillations and LTP. Science 252:722–724.

Merzenich, M. M., Recanzone, E. G., Jenkins, W. M., Allard, T. T., and Nudo, R. J. 1988. Cortical representational plasticity. In P. Rakic and W. Singer (eds.), Neurobiology of Neocortex. New York: Wiley, pp. 41–67.

Nicoll, R. A., Kauer, J. A., and Malenka, R. C. 1988. The current excitement in long-term potentiation. Neuron 1:97–103.

Pavlov, I. P. 1927. Conditioned Reflexes: An Investigation of the Physiological Activity of the Cerebral Cortex. G. V. Anrep (trans.) London: Oxford University Press.

Silva, A. J., Stevens, C. F., Tonegawa, S., and Wang, Y. 1992. Deficient hippocampal long-term potentiation in α-calcium–calmodulin kinase II mutant mice. Science 257:201–206.

Schuman, E. M., and Madison, D. V. 1994. Locally distributed synaptic potentiation in hippocampus. Science 263:532–536.

Schuman, E. M., and Madison, D. V. 1991. A requirement for the intercellular messenger nitric oxide in long-term potentiation. Science. 254:1503–1506.

Silva, A. J., Paylor, R., Wehner, J. M., and Tonegawa, S. 1992. Impaired spatial learning in α-calcium–calmodulin kinase II mutant mice. Science 257:206–211.

Stevens, C., and Wang, Y. 1994. Changes in reliability of synaptic function as a mechanism for plasticity. Nature 371:704–707.

Tsien, R. W., and Malinow, R. 1990. Long-term potentiation: Presynaptic enhancement following postsynaptic activation of Ca^{++}-dependent protein kinases. Cold Spring Harbor Symp. Quant. Biol. 55:147–159.

Weisskopf, M. G., Castillo, P. E., Zalutsky, R. A., and Nicoll, R. A. 1994. Mediation of hippocampal long-term potentiation by cyclic AMP. Science 265:1878–1882.

Yin, J. C. P., Wallach, J. S., Del Vecchio, M., Wilder, E. L., Zhou, H., Quinn, W. G., and Tully, T. 1994. Induction of a dominant negative CREB transgene specifically blocks long-term memory in *Drosophila*. Cell 79:49–58

Zalutsky, R. A., and Nicoll, R. A. 1990. Comparison of two forms of long-term potentiation in single hippocampal neurons. Science 248:1619–1624.

Figure Sources

Alberts, B., Bray, D., Lewis, J., Raff, M., Roberts, K., and Watson, J. D. 1994. Molecular Biology of the Cell, 3rd ed. New York: Garland.

Allen, G. I., and Tsukahara, N. 1974. Cerebrocerebellar communication systems. Physiol. Rev. 54:957-1006.

Almers, W., and Tse, F. W. 1990. Transmitter release from synapses: Does a preassembled fusion pore initiate exocytosis? Neuron 4:813-818.

Alsobrook, J. P. II, and Stevens, C. F. 1988. Cloning of calcium channel. Trends Neurosci. 11:1-2.

Anderson, O., and Providence, L. [Provided Figure 7-4B].

Bailey, C. H., and Chen, M. 1983. Morphological basis of long-term habituation and sensitization in *Aplysia*. Science 220:91-93.

Banerjee, U., and Zipursky, S. L. 1990. The role of cell-cell interaction in the development of the *Drosophila* visual system. Neuron 4:177-187.

Barre, J. J. [Provided Figure 1-1].

Beach, F. A., Noble, R. G., and Orndoff, R. K. 1969. Effects of perinatal androgen treatment on responses of male rats to gonadal hormones in adulthood. J. Comp. Physiol. Psychol. 68:490-497.

Bergen, J. R., and Julesz, B. 1983. Parallel versus serial processing in rapid pattern discrimination. Nature 303:696-698.

Berg, P., and Singer, M. 1993. Dealing with Genes. Mill Valley, Calif.: Blackwell Scientific Publications.

Berman, P. W., and Patrick, J. 1980. Experimental myasthenia gravis: A murine system. J. Exp. Med. 151:204-223.

Bermant. G., and Davidson, J. M. 1974. Biological Bases of Sexual Behavior. New York: Harper & Row.

Bershadsky, A. D., and Vasiliev, J. M. 1988. Cytoskeleton. New York: Plenum Press.

Bisiach, J. G., and Luzzatti, C. 1978. Unilateral neglect of representational space. Cortex 14:129-133.

Blakemore, C. 1977. Mechanics of the Mind. Cambridge, England: Cambridge University Press.

Bouchard, T. J., Jr., Lykken, D. T., McGue, M., Segal, N. L., and Tellegen, A. 1990. Sources of human psychological differences. The Minnesota twin study of twins reared apart. Science 250:223-228.

Boyd, I. A., and Martin, A. R. 1956. The end-plate potential in mammalian muscle. J. Physiol. (Lond.) 132:74-91.

Boyd, I. A., and Smith, R. S. 1984. The muscle spindle. In P. J. Byck, P. K. Thomas, E. H. Lambert, and R. Bunge (eds.), Peripheral Neuropathy, 2nd ed., Vol. 1. Philadelphia: Saunders, pp. 171-202.

Braun, M. 1992. Picturing Time: The Work of Etienne-Jules Marey (1830-1904). Chicago: University of Chicago Press.

Broca, Pierre Paul, portrait, courtesy of Michel Maumont, Maire de la Ville de Sainte-Foy-La-Grande.

Brown, R., and Herrnstein, R. J. 1975. Psychology. Boston: Little, Brown.

Bunge, M. P., Bunge R. P., and Ris, H. 1961. Ultrastructural study of remyelination in an experimental lesion in adult cat cord. J. Biophys. Biochem. Cytol. 10:677-94.

Burke, R. E. 1990. Spinal cord: Ventral horn. In G. M. Shepherd (ed.), The Synaptic Organization of the Brain, 3rd ed. New York: Oxford University Press, pp. 88-132.

Burke, R. E., Levine, D. N., Salcman, M., and Tsairis, P. 1974. Motor units in cat soleus muscle: Physiological, histochemical and morphological characteristics. J. Physiol. (Lond.) 238:503-514.

Carew, T. J., and Kandel, E. R. 1976. Two functional effects of decreased conductance EPSP's: Synaptic augmentation and increased electrotonic coupling. Science 192:150-153.

Carpenter, M. B., and Sutin, J. 1983. Human Neuroanatomy, 8th ed. Baltimore: Williams & Wilkins.

Castellucci, V. F., Carew, T. J., and Kandel, E. R. 1978. Cellular analysis of long-term habituation of the gill-withdrawal reflex of *Aplysia californica*. Science 202:1306-1308.

Catterall, W. A. 1988. Structure and function of voltage-sensitive ion channels. Science 242:50-61.

Clark, S. A., Allard, T., Jenkins, W. M., and Merzenich, M. M. 1988. Receptive fields in the body-surface map in adult cortex defined by temporally correlated inputs. Nature 332:444-445.

Coggeshall, R. E., and Mandriota, F. [Provided Figure 3-1B].

Colquhoun, D. 1981. How fast do drugs work? Trends Pharmacol. Sci. 2:212-217.

Conradi, S. 1969. Ultrastructure and distribution of neuronal and glial elements on the motoneuron surface in the lumbosacral spinal cord of the adult cat. Acta Physiol. Scand. [Suppl.] 332:5-48.

Constantin-Paton, M. 1981. Induced ocular-dominance zones in tectal cortex. In F. O. Schmitt, F. G. Worden, G. Adelman, S. D. Dennis (eds.), The Organization of the Cerebral Cortex: Proceedings of a Neurosciences Research Program Colloquium. Cambridge, Mass.: MIT Press, pp. 47-67.

Corbit, J. D. 1973. Voluntary control of hypothalamic temperature. J. Comp. Physiol. Psychol. 83:394-411.

Cole, K. S., and Curtis, H. J. 1939. Electric impedance of the squid giant axon during activity. J. Gen. Physiol. 22:649-670.

Dartnall, H. J. A., Bowmaker, J. K., and Mollon, J. D. 1983. Microspectrophotometry of human photoreceptors. In J. D. Mollon and L. T. Sharpe (eds.), Colour Vision: Physiology and Psychophysics. New York: Academic Press, pp. 69-80.

Davis, M. 1992. The role of the amygdala in fear and anxiety. Annu. Rev. Neurosci. 15:353-375.

Desmedt, J. E., and Godaux, E. 1977. Fast motor units are not preferentially activated in rapid voluntary contractions in man. Nature 267:717-719.

Deutsches Museum, Munich [Provided Figure 1-7].

DeYoe, E. A., and Van Essen, D. C. 1988. Concurrent processing streams in monkey visual cortex. Trends Neurosci. 11:219-226.

DiGirolamo, M., and Rudman, D. 1968. Variations in glucose metabolism and sensitivity to insulin of the rat's adipose tissue, in relation to age and body weight. Endocrinology 82:1133-1141.

Dogiel, A. S. 1908. Der Bau der Spinalganglien des Menschen und der Säugetiere. Jena: Fischer.

Doupe, A. J., Landis, S. C., and Patterson, P. H. 1993. Environmental influences in the development of the neural crest derivatives: Glucocorticoids, growth factors, and chromaffin cell plasticity. J. Neurosci. 5:2119-2142.

Dowling, J. E. 1979. Information processing by local circuits: The vertebrate retina as a model system. In F. O. Schmitt and F. G. Worden (eds.), The Neurosciences: Fourth Study Program. Cambridge, Mass.: MIT Press, pp. 163-181.

Dowling, J. E. 1992. Neurons and Networks: An Introduction to Neuroscience. Cambridge, Mass.: Harvard University Press.

Dowling, M. [Provided Figure 21-13B]

Drachman, D. B. 1983. Myasthenia gravis: Immunobiology of a receptor disorder. Trends Neurosci. 6:446-451.

Dudel, J. 1983. General sensory physiology. In R. F. Schmidt and G. Thews (eds.), Human Physiology. M. A. Biederman-Thorson (trans.) Berlin: Springer, pp. 177-192.

Dyck, P. J., Thomas, P. K., Lambert, E. H., and Bunge, R. (eds.) 1984. Peripheral Neuropathy, 2nd ed. Philadelphia: Saunders.

Eckert, R., Randall, D., and Augustine, G. 1988. Animal Physiology: Mechanisms and Adaptations, 3rd ed. New York: Freeman.

Elste, A., Koester, J., Shapiro, E., Panula, P., and Schwartz, J. H. 1990. Identification of histaminergic neurons in *Aplysia*. J. Neurophysiol. 64:736-744.

Escher, M. C. 1971. The Graphic Work of M. C. Escher. New Rev. and exp. ed. New York: Ballantine Books.

Evarts, E. V. 1968. Relation of the pyramidal tract activity to force exerted during voluntary movement. J. Neurophysiol. 31:14-27.

Fambrough, D. M., Drachman, D. B., and Satyamurti, S. 1973. Neuromuscular junction in myasthenia gravis: Decreased acetylcholine receptors. Science 182:293-295.

Fawcett, D. 1981. The Cell, 2nd ed. Philadelphia: Saunders.

Fernandez, J. M., Neher, E., and Gomperts, B. D. 1984. Capacitance measurements reveal stepwise fusion events in degranulating mast cells. Nature 312:453-455.

Fertuck, H. C., and Salpeter, M. M. 1974. Organization of acetylcholine receptor by ^{125}I-labeled α-bungarotoxin binding at mouse motor endplates. Proc. Natl. Acad. Sci. U.S.A. 71:1376-1378.

Finkel, A. S., and Redman, S. J. 1983. The synaptic current evoked in cat spinal motoneurones by impulses in single group Ia axons. J. Physiol. (Lond.) 422:203-225.

Fischer, J. M., Sossin, W., Newcomb, R., and Scheller, R. H. 1988. Multiple neuropeptides derived from a common precursor are differentially packaged and transported. Cell 54:813-822.

Forscher, P., and Smith, S. J. 1988. Actions of cytochalasins on the organization of actin filaments and microtubules in a neuronal growth cone. J. Cell Biol. 107:1505-1516.

Frank, E. [Provided Figure 3-2].

Frech, G. C., Van Dongen, A. M. J., Schuster, G. Brown, A. M., and Joho, R. H. 1989. A novel potassium channel with delayed rectifier properties isolated from rat brain by expression cloning. Nature 340:642-645.

Fremeau, T. T., Jr., Autelitano, D. J., Blum, M., Wilson, J., and Roberts, J. L. 1989. Intervening sequence-specific in situ hybridization: Detection of the proopiomelanocortin gene primary transcript in individual neurons. Mol. Brain Res. 6:197-201.

Frielle, T., Kobilka, B., Dohlman, H., Caron, M. G., and Lefkowitz, R. J. 1989. The β-adrenergic receptor and other receptors coupled to guanine nucleotide regulatory proteins. In S. Chien (ed.), Molecular Biology in Physiology. New York: Raven Press, pp. 79-91.

Frost, W. N., Castellucci, V. F., Hawkins, R. D., and Kandel, E. R. 1985. Monosynaptic connections from the sensory neurons participate in the storage of long-term memory in *Aplysia*. Proc. Natl. Acad. Sci. U.S.A. 82:8266-8269.

Fujisawa, H., Tani, N., Watanabe, K., and Ibata, Y. 1982. Branching of regenerating retinal axons and preferential selection of appropriate branches fr specific neuronal connections in the newt. Dev. Biol. 90:43-57.

Funahasi, S., Bruce, C. J., and Goldman-Rakic, P. S. 1989. Mnemonic coding of visual space in the monkey's dorsolateral prefrontal cortex. J. Neurophysiol. 61:331-349.

Furshpan, E. J., and Potter, D. D. 1957. Mechanisms of nerve-impulse transmission at the crayfish synapse. Nature 180:342-343.

Furshpan, E. J., and Potter, D. D. 1959. Transmission at the giant motor synapses of the crayfish. J. Physiol. (Lond.) 145:289-325.

Gardner, E. P. 1988. Somatosensory cortical mechanisms of feature detection in tactile and kinesthetic discrimination. Can. J. Physiol. Pharmacol. 66:439-454.

Gardner, E. L., and Lowinson, J. H. 1993. Drug craving and positive/negative hedonic brain substrates activated by addicting drugs. Semin. Neurosci. 5:359-368.

Gay, V. L. 1972. The hypothalamus: Physiology and clinical use of releasing factors. Fertil. Steril. 23:50-63.

Georgopoulos, A. P., Kalaska, J. F., Caminiti, R., and Massey, J. T. 1982. On the relations between the direction of two-dimensional arm movements and cell discharge in primate motor cortex. J. Neurosci. 2:1527-1537.

Georgopoulos, A. P., Lurito, J. T., Petrides, M., Schwartz, A. B., and Massey, J. T. 1989. Mental rotation of the neuronal population vector. Science 243:234-236.

Geschwind, N. 1979. Specializations of the human brain. Sci. Am. 241(3):180-199.

Geschwind, N., and Levitsky, W. 1968. Human brain: Left-right asymmetries in temporal speech region. Science 161:186-187.

Gilbert, P. F. C., and Thach, W. T. 1977. Purkinje cell activity during motor learning. Brain Res. 128:309-328.

Gilbert, C. D., and Wiesel, T. N. 1989. Columnar specificity of intrinsic horizontal and corticocortical connections in cat visual cortex. J. Neurosci. 9:2432-2442.

Gilbert, C. D., Wiesel, and Katz, unpublished experiments for Figure 25-5.

Gilula, N. [Provided Figure 11-5].

Gleitman, L. R., and Gleitman, H. 1981. Language. In H. Gleitman (ed.), Psychology. New York: Norton, chap. 10.

Gorski, R. A., Gordon, J. H., Shryne, J. E., and Southern, A. M. 1978. Evidence for a morphological sex difference within the medial preoptic area of the rat brain. Brain Res. 148:333-346.

Gross, C. G. 1981. [Will complete in galleys]

Groves, P. M., and Rebec, G. V. 1992. Introduction to Biological Psychology, 4th ed. Dubuque, Iowa: W. C. Brown.

Groves, P., and Schlesinger, K. 1979. Introduction to Biological Psychology. Dubuque, Iowa: W. C. Brown.

Gustafsson, B. and Wigström, H. 1988. Physiological mechanisms underlying long-term potentiation. Trends Neurosci. 11:156-162.

Hall, J. C. 1978. Courtship among males due to a male-sterile mutation in *Drosophila melanogaster*. Behav. Genet. 8:125-141.

Hall, J. C. 1985. Genetic analysis of behavior in insects. In Comparative Insect Physiology. Biochemistry and Pharmacology, Vol. 9, G. A. Kerkut, ed. Oxford: Pergamon Press, pp. 287-373.

Hall, J. C. 1990. Genetics of circadian rhythms. Annu. Rev. Genet. 24:659-697.

Hamburger, V. [Provided Figure 6-9].

Hardin, P. E., Hall, J. C., and Rosbash, M. 1990. Feedback of the *Drosophila* period gene produce on circadian cycling of its messenger RNA levels. Nature 343:536-540.

Harlow, H. F. 1958. Behavioral contributions to interdisciplinary research. In H. F. Harlow and C. N. Woolsey (eds.), Biological and Biochemical Bases of Behavior. Madison: University of Wisconsin Press, pp. 3-23.

Harvey, A. M., Lilienthal, J. L., Jr., and Talbot, S. A. 1941. Observations on the nature of myasthenia gravis: The phenomena of facilitation and depression of neuromuscular transmission. Bull. Johns Hopkins Hosp. 69:547-565.

Haymaker, W., and Schiller, F. 1970. The Founders of Neurology, 2nd ed. Springfield, Ill.: Thomas.

Hawkins, R. D., Abrams, T. W., Carew, T. J., and Kandel, E. R., 1983. A cellular mechanism of classical conditioning in Aplysia: Activity-dependent amplification of presynaptic facilitation. Science 219:400–406.

Hestrin, S., Nicoll, R. A., Perkel, D. J., and Sah, P. 1990. Analysis of excitatory synaptic action in pyramidal cells using whole-cell recording from rat hippocampal slices. J. Physiol. (Lond.) 422:203-225.

Heuser, J. E., and Reese, T. S. [Provided Figure 11-7].

Hikosaka, O., Tanaka, M., Sakamoto, M., and Iwamura, Y. 1985. Deficits in manipulative behaviors induced by local injections of muscimol in the first somatosensory cortex of the conscious monkey. Exp. Brain Res. 51:327-337.

Hille, B. 1984. Ionic channels of Excitable Membranes. Sunderland, Mass.: Sinauer.

Hille, B. 1992. G protein-coupled mechanisms and nervous signaling. Neuron 9:187-195.

Hirokawa, N., Pfister, K. K., Yorifuji, H., Wagner, M. C., Brady, S. T., and Bloom, G. S. 1989. Submolecular domains of bovine brain kinesin identified by electron microscopy and monoclonal antibody decoration. Cell 56:867-878.

Hochbergm J. E. 1968. Perception. Englewood Cliffs, N. J.: Prentice-Hall.

Hodgkin, A. L. 1992. Chance and Design: Reminiscences of science in Peah and War. Cambridge, England: Cambridge University Press.

Hodgkin, A. L., and Huxley, A. F. 1939. Action potentials recorded from inside a nerve fibre. Nature 144:710-711.

Hodgkin, A. L., and Huxley, A. F. 1945. Resting and action potentials in single nerve fibres. J. Physiol. (Lond.) 104:176-195.

Houk, J. C., Crago, P. E., and Rymer, W. Z. 1980. Functional properties of the Golgi tendon organs. In J. E. Desmedt (ed.), Spinal and Supraspinal Mechanisms of Voluntary Motor Control and Locomotion, Vol. 8. Progress in Clinical Neurophysiology. Basel: Karger, pp. 33-43.

Huang, K.-S., Radhakrishnan, R., Bayley, H., and Khorana, H. G. 1982. Orientation of retinal in bacteriorhodopsin as studied by cross-linking using a photosensitive analog of retinal. J. Biol. Chem. 257:13616-13623.

Hubel, D. M. 1988. Eye, Brain and Vision. New York: Scientific American Library.

Hubel, D. M., and Wiesel, T. W. Portrait. Courtesy of D. M. Hubel and T. W. Wiesel.

Hubel, D. H., and Wiesel, T. N. 1959. Receptive fields of single neurones in the cat's striate cortex. J. Physiol. (Lond.) 148:574-591.

Hubel, D. H., and Wiesel, T. N. 1962. Receptive fields, binocular interaction and functional architecture in the cat's visual cortex. J. Physiol. (Lond.) 160:106-154.

Hubel, D. H., and Wiesel, T. N. 1965. Binocular interaction in striate cortex of kittens reared with artificial squint. J. Neurophysiol. 28:1041-1059.

Hubel, D. H., and Wiesel, T. N. 1977. Ferrier Lecture: Functional architecture of macaque monkey visual cortex. Proc. R. Soc. Lond. [Biol.] 198:1-59.

Hubel, D. H., and Wiesel, T. N. 1979. Brain mechanisms of vision. Sci. Am. 241(3):150-162.

Hubel, D. H., Wiesel, T.N., and LeVay, S. 1977. Plasticity of ocular dominance columns in monkey striate cortex. Philos. Trans. R. Soc. Lond. [Biol.] 278:377-409.

Hubel, D. H., Wiesel, T. N., and Stryker, M. P. 1978. Anatomical demonstration of orientation columns in macaque monkey. J. Comp. Neurol. 177:361-379.

Hurvich, L. M. 1981. Color Vision. Sunderland, Mass.: Sinauer.

Huganir, R. L., and Greengard, P. 1990. Regulation of neurotransmitter receptor desensitization by protein phosphorylation. Neuron 5:555-567.

Hulliger, M. 1984. The mammalian muscle spindle and its central control. Rev. Physiol. Biochem. Pharmacol. 101:1-110.

Hunt, C. C. and Kuffler, S. W. 1951. Stretch receptor discharges during muscle contraction. J. Physiol. (Lond.) 113:298-315.

Huxley, A. F. [Provided Figure 10-2].

Ito, M. 1984. The Cerebellum and Neural Control. New York: Raven Press.

Iwamura, Y., Tanaka, M., Sakamoto, M., and Hikosaka, O. 1985. Comparison of the hand and finger representation in areas 3, 1, and 2 of the monkey somatosensory cortex. In M. Rowe and W. D. Willis, Jr. (eds.), Development, Organization, and Processing in Somatosensory Pathways. New York: Liss, pp. 239-245.

Iwamura, Y., Tanaka, M., Sakamoto, M., and Hikosaka, O. 1985. Vertical neuronal arrays in the postcentral gyrus signaling active touch: A receptive field study in the conscious monkey. Exp. Brain Res. 58:412-420.

Jan, Y. N., Jan, L. Y., and Kuffler, S. W. 1989. A peptide as a possible transmitter in sympathetic ganglia of the frog. Proc. Natl. Acad. Sci. U.S.A. 86:5159-5162.

Jen, J. and Stevens, C. F. [Provided Figures 13-10 and 13-13A].

Jenkins, W. M., Merzenich, M. M., Ochs, M. T., Allard, T., and Guic-Robles, E. 1990. Functional reorganization of primary somatosensory cortex in adult owl monkeys after behaviorally controlled tactile stimulation. J. Neurophysiol. 63:82-104.

Jessell, T. M., and Kandel, E. R. 1993. Synaptic transmission: A bidirectional and a self-modifiable form of cell-cell communication. Cell 72/Neuron 10 (Suppl.):1-30.

Julesz, B. 1983. Binocular depth perception in normal adults and its early development. Forschritte der Opthalmologie 80(5):378-383.

Julesz, B. 1986. Stereoscopic vision. Vision Res. 26:1601-1612.

Kaas, J. H., Merzenich, M. M., and Killackey, H.P. 1983. The reorganization of somatosensory cortex following peripheral nerve damage in adult and developing mammals. Annu. Rev. Neurosci. 6:325-356.

Kaas, J. H., Nelson, R. J., Sur, M., Lin, C.-S., and Merzenich, M. M. 1979. Multiple representations of the body within the primary somatosensory cortex of primates. Science 204:521-523.

Kandel, E. R., and Hawkins, R. S. 1992. The biological basis of learning and individuality. Sci. Am. 267:78-86.

Katz, B., and Miledi, R. 1967a. The study of synaptic transmission in the absence of nerve impulses. J. Physiol. (Lond.) 192:407-436.

Katz, B., and Miledi, R. 1967b. The timing of calcium action during neuromuscular transmission. J. Physiol. (Lond.) 189:535-544.

Keesey, R. E., Boyle, P. C., Kemnitz, J. W., and Mitchel, J. S. 1976. The role of the lateral hypothalamus in determining the body weight set point. In D. Novin, W.

Wrywicka, and G. A. Bray (eds.), Hunger: Basic Mechanisms and Clinical Implications. New York: Raven Press, pp. 243-255.

Keesey, R. E. 1989. Physiological regulation of body weight and the issue of obesity. Med. Clinics. N. Am. 7:371-418.

Kleiber, M. 1947. Body size and metabolic rate. Physiol. Rev. 27:511-541.

Kolb, B., and Whishaw, I. Q. 1990. Fundamentals of Human Neuropsychology, 3rd ed. New York: Freeman.

Konopka, R. J. and Benzer, S. 1971. Clock mutants of *Drosophila merganogaster*. Proc. Natl. Acad. Sci. USA 68:2112-2116.

Kuffler, S. W. 1953. Discharge patterns and functional organization of mammalian retina. J. Neurophysiol. 16:37-68.

Kuffler, S. W., Nicholls, J. G., and Martin, A. R. 1984. From Neuron to Brain: A Cellular Approach to the Function of the Nervous System, 2nd ed. Sunderland, Mass.: Sinauer.

Kuypers, H. G. J. M. 1985. The anatomical and functional organization of the motor system. In M. Swash and C. Kennard (eds.), Scientific Basis of Clinical Neurology. New York: Churchill Livingstone, pp. 3-18.

LeDoux, J. E. 1992. Brain mechanisms of emotion and emotional learning. Curr. Opin. Neurobiol., 2:191-197.

Lenneberg, E. H. 1967. Biological Foundations of Language. New York: Wiley.

LeVay, S., Wiesel, T. N., and Hubel, D. H. 1981. The postnatal development and plasticity of ocular-dominant columns in the monkey. In F. O. Schmitt, F. G. Worden, G. Adelman, and S. G. Dennis (eds.), The Organization of the Cerebral Cortex: Proceedings of a Neurosciences Research Program Colloquium. Cambridge, Mass.: MIT Press, pp. 29-45.

LeVay, S., Stryker, M. P., and Shatz, C. J. 1978. Ocular dominance columns and their development in layer IV of the cat's visual cortex: A quantitative study. J. Comp. Neurol. 178:223-244.

Levi-Montalcini, R. [Provided Figure 6-13].

Levi-Montalcini, R. 1972. The morphological effects of immunosympathectomy. In G. Steiner and E. Schonbaum (eds.), Immunosympathectomy. Amsterdam: Elsevier, pp. 55-78.

Levin, B. 1994. Genes, 5th ed. Oxford: Oxford University Press.

Leyton, A. S. F., and Sherrington, C. S. 1917. Observations on the excitable cortex of chimpanzee, orang-utan, and gorilla. Q. J. Exp. Physiol. 11:135-222.

Liley, A. W., 1956. The quantal components of the mammalian endplate potential. J. Physiol. (Lond.) 133:571-587.

Lindsay, P. H., and Norman, D. A. 1977. Human Information Processing, 2nd ed. New York: Academic Press.

Lindstrom, J. 1983. Using monoclonal antibodies to study acetylcholine receptors and myasthenia gravis. Neurosci. Comment 1:139-156.

Linstedt, A. D., and Kelly, R. B. 1991. Synaptophysin is sorted from endocytotic markers in neuroendocrine PC12 cells but not transfected fibrolasts. Neuron 7:309-317.

Lisak, R. P., and Barchi, E. L. 1982. Myasthenia Gravis. Philadelphia: Saunders.

Livingstone, M. Art, illusion and the visual system. Sci. Am. 258(1):78-85.

Livingstone, M. S., and Hubel, D. H. 1984. Anatomy and physiology of a color system in the primate cortex. J. Neuroscience 4:309-356.

Llinás, R. R. 1982. Calcium in synaptic transmission. Sci. Am (4):56-65.

Linás, R. R., and Heuser, J. E. 1977. Depolarization-release coupling systems in neurons. Neurosci. Res. Program Bull. 15:555-687.

Loewy, A. D., and Spyer, K. M. (eds.). 1990. Central Regulation of Autonomic Functions. New York: Oxford University Press.

Lund, J. S. 1988. Anatomical organization of macaque monkey striate cortex. Annu. Rev. Neurosci. 11:253-288.

Madison, D. V., and Nicoll, R. A. 1986. Cyclic adenosine 3'.5' monophosphate mediates β-receptor actions of noradrenaline in rat hippocampal pyramidal cells. J. Physiol. (Lond.) 372:245-259.

Makowski, L., Caspar, D. L. D., Phillips, W. C., Baker, T. S., and Goodenough, D. A. 1984. Gap junction structures. VI. Variation and conservation in connexon conformation and packing. Biophys. J. 45:208-218.

Malinow, R., and Tsien, R. W. 1990. Presynaptic enhancement shown by whole-cell recordings of long-term potentiation in hippocampal slices. Nature 346:177-180.

Marshall, W. H., Woolsey, C. N., and Bard, P. 1941. Observations on cortical somatic sensory mechanisms of cat and monkey. J. Neurophysiol. 4:1-24.

Martin, J. H. 1989. Neuroanatomy: Text and Atlas. New York: Elsevier.

Matthews, P. B. C. 1964. Muscle spindles and their motor control. Physiol. Rev. 44:219-288.

Matthews, P. B. 1972. Mammalian Muscle Receptors and Their Central Actions. London: Arnold.

McEwen, B. S. 1976. Interactions between hormones and nerve tissue. Sci. Am. 235:48-58.

McGuire, B. A., Gilbert, C. D., Rivlin, P. K., and Wiesel, T. N. 1991. Targets of horizontal connections in macaque primary visual cortex. J. Comp. Neurol. 305:370-392.

McMahan, U. J., and Kuffler, S. W. 1971. Visual identification of synaptic boutons on living ganglion cells and of varicosities in postganglionic axons in the heart of the frog. Proc. R. Soc. Lond. [Biol.] 177:485-508.

McMasters, R. [Provided Figure 5-1B].

Merzenich, M. M., Kaas, J. H., Wall, J., Nelson, R. J., Sur, M., and Felleman, D. 1983. Topographic reorganization of somatosensory cortical areas 3B and 1 in adult monkeys following restricted deafferentiation. Neuroscience 8:33-55.

Merzenich, M. M., Kaas, J. H., Wall, J. T., Sur, M., Nelson, R. J., and Felleman, D. J. 1983. Progression of change following median nerve section in the cortical representation of the hand in areas 3b and 1 in adult owl and squirrel monkeys. Neuroscience 10:639-665.

Miles, F. A. 1969. Excitable Cells. London: Heinemann.

Miller, K. D., Keller, J. B., and Stryker, M. P. 1989. Ocular dominance column development: Analysis and simulation. Science 245:605-615.

Miller, T. M., and Heuser, J. E. 1984. Endocytosis of synaptic vesicle membrane at the frog neuromuscular junction. J. Cell Biol. 98:685-698.

Milner, B. [Provided Figure 35-1]

Mogilmer, A., Grossman, J. A. I., Ribary, U., Joliot, M., Volkmann, J., Rapaport, D., Beasley, R. W., and Llinás, R. R. 1993. Somatosensory cortical plasticity in adult humans revealed by magnetoencephalography. Proc. Natl. Acad. Sci. U.S.A. 90:3593-3597.

Money, J., and Ehrhardt, A. A. 1972. Man & Woman, Boy & Girl. Baltimore: Johns Hopkins University Press.

Monster, A. W., and Chan, H. 1977. Isometric force production by motor units of extensor digitorum communis muscle in man. J. Neurophysiol. 40:1432-1443.

Mountcastle, V. 1974. Medical Physiology, 13th ed., Vol. I. St. Louis: Mosby.

Movshon, A. 1990. Visual processing of moving images. In H. Barlow, C. Blakemore, and M. Weston-Smith (eds.), Images and Understanding: Thoughts About Images; Ideas About Understanding. New York: Cambridge University Press, pp. 122-137.

Nathans, J., and Hogness, D. S. 1984. Isolation and nucleotide sequence of the gene encoding human rhodopsin. Proc. Natl. Acad. Sci. U.S.A. 81:4851-4855.

Nathans, J., Thomas, D., and Hogness, D. S. 1986. Molecular genetics of human color vision: The genes encoding blue, green, and red pigments. Science 232:193-202.

Nathans, J., Piantanida, T. P., Eddy, R. L., Shows, T. B., Hogness, D. S. 1986. Molecular genetics of inherited variation in human color vision. Science 232:203-210.

The National Gallery, London [Provided Figure 24-1].

Nauta, W. J. H., and Feirtag, M. 1986. Fundamental Neuroanatomy. New York: Freeman.

Nicoll, R. A., Kauer, J. A., and Malenka, R. C. 1988. The current excitement in long-term potentiation. Neuron 1:97-103.

Nicoll, R. A., Madison, D. V., and Lancaster, B. 1987. Noradrenergic modulation of neuronal excitability in mammalian hippocampus. In H. Y. Meltzer (ed.), Psychopharmacology: The Third Generation of Progress. New York: Raven Press, pp. 105-112.

Nieuwenhuys, R., Voogd, J., and van Huijzen, Chr. 1981. The Human Central Nervous System: A Synopsis and Atlas, 2nd ed. Berlin: Springer.

Nieuwenhuys, R., Voogd, J., and van Huijzen, Chr. 1988. The Human Central Nervous System: A Synopsis and Atlas, 3rd rev. ed. Berlin: Springer.

O'Brien, D. F. 1982. The chemistry of vision. Science 218:961-966.

Ochs, S. 1972. Fast transport of materials in mammalian nerve fibers. Science 176:2452-260.

Owens, J. L., and Kullberg, R. 1989. *In vivo* development of nicotinic acetylcholine receptor channels in *Xenopus* myotomal muscle. J. Neurosci. 9:1018-1028.

Pallis, C. A. 1955. Impaired identification of faces and places with agnosia for colours: Report of a case due to cerebral embolism. J. Neurol. Neurosurg. Psychiat. 18:218-224.

Patton, H. D. 1965. Reflex regulation of movement and posture. In T. C. Ruch and H. D. Patton (eds.), Physiology and Biophysics, 19th ed. Philadelphia: Saunders, pp. 181-206.

Patton, H. D., Sundsten, J. W., Crill, W. E., and Swanson, P. D. 1976. Introduction to Basic Neurology. Philadelphia: Saunders.

Pearson, K. 1976. The control of walking. Sci. Am. 235(6):72-86.

Penfield, W. (ed.) 1932. Cytology & Cellular Pathology of the Nervous System, Vol. 2. New York: Hoeber.

Penfield, W., and Rasmussen, T. 1950. The Cerebral Cortex of Man: A Clinical Study of Localization of Function. New York: Macmillan.

Peters, A., Palay, S. L., and Webster, H. deF. 1991. The Fine Structure of the Nervous System: Neurons and Their Supporting Cells, 3rd ed. New York: Oxford University Press.

Petersen, S. W., Fox, P. T., Posner, M. I., Minton, M., and Raichle, M. E. 1988. Positron emission tomographic studies of the cortical anatomy of single-word processing. Nature 331:585-589.

Phelps, M. E. and Mazziotta, J. C. Study of cerebral function with positron computer tomography. J. Cereb. Blood Flow Metab. 2:113-162.

Philadelphia Museum of Art, The William L. Elkins Collection.

Plomin, R., Owen, M. J., and McGuffin, P. 1994. The genetic basis of complex human behaviors. Science 264:1733-1739.

Price, C., MRC Cyclotron Unit, Hammersmith Hospital, London, England.

Purves, D. 1988. Body and Brain. A Trophic Theory of Neural Connections. Cambridge, Mass.: Harvard University Press.

Purves, D., and Lichtman, J. W. 1980. Elimination of synapses in the developing nervous system. Science 210:153-157.

Raibert, M. H. 1977. Motor control and learning by the state-space model. Cambridge, Mass.: MIT Press.

Rakic, P. 1971. Mode of cell migration to the superficial layers of the fetal monkey neocortex. J. Comp. Neurol. 145:61-83.

Ramachandran, V. S. 1987. Interaction between colour and motion in human vision. Nature 328:645-647.

Ramachandran, V. S. 1993. Behavioral and magnetoencephalographic correlates of plasticity in the adult human brain. Proc. Natl. Acad. Sci. U.S.A., 90:20.

Ramón y Cajal, S. 1933. Histology, 10th ed. Baltimore: Wood.

Ramón y Cajal, S. 1909. Histologie du Systéme Nerveux de l'Homme & des Vertebres, Vol. 1. L. Azoulay (trans.). Madrid: Instituto Ramón y Cajal, 1955.

Rasmussen, G. L. [Provided Figure 3-1D].

Readhead, C., Popko, B., Takahashi, N., Shine, H. D., Saavedra, R. A., Sidman, R. L., and Hood, L. 1987. Expression of a myelin basic protein gene in transgenic Shiverer mice: Correction of the dysmyelinating phenotype. Cell 48:703-712.

Ready, D. F., Hanson, T. E., and Benzer, S. 1976. Development of the *Drosophila* retina, a neurocrystalline lattice. Dev. Biol. 53:217-240.

Reichlin, S. 1978. Introduction. In S. Reichlin, R. J. Baldessarini, and J. B. Martin (eds.), The Hypothalamus. Res. Publ. Assoc. Res. Nerv. Ment. Dis. 56:1-14.

Robitaille, R., Adler, E. M., and Charlton, M. P. 1990. Strategic location of calcium channels at transmitter release sites of frog neuromuscular synapses. Neuron 5:773-779.

Rock, I. 1984. Perception. New York: Scientific American Books.

Roland, P. E., Larsen, B., Lassen, N. A., and Skinhut, E. 1980. Supplementary motor area and other cortical areas in organization of voluntary movements in man. J. Neurophysiol. 43:118-136.

Rosenkilde, C. E. 1979. Functional heterogeneity of the prefrontal cortex in the monkey: A Review. Behav. Neural Biol. 25:301-345.

Rossant, J. 1990. Manipulating the mouse genome: Implications for neurobiology. Neuron 4:323-334.

Rowland, L. P., Hoefer, P. F. A., and Aranow, H., Jr. 1960. Myasthenic syndromes. Res. Publ. Assoc. Res. Nerv. Ment. Dis. 38:548-600.

Rubin, E. [Provided Figure 21-3]

Sacks, O. and Wasserman, R. 1987. The Case of the Colorblind Painter. New York Review of Books, pp. 25-345.

Sakmann, B. [Provided Figures 5-5B, 10-11B, 10-11C, and 13-1B].

Satinoff, E. 1964. Behavioral thermoregulation in response to local cooling of the rat brain. Am. J. Physiol. 206:1389-1394.

Sawyer, D. B., Koeppe, R. E. II, and Andersen, O. S. 1989. Induction of conductance heterogeneity in gramicidin channels. Biochemistry 28:6571-6583.

Schiavo, G., Benfenati, F., Poulain, B., Rossetto, O., Polverino de Laureto, P., DasGupta, B. R., and Montecucco, C. 1992. Tetanus and botulinum-B neurotoxins block neurotransmitter release by proteolytic clevage of synaptobrevin. Nature 359:832-835.

Schmidt, R. F. 1985. Motor systems. In R. F. Schmidt and G. Thews (eds.), Human Physiology. M. A. Biderman-Thorson (trans.) Berlin: Springer.

Schnapp, B., and Reese, T. 1982. Cytoplasm structure in rapid-frozen axon. J. Cell Biol. 94:1667-1679.

Shepherd, G. M. 1990. The Synaptic Organization of the Brain, 3rd ed. New York: Oxford University Press.

Sherrington, C. 1947. The Integrative Action of the Nervous System, 2nd ed. New Haven: Yale University Press.

Silverman, A.-J., Carlos, L. D., Morrell, J. I. [Provided Figure 31-4]

Sossin, W. S., Fischer, J. M., and Scheller, R. H. 1989. Cellular and molecular biology of neuropeptide processing and packaging. Neuron 2:1407-1417.

Sperry, R. W. 1968. Mental unity following surgical disconnection of the cerebral hemispheres. Harvey Lect. 62:293-323.

Sperry, R.W. 1963. Chemoaffinity in the orderly growth of nerve fiber patterns and connections. Proc. Natl. Acad. Sci. U.S.A. 50:703-710.

Spurzheim, J. G. 1825. Phrenology, or the Doctrine of the Mind, 3rd ed. London: Knight.

Squire, L. R. 1987. Memory and Brain. New York: Oxford University Press.

Squire, L. R., Slater, P. C., and Chace, P. M. 1975. Retrograde amnesia: Temporal gradient in very long term memory following electroconvulsive therapy. Science 187:77-79.

Stevens, C. F. 1991. Making a submicroscopic hole in one. Nature 349:657-658.

Stryer, L. 1988. Biochemistry, 3rd ed. New York: Freeman.

Swaab, D. F., and Fliers, E. 1985. A sexually dimorphic nucleus in the human brain. Science 228:1112-1115.

Swett, J. E., and Schoultz, T. W. 1975. Mechanical transduction in the Golgi tendon organ: A hypothesis. Arch. Ital. Biol. 113:374-382.

Tomlinson, A. 1988. Cellular interactions in the developing *Drosophila* eye. Development 104:183-193.

Toran-Allerand, C. D. 1978. Gonadal hormones and brain development: Cellular aspects of sexual differentiation. Am Zool. 18:553-565.

Tosney, K. [Provided Figure 6-5C].

Toyoshima, C., and Unwin, M. 1988. Ion channel of acetylcholine receptor reconstructed from images of postsynaptic membranes. Nature 336:247-250.

Treisman, A. 1988. Features and objects: The Fourteenth Bartlett Memorial Lecture. J. Exp. Psychol. 40A:(2):201-237.

Ts'o, D., Gilbert, C., and Wiesel, T. 1990. Functional organization of primate visual cortex revealed by high-resolution optical imaging. Science 249:417-420.

Tully, T., and Hirsch, J. 1982. Behavior-genetic analysis of *Phormia regina*. I. Isolation of pure-breeding for high and low levels of the central excitatory state (CES) from an unselected population. Behav. Genet. 12:395-415.

Unwin, P. N. T., and Zampighi, G. 1980. Structure of the junction between communicating cells. Nature 283:545-549.

Vallbo, Å B. 1981. Basic patterns of muscle spindle discharge in man. In A. Taylor and A. Prochazka (eds.), Muscle Receptors and Movement. London: Macmillan, pp. 219-228, 263-275.

Wald, G. 1950. [Will complete in galleys]

Watson, J. D., Tooze, J., and Kurtz, D. T. (eds.) 1983. Recombinant DNA: A Short Course. San Francisco: Freeman.

Watson, J. D., Gilman, M., Witkowski, J., and Zoller, M. 1992. Recombinant DNA, 2nd ed. New York: Scientific American Books.

White, J. M. 1992. Membrane fusion. Science 258:917-924.

Williams, P. L., Warwick, R., Dyson, M., and Bannister, L. H. (eds). 1989. Gray's Anatomy, 37th ed. Edinburgh: Churchill Livingstone, pp. 859-919.

Wise, R. A., Bauco, P., Carlezon, W. A., Jr., and Trojniar, W. 1992. Self-stimulation and drug reward mechanisms. Ann. N.Y. Acad. Sci. 654:192-198.

Woolesy, C. N. 1958. Organization of somatis sensory and motor areas of the cerebral cortex. In: H. F. Harlow and C. N. Woolesy (eds.) Biological and Biochemical Bases of Behavior. Madison, WI: University of Wisconsin Press, pp. 63–81.

Wurtz, R. H., and Goldberg, M. E. (eds.) 1989. The Neurobiology of Saccadic Eye Movements, Reviews of Oculomotor Research, Vol. 3. Amsterdam: Elsevier.

Yarbus, A. L. 1967. Eye Movements and Vision. B. Haigh (trans.). New York: Plenum Press.

Young, R. W. 1970. Visual cells. Sci. Am. 22(4):80-91.

Zalutsky, R. A., and Nicoll, R. A. 1990. Comparison of two forms of long-term potentiation in single hippocampal neurons. Science 248:1619-1624.

Zeki, S. [Provided Figure 10-2].

Zerr, D., Hall, J. C., Rosbash, M., and Siwicki, K. K. 1990. Circadian fluctuations of period protein immunoreactivity in the CNS and the visual system of *Drosophila*. J. Neurosci. 19:2749-2762.

Zihl, J., Cramon, D. and Mai, N. 1983. Cerebral disturbances of movement vision. Brain, 106:313-340.

Glossary

Terms in *italic type* within definitions are key words and are themselves defined. The glossary is selective and does not include many basic terms in cell biology, biochemistry, or anatomy.

α-Bungarotoxin A peptide isolated from snake venom that binds strongly to *nicotinic acetylcholine receptors* at the vertebrate *neuromuscular junction*.

Acetylcholine (ACh) *Neurotransmitter* released at neuromuscular junctions and selective central *synapses*.

Acetylcholinesterase An enzyme that hydrolyzes the neurotransmitter *acetylcholine* (ACh) into choline and acetate, contributing to the rapid termination of the postsynaptic *current* and thereby to the rapid termination of the signal. It is densely concentrated in the *basal lamina* of the *junctional folds* in the muscle fiber. Much of the choline released by hydrolysis is recaptured for the *presynaptic terminal* by the choline *transporter*.

Action potential A large electrical signal of about 100 mV that is initiated at the initial segment or axon hillock of the neuron and propagated without failure, in an all-or-none fashion, to its presynaptic terminal.

Adequate stimulus The particular type of stimulus energy to which a sensory *receptor cell* is sensitive.

Adoptive twin studies Comparison of phenotypic traits ranging from various mental disorders (schizophrenia, depression, alcoholism) to less reliably characterized personality determinants (extroversion, conscientiousness, interest in sports) in twins raised apart by different adoptive parents. The phenotypic similarity of identical twins raised apart strongly suggests that genes make an important contribution to behavior and personality. Sharing of the particular phenotypic trait is called concordance, and has been found to be substantially higher in identical (or monozygotic) twins, whose genes are identical, than in fraternal (or dizygotic) twins, whose genes are only half identical.

Adrenergic transmission *Synaptic transmission* produced by a neuron that releases *norepinephrine* or *epinephrine*.

Afferent A neuron or *pathway* that sends signals to the *central nervous system* or a higher processing center. The term is sometimes used interchangeably with *sensory*; strictly speaking, however, the term "sensory" should be reserved for those neurons or pathways that contribute directly to perception. (cf. *Efferent*)

After-potential A transient *hyper-* or *depolarization* following the *action potential*.

Aggregate field theory A theory of *brain* function, which states that individual mental functions or behaviors are performed by the brain as a whole, rather than by discrete portions of the brain. (cf. *Localization*)

Agnosia Loss of knowledge. Inability to perceive objects through otherwise normally functioning sensory pathways, e.g., depth agnosia, movement agnosia, color agnosia, and prosopagnosia (impairment of face recognition).

Active zone Specialized site on the cytoplasmic surface of the terminal of the presynaptic cell membrane from which *neurotransmitter*, packaged in synaptic vesicles is released.

AMPA (alpha-amino 3 hydro 5 methyl 4 isoxazole propionic acid) A *glutamate* agonist that activates a class of non-NMDA receptors for glutamate (cf. *NMDA Receptor*)

Aminergic transmission *Synaptic transmission* produced by a *neuron* that releases a *biogenic amine*.

Amygdala Collection of several *nuclei* that lie deep in the temporal lobes of cerebral hemispheres and are reciprocally connected to the *hypothalamus*, hippocampal formation, and *thalamus*. The part of the *brain* most specifically concerned with emotion, especially anxiety. It coordinates *autonomic* and endocrine responses in conjunction with emotional states.

Aphasia A category of language disorders resulting from a lesion to specific structures in the brain. These include: Broca's aphasia, Wernicke's aphasia, *conduction aphasia, paraphasia*

Aphasia, conduction Impairment of neural pathways connecting perception to speech generation areas, resulting in the inability to repeat simple phrases while retaining the ability to understand spoken and written words.

Appetitive response The response of a subject to a positively reinforcing stimulus such as food or drink.

APV (2-amino-5-phosphonovaleric acid) A drug that blocks the *NMDA receptor* for *glutamate*.

Artificial intelligence A science that attempts to model human or animal intelligence using computers or robots.

Association area Region of the *cerebral cortex* that integrates diverse *sensory* or *motor* information for purposeful action. The association areas are connected to higher-order sensory or motor areas. There are three major association cortices: (1) the parietal-temporal-occipital association cortex, (2) the prefrontal association cortex, and (3) the limbic association cortex.

Astrocyte The most numerous of the three major types of glial cells. (cf. *Oligodendrocyte* and *Schwann cell*)

Ataxia A disorder of balance and gait that accompanies lesions of the *cerebellum* or the dorsal columns.

Attention Selective focusing in on a specific person, object, or event to the exclusion of other extraneous stimuli; a basis of consciousness.

Auditory system The system of neurons, beginning in the inner ear, through the *brain stem*, medial geniculate nucleus of the *thalamus*, and ending in the auditory *cortex*, that produces discriminatory hearing.

Autonomic division One of two major subdivisions of the *peripheral nervous system*. It is the motor system for the viscera, smooth muscles, and exocrine glands. The autonomic division consists of three spatially segregated subdivisions: the *sympathetic*, *parasympathetic*, and *enteric nervous systems*.

Axo-axonic synapse The synapse made by an axon terminal of one neuron onto the axon terminal of another. (cf. *Axo-dendritic* and *Axo-somatic synapse*)

Axo-dendritic synapse The synapse made by an *axon* terminal of one neuron onto the *dendrite* of another. (cf. *Axo-axonic* and *Axo-somatic synapse*)

Axolemma The plasma membrane of the *axon*.

Axon The output fiber of the neuron that ends in presynaptic terminals.

Axon hillock The specialized region of the neuron's *cell body* from which the cell's *axon* emerges.

Axonal pathfinding The process by which the *axon* of a developing neuron projects to an appropriate target cell. The factors considered important in axonal pathfinding include cell-cell and *cell adhesion molecules*, acting through mechanisms of chemorepulsion and *chemotropism*.

Axonal transport The active movement of a neuron's constituents within the *axon*. Membranous organelles are exported from the *cell body* to the axon terminal by anterograde axonal transport and from the axon terminal to the cell body by retrograde axonal transport. Vesicle movements are mediated by motor proteins: by kinesins in the anterograde direction; and by dyneins in the retrograde direction. Non-vesicular slow axoplasmic flow controls the anterograde movement of *filaments* and cytosol.

Axo-somatic synapse The synapse made by an *axon* terminal of one neuron on the *cell body* (soma) of another. (cf. *Axo-axonic* and *Axo-dendritic synapse*)

Barbiturates A class of hypnotic drugs that includes phenobarbital and secobarbital. Barbiturates bind to *postsynaptic receptors* for the *neurotransmitter GABA,* increasing chloride *current,* and thereby enhancing inhibitory synaptic transmission.

Basal ganglia Bilateral and deep-lying structures in both *cerebral hemispheres* including the putamen, caudate, globus pallidus, and substantia nigra. These ganglia participate in the regulation of motor performance.

Basal lamina A specialized layer of extracellular matrix, consisting of collagen, proteoglycans, and glycoproteins, that covers the surface of the entire muscle fiber. At the *junctional folds* of the muscle membrane the basal lamina contains *acetylcholinesterase*, the enzyme that breaks down the *neurotransmitter acetylcholine* in the *synaptic cleft*.

Battery An apparatus that produces an *electromotive force* as a result of a difference in chemical potentials.

Behaviorism The historical view, first developed at the beginning of the 20th century, that the only appropriate approach to the study of behavior is through observation of a subject's observable actions. "Mental function" in this view is regarded as unobservable. In recent decades behaviorism has been in decline, while cognitive and neurobiological approaches to the study of behavior have dominated.

Benzodiazepines A class of anti-anxiety drugs and muscle relaxants that include diazepam (Valium) and chlordiazepoxide (Librium). Benzodiazepines bind to *receptors* for the neurotransmitter GABA, and increase *chloride* current, thereby enhancing inhibitory synaptic transmission.

Bidirectional (= nonrectifying) Refers to current *(electrical synapse* or an *ion channel)* that flows in either of two directions.

Binding problem The problem of how unified perception arises from separate neural pathways processing different types of sensory information.

Biogenic amines A class of *neurotransmitters* which include serotonin, histamine, and the catecholamines.

Birthday (of a neuron) The time that a neuronal progenitor cell leaves the cell cycle and gives rise to a postmitotic neuron.

Blood brain barrier A specialized barrier formed by endothelial cells and *astrocytes* that prevents the access of large proteins and charged molecules to the central nervous system.

Brain The organ that mediates all mental functions and all behavior. Conventionally subdivided into seven main parts: *spinal cord, medulla oblongata, pons, cerebellum, midbrain, diencephalon*, and *cerebral hemispheres*.

Brain stem A collective term for three anatomical structures: the *medulla*, *pons*, and *midbrain*, all located rostral to the *spinal cord*. The brain stem processes sensation from the skin and joints in the head, neck, and face, as well as specialized senses, such as hearing, taste, and balance. The sensory input and motor output of the brain stem are carried by the *cranial nerves*.

Broca's area A region in the posterior part of the left frontal cortex that is critically involved in the expression of language. (cf. *Wernicke's area*)

Cadherins One of the major structural families of glycoproteins involved in neural *cell adhesion*. (cf. *Immunoglobulins, Integrins*)

Calcium (Ca^{2+}) An essential element in the function of the nervous system. The influx of Ca^{2+} *ions* controlled by membrane *channels* affects many metabolic processes within neurons, and triggers the release of *neurotransmitter*.

cAMP (cyclic adenosine 3',5'-monophosphate) This

nucleotide is synthesized by the enzyme adenylyl cyclase and triggers protein *phosphorylation* by cAMP-dependent protein *kinases*. It also can act by binding directly to certain ion *channels*, and in bacteria, to transcriptional activator proteins. Perhaps the best understood of the *second messengers*, cAMP research has influenced thinking on second-messenger mechanisms in general.

Capacitance (C) The ability of an insulating material, such as the cell membrane, to store electrical charges. Measured in farads.

Capacitor The insulating material between two conductors. In the neuron, the cell membrane (specifically the lipid bilayer) is the capacitor and the cytoplasm and extracellular fluid are *conductors*.

Carbon monoxide (CO) A *second messenger*, generated by heme oxygenase, which freely diffuses across cell membranes, making it a *transcellular messenger*.

Cell adhesion Proper adhesion of cells is essential for the organization of multicellular animals and for the formation of specific connections in the development and maintenance of the nervous system. Cell adhesion is mediated by protein molecules that belong to several gene families, for example, the *immunoglobulin* superfamily, the *cadherin* family, the *integrins* and the *selectins*.

Cell body The metabolic center of the neuron and the principal site of control of electric *signals* in most neurons. It gives rise to two types of processes, the *axon* and *dendrites*, both of which conduct electrical signals. Also called the soma.

Cell lineage The ancestry of a cell, defined by the sequences of cell divisions that give rise to the cell.

Cellular connectionism The historical view, originated by Ramón y Cajal at the beginning to the 20th century, that individual neurons are the signaling units of the *brain*, are generally situated together in functional groups, and are connected to one another in a precise fashion. (cf. *Localization*)

Central nervous system One of two anatomical divisions of the nervous system, the other being the *peripheral nervous system*. The central nervous system comprises the *brain* and *spinal cord*. Although anatomically separate, the central and peripheral nervous systems are interconnected functionally.

Central neuron A neuron in the *central nervous system*.

Central pattern generator A neuronal circuit that generates an oscillatory output used to drive behaviors such as breathing and locomotion. Some of these oscillators are located in the *brain stem*.

Cerebellum One of the seven main parts of the *brain*. It modulates the force and range of movement and is involved in the learning of motor skills.

Cerebral hemisphere One of the seven main parts of the *brain*. The bilateral pair of cerebral hemispheres are connected by the *corpus callosum*. The cerebral hemispheres comprise the *cerebral cortex* and three deep-lying structures: *basal ganglia*, *hippocampus*, and *amygdaloid nucleus*.

Cerebral cortex See *Cortex*

cGMP (cyclic guanosine 3',5'-monophosphate) This nucleotide is synthesized by the enzyme guanylyl cyclase and acts to modify proteins in two ways: (1) by binding directly to proteins such as cGMP-sensitive cation channels in the retina; and (2) by protein phosphorylation through the cGMP-dependent protein kinase.

Channel (= ion channel) A membrane-spanning protein that mediates the flow of *ions* into and out of all cells. In nerve cells, channels trigger the changes in *membrane potential* that alter excitability. Channels have three functional states: resting (closed and capable of being activated), active (open), and refractory (closed and not capable of being activated). Some channels behave like a simple *resistor*, others like a rectifier. The amplitude of the current through a single channel is relatively constant, whereas the duration of the opening of the channel is variable. Most channels are selective either for cations or for anions. See *Voltage- gated*, *Transmitter-gated*, *Mechanically-gated*, *Gap-junction*, and *Leakage channels*.

Channel, gap-junction A specialized protein structure that connects the cytoplasm of the *presynaptic* and *postsynaptic* cells at an *electrical synapse*. See *Connexon*.

Channel, leakage (see *Resting channel*)

Channel, mechanically-gated Specialized channels that transduce mechanical stimuli into electrical responses for sensing touch, vibration, and sound.

Channel, nongated (see *Resting channel*)

Channel, resting An *ion channel* in nerve cells that passively conducts ions (most commonly K^+) across the cell membrane. The *ionic current* that flows through these channels is responsible for the *resting membrane potential*. Also called leakage or nongated channels. (cf. *Gating*)

Channel, transmitter- (or ligand-) gated An *ion channel* whose opening and closing is regulated by the binding of a chemical messenger such as a *neurotransmitter*. The binding of transmitter either regulates the channel directly when the receptor is part of the channel protein (direct gating), or indirectly when the receptor is a separate protein, by activating a *second messenger* (indirect gating). The opening of the channels permits the flow of ions and thus changes the cell's *membrane potential*. Transmitter-gated channels are *excitatory* when they are permeable to *sodium* or *calcium* and inhibitory when they are permeable to *chloride* or *potassium*. (cf. *Voltage-gated channel*)

Channel, voltage-gated An *ion channel* that opens and closes in response to *depolarization* of the *membrane potential*. Voltage-gated channels in neurons are permeable to *sodium*, *potassium*, or *calcium*. The proportion of different types of voltage-gated channels in the neuron's *integrative* and *trigger zones* determines how a neuron responds to synaptic input. Voltage-gated channels generate the *action potential*.

Chemoaffinity hypothesis The hypothesis proposed by Roger Sperry that neurons form specific connections based on matching chemical affinities.

Chemotropism The hypothesis that developing *axons* are attracted toward diffusible factors secreted by particular groups of target cells.

Chloride (Cl^-) An essential constituent of the nervous system. The concentration of the negatively charged

Cl^- ions outside the resting neuron is higher than that inside the cell. Inward flux of Cl^- through *transmittergated* Cl^- *channels* is a common *inhibitory* mechanism.

Cholinergic transmission *Synaptic transmission* produced by a neuron that releases *acetylcholine*.

Circadian rhythm A biological activity (such as sleep) that occurs in approximately 24-hour periods or cycles. The "clock" resides in the suprachiasmatic nucleus of the brain.

Code, frequency A method of coding whereby the strength of the stimulus is coded by the firing rate of a neuron.

Code, labeled line A method for coding information based on specific neuronal pathways, or private lines that distinguish activity in one pathway (for example, pain) from that in another pathway (e.g., temperature).

Code, population A method of coding the properties of a stimulus strength or quality in terms of number of its responding neurons.

Coding region (of gene) (see *Regulatory region*)

Cognitive neural science The combination of methods designed to study the neural mechanisms that give rise to mental processes. These include systems neural science, brain imaging, behavioral neurology, cognitive psychology, and computer science.

Commissure A neural structure that contains only axonal *decussations*. The largest commissure is the *corpus callosum*.

Commisurotomy Surgical severing of the *corpus collosum* and anterior commissure in epileptics to prevent spread of epilepsy from one hemisphere side to the other. See *Split-brain animal*.

Competition, neuronal Excess innervation of target cells during development leads to competition between inputs. Weaker synapses are eliminated because other synapses more successfully maintain input to the *postsynaptic cell*.

Computerized tomography (= CT scan) Computer-enhanced X-ray images that reveal subtle differences and fine structure in tissue that cannot be detected by conventional X-ray techniques. (cf. *Positron emission tomography* and *Magnetic resonance imaging*).

Concordance Similarity of phenotypic traits in twins (see *Adoptive twin studies*).

Conditioning, classical A form of implicit associative learning discovered by Ivan Pavlov in which a subject learns the relationship between two stimuli, a conditioned stimulus (CS) and an unconditioned stimulus (US).

Conditioning, operant A form of implicit associative (or instrumental) learning discovered by Edward Thorndike in which an animal learns about the relationship between a stimulus and a response.

Conductance (G) A measure of the ease with which ions flow through a channel, measured in siemens.

Conductance, leakage (g_l) The total conductance of a population of *resting ion channels*.

Conductance, nonsynaptic Current flow through an area of *postsynaptic* membrane that surrounds a region containing synaptically activated ion channels.

Conduction, electrotonic The passive spread of voltage changes between different regions of the neuronal membrane.

Conduction, saltatory A form of conduction found in myelinated nerves, whereby the action potential jumps rapidly from one *node of Ranvier* to another.

Conductor Something that can carry *current*. Cytoplasm and extracellular fluid are conductors, because both are solutions of *ions*. (cf. *Resistor*)

Connectional specificity The principle that neurons form specific functional interconnections, based on three anatomical observations: (1) there is no cytoplasmic continuity between nerve cells; (2) neurons do not connect indiscriminately to one another or form random networks; and (3) each cell communicates with specific *postsynaptic cells* but not with others, and always at specialized sites (*synapses*).

Connectionist model A model of neural function with *parallel* and *distributed processing* components. Also known as neural net model. (cf. *Serial processing*)

Connexin The protein subunit of the *connexon*. Connexins from different tissues all appear to belong to one large gene family.

Connexon (= hemi-channel) A specialized transmembrane protein in neurons that forms an *electrical synapse*. Two connexons, one each in the *presynaptic* and *postsynaptic cells*, form continuous channels between the cytoplasms of the communicating cells termed the *gap junctions*. The channels open and close, presumably by the rotation of the protein subunits or *connexins*. When open they permit *ionic current* as well as non-ionic small molecules to flow directly between the two cells.

Convergence A pattern of connections between neurons in which several *presynaptic neurons* make a synaptic connection onto a common *postsynaptic cell*. (cf. *Divergence*)

Corpus callosum The largest *commissure* in the nervous system, connecting the two *cerebral hemispheres*.

Cortex, association See *Association area*

Cortex, cerebral One of the major components of the *cerebral hemispheres*. It is divided into four lobes (frontal, parietal, temporal, occipital). The cortex contains the primary, secondary, and tertiary (sensory or motor) areas.

Cortex, higher-order Any of several regions of cerebral cortex that process information from a *primary* (sensory or motor) area. There are higher-order motor cortices (i.e., the *premotor cortex* and the *supplementary motor cortex*) and several higher-order sensory regions. These in turn interconnect and project to the three *association areas*.

Cortex, premotor The higher-order motor area of the *cerebral cortex*; includes the *supplementary motor area* and premotor area. Important in the initiation of action. Also called the lateral premotor area.

Cortex, posterior parietal An area of cortex concerned with personal and extrapersonal space. Lesions give rise to the *neglect syndrome*.

Cortex, primary motor A region of the cerebral cortex that carries out the initial processing of motor information in the brain. (cf. *Association area*)

Cortex, primary sensory Specialized regions in the cerebral cortex that carry out the initial processing of specific types of *sensory* information in the brain: primary somatosensory cortex, primary visual cortex, and primary auditory cortex. (cf. *Primary motor cortex, Association areas*)

Co-transmission Co-release of more than one type of chemical messenger (e.g., peptides and small-molecule *neurotransmitters*) from the same *presynaptic cell*, resulting in the activation of appropriate *postsynaptic receptors* (see *Synaptic transmission*).

Cranial nerves The neurons that originate from sensory *ganglia* and motor nuclei in the *hindbrain* and *midbrain*.

Critical period A period in development, which varies in duration for different brain regions, when the genetically determined patterns of brain circuitry are particularly subject to environmental refinement. The critical period, therefore, provides a well studied example of the interactions of genes and environment during development. In particular, many mammals have a well- established critical period for the development of cells with binocular responses in their visual cortex.

Current (I) Net flow of positive charge per unit time, $\Delta Q/\Delta t$. In nerve and muscle cells current is carried by positive and negative *ions* in solution. Measured in amperes (A). One ampere of current represents the movement of 1 coulomb of charge per second.

Current, capacitive (I_c) The current carried by *ions* that change the net electrical charge stored on the neuronal membrane. Capacitive currents are responsible for changing the *membrane potential*.

Current, end-plate The *ionic current* at the *end-plate* or postsynaptic region of the *neuromuscular junction*. The normal end-plate current is the sum of the opening of more than 200,000 *ion channels*.

Current, ionic (I_i) The *current* carried by ions flowing through *ion channels* across the neuronal membrane. Also called resistive current. The ionic and *capacitive currents* comprise the *total membrane current*.

Current, leakage (I_l) The steady current that flows through the *resting ion channels* of the neuronal membrane.

Cytoskeleton The network of microfilaments, microtubules, neurofilaments and their associated proteins within the cell. The cytoskeleton is not a fixed structure but rapidly changes during development, growth, and injury. The filaments that make up the cytoskeleton are in a dynamic state and are influenced by activity. In addition to maintaining cell shape, they mediate the movement of organelles, for example, through fast axonal transport. The *filaments* also serve as substrates for the operation of *cell adhesion* molecules, many of which are membrane-spanning proteins anchored within the cell to the cytoskeleton. Cytoskeleton-associated proteins that regulate the dynamic state of filaments include a large number of actin-binding proteins and microtubule associated proteins (MAP). Motor proteins, dynein, myosin, and kinesins are associated with cytoskeletal *filaments*, the microtubules. In addition to *cell adhesion* proteins (for example, integrins and cadherins), several cytoskeletal proteins have been identified that anchor extracellular proteins to the cell (for example, adducins and ankyrins).

Decerebrate preparation Experimental animal in which the *brain stem* is cut at the level of the *midbrain*. Used to study *spinal reflexes*.

Decussation The crossing of a neural pathway in the *brain stem* or *spinal cord* from one side of the body to the contralateral side. The most prominent decussation is that of the human *visual system*. (cf. *Commissure*)

Demyelination Degeneration of the insulating *myelin sheath* around the axon due to disease.

Dendrite The receptive element of most nerve cells. The two major sites for synaptic input on the dendrite are the *spine* and shaft. (cf. See *Dendritic tree*)

Dendritic tree The branching pattern of a neuron's *dendrites*. Dendritic branching permits a neuron to receive many synaptic inputs from neurons over a large area. The *dendritic tree* can be elaborate; *central neurons* have 20 to 40 main dendrites.

Depolarization A decrease in the *membrane potential* of a cell. Depolarization increases the likelihood a neuron will generate an *action potential* and is therefore *excitatory*. (cf. *Hyperpolarization*)

Desensitization The process whereby the binding of a ligand to a receptor leads to the decreased effectiveness of the receptor.

Diacylglycerol-inositol polyphosphate system A *second-messenger* system activated by transmitters in nerve cells.

Diencephalon One of the seven main parts of the brain. It comprises two structures, the *thalamus* and *hypothalamus*. Also known as the "between brain" since it lies between the *midbrain* and the *cerebral hemispheres*.

Diffusion constant A measure of the rate at which a particle moves in solution. The diffusion constant of an *ion* is not simply a function of its size but also including its *waters of hydration*.

Distributed processing Routing of information to a number of different areas in the *brain*.

Divergence A pattern of connections between neurons in which a single presynaptic neuron forms synapses on several different *postsynaptic cells*. (cf. *Convergence*)

DNA (deoxyribonucleic acid) DNA, contained in the chromosomes within the cell's nucleus, contains the information for the synthesis of proteins. A greater portion of the total genetic information encoded in DNA is expressed in the brain than in any other organ of the body.

Docking Attachment of *transmitter*-containing *synaptic vesicles* to release sites at *active zones* in the *presynaptic terminal*.

Dorsal roots The *spinal nerves* that carry somatic *sensory* information into the *spinal cord* from sensory *ganglia*. (cf. *Ventral roots*)

Dysmetria Error in the range and force of movement.

Dynamic polarization The principle that information within a neuron flows in a predictable and consistent direction.

Efferent Indicates that a neuron or pathway sends signals from the *central nervous system* to the periphery or to a lower processing center. (cf. Afferent)

Electromotive force A source of electrical *potential*, such as an ionic gradient.

Electrotonic (or **electrical) transmission** *Synaptic transmission* at *bidirectional electrical synapses*, so called because the mechanism of transmission is similar to the passive (electrotonic) propagation of sub*threshold* electrical signals.

End-plate Region in muscle cells where nerve endings contact and form synapses. The end-plate region of vertebrate muscle fiber contains a significant number of *nicotinic acetylcholine receptors*.

Enhancer region (of gene) The portion of the regulatory region of a gene that binds proteins and controls in which cells and at what time the gene will be expressed.

Enteric nervous system One of three major subdivisions of the *autonomic division* of the *peripheral nervous system*. It controls the function of smooth muscle of the gut. (cf. *Sympathetic* and *Parasympathetic nervous systems*)

EPSP (= excitatory postsynaptic potential) The *excitatory synaptic potentials* that increase the probability that a cell will fire an *action potential*.

Equilibrium potential See *Nernst potential*

Equivalent circuit Representation of the electrical properties of individual neurons or groups of neurons in a conventional electrical circuit consisting only of *conductors* (or *resistors*), *batteries*, and *capacitors*.

Excitation The *depolarization* of a *postsynaptic cell*, increasing the likelihood that an *action potential* will be generated.

Excitatory Said of a *neuron*, *synapse*, or *synaptic potential* that produces *depolarization* of the *postsynaptic cell*. (cf. *Inhibitory*)

Exocytosis The process by which *synaptic vesicles* release *neurotransmitters* by fusing with the cell's plasma membrane. At *active zones* exocytosis is facilitated by proteins associated with both the *synaptic vesicles* (*vesicle-associated proteins*) and the nerve terminals.

Extrafusal fiber Main contractile element of a muscle. (cf. *Intrafusal fiber*)

Facilitation, presynaptic Action by a presynaptic neuron at an *axo-axonic synapse* increasing the amount of *neurotransmitter* released by the *postsynaptic cell* onto a third cell. (cf. *Presynaptic inhibition*)

Fiber When used alone, the term means an *axon*. When referring to a muscle cell, the term"muscle fiber" is always used.

Filaments Long polymers of similar subunit proteins (monomers) that constitute the *cytoskeleton*. In neurons, the the most abundant are microfilaments (F-[filamentous] *actin*), composed of G(globular) *actins*; microtubules, composed of heterodimers of α- and β-*tubulins*, and neurofilaments (neuronal intermediate filaments) whose components are the *neurofilament* proteins.

Flexure A characteristic curve in the longitudinal axis of the nervous systems of higher vertebrates, occurring during development. Principal flexures are the cervical flexure (the junction of the *spinal cord* and *hindbrain*), the cephalic flexure (junction of the *hindbrain* and *midbrain*), and the pontine flexure which lies between the cervical and cephalic flexure and bends the neural tube in the opposite direction.

Forebrain One of three primary embryonic vesicles of the nervous system. It gives rise to the *telencephalon* and *diencephalon*.

Fusion pore Narrow cytoplasmic bridge that unites the membrane of a *synaptic vesicle* with the plasma membrane at the *active zone* of a neuron and thus connects the lumen of the vesicle with extracellular space.

Ganglion (pl. ganglia) A cluster of functionally related neuron cell bodies in the *peripheral nervous system*. Examples are the segmentally organized dorsal root ganglia that mediate sensory functions, and the sympathetic and parasympathetic ganglia of the *autonomic division*.

Gap junction The zone of apposition between two neurons at an *electrical synapse*. (cf. *Synaptic cleft*)

Gating The active transition of an *ion channel* between open and closed states. Gating can be direct or indirect. (See *Transmitter-gated*, *Voltage-gated*, *Mechanically gated*, and *Resting channels*)

Gestalt psychology A school of psychology that particularly focused on visual perception and emphasized the fact that perception is based on an analysis of the relationship between an object and its surrounding background.

Glial cells One of two major types of cells in the nervous systems, the other being the *neuron*. Glial cells are between 10 and 50 times more numerous than neurons in the vertebrate *central nervous system*. They are not thought to be essential in processing information in the nervous system, but function as supporting elements, providing the *myelin* sheath around neurons, and as scavengers of cellular debris following injury or neuronal death. (See *Oligodendrocyte*, *Schwann cell*, and *Astrocyte*.)

Glutamate A common amino acid, L-glutamate is the major excitatory transmitter in the brain and spinal cord. L-glutamate interacts with three major classes of receptors: two classes of *ionotropic receptors* (the *NMDA* and the *non-NMDA receptors*) and a class of *metabotropic receptors*.

Goldman equation A formula for quantifying the *membrane potential* when it is determined by the flux of two or more *ion* species across the cell membrane.

Golgi tendon organ Sensory receptor in the muscle, sensitive to changes in tension. (cf. *Muscle spindle*)

G-protein A membrane protein that couples the extracellular receptor for a *neurotransmitter* to an intracellular enzyme. Subunits of G-proteins can also act by directly binding to *ion channels*. (cf. *Second messenger*)

Growth cone A specialized structure at the tip of *axons* of developing neurons that is involved in sensing cues in the environment and in providing the motive force for axon extension. (cf. *Axonal pathfinding*)

Growth factors Highly specialized proteins that appear to play a crucial role in the proliferation, differentation, and survival of neuron and *glial cells*.

Gustatory system The system of neurons, from the mouth to the cortex, that produces the perception of taste.

Gyrus (pl. gyri) The crest of a convolution in the *cerebral cortex*. Many of the gyri are invariant in location and help identify the surface areas of the cortex. The groove between two gyri is called a *sulcus*.

Habituation A simple nonassociative form of learning in which a subject learns about the properties of a single, innocuous stimulus. The subject learns to ignore the stimulus and to decrease its response to it.

Higher-order cortex See *Cortex*

Hindbrain One of three primary embryonic vesicles in the nervous system. It gives rise to the *metencephalon* and *myelencephalon*.

Hippocampus A deep-lying structure in the temporal lobe of the *cerebral hemispheres*. It is involved with aspects of memory storage.

Homeobox A conserved 180 base pair DNA sequence found in many developmentally important transcription factors in vertebrate and invertebrate embryos. The homeobox encodes a 60 amino acid DNA binding protein motif called the homeodomain. Proteins containing homeodomains function as transcription factors.

Hypercolumn A set of columns in the *cerebral cortex* that includes one set of ocular dominance columns for each eye and a complete set of orientation columns, covering 180° of orientation.

Hyperpolarization An increase in the *membrane potential* of the cell. Hyperpolarization makes it less likely a neuron will generate an *action potential* and is therefore *inhibitory*. (cf. *Depolarization*)

Hypothalamus One of two structures of the *diencephalon*. It regulates autonomic, endocrine, and visceral functions.

Immediate-early genes (Immediate-response genes) Genes that are rapidly and transiently activated in response to a stimulus.

Immunoglobulins A superfamily of glycoproteins, one of the major families of structural glycoproteins involved in neural *cell adhesion*: the immunoglobulin-related cell adhesion molecules. (cf. *Cadherins, Integrins*)

Inhibition, feedback (= lateral inhibition) An *inhibitory* circuit in which a neuron excites an inhibitory *interneuron* that makes a feedback connection onto the first neuron. This type of circuit is a form of self-regulation. (cf. *Feed-forward inhibition*)

Inhibition, feed-forward (= reciprocal inhibition) An *inhibitory* circuit in which a neuron both directly excites a target and indirectly inhibits it by exciting an interposed inhibitory neuron. This type of circuit, common in the reciprocal innervation activated by *monosynaptic reflex* systems, coordinates competing behaviors, e.g., the excitation of a flexor muscle and the inhibition of an extensor muscle.

Inhibition, lateral See *Feedback inhibition.*

Inhibition, postsynaptic The *hyperpolarization* of a *postsynaptic cell*, reducing the likelihood of or preventing an *action potential* in the postsynaptic cell. (See *Feed-forward inhibition, Feedback inhibition*)

Inhibition, presynaptic Action by a *presynaptic cell* at an *axo-axonic synapse* to reduce the amount of *neurotransmitter* released by the *postsynaptic cell* onto a third cell. (cf. *Presynaptic facilitation*)

Inhibition, reciprocal See *Feed-forward inhibition*

Inhibitory Said of a neuron or *synapse* that activates *ion channels* with a *reversal potential* more negative than threshold. (cf. *Excitatory*)

Integrative zone The initial segment of the *axon* where the decision is made to initiate an *action potential.*

Integrins One of the major structural families of membrane-spanning glycoproteins involved in neural *cell adhesion*. Integrins are membrane-spanning proteins. (cf. *Cadherins, Immunoglobulins*)

Interneuron One of three major functional types of neurons. There are two types of interneurons: *relay* (or projection) and *local*. (cf. *Motor neuron* and *Sensory neuron*)

Interneuron, local An interneuron with a short axon that carries information within a specific region of the brain. (cf. *Relay interneuron*)

Interneuron, relay An interneuron with a long *axon* that relays information to a distant target neuron. Also called projection interneuron. (cf. *Local interneuron*)

Intrafusal fiber Specialized muscle fiber that lies within the *muscle spindle* and whose contraction activates spindle afferents. (cf. *Extrafusal fiber*)

Ion An atom or molecule having a net positive or negative charge. The major ions found on the inside or outside of the neuronal membrane are *potassium*, sodium, *chloride, calcium, magnesium,* and *organic ions* such as amino acids.

Ion channel See *Channel*

IPSP (Inhibitory postsynaptic potential) Voltage change in a *postsynaptic* neuron in response to release of neurotransmitter from an *inhibitory presynaptic terminal.*

Junctional fold The specialized membrane in-folding at the *end-plate* of a muscle fiber in the region of the motor nerve terminal. (cf. *Basal lamina*)

Kinase (protein) An enzyme that uses ATP to catalyze the *phosphorylation* of other proteins, thereby modifying their function. Important examples are the cAMP-dependent protein kinase, Ca^{2+} calmodulin-dependent protein kinase, protein kinase C, and tyrosine *kinase*.

Learning, explicit A class of learning that requires conscious participation and is concerned with acquiring information about people, places, and things.

Learning, implicit A class of learning that does not require conscious participation and is concerned with perceptual strategies, motor skills, and habits

Length constant See *Membrane length constant*.

Ligand A substance that binds; typically refers to a *transmitter,* hormone, drug, or other agent that binds to a *postsynaptic receptor* or an *ion channel*.

Limbic lobe A ring of phylogenetically primitive cortex around the rostral *brain stem* and *diencephalon*, proposed by James Papez (1937) as the cortical substratum for emotion. (cf. *Limbic system*)

Limbic system A subdivision of the *brain*, it is not really a distinct area but consists of the *limbic lobe*, the medial portions of the frontal, parietal, and temporal lobes that form a continuous band overlying the rostral *brain stem* and *diencephalon*, as well as the *hippocampus* and the *amygdala*. The limbic system contains neurons that form complex circuits that play an important role in learning, memory, and emotion.

Lobe The overlying cortex of each of the *brain's* cerebral hemispheres is divided into four anatomically distinct lobes: frontal, parietal, occipital, and temporal.

Local sign Fixed spatial relationship between the locus of a stimulus for a reflex response and particular muscles that contract in response to the stimulus.

Localization Theory that specific functions are carried out by specialized parts of the nervous system. (cf. *Cellular connectionism* and *Aggregate field theory*)

Magnetic resonance imaging (MRI) A computer-assisted technique for imaging features of the brain in living subjects; used to explore function as well as structure. See *Computerized tomography*.

Matrix proteins, extracellular Glycoproteins (agrin, laminin, etc.) sequestered in the extracellular matrix that surrounds most neural cells.

Medulla One of the seven main parts of the *brain*. It is the direct rostral extension of the *spinal cord*, which it resembles in both organization and function. The medulla includes several centers responsible for such vital *autonomic* functions as digestion, breathing, and control of heart rate. The word "medulla" means "marrow" and is also the name of the central part of the adrenal gland.

Membrane length constant (λ) The distance at which a local change in *membrane potential* has decayed by 63 percent of its initial value. Typical values are in the range of 0.1 to 1.0 mm.

Membrane-spanning proteins Also called integral membrane proteins. A protein that is inserted across the cell membrane. Includes *receptors, ion channels*, and active transport pumps.

Membrane time constant (τ) The time it takes the *membrane potential* to move 63 percent of the way toward its final value in response to step change in current. Typical values are in the range of 1 to 20 ms.

Memory The storage of learned information. Memory exists in at least two stages: short-term (minutes to hours) and long- term (days to weeks). It also has two forms: *explicit* and *implicit*.

Memory, explicit The storage of consciously recalled information about people, places, and things.

Memory, implicit The storage of unconsciously recalled information, usually about habits, perceptual or motor strategies, and associative and nonassociative conditioning.

Memory, working Temporary storing of information used to guide future action.

Mesencephalon The primitive *midbrain* region in the *neural tube*. It gives rise to the inferior and superior colliculi, the red *nucleus*, and the substantia nigra.

Metencephalon Derived from the primitive *hindbrain* in the embryonic *neural tube*. It gives rise to the *pons* and *cerebellum*. (Also called the "afterbrain").

Midbrain One of the seven main parts of the *brain*. It lies rostral to the *pons*, and controls many sensory and motor functions, including eye movements and the coordination of visual and auditory *reflexes*. The term "midbrain" also refers to one of three primary embryonic vesicles in the nervous system.

Modality The class of stimulus used to evoke a sensory response.

Motivational system The system in the brain involved in the coordinated response to bodily needs, such as hunger, sex, and thirst. This system has an important dopaminergic component that is a target for drugs of abuse such as cocaine.

Motor map The neural representation of movement in the cerebral cortex. (cf. *Neural map*)

Motor plant The combined mechanical properties of the muscles, bones, and joints.

Motor proteins Kinesins, dynein, myosin, and other cytoskeletal proteins that generate force.

Motor system These systems mediate movement and other effector functions (refers to autonomic, fusimotor, skeletomotor systems).

Motor unit A single motor neuron and all fibers of the several muscles that it innervates.

Muscle spindle The large stretch receptor in skeletal muscles. An encapsulated structure that contains small (*intrafusal*) muscle fibers that are innervated by the *afferent nerve* endings of stretch receptor *sensory neurons* and by *gamma motor neurons*.

Mutagenesis, targeted Generation of specific changes in genes by homologous recombination.

Myelin sheath A insulating sheath or proteolipid surrounding the *axon*. The myelination of central neurons is performed by *oligodendrocytes*; myelination of peripheral neurons is performed by *Schwann cells*. (cf. *Demyelination*)

Mylencephalon A component of the primitive *hindbrain* in the *neural tube,* and comprising the *medulla.*

Neglect syndrome A syndrome associated with lesions of the *posterior parietal cortex,* manifested by a neglect of the opposite side of the body.

Nernst potential (= equilibrium potential) The specific membrane potential for a given *ion* species at which the electrical force, due to the charge on the membrane, is equal and oppositely directed to the chemical driving force resulting from the concentration gradient, so that no net movement of charge occurs. This potential is given for each ion by the Nernst equation, $E_{ion} = (RT/ZF) \ln \left(\frac{[\text{ion}]\ \text{outside}}{[\text{ion}]\ \text{inside}} \right)$

Nerve, dorsal root Peripheral nerves that carry incoming information from muscle and skin to the *spinal cord.* (cf. *Ventral root nerve*)

Nerve, spinal The nerves of the *peripheral nervous system* formed by the joining of the *dorsal* and *ventral roots.*

Nerve, ventral root Peripheral nerves carrying outgoing motor axons that innervate muscles, as well as axons of the *sympathetic* and *parasympathetic nervous systems.* (cf. *Dorsal root nerve*)

Nerve cell (see *Neuron*)

Nerve-muscle synapse (= *Neuromuscular junction*)

Neural crest (see *Neural crest cells*)

Neural crest cells Specialized cells that migrate from the neural tube soon after it has closed and differentiate into a wide variety of cell types in the peripheral nervous system.

Neural encoding Process by which the neuron translates the physical energy of a stimulus into electrical activity representing features of the stimulus.

Neural epithelium The cells that line the walls of the *neural tube,* and which generate all the *neurons* and *glial* cells of the *central nervous system.*

Neural map The orderly (topographic) arrangement of neurons at successive levels of information processing in a pathway, such that the spatial relationships of stimuli are preserved from the organ of perception to the *brain.* There is a similarly ordered *motor map* for movement.

Neural plate Columnar epithelium of undifferentiated neural cells.

Neural tube Tubular group of neural epithelial cells which develops out of the *neural plate* prior to neuronal differentiation.

Neuroblast A neuronal progenitor cell.

Neurology The classic field of medicine concerned with the nervous system, both normal and diseased. Clinical neurology is concerned with the diagnosis and treatment of nervous disorders, usually disorders that do not primarily affect mental processes. Neurology has posed many of the critical questions that cognitive neural science has attempted to address. By contrast, *psychiatry* attempts to address disorders of the *brain* that affect mental processes.

Neuromuscular junction The site of innervation of muscle. The neuromuscular junction is the most completely characterized and best understood of all *synapses.* Also called the nerve–muscle synapse.

Neuron One of two major classes of cells in the nervous system (the other is the *glial cell*). There are three major functional types of neurons: *sensory, motor,* and *interneurons.* Neurons are the fundamental signaling units of the nervous system. The human *brain* contains about 10^{11} neurons. The typical neuron forms about 1,000 *synapses* and receives even more. Neurons are similar to other cells in the body in having a common molecular machinery for cellular function. However, neurons have the unique ability to communicate rapidly with one another over great distances and with great precision.

Neuron, bipolar Neuron with two separate cell processes with different signaling functions that emerge from the opposite poles of the neuron. These cells have been particularly well studied in sensory systems where a *dendrite* conveys information from the periphery of the body, and an *axon* carries information toward the *central nervous system.* (cf. *Unipolar neuron* and *Multipolar neuron*)

Neuron, motor One of three major functional types of neurons. Motor neurons form synapses with muscle cells, conveying information from the central nervous system and converting it into muscular action. (cf. *Interneuron* and *Sensory neuron*)

Neuron, motor, alpha Alpha motor neurons regulate the length of the *intrafusal fibers* in muscle. (Also called the skeletomotor system.)

Neuron, motor, gamma Gamma motor neurons regulate the *extrafusal fibers* in muscle. (Also called the fusimotor system.)

Neuron, multipolar Neuron with a single *axon* and one or more *dendrites* that typically emerge from all parts of the *cell body.* Multipolar neurons predominate in the vertebrate nervous system. (cf. *Bipolar neuron* and *Unipolar neuron*)

Neuron, sensory One of three major functional types of neurons. This type of neuron is involved in the *afferent* transmission of environmental stimuli. The sensory receptor is the first neuron in each sensory pathway. (cf. *Interneuron* and *Motor neuron*)

Neuron, unipolar Neuron with a single cell process, usually with many branches. One branch serves as the *axon,* other branches serve as *dendrites.* Unipolar cells have no dendrites emerging from the *cell body.* They are the simplest anatomical type of neuron and predominate in invertebrate nervous systems. (cf. *Bipolar neuron* and *Multipolar neuron*)

Neuron doctrine The theory that individual neurons are the fundamental signaling elements of the nervous system.

Neuronal cell death Many neurons that are generated early in development die by a genetically programmed cell death process that is part of normal development.

Neuronal integration Process by which the *postsynaptic cell* sums the inputs from several *presynaptic cells* and decides whether or not to fire an *action potential.*

Neurotransmitter (= transmitter) A chemical substance that is released by a *presynaptic neuron* and binds to *postsynaptic receptors*, and alters *current* flow in the postsynaptic cell. The action (e.g., *excitatory* or *inhibitory*) of a *transmitter* in the *postsynaptic* cell does not depend on the chemical properties of the transmitter, but rather on the properties of the receptors. *Transmitters* fall into two categories: (1) small molecular *transmitters* like *glutamate*, GABA, ACh, and glycine, which are mostly related to amino acids; and (2) peptide transmitters, such as enkephalin, calcitonin- related peptides, VIP, and choleycystokinin. See specific *transmitter* names in the index, and Tables 16-1, 16-2, and 16-3 of the textbook for lists of transmitters.

Neurulation The folding of the neural plate to form a *neural tube*.

Nitric oxide (NO) A short-lived gas that readily passes through cell membranes to activate the synthesis of *cGMP* in a nearby cell. Since *cGMP* can modulate ion channels both directly and by protein *phosphorylation* through the *cGMP*- dependent *protein kinase*, NO is involved in a variety of intercellular signaling processes in the nervous system. (See *Transcellular chemical messengers*.)

Nociceptor A sensory receptor cell sensitive to noxious (tissue damaging) stimuli.

Nodes of Ranvier Interruptions in the *myelin sheath* at regular intervals (typically 1-2 mm) along the *axon* that allow active propagation of the *action potential* by *saltatory conduction*. The *voltage-gated* Na^+ *channels* in the exposed axon membrane at each node provide a source of current that helps propagate the *action potential*.

Nucleus (pl. nuclei) A cluster of functionally related neuron cell bodies in the *central nervous system*. In the *peripheral nervous system*, groups of *neurons* are arranged in *ganglia*.

Olfactory system The system of interconnections of neurons, from the nose to the cortex, that mediates the perception of odors.

Oligodendrocyte One of three major types of *glial cells*. Oligodendrocytes create the *myelin sheath* surrounding neurons in the central nervous system. (cf. *Astrocyte* and *Schwann cell*)

Organic ions (A^-) Charged molecules, including some amino acids and proteins. Of the most abundant ions inside nerve cells, only organic ions are unable to permeate the cell membrane.

Organizer region A region in the early gastrula embryo that provides signals for the induction of neural ectoderm and formation of the *neural plate*.

Paraphasia Impairment of the neural pathways involved in speech, resulting in incorrect use of words.

Parallel processing The deployment of several parallel neuronal pathways to convey similar information or to process different components of a common sensory modality such as the parallel processing of form and motion in vision. (cf. *Serial processing*)

Parasympathetic nervous system One of three subdivisions of the *autonomic division* of the *peripheral nervous system*. It acts to conserve the body's resources and restore the equilibrium of the resting state. (cf. *Sympathetic* and *Enteric nervous systems*)

Patch-clamp technique A method for recording the electrical activity of neurons or small patches of membrane by forming a tight seal between electrode and membrane.

Pathway A series of connections between neurons that constitute a functional circuit.

Peripheral nervous system One of two major anatomical components of the nervous system, the other being the *central nervous system*. The peripheral system comprises the *ganglia* and peripheral nerves outside the *brain* and *spinal cord*, and has two major divisions: *somatic* and *autonomic*. Although anatomically separate, the peripheral and central nervous systems are functionally interconnected.

Phantom limb sensation Sensations relating to a lost, amputated limb, that is characteristically experienced at the site of the amputation.

Phosphatase Enzyme that removes a phosphate group from a molecule such as a protein or lipid.

Phospholipid A small amphipathic molecule constructed from fatty acids and glycerol. Each phospholipid molecule has a hydrophobic tail, composed of the fatty acid chains, and a hydrophilic polar head group, where the phosphate is located. Phospholipids form the lipid bilayer which is the structural basis of all cell membranes.

Phosphorylation The covalent addition of a phosphate group to a protein by a protein *kinase*, which changes the activity of the recipient protein.

Plasticity, synaptic Increase or decrease of synaptic strength, for short or long periods, following activity. Thought to be critically involved in learning.

Pons One of the seven main parts of the *brain*. It is located rostral to the *medulla*, and conveys information about movement from the *cerebral hemisphere* to the *cerebellum*.

Positron emission tomography (= PET scan) A technique for producing images of live, real-time brain functioning. The technique combines the principles of *computerized tomography* and radioisotope imaging.

Postsynaptic cell The *neuron* whose excitability is affected by signals (electrical or chemical) from a *presynaptic cell* which communicates with it at a *synapse*.

Potassium (K^+) An essential element in the functioning of the nervous system. Potassium concentrations inside the resting neuron are higher than those outside the cell. The preponderance of *resting channels* in *glial cells* are permeable only to K^+.

Potential, action All-or-none transient electrical *depolarization* which actively propagates along neuronal *axons* and some *dendrites*. This electrical signal carries information from the *integrative* and *trigger zones* of a neuron to the axon terminals. Its amplitude, which is all-or-none at the site of initiation, is maintained by a regenerative mechanism that allows it to propagate without decrement. A *compound action potential* is total electrical activity resulting from action potentials in a population of nerve or muscle fibers.

Potential, electrotonic A passive response of the *membrane potential* of a neuron to *current* flow in the neuron, determined by the membrane's *length* and *time instants*.

Potential, end-plate The *excitatory synaptic potential* in muscle cells produced by transmitter release from a *motor* neuron and due to the simultaneous flow of Na^+ into and K^+ through the ACh receptor of the muscle cell. The amplitude of the end-plate potential is unusually large; a single motor neuron produces a synaptic potential of about +70 mV. (cf. *Miniature end-plate potential*)

Potential, equilibrium (*E*) See *Nernst potential*

Potential, evoked Electrical signal recorded with external scalp electrodes that represents the summed activity of thousands of neurons that are activated by a discrete sensory stimulus.

Potential, excitatory postsynaptic (= EPSP) See *Postsynaptic potential*

Potential, generator The graded potential produced in a *receptor cell* by an appropriate physical stimulus, e.g., the *depolarization* produced by a touch receptor or a stretch receptor in the afferent fiber.

Potential, inhibitory postsynaptic (= IPSP) See *Postsynaptic potential*

Potential, membrane (V_m) Electrical potential difference across the cell membrane resulting from the separation of electrical charge across the cell membrane. This electrical potential, measured in mV, together with the concentration gradients, determines the electrochemical potential that provides one driving force for current flow across the membrane. Membrane potential is related to current carried by an *ion*, *i*, by the equation $I_i = g_i\ (V_m - E_i)$, where E_i is the Nernst potential for the *ion*.

Potential, miniature end-plate The presynaptic motor *axon* releases the contents of *synaptic vesicles* spontaneously (in the absence of *action potentials*). These quantal release events give rise to spontaneous miniature *synaptic potentials*. Miniature *end-plate potentials* have all the properties of normal *synaptic (end-plate) potentials* except that they are much smaller and result from the release of a single vesicle. The synaptic potential evoked by an *action potential* results from the synchronous occurrence of many miniature *synaptic potentials*.

Potential, receptor Local change in membrane potential of the *primary sensory neuron* that is produced by a stimulus which spreads passively.

Potential, resting membrane (V_R) The *membrane potential* of a cell at rest. At the resting membrane potential the cell is not in equilibrium, but rather in a steady state, since metabolic energy is used by the *sodium-potassium pump* to maintain the gradients of *ions* inside and outside the cell membrane. The resting potential of neurons ranges from about −40 to −80 mV and in most mammalian nerve cells it is −65.

Potential, reversal The *membrane potential* at which the amplitude of the *synaptic potential* is zero. At the reversal potential there is no net current because an equal amount of charge flows in and out of the cell through the synaptic channels. For channels which are selectively permeable to only one *ion* species (such as the Na^+ or K^+ channels), the reversal potential approximates the *equilibrium potential* for that ion species.

Potential, unit synaptic *Postsynaptic potential* of fixed size produced by the release of *neurotransmitter* from a single *presynaptic cell*.

Potentiation Increase in size of *postsynaptic potentials* during *tetanic stimulation*. (cf. *Posttetanic potential*)

Potentiation, long-term A persistent increase (lasting hours to days) in the size of the *postsynaptic potential* following presynaptic *tetanic stimulation*.

Potentiation, posttetanic Increase in size of the postsynaptic potentials that persists for several minutes following presynaptic *tetanic stimulation*.

Presynaptic cell One of two cells, the postsynaptic cell being the other, needed to form a synapse.

Promoter region (of gene) The *regulatory region* of a gene, usually just upstream from the beginning of the coding region, to which the enzyme RNA polymerase binds in order to transcribe the gene.

Proprioception Sensation arising from stimuli within the organism itself, such as limb position.

Pruning Elimination of *axons* during development to remove projections to inappropriate targets and to refine the specificity of axonal projections.

Psychiatry The field of medicine concerned with mental disorders.

Psychophysics The physics of sensation and movement. Psychophysics is concerned with the perception of specific sensory stimuli and with the properties of movement.

Quantal transmission The units (quanta) that make up transmission across a chemical synapse. Each quantum of transmitter is thought to represent about 5,000 molecules of transmitter released from a single presynaptic vesicle. (See *Synaptic transmission, synaptic potential, miniature end-plate potential*)

Radial glial cell Elongated glial cells found in cortical structures that provide a substrate for the radial migration of neuronal progenitors from the ventricular zone. See (*Glial cell*)

Receptive field Area of the receptor sheet (e.g., retina or skin) that excites a neuron.

Receptor cell A sensory cell specialized to respond to a particular physical property such as touch, light, or temperature.

Receptor–channel (see *Receptor, ionotropic*)

Receptor, ionotropic A *postsynaptic receptor* consisting of several subunits (often 5) that includes both the recognition element (e.g., transmitter-binding site) and the *ion channel*. These receptors belong to a large family of proteins stemming from a common ancestor, are composed of subunits that span the membrane to form a water-filled pore, the *ion channel*. A prime example is the *nicotinic acetylcholine receptor*. In contrast to *metabotropic receptors* that gate ion channels indirectly, by means of intracellular messengers, ionotropic receptors gate ion channels directly.

Receptor, kainate A *non-NMDA receptor* for *glutamate*. One of three types of *postsynaptic* receptors for *glutamate*, the others being the *NMDA receptors*, and the *metabotropic glutamate receptors*.

Receptor, metabotropic A *postsynaptic* receptor consisting of a single subunit that is displayed across the membrane seven times and that couples to *G-proteins* which activate intracellular second messengers that engage *ion channels* and other targets within the cell. Prime examples are the *muscarinic receptors* for *acetylcholine*, the *metabotropic receptors* for *glutamate*, serotonin, and the β-adrenergic receptor.

Receptor, metabotropic glutamate A type of *metabotropic receptors* for *glutamate* that engages various intracellular *second-messenger* pathways. The metabotropic receptors are activated by the drug ACPD. The other two types of receptors for glutamate are ionotropic: the *NMDA receptor* and the *non- NMDA receptor*.

Receptor, muscarinic ACh One of two types of *postsynaptic receptors* that bind *acetylcholine*; so called because the drug muscarine binds exclusively to this type of receptor and mimics the actions of *ACh*. The *muscarinic ACh receptor* belongs to a large family of *metabotropic receptors* that activate *second-messenger* pathways in the postsynaptic neuron (indirect gating). (cf. *Nicotinic ACh receptor*)

Receptor, nicotinic ACh One of two types of *postsynaptic receptors* that bind *acetylcholine*; so called because nicotine binds exclusively to this type of receptor and mimics the actions of ACh. It is a prime example of an *ionotropic*, directly gated receptor. (cf. *Muscarinic ACh receptor*)

Receptor, NMDA One of three types of *postsynaptic receptors* for *glutamate*, it is selectively activated by the drug NMDA (N- methyl-D-aspartate) The *NMDA receptor* has been found to play a critical role in neurodevelopment and long-term potentiation, presumably because it is permeable to Ca^{2+}. (cf. *non-NMDA Receptor, non-NMDA* and *Receptor, metabotropic glutamate*)

Receptor, non-NMDA A type of ionotropic receptors that respond to *glutamate*. These receptors are activated by the drugs *AMPA*, kainate, and quisqualate, and are blocked by CNQX.

Receptor, postsynaptic A specialized protein macromolecule in the *postsynaptic cell* that binds the *neurotransmitter* released by the *presynaptic cell*. All receptors for chemical transmitters have two biochemical features in common: they have a recognition element for the *ligand* which they bind and they carry out an effector function within the cell, e.g., they are involved in *gating ion channels*, or in activating *second-messenger* cascades. Receptors fall into several major categories, including: *ionotropic, metabotropic*, and *receptor tyrosine kinases*.

Reflex, monosynaptic A reflex mediated by a single set of synaptic connections between two types of neurons (*sensory neurons* and *motor neurons*) in the *spinal cord*. (cf. *Polysynaptic reflex*)

Reflex, polysynaptic A reflex mediated by more than one synaptic connection in the CNS. (cf. *Monosynaptic reflex*)

Reflex, spinal Involuntary movement triggered by sensory input and produced by neural circuitry limited to the spinal cord.

Reflex, stretch *Monosynaptic reflex* mediated by *muscle spindles* and *gamma motor neurons* that causes *intrafusal fibers* to contract when a muscle is stretched.

Refractory state The state of *ion channels* and neurons following excitation, during which the membrane as a whole and many of its individual Na^+ channels cannot respond to another stimulus because they are still in an inactivated state.

Regulatory region (of gene) All genes are divided into two regions: a *coding region* transcribed into messenger RNA that is then translated into proteins, and a *regulatory* region that determines the circumstances during which the gene is transcribed. The regulatory region in turn divides into *promoter* and *enhancer regions*, each of which binds specific regulatory proteins essential for activating and repressing transcription of the gene.

Representation, internal A pattern of activity in one or more regions of the *brain* that represents an external stimulus, an action, a feeling, or a thought.

Resistance (*R*) The inverse of *conductance*, $1/G$. Measured in ohms.

Resistor An object that has resistance and opposes the flow of current. From Ohm's law, $R = V/I$.

Restriction fragment length polymorphism The difference between individuals (a polymorphism) in the length of the *DNA* fragment produced by different DNA cutting (restriction) enzymes. This method allows for the detection of stable genetic differences between individuals.

Retinotopic map The orderly projection of retinal axons onto the surface of the optic tectum.

Rhombomeres Conspicuous but transient swellings in the early *hindbrain* that are involved in the segmental organization of individual cranial motor and sensory nuclei and of other developing *hindbrain neurons*.

Schwann cell One of three types of *glial cells*. Schwann cells provide the myelin sheath surrounding neurons in the peripheral nervous system. (cf. *Astrocyte* and *Oligodendrocyte*)

Second messenger Substances produced within a cell subsequent to hormone or transmitter binding to external receptors. In neurons a *second messenger* is produced when a *neurotransmitter* (= first messenger) binds to a member of a large receptor family containing seven membrane-spanning domains (for example, the β-adrenergic receptor, the muscarine *acetylcholine* receptor, receptors for neuropeptides, and many others), initiating a cascade of molecular interactions involving GTP-binding proteins (*G-proteins*) that stimulate the enzymatic synthesis or release of the second messenger (for example, *cAMP* by adenylyl cyclase, *cGMP* by guanylyl cyclase, IP_3 and *diacyl glycerol* by phospholipase C and arachidonic acid by phospholipase A_2. These *second messengers* can act in a variety of ways: for example, *cAMP, cGMP,* and IP_3 can act directly by binding to an *ion channel* or other intracellular receptors, or they can act indirectly by protein *phosphorylation* through specific kinases such as the cyclic AMP-dependent protein kinase (PKA), and the cGMP-dependent *protein kinase C* (see *cAMP, cGMP*). Calcium (Ca^{2+}), which enters the neuron through *voltage-* or *ligand-gated channels*, is often called a *second messenger* because it triggers a sequence or cascade of molecular events, which include the activation of Ca^{2+}/calmodulin-dependent *protein kinase* and certain isoforms of *protein kinase* C.

Secretion The release of materials contained in vesicles by *exocytosis*. During this process the membrane of the vesicles is incorporated into the cell's external membrane. There are two secretory pathways, constitutive and regulated. In cells, vesicles fuse continuously, replacing surface membrane constituents by the constitutive pathway. In all secretory cells, secretory vesicles fuse and release their contents only when appropriately stimulated. On *depolarization, synaptic vesicles* fuse with the membrane of the neuron's axon terminal to release *neurotransmitters*, peptides, and other chemical messengers.

Selectivity filter A narrow region in the *ion channel* that acts as a molecular sieve, selecting which ions will pass by the criteria of charge and size.

Sensation Touch, pain, sight hearing, smell, taste. Each of these sensations is characterized by four attributes: modality, intensity, duration, and location.

Serial processing The relay of sensory or motor information along a single pathway comprising a fixed series of relay *nuclei*. (cf. *Parallel processing*)

Short circuiting An effect of *inhibition*. An *inhibitory postsynaptic potential* increases the *conductance* of the *postsynaptic cell*, thereby reducing the amplitude of an *excitatory postsynaptic potential* (since $V_{epsp} = I_{epsp}/G$).

Signal A change in the *membrane potential* of a postsynaptic neuron as a result of synaptic inputs from a presynaptic neuron or from activation of a sensory receptor. Signals between neurons are highly stereotyped throughout the nervous system; the "message" conveyed by a signal depends entirely on the pathway in which the active neuron is located. Within a neuron, four types of signals are processed at different sites: an input signal, integration signal, conducting signal, and output signal. Signals can be local (*receptor* and *synaptic potentials*) or propagated (the *action potential*).

Size principle Order of recruitment of motor neurons by the size of the *cell body*. Neurons with the largest cell body are recruited first.

Sodium-potassium pump A *membrane-spanning protein* that moves Na^+ and K^+ against their electrochemical *gradients*. It extrudes Na^+ from the cell and takes in K^+, using energy from the hydrolysis of ATP.

Somatic division One of two major divisions of the *peripheral nervous system*. It provides sensory information about muscle and limb position and about the environment outside the body to the *central nervous system*.

Somatic sensory system The sensory system concerned with sensation from the skin at the body surface: touch, pressure, pain, and position sense.

Somatotopic map The orderly projection of the *afferent axons* of neurons in the *somatosensory system* to their targets in the *thalamus* or *cortex*.

Spinal animal Experimental animal in which the *spinal cord* is severed to isolate the spinal circuits that are below the level of the cut from the descending influence of cortical signals.

Split-brain animal An experimental animal in which the two structures that connect the two *cerebral hemispheres*, the *corpus callosum* and anterior commissure, have been severed to test the independent functions of the hemispheres.

Spinal cord One of the seven main parts (the most caudal) of the *central nervous system*. It controls movements of the limbs and trunk, and processes sensory information from the skin, joints, and muscles of the limbs and trunk. Controls *autonomic* function.

Spine A tiny bud off of a dendritic shaft, it is specialized to receive synaptic input.

Stereognosis Perception of three-dimensional shapes of objects.

Stimulus Four attributes: modality (pathway), intensity (frequency and population codes), duration, and location (topographic organization of inputs).

Sulcus (pl. sulci) The groove between two *gyri* in the *cerebral cortex*.

Summation, spatial The process by which *synaptic potentials* generated in different regions of a neuron are added together at the *trigger zone*.

Summation, temporal The overlapping of *synaptic potentials* in the *postsynaptic cell*. Temporal summation is more prevalent in neurons with a long *membrane time constant*.

Supplementary motor area Part of the *Premotor cortex*. Also called the *Lateral premotor area*.

Sympathetic nervous system One of three major subdivisions of the *autonomic* division of the *peripheral nervous system*. It participates in the response of the body to stress. (cf. *Parasympathetic* and *Enteric nervous systems*)

Synapse The specialized site of communication between two neurons. Synapses can be categorized as *chemical* or *electrical*, depending on the mechanism of *synaptic transmission*.

Synapse, chemical A synapse at which the *presynaptic neuron* releases a *neurotransmitter* that binds to *receptors* on the *postsynaptic neuron* and in this way influences the excitability in the *postsynaptic cell*. The two cells are separated by the *synaptic cleft*. Chemical synapses can mediate either *excitatory* or *inhibitory* actions.

Synapse, electrical A *synapse* at which *current* flows directly from the *presynaptic cell* to the *postsynaptic cell*. At an electrical synapse the two cells are structurally connected by *gap-junction channels*. Electrical synapses are not unique to the nervous system. In the *brain* they are characterized by rapid passage of ions and other small molecules and by rather stereotyped action. Neurons that are interconnected by electrical synapses are able to fire synchronously to produce "explosive" behaviors, e.g., escape. Electrical synapses are thought to be important for transmitting developmental signals between neurons. They typically generate brief *excitatory* changes in the postsynaptic cell. They can be either *unidirectional* or *bidirectional*. (cf. *Chemical synapse* and *Synaptic transmission*)

Synaptic bouton Knob-like enlargements of the *axon* terminal from which *neurotransmitter* is released.

Synaptic cleft The extracellular gap between the *presynaptic cell* and *postsynaptic cell* at the *synapse*. It is typically 20–40 nm, somewhat wider than the separation of the cells at regions of the membranes without *synapses*. (cf. *Gap junction*)

Synaptic integration The process by which a central neuron adds up all incoming signals and determines whether an *action potential* will be generated. (cf. *Spatial summation, Temporal summation*)

Synaptic potential Also called *Postsynaptic potential*. A graded change in the *membrane potential* of a cell produced by a synaptic input. A *postsynaptic potential* can be either excitatory (EPSP) or inhibitory (IPSP). (cf. *Unit synaptic potential*)

Synaptic transmission The mechanism by which one neuron influences the excitability in another. The mechanisms of transmission can be *chemical* or *electrical*. *Electrical synaptic transmission* is mediated by current flow through a *gap junction* joining two neurons. *Chemical synaptic transmission* is mediated by the release of a chemical *neurotransmitter* from the *presynaptic* cell acting on receptors in the *postsynaptic* cell.

Synaptic vesicle Membranous organelles containing high concentrations of *neurotransmitters* to be released by *exocytosis* during *synaptic transmission*. Since *exocytosis* is triggered by rapid influx of Ca^{2+}, this process is a specialized form of the *regulated secretory pathway*. These vesicles, which are highly concentrated at *axon terminals*, exist in two size classes, small and large. Small (40–50 nm in diameter) typically are clustered at membrane specializations in terminals called *active zones*, and contain small molecule *neurotransmitters*, ATP, and other negatively charged substances. These vesicles contain *transporters* for concentrating *neurotransmitter*, in addition to other *membrane-spanning proteins* (synaptophysin, synaptotagmin, and synaptobrevin) that may be important for vesicle fusion and *exocytosis*. The large (90–200 nm) vesicles contain one or more types of neuropeptide, ATP, and, in noradrenergic cells, a small molecule neurotransmitter as well. These large vesicles typically do not fuse with the synaptic terminal membrane at regions outside the active zones.

Telencephalon A component of the primitive *forebrain* in the embryonic *neural tube* which gives rise to the constituents of the *cerebral hemispheres*. Also called the "endbrain."

Terminal, presynaptic The component of the neuron from which *synaptic vesicles* are released by *exocytosis* (*chemical synapses*) or which connects via *gap junctions* to the *postsynaptic cell* (*electrical synapses*).

Tetanic stimulation High-frequency stimulation of a *presynaptic neuron*, as much as 50 to 100 *action potentials* per second.

Tetanus Summation of forces produced by successive *action potentials* in a muscle without interruption.

Thalamus One of two major structures of the *diencephalon*. It processes most of the information reaching the *cerebral cortex* from the rest of the *central nervous system*.

Threshold (V_T) The critical level of *depolarization* of the neuron's membrane at which the cell can actively generate an *action potential*. It is the specific value of the *membrane potential* at which the net *ionic current* changes from outward to inward, depositing positive charge on the inside of the cell membrane.

Threshold, sensory Lowest stimulus intensity a subject can detect; can vary in one individual depending on such factors as fatigue and context.

Time constant See *Membrane time constant*

Topographic map Consistent spatial relationship between neurons at their targets, established by the orderly projection of *axons* to a target field.

Transcellular messenger Molecules produced through receptor activation in one neuron that can affect the activity of a neighboring neuron by diffusing through the membranes of both cells. Prime examples are the gases *nitric oxide* and *carbon monoxide*, and the lipid, arachidonic acid and its metabolites. These substances are also potential retrograde messengers.

Transcriptional activators Proteins that bind to regulatory regions, usually *enhancer regions* of genes, to activate transcription.

Transduction, stimulus Conversion of stimulus energy (e.g., light waves or mechanical pressure) into electrical activity in a *primary sensory neuron*. (cf. *Neural encoding*)

Transgene A foreign gene introduced into an animal.

Transmitter (see *Neurotransmitter*)

Transporters Membrane-spanning proteins that facilitate the passage of a small molecule (for example, glucose and *neurotransmitters*) across membranes. Operating as carriers, they bind the molecule to be transported and then bring the molecule across the membrane. This conformational change requires a driving force, which can be supplied by co- transport of ions, often Na^+, when *neurotransmitters* are taken up into the *presynaptic terminal*. There are also transporters for *transmitters* in the membrane of *synaptic vesicles,* which require a proton gradient and transport of Cl^- as a driving force.

Trigger zone A specialized component of the neuron, usually located in the initial segment of the *axon* where the *action potential* is initiated.

Trophic factors Proteins that help maintain the survival of neurons during development.

Tuning curve Quantification of sensitivity of a *sensory receptor cell* to a particular stimulus energy.

Tyrosine kinase pathway A pathway whereby a *ligand,* often a growth factor, activates a receptor tyrosine *kinase,* which in turn recruits to the membrane a number of proteins including other protein *kinases.*

Unidirectional (5 rectifying) Said of *current* that flows in one direction only, at an *electrical synapse* or in an *ion channel.* (cf. *Bidirectional*)

Varicosity The *axon* terminal portion of certain neurons that contain release sites for *neurotransmitters.*

Ventral root nerve The nerve formed by the projection of motor *axons* from the ventral *spinal cord.* (See *Nerve*)

Ventricles The two lateral ventricles and the third and fourth ventricles form part of a cavernous system within the *brain* and *spinal cord,* which is filled with cerebral spinal fluid.

Vesicle Small subcellular organelles that carry small molecule *transmitters* (small clear vesicles) and peptide transmitters (large electron-dense vesicles) to the active zone of the *presynaptic terminals* for the exocytotic release of *transmitter.* (cf. *Synaptic vesicle*)

Vesicle-associated proteins Synaptotagmin, syntaxin, VAMP, synapsin.

Visual system The system of interconnections of neurons, from the retina to the *cortex,* that produces vision.

Visuospatial task A task requiring the subject to observe the movement of an object in space.

Voltage-clamp An experimental device for preventing the opening and closing of *voltage-gated channels* from influencing the *membrane potential.* A direct measure of membrane *current* can be obtained with this technique by recording the current that must be generated by the voltage clamp to keep the membrane potential from changing.

Voluntary movement Movement that is planned according to some goal and shaped to specific features of the environment. (cf. *Reflex*)

Waters of hydration A cloud of water molecules that surrounds an *ion* in solution.

Wernicke's area The portion of the left parietal lobe concerned with comprehension of language. (cf. *Broca's area*)

Index

A

Absolute refractory period, 169
Accommodation, 523
Acetylcholine, 294–295
  effect on potassium current in sympathetic ganglion cells, 257
  influence of calcium concentration on amount released, 274–276
  neurotransmitter, 192
  release at neuromuscular junction, 50
  release by action potential in presynaptic nerve terminal, 200
  removal from synaptic cleft to terminate synaptic transmission, 302–303
Acetylcholine receptors
  antibodies to as cause of myasthenia gravis, 310–314
  changes in after muscle innervation, 104, 106
  channels
    complex, electron microscopic image, 200f
    functional model, 211f
    nicotinic, as membrane-spanning protein, 209–210
    subunit structure, 210f
  deducing secondary structure from amino acid sequence, 122–123f
  density of in muscle fibers in myasthenia gravis, 309f
  and ion conductance, 118
  labeling, 199f
  main immunogenic region, 312
  muscarinic, 198
  nicotinic, 198
    antibodies against produced in myasthenia gravis, 307
    messenger RNA sequence of alpha-subunit, 111–112f
    receptor-channel, 209–210
  postsynaptic, 201f
  rate of turnover in myasthenia gravis, 313f
  on surface of muscle fibers, 104, 106
  types, 198
  see also Slow-channel syndrome
Acetylcholinesterase
  lack of, 314
  and reversal of symptoms of myasthenia gravis, 308
ACh, see Acetylcholine
Achromatopsia, 398, 646
Action potential, 23, 32
  abolishes ion balance of resting membrane potential, 139–140
  accompanied by net increase in ionic conductance of membrane, 162f
  affected by myelination, 158f
  all-or-none nature, 34, 35–36, 169
  as conducting signal, 35–36
  basic mechanism same in all neurons, 170
  caused by sequential opening of voltage-gated sodium and potassium channels, 168f
  compound, 308
  digital versus analog signal, 37
  effect of intracellular ion concentrations, 171
  electrical equivalent circuit of spread, 157f
  followed by afterpotential, 169
  followed by refractory period, 169
  generated by sensory receptors, 376–377
  generation by flow of ions through voltage-gated channels, 161–162
  Hodgkin-Huxley model, 162, 163, 169–170
  nodes of Ranvier boost amplitude of, 157
  number of transmitter vesicles released modulated by calcium influx, 286–290
  passive conduction of depolarizing charge along axon and propagation of, 156f
  as propagated
    signal, 33
    signaling, 161–177
  propagation
    and electrotonic conduction, 155–156
    velocity affected by passive membrane properties and axon diameter, 156–159
  reconstruction from known electrical properties of neuron, 168–170
  refractory period following, 169
  and release of ACh into synaptic cleft at nerve-muscle synapse, 200
  saltatory conduction of in myelinated axon, 157
  as sequential current flow of sodium into the cell and potassium out, 163f
  slower conduction in demyelinating diseases, 157
  stereotyped throughout nervous system, 35
  trigger zone, 35
  triggering of at axon hillock, 157
  variations, 170–171
  voltage-gated channels and, 155–177
Active zone
  in chemical synapse, 192
  in neuromuscular junction, 198–200
  structure, 279f
  synapse, 276
Activity dependence, of presynaptic facilitation, 677
Adaptation
  and sensation, 372
  of sensory receptors, 378–379
Addiction, drug, and role of dopamine pathways, 626
Adenohypophyseal system, 603f
Adenosine, as neurotransmitter, 297
Adequate stimulus, and sensitivity of sensory receptors, 373
Adrenergic, 293
Afterpotential, following action potential, 169
Aggregate field view, of brain, 8
Agnosias, 340, 398, 399t, 646
Agraphia, 646

Alexia, 644, 646
All-*trans*-retinal
  role in vision, 411–412
  structure, 413f
Alpha motor neurons, innervation of extrafusal fibers, 510
Alpha-bungarotoxin, and study of ACh receptors, 309
Alpha-fetoprotein, and fetal protection from maternal estrogen, 585
Alpha-gamma coactivation, 510, 512f
Amino acid sequence, deducing ion channel structure from, 121–122, 122–123f
Amino acid transmitters, 296–297
Amnesia
  learning in, 654
  memory in, 654f, 655f
  retrograde versus anterograde, 664
Amputation, and phantom limb syndrome, 333–334
Amygdala
  and basolateral complex, 608–611
  of cerebral cortex, 83
  connections and projections, 608
  connections of central nucleus with hypothalamic and brain stem targets, 610f
  effect on output of autonomic nervous system, 600
  and emotions, 607–611
  role in implicit memory, 664
  sex difference in, 589
Androgen insensitivity syndrome, 581, 581f
Androgens, fetal source of, 583
Androstenedione, and masculinization of brain, 584
Angiotensin, role in water regulation, 623
Anterior commissure, 355
Anterior parietal lobe
  location of somatic sensory cortex, 324–325f
  see also Somatic sensory cortex
Anxiety, measures of in humans, 597t
Aphasias, 340
  anatomical basis of, 12–14
  classification, 642–643
  conduction, 14
  disorders of language, 639–644
  major, 640–641b
  versus motor disorders of speech, 642
Aplysia californica
  classical conditioning of gill-withdrawal reflex, 677–680, 678f
  electric synaptic transmission and release of ink, 187, 188f
  growth cone, 101f
  habituation, 668–671, 669f
  long-term sensitization, 673f
  short-term sensitization and presynaptic facilitation, 671–673f
Apoptosis, see Programmed cell death
Apparent motion, versus actual motion, 447, 447f
Apprehension, and amygdala, 608
Aprosodias, 644
Arachidonic acid
  generation of, 253f
  pathway, 253–254, 253f
Arousal, 614
  role of reticular formaton in sensory regulation, 371
  and use of sensory information, 371
Arrestin, role in light response, 414
Artificial intelligence
  parallel and distributed components, 40
  simulation of cognitive function, 359–362
Association areas, cerebral cortex, 82f, 83, 347–363, 348–350, 349t,f
  association areas involved in different cognitive functions, 348–350
  projections of central nucleus to, 611
Astereognosia, 340
Astrocytes, 27–28
  development of in rat optic nerve, 99
Ataxia, 544
ATP, as neurotransmitter, 297
Attention
  affected by lesions in posterior parietal cortex, 344
  effect on visual perception, 401–402
  and response of neurons in posterior parietal cortex to stimulus, 345f
  spotlight of, 402
  visual
    and conscious awareness, 403–404
    study at cellular levels, 449–450
Auditory system
  location of sound sources, 385
  neural map in cerebral cortex, 86–87
Autoimmune disease, myasthenia gravis as, 308–314
Autonomic motor system, 77
  see also Hypothalamus
  and voluntary movement, 83–84
Autonomic nervous system
  divisions, 598–600, 599f
  role in emotional states, 597–600
  role in temperature regulation, 616–618
  role of hypothalamus in controlling output, 600–602
  targets of innervation, 599f
Autophosphorylation, 250
Awareness, conscious, and visual attention, 403–404
Axolemma, 48
Axon
  active transport of membranes and secretory proteins in, 62–65
  competition during development and elimination, 106–107
  covered with myelin in sensory and motor neurons, 51–53
  crossing in visual system, 87
  cytoskeletal elements, 49f
  developing
    growth cone guidance to target in migration, 101
    guided by chemical cues, 101–104
  diameter, effect on propagation of action potential, 156–159
  growth, effect of hormones, 589–590
  hillock, 48–49, 50f
    nerve cell, 22, 23
    triggering of action potential at, 157
  motor, chemical guidance cues to targets in plexus, 105f
  motor neuron, 48–49
  myelinated, and saltatory conduction of action potential, 157
  nerve cell, 23
  from peripheral nervous system, 49f
  pathfinding, selection of peripheral pathways by developing of motor neurons, 102, 104
  pathways, formed in response to local guidance cues, 99–104
  regeneration
    and chemical cues, 102
    retinal ganglion neurons, 103f
  targeting, 108

terminals, modulatory synapses on, 226
transport, fast, of organelles, 62–64
unit length represented as electrical equivalent circuit, 154f
Axoplasmic flow, slow, transport of cytosol, 64–65

B
Back-propagation, in parallel distributed processing, 360
Bacteriorhodopsin, membrane protein, structure, 121–122
Balance, maintained by medial pathways of brain stem, 494
Basal ganglia
of cerebral cortex, 83
of cerebral hemispheres, 79
and cognitive behavior, 547–549
components, 544
connections between, 545, 546f
control of cortical and brain stem motor systems, 499
diseases of, 544
Parkinson's disease, 547
feedback circuits with cerebral cortex, 547–549, 548f
inputs and outputs, 545–547, 549
integrate information from areas of cortex, 544–549
relationship to cerebral hemispheres, 545f
somatotopic arrangement of motor portion, 547–549
specific behavioral functions of nuclei, 545
Basement membrane, 198f
Basolateral complex, receives sensory inputs to amygdala, 608–611
Batrachotoxin, for study of ion channels, 172
Battery
ion channels represented as, 141–142, 142f
potassium ion-selective ion channel represented as, 141f
Behavior, 551–628
brain, 5–19
complex, polygenic nature, 562
correlations between monozygotic and dizygotic twins, 558f
courtship, genetic determination, 571
critical periods in development of, 482–483
dimorphism, 582–583
effect of perception on, 323
emotional, 553–554
and genes, 555–577
genetic systems for study, 556–557b
increased blood flow in cerebral cortex indicates areas participating, 535f
influence of sex differences in brain, 590–592
inherited components, 555, 558–565
instinctive and learned, 553
and interaction of sensory, motor, and motivational systems, 84
and manipulation of gene expression in animals, 565–569
mesolimbic dopamine pathways in reinforcement and drugs of abuse, 625–626
motivated, 554
and nerve cells, 21–40
nerve cells as signaling units for, 29–31
and organization of the brain, 3
organization of by genes, 565–576
reinforcement by electrical stimulation of brain, 625
reproductive, 579–593
sexually dimorphic nature, 581–582
sex-linked, and perinatal hormones, 587–588
sexual, 553
single genes, 558–559
Behaviorism
as approach to study of cognition, 322
behaviorist psychology, 370
Binding problem, 368
and parallel processing pathways, 387
Binocular vision, and ocular dominance columns, 442, 442f
Binocular zone, 427
Biofeedback, as psychotherapeutic technique, 663
Biogenic amine transmitters, 295–296
Biological clock mechanisms, and behavioral responses, 624–625
Biological rhythms, and oscillations in gene expression, 571–576
Bipolar affective disorder, polygenic nature, 564–565
Bipolar nerve cells, 26, 27
bridge plexiform layers in retina, 420
excitatory connections with corresponding classes of ganglion cells, 423
receptive fields, 420, 421–422b
Birthday, neuron, 99–101
Blind spot
on retina, 427, 427f
in visual field, 409
Blobs
regions, primary visual cortex, 396f, 433, 442f
staining regions of primary visual cortex, 395
Blocking phenomenon, 660
Body surface
cortical representation, 326–328
neural map in cerebral cortex, 86
orderly arrangement of somatic sensory and motor projections from and to in cerebral cortex, 328f
and receptive fields, 335–339
represented by more than one sensory map in primary somatic sensory cortex, 689f
Body weight
regulation around set point, 618–620
set point altered by hypothalamic lesions, 621, 620f
boss gene, and photoreceptor identity in *Drosophila* eye, 91–94
Brain stem
as control level of motor system, 492–493
computer image during auditory stimulation, 1-2f
connections with central nucleus of amygdala, 610f
descending pathways, 494–496, 495f
lateral pathways, 494, 495f, 496
medial pathways, 494, 495f
medial and lateral systems, 493
modulation of motor neurons and interneurons, 494–496
pathways, and indirect action by motor cortex on spinal motor neurons, 496–498
role in auditory system, 385
role in motor function, 488
see also Climbing fibers; Cranial nerves; Medulla; Midbrain; Mossy fibers; Pons; Reticular formation
serotonergic neurons in, 296
structure, 79
Brain-reward circuits, rat, 627f
Broca's aphasia, 640–641b
Broca's area
and language, 642f, 643f
interconnection with other language areas of brain, 644

C
Cadherins, and axon guidance, 104
Caenorhabditis elegans
cell lineage in, 92f

Caenorhabditis elegans (cont.)
  example of neural differentiation in, 90, 91
  genetic system for study of behavior, 556b
  programmed cell death studies in, 107
Calcintonin gene-related peptide, co-release with ACh, 301
Calcium
  concentration
    influence on amount of ACh released, 274–276
    at presynaptic terminal regulated by intrinsic cellular mechanisms, 287–288
  control of synaptic vesicle mobilization, docking, and function, 284f
  and persistent enhancement of synaptic transmission in long-term potentiation, 684–685
  important in regulation of neuronal function, 290–291
  influx
    affects time course of synaptic transmission in presynaptic cell, 274f
    controls synaptic vesicle docking, fusion, and exocytosis, 282–285
    modulates number of transmitter vesicles released by action potential, 286–290
    triggers neurotransmitter release, 271–273
  ion channels, see Ion channels, calcium
  ion concentration, effect on action potential, 171
  ions, entry through NMDA-activated ion channels, 230
  intracellular, regulated by synaptic connections on presynaptic terminals, 288–290
  and light adaptation, 416b
  release of neurotransmitter is proportional to influx into presynaptic terminal, 273f
  required for synaptic release of neurotransmitters, 290–291
Calcium-calmodulin, role in activity-dependent presynaptic facilitation, 677
Calcium-calmodulin-dependent protein kinase, generation of, 252f
Capacitance
  membrane
    effect on rate of change of membrane potential in response to current flow, 152f
    effect on time course of electrical signals, 150–153
  passive electrical property of nerve cell, 143–144
Carbon monoxide, second-messenger pathway, 254–255
Carrier theory, movement of ions through channels, 117–118
Catechol-O-methyltransferase, 302
Catecholamine transmitters, 295–296
  and mental and neurological dysfunctions, 296
Caudate nucleus
  of basal ganglia, 544
  left, role in language, 644
Cell adhesion molecules, and axon guidance, 104
Cell body
  nerve cell, 22, 23
  role in motor and sensory neurons, 53–55
Cell lineage
  in *C. elegans*, 91, 92f
  and control of transcriptional regulatory proteins, 90
Cell membrane
  conductance, 138, 141
  electric potential difference across and open ion channels, 141–142
  equivalent circuit model of, 142–144
  glial cell, flux of potassium ions across, 136f
  ion permeability and generation of action potential, 162
  lipid bilayer as insulator, 144
  lipid bilayer structure, 116f
  resting, distribution of ions across, 134t
  separation of charge across and resting membrane potential, 134
Cell recognition proteins, see Myelin-associated glycoprotein
Cell–cell interactions
  and control of transcriptional regulatory proteins, 90
  see also Growth factors; Signaling
Cellular connectionist view, of brain, 8
Cellular differentiation, development of concepts, 90
Center-surround receptive field, retinal bipolar cells, 421–422b
Central nucleus
  projections to association areas, 611
  projects to areas of cerebral cortex concerned with representations of emotions, 611
Central pattern generators, spinal cord circuits, 524
Cerebellum, 10, 79
  anatomy, 536
  control of cortical and brain stem motor systems, 499
  cortex
    climbing fibers as excitatory input, 542
    mossy fiber afferent input to, 542
    Purkinje cell modulation of excitatory input to cerebellar nuclei, 542, 542f
  diseases and disorders of coordination, 544
  functional components, 539–541, 539f
  gross anatomy, 538f
  indirect regulation of movement, 536–544
  inputs and outputs, 549
  layers of cortex, 541–542, 541f
  lobes, 538
  outputs, 536
  parallel transverse convolutions of surface, 538
  role in implicit memory, 664
  role in motor learning, 542–544, 543f
  types of nerve cells in, 45, 47
Cerebral cortex
  alteration of structure by sensory deprivation, 471
  and autonomic nervous system, 600
  association areas, 347–363, 349t,f; see also Limbic association area; Parietal-temporal-occipital association area; Parietal association area
  and cognition, 347–363
  contralateral association with body, 9–10
  emotions, areas concerned with receive central nucleus projections, 611
  and emotional experience, 596–597
  higher-order sensory areas, 348–350, 349t,f
  increases in blood flow indicate areas of motor cortex participating in behavior, 535f
  intercortical connections, 350f
  and language, 12
  lesions in specific regions disrupt specific language functions, 639
  lobes of, 9, 81–83, 82f
  map for each sensation, 328–329
  motor areas, 348–350, 349t,f, 493
    as control level of motor system, 492–493
    and corticocortical connections, 534f
  orderly arrangement of somatic sensory and motor projections from and to body surface, 328f
  premotor areas, role in voluntary movement, 534–536
  primary areas, 348–350, 349t,f
  sensory and motor areas, 82f
  somatotopic arrangement of motor areas, 498
  somatic sensory centers in, 324–325f
  vision, 394f
Cerebral hemispheres, 10
  folding of, 81f

relationship of basal ganglia to, 545f
sexual dimorphism of asymmetry in humans, 590–592
Cerebrocerebellum, functional component of cerebellum, 539–541, 539f
CGRP, see Calcitonin gene-related peptide
Chaos theory, and neural networks, 360
Chemical cues, guide developing axons, 101–104, 105f
Chemical transmission, in nervous system, 183–184
Chemoaffinity hypothesis, and nerve cell development, 102
Chemotropism, 104
Chloride ions, passive distribution, 139
Chloride pump, 139
Chloride, receptor–channels selective for mediate inhibitory synaptic action in central neurons, 232–234
Chloride-selective receptor–channels, mediate inhibitory actions at chemical synapses, 233f
Cholecystokinin, effect on feeding behavior, 622–623
Cholinergic, 293
Chromaffin cells, 98–99, 98f
Chromatic abberation, 458
Chromosomes
function in nerve cells, 54–55
normal human, maps, 560f
Ciliary neurotrophic factor, role in neuron survival during development, 107
Cingulate gyrus, subarea of limbic association cortex, 354
Circadian rhythm
and behavioral responses, 624–625
Drosophila, and *per* gene, 571–576, 573f
Climbing fibers
excitatory input to cerebellar cortex, 542
role in learning, 543–544
Co-transmission, 301
Cocaine, effect on rate of intracranial self-stimulation, 626f
Coextensive single-opponent cells, retina and lateral geniculate nucleus, 460
Cognition
association areas involved in different functions of, 348–350
and cerebral cortex, 347–363
localization within cerebral cortex, 9–16
methods of study, 321–346
neural network simulation, 359–362
role of basal ganglia, 547–549
sexual dimorphism of in monkeys, 590
single-gene defects, 561–562
strategic planning for higher motor actions, location at principal sulcus, 351–353
Cognitive neural science, 317–363, 631
Cognitive psychology, as approach to study of cognition, 322–323
Cold-sensitive neurons, hypothalamic, 617
Color, 453–468
Color blindness, 464, 465–467b, 560
Color constancy, 453, 459–464
Color contrast, 459–464, 458f, 463f
Color opponency, 459–464, 458f
Colors, equiluminant, 396–398
Columnar organization
association areas, 350
of inputs by submodality in somatic sensory cortex, 336–339
of primary visual cortex, 440–444
spinal cord, 493
Commissures, 87
hemispheric, function, 355–357
see also Corpus callosum
Competition, between afferent fibers of two eyes, and establishment of ocular dominance columns, 475f
Complex cells, primary visual cortex
and specific orientation of image in visual field, 437f
end-stopping of, 438–440
receptive fields of, 436
Component direction-selective neurons, 448
Compound action potential, 308
Computational neurobiology, 40
Computer image, brain during auditory stimulation, 1–2f
Computer science, use of techniques in cognitive neural science, 323–324
Computerized tomography, 72–76b
Concentric broad-band cells, retina and lateral geniculate nucleus, 460
Concentric double-opponent cells, cortical, and color contrast, 462–464, 463f
Concentric single-opponent cells, retina and lateral geniculate nucleus, 460
Conditioning
appetitive versus defensive, 659
classical
of *Aplysia* gill-withdrawal reflex, 677–680, 678f
and associative enhancement of presynaptic facilitation, 677–680
contingency in, 660
importance of timing, 677
and learning predictive relationships between stimuli, 660–661
molecular model of synaptic action underlying, 679f
versus operant, 659
food-aversion, 662–663
operant, association of specific behavior with reinforcing event, 661–662
and psychotherapeutic techniques, 663
Conductance
membrane, 138, 141
calculation from voltage-clamp data, 167b
passive potassium ion channels, 142f
potassium, calculation from potassium current, 166–168
sodium, calculation from sodium current, 166–168
synaptic, 215f
Conduction
aphasia, 14, 641b
saltatory, of action potential in myelinated axon, 157
signal, efficiency affected by membrane and axoplasmic resistance, 153–156
velocity, effect of myelination on, 156–159
Conductor
ion channels as, 141–142
potassium channel represented as, 141f
Cone opsin, 412
Cone pigments, 457b
Cones
and color discrimination, 456–458
differences from rods, 409t
different systems response to different wavelengths of visible light, 455–456, 456f
structure, 411f
Connectional specificity, principle of, 25
Connectionist approach, cellular basis, 25
Connexin
in gap-junction channels, 189
membrane-spanning regions, 190f
Connexon, 189
three-dimensional structure, 190f
Contralateral projection, 355f
Contrast, visual, detection of, 419, 419f
Contrast information, from retina, not altered significantly in lateral geniculate nucleus, 431–432

Cooperation, and development of ocular dominance columns, 478–481
Corpus callosum, 87, 355
  and awareness of self, 18
  effect of lesions in on reading ability, 646
  effects of cutting, 355–357
Cortical representation, see Internal representation
Corticobulbar fibers, descending tract of motor cortex, 496
Corticospinal fibers, descending tract of motor cortex, 496
Corticospinal tract, and direct action by motor cortex on spinal motor neurons, 496–498
Corticostriate pathway, and motor pathway of basal ganglia, 547–548
Courtship behavior, *Drosophila, per* gene control, 574f
Cranial nerves, of brain stem, 79
Creativity, distinguishing feature of language, 635
Cross-talk, second-messenger pathways, 255–257
Current, see also Ion Channels
  capactive, 163
  flow
    in electrical versus chemical synapses, 185, 185f
    along axon and dendrite, 154–155
    passive, in nerve cell, modeled as electrical equivalent circuit, 143f
  flow through NMDA receptor–channel, 230f
  generator, in equivalent circuit model of nerve cell, 142–144
  depolarizing, degree of decrease and length constant of nerve cell, 223
  end-plate
    and rapid opening and closing of ACh-activated ion channels, 200
    calculation from equivalent electrical circuit, 213–216
    factors determining, 207
  gating, 173
  inhibitory, impact from synapses at two different sites in postsynaptic neuron, 225f
  injected, follows path of least resistance in neuronal process, 155f
  measuring in ion channels, 119–121, 119f
  recording in single ion channels, 120b
  relationship to ion concentation, 124
  saturation in ion channels, 124
  single-channel
    effect of membrane potential on, 208f
    reversal potential of those activated by glutamate and GABA, 235f
    same reversal potential as total end-plate current, 206f
    unitary nature, 204
  sodium and potassium, calculation of conductances from, 166–168
  through voltage-gated
    potassium channel, 163–166
    sodium channel, 163–166
  use of patch-clamp experiments to study flow through single ion channels, 204–208
Current-voltage relationships, nerve cells, 151f
Cyclic adenosine monophosphate
  as second messenger in synaptic transmission, 244–245
  cycle, 248–249f
  -dependent protein kinase, and gene activation in long-term memory for sensitization, 674, 674–675f
  pathway
    as neuronal second-messenger pathway, 249f
    role of cytoplasmic messenger, 247–250
  role in learning, 677
  role in sensitization, 671
Cyclic AMP, see Cyclic adenosine monophosphate
Cyclic AMP-dependent protein kinase, role in sensitization, 671
Cyclic GMP, see Cyclic guanosine 3′,5′-monophosphate
Cyclic guanosine 3′,5′-monophosphate
  concentration affected by light-activated visual pigments, 412–414
  role in phototransduction, 410–414
  second-messenger pathways, 254–255
Cytoarchitectonic method, functional areas of cerebral cortex, 14
Cytokeratins, cytoskeletal protein, 65
Cytoskeleton
  axon, 49f
  components, 66f
  nerve cell, 65–67
  see also Microtubules
Cytosol, transport, 64–65

D

Dark current, 415b
Decussations, 87
Delayed-response task, 351, 353f
Demyelination
  in central nervous system, 51–52
  diseases that cause, 157
Dendrites
  of motor neuron, 48
  nerve cell, 23
  spine, excitatory synapses on, 226
  unit length represented as electrical equivalent circuit, 154f
Dentate nuclei, outputs of cerebellum, 536
Depolarization, 32, 134
  enhances calcium influx at presynaptic terminal, 287
  local, and propagation of action potential, 155–156
  passive conduction along axon contributes to propagation of action potential, 156f
  signals, sent by electrical synapses, 184
Depth agnosia, 398
Depth perception, and magnocellular-thick stripes pathway, 396
Desensitization
  ion channels, 126–127
  and second-messenger action, 260–261
Development
  changes in ion channel functional properties over, 130
  cognitive, sexual dimorphism of in monkeys, 590
  critical periods of in brain, 482
  elimination of synpases during, 106–107
  gonadal
    hormones involved, 580–582
    single gene, 580
  human nervous system, 78f
  of nervous system, 43–110, 89–110
    effect of gonadal hormones, 582–586
  ocular dominance columns, 480–481f
  perinatal, critical periods of steroid hormone influence, 583–584
  role of signaling molecules, 107–108
  social competence, 482–483
  stages in language acquisition, 638t
  and survival of neurons, 107
Diacylglycerol-inositol triphosphate system, 251, 253f
Dichlorodiphenyl trichloroethane, and masculinization of brain, 584
Diencephalon, 10, 79
Diethylstilbestrol, and masculinization of brain, 584

Diffusion, and removal of neurotransmitter from synaptic cleft, 302
Diffusion constant, and ion channel, 117
Dimorphism, behavioral, 582–583
Disease
of basal ganglia, 544
Parkinson's disease, 547
cerebellar, 544
effect on language, 636
genetic and acquired, genetic component, 693f
ion channels as targets, 124
of motor systems, 488
see also Disorders
single-gene, 561–562
Dishabituation, 659
Dissociation constant, and ion channel, 124
Distributed processing theory, of brain, 13
Docking, synaptic vesicle, molecular machinery for, 285f
Dopamine
loss of in cells of substantia nigra leads to Parkinson's disease, 547
as neurotransmitter, 295–296
pathways, and drug addiction, 626
role in regulation of self-stimulation and pleasure, 626
Dorsal column-medial lemniscal system, organization, 324–325f
Dorsal root, spinal nerve, 79
Dorsal root ganglia cells, 46f, 47
Dorsal root ganglion, 30
Double-opponent cells, in primary visual cortex, 462
Drives, see Motivational states
Drosophila
cycling of RNA and protein products of period gene, 575f
differentiation of photoreceptors in compound eye, 94f
genetic system for study of behavior, 556b
manipulation of gene expression to study behavior, 566–567b
nerve cell identity in, 90, 91–94
period gene control of behavior, 571–576
role of cyclic AMP in learning, genetic studies, 677
sites of expression of *period* gene, 576f
Dynamic polarization
physiological evidence, 31
principle of, 25
Dynein, motor molecule for transport in nerve cell, 64
Dysarthria, 642
Dyslexia, 644, 646–648
developmental versus acquired, 646–648
Dysmetria, 544
Dysphonia, 642

E

11-cis-retinal
in cones, 412
structure, 413f
Ectoderm, and development of neural plate, 94–95
Electrical equivalent circuit
calculation of end-plate current, 213–216
deriving equation for resting membrane potential from, 144–147
effects of membrane capacitance on rate of change of membrane potential in response to current flow, 152f
end-plate potential, 214f
functional properties of nerve cell represented as, 140–144
model, of cell membrane, 142–144
nerve cell under voltage-clamp conditions, 167f
passive current flow in nerve cell modeled as, 143f
representation of axon and dendrite unit lengths as, 154
resting membrane potential as, 144f
sodium-potassium pump modeled as, 143f
spread of action potential, 157f
Electrical properties, passive, of neuron, and local signaling, 149–159
Electrical synapses, see Synapses
Electrical transmission, in nervous system, 183–184
Electrochemical driving force, 124, 138
Electrochemical energy
coversion of stimulus energy into, 375–376
transformation of sensory stimulus into, 372–373
Electroconvulsive treatment, effect on memory, 665, 665f
Electromotive force, 141
Electrotonic conduction, 155
action potential, 155–156
in electrical synapses, 186–187
Emotion, 595–612
amygdala, 610
hypothalamus, 605–607
learned, 608–611
neuroactive peptides, 299
pathways for processing, 609f
peripheral component, 596, 597
physiological changes accompanying, 597–600
proposed neural circuit, 607f
relationship of cognitive and physiological states, 596–597
role of autonomic nervous system, 597–600
role of hypothalamus in regulation, 597–607
role of temporal association areas, 354–355
End-plate, 198f
amplitude of synaptic potential at, 202f
current
and rapid opening and closing of ACh-activated ion channels, 200
calculation from equivalent electrical circuit, 213–216
factors determining, 207
total, effect of membrane potential on, 208f
morphology in myasthenia gravis, 311, 312f
muscle, ion channel permeable to sodium and potassium at, 203–204
potential
acetylcholine-activated, effect of membrane potential on, 208f
equivalent electrical circuit, 214f
excitatory postsynpatic potential in muscle cell, 200
factors determining time course, 215f
ion channels responsible for, 203f
miniature, 273
neuromuscular junction, transmitter-gated channel activated at, 211
produced by simultaneous sodium and potassium ion flow, 203f
slow change in, 202f
spontaneous, 273; see Miniature end-plate potentials
study by pharmacological isolation, 201f
synaptic potential at, 200–202
End-stopped cells, receptive fields, 438–440, 439f
Endocrine system
control by hypothalamus, 602
role in temperature regulation, 616–618
Endoplasmic reticulum, role in protein synthesis, 59f
Enteric nervous system, 77
division of autonomic nervous system, 598–600, 599f
Entorhinal cortex, role in memory, 657
Environment, influence on behavior, 553

Enzymatic degradation, and removal of neurotransmitter from synaptic cleft, 302
Epigenetic influences, nervous system development, 89–91
Epilepsy, and study of function of corpus callosum, 355
Epinephrine, as neurotransmitter, 295–296
Equilibrium potential, calculation, 137
Equiluminant stimuli, and study of parvocellular system, 396–398
Esterase, and degradation of ACh to terminate synaptic transmission, 302–303
Estradiol, and masculinization of brain, 584
Estradiol-sensitive neurons, in rat brain, 586f
Estrogen
  binding protects female fetuses from masculinization, 584–585
  induction of sexual receptivity, 582f
  sex differences in effect on regulation of progesterone receptors of hypothalamic ventromedial nucleus, 586–587
Ethology, 553
Evoked potentials, 326
Evolution, divergent versus convergent, 299
Excitatory postsynaptic potential
  calculation of reversal potential, 206b
  contribution of NMDA-activated channels to late phase, 231, 231f
  end-plate potential in muscle cells, 200
  interaction with inhibitory potential in nerve cell, 226f
  measurement of reversal potential for, 227
  mediated by transmitter-gated channels selective for sodium and potassium, 227–232
  reversal potential for, 228–229f
  see also Synaptic excitation
  in stretch reflex, 221
Excitatory synaptic action
  in central neuron, 221
  effect of norepinephrine, 257f
  transmitter and second-messenger-initiated, 256f
Exocytosis
  and changes in membrane surface area, 281, 281f
  and discharge of neurotransmitter from synaptic vesicle, 276, 279–282
  formation of fusion pore during, 284–285
  at presynaptic terminal, 280f
  synaptic vesicle, molecular machinery for, 285f
Experience
  and modification of somatotopic map in brain, 688–690
  can modify internal representation of personal space, 329–334
Experimental psychology, as approach to study of cognition, 322
Extinction, 660
  as psychotherapeutic technique, 663
Extrastriate cortex, 394f
  lesions and color perception, 646
  staining regions, 395
Eye
  closure, effect on development of ocular dominance columns, 476f
  compound
    development of in *Drosophila,* 93f
    identity of photoreceptors in *Drosophila,* 91–94

F

5-Hydroxytryptamine, see Serotonin
Faces, representation in inferotemporal cortex, 445
Facilitating neuromuscular block, 315
Facilitation, presynaptic, 288
  activity dependence, 677
  classical conditioning involves associative enhancement of, 677–680
  and short-term sensitization in *Aplysia,* biochemical steps, 672–673f
Fast fatigable, motor unit, 503, 504f
Fast fatigue-resistant, motor unit, 503, 504f
Fastigial nuclei, outputs of cerebellum, 536
Fear
  and amygdala, 608
  measures of in animal models, 597t
Feature maps, in brain, 402, 403f
Feedback signals, for water regulation, 623
Feedback system, and regulation of body weight, 619
Feeding behavior
  alteration by chemical stimulation of hypothalamus, 622
  aversion conditioning, 662–663
  control of food intake, 620–621
  and motivational states, 614–615
  regulation by short- and long-term cues, 622–623
  regulation, 618–623
Fibrillar proteins, of cytoskeleton, 65–67
Filopodia, 101
Fixed-action patterns, 571
Flexion
  reflexes, 521
  withdrawal reflex, circuits, 522f
Flexor-extensor rule, motor neurons, 493
Flight-or-fight reaction, 598
Forebrain, 10
Fovea, 408
  inputs to lateral geniculate nucleus, 430
Foveola, 409
Freeze-fracture technique, structural study of synaptic membranes, 278b
Frequency code
  and perception of stimulus strength, 379f
  conveys information about stimulus strength, 377–378
Frontal association area, see Prefrontal association cortex; Orbitofrontal cortex
  proportion of brain, 351f
  subdivisions, 351, 351f
Frontal cortex, lobotomy, 611
Frontal lobe, cerebral cortex, 81,82f
  association areas, 351–353
  function, 9
  sexual dimorphism of development in monkeys, 590
Fruit fly, see *Drosophila*
Fusimotor system, see Gamma motor neurons
Fusion pore, formation during exocytosis, 284–285
Fusion, synaptic vesicle, molecular machinery for, 285f

G

G-proteins, role in cyclic AMP pathway, 247–250
  action on different effectors, 260f
  control of synaptic vesicle mobilization, docking, and function, 284f
  -coupled receptors, 243–244
  direct action on ion channels, 258–259
  role in ion channel activity, 193f, 194
  role in sensitization, 671
  role in photoactivation, 414
  transducing, 246
GABA, see Gamma-amino butyric acid
Galanin, effect on feeding behavior, 622
Gamma motor neurons

alpha-gamma coactivation, 512f
control sensitivity of muscle spindles, 510–512, 511f,
Gamma-amino butyric acid, as neurotransmitter, 297
inhibitory neurotransmitter, 234
receptors, as transmembrane proteins, 236–237
reversal potential of single-channel current activated by, 235f
Ganglia, of peripheral nervous system, 77
Ganglion cells
detection of contrasts and rapid changes in visual image, 419
excitatory connections with corresponding classes of bipolar cells, 423
nuclear layer of retina, 417f
on-center versus off-center, 418f, 419
output of retina to, 414, 416
receptive field, 416–419
retina, types of, 460, 460–461f
signals from photoreceptors relayed by interneurons, 420–423
specialization, 419–420
Gap junction, 181
Gap junction channels, 189f
and electrical synapses, 189
molecular structure, 238f
second-messenger flow through, 187
structure, 189
three-dimensional structure, 190f
Gating
charge, redistribution within voltage-gated sodium channels when they are open, 172–173
current, 173
effect of exogeneous factors on, 128–129, 129f
ion channels, 125
of ion channels by postsynaptic receptors, 193–194
of sodium ion channel, 176f
of voltage-sensitive ion channels, influenced by intracellular ion concentrations, 170–171
Gender identity, development, and hormones, 581, 582–583
Genes
activation, and cellular differentiation, 90
and behavior, 555–577
cloning and sequencing, and determination of ion channel structure, 121–124
common ancestral, for potassium, sodium, and calcium ion channels, 173–177
encoding alpha subunits of voltage-gated ion channels, 177
engineering, method to deduce ion channel structure, 122, 124
expression
alterations in persistent mental states, 691
effects of second-messenger on, 261–265
manipulation of in animals to study behavior, 565–569
regulation, 263–265b
families of those encoding ion channels, 130
knockouts, study of LTP, 686–687b
and organization of behavior, 565–576
polymorphisms, 562–563b
single
and cognitive defect, 561–562
and color vision, 559–561
importance in behavioral traits, 558–559
mutation producing motor and cognitive defects, 561–562
switches initial development of gonad from male to female, 580
Genetic systems, for study of behavior, 556–557b
Gestalt psychology, 370
and creative nature of perception, 388
Gill withdrawal reflex, *Aplysia*
biochemical steps, 672–673f
classical conditioning, 677–680, 678f
and habituation, 668–671, 669f
Glial cells, 27–29
development of in rat optic nerve, 99
fibers, nerve cell migration along, 100f, 101
resting ion channels selective for postassium in, 136–137
Globus pallidus, of basal ganglia, 544
Glucoreceptors, in hypothalamus, 622
Glutamate
excitotoxicity, 232
excitatory transmitter in brain, 227–232
as neurotransmitter, 297
receptors, 227–232
classes, 229f
as transmembrane proteins, 236–237
reversal potential of single-channel current activated by, 235f
role in long-term potentiation, 681, 682f, 683
transmitter versus metabolic, 297
Glycine
inhibitory neurotransmitter, 234
as neurotransmitter, 297
receptors, as transmembrane proteins, 236–237
Glycoproteins, in cell membrane, 116f
Goldman equation, 140
Golgi apparatus, role in protein synthesis, 59f
Golgi tendon organs
functional differences from muscle spindles, 509
as muscle receptor, 506, 507f, 508f
Gonads
hormones from mother and fetus regulate continued development, 580–582
secretions are sexually dimorphic, 582
sexual development, 580
Granular layer, of cerebellar cortex, 541–542, 541f
Gray types I and II synaptic connections, 234–236
Growth cone
Aplysia, 101f
axon, arrival at skeletal muscle fiber target, 104
role in axon guidance to target during development, 101
Growth factors
gonadal steroid hormones as, 591f
role in differentiation of nerve cells, 95–99
Guanosine nucleotide-binding protein, receptor class, 243–244; see G-proteins
Guillain-Barre syndrome, demyelinating disease, 157
Gyri, cerebral cortex, 81, 82f

H

Habituation, 659
and depression of synaptic transmission, 668–671
and homosynaptic depression, 671
long-term
in *Aplysia,* 670f
and structural changes in presynaptic terminals of sensory neurons, 676f
Half-center model, of central neural circuit, 523, 524f
Handedness, and hemispheric specialization, 358, 358t
Head and eye movements, coordinated by tectospinal stract, 494
Hebb's rule, 681
Hemispheres, cerebral, 79, 81
anatomical asymmetry of, 358–359, 359f

Hemispheres, cerebral *(cont.)*
and handedness, 358
different cognitive capacities, 355–359
specialization, 359
Hindbrain, 10
segmental organization of in vertebrates, 95
Hippocampal formation, 607
of cerebral hemispheres, 79, 81
Hippocampus
afferent pathways in, 680–684, 680f
of cerebral cortex, 83
cooperative activity in production of long-term potentiation, 680, 681f
dendrite arrays from pyramidal cells, 629–630f
dorsal, sex difference in, 589
extent of removal in patient H.M., 653f
and long-term memory, 354
long-term potentiation in, 680f
associative in CA1 region, 681–684, 681f
and storage of explicit memory, 680–685
role in memory, 657–658
see also Long-term potentiation
storage of long-term memory in, 657f
Histamine, 296
Hodgkin-Huxley model, of action potential, 162–163, 169–170
Homeobox genes, and segmental organization of vertebrate hindbrain, 95
Homeostasis
anticipatory nature, 624–625
correspondence to motivational states, 614–615
role of hypopthalamus in, 597
Hormones
balance, alteration, effects on food intake, 621
and development of gender identity, 581, 582–583
effect on axonal growth, 589–590
and gonadal development, 580–582
gonadal, effect on developing nervous system, 582–586
gonadal steroids, growth-promoting properties, 591f
hypothalamic-releasing, 604t, 605f
male, fetal exposure causes pseudohermaphroditism in genetic females, 583
and masculinization of brain, 584–586
perinatal, and sex-linked behaviors, 587–588
receptors
as transcriptional regulators, 586–587
gonadal steroid, location in nerve cell, 585–586, 585f
sexually dimorphic secretion by gonads, 582
steroid, critical periods of influence on perinatal development, 583–584
Huntington's disease
inheritance, 561–562, 561f
role of glutamate toxicity, 232
single gene defect, 561–562
Hypercolumn, 442–443
Hyperpolarization, 32, 134
decreases calcium influx at presynaptic terminal, 287
Hypothalamus, 79
chemical stimulation of alters feeding behavior, 622
connections with central nucleus of amygdala, 610f
control of endocrine system, 602
control of pituitary, 603–605, 604f, 604t
lesions in alter set point for body weight, 621, 620f
location and structure, 600f
modulator of autonomic motor system, 84
nuclei of, 601f
possible role in biological clock mechanisms, 624
projections, 600, 602
regions concerned with heat conservation and heat dissipation, 616f
regions controlling body temperature, 616–618
and regulation of emotion, 597–607
regulation of water balance, 623–624
role in controlling output of autonomic nervous system, 600–602
stimulation of and selective eliciting of emotional states, 605–607
structure, 601f
summation of peripheral and central information on temperature, 617f

I

Illusions, and visual perception, 391–392, 391f, 392f
Imaging techniques, use of in cognitive neural science, 323
Immune response, elicited by myelin basic proteins, 51–52
Imprinting, 482
Inactivation, refractory state of ion channels, 127
Incentive stimuli, 615
Individuality, biological expression and learning-induced changes in somatotopic map, 688–690
Inferior prefrontal convexity, lesions in interfere with appropriate motor response, 353
Inferior temporal cortex
lesions affect speed of learning visual tasks, 354
and perception of form and color, 404
representation of complex forms in, 445
Inhibition
distal, in sensory relay nuclei, 383f
feed-forward versus feedback, 38
lateral,
inhibition
and discrimination between stimuli, 380–383
function in sensory pathways, 382f
sensory stimulus, 375
presynaptic, 288
synaptic, 221, 221f
see also Synaptic inhibition
types in sensory relay nuclei, 383f
types in sensory systems, 380–381
Inhibitory potential, interaction with excitatory potential in nerve cell, 226f
Inhibitory signals, in central neuron, 221
Inhibitory synaptic action, mediated by chloride-selective receptor–channels in central neurons, 232–234
Innervation, muscle
and change in distribution and stability of ACh receptors, 104, 106
and change in properties of ACh receptors after, 104, 106
Innervation ratio, 502
Inositol polyphosphates, as second messenger in synaptic transmission, 244–245
Input resistance, 150
Input signal, 31, 34
Insular cortex, of cerebral cortex, 81
Insulator, cell membrane lipid bilayer as, 144
Integral proteins, 59
Integration
neuronal, 222
signal, 31
synaptic, 219–241
of visual image, 404
Integrins
and axon guidance, 104
transmembrane protein, 67
Interblob region, primary visual cortex, 395

Internal representation
of body surface, 326–328
enlargement of in somatic sensory cortex by increased use, 331f
hand, 326f
modification of, 333–334
and mental rotation, 317–318f
and motor processing, 490
and phantom limb syndrome, 333–334
sensory representation, 370
of stimulus on skin, 336f
Interneurons
between photoreceptors and ganglion cells, 416
inhibitory
feed-forward versus feedback inhibition, 38
and negative feedback system for muscle tension, 520f
local and projection, 27, 84
mediation of action potential, 222
and motor neurons, modulation by brain stem, 494–495
nuclear layer of retina, 417f
projection, 27
receptive field, 374f
of relay nuclei, 374
relay signals from photoreceptors to ganglion cells, 420–423
role in polysynaptic spinal reflexes, 518–520
role in spinal circuits, 526
synaptic contact of motor neuron with, 49, 50f
types, 27
Interposed nuclei, outputs of cerebellum, 536
Interstripe region, extrastriate cortex, 395
Intracranial stimulation, of hypothalamus, as reinforcement for behavior, 625
Introspection, as approach to study of cognition, 322
Ion channel
acetylcholine-activated
nicotinic, 209–210, 209f
number open and end-plate current, 207f
complexed with ACh receptor, electron microscopic image, 200f
all-or-none behavior, 119f
calcium in presynaptic nerve terminal, 201f
calcium-activated potassium, 170
calcium voltage-gated
at chemical synapse, 191f
close relationship with sodium channels, 173–177
molecular structure of alpha subunit, 174f
role in action potential, 170
tertiary structure, 175f
cGMP-gated, 414
in photoreceptors, 410–411
changes in functional properties over development, 130
chloride
gating by GABA receptors and glycine, 234
in IPSP, 232–234
voltage-gated, 170
and inhibitory actions at nerve cell body, 224
common characteristics, 124–130
comparison of synaptic excitation produced by opening and closing, 259t
as conductor and battery, 141–142
conformational changes in opening and closing, 125
delayed rectifier, 170
direct action of second messengers and G-proteins on, 258–259
direct and indirect action of G-proteins on, 260f
direct versus indirect gating by neurotransmitter receptors, 243–245, 244f
dissociation constant for ion binding in, 124
drugs that block selectively, 165
effect of exogenous factors, 128–129, 129f
electrical potential across cell membrane and opening, 141
G-proteins in activity of, 193f, 194
GABA-activated, functional domains, 236–237
gating, 116, 125, 133
by postsynaptic receptors, 193–194
genes that encode, 130
investigation of structure and function, 118–124
ligand-gated, 127f
measuring current in, 119–121, 119f
mechanisms of neurotransmitter action on, 193f
mechanisms for refractory state, 128f
as membrane proteins, 116–118
molecular diversity, 39
molecular properties, 171–177
muscle end-plate, permeable to sodium and potassium, 203–204
N-type calcium, reduction/inactivation in habituation, 669
occlusion, 125
open, and electric potential difference across cell membrane, 141–142
opened or closed by second-messenger action through protein phosphorylation, 257–259
opening and closing, physical models for, 126f
passive flux of ions through, 124–125
patch-clamp experiments to study current flow through, 204–208
phosphorylation-gated, 127f
potassium
fast transient, 170
ion selective, electromotive force in, 141f
in IPSP, 232–234
reduction in sensitization, 671
properties, 115–116
refractory state, 127–128, 128f
resting, 116
and maintenance of membrane potential, 133, 134–139
selective for potassium in glial cells, 136–137
selective for several ion species in nerve cells, 137–138
second-messenger cascade in activity of, 193f, 194
saturation of current, 124
selective for sodium, potassium, or chloride, 142f
selectivity of, 117–118, 124
as simple resistor, 119, 121
single, recording current in, 120b
stimuli that control opening and closing, 127f
stretch or pressure-gated, 127f
structural features shared by different types, 237–239, 238f
structure, 121–124, 121f
as targets of disease, 124
transmitter-gated, 126–128
all-or-none activity, 204, 204–205f
at nerve-muscle synapse, 200–202
chloride, mediate IPSP in central neurons, 232–234
conduct unitary current, 204
ion flow through, 212f
molecular structure, 238f
sodium and potassium channels mediate EPSP, 227–232
versus voltage-gated, 210–212
types, 133
variations in different tissues, 130
voltage-dependent calcium, effect of depolarization on, 271
voltage-dependent potassium
closing of and excitation, 258

Ion channel *(cont.)*
M-type, 257
S-type, 257
voltage-gated, 117f, 126–128, 127f
all-or-none behavior, 172
calcium, in neuromuscular junction, 279
common ancestral gene, 173–177
genes encoding alpha subunits, 177
molecular structure, 238f
role in generation and propagation of action potential, 161–177
S4 membrane-spanning region, 174–175
sodium
basis for selection of sodium ions, 173
calcium, and potassium, molecular structure of alpha-subunit, 174f
at chemical synapse, 191f
distribution, 172
gating, 176f
and neurotransmitter release, 270f
molecular structure, 173
redistribution of charge when open, 172–173
tertiary structure, 175f
variety of in nervous system, 170
voltage-clamp experiment demonstrating, 166f
voltage-sensitive
gating influenced by intracellular ion concentrations, 170–171
potassium, blocking effect on presynaptic action potential, 272f
sodium, 35
effect of blocking on presynaptic action potential, 272f
Ion conductance, theories, 117–118
Ionic conductance, net increase in accompanies action potential, 162f
Ionic membrane current, study with voltage clamp, 162–168
Ionotropic receptors, 182
Ions
active pumping across cell membrane, 138–139
determination of contribution to resting membrane potential, 140
distribution across neuronal membrane at rest, 134t
flow through voltage-gated channels generates action potential, 161–162
gradient and resting membrane potential, 136–137
movement through cell membrane, 116–118
passive flux through ion channels, 124–125
rate of flow through channels, 115
relationship between concentration and current, 124
and waters of hydration, 117

J
Jacksonian seizure, 328
James-Lange hypothesis, emotion, 596
Jitter, 308
Junctional folds, 198f
Just noticeable difference, stimulus, 371

K
Kaniza triangle, 391f
Kinesin
model for transport, 64f
motor molecule for transport in nerve cell, 63
Kluver-Lacy syndrome, role of amygdala, 608
Knee jerk reflex
anatomy, soleus muscle, 48f
behavior, anatomical components, 29–31
neurons forming, 46f

L
L-Dopa, treatment of Parkinson's disease, 547
Labeled line code, 380
apply to smell and taste, 385
Language, 633–649
Lateral corticospinal tract, 497f
Lateral geniculate body, in thalamus, in visual system, 393f
Lateral geniculate nucleus
color coding ink, 459–462
neural map of retina in, 430
no significant alteration of contrast information from retina, 431–432
parallel pathways for visual perception, 395f
projection of retina to, 428–431, 429f
retinotopic representation of contralateral visual field in, 430
segregation of eye inputs, 478
structure, 430f
types of cells, 460, 460–461f
Lateral premotor area, 493, see Premotor cortex
Law of effect, and voluntary behavior, 662
Layered network, in parallel distributed processing, 360
Leakage conductance, 163
Leakage current, 163
Leakiness, of nerve cell, 159, 163
Learning, 651–666
in amnesic patients, 654
cellular mechanisms, 667–694
and classical versus operant conditioning, 659
implicit
and changes in effectiveness of synaptic transmission, 668–671
nonassociative versus associative, 658–660
measurement, 658b
memory processes during, 664f
molecular alphabet, 685
motor, role of cerebellum, 542–544, 543f
neuronal changes associated with provide insights into psychiatric disorders, 690–692
and relationship between implicit and explicit memory, 663
role of mossy and climbing fibers in, 543–544
Lesions, temporal lobe, effect on memory, 354
Light
adaptation, 414
absorption and activation in photoreceptors, 410–412
and calcium, 416b
intensity, slow adaptation of photoreceptors to, 414
wavelengths, 455f
Limb proprioception, as body sense, 383
Limbic association cortex, 83
functional areas, 350
orbitofrontal cortex as subdivision, 351, 351f
subareas, 354
Limbic lobe, 606f, 607
of cerebral cortex, 83
Limbic system, 606f, see also Amygdala; Hippocampal formation; Limbic lobe; Motivational system; Parahippocampal gyrus
Linguistics, Chomskian, 639f
Lipid bilayer, structure of cell membrane, 116f
Lobes, of cerebral cortex, 81–83
Lobotomy

frontal cortex, 611
prefrontal, 611
Local sign, 521
Localization, of complex functions of brain, 362
Locomotion
control by spinal circuits, 523–526, 525f
goal-directed, role of higher motor centers, 526
human, 527
Long-term potentiation, 288
associative
in CA1 region of hippocampus, 680–684, 681f
and spatial memory, 684
in hippocampus, 680f
and storage of explicit memory, 680–685
maintenance and increase in presynaptic transmitter release, 684f
model for induction, 680–684, 684f
nonassociative in CA3 hippocampal region, 685, 688f
study of using gene knockouts, 686–687b
Lordosis
and androgenized brain, 587
sexually dimorphic reproductive behavior, 582
Luteinizing hormone, secretion in male versus female brains, 586
Lysosome, role in protein synthesis, 59f

M

M cells, retinal ganglion cell type, 420
M-type potassium ion channel, 170
Macromolecules
synthesis of by spinal motor neuron, 54f
synthesized by cell body in sensory neuron, 53–55
Magnesium ions, and NMDA-activated ion channels, 230
Magnetic resonance imaging, use to determine structure of nervous system, 71
Magnocellular
cells, retina, projections, 394
layers, of lateral geniculate nucleus, 430
pathway, 396, 397f, 432
Main immunogenic region, ACh receptor, 312
Manic-depressive disorder, see Bipolar affective disorder
Masculinization, of brain, 584–586
Master map, in brain, 402, 403f
Mauthner's cell, and electric synaptic transmission, 187
Mechanoreceptors, slowly versus rapidly adapting, 380f
Medial lemniscus, role in sensory circuit, 372f
Medulla, 79
Medulla oblongata, 10
Medullary pyramid, 496
Membrane potential, 133–147
effect of membrane capacitance on rate of change in response to current flow, 152f
effect on ACh-activated end-plate potential, total end-plate current, and single-channel current, 208f
maintenance by resting ion channels, 133
maintenance with sodium-potassium pump, 139
methods of recording, 135b
resting, 31–32
in cell with resting sodium and potassium channels, 138f
determination of contribution of ions to, 140
as electrical equivalent circuit, 144f
equation for derived from electrical equivalent circuit, 144–147
ion balance of abolished during action potential, 139–140
and ionic gradient, 136–137
and separation of charge across cell membrane, 134
role of resting ion channels, 134–139
Membrane properties, passive, effect on propagation of action potential, 156–159
Membrane resistance
calculation, 154
effect on magnitude of electrical signals, 150
Membrane systems
in nerve cells, role in protein synthesis, 58
topological continuity in nerve cell, 58f
Membrane-spanning protein, deducing secondary structure from amino acid sequence, 122–123f
Memory
affected by lesions of temporal lobe, 354
in amnesic patients, 654f, 655f
cellular mechanisms, 667–694
classified as implicit and explicit, 653–658
effect of lesions of posterior parietal cortex on, 343f
and electroconvulsive treatment, 665, 665f
explicit
role of medial temporal lobe system, 656–658
storage involves long-term potentiation in hippocampus, 680–685
transformation, 656b
implicit
involves various perceptual and reflex pathways, 658
roles of amygdala and cerebellum, 664
localization
in different places throughout the nervous system, 665–666
of functions in brain, 651–653
long-term
for sensitization in *Aplysia,* changes in sensory neurons, 674–675f
requirement of new protein synthesis and growth of new synaptic connections, 671–676
storage in hippocampus, 657f
nerve cell, and long-term potentiation, 288
neural basis, 664–666
relationship between explicit and implicit in learning, 663
role
of entorhinal cortex, 657
of hippocampus, 657–658
of temporal association areas, 354–355
short-term and long-term storage, 664–665
spatial, and associative long-term potentiation, 684
of visual task, role of inferior temporal region, 354
working, effects of lesions of principal sulcus on, 352
Mental dysfunctions, role of catecholamine transmitters, 296
Mental functions, psychological, localizable internal representations, 632
Mental illness, organic and functional, 690
Mental processes, localization by their elementary operations, 16–19
Mental rotation, reaction time and shift in cortical representation, 317–318f
Mesaxon, 51
Mesoderm, signals from induce neural plate, 94–95
Mesolimbic dopamine pathways
acted on by some drugs of abuse, 625–626
role in reinforcement of behavior, 625–626
Messenger RNA, role in protein synthesis, 58
Metabolic signals, transmitted by electric synapses, 187
Metabotropic receptors, 182
Metarhodopsin II, 411–412
Microfilaments, structure, 66f
Microtubules
cytoskeletal element, 65

Microtubules *(cont.)*
role in fast anterograde transport, 63–64
and role of kinesin, 64f
structure, 66f
Midbrain, 10, 79
Middle temporal area
and perception of motion direction, 449
representation of motion in, 447–449
Migration
growth cone guidance of developing axon to target, 101
nerve cell
along glial cell fibers, 100f, 101
and establishing basic plan of central nervous system, 99
neural crest cells during development, 97
Mitchondria, in nerve cells, 55
Modality
sensory, 371
stimulus, and labeled line code, 380
Modeling, simulation of cognitive function with parallel processing, 359–362
Monoamine oxidase inhibitors, 302
Monocular zone, 427
Monosynaptic circuit, controlling knee-jerk stretch reflex, 46f
Monosynaptic circuits, 29–30
Mood, effect of sodium amytal injection on, 358
Mossy fibers
afferent input to cerebellar cortex, 542
afferent pathway in hippocampus, 680–684, 680f
role in learning, 543–544
Motion
agnosia case, 399b
detection of and magnocellular-thick stripes pathway, 396
direction, perception influenced by cells in middle temporal area, 449
global, method of computation, 448
information analyzed in two stages, 448
perception, and equiluminance, 398
representation in middle temporal and medial superior temporal areas, 447–449
visual perception of, 425–451; see also Magnocellular system
Motivational states
correspondence of homeostatic processes to, 614–615
functions, 614
as inferred internal states, 614
regulation by factors other than tissue needs, 624–625
Motivational system
and regulation of somatic and autonomic motor systems, 83–84
relay centers, 84
role in physical actions, 86f
distinct pathways, 85
Motor areas, see also Posterior parietal cortex; Prefrontal cortex; Premotor cortex; Primary motor cortex; Somatic sensory cortex; Supplementary motor area
cerebral cortex, 82f, 83
corticocortical connections, 534f
major, cerebral cortex, 348–350, 349t,f
Motor behaviors, types, 488
Motor circuit, of basal ganglia, 547–548, 548f
Motor control, and use of sensory informaton, 371
Motor coordination, 515
Motor cortex
control of movement achieved late in phylogeny, 496, 498
descending pathways, 497f
direct action on spinal motor neurons via the corticospinal tract, 496–498
indirect action on spinal motor neurons via brain stem pathways, 496–498
see also Posterior parietal cortex; Prefrontal cortex; Premotor cortex; Primary motor cortex; Somatic sensory cortex; Supplementary motor area
somatotopic arrangement, 498
Motor defects, produced by mutation in single gene, 561–562
Motor equivalence, 490, 490f
Motor function, 485–550
role of cerebellum and basal ganglia, 499
Motor learning, role of cerebellum, 543–544, 543f
Motor map, 498, 499f
in cereral cortex, 86–87
Motor molecules, see Dynein; Kinesin
Motor neurons, 27
alpha, innervation of extrafusal fibers, 510
alpha-gamma coactivation, 512f
as final common path, 493
axon, 48–49
branches, 50f
route from spinal cord, 50
cell bodies in spinal cord, 493–494
connections with sensory neurons, 47f
dendrites, 48
developing, selection of peripheral pathways by axons of, 102, 104
EPSP in, 221
gamma, control sensitivity of muscle spindles, 510–512, 511f
grouping in spinal cord, 494f
inductive interactions with muscle cell to form synapse at neuromuscular junction, 104
and interneurons, modulation by brain stem, 494–495
mediation of action potential, 222
and muscle fiber as motor unit, 501–502
myelin covering of axon, 51–53
myelin sheath of axon, 49f, 52f
and neuromuscular junction, 50
one innervates many muscle fibers, 504f
pools, spinal cord, 493
projection of primary sensory afferents from embryonic spinal cord, 41–42f
rate of firing and active muscle tension, 502f
recurrent branches of axon, 49, 50f
regions of termination in spinal cord according to species, 498f
role of cell body, 53–55
signals received by, 49–50
spinal, 46f
synthesis of macromolecules, 54f
structure, 48–51
synaptic inputs, 50
Motor processing, and internal representation, 490
Motor sequences, programmed by supplementary motor area, 534–536
Motor strategies and planning, role of prefrontal association cortex, 351–353
Motor systems, 85f, see Autonomic nervous system; Somatic nervous system; Voluntary nervous system
autonomic and somatic, comparison, 598f
distinct pathways, 85
disturbance by disease, 488
hierarchical and parallel arrangement, 492–493, 492f
redundancy in, 500
relay centers, 84
role in physical actions, 86f
role of cerebellum as comparator, 536
somatotopic maps in each component, 493

somatic and autonomic, regulated by motivational system, 83–84
Motor tasks, lesions of principal sulcus interfere with, 351–353
Motor unit, 501–502
increased firing rate and increased force output, 502–503
recruitment, 503–505, 503f, 505f
types, 503, 504f
Mounting
behavior, and androgenized brain, 587
sexually dimorphic reproductive behavior, 582
Mouse
manipulation of gene expression in to study behavior, 566–567b
targeted mutations in for study of behavior, 568–570b
Movement, 489–500
agnosia, 398, 399b
bilateral, programmed by supplementary motor area, 534–536
categories of, 491–492
complex sequences, programmed by supplementary motor area, 534–536
consequences of reported to neurons of primary motor cortex, 533–534
eye, role of basal ganglia, 547–549
fine, control by brain stem lateral pathways, 496
limb, role of basal ganglia, 547–549
perception of in visual field, 446–449, 446f
proximal, control by premotor cortex, 536
psychophysics of, 490–491
skilled, and somatic sensory cortex, 339
voluntary, 529–550
force and direction of encoded by nerve cells of primary motor cortex, 530–534
indirect regulation by cerebellum, 536–544
preparation for, neurons active in premotor cortex, 537f
regulation by motivational system, 83–84
role of basal ganglia in planning and triggering, 549
role of cerebellum in initiation, 540f
role of premotor areas of cerebral cortex, 534–536
walking, control by spinal cord, 523–526, 525f
Muller-Lyer illusion, 391f
Mullerian duct-inhibiting substance, and gondal development, 580
Multiple sclerosis, demyelinating disease, 52, 157
Multipolar nerve cells, 26, 27
Muscarinic acetylcholine receptor, 198
Muscimol, effect on synaptic transmission in somatic sensory cortex, 339f
Muscle
alternating control, 491
contraction, grading of force by nervous system, 502–505
distal, control by brain stem lateral pathways, 495f
Muscle cell
end-plate potential, 200
inductive interactions with motor neuron to form synapse at neuromuscular junction, 104
in neuromuscular junction, 198–200
passive membrane properties and time course of end-plate potential, 215f
Muscle fibers
ACh receptors on surface, 104, 106
innervation ratio, 502
many innervated by one motor neuron, 504f
Muscle spindle
excitation initiates stretch reflex, 517f
firing properties, 509–510, 509f
functional differences from Golgi tendon organs, 509
as muscle receptor, 506, 507f, 508f
sensitivity controlled by CNS through gamma motor neurons, 510–512, 511f
stretch receptor, 48, 48f
Muscle tension, and rate of firing of motor neurons, 502f
Muscle tone, regulation by stretch reflex, 518
Muscle twitch, 502
Muscles, 501–515
action around joint coordinated by inhibitory interneurons, 518–520, 519f
anatomical components of stretch reflex, 29–31
antagonist
co-contraction, 519
inhibited by group Ia inhibitory interneurons, 519–520
axial and proximal, controlled by brain stem medial pathway, 494
axial, proximal, and distal, 493
complex reflexes, initiated by stimulation of skin, 520–523
connection with dorsal root ganglion fiber, 47
direction of movement controlled by nerve cells of primary motor cortex, 530–534, 532f
flexion withdrawal, 521
force exerted by controlled by firing rate of neurons in primary motor cortex, 530–534, 531f
innervation and change in ACh receptors, 106
primary sensory nerve endings in, 48f
proximal and distal, method of control by supplementary motor area, 534
receptors, 501–515
sets, 491
skeletal, structure of muscle spindles and Golgi tendon organs, 507f
specialized receptors, 506–510
stretch, phases, 509–510
tension, negative feedback system for, 520f
Mutations
behavioral, 556–557b
cAMP and learning in *Drosophila,* 677
per gene, *Drosophila,* 571–576, 573f
reeler, mouse, 100
sevenless, Drosophila, 91–94
shiverer, mouse, 52–53, 53f
targeted, 568–570b
types, 560
Myasthenia gravis, 307–315
Myelin, 25, 48
covers axons of sensory and motor neurons, 51–53
Myelin basic protein
immune response elicited by, 51–52
transfection of gene to partially cure *shiverer* mutant mice, 53f
Myelin sheath
development in peripheral nervous system, 52f
diseases affecting, 51–53
of motor and sensory neuron axons, 52f
Myelin-associated glycoprotein, expressed by Schwann cells, 51
Myelination
affect on action potential, 158f
effect on conduction velocity, 156–159
genetic disorder of in mice, 53f
Myotactic unit, reflex pathways controlling group of muscles around joint, 519

N

N-ethyl-malemide sensitive fusion protein, role in synaptic vesicle fusion, 282, 284

N-methyl-D-aspartate, role in long-term potentiation, 681, 682f, 683; see also NMDA
Neglect syndrome, 340
  visual, 341–342f
Neocortex, projection of olfactory pathways to, 384
Neostigmine, compensation for reduced ACh activity in myasthenia gravis, 309f
Neostriatum
  of basal ganglia, 544
  receives cerebral cortex and thalamus inputs, 545
Nernst potential, 137
Nerve cells
  active transport in, 62–65
  and behavior, 21–40
  basic mechanism of action potential same in all, 170
  basic plan, 43
  bilayer structure of cell membrane, 116f
  body, inhibitory synapses on, 224–226
  calculation of membrane potential using equivalent circuit model, 142–144
  capacitance, 143–144
  central
    features of different types of electrical signals, 240t
    integration of excitatory and inhibitory signals into single response, 222–223
    receptive fields, 335–339
    synapses onto grouped according to function, 224–226
  chromosome function in, 54–55
  classes of proteins synthesized, 58–59
  complexity of interconnection coveys unique information, 39–40
  cytoskeleton, 65–67
  differentiation controlled by diffusible growth factors, 95–99
  establishing connections between, 89–91
  functional, properties represented as electrical equivalent circuit, 140–144
  importance of calcium in regulation of function, 290–291
  in neuromuscular junction, 198–200
  inhibitory and neurotransmitter release, 289f
  interaction of excitatory and inhibitory potentials, 226f
  leakiness of, 159, 163
  local signaling and passive electrical properties of, 149–159
  mechanisms for releasing substances, 281–282
  membrane systems within, 58, 58f, 62f
  mitochondrial function in, 55
  molecular diversity, 38–39
  as neuro-endocrine transducers, 602
  neuronal versus hormonal transmission, 192–193
  organization of signaling in, 31–38
  passive electrical properties. See Capacitance; Conductance; Electromotive force properties. See Resting membrane resistance; Membrane capacitance; Intracellular axial resistance; Extracellular axial resistance
  proteins of, 60b
  reconstruction of action potential from known electrical properties, 168–170
  regions of synaptic contact, 225f
  see also Receptive fields
  set-related, 536
  sexual differentiation and structure of certain, 588–589, 589f
  shape due to fibrillar proteins, 65–67
  as signaling units for behavioral responses, 29–31
  signaling between, 91–94, 133, 179–315
  signaling within, 112–177
  storage of biochemical information in, 288
  structure, 22, 23–27
  survival during development, 107
  trigger zone, 35
  types, 25–27
  variation in excitability properties, 171
Nerve growth factor
  and development of sympathetic ganglia, 109f
  role in neuron survival during development, 107
Network properties, computer analysis of in cognitive neural science, 323–324
Neural circuits
  development altered by early sensory deprivation, 470–471
  proposed for emotion, 607f
Neural crest cells
  differentiation, 95–99, 98f
  migration during development, 97
Neural encoding, sensory stimulus, 375
Neural induction, and organizer region, 96f
Neural maps
  auditory, 86–87
  body surface, 689f
  in cerebral cortex, 86–87
  feature, 402, 403f
  of receptive sheet, conveyed by sensory pathways to brain, 374f
  retinal, in lateral geniculate nucleus, 430
  somatotopic, 493
Neural networks, artificial
  simulation of cognitive function with, 359–362
  types, 361f
Neural pathways, central nervous system, cross body's midline, 87
Neural plate, induced by mesodermal signals, 94–95
Neural representation, of hand in somatic sensory cortex, 326f
Neural tube. See also Neural crest cells
  development of human nervous system from, 78f
  migration of neural crest cells from during development, 97
Neuroactive peptides, 298–301, 298t
  in brain, 298t
  co-release with small-molecule neurotransmitters in same neuron, 301–302
  families, 299t
  processing of from polyprotein, 299–301
  slow removal of from synaptic cleft, 302–303
Neuroanatomy, modern methods, 88
Neuroblasts, migration of to form basic plan of central nervous system, 99
Neuroendocrine cells, secretion by, 603f
Neurofilaments
  cytoskeletal element, 65, 67
  structure, 66f
Neurohypophyseal system, 603f
Neurology, behavioral, study of in cognitive neural science, 323
Neuromodulators, 623
Neuromuscular junction, 50, see also Acetylcholine receptors; End-plate current; End-plate potential
  effect of myasthenia gravis on, 307–315
  failure of synaptic transmission at in myasthenia gravis, 311f
  formation of receptor zone at, 104
  recyling of synaptic vesicles at, 285–286, 286f
  removal of neurotransmitter from to terminate synaptic transmission, 302–303
  structure, 198, 198f
  study of directed gated synaptic transmission at, 198–200
  synapse formation at, 104–107

synaptic
excitation at involves transmitter-gated ion channels, 200–202
transmission at, 197–217
topography of neurotransmitter release site at, 278f
transmitter-gated versus voltage-gated ion channels, 210–212
Neurons, 45–55. See also Nerve cell
cell body, 45
cerebral cortical, birthday defines eventual position and properties, 99–101
dendrites, 45
diversity, 45–47
doctrine, 25
excitability, 45
mediate stretch reflex, 47–51
motor, 46f
proteins synthesized by, 57–69
reflex, 47–51
sensory, 46f
sensory, transformation of physical stimulus into electrical activity, 34
Neuronal connectivity, and higher brain function, 87–88
Neuronal integration, 222
Neuronal precursors, see Neuroblasts
Neuropeptide Y, effect on feeding behavior, 622
Neurosecretion, 602
Neurotoxins, for study of ion channels, 172
Neurotransmitter, 293–306
binding to postsynaptic receptors, at chemical synapse, 192–193
calcium required for synaptic release, 290–291
calculation of probability of release, 277b
chemical
molecular diversity, 39
can produce synaptic action with different time courses, 262f
released by output component of signal, 36
requirements for, 293–294
discharged from synaptic vesicle by exocytosis, 276, 279–282
each quantum stored in synaptic vesicle, 276
effect of presynaptic inhibition and facilitation on amount released, 288–290
effect on feeding behavior by stimulation of hypothalamus, 622
and excitatory synaptic action, 256f
exocytosis at presynaptic terminal, 280f
histochemical detection, 304–305b
mechanisms of action on ion channels, 193f
receptors, direct versus indirect gating of ion channels, 243–245, 244f
release of, 269–291
inhibited or facilitated by presynaptic contact at axon terminal of another cell, 289f
not controlled by sodium influx or potassium efflux, 269–271
presynaptic, increase of and maintenance of long-term potentiation, 684f
proportional to calcium influx into presynaptic terminal, 273f
in quantal units, 273–276, 275f
triggered by calcium influx, 271–273
removal from synaptic cleft terminates synaptic transmission, 302–303
reuptake and inactivation, 303
role of voltage-gated sodium channels in release, 270f
see also Acetylcholine; Gamma-aminobutyric acid; Glutamate; Glycine; Synaptic vesicles
small-molecule, 294–298, 294t
co-release with neuropeptides in same neuron, 301–302
compared with peptides, 301
transport into synaptic vesicles, 297–298
therapeutic adminstration of in Parkinson's disease, 547
topography of release site at neuromuscular junction, 278f
vesicles, number release by action potential modulated by calcium influx, 286–290
Neurotrophic factors
and programmed cell death, 107
see also specific names; Ciliary neurotrophic factor; Nerve growth factor; Neurotrophin 3
Neurotrophin 3, role in neuron survival during development, 107
Neurotropins, effect on voltage-gated versus transmitter-gated channels, 211–212
NGF, see Nerve growth factor
Nicotine, effect on rate of intracranial self-stimulation, 626f
Night vision, rods specialized for, 409
Nissl substance, 46f
Nitric oxide
second-messenger pathway, 254–255
as second messenger in synaptic transmission, 245
NMDA receptor–channel
contribution to late phase of EPSP, 231, 231f
current flow through, 230f
as doubly gated channel in hippocampus, 681
properties, 227–232
Nociception, 324
Nociceptors, 384
Nodes of Ranvier, 25, 49f
boost amplitude of action potential, 157
effect on action potential, 158f
Nonpyramidal cells, primary visual cortex, 433
Nonsense words, processed differently from meaningful words, 644
Norepinephrine
amount in postsynaptic target regulated by cholinergic presynaptic neuron, 262
effect on excitatory synaptic action, 257f
as neurotransmitter, 295–296
Noxious stimuli, and flexion withdrawal, 521
NPY, see Neuropeptide Y
NSF, see N-ethyl-malemide sensitive fusion protein
Nucleus of solitary tract, projection of hypothalamus to, 600, 601f

O

Occipital cortex, left, effect of lesions on reading ability, 646
Occipital lobe, cerebral cortex, 81, 82f
function, 9
Ocular dominance columns
development, 480–481f
effect of vision deprivation during later phase of development, 477f
effect of cooperation on development, 478–481
effect of eye closure on development, 476f
established by competition between afferent fibers of two eyes, 475f
experimental induction, 478f
as model for fine tuning of visual circuitry by activity, 471–481
in primary visual cortex, 441–442, 442f
postnatal development of in visual cortex, 474f
segregation of afferent inputs, 474–478
visual sensory deprivation alters size, 473f

Ohm's law, 203
Olfactory bulb, 384
Olfactory system, and population codes, 384–385
Oligodendrocytes, 27–28, 51
  development in rat optic nerve, 99
Ommatidia, *Drosophila* compound eye, 93f
Opsin, 411, 413f
Opsin kinase, role in light response, 414
Optic chiasm, 428
Optic disc, blind spot, 427
Optic nerve, 393, 428
  experiments establishing methods of retinal axon guidance, 101–102
  rat, development of glial cells, 99
Optic tracts, 428
Orbitofrontal cortex
  subarea of limbic association cortex, 354
  subdivision of limbic association cortex, 351
Organelles, role of kinesin in transport, 64f
Organizer, of Spemann
  and development of nervous system, 95
  and neural induction, 96f
Orientation, of visual image in visual field, primary visual cortex, 437f
Orientation columns, in primary visual cortex, 441f
  horizontal connections, 443, 444f
  structure, 440–444
Oscillators, and control of biolgoical clock mechanisms, 624–625
Osmoreceptors, in hypothalamus, 623
Output signal, 31

P

P cells, retinal ganglion cell type, 420
Pain, see also Nociception
  as body sense, 384
Pallidum, see Globus pallidus
Parahippocampal gyrus, 607
Parallel distributed processing, simulation of cognitive function with, 359–362
Parallel processing
  binding problem, 387
  in learning, 665–666
  motor system, 492–493
  organizational principle of brain, 8–9
  retinal ganglion cells, 419–420
  of sensory information, 367–368
  in visual system, 393–402, 393f
Parasympathetic division, of autonomic nervous system, 598–600, 599f
Parietal association areas, role in higher sensory functions and language, 354
Parietal lobe, cerebral cortex, 81, 82f
  function, 9
  lesions result in aphasia and agnosia, 354
Parietal-temporal-occipital association cortex, 83
  functional areas, 350
  lesions in angular or supramarginal gyrus affect reading and writing, 646
Pars compacta, of basal ganglia, 544
Pars reticula, of basal ganglia, 544
Parvocellular
  blob pathway, and color perception, 395
  cells, retina, projections, 394
  interblob pathway, and shape perception, 395
  layers, of lateral geniculate nucleus, 430
  pathway, 432
  system, study of with equiluminant stimuli, 396–398
Parvocellular-blob
  pathway, information carried by, 397f
  system, processing of color information, 462
Parvocellular-interblob
  pathway, information carried by, 397f
  system, perception of form, 445
Patch-clamp experiments
  demonstrating all-or-none behavior of voltage-gated ion channels, 172
  single-channel recording, 120b, 121
  to study current flow through single ion channels, 204–208
Pattern direction-selective neurons, 448
Peptides
  compared with small-molecule neurotransmitters, 301
  neuroactive, 298–301, 298t
    in brain, 298t
    families, 299t
    processing of from polyprotein, 299–301
Perception, 365–484. See Visual perception
  effect on behavior, 323
  integration of as multistage process, 404
  Kant's idealism versus experimental pragmatism, 370
  selective, 381
  spatial, role of posterior parietal cortex, 404
  and use of sensory information, 371
Perceptions, internal construction, 370
Perforant pathway, afferent pathway in hippocampus, 680–684, 680f
period gene, *Drosophila*
  and control of biological rhythms, 571–576, 573f
  cycling of RNA and protein products, 575f
  sites of expression, 576f
Peripheral nerves, of peripheral nervous system, 77
Peripheral nervous system, 77
  axon from, 49f
  development of myelin sheath, 52f
  rearrangement of synaptic connections during early postnatal life, 106f
Personal space
  integration of information from mechanoreceptors by somatic sensory cortex, 339
  internal representation
    is modifiable by experience, 329–334
    study at cellular level, 335–339
  orderly representation of in brain, 324–328
  real, imagined, and remembered, representation of in posterior parietal association areas, 339–344
Personality, correlation between monozygotic and dizgotic twins, 558f
Phantom limb syndrome, and cortical representation, 333–334
Phasic control, muscle, 491
Phosphatidyl inositol, and second-messenger cascades, 250
Phospholipids
  in cell membrane, 116f
  hydrolysis of generates second messengers, 252f
  second messengers generated by hydrolysis of, 250–254
Phosphorylation
  cyclic AMP-dependent protein, 250
  and modulation of receptors that gate ion channels directly, 261
  protein, 246
    action of second messengers through to open or close ion channels, 257–259
    and transmitter receptor densensitization, 260–261
  sequence, 250

Photoreceptors
in compound of eye of *Drosophila*, identity, 91–94
differentiation in *Drosophila* eye, 94f
genetic defects in as cause for color blindness, 464, 465–467b
interneurons relay signals to ganglion cells, 420–423
ion channels in, 410–411
nuclear layer of retina, 417f
phototransduction in, 410–414
retinal, 408
slow adaptation to changes in light intensity, 414
structure, 411f
two types, 409–410; see also Cones; Rods
at least two types required for color discrimination, 456–458
visual pigments in, 410, 411–412
Phototransduction, as result of cascade of biochemical events in photoreceptors, 410–414
Physiology, regulation by autonomic motor system, 84
Pigment epithelium, of eye, 408
Pigments, cone, 457b
Pituitary, hypothalamic control, 603–605, 604f, 604t
Place conditioning, role of amygdala, 610
Place preference, role of amygdala, 610
Plantar reflex, 521
Plasticity, synaptic, 694
chemical synpases, 184
hippocampal neurons, 680
processes that modify, 286–287
Pleasure, and motivation of behavior, 625
Plexiform layers, outer and inner, of retina, 417f
Plexus, chemical cues guide motor axons to targets in, 105f
Polaroid camera, retinex method of color prediction, 463–464
Polymerization, protein, dynamics of, 68b
Polymorphism, in red pigment gene, 560
Polymorphisms, genetic, 562–563b
Polyprotein, processing of functional neuroactive peptides from, 299–301
Polyribosomes, role in protein synthesis, 59
Pons, 10, 79
Population codes
covey informaton about stimulus strength, 377–378
role in coding stimulus quality in taste and smell, 384–385
Population vector, of directionally tuned neurons in primary motor cortex, 532–533, 533f
Positional invariance, 436
Positron emission tomography
identification regions of cerebral cortex involved in recognition of written or spoken word, 17
study of brain function, 16
use to determine structure of nervous system, 71
Posterior parietal cortex
association cortex
effects of lesions in on attention, 344
functions with dorsolateral prefrontal association cortex, 352–353
representation of real, imagined, and remembered space, 339–344
effects of lesions in on memory, 343f
integration of somatic sensory input with visual and auditory input, 340
lesions of and neglect syndrome, 340, 341–343f
location of somatic sensory cortex, 324–325f
role in spatial perception, 404
Postinhibitory rebound, 523
Posttetanic potentiation, effect on synaptic effectiveness, 287, 288f
Postural adjustments, as category of movement, 491
Posture
indirect regulation by cerebellum, 536–544
maintained by medial pathways of brain stem, 494
maintenance by crossed extension reflex, 521–523
role of basal ganglia in adjustments, 549
Potassium, transmitter-gated channels selective for mediate EPSP, 227–232
Potassium ion channel
fast transient, 170
M-type, 170
passive, as battery in series with conductor, 142f
represented as conductor or resistor, 141f
voltage-gated
isolation of current through, 163–166
role in action potential, 162
sequential opening and action potential, 168f, 169–170
molecular structure of alpha subunit, 174f
Potassium ions
channels at muscle end-plate permeable to, 203–204
flux across glial cell membrane, 136f
movement through ion channel, 117–118
passive flux balanced by active pumping of ions, 138–139
resting ion channels selective for in glial cells, 136–137
role in end-plate potential, 203f
voltage-gated channel selective for, 117f
Potential. See also Action potential; End-plate potential; Equilibrium potential; Membrane potential; Receptor potential; Reversal potential
inhibitory, in central neurons, 232–234
postsynaptic
at end-plate, amplitude, 202f
fixed size according to quantol units of neurotransmitters released, 273
as local signal, 33
spatial decay, 223f
summation, 222
temporal and spatial summation by central neuron, 224f
Potentiation, effect on synaptic effectiveness, 287, 288f
Pre-attentive process, 401–402, 401f
Precursor cells, neural, 90
Precursors, hormone and neuropeptide, 300f
Prefrontal association cortex, 83, 351f
dorsolateral, functions with posterior parietal association cortex, 352–353
input from higher-order sensory cortices, 348
role in motor strategies and motor planning, 351–353
Prefrontal lobes, see also Principal sulcus
lobotomy, 611
size in schizophrenia, 353
Prefrontal convexity, inferior, lesions in interfere with appropriate motor response, 353
Preknowledge, 370
Premotor cortex
control of proximal movements projecting arm to target, 536
input from higher-order sensory cortices, 348
neurons active during preparation for voluntary movement, 537f
role in voluntary movement, 534
Preoptic area, sexually dimorphic nucleus of, 588–589, 589f
Presynaptic cell, 25
Presynaptic facilitation, see Facilitation, presynaptic
Presynaptic inhibition, see Inhibition, presynaptic
Presynaptic terminal, 25
effects of long-term sensitization and habituation, 676f
exocytosis at, 280f
regulation of calcium concentration at, 287–288

Presynaptic terminal *(cont.)*
- and release of neurotransmitters, 269–291
- synaptic connections on regulate intracellular free calcium, 288–290

Pretectum, projection of retina to, 429f
Primary auditory cortex, cerebral cortex, 83
Primary motor cortex, 493
- broad tuning of neurons to direction of movmement, 532f
- cerebral cortex, 83
- firing rate of neurons controls force exerted by innervated muscle, 531f
- nerve cells in encode force and direction of voluntary movements, 530–534
- neurons in informed of consequences of movements, 533–534
- population vector of directionally tuned neurons, 532–533, 533f

Primary sensory neuron, 372f
- as sensory receptor, 373
- receptor potential, 377f

Primary somatic sensory cortex, 324–328
- body surface represented by more than one sensory map in, 689f
- cerebral cortex, 83

Primary visual cortex
- analysis of form in, 433–440
- blob regions, 396f
- blobs, 442f
- cerebral cortex, 83
- end-stopping of simple and complex cells, 438–440
- horizontal connections between orientation columns, 443, 444f
- input from two eyes, 471f
- layers, 432f
- ocular dominance columns, 441–442, 442f
- organization, 440–444
- orientation columns, 441f
- processing of color information in, 462
- progressive convergence in and extraction of linear features of visual image, 440
- projection of retina to, 429f
- receptive fields of complex cells of, 436
- rectilinear receptive fields of simple cells of, 434–436, 434f, 435f
- specific orientation of image in visual field, 437f
- staining regions, 395
- visual fields in, 431f

Priming, 654
Principal sulcus
- firing of neuron in during delayed-response task, 353f
- lesions of interfere with specific motor tasks, 351–353

Programmed cell death, 107
Propagation, action potential, velocity affected by passive membrane properties and axon diameter, 156–159
Proprioception, 324
- information, monitored by muscle receptors, 501

Prosopagnosia, 398, 400b
Protective reflexes, 521
Protein kinase, 246
- calcium-calmodulin-dependent, 252f

Protein kinases
- characteristics, 251f
- protein kinase C, 251–253
- serine and threonine, 250

Protein phosphorylation, 246
- action of second messengers through to open or close ion channels, 257–258
- and transmitter receptor desensitization, 260–261

Protein synthesis
- mechanism, 58–59
- new, requirement for long-term memory, 671–676

Proteins
- gap-junction channels, 189
- integral, 59
- membrane-spanning
  - deducing secondary structure from amino acid sequence, 122–123f
  - membrane-spanning, receptors as, 193

Proximal-distal rule, motor neurons, 493
Proximity, and visual perception, 389
Pseudo-unipolar nerve cells, 26, 27
Pseudohermaphroditism, caused by fetal exposure of genetic females to male hormones, 583
Psychiatric disorders, insights provided by neuronal changes associated with learning, 690–692
Psychoactive drugs, and increased level of dopamine in neurons of ventral tegmental area, 626
Psychology, experimental, 369
Psychophysics
- of movement, 490–491
- and study of stimulus and perception, 370

Psychotherapeutic techniques, use of conditioning, 663
Psychotic illnesses, heritable susceptibility to, 691
Purkinje cells, cerebellar cortex, 541–543
Putamen, of basal ganglia, 544
Pyramidal cells, 26
- primary visual cortex, 433

Pyramidal decussation, 496

Q

Quantam, unit of neurotransmitter release, 273–276

R

Rab proteins, role in targeting synaptic vesicles to docking sites, 282
Reading disorders, 644–648
Receptive field, 335
- with axes of orientation, in primary visual cortex, 440
- of complex cells of primary visual cortex, 436
- concentric double-opponent cells, 462
- of end-stopped cells in primary visual cortex, 438–439f, 439
- ganglion cells, 461–419
- higher-order sensory neuron, 381f
- enlargement with successive stages of information processing, 339
- linear features of in cells of primary visual cortex, 433–440
- rectilinear in simple cells of primary visual cortex, 434–436, 434f
- of retinal bipolar cells, 420, 421–422b
- and sensory neurons in CNS, 374f
- sensory receptors and sensory neurons in CNS, 373–375
- size according to covergent input, 338f

Receptive sheet, 374f; see also Retina; Skin
Receptor ending, in sensory system, 372f
Receptor potential
- characteristics, 377f
- as local signal, 33
- stimulus, 375

Receptor–channels, see Ion channels
Receptors
- acetylcholine, 209–210
  - antibodies to as cause of myasthenia gravis, 310–314
  - change in functional properties after innervation, 106

changes in after muscle innervation, 104, 106
complexed with ion channel, electron microscopic image, 200f
deducing secondary structure from amino acid sequence, 122–123f
density of in muscle fibers in myasthenia gravis, 309f
and ion conductance, 118
labeling of, 199f
main immunogenic region, 312
on surface of muscle fibers, 104, 106
rate of turnover in myasthenia gravis, 313f
types, 198
G-protein-coupled, 243–244
structure, 246f
estrogen, and development of female brain, 584–585
GABA, as transmembrane proteins, 236–237
glutamate, 227–232
as transmemebrane proteins, 236–237
classes of, 229f
glycine, as transmembrane proteins, 236–237
gonadal steroid, location in nerve cell, 585–586, 585f
hormone, as transcriptional regulators, 586–587
idea of, 192–193
in somatic sensory syste, 383–384
ionophoric, 182
metabotropic, 182
modulation by phosphorylation, 261
molecular diversity, 39
muscle, 501–515
and neural cell differentiation, 90
neurotransmitter, direct versus indirect gating of ion channels, 243–245, 244f
nicotinic acetylcholine
antibodies against produced in myasthenia gravis, 307
messenger RNA sequence of alpha-subunit, 111–112f
NMDA, properties, 227–232
postsynaptic
ACh, 201f
at chemical synapse, 191f
binding of neurotransmitters to at chemical synapse, 192–193
gating of ion channels, 193–194
ionophoric, 193
metabotropic, 194
nerve-muscle synapse, and ACh, 200
sensory, 372
action potentials generated by, 376–377
adaptation of, 378–379
and receptive field, 373–375
specificity and stimulus line code, 380
and tuning curve, 373, 374f
types, 373
specialized in muscles, 506–510
transmitter, desensitization, 260–261
tyrosine kinase, 344
tyrosine kinase family, 194
Reciprocal inhibition, in sensory systems, 381
Recuitment, and force of muscle contraction, 502
Recurrent network, 360
reeler mutation, mouse, 100
Referred sensations, remapping in somatic sensory cortex, 334, 335f
Reflex, 515
as category of movement, 491
crossed extension, and maintenance of posture, 521–523
protective, 521
spinal, 515–527; see Spinal reflexes
scratch, 521–523
stretch, 29–31, 516–518
anatomical components, 29–31
combination of excitatory and inhibitory synaptic connections mediating in central nervous system, 220f
components, 515
inhibitory neurons in, 38
initiated by excitation of muscle spindles, 517f
as monosynaptic circuit, 516–518
neural circuit for, 516–518
neurons mediating, 47–51
and regulation of muscle tone, 518
synapses mediating, 221
testing strength of, 518
Reflex action, sequence of signals that produces, 37
Reflex systems, and implicit memory, 658
Regeneration
axonal, and chemical cues, 102
axons of retinal ganglion neurons, 103f
Relay centers, of sensory, motor, and motivational systems, 84
Relay nuclei, 84, 601f, see also Interneurons
sensory, types of inhibition, 383f
transmission of sensory information, 374
Representation, see Internal representation
Reproductive behavior, 579–593
sexually dimorphic nature, 581–582
Resistance, membrane
axial, calculation, 153–154
and axoplasmic, affect efficiency of signal conduction, 153–156
effect on magnitude of electrical signals, 150
Resistor
behavior of ion channel as, 119, 121
potassium channel represented as, 141f
Response, conditioned versus unconditioned, 659
Rest and digest reaction, 598
Retention, of visual task, role of inferior temporal region, 354
Reticular formation
of brain stem, 79
effect on output of autonomic nervous syste, 600
role in sensory regulation of arousal, 371
Reticulospinal tract, of brain stem medial pathway, 494, 495f
Retina, see Photoreceptors; Visual perception
Retinal, 411, 413f
Retinex method, and color prediction with Polaroid camera, 463–464
Retinotopy, 375
Reversal potential
for EPSP, 228–229f
of excitatory postsynaptic potential, calculation, 206b
measurement for EPSP, 227
of single-channel current, 206f
activated by glutamate and GABA, 235f
for synaptic action at muscle membrane, 203
Rhodopsin
structure, 413f
visual pigment in rod cells, 411
Rhombomeres, 95
Rhythmic
motor patterns, as category of movement, 491
movements, spinal reflexes consisting of, 521–523
pattern generation, in newborn humans, 527
Rhythms, biological, and oscillations in gene expression, 571–576
Ribosomes, role in protein synthesis, 59f
RNA, messenger, role in protein synthesis, 58
Rods, 409–410
differences from cones, 409t
structure, 411f

Rostral vental medulla, projection of hypothalamus to, 602
Round worm, see *Caenorhabditis elegans*
Rubrospinal tract, of brain stem lateral pathway, 495f

S

Saccades, 446f
Saliency map, in brain, 402
Saltatory conduction, of action potential in myelinated axon, 157
Satiation, 614
Saxitoxin, for study of ion channels, 172
Scanning, 446f
Schaffer collaterals, afferent pathway in hippocampus, 680–684, 680f
Schizophrenia
  lifetime risks of developing, 564f
  polygenic nature, 564–565
  prefrontal lobe size in, 353
Schwann cells, 27–28, 49f
  along axons, 51
Scorpion venom, for study of ion channels, 172
Scratch reflex, 521–523, 523f
Sculpturing, effect of synaptic inhibition, 221f
Second messengers
  action through protein phosphorylation to open or close ion channels, 257–258
  direct action on ion channels, 258–259
  effects on gene expression, 261–265
  and excitatory synaptic action, 256f
  generated by hydrolysis of phospholipids, 250–254, 252f
  long-lasting consequences on synaptic transmission, 261–263
  pathways
    interaction, 255–257
    molecular logic, 245–255
    neuronal, cyclic AMP pathway as typical, 249f
    specific, 254–255
  role in phototransduction, 410–414; see also Cyclic guanosine 3',5'-monophosphate
  substances acting as in synaptic transmission, 244–245
  synaptic, pathways, 245f
  system
    affects other types of receptors, 261f
    modulation of synaptic transmission, 243–267
  and transmitter receptor desensitization, 260–261
Second-messenger cascade, role in ion channel activity, 193f, 194
Second-messenger molecules, flow through gap-junction channels, 187
Secondary somatic sensory cortex, 324–325f
Segmental organization, of hindbrain and spinal cord in vertebrates, 95
Segregation, eye inputs to lateral geniculate nucleus, 478
Selective perception, 381
Self-stimulation
  intracranial, rat brain, 627f
  role of dopamine in regulation, 626
Sensation, 371
  body senses, 383–384
  map for each in cerebral cortex, 328–329
  thermal, 384
Sensitization, 659
  biochemical steps, 671
  and enhancement of synaptic transmission, 671
  long-term
    Aplysia, 673f
    and structural changes in presynaptic terminals of sensory neurons, 675–676, 676f
Sensory afferents, primary, projection from embryonic spinal cord to dendrites of motor neurons, 41–42f
Sensory areas
  cerebral cortex, 82f, 83
  primary and higher-order, cerebral cortex, 348–350, 349t,f
Sensory cortices, higher-order, input to premotor and prefrontal association cortex, 348
Sensory deficits, effects on food intake, 620–621
Sensory deprivation
  and alteration of structure of cerebral cortex, 471
  early
    alters development of neural circuits, 470–471
    effect on visual perception, 470
  visual, alters size of ocular dominance columns, 473f
Sensory experience, and formation of visual circuits, 469–484
Sensory functions, higher, role of parietal association areas, 354
Sensory information
  parallel processing of, 367–368
  processed by thalamus, 84
  uses of, 371
Sensory modalities, 371; see also Submodalities
Sensory nerve endings, primary, in muscle, 48f
Sensory neurons, 27
  changes in accompanying long-term memory for sensitization in Aplysia, 674–675f
  connections with motor neurons, 47f
  higher-order, receptive field, 381f
  intensity and frequency of discharge and intensity of stimulus, 378f
  long-term sensitization and habituation and structural changes in presynaptic terminals of, 675–676, 676f
  myelin sheath of axon, 51–53, 52f
  projections to motor neurons, 48
  and receptive field, 373–375
  role of cell body, 53–55
  structure, 47–48
  transformation of physical stimulus into electical activity, 34
Sensory stimulus, methods of study, 370
Sensory systems, 85f, 369–386
  common anatomical plan, 372–375
  distinct pathways, 85
  general organization, 372f
  mammalian, 373t
  parallel pathways, 375
  relay centers, 84
  role in physical actions, 86f
  sensitivity to differences in stimulus strength, 371–372
  types of inhibition in, 380–381
Serotonin
  as neurotransmitter, 295–296
  role in sensitization, 671
Servomechanisms, analysis of homeostatic processes in terms of, 614–615, 615f
Set point, and regulation of body weight, 618–620
Set-related neurons, 536
sevenless gene, and photoreceptor identity in *Drosophila* eye, 91–94
Sex differences
  brain
    cellular mechanisms, 589–590, 589f
    influence on behaviors, 590–592
  in nervous system, 588–589
Sex-determining region of Y gene, 580
Sex-linked behaviors, and perinatal hormonal exposure,

587–588
Sex-specific development, *Drosophila* brain, 572f
Sexual behavior, and testosterone replacement therapy, 587f
Sexual differentiation, 579–593
brain, 584–586
different physiological properties and behavioral tendencies, 586–590
of nervous system, and courtship behavior, 571
and structure of certain neurons, 588–589, 588f
Sexual dimorphism, 579
asymmetry of cerebral hemispheres in humans, 590–592
cognitive development in monkeys, 590
of reproductive behavior, 581–582
of secretions of gonads, 582
Shape perception, and parvocellular interblob pathway, 395
shiverer mutant, 52–53
Short circuit, in equivalent circuit model of nerve cell, 142–144
Signal, see also Action potential
amplification at chemical synapses, 192
attenuation, factors affecting, 153–156
between neighboring cells and control of neuronal identity, 91–94
between nerve cells, 179–315
chemical, common features of those produced by transmitter-gated and voltage-gated channels, 239
complexity of between nerve cells, 219–221
conduction, efficiency affected by membrane and axoplasmic resistance, 153–156
depolarizing, sent by electrical synapses, 184
electrical
and chemical in nerve cells, 133
features of different types in central neurons, 240t
local
versus propagated, 33–34
and passive electrical properties of the neuron, 149–159
mesodermal and induction of neural plate, 94–95
metabolic, transmitted by electric synapses, 187
molecules, role in nerve cell differentiation, 107–108
and neural cell differentiation, 90
organization of in nerve cells, 31–38
output component, and release of chemical transmitter, 36
propagated, 161–177
propagation and functional design of neuron, 159
see also Cell–cell interactions; Electrical signals; Ion channels; Synaptic integration
sequence to produce reflex action, 37
simplification by chemical synapses, 189–194
stereotyped throughout nervous system, 35–36
transcellular, 255f
transduction pathways, synaptic second messengers, 245f
transformation as it passes from one component to the next, 36–37
transmission, instantaneous in electrical synapses, 185–189
within nerve cells, 112–177
Similarity, and visual perception, 389
Simple cells, primary visual cortex
end-stopping of, 438–440
rectilinear receptive fields of, 434–436, 434f, 435f
Single-channel recording, 119–121, 119f, 120b
Size, judgment of, 392f
Skeletomotor system
role in temperature regulation, 616–618
see also Alpha motor neurons
Skilled movements, and somatic sensory cortex, 339
Skin
internal representation of stimulus on, 336f
stimulation of and initiation of complex muscle reflexes, 520–523
Slow-channel syndrome, 314–315
Small-molecule transmitters, 294–298, 294t
Smell
role of population codes in coding stimulus quality in, 384–385
as sensory modality, 371
Snake venoms, see Alpha-bungarotoxin
Social competence, 482–483
Sodium amytal, effect on speech and mood, 358
Sodium ion channel, transmitter-gated channels selective for mediate EPSP, 227–232
Sodium ion channel, voltage-gated
basis for selection of sodium ions, 173
close relationship with calcium channels, 173–177
distribution, 172
gating, 176f
inactivation after end of activating pulse, 169f
isolation of current through, 163–166
molecular structure, 173
of alpha subunit, 174f
redistribution of charge with open, 172–173
role
in action potential, 162
in depolarization and propagation of action potential, 155–156
sequential opening and action potential, 168f, 169–170
tertiary structure, 175f
Sodium ions
channels at muscle end-plate permeable to, 203–204
movement through ion channel, 117–118
nature of flux through at neuromuscular junction, 211, 212f
passive flux balanced by active pumping of ions, 138–139
role in end-plate potential, 203f
Sodium-potassium pump, 32
and maintenance of membrane potential, 139
modeled as electrical equivalent circuit, 143f
Solitary tract, nucleus of, projection of hypothalamus to, 600
Somatic division, of peripheral nervous system, 77
Somatic motor system, and voluntary movement, 83–84
Somatic sensory cortex, 324–328
in anterior parietal lobe, 354
cells with complex feature-detecting properties, 339
columnar organization of inputs by submodality, 336–339
convergent input and size of receptive fields, 338f
in different species, 329f
disruption of coordination by injection of muscimol, 339f
enlargement of representation by increased use, 331f
modification of representation of personal space in, 329–334
neural representation of specific areas of body, 326f
posterior parietal association areas, representation of real, imagined, and remembered personal space, 339–344
regions of, 330f
remapping of referred sensations, 334, 335f
representation of hand in determined by magnetoencephalography, 333f
and skilled movements, 339
Somatic sensory system, see also Body senses; Proprioception; Touch
and orderly representation of personal space, 324–328
maps, in cerebral cortex, 328–329
mediation of several different sensations, 383–384
neural architecture, 324–328, 324–325f
synaptic relays between sensory neurons and cerebral cortex, 324

Somatotopy, 375
  arrangement of motor cortex, 498
  map
    cellular mechanisms for changes, 688–690
    in each component of motor system, 493
  in motor portion of basal ganglia, 547–549
Spatial
  relationships, detection of and magnocellular-thick stripes pathway, 396
  resolution, provided by cones, 410
  summation, 155, 223
Specific membrane resistance, 150
Speed-accuracy trade-off, 491
Spike threshold, 34
Spinal cats, locomotion in, 524, 525f
Spinal cord, 10, see also Central nervous system; Climbing fibers; Mossy fibers; Spinal nerves
  anatomical regions of central nervous system, 79
  autonomic outflow, projection of hypothalamus to, 602
  circuits, as central pattern generators, 524
  columnar organization, 493
  connection with dorsal root ganglion fiber, 47
  control of walking movements, 523–526, 525f
  as control level of motor system, 492–493
  descending pathways of motor cortex to, 497f
  embryonic, projection of primary sensory afferents to dendrites of motor neurons, 41–42f
  glutamate as excitatory transmitter in, 227–232
  grouping of motor neurons, 494f
  motor neuron cell bodies in, 493–494
  principles governing organization, 84–88
  regions of termination of motor cortical neurons according to species, 498f
  role in motor function, 487–488
  segmentation, 79
  segmental organization of in vertebrates, 95
  sex difference in, 589
  and termination of corticospinal tract, 496
Spinal
  motor neurons, 26, 46f
  IPSP in, 232–234
  nerves, 79, see also Dorsal roots; Ventral roots
  reflexes, 29, 487, 493, 515–527
    consisting of rhythmic movements, 521–523
    monosynaptic, 516–518
    polysynaptic, 518
Spines, dendritic, 33
Spinocerebellum, functional component of cerebellum, 539–541, 539f
Split-brain studies, 355–357
Spotlight of attention, 402
Status epilepticus, role of glutamate toxicity, 232
Stellate cells, in primary visual cortex, 433
Stereognosia, 384
Stimulus
  capable of driving behavior, 615
  conditioned versus unconditioned, 659
  contrast enhanced by lateral inhibition, 382f
  discrimination of one from another, 380–383
  duration information encoded in discharge patterns of rapidly and slowly adapting receptors, 378–379
  energy, conversion into electrochemical energy, 375–376
  frequency and population codes convey information about strength, 377–378
  intensity and frequency of discharge of sensory neuron, 378f
  learning predictive relationships between in classical conditioning, 660–661
  modality and labeled line code, 380
  physical, transformation into electrical activity by sensory neuron, 34
  and process of perception, 388
  processing of information, 375–383
  receptor potential, 375
  representation of attributes in primary sensory neuron signals, 375, 376–377
  sensitivity of sensory system to differences in strength, 371–372
  sensory, transformation into electrochemical energy, 372–373
  on skin, internal representation, 336f
  strength, and frequency code, 379f
  transduction
    molecular mechanisms, 376f
    somatic sensory system, 375–376
Strabismus, treatment and development of visual perception, 484
Stress, and second-messenger regulaton of synaptic transmission, 263
Stretch
  muscle spindles as receptor for, 506, 507f, 508f
  receptor, muscle spindle, 48, 48f
  reflex, see Reflex
Stria terminalis, projection of amygdala, 608
Striate cortex, see Primary visual cortex
Stripes, staining region of extrastriate cortex, 395
Stroke, and possible role of glutamate toxicity, 232
Subfornical organ, role in water regulation, 623
Submodalities, 371
  columnar organizaton of somatic sensory cortex according to, 336–339, 337f
  integration in somatic sensory cortex by cells that have complex feature-detecting properties, 339
Substanatia nigra, of basal ganglia, 544
  loss of dopamine in cells leads to Parkinson's disease, 547
Subthalamic nucleus, of basal ganglia, 544
Sulci, cerebral cortex, 81, 82f
Summation
  of synaptic potentials, 222, 224f
  spatial, 223
Superior cervical ganglion, sex difference in, 589
Superior colliculus, projection of retina to, 429f
Superior temporal gyrus, stimulation produced altered perception of sounds and auditory illusions and hallucinations, 354
Supplementary motor area, 493
  role in voluntary movement, 534
Suprachiasmatic nucleus, possible role in biological clock mechanisms, 624–625, 625f
Surround, see Receptive field
Sympathetic division, of autonomic nervous system, 598–600, 599f
Sympathetic ganglia
  role of nerve growth factor in development of, 109f
  neurons, and covergent excitatory connections from two different neurons, 259f
Sympathetic neurons, development of from neural crest, 98f
Sympathetic system, 77
Sympathoadrenal lineage nerve cells, development, 98–99, 98f
Synapse formation, at neuromuscular junction, factors involved, 104–107
Synapses
  active zone, 276
  axo-axonic, control of calcium influx into presynaptic terminals, 288–290

onto central nerve cell, grouped according to function, 224–226
chemical
amplification at, 192
current flow in, 185, 185f
excitatory actions at, 228–229f
inhibitory actions mediated by chloride-selective receptor–channels, 233f
mediate excitatory and inhibitory actions, 184
and quantal transmission, 276
reason for synpatic delay in, 191
simplification of signals by, 189–194
size of synaptic cleft, 191
steps in transmission at, 191f
connections
brain, morphological types, 234–236
combination of excitatory and inhibitory in central nervous system, 220f
new, requirement for long-term memory, 671–676
number in brain, 183
peripheral nervous system, rearrangement during early postnatal life, 106f
plasticity and short-term memory in habituation, 670
on presynaptic terminals, regulation of intracellular calcium, 288–290
visual cortex, mechanism for cooperation and competition for fine tuning, 481f
contacts
formation, 108–109
retinal, 420
electrical, 189f
and gap-junction channels, 189
current flow in, 185, 185f
electrotonic transmission, 186–187
send simple depolarizing signals, 184
and transmission of metabolic signals, 187
electrical versus chemical, 184–185
distinguishing properties, 184t
elimination of during development, 106–107
excitatory on dendritic spine, 226
formation and plasticity, 694
inhibitory on nerve cell body, 224–226
modulatory on axon terminals, 226
nerve cell, 22
nerve-muscle, synaptic transmission at, 197–217
regions of contact on nerve cell, 225f
types in central nervous system, 236f
Synapsins, proteins that anchor synaptic vesicles, 282
Synaptic action
different times courses produced by single chemical transmitter, 262f
inhibitory, in central neurons, mediated by chloride-selective receptor–channels, 232–234
molecular model of undelying classical conditioning, 679f
Synaptic boutons, 46f, 49, 198f
Synaptic cleft, 25, 184, 198f
size in chemical synapses, 191
structure, 191f
Synaptic input, level of response of nerve cell to, 171
Synaptic potential, see Potential, postsynaptic
Synaptic relays, 84
Synaptic signals, transmission by cell body in motor neuron, 53
Synaptic transmission, 179–315
changes in with implicit learning, 668
chemical, modifiabilty, 194
chemical synapses, steps, 191f
chemical versus electrical, 181–182
clinical example of disruption, myasthenia gravis, 307–315
depression of in habituation, 668–671
directly gated, study of at neuromuscular junction, 198–200
effect of muscimol on in somatic sensory cortex, 339f
electrical
all-or-none behavior of neurons, 187, 188f
allows rapid and synchronous firing of interconnected cells, 187
demonstration, 186f
enhancement in sensitization, 671
failure at neuromuscular junction in myasthenia gravis, 311f
long-lasting consequences of second messengers, 261–263
modulation by second-messenger systems, 243–267
at nerve-muscle synapse, 197–217
substances acting as second messengers, 244–245
terminated by removal of neurotransmitters from synaptic cleft, 302–303
time course affected by calcium influx in presynaptic cell, 274f
and transcellular messengers, 255f
transmitting and receptive steps, 192–193
Synaptic vesicles, 191, 198
in active zone, 279f
chemiosmotic mechanism, 297–298
concentration of transmitter into, 297–298
control of mobilization, docking, and function by GTP-binding proteins and calcium, 284f
discharge of neurotransmitter from by exocytosis, 276, 279–282
docking, fusion, and exocytosis controlled by calcium influx, 282–285
exocytosis of and changes in membrane surface area, 281, 281f
molecular machinery for fusion and exocytosis, 285f
proteins, receptors and functions, 283f
quantum of neurotransmitter stored in, 276
recycling, 62f, 285–286, 286f
role of proteins in docking, fusion, and exocytosis, 282–285
Synaptophysin, synaptic vesicle protein, 284
Synaptotagmin, role in synatpic vesicle exocytosis, 284
Syndactyly, cortical representation, 333, 334f

T

Taste
as sensory modality, 371
role of population codes in coding stimulus quality in, 384–385
Taste buds, 384
Tectospinal tract, of brain stem medial pathway, 494, 495f
Temperature regulation
correspondence to motivational states, 614–615
and integration of autonomic, endocrine, and skeletomotor responses, 616–618
regions of hypothalamus concerned with, 616f
role of hypothalamus in controlling, 616–618
Temporal
pole, 350
resolution, by cones, 409, 410
summation, of synaptic potentials, 222
Temporal lobe
association areas, 354–355
cerebral cortex, 81, 82f
function, 9
effect of lesions in on memory, 654

Temporal lobe *(cont.)*
  medial, role in explicit memory, 656–658
  medial superior, representation of motion in, 447–449
  see also Medial temporal lobe
Tension, muscle, Golgi tendon organs as receptor for, 506–508, 507f, 508f
Testis-determining factor gene, 580
Testosterone
  fetal exposure of genetic females to causes pseudohermaphroditism, 583
  and gonadal development, 580
  and masculinzation of brain, 584
  metabolic pathways, 584f
  replacement therapy and sexual behavior, 587f
Tetanic stimulation, effect on synaptic effectiveness, 287, 288f
Tetanus, 503
Tetraethylammonium, potassium ion channel blocker, 165
Tetrodotoxin
  for study of ion channels, 172
  sodium ion channel blocker, 165
Thalamus, see also Lateral geniculate body
  left, role in languagae, 644
  receive basal ganglia projections to cortex, 545–547
  as relay for sensory pathways, 374
  role
    in processing sensory information, 84
    in sensory system, 372f
Thirst
  correspondence to motivational states, 614–615
  regulation, 623–624
Tissue osmolality, and regulation of thirst, 623–624
Tonic control, muscle, 491
Tonotopy, 375
Touch, 324
  as body sense, 383
  responses to in cerebral cortex, 327f
  as sensory modality, 371
Toxins, effects on ion channels, 128–129, 129f
Training, and expansion of afferent inputs in cortex, 691f
Transcellular signaling, 255f
Transcription, and regulation of gene expression, 263–265b
Transcription factors, see Homeobox genes
Transcriptional regulatory proteins, and cellular differentiation, 90
Transducin, role in photoactivation, 412
Transformation, stages as signal passes from one component to the next, 36–37
Transgenic animals, for study of behavior, 566–567b
Transmission, neuronal versus hormonal in nerve cell, 192–193
Transmitters, see Neurotransmitters
Transport
  active in nerve cell, 62–65
  calculating velocity of, 63f
  role of kinesin, 64f
Transporter molecules, and reuptake of neurotransmitters, 303
Trigger zone
  and integration of action potentials, 222
  as site of impule initation for action potential, 35
Trivariancy, of color vision, 456
Tuning curve, and sensitivity of sensory receptors, 373, 374f
Turner's syndrome, 581
Twins
  correlation between monozygotic and dizygotic for personality, 558f
  and inheritance of polygenic disorders of behavior, 564–565
Two-point threshold, and sensation, 372
Tyrosine kinase
  family receptors, 194, 344
  pathways, uses receptor and cytoplasmic kinases, 253

U

Unipolar nerve cells, 25, 26
Unit synaptic potential, 273
Universal grammar, and language acquisition, 638
Use, distinguishing feature of language, 635–636
Uterus, position of fetus and effect of hormone exposure on sexual development and behavior, 587–588

V

Vascular volume, and regulation of thirst, 623–624
Vasoactive intestinal peptide, co-release with ACh, 301
Vasopressin
  release as antipyretic response, 618
  role in water regulation, 623
Velocity
  action potential, effects of passive membrane properties and axon diameter on, 156–159
  sensitivity of muscle spindles, 510
Ventral
  amygdalofugal pathway, projection of amygdala, 608
  corticospinal tracts, 496, 497f
  horns, spinal cord, 493
  root, spinal nerve, 79
  tegmental area, activation and increase in dopamine levels, 626
Ventricular zone, cerebral cortex, 99
Vestibulocerebellum, functional component of cerebellum, 539–541, 539f
Vestibulospinal tract, of brain stem medial pathway, 494, 495f
Visceral afferents, processing of information as immediate reflex reponse or central autonomic circuit, 602f
Visible light
  separate cone systems respond to different wavelengths, 455–456, 456f
  wavelengths, 455f
Vision, see also Visual Perception; Visual System
  color, trivariancy of, 456
  deprivation, effect during later phase of ocular dominance column development, 477f
  effect of split-brain studies, 355–357
  form
    analysis of in primary visual cortex, 433–440
    distinguishing feature of language, 635
    judgement of, 392f
    perception, role of inferior temporal cortex, 404
    representation in inferotemporal cortex, 445
    visual perception of, 425–451; see also Parvocellular-interblob system
  regions of cerebral cortex dedicated to, 394f
  as sensory modality, 371
Visual attention
  and conscious awareness, 403–404
  study at cellular level, 449–450
Visual cortex
  convergence of input from eyes, 472f
  mechanism for cooperation and competition among neurons for fine tuning of synaptic connections, 481f
  postnatal development of ocular dominance columns, 474f

Visual deprivation, effect on architecture of ocular dominance columns, 471
Visual fields, 425–428
  binocular and monocular zones, 426f
  contralateral, retinotopic represenation in lateral geniculate nucleus, 430
  functional columns in primary visual cortex analyze, 443f
  neural map in cerebral cortex, 86
  perception of movement, 446–449, 446f
  in primary visual cortex, 431f
Visual image
  construction of, 387–405
  extraction of linear features of in primary visual cortex, 440
  inverted by eye lens on retina, 428f
Visual information, parallel processing supported by clinical evidence, 398–400
Visual pathways, coordination between facilitated by attention, 401–402
Visual perception
  as creative process, 388–393
  defects, 398–400
  development requires sensory experience, 485–486
  effect of attention, 401–402
  effect of early sensory deprivation, 470
  figure versus ground, 389f, 390f
  filling-in process, 391, 391f
  of form and motion, 425–451
  and illusons, 391–392, 391f, 392f
  laws, 389–393
  parallel pathways arising from lateral geniculate nucleus, 395f
  role of retina, 407–423
  stages in, 403f
  winner-take-all strategy, 389
Visual pigments
  activation by light in photoreceptors, 411–412
  genes for on X chromsome, 464f
  in photoreceptors, 410
  light activation affects cGMP concentration, 412–414
Visual system
  crossing of axons in, 87
  interactive integration of information, 404
  laws of perception, 389–393
  parallel pathways of, 393–402, 393f
  separate pathways carry different information, psychological evidence, 396–398, 397f
Visual tasks, speed of learning affected by lesions of inferior temporal region, 354
Voltage
  across cell membrane, 134
  passive spread of changes along neuron, 155
Voltage-clamp experiment, 165b
  calculation of membrane conductances from, 167b
  conditions, electrical equivalent circuit of nerve cell under, 167f
  demonstrating that sodium channels turn on and off more rapidly than potassium channels, 166f
  demonstrating two types of voltage-gated channels, 166f
  and study of action potential, 162–168
  technique, 164–165b
Voltage-gated ion channel, see Ion channels, voltage-gated
Voluntary movement, 529–550
  as category of movement, 491
  regulation by motivational system, 83–84
  role of spinal reflexes, 515

W

Wada test, demonstrates hemispheric differences regarding speech, 358
Walking movements, controlled by spinal circuits, 523–526, 525f
Warm-sensitive neurons, hypothalamic, 617
Water balance, regulation by hypothalamus, 623–624
Waters of hydration, 117
Weber's law, 371
Wernicke's aphasia, 640b
Wernicke's area
  interconnection with other language areas of brain, 644
  and language, 642f, 643f
Wernicke-Geschwind model
  and cortical connections, 644
  and distinguishing damage to language regions of brain, 642–643
Winner-take-all strategy, in visual perception, 389
Word blindness, 646
  with writing impairment, 646
Working memory, effects of lesions of principal sulcus on, 352
Writing, impairment, 646

X

X chromosome, 559, 560
  arrangement of red and green pigment genes on, 464f
X-linked genes, 560